Geohazards: Risk Assessment, Mitigation and Prevention

Geohazards: Risk Assessment, Mitigation and Prevention

Editors

Ricardo Castedo
Miguel Llorente Isidro
David Moncoulon

MDPI

Basel • Beijing • Wuhan • Barcelona • Belgrade • Novi Sad • Cluj • Manchester

Editors
Ricardo Castedo
Department of Geological and Mining Engineering
Universidad Politécnica de Madrid
Madrid
Spain

Miguel Llorente Isidro
Department of Climate Change and Geological Risks
Spanish Geological Survey
Madrid
Spain

David Moncoulon
R&D and Modeling
Caisse Centrale de Réassurance (CCR Group)
Paris
France

Editorial Office
MDPI
St. Alban-Anlage 66
4052 Basel, Switzerland

This is a reprint of articles from the Special Issue published online in the open access journal *Applied Sciences* (ISSN 2076-3417) (available at: www.mdpi.com/journal/applsci/special_issues/geohazards_risk).

For citation purposes, cite each article independently as indicated on the article page online and as indicated below:

Lastname, A.A.; Lastname, B.B. Article Title. *Journal Name* **Year**, *Volume Number*, Page Range.

ISBN 978-3-0365-9039-4 (Hbk)
ISBN 978-3-0365-9038-7 (PDF)
doi.org/10.3390/books978-3-0365-9038-7

Contents

About the Editors

Ricardo Castedo

Ricardo Castedo is an Associate Professor at the Geological and Mining Engineering Department (School of Mines and Energy; Universidad Politécnica de Madrid). He obtained his B.Sc. (2008) in Geological Engineering, M.Sc. (2014) in Teaching and Education, and Ph.D. (2012) in Engineering (Universidad Politécnica de Madrid). He had a one-year research stay at New Mexico Tech (USA) and three research stays at the University of Leeds (UK), the University of Windsor (Canada), and Politecnico di Torino (Italia). He is the author of more than 35 publications in scientific journals and more than 40 contributions to national and international conferences. Reviewer and editor of scientifically indexed (JCR) journals and international conferences. His research interests include rock mechanics, geomorphology, modeling, the dynamic behavior of structures, safety and security engineering, testing in structures with explosives, explosives behavior and characterization, and the safety and modeling of explosives.

Miguel Llorente Isidro

Miguel Llorente is Head of Scientific and Technical Projects in the Department of Climate Change and Geological Risks (Spanish Geological Survey, IGME-CSIC). He obtained his B.Sc. (2004) in Geology (University of Salamanca), M.Sc. (2005) in General and Applied Hydrology, and Ph.D. (2015) in Engineering (Universidad Politécnica de Madrid). He has two decades of work experience at the Head Research Unit in Galicia, as a Head of the Group in Applied Geosciences Research Unit, Director of Geology with the Camara Minera del Perú, EU Expert for Geohazards, and as a Tsunami Expert for the National Plan on Volcanic, Seismic, and Other Geophysical Process Monitoring. He is the author of more than 70 publications and contributions to national and international conferences and journals. He is also an Ad Honorem Profesor (Universidad Politécnica de Madrid). He is a reviewer and editor of scientifically indexed (JCR) journals and international conferences. His research interests include geohazards, geological risks, numerical modeling, floods, tsunamis, earthquakes, volcanoes, engineering, mapping, and GIS.

David Moncoulon

David Moncoulon was head of R&D and modeling at Caisse Centrale de Réassurance (CCR Group) until his death in July 2023, when he was 50 years old. He obtained his B.Sc. (1998) in Agronomical Engineering at L' Institut Agro Montpellier and his Ph.D. (2014) in hydrology and risk modeling at the Université Paul Sabatier Toulouse III. He worked three years at SEM Agriculture, four years at the French National Centre for Scientific Research (CNRS), and more than fifteen years at CCR. He is the author of several publications in scientific journals and at national and international conferences. His research interests include hydrological modeling, agricultural yield loss modeling, damage modeling, climate modeling, flood modeling, prevention, risk management, and insurance.

Preface

May this preface be a tribute to our colleague Dr. David Moncoulon, our fellow co-editor for this Special Issue (SI). He passed away too soon, too young, and still had much to contribute to research and development. We join his family, friends, and colleagues in mourning his loss.

This SI aims to provide a common forum to share knowledge about the growing threat of geohazards such as landslides, earthquakes, coastal changes, and drought, among many others, due to the expanding size of cities and urban areas, the critical use of agricultural lands, and the effects of climate change. A total of thirty-one papers (twenty-eight research papers, one review paper, and one technical note) on different topics like coastal changes, landslides, earthquakes, and others are presented in this reprint. The published works have been carried out in different parts of the world, such as Europe (i.e., Spain, France, Italy, etc.), Asia (i.e., China, South Korea, India, etc.), or America (i.e., Canada and Mexico).

Among the diverse research that the reader will find in this SI related to climate change and human intervention over coastal areas are the works of Fernández-Hernández et al. [1], Bio et al. [2], or Oh et al. [3]. These works use different data, such as aerial photographs (in some cases, more than 90 years ago), digital elevation models (DEMs), and video monitoring systems.

The contributions to this SI are mainly related to landslides (Saha et al. [4], Affandi et al. [5], Leonardi et al. [6], Gao and Zhang [7], Wu et al. [8], Liu et al. [9], Franco et al. [10], Cobos et al. [11], Niu et al. [12], and Gutiérrez-Martín et al. [13]), especially in populated areas, and focus on the development of prediction models, susceptibility maps, analysis of critical factors (e.g., rainfall, soil types, slope, etc.) or remediation techniques. These works include the use of geographic information systems, geotechnical flow modeling software, or electrical resistivity tomography.

Additionally, the contributions on earthquakes or seismic risks, including quantification of effects, stochastic generators, or analysis for insurance purposes (Cao et al. [14], Szabo et al. [15], Fidani [16], Gouache et al. [17], Hui et al. [18], Issadi et al. [19], Issadi et al. [20], and Pérez-Moreno et al. [21]), are noteworthy.

The reader will also find different laboratory studies on the influence of rainfall on slope erosion, the effect of dry-wet cycles on loess, or debris flow mobility (Tian et al. [22], Liu et al. [23], and Chang et al. [24]). To finish, readers will find articles on the effects of extreme droughts, forest fires, floods, or marine gas hydrate (Kapsambelis et al. [25], Gualdi et al. [26], Garrote et al. [27], Lee and Kim [28], León et al. [29], Jami et al. [30], and Li et al. [31]).

Finally, the editors would like to thank all the authors and reviewers, as well as the MDPI staff (especially Amy An), for their valuable contributions to this reprint.

Ricardo Castedo, Miguel Llorente Isidro, and David Moncoulon
Editors

applied sciences

MDPI

Article

Anthropic Action on Historical Shoreline Changes and Future Estimates Using GIS: Guadarmar Del Segura (Spain)

Marta Fernández-Hernández [1,*], Almudena Calvo [2], Luis Iglesias [1], Ricardo Castedo [1], Jose J. Ortega [1], Antonio J. Diaz-Honrubia [3], Pedro Mora [1] and Elisa Costamagna [4]

1 Escuela Técnica Superior de Ingenieros de Minas y Energía, Department Ingeniería Geológica y Minera, Universidad Politécnica de Madrid, 28220 Madrid, Spain; luis.iglesias@upm.es (L.I.); ricardo.castedo@upm.es (R.C.); josejoaquin.ortega@upm.es (J.J.O.); pedro.mora@upm.es (P.M.)
2 Euroconsult Nuevas Tecnologías, 28703 Madrid, Spain; acalvoc@euroconsultnt.es
3 Escuela Técnica Superior de Ingenieros Informáticos, Universidad Politécnica de Madrid, 28660 Madrid, Spain; antoniojesus.diaz@upm.es
4 Department of Environment, Land and Infrastructure Engineering (DIATI), Politecnico di Torino, 10129 Turin, Italy; elisa.costamagna@polito.it
* Correspondence: marta.fernandezh@upm.es; Tel.: +34-910676452

Abstract: A good understanding of historical change rates is a key requirement for effective coastal zone management and reliable predictions of shoreline evolution. Historical shoreline erosion for the coast of Guardamar del Segura (Alicante, Spain) is analyzed based on aerial photographs dating from 1930 to 2022 using the Digital Shoreline Analysis System (DSAS). This area is of special interest because the construction of a breakwater in the 1990s, which channels the mouth of the Segura River, has caused a change in coastal behavior. The prediction of future shorelines is conducted up to the year 2040 using two models based on data analysis techniques: the extrapolation of historical data (including the uncertainty of the historical measurements) and the Bruun-type model (considering the effect of sea level rises). The extrapolation of the natural erosion of the area up to 1989 is also compared with the reality, already affected by anthropic actions, in the years 2005 and 2022. The construction of the breakwater has accelerated the erosion along the coast downstream of this infrastructure by about 260%, endangering several houses that are located on the beach itself. The estimation models predict transects with erosions ranging from centimeters (±70 cm) to tens of meters (±30 m). However, both models are often overlapping, which gives a band where the shoreline may be thought to be in the future. The extrapolation of erosion up to 1989, and its subsequent comparison, shows that in most of the study areas, anthropic actions have increased erosion, reaching values of more than 35 m of shoreline loss. The effect of anthropic actions on the coast is also analyzed on the housing on the beach of Babilonia, which has lost around 17% of its built-up area in 40 years. This work demonstrates the importance of historical analysis and predictions before making any significant changes in coastal areas to develop sustainable plans for coastal area management.

Keywords: climate change; beach erosion; dune cliff; land management; data analysis

Citation: Fernández-Hernández, M.; Calvo, A.; Iglesias, L.; Castedo, R.; Ortega, J.J.; Diaz-Honrubia, A.J.; Mora, P.; Costamagna, E. Anthropic Action on Historical Shoreline Changes and Future Estimates Using GIS: Guadarmar Del Segura (Spain). *Appl. Sci.* **2023**, *13*, 9792. https://doi.org/10.3390/app13179792

Academic Editor: Athanasios Sfetsos

Received: 2 August 2023
Revised: 18 August 2023
Accepted: 28 August 2023
Published: 30 August 2023

1. Introduction

Coasts are valuable and vulnerable environments that provide numerous ecosystem services and support human populations in different ways [1]. It is estimated that sandy beaches make up approximately one-half of the world's ice-free coastline [2]. According to the United Nations, approximately 10% of the world's population currently lives in coastal areas that are less than 10 m above sea level, and about 40% lives within 100 km of the coast [3]. These areas are residentially, touristically and economically attractive; thus, they are becoming densely populated, but the concentration of infrastructures due to intense anthropic activity (i.e., ports, cities and resorts) puts additional stress on the coastal environment, sometimes pressing it to its limits [4].

Coastal zones are particularly vulnerable to natural erosive processes due to factors like wave action or storm events [5]. These processes can lead to the loss of land, degradation of coastal ecosystems, and increased risks for human settlements and infrastructure. The problem is emphasized when anthropic constructions modify coastal dynamics in some way, even though their primary objective is to protect the coast from erosion due to natural phenomena [6–8]. While these structures (i.e., harbors, channels or breakwaters) may provide immediate protection against erosion, they can disrupt the natural movement of sediment along the coastline, leading to unintended consequences [9,10]. Many of these hard structures were built from the 1950s to the 2000s, without accurate studies of the effects on coastal sediment dynamics and flora and fauna [1]. In contrast, in the last two or three decades, management strategies have focused on soft methods such as artificial beach nourishment, revegetation of dunes or bioengineering [11,12]. For example, some researchers have found a relationship between increased shoreline erosion in areas around the mouths of dammed rivers and in areas with high population densities in the Gulf of California [13]. Research conducted on the East Coast of South Korea found that the artificial structures constructed along the coast have not completely solved or stopped erosion but shifted it from one location to another [14]. On the contrary, work by Skilodimou et al. [15] shows how coastlines have been enhanced by 40% over the last 76 years due to human interventions, in this case, earth filling.

The effects of climate change on coastal regions are critical due to their exposure to rising sea levels, increased storm intensity and changes in oceanic and atmospheric conditions [16]. In general, rising sea levels can be a significant factor in shoreline changes [17]. Higher water levels mean that waves and storm surges can reach farther inland, eroding coastlines and flooding low-lying areas. This can result in the loss of beaches, dunes and other coastal landforms, which can lead to feedback loops, e.g., as coastal areas erode, sediment may be lost, reducing natural coastal barriers and exacerbating erosion. This, in turn, can increase vulnerability to coastal hazards. Changes in ocean wave characteristics, including alterations in wave height, period and direction, can further impact coastal erosion, sediment transport and stability. In addition to this, storms, including hurricanes and tropical cyclones, can cause significant changes due to intense erosion and the transportation of large amounts of sediment along the coastline [18]. An example is the work of Simeone et al. [19], who demonstrate that the beaches studied in Western Sardinia (Italy) show greater changes when faced with consecutive storms and not with an isolated one. These climate change impacts can lead to the loss of coastal land, damage to infrastructure and the displacement of communities. For these reasons, the analysis of the long-term effects of anthropic actions in past cases, in combination with information on the adverse effects of climate change (both at the global and regional levels), is fundamental to ensure the long-term sustainability and resilience of the new actions to be taken.

The aims of this work carried out in Guardamar del Segura, Alicante (Spain) are as follows: (1) analyze the historical evolution of the shoreline from 1930 to 2022, taking into account different anthropic constructions; (2) study the effects of these constructions on the erosion rates of the shoreline and a cliff dune; (3) evaluate the loss of built surface (houses) due to changes in coastal dynamics; (4) estimate the shoreline situation for the year 2040 with the information from available data and the use of two prediction models based on historical recession rates and changes in sea level; and (5) compare, with the use of these computational models, the positions that the coastline would have had in the years 2005 and 2022 without the constructions carried out in the area. The results and methodology provided by this work can be used as the basis for coastal planning and management, especially in areas with key infrastructures.

2. Study Area

This work focuses on the coastline of Guardamar del Segura located in the southeast of Alicante, Spain, on the Mediterranean coast (see Figure 1). The area has a population of more than 16,000 inhabitants, and the main economic activity in the region is summer tourism. The studied coast includes the beaches of Los Viveros (corresponding to zone B in Figure 1c) and Babilonia (zone C in Figure 1c). These two beaches have an actual length of 2.7 km and an average width of 12 m, reaching a maximum of 35 m. They are open beaches of golden sand with a grain size of around 0.27 mm [20], which is between fine and coarse sand.

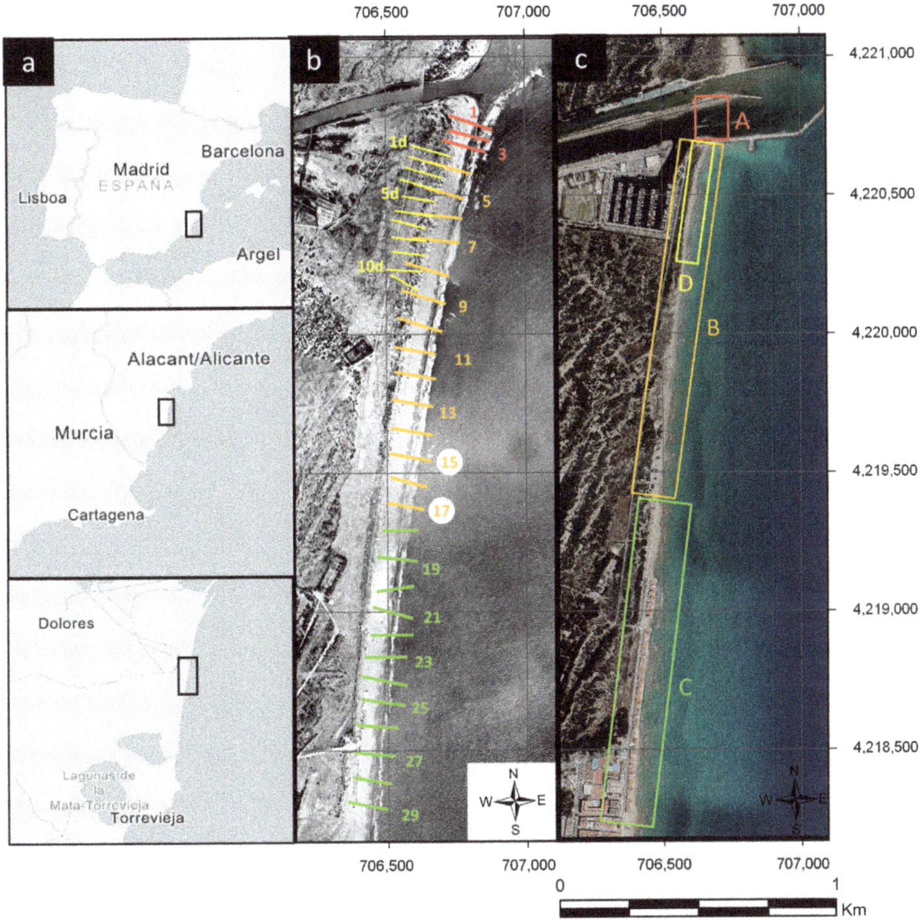

Figure 1. (**a**) Location of the study area. (**b**) Orthomosaic of the coast in 1930 and details of the transects used in the study. (**c**) Orthophoto of the coast in 2022. Note that the coordinates used are UTM ETRS89 H30N.

Guardamar del Segura coast is a fetch-limited environment with a predominance of sea waves. This coast is an environment with small astronomical tides (as is common in the Mediterranean) that oscillate about 0.3 m (SIMAR data from Puertos del Estado [21]). However, the so-called meteorological tides can reach up to 0.45 m. The dominant wave direction is northeast, creating a net N-S-oriented coastal current. Significant wave heights in this area, from 1958 to 2023, are 50.80% of the time between 0 and 0.5 m, 37.52% between 0.5 and 1 m, 8.7% between 1.0 and 1.5 m, 2.06% between 1.5 and 2.0 m and the remaining 0.92% between 2.0 and 3.5 m [21]. It should be noted that according to the available data [21], the maximum monthly values up to 1970 did not exceed 3 m of significant height. In the 1980s, there were already some peaks that exceeded 4 m, a trend that remained constant with peaks between 2.5 m and 4 m until 2010, but with an average value of around 2 m.

However, in the last 13 years, the average has risen around 3 m with some peaks in 2019 of almost 7 m. The latter peak is due to the large Isolated Depression at High Levels (DANA in Spanish) that occurred in the area in 2019 [22]. As for the peak period, the histogram is centered on the 5–6 s interval (23.94% of the time), distributed on the 2–4 s interval (29.12%) and then on the 4–5 s interval (19.59%), 6–7 s interval (17.89%) and 7–10 s interval (17.43%).

According to the latest data published by the Intergovernmental Panel on Climate Change (IPCC), the trend of accelerating sea level rise in the Mediterranean is consistent [23]. However, there are important differences depending on the methods and time horizons used in the analyses. In general, the accepted sea level rise referred to in the 20th century was 1.4 ± 0.2 mm yr^{-1} [24]. However, by 2050, it is estimated that the sea level may rise between 0.52 and 1.22 m above the mean relative sea level between 1996 and 2014. The relative sea level rise from 1927 to 2012 in Alicante was 1.28 ± 0.5 mm yr^{-1} [25].

This coastline has a long history of anthropogenic actions since the 18th century. The dune system in this area was left unfixed due to the massive felling of trees for the fishing industry during the 18th century. This meant that for a long time, the town of Guardamar was "threatened" by the movements of the dunes. From 1900 to 1930, the engineer Mira y Botella directed the reforestation of the area, which reduced the mobility of the dunes and, therefore, prevented the burial of the town. The houses of Babilonia (zone C in Figure 1) did not respond to the predatory model of the urbanism of the second half of the last century around the Spanish Levante. On the contrary, they had an environmental function and fit perfectly with the coastal physiognomy of the municipality in the 1930s. The number of houses increased until the end of the 1970s and remained constant until the 2000s, but since then, houses started to naturally and progressively collapse due to marine action. See Figure 2 for some photographs of the area since 2010.

Figure 2. (**a**) View from north of Babilonia Beach on 2 April 2010. (**b**) View of Los Viveros artificial dune on 2 April 2010. (**c**) View from north of Babilonia Beach on 5 May 2021. (**d**) View from north of Los Viveros Beach and artificial dune on 13 August 2021.

The main period of anthropic actions began in the mid-1970s and lasted about 20/25 years. In the 1970s, a small groin was built in the southern part of the mouth of the Segura River, which was completed in the mid-1980s with the construction of a larger groin in the northern part. In the period between 1987 and 1988, a large dredging of the mouth of the Segura River was carried out. Between 1990 and 1994, to reduce the risk of flooding, its mouth was channeled by means of a 525 m breakwater with an east and northeast orientation (zone A in Figure 1c), the opposite of all those built in the Spanish Levante [26]. This orientation stops the longitudinal transport of sediments from the north and blocks the outflow of sediments transported by the Segura River [7]. This forces the "Confederación Hidrográfica del Segura (CHS)" to schedule periodic dredging due to sediment accumulation that, if not executed, could cause breakage of the breakwater [27]. A volume of 14,000 m^3 of material extracted in the works of the early 1990s was used to raise Viveros beach (zone B in Figure 1c), and the rest of the material was used to raise other beaches to the south of the area of this study. Between 1996 and 1999, the Guardamar marina was built. In this project, a volume of sludge of approximately 200,000 m^3 was obtained, basically sand, with a low proportion of fines, very similar to that which makes up the current dune system. This sediment was dumped on a foredune originating an artificial dune (with the appearance of a small cliff) about 500 m long and of variable height (zone D in Figure 1) with a trapezoidal section, ranging from 8 m in the sector closest to the riverbed to 4.5 m in the southern part [6,28]. From 2002 to 2011, an attempt was made to eliminate the dumped anthropogenic materials, leaving the existing dune sands uncovered, and native species were planted to fix them [29]. Currently, this small cliff dune has areas up to 10 m high and slopes of up to >45° [30].

3. Methodology

The explanation of the methodology is divided into three blocks: shoreline evolution data, historical data analysis using the Digital Shoreline Analysis System (DSAS) [31] and prediction of future shorelines.

3.1. Dataset Preparation

Twenty aerial images covering a period of 93 years were used, with images in the following years: 1930, 1946, 1956, 1977, 1985, 1989, 1997, 1999, 2003, 2005, 2007, 2009, 2012, 2014, 2017, 2018, 2019, 2020, 2021 and 2022 (see Table 1 for details). They were obtained from different sources: National Geographic Institute (Instituto Geográfico Nacional (IGN) [32]) and Valencian Cartographic Institute (Institut Cartogràfic Valencià (ICV) [33]) photo libraries. The used reference system is ETRS89, UTM projection zone 30. In all images, the sea is calm, and therefore, all available data can be used for a comparative study.

Table 1. Main information from the aerial imagery used in this research. Note that GSD is the ground sampling distance: the distance between two consecutive pixel centers measured on the ground. This concept is equivalent to spatial resolution in remote sensing.

Year, Source	Flight Type	Date	GSD (m)	Photogrammetric Processing
1930 Ruiz de Alda	Analogical B/W	1929/1930	0.75	Orthomosaic
1946 American Flight Series A	Analogical B/W	1946	0.43	Rectification
1956 American Flight Series B	Analogical B/W	1956	0.63	Rectification
1977 Interministerial Flight	Analogical B/W	February 1977	0.45	Rectification
1985 National Flight	Analogical B/W	March 1985	0.75	Rectification
1989 Coastal flights	Analogical Color	March 1989	0.12	Rectification
1997 Flight OLISTAT	Analogical B/W	October 1997	1	Orthomosaic
1999 Five-Year Flight	Analogical Color	27 August 1999	1	Rectification
2002 Institut Cartogràfic Valencià	Analogical Color	12 May 2002	0.50	Orthomosaic
2005 Institut Cartogràfic Valencià	Analogical Color	30 September 2005	0.50	Orthomosaic
2007 Institut Cartogràfic Valencià	Digital Color	23 August 2007	0.50	Orthomosaic
2009 Institut Cartogràfic Valencià	Digital Color	5 August 2009	0.25	Orthomosaic
2012 Institut Cartogràfic Valencià	Digital Color	16 June 2012	0.50	Orthomosaic
2014 Institut Cartogràfic Valencià	Digital Color	21 June 2014	0.22	Rectification
2017 Institut Cartogràfic Valencià	Digital Color	18 August 2017	0.25	Orthomosaic
2018 Institut Cartogràfic Valencià	Digital Color	13 June 2018	0.25	Orthomosaic
2019 Institut Cartogràfic Valencià	Digital Color	14 May 2019	0.25	Orthomosaic
2020 Institut Cartogràfic Valencià	Digital Color	2 May 2020	0.25	Orthomosaic
2021 Institut Cartogràfic Valencià	Digital Color	13 May 2021	0.25	Orthomosaic
2022 Institut Cartogràfic Valencià	Digital Color	8 May 2022	0.25	Orthomosaic

The photographs of the first flight used (Ruiz de Alda 1929–1930) were treated with AgiSoft Metashape to obtain an orthomosaic of the study area since the photos were not taken systematically [34]. Control points identified on the 2022 orthomosaic were used to make the 1929–1930 orthomosaic. It seems clear that, although the 1930 orthomosaic has an error (Table 1), it can be considered within acceptable limits when working with data 50 years old or more [35], especially when you consider that it provides an important piece of information that is worth considering. The 1946, 1956, 1977, 1985, 1989, 1999 and 2017 aerial photographs were rectified using easily identified points on the 2022 orthomosaic, which was taken as a reference for the entire work. The remaining used images were orthomosaics produced by the official cartographic agencies and downloaded from their data servers.

Once the 20 mosaics with the orthophotos of all the years of study were uploaded to GIS, the next step was the vectorization of the shorelines, the cliff top and the houses of Babilonia Beach. The shoreline was drawn manually by the same operator instead of using automatic techniques more commonly used in remote sensing [36], following the last wet tide mark on the beach profile (see Figure 3 as an example). This line defines the boundary with the backshore and can be visually identified in orthophotos [37,38]. This methodology can be used in sedimentary littoral formations exposed to the open sea (i.e., beaches). To profile the cliff dune, the top (crest) of the slope is used, which is a visually perceptible feature [39]. The calculation of the number of dwellings and occupied area is based on the most current 2011 cadastral data [40]. On these cadastral data, the remaining 19 layers were manually modified to adapt them to each orthophoto from 1930 to 2022. All this work was carried out by the same operator to ensure a uniform criterion in the choice of geomorphological features.

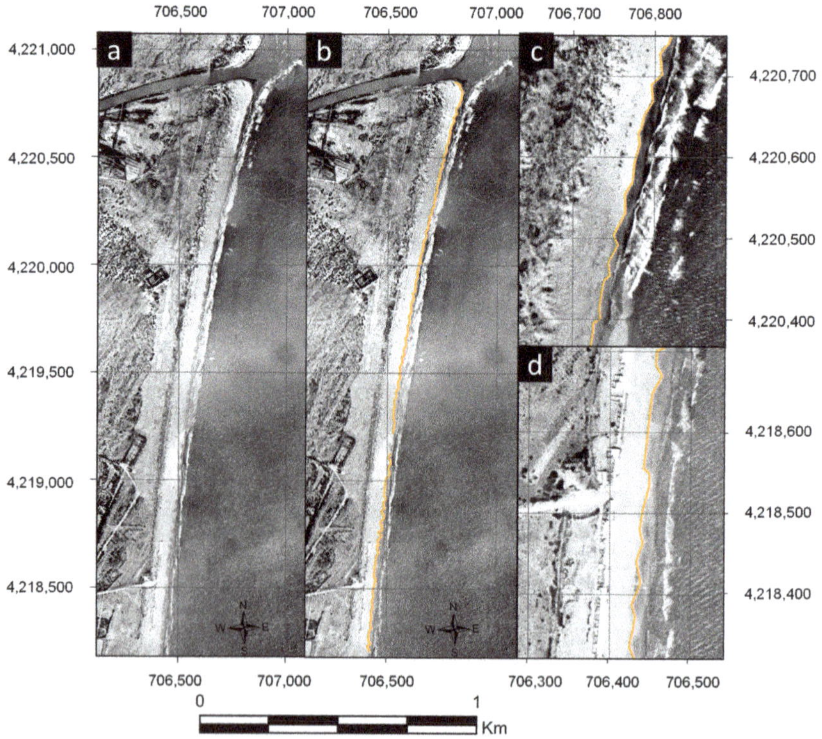

Figure 3. (**a**) Orthomosaic of the coast in 1930. (**b**) Orthomosaic of the coast in 1930 with the drawing of the shoreline. (**c**) Detail of the shoreline in the area of transects 4 and 8 (north of zone B in Figure 1). (**d**) Detail of the shoreline in the area of transects 25 and 29 (south of zone C in Figure 1).

3.2. DSAS—Historical Shoreline Analysis

Digital Shoreline Analysis System (DSAS) version 5.1 [31] was used to analyze the historical evolution of the coastline. For its analysis, the study region was divided into three parts, in addition to the cliff area (see Figure 1c). In total, there are 40 transects, 3 of which are in zone A (numbered 1 to 3), 14 are in zone B (numbered 4 to 17), 12 are in zone C (numbered 18 to 29) and 11 are in the zone D or cliff zone (numbered 1d to 11d). The transects are spaced 50 m apart in zones A and D, while in zones B and C, they were spaced 100 m apart. Each transect is a line perpendicular to the coastline, which joins the intersection points of the coastline of each of the available images (dates) with said perpendicular line.

The first analysis of the historical data was performed considering the complete time series from 1930 to 2022, except for zones A and D. In zone A, the analysis was conducted up to 1989, as this area was removed during the construction of the north breakwater and the channeling of the Segura River. Zone D was analyzed from 1999 to 2022 after human intervention. In addition, to improve the analysis and interpretation of the results and consider the important anthropic actions in zones B and C, an analysis of two-time intervals was also carried out: 1930–1989 (before the channeling of the Segura River) and 1997–2022 (after the construction of the marina and artificial dune).

The classical linear regression model was selected to estimate the rates of shoreline change. This technique is based on accepted statistical concepts, includes all the data of the series and provides the necessary data for the prediction models used in this research. In detail, the data provided by the DSAS software with this calculation method are as follows:

- Net shoreline movement (NSM): Maximum displacement of each transect between the first and the last available data. Measured in meters.
- Linear regression rate (LRR): A least-squares linear regression is fitted to all points of a transect. The historical rate of change in the shoreline is the slope of the fitted linear regression. Measured in meters per year (m/yr).
- Standard error of the slope (LCI): This estimator is often referred to as the standard deviation and describes the uncertainty (or variability) associated with the calculation of the LRR at a given confidence interval. In this case, it was calculated at 99%, thus ensuring that 99% of the data is within the LRR $\pm$ LCI. Measured also in meters per year (m/yr).
- R-squared of linear regression (LR2): Also known as the coefficient of determination, it is the percentage of variance in the data that is explained by a regression. Values close to 1 imply that the regression fits the data well, while values close to 0 imply that the regression does not fit as well.

3.3. Prediction of Future Shoreline

While it is challenging to make precise projections on future shorelines, researchers use available data and models to understand potential scenarios [41,42]. Calculated data on historical shoreline changes generally serve as the first, and perhaps most important, input for prediction models. However, these data alone are valid for extrapolation, if in the future it is assumed that the system remains in the same dynamic equilibrium as in the historical period analyzed. Otherwise, some variability must be added to the historical data (i.e., standard deviation) or, in more complex models, the elements in each zone that are most likely to be affected by future changes must be included.

In this paper, two simple but well-known models were used to estimate the position of the shoreline in the future: Lee–Clark [43,44] and Leatherman [45–47] models. The Lee–Clark model is based on the extrapolation of historical data (LRR). In this model, the standard deviation (LCI) of the measured erosions is incorporated to consider the variability in the rates over time. Erosion calculation is simple, and the result in year T would be: LCe = (LRR $\pm$ LCI) $\times$ T years. The Leatherman model (also known as historical trend analysis) is basically based on the "Bruun rule" to estimate the potential shoreline retreat (R_2 measured in m/yr) resulting from a rise in sea level as: $R_2 = S_2(LRR/S_1)$, where S_1 is the sea level rise rate (m/yr) during the analyzed period and S_2 is the projected sea level rise rate (m/yr) in the study area.

4. Results and Discussion

The results and their discussion were structured by zones, starting with the evolution of the shoreline, then the cliff dune and ending with a section dedicated to the comparison of the coastline in 2005 and 2022 with and without the effects of anthropogenic actions.

4.1. Shoreline Evolution (Past and Future)

Table 2 shows the rates of coastline change for the three periods analyzed (1930–1989; 1997–2022; and 1930–2022) in the three areas (A, B and C) that have or have had shorelines. In zone A, between 1930 and 1989, the erosion is important, with an average rate in the zone of 1.13 m/yr (Figure 4). As can be seen in Figure 5, most of this erosion is due to the period 1930–1956. Even in transect 1, the 1977 shoreline was ahead of the 1956 shoreline, reflecting a small accretion in the area, probably due to the construction of the small groin on the south side of the river mouth. It seems evident that in this area, there was some erosion near the natural mouth of the Segura River before it was channeled.

Table 2. Erosion rates for profiles 1 to 29, including zone A, B and C for three different study periods: 1930–1989, 1997–2023, 1930–2022. Negative sign means erosion, while positive sign means accretion for NSM and LRR.

		Period 1930–1989				Period 1997–2022				Period 1930–2022			
ZONE	TRANSECT	NSM (m)	LRR (m/yr)	LCI (m/yr)	LR2	NSM (m)	LRR (m/yr)	LCI (m/yr)	LR2	NSM (m)	LRR (m/yr)	LCI (m/yr)	LR2
	1	−78.70	−1.08	1.38	0.76	-	-	-	-	-	-	-	-
A	2	−75.33	−1.16	0.74	0.93	-	-	-	-	-	-	-	-
	3	−78.24	−1.16	0.88	0.90	-	-	-	-	-	-	-	-
	4	−69.08	−1.02	0.87	0.88	−41.80	−1.25	0.96	0.60	−128.64	−1.16	0.22	0.93
	5	−64.72	−0.94	0.67	0.91	−35.26	−1.31	1.25	0.53	−131.03	−1.23	0.21	0.94
	6	−59.41	−0.91	0.50	0.95	−47.14	−1.24	0.95	0.60	−126.75	−1.15	0.20	0.94
	7	−58.93	−0.90	0.46	0.95	−44.07	−1.49	1.24	0.59	−121.82	−1.12	0.20	0.94
	8	−46.07	−0.68	0.47	0.92	−43.63	−1.17	0.70	0.68	−111.34	−1.05	0.20	0.93
	9	−35.72	−0.62	0.38	0.94	−36.86	−1.25	0.69	0.74	−100.45	−1.03	0.20	0.93
B	10	−32.51	−0.43	0.37	0.88	−43.60	−1.18	0.49	0.82	−90.12	−0.93	0.21	0.90
	11	−26.43	−0.33	0.45	0.74	−35.38	−1.23	0.49	0.85	−88.18	−0.94	0.24	0.87
	12	−21.07	−0.24	0.41	0.64	−38.48	−1.21	0.61	0.77	−79.76	−0.84	0.25	0.84
	13	−16.49	−0.21	0.31	0.71	−30.41	−1.22	0.40	0.89	−69.21	−0.81	0.23	0.85
	14	−10.70	−0.05	0.41	0.08	−34.21	−1.15	0.50	0.84	−68.15	−0.73	0.27	0.77
	15	−6.42	0.04	0.57	0.03	−33.85	−1.15	0.55	0.79	−60.27	−0.64	0.28	0.71
	16	−3.40	0.01	0.67	0.00	−22.74	−1.40	0.73	0.79	−60.59	−0.64	0.29	0.69
	17	−0.70	0.15	0.69	0.19	−36.04	−1.15	0.72	0.72	−57.08	−0.54	0.30	0.59
	18	−4.30	0.06	0.77	0.03	−19.51	−0.77	0.31	0.93	−35.21	−0.37	0.30	0.55
	19	3.68	0.17	0.52	0.55	−21.66	−0.98	0.59	0.71	−35.72	−0.60	0.29	0.68
	20	−5.60	−0.03	0.62	0.01	−5.24	−0.33	0.54	0.40	−35.03	−0.32	0.18	0.65
	21	−7.69	−0.05	0.47	0.06	−4.37	−0.17	0.73	0.16	−30.94	−0.32	0.15	0.70
	22	−18.75	−0.24	0.77	0.34	−11.56	−0.35	0.35	0.53	−40.14	−0.31	0.13	0.75
C	23	−19.27	−0.23	0.68	0.38	−4.73	−0.26	1.44	0.28	−45.87	−0.39	0.15	0.78
	24	−7.33	−0.09	0.55	0.11	−15.63	−0.48	0.39	0.63	−29.02	−0.33	0.15	0.73
	25	−7.44	−0.12	0.41	0.31	−13.64	−0.39	0.31	0.69	−31.96	−0.33	0.12	0.79
	26	−5.27	−0.08	0.32	0.26	−13.76	−0.43	1.25	0.38	−32.85	−0.28	0.16	0.60
	27	−9.48	−0.14	0.27	0.59	−10.27	−0.09	0.55	0.06	−29.36	−0.19	0.14	0.48
	28	−15.42	−0.24	1.04	0.73	−13.29	−0.37	1.07	0.39	−31.99	−0.20	0.16	0.45
	29	−7.77	−0.08	2.00	0.08	−11.39	−0.27	0.71	0.43	−28.25	−0.17	0.17	0.34

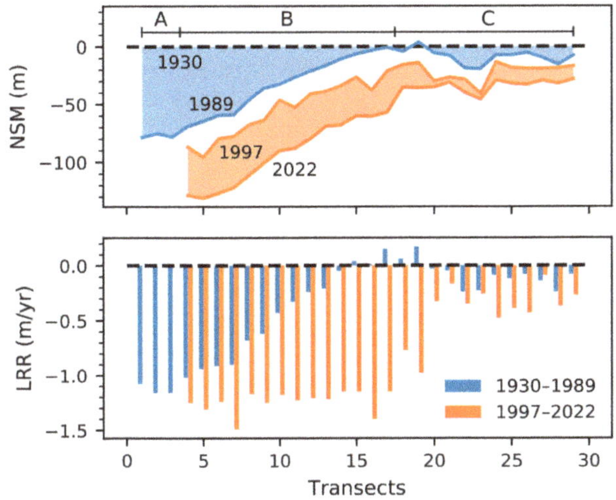

Figure 4. The upper graph shows the NSM of all transects in zones A, B and C in the periods 1930–1989 and 1997–2022. The lower graph shows the LRR values of each transect for the same zones and periods.

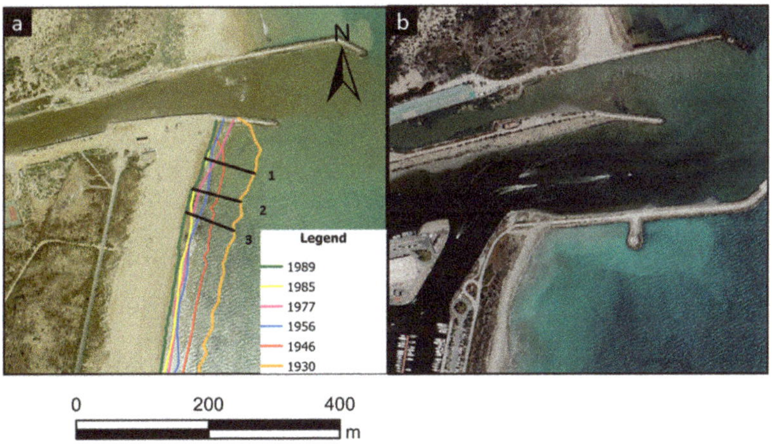

Figure 5. (**a**) Shorelines and transects 1 to 3 from 1930 to 1989 (background image from 1989). (**b**) Image from 2022 of the same area.

In the northern area of zone B, near the breakwater (transects 4–8, Table 2 and Figure 6), the erosion during the period 1930–1989 is significant, with an average of 0.89 m/yr. It is worth remembering that between 1990 and 1992, the beach was regrown by about 50 m due to the creation of the marina [4]. As shown in Figure 4, this material added to the beach eroded rapidly, as the 1997 shoreline was behind the 1989 one. However, for the period 1997–2022, the values shot up to an average of 1.29 m/yr. It seems more than evident that the construction of the breakwater and the channelization of the river has accelerated a natural process (Figure 4). From transects 9 to 13, erosion rates only reach 0.36 m/yr for the period 1930–1989. This indicates that this zone, already farther from the mouth of the river, was less affected by the natural shoreline erosion. In contrast, after construction works in the area, the rate skyrockets to 1.22 m/yr, almost one meter more in a much shorter period of only 25 years. The erosion data for the area between transects 14 and 17 present a low LR2 for the first epoch analyzed. This is because the shorelines in this area are very close together, having accretion and erosion epochs, which makes it difficult to fit the linear regression [48]. In any case, the net movement of shoreline erosion in these transects was 5 m for the 60-year period (1930–1989). However, for the period 1997–2022, the mean NSM for these transects is 31.71 m, which seems to confirm the increase in erosion south of the breakwater. Looking at the entire period of 92 years, the mean erosion of all transects in zone B is 0.92 ± 0.24 m/yr. It happens that throughout zone B, erosion rates have increased after anthropic actions, reflecting their negative effects, especially in the southern part (Figure 4).

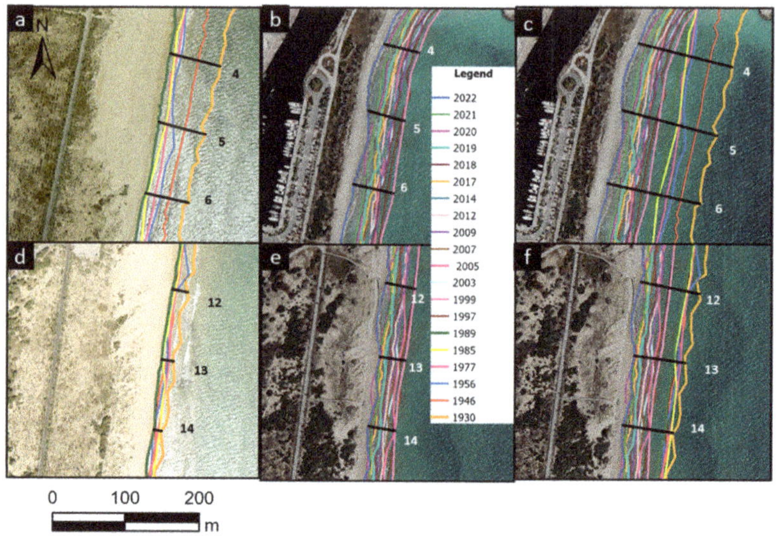

Figure 6. Zone B. (**a**,**d**) Shorelines and transects from 1930 to 1989 (background image from 1989). (**b**,**e**) Shorelines and transects from 1997 to 2022 (background image from 2022). (**c**,**f**) Shorelines and transects from 1930 to 2022 (background image from 2022).

In zone C, for the period 1930–1989, something similar to zone B occurs with some of the linear regression, where the LR2 value is low. In this zone, some of the transects show a net accretion for this period (see Figures 4 and 7a). Erosion at Babilonia Beach was very small until 1989, with only 8.72 m in 60 years, whereas in the last 25 years, 12.09 m, about 40% more, eroded in less than half the time. This has caused the loss of beaches in front of houses and, therefore, has left them directly exposed to sea waves (see Figure 2). However, as can be seen in Figure 7, the houses of Babilonia Beach act as an artificial barrier (or artificial shoreline), and therefore, there can be no further erosion except for the destruction of the houses themselves. If zone C is analyzed for the entire available time set, it can be observed that the fits are better and that the NSM value amounts to 33.86 m (0.32 m/yr). As in zone B, Figure 4 shows how erosion rates have increased after anthropogenic actions.

The effect of erosion on the houses of Babilonia Beach was also analyzed. During the 1930s and up to the 1970s, the built-up area increased, with 8533 m^2 (55 houses) in 1930; 12,277 m^2 (80 houses) in 1946; 17,976 m^2 (117 houses) in 1956; and 19,463 m^2 (126 houses) in 1977. This last value remained stable until 1999. From then on, and as a natural response to the values already discussed, the built-up area has been in continuous decline. Until 2005 the loss of floor area was a mere 131 m^2 (one house). But from 2005 to 2012, the loss increased by 2218 m^2 (14 houses) and has not stopped until 2022 (last data available), with a loss of another 1003 m^2 (6 houses). One should consider the numerous attempts to protect these houses with the use of rock dikes, Hesco barriers filled with rock and sand, etc., that have managed to somewhat slow down the destruction of these houses [49].

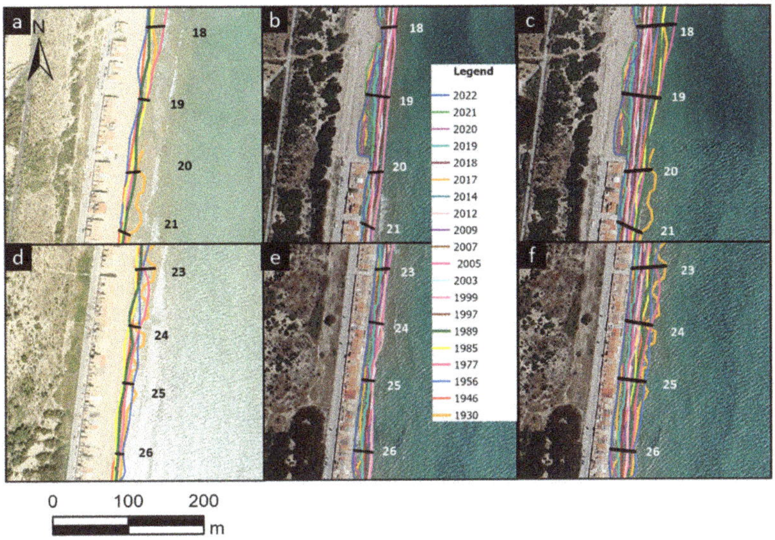

Figure 7. Zone C. (**a**,**d**) Shorelines and transects from 1930 to 1989 (background image from 1989). (**b**,**e**) Shorelines and transects from 1997 to 2022 (background image from 2022). (**c**,**f**) Shorelines and transects from 1930 to 2022 (background image from 2022).

Having historical erosion data in different transects, it is possible to apply the prediction models explained in Section 3.3. An adequate time horizon, preferably shorter than the available observation period (92 years), should be chosen to consider the data variability and, at the same time, to reduce the impact of data uncertainty on the quality of the results [39]. Therefore, the simulation is extended for 18 years, up to the year 2040 (see Table 3). Uncertainties have been incorporated for each model, including lower/upper bounds, having a surface area that delimits the possible solution (Figure 8). In the Lee–Clark model, uncertainty in the calculation of future erosion (LCe) is incorporated by adding (upper bound) and/or subtracting (lower bound) the LCI to the LRR value for each transect (see Table 2). For the Leatherman model, the limits are introduced using a minimum and maximum estimate of the rate of sea level rise in the future (S_2). According to local estimates, the future rates will be a maximum (upper bound) of 0.5 mm/yr, with a reasonable minimum (lower bound) value of 0.25 mm/yr [25]. As mentioned in Section 2, the historical relative sea level rise (S_1) has been 1.28 mm/yr.

As can be seen in Figure 8 and Table 3, the prediction models overlap in most cases. In this situation, it can be said that the shoreline is most likely to be located at some intersection surface of the two models. The Lee and Clark model has a larger difference between the lower and upper limits than the Leatherman. This is probably because the variability in historical erosion rates is greater than the effect of sea level changes for this relatively short simulation period. In the northern part of zone B, the shoreline is predicted to reach the area currently occupied by the cliff dune that, therefore, would also be eroded. In zone C, the houses continue to act as a barrier, although in the northern zone, erosion will increase and the loss of built-up area is likely to increase following the actual trend.

4.2. *Cliff Dune Evolution (Past and Future)*

The historical evolution of the cliff dune shows episodic erosion with major events between 2005 and 2007 and, more recently, between 2021 and 2022 (see Table 4 and Figure 9). In general, this is due to periods of major storms and an increase in wave height in the region [4,50]. In any case, the mean erosion value (including the 11 transects) for this area is 1.12 ± 0.66 m/yr. In this part of zone B, erosion data are 1.11 ± 0.20 m/yr for the period

1930–2022 and a little bit higher, 1.31 ± 0.77 m/yr, for the period 1999–2022. This shows, as is logical, that both cliff dunes and shorelines are closely related.

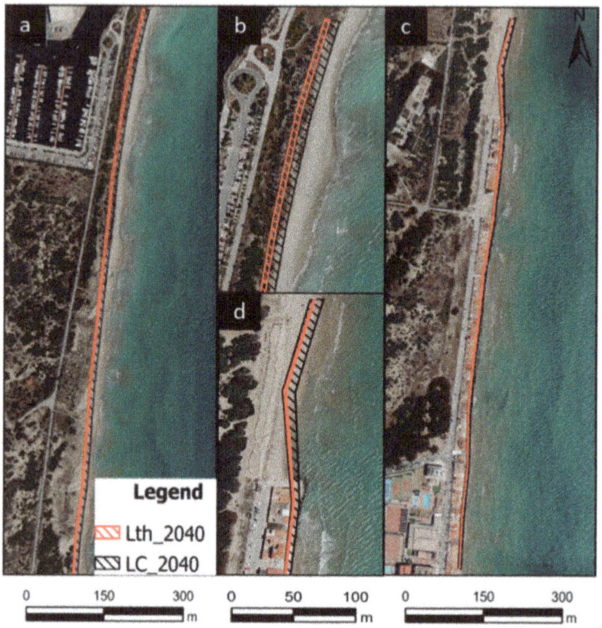

Figure 8. Shoreline model predictions (Lth—Leatherman; LC—Lee and Clark) by year 2040: (**a**) general view of zone B; (**b**) detailed view of zone B; (**c**) detailed view of zone C; (**d**) general view for zone C.

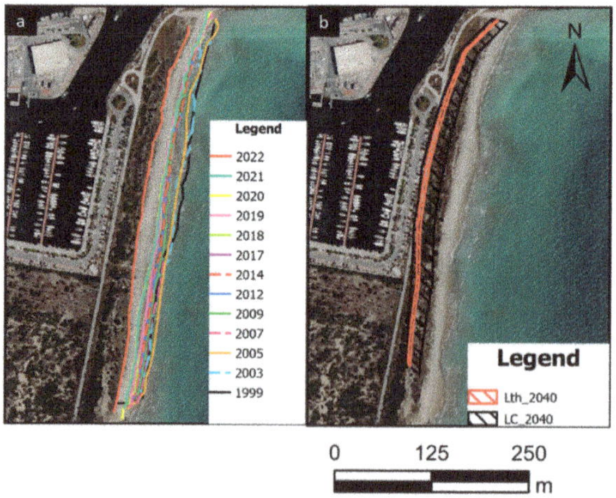

Figure 9. (**a**) Cliff dune erosion since 1999, zone D. (**b**) Cliff dune erosion predictions (Lth—Leatherman; LC—Lee and Clark) by the year 2040.

The behavior of the Lee–Clark and Leatherman model prediction results are very similar to those obtained for the shoreline, with greater variability in the LC model and the Leatherman model mostly bordering the upper bound of the LC model. With both models, the prediction for the year 2040 is that erosion will be such that the paved road that parallels part of the marina will be eroded.

Table 3. Total shoreline erosion by 2040 in meters.

Zone	Transect	Lee–Clark		Leatherman	
		Lower Bound	Upper Bound	Lower Bound	Upper Bound
B	4	−17.02	−24.79	−24.98	−29.07
	5	−18.21	−25.93	−26.38	−30.69
	6	−17.12	−24.44	−24.84	−28.90
	7	−16.69	−23.76	−24.18	−28.13
	8	−15.22	−22.50	−22.54	−26.23
	9	−14.95	−22.04	−22.10	−25.72
	10	−13.02	−20.63	−20.11	−23.40
	11	−12.52	−21.25	−20.18	−23.48
	12	−10.70	−19.71	−18.17	−21.14
	13	−10.31	−18.75	−17.37	−20.21
	14	−8.25	−17.96	−15.67	−18.23
	15	−6.53	−16.63	−13.84	−16.10
	16	−6.36	−16.81	−13.85	−16.11
	17	−4.20	−15.12	−11.55	−13.44
C	18	−1.34	−12.09	−8.03	−9.34
	19	−5.58	−15.94	−12.86	−14.96
	20	−2.59	−9.05	−6.95	−8.09
	21	−2.96	−8.46	−6.83	−7.94
	22	−3.30	−8.04	−6.77	−7.88
	23	−4.27	−9.73	−8.37	−9.74
	24	−3.31	−8.53	−7.08	−8.24
	25	−3.73	−7.99	−7.01	−8.15
	26	−2.17	−7.90	−6.02	−7.01
	27	−1.02	−5.92	−4.15	−4.83
	28	−0.70	−6.58	−4.35	−5.06
	29	−0.05	−6.24	−3.70	−4.30

Table 4. Erosion rates for profiles in zone D (cliff area) for the period: 1999-2022. Negative sign means erosion, while positive sign means accretion for NSM and LRR. Also, the total shoreline erosion with both models by 2040 in meters is shown.

Transect	NSM (m)	Period 1999–2022			Lee–Clark		Leatherman	
		LRR (m/yr)	LCI (m/yr)	LR2	Lower Bound	Upper Bound	Lower Bound	Upper Bound
1d	−11.61	−0.64	0.57	0.81	−1.36	−21.71	−13.79	−16.04
2d	−33.56	−0.92	0.59	0.68	−6.02	−27.13	−19.81	−23.05
3d	−37.87	−1.08	0.64	0.74	−7.97	−31.00	−23.29	−27.09
4d	−42.91	−1.34	0.72	0.78	−11.16	−37.23	−28.92	−33.65
5d	−48.39	−1.36	0.77	0.76	−10.58	−38.46	−29.31	−34.10
6d	−46.96	−1.34	0.67	0.78	−11.93	−36.16	−28.74	−33.44
7d	−42.83	−1.34	0.76	0.73	−10.41	−37.81	−28.82	−33.53
8d	−32.11	−0.99	0.76	0.63	−4.29	−31.51	−21.40	−24.90
9d	−29.69	−1.08	0.62	0.73	−8.25	−30.54	−23.18	−26.97
10d	−30.31	−1.05	0.51	0.78	−9.57	−28.07	−22.50	−26.18
11d	−36.85	−1.19	0.65	0.77	−9.68	−33.12	−25.58	−29.76

4.3. What Would Have Happened in the Area without Large-Scale Anthropogenic Actions?

With the data obtained in the previous sections, it is possible to simulate where the shoreline would be without the anthropic actions and compare it with the real situation. For this purpose, the Lee–Clark model is used in zones B and C for the pre-works period, i.e., from 1930 to 1989, as shown in Table 2.

In zone B (Figure 10), it can be seen how erosion has been advancing faster than natural erosion would have. This effect is especially noticeable in 2022 (Figure 10c,d), about 32 years after the start of the main works. At some points, such as transect 6, there was a difference of up to 10.85 m in 2005 (measured from the upper bound of the prediction to the shoreline), while in 2022, that difference increased to 20.85 m. However, this effect is felt more in the central sector of zone B; at transect 12, there was a difference of 23 m in 2005 and of 36 m in 2022. It appears evident that the effect of upstream constructions has accelerated the erosive process on the shoreline, especially in the central section of zone B (Figure 10b,d).

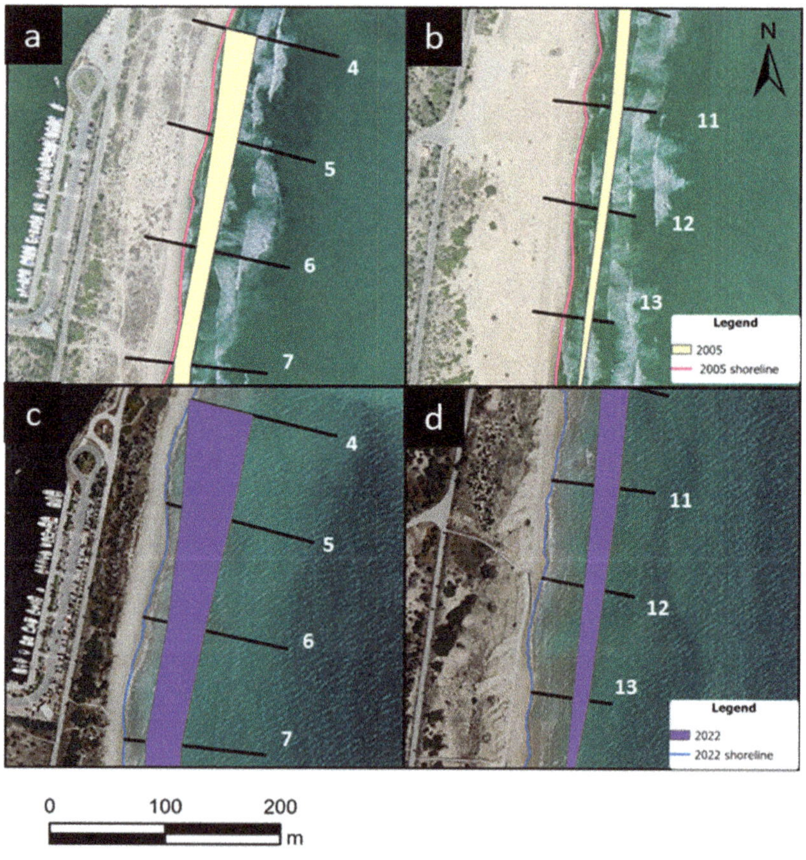

Figure 10. Shoreline and area of possible erosion predicted with the LC model: (**a**) north of zone B by 2005; (**b**) middle sector of zone B by 2005; (**c**) north of zone B by 2022; (**d**) middle sector of zone B by 2022.

In Babilonia Beach, the erosion behavior is strongly conditioned by houses. An important characteristic in zone C is that in 2005 (Figure 11a,b), the houses had an important beach area that protected them from the direct action of the sea. In the 2022 images (Figure 11c,d), this beach does not exist anymore. Figure 11a,c show how a significant number of buildings have been destroyed from 2005 to 2022 and how these houses have been replaced by sand. This has left the way open for erosion, which has been increasing in contrast to what the model predicts (transect 19). According to the projections of the model used, erosion in some areas would have reached the housing area even without the works carried out, but these have undoubtedly had an important effect on the acceleration of this phenomenon.

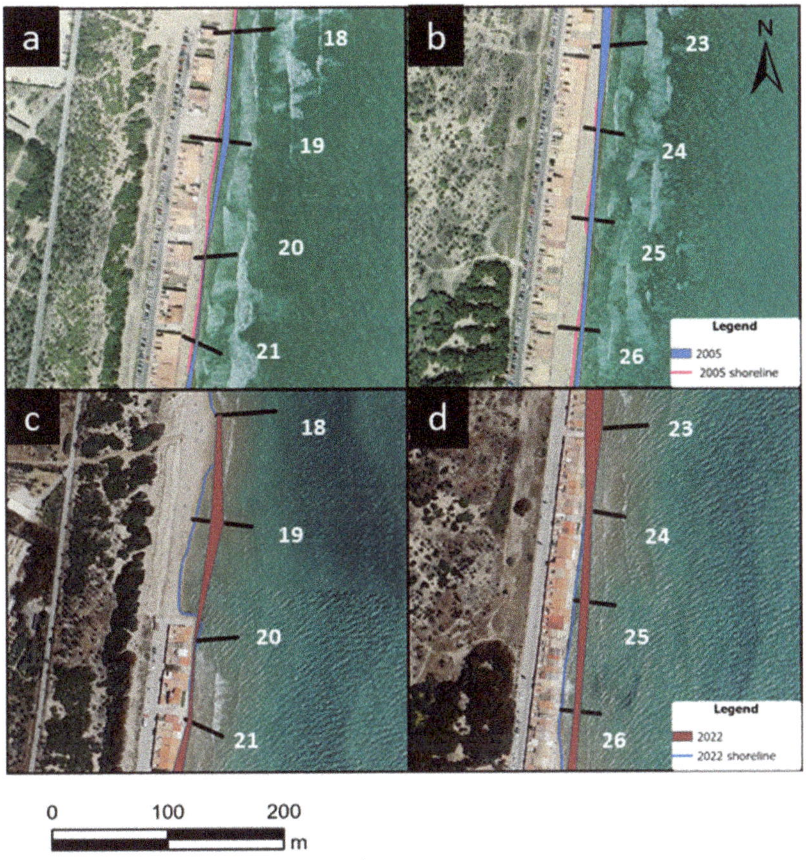

Figure 11. Shoreline and area of possible erosion predicted with the LC model: (**a**) north of zone C by 2005; (**b**) middle sector of zone C by 2005; (**c**) north of zone C by 2022; (**d**) middle sector of zone C by 2022.

5. Conclusions

In general, the results presented and the methodology shown can be used to improve short- and medium-term decisions in areas where major construction is to be carried out, as well as where and when to intervene in coastal erosion processes. The conclusions in response to the objectives set out in the study are as follows:

1. Erosion changes from one point of the coast to another and according to the period analyzed (1930–1989 or 1997–2022), which demonstrates the complex effect of anthropic actions that have accelerated the natural erosion of the area.
2. The erosion of the cliff dune (zone D) shows episodic erosion due to strong storms but with rates in meters/year like those of the shoreline that protects it (zone B).
3. The breakwaters at the mouth of the Segura River caused a loss of 3353 m^2 of housing and an increment in shoreline erosion rates until 2022.
4. The prediction models used overlap with each other; therefore, the predicted area would be the most likely for the shoreline situation in the year 2040. The Lee–Clark model (based on learning a model created on the derived dataset that is used for extrapolation of historical data) has greater variability than the Leatherman one (based on rates of sea level change), although the latter is always close to the maximum erosion limit of the former. In this area, the effect of the rate of sea level change does

not seem to be very important in shoreline erosion since this value was already quite high during the last century.

5. Comparing the actual shorelines in the years 2005 and 2022 with those simulated using the Lee–Clark model, it can be said that anthropogenic actions exhibit their effect by increasing erosion rates. This effect is more noticeable in the central zone of this study (southern part of zone B and northern part of zone C) and more limited in the marina zone (northern part of zone B) and in the houses on Babilonia Beach (southern part of zone C).

Author Contributions: Conceptualization, M.F.-H. and R.C.; methodology, A.C. and R.C.; software and data-based model learning, L.I., A.C. and A.J.D.-H.; validation, L.I., A.C. and M.F.-H.; formal analysis, R.C., J.J.O. and E.C.; investigation, M.F.-H., A.C. and P.M.; data curation, A.C. and L.I.; writing—original draft preparation, E.C., M.F.-H., R.C., J.J.O. and A.J.D.-H.; writing—review and editing, M.F.-H., A.C., L.I. and R.C.; visualization, A.C. and M.F.-H.; supervision, M.F.-H.; funding acquisition, R.C. All authors have read and agreed to the published version of the manuscript.

Funding: This work has been partially funded by the Universidad Politécnica de Madrid with grants to PDI and PhD researchers for international research stays of one month or more (resolution of 30 May 2022, of the Rector of the Universidad Politécnica de Madrid).

Institutional Review Board Statement: Not applicable.

Informed Consent Statement: Not applicable.

Data Availability Statement: The data presented in this study are available on request from the corresponding author. The data are not available to the public, as some of the images used must be requested from the Cartographic and Photographic Center of the Spanish Air Force and cannot be disseminated without permission.

Acknowledgments: The authors of this work would like to thank all the institutions that provide access to their material through official websites. The authors would like to thank the two anonymous reviewers for their contributions that have improved the final manuscript.

Conflicts of Interest: The authors declare no conflict of interest.

References

1. Bio, A.; Gonçalves, J.A.; Iglesias, I.; Granja, H.; Pinho, J.; Bastos, L. Linking short-to medium-term beach dune dynamics to local features under wave and wind actions: A Northern Portuguese case study. *Appl. Sci.* **2022**, *12*, 4365. [CrossRef]
2. Defeo, O.; McLachlan, A.; Armitage, D.; Elliott, M.; Pittman, J. Sandy beach social–ecological systems at risk: Regime shifts, collapses, and governance challenges. *Front. Ecol. Environ.* **2021**, *19*, 564–573. [CrossRef]
3. United Nations. Available online: https://www.un.org/esa/sustdev/natlinfo/indicators/methodology_sheets/oceans_seas_coasts/pop_coastal_areas.pdf (accessed on 17 August 2023).
4. Pagán, J.I.; López, I.; Aragonés, L.; Garcia-Barba, J. The effects of the anthropic actions on the sandy beaches of Guardamar del Segura, Spain. *Sci. Total Environ.* **2017**, *601*, 1364–1377. [CrossRef] [PubMed]
5. Trenhaile, A.S. Climate change and its impact on rock coasts. *Geol. Soc. Lond. Mem.* **2014**, *40*, 7–17. [CrossRef]
6. Syvitski, J.P.M.; Vörösmarty, C.J.; Kettner, A.J.; Green, P. Impact of humans on the flux of terrestrial sediment to the global coastal ocean. *Science* **2005**, *308*, 376–380. [CrossRef] [PubMed]
7. Aragonés, L.; Pagán, J.I.; López, M.P.; García-Barba, J. 2016. The impacts of Segura River (Spain) channelization on the coastal seabed. *Sci. Total Environ.* **2016**, *543*, 493–504. [CrossRef]
8. Biondo, M.; Buosi, C.; Trogu, D.; Mansfield, H.; Vacchi, M.; Ibba, A.; Porta, M.; Ruju, A.; De Muro, S. Natural vs. Anthropic Influence on the Multidecadal Shoreline Changes of Mediterranean Urban Beaches: Lessons from the Gulf of Cagliari (Sardinia). *Water* **2020**, *12*, 3578. [CrossRef]
9. Anthony, E.J.; Almar, R.; Aagaard, T. Recent shoreline changes in the Volta River delta, West Africa: The roles of natural processes and human impacts. *Afr. J. Aquat. Sci.* **2016**, *41*, 81–87. [CrossRef]
10. Villate, D.A.; Portz, L.; Manzolli, R.P.; Alcántara-Carrió, J. Human disturbances of shoreline morphodynamics and dune ecosystem at the Puerto Velero spit (Colombian Caribbean). *J. Coast. Res.* **2020**, *95*, 711–716. [CrossRef]
11. de Schipper, M.A.; Ludka, B.C.; Raubenheimer, B.; Luijendijk, A.P.; Schlacher, T.A. Beach nourishment has complex implications for the future of sandy shores. *Nat. Rev. Earth Environ.* **2021**, *2*, 70–84. [CrossRef]
12. Lee, J.T.; Shih, C.Y.; Hsu, Y.S. Root biomechanical features and wind erosion resistance of three native leguminous psammophytes for coastal dune restoration. *Ecol. Eng.* **2023**, *191*, 106966. [CrossRef]

13. Franco-Ochoa, C.; Zambrano-Medina, Y.; Plata-Rocha, W.; Monjardín-Armenta, S.; Rodríguez-Cueto, Y.; Escudero, M.; Mendoza, E. Long-term analysis of wave climate and shoreline change along the Gulf of California. *Appl. Sci.* **2020**, *10*, 8719. [CrossRef]
14. Yum, S.G.; Park, S.; Lee, J.J.; Adhikari, M.D. A quantitative analysis of multi-decadal shoreline changes along the East Coast of South Korea. *Sci. Total Environ.* **2023**, *876*, 162756. [CrossRef] [PubMed]
15. Skilodimou, H.D.; Antoniou, V.; Bathrellos, G.D.; Tsami, E. Mapping of coastline changes in Athens Riviera over the past 76 year's measurements. *Water* **2021**, *13*, 2135. [CrossRef]
16. Griggs, G.; Reguero, B.G. Coastal adaptation to climate change and sea-level rise. *Water* **2021**, *13*, 2151. [CrossRef]
17. Zhu, X.; Linham, M.M.; Nicholls, R.J. *Technologies for Climate Change Adaptation. Coastal Erosion and Flooding*, 1st ed.; UNEP Risø Centre on Energy, Climate and Sustainable Development: Roskilde, Denmark, 2010.
18. Bacopoulos, P.; Clark, R.R. Coastal erosion and structural damage due to four consecutive-year major hurricanes: Beach projects afford resilience and coastal protection. *Ocean. Coast. Manag.* **2021**, *209*, 105643. [CrossRef]
19. Simeone, S.; Palombo, L.; Molinaroli, E.; Brambilla, W.; Conforti, A.; De Falco, G. Shoreline response to wave forcing and sea level rise along a geomorphological complex coastline (western Sardinia, Mediterranean Sea). *Appl. Sci.* **2021**, *11*, 4009. [CrossRef]
20. Pagán, J.I.; López, M.; López, I.; Tenza-Abril, A.J.; Aragonés, L. Causes of the different behaviour of the shoreline on beaches with similar characteristics. Study case of the San Juan and Guardamar del Segura beaches, Spain. *Sci. Total Environ.* **2018**, *634*, 739–748. [CrossRef]
21. Puertos del Estado. Available online: https://www.puertos.es/es-ES/Paginas/avisolegal.aspx (accessed on 5 June 2023).
22. Martí Talavera, J.; Amor Jiménez, J.A.; Giménez García, R.; Ruiz Álvarez, V.; Biener Camacho, S. Episodio de lluvias torrenciales del 11 al 15 de septiembre de 2019 en el sureste de la Península Ibérica: Análisis meteorológico y consecuencias de las transformaciones en los usos del suelo. *Finisterra* **2021**, *56*, 151–174. [CrossRef]
23. Ali, E.; Cramer, W.; Carnicer, J.; Georgopoulou, E.; Hilmi, N.J.M.; Le Cozannet, G.; Lionello, P. Cross-Chapter Paper 4: Mediterranean Region. In *Climate Change 2022: Impacts, Adaptation and Vulnerability. Contribution of Working Group II to the Sixth Assessment Report of the Intergovernmental Panel on Climate Change*, 1st ed.; Pörtner, H.O., Roberts, D.C., Tignor, M., Poloczanska, E.S., Mintenbeck, K., Alegría, A., Craig, M., Langsdorf, S., Löschke, S., Möller, V., et al., Eds.; Cambridge University Press: Cambridge, UK; New York, NY, USA, 2022; pp. 2233–2272. [CrossRef]
24. Wöppelmann, G.; Marcos, M. Coastal Sea level rise in southern Europe and the nonclimate contribution of vertical land motion. *J. Geophys. Res.* **2012**, *117*, C01007. [CrossRef]
25. Zerbini, S.; Raicich, F.; Prati, C.M.; Bruni, S.; Del Conte, S.; Errico, M.; Santi, E. Sea-level change in the Northern Mediterranean Sea from long-period tide gauge time series. *Earth-Sci. Rev.* **2017**, *167*, 72–87. [CrossRef]
26. Alfosea, F.J.T.; Cantos, J.O. Incidencia de los temporales de levante en la ordenación del litoral alicantino. *Papeles Geogr.* **1997**, *26*, 105–132.
27. CHS—Confederación Hidrográfica del Segura. Available online: https://www.chsegura.es/es/ (accessed on 8 June 2023).
28. Matarredona Coll, E.; Marco Molina, J.A.; Prieto Cerdán, A. La configuración física del litoral sur alicantino. In *Guía de Campo de las XXI Jornadas de Geografía Física*, 1st ed.; Giménez-Font, P., Marco-Molina, J.A., Matarredona-Coll, E., Padilla-Blanco, A., Sánchez-Pardo, A., Eds.; Universitat d'Alacant, Instituto Universitario de Geografía: Alicante, Spain, 2006; pp. 35–47.
29. MAPAMA. *Recuperación del Ecosistema Dunar de Guardamar del Segura, Tramo Casas de Babilonia—Desembocadura del río Segura, Alicante*; Costas, D.G.D., Ed.; Ministerio de Agricultura, Pesca y Alimentación: Madrid, Spain, 2001.
30. Pagán, J.I.; Bañón, L.; López, I.; Bañón, C.; Aragonés, L. Monitoring the dune-beach system of Guardamar del Segura (Spain) using UAV, SfM and GIS techniques. *Sci. Total Environ.* **2019**, *687*, 1034–1045. [CrossRef] [PubMed]
31. Himmelstoss, E.A.; Henderson, R.E.; Kratzmann, M.G.; Farris, A.S. *Digital Shoreline Analysis System (DSAS) Version 5.1 User Guide: U.S. Geological Survey Open-File Report 2021–1091*; U.S. Geological Survey: Reston, VA, USA, 2021; p. 104. [CrossRef]
32. Centro de Descargas—Centro Nacional de Información Geográfica. Available online: https://centrodedescargas.cnig.es/CentroDescargas/index.jsp (accessed on 8 February 2023).
33. Institut Cartogràfic Valencià. Available online: https://icv.gva.es/es/ (accessed on 14 February 2023).
34. AgiSoft PhotoScan Professional. Available online: http://www.agisoft.com/downloads/installer/ (accessed on 2 February 2023).
35. Longley, P.A.; Goodchild, M.F.; Maguire, D.J.; Rhind, D.W. *Geographic Information Science and Systems*, 4th ed.; John Wiley & Sons, Ltd.: Chichester, England, 2015; p. 480.
36. Tsiakos, C.A.D.; Chalkias, C. Use of Machine Learning and Remote Sensing Techniques for Shoreline Monitoring: A Review of Recent Literature. *Appl. Sci.* **2023**, *13*, 3268. [CrossRef]
37. Ojeda Zújar, J.; Fernández Núñez, M.; Prieto Campos, A.; Pérez Alcántara, J.P.; Vallejo Villalta, I. Levantamiento de líneas de costa a escalas de detalle para el litoral de Andalucía: Criterios, modelo de datos y explotación. In *Tecnologías de la Información Geográfica: La Información Geográfica al Servicio de los Ciudadanos*, 1st ed.; Ojeda, J., Pita, M.F., Vallejo, I., Eds.; Secretariado de Publicaciones de la Universidad de Sevilla: Seville, Spain, 2010; pp. 324–336.
38. Pagán, J.I.; Aragonés, L.; Tenza-Abril, A.J.; Pallarés, P. The influence of anthropic actions on the evolution of an urban beach: Case study of Marineta Cassiana beach, Spain. *Sci. Total Environ.* **2016**, *559*, 242–255. [CrossRef]
39. Castedo, R.; de la Vega-Panizo, R.; Fernández-Hernández, M.; Paredes, C. Measurement of historical cliff-top changes and estimation of future trends using GIS data between Bridlington and Hornsea–Holderness Coast (UK). *Geomorphology* **2015**, *230*, 146–160. [CrossRef]
40. Sede Electrónica del Catastro. Available online: https://www.sedecatastro.gob.es/ (accessed on 23 March 2023).

41. Islam, M.S.; Crawford, T.W. Assessment of spatio-temporal empirical forecasting performance of future shoreline positions. *Remote Sen.* **2022**, *14*, 6364. [CrossRef]
42. Farris, A.S.; Long, J.W.; Himmelstoss, E.A. Accuracy of shoreline forecasting using sparse data. *Ocean Coast. Manag.* **2023**, *239*, 106621. [CrossRef]
43. Lee, E.M.; Clark, A.R. *Investigation and Management of Soft Rock Cliffs*; Thomas Telford: London, UK, 2002.
44. Castedo, R.; Paredes, C.; de la Vega-Panizo, R.; Santos, A.P. The modelling of coastal cliffs and future trends. In *Hydro-Geomorphology–Models and Trends*, 2nd ed.; Shukla, D.P., Ed.; IntechOpen: Rijeka, Croatia, 2017; pp. 53–78.
45. Leatherman, S.P. Modeling shore response to sea-level rise on sedimentary coasts. *Prog. Phys. Geogr.* **1990**, *14*, 447–464. [CrossRef]
46. Athanasiou, P.; van Dongeren, A.; Giardino, A.; Vousdoukas, M.I.; Ranasinghe, R.; Kwadijk, J. Uncertainties in projections of sandy beach erosion due to sea level rise: An analysis at the European scale. *Sci. Rep.* **2020**, *10*, 11895. [CrossRef]
47. Bagheri, M.; Ibrahim, Z.Z.; Mansor, S.; Abd Manaf, L.; Akhir, M.F.; Talaat, W.I.A.W.; Wolf, I.D. Hazard Assessment and Modeling of Erosion and Sea Level Rise under Global Climate Change Conditions for Coastal City Management. *Nat. Hazards Rev.* **2023**, *24*, 04022038. [CrossRef]
48. Frost, J. *Regression Analysis: An Intuitive Guide for Using and Interpreting Linear Models*, 1st ed.; Jim Publishing: Pennsylvania, PA, USA, 2020; p. 355.
49. NIUS Diario. Available online: https://www.niusdiario.es/espana/valencia/20230406/colonia-babilonia-playa-ruinas-guardamar-segura_18_09202722.html (accessed on 29 June 2023).
50. Oliva Cañizares, A.; Olcina, J. Temporales marítimos, cambio climático y cartografía de detalle de ocupación de la franja costera: Diagnóstico en el sur de la provincia de Alicante (España). *Doc. D'anàlisi Geogràfica* **2022**, *68*, 107–138. [CrossRef]

applied sciences

MDPI

Article

Linking Short- to Medium-Term Beach Dune Dynamics to Local Features under Wave and Wind Actions: A Northern Portuguese Case Study

Ana Bio [1,*], José Alberto Gonçalves [1,2], Isabel Iglesias [1], Helena Granja [1], José Pinho [3] and Luísa Bastos [1,2]

1 Interdisciplinary Centre of Marine and Environmental Research (CIIMAR/CIMAR), University of Porto, 4450-208 Matosinhos, Portugal; jagoncal@fc.up.pt (J.A.G.); iiglesias@ciimar.up.pt (I.I.); helenapgranja@gmail.com (H.G.); lcbastos@fc.up.pt (L.B.)

2 Department of Geosciences Environment and Spatial Planning, Faculty of Sciences, University of Porto, 4169-007 Porto, Portugal

3 Centre of Territory, Environment and Construction (CTAC), Department of Civil Engineering, University of Minho, 4800-058 Guimarães, Portugal; jpinho@civil.uminho.pt

* Correspondence: author: anabio@ciimar.up.pt

check for updates

Citation: Bio, A.; Gonçalves, J.A.; Iglesias, I.; Granja, H.; Pinho, J.; Bastos, L. Linking Short- to Medium-Term Beach Dune Dynamics to Local Features under Wave and Wind Actions: A Northern Portuguese Case Study. *Appl. Sci.* **2022**, *12*, 4365. https://doi.org/10.3390/app12094365

Academic Editors: Miguel Llorente Isidro, Ricardo Castedo and David Moncoulon

Received: 31 March 2022
Accepted: 22 April 2022
Published: 26 April 2022

Publisher's Note: MDPI stays neutral with regard to jurisdictional claims in published maps and institutional affiliations.

Abstract: Many coasts suffer from prevailing erosion, with them being particularly vulnerable to predicted climate change impacts, threatening coastal ecosystems, their services, infrastructures and populations. Understanding coastal morpho-sedimentary dynamics is thus essential for coastal management. However, coastal vulnerability may differ locally, depending on exposure/protection and local geological and morpho-hydrodynamical features, suggesting that a local approach to erosion risk assessment is needed to identify and understand local patterns. Digital elevation models of a 14 km long coastal stretch in northern Portugal that were extracted from aerial surveys obtained between November 2008 and February 2019 were analysed to quantify changes in shoreline position and sediment budgets, both for the whole study area and for distinct beach segments. The observed dynamics were subsequently analysed by considering prevailing wave and wind intensities and directions. Overall and during the decade analysed, the beach–dune system of the studied stretch slightly increased in volume (0.6%), although the shoreline retreated (by 1.6 m on average). Temporal variability in coastal dynamics was observed at all of the temporal scales considered—from seasons to 5-year periods—with them being related to variability in ocean and wind patterns. There was a trend from accretional to erosional conditions, with the first 5-year period showing a mean increase in the beach–dune system's volume of 0.6% and a mean shoreline progradation of 1.5 m, followed by 5-years with 0.0% volume change and 3.1 m shoreline retreat. Locally, the dynamics were very variable, with shoreline dynamics ranging from 24.0 m regression to 51.5 m progradation, and sediment budgets from 213.8 m^3 loss to 417.0 m^3 gain, per segment and for the decade. Stretches with relatively stable morphologies and others with erosional or accretional trends were found, depending on the beach type, shoreline orientation and the presence of defence structures. Rocky beaches were the least dynamic and sandy beaches the most dynamic, with mean shoreline position changes of 0.0 m and −3.4 m, respectively, and mean sediment budgets of −1.1 m^3 and −2.9 m^3 per linear meter of coastline, respectively, for the studied decade. The observed dynamics showed how local conditions interacted with meteo-ocean conditions in shaping local morpho-sedimentary dynamics, stressing the importance of a local approach to coastal erosion monitoring and risk assessment.

Keywords: coastal morpho-sedimentary dynamics; erosion/accretion patterns; remote sensing; beach types; meteo-ocean effects

1. Introduction

Coasts are land–ocean interfaces of high environmental and economic value; they provide important ecosystem services, from coastal buffering and inland protection to nutrient cycling, biodiversity, and recreational and cultural environments. As a consequence, coasts are often densely populated and modified, suffering increasing anthropogenic pressure [1]. This raises concerns about their increasing vulnerability [2], particularly in the light of potential climate change impacts, such as sea-level rise and changes in ocean wave characteristics [3,4]. Furthermore, many coasts suffer from erosion due to a sediment deficit, mostly caused by human actions that disrupt sediment fluxes and retain sediments in reservoirs and constructions [5]. More than 20% of the European and 30% of the Portuguese coastline are estimated to suffer from coastal erosion [6]. Erosion causes land loss, threatens ecosystems and infrastructures through increased exposure to wave impacts, and increases flood risk [7], and it is expected to be aggravated by climate change [8]. For Portugal, for instance, the projected sea-level rise is 1.14 m (ranging between 0.39 m and 1.89 m, with a 95% probability) by 2100 [9]. Erosion and shoreline retreat are therefore likely to become an even more relevant problem in the coming decades. There is hence an urgent need for coastal zone management and maritime and coastal planning.

The retreat of coastlines can be prevented by hard structures, such as groynes, seawalls and breakwaters, or through soft methods, such as beach nourishment, dune stabilization and bioengineering [10–12]. Given the high implementation and maintenance costs of hard defence structures and their negative impacts on ecosystems, as well as on neighbouring hydro-morphodynamics, soft solutions have gained in popularity [12–14]. This trend is also visible in Portugal, where hard defence structures were implemented from the late 1950s onward, often producing unexpected or unwanted effects and exacerbating downdrift erosion problems [15,16]. Over the past two decades, management strategies have therefore favoured soft defence approaches through artificial beach nourishment, placement of fences, construction of footbridges and revegetation of dunes [17,18].

Furthermore, also after 1950, important dams were built in the main river basins of the Iberian Peninsula, and large harbours were constructed at the river mouths. These two anthropic interventions, next to other significant soil-use changes in the river basins, greatly contributed to diminishing the volumes of sediments that are transported to the coastal platform or feed the alongshore dominant drift. The consequent sediment starvation at the coast implies that the design principles of groynes and breakwaters are not satisfied, making these defence solutions inefficient in most coastal environments.

Sustainable coastal management requires an integrated evaluation of coastal protection through hard defence structures or soft interventions, such as beach nourishment. Long-term cost-benefits and feasibility depending on local conditions, vulnerabilities and values, as well as sediment availability, need to be assessed to decide on mitigation measures or the alternative of a managed retreat, sacrificing areas that are considered less valuable [11].

Coastal morphology reflects the local natural and man-made conditions. Beach morphological changes occur due to the influence of natural phenomena, such as ocean waves and coastal currents, wind and river-flow effects, as well as due to human activities, such as urbanisation, infrastructures and coastal defence intervention [19–21]. Coastal vulnerability may therefore differ locally, depending on exposure/protection and local geological features, as well as local hydrodynamic and meteorological conditions, suggesting that a local approach to monitoring and management is needed [22]. Spatial and temporal fine-scale coastal sedimentary morphodynamics studies showed that the trends observed in large-scale studies may be the opposite of those observed in certain beach segments [23]. The consideration of local specificities, such as coastline orientation, exposure and natural or man-made structures, is thus crucial for the assessment of coastal risk and decision-making regarding protective measures for coastal populations and infrastructures.

Analogously, the characteristics of forcing variables, such as the prevailing wave and wind directions and intensities, have to be considered, as these are likely to influence the local impacts [24,25]. The importance of wave direction for the variability of coastal morphodynamics has been widely studied and acknowledged, with wave direction affecting sediment transport, beach rotation and morphology [26]. While, on open-coast, non-embayed beaches, morphology is considered to be primarily controlled by wave direction and energy, sediment transport will vary locally along a non-straight coastline due to uneven wave attenuation [27]. Lemke and Miller [22], who studied the impact of storm erosion on beach morphology, furthermore concluded that due to the propensity for beach conditions to change over short spatial scales, it is important to assess impacts on a local scale.

Another important parameter that affects beach morphology is the wind. Winds affect wave formation, as well as sediment transport on land, with this being particularly important for dune formation, as well as erosion. The aeolian sediment transport, which can be considered as the primary driver of the sediment flux not influenced by oceanic transport, is dependent on the wind velocity [28]. Wind speed needs to exceed a certain threshold value to start the sediment particle movement [29]. However, the wind direction and the angle of wind approach are also important for the transport of sediments, particularly regarding dune formation or erosion.

Other parameters that control sediment transport are the moisture content because moisture increases the inter-particle cohesion and, consequently, reduces the overall rate of transport [30]; vegetation, which helps to fix the sediments; and the availability of sediments for dune formation, which is also dependent on the sediment supply from waves, coastal drifts and rivers [29].

For sound coastal erosion management and mitigation, an analysis of the relationship between local conditions and forcing variables, as well as local morphodynamics, is thus needed to quantify and, ultimately, model these effects, allowing for much-needed prediction of future erosion trends and likely changes, for instance, following anthropogenic interventions (e.g., implementation of coastal defence structures such as breakwaters and groynes, or structures, such as promenades and buildings) or as a result of the predicted climate change impacts.

This can only be achieved through monitoring at temporal and spatial scales that are adequate to capture relevant variabilities and trends. However, high-temporal- and high-spatial-resolution studies of coastal morphology are rare, particularly in countries with a lower budget available for regular monitoring, although technological developments in remote sensing systems and methodologies nowadays allow for regular surveys at adequate spatial resolutions and affordable costs [31]. The present work aimed to show how a combination of digital elevation models, extracted from aerial photography, and in situ data of beach characteristics could provide the necessary information to characterize beach–dune morphodynamics at a local scale. This constituted a novel, local and comprehensive approach to the study of coastal sedimentary morphodynamics.

A case study is presented, where the morpho-sedimentary dynamics of a coastal stretch were analysed both temporally and locally and related to local features, as well as to the regional meteo-ocean conditions. Digital elevation and terrain models of the Vila Nova de Gaia coast in Northern Portugal, extracted from aerial photographic surveys carried out between November 2008 and February 2019, were analysed to quantify changes in shoreline position and sediment budgets for the study area as a whole and distinct beach segments. Changes were analysed at seasonal, yearly, five-yearly and decadal time scales. The observed dynamics were subsequently related to relevant parameters, including beach type/geology, the presence of a groyne and a breakwater, and dominant wave and wind directions and intensities.

2. Materials and Methods

2.1. Study Area

The study area covers a coastal stretch of about 14 km length in northern Portugal, between the Douro river mouth in the north and the city of Espinho in the south, corresponding to the coastline of the Vila Nova de Gaia municipality (Figure 1). This municipality is the third most densely populated in Portugal, with a population density in the coastal zone (up to 5 km from the coastline) of about 2500 inhabitants per square kilometre (data from the National Institute of Statistics, referring to 2011; www.ine.pt, accessed on

15 March 2022). Consequently, its coastline is densely occupied and urbanised. The coast is characterised by rocky beaches in the north (Figure 1a) and mostly sandy beaches with rocky outcrops in the centre and south. Being exposed to the high-energy ocean climate of the North Atlantic and suffering from sediment depletion, the Portuguese coast has several sectors that are prone to erosion [18,32]. Therefore, about 14% of the Portuguese coast is currently defended by artificial structures [33], although these structures, which mitigate sediment loss locally, often increase downdrift erosion [18]. The studied coastal stretch comprises a groyne (Figure 1b), which was built to retain sediments and to fix a submarine outfall, as well as a breakwater (Figure 1c) that was meant to be detached but developed a tombolo during construction, connecting it to the coast [15]. The beaches near Espinho have been managed with repeated artificial beach nourishment [34].

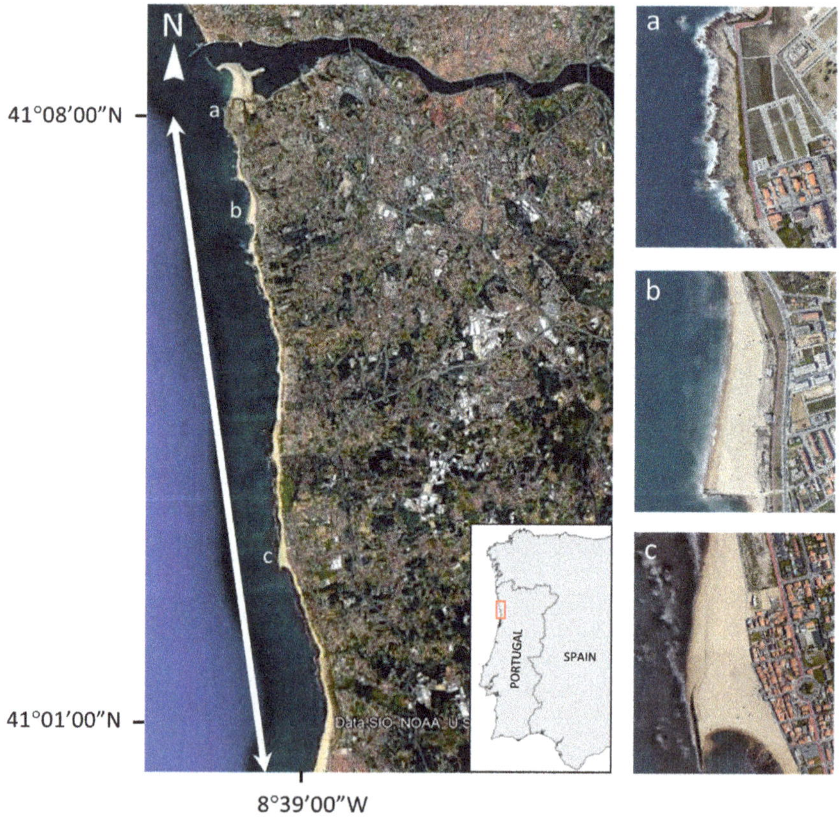

Figure 1. The Vila Nova de Gaia coast (left panel) and its location in Northern Portugal (red rectangle), and closeups of the rocky sector in the North (**a**), the groyne at Praia de Canide (**b**) and the breakwater at Aguda beach (**c**) (image: GoogleEarth, Data SIO, NOAA, U.S. Navy, NGA, GEBCO).

The north-western Portuguese Atlantic coast is highly energetic, presenting mean significant wave heights of 2–3 m offshore, and mean wave periods of 8–12 s [35,36]. Waves usually come from the NW, which induces a longshore drift current from north to south. This current is in some areas inverted due to the presence of obstacles (such as breakwaters, jetties, groynes, ebb tidal deltas and bars) that promote wave diffraction. The continental shelf is 30–40 km wide [37] and the local tides are dominated by a semidiurnal regime [38], with amplitudes ranging between 2.5 m and 3.8 m.

The regional atmospheric climate varies seasonally. Mediated by the Azores Anticyclone [39], predominant winds from northerly and north-westerly directions occur during the whole year, but with the highest amplitude during the summer months [40]. During autumn and winter, southerly, south-westerly and westerly winds become dominant, though northerly winter winds continue to occur.

2.2. Data

2.2.1. Surveys and Elevation Models

Aerial photos from surveys commissioned in the scope of several research projects were used to obtain orthomosaics and digital elevation models (DEMs) for the study area. Series of overlapping photos were taken from a small manned airplane carrying a digital photogrammetric camera (ZI-DMC, 7680 × 13,824 pixels, for the 2008–2010 surveys, and Vexcel UltraCam Falcon, 9420 × 14,430 pixels, for the 2018–2019 surveys) and flying at about 1000 m and 1600 m heights for the 2008–2010 and the 2018 and 2019 surveys, respectively, to provide high-resolution images with 10 cm and 12–13 cm ground-sampling distance, respectively. Images were directly georeferenced using an on-board GNSS/INS system. For postprocessing, the GNSS relative positioning mode was used. For each image, the position of the camera projection centre and the attitudinal angles were obtained, and a boresight alignment with ground control points was carried out to correct slight systematic effects in the attitude angles of about 0.02 degrees [31,41,42]. In situ measurements in built-up areas were used for calibration, as these contain features that can be used as accurate ground control points.

Surveys were taken on 14 November 2008, 23 April and 11 November 2009, 5 May 2010, 17 May 2018 and 20 February 2019 during spring low tides. DEMs with 1 m resolution were computed from the pairs of stereoscopic aerial images after extracting correlated points through stereo-matching using the Agisoft software [43]. Final DEM accuracies were in the order of 10 cm and, therefore, close to the image resolution [42].

Additionally, a Digital Terrain Model (DTM) from 2014 (kindly provided by the Direção Geral do Território, available at http://mapas.dgterritorio.pt/inspire/atom/downloadservice.xml (accessed on 15 March 2022)) with a 2 m resolution was used to complement the elevation data.

2.2.2. Wave Conditions

Wave data were obtained from a DATAWELL directional wave buoy located about 32 km NW of the study area, off the coast of Leixões harbour (41°19.00′ N, 008°59.00′ W), at 83 m water depth (processed data supplied by the Portuguese Hydrographical Institute—IH; https://www.hidrografico.pt (accessed on 15 March 2022)). Average significant heights (Hs) and wave peak directions for 3 h intervals were used (Figure 2). Wave roses were produced to show the wave direction distribution depending on the wave height (Figures 3–5), using 16 cardinal directions and frequencies for 3 significant-wave-height classes: above 4.5 m, between 3 m and 4.5 m, and above 4.5 m. These classes were based on the 85th and 95th percentiles of the mean significant wave heights (Hs) recorded during the study period, which were 3.1 m and 4.4 m, respectively. Three-hourly wave data were available for 89.5% of the total study period, with data missing for January 2010, January 2016, May and June 2018, and February 2019 due to equipment failure (mostly during winter) or maintenance (in spring/summer). Notice, that failures due to the damage or breakdown of the equipment during storm events meant that some data of extreme conditions, which are relevant for sediment transport, were missed.

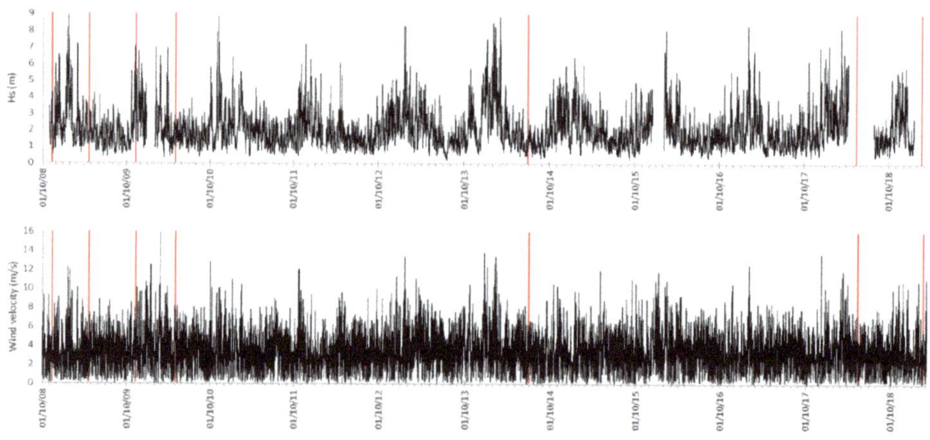

Figure 2. Significant wave heights (Hs) measured at the Leixões wave buoy and estimated wind velocities for the study area during the study period; survey dates are marked with red vertical lines.

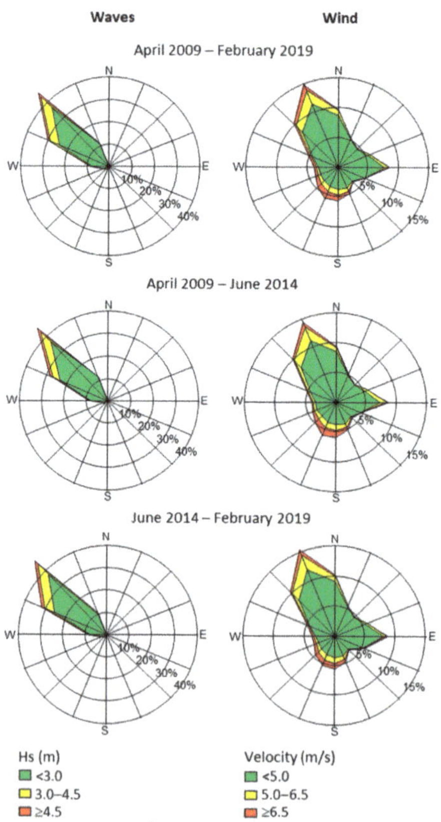

Figure 3. Distribution of 3-hourly wave directions for three significant-wave-height classes (**left plots**) and of hourly wind directions for three velocity classes (**right plots**) for the last decade and the two 5-year periods.

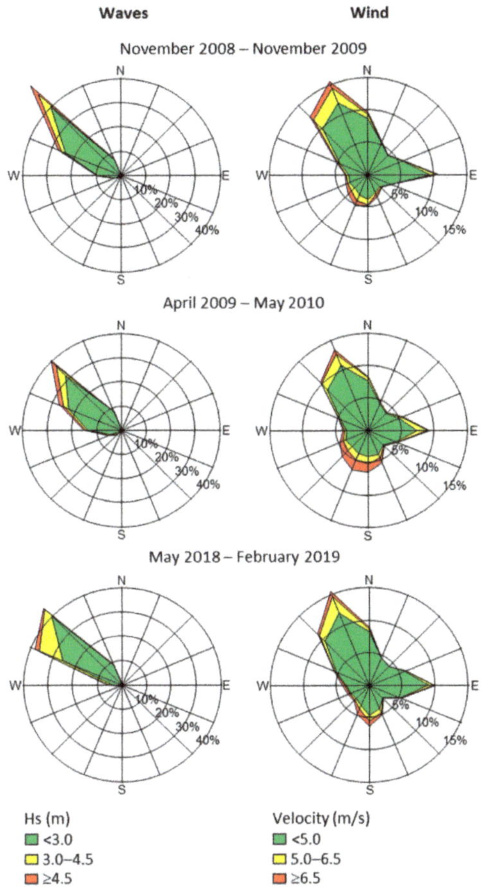

Figure 4. Distribution of 3-hourly wave directions for three significant-wave-height classes (**left plots**) and of hourly wind directions for three velocity classes (**right plots**) for the three yearly periods.

2.2.3. Wind Conditions

Wind velocity and direction were obtained from the ERA5-Land reanalysis dataset [44]. ERA5-Land provides hourly high-resolution gridded climate variable estimations from 1950 to present with a resolution of 0.1° × 0.1° (i.e., about 11.1 km in the N–S direction and 8.3 km in the E–W direction at 41.1° N latitude), with the closest available location for the study region being at 41.1° N, 8.6° W. Hourly wind velocities (in m/s) were extracted and wind directions were computed from the respective estimates of the horizontal speed of air moving towards the east (u-component) and toward the north (v-component) at a height of ten metres above the surface of the Earth (Figure 2). Analogously to the processing of the wave data, wind roses were produced to show the wind direction distribution depending on the wind velocity, using 16 cardinal directions and frequencies for 3 wind-speed classes: more than 6.5 m/s, between 5 m/s and 6.5 m/s, and less than 5 m/s (Figures 3–5). The classes were based on the 85th and 95th percentiles of the hourly wind velocities recorded during the total study period, which were 5.2 m/s and 6.6 m/s, respectively. Notice that these data were model simulations and, although ERA5-Land wind data may be generally acceptable for coastal locations with moderate variability in topography, they should be used with caution for coastal zones because the ocean–land discontinuity affects model stability conditions [45].

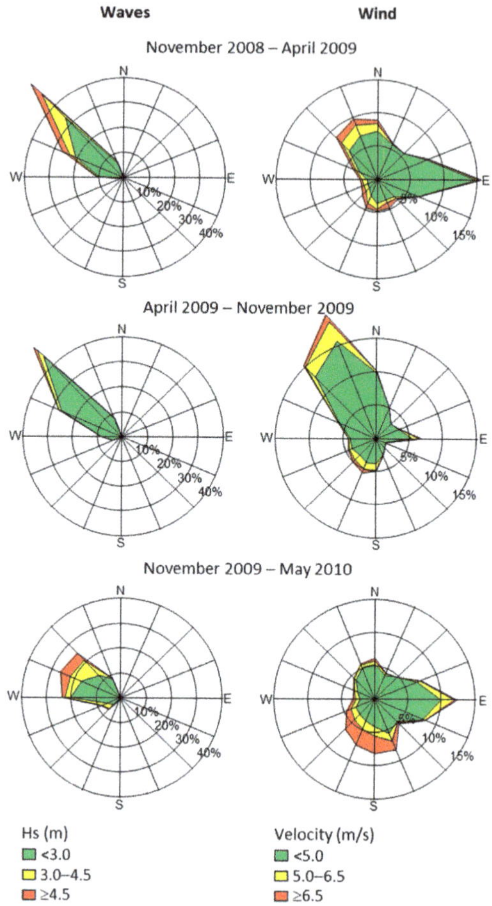

Figure 5. Distribution of 3-hourly wave directions for three significant-wave-height classes (**left plots**) and hourly wind directions for three velocity classes (**right plots**) for the 3 seasons monitored (from top to bottom: winter, summer, winter).

2.3. Analyses

The adopted procedure is schematised in Figure 6. The DEMs and DTM, on a regular 1 × 1 m grid and 2 × 2 m grid, respectively, were mapped and analysed in a GIS tool (using ArcGIS 10.6 and its Spatial Analyst and 3D Analyst modules). Differences between elevation models were computed to quantify sediment budgets and changes in shoreline position for different periods. The data series allowed for assessing changes that occurred during the following periods: an approximate decade, between April 2009 and February 2019; two approximate 5-year periods, between April 2009 and 2014 and between 2014 and February 2019 (we assumed that the 2014 surveys took place on the 30th of June, but precise dates are unknown); three periods of a year, between November 2008 and November 2009, between April 2009 and May 2010 and between May 2018 and February 2019 (not a complete year); and three seasons, between November 2008 and April 2009 (winter), between April 2009 and November 2009 (summer) and between November 2009 and May 2010 (winter) (Table 1).

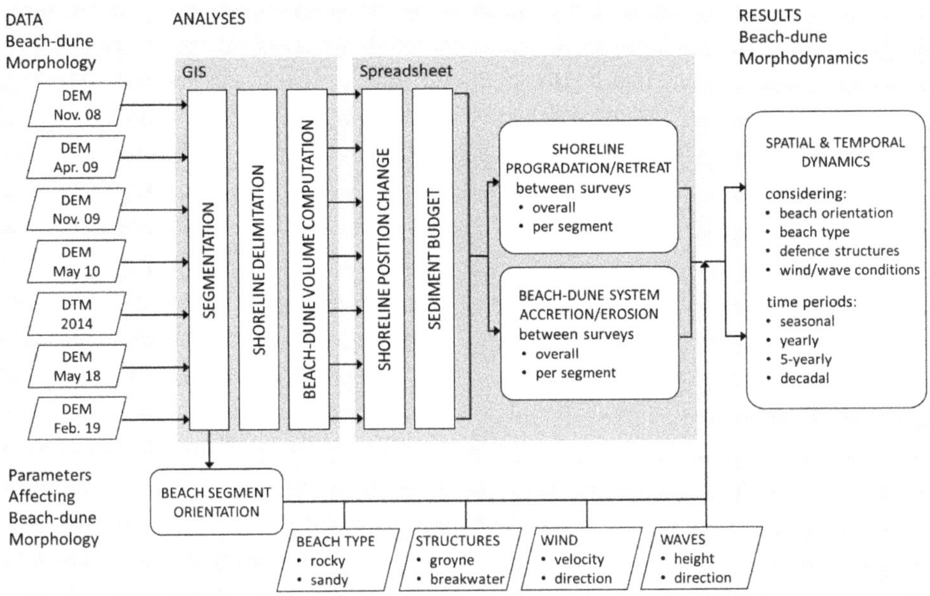

Figure 6. Scheme of the adopted procedure, where digital elevation models (DEMs) and the digital terrain model (DTM) were analysed in terms of the spatial and temporal morphodynamics, which were subsequently related to local features and wave and wind patterns.

Table 1. Changes in volume of the beach–dune system and in the shoreline position (positive and negative values indicate seaward and landward movements, respectively) that were observed between subsequent surveys (upper rows), and for the yearly, 5-yearly and the decadal periods of the time series analysed.

Dates	Period	Δ Volume (m^3)		Δ Shoreline (m)
November 2008–April 2009	Season: winter	223,094	+2.9%	2.7
April 2009–November 2009	Season: summer	108,848	+1.4%	0.7
November 2009–May 2010	Season: winter	−6889	−0.1%	1.2
May 2010–June 2014		−54,483	−0.7%	−0.4
June 2014–May 2018		90,981	1.2%	−3.3
May 2018–February 2019	1 year (approx.)	−89,372	−1.1%	0.2
November 2008–November 2009	1 year	331,942	+4.4%	3.3
April 2009–May 2010	1 year	101,959	+1.3%	1.9
April 2009–June 2014	5 years	47,476	+0.6%	1.5
June 2014–February 2019	5 years	1608	0.0%	−3.1
April 2009–February 2019	Decade	49,084	+0.6%	−1.6

To assess the local patterns, as well as the effect of beach exposure to wind and wave impacts, the studied coastal stretch was divided into segments. Therefore, the contour line of 1 m above the mean sea level (MSL) of the first survey (November 2008) was drawn and generalised (simplified) using the Douglas–Peucker simplification algorithm [46] with a specified maximum offset tolerance of 30 m. Each segment of this generalized line represented a beach segment, and for each segment, its length and orientation (facing direction) were extracted.

Considering the isoline 1 m above MSL as representative of the shoreline, shoreline dynamics were obtained by calculating the difference between surveys in 2D area per linear meter, using the simplified isoline to determine the coastal stretch or segment length. Given that the inside limit of the study area (and of the segments) is fixed, the change in area per linear meter corresponded to the mean change in shoreline position.

Volumes, differences in volume and differences in shoreline position between surveys were computed for the beach–dune system of the entire study area and per segment. For the segments, volumes were obtained using a buffer for each segment, with straight limits perpendicular to the beach line (Figure 7). To obtain comparable results for the different beach segments, which vary in length, changes in volume were presented per linear meter of segment length. Furthermore, the slope of the more dynamic part of the beach was computed per segment by considering the beach zone with elevations between 1 and 4 m only, approximately corresponding to the upper beach face. Mean slopes per segment were obtained based on the segment area for this elevation range and the segment length. Finally, the results were analysed by considering beach types, beach orientation and local wind and wave climates.

Figure 7. Segmentation of the coastal stretch for local analysis (only part of the stretch is shown): November 2008 DEM (**a**); contours (1 m above MSL contour in red) (**b**); and the simplified contour line (in black) providing the segments used for analysis, with their length and facing direction (**c**).

3. Results

3.1. Overall Morphodynamics

The beach and dune system of the studied coastal stretch had an approximate area of 1,400,000 m^2 and a volume of 790,000 m^3 above the MSL. Overall, the analysed time series started with a year showing marked accretion and shoreline progradation (Table 1), followed by less accentuated erosional and accretional periods. Shoreline dynamics did not always reflect sedimentary budgets, with some periods showing landward migration when the volume increased and seaward movements when the volume decreased.

Looking at the morphodynamics for approximate seasons, years, 5-year periods and decade of the data series (Table 1), different behaviours were observed for comparable periods. Seasonal changes, which could only be studied for the first one and a half years, showed accretion during the first winter and the summer period, followed by a nearly stable second winter. There was also inter-annual variability, with the early years (2008/2009, 2009/2010) showing increases in volume and shoreline progradation, whereas, during the last (approximate) year (2018/2019), volume was lost, though the shoreline position remained on average stable. Considering the changes observed during 5-year periods, the first (2009/2014) showed a slight increase in volume and shoreline progradation, while the second (2014/2019) showed stable volume and shoreline regression. For the approximate decade (2009/2019), there was a slightly accretional trend in terms of volume but an erosional trend in terms of the shoreline position, which retreated 1.6 m on average.

3.2. Local Morphodynamics

Of the 52 beach segments obtained through beach line simplification (Figure 8), 3 were not analysed because they represented hard defence structures without a beach (i.e., the southern face of the Praia de Canide groyne (Figure 1b), segment 14, and the two faces of the Aguda breakwater (Figure 1c, segments 41 and 42). In the northern part, segments 1 to 7 were predominantly rocky (Figure 1a), with slopes between 7.6° and 20.9°, with a mean slope of 10.6°. To the south, from segment 8 to segment 52, the beaches were sandy with rocky outcrops and slopes between 2.1° and 7.9°, with a mean slope of 5.1°.

Considering a decade (Table 1), sediment budgets and shoreline dynamics varied spatially (Figure 9). Segments showed sedimentary budgets ranging from losses of 213.8 m^3 to gains of 417.0 m^3 per linear meter of coast and shoreline dynamics ranging from 24.0 m regression to 51.5 m progradation. Stretches with relatively stable morphology and others with erosional or accretional trends were found, depending on beach type, shoreline orientation and the presence of defence structures. Rocky beaches (segments 1–7) were the least dynamic. Sandy beaches were the most dynamic, with mean shoreline position changes of 0.0 m and −3.4 m, respectively, and mean sediment budgets of −1.1 m^3 and −2.9 m^3, respectively, per linear meter of coastline and for the studied decade. However, two zones of sandy segments could be distinguished as segments with varying erosional trends in the north and centre of the study area (8–39) and segments with accretional trends in the south (40–52). Patterns differed, however, for the two 5-year periods comprising the decade, except for the northern rocky segments, which remained relatively stable throughout. The accretion in the most southern segments (40–52) and the erosion in the northern sandy segments (12–20) took place during the first 5 years. The erosion of the central-southern segments (24–40) occurred during the second 5 years.

The three yearly periods analysed showed a spatial variability similar to the decadal period, with sediment budgets between −243.7 m^3 and 377.9 m^3 per linear meter and shoreline changes between −29.3 m and 30.7 m. There was a marked interannual variability (Figure 10). The first year (2008/2009) showed accretion for most segments, except for the rocky northern segments and the two most southern segments, which lost volume and retreated. The behaviour in terms of volume was not always analogous to the shoreline change. The second (2009/2010) and third (2018/2019) years analysed showed patterns that were more similar to the decadal pattern. The northern rocky segments were followed by predominantly eroding segments and a few accretionary southern segments, although there was some variability in the central part during the second year.

The seasonal analysis (Figure 11) showed a first overall accretional winter with a pattern similar to that of the first year, followed by a rather stable summer season, marked by accretion in some southern segments (40–44, 50–52). The second winter, however, displayed much more variability, with some central-zone segments losing a lot of volume. Seasons showed segments with sediment budgets between −264.3 m^3 and 301.6 m^3 per linear meter and shoreline changes between −26.1 m and 27.1 m.

Analysis of the segments per beach type and segment orientation (Figures 12–14), confirmed that rocky beaches presented less sedimentary dynamics in general than sandy beaches, as expected, independently of their orientation. Over the 10-year period and on average, rocky segments lost 1.1 m^3 of volume per linear meter (SD = 1.7) and did not change their shoreline position (change = 0.0 m, SD = 1.7). Sandy segments lost 2.9 m^3 of volume per linear meter (SD = 117.6) and retreated 3.4 m (SD = 10.7). Segments neighbouring defence structures, particularly the segment to the north of the Aguda breakwater, showed variable behaviour. For the decadal period (Figure 12), erosion and shoreline retreat tended to be higher for westward oriented segments compared to north-westward oriented segments, although some segments (particularly the most southern segments of the study area) displayed marked accretion despite facing approximately west. The behaviours of the two 5-year periods were distinct, with a second 5-year period characterized by more intense erosion and shoreline retreat, which seems to become more severe as the segments tended towards a western orientation. Yearly patterns (Figure 13) showed accretion in most segments in the first year, independent of their orientation, and higher erosion in the second year, particularly in segments oriented towards the W. For the third analysed yearly period (corresponding to the last year of the data series), the shoreline dynamics were apparently related to the segment's orientation, similar to the behaviour observed for the last 5-year period. Seasonal patterns (Figure 14) were also distinct. Segment orientation seemed of little importance for the first winter period, which was marked by accretion, and for the following summer period, which showed less intense dynamics. The second winter, on the other hand, showed again a tendency towards higher erosion in the segments oriented towards the west. As seen earlier, volumetric changes did not always correspond to shoreline changes.

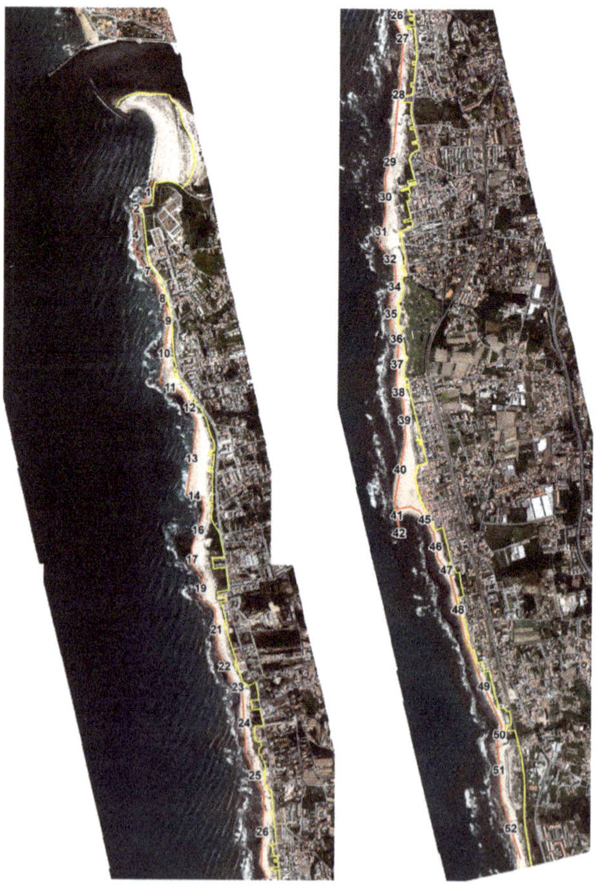

Figure 8. Orthomosaic of the aerial photographs, with the analysed coastal segments (simplified 1 m isoline in red, segments labelled with their ID numbers) and the delimitation of the beach–dune system (yellow line); segment 14 marks the Canide groyne, segments 41 and 42 mark the Aguda breakwater.

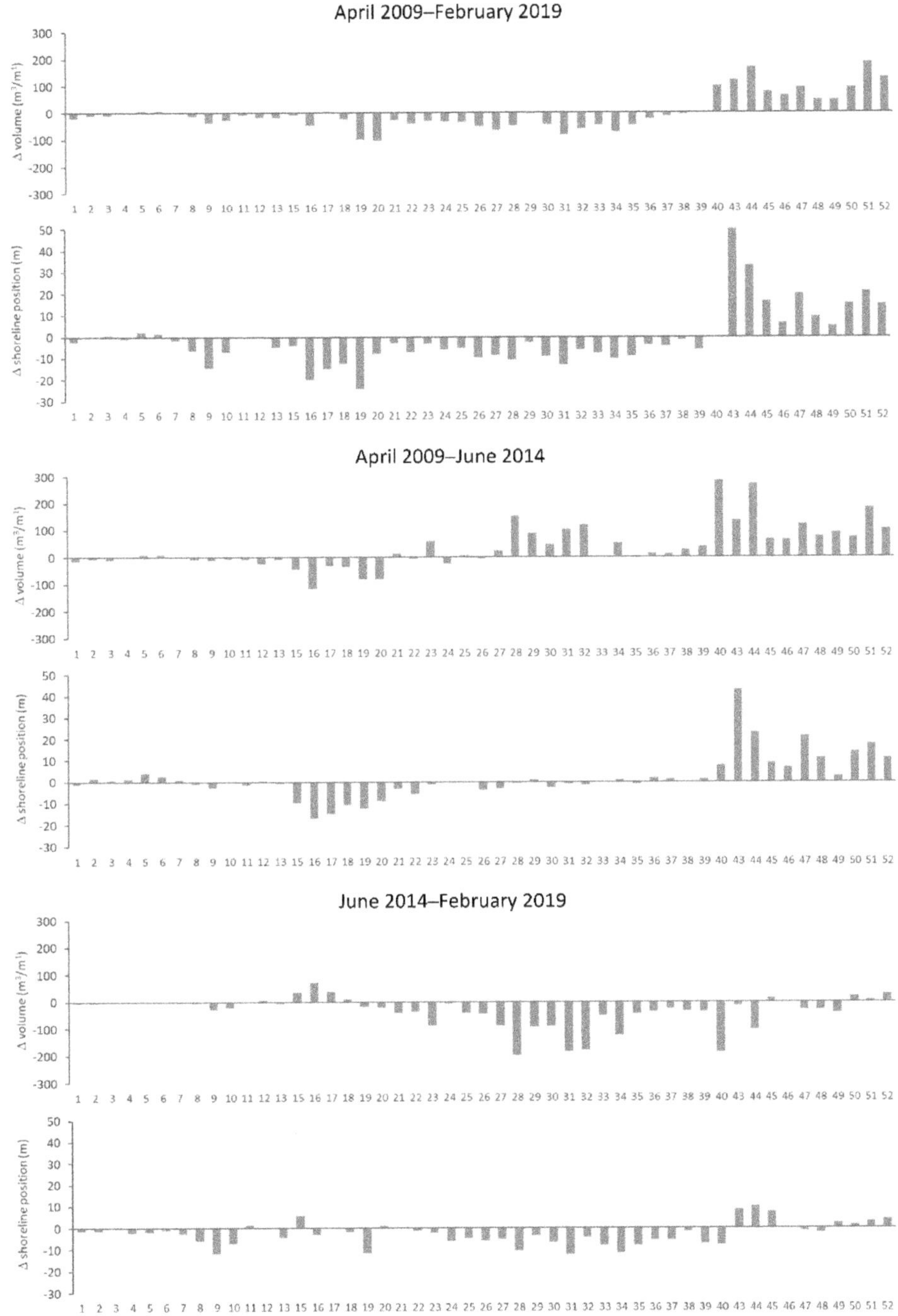

Figure 9. Decadal and 5-yearly changes in volume and shoreline position per segment, presented from north (segment 1) to south (segment 52); segment IDs are presented below the graphs; positive shoreline change values represent seaward migration and negative values represent landward migration.

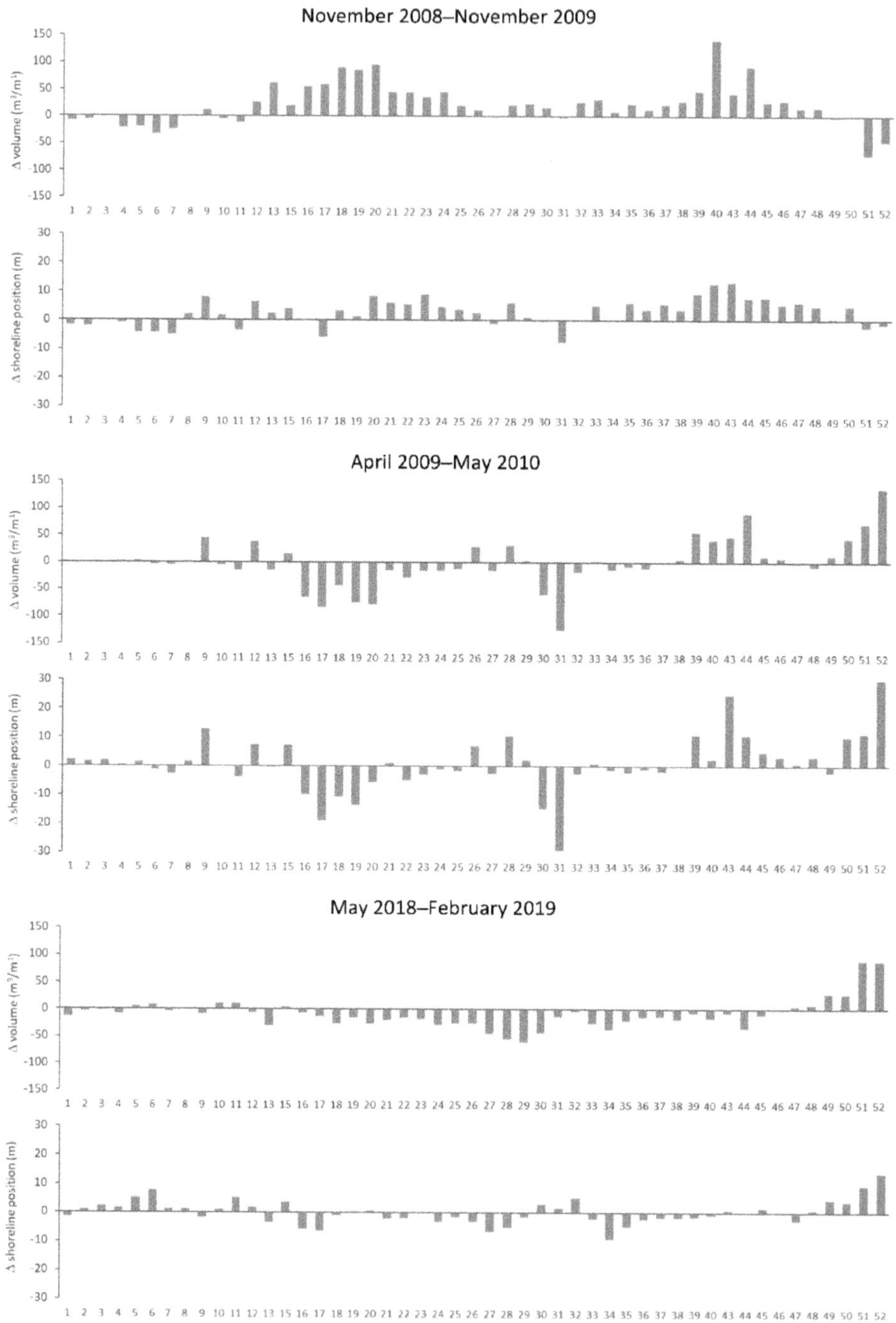

Figure 10. Yearly changes in volume and shoreline position per segment, presented from north (segment 1) to south (segment 52); segment IDs are presented below the graphs; positive shoreline change values represent seaward migration and negative values represent landward migration.

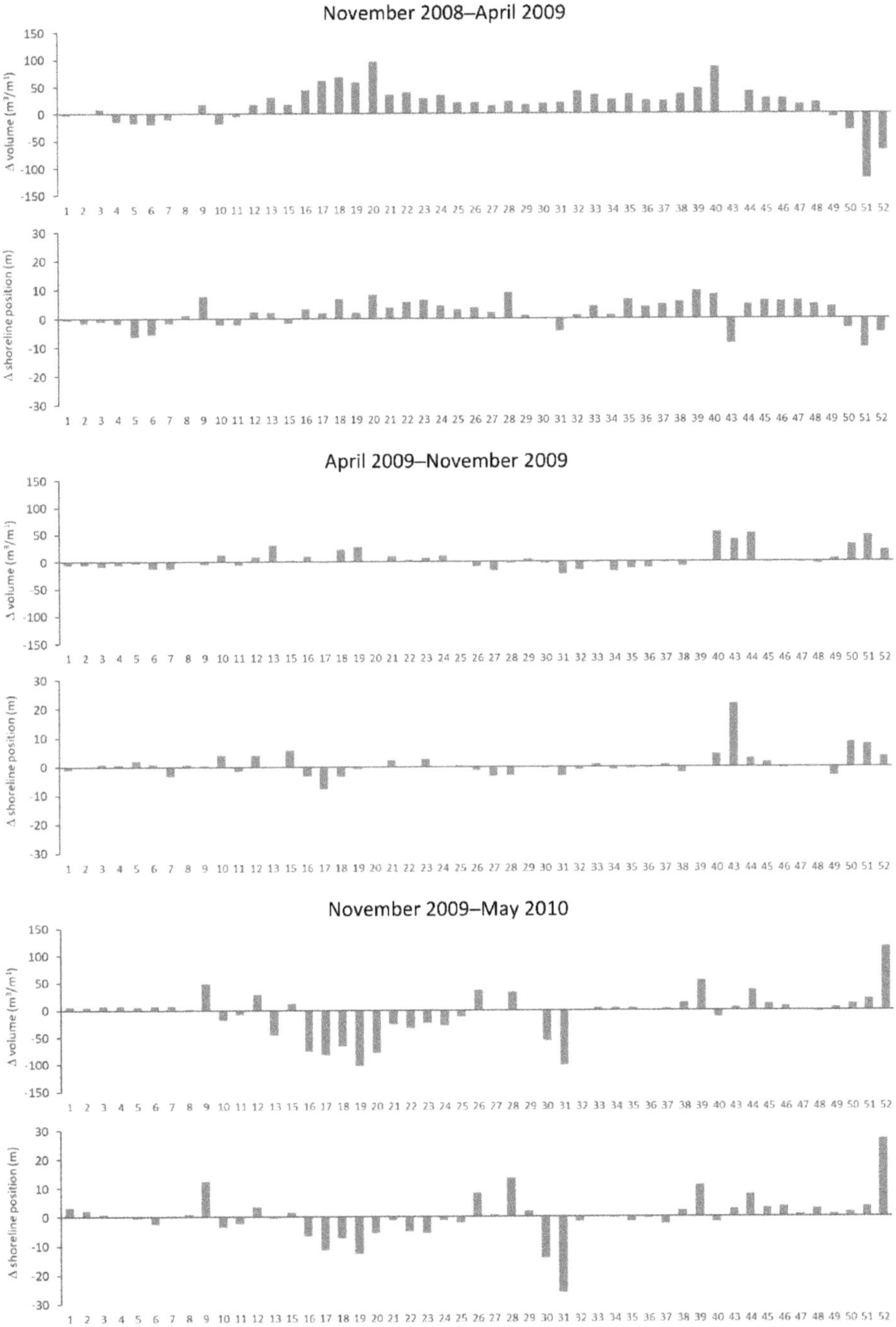

Figure 11. Seasonal changes in volume and shoreline position per segment, presented from north (segment 1) to south (segment 52); segment IDs are presented below the graphs; positive shoreline change values represent seaward migration and negative values represent landward migration.

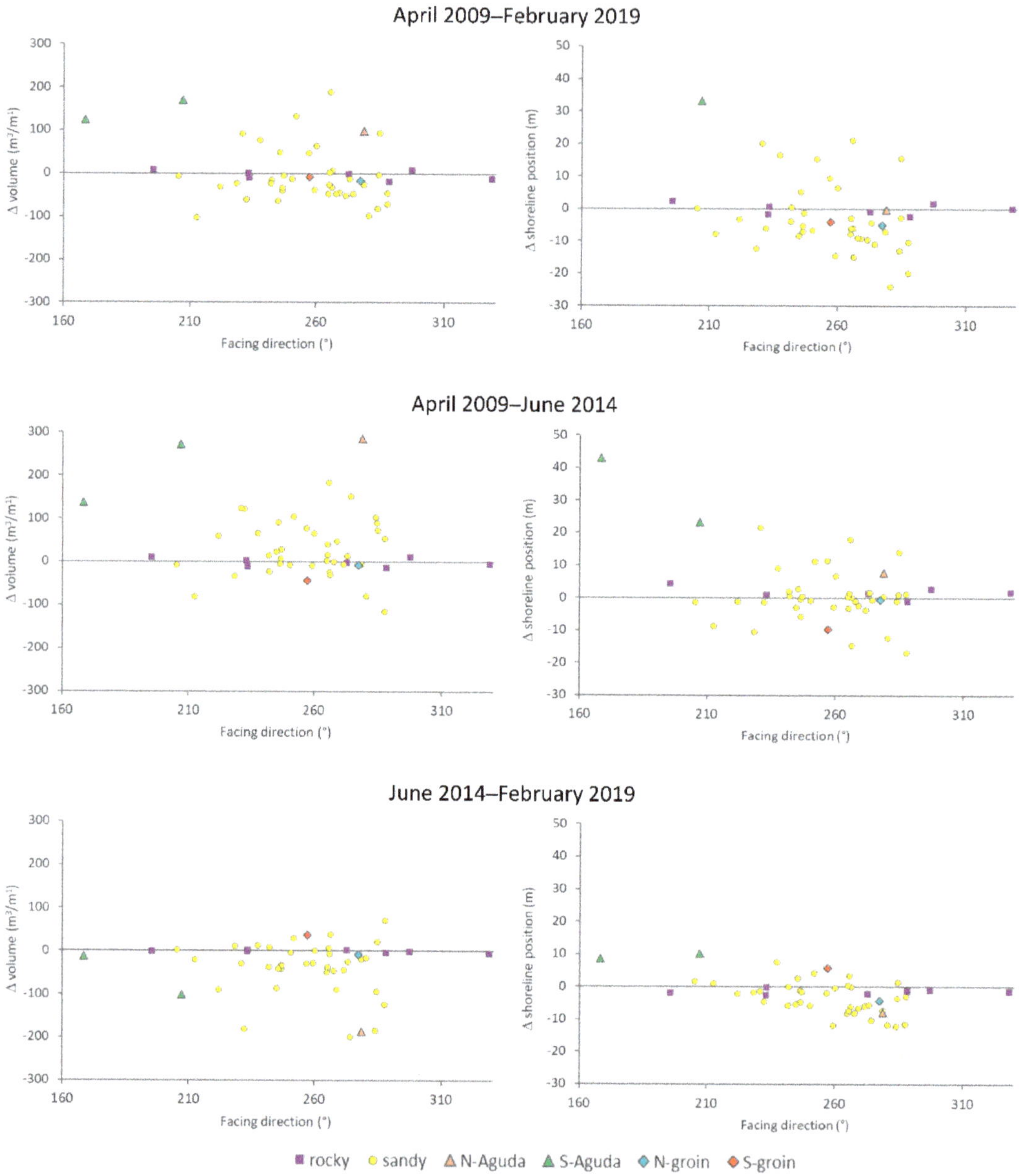

Figure 12. Decadal and 5-yearly changes in volume and shoreline position per segment presented according to segment orientation and marked according to the segment type or its location (N-Aguda: segment 40 north of the Aguda breakwater, S-Aguda: segments 43 and 44 south of the Aguda breakwater, N-groyne: segment 13 north of the groyne, S-groyne: segment 15 south of the groyne).

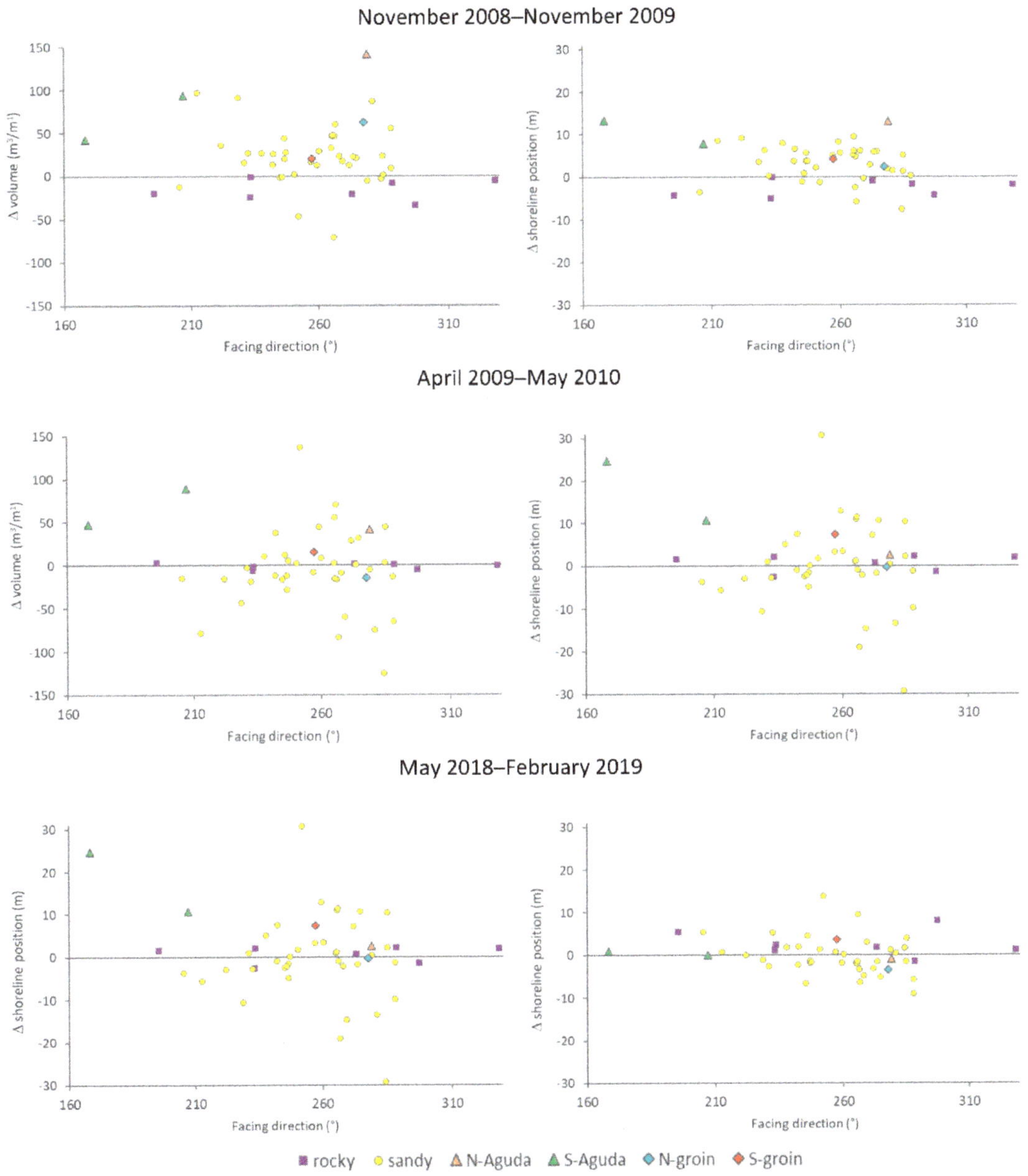

Figure 13. Yearly changes in volume and shoreline position per segment presented according to the segment orientation and marked according to the segment type or its location (N-Aguda: segment 40 north of the Aguda breakwater, S-Aguda: segments 43 and 44 south of the Aguda breakwater, N-groyne: segment 13 north of the groyne, S-groyne: segment 15 south of the groyne).

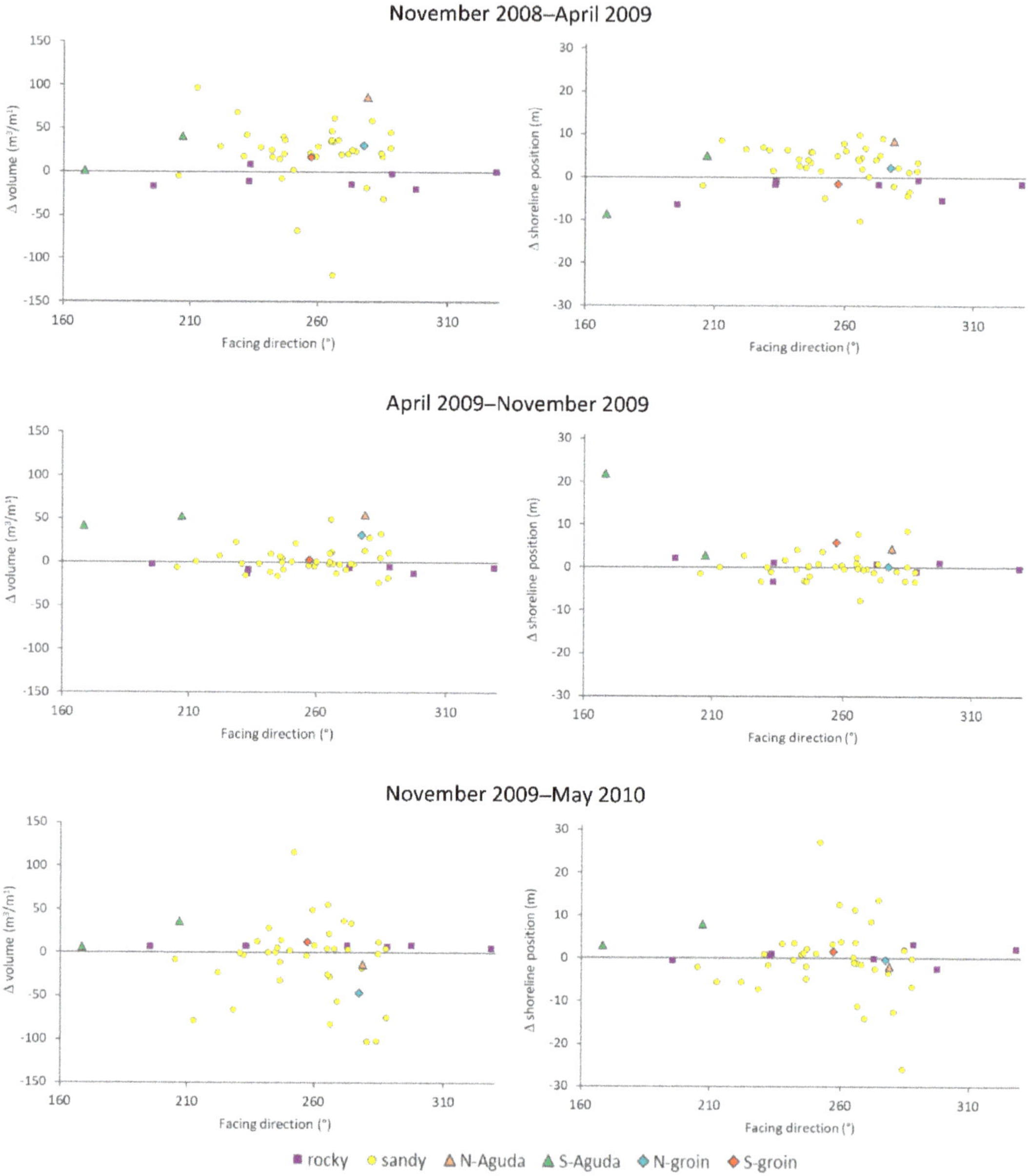

Figure 14. Seasonal changes in volume and shoreline position per segment presented according to segment orientation and marked according to the segment type or its location (N-Aguda: segment 40 north of the Aguda breakwater, S-Aguda: segments 43 and 44 south of the Aguda breakwater, N-groyne: segment 13 north of the groyne, S-groyne: segment 15 south of the groyne).

4. Discussion

4.1. Overall Morphodynamics

For the whole study period (November 2009–February 2019), the beach–dune system studied gained 3.6% in volume and 1.1% in area, with the shoreline moving, on average,

1.1 m seaward (Table 1). However, most of this accretion took place during the first year. After that, the beach volume and shoreline position stabilized. Considering the two analysed 5-year periods (2009/2014 and 2014/2019), the first showed a slight increase in volume and shoreline progradation, whereas the second showed a stable volume but a marked shoreline retreat of 3.1 m, despite protection measures (such as the groyne and breakwater) and artificial beach nourishments. The northwest coast of Portugal has been suffering from a scarce sediment input and a high-energy wave climate, which turns this coast into one of the most active in terms of sediment transport fluxes, with the segment just south of the study area between Espinho and Torreira considered one of the most exposed and vulnerable areas to erosion in the country [18]. Erosion hotspots with 2.5 m shoreline retreat have been found in that region at Vagueira beach [25], despite frequent artificial beach nourishments. A retreat of 1–5 m per year was estimated if no mitigation measures were taken [14]. Therefore, although the studied stretch seems to be rather stable in the medium term (decade), there seems to be a trend towards erosion in recent years (though longer time series are needed to confirm whether this trend is persistent). Furthermore, this trend, if confirmed, in combination with the projected sea-level rise [9], suggests that erosion and shoreline retreat is likely to become an even more serious problem in the future.

There are several possible reasons for these differences between early and later periods of the analysed time series, ranging from differences in forcing variables, such as wave and wind intensities and directions, to differences in sediment availability (dependent on river flows, ocean currents, sediment extraction and beach nourishments). Wave and wind conditions during the period between surveys were therefore analysed to look for patterns that may explain the observed morphodynamics. Offshore wave and onshore wind conditions were quite similar for these two 5-year periods (Figure 3), but the strongest and most damaging winter of the time series occurred in 2013/2014. This winter caused widespread and intense erosion on Atlantic coasts, which took months to some years to recover [47,48]. This may have contributed to the sediment loss between May 2010 and June 2014 and to the recovery between June 2014 and May 2018. Nonetheless, the second 5-year period showed a remarkable and concerning average shoreline retreat.

The difference between early and later periods of the data series, representing more accretional and more erosional behaviours, respectively, was also reflected in the marked inter-annual variation found. During the first year (November 2008 to November 2009), the beach–dune system volume increased by more than 4% and the shoreline moved, on average, more than 3 m seaward. Accretion was less intense for the second yearly period analysed (April 2009 to May 2010), and the (approximate) last year (May 2018 to February 2019) showed erosion instead. Looking at the wave and wind conditions during the yearly periods, patterns diverged, reflecting the general medium-term pattern during the first year, with dominant wave directions from the NW and wind directions from NNW, showing more distributed wave directions and a strong southern wind component during the second year and revealing a more frequent wave direction component from the WNW in the third year. Given that the coastal stretch faced roughly WSW, wave incidence in the last year was less oblique for most segments, possibly reducing sediment transport and deposition via the induced N–S littoral drift.

Seasonal analyses showed two very different winter periods (November 2008 to April 2009 and November 2009 to May 2010, respectively), the first presenting overall accretion and shoreline progradation, while the second presented erosion (though with a progressing shoreline). Looking at the corresponding wave and wind climates, the first winter was characterized by NW–WNW waves and winds distributed from N to W and S, whereas the second winter showed dominant W–NW waves and southern winds, suggesting, again, that more oblique wave incidence promoted accretion and less oblique wave incidence promoted erosion.

However, notice that the temporal changes were analysed based on non-regular surveys. Hence, seasonal and annual analyses do not cover the whole time series, as would be desirable. Moreover, the wave data present gaps caused by equipment failure due to extreme weather and wave conditions, meaning that extreme events may be underrepresented in the data. In terms of wind effects, no clear relationship between wind directions and the morpho-sedimentary dynamics could be identified, possibly because the ERA5-Land wind parameters used are model simulations that may not represent local coastal patterns well [45]. Furthermore, morphodynamics results for the higher beach–dune areas, where wind effects are expected to be most noticeable, are likely to be less precise, particularly in the presence of vegetation. Analyses were based on six DEMs, where dune volume reflects sediments, as well as vegetation, and on one DTM, where vegetation was probably filtered out. In vegetated dunes, a DEM will therefore tend to produce higher volumes than a DTM. Furthermore, dune grass and shrub vegetation will increase in volume during the growing season, affecting seasonal budgets, and in some of the dunes, trees were also found (which may cause interannual, 5-yearly or decadal differences due to growth or felling). Detailed analysis of rocky outcrops showed very similar altitudes in the DEM and the DTM, suggesting that they were comparable. However, the fact that the models may have been processed differently and the lower resolution of the DTM have to be considered as potential error sources.

4.2. Local Morphodynamics

Local morphodynamics often diverged from the general pattern of the coastal stretch as a whole, highlighting the importance of local approaches to erosion monitoring. Morpho-sedimentary dynamics were found to vary temporally, as well as spatially, with patterns depending on beach type and exposure. This was consistent with findings that sediment transport and budget will vary locally along a non-straight coastline due to uneven wave attenuation [27]. Variability between segments was high and the magnitude of change was of a similar order for the different periods analysed, from seasonal to decadal, showing how rapidly coastal morphology can change and recover.

Over the last decade and from north to south, three zones could be distinguished: the rocky segments in the north, which were morphologically stable, as expected; a central zone of sandy beaches with rocky outcrops, displaying variable patterns but a tendency for erosion; and a southern part covering the beaches close to the city of Espinho that tended towards accretion (Figures 7–12). This pattern also occurred during the second winter (2009/2010), the second year (2010), the second 5-year period (2014–2019) and the last year (2018/2019). The first season and first year showed overall accretion in most segments (12–48), and erosion in some rocky segments (4–7) and the most southern segments (51–52), contradicting the pattern of the later periods.

The observed sedimentary dynamics of the rocky shores were unexpected. Detailed analysis showed that differences were found on some sandy patches in the first segments, as well as between rocks, suggesting that there may be artefacts in the DEM due to shadow effects, which are particularly strong in the crevices between rocks (notice that surveys were always done during spring low tide, which may occur at different times of the day, causing more or less shadow) but also near the shoreline, where wave breaking and runup between the rocks and in crevices may have affected DEM precision.

During the first 5-year period, the central part of the study area already showed erosion, but mostly in the upper central segments (16–23). The segments to the south of the Aguda breakwater (43–52) showed accretion, which could be (at least partly) attributed to a protective effect of the breakwater. In the second 5-year period, nearly all of the central segments showed erosion and shoreline retreat. In fact, comparing the two 5-year periods, many segments showed the opposite behaviour, particularly in terms of shoreline position (e.g., 28–31, 34, 40, 44). This suggested a change in the spatial pattern between 2008 and 2019, with a trend towards erosion in the central zone and a trend towards accretion for the southern zone of the studied coastal stretch. Notice, that there is a large groyne just south of the study area at Espinho, which may have contributed to the observed updrift accretion. However, this groyne has been there for decades with more or less success in retaining updrift sediments, suggesting that a change in hydrodynamics and/or sediment supply through increased artificial beach nourishments [34] supported the accretion in recent years.

The temporal variability observed was likely a result of the different wave and wind patterns, which affected segments differently depending on the local conditions—i.e., geology, exposure, and updrift and downdrift structures. The wave incidence direction and wave impact will depend on the direction of the waves and the segment's orientation. Murray and Asthon [49] evaluated the alongshore sediment flux for different wave angles and concluded that coastline segments with different orientations experience different alongshore sediment fluxes. Their results show that the wave angle that leads to the highest value of sediment transport is not necessarily the most oblique wave in shallow water. However, wave incidence will also depend on local hydrodynamics. The wave data used provide offshore wave heights and directions, but onshore, breaking wave characteristics are determined by wave propagation phenomena, such as shoaling and refraction, which depend on the coastal morphology and setting.

Nonetheless, analyses of the erosion/accretion indicators (sediment budget and shoreline dynamics) in relation to beach orientation (Figures 12–14), showed: (i) the general temporal trends, with periods of more accretion or erosion, that were also seen in the overall and the spatial analyses; (ii) differences between beach types, with rocky beaches being relatively stable, independent of orientation, and sandy beaches showing marked variability; and, (iii) an apparent tendency towards more erosion/shoreline retreat for segments facing more western directions (despite the scatter). The latter was particularly the case for the periods of the last 5 years and the last year, suggesting that shorelines with less obliquely incident waves presented a higher erosion risk. This could be explained by the dominance of cross-shore transport towards the subaerial beach during wave storms, with the sand deposits occurring at or behind the rocky outcrops. The most westward-facing segments tended to have the highest volume losses and shoreline retreats, and erosion/shoreline retreat increased with more western-dominant wave directions, as was the case in the last year. For the 5-year periods, the wave roses seemed rather similar (as could be expected for such a long period), but the WNW component was higher for the second 5-year period than for the first.

In general, more exposed segments (e.g., segments 17, 19, 31; Figure 5) showed frequent sediment loss and shoreline retreat, yet many apparently protected (embayed) segments retreated too (e.g., 18, 20, 23). This may have been due to wave diffraction. The main wave crest orientation was from the NW, inducing a drift current from north to south, which was in some areas inverted due to the presence of obstacles that promoted wave diffraction, causing downdrift erosion. Shoreline retreat was also found downdrift the Canide groyne, which is a typical effect of these structures that are intended to trap sediments in updrift areas. However, even the segments above the groyne did retreat, though less than those downdrift. Segments north (39, 40), as well as south, of the Aguda breakwater (43, 44) moved from accretional during the first years of the survey series to erosional during the last 5 years. This may have been due to changes in wave action, as explained above, but also due to sand beach management operations that removed sediments from north of the breakwater to nourish the beaches of the city Espinho in the south.

There were two flood events in the Douro river during the study period, one in February 2010 and another in March 2018, with peak river flows of more than 6000 m^3/s and 4000 m^3/s, respectively. These may have affected longshore sediment transport to the sectors south of the river outlet, as the periods comprising these flood events showed consistent erosion in the intermediate segments (16–25) of the study area. However, a detailed hydrodynamic analysis would be needed to confirm this hypothesis.

An interesting issue was that volumetric changes did not always correspond to shoreline dynamics, suggesting sediment transport from the beach to the dunes. There are different modes of beach dynamics—shoreline advance/retreat (translation modes) and beach steepening/flattening (rotation modes)—that can occur simultaneously and are often linked to sediment redistribution within the beach–dune system [50]. For instance, segments 39 and 40 (located to the north of the Aguda breakwater) presented shoreline regression and stability (−6.1 m and −0.2 m, respectively) but volume stability and accretion (2.1 m^3/m and 265.4 m^3/m, respectively) for the decade. Simultaneously, a steepening in the profiles was observed, with the beach slope increasing from 6.5 to 7.7° and from 3.7 to 6.6°, respectively. Notice that the enormous difference in volume in segment 40 was also due to the beach width. Larger segments will have more surface and hence the capacity to present larger sediment budgets.

5. Conclusions

Concluding, despite the limitations of the data time series, the present study demonstrated how local conditions interact with meteo-ocean conditions in shaping local morpho-sedimentary dynamics, stressing the need for local approaches to monitoring and erosion risk analyses. Dynamics varied markedly in space, with patterns depending on beach type and exposure, as well as on the presence of coastal defence structures that alter local hydrodynamics and, therefore, sediment transport and deposition.

The analysed coastal stretch suffered an average 1.6 m retreat of its shoreline in a decade with a slight increase in its volume (0.6%). However, analyses of shorter periods (annual and five years) revealed greater shoreline retreat and sediment budget values, demonstrating the high temporal dynamics of this coastal stretch and stressing the need for longer monitoring periods (more than 10 years) to assess the main trends of coastal morphodynamics.

Although the studied stretch seemed to be rather stable in the medium term (decade), the northwest coast of Portugal has been suffering from a scarce sediment input and a high-energy wave climate, which turns this coast into one of the most active in terms of sediment transport fluxes. If a projected mean sea-level rise of the order of 1 m by 2100 is confirmed, shoreline retreat will become a major problem in the next few decades.

Coastal management should therefore be based on structural monitoring programs, with surveys at adequate temporal and spatial scales to understand local dynamics and to be able to apply adequate erosion mitigation measures in the right places. Furthermore, given that many coastal systems show sediment deficits, mainly due to anthropogenic interventions in rivers, such as dams, and sediment extraction, sediment budgets should play a central role in the development of coastal defence strategies.

Author Contributions: Conceptualization, H.G., J.P. and A.B.; methodology, A.B., J.A.G. and J.P.; software, J.A.G. and A.B.; validation, J.A.G.; formal analysis, J.A.G., A.B. and I.I.; investigation, H.G., J.P., J.A.G., A.B. and I.I.; resources, H.G., J.P. and J.A.G.; data curation, J.A.G.; writing—original draft preparation, A.B.; writing—review and editing, A.B., J.A.G., I.I., H.G., J.P. and L.B.; visualization, A.B.; supervision, L.B.; project administration, H.G., J.P., L.B. and A.B.; funding acquisition, H.G., J.P., L.B. and A.B. All authors have read and agreed to the published version of the manuscript.

Funding: This research was partially supported by the Strategic Funding UIDB/04423/2020 and UIDP/04423/2020 through national funds provided by the FCT—Foundation for Science and Technology and European Regional Development Fund (ERDF). This work was further funded by the projects ATLANTIDA (ref. NORTE-01-0145-FEDER-000040) and Ocean3R (ref. NORTE-01-0145-FEDER-000064), both of which were supported by the North Portugal Regional Operational Programme (NORTE2020) under the PORTUGAL 2020 Partnership Agreement and through the European Regional Development Fund (ERDF).

Institutional Review Board Statement: Not applicable.

Informed Consent Statement: Not applicable.

Acknowledgments: Data were supplied by the European Union MarRISK project: Adaptación costera ante el Cambio Climático: conocer los riesgos y aumentar la resiliencia (0262_MarRISK_1_E) through EP INTERREG V A España-Portugal (POCTEP) program and the project INNOVMAR-Innovation and Sustainability in the Management and Exploitation of Marine Resources (NORTE-01-0145-FEDER-

000035, within Research Line ECOSERVICES), supported by NORTE 2020, under the PORTUGAL 2020 Partnership Agreement, through the ERDF.

Conflicts of Interest: The authors declare no conflict of interest.

References

1. Nel, R.; Campbell, E.E.; Harris, L.; Hauser, L.; Schoeman, D.S.; McLachlan, A.; du Preez, D.R.; Bezuidenhout, K.; Schlacher, T.A. The status of sandy beach science: Past trends, progress, and possible futures. *Estuar. Coast. Shelf Sci.* **2014**, *150*, 1–10. [CrossRef]
2. Newton, A.; Weichselgartner, J. Hotspots of coastal vulnerability: A DPSIR analysis to find societal pathways and responses. *Estuar. Coast. Shelf Sci.* **2014**, *140*, 123–133. [CrossRef]
3. Jones, A.; Phillips, M. *Disappearing Destinations: Climate Change and Future Challenges for Coastal Tourism*; CABI: Oxfordshire, UK, 2010; p. 273.
4. Lobeto, H.; Menendez, M.; Losada, I.J. Future behavior of wind wave extremes due to climate change. *Sci. Rep.* **2021**, *11*, 7869. [CrossRef] [PubMed]
5. Syvitski, J.P.M.; Vörösmarty, C.J.; Kettner, A.J.; Green, P. Impact of humans on the flux of terrestrial sediment to the global coastal ocean. *Science* **2005**, *308*, 376–380. [CrossRef] [PubMed]
6. European Communities. *Living with Coastal Erosion in Europe–Sediment and Space for Sustainability*; Office for Official Publications of the European Communities: Luxembourg, 2004.
7. Pollard, J.A.; Spencer, T.; Brooks, S.M. The interactive relationship between coastal erosion and flood risk. *Prog. Phys. Geogr.* **2019**, *43*, 574–585. [CrossRef]
8. Vousdoukas, M.I.; Ranasinghe, R.; Mentaschi, L.; Plomaritis, T.A.; Athanasiou, P.; Luijendijk, A.; Feyen, L. Sandy coastlines under threat of erosion. *Nat. Clim. Chang.* **2020**, *10*, 260–263. [CrossRef]
9. Antunes, C. Assessment of sea level rise at West Coast of Portugal Mainland and its projection for the 21st century. *J. Mar. Sci. Eng.* **2019**, *7*, 61. [CrossRef]
10. Turner, R.K.; Burgess, D.; Hadley, D.; Coombes, E.; Jackson, N. A cost-benefit appraisal of coastal managed realignment policy. *Glob. Environ. Chang.* **2007**, *17*, 397–407. [CrossRef]
11. Williams, A.T.; Rangel-Buitrago, N.; Pranzini, E.; Anfuso, G. The management of coastal erosion. *Ocean Coast. Manag.* **2018**, *156*, 4–20. [CrossRef]
12. Lima, M.; Coelho, C.; Veloso-Gomes, F.; Roebeling, P. An integrated physical and cost-benefit approach to assess groins as a coastal erosion mitigation strategy. *Coast. Eng.* **2020**, *156*, 103614. [CrossRef]
13. Rangel-Buitrago, N.; Williams, A.T.; Anfuso, G. Hard protection structures as a principal coastal erosion management strategy along the Caribbean coast of Colombia. A chronicle of pitfalls. *Ocean Coast. Manag.* **2018**, *156*, 58–75. [CrossRef]
14. Stronkhorst, J.; Huisman, B.; Giardino, A.; Santinelli, G.; Santos, F.D. Sand nourishment strategies to mitigate coastal erosion and sea level rise at the coasts of Holland (The Netherlands) and Aveiro (Portugal) in the 21st century. *Ocean Coast. Manag.* **2018**, *156*, 266–276. [CrossRef]
15. Granja, H.; Bastos, L.; Pinho, J.; Gonçalves, J.; Henriques, R.; Bio, A.; Magalhães, A. *Small Harbours Risks: Lowering for Fishery and Increasing Erosion. The Case of Portinho da Aguda (NW Portugal)*; Littoral 2010; EDP Sciences: Les Ulis, France, 2011; p. 09003. [CrossRef]
16. Vieira, B.F.V.; Pinho, J.L.S.; Barros, J.A.O.; Antunes do Carmo, J.S. Hydrodynamics and morphodynamics performance assessment of three coastal protection structures. *J. Mar. Sci. Eng.* **2020**, *8*, 175. [CrossRef]
17. Semeoshenkova, V.; Newton, A. Overview of erosion and beach quality issues in three Southern European countries: Portugal; Spain and Italy. *Ocean Coast. Manag.* **2015**, *118*, 12–21. [CrossRef]
18. Marinho, B.; Coelho, C.; Hanson, H.; Tussupova, K. Coastal management in Portugal: Practices for reflection and learning. *Ocean Coast. Manag.* **2019**, *181*, 104874. [CrossRef]
19. Guillén, J.; Stive, M.J.F.; Capobianco, M. Shoreline evolution of the Holland coast on a decadal scale. *Earth Surf. Process. Landf.* **1999**, *24*, 517–536. [CrossRef]
20. Van Rijn, L.C. Coastal erosion and control. *Ocean Coast. Manag.* **2011**, *54*, 867–887. [CrossRef]
21. Casella, E.; Rovere, A.; Pedroncini, A.; Stark, C.P.; Casella, M.; Ferrari, M.; Firpo, M. Drones as tools for monitoring beach topography changes in the Ligurian Sea (NW Mediterranean). *Geo-Mar. Lett.* **2016**, *36*, 151–163. [CrossRef]
22. Lemke, L.; Miller, J.K. Role of storm erosion potential and beach morphology in controlling dune erosion. *J. Mar. Sci. Eng.* **2021**, *9*, 1428. [CrossRef]
23. Hinestroza-Mena, K.M.; Toro, V.G.; Londoño-Colorado, G.S.; Chávez, V.; García-Blanco, J.K.; Silva, R. Fine spatial scale, frequent morphological monitoring of urbanised beaches to improve coastal management. *J. Mar. Sci. Eng.* **2021**, *9*, 550. [CrossRef]
24. Flor-Blanco, G.; Alcántara-Carrió, J.; Jackson, D.W.T.; Flor, G.; Flores-Soriano, C. Coastal erosion in NW Spain: Recent patterns under extreme storm wave events. *Geomorphology* **2021**, *387*, 107767. [CrossRef]
25. Ferreira, A.M.; Coelho, C. Artificial nourishments effects on longshore sediments transport. *J. Mar. Sci. Eng.* **2021**, *9*, 240. [CrossRef]

26. Murray, T.; Strauss, D.; Da Silva, G.V.; Wharton, C. Spatial variability in beach morphology with respect to wave exposure along a zeta-shaped coastline. In Proceedings of the 9th International Conference on Coastal Sediments 2019, Tampa/St. Petersburg, FL, USA, 27–31 May 2019; pp. 632–645. [CrossRef]
27. Vieira Da Silva, G.; Murray, T.; Strauss, D. Longshore wave variability along non-straight coastlines. *Estuar. Coast. Shelf Sci.* **2018**, *212*, 318–328. [CrossRef]
28. Bauer, B.O.; Davidson-Arnott, R.G.D.; Walker, I.J.; Hesp, P.A.; Ollerhead, J. Wind direction and complex sediment transport response across a beach-dune system. *Earth Surf. Process. Landf.* **2012**, *37*, 1661–1677. [CrossRef]
29. Sloss, C.R.; Hesp, P.; Shepherd, M. Coastal Dunes: Aeolian Transport. *Nat. Educ. Knowl.* **2012**, *3*, 21. Available online: https://www.nature.com/scitable/knowledge/library/coastal-dunes-aeolian-transport-88264671/ (accessed on 15 April 2022).
30. Bauer, B.O.; Davidson-Arnott, R.G.D.; Hesp, P.A.; Namikas, S.L.; Ollerhead, J.; Walker, I.J. Aeolian sediment transport on a beach: Surface moisture, wind fetch, and mean transport. *Geomorphology* **2009**, *105*, 106–116. [CrossRef]
31. Bio, A.; Bastos, L.; Granja, H.; Pinho, J.L.S.; Goncalves, J.A.; Henriques, R.; Madeira, S.; Magalhaes, A.; Rodrigues, D. Methods for coastal monitoring and erosion risk assessment: Two Portuguese case studies. *J. Integr. Coast. Zone Manag.* **2015**, *15*, 47–63. [CrossRef]
32. Taveira Pinto, F. The practice of coastal zone management in Portugal. *J. Coast. Conserv.* **2004**, *10*, 147–158. [CrossRef]
33. APA; ARH Norte; ARH Centro; ARH Tejo e Oeste; ARH Alentejo; ARH Algarve. *Plano de Ação Litoral XXI*; Agência Portuguesa do Ambiente: Amadora, Portugal, 2017; pp. 1–128.
34. Pinto, C.A.; Silveira, T.M.; Teixeira, S.B. Beach nourishment practice in mainland Portugal (1950–2017): Overview and retrospective. *Ocean Coast. Manag.* **2020**, *192*, 105211. [CrossRef]
35. Cruz, J. *Ocean Wave Energy: Current Status and Future Perspectives*; Green Energy and Technology; Springer: Berlin/Heidelberg, Germany, 2008. [CrossRef]
36. Viitak, M.; Avilez-Valente, P.; Bio, A.; Bastos, L.; Iglesias, I. Evaluating wind datasets for wave hindcasting in the NW Iberian Peninsula coast. *J. Oper. Oceanogr.* **2021**, *14*, 152–165. [CrossRef]
37. Relvas, P.; Barton, E.D.; Dubert, J.; Oliveira, P.B.; Peliz, Á.; da Silva, J.C.B.; Santos, A.M.P. Physical oceanography of the western Iberia ecosystem: Latest views and challenges. *Prog. Oceanogr.* **2007**, *74*, 149–173. [CrossRef]
38. Benavent, M.; Arnoso, J.; Montesinos, F.G. Regional ocean tide loading modelling around the Iberian Peninsula. *J. Geodyn.* **2009**, *48*, 132–137. [CrossRef]
39. Lorenzo, M.N.; Taboada, J.J.; Gimeno, L. Links between circulation weather types and teleconnection patterns and their influence on precipitation patterns in Galicia (NW Spain). *Int. J. Climatol.* **2008**, *28*, 1493–1505. [CrossRef]
40. Ramos, A.M.; Ramos, R.; Sousa, P.; Trigo, R.M.; Janeira, M.; Prior, V. Cloud to ground lightning activity over Portugal and its association with circulation weather types. *Atmos. Res.* **2011**, *101*, 84–101. [CrossRef]
41. Gonçalves, J.; Bastos, L.; Pinho, J.; Granja, H. Digital aerial photography to monitor changes in coastal areas based on direct georeferencing. In Proceedings of the 5th EARSeL Workshop on Remote Sensing of the Coastal Zone, Prague, Czech Republic, 1–3 June 2011. Available online: http://www.conferences.earsel.org/abstract/show/2689 (accessed on 12 June 2014).
42. Gonçalves, J.A.; Bastos, L.; Madeira, S.; Magalhães, A.; Bio, A. Three-dimensional data collection for coastal management–efficiency and applicability of terrestrial and airborne methods. *Int. J. Remote Sens.* **2018**, *39*, 9380–9399. [CrossRef]
43. Agisoft LLC. Agisoft Metashape User Manual Professional Edition, Version 1.8; Agisoft Metashape: 2022. Available online: https://www.agisoft.com/pdf/metashape-pro_1_8_en.pdf (accessed on 15 April 2022).
44. Muñoz Sabater, J. *ERA5-Land Hourly Data from 1981 to Present*; Copernicus Climate Change Service (C3S) Climate Data Store (CDS): Sherfield, UK, 2019. [CrossRef]
45. Gualtieri, G. Reliability of era5 reanalysis data for wind resource assessment: A comparison against tall towers. *Energies* **2021**, *14*, 4169. [CrossRef]
46. Douglas, D.H.; Peucker, T.K. Algorithms for the reduction of the number of points required to represent a digitized line or its caricature. *Cartogr. Int. J. Geogr. Inf. Geovis.* **1973**, *10*, 112–122. [CrossRef]
47. Dodet, G.; Castelle, B.; Masselink, G.; Scott, T.; Davidson, M.; Floc'h, F.; Jackson, D.; Suanez, S. Beach recovery from extreme storm activity during the 2013–14 winter along the Atlantic coast of Europe. *Earth Surf. Process. Landf.* **2019**, *44*, 393–401. [CrossRef]
48. Nicolae-Lerma, A.; Ayache, B.; Ulvoas, B.; Paris, F.; Bernon, N.; Bulteau, T.; Mallet, C. Pluriannual beach-dune evolutions at regional scale: Erosion and recovery sequences analysis along the Aquitaine coast based on airborne LiDAR data. *Cont. Shelf Res.* **2019**, *189*, 103974. [CrossRef]
49. Murray, A.; Asthon, A. Instability and finite-amplitude self-organization of large-scale coastline shapes. *Philos. Trans. R. Soc. A* **2013**, *371*, 20120363. [CrossRef]
50. Montaño, J.; Coco, G.; Chataigner, T.; Yates, M.; Le Dantec, N.; Suanez, S.; Cagigal, L.; Floc'h, F.; Townend, I. Time-Scales of a Dune-Beach System and Implications for Shoreline Modeling. *J. Geophys. Res. Earth Surf.* **2021**, *126*, e2021JF006169. [CrossRef]

applied sciences

MDPI

Article

Monitoring of Recovery Process at Yeongildae Beach, South Korea, Using a Video System

Jung-Eun Oh [1], Weon-Mu Jeong [1], Kyong-Ho Ryu [1], Jin-Young Park [2] and Yeon-S. Chang [1,]*

[1] Maritime ICT R&D Center, Korea Institute of Ocean Science and Technology, 385 Haeyang-ro, Yeongdo-gu, Busan 49111, Korea; jungeunoh@kiost.ac.kr (J.-E.O.); wmjeong@kiost.ac.kr (W.-M.J.); ryukh75@kiost.ac.kr (K.-H.R.)

[2] Department of Coastal Management, GeoSystem Research Corporation, 172 LS-ro, Gunpo 15807, Korea; pjynuri@geosr.com

* Correspondence: yeonschang@kiost.ac.kr

Featured Application: The video monitoring system used in this study can be applied for monitoring coastal processes of erosion and recovery and thus can be useful in designing beach protection/prevention plans for damage by storm waves.

Abstract: Once a beach is eroded by storm waves, it is generally recovered under milder wave conditions. To prevent or reduce damage, it is therefore important to understand the characteristics of the site-specific recovery process. Here, we present the results, based on a data set from a video monitoring system and wave measurements, of the recovery process in a pocketed beach located inside a bay where the shoreline retreated harshly (~12 m, on average, of beach width) during Typhoon TAPAH (T1917) in September 2019. It took about 1.5 years for the beach to be recovered to the level before the typhoon. During this period, the erosion and accretion were repeated, with the pattern highly related to the wave power (P_w); most of the erosion occurred when P_w became greater than 30 kWatt/m, whereas the accretion prevailed when P_w was no greater than 10 kWatt/m. The recovery pattern showed discrepancies between different parts of the beach. The erosion during storm events was most severe in the southern part, whereas the northern shoreline did not significantly change even during TAPAH (T1917). In contrast, the recovery process occurred almost equally at all locations. This discrepancy in the erosion/accretion process was likely due to human intervention, as a shadow zone was formed in the northern end due to the breakwaters, causing disequilibrium in the sediment transport gradient along the shore. The results in this study could be applied in designing the protection plans from severe wave attacks by effectively estimating the size of coastal structures and by correctly arranging the horizontal placement of such interventions or beach nourishment. Although the application of these results should be confined to this specific site, the method using wave energy parameters as criteria can be considered in other areas with similar environments, for future planning of beach protection.

Keywords: beach erosion; storm waves; recovery process; video monitoring; pocket beach

Citation: Oh, J.-E.; Jeong, W.-M.; Ryu, K.-H.; Park, J.-Y.; Chang, Y.-S. Monitoring of Recovery Process at Yeongildae Beach, South Korea, Using a Video System. *Appl. Sci.* **2021**, *11*, 10195. https://doi.org/10.3390/app112110195

Academic Editors: Ricardo Castedo, Miguel Llorente Isidro and David Moncoulon

Received: 30 September 2021
Accepted: 28 October 2021
Published: 30 October 2021

1. Introduction

Beaches are important for nearby communities, as they provide areas for recreational activities. They are sources of tourism, and thus well-groomed beaches with a nice coastal environment attract many visitors, even from abroad. Therefore, maintaining these beaches in good condition is necessary to improve their value. Apart from economic reasons, beaches are also environmentally important because they are buffer zones between the ocean and land, protecting on-land facilities by mitigating wave energy. Because of this buffer role, beaches themselves are exposed to attacks of extreme waves under storms such as tropical cyclones. When the wave energy is released in the surf zone due to the breaking of waves, sediments in the beach face and the nearshore seabed are dynamically transported. Therefore, one of the serious types of damage as a result of storm attacks is beach erosion.

Under energetic wave conditions, nearshore sediments can move offshore rapidly due to wave-induced underwater currents [1], and the shoreline can significantly retreat, even for a short time of hours or days [2,3]. The volume of erosion is directly related to the maximum wave height during the storm, as the near-bed shear stress initiates sediment motion proportional to the square of the wave height, and more sediment moves under higher shear stress once it exceeds a critical value. In addition, the wave period that contributes to the power of storm waves and the wave direction that influences the distribution of wave energy along the shore are important factors in the amount of eroded sediment. In addition to the wave parameters, the duration of a storm [4] or a sequence of consecutive storms [5,6] could play a significant role in the erosional process.

Once a storm event ends, onshore sediment motion prevails, under mild wave conditions, due to wave nonlinearity such as skewness and asymmetry [7], which results in the natural process of shoreline recovery following storm events. Therefore, studies on the recovery process after storms are as important as studies on the erosional process during storms [8]. The time scales of the post-storm nearshore morphological recovery are varying, from days [5,8] to months [9]. However, large storms can cause rapid erosion locally, from which recovery may take many years or even decades if the impacts are sufficiently large [10]. In extreme cases, the damage can be irreversible if the coastal structures that are built to protect the shore are destroyed by the storms [11].

In the recovery process, the interval between multiple storms is important, as it is inversely related to the storm damage. If the interval between two consecutive storms is less than the recovery time of the beach, the next storm could cause greater damage to the beach [12]. For example, storm clusters with small return periods could impose similar erosion impacts to a single storm that had a much longer return period (i.e., much greater storm). If a beach reached equilibrium after the attack of the first storm, additional erosion would occur only when the intensity of the following storms exceeded that of the first one [5]. Therefore, the recovery after storm events is a complex process, not only depending on the post-storm recovery time but also depending on the intensity of the following storms.

The studies on beach recovery after storms have used various approaches. One of the traditional methods is the direct measurement of the elevation and water depth along the lines set perpendicular to the coastline, which is called profiling. The profiling would be useful for long-term monitoring for beach processes with a low initial cost [13]. The profile data could be also used to validate simulation results calculated from numerical models [14]. The model approach is usefully applied for the prediction of beach recovery for either short-term or long-term processes. Short-term prediction models such as XBeach [15] calculate the rapid changes in the seabed and dunes during storms and the post-storm recovery process in the time scale of months [9]. Due to the high computational cost, however, the application of process models is still restricted for year-scale predictions. For such long-term predictions, line models [16] based on the one-line theory [17] or statistical models such as Monte Carlo have been applied [18].

Another tool that is useful for the long-term observation of the beach process is remote sensing. The satellite imagery can be regularly collected over a long period, and it has been used to derive shorelines and identify their spatial patterns [19,20]. However, it is sensitive to cloud cover and has a limited temporal resolution. Similarly, aerial photos are useful to detect geographical changes over decades, although they cannot provide detailed information on detected variation [21]. On the other hand, airborne and in situ LiDAR also provides detailed information on the beach geography, for comparison and validation of other measurements [22,23]. Despite their advantages, the application of aerial photos and LiDAR data is restricted in measuring the beach recovery process continuously due to the high cost. For this reason, Video Monitoring Systems (VMSs) installed at beaches have been successfully applied to observe the recovery process in specific sites [8,9]. The VMS is not only useful for long-term monitoring [24] but also for short-term observation of the recovery process after storms [5] and for estimation of the wave parameters [25] as well as wave-induced currents [26].

In this study, we also investigated a recovery process after an attack of storm waves, when a severe erosion occurred, by analyzing the shoreline variations in a pocketed beach located inside a bay in the southeast coast of Korea. Our study focused on the conditions that determined the erosional or depositional process at the beach using data sets from a VMS and wave measurements. As previously described, it is a natural process that beaches undergo erosion by attacks of storm waves, and then they are slowly recovered under milder conditions. However, the criteria that decide the shoreline erosion and accretion are still unclear. Although such information obtained at a beach is site-specific and thus cannot be generally applied for other beaches, the analytical methods may still be applied to beaches with different conditions. In addition, if a criterion is provided for a specific beach, then plans can be designed to protect the beach from severe damage that can lead to a loss of its value.

Another focus was examination of the locality in the recovery pattern between different locations along the beach. Previous studies using the VMS generally focused on the time scale of the recovery process [8]. However, it is also important to analyze the discrepancy in the recovery process between different locations, if it exists, because such disequilibrium may lead to the concentration of damage at specific locations if subsequent storms attack the site before fully recovered. The results of the present study are, therefore, useful not only to understand the criteria for erosion/accretion processes at the study site, but also to analyze the causes of disequilibrium during the recovery process. These outcomes are then applied to suggest measures for conservation of the beach, considering the characteristic erosion and recovery pattern, generally or locally.

The paper is organized as follows. The information of the study site and the damage as a result of Typhoon TAPAH (T1917) are described in Section 2.1. The VMS used in this study and its data are introduced in Section 2.2. The general trend of the beach process is described in Section 3.1, and the locality (discrepancies in the process between areas of the beach) is analyzed in Section 3.2. The discussion on the results is provided in Section 4, and the conclusion of the study in Section 5.

2. Materials and Methods

2.1. Study Sites

Yeongildae Beach is in the southeast of the Korean Peninsula (Figure 1a), inside Yeongil Bay, which faces northeast with a mouth width of ~10 km (Figure 1b). Due to its location in the west corner of the bay, the beach has been protected from severe erosion because the wave energy is generally attenuated when reaching the site. Yeongildae Beach is a ~1.7 km long sandy beach where the sediment size ranges from about 0.15 to 0.35 mm. It is a pocketed beach facing ESE (East-Southeast), bounded by two ports—Pohang Port at the southern end and Duho Port at the northern end. Due to the breakwaters that were built for these ports, the sediments were expected to remain within the coastal cell without loss in the longshore direction. However, the beach has been classified as an erosive beach because input sources of sediment to the beach have been lost due to the construction of the ports. In addition, after the breakwaters of Duho Port were constructed, a shadow zone could have been formed in the lee area of the breakwaters, reducing the wave energy by wave diffraction [27]. The shadow zone has caused redistribution of sediments along the beach shore so that the shoreline width has become narrower near the Yeongil Bridge area but thicker at both ends of the beach near the two ports (Figure 1b). The aerial photograph and satellite image in Figure 2 compare the shorelines measured in 1977 (before the construction of the Duho Port breakwaters) and in 2019 (after construction). It shows that the shoreline severely retreated in the middle of the beach, whereas it accredited at both ends in 2019, compared to 1977, indicating the redistribution of sediment to reach an equilibrium by the changed wave energy field after the construction of Duho Port.

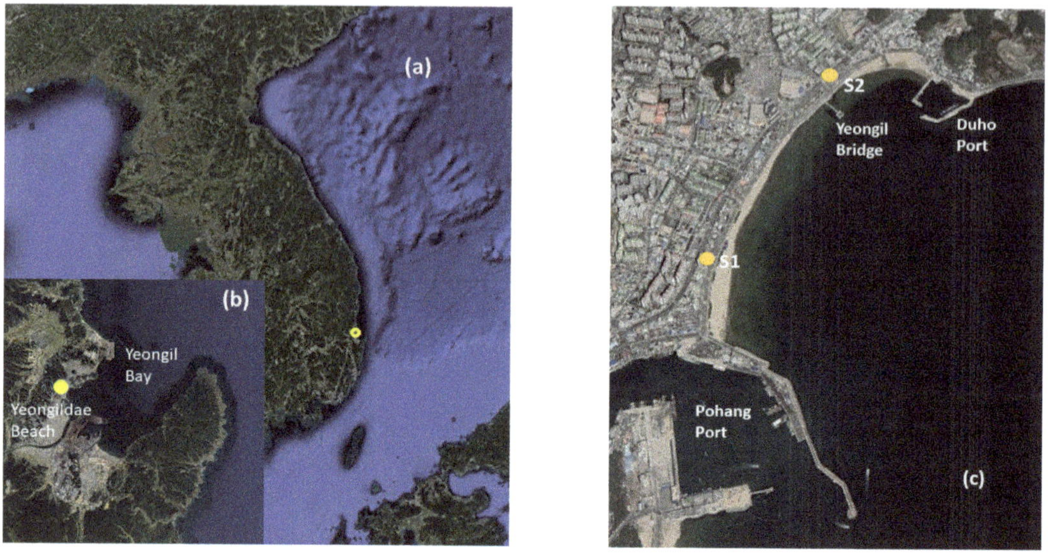

Figure 1. (**a**,**b**) Location of Yeongildae Beach, the Republic of Korea, (**c**) map of Yeongildae Beach with the location of the video monitoring system (S1 and S2). The pictures are captured from Google Earth.

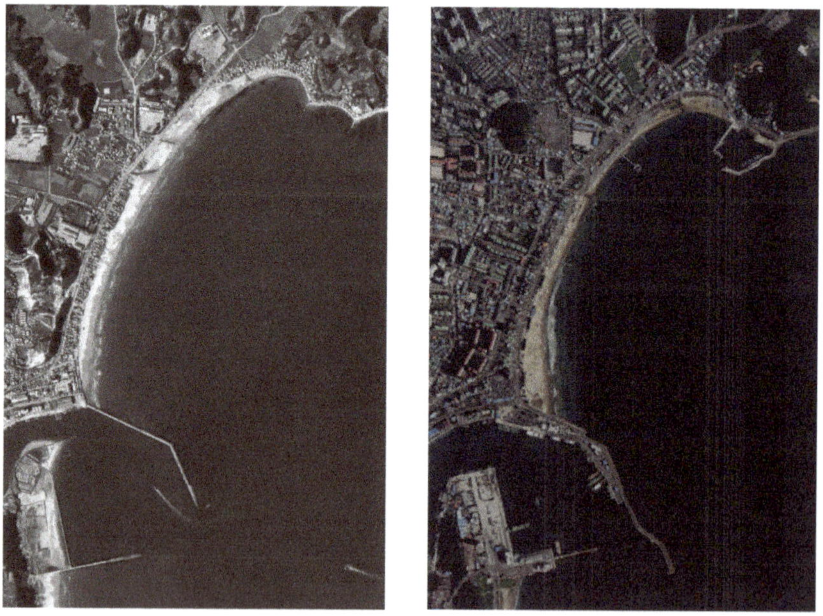

(**a**) (**b**)

Figure 2. Comparison of shoreline positions of Yeongildae Beach: (**a**) aerial photograph taken in 1977 and (**b**) satellite image measured in 2019.

The wave data were measured at two locations, M1 and M2, as shown in Figure 3a. At M1, a pressure transducer was moored at the bottom, at an 8.5 m depth, ~1 km away from the shore of the Yeongildae Beach, measuring the wave height (H_s) and period (T_p).

Because the pressure transducer could not provide information on the wave propagation (D_p), it was measured at M2, where an Acoustic Wave and Current Profiler (AWAC) was moored at the bottom, at a 21.6 m depth, located in nearly the center of Yeongil Bay, ~7 km away from Yeongildae Beach. The tidal range in the study area was no greater than 0.3 m, and the tide-induced current was no greater than 0.2 m/s. Therefore, the effect by the tide on the beach processes may not be significant, and thus was not considered in this study.

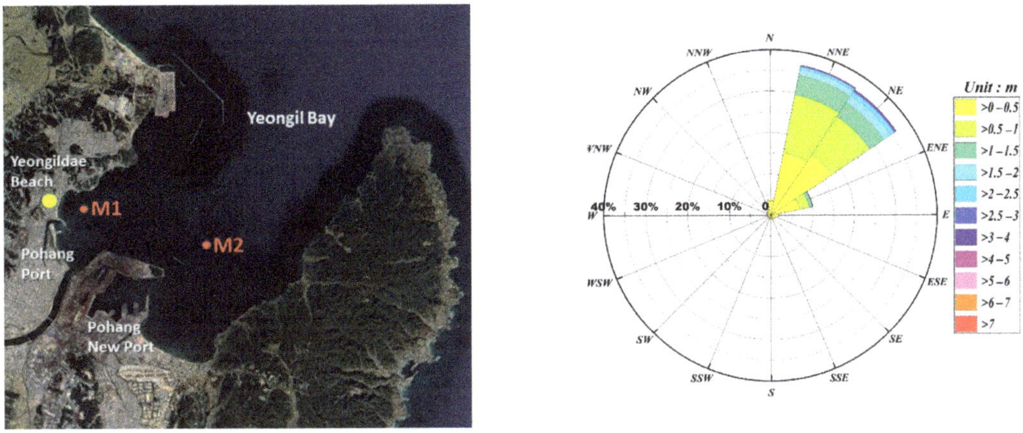

(**a**) (**b**)

Figure 3. (**a**) Map of the two locations of the wave measurements. Wave height and period were measured at M1 using a pressure transducer moored at water depth 8.5 m in the nearshore area of Yeongildae Beach. D_p was measured at M2 using an AWAC moored at water depth 21.6 m in the middle of Yeongil Bay, (**b**) a rose diagram for wave conditions measured at M2 for 2.5 years (from 21 May 2018 to 26 November 2020).

Figure 3b shows a wave rose diagram that represents the wave conditions at M2. Inside Yeongil Bay, the waves mainly propagate from the NE direction, which is similar to the wave propagation outside the bay. Wave conditions in M2 are moderate because the significant wave heights (H_s) are generally less than 2.5 m, except for several extreme storm cases. The dominant wave directions in M2 are NE and NNE, which indicates that the wave propagates inside the Yeongil Bay generally in the normal direction to the mouth of the bay.

Due to the geographical characteristic of Yeongildae Beach being located at the northwest end, inside Yeongil Bay (Figure 1b), the impact of storm waves has not been significant compared to other beaches located outside the bay. In September 2019, however, Typhoon TAPAH (T1917) attacked the study site. Although TAPAH (T1917) was a Category 1 typhoon (max. wind speed 35 m/s), it caused serious damage to many beaches located on the southeast coast of Korea due to the proximity of its path (Figure 4). The damage was also significant at Yeongildae Beach because the beach face was rapidly eroded during the attack of TAPAH (T1917), which provided the motivation of the present study to investigate the characteristic pattern of its recovery process.

2.2. Video Monitoring System

At Yeongildae Beach, two VMSs were constructed at S1 and S2, as shown in Figure 1c. The monitoring by the VMS was initiated in November 2018 and in February 2019 at S1 and S2, respectively. The VMSs were built at the top of buildings in both locations to save the cost of constructing new towers to mount the VMSs. At S1, four video cameras were mounted to cover the central and southern part of the beach, and two additional cameras were mounted at S2 to cover the resting northern end of the beach.

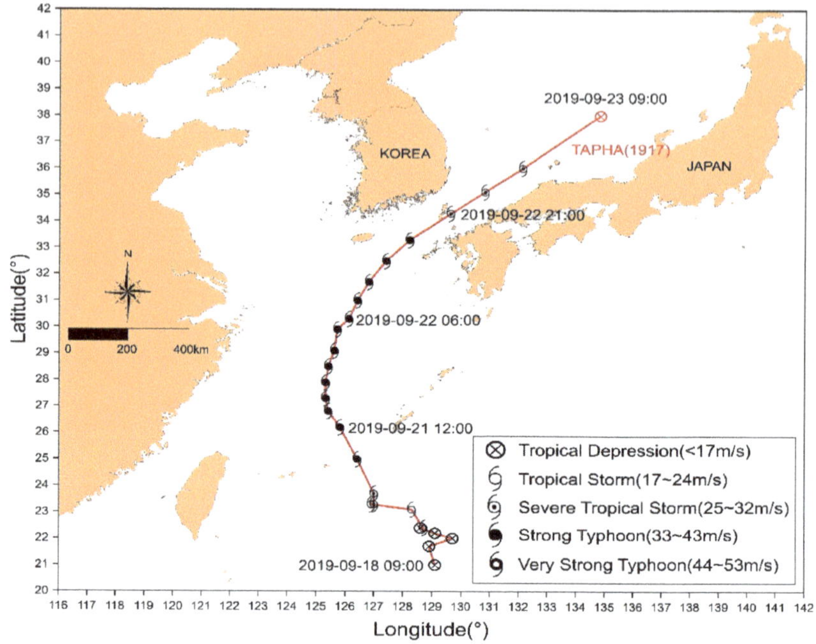

Figure 4. Track of Typhoon TAPAH (T1917).

The VMS consists of the cameras, camera controller, data processing, and data transfer systems that send processed data (averaged images) to the main server. Every 30 min from 7:00 a.m. to sunset, each camera took snapshot pictures for 3 min every half-second. In Figure 5a,b, examples of the snapshot images taken at S1 and S2 are shown. The 180 snapshot images collected for 3 min were then averaged to produce an 'averaged image'; in this way, the averaged images were provided twice per hour. The averaged images stored in the controller were transferred to the main server one time per day in the nighttime, when the internet network was not busy. The snapshot images were not saved in the system unless specified to do so, to save the cost of running the VMS. Once the averaged-image data were transferred to the main server, the locations of the images were corrected in the x-y coordinate system. Because the images were obliquely measured with angles that were different between the cameras, the data needed to be converted into orthogonal images to the ground. After this, the images from each camera were combined into one-image data that covered the whole beach. Figure 5c shows the data in which the orthogonal images are combined into one.

The shoreline positions and beach width were estimated from the image data of the whole beach. For this process, a set of baselines needed to be established, as shown in Figure 5d. The baselines started from the inner line set, along the landward end of the backshore of the beach. Each baseline was set perpendicular to the inner line to be ended in the water so that the position of the shore could be determined manually by comparing the color of the pixels along the baseline. Once the shoreline positions were determined, the beach width could be calculated by connecting the shoreline positions of a baseline to that of the adjacent baselines. Similarly, the position where a baseline crossed the inner line could be connected to the adjacent positions to form a polygon, as marked with the blue solid lines in Figure 5. The beach width was then obtained by calculating the area inside the polygon. To measure the time variation of the beach width, the inner line and baselines were fixed once determined. In the case of Yeongildae Beach, a total of 34 baselines were established.

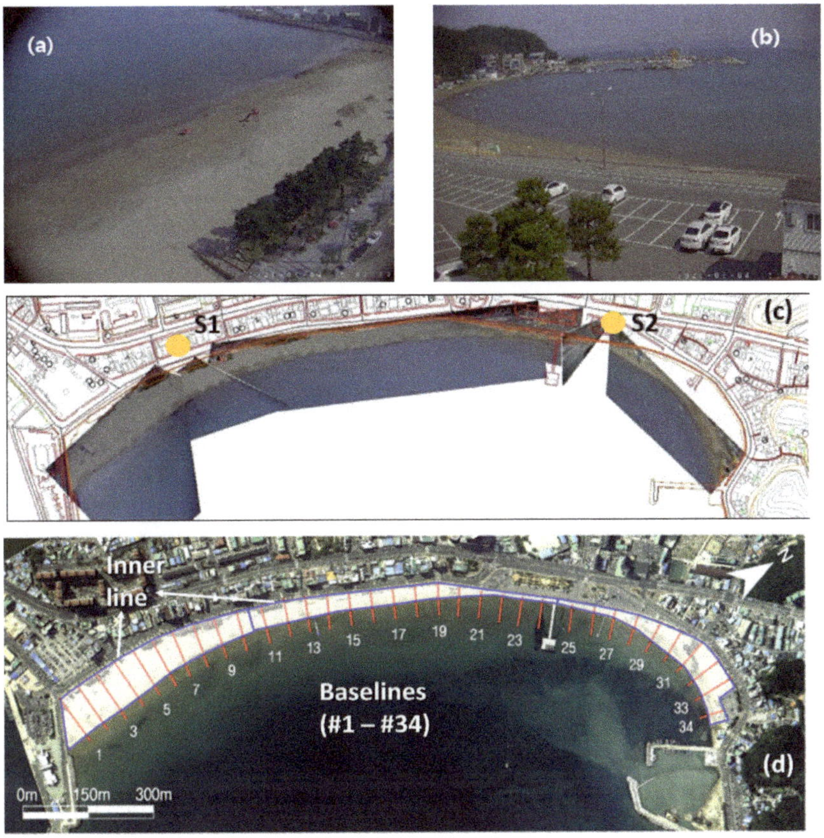

Figure 5. (**a**,**b**) Snapshots of VMS images captured from one of the cameras on S1 and S2, (**c**) orthogonal image in which the pieces of image data captured by the cameras in the VMS system were combined, (**d**) the baseline system to estimate the shoreline positions and beach width. The inner line is set along the landward end of the backshore of the beach, and the baselines are set to start from the inner line and extend seaward perpendicular to the shoreline. A total of 34 baselines are set in Yeongildae Beach, as marked with red solid lines. The shoreline positions are determined by comparing the color of image pixels along each baseline, and the beach width can be calculated from the area of the polygon (marked with blue lines in the figure), which is composed by connecting the crossing points of the baselines to the inner line and shoreline positions horizontally.

3. Results

3.1. General Pattern of Shoreline Recovery

In Figure 6, time variations of the four parameters measured in the site are compared for about 33 months, since the initiation of the VMS measurement in November 2018. The significant wave height, H_{m0}, and the significant wave period, T_s, were measured at M1 (Figure 3). The time variations of H_{m0} and T_s in Figure 6a show complex patterns, as they highly fluctuate, which may cause difficulties in understanding the impact of these wave parameters on the shoreline change. Therefore, another parameter, wave power ($P_w = \frac{\rho g^2}{64\pi} H_{m0}^2 T_s$) was calculated (Figure 6b). P_w is an index that directly measures the wave energy flux and can be used as a parameter that combines the effects of the wave height and period. The time variation of P_w is more clearly distinct from those of H_{m0} and T_s. Its magnitude is usually less than 10 kWatt/m, with an average value over the

33 months of ~0.9 kWatt/m. However, there were times when P_w sharply increased, exceeding 20 kWatt/m, with its maximum value reaching ~75 kWatt/m.

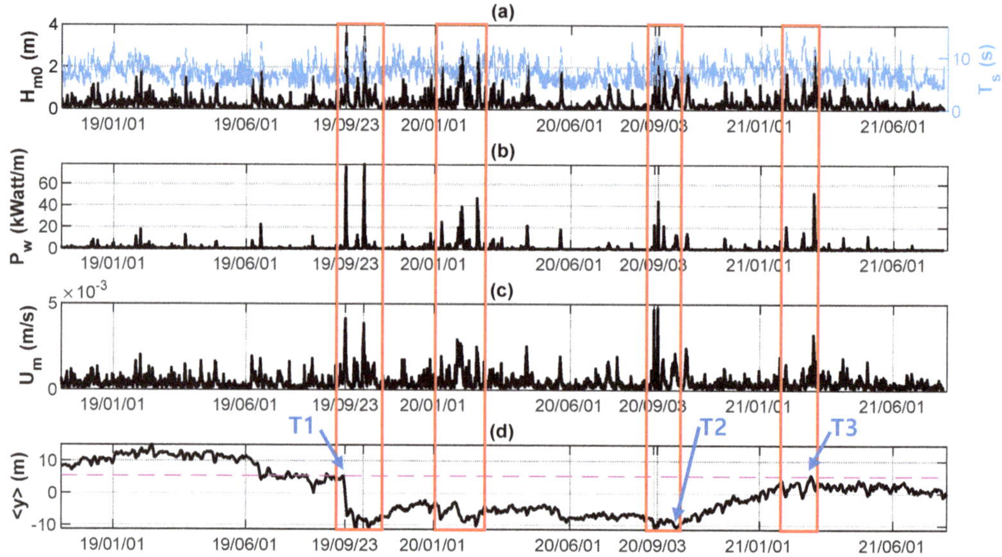

Figure 6. Time variations of (**a**) the significant wave height, H_{m0}, and the significant wave period, T_s, (**b**) the wave power, P_w, (**c**) the amplitude of the near-bed velocity, U_m, and (**d**) the mean beach width, $\langle y \rangle$, averaged over the 34 baselines. The red rectangles mark the 4 time periods when P_w exceeded 20 kWatt/m and $\langle y \rangle$ was decreased, indicating the high correlation between P_w and $\langle y \rangle$. The T1 in (**d**) denotes the time just before the beach was severely eroded by Typhoon TAPAH (T1917) in September 2019, T2 denotes the time when the recovery process started in October 2020, and T3 denotes the time when $\langle y \rangle$ was recovered to the level of T1 in March 2021.

The mean beach width, $\langle y \rangle$, was estimated by integrating the beach width, y, along each baseline, from baseline #1 to #34. Here, the y magnitude of each baseline was obtained by subtracting the average value during the first 2 years from the original VMS beach width (i.e., y is the average removed beach width). As shown in Figure 6d, the beach was significantly eroded when TAPAH (T1917) attacked the site in September 2019. During the period of ~2 days, $\langle y \rangle$ was reduced by ~12 m. In this study, the time of 23 September 2019 was set as T1, as it was the initial time for the erosion to occur due to the typhoon. Since the severe erosion, the beach width hardly recovered, as $\langle y \rangle$ could not reach the original value in T1 (marked with the dashed magenta line in Figure 6d) until 1 March 2021, and this time was set as T3 as an indication of the time of shoreline recovery, although $\langle y \rangle$ touched the original value only for a short time at T3 and decreased again sharply after that. The time variation of $\langle y \rangle$ from T1 and T3 shows an interesting pattern. Once the severe erosion occurred in T1, $\langle y \rangle$ tended to increase slowly (i.e., with much lower speed than the erosion in T1). However, there were times when $\langle y \rangle$ decreased back to the level when the most severe erosion occurred in T1. This accretion and erosion pattern was repeated until 5 October 2020 (this time was set as T2). After T2, $\langle y \rangle$ continued to increase gradually (without severe erosion) until T3, when $\langle y \rangle$ recovered to the level before the erosion in T1. Once it reached the maximum level in T3, $\langle y \rangle$ decreased again, and the gradually increasing process stopped. After this, $\langle y \rangle$ remained level, without showing a clear pattern of erosion/accretion until July 2021, the time of the most recent data.

The pattern of $\langle y \rangle$ shows a high correlation with that of P_w, especially when the erosions occurred from T1 to T3. To increase visibility, four red rectangles are marked to show the time periods when $\langle y \rangle$ and P_w are significantly correlated. In these four periods, the beach width showed a significant decrease (or the increasing trend in $\langle y \rangle$ was halted, as shown in T3), and the maximum magnitude of P_w was observed to exceed 20 kWatt/m, which was significantly higher than observed at other times. In addition to the wave power, the amplitude of the near-bed velocities, U_m, was estimated using the formula ($U_m = 0.0116 H_{m0}/T_s$) by Soulsby (1986) [28]. U_m could be an important factor for sediment motion because cross-shore sediment fluxes are calculated from bed shear stresses that are based on near-bed velocities [29]. As shown in Figure 6c, the time variation pattern of U_m was similar to that of P_w; hence, U_m increasing with increasing P_w might trigger the sediment motions. However, the correlation between $\langle y \rangle$ and U_m is less relevant compared to that between $\langle y \rangle$ and P_w; that is, U_m peaked on 3 September 2020, not on 23 September 2019, when $\langle y \rangle$ declined considerably. Furthermore, U_m was an indirect parameter estimated from the empirical formula using the measured wave heights and periods from the wave sensor. Therefore, it may not be appropriate to apply U_m to directly measure the erosion/accretion processes in the beach face. In contrast, P_w was directly estimated from the wave measurements and thus can be suggested as a controlling indicator in this study.

The correlation between the beach width and wave power can be more clearly investigated in Figure 7, in which the time variation of P_w is also compared with that of $\langle y \rangle$ for the same observation period. To increase the visibility, however, the range of P_w in the y-axis is adjusted so that its upper limit is set to 30 kWatt/m (Figure 7a). In addition, four green rectangles are added in the figure to mark the times when the magnitude of $\langle y \rangle$ increased (i.e., when the shoreline was advanced). Red/green rectangles are marked over the whole observational period of 33 months so that the erosion/accretion trend can be examined even during the time before Typhoon TAPAH (T1917). Compared to the pattern in the red rectangles, the times of shoreline accretion are also closely related to the wave power in the green rectangles. The maximum magnitude of P_w was generally higher than 20 kWatt/m in the periods of five red rectangles, whereas it was less than 10 kWatt/m in the periods of four green rectangles. In particular, at four out of five periods of the red rectangles, the maximum magnitude of P_w exceeded 30 kWatt/m and could reach up to ~75 kWatt/m. During the longest time of the gradual shoreline accretion for ~4 months, which started from T2, the maximum magnitude of P_w was generally lower than 10 kWatt/m. This indicates that a continuation of low-powered wave conditions is necessary for a recovery process to occur consistently. In contrast, the erosion process occurred in relatively shorter periods, when high-powered waves attacked the site. Specifically, it is noted that the gradual increase of $\langle y \rangle$ stopped in T3, when other high-powered waves with P_w of ~50 kWatt/m (Figure 6b) attacked the site, which indicates that the high-powered waves not only caused the rapid erosion but also could change the accretion trend and bring it to a halt.

Following the time variation of the total mean beach width $\langle y \rangle$ in Figure 7b, c displays the time variations of the beach widths in three different parts of the shoreline. In the figure, the solid blue line shows the time variation of averaged beach width for the baselines #2–#5, which is symbolled as $\langle y \rangle_s$ as it represents the pattern in the southern end area. Similarly, the orange line, $\langle y \rangle_m$, represents the variation in the middle of the beach as averaged for the baselines #15–#17. The yellow line, $\langle y \rangle_n$, represents the variation in the northern end as averaged for the baselines #30–#33. The variations of each group of beach widths show different changes from time to time. However, all the time variations of each group of beach widths follow the trend of the total mean beach width $\langle y \rangle$ in T1, T2, and T3. That is, $\langle y \rangle_s$, $\langle y \rangle_m$, and $\langle y \rangle_n$ started to decline rapidly at T1 and formed troughs around T2, and then they all gradually increased until around T3.

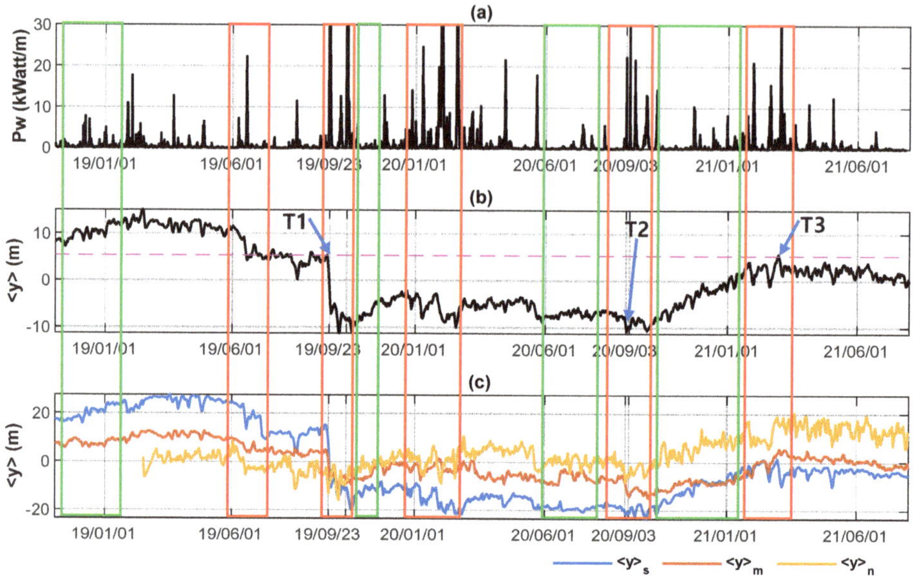

Figure 7. Comparison of time variations between P_w and $\langle y \rangle$ as shown in Figure 6. (**a**) the wave power, P_w, the range of the y-axis for P_w is adjusted from 0 to 30 kWatt/m to increase the visibility of the correlation between the two parameters. (**b**) the mean beach width, $\langle y \rangle$, averaged over the 34 baselines, and (**c**) the mean beach widths of the southern end, the middle, and the northern end of the beach, $\langle y_s \rangle$, $\langle y_m \rangle$, and $\langle y_n \rangle$. T1, T2, and T3 are the same time steps marked in Figure 6. The red rectangles mark the period when the maximum magnitude of P_w became higher than 20 kWatt and $\langle y \rangle$ decreased. The green rectangles mark the periods when the maximum magnitude of P_w was no greater than 10 kWatt/m and $\langle y \rangle$ generally increased.

3.2. Locality in the Erosion/Accretion Process

The time variation in Figures 6 and 7 shows the general pattern of the erosion and recovery processes of the averaged beach width of the 34 baselines. As described in Section 2.1 with Figures 1c and 2, however, the beach widths show high locality—i.e., a discrepancy in y between different baselines. Due to the construction of Duho Port, a shadow zone was formed in the area near the baselines #29–#34 (Figure 5), where the shoreline was advanced. In contrast, the shoreline at the area near baselines #21–#27 severely retreated even before Typhoon TAPAH (T1917) attacked the site. Similarly, the beach widths in the southern end (near baselines #1–#10) were thicker compared to those in the middle of the beach (near baselines #11–#20). This high locality in the beach width between different areas of the beach might affect the erosion and recovery processes, which is analyzed in this section.

Figure 8 shows the time variation of parameters that are related to the locality in terms of wave power. In Figure 8a, the colored contours represent the beach width (y) variations of the 34 baselines, as the numbers in the y-axis denote the baseline numbers marked in Figure 5. Here, it is noted that the colors in the contours show the variation in y along each baseline but cannot be used to compare the magnitude between the baselines because each y is the average removed beach width using the first 2-year average. It is clear that the y values rapidly dropped after TAPAH (T1917). Compared to the northern area (# of baseline >25), however, the y values decreased more severely in the southern end (# of baseline <6). The locality was also observed after the typhoon, as y values were even reduced in the southern end near baselines < #5 until the recovery process prevailed, starting in November 2020. In contrast, y values slightly increased in the northern end (# of baseline >28) by that time. This pattern continued during the recovery process after November

2020, because the y values in the northern end became even greater than those before the typhoon since January 2021, whereas they were smaller than those before the typhoon in the southern end. This pattern of locality can be more clearly quantified through $\langle y \rangle_s$, $\langle y \rangle_m$, $\langle y \rangle_n$, and $\langle y \rangle$ in Figure 8c, in which the time variations of the beach widths calculated for three different groups are compared to the mean beach width. In the southern end, the $\langle y \rangle_s$ rapidly decreased ~30 m after TAPAH (T1917) and reached a minimum on 3 September 2020. The $\langle y \rangle_s$ magnitude could not be recovered even during the recovery process, since November 2020, as the maximum $\langle y \rangle_s$ observed in 2021 was still ~10 m smaller than that before the typhoon. In the northern end, $\langle y \rangle_n$ was not significantly reduced by the typhoon and gradually increased after that time as it became greater than the value before the typhoon during the recovery process, since November 2020. In the middle of the beach, $\langle y \rangle_m$ shows the middle course between the two previous cases. It is also noted that the pattern of $\langle y \rangle_m$ is similar to that of $\langle y \rangle$, indicating the time variation in the middle of the beach can represent the pattern for the whole beach.

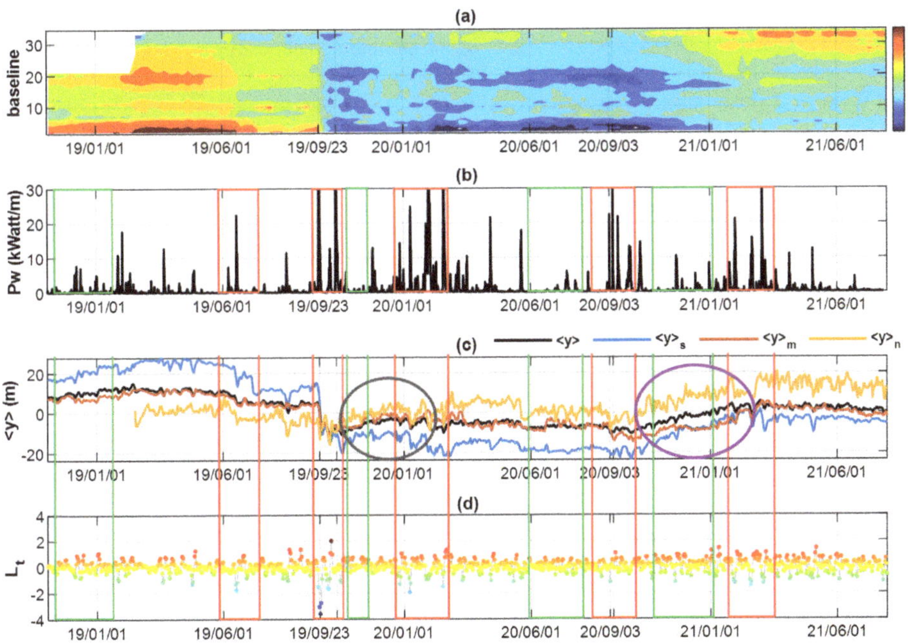

Figure 8. Time variation of (**a**) beach widths of all 34 baselines, where the y-axis represents the baseline numbers, (**b**) wave power of Figure 7a, (**c**) the beach width, $\langle y \rangle_s$, averaged for baselines #2–#5, representing the southern end area (blue), $\langle y \rangle_m$, averaged for baselines #15–#17, representing the middle part of the beach (orange), $\langle y \rangle_n$, averaged for baselines #30–#33, representing the northern end area (yellow), and $\langle y \rangle$, averaged for the all 34 baselines (black), (**d**) $L_t = \int (\partial y/\partial t) dx$, the parameter that indicates the locality of erosion/accretion pattern. The colors of the dots in (**d**) represent the range of L_t magnitude, with reds for positive and blues for negative values. The red and green rectangles in the figure are marked the same as those in Figure 7. (**c**) The grey and purple circles mark the periods when the time variation pattern of the beach width shows locality between the two ends of the beach.

In Figure 8c, the beach width change pattern is compared at two additional time periods, marked with circles. The grey circle covers the period from October 2019 to January 2020, when the beach settled down after TAPAH (T1917). The purple circle covers the recovery period from October 2020 to February 2021. These two periods were chosen because the beach width change pattern shows a discrepancy between the southern and northern ends. During the period of the grey circle, the magnitude of $\langle y \rangle_s$ decreased, whereas that of $\langle y \rangle_n$ increased, showing that the shoreline retreated in the southern end but advanced in the northern end. These results may be interpreted as an indication of a long sediment movement from the southern area toward the northern area. In contrast, both $\langle y \rangle_s$ and $\langle y \rangle_n$ increased during the recovery process, as marked with the purple circle, which indicates that the shoreline was advanced in both areas during the period.

For a better understanding of the locality in the shoreline variation pattern, a parameter was developed using the formula $L_t = \int (\partial y/\partial t) dx$, as plotted in Figure 8d. The dots in the figure show the magnitude of L_t at each time step, and their colors also indicate the range of L_t magnitude, with reds for positive and blues for negative values. In the formula of L_t, x is the alongshore direction toward the northern end. Therefore, the negative value of L_t indicates that $\partial y/\partial t$ magnitude decreased in the positive x-direction. In other words, the positive/negative values of L_t indicate the beach width change rate ($\partial y/\partial t$) increased/decreased toward the northern end of the beach. For example, the minimum L_t value (~−4) during the time of TAPAH (T1917) indicates that the erosion rate was greater in the southern area and decreased in the increasing x-direction so that beach might be most severely eroded in the southern end. Figure 9a shows the distribution of L_t in terms of wave power. Although its probability of occurrence is low (Figure 9b), L_t usually became negative with increasing magnitude when P_w was greater than 10 kWatt/m. In particular, the extreme L_t values (<−3) occurred when TAPAH (T1917) afflicted the beach. This pattern of L_t distribution indicates that the shoreline was more severely eroded in the southern part of the beach, compared to that in the northern part, when storm waves attacked the site. In Figure 9c, the distribution of wave direction, D_p, is also plotted in terms of L_t. The x-axis denotes the wave propagation angles, which increase clockwise from the origin (0°) at the north. Therefore, 45° and 90° denote the NE and E, respectively. The high-powered waves generally approached the beach from ~NE. In the case of Typhoon TAPAH (T1917), the wave propagation direction ranged between 50° and 60°. Considering the shoreline in Figures 1c and 3a, the storm waves during the typhoon directly attacked the southern part, with a slight slope from the normal direction, whereas a shadow zone was likely formed in the northern part of the beach due to the breakwaters of Duho Port.

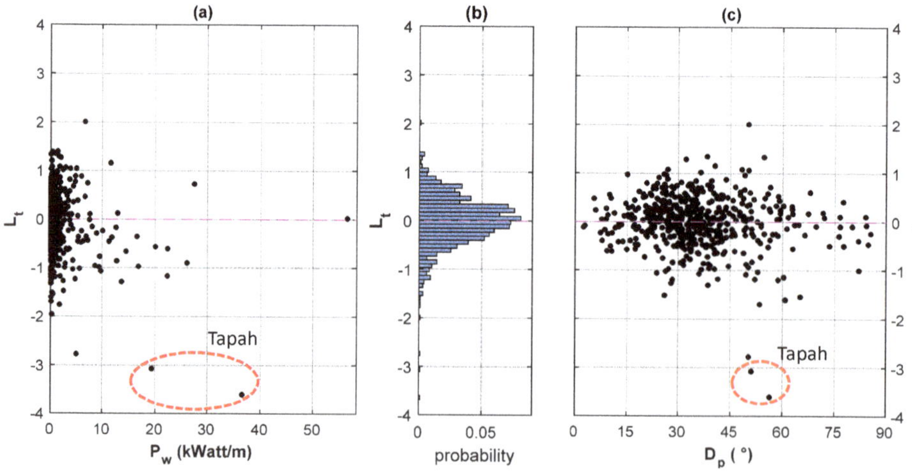

Figure 9. Distribution of (**a**) wave power, (**b**) probability of occurrence, and (**c**) wave propagation direction in terms of $L_t = \int (\partial y/\partial t) dx$. The red circles in (**a**,**c**) mark P_w and D_p for Typhoon TAPAH (T1917), respectively.

The discrepancy in the erosion/accretion pattern between the southern and northern parts of the beach can be also observed in the plane view of shoreline changes. In Figure 10, the beach widths of all 34 baselines are compared at three different times, as they correspond to the times of T1, T2, and T3 in Figures 6 and 7. The figure clearly shows that the reduction in the beach width from T1 to T2 was greatest in the southern end (# of baseline <6), where the maximum reduction distance along baseline #2 was ~40 m. In contrast, the beach width in the northern end (# of baseline >28) was not significantly changed from T1 to T2, as the maximum change in the beach width was no greater than 5 m along baseline #29. At the baselines higher than #30, the change was minimal and not observed. During the recovery process from T2 to T3, the beach width increased at most of the baselines. However, the locality is also clearly observed, as its magnitude at T3 was smaller than that at T1 for the baselines #2–#20, indicating the shoreline was not fully recovered to the level before the typhoon attack in the southern part of the beach. In contrast, the beach width at T3 was greater than that at T1 for the baselines higher than #28. This result indicates that the shoreline in this northern area was not eroded significantly during the attacks of storm waves but advanced under milder wave conditions, leading to an accretion of the shoreline in general.

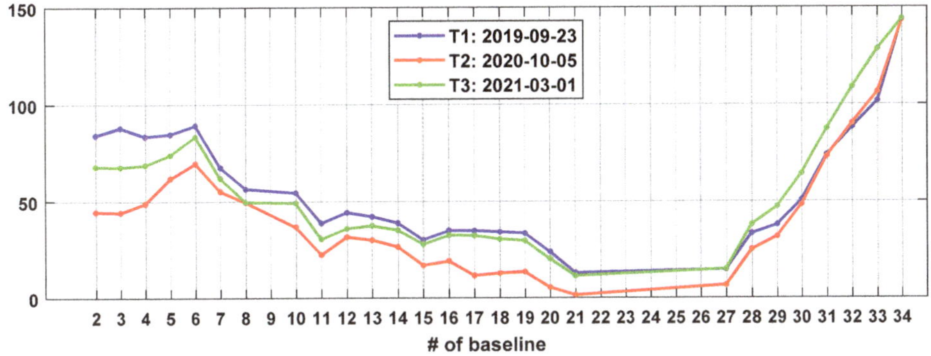

Figure 10. Beach width variation along the baseline at three different times. Data was calculated by averaging the beach widths over one week, as their centered times were on 23 September 2019 (navy), 5 October 2020 (red), and 1 March 2021 (green). These three times correspond to T1, T2, and T3, as marked in Figures 6 and 7.

In the present study, the wave data were measured in location M1, which led to the assumption that the wave conditions were uniform along the coast of the beach. Therefore, alongshore variation of sediment transport caused by the spatial difference in the wave conditions could not be detected using the present data sets. As described, the locality in the results was induced by the shadowing by breakwaters and the curved shoreline, which might produce alongshore discrepancies in the shoreline evolution pattern, as implied from the locality parameter, L_t, in Figure 8d. The impacts of alongshore variation of wave conditions can be established, in future studies, by employing an array of wave gauges in the longshore direction.

4. Discussion

From the shoreline erosion and accretion patterns shown in Figures 6 and 7, a consistent description can be provided for the recovery process, after the severe erosion by TAPAH (T1917) occurred in T1. The beach width decreased under high wave conditions and increased under low conditions, which corresponds to the general understanding of the shoreline evolution as previously described in Section 1. However, the criteria that determine the range of accretion and erosion are still unclear. The results in the present study can provide these criteria, although their application should be only confined to this study site.

Once the severe erosion by TAPAH (T1917) occurred in T1, the recovery process mainly started in T2, about 1 year after the erosion, and continued until T3, about 5 months after T2. The factors that affected short-term (days) processes can be further examined by posing a question about the recovery process—why didn't the recovery process start earlier rather than starting ~1 year, until T2, after the typhoon? The answer may be simple: there were earlier times in which the recovery actually occurred under mild wave conditions. However, it could not continue because another set of high waves attacked the site, restraining the shoreline accretion. Some of the high waves were powerful enough to reverse the accretion trend into erosion. Therefore, the recovery process could not continue further, while the accretion and erosion processes were repeated until T2. During the period of ~5 months from T2 to T3, the wave power remained low so that the recovery process could continue without severe disturbances.

The next question on the process is which wave condition determines the trend of erosion or accretion? Based on the results of this study, a rough estimation can be provided: the shoreline was advanced when the wave power was no greater than 10 kWatt/m. In contrast, the erosion likely occurred when the wave power became greater than 20 kWatt/m. In fact, most rapid erosions occurred when the wave power exceeded 30 kWatt/m. During the total observational period of 33 months, the maximum wave power reached up to 75 kWatt/m, whereas the average wave power during the period was ~0.9 kWatt/m. When the wave power was too low in this site, the sediments did not move, and neither erosion nor accretion could occur. However, the criteria for this 'no sediment motion' condition could not be determined based on the results of this study.

It should be noted that there are critical cases in which the criteria may not be applicable. For example, once the beach width reached the minimum value at T1, its magnitude did not fall below this value during the rest of the observation period. This indicates that the erosional status reached a critical limit with this minimum width, and thus further erosion was prevented unless more extreme events occurred to beak the equilibrium again. On the other hand, there was a period of ~3 months from mid-February to mid-May 2019 (between the first green rectangle and the first red rectangle in Figure 7) when the wave power was generally no greater than 10 kWatt/m, but the beach width did not increase. This is likely because the beach width was thicker than the other period, and thus additional accretion of the shoreline might be disturbed, i.e., the beach width already reached saturation.

The results from locality analysis on the erosion/accretion process showed that the southern part of the beach was more severely eroded, especially when high-powered waves afflicted the site, whereas the shoreline in the northern part of the beach was relatively well protected. During the recovery process after October 2020, however, the imbalance between the two ends was reduced, and the shoreline was advanced almost equally at all parts—southern, middle, and northern—of the beach. The reason for this is still unclear, but the northern end could be protected from erosion due to the shadow zone formed by the breakwaters of Duho Port, in which the wave energy was attenuated, especially when the high-powered storm waves attacked the site.

In the southern end, the breakwater of Pohang Port does not form a shadow zone, considering that most of the waves approached from the NE or NNE, as shown in Figure 3b. Figure 9c also shows that the propagating direction of waves during Typhoon TAPAH (T1917) or other storm waves ranged from 50° to 60°, which confirms that the damage by storm waves could be focused on the southern part of the beach. In contrast, the recovery process occurred under milder wave conditions, by transporting the sediments equally at all parts of the beach. However, there was a possibility of alongshore sediment movement from the southern part to the middle and northern areas. As shown in Figure 8c, $\langle y \rangle_s$ decreased but $\langle y \rangle_m$ and $\langle y \rangle_n$ increased during the period just after TAPAH (T1917), as marked with the grey circle. This locality of different shoreline evolution processes implies that the sediments could be transported in the longshore direction from the south to the north.

The results of this study can be usefully applied in designing the protection/prevention plans against damage to the beach and nearby coastal areas, not only from the disastrous storm events but also from the processes that may break the equilibrium in the beach status. For example, the suggested criteria that could determine the erosion/accretion conditions could be useful in planning coastal structures. The size of these structures can be effectively estimated if information on the wave conditions that affect the erosion or accretion in this specific site is known in advance. If the size of such a structure is too large, it may be cost-ineffective. In addition, the beach recovery process could be disturbed if the structure reduces too much wave energy; in this case, the minimum power required for accretion cannot be met. Similarly, the structure would not correctly function if its size is too small to reduce the wave power below the criteria for erosion. The information on the locality in the shoreline evolution process could be also useful in planning hard and soft coastal structures. If the longshore transport from the southern to the northern part is quantitatively understood (although it has not been done in this study), structures such as groins can be designed to reduce the loss of sediment in specific areas. In addition, the results of the locality analysis in this study would be useful in planning beach nourishment—in determining where to put the sand to maximize the effect of the beach fill by considering the imbalance in the beach process. For example, beach nourishment and placing some coastal structures have been planned in the study area, although they have not been finalized. The results of this study are, therefore, expected to provide useful information in determining such plans, to maximize the effect of the structures and nourishment.

It should be noted that the results from this study are confined only to this specific site. This small beach is characterized as a pocketed beach, where the sediments cannot be added from or lost to other coastal cells. In addition, the beach is located inside a bay, so the waves are usually attenuated when reaching the study site such that the shoreline can reach equilibrium under lower wave energy compared to beaches on the open coast. Figure 11 compares the wave power between Yeongildae Beach (8.5 m depth) and Hwajin Beach (32.0 m depth), located ~20 km north of Yeongildae Beach. Because Hwajin Beach is located on the open coast outside Yeongil Bay, its wave force is much greater than that observed at Yeongildae Beach despite, the difference in water depth.

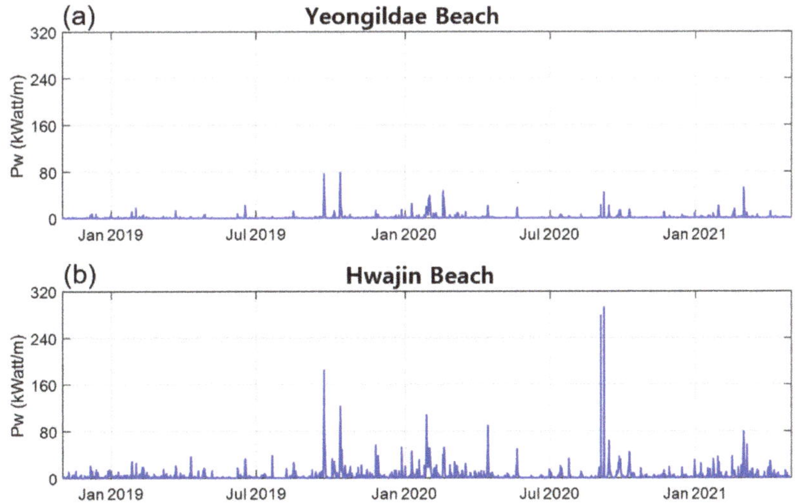

Figure 11. Comparison of the observed wave power between the two beaches of (**a**) Yeongildae Beach (this study site) and (**b**) Hwajin Beach, located in the open coast outside Yeongil Bay. Although the distance between the two beaches is only ~20 km, the wave power in Hwajin Beach is much greater than that in Yeongildae Beach.

As a result, the sediments in this site could respond more easily at lower wave energy levels, which might lead to a different pattern of erosion and recovery process. For example, the sand size in the study site ranged from 0.15 to 0.35 mm, as it is categorized as fine to medium sand, which was smaller than the sand size measured in the two beaches located in the open coast outside the bay, where it was ~0.5 mm at both beaches (categorized as medium to coarse sand) [30,31]. The criteria for erosion/accretion suggested in this study can hardly be determined in other beaches on the open coast, where nearshore crescentic sandbars are actively developed. Because the horns of the sandbars are coupled with the shoreline pattern [32], the erosion/accretion of the shoreline has a strong locality rather than showing the general trend as observed in this study site, where no crescentic sandbars developed. For this reason, the information on the locality from these study results should be also confined to this specific site.

5. Conclusions

In this study, we employed a data set of VMS and wave parameters measured over 33 months to investigate the erosional and depositional processes in Yeongildae Beach, located inside Yeongil Bay on the southeastern coast of the Korean Peninsula. The beach was severely eroded when Typhoon TAPAH (T1917) hit the site in late September 2019, during which the beach width retreated ~12 m on average; it took about 1.5 years for the beach width to be recovered to the level before the attack of the typhoon. The study then analyzed the recovery process, during which shoreline erosion and accretion repeated, corresponding to the wave conditions.

The study site is a pocketed beach, as the breakwaters of two ports were built at both ends of the beach, blocking the input or loss of sediments in the longshore direction crossing the lateral boundary of the beach. The beach is also located at the west corner, inside the bay, and the distance from the beach to the mouths of the bay is 8–15 km. Due to its location on the beach, the wave energy was lower and the sediment size was finer, compared to those observed on the open coast outside the bay, because the waves were attenuated when reaching the site.

The results of the analysis corresponded to the general understanding of beach processes—that beach widths decreased/increased under high/low wave energy conditions. In this study site, however, the pattern of shoreline evolution was found to be highly correlated with P_w, the wave power that combined the impacts of wave height and period, rather than the wave propagating direction, which was observed to be similar for high-powered waves. In particular, the beach width generally increased when P_w was no greater than 10 kWatt/m and tended to decrease under the condition of P_w greater than 20 kWatt/m. However, most erosional events occurred when P_w exceeded 30 kWatt/m. Especially after TAPAH (T1917), it took a long time (~1.5 years) for the beach width to be fully recovered to the previous level because high-powered waves occasionally disturbed the recovery trend. Therefore, the full recovery of the beach width could be reached over the last five months in the recovery period, during which P_w was kept continuously lower than 10 kWatt/m.

The locality of beach recovery process was also analyzed by examining the discrepancy in the erosion/accretion pattern between different locations within the beach. It was observed that the erosion by TAPAH (T1917) was most severe in the southern part of the beach, whereas the change in the beach width was minimal in the northern part during the same period. This pattern of locality was similarly observed under other storm conditions. Considering the wave propagation direction was similar (~NE) for the storm waves, it is likely that a shadow zone was formed in the northern end due to the breakwaters of Duho Port, such that the sediments in the lee area of the breakwaters were protected because the wave energy was attenuated. On the contrary, the breakwater of Pohang Port located in the southern end did not form a shallow zone, and thus the storm waves directly attacked the southern part, causing erosion. In the phase of recovery under milder wave conditions, the shadow zone did not play an important role, and the shoreline was advanced almost equally at all parts of the beach. However, it was also observed that there was a period just after TAPAH (T1917) when erosion occurred in the southern part while accretion occurred in the northern part, indicating a possible alongshore sediment movement from the south to the north.

The results from this study can be usefully applied to design the protection and prevention plans of the beach from storm damage, considering the general and local erosion and recovery patterns. For example, the criteria that determined the trends of erosion and accretion based on wave power can be used to effectively estimate the size of coastal structures, such as detached breakwaters for shore protection. In addition, the results of locality analysis can be useful to plan the horizontal arrangement of such hard structures or beach nourishment, because they can provide information on where to place the structure or sand to maximize their ability to stabilize the beach. It is noted that the application of these results should be confined only to this specific site and may not be directly applicable to other beaches on the open coasts. For example, the suggested criteria are valid where the wave energy is relatively lower than that on the open coasts. The conditions in this specific beach might result in characteristic shoreline responses. In the case of other beaches located on the open coasts, where various nearshore sandbars develop, the shoreline could be coupled with sandbar positions, and the erosion/accretion pattern would be more complex than that observed in this study site. Regardless of these limitations, the method in this study can be usefully applied in areas with a similar environment for designing prevention plans for disasters due to storms.

Author Contributions: Conceptualization, J.-E.O., K.-H.R., W.-M.J. and Y.-S.C.; methodology, J.-E.O., K.-H.R. and J.-Y.P.; formal analysis, J.-E.O.; investigation, J.-E.O., W.-M.J. and Y.-S.C.; resources, J.-E.O. and J.-Y.P.; data curation, J.-E.O.; writing—original draft preparation, J.-E.O. and Y.-S.C.; writing—review and editing, J.-E.O. and Y.-S.C.; visualization, J.-E.O. and Y.-S.C.; project administration, K.-H.R.; funding acquisition, W.-M.J. All authors have read and agreed to the published version of the manuscript.

Funding: This research was funded by the Korea Institute of Ocean Science and Technology (KIOST), grant number PE99932, and the Ministry of Oceans and Fisheries, grant number PG52320.

Conflicts of Interest: The authors declare no conflict of interest.

References

1. Russell, P.E. Mechanisms for beach erosion during storms. *Cont. Shelf Res.* **1993**, *13*, 1243–1265. [CrossRef]
2. Davidson-Arnott, R.; Bauer, B.; Houser, C. *Introduction to Coastal Processes and Geomorphology*; Cambridge University Press: Cambridge, UK, 2019.
3. Harter, C.; Figlus, J. Numerical modeling of the morphodynamic response of a low-lying barrier island beach and foredune system inundated during Hurricane Ike using XBeach and CSHORE. *Coast Eng.* **2017**, *120*, 64–74. [CrossRef]
4. Cox, J.; Pirrello, M. Applying joint probabilities and cumulative effects to estimate storm-induced erosion and shoreline recession. *Shore Beach.* **2001**, *69*, 5–7.
5. Vousdoukas, M.I.; Almeida, L.P.M.; Ferreira, Ó. Beach erosion and recovery during consecutive storms at a steep-sloping, meso-tidal beach. *Earth Surf. Process. Landf.* **2012**, *37*, 583–593. [CrossRef]
6. Lee, G.-H.; Nicholls, R.J.; Birkemeier, W.A. Storm-driven variability of the beach-nearshore profile at Duck, North Carolina, USA, 1981–1991. *Mar. Geol.* **1998**, *148*, 163–177. [CrossRef]
7. Hsu, T.J.; Hanes, D.M. Effects of wave shape on sheet flow sediment transport. *J Geophys. Res. Ocean.* **2004**, *109*. [CrossRef]
8. Ranasinghe, R.; Holman, R.; De Schipper, M.; Lippmann, T.; Wehof, J.; Duong, T.M.; Roelvink, D.; Stive, M. Quantifying nearshore morphological recovery time scales using argus video imaging: Palm Beach, Sydney and Duck, North Carolina. In Proceedings of the 33rd International Conference on Coastal Engineering, Santander, Spain, 1–6 July 2012.
9. Morales-Márquez, V.; Orfila, A.; Simarro, G.; Gómez-Pujol, L.; Álvarez-Ellacuría, A.; Conti, D.; Galán, Á.; Osorio, A.F.; Marcos, M. Numerical and remote techniques for operational beach management under storm group forcing. *Nat. Hazards Earth Syst. Sci.* **2018**, *18*, 3211–3223. [CrossRef]
10. Forbes, D.L.; Parkes, G.S.; Manson, G.K.; Ketch, L.A. Storms and shoreline retreat in the southern Gulf of St. Lawrence. *Mar. Geol.* **2004**, *210*, 169–204. [CrossRef]
11. Do, J.D.; Jin, J.-Y.; Jeong, W.M.; Lee, B.; Choi, J.Y.; Chang, Y.S. Collapse of a Coastal Revetment Due to the Combined Effect of Anthropogenic and Natural Disturbances. *Sustainability* **2021**, *13*, 3712. [CrossRef]
12. Karunarathna, H.; Pender, D.; Ranasinghe, R.; Short, A.D.; Reeve, D.E. The effects of storm clustering on beach profile variability. *Mar. Geol.* **2014**, *348*, 103–112. [CrossRef]
13. Suanez, S.; Cariolet, J.-M.; Cancouët, R.; Ardhuin, F.; Delacourt, C. Dune recovery after storm erosion on a high-energy beach: Vougot Beach, Brittany (France). *Geomorphology* **2012**, *139*, 16–33. [CrossRef]

14. Callaghan, D.; Nielsen, P.; Short, A.; Ranasinghe, R. Statistical simulation of wave climate and extreme beach erosion. *Coast Eng.* **2008**, *55*, 375–390. [CrossRef]
15. Roelvink, D.; Reniers, A.; Van Dongeren, A.; De Vries, J.V.T.; McCall, R.; Lescinski, J. Modelling storm impacts on beaches, dunes and barrier islands. *Coast Eng.* **2009**, *56*, 1133–1152. [CrossRef]
16. Chang, Y.S.; Huisman, B.; de Boer, W.; Yoo, J. Hindcast of Long-term Shoreline Change due to Coastal Interventions at Namhangjin, Korea. *J. Coast. Res.* **2018**, *85*, 201–205. [CrossRef]
17. Pelnard-Considère, R. Essai de theorie de l'evolution des formes de rivage en plages de sable et de galets. *Journées De L'hydraulique* **1957**, *4*, 289–298.
18. Davidson, M.A.; Turner, I.L.; Splinter, K.D.; Harley, M.D. Annual prediction of shoreline erosion and subsequent recovery. *Coast Eng.* **2017**, *130*, 14–25. [CrossRef]
19. Konlechner, T.M.; Kennedy, D.M.; O'Grady, J.J.; Leach, C.; Ranasinghe, R.; Carvalho, R.C.; Luijendijk, A.P.; McInnes, K.L.; Ierodiaconou, D. Mapping spatial variability in shoreline change hotspots from satellite data; a case study in southeast Australia. *Estuar. Coast. Shelf Sci.* **2020**, *246*, 107018. [CrossRef]
20. Castelle, B.; Masselink, G.; Scott, T.; Stokes, C.; Konstantinou, A.; Marieu, V.; Bujan, S. Satellite-derived shoreline detection at a high-energy meso-macrotidal beach. *Geomorphology* **2021**, *383*, 107707. [CrossRef]
21. Brooks, S.M.; Spencer, T.; Christie, E.K. Storm impacts and shoreline recovery: Mechanisms and controls in the southern North Sea. *Geomorphology* **2017**, *283*, 48–60. [CrossRef]
22. Héquette, A.; Ruz, M.-H.; Zemmour, A.; Marin, D.; Cartier, A.; Sipka, V. Alongshore variability in coastal dune erosion and post-storm recovery, northern coast of France. *J. Coast. Res.* **2019**, *88*, 25–45. [CrossRef]
23. Do, J.D.; Jin, J.-Y.; Kim, C.H.; Kim, W.-H.; Lee, B.-G.; Wie, G.J.; Chang, Y.S. Measurement of Nearshore Seabed Bathymetry using Airborne/Mobile LiDAR and Multibeam Sonar at Hujeong Beach, Korea. *J. Coast. Res.* **2020**, *95*, 1067–1071. [CrossRef]
24. Davidson, M.; Aarninkhof, S.; Van Koningsveld, M.; Holman, R. Developing coastal video monitoring systems in support of coastal zone management. *J. Coast. Res.* **2006**, 49–56.
25. Holman, R.A.; Stanley, J. The history and technical capabilities of Argus. *Coast. Eng.* **2007**, *54*, 477–491. [CrossRef]
26. Chickadel, C.; Holman, R. Measuring longshore current with video techniques. In Proceedings of the EOS Trans, Fall Meeting, San Francisco, CA, USA, 10-14 December 2001; American Geophysical Union: Washington, DC, USA, 2002; Volume 82.
27. Do, J.D.; Jin, J.-Y.; Hyun, S.K.; Jeong, W.M.; Chang, Y.S. Numerical investigation of the effect of wave diffraction on beach erosion/accretion at the Gangneung Harbor, Korea. *J. Hydro-Environ. Res.* **2020**, *29*, 31–44. [CrossRef]
28. Soulsby, R.; Smallman, J. *A Direct Method of Calculating Bottom Orbital Velocity under Waves*; Hydraulics Research Limited: Wallingford, Oxfordshire, UK, 1986.
29. Ribberink, J.S. Bed-load transport for steady flows and unsteady oscillatory flows. *Coast. Eng.* **1998**, *34*, 59–82. [CrossRef]
30. Do, J.D.; Chang, Y.S.; Jin, J.-Y.; Jeong, W.M.; Lee, B.; Ha, H.K. Hydrodynamic Measurements of Propagating Waves at Different Nearshore Depths in Hujeong Beach, Korea. *J. Mar. Sci. Eng.* **2020**, *8*, 690. [CrossRef]
31. Oh, J.-E.; Chang, Y.S.; Jeong, W.M.; Kim, K.H.; Ryu, K.H. Estimation of longshore sediment transport using video monitoring shoreline data. *J. Mar. Sci. Eng.* **2020**, *8*, 572. [CrossRef]
32. Chang, Y.S.; Jin, J.-Y.; Jeong, W.M.; Kim, C.H.; Do, J.-D. Video Monitoring of Shoreline Positions in Hujeong Beach, Korea. *Appl. Sci.* **2019**, *9*, 4984. [CrossRef]

applied sciences

Article

A Multi-Criteria Decision Analysis (MCDA) Approach for Landslide Susceptibility Mapping of a Part of Darjeeling District in North-East Himalaya, India

Abhik Saha [1], Vasanta Govind Kumar Villuri [1,*], Ashutosh Bhardwaj [2] and Satish Kumar [1]

1 Department of Mining Engineering, Indian Institute of Technology, Indian School of Mines, Dhanbad 826004, India; abhik.17dr000597@me.ism.ac.in (A.S.); satish.19pr0049@me.iitism.ac.in (S.K.)
2 Research Project Monitoring Department, Indian Institute of Remote Sensing, 4, Kalidas Road, Dehradun 248001, India; ashutosh@iirs.gov.in
* Correspondence: vgkvilluri@iitism.ac.in; Tel.: +91-9471192205

Abstract: Landslides are the nation's hidden disaster, significantly increasing economic loss and social disruption. Unfortunately, limited information is available about the depth and extent of landslides. Therefore, in order to identify landslide-prone zones in advance, a well-planned landslide susceptibility mapping (LSM) approach is needed. The present study evaluates the efficacy of an MCDA-based model (analytical hierarchy process (AHP)) and determines the most accurate approach for detecting landslide-prone zones in one part of Darjeeling, India. LSM is prepared using remote sensing thematic layers such as slope, rainfall earthquake, lineament density, drainage density, geology, geomorphology, aspect, land use and land cover (LULC), and soil. The result obtained is classified into four classes, i.e., very high (11.68%), high (26.18%), moderate (48.87%), and low (13.27%) landslide susceptibility. It is observed that an entire 37.86% of the area is in a high to very high susceptibility zone. The efficiency of the LSM was validated with the help of the receiver operating characteristics (ROC) curve, which demonstrate an accuracy of 96.8%, and the success rate curve showed an accuracy of 81.3%, both of which are very satisfactory results. Thus, the proposed framework will help natural disaster experts to reduce land vulnerability, as well as aid in future development.

Keywords: landslide; multi-criteria decision analysis (MCDA); analytical hierarchy process (AHP); receiver operating characteristics; area under the curve (AUC)

Citation: Saha, A.; Villuri, V.G.K.; Bhardwaj, A.; Kumar, S. A Multi-Criteria Decision Analysis (MCDA) Approach for Landslide Susceptibility Mapping of a Part of Darjeeling District in North-East Himalaya, India. *Appl. Sci.* **2023**, *13*, 5062. https://doi.org/10.3390/app13085062

Academic Editors: Miguel Llorente Isidro, Ricardo Castedo and David Moncoulon

Received: 30 March 2023
Revised: 11 April 2023
Accepted: 12 April 2023
Published: 18 April 2023

1. Introduction

Disastrous natural hazards such as landslides have brought enormous casualties and economic losses in the past. In hilly areas, the essential requirement for ensuring people's safety is the identification of high-risk landslide-prone zones where development should not be conducted, and where soil stabilization works should be conducted. Nearly 12.60% of landslides have occurred in the Indian Himalayan region, and all areas can be detected as being landslide-prone. In developing countries such as India, various terrain hills are experiencing increasing population density and rapid infrastructural development, which may increase the vulnerability to potential landslides and thus socio-economic losses. The study area is situated in the Himalayan region, where landslides are a common occurrence due to the area's topography and geology. The region encompasses several major cities, including Darjeeling, Ghum, Rangbull, and Sonada, which attract a large number of tourists daily. The population in the area is dense, and significant land-use changes have taken place over the years, further exacerbating the risk of landslides. By conducting landslide susceptibility mapping in these areas, it is possible to identify high-risk zones and inform land-use planning decisions to mitigate the impact of landslides on the local population and infrastructure. Quantitative and qualitative approaches are used to classify landslide susceptibility mapping (LSM) [1–4]. To prevent these losses and improve the efficiency of LSM, new methods and different hybrid models are introduced every day by researchers [5–10]. Machine learning approaches have also been introduced in this field in recent years [11–14]. Different ensemble methods and hybrid models are also used to assess and identify LSM [8,15]. The overall incidences of landslides are increasing, and the factors causing the landslides are the growth of urbanization, and development in landslide-prone areas.

Changing climate patterns, increased local rainfall, etc., are the reasons behind constant changes in landslide-prone zones [16,17]. Most losses would be avoided by identifying the problematic areas, land deforestation, etc. Hence, recognizing the existing and potentially unstable slopes is most significant task. Due to mass changes in urbanization and growing awareness of its socio-economic impact in the hilly region, significant attention is focused on the study of landslides [18]. Hence, LSM of any terrain helps to classify areas into different classes concerning the degree of potential hazards. Recently, various GIS tools have helped to identify conditioning factors that may influence an area's vulnerability to landslides. Rainfall and earthquakes significantly trigger landslides due to their external and temporal aspects [19]. For effective disaster management and future improvement, developers and engineers use landslide susceptibility maps [12,20,21]. The identification of risk areas, coupled with a proper landslide assessment plan, can help disaster management planners and developers to minimize accidents and losses. Many parts of the Himalayas are not physically accessible but geospatially can be accessible everywhere, so assessing landslides using the geospatial technique is easy [22,23]. Geospatial and geotechnical modeling has been used by different researchers in the last few years for landslide susceptibility and hazard assessment using different parameters [24–26]. Land-use change is a major contributor to landslides, causing significant environmental and socioeconomic impacts [27]. In order to achieve sustainability goals and reduce landslide risks, understanding the spatiotemporal evolution and influencing factors of land-use change is essential [28–30]. Land-use change can create slope instability, increasing the susceptibility of an area to landslides. Deforestation, urbanization, and mining are examples of land-use changes that can alter soil properties and destabilize slopes [31]. The impacts of land-use change on landslide susceptibility are complex and influenced by factors such as topography, geology, climate, land-use policies, and population growth [26]. A comprehensive approach is necessary to achieve sustainability goals, which includes considering the impacts of land-use change on landslide susceptibility. Sustainable land-use planning can play a critical role in reducing landslide risks by preserving natural ecosystems, restoring degraded land, and promoting low-impact development practices [32,33]. Accurate landslide susceptibility mapping is crucial for identifying at-risk areas and making informed land-use planning decisions. With advances in remote sensing and geographic information systems (GIS) technology, sophisticated landslide susceptibility models have been developed to capture the complex interactions between land-use change and landslide susceptibility [23,34,35]. Understanding the spatiotemporal evolution and influencing factors of land-use change is the key to achieving sustainability goals and mitigating landslide risks. Sustainable land-use planning and effective landslide susceptibility mapping can contribute to promoting a low-carbon, resilient future. The LSM has multiple uses, identifying unstable areas in advance, disallowing new construction in hazard-prone areas, relief operations, etc. An analytical hierarchy process (AHP)-based method has been attempted in the study area to prepare LSM due to its simplifying method of assigning ranks and weights [21,29,36]. Different natural hazards such as floods, gully erosion, earthquakes, and liquification can also be classified through AHP methodology [37–39]. It is a method to derive ratio scales from paired comparisons from the principal eigenvectors [29].

The study of landslide susceptibility mapping in Darjeeling is due to a combination of various factors in the area. A high-resolution DEM is used for the delineation of various thematic layers. With a high population density, significant land-use changes, and challenging topography, the study region provides an interesting case study for analyzing the complex interactions between topography, land use, and population growth contributing to landslide risk in the Himalayan region. This study has the potential to provide valuable insights for informing land-use planning decisions and reducing the impact of landslides on local communities. The current study focuses on identifying the possible landslide areas for stabilizing the citizens' safety and the area's future development. To aid in the development of more accurate and reliable landslide susceptibility models that can be applied to other regions with similar challenges, the AHP method is adopted to recognize the landslide prone areas. The published result can be utilized by West Bengal tourism for future development and controlled habitation in hazard-prone areas.

2. Materials and Methods

2.1. Study Area

The study region is located in the Darjeeling district of West Bengal, India, in the Eastern Himalayan mountain environment (Figure 1). The area is situated in latitudes 88°171′ E to 88°343′ E, 26°933′ N to 27°082′ N and encompasses around 256 Km^2 of the region. The maximum elevation in this region is 2600 m from mean sea level (MSL). The most dreaded and widespread landslide event in the Darjeeling Himalayas occurred in 1968, triggering many shallow and deep-seated landslides [40]. It is believed that many prominent and large landslides in the Darjeeling hills are still active today [41]. Thus, the region requires a systematic study for LSM. The landslide's study area induces inherent factors such as highly dissected slopes, barren land, etc. Tea plantation is the primary land use operation in this area. The study region had a population density of 586 per sq·Km, while the Darjeeling district's population was 1,846,823 as per the 2011 census.

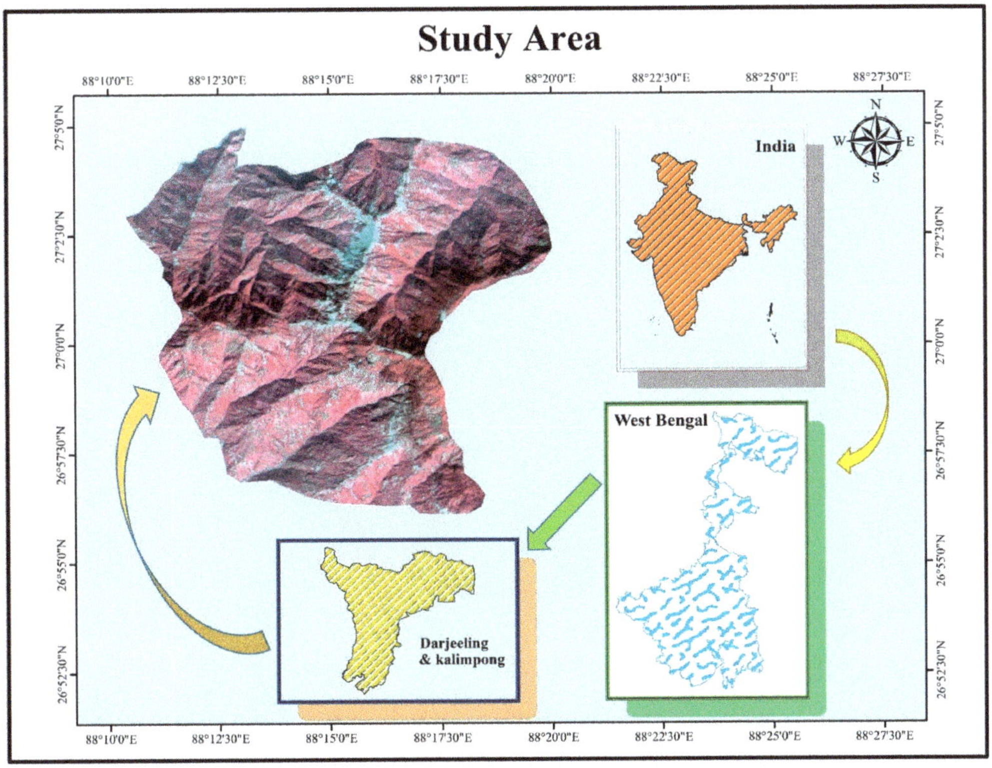

Figure 1. The study area shows a part of the Darjeeling district.

2.2. Datasets Used

Thematic layers significant to the causative factors presented in Table 1 were prepared using various remotely sensed data from multiple sources. The GIS software (ArcGIS 10.4) was used for preparing layers such as slope, aspect, drainage density, and lineament density from ALOS PALSAR DEM. The soil and geomorphology maps of West Bengal were collected from the National Bureau of Soil Survey and Land Use Planning (NBSS and LUP), Government of India. A geology map was compiled from the Geological Survey of India (GSI). The Catrosat-2 series (MX) dataset generated a land use and land cover (LULC) map. The drainage density and lineament density maps were made from the ALOS PALSAR DEM data. Data on rainfall were gathered from the Current Research Unit (CRU). The inverse distance weighting (IDW) method for interpolation was used to prepare a rainfall map. An earthquake map was prepared from the data procured from the National Centre for Seismology, New Delhi, India.

Table 1. Data used for generating landslide conditioning factor.

Sl. No.	Datasets Name	Purpose	Resolution (m)
1	ALOS PALSAR DEM	To prepare aspect, slope, lineament density, drainage density	12.5
2	Catrosat-2 series (MX)	To extract LULC	1.6
3	CRU	To generate a rainfall map	Numerical
4	Soil	To generate a soil map	-
5	Geology and geomorphology	To extract geology and geomorphology map	-
6	Earthquake data	Earthquake map generations	Point data

2.3. Methodology

The basic methodology adopted for preparing LSM is shown in Figure 2. The initial stage of a susceptibility assessment involves gathering all available information and data about the study area. This stage is crucial as it lays the foundation for further analysis of the relationship between landslide occurrence and conditioning parameters. However, assessing the cause–effect relationships can be challenging since landslides are typically caused by multiple factors. To create a landslide susceptibility map for the Darjeeling region of West Bengal, ten inputs were chosen based on the most significant factors influencing landslides, including slope, aspect, earthquake activity, drainage density, rainfall, land use/land cover (LULC), lineament density, geology, geomorphology, and soil type [12,29]. Our study expanded upon the number of factor classes considered compared to prior research. We reviewed the literature by other authors and sought to assign non-homogeneous weights to each factor class, in contrast to the uniform weighting used in previous studies. The various layer classes are rated in the range between 0 and 9, where higher ratings represent more decisive landslide influence. After preparing the final LSM, the study area was differentiated into different susceptibility zones, such as low, high, moderate, and very high.

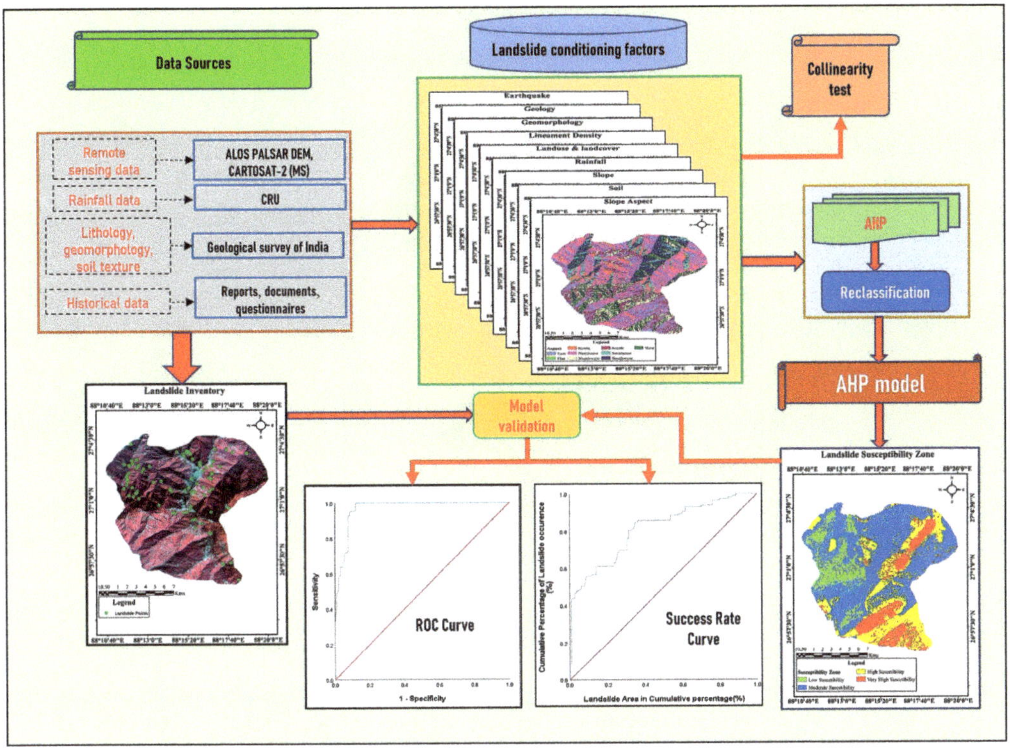

Figure 2. Graphical representation of overall methodology of landslide susceptibility mapping.

2.4. Landslide Inventory

The position and extent of the landslide must be appropriately identified when creating the maps of landslide susceptibility. Landslide inventory is a critical component of all types of landslide zoning classification, including hazard, susceptibility, and risk zoning [42]. It has to deal with the time, location, sort, amount, and travel distance of land sliding in a specific area. Different methods are available for identifying landslides. Field observations, satellite photographs, book studies for details on previous landslides, and aerial photography are a few of them [43]. The landslide inventory map was produced using a combination of visual analysis of aerial imagery, field data, and satellite photography [14]. The whole data on landslides in India were made available by the Geological Survey of India (GSI). Resampled public data from the GSI were used to create the inventory map for the study area. One hundred and fourteen landslides have been found by this study using the available data (Figure 3). Most of the landslides recorded in this area have cliffs, highways, and banks of rivers. Erosion, steep terrain, and a lack of flora are often the causes of landslides along cliffsides.

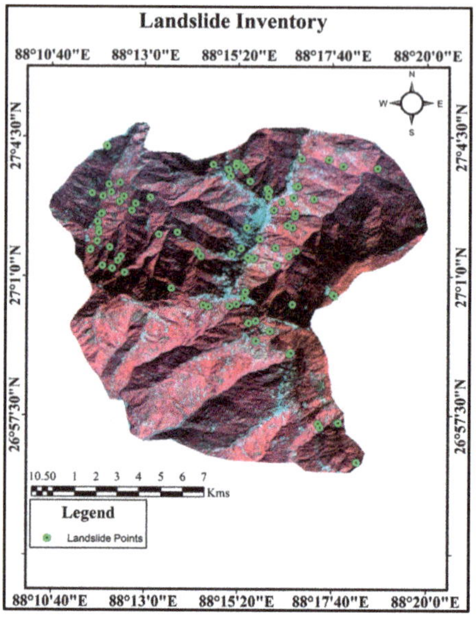

Figure 3. Map of landslide inventory of study area.

2.5. Landslide Conditioning Factor (LCF)

2.5.1. Slope

The slope is one of the triggering factors for landslides in any hilly region. Steeper slopes are more hazard-prone to instability compared with lesser slopes. Shear stress in soil and other uncemented materials gradually rises as the slope angle increases [44,45]. In a steep slope, the gravitational force works more than in a moderate slope, though the shear stress is the same [31]. Depending on the slope angle, it is classified into five different classes with the help of a natural break classification technique: <15°, 14.6–22.45°, 22.46–29.74°, 29.75–37.88°, >37.89°. (Figure 4a). Landslides usually happen in areas with high slopes. Hence, areas with high and extremely high slopes have a higher likelihood of experiencing landslides. On the other hand, landslide incidence is more or less stable in areas with a moderate to low slope angle [46].

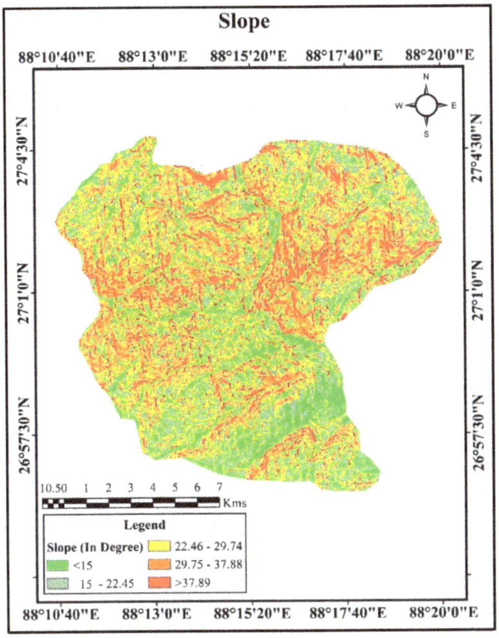

(a)

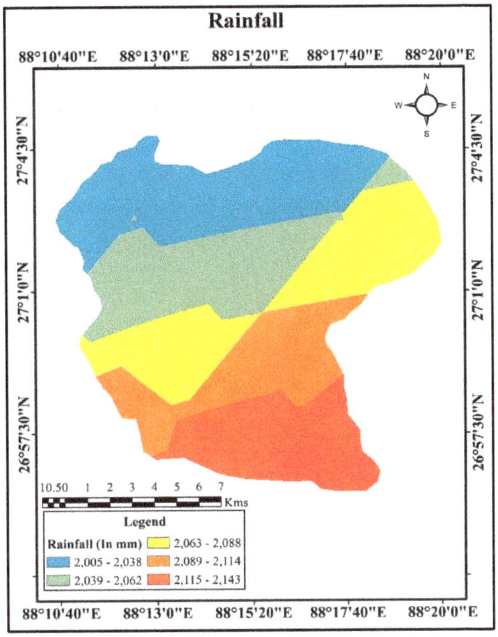

(b)

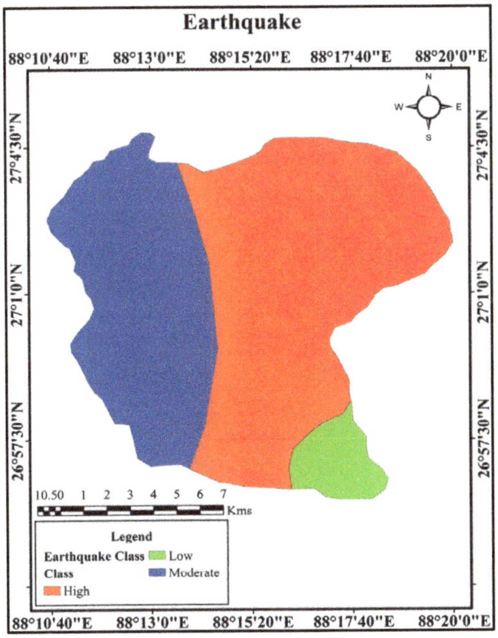

(c)

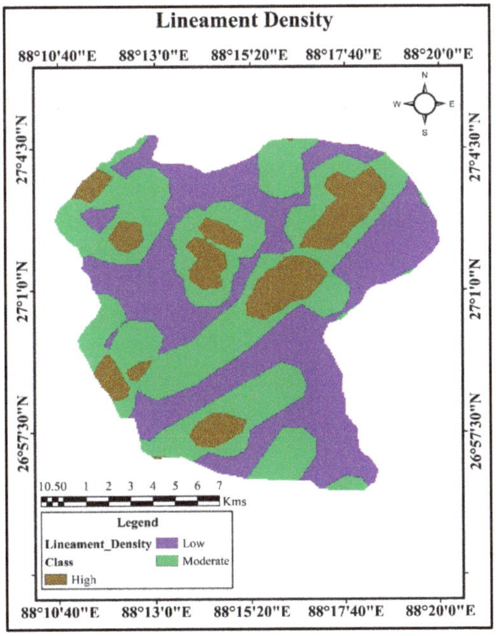

(d)

Figure 4. *Cont.*

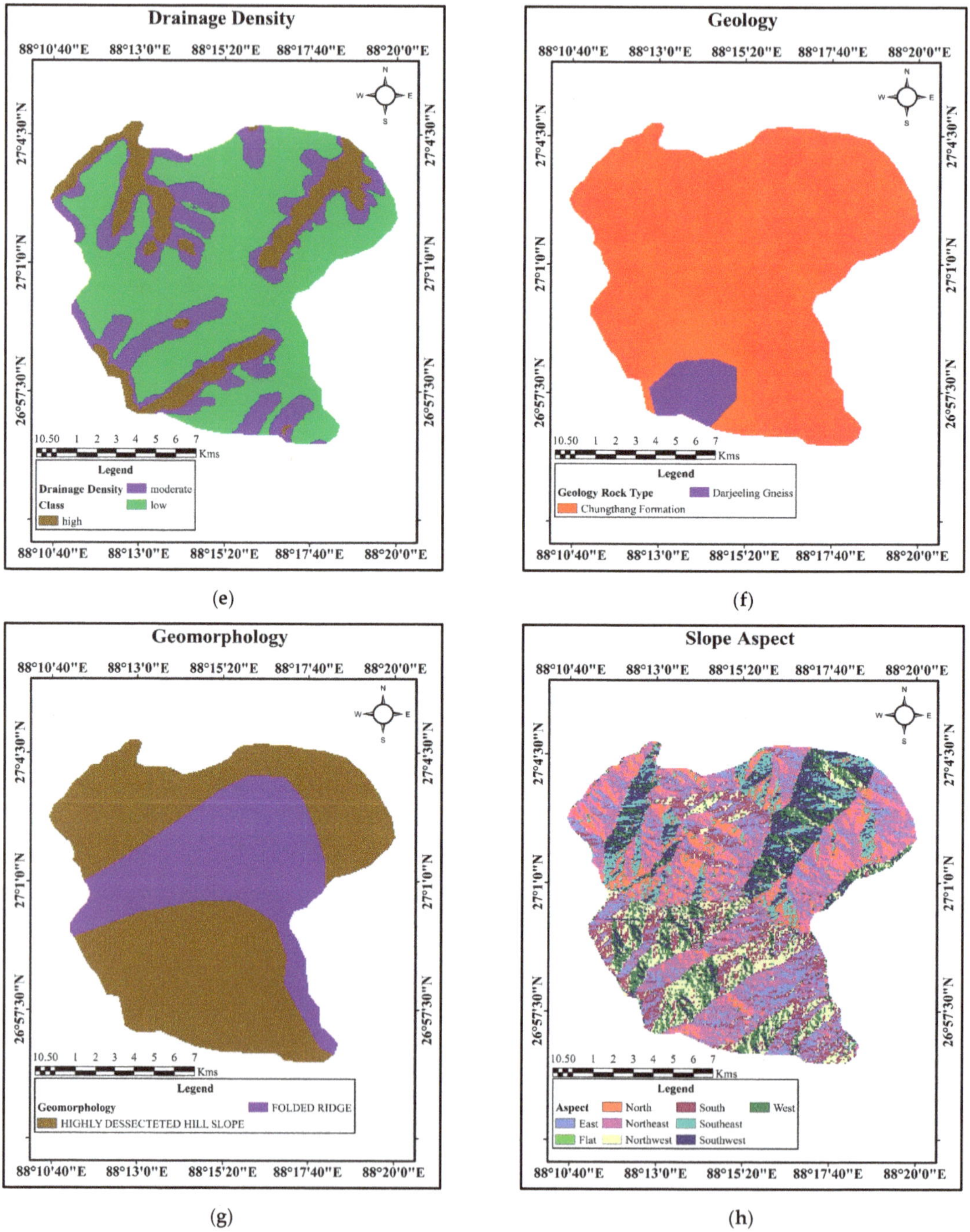

Figure 4. *Cont.*

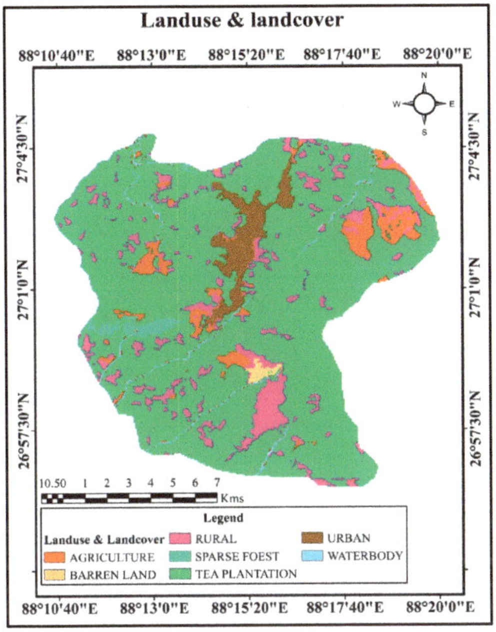

(i)

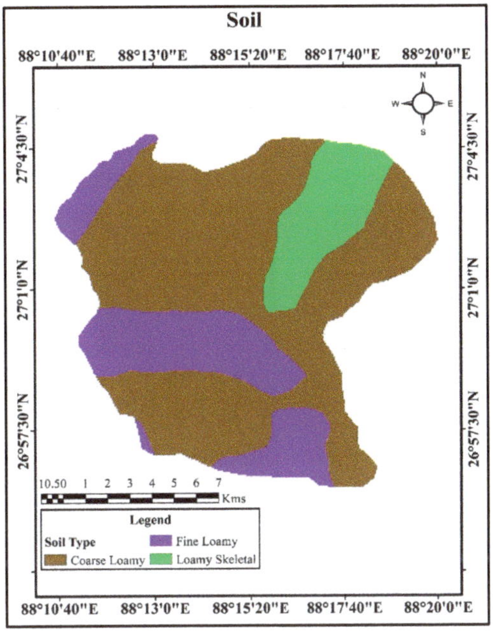

(j)

Figure 4. Landslide conditioning factor: (**a**) slope, (**b**) rainfall, (**c**) earthquake, (**d**) lineament density, (**e**) drainage density, (**f**) geology, (**g**) geomorphology, (**h**) aspect, (**i**) LULC, (**j**) soil.

2.5.2. Rainfall

As rainfall is one of the triggering factors for landslides, May to July is the most vulnerable period [47,48]. The inverse distance weighting (IDW) method for interpolation is used to prepare a rainfall map (Figure 4b). For the rainfall map, the last ten years of data from the Darjeeling and Kalimpong areas were collected from the Climate Research Unit (CRU). The range of rainfall data are 2005–2143 mm, and are further divided into five classes: 2005–2038 mm, 2039–2062 mm, 2063–2088 mm, 2089–2114 mm, and 2115–2143 mm.

2.5.3. Earthquake

A landslide due to an earthquake can be triggered either by increased shear stress due to acceleration horizontally, or decreased strength of materials [49]. When any hilly area is seismically activated, there is potential for an earthquake to occur [49]. The Eastern Himalayas region is very active seismically. A map of earthquakes has been created using point data from the National Centre for Seismology in New Delhi, which spans more than 200 years. The earthquake map (Figure 4c) has been created using the Inverse Distance Weighted tool in a GIS environment, and is divided into three classes.

2.5.4. Lineament Density

A study of landslide density helps us to understand the causative elements of landslides [50]. From the literature, it is understood that in the Himalayan region, landslide circumstances are generally very close to the local geological lineament [50–53]. Landslides are further vulnerable in the joint, fractured, and faulted zones [28]. The thematic map lineament density (Figure 4d) was generated from ALOS PALSAR DEM images after interpretation in a GIS environment. Rectilinear inclinations of morphological features, linear stream courses, structural alignments, and tonal contrast aid in interpreting the lineaments.

Despite discovering huge lineaments, no substantial thrusts or faults have been observed in the study area [12].

2.5.5. Drainage Density

Drainage density (DD) is a triggering factor for landslides, where drainage controls the landslide, and its densities correspond to the nature of geotechnical features as well as soil properties [40]. River Tista is the biggest river in the study area. Almost all streamlines initiated from high ridges flow down by making valleys and meet the Tista river. Infiltration is inversely related to drainage density [54]. Equation (1) is used to generate drainage density:

$$DD = \frac{L_d}{S_d} \quad (1)$$

where L_d is the measurement term of the drainage system and S_d represents drainage basin size. The Euclidean distance approach generates drainage density maps in GIS and is classified into three classes (Figure 4e).

2.5.6. Geology

The structure and content of different geology determine the strength of the rock [14]. Compared to soft rocks, the more substantial rocks provide excellent protection from the main thrusts and are less prone to landslides [55,56]. Geological rock properties represent texture, color, grain size, or properties. The geology map (Figure 4f) was prepared using the geology of the study area published by GSI.

2.5.7. Geomorphology

Geomorphology plays a significant role in the command of landslides. Inflated altitude regions are more vulnerable to landslides than lower altitude regions. The geomorphology map (Figure 4g) was extracted from the published map from the BHUKOSH portal (https://www.bhukosh.gsi.gov.in (accessed on 31 May 2022)) in the GIS environment.

2.5.8. Aspect

The slope aspect represents the orientation of the slope angle [57]. The south-east direction is more vulnerable than the rest of the directions [56]. East- and south-facing slopes are more likely to have landslides, according to the distribution of landslides [58]. Generally, north-facing slopes have more vegetation density than south-facing slopes [40]. Due to sun rotation, in the afternoon, west-facing slopes experience the hottest time of the day [59]. The slope aspect map (Figure 4h) is classified into 1. flat, 2. north, 3. north-east, 4. east, 5. south-east, 6. south, 7. south-west, 8. west, 9. north-west, and this map was generated in the GIS platform. The slope facing south and east makes up a significant component of the study area. The west-facing slope covers the lower part of the area, while a slope angle pointing north comprises a moderate to low slope zone.

2.5.9. Land Use and Land Cover

A key element contributing to the incidence of landslides is land use and land cover. Barren regions are more prone to landslides than lush trees. Deforestation is another essential factor that causes landslides. Areas covered with vegetation have seen that the big woody trees with long root systems have helped to improve slope stability in the area. Vegetation density is inversely proportional to landslides [55]. The LULC map (Figure 4i) was extracted using CARTOSAT-2 (MX) series imagery. The LULC map is organized into many categories, including rural and urban areas, sparse forests, tea plantations, agricultural land, barren land, and waterbodies. The southern portion of the area is where you will find agricultural land and developed regions with excellent road access. The bulk of the basin is covered in forest. In the southern zone, human intervention is more common than in the northern region, owing to accessibility issues.

2.5.10. Soil

The occurrence of landslides impacts topsoil cover on a slope. Regarding soil texture, rocky and sandy loam have higher landslide potential than silt loam, fine sandy loam, gravelly silt loam, and loam [22,60]. The soil map (Figure 4j) for the study area was created from a local soil map made by NBSS and LUP data on a scale of 1:500,000. Three different textural groups: fine loamy, coarse loamy, and loamy soil skeletal, are found in this region.

2.6. Method

2.6.1. Analytical Hierarchy Process (AHP)

The multi-criteria decision analysis (MCDA) method, AHP, is a systematic way to arrange and evaluate difficult mathematical choices. Since then, it has undergone substantial research and development. The factors affecting landslide susceptibility must be added while creating the susceptibility map. The layers have been weighted for this based on their significance. In AHP, the pair-wise comparison matrix method is implemented for the MCDM structure. Qualitative (subjective) and quantitative (objective) decision-making analysis is performed using the AHP method. The number of columns and rows in the comparison matrix is equal, where value 1 is placed on the diagonal of the matrix, and one side of the diagonal stores the scores. Each layer must rate against the other to construct a pair-wise confusion matrix. The rating value ranges from 1 to 9 (Table 2). Each pair-wise comparison matrix value represents the importance of two factors. It was thought that if attribute A is given a score of 9, meaning it is more important than attribute B, then B must receive a score of 1/9, meaning it is less important than A.

The consistency ratio (CR) and consistency index (CI) need to be calculated to validate the pair-wise comparison matrix [36,61,62]. We need to know the value of the 1970 Saaty-projected random consistency index (RI) (Table 3) to compute the CR [59]. If the value of CR < 0.10, we will accept the pair-wise comparison matrix. The following formula used for calculating CI and CR [63]:

$$\text{Consistency Index(CI)} = \frac{\lambda_{max} - n}{n - 1} \quad (2)$$

Table 2. The scale of weightage for AHP given by Saaty [63].

Scale	Definition	Explanation
1	Equally Important	The goal is equally affected by two elements.
3	A little more significant	One somewhat edges out the other in favor of experience and judgment.
5	Far more important	Judgment and experience greatly favor one another.
7	Very much more important	Experience and judgment outweigh each other quite significantly.
9	Definitely more important	The evidence favoring one over the other may have the most potential validity.
2, 4, 6, 8	Intermediate values	This is when the compromises are needed.

Table 3. RI value for calculating CR by Saaty [63].

NO.	1	2	3	4	5	6	7	8	9
RI	0	0	0.58	0.90	1.12	1.24	1.32	1.42	1.45

The principle eigenvalue is λ_{max} and n is the number of factors.

$$\text{Consistency Ratio(CR)} = \frac{\text{Consistency index(CI)}}{\text{Random Consistency Index(RI)}} \quad (3)$$

2.6.2. Multi-Collinearity Analysis of LFC

In the least squares regression analysis, the variance inflation factor (VIF) is used to assess the level of multi-collinearity. The exponent shows the multicollinearity-based increases in the coefficient [64]. The amount of multi-collinearity may be evaluated using the VIF's value. An observational rule states that multi-collinearity is high if the VIF value exceeds 5. The second approach to investigating multi-collinearity uses the tolerance (T) margin of error. Tolerance is a comparatively universal kind of multiple correlation coefficient [65]. Fully multi-collinear variables have no margin of error since they are entirely predictable from all independent variables. If a variable's tolerance value is one, it has zero correlation with any of the independent variables [33,65]. The criteria VIF and T revealed that the study was multi-collinear. It is shown below.

$$\text{Tolerance} = 1 - {R_j}^2 \quad (4)$$

$$\text{VIF} = \frac{1}{\text{Tolerance}} \quad (5)$$

Here, ${R_j}^2$ is the coefficient of determination (R-squared) of the model of the descriptive variable j as the response variable, and the other explanatory variables as the independent variable.

2.6.3. Model Accuracy Evaluation Method

A range of statistical indicators may be used to evaluate the efficacy of landslide hazard zonation models. To assess the performance of the prediction model, many validation model evaluation approaches were employed in this study, including sensitivity, receiver operating characteristics (ROC), specificity, area under curve (AUC), and accuracy. The effectiveness of landslide susceptibility prediction has recently been extensively studied using the ROC_AUC technique [26,66]. The inputs used to create the ROC curve were true positive, which refers to a landslide that was correctly anticipated on the axis-"X", and false positive, which refers to a landslide that was not correctly predicted on the axis-"Y". The models were statistically compared using AUC, a measure of the inclusive effectiveness of the models. ROC-AUC's accuracy is assessed as poor when the value is between 0.5 and 0.6, moderate when the value is between 0.6 and 0.7, high when the value is between 0.7 and 0.8, and exceptional when the value is between 0.8 and 0.9 [67]. Consequently, AUC values may be used to evaluate the accuracy of a prediction model.

The receiver operating characteristics (ROC) curve was used to assess the map and the model's precision [38,68–70]. The ROC curve was shown concerning the axis-"X" and axis-"Y", where the axis-"X" is the false positive rate (1-specificity) and the "Y" axis is the true positive rate (sensitivity) [25,31,70,71].

$$\text{X} - \text{axis} = \frac{\text{TN}}{\text{TN} + \text{FP}} = 1 - \text{Specificity} \quad (6)$$

$$\text{Y} - \text{axis} = \frac{\text{TP}}{\text{TP} + \text{FN}} = \text{Senstivity} \quad (7)$$

where TP = true positive, FP = false positive, TN = true negative, FN = false negative. A ROC curve was used for the assessment of the hazard map.

3. Result

3.1. Multi-Collinearity Analysis of LCF

Table 4 displays the findings of the multi-collinearity connection between the ten independent components employed in this investigation. The lineament density has the highest value (VIF = 1.423), according to the findings of multi-collinearity between these variables. In contrast, the land use and land cover component variance factor has the lowest value (VIF = 1.029). In addition, all variables have VIF values less than five, and all factors have high T values. Since each independent variable uniquely impacts the dependent variable, there is no multi-collinearity among the variables used in this study. Hence, with less collinearity, every layer is used to compute the final LSM map.

Table 4. Multi-collinearity analysis of landslide conditioning factors.

Sl. No.	Class	T	VIF
1	Aspect	0.751	1.331
2	Soil	0.711	1.406
3	Rainfall	0.703	1.422
4	Slope	0.932	1.073
5	Geology	0.844	1.185
6	Land use and land cover	0.972	1.029
7	Lineament Density	0.703	1.423
8	Geomorphology	0.797	1.255
9	Drainage Density	0.762	1.312
10	Earthquake	0.711	1.406

3.2. Assessment with AHP for Generating Landslide Susceptibility Mapping

The LSM map was prepared by integrating all ten layers: drainage density, rainfall, lineament density, slope, geology, LULC, geomorphology, soil, aspect, and earthquake. Each sub-factor of LCF is assigned weights for generating LSM [72]. Table 5 represents the weights of each sub-factor of LCFs. LSM is created by the AHP method in the GIS environment with ten selected LCFs by the formula

$$\mathrm{LSM} = \sum_{i=1}^{n} w_i * l_i \tag{8}$$

where l are the individual layers, w are their corresponding weights, and n is the number of layers (Table 6).

Table 5. LSM weights and rating system of different landslide conditioning factors.

Layer	Weight	Sub-Layer		Rating
		Class	Weight	
Slope (in degree)	0.288	<15°	0.034	1
		15–22.45°	0.067	3
		22.46–29.74°	0.134	5
		29.75–37.88°	0.260	7
		>37.89°	0.502	9
Rainfall (in mm)	0.210	<2038	0.034	1
		2039–2062	0.067	3
		2063–2088	0.134	5
		2089–2114	0.259	7
		>2115	0.503	9

Table 5. *Cont.*

Layer	Weight	Sub-Layer		Rating
		Class	Weight	
Earthquake	0.151	Low	0.106	1
		Moderate	0.260	3
		High	0.633	5
Lineament Density	0.109	Low	0.083	1
		Moderate	0.193	3
		High	0.723	7
Drainage Density	0.078	Low	0.083	1
		Moderate	0.193	3
		High	0.723	7
Geology	0.056	Chungthang Formation	0.750	1
		Darjeeling Gneiss	0.249	3
Geomorphology	0.040	Highly Dissertated Hill Slope	0.750	1
		Folded Ridge	0.249	3
Soil type	0.015	Coarse Loamy	0.103	1
		Fine Loamy	0.174	2
		Loamy Skeletal	0.722	6
Aspect	0.028	Flat	0.028	1
		North	0.071	3
		North-east	0.189	5
		East	0.071	3
		Southeast	0.071	3
		South	0.071	3
		South-west	0.353	7
		West	0.071	3
		North-west	0.071	7
Land use and Land Cover	0.020	Agriculture	0.138	7
		Barren Land	0.138	7
		Rural	0.030	2
		Sparse Forest	0.072	5
		Tea Plantation	0.298	9
		Urban	0.298	9
		waterbody	0.022	7

Table 6. The final pair-wise comparison matrix of landslide conditioning factors.

	SL	RF	EQ	LD	DD	GEO	GEM	AS	LULC	SO	Weight
Slope	1	2	3	4	5	6	7	8	9	9	0.288
Rainfall	1/2	1	2	3	4	5	6	7	8	9	0.210
Earthquake	1/3	1/2	1	2	3	4	5	6	7	8	0.151
Lineament Density	1/4	1/3	1/2	1	2	3	4	5	6	7	0.109
Drainage Density	1/5	1/4	1/3	1/2	1	2	3	4	5	6	0.078
Geology	1/6	1/5	1/4	1/3	1/2	1	2	3	4	5	0.056
Geomorphology	1/7	1/6	1/5	1/4	1/3	1/2	1	2	3	4	0.040
Aspect	1/8	1/7	1/6	1/5	1/4	1/3	1/2	1	2	3	0.028
LULC	1/9	1/8	1/7	1/6	1/5	1/4	1/3	1/2	1	2	0.020
Soil	1/9	1/9	1/8	1/7	1/6	1/5	1/4	1/3	1/2	1	0.015

Where SL = Slope, Rainfall = RF, Earthquake = EQ, LD = Lineament Density, DD = Drainage Density, GEO = Geology, GEM = Geomorphology, AS = Aspect, LULC = land use and land cover, SO = Soil.

From the outcome of the AHP model, the results were utilized to generate the final LSM and classified into low susceptibility, moderate susceptibility, high susceptibility, and very high susceptibility zones (Figure 5). Susceptibility mapping for the study area shows that the very high susceptibility zone covers 14.40% (19.312 km^2), the high susceptibility zone covers 32.20% (43.180 km^2), and the moderate susceptibility zone covers 36.77% (49.321 km^2). The low susceptibility zone covers 16.63% (22.303 km^2) of the total area. So, we observed that 83.37% of the total area comes under the moderate to very high zone.

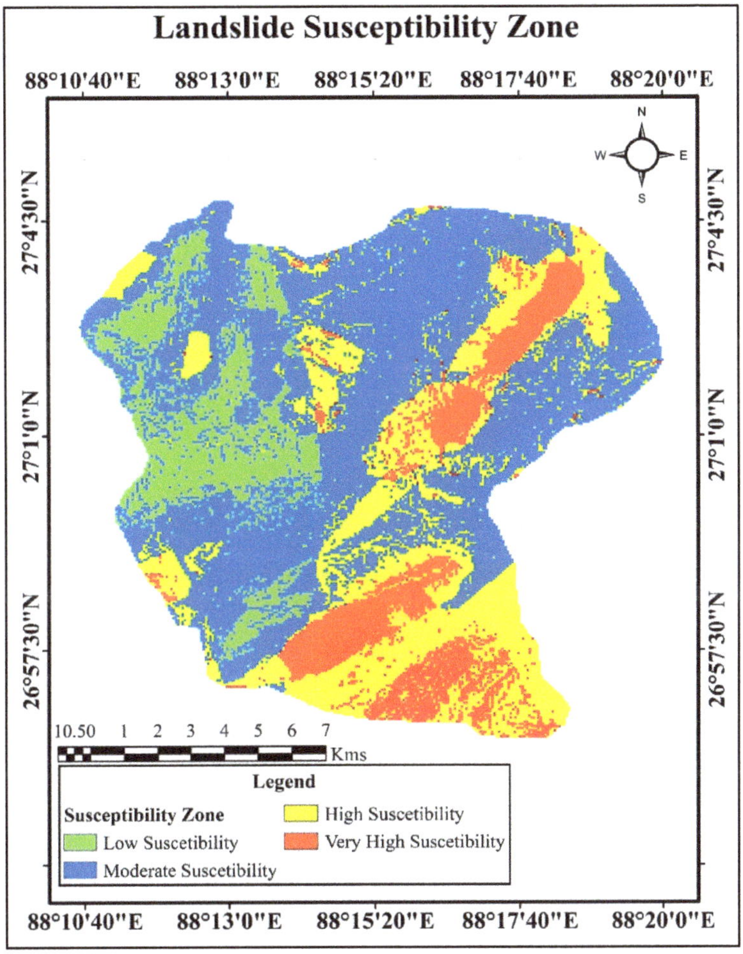

Figure 5. Final landslide susceptibility map of the part of the Darjeeling district.

3.3. Validation of LSM and Comparison with Field Data

AHP prepares an LSM of the study area. The AUC curve is drawn using a validating dataset. The dataset is divided into training (70%) and testing (30%) datasets. The testing dataset is used for the validation procedure. The AUC value obtained for this study region is 96.8%. An acceptable model has an AUC value from 0.70 to 0.8, a good hazard model has an AUC from 0.8 to 0.9, and an outstanding model has an AUC greater than 0.9. Our findings show that our model had an AUC of 96.8% (Figure 6), suggesting an extradentary model for susceptibility and functioning [57].

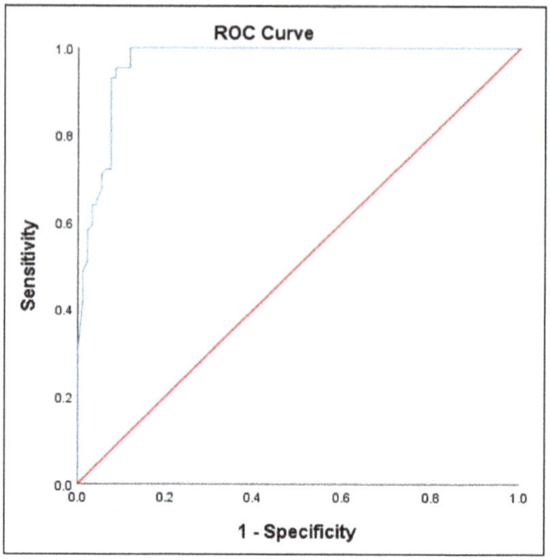

Figure 6. Overall model's accuracy is represented in AUC-ROC curve.

To evaluate the accuracy of a predictive model, it is important to compare the model's predictions to real-world data. In the case of a landslide susceptibility map developed using the analytic hierarchy process (AHP) model, one method of validation is to use a success rate curve. The success rate curve is constructed by plotting the cumulative percentage of observed landslide occurrence against the cumulative percentage of the study area in decreasing the landslide susceptibility index (LSI) values [21,73]. The area under the success rate curve can be used to assess the accuracy of the prediction [53]. To validate the AHP model's landslide susceptibility map, 30% of the data were randomly selected and used for model validation. The results of the success rate curve analysis for the AHP model showed an area under the curve of 0.813, corresponding to a prediction accuracy of 81.30% (Figure 7). The results show good accuracy of the model.

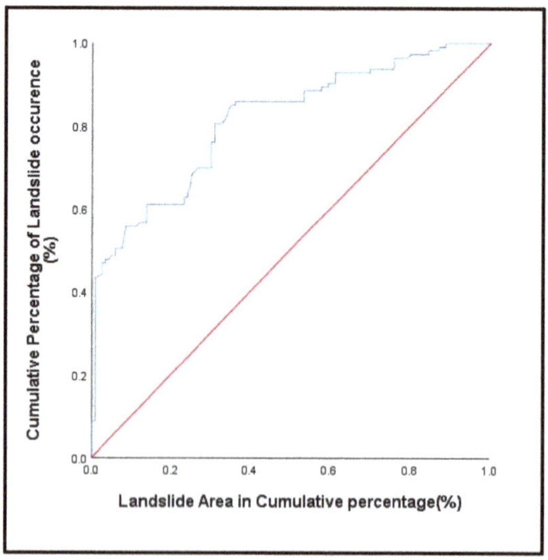

Figure 7. Accuracy of the model represented in a success rate curve.

4. Discussion

Since hilly and mountainous areas are more vulnerable to natural dangers, planning and development must be performed cautiously. One of these dangers is the possibility of a landslide. Landslides are likely to occur under similar conditions as those observed in previous studies [74,75]. Landslide susceptibility mappings are crucial in these areas because they provide decision-makers and planners with a better understanding of the process, allowing them to take the first line of defensive action [76]. It is challenging to create an accurate LSM that can be used to designate areas prone to landslides [77]. As a consequence, several solutions are being developed daily all around the globe to handle these challenges of accuracy and reliability [78]. Due to the exhaustive research of LSM, new methodologies have been created. Our research employs MCDA algorithms to develop an accurate model of landslide hazard zonation. LCFs help to initiate landslides. It is critical to choose the appropriate LCFs when developing an accurate landslide susceptibility model. LSM models provide strong predictive capability with less inaccuracy as a consequence. There are numerous LCFs, and they vary depending on the characteristics locally and globally. These landslide indicators are linked to the land use, climatic, geological, and geomorphological variables that regulate landslides. The area's edge is dominated by the convex slope, the middle section by the straight slope, and the bottom portion by the concave slope. There are no standards for selecting LCFs based on the variety of indicators and the characteristics of the region [79]. Much effort has been spent in choosing the most appropriate and stressful factors. The multi-collinearity test helps us to find the correlation between LCFs that may impact the overall accuracy of the models. Ten (10) LCFs were chosen as independent layers in the present research to assess the susceptibility of the area to landslides. The VIF was used to evaluate the LCFs' multi-collinearity. Our findings show that the layers have no multi-collinearity between them. As a result, all variables were included in this model. According to the study's results, one of the reliable approaches for LSM is AHP, which describes the weighted-overlay analysis technique with a multi-criteria decision approach. To classify the area into different susceptibility zones, ten (10) thematic layers were used, including rainfall, slope, earthquake, aspect, lithology, drainage density, lineament density, LULC, soil, and geomorphology. Figure 5 depicts a statistically generated landslide susceptibility map interpreted using the landslide hazard index (LHI) value. The model's min and max LHI values were 1393 and 5391, respectively, with a mean of 2606.88 and a standard deviation (s.d.) of 584.93 (Figure 8). Since this histogram showed that the values were unevenly distributed, the natural-break categorization approach was employed for zonation mapping [80]. As a result, four landslide hazard zones were identified and mapped: low susceptibility, moderate susceptibility, high susceptibility, and extremely high susceptibility (Figure 5). The analytical area % of the extremely high susceptibility zone for AHP is 11.68. (Figure 9). According to the hazard area, AHP falls within the high to extremely high zone with 37.86 percent (Figure 10). As a result, we may assume that the whole study area is in the high-risk zone. The AHP methodology is widely used in landslide susceptibility mapping. However, it has some limitations in the context of the selected area. One limitation of using AHP methodology for landslide susceptibility mapping in Darjeeling is that it relies heavily on expert opinion, which may be subjective and may vary from person to person. The quality and availability of data can also impact the accuracy of the AHP model, especially in areas where data are limited. Another limitation of AHP methodology for landslide susceptibility mapping in Darjeeling is that it requires extensive data inputs, which can be challenging to obtain and verify in remote and mountainous regions. The AHP method assumes that the weights assigned to each criterion remain constant over time, which may not be the case in dynamic and rapidly changing environments such as Darjeeling. The AHP method does not consider the temporal and spatial variability of landslide triggering factors, which can vary significantly depending on the location and time of year. The AHP method does not account for the complex interactions and feedback mechanisms between factors, which can lead to underestimating or overestimating the risk of landslides in certain areas. Moreover, the trustworthiness of the LSM map is dependent on the findings obtained from the ROC-AUC and success rate curve. The GSI's released data were used to create a landslide inventory map. A total of 114 landslides were recorded, with 79 (70%) serving as training data and 35 (30%) serving as testing data. The AHP model generated maps with AUC values of 96.8 percent. The success rate curve was also plotted for determining the accuracy of the model, which was found to be 81.30 percent. Since time is an important factor in hazard research, this finding is useful in an emergency. One may conclude that the models' accuracy is comparable. We recommend using the AHP model in landslide investigations because it can generate excellent and reliable landslide hazard maps for risk reduction and management planning.

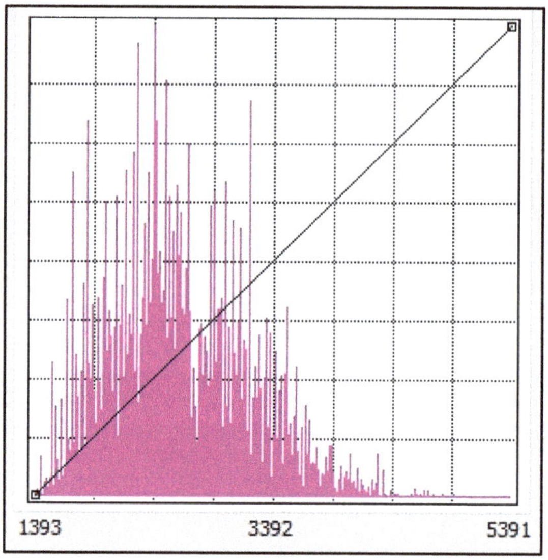

Figure 8. Histogram analysis of the LSM model.

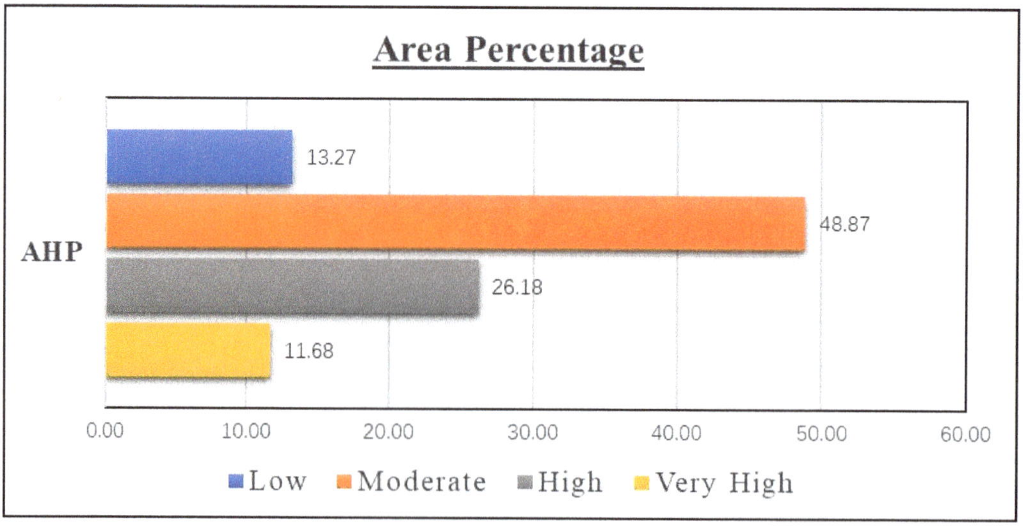

Figure 9. The graph shows the area percentage of different landslide susceptibility classes.

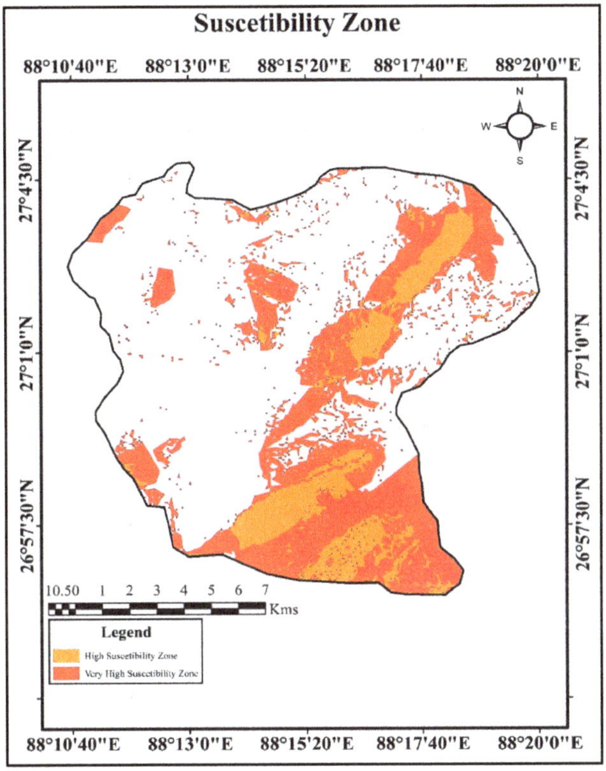

Figure 10. The map shows the high to very high susceptibility zone of the study area.

5. Conclusions

One of the deadliest natural dangers in the mountainous terrain is a landslide. Due to man-made and natural events, the highly damaged mountainous landscape is vulnerable to landslides (such as earthquakes, climate change, and human intervention, respectively). Because of these threats, communities in West Bengal's northern Himalayan highlands live in a never-ending nightmare, leading to socio-economic losses. To protect property and livelihood, interim and long-term solutions to reduce the risk of landslides in this region are required. To aid in infrastructural development and socio-economic development planning, we must identify and map the most vulnerable areas. LSM may be a critical tool for assessing risk management in rugged terrain. For many years, the application of MCDA models for landslide hazard assessment has produced amazingly efficient and exact results. The primary goal of this study is to evaluate the efficacy of MCDA models AHP and to determine the most accurate and helpful approach for detecting landslide-prone places in the research region. The result is based on AHP, classified into four landslide susceptibility classes. The classes are very high, high, moderate, and low landslide susceptibility. The results show that 26.18% of the total area falls under high landslide susceptibility zones, and 11.68% falls into very high susceptibility areas. So, the entire 37.86% area is in a high to very high susceptibility zone. Our findings (based on LSM and AUC) show that the usual strategy is successful. The AHP MCDA algorithm achieved 96.8 percent for the AUC-ROC curve and 81.3 percent for the success rate curve. From the result, we can identify that most areas which fall under the high to very high susceptibility classes are in the eastern, south-eastern, and southern parts of the area. The pattern of landslide sites reveals that landslides often occur on heavily dissected slopes. Less debris-filled vegetation areas can increase the risk of landslides. The area's combined landslip categorization is debris slide. Hence, it can be concluded that the most likely place for a landslide to occur in this study area is a severely dissected hill with debris and low to moderate vegetation, usually sparse forest, barren land, or a tea plantation area. This research successfully identified the landslide-prone locations, and the best location for planners for development in the hilly region of Darjeeling, West Bengal, India. Finally, the LSM established in this research may be used by decision-makers, land-use planners, and other governmental and non-governmental entities as an efficient tool for optimizing resource management, infrastructure development, and human activity in the studied area.

Author Contributions: Conceptualization, A.S.; methodology, A.S., V.G.K.V. and A.B.; software, A.S.; validation, A.S., V.G.K.V. and A.B.; formal analysis, A.S.; investigation, A.S., V.G.K.V., A.B. and S.K.; resources, A.S., V.G.K.V. and A.B.; writing—original draft preparation, A.S.; writing—review and editing, A.S., V.G.K.V., A.B. and S.K.; visualization, A.S., V.G.K.V., A.B. and S.K.; supervision, V.G.K.V. and A.B. All authors have read and agreed to the published version of the manuscript.

Funding: This research received no external funding.

Acknowledgments: The authors are thankful to the Department of Mining Engineering, Indian Institute of Technology (Indian School of Mines), Dhanbad, for providing an adequate environment for the research work. Authors thanks Lakshya Tripathi for the support during the study.

Conflicts of Interest: The authors declare no conflict of interest.

References

1. Shano, L.; Raghuvanshi, T.K.; Meten, M. Landslide susceptibility evaluation and hazard zonation techniques—A review. *Geoenvironmental Disasters* **2020**, *7*, 18. [CrossRef]
2. Kalantar, B.; Ueda, N.; Saeidi, V.; Ahmadi, K.; Halin, A.A.; Shabani, F. Landslide Susceptibility Mapping: Machine and Ensemble Learning Based on Remote Sensing Big Data. *Remote Sens.* **2020**, *12*, 1737. [CrossRef]
3. Karagianni, A.; Lazos, I.; Chatzipetros, A. Remote sensing techniques in disaster management: Amynteon mine landslides, Greece. In *Intelligent Systems for Crisis Management: Gi4DM 2018*, 1st ed.; Springer: Berlin/Heidelberg, Germany, 2019; pp. 209–235.
4. Liu, P.; Wei, Y.; Wang, Q.; Chen, Y.; Xie, J. Research on Post-Earthquake Landslide Extraction Algorithm Based on Improved U-Net Model. *Remote Sens.* **2020**, *12*, 894. [CrossRef]
5. Huang, F.; Yin, K.; Huang, J.; Gui, L.; Wang, P. Landslide susceptibility mapping based on self-organizing-map network and extreme learning machine. *Eng. Geol.* **2017**, *223*, 11–22. [CrossRef]
6. Lagomarsino, D.; Tofani, V.; Segoni, S.; Catani, F.; Casagli, N. A Tool for Classification and Regression Using Random Forest Methodology: Applications to Landslide Susceptibility Mapping and Soil Thickness Modeling. *Environ. Model. Assess.* **2017**, *22*, 201–214. [CrossRef]
7. Legorreta Paulín, G.; Bursik, M.; Hubp, J.L.; Mejía, L.M.P.; Aceves Quesada, F. A GIS method for landslide inventory and susceptibility mapping in the Río El Estado watershed, Pico de Orizaba volcano, México. *Nat. Hazards* **2014**, *71*, 229–241. [CrossRef]
8. Shirzadi, A.; Bui, D.T.; Pham, B.T.; Solaimani, K.; Chapi, K.; Kavian, A.; Shahabi, H.; Revhaug, I. Shallow landslide susceptibility assessment using a novel hybrid intelligence approach. *Environ. Earth Sci.* **2017**, *76*, 60. [CrossRef]
9. Tsangaratos, P.; Ilia, I.; Hong, H.; Chen, W.; Xu, C. Applying Information Theory and GIS-based quantitative methods to produce landslide susceptibility maps in Nancheng County, China. *Landslides* **2017**, *14*, 1091–1111. [CrossRef]
10. Pham, B.T.; Jaafari, A.; Prakash, I.; Bui, D.T. A novel hybrid intelligent model of support vector machines and the MultiBoost ensemble for landslide susceptibility modeling. *Bull. Eng. Geol. Environ.* **2019**, *78*, 2865–2886. [CrossRef]
11. Wang, H.; Zhang, L.; Yin, K.; Luo, H.; Li, J. Landslide identification using machine learning. *Geosci. Front.* **2020**, *12*, 351–364. [CrossRef]
12. Saha, A.; Villuri, V.G.K.; Bhardwaj, A. Development and Assessment of GIS-Based Landslide Susceptibility Mapping Models Using ANN, Fuzzy-AHP, and MCDA in Darjeeling Himalayas, West Bengal, India. *Land* **2022**, *11*, 1711. [CrossRef]
13. Zhang, S.; Bai, L.; Li, Y.; Li, W.; Xie, M. Comparing Convolutional Neural Network and Machine Learning Models in Landslide Susceptibility Mapping: A Case Study in Wenchuan County. *Front. Environ. Sci.* **2022**, *10*, 886841. [CrossRef]
14. Selamat, S.N.; Majid, N.A.; Taha, M.R.; Osman, A. Landslide Susceptibility Model Using Artificial Neural Network (ANN) Approach in Langat River Basin, Selangor, Malaysia. *Land* **2022**, *11*, 833. [CrossRef]
15. Merghadi, A.; Yunus, A.P.; Dou, J.; Whiteley, J.; Thaipham, B.; Bui, D.T.; Avtar, R.; Abderrahmane, B. Machine learning methods for landslide susceptibility studies: A comparative overview of algorithm performance. *Earth Sci. Rev.* **2020**, *207*, 103225. [CrossRef]
16. Dikshit, A.; Satyam, D.N.; Towhata, I. Early warning system using tilt sensors in Chibo, Kalimpong, Darjeeling Himalayas, India. *Nat. Hazards* **2018**, *94*, 727–741. [CrossRef]

17. Borgatti, L.; Soldati, M. Landslides as a geomorphological proxy for climate change: A record from the Dolomites (northern Italy). *Geomorphology* **2010**, *120*, 56–64. [CrossRef]
18. Aleotti, P.; Chowdhury, R. Landslide hazard assessment: Summary review and new perspectives. *Bull. Eng. Geol. Environ.* **1999**, *58*, 21–44. [CrossRef]
19. Mallick, J.; Singh, R.K.; AlAwadh, M.A.; Islam, S.; Khan, R.A.; Qureshi, M.N. GIS-based landslide susceptibility evaluation using fuzzy-AHP multi-criteria decision-making techniques in the Abha Watershed, Saudi Arabia. *Environ. Earth Sci.* **2018**, *77*, 276. [CrossRef]
20. Kanwal, S.; Atif, S.; Shafiq, M. GIS based landslide susceptibility mapping of northern areas of Pakistan, a case study of Shigar and Shyok Basins. *Geomat. Nat. Hazards Risk* **2017**, *8*, 348–366. [CrossRef]
21. Das, S.; Sarkar, S.; Kanungo, D.P. GIS-based landslide susceptibility zonation mapping using the analytic hierarchy process (AHP) method in parts of Kalimpong Region of Darjeeling Himalaya. *Environ. Monit. Assess.* **2022**, *194*, 234. [CrossRef]
22. Anbalagan, R.; Kumar, R.; Parida, S.; Lakshmanan, K. GIS Based Post Earthquake Landslide Hazard Zonation Mapping of Lachung Basin, Sikkim. *Int. J. Emerg. Technol. Adv. Eng.* **2014**, *4*, 431–441.
23. Saha, A.; Villuri, V.G.K.; Bhardwaj, A. Assessment and preparation of landslide susceptibility zonation map by geospatial method using remote sensing and GIS. In Proceedings of the ACRS 2020—41st Asian Conference on Remote Sensing, Deqing, China, 9–11 November 2020; Volume 1–3, pp. 461–470.
24. Sarma, C.P.; Dey, A.; Krishna, A.M. Influence of digital elevation models on the simulation of rainfall-induced landslides in the hillslopes of Guwahati, India. *Eng. Geol.* **2020**, *268*, 105523. [CrossRef]
25. Huang, W.; Ding, M.; Li, Z.; Zhuang, J.; Yang, J.; Li, X.; Meng, L.; Zhang, H.; Dong, Y. An Efficient User-Friendly Integration Tool for Landslide Susceptibility Mapping Based on Support Vector Machines: SVM-LSM Toolbox. *Remote Sens.* **2022**, *14*, 3408. [CrossRef]
26. Zhao, P.; Masoumi, Z.; Kalantari, M.; Aflaki, M.; Mansourian, A. A GIS-Based Landslide Susceptibility Mapping and Variable Importance Analysis Using Artificial Intelligent Training-Based Methods. *Remote Sens.* **2022**, *14*, 211. [CrossRef]
27. Pal, S.C.; Das, B.; Malik, S. Potential Landslide Vulnerability Zonation Using Integrated Analytic Hierarchy Process and GIS Technique of Upper Rangit Catchment Area, West Sikkim, India. *J. Indian Soc. Remote Sens.* **2019**, *47*, 1643–1655. [CrossRef]
28. Basu, T.; Pal, S. RS-GIS based morphometrical and geological multi-criteria approach to the landslide susceptibility mapping in Gish River Basin, West Bengal, India. *Adv. Space Res.* **2019**, *63*, 1253–1269. [CrossRef]
29. Yunis, C.R.C.; López, R.S.; Cruz, S.M.O.; Castillo, E.B.; López, J.O.S.; Trigoso, D.I.; Briceño, N.B.R. Land Suitability for Sustainable Aquaculture of Rainbow Trout (Oncorhynchus mykiss) in Molinopampa (Peru) Based on RS, GIS, and AHP. *ISPRS Int. J. Geo-Inf.* **2020**, *9*, 28. [CrossRef]
30. Ma, J.; Liu, X.; Niu, X.; Wang, Y.; Wen, T. Forecasting of Landslide Displacement Using a Probability-Scheme Combination Ensemble Prediction Technique. *Int. J. Environ. Res. Public Health* **2020**, *17*, 4788. [CrossRef] [PubMed]
31. Ahmed, B. Landslide susceptibility mapping using multi-criteria evaluation techniques in Chittagong Metropolitan Area, Bangladesh. *Landslides* **2015**, *12*, 1077–1095. [CrossRef]
32. Calligaris, C.; Poretti, G.; Tariq, S.; Melis, M.T. First steps towards a landslide inventory map of the Central Karakoram National Park. *Eur. J. Remote Sens.* **2013**, *46*, 272–287. [CrossRef]
33. Arabameri, A.; Pradhan, B.; Rezaei, K.; Lee, S.; Sohrabi, M. An ensemble model for landslide susceptibility mapping in a forested area. *Geocarto Int.* **2020**, *35*, 1680–1705. [CrossRef]
34. Arabameri, A.; Rezaei, K.; Cerda, A.; Lombardo, L.; Rodrigo-Comino, J. GIS-based groundwater potential mapping in Shahroud plain, Iran. A comparison among statistical (bivariate and multivariate), data mining and MCDM approaches. *Sci. Total Environ.* **2019**, *658*, 160–177. [CrossRef] [PubMed]
35. Mahalingam, R.; Olsen, M.J. Evaluation of the influence of source and spatial resolution of DEMs on derivative products used in landslide mapping. *Geomat. Nat. Hazards Risk* **2016**, *7*, 1835–1855. [CrossRef]
36. Feizizadeh, B.; Blaschke, T. GIS-multicriteria decision analysis for landslide susceptibility mapping: Comparing three methods for the Urmia lake basin, Iran. *Nat. Hazards* **2013**, *65*, 2105–2128. [CrossRef]
37. Radwan, F.; Alazba, A.A.; Mossad, A. Flood risk assessment and mapping using AHP in arid and semiarid regions. *Acta Geophys.* **2019**, *67*, 215–229. [CrossRef]
38. Arabameri, A.; Pradhan, B.; Rezaei, K.; Conoscenti, C. Gully erosion susceptibility mapping using GIS-based multi-criteria decision analysis techniques. *Catena* **2019**, *180*, 282–297. [CrossRef]
39. Rawat, A.; Kumar, D.; Chatterjee, R.S.; Kumar, H. A GIS-based liquefaction susceptibility mapping utilising the morphotectonic analysis to highlight potential hazard zones in the East Ganga plain. *Environ. Earth Sci.* **2022**, *81*, 358. [CrossRef]
40. Singh, P.; Gupta, A.; Singh, M. Hydrological inferences from watershed analysis for water resource management using remote sensing and GIS techniques. *Egypt. J. Remote Sens. Space Sci.* **2014**, *17*, 111–121. [CrossRef]
41. Ghosh, S.; van Westen, C.J.; Carranza, E.J.M.; Jetten, V.G.; Cardinali, M.; Rossi, M.; Guzzetti, F. Generating event-based landslide maps in a data-scarce Himalayan environment for estimating temporal and magnitude probabilities. *Eng. Geol.* **2012**, *128*, 49–62. [CrossRef]
42. Varnes, D.J. *Landslide Hazard Zonation: A Review of Principles and Practice*; Amazon: Charlotte, NC, USA, 1984.
43. Fell, R.; Whitt, G.; Miner, T.; Flentje, P. Guidelines for landslide susceptibility, hazard and risk zoning for land use planning. *Eng. Geol.* **2008**, *102*, 83–84.

44. Achour, Y.; Boumezbeur, A.; Hadji, R.; Chouabbi, A.; Cavaleiro, V.; Bendaoud, E.A. Landslide susceptibility mapping using analytic hierarchy process and information value methods along a highway road section in Constantine, Algeria. *Arab. J. Geosci.* **2017**, *10*, 194. [CrossRef]
45. Roy, J.; Saha, S. Landslide susceptibility mapping using knowledge driven statistical models in Darjeeling District, West Bengal, India. *Geoenvironmental Disasters* **2019**, *6*, 11. [CrossRef]
46. Tanyaş, H.; van Westen, C.J.; Allstadt, K.E.; Jibson, R.W. Factors controlling landslide frequency–area distributions. *Earth Surf. Process Landforms* **2019**, *44*, 900–917. [CrossRef]
47. Chikalamo, E.E.; Mavrouli, O.C.; Ettema, J.; van Westen, C.J.; Muntohar, A.S.; Mustofa, A. Satellite-derived rainfall thresholds for landslide early warning in Bogowonto Catchment, Central Java, Indonesia. *Int. J. Appl. Earth Obs. Geoinf.* **2020**, *89*, 102093. [CrossRef]
48. Sengupta, A.; Gupta, S.; Anbarasu, K. Rainfall thresholds for the initiation of landslide at Lanta Khola in north Sikkim, India. *Nat. Hazards* **2010**, *52*, 31–42. [CrossRef]
49. Farooq Ahmed, M.; Rogers, J.D.; Ismail, E.H. A regional level preliminary landslide susceptibility study of the upper Indus river basin. *Eur. J. Remote Sens.* **2014**, *47*, 343–373. [CrossRef]
50. Ramli, M.F.; Yusof, N.; Yusoff, M.K.; Juahir, H.; Shafri, H.Z.M. Lineament mapping and its application in landslide hazard assessment: A review. *Bull. Eng. Geol. Environ.* **2010**, *69*, 215–233. [CrossRef]
51. Mathew, J.; Jha, V.K.; Rawat, G.S. Application of binary logistic regression analysis and its validation for landslide susceptibility mapping in part of Garhwal Himalaya, India. *Int. J. Remote Sens.* **2007**, *28*, 2257–2275. [CrossRef]
52. Salui, C.L. Methodological Validation for Automated Lineament Extraction by LINE Method in PCI Geomatica and MATLAB based Hough Transformation. *J. Geol. Soc. India* **2018**, *92*, 321–328. [CrossRef]
53. Kumar, A.; Sharma, R.K.; Bansal, V.K. GIS-based comparative study of information value and frequency ratio method for landslide hazard zonation in a part of mid-Himalaya in Himachal Pradesh. *Innov. Infrastruct. Solut.* **2019**, *4*, 28. [CrossRef]
54. Bera, A.; Mukhopadhyay, B.; Das, D. Morphometric Analysis of Adula River Basin in Maharashtra, India using GIS and Remote Sensing techniques. *Geo-Spat. Data Nat. Resour.* **2018**, *2018*, 13–35. [CrossRef]
55. Kanungo, D.P.; Arora, M.K.; Sarkar, S.; Gupta, R.P. A comparative study of conventional, ANN black box, fuzzy and combined neural and fuzzy weighting procedures for landslide susceptibility zonation in Darjeeling Himalayas. *Eng. Geol.* **2006**, *85*, 347–366. [CrossRef]
56. Sarkar, S.; Roy, A.K.; Martha, T.R. Landslide susceptibility assessment using Information Value Method in parts of the Darjeeling Himalayas. *J. Geol. Soc. India* **2013**, *82*, 351–362. [CrossRef]
57. Youssef, A.M.; Pradhan, B.; Dikshit, A.; Al-Katheri, M.M.; Matar, S.S.; Mahdi, A.M. Landslide susceptibility mapping using CNN-1D and 2D deep learning algorithms: Comparison of their performance at Asir Region, KSA. *Bull. Eng. Geol. Environ.* **2022**, *81*, 165. [CrossRef]
58. Dhakal, A.S.; Amada, T.; Aniya, M. Landslide hazard mapping and its evaluation using GIS: An investigation of sampling schemes for a grid-cell based quantitative method. *Photogramm. Eng. Remote Sens.* **2000**, *66*, 981–989.
59. Bera, A.; Mukhopadhyay, B.P.; Das, D. Landslide hazard zonation mapping using multi-criteria analysis with the help of GIS techniques: A case study from Eastern Himalayas, Namchi, South Sikkim. *Nat. Hazards* **2019**, *96*, 935–959. [CrossRef]
60. Lee, S.; Hong, S.-M.; Jung, H.-S. A Support Vector Machine for Landslide Susceptibility Mapping in Gangwon Province, Korea. *Sustainability* **2017**, *9*, 48. [CrossRef]
61. Chen, Y.; Khan, S.; Paydar, Z. To retire or expand? A fuzzy GIS-based spatial multi-criteria evaluation framework for irrigated agriculture. *Irrig. Drain.* **2010**, *59*, 174–188. [CrossRef]
62. Kolat, C.; Ulusay, R.; Suzen, M.L. Development of geotechnical microzonation model for Yenisehir (Bursa, Turkey) located at a seismically active region. *Eng. Geol.* **2012**, *127*, 36–53. [CrossRef]
63. Wind, Y.; Saaty, T.L. Marketing applications of the analytic hierarchy process. *Manag. Sci.* **1980**, *26*, 641–658. [CrossRef]
64. Arabameri, A.; Pradhan, B.; Rezaei, K. Gully erosion zonation mapping using integrated geographically weighted regression with certainty factor and random forest models in GIS. *J. Environ. Manag.* **2019**, *232*, 928–942. [CrossRef]
65. Arabameri, A.; Pradhan, B.; Rezaei, K.; Sohrabi, M.; Kalantari, Z. GIS-based landslide susceptibility mapping using numerical risk factor bivariate model and its ensemble with linear multivariate regression and boosted regression tree algorithms. *J. Mt. Sci.* **2019**, *16*, 595–618. [CrossRef]
66. Sengupta, A.; Nath, S.K. GIS-Based Landslide Susceptibility Mapping in Eastern Boundary Zone of Northeast India in Compliance with Indo-Burmese Subduction Tectonics. In *Geospatial Technology for Environmental Hazards*; Springer: Berlin/Heidelberg, Germany, 2022; Volume 1, pp. 19–37. [CrossRef]
67. Tien Bui, D.; Tuan, T.A.; Klempe, H.; Pradhan, B.; Revhaug, I. Spatial prediction models for shallow landslide hazards: A comparative assessment of the efficacy of support vector machines, artificial neural networks, kernel logistic regression, and logistic model tree. *Landslides* **2016**, *13*, 361–378. [CrossRef]
68. Anbalagan, R.; Kumar, R.; Lakshmanan, K.; Parida, S.; Neethu, S. Landslide hazard zonation mapping using frequency ratio and fuzzy logic approach, a case study of Lachung Valley, Sikkim. *Geoenvironmental Disasters* **2015**, *2*, 6. [CrossRef]
69. Samui, P. Slope stability analysis: A support vector machine approach. *Environ. Geol.* **2008**, *56*, 255–267. [CrossRef]
70. Avand, M.; Moradi, H.; Lasboyee, M.R. Spatial modeling of flood probability using geo-environmental variables and machine learning models, case study: Tajan watershed, Iran. *Adv. Space Res.* **2021**, *67*, 3169–3186. [CrossRef]

71. Youssef, A.M.; Pradhan, B.; Dikshit, A.; Mahdi, A.M. Comparative study of convolutional neural network (CNN) and support vector machine (SVM) for flood susceptibility mapping: A case study at Ras Gharib, Red Sea, Egypt. *Geocarto Int.* **2022**, *37*, 11088–11115. [CrossRef]
72. Das, S.; Sarkar, S.; Kanungo, D.P. *A Critical Review on Landslide Susceptibility Zonation: Recent Trends, Techniques, and Practices in INDIAN Himalaya*; Springer: Dordrecht, The Netherlands, 2023; Volume 115, ISBN 0123456789.
73. Mahdadi, F.; Boumezbeur, A.; Hadji, R.; Kanungo, D.P.; Zahri, F. GIS-based landslide susceptibility assessment using statistical models: A case study from Souk Ahras province, N-E Algeria. *Arab. J. Geosci.* **2018**, *11*, 476. [CrossRef]
74. Kranjčić, N.; Medak, D.; Župan, R.; Rezo, M. Support Vector Machine Accuracy Assessment for Extracting Green Urban Areas in Towns. *Remote Sens.* **2019**, *11*, 655. [CrossRef]
75. Kumar, D.; Thakur, M.; Dubey, C.S.; Shukla, D.P. Landslide susceptibility mapping\& prediction using support vector machine for Mandakini River Basin, Garhwal Himalaya, India. *Geomorphology* **2017**, *295*, 115–125.
76. Roccati, A.; Paliaga, G.; Luino, F.; Faccini, F.; Turconi, L. GIS-Based Landslide Susceptibility Mapping for Land Use Planning and Risk Assessment. *Land* **2021**, *10*, 162. [CrossRef]
77. Mandal, K.; Saha, S.; Mandal, S. Applying deep learning and benchmark machine learning algorithms for landslide susceptibility modelling in Rorachu river basin of Sikkim Himalaya, India. *Geosci. Front.* **2021**, *12*, 101203. [CrossRef]
78. Prakash, N.; Manconi, A.; Loew, S. A new strategy to map landslides with a generalized convolutional neural network. *Sci. Rep.* **2021**, *11*, 1–15. [CrossRef] [PubMed]
79. Park, S.; Kim, J. Landslide Susceptibility Mapping Based on Random Forest and Boosted Regression Tree Models, and a Comparison of Their Performance. *Appl. Sci.* **2019**, *9*, 942. [CrossRef]
80. Constantin, M.; Bednarik, M.; Jurchescu, M.C.; Vlaicu, M. Landslide susceptibility assessment using the bivariate statistical analysis and the index of entropy in the Sibiciu Basin (Romania). *Environ. Earth Sci.* **2011**, *63*, 397–406. [CrossRef]

applied sciences

MDPI

Article

Revalidation Technique on Landslide Susceptibility Modelling: An Approach to Local Level Disaster Risk Management in Kuala Lumpur, Malaysia

Elanni Affandi [1], Tham Fatt Ng [1,*], Joy J. Pereira [2], Ferdaus Ahmad [3] and Vanessa J. Banks [4]

1 Department of Geology, University of Malaya, Kuala Lumpur 50603, Malaysia
2 Southeast Asia Disaster Prevention Research Initiative (SEADPRI-UKM), Universiti Kebangsaan Malaysia, Bangi 43600, Malaysia
3 Technical Services Division, Department of Mineral and Geoscience Malaysia, Jalan Sultan Azlan Shah, Ipoh 31400, Malaysia
4 British Geological Survey, Nicker Hill, Keyworth, Nottingham NG12 5GG, UK
* Correspondence: ntf@um.edu.my; Tel.: +60-3-79674153

Abstract: Landslide susceptibility modelling in tropical climates is hindered by incomplete inventory due to rapid development and natural processes that obliterate field evidence, making validation a challenge. Susceptibility modelling was conducted in Kuala Lumpur, Malaysia using a new spatial partitioning technique for cross-validation. This involved a series of two alternating east-west linear zones, where the first zone served as the training dataset and the second zone was the test dataset, and vice versa. The results show that the susceptibility models have good compatibility with the selected landslide conditioning factors and high predictive accuracy. The model with the highest area under curve (AUC) values (SRC = 0.92, PRC = 0.90) was submitted to the City Council of Kuala Lumpur for land use planning and development control. Rainfall-induced landslides are prominent within the study area, especially during the monsoon period. An extreme rainfall event in December 2021 that triggered 122 landslides provided an opportunity to conduct retrospective validation of the model; the high predictive capability (AUC of PRC = 0.93) was reaffirmed. The findings proved that retrospective validation is vital for landslide susceptibility modelling, especially where the inventory is not of the best quality. This is to encourage wider usage and acceptance among end users, especially decision-makers in cities, to support disaster risk management in a changing climate.

Keywords: landslide susceptibility; validation; predictive capability; disaster risk; tropical climate; Malaysia

check for updates

Citation: Affandi, E.; Ng, T.F.; Pereira, J.J.; Ahmad, F.; Banks, V.J. Revalidation Technique on Landslide Susceptibility Modelling: An Approach to Local Level Disaster Risk Management in Kuala Lumpur, Malaysia. *Appl. Sci.* **2023**, *13*, 768. https://doi.org/10.3390/app13020768

Academic Editors: Ricardo Castedo, Miguel Llorente Isidro and David Moncoulon

Received: 2 December 2022
Revised: 21 December 2022
Accepted: 28 December 2022
Published: 5 January 2023

1. Introduction

Landslides triggered by rainfall have resulted in the highest number of fatalities in Asia, and whilst its attribution to climate change is limited by deficient records, landslide occurrence is expected to increase in many regions [1,2]. Development in steep hills and unstable slopes, in combination with seasonally dry periods followed by excessive rainfall are contributing factors for frequent landslides in Southeast Asia [3]. The number of people in cities and urban settlements that will be exposed to landslides is expected to increase in Southeast Asia due to increased monsoon rainfall as a result of climate change [1,4]. The expected increase of landslides calls for reliable demarcation of areas susceptible to this hazard for supporting disaster risk management in cities.

Quantitative techniques predominate in landslide susceptibility modelling. The techniques could be deterministic (analytically driven), heuristic (knowledge-driven), or data-driven statistics, encompassing bivariate, multivariate, and machine learning approaches [5,6]. The use of statistical approaches in actual practice is dependent on a reliable inventory that reflects the actual factors that cause a slope to fail [7,8]. Statistical approaches are reliable for shallow landslides when drawing on inventories that integrate field investigation, high resolution digital elevation models (DEM) and images, combined with expert inputs for more nuanced representation of local conditions [9–12].

Validation is critical for susceptibility modelling to ensure the quality and reliability of outputs for wider usage among the decision makers [13–15]. Model validation generally involves two aspects, evaluating compatibility and assessing predictive capability [5,16,17]. The area under curve (AUC) is computed from the receiver operating characteristics (ROC) curve to report results. The evaluation of compatibility (i.e., agreement between the outputs and observed data) is from success rate curves (SRC) while the assessment of predictive capability is through prediction rate curves (PRC). Partitioning of the inventory into test and training datasets is the basis for validation [15,18].

Partitioning of the inventory is handled using three types of techniques. These are random, spatial, and temporal techniques [13,14,19,20]. Temporal validation uses dataset of landslides that occurred at different time periods to assess the predicted model capability. Random partitioning involves separation of the landslide inventory into two subsets, where one subset is arbitrarily removed to assess predictability. Spatial partitioning is a comparison of two mutually exclusive sub-regions of the area under investigation, or to another area with similar setting. This technique has limitations in lithologically heterogeneous areas, where some parameters that influence landslides may be absent in one of the sub-regions, leading to low predictive capability [15,21]. Retrospective validation has recently been introduced to assess the actual accuracy of prediction from susceptibility modelling [18]. The application of this approach is expected to increase the reliability, transparency, and acceptance of susceptibility modelling.

Malaysia has tropical climate with mean daily temperatures of between 26 °C to 28 °C and annual average rainfall ranging from 2000 mm to 4000 mm, with two distinct monsoon seasons [22]. The hot and humid conditions have resulted in deep weathered soil profiles that are prone to landslides. Rainfall-induced shallow landslides are common in cities primarily due to land use change and other human activity, and several have been observed to be reactivated failures [23–25]. In some cases, failures have been linked to inadequate engineering design during construction and poor slope maintenance [23]. Landslide mapping is a challenge due to vegetation growth, active erosion, and rapid development that obliterates evidence of landslide occurrence. This is compounded by inadequate public records, which is a prevailing issue for many developing countries due to limited resources [26].

Landslide susceptibility modelling has been conducted sporadically in Malaysia over the past decade. Statistical approaches are prevalent in areas where inventories are available such as Cameron Highlands, Penang, Putrajaya, and eastern Selangor [27–36]. Generally, the inventories are limited, and the best available data are utilised. More recently, a bivariate statistics approach combined with expert inputs was used to improve the susceptibility model [9]. In many cases, model validation is not mentioned [28–30,32,33], while in others only the success rates are reported [9,34]. Investigations that report on predictive capability have relied on random partitioning and spatial correlation techniques [27,31,35,36]. All the investigations have been conducted for academic purposes and retrospective validation has not been reported.

Kuala Lumpur, the capital city of Malaysia, suffers from numerous slope failures especially after a severe rainfall episode, where residential infrastructure and business complexes are most affected [23]. Many of the incidents occur in cut and fill slopes, road embankments, highways, and hill slope development projects [28,37,38]. The situation is expected to worsen with climate change and continuous expansion of the city. This paper highlights a new spatial partitioning technique for landslide susceptibility modelling that was conducted in Kuala Lumpur, Malaysia. The technique is introduced for assessing the predictive capability of landslides, to overcome the limitations associated with lithological heterogeneity. The ensuing landslide susceptibility map was then formally submitted to the City Council of Kuala Lumpur (DBKL) to support decision-making. An extreme rainfall event with a recorded precipitation of more than 250 mm occurred in December 2021 exceeding the monthly rainfall in over a day. The unusual phenomenon was caused by the northeast monsoon flow factors and a low-pressure weather system which caused continuous heavy rain over the west coast of peninsula regions besides the usually affected central east coast and northern peninsula regions. The event caused massive flooding of 100-year return period and triggered 122 landslides in the city. This provided an opportunity to conduct retrospective validation to confirm the robustness of the partitioning technique and increase confidence in the landslide susceptibility map.

2. Materials and Methods

2.1. Study Area

Kuala Lumpur covers an area of 243 km^2 and is the most urbanised and densely populated territory in Malaysia. The city is located in west-central Peninsular Malaysia, underlain by a variety of rock types, geological structures and geomorphological features [39]. The Dinding Schist present in northeastern Kuala Lumpur consists of meta-volcanics and quartz-mica schist (Figure 1). It is overlain by the Hawthornden Schist comprising mainly of graphitic quartz-mica schist. The schistosity of these two rock units is trending approximately north-south. The northern and east-central area is underlain by the Kuala Lumpur Limestone. The limestone is fully covered by alluvium, fill material, and mine tailing with thickness up to 66 m. Kenny Hill Formation is mainly found in the southern and central parts of the city. It comprises interbedded phyllite and quartzite, and the strata are mainly trending north–south. All the above metasedimentary sequences were intruded by granite, which is present mainly in the western and southeastern Kuala Lumpur. Soils developed over the schists, Kenny Hill Formation, and granite average 13 m, 9 m, and 15 m thick, respectively. The ground elevation obtained from the digital terrain model (DTM) of Kuala Lumpur ranges from 9 m to 320 m; about 90% of the city is below 100 m. Flat alluvial plains occur along the Kelang River and its tributaries. The alluvium has been largely mined for tin in the past. The alluvial plain is flanked by hills at the east and west. The hills in the Kenny Hill Formation areas have low relief (~50 m) and are generally elongated in the north–south direction. The hills in the granite and schist areas have higher relief (~100 m) and are also elongated in the north–south direction.

2.2. Input Data

The landslide inventory consists of 650 landslide points sourced from Department of Mineral and Geoscience Malaysia (JMG). It is the sole official landslide repository that was established in 2014 through field mapping and includes only rainfall-triggered landslides. Although it contains only partial information on location and type of slope failure (man-made or natural), the original landslide points were refined through quality assessment to ensure the evidence of landslides location and its accuracy. The points were overlain on the DTM, orthorectified aerial photographs and satellite imageries from Google Earth, and repositioned at the crown of the landslides.

The DTM was derived from LiDAR data from DBKL. The 2014 LiDAR data have a resolution of 1 m, and they were resampled to a pixel size of 5 m for the study. Aerial photographs were acquired from the Department of Survey and Mapping Malaysia (JUPEM) and DBKL. Topographic maps used were at the scales of 1:50,000, 1:25,000 and 1:10,000 published by JUPEM. Bedrock geology maps obtained from JMG were updated to incorporate information on surface geology and geological structures. The additional data were obtained from boreholes and field visits as well as interpretation from aerial photographs and topographic maps.

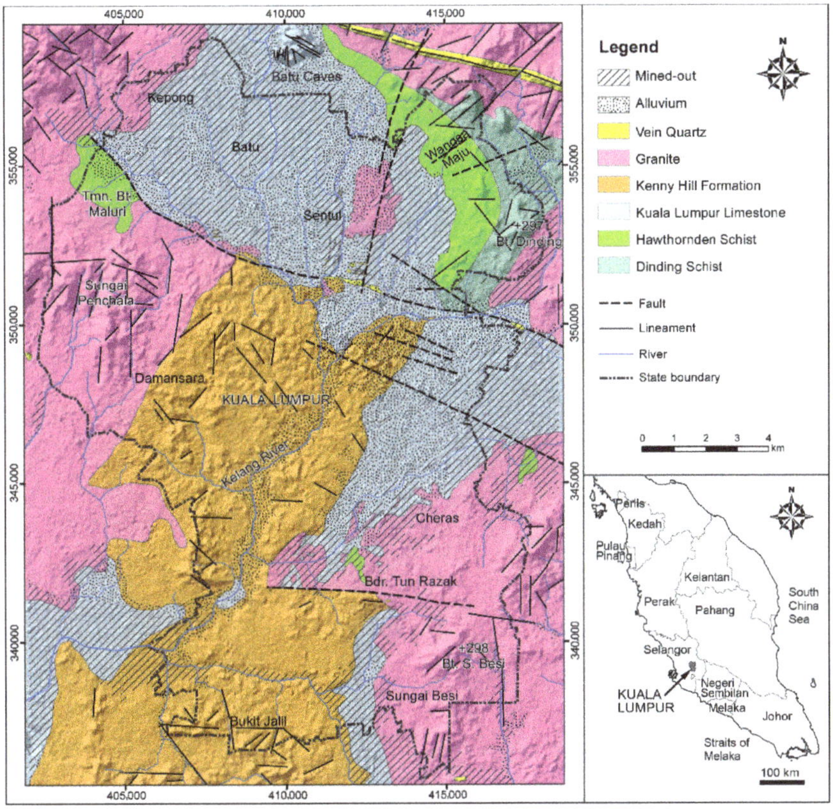

Figure 1. Geological map of the Kuala Lumpur area. The city of Kuala Lumpur is located within the Selangor state, in central-west Peninsular Malaysia.

2.3. Landslide Conditioning Factors

A total of fourteen intrinsic landslide conditioning factors were prepared from the geological maps, DTM, topographic maps, and satellite imageries. All the geoprocessing was done using the software ArcGIS 10.5. Bedrock geology and surface geology were extracted from the geological data. Geomorphological factors such as slope gradient, slope aspect, and slope curvature were derived directly from the DTM. The standard deviation of elevation calculated from a 10 m by 10 m moving window was used to derive the surface roughness [40]. The topographical position index (TPI) [41,42] was calculated as the difference between the elevation at a point and the mean elevation within a 250 m radius circular moving window.

The topographic wetness index (TWI) is a contributing factor that affects the topography and used to quantify the topographic control on hydrological processes, expressed by Equation (1), whereas the stream power index (SPI) measures the erosive power of water flow as defined by Equation (2). The value α is calculated by multiplying the flow accumulation with the cell area. The distance to stream, distance to road and distance to lineament maps were created by making multiple ring buffer around the streams, roads and lineament, respectively. Landsat 8 imagery was used to produce the normalised difference vegetation index (NDVI) map. The value can be calculated using Equation (3).

$$\mathrm{TWI} = \ln\left(\frac{\alpha}{\tan\beta}\right) \tag{1}$$

$$\mathrm{SPI} = \ln(\alpha . \tan\beta) \quad (2)$$

where α is the local upslope contributing area, and β is the slope [43].

$$\mathrm{NDVI} = \frac{NIR - Red}{NIR + Red} \quad (3)$$

where *NIR* is the near-infrared band, and *Red* is the red band.

An expert consultation process was carried out to select the best landslide conditioning factors for the area [9]. The conditioning factors went through an iterative selection and elimination procedure by experts with experience working on landslides in tropical terrain. The evaluation was based on redundancy and relevance of the factors to the local geological and terrain conditions, as well as the initial results of bivariate statistical analysis. Only seven of the original conditioning factors were selected as shown in Table 1. They are surface geology, slope gradient, elevation, distance to lineament, distance to road, surface roughness, and TPI (Figures 2 and 3).

Most of the ground is covered by surface geology and shows a strong influence towards landslide initiation compared to bedrock geology, caused by the deep weathering profile which generates thick residual soil over most area [44]. The residual soil of Kenny Hill formation holds the highest weightage of 1.00 with 68.7% of total landslide, due to the interlayering of metasedimentary rocks of variable strength and properties while the granite represents 127 (20%) landslides with a weightage value of −0.18. Slope gradient is usually regarded as the primary causative factor for the onset of slope failure associated to shear strength of the material on slope [45]. The analysis of the parameter indicates that the landslide density and weightage values increase with increasing slope gradient. Elevation of the slope is not a conditioning factor by itself but affects the overall surface of the terrain and topographic features which controls the vegetation distribution [31]. The highest weightage is represented by the 100–150 m class with a value of 0.55 and has 10% of the total landslides while 50–100 m class represents 68.8% of the total landslides with a lower weightage of 0.49 due to the larger area. The correlation between distance to lineament parameter and landslides is explained by weakened earth material by deformation along faults. The structural stability of the surrounding area is reduced by induced regional perturbations in the fracturing occurrence and enhanced weathering in the rocks [46]. The distance to lineament factor and landslide shows direct correlation with the highest weightage (0.93) is within 0–250 m proximity. Slope which has closer proximity to roads especially near cut and fill slope could affect the stability by increase of stress in its base and accumulation of water from nearby slope [47,48]. The highest weightage of 0.18 produced from the 25–50 m class interval representing 25% of total landslides while the majority of landslides (47%) occurs within 0–25 m distance from a roadway yields a low weightage of 0.02 due to larger class area. Surface roughness generally correlates to excavation, surface erosion and the density of vegetation cover and could signify the type of land cover of that particular area and has been classified into high, moderate and low values [46]. Half of the landslides (53%) are categorised as having moderate surface roughness with a weightage of 0.07, however the highest weightage of 0.29 represented by slope with high roughness due to smaller areal class. TPI gives an indication to where the landslide point is located with reference to the topographical position or sometimes expressed as landscape position [49,50]. The highest percentage and weightage of landslides is within the middle slope with a value of 61.17% and 1.13, respectively, with the upper slope class having a weightage of −0.01 with 15.08% of total landslides.

Table 1. Landslide conditioning factors and their definition as used in the bivariate statistical analysis.

No.	Landslide Conditioning Factors	Definition	Source
1	Topographic Position Index (TPI)	Relative slope position or landform category based on relative relief values	DTM
2	Slope Gradient	Degree of inclination of the slope	DTM
3	Distance to Lineament	Distance to fault and lineament in metres	DTM, air photo
4	Distance to Road	Distance to major roads in metres	DBKL road map
5	Surface Geology	Recent material deposit on the ground	Topographic map, geological map, air photo and borehole
6	Elevation	Height above sea level	DTM
7	Surface Roughness	Degree of variation of surface elevation	DTM

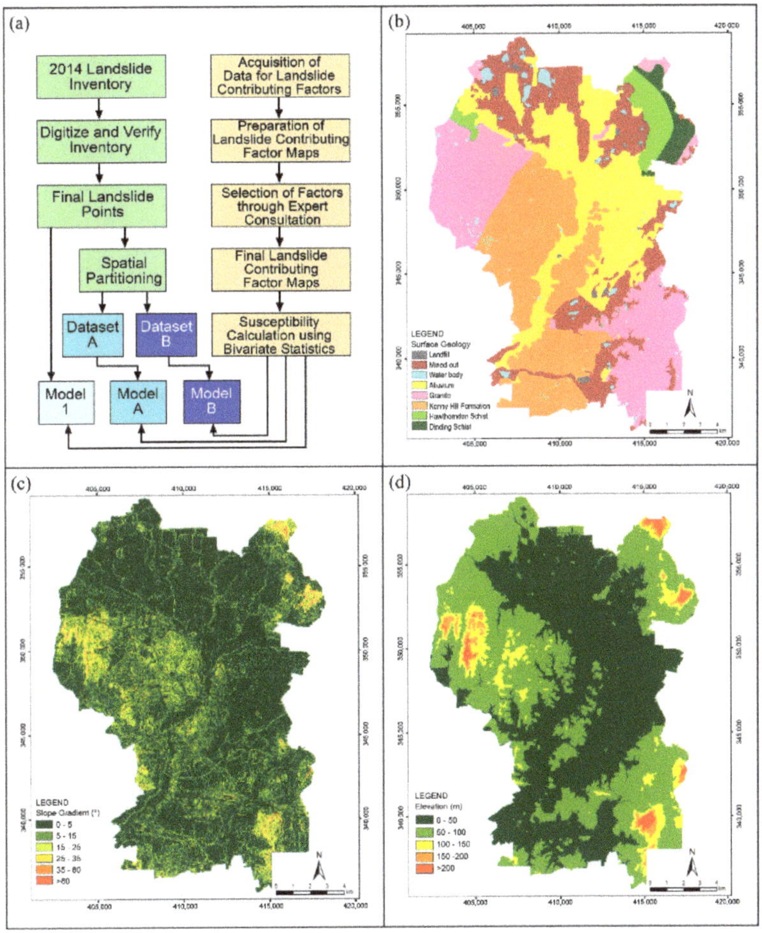

Figure 2. Seven landslide conditioning factors were selected according to the steps highlighted in the flow chart (**a**). The factors are surface geology (**b**), slope gradient (**c**), and elevation (**d**).

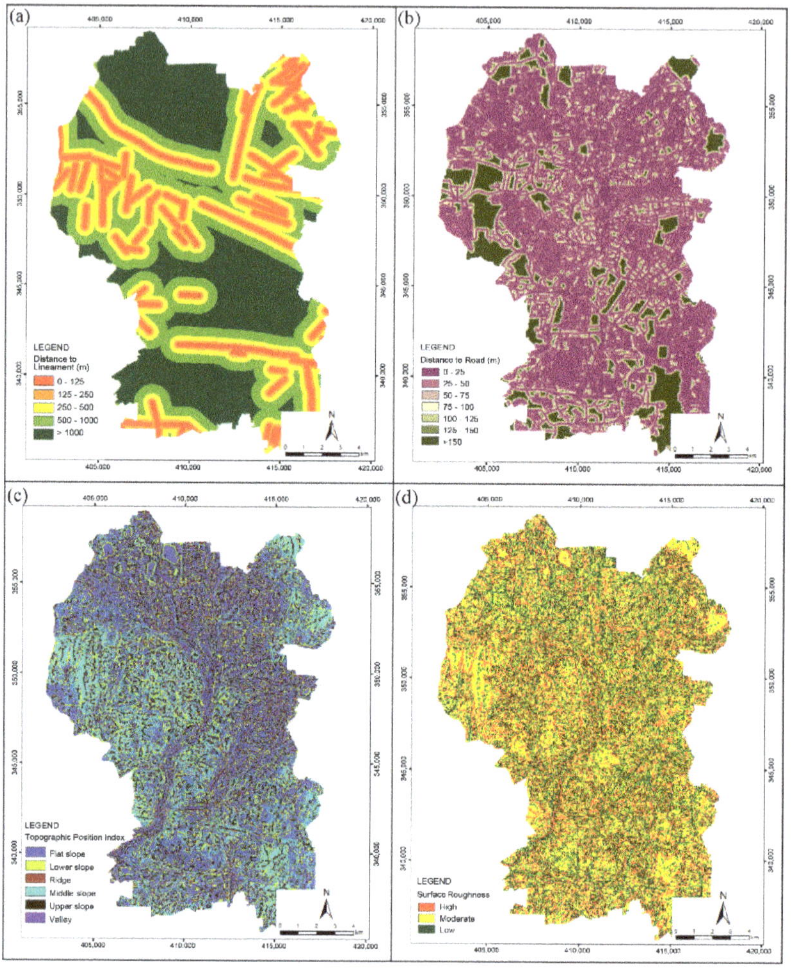

Figure 3. Map of the landslide conditioning factors on distance to lineaments (**a**), distance to roads (**b**), topographic position index (**c**), and surface roughness (**d**).

2.4. Bivariate Statistical Analysis

Landslide susceptibility modelling was conducted using the bivariate statistical approach that relies on the correlation between landslide population and the landslide conditioning factors. Each conditioning factor is subdivided into several classes, and a weight is calculated for each class from its statistical spatial relationship with landslide distribution [5,12,51,52].

The method proposed by [53] involves calculation of the natural logarithm of landslide density of each class within each factor divided by the overall landslide density in the area expressed as the Equation (4).

$$W_i = \ln\left(\frac{Densclass}{Densmap}\right) = \ln\left(\frac{\frac{Npix(S_i)}{Npix(N_i)}}{\frac{\sum Npix(S_i)}{\sum Npix(N_i)}}\right) \quad (4)$$

where, W_i = the weight given to a certain factor class. *Densclass* = the landslide density within the factor class. *Densmap* = the landslide density within the entire map.

$Npix(S_i)$ = the number of pixels with landslide occurrence in a certain factor class. $Npix(N_i)$ = total number of pixels in a certain factor class.

The weight indicates the correlation of the class factor with the landslide occurrence and refers to landslide density of each factor class. Negative weights indicate that the landslide density is lower than average, while a positive value signifies that it is higher. A zero weight is obtained when there is an absence of landslide occurrences within a parameter class [53]. Since the analysis is based on statistical calculation of the pixels in a map, raster maps are required for the processing. Vector maps in the form of polygon shapefiles were converted into raster files for further processing.

The raster factor maps were reclassified by assigning the weight to all the factor classes. The landslide susceptibility is a summation of the reclassified factor maps. The susceptibility classes can be defined based on the landslide density [54,55]. The classification was carried out manually based on the percentage of landslide points within each class. The classes are very high (>50%), high (20–50%), moderate (10–20%), low (2–10%) and very low (<2%) susceptibility, respectively.

2.5. Model Evaluation and Validation

The study employed the ROC method for model evaluation and validation. The AUC is used as the metric to evaluate the quality of the model. The susceptibility map was divided into 25 equal area classes with decreasing susceptibility and the percentage of landslide in each of class is calculated. The cumulative landslide percentage was plotted against the cumulative area to derive the ROC curve.

The landslide inventory was partitioned into two groups, i.e., training dataset and test dataset. The SRC is acquired when ROC curve of the same susceptibility map is plotted against the training set. The PRC is generated from the ROC curve derived from the susceptibility map generated from the training set against landslide points in the test set. The performance of the model is represented by a value range under the curve of 0.5 to 1.0, where 0.5 represents a test with accuracy no better than chance while value close to 1.0 which suggests an ideal model with a perfect fit [19].

The whole 2014 landslide inventory was used to generate a landslide susceptibility map and the SRC is plotted to assess the model's compatibility. Spatial cross-validation was done on the same inventory. The study area was partitioned into alternating linear zones (zone A and zone B) measuring 1 km in width as to represent the smallest geological unit in Kuala Lumpur known as schist. Another reason for the selection of the dimension is to ensure consistency since the topographic map of Malaysia has a 1 km grid. The zones are orientated east–west, perpendicular to the north-south geological and geomorphological trend. This is to ensure the inclusivity of the factors involved in both sub-regions. The validation was done twice. First zone A was used as the training set and zone B as the test set, and in second validation, the role of zones A and B is reversed. The SRC and PRC were plotted for both cases.

2.6. Retrospective Validation

In December 2021, an extreme rainfall event occurred in Kuala Lumpur and its surrounding areas causing massive flood events. The event also triggered widespread landslides. Fieldwork was carried out in January 2022 to record the landslides. The research team mapped 122 landslides more precisely than the older events in the original inventory. The position of the landslides was recorded directly in the field using the application Avenza Maps, where the DTM used as the base map. Retrospective validation was conducted using a quantitative approach. The susceptibility maps and the landslide points were compared by calculating the AUC values of PRC to evaluate the performance of the landslide susceptibility maps that was generated from the original inventory.

3. Results

3.1. Landslide Inventories

The 2014 landslide inventory comprising 650 events is represented mainly by small rotational landslides that occurred in the Kuala Lumpur granite and rotational to translational failures observed in the Kenny Hill formation. The mechanism of the landslides is recognised as similar as they often occur together and caused sliding of material, but data for each location are not available in the inventory. The landslides within the area are rainfall-induced which occurred on both natural and man-made slope. About 70% of the landslide are less than 10 m in length, 27% between 10 m and 20 m, and only 3% are more than 20 m. The landslides occurred mainly in areas under the Kenny Hill Formation (70%) and granite (20%). Landslides are common in the hilly areas on slope with gradient of 15–25° (35%) and 25–35° (31%), in the west-central, south, and northeast Kuala Lumpur. The distribution of the 2021 landslides is similar, but there is a noticeable increase of landslides in granite (31%). Spatially partitioning the 2014 inventory for model validation revealed that 58% of landslides occurred in the first subset (zone A) while the remaining 42% were in the second subset (zone B).

3.2. Landslide Susceptibility Models

Three landslide susceptibility models were produced from bivariate statistical analysis using 7 landslide conditioning factors (Figures 2 and 3). The weights calculated for these three models are listed in Table S1. In the first model (model 1), all the landslide points from the 2014 inventory were used and there was no spatial partitioning of the dataset (Figure 4a). The AUC of the SRC of this model is 0.91 (Figure 5a). The areas classified as very high and high susceptibility cover 7% and 9% of the city, respectively. These two susceptibility classes are concentrated in the northeastern, west-central, and southern parts of Kuala Lumpur; much of this area is not developed. About 7% of the area is classified as moderately susceptible. The moderately susceptible areas are distributed mainly in the vicinity of the two earlier classes. The low susceptibility areas (20%) are scattered around the earlier classes and within the very low susceptibility class. Covering 57% of the area, the very low susceptibility class is the largest. It is distributed mainly in the northern and eastern Kuala Lumpur, in relatively flat areas underlain by alluvium and mine tailings.

The next two models were used for spatial cross-validation, where the study area is partitioned into a series of two alternating linear zones (zones A and B) trending east-west. In the initial partitioned model (model A), the susceptibility map is generated using landslide points in zone A as the training dataset and landslide points in zone B as the test dataset (Figure 4b). The role of landslide points in zones A and B are reversed in the next model (model B) (Figure 4c). The distribution of the very high to moderate susceptibility classes of these two models are similar to the unpartitioned model (model 1), which obtained using the entire dataset of the inventory (Figure 4a–c). Compared to model 1, the low susceptibility areas of both models A and B are higher (27%), and the very low susceptibility areas are lower (48%).

3.3. Model Evaluation and Validation

The spatial cross-validation produced results that are similar. The AUC of the SRC for model A and model B is 0.92 and 0.89, respectively (Figure 5a). The high AUC values of these two models and the model 1 indicate good compatibility of the models derived from the 7 selected landslide conditioning factors. It also indicates a high level of inclusivity where all the parameter classes are well represented in the three models. Model 1 and model A produced similar high SRC values, which indicates excellent model fitness. Although the 0.01 difference may not be statistically significant, a possible explanation for this finding is the higher distribution of landslides within zone A subset (58%) generated a model with the closest SRC value to that of model 1.

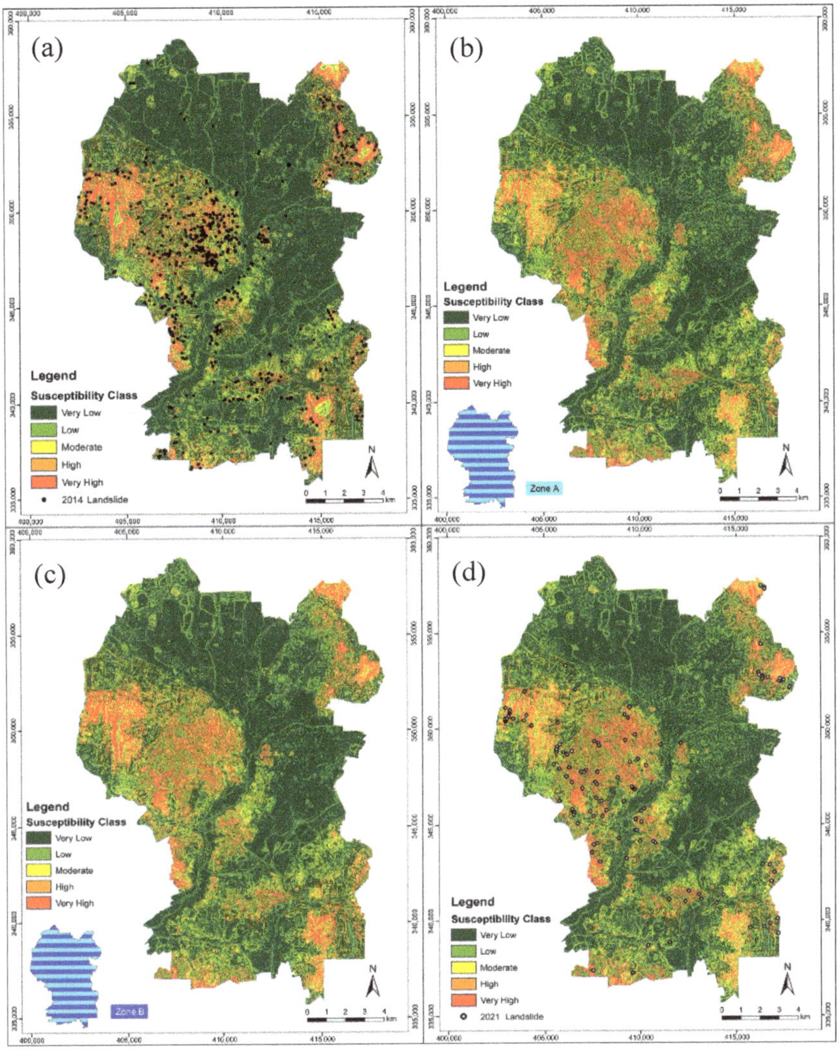

Figure 4. (**a**) Landslide susceptibility map of model 1, where all the landslide points in the 2014 inventory were used to generate the model. (**b**) Landslide susceptibility map of model A, where landslides in zone A were used as training dataset and landslides in zone B as test dataset. (**c**) Landslide susceptibility map of model B where landslides in zone B were used as training dataset and landslides in zone A as test dataset. (**d**) Landslides occurred in 2021 that were used for retrospective validation are plotted on model A.

The PRC for model A and model B also produced AUC values of 0.90 and 0.89, respectively, indicating that both models have similarly high predictive capability (Figure 5a). The similar AUC values show that the spatial partitioning method used is suitable for model validation. The AUC of model A is marginally better than model B. In 2020, this model was selected to be formally submitted to DBKL, the local council of Kuala Lumpur, to support decision-making on land use planning and development control.

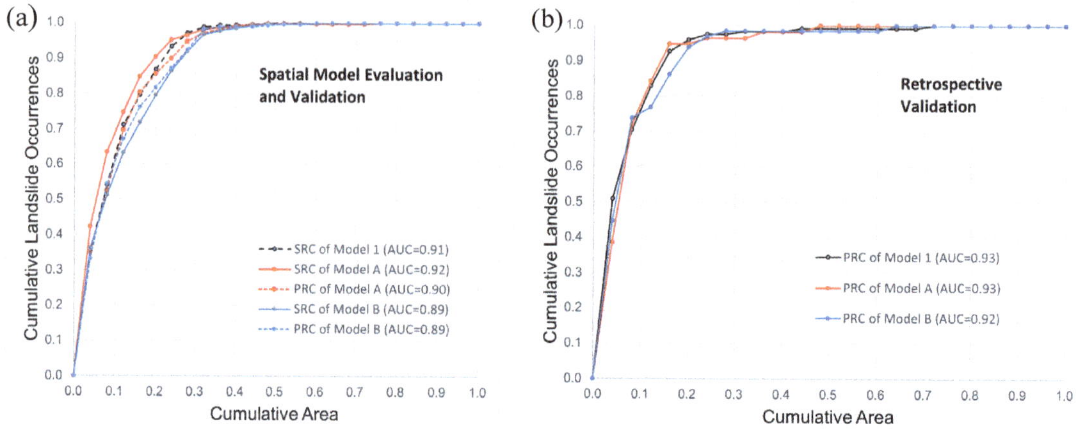

Figure 5. (**a**) ROC curves of spatial model evaluation and validation with associated AUC values for model 1, model A, and model B. (**b**) ROC curves of retrospective validation with associated AUC values for the same models.

The retrospective validation of all the three models using the 122 landslide events that occurred in 2021 showed a slight increase in confidence level (Figure 4d). The AUC of the PRC for model 1, model A and model B is 0.93, 0.93, and 0.92, respectively (Figure 5b). The values proved to be marginally higher than the spatial partitioning method. Comparing the PRC to that of model 1, the same AUC values could be attributed to the small sample size of the 2021 landslide events and similar landslide distribution in both models. The revalidation using the retrospective method reaffirms the high predictive ability of all three landslide susceptibility models. This indicates that the product that was submitted to the city council is reliable for decision-making purposes.

4. Discussion

The sole official landslide inventory established in 2014 contained only 650 landslide points that could be verified despite the numerous slope failures have been reported in Kuala Lumpur for decades, especially after a severe rainfall. This is primarily because much of the evidence in the field was obliterated. In comparison, after one unusually heavy rainfall incident, 122 events were recorded by the research team. This indicates that the number of historical landslides is most likely higher and has gone unrecorded. Furthermore, information in the original 2014 landslide inventory was limited to event locations, while more complete information was obtained from the fresh landslide scars in 2021; indicating the need for urgency in field data collection. Though the 2014 inventory was manually improved to correctly position and interpret the type and size of landslides using DTM and orthorectified aerial photographs; much time and expertise on local conditions was required. An automated approach has been advocated for the identification of landslides using high resolution DEM [7]. However, the accuracy of this method in cities is yet to be reported. In order to obtain complete and better-quality datasets, landslide inventories in tropical cities need to be routinely updated after heavy rainfalls.

Landslide inventories with information on landslide masses reportedly produced more accurate susceptibility maps compared to the use of landslide mass centre point [7]. However, it has also been reported that the scarp has higher predictive accuracy than the mass centre as it closely represents the pre-failure conditions [56]. The 2014 inventory utilised the position of the landslide scarp centre as it was a more prominent feature that was more readily identified. In most cases, the body and toe of the landslide is not clear due to erosion and vegetation. In this study, the inventory sourced from JMG was not of the best quality with respect to data on the type and extent of the landslides. Notwithstanding, a consistent sampling technique using information on landslide scarp resulted in sufficiently accurate susceptibility models.

The landslide conditioning factors used for modelling were restricted to 7, although a total of 14 factors were initially produced. This smaller selection was considered appropriate through expert consultation and resulted in a susceptibility model that had high prediction accuracy. In this investigation, the high resolution (1 m) Light Detection and Ranging (LiDAR) derived DTM was resampled to 5 m, to reduce the noise, artifacts, and processing time. This was also considered sufficient as the topographic data (DTM) was of relatively high resolution. It has been demonstrated that finer resolution of topographic data leads to more accurate and precise susceptibility models. This study supports recent findings that as long as suitable input data and techniques are selected based on the data quality and purpose, most landslide susceptibility models result in sound overall prediction accuracy regardless of the approach [7].

Retrospective validation conducted in this study indicates that the spatial partitioning technique along alternating linear zones trending east–west that was used to cross validate the susceptibility models is able to resolve the issue of heterogeneity caused by north–south trending geological and geomorphological features. All the landslide conditioning factor classes were represented in both zones A and B (Supplementary, Table S1). The difference in the area for 70% of the factor classes between the two zones is less than 10% (Supplementary Table S2). It appears that spatial partitioning techniques that are context specific, taking into account local geological and geomorphological features, result in high confidence susceptibility maps. Nevertheless, additional analysis is required using linear zones trending in other directions to confirm this outcome. Other random partitioning and spatial correlation techniques that have been previously reported in the country should also be investigated [27,31,35,36].

Notwithstanding, the landslide susceptibility map that was submitted to DBKL is reliable for decision-making as indicated by the high predictive value (AUC = 0.93) from retrospective validation. The map indicates that that 16% of the Kuala Lumpur area has very high to high susceptibility. Fortunately, much of this comprises rugged terrain that is currently not developed. The population density map of district level produced using the 2020 census data indicates the highest population per km^2 is distributed at the northwest region found in Wangsa Maju followed by the second highest class represented by Setiawangsa, Batu and Seputeh in southwest region (Figure 6) [57]. With reference to the landslide susceptibility model, we recognise that part of the hilly areas in Wangsa Maju, Setiawangsa and Seputeh located in high to very high susceptibility class, requiring urgent effort in hazard mitigation, disaster preparedness, and stringent actions for development. In the current scenario, targeted disaster risk management strategies could be formulated by DBKL to manage the zones that have already been developed and the vulnerable areas. This could include investing in monitoring and early warning measures and conducting preparedness programs which aim to build the resilience of communities. The undeveloped areas could be either be gazetted for protection or stringent development, where developers are required to conduct detailed geological and geotechnical investigation, with provision of appropriate disaster mitigation measures.

Resource constraints are a contributing factor to the lack of effort to maintain and regularly update the inventory in Kuala Lumpur. A complete and frequently updated inventory would facilitate further investigation including the development of hazard maps to determine the probability of landslide events. To evaluate the accuracy of bivariate statistical model in comparison with other established methods, future work using a semi-quantitative approach, namely the analytic hierarchy process (AHP), is recommended to further improve the findings. In addition, the determination of rainfall threshold values that trigger landslides could be used to establish early warning for disaster risk management [58,59]. Further investigation is also required in Kuala Lumpur to understand the significant increase of landslides in areas underlain by granites in 2021, whether this is due to unevenness in rainfall distribution or inherent characteristics of the terrain. This is critical in light of the worsening situation expected with climate change and continuous expansion of the city. Innovative partnerships could be considered, encompassing academia, the government agencies, and the local council, with involvement of the insurance and banking sectors, to maintain and conduct further investigation of landslide risks in Kuala Lumpur. Such public–private partnerships could be scaled up to cover more cities in the country, if successful.

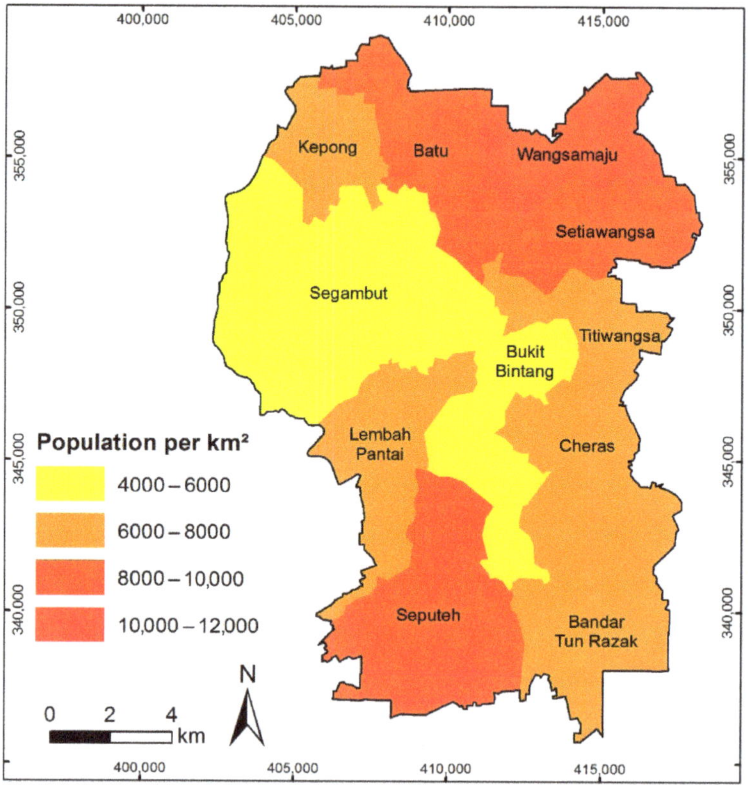

Figure 6. Population density map of Kuala Lumpur at district level based on 2020 census data.

5. Conclusions

The projected intensification of monsoon rainfall over Southeast Asia and continuous growth of the city is expected to increase landslide incidents in Kuala Lumpur, Malaysia. This calls for reliable demarcation of areas within the city that are susceptible to landslides to support disaster risk management. Kuala Lumpur is underlain by heterogeneous lithology and north–south trending geological structures and geomorphological features, which limits the effectiveness of conventional spatial partitioning techniques for validation in landslide susceptibility modelling. This limitation was overcome by employing a new spatial partitioning technique using a series of two alternating east–west linear zones, where the first zone served as the training dataset and the second zone was the test dataset, and vice versa. The portioning resulted in an inclusive distribution of landslide conditioning factor classes; the difference in the area of factor classes between the two zones is mainly less than 10%. The ROC curves show that the susceptibility models have good compatibility with the selected landslide conditioning factors and have high predictive accuracy. Model A with the highest AUC values (SRC = 0.92, PRC = 0.90) was submitted to DBKL for land use planning and development control. Retrospective revalidation reaffirmed the high predictive ability (AUC of PRC = 0.93) of the landslide susceptibility model. The findings indicates that the product is reliable for decision-making purposes, supporting local level hazard and risk mitigation efforts and better ground prediction in future city development The authors suggest future work to be done using other established methods where comparison of the findings can be done, as data acquisition of landslide occurrences and precipitation data within the study area improves.

Supplementary Materials: The following supporting information can be downloaded at: https://www.mdpi.com/article/10.3390/app13020768/s1, Table S1: (a) The weights of the 14 landslide conditioning factors. Only the first seven factors were used in model 1. (b) The weights for model A where all the 377 landslide points in zone A were used as the training dataset. (c) The weights for model B where all the 274 landslide points in zone B were used as the training dataset. Table S2: Comparison of the area extent of the landslide conditioning factor classes in zone A and zone B.

Author Contributions: Funding acquisition, J.J.P.; resources, J.J.P.; conceptualisation, T.F.N., E.A., V.J.B. and F.A.; data collection (2014 landslide inventory), F.A.; field investigation, E.A. and T.F.N.; data analysis, E.A.; writing, E.A., T.F.N. and J.J.P.; editing, T.F.N., J.J.P. and V.J.B. All authors have read and agreed to the published version of the manuscript.

Funding: The authors would like to thank the Newton-Ungku Omar Fund (Application ID: 59348-455144, UM Project No. IF002–2017, UKM Project No. XX-2017-002) for supporting this research.

Institutional Review Board Statement: Not applicable.

Informed Consent Statement: Not applicable.

Data Availability Statement: The datasets generated during and analysed are available from the corresponding author on reasonable request.

Acknowledgments: The authors would like to thank the Editor for the opportunity to be part of this special issue. The assistance of project members from multiple institutions is gratefully acknowledged.

Conflicts of Interest: The authors declare no conflict of interests.

References

1. IPCC. Summary for Policymakers. In *Climate Change 2021: The Physical Science Basis. Contribution of Working Group I to the Sixth Assessment Report of the Intergovernmental Panel on Climate Change*; Masson-Delmotte, V., Zhai, P., Pirani, A., Connors, S., Péan, C., Berger, S., Caud, N., Chen, Y., Goldfarb, L., Gomis, M., et al., Eds.; Cambridge University Press: Cambridge, UK, 2021.
2. Froude, M.J.; Petley, D.N. Global fatal landslide occurrence from 2004 to 2016. *Nat. Hazard Earth Sys.* **2018**, *18*, 2161–2181. [CrossRef]
3. Douglas, I. The urban environment in southeast Asia. In *The Physical Geography of Southeast Asia*; Gupta, A., Ed.; Oxford University Press: Oxford, UK, 2005; pp. 314–334.
4. Gariano, S.L.; Guzzetti, F. Landslides in a changing climate. *Earth Sci. Rev.* **2016**, *162*, 227–252. [CrossRef]
5. Reichenbach, P.; Rossi, M.; Malamud, B.D.; Mihir, M.; Guzzetti, F. A review of statistically-based landslide susceptibility models. *Earth Sci. Rev.* **2018**, *180*, 60–91. [CrossRef]
6. Soeters, R.; Van Westen, C. Slope instability recognition, analysis and zonation. In *Landslides: Investigation and Mitigation. Special Report No. 247. Transportation Research Board National Research Council*; Turner, K., Schuster, R., Eds.; National Academy Press: Washington, DC, USA, 1996; Volume 247, pp. 129–177.
7. Gaidzik, K.; Ramírez-Herrera, M.T. The importance of input data on landslide susceptibility mapping. *Sci. Rep.* **2021**, *11*, 19334. [CrossRef] [PubMed]
8. Hearn, G.; Hart, A. Landslide susceptibility mapping: A practitioner's view. *Bull. Eng. Geol. Environ.* **2019**, *78*, 5811–5826. [CrossRef]
9. Daniel, M.T.; Ng, T.F.; Kadir, A.; Farid, M.; Pereira, J.J. Landslide Susceptibility Modeling Using a Hybrid Bivariate Statistical and Expert Consultation Approach in Canada Hill, Sarawak, Malaysia. *Front. Earth Sci.* **2021**, *9*, 71. [CrossRef]
10. Bordoni, M.; Galanti, Y.; Bartelletti, C.; Persichillo, M.G.; Barsanti, M.; Giannecchini, R.; Avanzi, G.D.A.; Cevasco, A.; Brandolini, P.; Galve, J.P.; et al. The influence of the inventory on the determination of the rainfall-induced shallow landslides susceptibility using generalized additive models. *Catena* **2020**, *193*, 104630. [CrossRef]
11. Merghadi, A.; Yunus, A.P.; Dou, J.; Whiteley, J.; ThaiPham, B.; Bui, D.T.; Avtar, R.; Abderrahmane, B. Machine learning methods for landslide susceptibility studies: A comparative overview of algorithm performance. *Earth Sci. Rev.* **2020**, *207*, 103225. [CrossRef]

12. Van Westen, C.; Rengers, N.; Soeters, R. Use of geomorphological information in indirect landslide susceptibility assessment. *Nat Hazards* **2003**, *30*, 399–419. [CrossRef]
13. Chung, C.-J.; Fabbri, A.G. Validation of spatial prediction models for landslide hazard mapping. *Nat Hazards* **2003**, *30*, 451–472. [CrossRef]
14. Chung, C.-J.; Fabbri, A.G. Predicting landslides for risk analysis—Spatial models tested by a cross-validation technique. *Geomorphology* **2008**, *94*, 438–452. [CrossRef]
15. Remondo, J.; Gonzalez, A.; De Teran, J.R.D.; Cendrero, A.; Fabbri, A.; Chung, C.J.F. Validation of landslide susceptibility maps; Examples and applications from a case study in northern Spain. *Nat Hazards* **2003**, *30*, 437–449. [CrossRef]
16. Corominas, J.; van Westen, C.; Frattini, P.; Cascini, L.; Malet, J.P.; Fotopoulou, S.; Catani, F.; Van Den Eeckhaut, M.; Mavrouli, O.; Agliardi, F.; et al. Recommendations for the quantitative analysis of landslide risk. *Bull. Eng. Geol. Environ.* **2014**, *73*, 209–263. [CrossRef]
17. Frattini, P.; Crosta, G.; Carrara, A. Techniques for evaluating the performance of landslide susceptibility models. *Eng. Geol.* **2010**, *111*, 62–72. [CrossRef]
18. Fleuchaus, P.; Blum, P.; Wilde, M.; Terhorst, B.; Butscher, C. Retrospective evaluation of landslide susceptibility maps and review of validation practice. *Environ. Earth Sci.* **2021**, *80*, 485. [CrossRef]
19. Deng, X.; Li, L.; Tan, Y. Validation of spatial prediction models for landslide susceptibility mapping by considering structural similarity. *ISPRS Int. J. Geo-Inf.* **2017**, *6*, 103. [CrossRef]
20. Lee, S.; Ryu, J.-H.; Kim, L. Landslide susceptibility analysis and its verification using likelihood ratio, logistic regression, and artificial neural network models: Case study of Youngin, Korea. *Landslides* **2007**, *4*, 327–338. [CrossRef]
21. Dornik, A.; Drăguţ, L.; Oguchi, T.; Hayakawa, Y.; Micu, M. Influence of sampling design on landslide susceptibility modeling in lithologically heterogeneous areas. *Sci. Rep.* **2022**, *12*, 2106. [CrossRef]
22. NC3 Malaysia. Third National Communication and Second Biennial Update Report to the UNFCCC. Available online: https://unfccc.int/documents/182748 (accessed on 25 June 2022).
23. Majid, N.; Taha, M.; Selamat, S. Historical landslide events in Malaysia 1993–2019. *Indian J. Sci. Technol.* **2020**, *13*, 3387–3399. [CrossRef]
24. Alnaimat, A.; Choy, L.K.; Jaafar, M. An assessment of current practices on landslides risk management: A case of Kuala Lumpur territory. *Geografia* **2017**, *13*, 1–12.
25. Saadatkhah, N.; Kassim, A.; Lee, L.M.; Rahnamarad, J. Spatiotemporal regional modeling of rainfall-induced slope failure in Hulu Kelang, Malaysia. *Environ. Earth Sci.* **2015**, *73*, 8425–8441. [CrossRef]
26. Gariano, S.L.; Rianna, G.; Petrucci, O.; Guzzetti, F. Assessing future changes in the occurrence of rainfall-induced landslides at a regional scale. *Sci. Total Environ.* **2017**, *596–597*, 417–426. [CrossRef] [PubMed]
27. Sameen, M.I.; Pradhan, B.; Bui, D.T.; Alamri, A.M. Systematic sample subdividing strategy for training landslide susceptibility models. *Catena* **2020**, *187*, 104358. [CrossRef]
28. Sulaiman, M.S.; Nazaruddin, A.; Salleh, N.M.; Abidin, R.Z.; Miniandi, N.D.; Yusoff, A.H. Landslide occurrences in Malaysia based on soil series and lithology factors. *Int. J. Adv. Sci. Technol.* **2019**, *28*, 1–26.
29. Althuwaynee, O.F.; Pradhan, B.; Mahmud, A.R.; Yusoff, Z.M. Prediction of Slope Failures Using Bivariate Statistical Based Index of Entropy Model. In Proceedings of the 2012 IEEE Colloquium on Humanities, Science & Engineering Research (CHUSER), Kota Kinabalu, Malaysia, 3–4 December 2012; pp. 362–367.
30. Althuwaynee, O.F.; Pradhan, B. Semi-quantitative landslide risk assessment using GIS-based exposure analysis in Kuala Lumpur City. *Geomat. Nat. Haz. Risk.* **2017**, *8*, 706–732. [CrossRef]
31. Saadatkhah, N.; Kassim, A.; Lee, L.M.; Yunusa, G.H. Quantitative hazard analysis for landslides in Hulu Kelang area, Malaysia. *Jurnal Teknologi* **2015**, 73. [CrossRef]
32. Elmahdy, S.I.; Mostafa, M.M. Natural hazards susceptibility mapping in Kuala Lumpur, Malaysia: An assessment using remote sensing and geographic information system (GIS). *Geomat. Nat. Haz. Risk.* **2013**, *4*, 71–91. [CrossRef]
33. Mahmud, A.R.; Awad, A.; Billa, R. Landslide susceptibility mapping using averaged weightage score and GIS: A case study at Kuala Lumpur. *Pertanika J. Sci. Technol.* **2013**, *21*, 473–486.
34. Sezer, E.A.; Pradhan, B.; Gokceoglu, C. Manifestation of an adaptive neuro-fuzzy model on landslide susceptibility mapping: Klang valley, Malaysia (vol 38, pg 8208, 2011). *Expert Syst. Appl.* **2013**, *40*, 2360. [CrossRef]
35. Pradhan, B.; Lee, B.; Buchroithner, M.F. Remote Sensing and GIS-based Landslide Susceptibility Analysis and its Cross-validation in Three Test Areas Using a Frequency Ratio Model. *Photogramm. Fernerkun.* **2010**, *1*, 17–32. [CrossRef]
36. Pradhan, B. Remote sensing and GIS-based landslide hazard analysis and cross-validation using multivariate logistic regression model on three test areas in Malaysia. *Adv. Space. Res.* **2010**, *45*, 1244–1256. [CrossRef]
37. Jamaluddin, T. Human factors and slope failures in Malaysia. *Bull. Geol. Soc. Malays.* **2006**, *52*, 75–84. [CrossRef]
38. Tan, B.K.; Komoo, I. Urban geology: Case study of Kuala Lumpur, Malaysia. *Eng. Geol.* **1990**, *28*, 71–94. [CrossRef]
39. Leslie, A.G.; Dobbs, M.R.; Fatt, N.T.; Rosle, Q.A.; Noh, M.R.M.; Dodd, T.J.; Gillespie, M.R. The Ukay Perdana Shear Zone in Kuala Lumpur: A crustal-scale marker of early Jurassic orogenic deformation in Peninsular Malaysia. *Bull. Geol. Soc. Malays.* **2020**, *69*, 135–147. [CrossRef]
40. Tian, B.; Wang, L.; Koike, K. Spatial statistics of surface roughness change derived from multi-scale digital elevation models. *Procedia Environ. Sci.* **2011**, *7*, 252–257. [CrossRef]

41. De Reu, J.; Bourgeois, J.; Bats, M.; Zwertvaegher, A.; Gelorini, V.; De Smedt, P.; Chu, W.; Antrop, M.; De Maeyer, P.; Finke, P.; et al. Application of the topographic position index to heterogeneous landscapes. *Geomorphology* **2013**, *186*, 39–49. [CrossRef]
42. Gallant, J.; Wilson, J. Primary topographic attributes. In *Terrain Analysis: Principles and Applications*; Wilson, J., Gallant, J., Eds.; Wiley: New York, NY, USA, 2000; pp. 51–85.
43. Beven, K.J.; Kirkby, M.J. A physically based, variable contributing area model of basin hydrology. *Hydrol. Sci. B* **1979**, *24*, 43–69. [CrossRef]
44. Paramananthan, S.; Muhamad, N.; Pereira, J.J. Soil related factors controlling erosion and landslides in Malaysia. *Bull. Geol. Soc. Malays.* **2021**, *72*, 165–175.
45. Dai, F.C.; Lee, C.F. Landslide characteristics and slope instability modeling using GIS, Lantau Island, Hong Kong. *Geomorphology* **2002**, *42*, 213–228. [CrossRef]
46. Yusof, N.M.; Pradhan, B. Landslide susceptibility mapping along PLUS expressways in Malaysia using probabilistic based model in GIS. *IOP Conf. Ser. Earth Environ. Sci.* **2014**, *20*, 012031. [CrossRef]
47. Yalcin, A. GIS-based landslide susceptibility mapping using analytical hierarchy process and bivariate statistics in Ardesen (Turkey): Comparisons of results and confirmations. *Catena* **2008**, *72*, 1–12. [CrossRef]
48. Althuwaynee, O.F.; Pradhan, B.; Lee, S. Application of an evidential belief function model in landslide susceptibility mapping. *Comput. Geosci.* **2012**, *44*, 120–135. [CrossRef]
49. Jebur, M.N.; Pradhan, B.; Tehrany, M.S. Optimization of landslide conditioning factors using very high-resolution airborne laser scanning (LiDAR) data at catchment scale. *Remote Sens. Environ.* **2014**, *152*, 150–165. [CrossRef]
50. Pourghasemi, H.; Moradi, H.; Fatemi Aghda, S.; Gokceoglu, C.; Pradhan, B. GIS-based landslide susceptibility mapping with probabilistic likelihood ratio and spatial multi-criteria evaluation models (North of Tehran, Iran). *Arab. J. Geosci.* **2014**, *7*, 1857–1878. [CrossRef]
51. Guzzetti, F.; Carrara, A.; Cardinali, M.; Reichenbach, P. Landslide hazard evaluation: A review of current techniques and their application in a multi-scale study, Central Italy. *Geomorphology* **1999**, *31*, 181–216. [CrossRef]
52. Carrara, A. Multivariate models for landslide hazard evaluation. *J. Int. Assoc. Math. Geol.* **1983**, *15*, 403–426. [CrossRef]
53. Van Westen, C. Statistical landslide hazard analysis, ILWIS 2.1 For Windows Application Guide. *ITC Publ. Enschede* **1997**, *2*, 73–84.
54. Fell, R.; Corominas, J.; Bonnard, C.; Cascini, L.; Leroi, E.; Savage, W.Z.; Eng, J.-J.T.C.L. Guidelines for landslide susceptibility, hazard and risk zoning for land-use planning Commentary. *Eng. Geol.* **2008**, *102*, 99–111. [CrossRef]
55. Flentje, P.N.; Miner, A.; Whitt, G.; Fell, R. Guidelines for landslide Susceptibility, Hazard and Risk Zoning for Land Use Planning. Available online: https://ro.uow.edu.au/engpapers/2823 (accessed on 25 January 2021).
56. Regmi, N.R.; McDonald, E.V.; Giardino, J.R.; Vitek, J.D. A comparison of logistic regression-based models of susceptibility to landslides in western Colorado, USA. *Landslides* **2014**, *11*, 247–262. [CrossRef]
57. Department of Statistics Malaysia. *Key Findings: Population and Housing Census of Malaysia 2020*; Department of Statistics Malaysia: Putrajaya, Malaysia, 2020; pp. 86–87.
58. Wu, Y.-m.; Lan, H.-x.; Gao, X.; Li, L.-p.; Yang, Z.-h. A simplified physically based coupled rainfall threshold model for triggering landslides. *Eng. Geol.* **2015**, *195*, 63–69. [CrossRef]
59. Luca, D.L.D.; Versace, P. A General Formulation to Describe Empirical Rainfall Thresholds for Landslides. *Procedia Earth Planet. Sci.* **2016**, *16*, 98–107. [CrossRef]

applied sciences

MDPI

Article

GIS-Multicriteria Analysis Using AHP to Evaluate the Landslide Risk in Road Lifelines

Giovanni Leonardi [1,*], Rocco Palamara [1], Francesco Manti [2] and Antonio Tufano [3]

[1] DICEAM Department, University of Reggio Calabria, Via Graziella, Loc. Feo di Vito, 89122 Reggio Calabria, Italy; rocco.palamara@unirc.it

[2] Civil Protection Department of the Calabria Region, Via Palazzo Cedir snc, 89121 Reggio Calabria, Italy; f.manti@regione.calabria.it

[3] Department of Engineering, Telematic University Pegaso, Centro Direzionale Napoli Isola F2, Pegaso Tower, 80143 Napoli, Italy; antonio.tufano@unipegaso.it

* Correspondence: giovanni.leonardi@unirc.it

Abstract: The present paper proposes a new methodology to characterize the landslide susceptibility of the Reggio Calabria metropolitan area. For this purpose, various factors were used, such as land use, slope, rainfall, elevation, lithology, distance from roads and rivers, and thanks to the use of GIS devices and the AHP method, the landslide risk was defined for the whole territory. The values obtained were classified into four categories: low, moderate, high, and very high. They were then exported into the GIS environment to produce a landslide susceptibility map. The study carried out demonstrates the fragility of the Calabrian territory. From the results obtained, in fact, 66% of the metropolitan territory of Reggio Calabria appears to have a medium–high landslide risk.

Keywords: landslide risk; natural phenomena; Reggio Calabria; GIS; AHP method

check for updates

Citation: Leonardi, G.; Palamara, R.; Manti, F.; Tufano, A. GIS-Multicriteria Analysis Using AHP to Evaluate the Landslide Risk in Road Lifelines. *Appl. Sci.* **2022**, *12*, 4707. https://doi.org/10.3390/app12094707

Academic Editors: Miguel Llorente Isidro, Ricardo Castedo and David Moncoulon

Received: 15 March 2022
Accepted: 3 May 2022
Published: 7 May 2022

1. Introduction

Landslides are natural phenomena as well as man-made disasters that frequently lead to loss of human life and property, as well as causing serious damage to natural resources throughout the world. These occurs because of different associated natural hazards or anthropogenic activities. Natural phenomena include meteorological changes, such as intense rainfall, prolonged rainfall, or snowmelt and again earthquakes or volcanic eruptions. Human disturbances instead include land use alteration, deforestation, carrying out excavations, changes in the slope profile, and infrastructure constructions [1].

The effects associated with the occurrence of landslides include injuries, loss of human life, significant damage to natural resources and infrastructure, such as roads, bridges, and communication lines. To limit this, the realization of the landslide susceptibility map is fundamental to identify possible preventive measures and the planning of evacuation strategies to avoid significant damage and victims. Moreover, landslide susceptibility, hazard and risk maps are of great help to planners for selecting suitable areas to implement development schemes in any area [2,3].

Despite tremendous progress in science and technology, landslides considerably affect the socio-economic conditions of all regions of the globe. The susceptibility to landslides can be defined as the probability of the spatial occurrence of a landslide event based on the relationships between the occurrence distribution and a set of predisposing factors in a given area, such as geo-environmental thematic variables [4]. Landslides cannot be completely prevented but an accurate prediction of the landslide-susceptible areas is of particular importance since this can lead to saving natural resources as well as human lives.

Landslides occur and are controlled by one or more conditioning factors, such as topography, slope, failure mechanism, intensity of rainfall, land cover, geological formation, strength of rocks, and many more [5–7].

The landslide susceptibility assessment can be tricky because of the difficult evaluation of both the spatial and temporal distribution of past events for large areas, mainly due to the limitations and gaps of historical records and geographic information [8].

Many different types of landslide risk assessment zonation techniques have been developed over the last decades. These methods can typically be labeled as being either qualitative or quantitative procedures, also called direct and indirect methods [9–11].

The qualitative approaches are inventory-based and knowledge-driven methods such as distribution analysis or inventory [12], geomorphic mapping [13], and map integration models [14]. They yield direct and countable results but are limited by data availability. By contrast, the qualitative approach is less complex, but the subjective estimation of experts renders its degree of confidence and the accuracy of its results questionable [15].

Direct mapping has the purpose of determining the degree of landslide susceptibility based on a qualitative approach and provides an estimate of "where" landslides are expected.

In this approach, the experience, knowledge of the ground boundary conditions, and the evaluation of pre-existing terrain maps are taken into consideration. Precisely for these reasons, direct methods present results that are subjective since they are linked to the experience and knowledge of the personnel and present many difficulties in transcribing the phenomena due to the occurrence of particular environmental changes in certain areas [16]. This methodology usually uses factors characteristic of the locations where landslides have occurred in the past, together with an assessment of the areas with a potential to experience land sliding in the future, but with no assessment of the landslide's occurrence frequency.

The reliability of landslide-susceptibility maps mainly depends on the scale of work, the quality of the data and the analysis methodology.

On the other hand, indirect mapping is based on quantitative models, such as statistical and deterministic ones, which aim to find numerical correlations between the various factors that favor the occurrence of landslides and their distribution on the earth's surface and based on information obtained from the interrelation between landslide conditioning factors and the landslide distribution. For large areas, statistical methods are the most widely used [17]. The statistical methods include bivariate statistical analysis [18], logistic regression [19–21], multivariate regression [22,23]; multivariate adaptive regression spline [24], statistical index [25,26], analytical hierarchy process analytical hierarchy process [27,28], weight of evidence [29,30], and evidential belief function [31,32].

In recent times, quantitative approaches have been widely used for landslide susceptibility and hazard evaluation. However, it is mostly very difficult to include temporal probability in this analysis of large areas, due to the heterogeneity of subsoil conditions, the numerous conditioning factors, and the absence of a complete historical record of the events that can determine the occurrence of landslides (for example precipitation, earthquakes, and landslides themselves).

Furthermore, these models, however, show a high sensitivity of the results to the quality and accuracy of the thematic data and the specificity of some sensitive factors that could end up not being taken into consideration.

Landslide susceptibility and hazard assessments often use multi-criteria decision analysis (MCDA) techniques since in most cases, the types and format of data that are available are qualitative and quantitative, thereby requiring a semi-quantitative method that incorporates both types of data [33].

The semi-quantitative landslide assessment methods can be considered an effective expert's tool for weighting and ranking the chosen factors, which represent the main causes for landslide susceptibility of the study area, in an objectively optimal and simple way.

Some qualitative methods become semi-quantitative by incorporating ranking and weighting [34,35], as is the case of the analytic hierarchy process (AHP), a multi-objective and multi-criteria decision-making methodology which has been widely applied for the solution of decision problems [36].

This method is based on the analytical hierarchy of involved factors and the comparison between the various pairs of them in order to enable the assignment of a relevant ratio for each factor. Thanks to this method, it is therefore possible to estimate the weight of each factor considered, through the linear correlation of each factor with respect to the others. The ability to correlate different factors has made this method a valuable tool for many researchers in compiling landslide susceptibility maps, which are obtained by correlating and comparing a large number of factors [37].

The management of a large number of factors to be correlated and estimated for the definition of landslide susceptibility is therefore carried out, in most cases, through the use of the geographical information system (GIS) [35].

The increasing popularity of GIS and its ease of use over the last decades has led many studies to use indirect susceptibility mapping approaches.

GIS, which uses data-integration techniques, is a very suitable tool for landslide susceptibility mapping. In fact, with the increasing availability of high-resolution spatial data sets, GIS, remote sensing, and computers with large and fast processing capacity, it has becoming possible to partially automate the evaluation of landslide hazard and susceptibility mapping process and thus minimize fieldwork.

The reliability of these maps depends mostly on the applied methodology as well as on the available data used for the hazard risk estimation. Moreover, GIS is an excellent and useful tool for mapping the susceptibility of an area prone to landslide manifestation.

This study focused on producing a landslide susceptibility map of the Reggio Calabria metropolitan area (Italy) by combining GIS techniques and the AHP method.

The main objectives of the present work can be synthesized to create a landslide susceptibility map for the province of Reggio Calabria, using the GIS software by means of the weighted combination of various factors, such as the slope, lithology, elevation, rainfall, land use, distance of the road and river. From the obtained map, it is possible to evaluate the landslide susceptibility in the areas that are particularly relevant because of the connections between the internal urban areas and the main towns and services located along the coasts and to analyze the degree of landslide risk of the connecting infrastructures (lifelines). This method could be used in other areas of the Calabria Region for the realization of the landslide susceptibility map in order to be able to use the maps obtained to plan the safety of the slopes or, in particular, identify the main road infrastructures more at landslides risk, comparing the results obtained with the various inventories of landslides that occurred in the past.

2. Study Area

The Reggio Calabria metropolitan city, in the Calabria region in the south of Italy, was selected as the study area because the territory is subject to phenomena of hydrogeological and seismic risk; the present study area has suffered from every landslide type.

The territory of Calabria is geologically young and often subject to natural modifications, so hydro-geological disaster (landslides and floods) is one of the risk factors to which Calabria is exposed. This is due, among other conditions, to the physical conformation of the region and the climatic conditions.

Landslides are the most common consequence of the soil instability due to the seismic activity and the hydrogeological problems in Reggio Calabria metropolitan city. Extraordinary conditions, such as earthquakes, and more ordinary meteorological conditions, such as rainfalls, can both induce landslides, rock fall, or debris flow.

The considered study area with 97 cities and towns and with a total population of 530,000 inhabitants is the area of Calabria with the highest population density equal to 165.57 residents/km^2.

The Reggio Calabria metropolitan city covers an area of 3183 km^2, of which 1685 km^2 (52.95%) is represented by hilly terrain, 1.275 km^2 (40.07%) is mountainous and the remaining 223 km^2 (6.97%) are represented by land lowland.

In regions where urban residential areas coincide with mountainous terrains, like in this area, the risk is higher for people, and the economic costs include relocating communities and repairing physical structures.

The Autorità di Bacino of Calabria Region, considering the various mentioned problems, has made available several thematic maps for a complete hydrogeological balance (PAI—Piano di Assetto Idrogeologico, 2001) [38]. These maps, updated in 2016, are accessible through Quantum—GIS, and they describe the landslide hazard and risk areas in the whole region, but they do not give any information about the landslide risk assessment in the strategical infrastructures identified as lifelines [2]. Furthermore, these assessments were carried out without taking into consideration the main connecting and emergency road infrastructures, which, however, play a role particularly relevant during an emergency to allow rapid and efficient access, assistance, and rescue, guaranteeing evacuation and, more in general, maintaining access for all emergency services [2,3].

Therefore, starting from the data provided by the PAI, containing the landslides that have occurred in the past together with an assessment of the areas with a potential to experience land-sliding in the future, and with the integration of factors of particular importance, such as distance from roads and from the rivers, the objective of the present work is to create a landslide susceptibility map for the metropolitan area of Reggio Calabria by means of the weighted combination of various indices using the AHP method.

3. Materials and Method

The first step in every susceptibility assessment consists of collecting all available information and data for the study area, and this stage may be the most important part.

Evaluating the relationship between the landslide occurrence and the conditioning parameters becomes very important for the landslide susceptibility mapping; further analysis of the cause–effect relationships is not always simple, as a landslide is seldom linked to a single cause. So, to produce the landslide susceptibility map in the metropolitan area of Reggio Calabria, a total of 7 inputs were selected for the model, considering the main influencing factors of landslides: the slope, the lithology, the elevation, the rainfall, the land use, the distance of the road, and the river.

In the study carried out, the number of factor classes was increased compared to that of previously published studies, analyzing what was done in the literature by other authors and wanting to obtain a non-homogeneous weight of each class of factor, unlike what was obtained in the previous studies.

3.1. Slope Factor

The slope and the aspect of the slopes play an important role in the occurrence of landslides because they represent the result of the combined influence of many agents. Slope is an important factor in the analysis of landslide; in fact, as the slope increases, the probability of the occurrence of landslide increases because as the slope angle increases, the shear stress of the soil increases.

A digital elevation model (DEM) was utilized using 3D extension of ArcGIS, and the slope angle was extracted from it, using contours with 5 m intervals digitized from topographic sheets and saved as a line layer.

Six slope categories were used as factor classes for the analysis of landslides risk as shown in Table 1.

Table 1. Weight of the individual classes of landslide hazard factors.

	Factor Class	$W_{i,L}$
Slope degree [°]	0–8	0.21
	8.01–18	0.15
	18.01–28	0.13
	28.01–37	0.09
	37.01–47	0.09
	>47	0.33
Land Use	Artificial and waterproof surfaces	0.04
	Arable agricultural areas with sparse vegetation	0.26
	Agroforestry areas with scattered trees and shrubs	0.42
	Wooded areas or areas with important vegetation	0.28
Elevation [m]	<100	0.19
	101–300	0.25
	301–500	0.15
	501–1000	0.27
	>1001	0.13
Rainfall [mm]	0–600	0.02
	600–900	0.19
	900–1200	0.36
	1200–1500	0.25
	<1500	0.18
Lithology	Sedimentary deposits	0.33
	Clays and sandstones	0.21
	Rocks	0.45
River Distance [m]	0–50	0.10
	50.01–100	0.10
	100.01–200	0.18
	200.01–250	0.08
	<250	0.54
Road Distance [m]	0–100	0.10
	101–200	0.08
	201–300	0.07
	301–400	0.06
	401–500	0.06
	>500	0.62

3.2. Lithology Factor

Lithology is a further important parameter with regard to landslide manifestation. It is widely recognized that geology greatly influences the occurrence of landslides because lithological and structural variations often lead to a difference in the strength and permeability of rocks and soils.

For the study area, three factor classes were determined by grouping the different geological formations according to their geological engineering behavior and according to their physical and mechanical characteristics.

Thus, lithology includes three classes as follows: sedimentary deposits, clays and sandstones and rocks.

3.3. Elevation Factor

The altitude does not contribute directly to landslide manifestation, but in relation to the other parameters, such as precipitation, the altitude contributes to landslide manifestation and influences the whole system.

Elevation is useful to classify the local relief and to locate points of maximum and minimum heights within terrains. The elevation is considered the fourth important parameter in the classification of landslide risk. The grid maps of the altitude with cell size 5×5 m were produced from the DEM. The separation of the altitude into 5 classes was as follows: <100 m, 101–300 m, 301–500 m, 250–500 m, and >1000 m. The highest density corresponds to the class with an elevation range of 500–1000 m, representing hilly areas.

3.4. Rainfall Factor

As it is well known, precipitation is among the most usual triggering factors for landslide manifestation. For the necessities of this study, the precipitation map was produced, using the annual average precipitation data of the main meteorological stations well distributed in the study area.

This map was separated into 5 classes as shown in Table 1.

3.5. Land Use Factor

The data for the land use in an area are a parameter that seriously affects the slope failures, as slope stability is very sensitive to changes in vegetation. For the necessities of this study, the land use, which reflects the vegetation covering, was classified into 4 categories as follows: artificial and waterproof surfaces, arable agricultural areas with sparse vegetation, agroforestry areas with scattered trees and shrubs, wooded areas, or areas with important vegetation.

3.6. Distance of Road and River Factor

As it is obvious, the artificial and natural parts of the slopes around a road are more sensitive in landslide manifestation. In addition, some lifeline roads have to guarantee effectiveness and efficiency in the immediate aftermath of a natural disaster; in some cases, they have to be maintained also during the event to allow rapid and efficient access and assistance and rescue, guaranteeing evacuation and, more in general, maintaining access for all emergency services. Therefore, the road network was chosen as a principal parameter that consists of the road network of the province of Reggio Calabria, composed of 1850 km of roads.

Buffer zones were created around road lifelines at distances of 100, 200, 300, 400 and 500 m.

Similar to the road network, the hydrographic network was digitized and saved as line layers in the GIS database, using the topographic sheets as a data source, and buffer zones were created around the bed of the rivers and the streams of the area, at distances of 50, 100, 200, and 250 m. These distances of river are measured from the river's bed boundaries from both sides.

Each factor was characterized in classes whose weight ($\mathbf{W}_{\mathbf{i,L}}$) was determined on the basis of the portion of the territory occupied by the class of the factor in the entire metropolitan area of Reggio Calabria (Table 1).

This method was used to identify the amount of territory that the factor class occupies to then be put into relation with the weights of the same calculated with other methods ($\mathbf{W}_{\mathbf{i,PAI}}$, $\mathbf{W}_{\mathbf{i,AHP}}$) and to be able to determine the degree of landslide susceptibility.

The analysis was carried out by means of GIS devices, which allow to easily analyze a considerable amount of data and to convert the map pixels into data sets.

The liability and accuracy of the collected data also influence the success of the applied methodology. For this reason, raster images were used with an accuracy of a pixel equal to 5 m. The images obtained are shown in Figure 1.

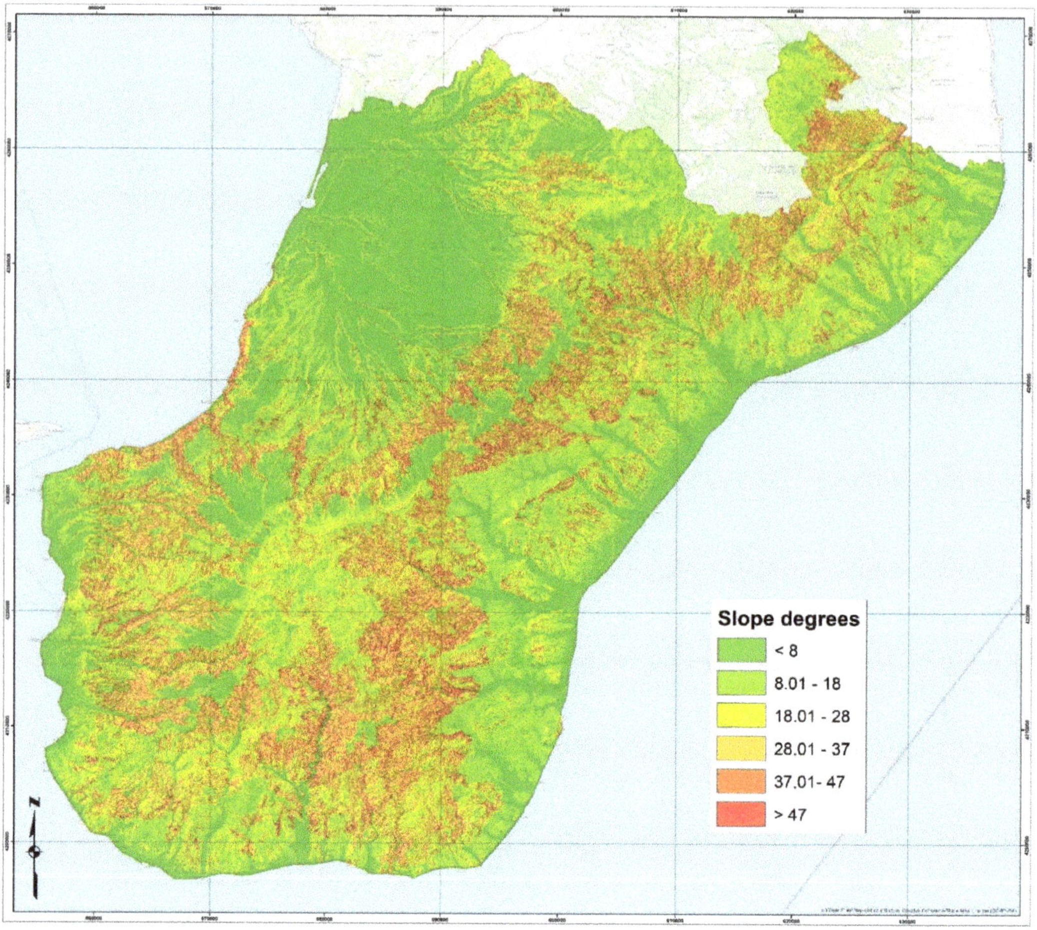

(a)

Figure 1. *Cont.*

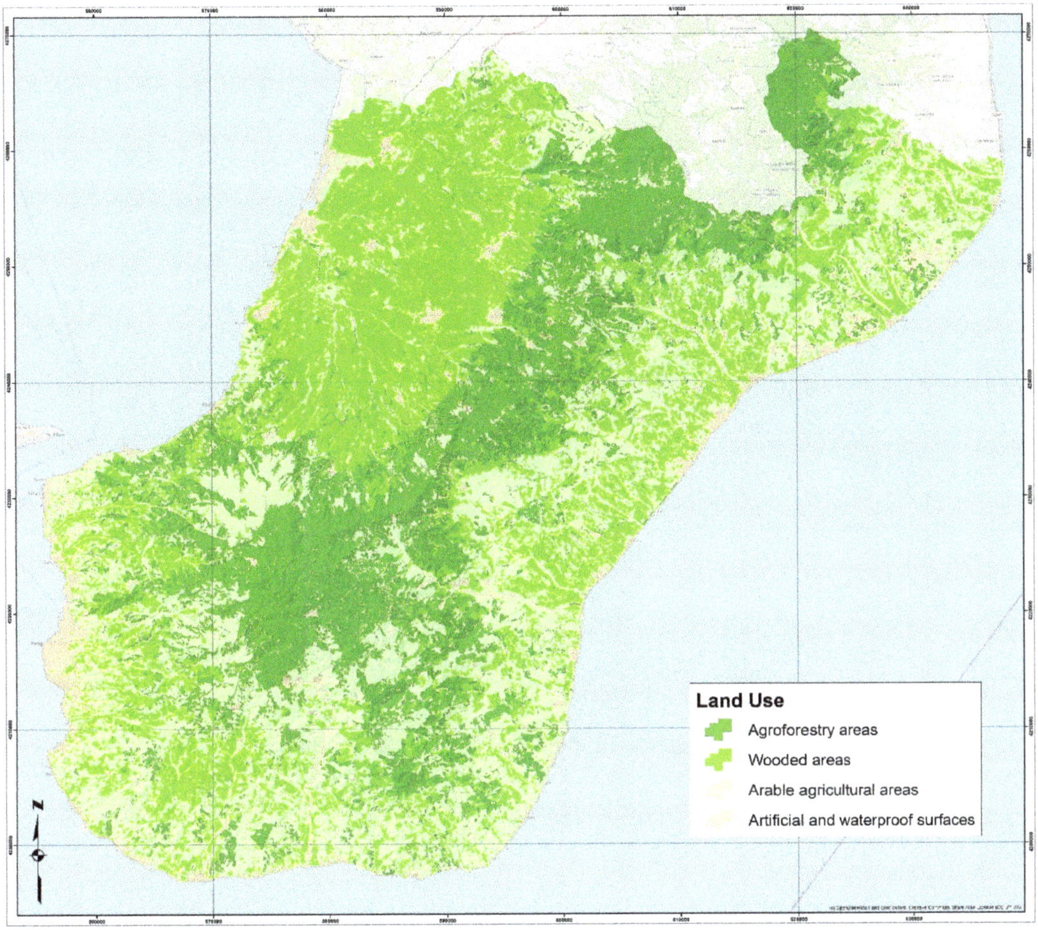

(b)

Figure 1. *Cont.*

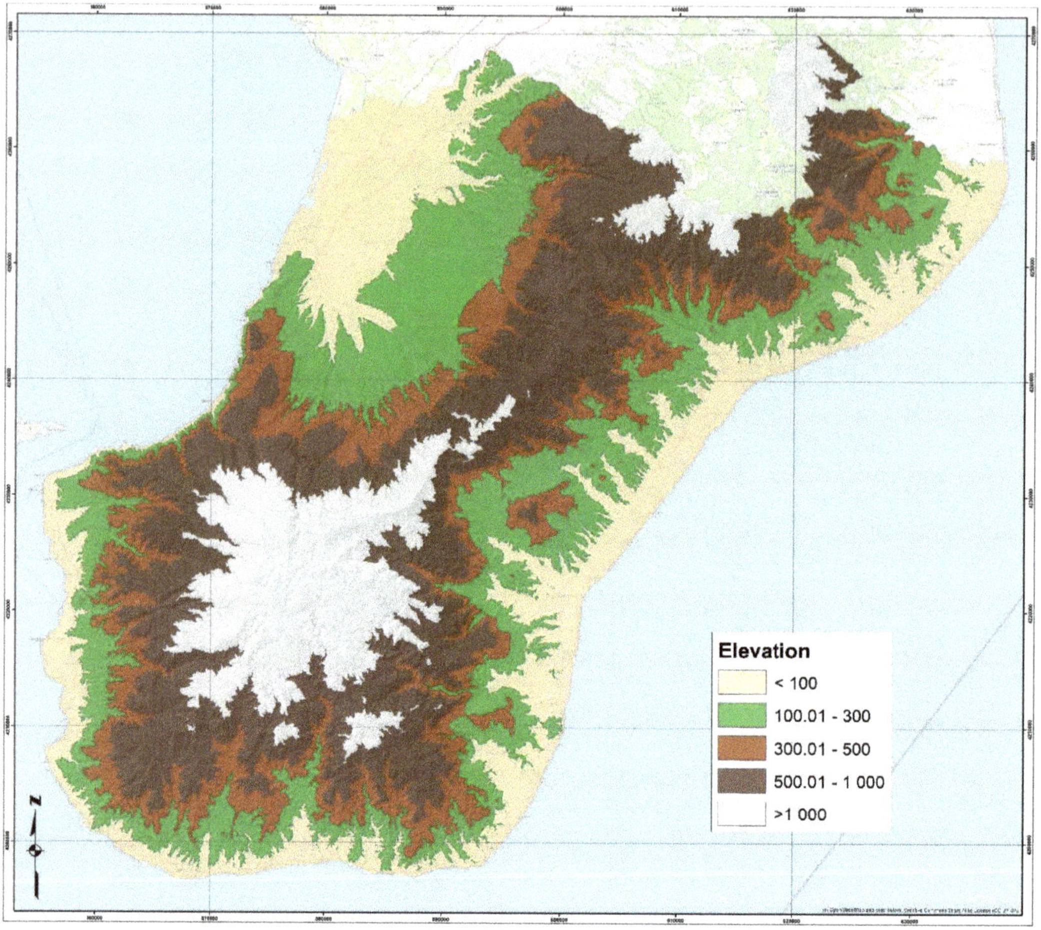

(c)

Figure 1. *Cont.*

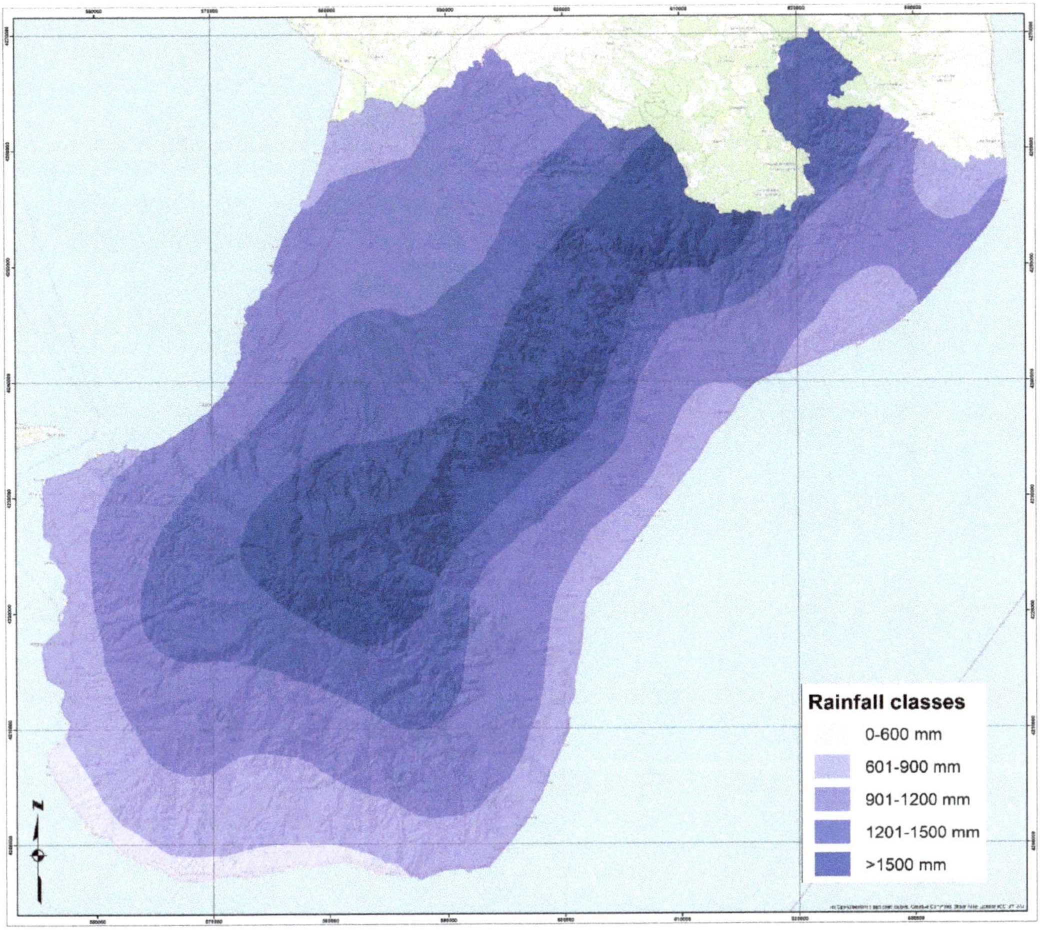

(d)

Figure 1. *Cont.*

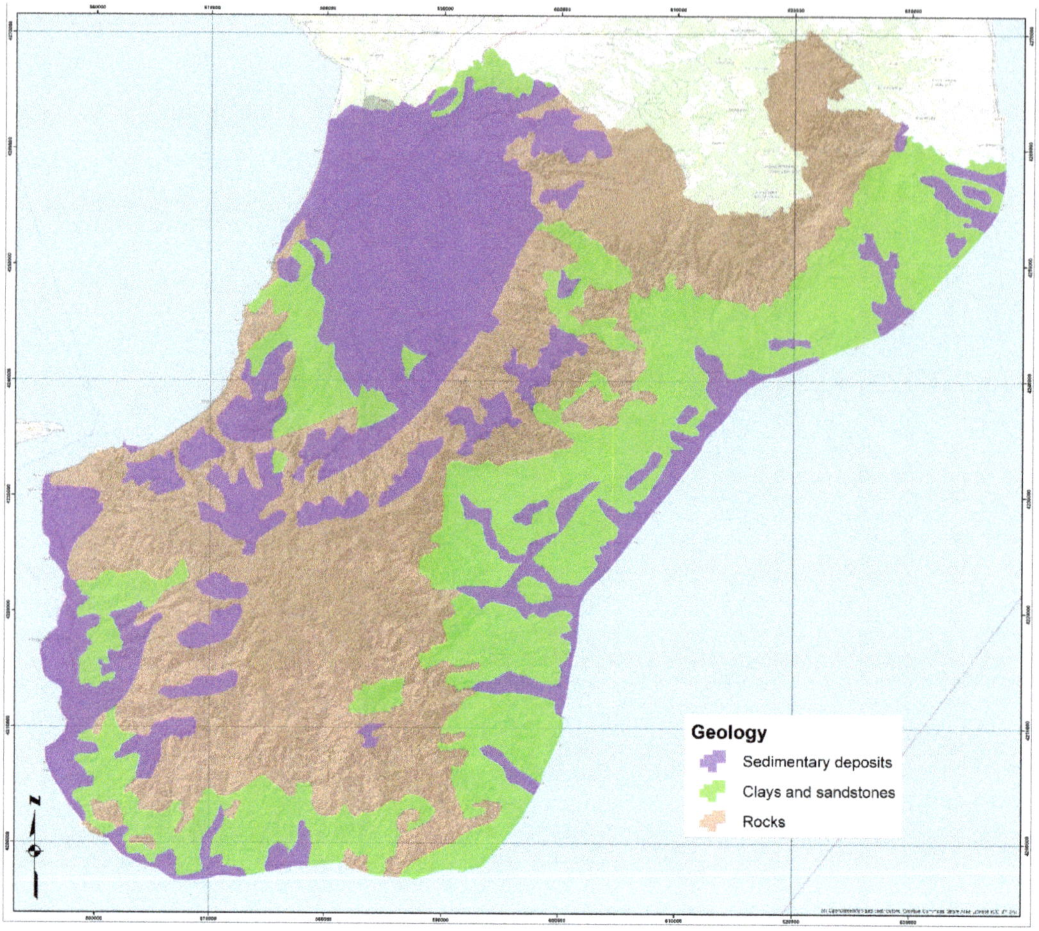

(e)

Figure 1. *Cont.*

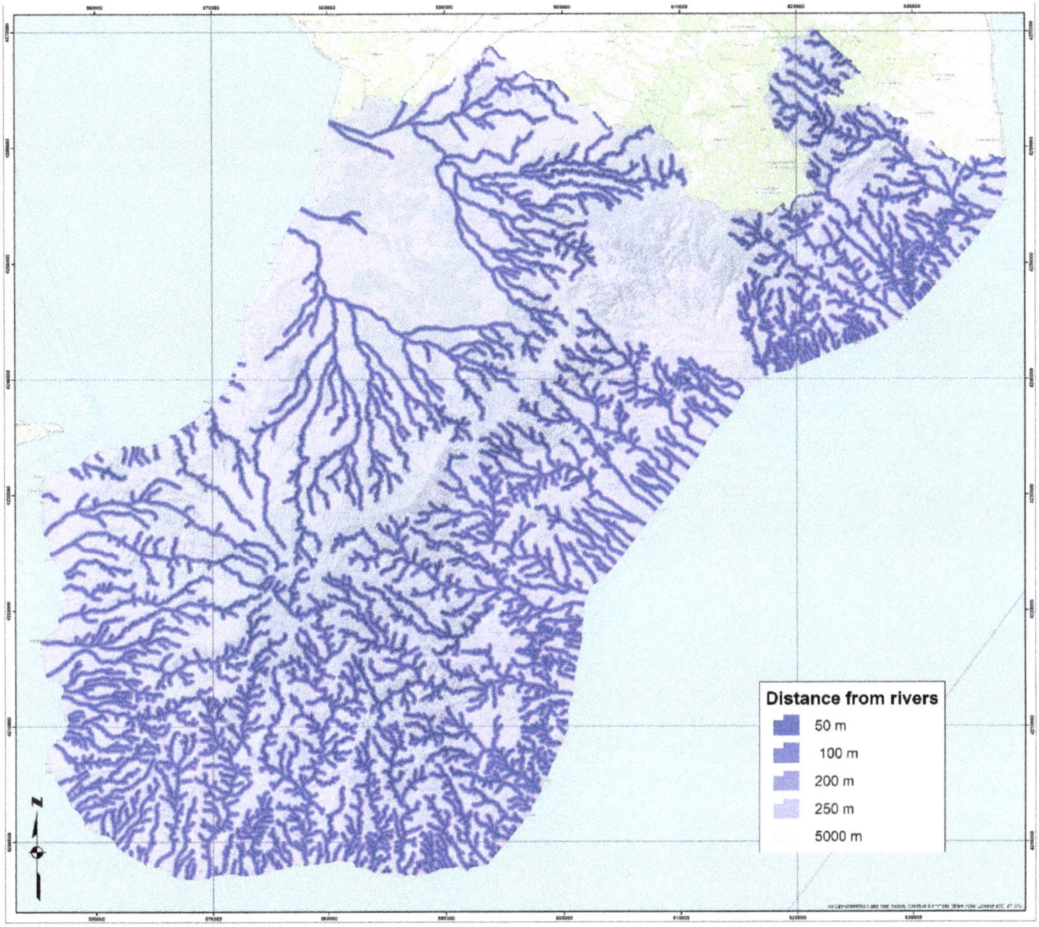

(f)

Figure 1. *Cont.*

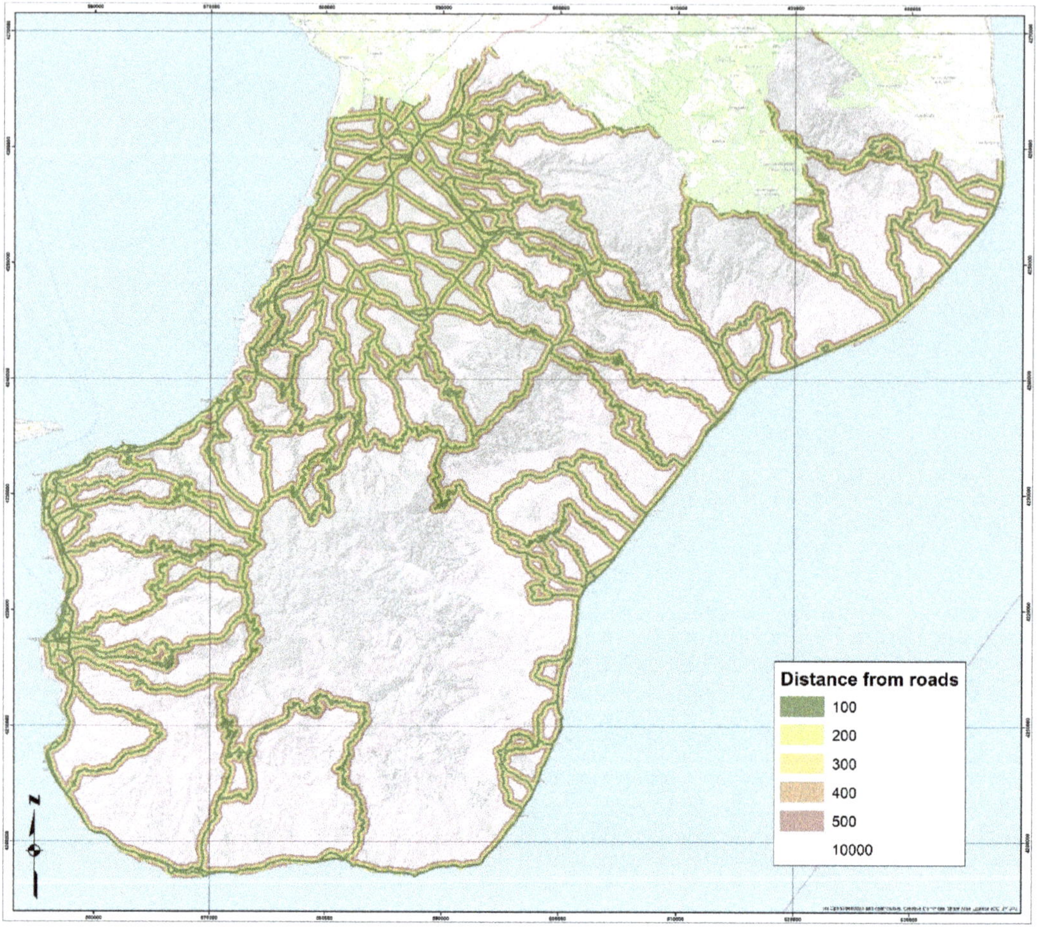

(g)

Figure 1. *Cont.*

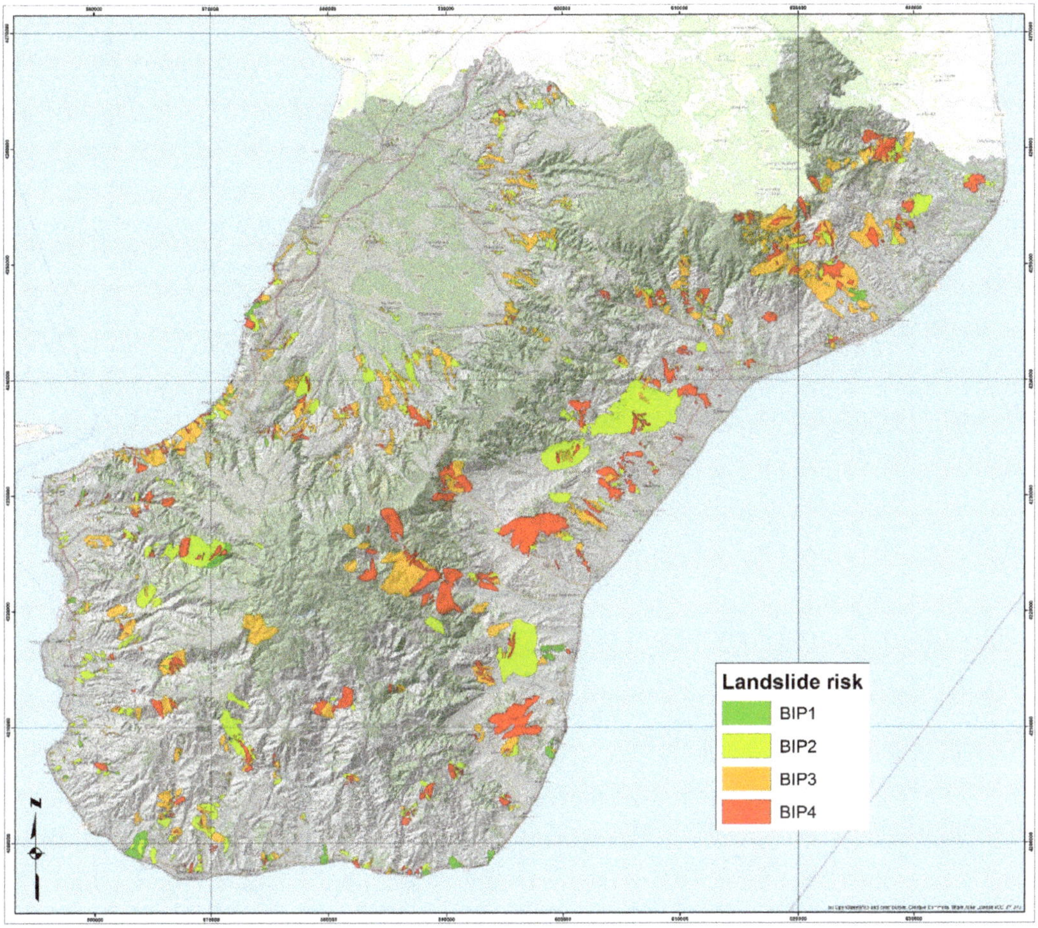

(**h**)

Figure 1. Topographical parameter maps: (**a**) slope degree; (**b**) land use; (**c**) elevation; (**d**) rainfall; (**e**) geology; (**f**) river distance; (**g**) road distance, (**h**) PAI landslide risk areas.

So, the weight of each class was determined by the examination of the value attained by each factor class in the landslide areas identified by the PAI, with respect to the past landslide events ($\mathbf{W_{i,PAI}}$) that occurred in the territory of the metropolitan area (Table 2). This parameter allows to determine the importance of each individual class based on historical data relating to landslides that occurred in the past.

Table 2. Weight of classes of factor in the landslide areas identified by the PAI.

	Factor Class	$W_{i,L}$	$W_{i,PAI}$
Slope degree [°]	0–8	0.21	0.07
	8.01–18	0.15	0.15
	18.01–28	0.13	0.17
	28.01–37	0.09	0.11
	37.01–47	0.09	0.10
	<47	0.33	0.39
Land Use	Artificial and waterproof surfaces	0.04	0.04
	Arable agricultural areas with sparse vegetation	0.26	0.37
	Agroforestry areas with scattered trees and shrubs	0.42	0.38
	Wooded areas or areas with important vegetation	0.28	0.21
Elevation [m]	<100	0.19	0.10
	101–300	0.25	0.40
	301–500	0.15	0.23
	501–1000	0.27	0.21
	>1001	0.13	0.06
Rainfall [mm]	0–600	0.02	0.02
	600–900	0.19	0.12
	900–1200	0.36	0.46
	1200–1500	0.25	0.29
	<1500	0.18	0.11
Lithology	Sedimentary deposits	0.33	0.26
	Clays and sandstones	0.21	0.34
	Rocks	0.45	0.46
River Distance [m]	0–50	0.10	0.11
	50.01–100	0.10	0.10
	100.01–200	0.18	0.18
	200.01–250	0.08	0.08
	<250	0.54	0.53
Road Distance [m]	0–100	0.10	0.14
	101–200	0.08	0.11
	201–300	0.07	0.09
	301–400	0.06	0.07
	401–500	0.06	0.06
	>500	0.62	0.53

Later using the AHP method, we were able to obtain a relative significance of the relevant factors after the pairwise comparison matrix was constructed.

The pairwise comparison method was developed by Saaty in the context of the analytical hierarchy process. The AHP approach allows assessing the relative weight of multiple elements in an initiative manner [39].

After defining the weight of each factor, with respect to the past landslide events identified by the PAI, its relative weight was calculated with the AHP method. The rating score of relative significance was set from 1 to 9, indicating less important to much more important factors.

The pairwise comparison matrix is shown in Table 3 using a 7 × 7 matrix, where diagonal elements are equal to 1.

The row describes the importance of factor, and the values of each row are compared with each column to define the relative importance to obtain a rating score. For example, the slope is significantly more important than the elevation and therefore assigned the value 8. Conversely, the importance of the elevation with respect to the slope is equal to 1/8.

Table 3. Pairwise comparison matrix.

	Slope	Land Use	Elevation	Rainfall	Lithology	River Distance	Road Distance
Slope degree	1.00	7.00	8.00	2.00	7.00	4.00	9.00
Land Use	0.14	1.00	3.00	0.20	4.00	0.33	5.00
Elevation	0.13	0.33	1.00	0.20	4.00	0.20	8.00
Rainfall	0.50	5.00	5.00	1.00	3.00	3.00	7.00
Lithology	0.14	0.25	0.25	0.33	1.00	0.17	3.00
River Distance	0.25	3.00	5.00	0.33	4.00	1.00	5.00
Road Distance	0.20	0.20	0.13	0.14	0.33	0.20	1.00

After evaluating and confirmed the consistency of the created eigenvector matrix for the AHP method, the weight for each factor using this method was obtained (Table 4).

Table 4. Weight of normalized landslide hazard factors through AHP method.

	$\mathbf{W_{i,AHP}}$
Slope	0.403
Land Use	0.086
Elevation	0.061
Rainfall	0.242
Geology	0.037
Road distance	0.148
River distance	0.023

The landslides index for each individual pixel was finally obtained by the weighted sum of the weight determined on the basis of the portion of territory occupied by the factor class ($\mathbf{W_{i,L}}$), the weight reached by each factor class in the landslide areas identified by the PAI ($\mathbf{W_{i,PAI}}$), and the weight of the same factor obtained from the AHP method ($\mathbf{W_{i,AHP}}$).

$$\mathbf{LI} = \sum_{i=1}^{n} \mathbf{W_{i,L}} + \mathbf{W_{i,PAI}} + \mathbf{W_{i,AHP}}$$

The values thus obtained for the assessment of susceptibility were classified into five ranges; in particular, values below 0.17 represent the portions of territory with low risk, values between 0.1701 and 0.25 represent the portions of territory with moderate risk, values between 0.2501 and 0.3 represent the portions of territory with high risk, and values greater than 0.301 represent the portions of the territory with very high risk.

The above-mentioned results highlight that 61% of the whole territory of the metropolitan area is affected by a high landslide susceptibility value. In particular, the surface at low risk of landslide is equal to 7% of the entire metropolitan area, at moderate risk is equal to 32%, at high risk is equal to 36%, and at very high risk is equal to 25%.

The values obtained for the assessment of susceptibility were classified into five ranges in such a way that, having identified the lower and higher values in each range considered, the number of pixels with respect to the total was equalized between the various classes. This ensures that each class range has approximately the same number of values in each class and that the change between intervals is fairly consistent [40].

From the obtained map (Figure 2), it is possible to notice that the highly and very highly susceptible zones are the hills, i.e., the areas between the coasts and the central part of the province. These areas are particularly relevant because of the connections within the various internal urban areas and the main towns and services located along the coasts. Further, the landslide susceptibility evaluation procedure adopted resulted to be in agreement with the work performed by the Calabria Region Basin Authority: 80% of the highly and very highly susceptible areas coincide with the landslide areas identified by the PAI maps.

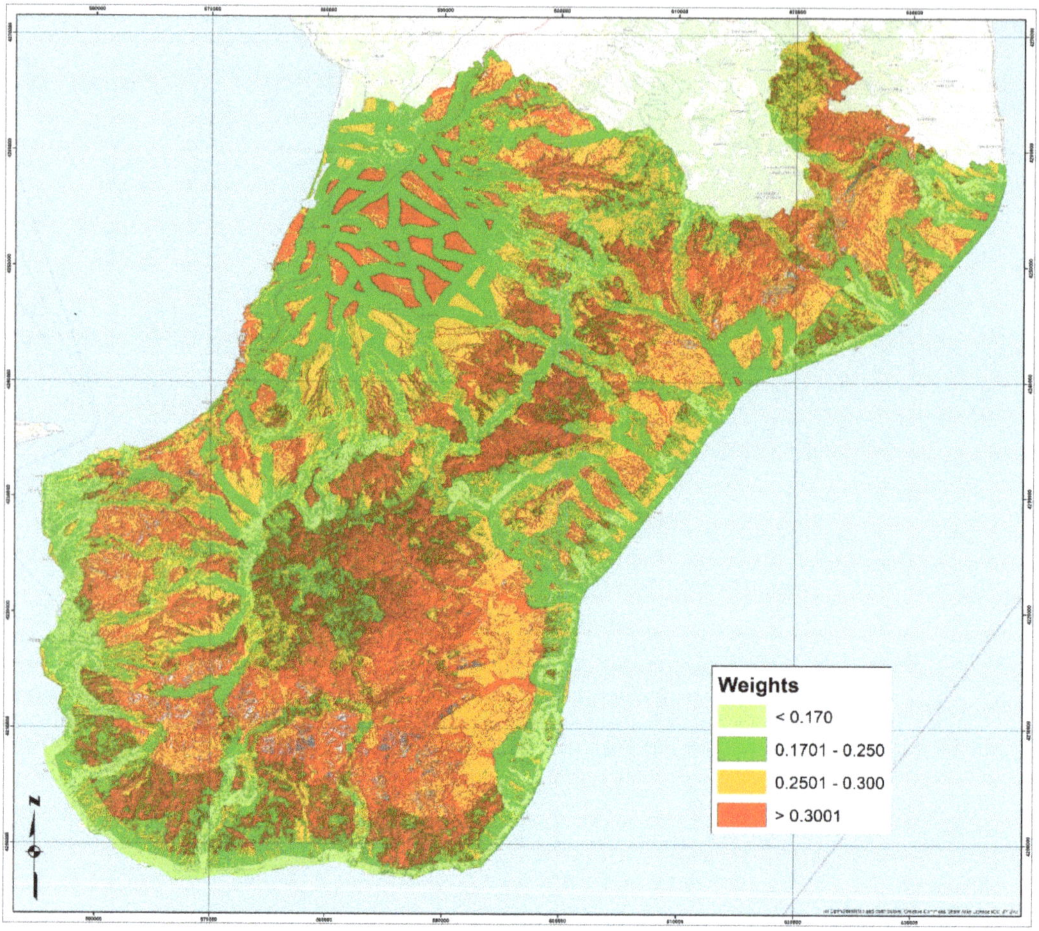

Figure 2. Landslides susceptibility map.

4. Results

The study carried out demonstrates the fragility of the Calabrian territory. From the results obtained, in fact, 38% of the metropolitan territory of Reggio Calabria appears to have a medium landslide risk and 28%, a high landslide risk. The results obtained from the model are congruent with the landslide risk areas identified by the PAI and are more reliable than previous studies [5]. In particular, the model produces an overestimation of only 8% of the areas at risk of medium–high landslide, compared to the areas surveyed by the PAI, thus also managing to determine the risk of landslides in those areas not classified by the PAI due to the absence of exposed elements. These areas, however, are still of fundamental importance, as they are crossed by connecting infrastructures.

This method could be used in other areas of the Calabria Region for the realization of the landslide susceptibility map in order to be able to use the maps obtained to plan the safety of the slopes or, in particular, identify the main road infrastructures on which to carry out work risk mitigation (lifelines). The results of this preliminary analysis can give important information on the relative criticality of the different road sectors, thereby allowing attention and economic budgets to be shifted toward the most critical aspects, where structural and non-structural mitigation measures could be implemented. In such a sense, they supply useful information regarding the intervention priorities and, above all, a scale of inspection surveys and analyses of a structural nature which can find the effective level of susceptibility of the infrastructure, as well as the most opportune mitigation actions for the reduction of risk levels. This latter aspect will be the subject of future studies.

Author Contributions: Data curation, F.M.; Methodology, G.L., R.P., F.M. and A.T.; Writing—original draft, G.L., R.P. and A.T. All authors have read and agreed to the published version of the manuscript.

Funding: This research received no external funding.

Conflicts of Interest: The authors declare no conflict of interest.

References

1. Zaruba, Q.; Mencl, V. *Landslides and Their Control*; Elsevier: New York, NY, USA, 2014.
2. Cirianni, F.; Fonte, F.; Leonardi, G.; Scopelliti, F. Analysis of lifelines transportation vulnerability. *Procedia Soc. Behav. Sci.* **2012**, *53*, 29–38. [CrossRef]
3. Cirianni, F.; Leonardi, G.; Scopelliti, F. A Methodology for Assessing the Seismic Vulnerability of Highway Systems. *AIP Conf. Proc.* **2008**, *1020*, 864–871.
4. Guzzetti, F.; Carrara, A.; Cardinali, M.; Reichenbach, P. Landslide hazard evaluation: A review of current techniques and their application in a multi-scale study, Central Italy. *Geomorphology* **1999**, *31*, 181–216. [CrossRef]
5. Leonardi, G.; Palamara, R.; Suraci, F. A fuzzy methodology to evaluate the landslide risk in road lifelines. *Transp. Res. Procedia* **2020**, *45*, 732–739. [CrossRef]
6. Pellicani, R.; Argentiero, I.; Spilotro, G. GIS-based predictive models for regional-scale landslide susceptibility assessment and risk mapping along road corridors. *Geomat. Nat. Hazards Risk* **2017**, *8*, 1012–1033. [CrossRef]
7. Chen, W.; Xie, X.; Peng, J.; Shahabi, H.; Hong, H.; Bui, D.T.; Duan, Z.; Li, S.; Zhu, A.-X. GIS-based landslide susceptibility evaluation using a novel hybrid integration approach of bivariate statistical based random forest method. *Catena* **2018**, *164*, 135–149. [CrossRef]
8. Leonardi, G.; Palamara, R.; Cirianni, F. Landslide susceptibility mapping using a fuzzy approach. *Procedia Eng.* **2016**, *161*, 380–387. [CrossRef]
9. Park, H.J.; Lee, J.H.; Woo, I. Assessment of rainfall-induced shallow landslide susceptibility using a GIS-based probabilistic approach. *Eng. Geol.* **2013**, *161*, 1–15. [CrossRef]
10. Peethambaran, B.; Anbalagan, R.; Shihabudheen, K.V. Landslide susceptibility mapping in and around Mussoorie Township using fuzzy set procedure, MamLand and improved fuzzy expert system—A comparative study. *Nat. Hazards* **2019**, *96*, 121–147. [CrossRef]
11. Kamran, K.V.; Feizizadeh, B.; Khorrami, B.; Ebadi, Y. A comparative approach of support vector machine kernel functions for GIS-based landslide susceptibility mapping. *Appl. Geomat.* **2021**, *13*, 837–851. [CrossRef]
12. Wright, R.H.; Campbell, R.H.; Nilsen, T.H. Preparation and use of isopleth maps of landslide deposits. *Geology* **1974**, *2*, 483–485. [CrossRef]
13. Humbert, M. Risk Mapping of areas exposed to Movements of Soil and Sub-soil: French "Zermos" maps. *Bull. Int. Assoc. Eng. Geol.* **1977**, *16*, 80–82. [CrossRef]
14. Brabb, E.E. The world landslide problem. *Epis. J. Int. Geosci.* **1991**, *14*, 52–61. [CrossRef]
15. Althuwaynee, O.F.; Pradhan, B. Semi-quantitative landslide risk assessment using GIS-based exposure analysis in Kuala Lumpur City. *Geomat. Nat. Hazards Risk* **2017**, *8*, 706–732. [CrossRef]
16. Van Westen, C.; Seijmonsbergen, A.; Mantovani, F. Comparing landslide hazard maps. *Nat. Hazards* **1999**, *20*, 137–158. [CrossRef]
17. Shano, L.; Raghuvanshi, T.K.; Meten, M. Landslide susceptibility evaluation and hazard zonation techniques—A review. *Geoenviron. Disasters* **2020**, *7*, 18. [CrossRef]
18. Constantin, M.; Bednarik, M.; Jurchescu, M.C.; Vlaicu, M. Landslide susceptibility assessment using the bivariate statistical analysis and the index of entropy in the Sibiciu Basin (Romania). *Environ. Earth Sci.* **2011**, *63*, 397–406. [CrossRef]
19. Mousavi, S.Z.; Kavian, A.; Soleimani, K.; Mousavi, S.R.; Shirzadi, A. GIS-based spatial prediction of landslide susceptibility using logistic regression model. *Geomat. Nat. Hazards Risk* **2011**, *2*, 33–50. [CrossRef]
20. Shahabi, H.; Hashim, M. Landslide susceptibility mapping using GIS-based statistical models and Remote sensing data in tropical environment. *Sci. Rep.* **2015**, *5*, 9899. [CrossRef]
21. Wang, L.-J.; Guo, M.; Sawada, K.; Lin, J.; Zhang, J. A comparative study of landslide susceptibility maps using logistic regression, frequency ratio, decision tree, weights of evidence and artificial neural network. *Geosci. J.* **2016**, *20*, 117–136. [CrossRef]
22. Akgün, A.; Türk, N. Mapping erosion susceptibility by a multivariate statistical method: A case study from the Ayvalık region, NW Turkey. *Comput. Geosci.* **2011**, *37*, 1515–1524. [CrossRef]
23. Guzzetti, F.; Galli, M.; Reichenbach, P.; Ardizzone, F.; Cardinali, M. Landslide hazard assessment in the Collazzone area, Umbria, Central Italy. *Nat. Hazards Earth Syst. Sci.* **2006**, *6*, 115–131. [CrossRef]

24. Felicísimo, Á.M.; Cuartero, A.; Remondo, J.; Quirós, E. Mapping landslide susceptibility with logistic regression, multiple adaptive regression splines, classification and regression trees, and maximum entropy methods: A comparative study. *Landslides* **2013**, *10*, 175–189. [CrossRef]
25. Razavizadeh, S.; Solaimani, K.; Massironi, M.; Kavian, A. Mapping landslide susceptibility with frequency ratio, statistical index, and weights of evidence models: A case study in northern Iran. *Environ. Earth Sci.* **2017**, *76*, 499. [CrossRef]
26. Zhang, G.; Cai, Y.; Zheng, Z.; Zhen, J.; Liu, Y.; Huang, K. Integration of the statistical index method and the analytic hierarchy process technique for the assessment of landslide susceptibility in Huizhou, China. *Catena* **2016**, *142*, 233–244. [CrossRef]
27. Rahim, I.; Ali, S.M.; Aslam, M. GIS Based landslide susceptibility mapping with application of analytical hierarchy process in District Ghizer, Gilgit Baltistan Pakistan. *J. Geosci. Environ. Prot.* **2018**, *6*, 34–49. [CrossRef]
28. Kumar, R.; Anbalagan, R. Landslide susceptibility mapping using analytical hierarchy process (AHP) in Tehri reservoir rim region, Uttarakhand. *J. Geol. Soc. India* **2016**, *87*, 271–286. [CrossRef]
29. Tahmassebipoor, N.; Rahmati, O.; Noormohamadi, F.; Lee, S. Spatial analysis of groundwater potential using weights-of-evidence and evidential belief function models and remote sensing. *Arab. J. Geosci.* **2016**, *9*, 79. [CrossRef]
30. Ilia, I.; Tsangaratos, P. Applying weight of evidence method and sensitivity analysis to produce a landslide susceptibility map. *Landslides* **2016**, *13*, 379–397. [CrossRef]
31. Pradhan, B.; Abokharima, M.H.; Jebur, M.N.; Tehrany, M.S. Land subsidence susceptibility mapping at Kinta Valley (Malaysia) using the evidential belief function model in GIS. *Nat. Hazards* **2014**, *73*, 1019–1042. [CrossRef]
32. Bui, D.T.; Pradhan, B.; Revhaug, I.; Nguyen, D.B.; Pham, H.V.; Bui, Q.N. A novel hybrid evidential belief function-based fuzzy logic model in spatial prediction of rainfall-induced shallow landslides in the Lang Son city area (Vietnam). *Geomat. Nat. Hazards Risk* **2015**, *6*, 243–271. [CrossRef]
33. Tsangaratos, P.; Rozos, D.E. Producing landslide susceptibility maps by applying expert knowledge in a GIS-based environment. *Bull. Geol. Soci. Greece* **2013**, *47*, 1539–1549. [CrossRef]
34. Ayalew, L.; Yamagishi, H.; Marui, H.; Kanno, T. Landslides in Sado Island of Japan: Part II. GIS-based susceptibility mapping with comparisons of results from two methods and verifications. *Eng. Geol.* **2005**, *81*, 432–445. [CrossRef]
35. El Jazouli, A.; Barakat, A.; Khellouk, R. GIS-multicriteria evaluation using AHP for landslide susceptibility mapping in Oum Er Rbia high basin (Morocco). *Geoenviron. Disasters* **2019**, *6*, 3. [CrossRef]
36. Saaty, T.L. *Decision Making for Leaders: The Analytic Hierarchy Process for Decisions in a Complex World*; RWS Publications: Pittsburgh, PA, USA, 1990.
37. Saaty, T.L. How to make a decision: The analytic hierarchy process. *Eur. J. Oper. Res.* **1990**, *48*, 9–26. [CrossRef]
38. Calabria, A.d.B.R. *Piano Stralcio per l'Assetto Idrogeologico (PAI)*; Regione Calabria, Specifiche Tecniche: Catanzaro, Italy, 2001.
39. Saaty, R.W. The analytic hierarchy process—What it is and how it is used. *Math. Model.* **1987**, *9*, 161–176. [CrossRef]
40. Orhan, O.; Bilgilioglu, S.S.; Kaya, Z.; Ozcan, A.K.; Bilgilioglu, H. Assessing and mapping landslide susceptibility using different machine learning methods. *Geocarto Int.* **2020**, 1–26. [CrossRef]

applied sciences

MDPI

Article

Landslide Susceptibility Assessment Considering Landslide Volume: A Case Study of Yangou Watershed on the Loess Plateau (China)

Hang Gao [1] and Xia Zhang [2,*]

[1] Faculty of Infrastructure Engineering, Dalian University of Technology, Dalian 116024, China; hgao_muriel@foxmail.com
[2] State Key Laboratory of Eco-Hydraulics in Northwest Arid Region, Xi'an University of Technology, Xi'an 710054, China
* Correspondence: zhangxia20002000@163.com

Abstract: Because of the special geological conditions on the Loess Plateau, Landslide erosion is not only the main goal of prevention and control of geological disasters, but also an important erosion mode of soil and water loss in the basin. Thus, landslide susceptibility assessment before only considering landslide frequency is not far enough for a decision-maker. The study aims to consider both frequency and scale of landslides for a better landslide susceptibility evaluation. Taking the Yangou small watershed as an example, this study used a VR model, RIRA method, and the GIS method to comprehensively consider frequency and scale to analyze landslide susceptibility of the small watershed. Based on the detailed analysis of the existing literature, slope, elevation, NDVI, land-use, lithology, amount distant to road, amount distant to river, profile curvature, and rainfall as landslide are selected as the conditioning factors (CFs) of the landslide, to draw the sensitivity map. The map of landslide susceptibility was classified into five zones: very low, low, medium, high, and very high, and the cover areas occupy 6.90, 12.81, 12.83, 9.42, and 5.87 km^2, respectively. A total of 60% of the landslide occurred in the zones of high and very high susceptibility, accounting for 87% of the total volume in the study area. The very high susceptibility is the area with a larger relief and along the river and road. The findings will help decision makers to formulate scientific comprehensive policies that take into account disaster prevention and soil conservation measures in specific regions.

Keywords: landslide susceptibility; volume ratio; reserve increase-rate-analysis method; GIS; Loess Plateau

Citation: Gao, H.; Zhang, X. Landslide Susceptibility Assessment Considering Landslide Volume: A Case Study of Yangou Watershed on the Loess Plateau (China). *Appl. Sci.* **2022**, *12*, 4381. https://doi.org/10.3390/app12094381

Academic Editors: Miguel Llorente Isidro, Ricardo Castedo and David Moncoulon

Received: 24 March 2022
Accepted: 24 April 2022
Published: 26 April 2022

1. Introduction

Landslides, a mass failure on steep slopes, facilitated by gravity are an important process controlling the sedimentary structures and growth patterns of the steep slope [1], and major natural disasters that often cause human and economic losses due to natural forces and human actions [2]. There are mainly three manifestations such as falling, sliding, or flowing [3]. China is covered by a vast area that is considered to be one of the most landslide-prone regions in the world, while landslide phenomena cause an estimated 700 to 1000 deaths every year and more than RMB 10 billion annually in infrastructure and property damage [4]. Landslides, especially on the Loess Plateau, have been widespread due to their topography and geographical location. In order to prevent and mitigate the damage, it is necessary to need comprehensive information regarding landslides, such as occurrence, spatial distribution, and susceptibility. Landslides also deliver huge amounts of sediment to rivers in the mountainous and hilly watersheds [5,6]. In particular, most landslides occur on slopes, whether man-made or natural, where loose material is always transported to the toe and downhill, which may have an out-of-field effect on sediment transport through river flows to the downhill [7]. On the Loess Plateau, the phenomenon is particularly serious, which is one of the reasons for the poor soil erosion in the region due to unmanageable, heavily containing sediment flow [8]. It can be seen that landslides have a great effect on water and soil loss in the basin. Older land use policies may not always reflect optimal planning for land use prone to landslides [9]. A reasonable zoning prediction of regional landslides can therefore assist planners and engineers in making decisions on the use of such lands [6].

Landslide susceptibility is recognized as the propensity of an area to generate landslides [10]. Generally, landslide occurrence in each region is a function of various factors. Each of the factors and function has a different influence. In order to assess susceptibility from gravity, it is, therefore, necessary to identify and analyze the factors leading to gravity [11]. The occurrence of landslides is controlled both by a series of predisposing factors, e.g., geological, geomorphological, climatic and hydro-geological, and triggering factors, including seismicity, heavy rainfall, human activities, and freeze-thaw [12]. Although it is still difficult to predict a landslide event in space and time, a region may be divided into sub-regions with homogenous properties, which are classified and ranked according to the potential hazard degrees of group movement considering the prerequisite factors explained above. A landslide susceptibility map is considered as an effective solution, and it is also an important task for decision making on regional planning and protection.

A variety of methods can be employed to develop the landslide susceptibility assessment. There are two main types: a "knowledge-driven model" that uses relevant domain knowledge for analysis and a "data-driven model" that uses intelligent methods to automatically find patterns from data. The former methods include fuzzy logic [13], fuzzy comprehensive evaluation [14], and the analytic hierarchy process [15]. The latter methods include logistic regression analysis [16], decision tree [17], random forest [18], artificial neural network [19], and support vector machine [20]. In general, the knowledge-driven model is more subjective while the data-driven model is relatively objective and can accurately reflect the correlation between landslide susceptibility and its basic environmental factors. However, the data-driven model is complex in the model training and testing process and requires a large number of training samples. Both types of models can get a more reasonable result, however; they mainly provide information on the probability of landslide occurrence, but no information on the size of a potential landslide once occurring. In other words, the landslide susceptibility map applied by those methods only reflected the possibility of landslide occurrence, while it cannot express the probability of a landslide. Research has shown that there is a significant difference in spatial variation between the susceptibility and erosion risk of a mass movement in the watershed [21]. Thus, the results only consider landslide frequency as incomplete for the decision-makers to implement land planning or other policy in the region. Therefore, it will also be meaningful to understand the sediment yield and prevent soil erosion in the catchment when both considering the scale and the frequency of the landslide.

If the volume of the landslide is considered in the assessment of landslide susceptibility, it is necessary to choose a suitable method. Whatever the relationship between landslide and the landslide-related factors, it can be inferred from the relationship between landslide and the landslide-related factors that had occurred. Thus, volume ratio (VR), which reflected the relationship between the volume of the landslide and each landslide conditioning factors, is used in the study, and the revised increase-rate-analysis method (RIRA) method is also a suitable way for quantitative analysis of the relationship between landslide scale and various factors. The RIRA method is an improvement on the increase-rate-analysis method (IRA), which is used to evaluate the sensitivity of the landslide to a single variable [1]. On this basis, Lucas [22] has proposed a revised increase-rate-analysis method (RIRA) based on the importance ranking of influencing factors and sensitivity analysis for a large number of data-influencing factors. The method has been successfully applied in the analysis of the influencing factors of the silt dam conditions in the Culiacan basin in northwestern Mexico. Recently, as a basic analysis tool for landslide hazards,

ArcGIS can effectively perform spatial data management and manipulation for the analysis and has been used in many studies [16]. It is also an indispensable tool for landslide susceptibility mapping.

The objective of this study included: (a) applying the VR model, the RIRA method, and GIS in the spatial event prediction based on considering landslide scale; (b) creating a landslide susceptibility map for the Yangou small watershed on the Loess Plateau of China; (c) assessing the landslide susceptibility and providing suggestions for landslide prevention and soil erosion of the small catchment.

2. Study Area

The study focuses on the Yangou watershed, a small catchment located in Yanan city in the north Shanxi province of China (Figure 1). It is enclosed within the latitudes 36°28′ N and 36°34′ N and longitudes 109°27′ E and 109°33′ E, covering an area of about 48 km^2. The catchment is typical of the gullied-hill zone and is located in the central area of the Loess Plateau, which is a sub-watershed in the Yanhe basin. The river is a secondary branch of the Yan River and flows from southeast to northwest. The watershed exhibits complex topographic variations, with a gully density of 4.8 km km^{-2} and an elevation ranging from 988 m to 1404 m. The terrain gradient of the basin is mostly composed of steep hill slopes. The slope analysis showed that the slope distribution in the study area ranged from 0° to 51°. The climate belongs to the transitional zone from warm semi-dry to semi-wet. The average annual precipitation is 550 mm, and 70% of the annual precipitation occurs from June to September. Generally, the southern section had higher values than the north. The geotechnical medium in the area can be divided into rock mass and soil mass, mainly sandstone and loess. The saturated bulk density (γ), Poisson's ratio, cohesion (C), and friction angle (φ) of sandstone are 2.68 g/cm^3, 0.22, 4.23 MPa, and 41.6°, respectively. The moisture content (ω), dry bulk density (γ), cohesion (C), and friction angle (φ) of loess are 15.6%, 1.51 g/cm^3, 39.48 MPa, and 26.95°.

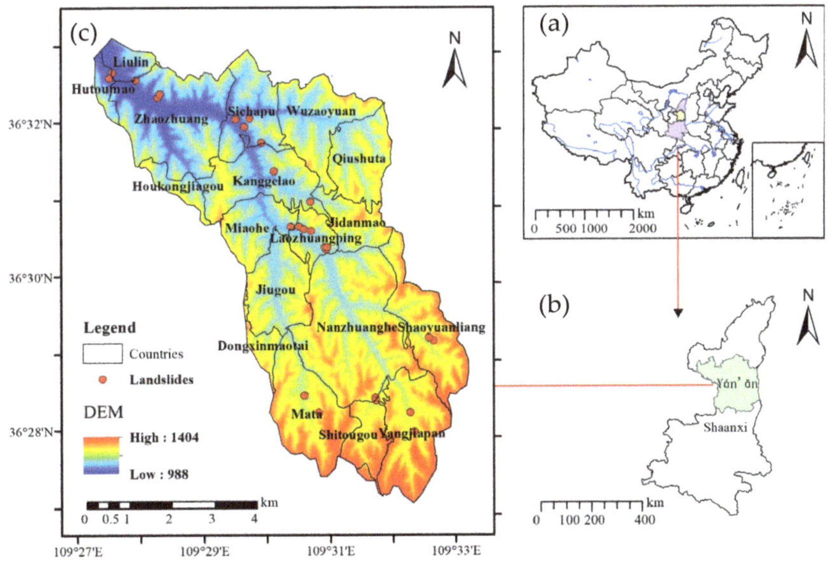

Figure 1. Study area in China. (**a**) Shaanxi province in China; (**b**) Yan'an city in the Shaanxi province; (**c**) Yangou watershed in the Yan'an city.

3. Methods

The landslide susceptibility maps were prepared using VR and RIRA models to assess the landslide susceptibility in the Yangou catchment. Figure 2 showed the flow chart of the application method, including the data collection, gathering of the landslide conditioning factors, determination of relationship between each condition factors' subclasses, and landslide inventory map using VR, application of the RIRA method in the landslide sensibility, landslide susceptibility assessment of numerical data integration, and landslide susceptibility mapping.

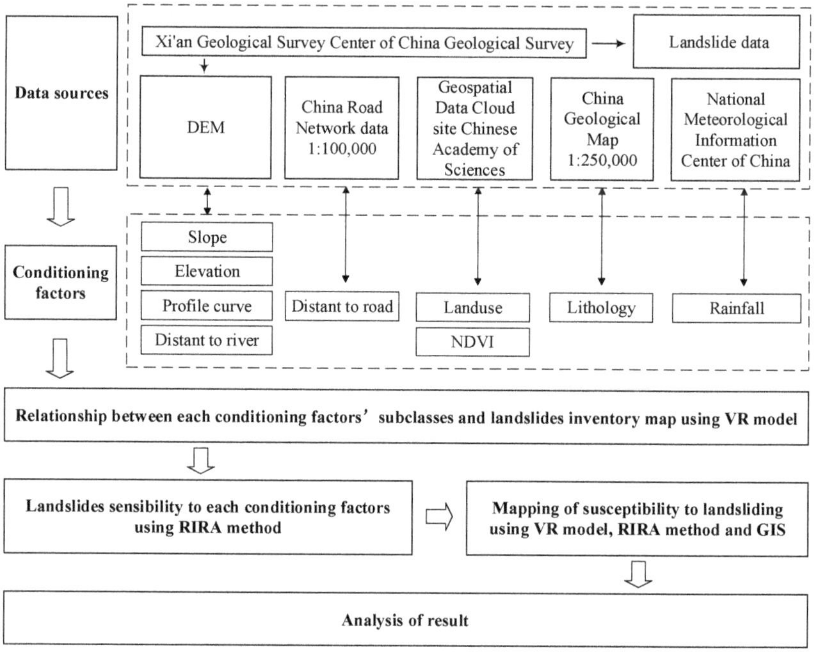

Figure 2. Flowchart of the used methodology.

The landslide inventory database, which included 27 locations, volumes, and types were provided by the Xi'an Geological Survey Center of the China Geological Survey. The landslide is divided into four sections: $\leq 1 \times 10^4$, 1×10^4–10×10^4, 10×10^4–100×10^4, and $\geq 100 \times 10^4$ m^3, corresponding to tiny, small, medium, and large landslides. The amount of the landslide and its percentage in each interval were counted. In this study, nine parameters were chosen according to the literature review and the general environment of the study area. These conditioning factors are land-use, slope, profile curvature, altitude/elevation, distance to roads, distance to rivers, lithology, normalized difference vegetation index (NDVI), and rainfall. The altitude/elevation, slope, and profile curvature were extracted from the Digital Elevation Model (DEM) (cell size: 25 m $\times$ 25 m), which is converted from a topographic map with a scale of 1:100,000. They were also collected from the "Xi'an Geological Survey Center of China Geological Survey". The NDVI was extracted from the Landsat 8/OLI data set, which is provided by the Geospatial Data Cloud site, Computer Network Information Center [23]. The computational formula of NDVI is

$$\text{NDVI} = (NIR - R)/(NIR + R) \tag{1}$$

where NIR is the reflection value of the near-infrared band, and R is the reflection value of the red band. The landuse data were provided by the Resource and Environment Science and Data Center [24]. It was decided that the study area was mainly covered by eight land-use types, namely water, wetland, farmland, resident land, bareland, forest, shrubland, and grassland. The lithology of the Yangou Watershed was delineated using the China Geological Map at a 1:250,000-scale. The distance to rivers was extracted using hydrological analysis of DEM and neighborhood statistic in the ArcGIS. The distance to roads was developed using China Road Network data at a 1:100,000-scale. The rainfall data were from the China meteorological background data set of a 500 m × 500 m pixel size, which is provided by the Resource and Environment Data Cloud Platform [25]. Then, these nine conditioning factors were selected in the present study and were standardized to the same size of 25 m × 25 m for further analyses. The study area covers 355 columns and 447'rows. The spatial analysis tool in the ArcGIS software was used to count the corresponding values of each factor at the landslide location. Also, the occurrence frequency ratio of the different classes was calculated in the land use and geology as the corresponding values of the landslide.

To show the relationship between each factor's subclasses and landslide volumes, the VR approach is proposed. It is a variant of the probabilistic method that is based on the observed relationships between the volume of the landslide and each landslide conditioning factor. The volume ratios for the classes or types of each conditioning factor were calculated by dividing the landslide volume by the typical value ratio of the study area. Each factor's ratio value was calculated using Equation (2):

$$VR_{jk} = \frac{V_{jk}^*}{V_{jk}} / \frac{T^*}{T} \tag{2}$$

VR_{jk} = volume ratio of each factor; $V_{jk}{}^*$ = volume of observed landslide of class "k" of factor "j"; V_{jk} = volume of class "k" of factor "j"; T = typical value of the map, such as the total area of the map, the average distance to roads or rivers in each class, the average elevation of each class, and the average rainfall of each class. T^* = typical value of the observed landslide of the map, such as the total area of each class, the average distance of the landslide to roads or rivers, the average elevation of the landslide, and the average rainfall of the landslide. A value of 1 is the neutral value, and the values higher than 1 show a high positive correlation with a certain factor of the landslide [26]. A higher VR value shows that a higher probability of the landslide volume occurs.

The RIRA is a method of sensitivity analysis used by calculating the relationship between dependent variables and independent variables. RIRA can assess the relative contribution of each variable by placing greater emphasis on known variables contributing to landslides, and which are sorted by contribution [22]. In addition, each variable has a specific value in the method, which ensures the objectivity of the data calculation. To eliminate the dimensional impact between the independent variables, all data of the factors were standardized using the z-scores method before calculation. The dependent variables (such as the amount of landslide) and the independent variables (such as profile curvature, slope, altitude, NDVI, land-use, geology, distance to rivers, distance to roads, and rainfall) were listed separately. The variables were sorted in order from small to large, and the sensitivity coefficients of each influencing factor were calculated. In order to quantify the impact of landslide-related factors on the landslide, the calculation steps of RIRA were as follows: first, the landslide increase rate (%) was calculated using Equation (3):

$$RP_i = \frac{2(P_i - P_{i-1})}{P_i + P_{i-1}} \tag{3}$$

where P_i is the amount of the i-th landslide (104 m^3). Then, the increase rate (%) of each impact factor on the landslide was calculated as follows:

$$Rt_i = \frac{2(t_i - t_{i-1})}{t_i + t_{i-1}} \tag{4}$$

where t_i and t_{i-1} are the values of the explanatory variable that are sequentially connected. Rt_i is the increase rate calculated for two successively independent variables ordered by the values. Therefore, the Absolute Sensitivity Parameter (s), towards an independent variable, t, was calculated by its mean growth rate:

$$s_{aj} = \left|\overline{(RP_{i,j}/Rt_{i,j})}_{i=1,N}\right| \quad (5)$$

Being j the j-th independent variable, N the total number of landslide data, namely slope, aspect, altitude, NDVI, land use, geology, distance to rivers and distance to roads. A factor with a larger absolute value of the s represents the higher sensitivity of the amount of the landslide to the related factor. In the study, the revised increase-rate-analysis (RIRA) method is used to assess the relationship between the factors and landslide.

At last, the landslide susceptibility of each subclass was calculated by Equation (6):

$$S_{ak} = s_{aj} \times VR_{jk} \quad (6)$$

S_{ak} is the susceptibility of each *k-th* class. Then, each class susceptibility map can be obtained using the S_{ak} value in the ArcGIS software. At last, all thematic layers were combined using the S_{ak} value to derive the landslide susceptibility map of the Yangou watershed. Calculated susceptibility values were classified into areas of very low, low, medium, high, and very high susceptibility using the natural break method available within the ArcGIS software.

4. Results and Discussion

4.1. Landslide Inventory

Landslides in the study area primarily lie along the northwest and southeast direction. There are two types of landslides in the region: slides and falls. About 15 observed landslides belong to slides and 12 falls. The amount of erosion is 704×10^4 m^3 and 14.4×10^4 m^3, accounting for 98% and 2% of the total landslide, respectively. The slide is the main type of landslide in the region, which had an important contribution to the frequency of landslide and the total amount of slope movements in the region. The triggering mechanism of falls is different from that of slides. When the water in the gully scours and transports the lower accumulation, and the gravity moment generated by the upper steep wall under the action of gravity is greater than the tensile force moment of the soil, the upper soil layer loses its balance to fall [27]. There were many small falls in the study area. Most of the fall-bodies fell and broke, which were difficult to preserve for a long time. Most of them have been transported to form sand production, which is difficult to investigate [28].

The volume of the smallest landslide is approximately 0.2×10^4 m^3, the largest is around 270×10^4 m^3, and the average is estimated to be 26.6×10^4 m^3. Figure 3 showed the amount and frequency of landslides in each interval, which were tiny, small, medium, and large landslides. The total amount of erosion caused by large landslides of $\geq 100 \times 10^4$ m^3 was the largest, which was 503.6×10^4 m^3, accounting for 70.2% of the total erosion and followed by the erosion of the medium landslide, which was 174.7×10^4 m^3, accounting for 24.4% of total erosion. The small landslide was 37.1×10^4 m^3, accounting for 5.2% of total erosion, and the tiny landslide had the least landslide, at 2.7×10^4 m^3, accounting for 0.2% of total erosion. In the case of frequency of landslides, there were 19 small-scale erosions with a volume of fewer than 10×10^4 m^3 in the study area, accounting for 70% of the total frequency of erosion, that is, the occurrence of tiny-scale and small-scale landslides were higher in the area. While the amount of medium-sized and large-scale collapse accounted for 94.6% of the total, their frequencies of occurrence were only 5 and 3 times, accounting for 19% and 11% of the total frequency, respectively. It can be seen that the number of tiny–small landslides in the study area was more frequent. The frequency of medium and large-scale landslides was little, but they had a major contribution to the total amount of landslides. It can also be seen from the statistics that the contribution of frequency and the volume of landslides are not the same, and it is more meaningful to consider both effects in the study. The cumulative amount of landslide erosion in the study area reached 718.1 × 10^4 m^3. A large amount of landslide erosion has an important impact on sediment yield and sediment transport in the basin. The landslide-triggered erosion enters the channel and then flows into the river system by runoff, directly or indirectly transporting a large amount of sediment to the channel, which is one of the main processes of soil erosion and river sediment sources. Large slides also cause road damage and soil erosion. The volume of the falls is not large but their migration speed is fast. Because the occurrence to stop movement does not exceed 1 min, people are often unable to hide and suffer the risk of casualties; the harm is no less than the landslide. Between 1980 and 2015, there were 53 landslides over the Loess Plateau, killing and missing 717 people [29].

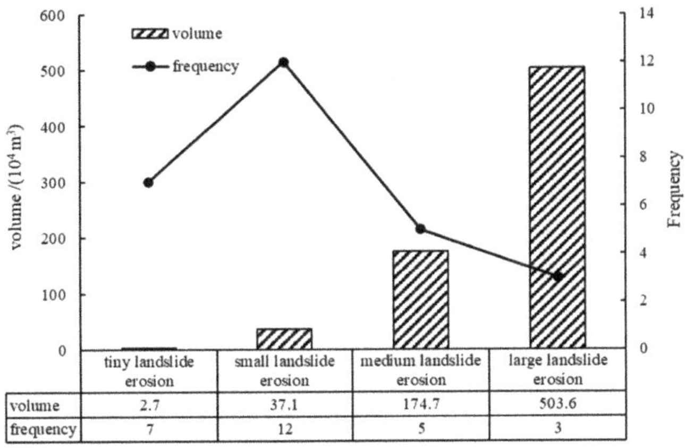

Figure 3. Amount and frequency of landslides in the Yangou watershed.

4.2. Application of the VR Model

In order to study, in detail, the relationship between each factor and the scale of the landslide, each factor is divided into several classes. The results of the spatial relationship between landslide volumes and the classes of landslide conditioning factors using the VR model are shown in Table 1.

Table 1. Spatial relationship between landslide conditioning factors and landslide by volume ratio model.

Independent Variable	Classes	V_{jk}^*/V_{jk}	T^*/T	VR_{jk}
Slope angle (°)	0–15	0.7294	0.8629	0.8453
	15–25	0.2427	0.1272	1.9089
	25–35	0.0259	0.0094	2.7538
	>35	0.0019	0.0006	3.5264
Elevation (m)	<1000	0.1448	0.9991	0.1450
	1000–1100	0.0947	0.9874	0.0959
	1100–1200	0.1886	0.9946	0.1896
	1200–1300	0.5719	0.9937	0.5755
	>1300	0.0000	0.0000	0.0000
NDVI	no cover	0.0000	0.0004	0.0000
	very low cover	0.1478	0.0263	0.1778
	low cover	0.7124	0.2459	0.3451
	medium cover	0.1398	0.5355	3.8301
	high cover	0.0000	0.1919	0.0000

Table 1. *Cont.*

Independent Variable	Classes	V_{jk}^*/V_{jk}	T^*/T	VR_{jk}
Land-use	Farmland	0.2981	0.3063	0.9734
	Forest	0.0199	0.4000	0.0498
	Grassland	0.4029	0.2345	1.7180
	Shrubland	0.0000	0.0021	0.0000
	Wetland	0.0000	0.0000	0.0000
	Water	0.0000	0.0006	0.0000
	Resident land	0.1313	0.0367	3.5780
	Bareland	0.1478	0.0198	7.4512
Lithology	Sandstone	0.4169	0.2848	1.4640
	Loess	0.5831	0.7152	0.8152
Distant to road (m)	<50	0.5289	0.4101	1.2897
	50–150	0.2454	0.9813	0.2501
	150–300	0.0100	0.8237	0.0122
	300–500	0.1824	0.9151	0.1994
	500–800	0.0333	1.1883	0.0280
	>800	0.0000	0.0000	0.0000
Distant to river (m)	<50	0.7368	1.1093	0.6642
	50–150	0.2543	0.9612	0.2645
	150–300	0.0089	0.8106	0.0110
	>300	0.0000	0.0000	0.0000
Profile curvature	Concave slope	0.5151	0.5384	0.9567
	Convex slope	0.4849	0.4616	1.0505
Rainfall	Less rainfall area	0.2051	0.9998	0.2052
	Medium rainfall area	0.3779	1.0002	0.3779
	More rainfall area	0.2415	0.9996	0.2416
	Heavy rainfall area	0.1755	0.9996	0.1755

In the case of slope angle, the results showed that the VR values also increased as the slope increased, indicating that the slope angle has a good positive correlation with the probability of the landslide volume. The slope is a major factor in landslide occurrence because it relates to drainage, fault line, and road networks [2]. In the Yangou watershed, the 0°–15° slope area constituted 20.34% of the total area, the 15°–25° area accounted for 35.49%, the 25°–35° area accounted for 36.57%, and slope areas of more than 35° were 7.61% of the total area. The findings revealed a wide range of slope changes across the whole research area. According to the survey, 78% of the landslides occurred on slopes greater than 15° in the study area. Generally, the mass movement shifts from sliding to slumping with an increasing angle [30,31]. Thus, landslides are created in a hilly region or mountainous areas with steep slopes. The slope affects the scale and intensity of surface material flow and energy conversion. For example, the slope affects the area and amount of rainfall on the slope, thereby affecting the magnitude of runoff, infiltration, and runoff kinetic energy [32].Thus, the greater the slope, the greater the gravity erosion [20,33].

There is no correlation between landslide volume and elevation rise in terms of elevation. There are many reasons for the influence of elevation on the development of landslide disasters, such as that different elevation ranges have different climatic characteristics and local depressions are due to slope differences, whether there is a sliding surface and an intensity of human activities in different elevation ranges [34]. Thus, elevation has no direct impact on the distribution of gravity erosion disaster points in the region. The 1200–1300 m area had the highest VR value of 0.5755, followed by 1100–1200 m (0.1896), and <1000 m (0.1450). The 1000–1100 m area has the lowest value of 0.959. In the Yangou watershed, 96.3% of the landslides occurred on the altitude ranging from 1000 m to 1300 m. This showed that landslides usually occurred at intermediate elevations. The results can also be found in the other places since slopes tend be covered by a layer of thin colluvium that is prone to landslides [35,36].

The NDVI is a measure of surface reflectance to give a quantitative estimate of biomass and vegetation growth [37]. The value of NDVI in the study area ranges from −0.27 to 0.59 in the summer. Generally, the larger the NDVI value is, the higher the vegetation coverage is. In the case of NDVI, there is no landslide in the classes of no cover and high cover. In the other classes, the VR value reflected the probability of landslide volume increased with the increase of the plant cover. In other words, the vegetation didn't play important roles in the decrease of the landslide amount. There are positive and negative effects of vegetation cover on landslides. Vegetation roots can reduce the creep of the soil layer by increasing the shear strength of soil, which is negatively correlated with the amount of landslide erosion. On the other hand, in the field investigation of small watersheds, it was found that with the increase of vegetation coverage, the runoff on the slope increased, thereby increasing the gully erosion of the watershed [38]. Plant root splitting may also promote the occurrence of landslide erosion.

Land-use is another factor used to consider the natural and man-made environmental impacts on the land surface. The main land use types in the study area were forestland and cropland, accounting for 40% and 31% of the total study area, respectively. Secondly, the grassland accounted for 23% of the total study area. Other classes accounted for a combined 6% of the total study area. The bareland had the highest VR value of 7.4512, followed by resident land (3.5780) and grassland (1.7180). The class of forest and farmland had the VR values of 0.9734 and 0.0498, respectively. In the shrubland, wetland, and water, the VR values are zero. The results indicated surface attachment plays an important role in slope stability compared with bareland, and in our case, the cultivated areas are less prone to landslides when compared to other land cover types. This may be due to positive impacts of land management with better water and soil control measures on slope stability [39].

Different geological formations have different compositions and structure, which contribute to the strength of the material. Stronger rocks are more resistant to the driving force of landslides and less prone to damage than weaker rocks [40]. The main lithology units of the study area consisted of Jurassic sandstone and Quaternary loess. Vertical joints, structural joints, collapse joints, and weathering joints of loess are generally developed, which are potential geological factors leading to geological disasters such as loess slides and falls. The deformation and failure caused by loess collapsibility provide a channel for precipitation collection and rapid infiltration, and often lead to geological disasters such as landslides. The vertical joints and weathered joints of sandstone in the slope zone are developed, and the development and expansion of joint cracks form the dangerous rock mass of slope is seen, which often leads to fall disasters [25]. The joints in all the study area are very developed. From the VR values, it is seen that the high landslide volume occurred in the lithological units of sandstone with the VR value of 1.4640, followed by the lithological unit of loess with the VR value of 0.8152. The Jurassic sandstone had a higher value because it is mainly in the valley area; as the drainage area, the hydrological conditions are more complicated.

The stability of the soil mass might be harmed by road cutting caused by slope excavations. As a result, one of the influencing elements for landslides is closeness to a road. In the case of distance to roads, distance to roads of 0–50 m had the highest VR values, while the VR value in the distance to roads of 300–500 m had saltation. The result may be that the population density is high in the study area [41], and thus the road net is also density. In the range of 300–500 m, large collapses had occurred. However, the overall trend is that the VR value decreased as the distance increased, indicating that the distance to roads has a high influence on landslide occurrence and scale. Actually, the amount of landslide occurrence within 150 m of the road accounted for more than 78% in the study area.

There are reverse correlations between landslide volume and parameters based on the VR value of the distance to river. In the study area, 89% of the landslide occurred within 150 m of streams. At the same time, the hydrographic axes of the third and fourth order streams are considered to be the main factor of the occurrence of landslides [28]. This is because the scouring effect of river flow will erode the slope. At the same time, the seepage effect of water flow makes the slope soil saturated. A large amount of free water is accumulated in collapsible loess, and the shear strength of soil is reduced, which has a negative impact on the slope stability. In addition, human activities are mostly distributed in the valley, which reduces the slope stability and is also the cause of frequent disasters and serious landslide erosion on both sides of the river.

Profile curvature measures the rate of change of slope, which controls erosion and deposition by affecting the acceleration and deceleration of the flow across the surface. Convex surface is positive and concave surface is negative [42]. For the profile curvature, 44% of the landslide occurred on the concave slopes, while 56% occurred on the convex slopes, and the VR value for convex slope (1.0505) is greater than that of the concave slope (0.9567), indicating that the convex slope has a little more effect on the volume of landslides. The concave slope easily collects rainwater, and rainwater infiltrates along the slope, which reduces the shear strength of soil and causes the instability and deformation of soil. At the same time, the potential energy of convex slope soil is closer to the limit value of mechanical equilibrium, and is also vulnerable to stronger weathering, which may also lead to the occurrence of slope instability. Stochastic statistics of 300 unstable slopes in China have shown that the instability of convex slopes is better than that of concave slopes [43]. Besides, in the experiment, it is verified the larger volume of landslides easily occurred in the convex slope [44]. The failure characteristics of loess soil slope are creep, and soil creep tends to produce convex terrain [45]. Therefore, the convex slope is dominant in the steep loess slope, and it is also prone to landslides [46].

The precipitation intensity is an important factor to affect the occurrence of landslides in the semi-arid region, although the interval between minimum and maximum rainfall is small [40]. In the case of rainfall, the medium rainfall area had the highest VR value of 0.3799, followed by more rainfall area (0.2416), less rainfall area (0.2052), and heavy rainfall area (0.1755), which indicated that there is no cear relationship between the amount of this factor and landslide volume. On the Loess Plateau, long-term and strong early effective rainfall both triggered clusters of shallow landslides in the previous study [47].

4.3. Application of the RIRA Method

The sensitivity coefficient of different related factors on landslides is quite different. The results of the sensitivity of the slope, elevation, NDVI, land-use, lithology, distance to road, distance to river, profile curvature, and rainfall on the volume of landslides performed by RIRA are reported in Table 2. The larger the sensitivity of an independent variable, the more prominently this variable influences the volumes of landslide. Profile curvature was found to be the most important influential factor, as it is the most sensitive parameter (0.3616), followed by the distance to river (0.2567) and distance to road (0.0908). Rainfall was the fourth in the ranking of the relative sensitivity, and the value is 0.0628. The relative sensitivity of elevation, slope, and NDVI are comparable to each other, although these variables are less important than the three variables motioned above (0.0523, 0.0345, and 0.0203, respectively). Finally, lithology and land use have relatively little relative sensitivity.

Table 2. Sensitivity analysis of landslides and related factors in the Yangou watershed (China).

Parameter	Independent Variables								
	Profile Curvature	Distant to River	Distant to Road	Rainfall	Elevation	Slope	NDVI	Lithology	Land-Use
Absolute Sensitivity	0.3616	0.2567	0.0908	0.0628	0.0523	0.0345	0.0203	0.0099	0.0097

In summary, the independent variables, profile curvature, distance to river, distance to road, rainfall, elevation, and slope, have an important impact on the triggering and scale of landslides. Especially, both the relative sensitivity of two variables, profile curvature, and the distance to river, are very high. Convex and concave slope profiles are common in natural hill slopes. In the Yangou watershed, it is the most import factor. That means landslides in areas with large undulating terrain are subject to key monitoring. Therefore, the influence of slope morphology should be taken into account when humans are carrying out activities. In the case of the relationship between landslides and distance to streams, the VR value rises as the distance to a stream decreases. The river has a strong erosive effect on the exposed area of the bedrock, which affects the stability of the valley slope. Especially in smaller valleys, both the down erosion and side erosion of flowing water exist, and the valley slopes on both sides are steep, which are in the erosion of flowing water and are high-risk areas of slides and falls. In addition, the most direct impact of long-term surface water accumulation is the rise of the groundwater level, which forms groundwater in a large range and affects slope stability through groundwater. The groundwater on both sides of the basin passes can aggravate the slope instability by softening and eroding the earth and soil, reducing the strength of the rock and soil and simultaneously generating dynamic and hydrostatic pressures and pore water pressures, exerting floating force on the rock and soil and increasing the weight of the rock and soil [25]. Thus, the distance to river has the second sensitivity of all factors. In the study area, most of the collapse of the loess hillside slope, or the initiation of loess landslides, after having been excavated during the construction of an expressway, was induced by excavation and rainfall [48]. The sensitivity of distance to road on the scale of the landslide is of third importance. In the construction of rural road projects, rough construction, such as cut-off or slope cutting and filling, is blindly carried out without the stability analysis of the slopes on both sides of the valley, which makes the slope steeper, loses support, forms a steep slope, and causes hidden dangers for the occurrence of landslide disasters [49]. Precipitation is the main recharge source for soil moisture in the hill and gully region of the Loess Plateau [50]. According to the survey, 40% of the catastrophic landslides are trigged by rainfall on the Loess Plateau of China [26]. It is not easy to infiltrate after precipitation, however, once the effective strength of the loess is reduced by rainfall, the landslide is triggered. There is a close correspondence between landslide occurrence and the pattern of precipitation, and 84.6% of landslides have happened in the monsoon in the Baota district since 1985 [51]. Therefore, rainfall is the fourth important factor in the study area. Elevation data reflected the location of the landslide, which means the volume of landslides has higher relative sensitivity on the variable of elevation in the study area. The slope is an important performance of the slope geometry, which determines the state and distribution of stresses in the slope mass and controls the stability and mode of the instability [51]. However, the slope has relatively little impact on the occurrence and scale of landslides in the study area. The increased vegetation coverage has a certain inhibitory effect on the amount of landslides, but considering the comprehensive effect of other factors, the influence of plants is not obvious. In addition, landslides with different scales occurred in different lithology. Thus, the influence of lithology on landslides is also relatively rare in the study area. The impact of environmental and human factors on the surface, such as construction activities, road cutting, and natural resources exploitation, may lead to landslides [34]. However, during the 9th Five-Year Plan period (1996–2000), Yangou watershed was selected as a demonstration region for integrated research for the development of ecological agriculture in the Loess Plateau, and the land use structure changed rapidly [41]. Nowadays, the land cover tends to be stable, so the absolute sensitivity of landslides to land use is lowest.

4.4. Results of the Landslide Susceptibility Mapping

Figures 4 and 5 showed the different space layers of conditioning factors and the landslide susceptibility map constructed by the VR model and RIRA method, respectively. In Figure 4, higher susceptibility values were in the layers with the convex slope and concave slope, as well as the distance of less than 50 m to the river, the distance of less than 50 m to the road, the slope of 25–35°, and the slope higher than 35°; the values were all more than 10. Among the nine related factors examined in the study, elevation, distance to river, lithology, and rainfall are hard to change, and the other factors, slope, profile curve, distance to road, land-use, and NDVI, can be modified to reduce the landslide susceptibility. Slope gradients can be reduced by converting slope land into terraces [21]. In terms of profile curve and distance to road, human activity and road construction should be carried out in a planned way. Especially, a preliminary reconnaissance survey should be implemented on the topography and geology of the area. In terms of land use, many landslides were observed on the bareland. Thus, the reduction of ground exposure may reduce landslide susceptibility. For the NDVI, it can be seen that soil bioengineering is not omnipotent and did not apply to all situations, e.g., deep-seated, rock, or extremely steep slopes [52]. Therefore, the comprehensive effect of the conditioning factors should be considered in the process of prevention and control of landslides. In the study area, reconstruction of the eco-environment had already been conducted since 1997. In the past decade, grasses or shrubs are the main types of plantings on the top of hills or upper part of hill slopes, while trees were mainly selected in the hilly slope with a moderate gradient, and most of the slope of cropland was converted to woodland or grassland [53]. Nevertheless, a quantitative assessment of the effectiveness of those comprehensive measures in inhabiting landslides is still lacking.

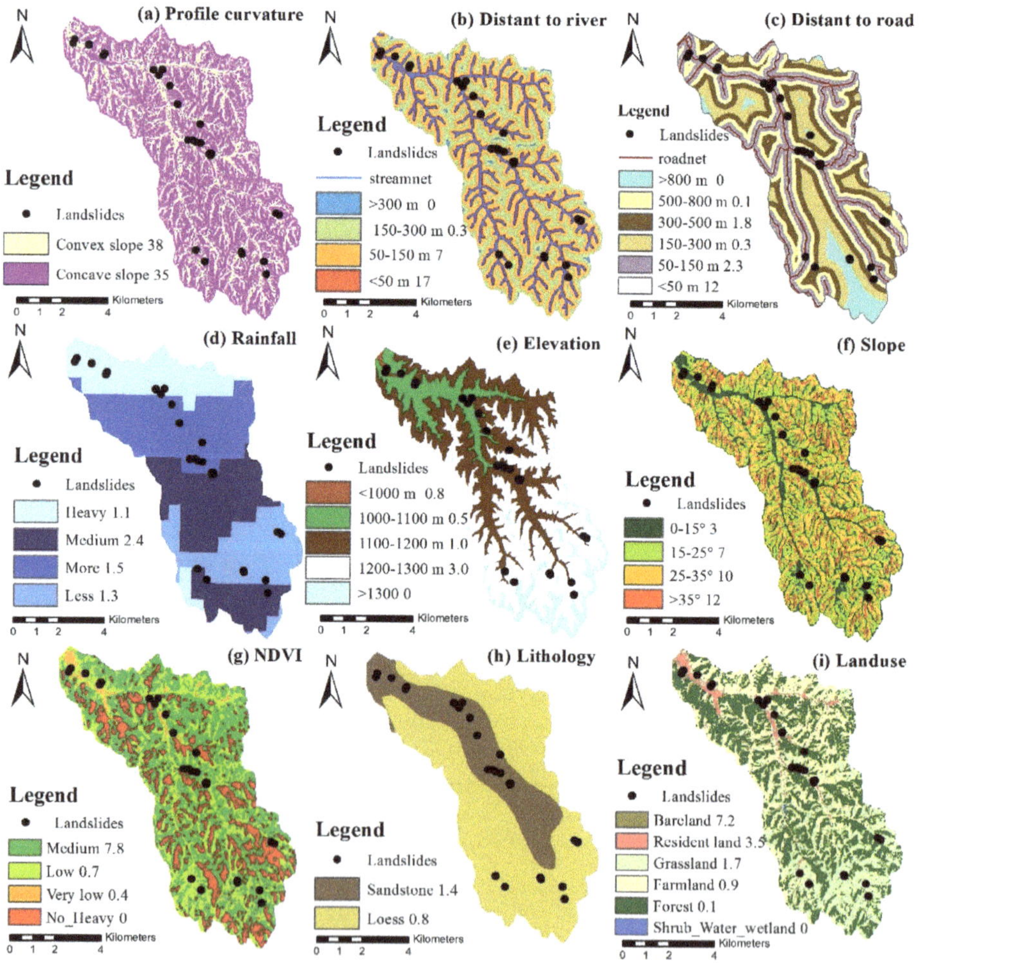

Figure 4. Susceptibility value of spatial layers influencing landslide occurrence in the study area: (**a**) profile curvature, (**b**) distance to river, (**c**) distance to road, (**d**) rainfall, (**e**) elevation, (**f**) slope, (**g**) NDVI, (**h**) lithology, and (**i**) land use.

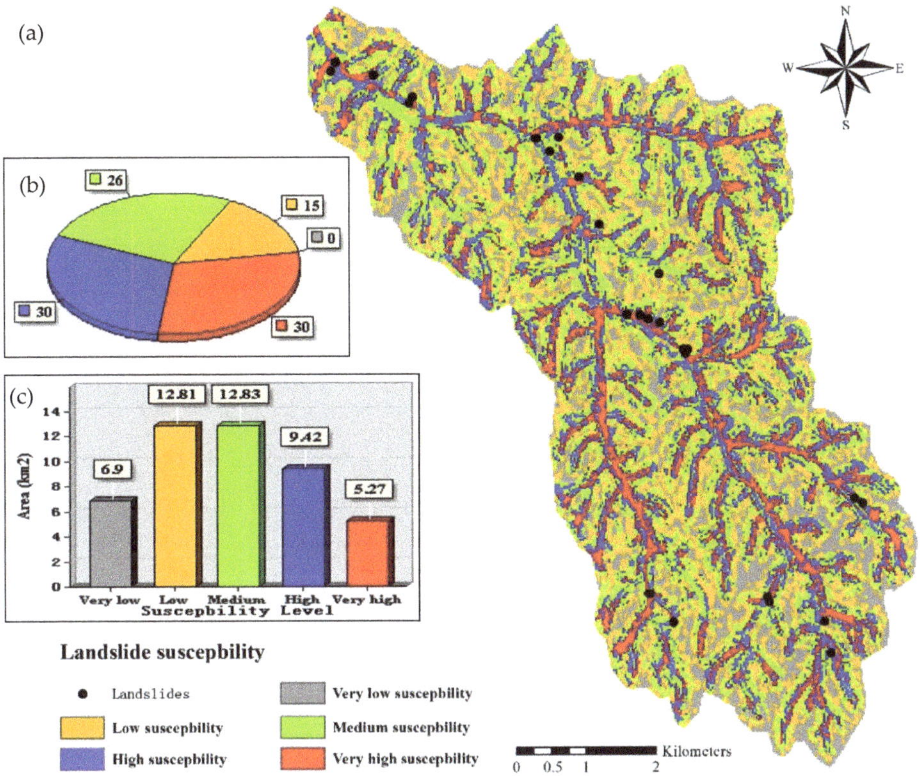

Figure 5. Landslide susceptibility map by the VR model and RIRA method: (**a**) landslide susceptibility classes in the Yangou watershed, (**b**) distribution of susceptibility over the landslide at susceptibility map (%), and (**c**) histogram of landslide susceptibility index value.

As shown in Figure 5c, a very low susceptibility zone occupies 6.90 km^2. Similarly, low, medium, high, and very high susceptibility zones occupy 12.81, 12.83, 9.42, and 5.87 km^2, respectively. As depicted in Figure 5a, the susceptibility levels decreased with distance from rivers and road. Very high susceptibility was in the region with a larger relief and along the river and road, which was also in line with the site investigation. In the whole province, landslides were mostly distributed on both sides of the roads and rivers. For example, collapses often take place on both sides of the Yan and the Fenchuan Rivers, especially along the roads and railways, [51]. This is due to the fact that Yan'an City and urban population centers are located in major river valleys and that intensive human engineering activities are often responsible for many landslides and related serious damages [51]. Therefore, in the road construction and other engineering works on the Loess Plateau, the engineers and researchers should be committed to the prevention and control of loess landslides [48], and, also, the road and the river located in the gullies. As a result, the down-cutting of channels during rainstorms, as well as the high content of soil moisture adjacent to a channel, make it more prone to landslides [21]. Moreover, in the regional susceptibility assessment of the Baota district, the specific area was extended along the Nanchuan River, and its tributaries to Shitougou village got a high susceptibility [28]. Figure 5b revealed that no landslide is located within the very low susceptibility areas; 15% of the landslide are situated within the low susceptibility areas; 26% within the moderate suitability areas; 30% within the high susceptibility areas; 30% within the very high susceptibility areas. The result of the landslide assessment highlighted that the majority of the occurred landslides

(60%), are situated within the zones of high and very high susceptibility, while the part of landslide volume was accounting for 87% of the total landslides. Landslide susceptibility is predicting the probability of landslides' occurrence regardless of their magnitude which is, however, a critical piece of information in mass movement erosion control. However, the previous study revealed small and medium mass movements often have much higher transport rate of debris than large ones [21]. Both the volume and frequency of landslides were considered in our study; the landslide mitigation and control areas can be further prioritized in the watershed from the map. The high and very high susceptibility area must receive higher priority in the watershed for large landslides, while the low susceptibility risk area can be excluded in the observation. Moreover, the area of low and medium susceptibility can be paid more attention to, as well as the soil erosion caused by landslides. The results of a zoning plan can also reduce the workload of field investigations and move the focus of the actual work in watershed management.

5. Conclusions

To understand the role of landslide disasters in small watersheds more comprehensively, a landslide susceptibility assessment that considers the amount of landslides is necessary. The applied analysis of the VR model, RIRA method, and ArcGIS within the framework of the present study achieved more comprehension for disaster prevention and land planning with respect to the location and volume of the past landslides. The relationship among landslide occurrence, landslide volume, and nine landslide conditioning factors such as slope, elevation, NDVI, land use, lithology, distance to road, distance to river, profile curvature, and rainfall are evaluated using the above models. Our analysis demonstrated that the sub-classes in the same conditioning factors had different VR values, as indicated that sub-regions with homogenous properties contributed differently to the landslide scales. In addition, the results showed that the various parameters of landslides had different sensitive values to the landslide volume and occurrence. Among those factors, profile curvature and distance to river were the two most highly sensitive, followed by distance to road, rainfall, elevation, slope, and NDVI. The lithology and land-use were the two lowest sensitive factors. The sensitive values are 0.3616, 0.2567, 0.0908, 0.0628, 0.0523, 0.0345, 0.0203, 0.0099, and 0.0097, respectively. Also, the landslide susceptibility map considering landslide volume classified the study area into five zones, with susceptibility degrees of very high, high, medium, low, and very low, which showed that about 60% of the landslides occurred in a high and very high susceptibility area, accounting for 87% of the total volume of landslides. Landslides such as slides and falls are important contents of geological disaster prevention and control, and this phenomenon is also the main form of soil erosion. Small landslide erosion is more likely to cause sediment yield. Thus, the spatial distribution of the existing landslide susceptibility can be analyzed in order to identify the components that are located in large landslide scale areas and small landslide scale areas, so as to set the priorities to prevent and control landslide disasters and soil erosion. Engineers, decision-makers, and environmental managers may implement the analysis that we implemented in the present study during new or existing planning projects and produce maps that will make possible the adoption of policies and strategies aimed at multi-hazard mitigation.

Author Contributions: Writing—original draft preparation, H.G.; writing—review and editing, X.Z. All authors have read and agreed to the published version of the manuscript.

Funding: This research was funded and supported by the National Natural Science Foundation of China (Grant No: 52009103, 42177346) and the Innovation Capability Support Program of Shaanxi (2019TD-040).

Institutional Review Board Statement: Not applicable.

Informed Consent Statement: Not applicable.

Data Availability Statement: Not applicable.

Acknowledgments: We are grateful to the anonymous reviewers for their valuable suggestions in improving the manuscript.

Conflicts of Interest: The authors declare no conflict of interest.

References

1. Xu, X.; Liu, Z.; Xiao, P.; Guo, W.; Zhang, H.; Zhao, C.; Yan, Q. Landslide on the steep loess slope: Behavior, trigger and sensitivity. *Catena* **2015**, *135*, 231–239. [CrossRef]
2. Hadmoko, D.S.; Lavigne, F.; Samodra, G. Application of a semiquantitative and GIS-based statistical model to landslide susceptibility zonation in Kayangan Catchment, Java, Indonesia. *Nat. Hazards* **2017**, *87*, 437–468. [CrossRef]
3. Varnes, D.J. Landslide types and processes. *Landslides Eng. Pract.* **1958**, *24*, 20–47.
4. Tsangaratos, P.; Ilia, I.; Hong, H.; Chen, W.; Xu, C. Applying Information Theory and GIS-based quantitative methods to produce landslide susceptibility maps in Nancheng County, China. *Landslides* **2017**, *14*, 1091–1111. [CrossRef]
5. Althuwaynee, O.F.; Pradhan, B. Semi-quantitative landslide risk assessment using GIS-based exposure analysis in Kuala Lumpur City. *Geomat. Nat. Hazards Risk* **2017**, *8*, 706–732. [CrossRef]
6. Regmi, A.D.; Poudel, K. Assessment of landslide susceptibility using GIS-based evidential belief function in Patu Khola watershed, Dang, Nepal. *Environ. Earth Sci.* **2016**, *75*, 743. [CrossRef]
7. Beek, R.V.; Cammeraat, E.; Andreu, V.; Mickovski, S.B.; Dorren, L. Hillslope processes: Mass wasting, slope stability and erosion. In *Slope Stability and Erosion Control: Ecotechnological Solutions*; Norris, J.E., Stokes, A., Mickovski, S.B., Cammeraat, E., van Beek, R., Nicoll, B.C., Achim, A., Eds.; Springer: Berlin/Heidelberg, Germany, 2008; pp. 17–64.
8. Xu, J. Sediment-heavily containing flows in slope-channel systems of gullied hilly area in the Loess Plateau (I): Influences of landforms and gravitational erosion. *J. Nat. Disasters* **2004**, *13*, 55–60. (In Chinese)
9. Highland, L.; Bobrowsky, P.T. *The Landslide Handbook: A Guide to Understanding Landslides*; US Geological Survey: Reston, VA, USA, 2008.
10. Guzzetti, F.; Reichenbach, P.; Ardizzone, F.; Cardinali, M.; Galli, M. Estimating the quality of landslide susceptibility models. *Geomorphology* **2006**, *81*, 166–184. [CrossRef]
11. Oh, H.-J.; Lee, S. Landslide susceptibility mapping on Panaon Island, Philippines using a geographic information system. *Environ. Earth Sci.* **2011**, *62*, 935–951. [CrossRef]
12. Conforti, M.; Pascale, S.; Sdao, F. Mass movements inventory map of the Rubbio stream catchment (Basilicata—South Italy). *J. Maps* **2015**, *11*, 454–463. [CrossRef]
13. Pourghasemi, H.R.; Pradhan, B.; Gokceoglu, C. Application of fuzzy logic and analytical hierarchy process (AHP) to landslide susceptibility mapping at Haraz watershed, Iran. *Nat. Hazards* **2012**, *63*, 965–996. [CrossRef]
14. Zhao, H.; Yao, L.; Mei, G.; Liu, T.; Ning, Y. A Fuzzy Comprehensive Evaluation Method Based on AHP and Entropy for a Landslide Susceptibility Map. *Entropy* **2017**, *19*, 396. [CrossRef]
15. Yoshimatsu, H.; Abe, S. A review of landslide hazards in Japan and assessment of their susceptibility using an analytical hierarchic process (AHP) method. *Landslides* **2006**, *3*, 149–158. [CrossRef]
16. Lee, S.; Sambath, T. Landslide susceptibility mapping in the Damrei Romel area, Cambodia using frequency ratio and logistic regression models. *Environ. Earth Sci.* **2006**, *50*, 847–855. [CrossRef]
17. Saito, H.; Nakayama, D.; Matsuyama, H. Comparison of landslide susceptibility based on a decision-tree model and actual landslide occurrence: The Akaishi Mountains, Japan. *Geomorphology* **2009**, *109*, 108–121. [CrossRef]
18. Youssef, A.M.; Pourghasemi, H.R.; Pourtaghi, Z.S.; Al-Katheeri, M.M. Landslide susceptibility mapping using random forest, boosted regression tree, classification and regression tree, and general linear models and comparison of their performance at Wadi Tayyah Basin, Asir Region, Saudi Arabia. *Landslides* **2016**, *13*, 839–856. [CrossRef]
19. Chen, W.; Pourghasemi, H.R.; Zhao, Z. A GIS-based comparative study of Dempster-Shafer, logistic regression and artificial neural network models for landslide susceptibility mapping. *Geocarto Int.* **2017**, *32*, 367–385. [CrossRef]
20. Kavzoglu, T.; Sahin, E.K.; Colkesen, I. Landslide susceptibility mapping using GIS-based multi-criteria decision analysis, support vector machines, and logistic regression. *Landslides* **2014**, *11*, 425–439. [CrossRef]
21. Zhu, T.; Xu, X.; Zhu, A. Spatial variation in the frequency and magnitude of mass movement in a semiarid, complex-terrain agricultural watershed on the Loess Plateau of China. *Land Degrad. Dev.* **2019**, *30*, 1095–1106. [CrossRef]
22. Lucas-Borja, M.E.; Zema, D.A.; Guzman, M.D.H.; Yang, Y.; Hernández, A.C.; Xiangzhou, X.; Carrà, M.G.; Nichols, M.; Cerdà, A. Exploring the influence of vegetation cover, sediment storage capacity and channel dimensions on stone check dam conditions and effectiveness in a large regulated river in México. *Ecol. Eng.* **2018**, *122*, 39–47. [CrossRef]
23. Geospatial Data Cloud Site, Computer Network Information Center, Chinese Academy of Sciences. Available online: http://www.gscloud.cn (accessed on 23 March 2022).
24. Resource and Environment Science and Data Center. Available online: https://www.resdc.cn/data.aspx?DATAID=335 (accessed on 23 March 2022).
25. Resource and Environment Data Cloud Platform. Available online: http://www.resdc.cn/Default.aspx (accessed on 23 March 2022).
26. Sassa, K.; Canuti, P.; Yin, Y. *Landslide Science for a Safer Geoenvironment: Volume 2: Methods of Landslide Studies*; Springer: Berlin/Heidelberg, Germany, 2014.

27. Chen, A.; Zhang, D.; Fan, J.; Liu, G. Mechanical mechanism and simulation experiment of the collapse of gully cliff in Yuanmou Dry-Hot Valley. *Sci. Soil Water Conserv.* **2012**, *10*, 29–35. (In Chinese)
28. Zhang, M. *Geological Disaster of Landslide and Collapse in Baota District, Yan'an*; Geological Publishing House: Beijing, China, 2008.
29. Xu, X.-Z.; Guo, W.-Z.; Liu, Y.-K.; Ma, J.-Z.; Wang, W.-L.; Zhang, H.-W.; Gao, H. Landslides on the Loess Plateau of China: A latest statistics together with a close look. *Nat. Hazards* **2017**, *86*, 1393–1403. [CrossRef]
30. Ashournejad, Q.; Hosseini, A.; Pradhan, B.; Hosseini, S.J. Hazard zoning for spatial planning using GIS-based landslide susceptibility assessment: A new hybrid integrated data-driven and knowledge-based model. *Arab. J. Geosci.* **2019**, *12*, 126. [CrossRef]
31. Skilodimou, H.D.; Bathrellos, G.D.; Chousianitis, K.; Youssef, A.M.; Pradhan, B. Multi-hazard assessment modeling via multi-criteria analysis and GIS: A case study. *Environ. Earth Sci.* **2019**, *78*, 47. [CrossRef]
32. Assouline, S.; Ben-Hur, M. Effects of rainfall intensity and slope gradient on the dynamics of interrill erosion during soil surface sealing. *Catena* **2006**, *66*, 211–220. [CrossRef]
33. De Rose, R.C. Slope control on the frequency distribution of shallow landslides and associated soil properties, North Island, New Zealand. *Earth Surf. Processes Landf.* **2013**, *38*, 356–371. [CrossRef]
34. Kavzoglu, T.; Sahin, E.K.; Colkesen, I. An assessment of multivariate and bivariate approaches in landslide susceptibility mapping: A case study of Duzkoy district. *Nat. Hazards* **2015**, *76*, 471–496. [CrossRef]
35. Dai, F.C.; Lee, C.F. Landslide characteristics and slope instability modeling using GIS, Lantau Island, Hong Kong. *Geomorphology* **2002**, *42*, 213–228. [CrossRef]
36. Dragićević, S.; Lai, T.; Balram, S. GIS-based multicriteria evaluation with multiscale analysis to characterize urban landslide susceptibility in data-scarce environments. *Habitat Int.* **2015**, *45*, 114–125. [CrossRef]
37. Chen, W.; Pourghasemi, H.R.; Naghibi, S.A. Prioritization of landslide conditioning factors and its spatial modeling in Shangnan County, China using GIS-based data mining algorithms. *Bull. Eng. Geol. Environ.* **2018**, *77*, 611–629. [CrossRef]
38. Liu, Y.; Fu, B.; Lü, Y.; Wang, Z.; Gao, G. Hydrological responses and soil erosion potential of abandoned cropland in the Loess Plateau, China. *Geomorphology* **2012**, *138*, 404–414. [CrossRef]
39. Tang, Q.; Bennett, S.J.; Xu, Y.; Li, Y. Agricultural practices and sustainable livelihoods: Rural transformation within the Loess Plateau, China. *Appl. Geogr.* **2013**, *41*, 15–23. [CrossRef]
40. Mansouri Daneshvar, M.R. Landslide susceptibility zonation using analytical hierarchy process and GIS for the Bojnurd region, northeast of Iran. *Landslides* **2014**, *11*, 1079–1091. [CrossRef]
41. Xu, Y.; Tang, Q. Land use optimization at small watershed scale on the Loess Plateau. *J. Geogr. Sci.* **2009**, *19*, 577–586. [CrossRef]
42. Thomas, J.; Joseph, S.; Thrivikramji, K.; Arunkumar, K. Sensitivity of digital elevation models: The scenario from two tropical mountain river basins of the Western Ghats, India. *Geosci. Front.* **2014**, *5*, 893–909. [CrossRef]
43. Lu, K.; Zhu, D. Theoretical and experimental study of effect of slope topography on its stability. *Chin. J. Rock Mech. Eng.* **2014**, *33*, 35–42. (In Chinese)
44. Xu, X.; Ma, Y.; Yang, W.; Zhang, H.; Tarolli, P.; Jiang, Y.; Yan, Q. Qualifying the influence of scar geometry on mass failure in loess gully sidewalls using laboratory experimentation. *Catena* **2020**, *187*, 104252. [CrossRef]
45. Derbyshire, E.; Dijkstra, T.; Smalley, I.; Li, Y. Failure mechanisms in loess and the effects of moisture content changes on remoulded strength. *Quat. Int.* **1994**, *24*, 5–15. [CrossRef]
46. Zhang, F.; Pei, X.; Chen, W.; Liu, G.; Liang, S. Spatial variation in geotechnical properties and topographic attributes on the different types of shallow landslides in a loess catchment, China. *Eur. J. Environ. Civ. Eng.* **2014**, *18*, 470–488. [CrossRef]
47. Liu, C.-C. Preparing a landslide and shadow inventory map from high-spatial-resolution imagery facilitated by an expert system. *J. Appl. Remote Sens.* **2015**, *9*, 96080. [CrossRef]
48. Wang, J.-J.; Liang, Y.; Zhang, H.-P.; Wu, Y.; Lin, X. A loess landslide induced by excavation and rainfall. *Landslides* **2014**, *11*, 141–152. [CrossRef]
49. Shang, J. *Study on the Stability of Loess Landslide in the Baota District of Yan'an City—A Case Study: Wohushan Landslide Chang'an University*; Chang'an University: Xi'an, China, 2011.
50. Wang, L.; Wei, S.; Horton, R.; Shao, M. Effects of vegetation and slope aspect on water budget in the hill and gully region of the Loess Plateau of China. *Catena* **2011**, *87*, 90–100. [CrossRef]
51. Zhang, M.; Liu, J. Controlling factors of loess landslides in western China. *Environ. Earth Sci.* **2010**, *59*, 1671–1680. [CrossRef]
52. Stokes, A.; Sotir, R.; Chen, W.; Ghestem, M. Soil bio- and eco-engineering in China: Past experience and future priorities. *Ecol. Eng.* **2010**, *36*, 247–257. [CrossRef]
53. Tian, J.; Liu, P.; Shimizu, H.; Inanaga, S. A case study on combating desertification at a small watershed in the hills-gully area of loess plateau, China. In *Plant Responses to Air Pollution and Global Change*; Springer: Tokyo, Japan, 2005; pp. 277–284.

applied sciences

MDPI

Article

Comparisons of Dynamic Landslide Models on GIS Platforms

Yuming Wu [1], Aohua Tian [1,2,3] and Hengxing Lan [1,*]

[1] State Key Laboratory of Resources and Environmental Information System, Institute of Geographic Sciences and Natural Resources Research, Chinese Academy of Sciences, Beijing 100101, China; wuym@lreis.ac.cn (Y.W.); tian1454565116@163.com (A.T.)
[2] College of Geoscience and Surverying Engineering, China University of Mining and Technology-Beijing, Beijing 100083, China
[3] College of Resources and Environment, University of Chinese Academy of Sciences, Beijing 100049, China
[*] Correspondence: lanhx@lreis.ac.cn; Tel.: +86-10-64888783

Abstract: Numerical simulation is one of the methods to assess landslide movement processes, which is beneficial for engineering design and urban planning. With the development of computer technology, GIS has gradually become the mainstream platform for landslide simulation due to data availability and algorithm integrability. However, the dynamic processes of landslides are complicated, which makes integration difficult on GIS platforms. Some assumptions are applied to simplify these dynamic processes and solve this problem. Generally, there are two main types of numerical models on GIS platforms: models based on the Eulerian description and models based on the Lagrangian description. Case studies show that Eulerian models are suitable for flow-like movement, and Lagrangian models are suitable for discrete rigid bodies movement. Different models face different problems: the Eulerian-based models show numerical diffusion and oscillation, and the Lagrangian-based model needs to consider complicated shear and collision processes. In addition, the 3-D model can describe more details in the z-direction, while the depth-averaged model can obtain a reasonable range of motion, depth, and speed quickly. From the view of numerical simulation, inappropriate models, assumptions, and numerical schemes will produce errors. The landslide type refers to several forms of mass wasting associated with a wide range of ground movements, which guides establishing dynamic models and numerical schemes on GIS platforms and helps us obtain results accurately.

Keywords: landslide dynamics; landslide classification; dynamic models; depth-averaged model; GIS

Citation: Wu, Y.; Tian, A.; Lan, H. Comparisons of Dynamic Landslide Models on GIS Platforms. *Appl. Sci.* **2022**, *12*, 3093. https://doi.org/10.3390/app12063093

Academic Editors: Ricardo Castedo, Miguel Llorente Isidro and David Moncoulon

Received: 18 February 2022
Accepted: 16 March 2022
Published: 17 March 2022

1. Introduction

Landslides cause a large number of casualties and property losses every year. Quantitative risk analysis (QRA) is essential for reducing property damage and loss of life. Numerical simulation is one of the methods for assessing the landslide movement process and potential hazard levels [1,2], which is part of quantitative risk analysis [3]. Many numerical models are used to obtain the landslide runout distance, thus providing a basis for urban planning [4–7].

From dynamics, all landslides follow Newton's second law [2] and, thus, mass conservation, momentum conservation, and energy conservation are the governing equations in numerical simulations. In terms of force, gravity is the primary driving force [8], and friction is the major resistance force during motion. However, different shear processes and collision processes make landslide dynamics complex.

There are two model types: discrete models and continuum models. The discrete models describe the interactions and shocks well, but the continuum models describe the flow well. The Lagrangian functions describe the translation and rotation of each discrete element for the discrete models. Eulerian functions are the best to calculate the field for the continuum models. With the development of computation, meshless methods such as

SPH are proposed to solve problems dominated by complex boundary dynamics for the continuum models. In the SPH method, finding neighboring nodes is difficult. Models with different accuracies are used to describe the processes [7]. These models include Rockfall Analyst [9], Rocfall [10], and Rockyfor3D [11] for rockfalls, and Massflow [12], Flo-2D [13], r.avaflow [14], and Titan2D [15] for mass flow.

In some simulations, different software programs yield different trajectories based on methods, assumptions, and numerical schemes [16]. Models may face various challenges. For example, fluid mechanics models may encounter numerical diffusions and oscillations [17]; rigid body models need to consider complicated shear and collision processes [18]. Therefore, a suitable model is a prerequisite for obtaining accurate results, and the physical insights provided by models can help us understand the landslide process.

GIS provides some functions that allow users to build numerical models [19]. It makes programming convenient and easy. There are two types of landslide simulations on GIS platforms: add-in programs and stand-alone programs. Add-in programs use the GUI and functions on GIS platforms to build a dynamic landslide model. This combination is easy to implement, but the GIS platform may limit some capacities. Stand-alone programs use open libraries such as GDAL. They are computationally efficient but require a lot of development time. The landslide dynamics programs on GIS platforms need to be simplified to accommodate the GIS file structure. Therefore, appropriate assumptions and suitable models are the keys to simulation on GIS platforms.

Landslide classification uses simple words to describe the characteristics of landslide phenomena, which may guide us to select a suitable model for simulation on GIS platforms. Several authors, including Heim [20], Zaruba and Mencl [21], and Sharpe [22], proposed different landslide classification systems. In most of these classifications, the name of the landslide is a combination of the material and motion type [23–26]. The most well-known classification system is Varnes' classification [27]. In this classification, material types include rock, debris, and earth, and motion types include falling, toppling, sliding, spreading flow, and complex. Subsequently, Hungr [23] updated the Varnes' classification to assess landslide phenomena, and the material types were more detailed than those in the original classification. Landslide types are often indicators of the critical movement process, and a suitable numerical model must consider the appropriate variables.

However, the relationship between the landslide movement process and the physical mechanism of landslides is not very clear. It results in difficulty in obtaining accurate runout zones on GIS platforms. In this paper, we compared numerical models on GIS platforms and established relationships between landslide types and the selection of numerical models on GIS platforms.

2. Models

2.1. Descriptions

Landslides contain two main motion processes: rigid body motion and flow-like motion. The rigid body motion assumes that the block does not deform or change shape [28]; the flow-like motion assumes that the material is continuous mass rather than discrete particles. The driving force is gravity, and the resistance force is mainly friction during the movement (Figure 1).

In dynamics, there are five major categories in the Varnes' movement type: falling, toppling, sliding, spreading, and flowing. The movement type is related to the critical processes that occur during motion. All the processes follow Newton's laws of motion [29,30], which are given by

$$\frac{dmv}{dt} = f \tag{1}$$

where m is the mass, v is the velocity, and f is the resultant force, which depends on the movement type. The left-hand side (LHS) of Equation (1) represents the motion process, and the right-hand side (RHS) is the resultant force.

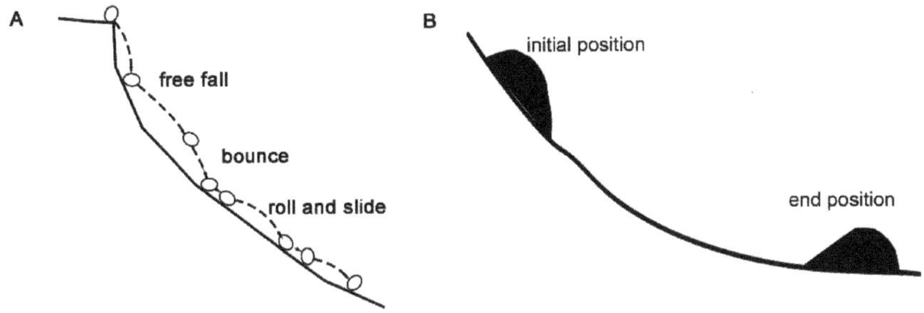

Figure 1. Two main motion processes: (**A**) the rigid body motion, and (**B**) the flow-like motion.

There are two common descriptions of motion in numerical models: the Eulerian description and the Lagrangian description [31] (Figure 2). In the Eulerian description, a model is not concerned about the location or velocity of any particular particle or columns, and the focus is on the grid. In the Lagrangian description, individual particles or columns are marked, and their positions and velocities are described as a function of time.

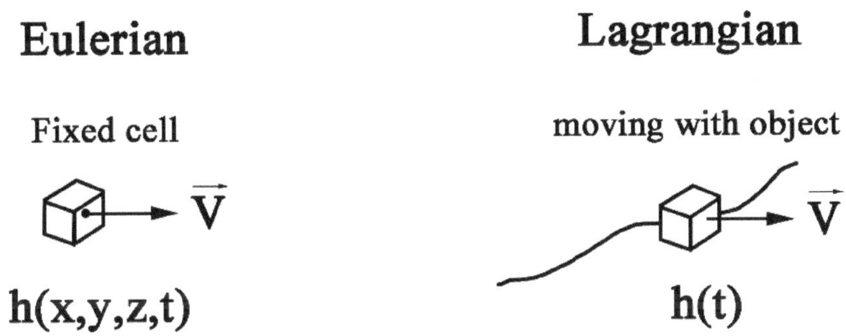

Figure 2. Eulerian and Lagrangian descriptions.

2.1.1. Models Based on the Lagrangian Description

The rigid body motion includes translation and rotation, which are six-degrees-of-freedom processes (three components for the translation process and three components for the rotation process). Using the center of mass and inertia matrix, the force and torque equations take the form:

$$\mathbf{F} = m\mathbf{a},\ \mathbf{T} = [I_R]\alpha + \omega \times [I_R]\omega \quad (2)$$

where **F** is the force, **T** is the torque, **a** is the acceleration, m is the mass, I_R is the moment of inertia matrix, ω is the angular velocity, and α is the angular acceleration. The equations of translation and rotation on the Lagrangian description are

$$\mathbf{x}_{t+\Delta t} = \mathbf{x}_t + \mathbf{v}_t\Delta t + \frac{1}{2}\mathbf{a}_t\Delta t^2 \quad (3)$$

$$\boldsymbol{\theta}_{t+\Delta t} = \boldsymbol{\theta}_t + \boldsymbol{\omega}_t\Delta t + \frac{1}{2}\boldsymbol{\alpha}_t\Delta t^2 \quad (4)$$

where $\mathbf{x}_t$ is the location at time t, $\mathbf{v}_t$ is the velocity at time t, $\mathbf{a}_t$ is the acceleration at t, $\boldsymbol{\theta}_t$ is the angle at time t, $\boldsymbol{\omega}_t$ is the angular velocity at time t, and $\boldsymbol{\alpha}_t$ is the angular acceleration at time t.

2.1.2. Models Based on the Eulerian Description

In the continuum systems, the Eulerian description is best for simulating motion, and the Navier–Stokes equations are the basis of modeling [32]:

$$\frac{\partial \rho}{\partial t} + \nabla \cdot (\rho \mathbf{v}) = 0 \tag{5}$$

$$\rho(\partial \mathbf{v}/\partial t + \mathbf{v} \cdot \nabla \mathbf{v}) = \mathbf{S_f} \tag{6}$$

where ρ is the mass density, t is time, $\mathbf{v}$ is the velocity, $\mathbf{S_f}$ is the field which includes gravity, friction, and other forces. The best way to solve these equations is through numerical analysis due to lacking a smooth analytical solution. Traditional CFDs such as Openfoam® use these equations to simulate the landslide dynamics. In these programs, the geometric modeling process is complex.

To describe flowing grain–fluid mixture, Iverson proposed a mixture theory [33], and the $\mathbf{S_f}$ is:

$$\mathbf{S_f} = -\nabla \cdot \left(\mathbf{T_s} + \mathbf{T_f} + \mathbf{T}'\right) + \rho \mathbf{g} \tag{7}$$

in which

$$\rho = \rho_s v_s + \rho_v v_v \tag{8}$$

$$\mathbf{v} = \left(\rho_s v_s \mathbf{v}_s + \rho_f v_f \mathbf{v}_f\right)/\rho. \tag{9}$$

Here, ρ is the mixture mass density, $\mathbf{v}$ is the mixture velocity, $\mathbf{v_s}$ is the velocity in the solid phase and $\mathbf{v_f}$ is the velocity in the fluid phase, v_s is the volume fraction of solid, v_f is the volume fraction of fluid, $\mathbf{g}$ is the gravitational acceleration, $\mathbf{T_s}$ is the solid stress, $\mathbf{T_f}$ is the fluid stress, and $\mathbf{T}'$ is a contribution to the mixture stress. The stress $\mathbf{T}'$ can be avoided by using an approximation suitable for many debris flows.

To reduce the difficulty of traditional CFDs, Savage and Hutter proposed the depth-averaged theory [34]. This model allows the GIS to integrate simulation codes. Depth averaging is one of the steps to eliminate the calculation in the z-direction. The averaged velocities and resultant forces are calculated on the x–y plane [32,34,35]. In this method, the assumption is that ρ is constant and the landslide depth is shallow. Thus, the equations [32,36] are

$$\frac{\partial h}{\partial t} + \frac{\partial h\overline{v}_x}{\partial x} + \frac{\partial h\overline{v}_y}{\partial y} = 0 \tag{10}$$

$$\rho\left[\frac{\partial(h\overline{v}_x)}{\partial t} + \frac{\partial(h\overline{v}_x{}^2)}{\partial x} + \frac{\partial(h\overline{v}_x\overline{v}_y)}{\partial y}\right] = -\int_0^h \left[\frac{\partial T_{xx}}{\partial x} + \frac{\partial T_{yx}}{\partial y} + \frac{\partial T_{zx}}{\partial z} - \rho g_x\right] dz \tag{11}$$

$$\rho\left[\frac{\partial(h\overline{v}_y)}{\partial t} + \frac{\partial(h\overline{v}_y{}^2)}{\partial y} + \frac{\partial(h\overline{v}_x\overline{v}_y)}{\partial x}\right] = -\int_0^h \left[\frac{\partial T_{xy}}{\partial x} + \frac{\partial T_{yy}}{\partial y} + \frac{\partial T_{zy}}{\partial z} - \rho g_y\right] dz. \tag{12}$$

where $\overline{v}_x = \frac{1}{h}\int_0^h v_x dz$, $\overline{v}_y = \frac{1}{h}\int_0^h v_y dz$. As shown in the equations, the velocity in the z-direction is averaged. Based on the depth-averaged theory, a depth-averaged mixture model was proposed by Iverson and Denlinger [32]. Subsequently, two-phase and multi-phase depth-averaged models were proposed to describe distinct mechanical responses and dynamic behaviors of material [35,37].

Some studies considered the erosion process, a mechanical process by which the bed material is mobilized by the flow and dominant mechanical processes in geophysical mass flows [38]. Erosion determines enhanced or reduced mobility but is not understood thoroughly. In the dynamic model, the erosion rate $E = -\partial b/\partial t$ and erosion velocity u_b are two important parameters, which change mass and momentum productions E and $u_b E$ [38], respectively. Subsequently, Pudashini and Fischer proposed a two-phase erosion model on this basis [39]. The emergence of these models has further developed the continuum model on GIS platforms.

2.2. Forces

Gravity, friction, collision force, and hydraulic pressure are the major forces during motion. Some forces have less influence and can be ignored during the movement.

2.2.1. Collision Force

Collision force is one of the factors affecting the process of landslide movement, especially during falls and topples. The collision process is complicated, which makes the calculation difficult. Traditionally, the spring–dashpot model is widely used to describe the nonlinear process, which is

$$F_n = F_{el} + F_{diss} \tag{13}$$

where F_n is the collision force, F_{el} is the elastic force (spring), and F_{diss} is the dissipative force (dashpot). The spring obeys Hooke's law, and the dashpot obeys Newton's law of viscosity [40,41]. To simplify the process, Evans and Hungr used a lumped mass model in ROCKFALL programs in 1993 [42]. The associated assumptions of the lumped mass model are as follows: (1) each rock is a small spherical particle; (2) rocks do not have any size, only mass. The lumped mass model uses one or two restitution coefficients to express this process to avoid calculating complex collision forces. The restitution coefficients are the ratio of the rebound velocity to the incident velocity, the impulse ratio, and the work ratio, which involves the square root of work performed. Most models use two restitution coefficients: the tangential restitution coefficient R_t and the normal restitution coefficient R_n; however, a few models use only one restitution coefficient to quantify dissipation in terms of velocity magnitude loss. Hybrid approaches were proposed on GIS platforms to simulate rockfall accurately. These models, such as CRSP [43], RocFall [44], and STONE [45], consider the influence of shape and nonlinear collision based on the lumped mass model. Hybrid approaches have been the mainstream models on GIS platforms. Therefore, there are three models for describing the collision force, a fully rigid body model, a lumped mass model, and a hybrid approach [46], on GIS platforms.

2.2.2. Friction Force

Friction is the force that resists the relative motion of solid surfaces, fluid layers, and material elements sliding against each other, which is important for the movement calculation. In this process, kinetic energy is converted to thermal energy when motion with friction occurs. There are two types of resistance: viscosity and Coulomb's friction.

For dry granular flow, Coulomb's friction is adopted, which is

$$\tau = \mu N \tag{14}$$

where τ is the unit base resistance and μ is the friction coefficient, and N is the normal force. Generally, the basal shear forces, obtained by simple infinite landslide models, are calculated by Coulomb's friction.

For fluid flow, viscosity is a measure of resistance to deformation at a given rate. It can be conceptualized as the internal frictional force that arises between adjacent layers of fluid that are in relative motion. There are three viscosity equations: Newtonian, Bingham, and quadratic fluids.

(1) The Newtonian fluid is:

$$\tau = \mu \gamma' \tag{15}$$

where μ is the shear viscosity of a fluid and γ' is the derivative of the velocity component.

(2) The Bingham fluid model is

$$\tau = \tau_0 + \mu \gamma' \tag{16}$$

where τ_0 is a constant yield strength, μ is the shear viscosity of the fluid, and γ' is the derivative of the velocity component.

(3) The quadratic fluid model is

$$\tau = \tau_0 + \mu\gamma' + \zeta\gamma'^2 \tag{17}$$

The first two terms are referred to as the Bingham shear stresses. The last term represents the dispersive and turbulent shear stresses. Fluid friction can be generalized as:

$$f = f_s + f_v + f_t \tag{18}$$

where f_v is the viscosity term, f_s is the constant term, and f_t is a turbulent term. Based on the depth-averaged theory, shear stress is depth-integrated, and the corresponding equation is

$$S = \frac{1}{h}\int f dz \tag{19}$$

where S is depth-integrated shear stress. Therefore, Equation (18) can be transformed to:

$$S_{fx} = S_\tau + S_v + S_{td}. \tag{20}$$

2.2.3. Other Forces

Hydraulic pressure in the depth-averaged model is the force imparted per unit area of liquid or flow-like materials on the surfaces, which can be expressed in the Eulerian description as:

$$F_i = \frac{\partial}{\partial x}\beta\frac{h^2}{2} \tag{21}$$

where β can be changed to β_s^x and β_f^x based on the form of the phase [37]. For fluid, the β is g^z. For solids, the force created by collisions among particles is simplified to an internal force based on soil mechanics, and the β is:

$$\beta_s^x = Kg^z\left(1 - \gamma_s^f\right) \tag{22}$$

$$K_{pas/act} = 2\sec^2\phi\left\{1 \pm \left(1 - \cos^2\phi\sec^2\delta\right)^{1/2}\right\} - 1 \tag{23}$$

where ϕ is the internal frictional angle and δ is the basal frictional angle. The collisions and frictions among particles are difficult to calculate during the motion.

There are two methods to describe the form of hydraulic pressure on the Lagrangian description. The first method is the particle-in-cell (PIC) method [47], which converts particles or columns to fields based on the volume in each cell. The other is smoothed-particle hydrodynamics (SPH) approximation [48–50], and they are:

$$\overline{P}_I = \frac{1}{2}\beta h_I{}^2 \tag{24}$$

$$H_p = \sum_J m_J\left(\frac{\overline{P}_I}{h_I{}^2} + \frac{\overline{P}_J}{h_J{}^2}\right) gradW_{IJ} \tag{25}$$

where $\overline{P}_I$ is an averaged hydraulic pressure term, W_{IJ} is the value of the SPH kernel function W_{IJ} centered at node I evaluated at node J. The weighting function or kernel W_{IJ} is a symmetric function of $x_I - x_J$. Additionally, m_J has no physical meaning. When the node moves, the material contained in a column of base Ω_I has entered it or will leave it as the column moves with an averaged velocity, which is not the same for all particles or columns in it [48].

In addition, buoyancy and drag forces in the two-phase and multi-phase flow can also influence the landslide motion. Buoyancy is an upward force exerted by a fluid that opposes the weight of a partially or fully immersed object, and it is a vertical force. Buoyancy can reduce the pressure at the basal surface. Therefore, buoyancy can reduce the resistance, especially in multi-phase flow. Drag forces are shear forces caused by different velocities and accelerations in different phases. In other words, solid particles may accelerate relative to fine solids or fluids [37].

3. Software

With the development of computer technology, GIS has gradually become a mainstream system for engineering design and urban planning. There are a lot of GIS programs, such as GRASS GIS, QGIS, and ArcGIS. GRASS GIS and QGIS are popular in program development due to being free and open source. ArcGIS is a mature commercial program and is widely applied in urban planning and engineering design. These programs provide a rich interface such as raster import, vector import, raster statistics, and vector analysis, and are convenient for users to call the functions and develop their programs. In addition, these programs provide a GUI to display the results calculated by their programs. More and more landslide simulation codes support GIS. At present, there are two kinds of programs: programs based on the rigid body model and programs based on the flow-like model.

3.1. Programs Based on the Rigid Body Model

On GIS platforms, there are several programs to simulate the discrete rigid body motions, as shown in Table 1. On GIS platforms, the format of input parameters of GIS data needs to be considered, especially DEM data. The choice of parameter expressions determines the model. There are three types of DEMs to express on GIS platforms: triangulated irregular networks (TINs), grid networks, and vector or contour-based networks. Relying on the algorithm, the selection of the DEM is also different. For example, Rockfall Analyst and Rockyfor3D select the grid networks. The lumped mass models are popular in early GIS platforms among these models. With the GIS technique development, the hybrid model is mainstream on GIS platforms. These programs include Hy-STONE, Rockyfor3D, and PICUS Rock'n'Roll.

Table 1. Rigid body programs on GIS platforms.

Software	Scheme	Platform	Format
STONE [45]	Lumped mass		ASCII
Hy-STONE [51]	Hybrid		ASCII
Rockfall Analyst [9]	Lumped mass	ArcGIS	All raster format
Rockyfor3D [11]	Hybrid		ASCII
PICUS Rock'n'Roll [52]	Hybrid	PICUS	
RAMMS::ROCKFALL [53]	Rigid body		ASCII
RockGIS [54]	Lumped mass		ASCII

3.2. Programs Based on the Flow-like Model

Based on the description, the grid networks of DEM are best for flow-like models on GIS platforms. The grid networks make the calculation simple and accurate. At present, there are a lot of codes on the GIS platforms, such as r.avaflow, LA, DA, Titan2D, and Massflow (Table 2). As for the model, the depth-averaged theory model is employed to obtain the depths and velocities in each cell. The simulation r.avaflow is popular around the world. The program has a built-in multi-phase flow method based on the depth-averaged theory and is applied in GRASS GIS and R. It is a very good cross-platform program, which means that we can use the program on Windows, Linux, and Mac OS [35,37,38].

Table 2. Flow-like programs on GIS platforms.

Software	Description	Scheme	Format	Platform
r.avaflow [14]	Eulerian	NOC	All raster format	GRASS
RAMMS [55]	Eulerian	1st/2nd order	ASCII, Geotiff	
Massflow [12]	Eulerian	TVD–MacCormack	ASCII	
Massmov2D [56]	Eulerian	2-step scheme	PCRaster	PCRaster
LA and DA [31,57]	Eulerian–Lagrangian	PIC-like	GeoTIFF	
Titan2D [15]	Eulerian	AMR	All raster format	GRASS
IMEX_SfloW2D [58]	Eulerian	Semi-discrete central scheme	ASCII	
Geo-Claw [59]	Eulerian	AMR	ASCII, NetCDF	
FLO-2D [13]	Eulerian	1st order	All raster format	QGIS
DAN3D [60]	Lagrangian	SPH	ASCII	
SPHERA [61]	Lagrangian	SPH	All raster format	QGIS

For flow-like programs, the numerical scheme is one of the factors affecting results. Most codes use 1st or 2nd order finite difference methods to solve the partial differential equations. In these programs, the numerical diffusion and numerical oscillation are the difficulties for the models in the Eulerian description. Some codes use the TVD method and adaptive mesh refinement (AMR) to improve precision. In addition, some methods use the Lagrangian or Eulerian–Lagrangian description to solve difficulties such as the material point method (MPM) [62], particle-in-cell (PIC) [47], and smoothed particle hydrodynamics (SPH) [63]. In these models, the computational cost of simulations per number of particles may be higher than the cost of grid-based simulations per number of cells. In some cases, these methods solve the numerical solution problem of differential equations to a certain extent.

4. Results and Discussion

Many factors affect the simulation results, such as models, algorithms, and descriptions. In this section, we show cases such as bilateral dam break, a rockfall example in the RA program, and the Yigong landslide to analyze the effect on models, algorithms, and descriptions.

4.1. Reason for Differences

4.1.1. Differences Caused by Models

Different models can produce different results due to assumptions. We used a 2-D bilateral dam break simulation to analyze the applicability of the traditional CFD model (two-phase model), the depth-averaged model, and the SPH model (Figure 3). In these simulations, we set the initial state to 1 m at [0, 1]. Under the action of gravity, the fluid moves downward. In this case, we applied the interfoam solver of OpenFOAM® to calculate the two-phase model [64], PySPH to calculate the SPH model [65], and the Lax–Friedrichs scheme to calculate the depth-averaged model. In OpenFOAM®, we select the area in the grid with water content greater than 0.6 to obtain the profile. In the SPH model, we set the ball to have a diameter of 0.03 m.

The results show that the range of movement observed in all models is similar at 0.5 s, spanning [−2, 3]. In the 2-D simulation, different models will have different results in terms of details. The depth-averaged model can obtain a smoother result than the other models. Additionally, the results of the two-phase model and the SPH model include more details in the z-directions than the depth-averaged model. Based on the above analysis, the model selection depends on the relevant requirements. When detailed information is required, we must select a complex model to calculate the results. When we focus on the range, we can use the depth-averaged model.

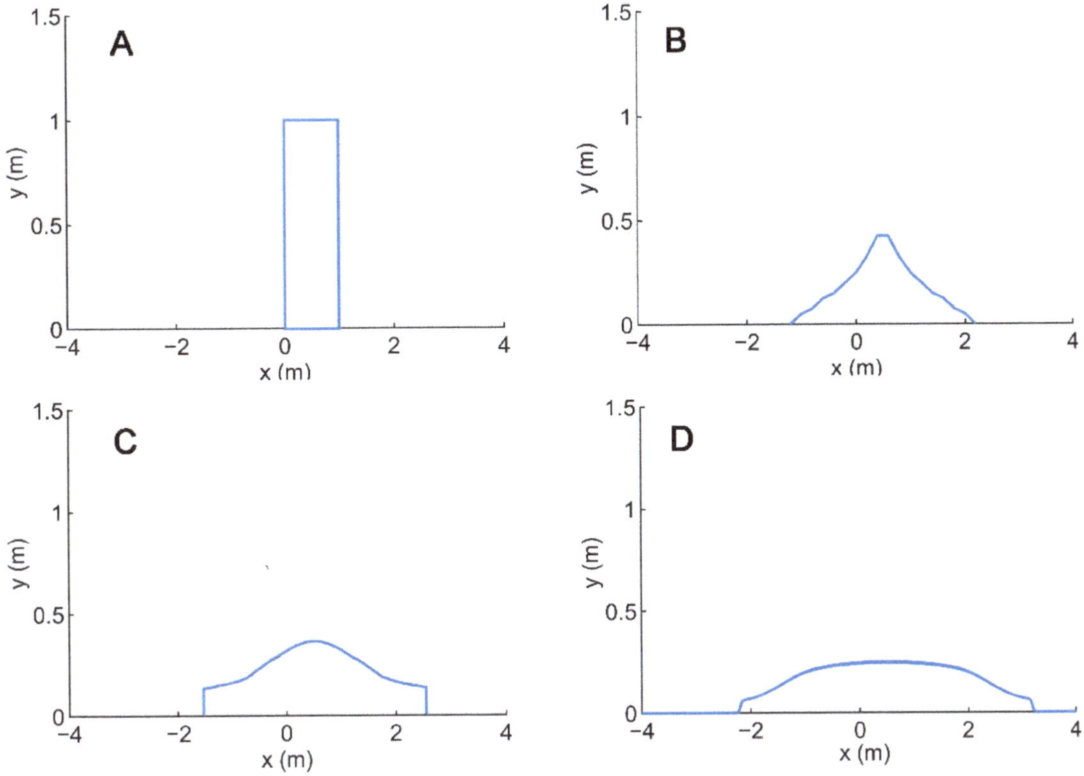

Figure 3. Bilateral dam break simulation with different models: (**A**) the initial state of the simulation, (**B**) obtained by OpenFOAM at 0.5 s, (**C**) obtained by PySPH at 0.5 s, and (**D**) obtained by the depth-averaged model at 0.5 s.

4.1.2. Differences Caused by Algorithms

Differences in parameters and numerical schemes can affect the runout zones significantly (Figure 4). In this case, we used Rockfall Analyst to analyze the influence of algorithms. The results obtained with ArcGIS 9.x and ArcGIS 10.x differ because of different point extraction algorithms (Figure 4). The small difference in the point extraction of the GIS module will affect the runout zone. The rock fell into the river in ArcGIS 9.x. However, the rock stops on the road in ArcGIS 10.x, although the DEM is the same. The results show that a small difference in the algorithm can produce large differences in trajectory.

In the Eulerian-based model, some numerical schemes can handle one property of ADEs but process badly with another one [17]. Therefore, the balance between numerical diffusion and numerical oscillation is key to obtaining a suitable physical solution. For the model in the Eulerian description, we simulated the uniform linear motion of a block with 1 m/s. In this simulation, the space interval is 0.1 m, and the time interval is 0.01 s. The Lax–Friedrichs scheme, a first-order in time and second-order in space method, shows numerical diffusion during motion (Figure 5A) but the Lax–Wendroff scheme, a second-order in both space and time method, shows numerical dispersion (Figure 5B). The errors are generally caused by neglecting high-order terms. However, higher-order linear schemes such as 3rd order, although more accurate for smooth solutions, are not TVD and tend to introduce spurious oscillations (wiggles) where discontinuities or shocks arise. Various high-resolution schemes use flux/slope limiters to maintain the TVD, thereby reducing the impact of numerical dissipation and numerical diffusion [66,67]. In these methods, the accuracy is high in the smooth area, and the flux/slope limiter method is used in the shock area to avoid producing nonphysical solutions.

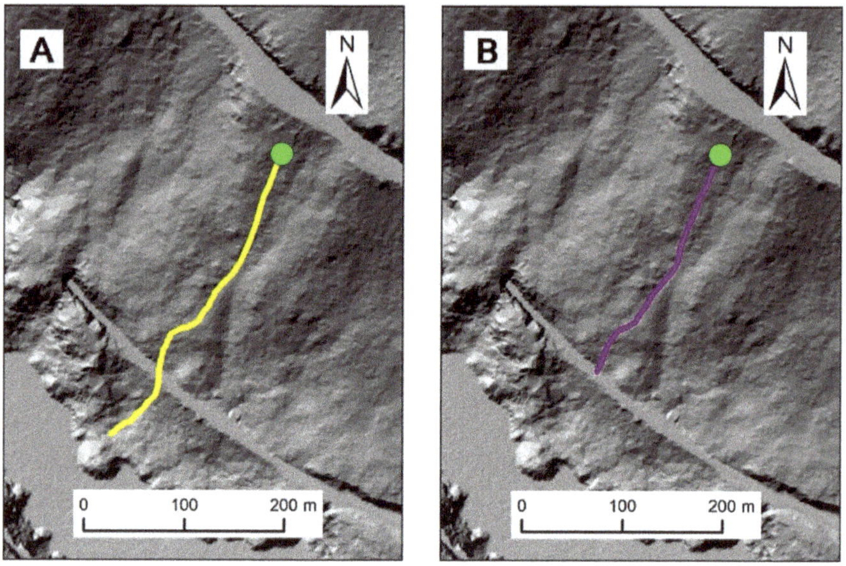

Figure 4. Rockfall simulations using RA: (**A**) in ArcGIS 9.x and (**B**) in ArcGIS 10.x.

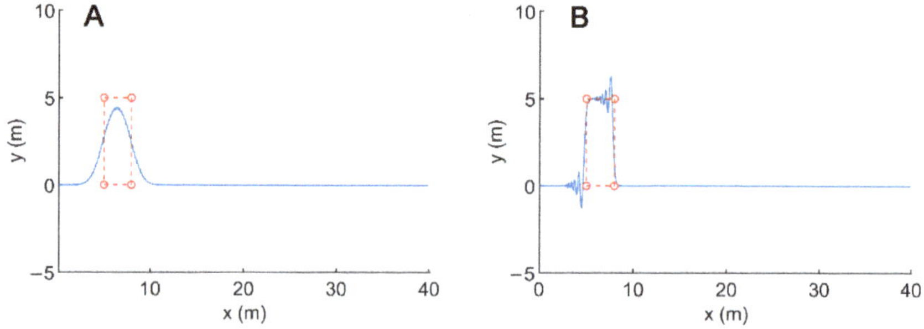

Figure 5. Numerical diffusion and numerical oscillation: (**A**) numerical diffusion obtained by the Lax–Friedrichs scheme, and (**B**) numerical oscillation obtained by the Lax–Wendroff scheme.

4.1.3. Differences Caused by Description

In this section, we show the 3-D sliding block used to analyze the influence of different descriptions. In this case, we assumed the acceleration in the x- and y-direction is 5 m/s^2, and that the x-direction and y-direction interval is 10 m. The location of the center of the block is (200, 200) (Figure 6A). In this motion process, the deformation of the block is zero, and the block can be considered as a rigid body. After 20 s, the block moves to (1200, 1200) based on Newton's second law. We selected the lumped mass model in the Lagrangian description and the depth-averaged model in the Eulerian description to simulate the same motion.

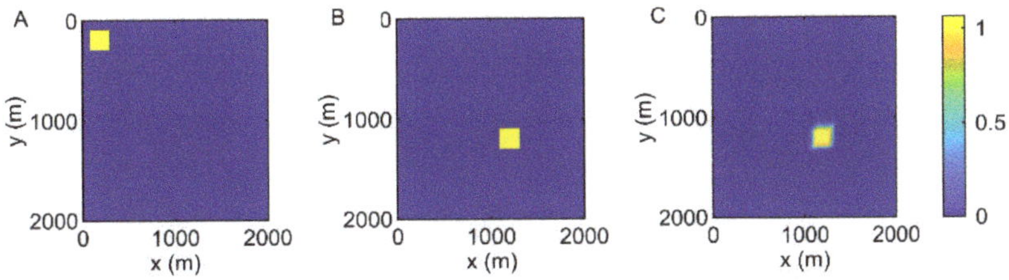

Figure 6. Simulations using different numerical schemes: (**A**) the original location, (**B**) the result in the Lagrangian description, and (**C**) the result in the Eulerian description.

In the Lagrangian description, the lumped mass model is simple (Figure 6B) because the block assumes a mass point in the model. Based on the lumped mass model, we can calculate the location and the result close to the observed one. In the Eulerian description, we used depth-averaged theory and the McCormack–TVD scheme. After 20 s, the block also moves to (1200, 1200) (Figure 6C). However, numerical diffusion (the green area) is notable at the boundary of the block, even though the TVD method is applied to reduce diffusion (Figure 6C). Additionally, the calculation efficiency is far lower than the method based on the Lagrangian description, when reaching the same precision. This case shows that proper model assumptions are a prerequisite for good results. Suitable assumptions can reduce some errors in the calculation process. The lumped mass model is simple and obtains results quickly and accurately. Therefore, the lumped mass model is more suitable for this motion.

In the 3-D landslide simulation, we selected the Yigong landslides to evaluate the influence of different numerical schemes. The Yigong landslide happened at the head of the Zhamulong gully (30.178° N; 94.940° E) on 9 April 2000 [68] and blocked the Yi Gong River at the foot of the slope with 3×10^8 m^3 of sediment that formed a 60 m high dam [69]. Aiming at the characteristics of long-distance and high-speed movement of the Yigong landslide, researchers use many numerical models to investigate it.

We selected different descriptions to simulate this process. Based on previous studies [70,71], the Voellmy model can obtain a suitable result [72]. In this case, the basal friction angle is 12°, the internal friction angle is 13°, and the turbulence assumes 1000 s^2/m (Table 3).

Table 3. Mechanical parameters of the Yigong landslide.

Parameters	Values
Basal friction angle	12°
Internal friction angle	13°
Turbulent coefficient	1000

In numerical schemes, we selected the NOC–TVD scheme in the Eulerian description, which is applied in r.avaflow, and the depth-averaged SPH method in the Lagrangian description to analyze the differences. The states of 50 s, 100 s, and 200 s using two methods are given in Figure 7. From the simulation, both methods obtain similar results. The Yigong landslide reached the foot of the mountain at 200 s. In detail, the process calculated in the Eulerian description is smoother than that in the Lagrangian description. The material in the Lagrangian description is concentrated in the channel. The maximum height is higher than the model in the Eulerian description. As for the maximum speed, the NOC–TVD scheme is 90.23 m/s, and the depth-averaged SPH is 97.20 m/s. The precision of the SPH model is related to the number of particles or the number of columns. The more particles or columns, the finer the description and the closer to the analytical solution of the fluid equation. Fewer particles or columns will result in lower interaction forces between columns or particles. Additionally, the computational cost of SPH simulations per number of particles or columns is significantly larger than the cost of grid-based simulations for flow-like motion.

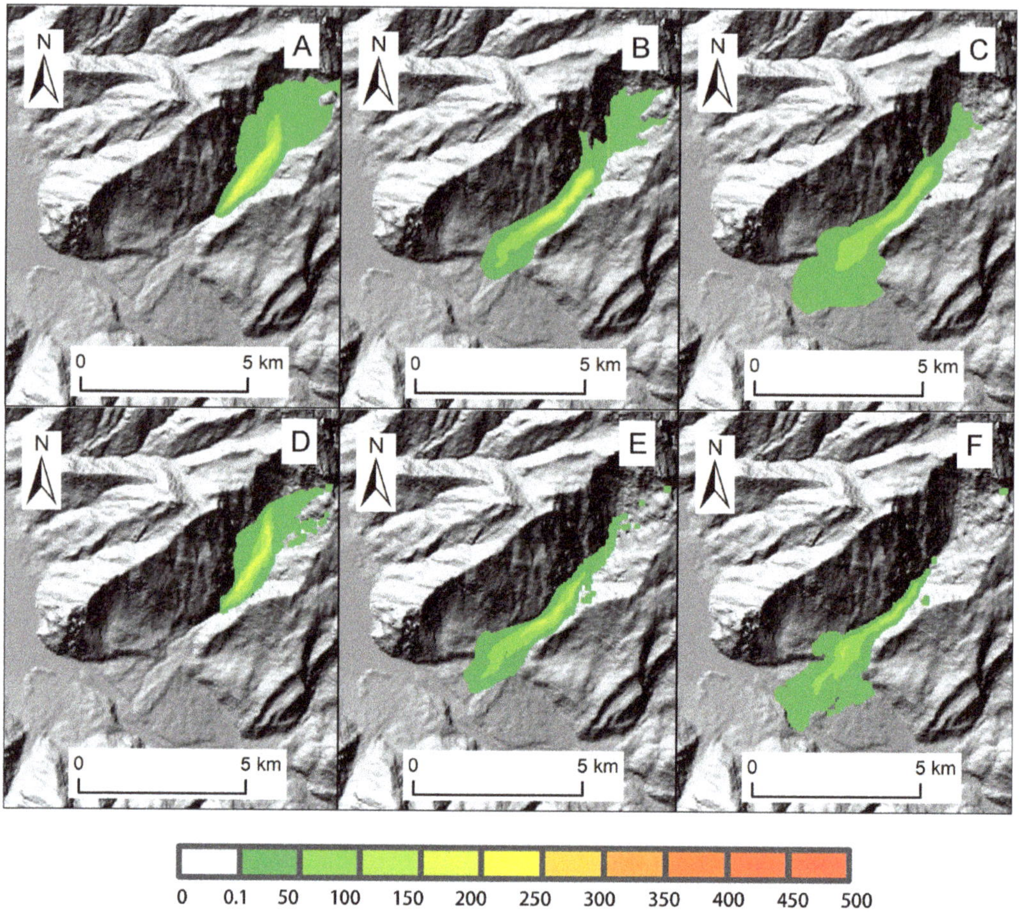

Figure 7. Yigong landslide simulation using different descriptions: (**A**) at 50 s using Eulerian method; (**B**) at 100 s using Eulerian method; (**C**) at 200 s using Eulerian method; (**D**) at 50 s using depth-averaged SPH; (**E**) at 100 s using depth-averaged SPH; (**F**) at 200 s using depth-averaged SPH.

Based on the above analysis, the Eulerian description is more suitable for flow-like motion, but the Lagrangian description is more suitable for the discrete rigid body motion. The description is one of the factors influencing the result.

4.2. Model Selection

Based on Varnes' classification [27], the materials include rock, debris (coarse soil), and earth (fine soil). Rock is a solid mass of geological materials, debris is scattered material (large rock fragments), and earth is a cohesive, plastic, clayey soil. These terms are neither geological nor geotechnical [32,34,35,73,74] but are related to the size, shape, quantity, and properties, which help us select a suitable description. A rock is a solid mass and an aggregate of minerals that can be considered a discontinuous rigid body in landslide dynamics due to its characteristics. "Earth" is neither a geological term nor a geotechnical term, and it describes construction material or agricultural soil [75]. Earth is defined as a material in which at least 80% of particles are smaller than 2 mm [27]. Debris is a mixture of large and small blocks of rock, and debris motion involves multi-phase flow (20% to 80% of particles >2 mm). The properties of debris encompass the characteristics of both rigid bodies and fluids. Particle size is one of the critical factors to consider in the description of motion. When the material volume is small, and the quantity of material is large, the Eulerian description including the depth-averaged model is generally suitable for describing these motions. The Lagrangian description is suitable for large-volume and small-quantity discrete rigid body movement. The Eulerian–Lagrangian method may be suitable for the rock and soil aggregate movement.

In addition, the motion types include falling, sliding, spreading, and flowing. In falling and toppling, collisions and shearing are the major contact processes, and shearing is the main contact process in sliding, flowing, and spreading. This indicates that the movement type determines the force model in a given situation. Collision and friction are critical forces to change the movement state for falling and toppling. However, dry friction and viscosity need to be considered in sliding, spreading, and flowing. Therefore, landslide classification can guide the selection of models of landslide dynamics (Table 4).

Table 4. Relationship between models and types.

Motion	Rock	Earth	Debris
Falling and toppling	Force: collision, friction Motion: Lagrangian	Force: collision, friction Motion: Eulerian	Force: collision, friction Motion: Eulerian/Lagrangian
Sliding, spreading, and flowing	Force: dry friction and viscosity Motion: Lagrangian or Eulerian	Force: dry friction and viscosity Motion: Eulerian	Force: dry friction and viscosity Motion: Eulerian/Lagrangian

During the movement, the movement forms transform into each other. Fragmentation is a key process in rock movement and influences the whole process. Fragmentation can cause changes in movement form from falling to flowing [76]. As the number of debris increases due to fragmentation, the interactions in the debris become increasingly complex. From an energy perspective, fragmentation can result in energy dissipation [62] and drag reduction during fragmentation. After fragmentation, small debris has a lubricating effect, and large debris can be transported for a long distance. Therefore, fragmentation in long-runout landslides is very complicated [77]. The movement process gradually transforms from falling and bouncing to sliding and flowing. Therefore, a single-phase model is not suitable for these processes. We should build an episode-based multi-phase model on GIS platforms to describe different states. The best method is a rigid body model in the initial stage and a flow-like model after fragmentation. Therefore, we may use various model forms to obtain more accurate results in landslide simulation.

5. Conclusions

Based on the above analysis, we can draw the following conclusions:

(1) A suitable model with proper assumptions can reduce uncertainties and simplify calculations. We must select suitable models to simulate different types of landslides. The proposed classification can provide some guidance for model selection. Landslide classification helps us understand landslide phenomena and select suitable dynamic models.

(2) Compared with the two different models, the 3-D model can describe more details of the moving process than the depth-averaged model. Landslide runout zone, height, and speed are critical parameters in engineering design. The depth-averaged model meets the actual needs and provides engineering parameters quickly. Therefore, we can use the depth-averaged model to obtain the runout zones. If we pay more attention to the details, such as surge waves, the 3-D landslide simulation model is better.

(3) A small difference in algorithms can produce a large difference in the runout zones. We should use as many algorithms as possible to obtain the trajectory for engineering design.
(4) The number of elements, property, and material size determine the model selection. For discrete rigid body motion, models based on the Lagrangian description are suitable; for flow-like motion, models based on the Eulerian description are proper; for materials with various properties, models based on the Eulerian–Lagrangian description are best.

Author Contributions: Conceptualization, H.L. and Y.W.; methodology, Y.W.; software, Y.W. and A.T.; validation, H.L. and Y.W.; formal analysis, Y.W.; writing—original draft preparation, Y.W.; writing—review and editing, Y.W.; visualization, Y.W.; supervision, Y.W.; funding acquisition, H.L. and Y.W. All authors have read and agreed to the published version of the manuscript.

Funding: This research was supported by the Strategic Priority Research Program of the Chinese Academy of Sciences (CAS) (Grant No. XDA23090301), National Natural Science Foundation of China (Grant No. 41941019, 42041006), and the Second Tibetan Plateau Scientific Expedition and Research (STEP) program (Grant No. 2019QZKK0904).

Institutional Review Board Statement: Not applicable.

Informed Consent Statement: Not applicable.

Conflicts of Interest: The authors declare no conflict of interest.

References

1. McDougall, S. 2014 Canadian Geotechnical Colloquium: Landslide runout analysis—Current practice and challenges. *Can. Geotech. J.* **2017**, *54*, 605–620. [CrossRef]
2. Lo, C.-M.; Feng, Z.-Y.; Chang, K.-T. Landslide hazard zoning based on numerical simulation and hazard assessment. *Geomat. Nat. Hazards Risk* **2018**, *9*, 368–388. [CrossRef]
3. Pastor, M.; Blanc, T.; Haddad, B.; Petrone, S.; Sanchez Morles, M.; Drempetic, V.; Issler, D.; Crosta, G.; Cascini, L.; Sorbino, G. Application of a SPH depth-integrated model to landslide run-out analysis. *Landslides* **2014**, *11*, 793–812. [CrossRef]
4. Pedrazzini, A.; Froese, C.R.; Jaboyedoff, M.; Hungr, O.; Humair, F. Combining digital elevation model analysis and run-out modeling to characterize hazard posed by a potentially unstable rock slope at Turtle Mountain, Alberta, Canada. *Eng. Geol.* **2012**, *128*, 76–94. [CrossRef]
5. Sun, D.; Xu, J.; Wen, H.; Wang, D. Assessment of landslide susceptibility mapping based on Bayesian hyperparameter optimization: A comparison between logistic regression and random forest. *Eng. Geol.* **2021**, *281*, 105972. [CrossRef]
6. Liu, J.; Wu, Y.; Gao, X.; Zhang, X. A Simple Method of Mapping Landslides Runout Zones Considering Kinematic Uncertainties. *Remote Sens.* **2022**, *14*, 668. [CrossRef]
7. Fell, R.; Corominas, J.; Bonnard, C.; Cascini, L.; Leroi, E.; Savage, W.Z. Guidelines for landslide susceptibility, hazard and risk zoning for land-use planning. *Eng. Geol.* **2008**, *102*, 99–111. [CrossRef]
8. Glade, T.; Crozier, M.J. The nature of landslide hazard impact. In *Landslide Hazard Risk*; Wiley: Chichester, UK, 2005; pp. 43–74.
9. Lan, H.; Martin, C.D.; Lim, C.H. RockFall analyst: A GIS extension for three-dimensional and spatially distributed rockfall hazard modeling. *Comput. Geosci.* **2007**, *33*, 262–279. [CrossRef]
10. Stevens, W.D. RocFall, a Tool for Probabilistic Analysis, Design of Remedial Measures and Prediction of Rockfalls. Ph.D. Thesis, University of Toronto, Toronto, ON, Canada, 1998.
11. Dorren, L.K. Rockyfor3D (v5. 2) revealed—Transparent description of the complete 3D rockfall model. *ecorisQ Pap.* **2015**, *32*, 1–33.
12. Ouyang, C.; He, S.; Xu, Q.; Luo, Y.; Zhang, W. A MacCormack-TVD finite difference method to simulate the mass flow in mountainous terrain with variable computational domain. *Comput. Geosci.* **2013**, *52*, 1–10. [CrossRef]
13. FLO-2D. *FLO-2D Reference Manual*; FLO-2D Software Inc.: Nutrioso, AZ, USA, 2017.
14. Mergili, M.; Fischer, J.-T.; Krenn, J.; Pudasaini, S.P. R.avaflow v1, an advanced open-source computational framework for the propagation and interaction of two-phase mass flows. *Geosci. Model Dev.* **2017**, *10*, 553–569. [CrossRef]
15. Sheridan, M.F.; Stinton, A.J.; Patra, A.; Pitman, E.; Bauer, A.; Nichita, C. Evaluating Titan2D mass-flow model using the 1963 Little Tahoma peak avalanches, Mount Rainier, Washington. *J. Volcanol. Geotherm. Res.* **2005**, *139*, 89–102. [CrossRef]
16. Scaringi, G.; Fan, X.; Xu, Q.; Liu, C.; Ouyang, C.; Domènech, G.; Yang, F.; Dai, L. Some considerations on the use of numerical methods to simulate past landslides and possible new failures: The case of the recent Xinmo landslide (Sichuan, China). *Landslides* **2018**, *15*, 1359–1375. [CrossRef]
17. Chang, Y.-S.; Chang, T.-J. SPH simulations of solute transport in flows with steep velocity and concentration gradients. *Water* **2017**, *9*, 132. [CrossRef]
18. Mishra, B.; Rajamani, R.K. The discrete element method for the simulation of ball mills. *Appl. Math. Model.* **1992**, *16*, 598–604. [CrossRef]

19. Rigaux, P.; Scholl, M.; Voisard, A. *Spatial Databases: With Application to GIS*; Morgan Kaufmann: Burlington, MA, USA, 2002.
20. Heim, A. *Landslides and Human Lives (Bergsturz und Menchenleben)*; Bi-Tech Publishers: Vancouver, BC, USA, 1932; p. 196.
21. Zaruba, Q.; Mencl, V. *Landslides and Their Control*; Elsevier: Amsterdam, The Netherlands, 2014.
22. Sharpe, C. *Landslides and Related Phenomena*; Columbia University Press: New York, NY, USA, 1938.
23. Hungr, O.; Leroueil, S.; Picarelli, L. The Varnes classification of landslide types, an update. *Landslides* **2014**, *11*, 167–194. [CrossRef]
24. Varnes, D.J. Landslide types and processes. *Landslides Eng. Pract.* **1958**, *24*, 20–47.
25. Cruden, D.; Varnes, D. Landslide types and processes. In *Landslides-Investigation and Mitigation*; National Research Council Transportation Research Board Special Report 247; Turner, K.A., Schuster, R.L., Eds.; Transportation Research Board: Washington, DC, USA, 1996.
26. WP/WLI. *UNESCO Working Party for World Landslide Inventory G 1993*; BiTech Publishers Ltd.: Richmond, BC, Canada, 1993.
27. Varnes, D.J. Slope movement types and processes. *Spec. Rep.* **1978**, *176*, 11–33.
28. Dorren, L.K. A review of rockfall mechanics and modelling approaches. *Prog. Phys. Geogr.* **2003**, *27*, 69–87. [CrossRef]
29. Iverson, R.M. Landslide triggering by rain infiltration. *Water Resour. Res.* **2000**, *36*, 1897–1910. [CrossRef]
30. Wang, Y.; Xu, G. Back-Analysis of Water Waves Generated by the Xintan Landslide. In *Landslide Disaster Mitigation in Three Gorges Reservoir, China*; Springer: Berlin/Heidelberg, Germany, 2009; pp. 433–445.
31. Wu, Y.; Lan, H. Debris flow analyst (DA): A debris flow model considering kinematic uncertainties and using a GIS platform. *Eng. Geol.* **2020**, *279*, 105877. [CrossRef]
32. Iverson, R.M.; Denlinger, R.P. Flow of variably fluidized granular masses across three-dimensional terrain: 1. Coulomb mixture theory. *J. Geophys. Res. Solid Earth* **2001**, *106*, 537–552. [CrossRef]
33. Iverson, R.M. The physics of debris flows. *Rev. Geophys.* **1997**, *35*, 245–296. [CrossRef]
34. Savage, S.B.; Hutter, K. The motion of a finite mass of granular material down a rough incline. *J. Fluid Mech.* **1989**, *199*, 177–215. [CrossRef]
35. Pudasaini, S.P. A general two-phase debris flow model. *J. Geophys. Res. Earth Surf.* **2012**, *117*, 1–28. [CrossRef]
36. Denlinger, R.P.; Iverson, R.M. Granular avalanches across irregular three-dimensional terrain: 1. Theory and computation. *J. Geophys. Res. Earth Surf.* **2004**, *109*, 1–16. [CrossRef]
37. Pudasaini, S.P.; Mergili, M. A multi-phase mass flow model. *J. Geophys. Res. Earth Surf.* **2019**, *124*, 2920–2942. [CrossRef]
38. Pudasaini, S.P.; Krautblatter, M. The mechanics of landslide mobility with erosion. *Nat. Commun.* **2021**, *12*, 6793. [CrossRef]
39. Pudasaini, S.P.; Fischer, J.-T. A mechanical erosion model for two-phase mass flows. *Int. J. Multiph. Flow* **2020**, *132*, 103416. [CrossRef]
40. Worgull, M. Chapter 3—Molding Materials for Hot Embossing. In *Hot Embossing*; Worgull, M., Ed.; William Andrew Publishing: Boston, MA, USA, 2009; pp. 57–112.
41. Zeng, S.; Migórski, S. A class of time-fractional hemivariational inequalities with application to frictional contact problem. *Commun. Nonlinear Sci. Numer. Simul.* **2018**, *56*, 34–48. [CrossRef]
42. Evans, S.; Hungr, O. The assessment of rockfall hazard at the base of talus slopes. *Can. Geotech. J.* **1993**, *30*, 620–636. [CrossRef]
43. Pfeiffer, T.J. Rockfall Hazard Analysis Using Computer Simulation of Rockfalls. Ph.D. Thesis, Colorado School of Mines, Golden, CO, USA, 1989.
44. Spadari, M.; Kardani, M.; De Carteret, R.; Giacomini, A.; Buzzi, O.; Fityus, S.; Sloan, S. Statistical evaluation of rockfall energy ranges for different geological settings of New South Wales, Australia. *Eng. Geol.* **2013**, *158*, 57–65. [CrossRef]
45. Guzzetti, F.; Crosta, G.; Detti, R.; Agliardi, F. STONE: A computer program for the three-dimensional simulation of rock-falls. *Comput. Geosci.* **2002**, *28*, 1079–1093. [CrossRef]
46. Li, L.; Lan, H. Probabilistic modeling of rockfall trajectories: A review. *Bull. Eng. Geol. Environ.* **2015**, *74*, 1163–1176. [CrossRef]
47. Tskhakaya, D.; Matyash, K.; Schneider, R.; Taccogna, F. The Particle-In-Cell Method. *Contrib. Plasma Phys.* **2007**, *47*, 563–594. [CrossRef]
48. Pastor, M.; Haddad, B.; Sorbino, G.; Cuomo, S.; Drempetic, V. A depth-integrated, coupled SPH model for flow-like landslides and related phenomena. *Int. J. Numer. Anal. Methods Geomech.* **2009**, *33*, 143–172. [CrossRef]
49. Lin, C.; Pastor, M.; Li, T.; Liu, X.; Qi, H.; Lin, C. A SPH two-layer depth-integrated model for landslide-generated waves in reservoirs: Application to Halaowo in Jinsha River (China). *Landslides* **2019**, *16*, 2167–2185. [CrossRef]
50. Longo, A.; Pastor, M.; Sanavia, L.; Manzanal, D.; Martin Stickle, M.; Lin, C.; Yague, A.; Tayyebi, S.M. A depth average SPH model including μ (I) rheology and crushing for rock avalanches. *Int. J. Numer. Anal. Methods Geomech.* **2019**, *43*, 833–857. [CrossRef]
51. Agliardi, F.; Crosta, G.B. High resolution three-dimensional numerical modelling of rockfalls. *Int. J. Rock Mech. Min. Sci.* **2003**, *40*, 455–471. [CrossRef]
52. Woltjer, M.; Rammer, W.; Brauner, M.; Seidl, R.; Mohren, G.; Lexer, M. Coupling a 3D patch model and a rockfall module to assess rockfall protection in mountain forests. *J. Environ. Manag.* **2008**, *87*, 373–388. [CrossRef]
53. Bartelt, P.; Bieler, C.; Bühler, Y.; Christen, M.; Christen, M.; Dreier, L.; Gerber, W.; Glover, J.; Schneider, M.; Glocker, C. *RAMMS: Rockfall User Manual v1. 6*; WSL Institute for Snow and Avalanche Research SLF: Davos, Switzerland, 2016.
54. Matas, G.; Lantada, N.; Corominas, J.; Gili, J.; Ruiz-Carulla, R.; Prades, A. RockGIS: A GIS-based model for the analysis of fragmentation in rockfalls. *Landslides* **2017**, *14*, 1565–1578. [CrossRef]
55. Christen, M.; Kowalski, J.; Bartelt, P. RAMMS: Numerical simulation of dense snow avalanches in three-dimensional terrain. *Cold Reg. Sci. Technol.* **2010**, *63*, 1–14. [CrossRef]

56. Molinari, M.E.; Cannata, M.; Begueria, S.; Ambrosi, C. GIS-based Calibration of MassMov2D. *Trans. GIS* **2012**, *16*, 215–231. [CrossRef]
57. Wu, Y.; Lan, H. Landslide Analyst—A landslide propagation model considering block size heterogeneity. *Landslides* **2019**, *16*, 1107–1120. [CrossRef]
58. de'Michieli Vitturi, M.; Esposti Ongaro, T.; Lari, G.; Aravena, A. IMEX_SfloW2D 1.0: A depth-averaged numerical flow model for pyroclastic avalanches. *Geosci. Model Dev.* **2019**, *12*, 581–595. [CrossRef]
59. Berger, M.J.; George, D.L.; LeVeque, R.J.; Mandli, K.T. The GeoClaw software for depth-averaged flows with adaptive refinement. *Adv. Water Resour.* **2011**, *34*, 1195–1206. [CrossRef]
60. Hungr, O.; McDougall, S. Two numerical models for landslide dynamic analysis. *Comput. Geosci.* **2009**, *35*, 978–992. [CrossRef]
61. Amicarelli, A.; Manenti, S.; Albano, R.; Agate, G.; Paggi, M.; Longoni, L.; Mirauda, D.; Ziane, L.; Viccione, G.; Todeschini, S. SPHERA v. 9.0. 0: A Computational Fluid Dynamics research code, based on the Smoothed Particle Hydrodynamics mesh-less method. *Comput. Phys. Commun.* **2020**, *250*, 107157. [CrossRef]
62. Li, X.; Tang, X.; Zhao, S.; Yan, Q.; Wu, Y. MPM evaluation of the dynamic runout process of the giant Daguangbao landslide. *Landslides* **2021**, *18*, 1509–1518. [CrossRef]
63. Crosta, G.; Imposimato, S.; Roddeman, D. Numerical modelling of entrainment/deposition in rock and debris-avalanches. *Eng. Geol.* **2009**, *109*, 135–145. [CrossRef]
64. Jasak, H.; Jemcov, A.; Tukovic, Z. OpenFOAM: A C++ library for complex physics simulations. In Proceedings of the International Workshop on Coupled Methods in Numerical Dynamics, Dubrovnik, Croatia, 19–21 September 2007; pp. 1–20.
65. Ramachandran, P. PySPH: A reproducible and high-performance framework for smoothed particle hydrodynamics. In Proceedings of the 15th Python in Science Conference, Austin, TX, USA, 11–17 July 2016; pp. 127–135.
66. Wang, J.-S.; Ni, H.-G.; He, Y.-S. Finite-difference TVD scheme for computation of dam-break problems. *J. Hydraul. Eng.* **2000**, *126*, 253–262. [CrossRef]
67. Ming, H.T.; Chu, C.R. Two-dimensional shallow water flows simulation using TVD-MacCormack scheme. *J. Hydraul. Res.* **2000**, *38*, 123–131. [CrossRef]
68. Delaney, K.B.; Evans, S.G. The 2000 Yigong landslide (Tibetan Plateau), rockslide-dammed lake and outburst flood: Review, remote sensing analysis, and process modelling. *Geomorphology* **2015**, *246*, 377–393. [CrossRef]
69. Guo, C.; Montgomery, D.R.; Zhang, Y.; Zhong, N.; Fan, C.; Wu, R.; Yang, Z.; Ding, Y.; Jin, J.; Yan, Y. Evidence for repeated failure of the giant Yigong landslide on the edge of the Tibetan Plateau. *Sci. Rep.* **2020**, *10*, 14371. [CrossRef]
70. Kang, C. Modelling Entrainment in Debris Flow Analysis for Dry Granular Material. *Int. J. Geomech.* **2016**, *17*, 04017087. [CrossRef]
71. Zhuang, Y.; Yin, Y.; Xing, A.; Jin, K. Combined numerical investigation of the Yigong rock slide-debris avalanche and subsequent dam-break flood propagation in Tibet, China. *Landslides* **2020**, *17*, 2217–2229. [CrossRef]
72. Voellmy, A. Die Zerstorungs-Kraft von lawinen, Sonderdruck aus der Schweiz. *Bauzeitung* **1955**, *73*, 159–165.
73. Morgenstern, N.R. The evaluation of slope stability—A 25 year perspective. In Proceedings of the Stability and Performance of Slopes and Embankments II, Berkeley, CA, USA, 29 June–1 July 1992; pp. 1–26.
74. Leroueil, S.; Locat, J.; Vaunat, J.; Picarelli, L.; Lee, H. Geotechnical characterization of slope movements. In Proceedings of the Landslides, Trondheim, Norway, 17–21 June 1996; pp. 53–74.
75. Bates, R.L.; Jackson, J.A. *Dictionary of Geological Terms*; Anchor Books: New York, NY, USA, 1984; Volume 584.
76. Ruiz-Carulla, R.; Corominas, J.; Mavrouli, O. A fractal fragmentation model for rockfalls. *Landslides* **2017**, *14*, 875–889. [CrossRef]
77. Davies, T.; McSaveney, M.; Hodgson, K. A fragmentation-spreading model for long-runout rock avalanches. *Can. Geotech. J.* **1999**, *36*, 1096–1110. [CrossRef]

applied sciences

MDPI

Article

The Effects of Rainfall, Soil Type and Slope on the Processes and Mechanisms of Rainfall-Induced Shallow Landslides

Yan Liu, Zhiyuan Deng and Xiekang Wang *

State Key Laboratory of Hydraulics and Mountain River Engineering, Sichuan University, Chengdu 610065, China; liuyan2021scu@gmail.com (Y.L.); dengzhiyuan2019@163.com (Z.D.)
* Correspondence: wangxiekang@scu.edu.cn

Abstract: Landslides are a serious geohazard worldwide, causing many casualties and considerable economic losses every year. Rainfall-induced shallow landslides commonly occur in mountainous regions. Many factors affect an area's susceptibility, such as rainfall, the soil, and the slope. In this paper, the effects of rainfall intensity, rainfall pattern, slope gradient, and soil type on landslide susceptibility are studied. Variables including soil volumetric water content, matrix suction, pore water pressure, and the total stress throughout the rainfall were measured. The results show that, under the experimental conditions of this paper, no landslides occurred on a 5° slope. On a 15° slope, when the rainfall intensity was equal to or less than 80 mm/h with a 1 h duration, landslides also did not happen. With a rainfall intensity of 120 mm/h, the rainfall pattern in which the intensity gradually diminishes could not induce landslides. Compared with fine soils, coarser soils with gravels were found to be prone to landslides. As the volumetric water content rose, the matrix suction declined from the time that the level of infiltration reached the position of the matrix. The pore water pressure and the total stress both changed drastically either immediately before or after the landslide. In addition, the sediment yield depended on the above factors. Steeper slopes, stronger rainfall, and coarser soils were all found to increase the amount of sediment yield.

Keywords: landslides; artificial rainfall; grain size; rainfall pattern; pore water pressure

check for updates

Citation: Liu, Y.; Deng, Z.; Wang, X. The Effects of Rainfall, Soil Type and Slope on the Processes and Mechanisms of Rainfall-Induced Shallow Landslides. *Appl. Sci.* **2021**, *11*, 11652. https://doi.org/10.3390/app112411652

Academic Editors: Ricardo Castedo, Miguel Llorente Isidro and David Moncoulon

Received: 3 November 2021
Accepted: 3 December 2021
Published: 8 December 2021

1. Introduction

Landslides refer to the geological phenomenon of rock and soil mass sliding along a slope, these are a type of natural hazard that is widely distributed throughout the world [1,2]. Landslides cause tens of billions of dollars of economic losses and serious casualties worldwide every year [3]. In areas with complex geological conditions, landslides occur more frequently [4–7].

In mountain regions, landslides are often triggered by intensive rainfall [8,9]. Rainfall and infiltration enhance moisture content, which further decreases the matrix suction and soil shear strength [10,11]. Rainfall characteristics are often used as criteria for landslide occurrence [12,13]. The rainfall's intensity and pattern influence the characteristics of landslides [14–16]. To date, there have been many studies on the effects of rainfall on landslides. Research methods include in situ experiments, laboratory experiments, and numerical simulations [17,18]. The artificial rainfall test is an effective method when used to study rainfall-induced landslides [19].

Slopes with different soil compositions respond differently to rainfall. Fine particle migration leads to pore blockage [20] and the soil composition of slopes is closely related to their landslide susceptibility. Research has showed that a prerequisite for landslides to occur is that the clay percentage of the soil is higher than 2.5% [21]. The soils in a region where earthquakes happen often have many coarse particles, such as gravel. The proportion of gravel in soils has a great influence on the density and void ratio, which determines the timing and type of landslides [22]. The failure mode is closely related to the grain size [23].

The type of landslide is usually a gully failure when small gravel content is present and usually a "layer-by-layer sliding" failure when large gravel content is present [24].

Rainfall and infiltration change the soil characteristics greatly [25]. Some physical qualities of the soil change during rainfall, such as the water content, matrix suction, pore water press, and total stress [26,27]. The volumetric water content is the ratio of the volume of water to the unit volume of soil and it increases during rainfall. Matrix suction is a sensitive parameter when unsaturated soils encounter rainfall and it is differently affected for soils with varying levels of rainfall infiltration [28,29]. The pore water press increases during rainfall and reduces the soil's shear strength. Slopes with greater inclination have larger pore water press [30]. Total stress is the basis of stability analyses that are used for calculating the factor of safety [31]. However, all of the above studies measured one or several of the physical quantities of the soil and lacked a comprehensive reflection of the changes in soil properties during landslides.

The landslide is a typical example of gravity-based erosion. The sediment yield of landslides varies with the presence of a number of factors [32,33]. The time scale of the impact of a landslide on the sediment yield in the basin is large. To comprehensively analyze sediment transport, it is necessary to investigate the landslide history of the basin for at least the preceding 100 years [34]. The landslide sediment yield of soil that is affected by an earthquake under rainfall has been studied and quantified [35]. One model that is used to express the contribution of shallow landslides to sediment yield as a rainfall characteristic function has been established [36]. Another model called SHEETRAN was established in order to analyze the impact of rainfall on landslides and sediment transport. It was applied to the Valsassina Basin, which is a wide glaciated valley with a U-shaped profile. The superficial deposits that are found on the valley's slopes consist of calcareous-dolomitic chaotic material with loose and sharp-edged fragments. [37–39]. The effects of rainfall on sediment yield are obvious and the sediment yield increases with an increase in the gravel percentage [40,41].

Therefore, the purpose of this paper is to make clear how each factor influences landslide occurrences. A series of physical model tests that were carried out by an artificial rainfall system was performed in order to investigate the process and mechanics of slope failure. The rainfall intensity, rainfall pattern, soil type, and slope gradients each play a unique role in slope stability. In the process, the main physical parameters of the soil, including volumetric water content, matrix suction, pore water pressure, and total stress, were all measured. Based on the detailed measured data, we were able to derive the relationship between these physical qualities and the timing of landslides. Moreover, the sediment yield of landslides was quantified and the dependence of the yield on the test variables was analyzed. The present experimental results contribute to improving the understanding of landslide mechanisms and mitigating landslide disasters.

2. Materials and Methods

2.1. Experimental System

The artificial rainfall test site that was used in this study is located in the State Key Laboratory of Hydraulics and Mountain River Engineering of Sichuan University. The system is composed of a reservoir, water pump, water delivery pipe, rain gauge, electromagnetic flowmeter, nozzle, and valve (Figure 1). It is controlled by computer software and can be self-adjusted automatically. When the automatic adjustment mode is turned on, the water pressure and valve opening can be adjusted automatically in order to hold or change the rainfall intensity. A water content sensor, tensiometer, earth press cell, and pore water pressure sensor were used in the test (Figure 2c). The flume that was used for the test was made of impermeable transparent polymer plastic material, with a length of 2 m, a width of 0.3 m, and a height of 0.8 m. The flume was placed horizontally. The three slope angles that were set in this research were 5°, 15°, and 30°. The flume, soil, and instruments that were used are shown in Figure 2.

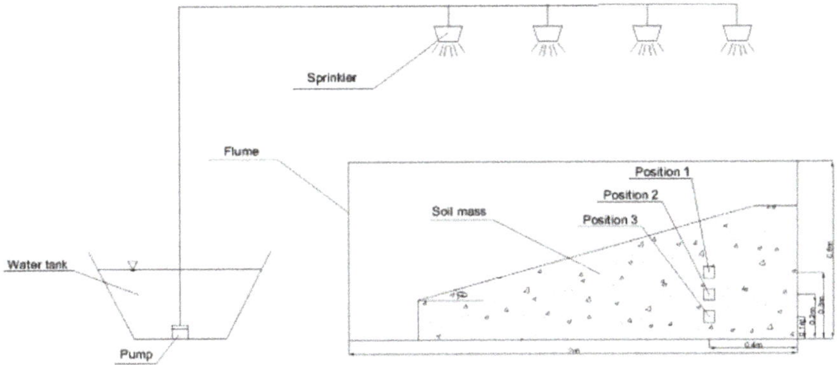

Figure 1. Sketch of experimental setup.

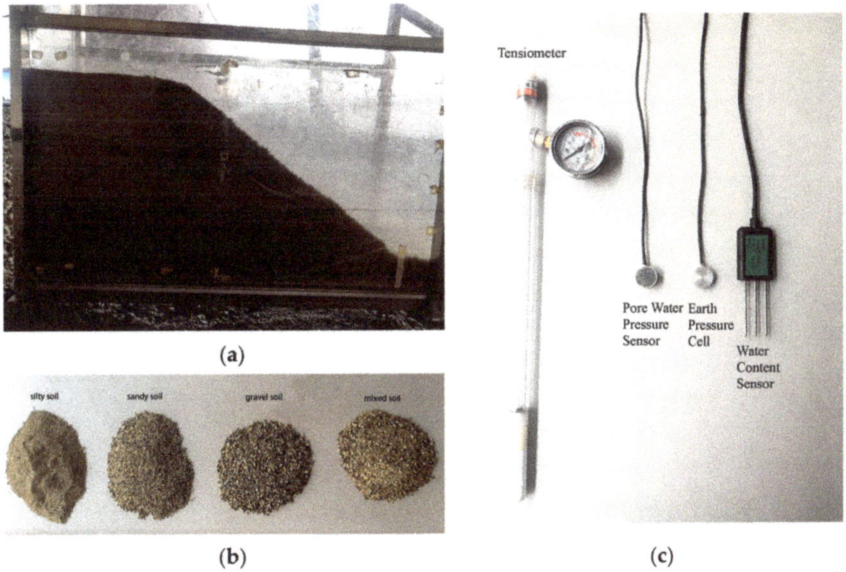

Figure 2. Tested samples and experimental apparatus: (**a**) side view of the slope; (**b**) soil samples; (**c**) measuring apparatuses.

2.2. Experimental Program

The experiment was set up with four variables: rainfall intensity, rainfall pattern, soil type, and slope gradient (Table 1). Each variable was changed only when the other variables were kept constant in order to study the relationship between that variable and landslide susceptibility. Four values of 40 mm/h, 80 mm/h, 120 mm/h, and 160 mm/h were taken for the rainfall intensity variable when the rainfall pattern was uniform (Tests no. 1–4). The rainfall pattern variable was set to I, II, III, and IV (test no. 5–7 and 3), these rainfall intensity change processes are shown in Figure 3.

In tests no. 8–10 and 3, four soil types were used. Three of the soil types were mainly composed of silt, sand, and gravel, respectively, and the mixed type was a 1:1:1 mixture of the aforementioned three. The soil compositions were set according to the common soil types that are seen in Min Jiang River basin in southwest China. The soil that was used for the tests was collected in nature through systemic screening and mixed. The soil compositions are listed in Table 2 and the grain size distribution curves of the soil types are shown in Figure 4.

Table 1. Test variables.

Test No.	Rainfall Intensity (mm/h)	Rainfall Pattern	Soil Type	Slope Gradient (°)
1	40	IV	Mixed soil	15
2	80	IV	Mixed soil	15
3	120	IV	Mixed soil	15
4	160	IV	Mixed soil	15
5	120	I	Mixed soil	15
6	120	II	Mixed soil	15
7	120	III	Mixed soil	15
8	120	IV	Silty soil	15
9	120	IV	Sandy soil	15
10	120	IV	Gravel soil	15
11	120	IV	Mixed soil	5
12	120	IV	Mixed soil	30

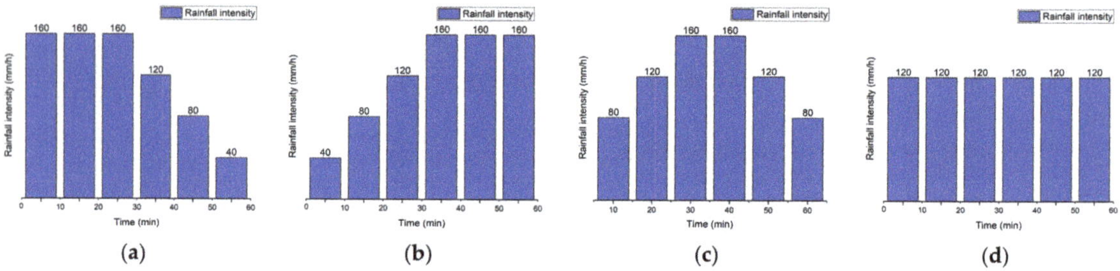

Figure 3. Rainfall patterns: (**a**) Pattern I; (**b**) Pattern II; (**c**) Pattern III; (**d**) Pattern IV.

Table 2. Soil composition.

Soil Type	Clay (%)	Silt (%)	Sand (%)	Gravel (%)	D_{50} (mm)	Initial Volume Moisture Content (%)	Initial Dry Density (g/cm^3)
Silty soil	10.2	68.6	21.2	0	0.021	9.1	1.88
Sandy soil	7.1	18.5	74.4	0	0.71	8.2	1.74
Gravel soil	8.8	16.9	18.6	55.7	2.54	6.5	1.50
Mixed soil	10.2	24.9	36.7	28.2	1.68	7.7	1.61

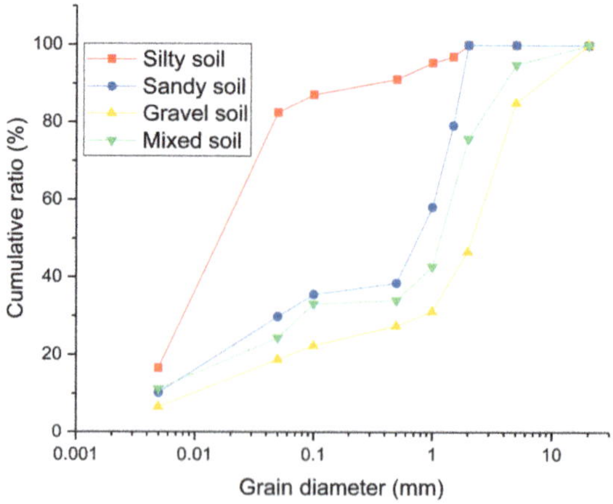

Figure 4. Grain size distribution curves.

2.3. Test Procedures

Before the test began, the soil was filled into the flume, with each layer being 10 cm deep. The soil was paved to the same thickness in each layer and then knocked evenly with wood blocks. Measuring instruments were embedded in the positions that are shown in Figure 1. After the preparation was completed, the test started with the commencement of the rainfall. Each rainfall lasted 1 h. At the end of a test, the amount of sediment that was yielded by the landslide was measured.

3. Results

3.1. Slope Instability Processes under Different Rainfall Intensities

When a slope encounters rainfall with different intensities, its water content, matrix suction, pore water pressure, and total stress exhibit different change processes. Water content is a basic parameter that is used to describe soil's properties. The volumetric water content of the tests with higher rainfall intensity were found to rise earlier and the matrix suction declined earlier, too. When the rainfall intensity was 160 mm/h, the volumetric water content reached its maximum at about 40 min. When the rainfall intensity was 120 mm/h, the volumetric water content reached its maximum at about 55 min. The water gradually penetrated downward from the soil's surface, so the volumetric water content deeper down within the soil was found to increase later than that of the position near the surface.

Matrix suction is an important parameter of the mechanical properties of unsaturated soils. The pores of unsaturated soil are filled with water and air. The water–air interface has surface tension. In unsaturated soil, through capillary action, the pore water pressure under the bent liquid surface is less than the pore air pressure. The shrinkage membrane is subjected to air pressure greater than the water pressure, and this pressure difference is called matrix suction. The processes of matrix suction that were observed at the different depths under the different rainfall intensities are shown in Figure 5b. It can be seen from the figure that the change curve of matrix suction was divided into three stages, namely the initial stage, steep fall stage, and stable stage. Taking, as an example, the matrix suction change process that was observed at position I when the rainfall intensity is 160 mm/h, it was noted that within 22 min of the beginning of rainfall, the change in the soil matrix suction was not obvious. At 22–42 min, the matrix suction of the slope soil decreased abruptly. As the infiltration of the rainfall continued to increase, the soil matrix suction decreased to the minimum and became stable.

Pore water pressure is the pressure of the groundwater that is present in soil or rock, which acts between particles or pores and is an important indicator of stress changes in the soil. The variation of pore water pressure that was observed at the different slope locations under the different rainfall intensities is shown in Figure 5c. The pore water pressure variation curves during the test were observed in three stages: the initial stage, surging stage, and slowly increasing stage. It was specifically noted that a greater intensity of rainfall led to a shorter duration of the initial stage. For example, the initial stage of pore water pressure at position I lasted about 8 min when the rainfall intensity was 160 mm/h, and about 24 min when the rainfall intensity was 40 mm/h.

The total stress is the total force per unit area that is acting within a mass of soil. It increases with the greater depth of the measurement point. Different rainfall intensities lead to different variation processes of total stress. When the rainfall intensity was small, the total stress started to increase a short time after the rainfall began and the increase process was relatively smooth. When the rainfall intensity was higher, the total stress started to increase at the beginning of the rainfall.

The intensity of the rainfall is closely related to the occurrence of a landslide. When the intensity of rainfall was 40 mm/h and 80 mm/h, no landslide occurred. For the case of a rainfall intensity of 120 mm/h, the time of the initial landslide occurrence was about 47 min. For the case of a rainfall intensity of 160 mm/h, the time of the initial landslide occurrence was about 40 min.

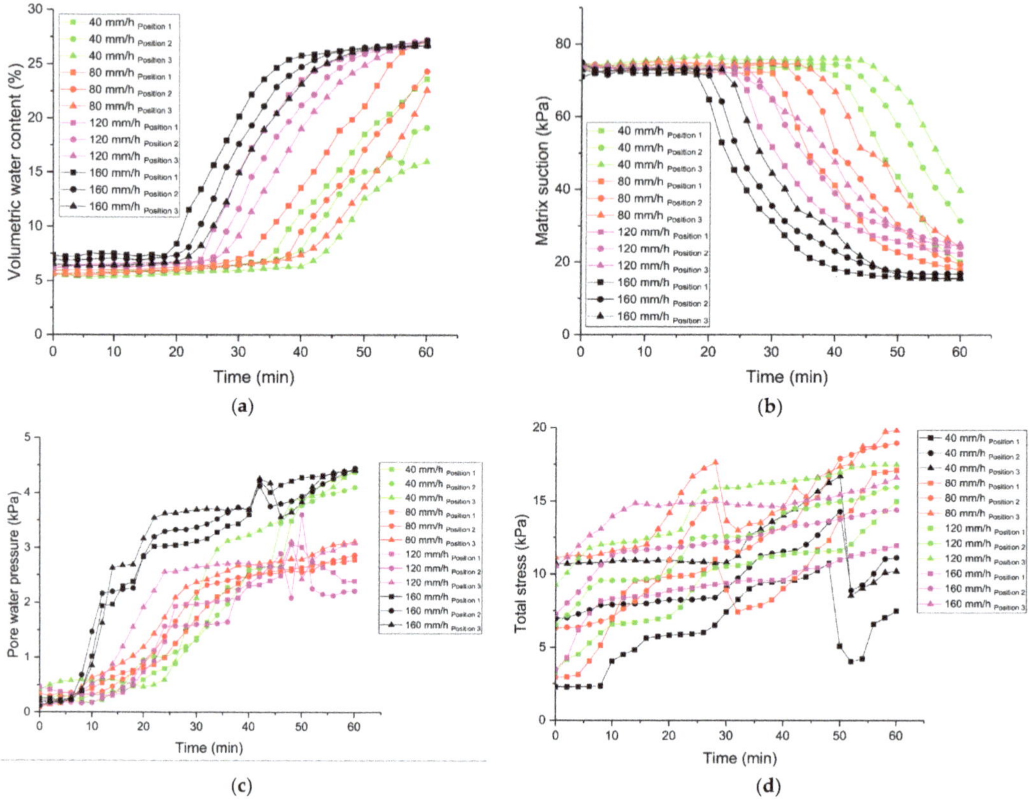

Figure 5. Variation in measured data when rainfall intensity is different: (**a**) water content; (**b**) matrix suction; (**c**) pore water pressure; (**d**) total stress.

3.2. Slope Instability Processes under Different Rainfall Patterns

The changes in the water content, matrix suction, pore water pressure, and total stress that were observed under the different rainfall patterns are shown in Figure 6. The volumetric water content that was measured at the measurement points did not change for a period of time after the onset of rainfall. The volumetric water content of rainfall pattern I started to increase at the earliest time point. Later, the volumetric water content of rainfall patterns IV and III started to increase. The volumetric water content of rainfall pattern II started to increase last. The rate of the increase in the volumetric water content in the rainfall pattern I test decreased gradually. The rate of increase in the rainfall pattern IV test remained constant. The rate of increase in rainfall pattern II and rainfall pattern III gradually increased.

The matrix suction of the tests with rainfall pattern I and rainfall pattern IV began to diminish earlier than the matrix suction of pattern II and pattern III. The matrix suction curve of the tests with rainfall pattern I and rainfall pattern IV began to enter the attenuation stage at about 20 min, while those of pattern II and pattern III began to enter the attenuation stage at about 35 min. The rainfall of patterns I and IV was relatively large in the initial stage. The attenuation processes of these tests were similar. In the stable stage, the matrix suction of the test with patterns I and IV was the smallest, and those of patterns II and III were relatively large.

In the changing process of pore water pressure, the slow-changing stages of the rainfall pattern I and IV tests lasted the shortest time. The rate of the curve was very large at the beginning. After the surge stage, it tended to be smooth. The slow change phase of rainfall patterns II and III lasted longer and the pore water pressure increased slowly throughout the test process. As the rainfall continued, the rainfall intensity of rainfall patterns II and III gradually increased and the pore water pressure changed into the surge stage. The pore water pressure of rainfall patterns I, III, and IV increased slightly in the stable stage.

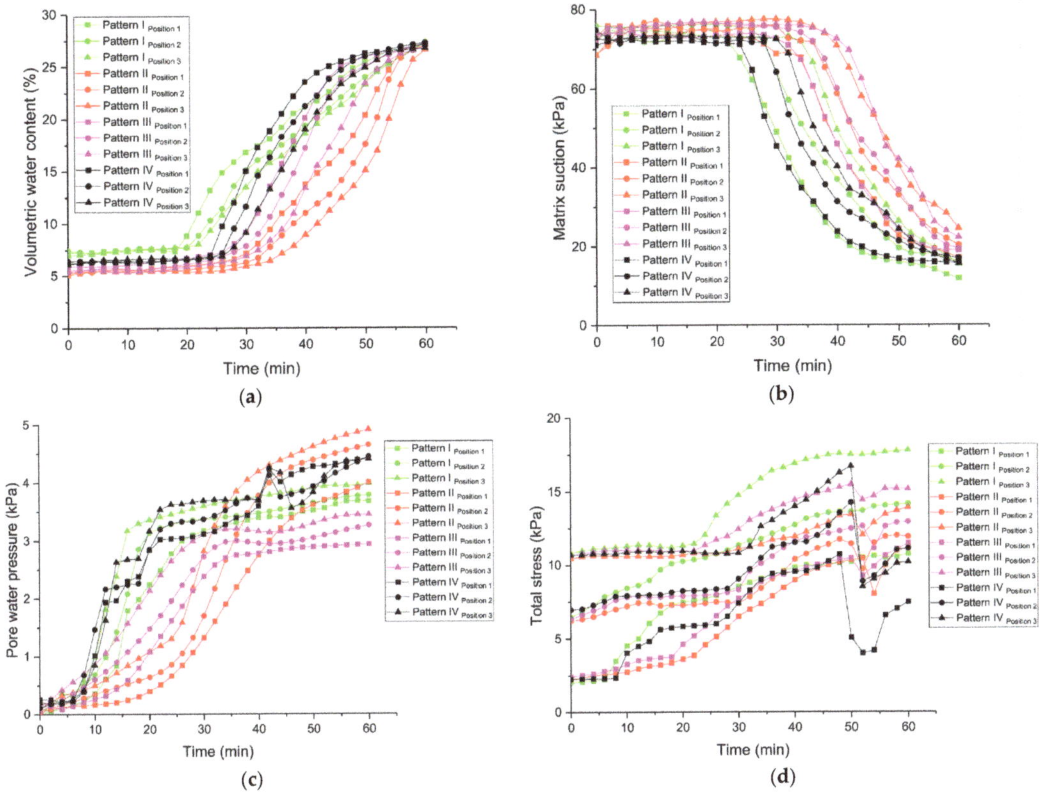

Figure 6. Variation in measured data when rainfall patterns are different: (**a**) water content; (**b**) matrix suction; (**c**) pore water pressure; (**d**) total stress.

Different rainfall patterns cause different response processes of total stress. The soil response under rainfall pattern I was rapid. The total stress began to rise at 5 min. The growth rate slowed down at about 20 min. The total stress of the tests with rainfall pattern II and rainfall pattern III began to increase by close to the 20 min mark. Compared with rainfall pattern I, these patterns displayed lag. The total stress of the tests with rainfall patterns II and IV suddenly decreased in the later rainfall stage, due to the landslide.

The occurrence of landslides varied under the different rainfall patterns: no landslides occurred in rainfall pattern I, while landslides did occur in rainfall patterns II, III, and IV. The first landslide of the test with rainfall pattern II occurred at about 42 min. The first landslide of the test with rainfall pattern III occurred at about 45 min. The first landslide of the test with rainfall pattern IV occurred at about 47 min.

3.3. Slope Instability Processes with Different Soil Types

Figure 7 shows the variation of water content, matrix suction, pore water pressure, and total stress of the slopes with different soil compositions. For the same rainfall intensity, there was no significant difference in the time at which the volumetric water content started to rise for the different soils. The volumetric water content for all of them started to increase at about 25 min. However, there were large differences in the final volumetric water content of the different soil types. The upper limit of the volumetric water content was the highest for the silty soil, at about 36%. The upper limit of the volumetric water content for the mixed soil was about 30%. The sandy soil had an upper volumetric water content of about 27%. The gravel soil had the lowest upper limit of volumetric water content, which was about 18%. The volumetric water content of the silty and mixed soils rose very quickly and reached their maximum water content at about 35 min. The volumetric water content of the sandy soil rose over a longer period of time and the rate gradually decreased. The volumetric water content of the gravel soil plateaued after a small increase.

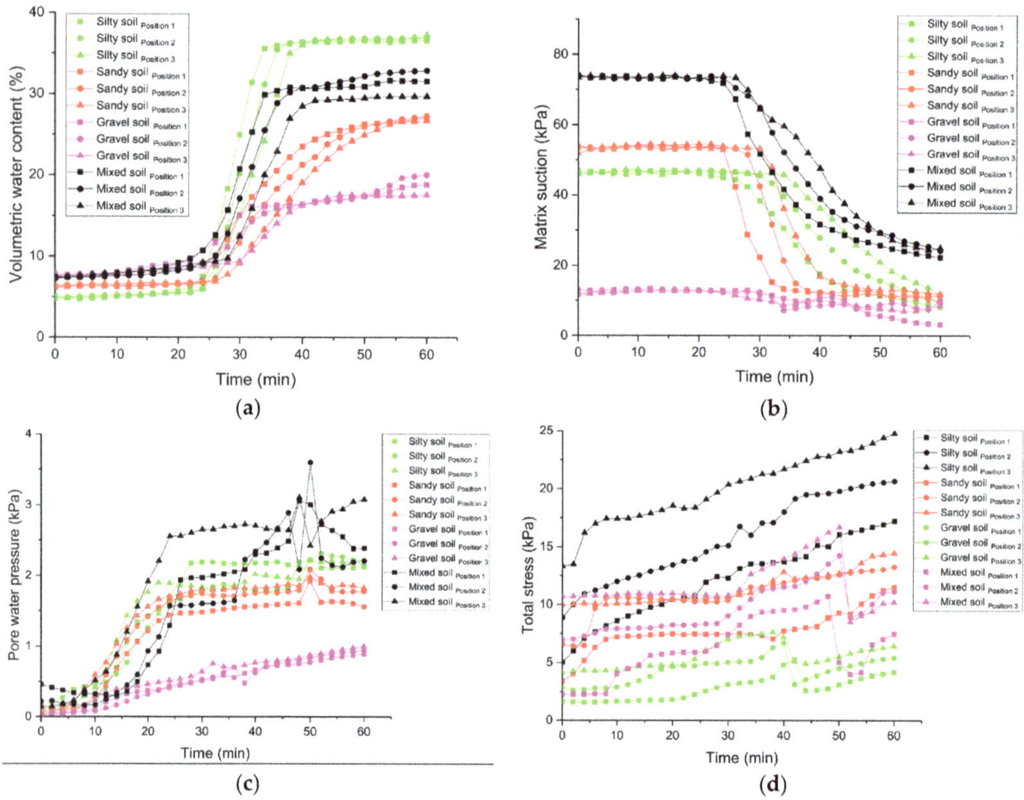

Figure 7. Variation in measured data when soil types are different: (**a**) water content; (**b**) matrix suction; (**c**) pore water pressure; (**d**) total stress.

The initial matrix suction of the different soils showed great differences. The initial matrix suction of the fine-grained soils was higher than that of the coarse-grained soils. The initial matrix suction of the silty soil was about 78 kPa, that of the sandy soil was about 52 kPa, that of the gravel soil was only about 14 kPa, and that of the mixed soil was about 48 kPa. The changing processes of matrix suction were also different. The matrix suction of the silty soil at location I entered the diminished stage at about 28 min, decreased sharply in the time period of 30 ~ 40 min, and finally stabilized at about 20 kPa. The matrix suction of the sandy soil and mixed soil decreased at about 25 min and the matrix suction of the sandy soil reached the stable stage at 35 min with a value of 18 kPa. The matrix suction of the mixed soil continued to decline and finally decreased to about 8 kPa. The matrix suction of the gravel soil was very small and did not change significantly in the first 30 min of rainfall. It entered the fluctuating stage at 30 min.

The soil type had a strong influence on the changing process of pore water pressure. The increasing process of pore water pressure was longer for the mixed soil. The peak occurred between 30 and 40 min and it reached about 4 kPa. The pore water pressure changes that were observed in the silty, sandy, and gravel soils can be divided into two stages: a rapid-increase stage and a stable stage. The increasing stage was seen in approximately the first 20 min of the test. When the increasing stage was over, the pore water pressure of the silty soil was greater than that of the sandy soil, and that of the sandy soil was greater than that of the gravel soil.

The different soil types had different total stress values. At position I, the total stress was highest in the silty soil. It increased slowly during the time period from 10 to 40 min and remained constant at about 10 kPa during the time period from 40 to 60 min. The total stress of the sandy soil varied little in the first 20 min, increased and fluctuated in the time period from 20–50 min, and decreased abruptly at about 50 min due to the landslide. The total stress of the gravel soil was the smallest and it increased only slightly during the rainfall and then stabilized around 4 kPa. The total stress variation pattern of the mixed soil was similar to that of the sandy soil.

Soil type is a key factor for determining the stability of slopes. Under the fixed rainfall process and slope conditions that were set in these tests, no landslide occurred when the soil type was silty or sandy. Landslides were generated when the soil type was gravelly or mixed. The initial landslide occurrence time was about 39 min for the gravel soil and about 47 min for the mixed soil.

3.4. Slope Instability Processes on Differently Angled Slopes

Figure 8 shows the variation process of matrix suction when the slope gradients were different. The different slope gradients had little influence on the variation of the volume water content, the variation of each test was generally similar. The rise speed rate of the 30° slope was less than that of the 5° slope and 15° slope.

The changing process of the matrix suction of the different gradient slopes was generally similar. For example, in a test with a 30° slope at position I it was observed that, in the first 25 min of rainfall, the change in the matrix suction was not obvious. At 26–46 min, the matrix suction at this location decreased abruptly. As the infiltration of the rainfall continued to increase, the soil matrix suction dropped to the minimum level and then stabilized. Since the precipitation on the steep slope was largely converted into surface runoff during the rainfall-infiltration process, the infiltration volume was smaller than that of the gentle slope at the same time. The matrix suction started to decrease earlier for the tests on the gentle slope than those on the steep slope.

The increasing and stable stages of pore water pressure on the slopes with different gradients had large differences. Taking location I as an example, when the slope was gentle (5°), the rising curve of the pore water pressure resembled a convex function. That is, the function slope was larger at the beginning and the period from 5–12 min accommodated the concentrated rising section of the pore water pressure process. When the slope was 15°, the growth of the pore water pressure was approximately linear. When the slope was 30°, the curve of the pore water pressure increase was similar to the concave function, which decreased slightly in the time period from 0–15 min and increased rapidly in the time period from 20–25 min. In the stable stage, the pore water pressure on the 15° slope was the largest, around 3.2 kPa. The pore water pressure on the 30° slope was about 2.5 kPa and the pore water pressure on the 5° slope was stable at about 2.8 kPa.

The slope had a significant effect on the changing process of the total stress. For the case of a 5° slope, the total stress at position I and position II increased linearly throughout the test. For the case of the 15° slope, the total stress at position II changed very little in the first 25 min, increased significantly in the time period from 25 to 45 min, and suddenly decreased at about 47 min due to a landslide. For the case of the 30° slope, the total stress changed little in the first 40 min at position II, increased continuously from 40 min, and decreased suddenly at around 50 min due to the landslide.

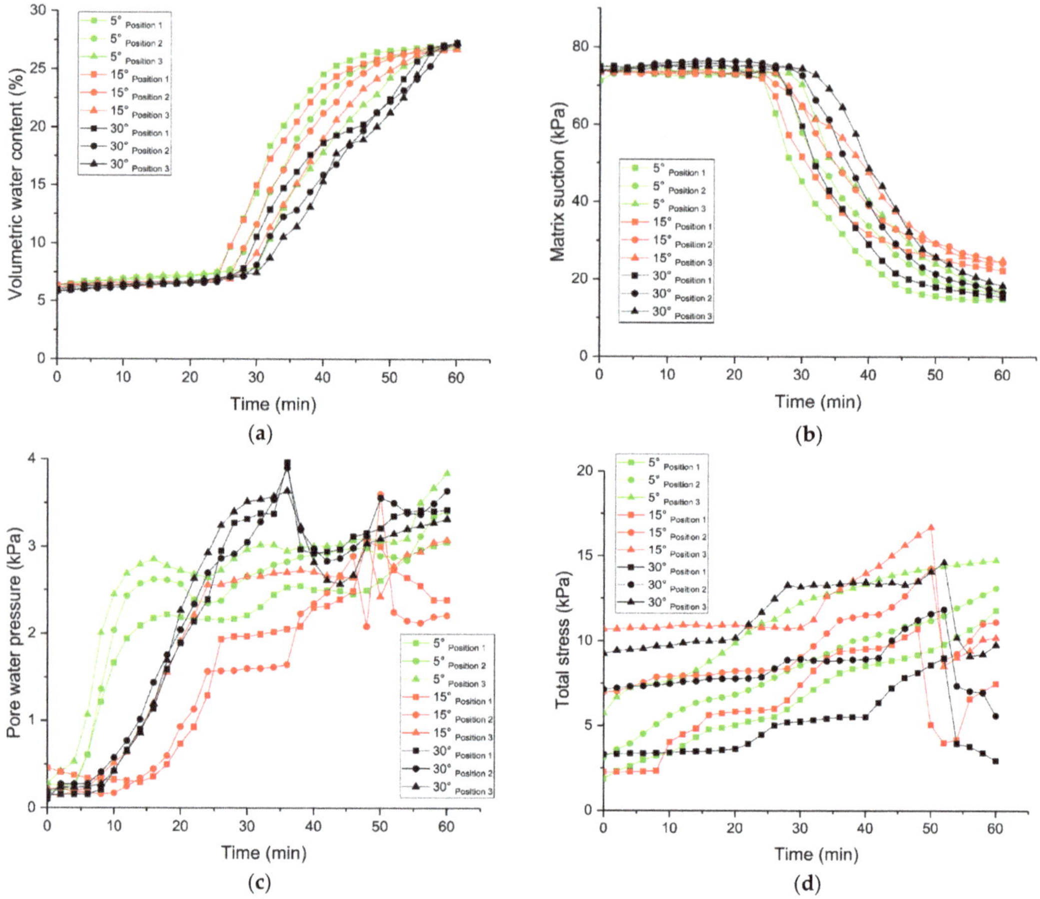

Figure 8. Variation in measured data when slope gradients are different: (**a**) water content; (**b**) matrix suction; (**c**) pore water pressure; (**d**) total stress.

The gradient of the slope was a decisive factor for the occurrence of landslides. No landslide occurred on the 5° slope. Landslides occurred on both the 15° and 30° slopes. The initial landslide occurred at about 47 min on the 15° slope and about 33 min on the 30° slope.

4. Discussion

A series of physical model tests for rainfall-induced shallow slides have been carried out then and reported on in this paper. The soil samples were made according to the natural soil in the Min Jiang River basin in southwestern China. The different soil types have unique characteristics. Silty soils are usually well-aggregated, but the aggregates break down rapidly when wetted, allowing non-aggregated soil particles to be easily transported [42]. For sandy soil, there is a clear linkage between landslides and sediment yield [43]. Seismic activity will generate a large amount of gravel soils and make the region susceptible to geohazards [22]. Therefore, the impact of the typical soil constitution on the landslides and sediment yield is analyzed in this paper. The sediment yield of the landslides that resulted from tests with different rainfall intensities, rainfall patterns, slope gradients, and soil particle compositions is shown in Table 3. All of the soil that slid down came from within 10 cm of the surface layer of the original mass, so the mass of the soil that slid down by landslide, as a percentage of the total mass of the upper 10 cm layer of the original slope, was used to measure the severity of the landslide.

Table 3. Sediment yield from landslides.

Test No.	Slope (°)	Rainfall Intensity (mm/h)	Rainfall Pattern	Soil Sample	Time of Initial Landslide (min)	Sediment Yield (%)
3	15	120	IV	Mixed soil	47	11.20
4	15	160	IV	Mixed soil	40	18.20
6	15	120	II	Mixed soil	42	14.6
7	15	120	III	Mixed soil	45	12.4
10	15	120	IV	Gravel soil	39	9.80
12	30	120	IV	Mixed soil	33	24.20

Some research on rainfall-induced shallow landslides has been conducted. A comprehensive physics-based Integrated Hydrology Model was set up, which is appliable to different rainfall characteristics [44]. In this paper, more factors are studied in order to investigate the mechanisms of rainfall-induced shallow landslides. The impact of the water content and pore water pressure was analyzed by monitoring a natural slope [45]. However, the variables of that study are not complete because matrix suction and total stress are lacking. A distributed one-dimensional modeling approach for predicting shallow-rainfall-induced landslides was established [46], but the actual landslides are complex. In this paper, more factors have been considered. The influence of soil depth on the occurrence of shallow landslides has been previously investigated [43], but evidence of how the rainfall and soil condition affect a landslide's occurrence is lacking.

The occurrence of landslides and their sediment yield is related to many factors. To investigate the effect of one factor and the following comparisons, the other factors were kept the same. Under a rainfall intensity of 120 mm/h (Test 3), the first landslide occurred at 39 min. Under a rainfall intensity of 160 mm/h (Test 4), the first landslide occurred at 47 min. This suggests that the higher the intensity of rainfall, the earlier the initial landslide occurs. The sediment yield of Test 4 was also larger than that of Test 3. When the rainfall intensity was 80 mm/h, there was no landslide. This indicates that the occurrence of landslides under the test conditions requires the rainfall intensity to exceed 80 mm/h. The rainfall pattern also affected the occurrence of landslides and the first landslide's time of occurrence. In the tests with rainfall patterns II, III, and IV (Test 6, 7, and 3), the first landslides were at 42 min, 45 min, and 47 min, respectively. In the test with rainfall pattern I, no landslide occurred. For the different soil types, the initial landslide occurrence times of the gravel soil and the mixed soil were 39 min and 47 min, and the sediment yields were 9.8% and 11.2%, respectively. No landslide occurred in the silty soil or the sandy soil. Soils composed of coarse grains were shown to be prone to landslides. The landslide on the 30° slope (Test 12) occurred earlier, at 33 min, than those which occurred during all of the tests on the 15° slope. The sediment yield was 24.2%, which was also higher than those on the 15° slope.

5. Conclusions

Landslides are a gravity-driven mass movement and induce an increase in sediment yield in the watershed. In this paper, a series of artificial rainfall tests were conducted in order to investigate rainfall-induced shallow landslides. Four impact factors, including rainfall intensity, rainfall pattern, soil type, and slope gradient, were set to be studied. The changing processes of volumetric water content, matrix suction, pore water pressure, and total stress during rainfall were analyzed. The conclusions were as follows.

The occurrence of rainfall-induced shallow landslides was related to the intensity and pattern of the rainfall, slope gradient, and soil composition. Landslides were triggered by rainfall of a certain intensity and, according to the results obtained from the present tests, the rainfall intensity must exceed 40 mm/h in order to trigger a landslide. The higher the rainfall intensity, the earlier the landslide occurs. The rainfall pattern also influenced landslide generation.

The variables had unique processes of change during the rainfall that occurred under the different combinations of impact factors. The process of matrix suction consisted of an initial stage, steep decline stage, and stable stage. With greater intensity of rainfall, the matrix suction diminished earlier. The pore water pressure continued to rise through the rainfall and it started to rise earlier when the rainfall intensity was greater. In the tests with different rainfall patterns, matrix suction began to diminish later when the rainfall intensity peaked in the middle or late stage (patterns II and III). On the steep slope, soil water content rose and matrix suction diminished later than on the gentler slopes.

When rainfall-induced shallow landslides occurred, there was a corresponding significant change in the physical parameters of the slope. The landslides occurred after the matrix suction entered the diminished stage. The different soil compositions had sequential landslide occurrence times, with the silty and sandy types occurring earlier than the gravelly and mixed types. The pore water pressure rose briefly before the landslide occurred and fell back down after the landslide occurred. As the landslides caused rapid local soil movement, total stress produced rapid changes during the landslides and, in most cases, obviously decreased.

The sediment yield from the landslides was influenced by various factors. When the intensity of the rainfall increased, the sediment yield increased. The sediment yield of the landslides with coarser particle composition was greater than that of the finer soils. The sediment yield of the 30° slope was significantly higher than that of the 15° slope.

Author Contributions: Conceptualization, Y.L. and X.W.; methodology, Y.L. and X.W.; software, Y.L. and Z.D.; validation, Z.D.; investigation, Y.L. and Z.D.; data curation, Y.L. and Z.D.; writing—original draft preparation, Y.L.; writing—review and editing, Y.L. and X.W.; funding acquisition, X.W. All authors have read and agreed to the published version of the manuscript.

Funding: This research is supported by the opening fund from the State Key Laboratory of Hydraulics and Mountain River Engineering, Sichuan University (grant no. SKHL1804 and SKHL 2009).

Institutional Review Board Statement: Not applicable.

Informed Consent Statement: Not applicable.

Data Availability Statement: The data presented in this study are available on request from the corresponding author.

Conflicts of Interest: The authors declare no conflict of interest.

References

1. Glade, T.; Anderson, M.G.; Crozier, M.J. *Landslide Hazard and Risk*; John Wiley & Sons: Hoboken, NJ, USA, 2006.
2. Ray, A.; Bharati, A.K.; Rai, R.; Singh, T.N. Landslide occurrences in Himalayan residual soil: A review. *Himal. Geol.* **2021**, *42*, 189–204.
3. Clague, J.J.; Stead, D. *Landslides: Types, Mechanisms and Modeling*; Cambridge University Press: New York, NY, USA, 2012.
4. Chen, Y. *Landslide Disaster*; Seismological Press: Beijing, China, 2020.
5. Yi, X.; Feng, W.; Bai, H.; Shen, H.; Li, H. Catastrophic landslide triggered by persistent rainfall in Sichuan, China: August 21, 2020, Zhonghaicun landslide. *Landslides* **2021**, *18*, 2907–2921. [CrossRef]
6. Wu, Y.; Zhang, M.; Yang, L.; Liu, T.; Zhang, T.; Sun, Q.; Wang, B.; Xie, X. Failure mechanisms and dynamics of the Shanzao rockslide in Yongjia County, China on 10 August 2019. *Landslides* **2021**, *18*, 2565–2574. [CrossRef]
7. Xia, M.; Ren, G.M.; Yang, X.L. Mechanism of a catastrophic landslide occurred on May 12, 2019, Qinghai Province, China. *Landslides* **2021**, *18*, 707–720. [CrossRef]
8. Dikshit, A.; Sarkar, R.; Pradhan, B.; Segoni, S.; Alamri, A.M. Rainfall Induced Landslide Studies in Indian Himalayan Region: A Critical Review. *Appl. Sci.* **2020**, *10*, 2466. [CrossRef]
9. Wang, K.; Zhang, S. Rainfall-induced landslides assessment in the Fengjie County, Three-Gorge reservoir area, China. *Nat. Hazards* **2021**, *108*, 451–478. [CrossRef]
10. Huang, G.; Zheng, M.; Peng, J. Effect of Vegetation Roots on the Threshold of Slope Instability Induced by Rainfall and Runoff. *Geofluids* **2021**, *2021*, 6682113. [CrossRef]

11. Garg, A.; Huang, H.; Kushvaha, V.; Madhushri, P.; Kamchoom, V.; Wani, I.; Koshy, N.; Zhu, H.-H. Mechanism of biochar soil pore–gas–water interaction: Gas properties of biochar-amended sandy soil at different degrees of compaction using KNN modeling. *Acta Geophys.* **2019**, *68*, 207–217. [CrossRef]
12. Chang, J.-M.; Chen, H.; Jou, B.J.-D.; Tsou, N.-C.; Lin, G.-W. Characteristics of rainfall intensity, duration, and kinetic energy for landslide triggering in Taiwan. *Eng. Geol.* **2017**, *231*, 81–87. [CrossRef]
13. Marin, R.J. Physically based and distributed rainfall intensity and duration thresholds for shallow landslides. *Landslides* **2020**, *17*, 2907–2917. [CrossRef]
14. Li, Q.; Huang, D.; Pei, S.; Qiao, J.; Wang, M. Using Physical Model Experiments for Hazards Assessment of Rainfall-Induced Debris Landslides. *J. Earth Sci.* **2021**, *32*, 1113–1128. [CrossRef]
15. Wu, L.Z.; Xu, Q.; Zhu, J.D. Incorporating hydro-mechanical coupling in an analysis of the effects of rainfall patterns on unsaturated soil slope stability. *Arab. J. Geosci.* **2017**, *10*, 386. [CrossRef]
16. Sagitaningrum, F.H.; Bahsan, E. Parametric study on the effect of rainfall pattern to slope stability. *MATEC Web Conf.* **2017**, *101*, 5005. [CrossRef]
17. Buscarnera, G.; Di Prisco, C.P. Soil stability and flow slides in unsaturated shallow slopes: Can saturation events trigger liquefaction processes? *Géotechnique* **2013**, *63*, 801–817. [CrossRef]
18. Li, W.C.; Dai, F.C.; Wei, Y.Q.; Wang, M.L.; Min, H.; Lee, M.L. Implication of subsurface flow on rainfall-induced landslide: A case study. *Landslides* **2015**, *13*, 1109–1123. [CrossRef]
19. Feng, C.; Tian, B.; Lu, X.; Beer, M.; Broggi, M.; Bi, S.; Xiong, B.; He, T. Bayesian Updating of Soil–Water Character Curve Parameters Based on the Monitor Data of a Large-Scale Landslide Model Experiment. *Appl. Sci.* **2020**, *10*, 5526. [CrossRef]
20. Cui, Y.-F.; Zhou, X.-J.; Guo, C.-X. Experimental study on the moving characteristics of fine grains in wide grading unconsolidated soil under heavy rainfall. *J. Mt. Sci.* **2017**, *14*, 417–431. [CrossRef]
21. Chen, N.; Zhou, W.; Yang, C.; Hu, G.; Gao, Y.; Han, D. The processes and mechanism of failure and debris flow initiation for gravel soil with different clay content. *Geomorphology* **2010**, *121*, 222–230. [CrossRef]
22. Liao, L.; Yang, Y.; Yang, Z.; Zhu, Y.; Hu, J.; Zou, D.H.S. Mechanical state of gravel soil in mobilization of rainfall-induced landslides in the Wenchuan seismic area, Sichuan province, China. *Earth Surf. Dyn.* **2018**, *6*, 637–649. [CrossRef]
23. Wang, G.; Sassa, K. Pore-pressure generation and movement of rainfall-induced landslides: Effects of grain size and fine-particle content. *Eng. Geol.* **2003**, *69*, 109–125. [CrossRef]
24. Huang, K.; Duan, H.; Yi, Y.; Yu, F.; Chen, S.; Dai, Z. Laboratory Model Tests on Flow Erosion Failure Mechanism of a Slope Consisting of Anqing Group Clay Gravel Layer. *Geofluids* **2021**, *2021*, 5559052. [CrossRef]
25. Pajalić, S.; Peranić, J.; Maksimović, S.; Čeh, N.; Jagodnik, V.; Željko, A. Monitoring and Data Analysis in Small-Scale Landslide Physical Model. *Appl. Sci.* **2021**, *11*, 5040. [CrossRef]
26. Rahardjo, H.; Lee, T.T.; Leong, E.C.; Rezaur, R.B. Response of a residual soil slope to rainfall. *Can. Geotech. J.* **2005**, *42*, 340–351. [CrossRef]
27. Hou, X.; Li, T.; Qi, S.; Guo, S.; Li, P.; Xi, Y.; Xing, X. Investigation of the cumulative influence of infiltration on the slope stability with a thick unsaturated zone. *Bull. Int. Assoc. Eng. Geol.* **2021**, *80*, 5467–5480. [CrossRef]
28. Chen, H.; Lee, C.F.; Law, K.T. Causative Mechanisms of Rainfall-Induced Fill Slope Failures. *J. Geotech. Geoenvironmental Eng.* **2004**, *130*, 593–602. [CrossRef]
29. Mahmood, K.; Kim, J.M.; Ashraf, M.; Ziaurrehman. The effect of soil type on matric suction and stability of unsaturated slope under uniform rainfall. *KSCE J. Civ. Eng.* **2016**, *20*, 1294–1299. [CrossRef]
30. Cui, P.; Guo, C.-X.; Zhou, J.-W.; Hao, M.-H.; Xu, F.-G. The mechanisms behind shallow failures in slopes comprised of landslide deposits. *Eng. Geol.* **2014**, *180*, 34–44. [CrossRef]
31. Zhan, L.-T.; Guo, X.-G.; Sun, Q.-Q.; Chen, Y.-M.; Chen, Z.-Y. The 2015 Shenzhen catastrophic landslide in a construction waste dump: Analyses of undrained strength and slope stability. *Acta Geotech.* **2021**, *16*, 1247–1263. [CrossRef]
32. Claessens, L.; Knapen, A.; Kitutu, M.; Poesen, J.; Deckers, J. Modelling landslide hazard, soil redistribution and sediment yield of landslides on the Ugandan footslopes of Mount Elgon. *Geomorphology* **2007**, *90*, 23–35. [CrossRef]
33. Al-Sheriadeh, M.S.; Malkawi, A.I.H.; Al-Hamdan, A.; Abderahman, N.S. Evaluating sediment yield at King Talal Reservoir from landslides along Irbid–Amman Highway. *Eng. Geol.* **2000**, *56*, 361–372. [CrossRef]
34. Koi, T.; Hotta, N.; Ishigaki, I.; Matuzaki, N.; Uchiyama, Y.; Suzuki, M. Prolonged impact of earthquake-induced landslides on sediment yield in a mountain watershed: The Tanzawa region, Japan. *Geomorphology* **2008**, *101*, 692–702. [CrossRef]
35. Chuang, S.-C.; Chen, H.; Lin, G.-W.; Lin, C.-W.; Chang, C.-P. Increase in basin sediment yield from landslides in storms following major seismic disturbance. *Eng. Geol.* **2009**, *103*, 59–65. [CrossRef]
36. Bovolo, C.I.; Bathurst, J.C. Modelling catchment-scale shallow landslide occurrence and sediment yield as a function of rainfall return period. *Hydrol. Process.* **2011**, *26*, 579–596. [CrossRef]
37. Bathurst, J.C.; Moretti, G.; El-Hames, A.; Beguería, S.; García-Ruiz, J.M. Modelling the impact of forest loss on shallow landslide sediment yield, Ijuez river catchment, Spanish Pyrenees. *Hydrol. Earth Syst. Sci.* **2007**, *11*, 569–583. [CrossRef]
38. Bathurst, J.C.; Moretti, G.; El-Hames, A.; Moaven-Hashemi, A.; Burton, A. Scenario modelling of basin-scale, shallow landslide sediment yield, Valsassina, Italian Southern Alps. *Nat. Hazards Earth Syst. Sci.* **2005**, *5*, 189–202. [CrossRef]
39. Burton, A.; Bathurst, J.C. Physically based modelling of shallow landslide sediment yield at a catchment scale. *Environ. Earth Sci.* **1998**, *35*, 89–99. [CrossRef]

40. Yang, S.-Y.; Jan, C.-D.; Yen, H.; Wang, J.-S. Characterization of landslide distribution and sediment yield in the TsengWen River Watershed, Taiwan. *Catena* **2019**, *174*, 184–198. [CrossRef]
41. Gan, F.; He, B.; Wang, T. Water and soil loss from landslide deposits as a function of gravel content in the Wenchuan earthquake area, China, revealed by artificial rainfall simulations. *PLoS ONE* **2018**, *13*, e0196657. [CrossRef]
42. Gilley, J.E. Erosion | Water-Induced. In *Encyclopedia of Soils in the Environment*; Hillel, D., Ed.; Elsevier: Oxford, UK, 2005; pp. 463–469.
43. Acharya, G.; Cochrane, T.; Davies, T.; Bowman, E. The influence of shallow landslides on sediment supply: A flume-based investigation using sandy soil. *Eng. Geol.* **2009**, *109*, 161–169. [CrossRef]
44. Ran, Q.; Hong, Y.; Li, W.; Gao, J. A modelling study of rainfall-induced shallow landslide mechanisms under different rainfall characteristics. *J. Hydrol.* **2018**, *563*, 790–801. [CrossRef]
45. Bordoni, M.; Meisina, C.; Valentino, R.; Lu, N.; Bittelli, M.; Chersich, S. Hydrological factors affecting rainfall-induced shallow landslides: From the field monitoring to a simplified slope stability analysis. *Eng. Geol.* **2015**, *193*, 19–37. [CrossRef]
46. Baum, R.L.; Godt, J.W.; Savage, W.Z. Estimating the timing and location of shallow rainfall-induced landslides using a model for transient, unsaturated infiltration. *J. Geophys. Res. Earth Surf.* **2010**, *118*, 1999. [CrossRef]

applied sciences

MDPI

Article

Geometry-Based Preliminary Quantification of Landslide-Induced Impulse Wave Attenuation in Mountain Lakes

Andrea Franco [1,*], Barbara Schneider-Muntau [2], Nicholas J. Roberts [3,4], John J. Clague [3] and Bernhard Gems [1]

1 Unit of Hydraulic Engineering, University of Innsbruck, Technikerstraße 13, 6020 Innsbruck, Austria; bernhard.gems@uibk.ac.at
2 Unit of Geotechnical Engineering, University of Innsbruck, Technikerstraße 13, 6020 Innsbruck, Austria; barbara.schneider-muntau@uibk.ac.at
3 Department of Earth Sciences, Simon Fraser University, 8888 University Drive, Burnaby, BC V5A 1S6, Canada; nickr@sfu.ca (N.J.R.); john_clague@sfu.ca (J.J.C.)
4 Mineral Resources Tasmania, Department of State Growth, 30 Gordons Hill Road, Rosny Park, TAS 7018, Australia
* Correspondence: andrea.franco@uibk.ac.at

Abstract: In this work, a simple methodology for preliminarily assessing the magnitude of potential landslide-induced impulse waves' attenuation in mountain lakes is presented. A set of metrics is used to define the geometries of theoretical mountain lakes of different sizes and shapes and to simulate impulse waves in them using the hydrodynamic software Flow-3D. The modeling results provide the 'wave decay potential', a ratio between the maximum wave amplitude and the flow depth at the shoreline. Wave decay potential is highly correlated with what is defined as the 'shape product', a metric that represents lake geometry. The relation between these two parameters can be used to evaluate wave dissipation in a natural lake given its geometric properties, and thus estimate expected flow depth at the shoreline. This novel approach is tested by applying it to a real-world event, the 2007 landslide-generated wave in Chehalis Lake (Canada), where the results match well with those obtained using the empirical equation provided by ETH Zurich (2019 Edition). This work represents the initial stage in the development of this method, and it encourages additional research and modeling in which the influence of the impacting characteristics on the resulting waves and flow depths is investigated.

Keywords: landslide-induced wave; lake-tsunami assessment; mountain lakes

Citation: Franco, A.; Schneider-Muntau, B.; Roberts, N.J.; Clague, J.J.; Gems, B. Geometry-Based Preliminary Quantification of Landslide-Induced Impulse Wave Attenuation in Mountain Lakes. *Appl. Sci.* **2021**, *11*, 11614. https://doi.org/10.3390/app112411614

Academic Editor: Ricardo Castedo

Received: 29 October 2021
Accepted: 2 December 2021
Published: 7 December 2021

1. Introduction

Landslide-induced impulse waves in lakes are gaining interest in the scientific community due to the hazards they pose to people living or recreating along their shorelines and to dams and other infrastructure [1]. Additionally, climate change is driving rapid geomorphological changes in high mountains, which may increase the likelihood of landslides into mountain lakes. Rapid thinning and retreat of glaciers and an increase in heavy rainfall events can destabilize slopes adjacent to water bodies [2–5].

In December 2007, a 3-Mm^3 rock avalanche entered the Chehalis Lake (Canada), generating an impulse wave that destroyed forest and campgrounds along the shoreline and achieved a maximum run-up of 37.8 m above average lake level [6,7]. In July 2014, a 10-Mm^3 rockslide collapsed into Askja Lake (Iceland) from its rimming caldera and generated a wave that propagated >3 km across the lake with localized run-ups of up to 60–80 m [8]. In October 2015, a 50-Mm^3 landslide collapsed into Taan Fiord (Alaska), inducing one of the largest landslide-induced waves ever recorded, with a maximum run-up of 193 m on the slope facing the landslide; the wave traveled >17 km down the fiord, devastating forests and eroding soil and sediments along its path [5,9]. In June 2017, an impulse wave generated by a 50-Mm^3 subaerial rockslide into Karrat Fiord on Greenland's west coast killed four people and destroyed 11 buildings in the village of Nuugaatsiaq, 32 km from its source, and flooded other settlements along the coast [10]. These events are just a small subset of all the subaerial landslides known to have generated impulse waves around the world [1].

Different methods, including scaled physical tests, field investigations, and numerical models are used to investigate aspects of this phenomenon, including landslide behavior, landslide–water interaction, and wave formation, propagation, and inundation. The effect of basin geometry on wave propagation has been experimentally and numerically investigated [11–13]. Researchers have also analyzed wave dispersion and related wave decay. Ruffini et al. [14] state that wave decay results from (i) frequency dispersion, (ii) bottom friction, (iii) lateral spreading of the water, and (iv) breaking of waves during generation and propagation. According to those authors, an increase in the lateral angle of the basin leads to a decay of solitary waves. The effect of lateral energy spread on wave amplitude is larger in a 3D basin-type geometry than the effect of frequency dispersion in a 2D flume-type geometry.

To virtually reproduce wave dynamics generated by an impact, researchers have used several different numerical approaches, notably non-linear shallow water equations (NSWE), Reynolds-averaged Navier–Stokes equations (RANS), and smoothed particle hydrodynamic methods (SPH) [13]. These approaches have been used to retrospectively investigate specific landslide-induced impulse wave events, typically with much success. However, such modeling approaches, when used in assessments of possible future events, are compromised by cost, computational time, the considerable amount of data required for model calibration and validation, and the absence of run-up and inundation evidence typically used to fine-tune model parameters

Other methods, such as generic empirical equations, can be used for preliminary assessment of the potential impact of landslide-generated waves and to guide decisions on the need for further investigations. Empirical relationships, commonly based on field-measured wave impacts of historical events or scaled physical experiments, are used to determine wave characteristics. The analytical equations of Heller et al. [15], Heller and Hager [16], and Evers et al. [17] were developed to assess impulse waves in artificial water basins impounded by dams but are also applicable to natural lakes. Their equations, and the 3D approach employed in them, provide estimates of wave characteristics while accounting for variable propagation angles and travel distances, thus covering a wide range of water-basin geometries. The main inputs into their set of equations are landslide properties and a representative lake depth. Their workflow provides the first estimate of impulse wave celerity and waves run-up.

Strupler et al. [18] suggest a classification of mountain lakes based on their impulse wave potential, which they derived from both subaerial and subaqueous mass movements at Swiss perialpine lakes >1 km^2. Their method relies on parameters calculated from digital elevation data (using geospatial software), for instance, topographic surfaces and bathymetry, together with seismologic data (e.g., the local acceleration), considered as an external factor for landslide initiation. They argue that lakes in the Alps have a high potential for impact by subaerial or subaqueous mass movements due to the surrounding steep slopes and fiord-like morphology of the lake basins compared to perialpine lakes. The latter lakes have a lower potential for subaerial mass movement, but a larger potential to inundate surrounding areas because of the typically lower relief surrounding them.

Despite the simplicity of diverse approaches available in the literature, there are some limitations, such as the technical knowledge required to properly use these tools and limited availability of high-resolution geospatial data—particularly bathymetry—which may necessitate the use of lower-quality digital elevation data. In this paper, a simple alternative method, based on lake shape and extent, for estimating impulse wave propagation in mountain lakes (Figure 1) is presented. The method has a short workflow based on analytical equations derived from the analysis of wave characteristics and estimated using hydrodynamic numerical models in Flow-3D for theoretical mountain lakes. A fixed impact volume is used to induce the impulse wave. In comparison to other approaches, the equations are easy to apply. The variables required are the geometric characteristic of lakes; no specifics of the impacting landslide are needed.

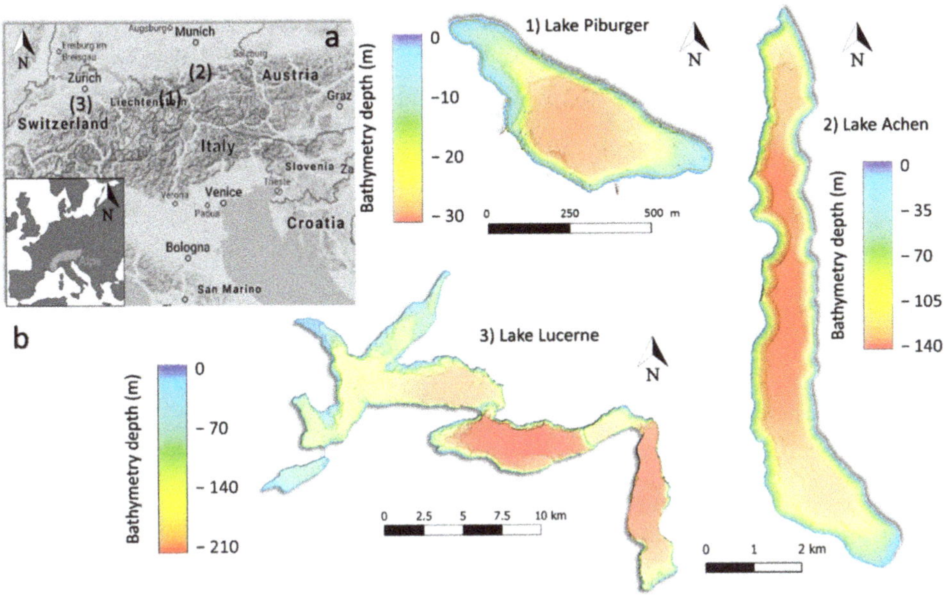

Figure 1. Examples of mountain lakes, differing in shape, bathymetry, and spatial extent. (**a**) Location of the presented lakes in the Alps region. (**b**) Examples of diverse lake geometries: Lake Piburger (1-Austria), Lake Achen (2-Austria), and Lake Lucerne (3-Switzerland). Bathymetric data provided by the Sedimentary Geology Working Group of the University of Innsbruck [19,20] and the Swiss Service-Federal Office of Topography swisstopo (https://www.swisstopo.admin.ch/en/geodata/height/bathy3d.html) (Accessed on 3 September 2021), [21,22]. The topographical background is taken from Microsoft Bing Maps—2021 Microsoft Corporation Earthstar Geographics Sio (Accessed on 26 July 2021).

We describe the near-shore impulse-wave magnitude in terms of flow depth at the shoreline. This novel approach enables rapid, resource-efficient, preliminary investigations that may be required to identify basins of particular concern or high risk and to justify more detailed data collection and analysis. An Excel spreadsheet, as a Computational tool, comprising the presented workflow is available in the Supplementary Materials.

2. Materials and Methods

Data used in this study are derived from 56 subjectively selected alpine and perialpine lakes in the Alps region (table in Supplementary Materials). All of the lakes are bordered by slopes that could potentially generate subaerial mass movements. Characteristics of most lakes—water volume (V_w), lake area (A), width (W), length (L), and mean and maximum water depth (d_w and D_w)—are already available from the literature and diverse online sources [23–26]. Bathymetric data for some of the lakes were provided by the Sedimentary Geology Working Group at the University of Innsbruck (see a listing in the table in the Supplementary Materials [19,20]) and the Swiss Service (Federal Office of Topography-swisstopo), enabling a more accurate geometric data calculation utilizing the Raster Layer Statistics and Raster Surface Volume tools in QGIS v3.16.

A classification scheme is developed to group and discriminate the geometries of the 56 lakes in the dataset (Section 3). Frequency analyses are performed to identify the most common lake metrics and to combine them to produce a representative range of theoretical lake-basin shapes for use in subsequent numerical modeling. These theoretical lake basins are generated as 3D solid bodies and exported to stereolithography (STL) files using Rhinoceros 6 software (see Supplementary Materials).

Numerical models are implemented in the finite-volume-based for computational fluid dynamics (CFD) software Flow-3D v11.1 [27–30]. This software simulates two-fluid problems, where all velocity components (u, v, w) are computed in the 3D domain using RANS equations [31] in combination with the volume-of-fluid method [29,32], and adopting the fractional-area/volume-obstacle-representation [33]. To compute turbulence and viscosity processes in Flow-3D, the renormalized group model-based k-epsilon turbulence model [34] is applied to create a fluid–fluid coupled model of the impulse wave. This model uses statistical formulations to compute the turbulent kinetic energy dissipation rate [35–37]. A Newtonian-like fluid (see Section 4), featuring a higher density compared to the still-water density, is adopted to simulate the impacting volume [38]. Modeled free-water-surface elevations, wave-crest elevations in open water, and flow depth at the lake shoreline are post-processed in FlowSight v11.1 [27]. Finally, correlation matrices between the numerical input parameters and the modeling results are generated to establish a relationship that can be used to assess potential impulse waves in natural mountain lakes.

3. Classification and Frequency Analysis of Alpine Lake Geometries

A classification of the considered alpine lakes is shown in Figure 2. The figure shows the relationships between water volume (V_w) and lake area (A) and between water volume and mean (d_w) and maximum (D_w) lake depth (each data point represents a single lake). The data are subjectively grouped and named based on ranges of V_w as follows: small-size lakes (V_w < 1 Mm^3); medium-size lakes (V_w 1–100 Mm^3); large-size lakes (V_w 100–10,000 Mm^3); and very large-size lakes (V_w > 10,000 Mm^3).

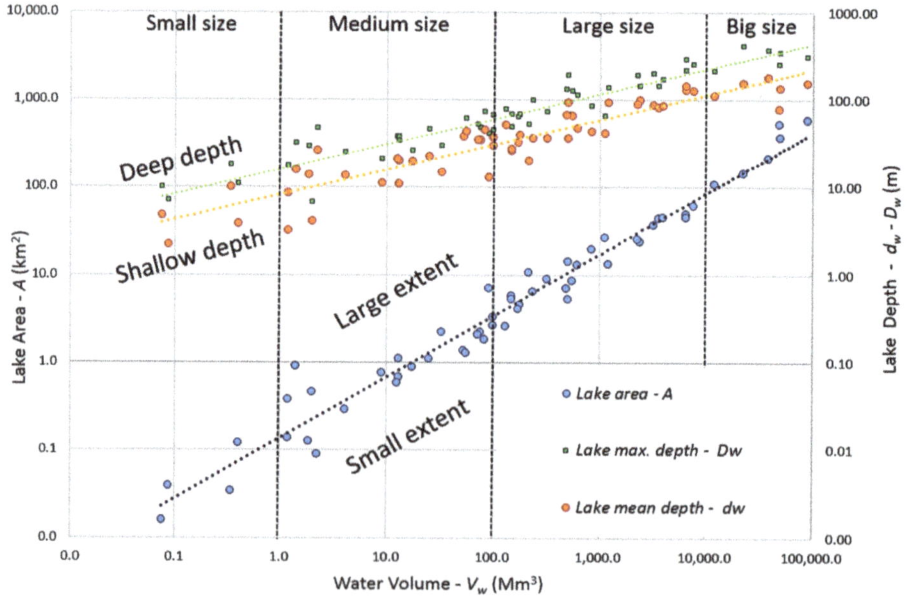

Figure 2. Classification of 56 alpine lakes based on water volume, the lake area, and mean and maximum lake depths: dotted blue line is V_w against A; dotted orange line is V_w against d_w; the dotted green line is V_w against D_w. Trendlines enable a qualitative description of the lakes in terms of their extent and depths relative to water volume. Axes are logarithmic (base 10.).

Trendlines in Figure 2 quantitatively describe lake extents and water depths relative to water volume. The higher a lake is above the trendline, the larger the volume-normalized extent or the deeper the volume-normalized depth. Conversely, increasing distance below trendlines indicates a smaller extent or shallower depth relative to lake volume. Among the 56 lakes considered, 42% and 58% have large- and small-volume-normalized extents, respectively; and 53% and 47% have deep and shallow-volume-normalized depths, respectively.

Considering a ratio between L and W, it is possible to differentiate elongate lake shapes ($L/W > 1.5$, about 60% of them; e.g., Lake Achen in Figure 1) from those that are approximately equidimensional ($1 < L/W < 1.5$, about 16%). Other lakes (the remaining 24%) have complex shapes, generally comprising sub-basins (e.g., Lake Lucerne in Figure 1). These are mostly large- and very large-size lakes. Given that sub-basins of complex lakes are themselves either medium or large size and that only a few small lakes were identified, the small-size and very large-size lake classes are not considered further in the analysis. Results of a frequency analysis of lake geometrical characteristics for the medium-size and large-size lake classes are shown in Figure 3.

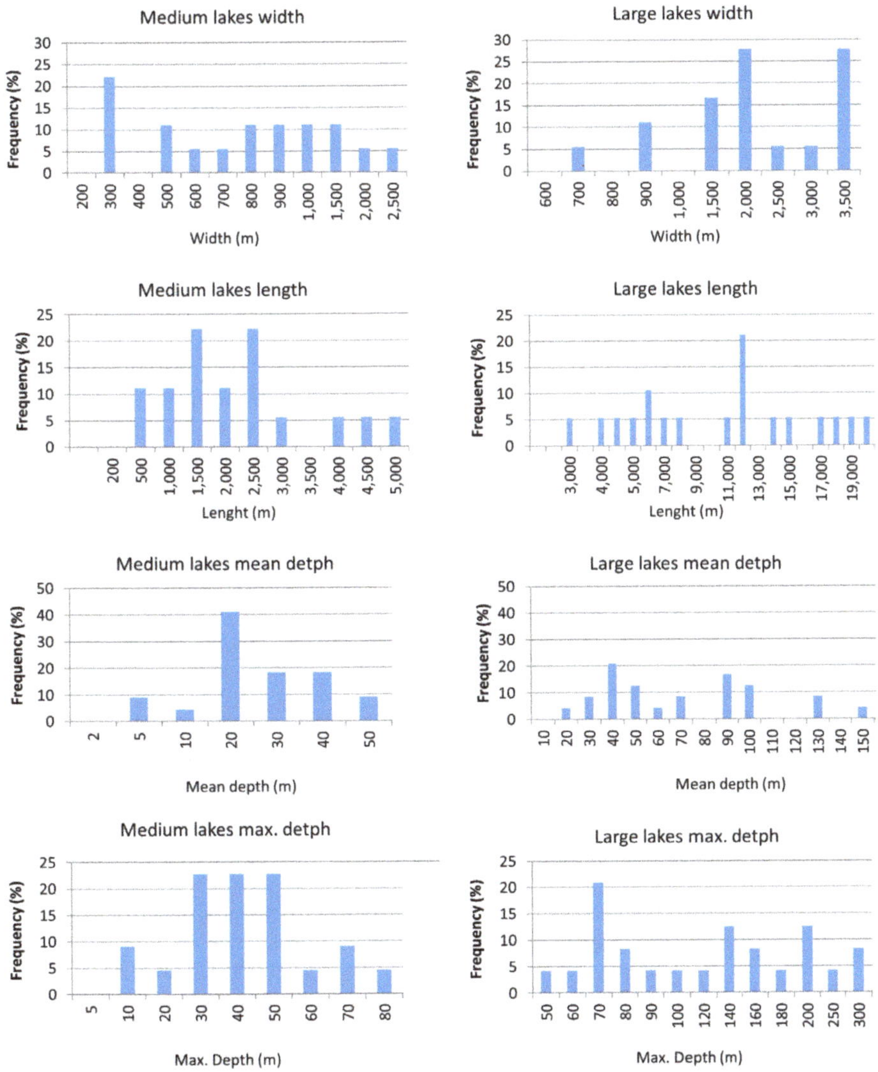

Figure 3. Frequency analysis and related histograms of geometrical characteristics (lake width and length, mean and maximum depth) of medium- and large-size lakes.

Based on the aforementioned classification, different combinations of parameters are subjectively defined to design theoretical lakes that cover the considerable variability of mountain lake configurations in the Alps. Table 1 shows the selected values for the two size classes, resulting in 51 combinations for medium-size lakes and 18 for large-size lakes. These 69 theoretical lakes form the basis for the numerical modeling part of this study (both a complete table with related information and STL files are available in the Supplementary Materials).

Table 1. Geometry parameters chosen for the theoretical lakes [lake width (W), length (L), and mean and maximum water depths (d_w and D_w)]. Permutations of these three variables yield a total of 69 lake-basin configurations.

Lake Class	d_w–D_w (m)	W (m)	L (m)
Medium size lakes	10–20 20–40 40–50	250 500 1000 1500	500 1000 1500 2000 2500
Large size lakes	40–70 80–150 100–200	1000 2000 3000	6000 12,000

4. Numerical Model Set-Up

Numerical models are designed to reflect conditions typical of mountain lakes in alpine settings, albeit with some necessary simplifications. Alpine basins, whether small or large, commonly fill topographic depressions with adjacent steep slopes that in many cases are the product of glacial erosion or deposition [39]. Steep slopes are particularly common along valley sides, making subaerial landslides capable of generating impulse waves more likely along lakesides than ends (e.g., [6,39,40]). Consequently, impacting volumes initiated in these environments enter the lake along its sides in all modeled scenarios.

Each numerical simulation considers that same impacting volume for triggering the wave. This lies initially on a sliding surface dipping 45° toward the lake and is represented by a 0.5 Mm^3 prismatic fluid body, 20 m thick, 208 m long, and 120 m wide, with a toe position directly above the lake surface (Figure 4). The volume is located mid-way along the side of each theoretical lake.

A bulk material density of 1620 kg m^{-3} is used for the landslide [38]. Additionally, an initial speed of 20 m s^{-1} at simulation time 0, is arbitrarily set. This, together with volume deformation during the sliding process, results in a maximum speed of 60 m s^{-1} for the center of mass at the impact in all models, regardless of lake geometry.

For each simulation, the model domain includes the slide area, the entire water body, and the air above it to 40 m above lake level (m a.l.l.). The origin of the system (x,y,z equal to 0) is located midway along the long side of each artificial lake at the middle of the landslide toe (0 m a.l.l.). The maximum depth D_w is located at the center of each theoretical lake (Figure 4e). A mesh block of 120 m × 140 m × 162 m, comprising 2 m × 2 m × 2 m mesh cells, includes the slope and the impacting volume (Figure 4a,e). To improve computational efficiency while also representing the complexity of nearfield landslide–water interaction, a finer mesh block within 500 m of the model origin and a coarser one beyond that is used. Medium-size lakes up to 500 m wide are modeled using uniform 2 m × 2 m × 2 m mesh cells in the nearfield and non-uniform 4 m × 4 m × 2 m cells beyond 500 m.

Medium-size lakes wider than 500 m are modeled using uniform cells that increase beyond 500 m from 2 to 4 m. For all large-size lakes, uniform cells (5 m) and non-uniform cells (10 m × 10 m × 5 m) are used, respectively, within and beyond 500 m of the origin.

In all models, the "wall" boundary condition of the mesh block at the lateral margins of the impacting volume is set in Flow-3D to simulate a continuous slope along the length of the lake, allowing wave reflection modeling. A "symmetry" boundary condition is set to allow the impacting volume to leave the mesh of the sliding area and enter the mesh comprising the lake. "Outflow" boundary conditions, at the lake borders, on the other edges of the model domain allow the flow to exit the domain without reflection to avoid wave interference and, thus, simplify the computation process. This simplification was implemented because wave characteristics of the first front arrival at the shorelines are of principal interest (Figure 4a,e). The water level at lake rest (0 m a.l.l. at z-axis in the domain system) is used as the initial condition for all model runs. Recording data intervals of 0.5 and 1 s are set for the medium-size and large-size lakes, respectively.

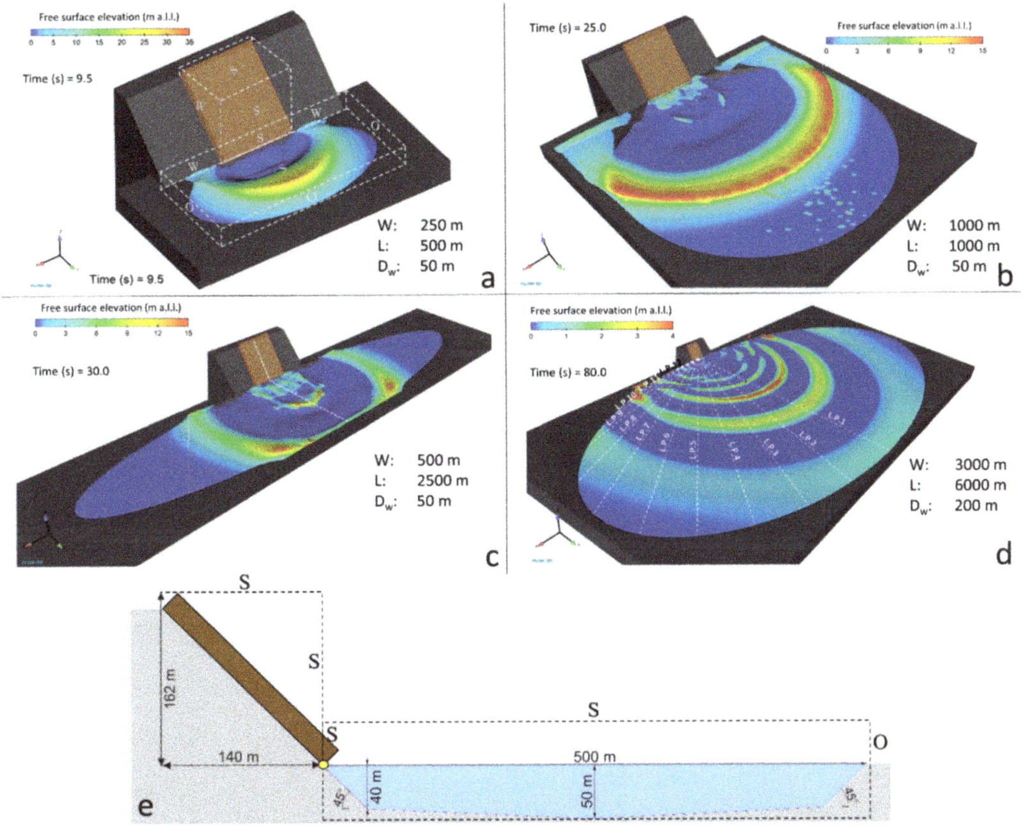

Figure 4. Examples of theoretical lakes and model set-ups showing simulations of a landslide-induced impulse wave. (**a**) Small elongated medium-size lake; in this example, the boundary conditions are shown (S-symmetry; W—wall; O—outflow). (**b**) Equidimensional medium-size lake. (**c**) Narrow elongated medium-size lake. (**d**) Elongated large-size lake (dotted white lines represent lines probes (L.P. 1–12) used to record water-surface elevations in FlowSight). (**e**) Example of longitudinal section for the theoretical lake in (**c**) along the slide direction (dotted white line); boundary conditions are shown. The yellow dot represents the origin of the system in the model domain.

The model set-ups are chosen to allow completion of each simulation in a reasonable amount of time (between 2 and 24 h per simulation) and to optimize the balance between output accuracy and output file size (4 to 109 Gb per simulation). The model environments have been designed to ensure that output files are within the processing capabilities of FlowSight, while also producing results suited for inter-model comparisons.

The computational resource and hardware components used for numerical modeling are the following:

- Processor: Intel® Core™ i7-3820 CPU 3.60 GHz;
- RAM: 32 GB;
- System type: 64-bit operating system;
- Number of core license tokens: 8 (Flow-3D parallel license code);
- Graphics card: GeForce GTX 6602 (Integrated RAMDAC, total available memory 4096 MB).

5. Results

5.1. Numerical Simulations and Wave Decay Potential Parameter

For each scenario, lake-surface elevations are estimated along with line probes (L.P.), which are lines along which the free-water-surface in the 3D domain is monitored over time following impact (Figure 4d). L.Ps are used to document and analyze maximum wave crest elevations in open water and flow depths at the shoreline. A set of 12 L.Ps extend from the domain origin (Section 4) to equal space-distributed points along the shoreline, allowing to qualitatively and quantitatively analyze the propagation of the impulse wave and how lake geometry affects the wave. An example is provided in Figure 5, which shows that a decrease in water depth results in progressively higher wave dissipation and a lower flow depth at the shore (compare Figure 5a–c). These trends reflect the increase in water volume that is mobilized by the impacting volume and reduced interaction and friction with the lake floor as the water depth increases. Conversely, the longer travel distances facilitated by incrementally wider lakes result in greater wave dissipation and, consequently, lower flow depth at the shoreline. The slight increase of flow depth along with the L.Ps in the proximity of the shoreline (Figure 5a,d,e) shows the wave deformation due to the interaction with the lake floor while approaching the shoreline.

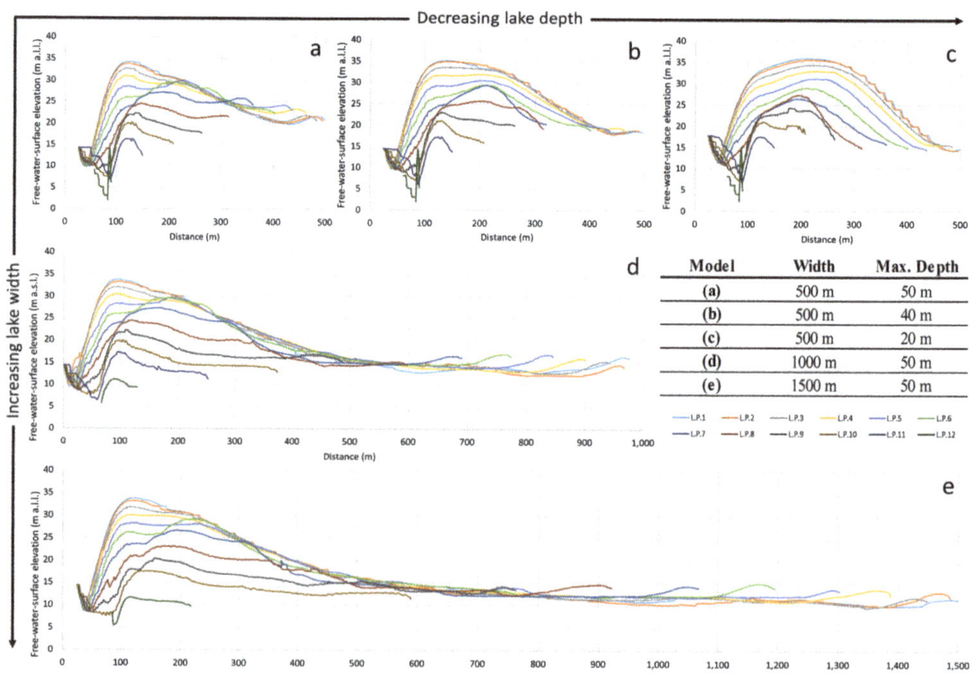

Figure 5. Free-water-surface elevation computed along with the L.Ps (e.g., Figure 4d) for typical circular artificial lakes with different widths and maximum depths (**a**–**e**). Each lake configuration influences the landslide–water interaction and wave dissipation, thus causing different impulse wave propagation patterns.

The same patterns are observed for all 69 scenarios. Generally, for the medium-size lakes, the maximum wave crest elevation ranges from 29 to 5.9 m a.l.l along with L.P. 1 (the midline representative of the landslide's travel direction). Considering all L.Ps, average wave crest elevations range from 15.4 to 27.5 m a.l.l. At the shoreline, the average flow depth is 4.1 to 19.7 m a.l.l. The highest values of flow depth (6 to 35.6 m a.l.l.) occur at the shoreline along L.P. 1. In the case of large-size lakes, maximum wave crest elevations range from 29.4 to 33.8 m a.l.l. along with L.P. 1, and the average for all L.Ps ranges from 15.4 to 24 m a.l.l. At the shoreline, the average flow depth is 4.1 to 7.4 m a.l.l., with the highest values on L.P. 1 (4.2 to 15.9 m a.l.l.).

The numerical analyses provide insight into potential impulse-wave threats for particular mountain lake configurations. A possible indicator of the threat level is the "dissipation power" of an impulse wave in a lake, which is a measure of the degree to which the wave attenuates as it moves away from the impact location. This metric is determined for each L.P. by calculating the ratio of the maximum wave crest elevation and the flow depth at the shoreline. A new parameter, defined as a single weighted-average decay ratio value and representative for the entire lake is henceforth termed the "wave decay potential parameter" (*WDPP*) is given by Equation (1):

$$WDPP = \frac{\sum_{i=1}^{n}(a_{mi})\left(\frac{a_{mi}}{f_{dci}}\right)}{\sum_{i=1}^{n} a_{mi}}\ (-) \tag{1}$$

where a_m and f_{dc} are, respectively, the maximum wave elevation and the flow depth at the shoreline location for each of the L.Ps. For medium-size lakes, values of *WDPP* range from 1.09 to 4.87, and for large-size lakes from 2.61 to 6.6. When a higher value of *WDPP* is obtained, a higher dissipation power of the lake is expected, and vice versa.

5.2. Correlation Analysis and the Shape-Product Approach

A correlation analysis is completed to better understand how the numerical results are related to the input parameters with matrices showing the resulting correlation coefficients (r) calculated by the Pearson function. The correlation matrix in Figure 6a considers all analyzed theoretical lakes. The maximum wave crest elevation does not correlate well with any input parameter, suggesting that it depends on the combination of all lake characteristics (and in a real situation also on the impacting landslide properties, which are not considered in this study). Similarly, no relevant correlations are evident for flow depth at the shoreline. In the case of the mean flow depth at the shoreline, r-values of 0.669 and 0.665 are obtained for lake width and length, respectively. Maximum and minimum flow depths yield r-values of 0.71 and 0.694 for lake width and length, respectively. Higher correlation coefficients are calculated between *WDPP* and lake characteristics: r-values of 0.835 and 0.786 for lake width and length, respectively; and r-values of 0.830 and 0.806 for lake area and volume, respectively.

By plotting *WDPP* against lake area and water volume (Figure 7) and considering the maximum water depth as an indicator, it is found that the theoretical lake basins form two groups. Lakes with a maximum D_w of 20 m (small red circles in Figure 7) can be separated from the others. This separation is supported by correlation analyses provided for the two lake geometry subgroups ($D_w \leq 20$ m and $D_w > 20$ m). In the case of lakes with maximum depths greater than 20 m, the correlations between *WDPP* and lake area and lake volume (see Figure 6b) have r values of 0.90 and 0.88, respectively.

These relationships can be approximated by hyperbolic (power-type) functions. The coefficients of determination, R^2, which express the variation of the dependent variable predicted by the independent variable and provide a measure of how well the curves fit the data (the closer R^2 is to 1, the better fit-Figure 7 and Table 2).

Lake basins with $D_w \leq 20$ m likely plot farther from deeper basins because their water depths are similar to or less than the thickness of the impacting volume. This may imply that numerical results for these scenarios are influenced by how the impacting volume (the dense fluid) enters the lake and propagates through the water body [41]. For example, the dense volume might induce an overestimation of the maximum wave crest in open water. Consequently, further analysis and discussion only consider model scenarios with maximum depths greater than 20 m. *WDPP* is well correlated to the lake area and water volume, thus a parameter that relates different lake characteristics and *WDPP* is formulated to improve rapid assessments of impulse waves in mountain lakes.

The new parameter is the "shape product" (herein labeled *ShpP*), which is defined using empirical Equation (2). Results are well correlated to *WDPP* with an *r*-value of 0.962 (Figure 6 and Table 2):

$$ShpP = \frac{V_w}{A} * \left(\frac{W * L}{d_w^3}\right)^{1/3} \ (\text{m}^{2/3}) \tag{2}$$

where V_w is lake volume, A is lake area, W and L are lake width and length, respectively, and d_w is mean lake depth.

The shape parameter does not represent any specific physical measurement or a dimensionless parameter. On the contrary, it is the measure of a lake's geometric characteristics most closely related to the *WDPP*.

A plot of *ShpP* against *WDPP* reveals a linear relationship with an R^2 of 0.9257 (Figure 8a and Table 2). Linear regression of *ShpP* against *WDPP* verifies the reliability of this relationship (Figure 8b), as it yields the same equation and R^2 proposed in the previous analysis (Figure 8a).

After calculating *WDPP* from a specific *ShpP*, the equations in Table 3 provide a rough estimate of the expected flow depth at the lakeshore. Plots of *WDPP* against maximum, mean, and minimum flow depth obtained from the numerical models are shown in Figure 9. The mean flow depth is strongly and negatively correlated to *WDPP* ($r = -0.849$, Table 3). The maximum and minimum flow depths are well negatively correlated to *WDPP* calculated at L.Ps 1 and 5 (r = −0.751 and −0.749, respectively; Figure 6b).

All lakes considered (tot. 69)	Width (m)	Length (m)	Mean depth (m)	Max. depth (m)	Lake area (m²)	Water volume (m³)	Max. wave crest elevation (m)	Mean flow depth at shoreline (m)	Max. flow depth at shoreline (m)	Min. flow depth at shoreline (m)	Wave decay in L.P. 1	Wave decay in L.P. 5	Wave decay potential parameter	Shape product (m^2/3)
Max. wave crest elevation (m)	-0.227	-0.110	-0.256	-0.222	-0.105	-0.164	1.000							
Mean flow depth at shoreline (m)	-0.669	-0.665	-0.451	-0.474	-0.617	-0.553	0.044	1.000						
Max. flow depth at shoreline (m)	-0.710	-0.397	-0.346	-0.322	-0.447	-0.371	-0.443	0.706	1.000					
Min. flow depth at shoreline (m)	-0.338	-0.694	-0.423	-0.501	-0.556	-0.529	0.312	0.665	0.218	1.000				
Wave decay in L.P. 1	0.936	0.575	0.541	0.500	0.760	0.679	0.423	-0.707	-0.767	-0.362	1.000			
Wave decay in L.P. 5	0.755	0.814	0.702	0.732	0.829	0.820	0.155	-0.814	-0.551	-0.718	0.766	1.000		
Wave decay potential parameter	0.835	0.786	0.668	0.692	0.830	0.806	0.288	-0.836	-0.676	-0.629	0.881	0.958	1.000	
Shape product (m^2/3)	0.867	0.931	0.801	0.788	0.930	0.866	0.066	-0.764	-0.611	-0.621	0.795	0.880	0.896	1.000

a

Excluding lakes with maximum depth of 20 m (tot. 52)	Width (m)	Length (m)	Mean depth (m)	Max. depth (m)	Lake area (m²)	Water volume (m³)	Max. wave crest elevation (m)	Mean flow depth at shoreline (m)	Max. flow depth at shoreline (m)	Min. flow depth at shoreline (m)	Wave decay in L.P. 1	Wave decay in L.P. 5	Wave decay potential parameter	Shape product (m^2/3)
Max. wave crest elevation (m)	-0.301	-0.136	-0.398	-0.328	-0.142	-0.215	1.000							
Mean flow depth at shoreline (m)	-0.750	-0.781	-0.665	-0.652	-0.729	-0.666	0.041	1.000						
Max. flow depth at shoreline (m)	-0.756	-0.487	-0.500	-0.443	-0.530	-0.457	-0.420	0.732	1.000					
Min. flow depth at shoreline (m)	-0.432	-0.799	-0.629	-0.677	-0.653	-0.623	0.295	0.690	0.267	1.000				
Wave decay in L.P. 1	0.972	0.650	0.682	0.602	0.843	0.765	0.372	-0.750	-0.751	-0.452	1.000			
Wave decay in L.P. 5	0.810	0.876	0.863	0.856	0.889	0.880	0.188	-0.833	-0.596	-0.749	0.820	1.000		
Wave decay potential parameter	0.871	0.862	0.841	0.828	0.899	0.882	0.283	-0.849	-0.683	-0.679	0.892	0.972	1.000	
Shape product (m^2/3)	0.862	0.932	0.806	0.779	0.931	0.868	0.112	-0.875	-0.686	-0.736	0.855	0.948	0.962	1.000

b

Figure 6. Correlation matrices show the correlation coefficients (*r*) between lake geometrical characteristics and the hydrodynamic modeling results. (**a**) Matrix for all lakes (69 models). (**b**) Matrix for all lakes with a maximum depth greater than 20 m (52 scenarios).

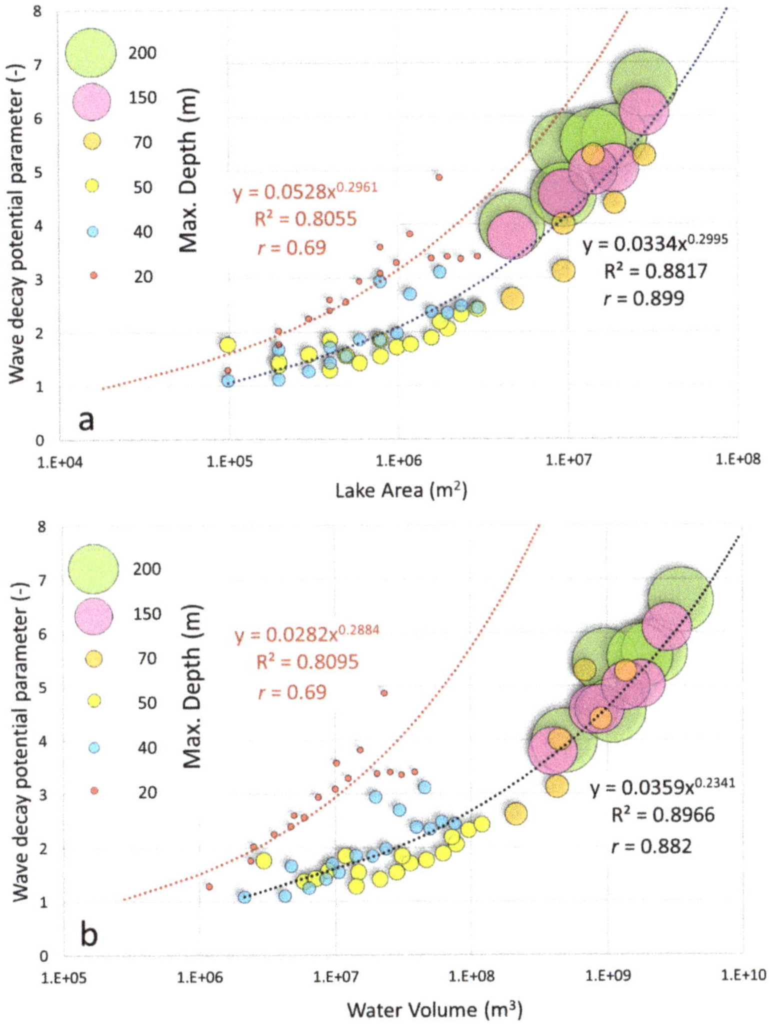

Figure 7. Hyperbolic relations between *WDPP* and (**a**) lake area and (**b**) water volume (X-axis in logarithmic scale, base 10). Circle size and color relate to lake depth. Related equations, coefficient of determination (R^2), and correlation coefficient (r) are also shown. The r-values labeled in red, which are not shown in Figure 6, are correlation coefficients specifically for medium-size lakes with a maximum depth of 20 m. The r-values labeled in black, shown in Figure 6b, are correlation coefficients for theoretical lakes with a maximum depth greater than 20 m.

Table 2. Correlation coefficients (r) between *WDPP* and A, V_w, and *ShpP*; equations; and corresponding coefficients of determination (R^2), where y is *WDPP* and x is A, V_w, or *ShpP*.

Wave Decay Potential Parameter	r	Equation	R^2
Depending on lake area (A)	0.899	(3) $y = 0.0334x^{0.2995}$	0.8817
Depending on water volume (V_w)	0.882	(4) $y = 0.0359x^{0.2341}$	0.8966
Depending on the shape product (*ShpP*)	0.962	(5) $y = 0.0152x + 0.3775$	0.9257

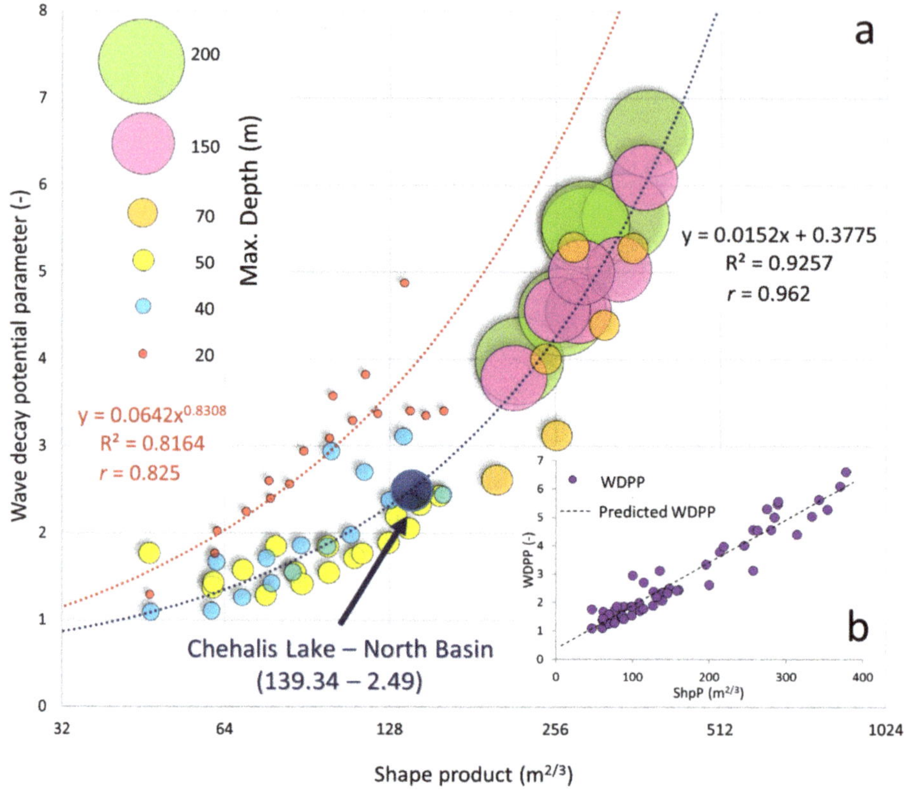

Figure 8. (**a**) Plot of *WDPP* against *ShpP* (x-axis in logarithmic scale, base 2), and related equations, coefficient of determination (R^2), and correlation coefficient (r). Data size and color relate to maximum lake depth. The blue circle refers to the 2007 Chehalis Lake landslide-generated wave case, used as a test of the proposed approach (Section 7). The r-values with red font, which are not shown in Figure 6, correspond only to the red dots. (**b**) Linear regression plot obtained for the relation between *ShpP* and *WDPP* (the resulting equation is the same as the one obtained with the trendline in Figure 8a).

Table 3. Relationships between *WDPP* and flow depths at the shoreline, including correlation coefficients (r), equations, and coefficients of determination (R^2). In the equations, y is flow depth and x is *WDPP* obtained from the equations in Table 2.

Flow Depth at the Shoreline (m)	r	Equation	R^2
Maximum	−0.751	(6) $y = 29.376x^{-0.696}$	0.8745
Mean	−0.849	(7) $y = 17.878x^{-0.779}$	0.9071
Minimum	−0.749	(8) $y = 8.3583x^{-0.583}$	0.6578

The Supplementary Materials include an Excel spreadsheet that summarizes the workflow processes, including the entire calculation procedure and includes informative charts to display the results.

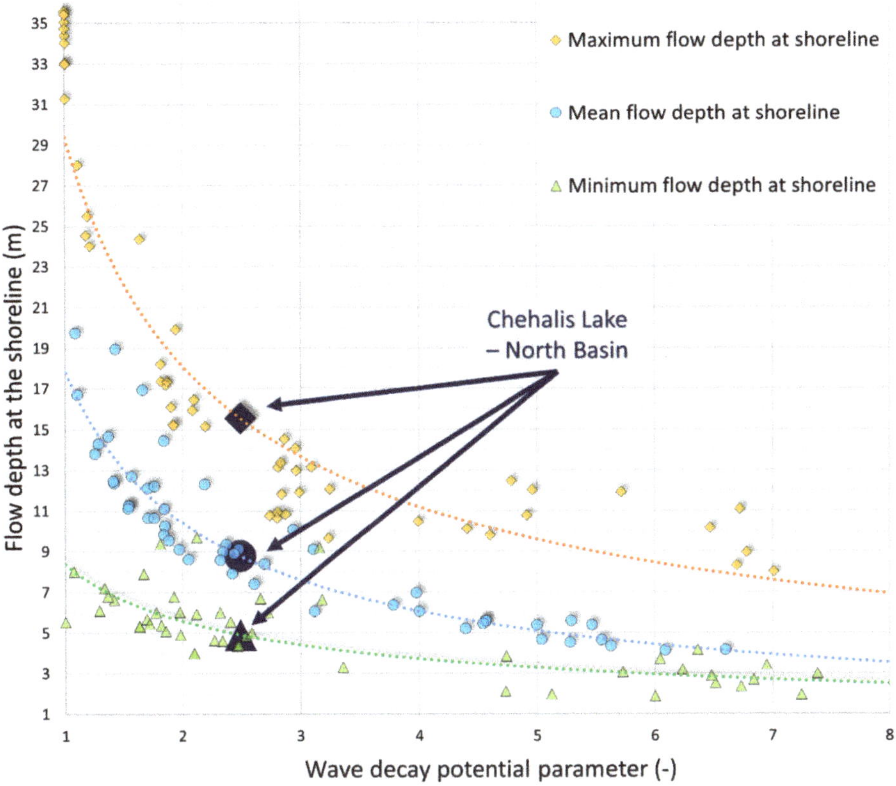

Figure 9. Hyperbolic relation between *WDPP* and flow depth at the shoreline (mean—blue data; maximum—orange data; minimum—green data). Flow depth values for the Chehalis Lake event (blue) are obtained using the 2019 ETH Zurich equations.

6. Example of Application of the Proposed Approach—The Chehalis Lake Landslide-Generated Wave

The landslide-induced wave in Chehalis Lake on 4 December 2007 (Figure 10, see Section 1) [6,7] is chosen as a test case of the proposed methodology for several reasons. It involves a representative mountain basin, and both the landslide [6] and lake characteristics [7] are well documented. In addition, multiple numerical modeling studies [6,42–44] provide insights into open-water hydrodynamics necessary to reproduce documented impacts. Bathymetric data acquired through a SONAR survey [7] are used in QGIS v3.16 to estimate the lake's characteristics (see Section 2). The volume of the initial rockslide is estimated to be about 3 Mm^3. The rock mass rapidly fragmented, transforming into a rock avalanche as it approached the shoreline. Approximately 2.2 Mm^3 of rock debris entered the lake and triggered the wave [6,7,44]. Although this volume is larger than the one employed in hydrodynamics models (see Section 4), the maximum impact speeds are similar (about 60 ms^{-1} [42,45], Figure 10), making the case study suitable for testing the method.

The case study data are first used as inputs in the empirical equations provided by Evers 2019 [46] (ETH-Zurich, 3D approach—Overland flow) to estimate flow depths at the Chehalis Lake shoreline. As this approach cannot be applied to the entire area of the lake, and because the lake is also divided into two sub-basins by a shallow subaqueous ridge, the test is limited to the north sub-basin (yellow rectangle in Figure 10). The sub-basins characteristics and landslide properties are summarized in Figure 10.

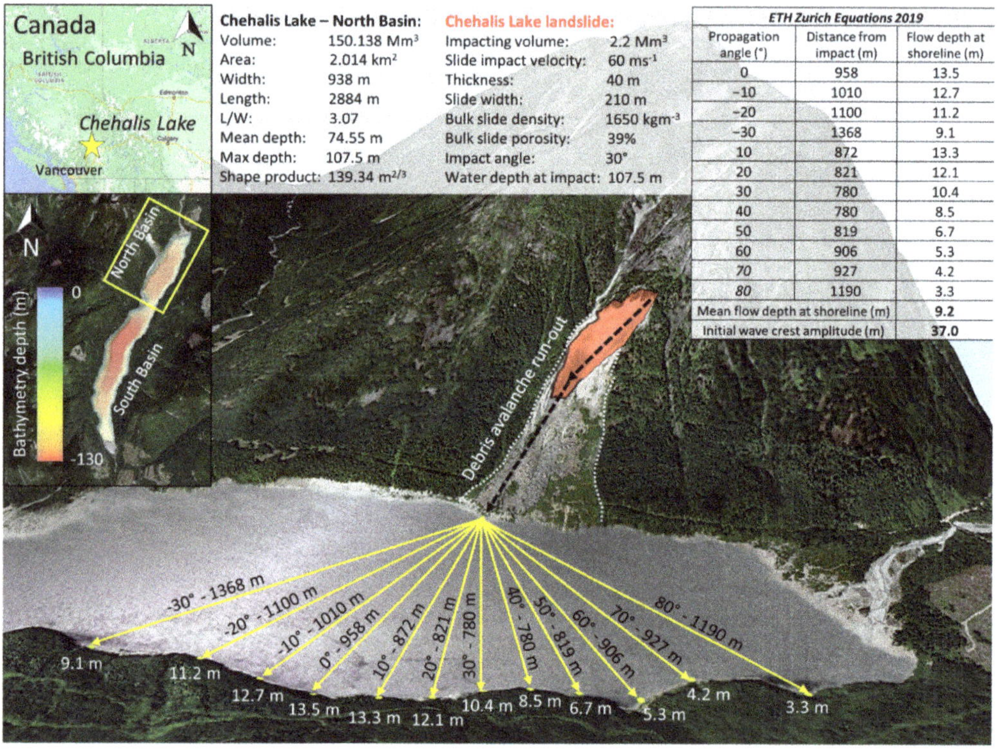

Figure 10. Chehalis Lake overview. Lake and landslide properties are shown (data from [6,7]). The distance and angle of wave propagation from the landslide impact point, which are useful inputs for the 2019 ETH-Zurich equations [46], are also shown. Flow depths at the shoreline in different locations are shown in white. Topographical background taken from Microsoft Bing Maps—2021 Microsoft Corporation Earthstar Geographics Sio (Accessed on 22 July 2021).

Flow depths are calculated at shoreline locations corresponding to wave propagation angles in 10° increments and ranging from −30° to +80° on both sides of the landslide midline (Figure 10). Calculations yield an open-water amplitude of the initial impulse wave of 37 m a.l.l., and shoreline flow depths ranging from 3.3 to 13.5 m, with a maximum opposite the slide source and a mean shoreline flow depth of 9.2 m (Figure 10).

A *ShpP* value of 139.34 $m^{2/3}$ and a *WDPP* value of 2.495 (Equation (5) in Table 2 and Figure 8a) are calculated for the north sub-basin. Appling the latter value in Equations (6)–(8) yields estimates of maximum, mean, and minimum flow depths of 15.5, 8.8, and 4.9 m, respectively. The maximum and minimum values obtained with Equations (6)–(8) are higher than the ones obtained with the ETH equations, but a close match is found for the mean flow depth (Figure 9). However, the ratio of the maximum wave amplitude (37 m a.l.l.) and the *WDPP* of 2.49 is 14.8 m, which is very close to the maximum flow depth calculated using the ETH equations. These results pertain only to the north sub-basin of the Chehalis Lake. Considering the whole lake, the *ShpP*-value is 196 $m^{2/3}$ and the *WDPP* value is 3.36, yielding a slightly lower estimate of maximum, mean, and minimum flow depths at the shoreline of 12.63, 6.95, and 4.12 m, respectively.

These results show the applicability of the proposed method and use of equations in Table 3, with specific values of *WDPP*, for a preliminary evaluation of potential landslide-induced waves in natural alpine lakes to provide reliable estimates of the flow depths at the lake shoreline. Comparisons between the resulting flow depths obtained with the proposed equations and the ones derived using the ETH equations suggest that the new methodology provides a reliable estimate of flow characteristics. Regardless of what was observed with the presented test, appropriate validation of the proposed method is still required, where a comparison with real data or observations would increase the reliability of this approach's application.

7. Comparison between Real and Theoretical Lakes

Based on this work, the *WDPP* is proposed as a suitable first-order descriptor of a lake basin's potential for dissipating an impulse wave. Figure 11 is a plot of *WDPP* values superposed on the alpine lake classification. It provides a screening tool to identify lakes that are most likely to disperse a wave (Section 3). For the medium-size lakes, *WDPP* generally increases as lake size increases and depth decreases. This is consistent with the numerical modeling results (Figure 5)—lower water depths and longer distances to shorelines enhance wave dissipation and thus decrease shoreline flow depths.

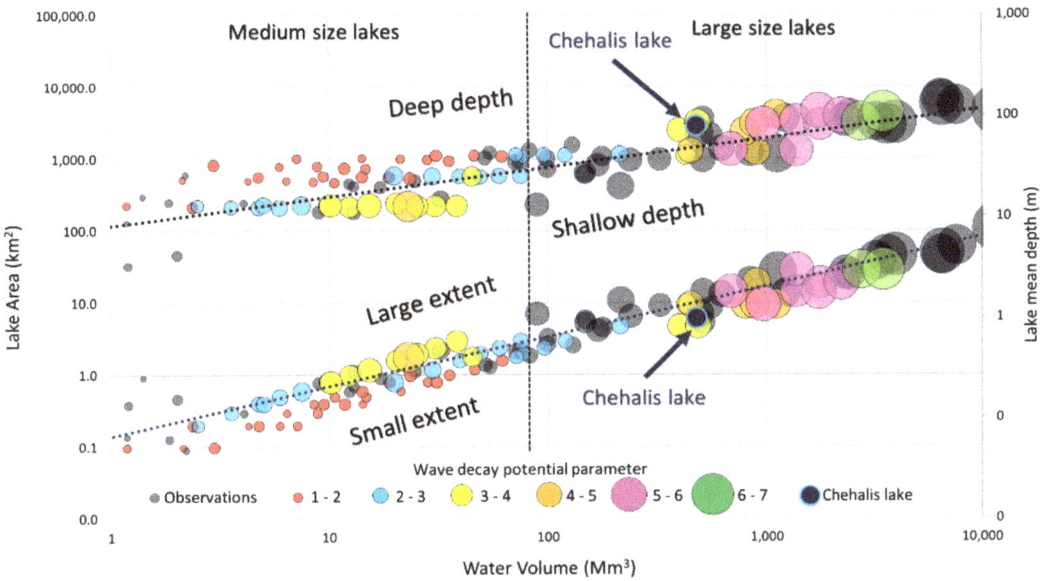

Figure 11. Comparison of modeling results (circle size and color based on *WDPP*; black circles are from the alpine lakes classification chart, see Section 3).

In the case of large-size lakes, *WDPP* provides no clear separation of geometrical characteristics of lake basins (lake extent or depth), although a larger *WDPP* appears to be favored by larger lake volumes.

Figure 12 shows expected *WDPP* values for all the natural lakes in the Alps examined in this study. Circles on the map are lakes for which *WDPP* has been derived from *ShpP* Equation (5). Still, *WDPPs* for lakes with complex shapes (diamonds in Figure 12), for which *ShpP* is not available, are here based, for instance, on the water volume (Table 2, Equation (4)). Generally, high values of *WDPP* are related to perialpine lakes with large areas and volumes, whereas lower *WDPP* values are associated with smaller alpine lakes.

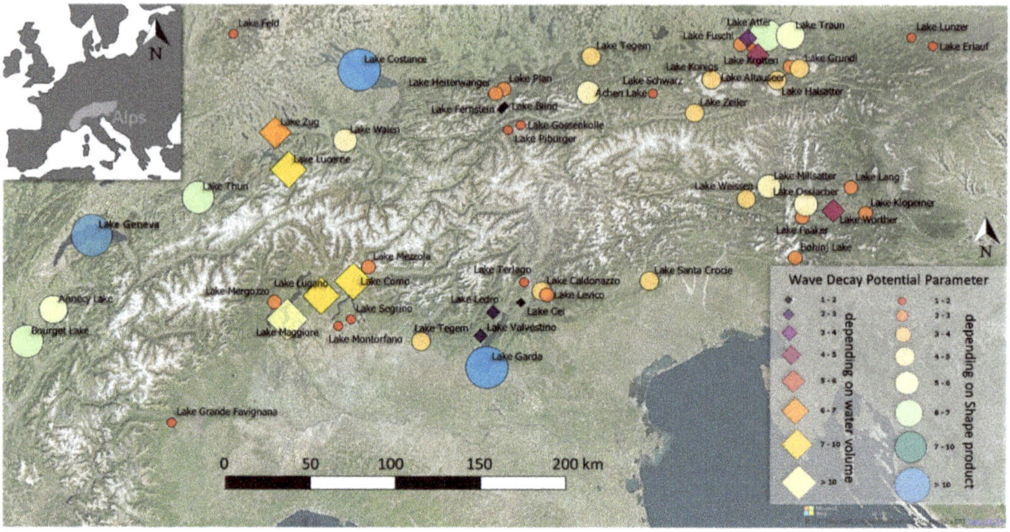

Figure 12. Alpine and perialpine lakes considered in this study and related *WDPP* values obtained using the *ShpP* relation (circles in the legend). Where *ShpP* is unavailable, *WDPP* was obtained using the water volume relation (squares in the legend, see also Figure 7b). Topographical background taken from Microsoft Bing Maps—2021 Microsoft Corporation Earthstar Geographics Sio (Accessed on 26 July 2021).

A critical issue in this work is the reliable range of lakes for which the new approach is applicable (see Table 1 and the complete table of theoretical lakes in the Supplementary Data). The data suggest that the approach has limitations when used for some lakes that are outside of this range, notably some large-size and very large-size lakes (i.e., *WDPP* > 7 in Figure 12). Further study is required to extend these investigations to alpine lakes outside of the stated range in this work to test the broader application of the proposed method.

8. Discussion and Further Required Research

8.1. Applicability of the Proposed Approach

The approach introduced in this work is intended to be a high-level screening tool for identifying the greatest landslide impulse wave threats and thus helping prioritize resources for a more detailed and thorough analysis. The study is based on 3D hydrodynamic modeling and considers a wide range of water body geometries. Results demonstrate the reliability of a lake geometry-based approach for a first-order assessment of the danger posed by potential landslide-triggered impulse waves in mountain lakes. The suggested method finds its applicability in situations where the characteristics of the possible impacting volume are unknown, as only the geometrical properties of lakes are required. Unlike other methods, this approach does not require a deep knowledge of wave theory or the need for additional software. A reliable constraint on wave attenuation, as is provided by this approach, means that the initial wave amplitude needs only to be generally approximated.

The method can be applied to lakes with diverse shapes and dimensions, subject to the limitation that the potential impacting volume has similar characteristics to the one adopted for the modeling in this work (Section 4). In the case of lakes with complex shapes, Equation (5) can be applied to sub-basins with specific geometrical properties to estimate the wave decay potential parameter *WDPP*, as has been done with the Chehalis Lake test case (Section 6). If Equation (5) is not applicable due to unknown lake-geometry parameters in Equation (2), *WDPP* can be estimated using Equation (3) or Equation (4), which consider lake area or volume respectively. The equations in Table 3 provide flow depths expected at the shoreline. However, because models with a maximum depth of less than 20 m were excluded from the later stages of the workflow, this method might overestimate the hazard potential for small shallow lakes.

The Chehalis Lake example shows that analyzing singular sub-basins provides a more conservative estimate of flow characteristics compared to an analysis of the entire lake. As waves propagating from one basin to the next may have already attenuated, sub-basin-specific estimations using this approach can be deemed as worst-case scenarios. It is worth noting that the subdivision of sub-basins does not take into account wave propagation from one basin to another. However, this limitation is a minor issue because a wave entering a second sub-basin has already significantly attenuated, implying that the highest threat is already identified by intra-basin events.

Nonetheless, wherever sub-basins join, complex variations in flow depth at the shoreline might be expected as impulse waves approach from the adjacent sub-basin. This is due to changes in bathymetry (e.g., a shallow sill as at Chehalis Lake, Figure 10), basin alignment (e.g., non-parallel valleys such as the western part of Lake Lucerne, Figure 1b), or both (e.g., the eastern part of Lake Lucerne, Figure 1b). In some instances, these effects might result in locally increased near-shore wave energy and run-up. As a consequence, although intra-basin wave generation gives a conservative assessment of near-shore wave threats, intra-basin waves may underestimate such threats where sub-basins join.

Given the numerical modeling results for the theoretical lakes (Section 5.1) and the results produced for the Chehalis Lake test (Section 6), it is concluded that the proposed approach is applicable for lakes that are flanked by steep unstable slopes, for which failures with volumes ranging from thousands to a few million cubic meters can be expected. However, this assertion requires validation with additional historical case studies from a variety of mountain regions, as well as fiords with basin characteristics similar to alpine lakes. Extending the dataset to other mountain regions for which good data are available (e.g., Canadian Cordillera or a select part of the Andes) to newly formed glacial water bodies [39] and artificial reservoirs would also improve the analysis and results.

8.2. Limitations, Uncertainties, and Further Potential Development

This work provides a workflow to estimate near-shore impulse-wave magnitude in terms of flow depth at the shoreline (see "Excel spreadsheet—Computational tool" in the Supplementary Materials). By contrast, most other studies report impacts in terms of inundation distance and run-up height, metrics that are not considered in this work. Nonetheless, shoreline water depth provides a preliminary characterization of run-up potential and inundation, although these are also heavily affected by the morphology and the steepness of the surrounding topography. As a result, observations of an impulse wave or post-event measurements of its impacts provide only a very general indication of the large-scale variability of flow depth at the shoreline. A more reasonable assessment of this new approach can be achieved using previously established relationships that independently estimate flow depth at the shoreline, as was done for the Chehalis Lake case study (Section 6).

A current disadvantage of this work is the lack of adequate validation. The Chehalis Lake case study and comparison of the results with those based on the ETH equations are insufficient to fully evaluate the validity of this approach. Proper validation would entail a comparison of the results derived using this method with real data collected in the field. However, flow depths at the shoreline caused by an impulse wave are not directly observable in the field and therefore must be calculated independently. Such calculation requires the use of detailed, site-specific, retrospective numerical simulations capable of accurately recreating a landslide-generated wave event.

Future extension and expansion of the proposed approach will involve two critical steps: (1) implementing numerical analyses on well documented, real-world events to validate the geometry-based strategy as a predictor of flow depth at the shoreline; and (2) developing rapid, low-computational-cost methods for estimating run-up and flooding potential from flow depth at the shoreline. This next phase of development will expand the range of outputs to include expected run-up, thereby extending the applicability of the proposed approach and enchaining opportunities to reliably validate it.

Another possible limitation of this study is its applicability to large lakes and a single impact volume. The size and properties of the impacting landslide substantially influence wave characteristics. For example, the maximum wave height resulting from an impact is dependent on landslide volume and competence, debris thickness and frontal width, impact speed, and slope angle and roughness [17,37]. Further research that addresses these issues is required. For example, the approach can be extended to consider diverse characteristics of the impacting volume to provide a relation for *WDPP* that considers landslide properties in addition to the geometrical characteristics of the lake (the *ShpP*). Incorporation of additional consideration into the workflow—including by varying impact processes, volumes, and velocities—might lead to the development of additional equations that can be applied to the full range of real-world situations. This can be accomplished in Flow-3D by adopting the concept of the impacting volume being a dense fluid, as done in the present study. Different volumes, shapes, impact velocities, and slope angles can be implemented in the model set-up. Scaling impact volumes relative to lake volume may assist in standardizing comparisons between lakes. In this work, the slide source is located halfway along the long side of the lake. Further development of the approach should consider different impacting locations along the lakeshore to extend the applicability of the suggested method. Moreover, it would be valuable to consider submerged slide sources, as subaqueous mass movements can also trigger lake impulse waves [47,48].

Further development of this methodology would also benefit from the inclusion of geotechnical properties of the materials that collapse and enter the lake. This would be especially interesting when studying mass movements made up of unlithified materials or heterogeneous rock masses, as well as studies of moraine-dammed lakes where large sediment masses can fail. Finally, landslide hazard and the erosional vulnerability of nearshore elements should be addressed in the context of risk analysis to complement the relative level of threat posed by the degree and pattern of impulse wave attenuation in a given basin.

9. Conclusions

A novel method for conducting a preliminary evaluation of the size and propagation of landslide-induced waves in mountain lakes is proposed. A set of equations is used within a numerical modeling framework to characterize the geometric characteristics of a lake and quickly assess the possible wave threat in terms of wave dissipation and expected flow depth along the shoreline (which is different from the expected run-up). Wave propagation controlled by lake geometry provides a general indication of where run-up or inundation would be the highest, although the behavior of the breaking wave is also heavily influenced by topography along the shoreline. The incorporation of landslide properties, behavior, and locations into the workflow will further increase the utility of this approach. The proposed method can be used as a first-order indicator for prioritizing lakes that appear to be particularly susceptible to landslide-induced impulse waves.

The findings of this study reveal that *WDPP* is a valid metric for the "dissipation power" (attenuation) of an impulse wave in a mountain lake and that it correlates well to lake geometrical characteristics, particularly area and water volume. The linear relationship between *WDPP* and *ShpP*, a metric that takes into account diverse lake properties, provides a valuable approach for estimating potential wave dissipation. If the *ShpP* is unavailable, for example in lakes with complex shapes, water volume can be used to estimate *WDPP*. Furthermore, *WDPP* can be used to calculate the expected flow depth at the lake shoreline. Results suggest that large perialpine lakes ($V_w > 10.000$ Mm^3) are more likely to disperse impulse waves; these attenuate less in smaller alpine lakes ($V_w < 10$ Mm^3).

This method can be applied worldwide to mountain lakes that differ in shape, extent, and volume. The minimum required input data are the geometrical characteristics of the lakes. Unlike other proposed methodologies, specific technical knowledge or additional inputs such as digital elevation data are not required, making the proposed method easy to use. However, inclusion of high-resolution digital bathymetric data would produce more accurate assessments.

The proposed method is tested with data available for the 2007 Chehalis Lake landslide-generated wave event. Results match well those obtained using the empirical equations published by ETH Zurich (2019 Edition—3D approach, [17,46]). Despite the positive results, one case study is insufficient to assess the overall reliability of our novel approach, and a validation considering real, observed data is still required. Indeed, additional historic events in which wave characteristics have been directly observed would be particularly valuable for further assessing the method. Furthermore, the applicability of impact volumes that differ from the one used in this study, together with a variety of impact characteristics, must be determined.

This study is an initial step in the development of this methodology; additional research is required to improve the method and its application, for example by taking into account properties of the potential impacting mass. Stability analyses of slopes bordering lakes with high hazards from impulse waves, including implications of external triggers such as intense rain events or earthquakes, are recommended to contribute to a proper estimate of the related hazard in a cascade effect context. Recently formed alpine lakes due to glacier retreat should also be included in these investigations.

Supplementary Materials: The following are available online at: https://zenodo.org/record/5569220 (accessed on 29 November 2021), Data set—Zenodo: https://doi.org/10.5281/zenodo.5569220, Data set (Excel spreadsheet): Alpine Lakes and Artificial Lakes; Data set (STL files): Alpine lakes. The following is available online at: https://zenodo.org/record/5733950 (accessed on 29 November 2021), Zenodo: https://doi.org/10.5281/zenodo.5733950, Computational tool (Excel spreadsheet): UIBK_Computational tool for Landslide_Induced Impulse Wave 2021.

Author Contributions: Conceptualization, A.F. and B.G.; data curation, A.F., N.J.R., J.J.C. and B.G.; formal analysis, A.F., B.S.-M. and N.J.R.; investigation, A.F. and B.S.-M.; methodology, A.F. and B.G.; software, B.G.; supervision, B.S.-M. and B.G; validation, N.J.R.; visualization, A.F., N.J.R. and J.J.C.; writing—original draft, A.F.; writing—review and editing, B.S.-M., N.J.R., J.J.C. and B.G. All authors have read and agreed to the published version of the manuscript.

Funding: This research received no external funding.

Data Availability Statement: Open source data set and free access tools that support the findings in this work are available in the links posted in the Supplementary Materials.

Acknowledgments: We thank the Sedimentary Geology Working Group at the University of Innsbruck for the useful data, especially Jasper Moernaut and Michael Strasser. We thank the anonymous referees for their constructive contribution in helping to improve this work.

Conflicts of Interest: The authors declare no conflict of interest. The funders had no role in the design of the study; in the collection, analyses, or interpretation of data; in the writing of the manuscript, or in the decision to publish the results.

References

1. Roberts, N.J.; McKillop, R.; Hermanns, R.L.; Clague, J.J.; Oppikofer, T. Preliminary Global Catalogue of Displacement Waves from Subaerial Landslides. In *Landslide Science for a Safer Geoenvironment*; Sassa, K., Canuti, P., Yin, Y., Eds.; Springer: New York, NY, USA, 2014; pp. 687–692.
2. Coe, J.A.; Bessette-Kirton, E.K.; Geertsema, M. Increasing rock-avalanche size and mobility in Glacier Bay National Park and Preserve, Alaska detected from 1984 to 2016 Landsat imagery. *Landslides* **2018**, *15*, 393–407. [CrossRef]
3. Kos, A.; Amann, F.; Strozzi, T.; Delaloye, R.; Von Ruette, J.; Springman, S. Contemporary glacier retreat triggers a rapid landslide response, Great Aletsch Glacier, Switzerland. *Geophys. Res. Lett.* **2016**, *43*, 12–466. [CrossRef]
4. Harrison, S.; Kargel, J.S.; Huggel, C.; Reynolds, J.; Shugar, D.H.; Betts, R.A.; Emmer, A.; Glasser, N.; Haritashya, U.K.; Klimeš, J.; et al. Climate change and the global pattern of moraine-dammed glacial lake outburst floods. *Cryosphere* **2018**, *12*, 1195–1209. [CrossRef]

5. Higman, B.; Shugar, D.H.; Stark, C.P.; Ekström, G.; Koppes, M.N.; Lynett, P.; Dufresne, A.; Haeussler, P.J.; Geertsema, M.; Gulick, S.; et al. The 2015 landslide and tsunami in Taan Fiord, Alaska. *Sci. Rep.* **2018**, *8*, 1–12. [CrossRef]
6. Brideau, M.A.; Sturzenegger, M.; Stead, D.; Jaboyedoff, M.; Lawrence, M.; Roberts, N.; Ward, B.; Millard, T.; Clague, J. Stability analysis of the 2007 Chehalis lake landslide based on long-range terrestrial photogrammetry and airborne LiDAR data. *Landslides* **2012**, *9*, 75–91. [CrossRef]
7. Roberts, N.J.; Mckillop, R.J.; Lawrence, M.S.; Psutka, J.F.; Clague, J.J.; Brideau, M.; Ward, B.C. Impacts of the 2007 Landslide-Generated Tsunami in Chehalis Lake, Canada. *Landslide Sci. Pract.* **2013**, *6*, 133–140. [CrossRef]
8. Gylfadóttir, S.S.; Kim, J.; Helgason, J.K.; Brynjólfsson, S.; Höskuldsson, Á.; Jóhannesson, T.; Harbitz, C.B.; Løvholt, F. The 2014 Lake Askja rockslide-induced tsunami: Optimization of numerical tsunami model using observed data. *J. Geophys. Res. Ocean.* **2017**, *122*, 2647–2651. [CrossRef]
9. Franco, A.; Moernaut, J.; Schneider-Muntau, B.; Strasser, M.; Gems, B. Triggers and consequences of landslide-induced impulse waves—3D dynamic reconstruction of the Taan Fiord 2015 tsunami event. *Eng. Geol.* **2021**, *294*, 106384. [CrossRef]
10. Paris, A.; Okal, E.A.; Guérin, C.; Heinrich, P.; Schindelé, F.; Hébert, H. Numerical Modeling of the June 17, 2017 Landslide and Tsunami Events in Karrat Fjord, West Greenland. *Pure Appl. Geophys.* **2019**, *176*, 3035–3057. [CrossRef]
11. Heller, V.; Moalemi, M.; Kinnear, R.D.; Adamns, R.A. Geometrical Effects on Landslide-Generated Tsunamis. *J. Waterw. Port Coast. Ocean Eng.* **2012**, *138*, 286–298. [CrossRef]
12. Ruffini, G.; Heller, V.; Briganti, R. Numerical characterisation and efficient prediction of landslide-tsunami propagation over a wide range of idealised bathymetries. *Coast. Eng.* **2021**, *167*, 103854. [CrossRef]
13. Zhang, Y.; Li, D.; Chen, L.; Yin, K.; Xiao, L.; Fu, X.; Glade, T.; Leo, C. Numerical analysis of landslide-generated impulse waves affected by the reservoir geometry. *Eng. Geol.* **2019**, *266*, 105390. [CrossRef]
14. Ruffini, G.; Heller, V.; Briganti, R. Numerical modelling of landslide-tsunami propagation in a wide range of idealised water body geometries. *Coast. Eng.* **2019**, *153*, 103518. [CrossRef]
15. Heller, V.; Hager, W.H.; Minor, H. *Landslide Generated Impulse Waves in Reservoirs: Basics and Computation*; ETH Zurich: Zurich, Switzerland, 2009.
16. Heller, V.; Hager, W.H. Impulse product parameter in landslide generated impulse waves. *J. Waterw. Port Coast. Ocean Eng.* **2010**, *136*, 145–155. [CrossRef]
17. Evers, F.; Heller, V.; Fuchs, H.; Hager, W.H.; Boes, R. *Landslide-Generated Impulse Waves in Reservoirs: Basics and Computation*, 2nd ed.; ETH Zurich: Zurich, Switzerland, 2019.
18. Strupler, M.; Evers, F.M.; Kremer, K.; Cauzzi, C.; Bacigaluppi, P.; Vetsch, D.; Boes, R.M.; Fäh, D.; Anselmetti, F.S.; Wiemer, S. A Simplified Classification of the Relative Tsunami Potential in Swiss Perialpine Lakes Caused by Subaqueous and Subaerial Mass-Movements. *Front. Earth Sci.* **2020**, *8*, 564783. [CrossRef]
19. Oswald, P.; Moernaut, J.; Fabbri, S.C.; De Batist, M.; Hajdas, I.; Ortner, H.; Titzler, S.; Strasser, M. Combined On-Fault and Off-Fault Paleoseismic Evidence in the Postglacial Infill of the Inner-Alpine Lake Achensee (Austria, Eastern Alps). *Front. Earth Sci.* **2021**, *9*, 438. [CrossRef]
20. Oswald, P.; Strasser, M.; Hammerl, C.; Moernaut, J. Seismic control of large prehistoric rockslides in the Eastern Alps. *Nat. Commun.* **2021**, *12*, 1059. [CrossRef] [PubMed]
21. Hilbe, M.; Anselmetti, F.S. Signatures of slope failures and river-delta collapses in a perialpine lake (Lake Lucerne, Switzerland). *Sedimentology* **2014**, *61*, 1883–1907. [CrossRef]
22. Hilbe, M.; Anselmetti, F.S.; Eilertsen, R.S.; Hansen, L.; Wildi, W. Subaqueous morphology of Lake Lucerne (Central Switzerland): Implications for mass movements and glacial history. *Swiss J. Geosci.* **2011**, *104*, 425–443. [CrossRef]
23. Sample, H. Carinthia Lakes and Reservoirs in Austria. In *Encyclopedia of Lakes and Reservoirs*; Bengtsson, L., Herschy, R.W., Fairbridge, R.W., Eds.; Encyclopedia of Earth Sciences Series; Springer: New York, NY, USA, 2012; pp. 121–175. [CrossRef]
24. Chinaglia, N.; Magni, D.; Premazzi, G. Alplakes. Survey between Land and Water. AlpLakes. Alpine Lakes Network. Available online: http://www.alpine-space.org/ (accessed on 4 December 2020).
25. Hellström, T. Alpine lakes. In *Encyclopedia of Earth Sciences Series*; Facts on File, Inc.: New York, NY, USA, 2012. [CrossRef]
26. Dokulil, M. European Alpine lakes. In *The Lakes Handbook*; O'Sullivan, P.E., Reynolds, C.S., Eds.; Blackwell: Oxford, UK, 2005; Volume 2.
27. *Flow Science Inc.FLOW-3D®Solver*; Version 11.1; Flow Science Inc.: Santa Fe, NM, USA, 2017.
28. Harlow, F.H.; Welch, J.E. A domain decomposition method for incompressible viscous flow of fluid with free surface. *Phys. Fluids* **1965**, *8*, 2182–2189. [CrossRef]
29. Nichols, B.D.; Hirt, C.W.; Hotchkiss, R.S. Volume of fluid (VOF) method for the dynamics of free boundaries. *J. Comput. Phys.* **1981**, *39*, 201–225.
30. Welch, J.E.; Harlow, F.H.; Shannon, J.P.; Daly, B.J. *The MAC Method-A Computing Technique for Solving Viscous, Incompressible, Transient Fluid-Flow Problems Involving Free Surfaces*; Los Alamos Scientific Laboratory Report LA-3425; Los Alamos Scientific Laboratory: Los Alamos, NM, USA, 1968. [CrossRef]
31. Hinze, J.O. Isotropic Turbulence. In *Turbulence*, 2nd ed.; McGraw-Hill: New York, NY, USA, 1975; pp. 165–204.
32. Rady, R. 2D-3D modeling of flow over sharp-crested weirs. *J. Appl. Sci. Res.* **2011**, *7*, 2495–2505.
33. Hirt, C.W.; Sicilian, J.M. A Porosity Technique for the Definition of Obstacles in Rectangular Cell Meshes. In Proceedings of the 4th International Conference on Numerical Ship Hydrodynamics, Washington, DC, USA, 24–27 September 1985; p. 19.

34. Yakhot, V.; Smith, L.M. The renormalization group, the ϵ-expansion and derivation of turbulence models. *J. Sci. Comput.* **1992**, *7*, 35–61. [CrossRef]
35. Harlow, F.H.; Nakayama, P.I. *Transport of Turbulence Energy Decay Rate*; Los Alamos Scientific Laboratory University California Report LA-3854; Los Alamos Scientific Laboratory University California: Los Alamos, NM, USA, 1968.
36. Chung, T.J. *Computational Fluid Dynamics*; Cambridge University Press: Cambridge, UK, 2010.
37. Ersoy, H.; Karahan, M.; Gelişli, K.; Akgün, A.; Anılan, T.; Sünnetci, M.O.; Yahşi, B.K. Modelling of the landslide-induced impulse waves in the Artvin Dam reservoir by empirical approach and 3D numerical simulation. *Eng. Geol.* **2019**, *249*, 112–128. [CrossRef]
38. Franco, A.; Moernaut, J.; Schneider-Muntau, B.; Strasser, M.; Gems, B. The 1958 Lituya Bay tsunami—Pre-event bathymetry reconstruction and 3D numerical modelling utilising the computational fluid dynamics software Flow-3D. *Nat. Hazards Earth Syst. Sci.* **2020**, *20*, 2255–2279. [CrossRef]
39. Haeberli, W.; Buetler, M.; Huggel, C.; Friedli, T.L.; Schaub, Y.; Schleiss, A.J. New lakes in deglaciating high-mountain regions—Opportunities and risks. *Clim. Chang.* **2016**, *139*, 201–214. [CrossRef]
40. Schnellmann, M.; Anselmetti, F.S.; Giardini, D.; McKenzie, J.A. 15,000 Years of mass-movement history in Lake Lucerne: Implications for seismic and tsunami hazards. *Eclogae Geol. Helv.* **2006**, *99*, 409–428. [CrossRef]
41. De Blasio, F.V. Landslides in Peculiar Environments. In *Introduction to the Physics of Landslides*; Springer: New York, NY, USA, 2011; pp. 238–261. [CrossRef]
42. Ghaïtanellis, A.; Violeau, D.; Liu, P.L.F.; Viard, T. SPH simulation of the 2007 Chehalis Lake landslide and subsequent tsunami. *J. Hydraul. Res.* **2021**, *59*, 863–887. [CrossRef]
43. Viard, T. Hydraulic 2D Model of Landslide Generated Wave–Cases of Chehalis Lake and Chambon Lake. In Proceedings of the XXIVth TELEMAC-MASCARET User Conference, Graz, Austria, 17–20 October 2017; pp. 113–118.
44. Si, P.; Aaron, J.; McDougall, S.; Lu, J.; Yu, X.; Roberts, N.J.; Clague, J.J. A non-hydrostatic model for the numerical study of landslide-generated waves. *Landslides* **2018**, *15*, 711–726. [CrossRef]
45. Wang, J.; Ward, S.N.; Xiao, L. Numerical simulation of the December 4, 2007 landslide-generated tsunami in Chehalis Lake, Canada. *Geophys. J. Int.* **2015**, *201*, 372–376. [CrossRef]
46. Evers, F.M. Computational Tool for 'Landslide-Generated Impulse Waves in Reservoirs—Basics and Computation' (2019). (1.0) Zenodo. Available online: https://zenodo.org/record/3492000#.YadWZtDMJPY (accessed on 29 November 2021). [CrossRef]
47. Strupler, M.; Hilbe, M.; Kremer, K.; Danciu, L.; Anselmetti, F.S.; Strasser, M.; Wiemer, S. Subaqueous landslide-triggered tsunami hazard for Lake Zurich, Switzerland. *Swiss J. Geosci.* **2018**, *111*, 353–371. [CrossRef]
48. Strupler, M.; Anselmetti, F.S.; Michael Hilbe, M.; Kremen, S.; Wiemer, S. A workflow for the rapid assessment of the landslide-tsunami hazard in peri-alpine lakes. *Geol. Soc. Lond. Spec. Publ.* **2020**, *500*, 81–95. [CrossRef]

applied sciences

MDPI

Article

A Case Study of a Large Unstable Mass Stabilization: "El Portalet" Pass at the Central Spanish Pyrenees

Guillermo Cobos [1,*], Miguel Ángel Eguibar [2], Francisco Javier Torrijo [1,3] and Julio Garzón-Roca [4]

[1] Department of Geological and Geotechnical Engineering, Universitat Politècnica de València, 46022 Valencia, Spain; meguibar@hma.upv.es
[2] Department of Hydraulic Engineering and Environment, Institute for Water and Environmental Engineering (IIAMA), Universitat Politècnica de València, 46022 Valencia, Spain; fratorec@trr.upv.es
[3] Research Centre PEGASO, Universitat Politècnica de València, 46022 Valencia, Spain
[4] Department of Geodynamics (GEODESPAL), Faculty of Geology, Complutense University of Madrid, 28040 Madrid, Spain; julgarzo@ucm.es
* Correspondence: gcobosc@trr.upv.es

Abstract: This case study presents the engineering approach conducted for stabilizing a landslide that occurred at "El Portalet" Pass in the Central Spanish Pyrenees activated due to the construction of a parking lot. Unlike common slope stabilization cases, measures projected here were aimed at slowing and controlling the landslide, and not completely stopping the movement. This decision was taken due to the slow movement of the landslide and the large unstable mass involved. The degree of success of the stabilization measures was assessed by stability analyses and data obtained from different geotechnical investigations and satellite survey techniques such as GB-SAR and DinSAR conducted by different authors in the area under study. The water table was found to be a critical factor in the landslide's stability, and the tendency of the unstable slope for null movement (total stability) was related to the water table lowering process, which needs more than 10 years to occur due to regional and climatic issues. Results showed a good performance of the stabilization measures to control the landslide, demonstrating the effectiveness of the approach followed, and which became an example of a good response to the classical engineering duality cost–safety.

Keywords: landslide; safety factor; Central Spanish Pyrenees; soft rocks; stabilization measures

Citation: Cobos, G.; Eguibar, M.Á.; Torrijo, F.J.; Garzón-Roca, J. A Case Study of a Large Unstable Mass Stabilization: "El Portalet" Pass at the Central Spanish Pyrenees. *Appl. Sci.* **2021**, *11*, 7176. https://doi.org/10.3390/app11167176

Academic Editors: Ricardo Castedo, Miguel Llorente Isidro and David Moncoulon

Received: 2 July 2021
Accepted: 2 August 2021
Published: 4 August 2021

1. Introduction

Slope stabilization is probably one of the most typical, ancient and challenging issues of civil engineering. Commonly, problematic landslides affecting buildings and infrastructures are solved by installing a series of measures that lead to completely and immediately stopping the ground movement. Those measures normally include rigid retaining walls, anchors or a substantial variation of the ground profile's geometry, and their implementation can give rise to a compromise between cost and safety [1–3]. However, when dealing with a great amount of ground material, stopping the movement absolutely and instantaneously may not be the optimal solution. The literature shows the possibility of using different techniques for reducing the mobility of landslides, including small excavations and adjustments of the slope profile, the implementation of flexible walls or piles and the installation of drainage [4–9]. The latter is especially important since landslides are often triggered by precipitations and the resulting change in the groundwater level. Water reduces ground strength and increases pore pressures, contributing to the instability of a slope [7,10–13]. Controlling the water table is essential when dealing with shallow and slow movements, as any rainfall may accelerate or even reactivate a landslide. The previous issues are particularly relevant in areas characterized by rainy seasons followed by dry periods, such as the Mediterranean region and, more precisely, the Iberian Peninsula.

The "El Portalet" Pass is located at the municipality of "Sallent de Gallego", Huesca province, in Spain, and belongs to the Central Spanish Pyrenees. In 2004, an excavation to create a parking lot next to a local road was performed in the area. Parking excavation was conducted and resulted in the activation of a large landslide. That landslide occurred on a hillside in the surroundings of Petrusos Peak, in an area of about 0.35 km^2, where superficial cracking and diverse instabilities at different movement rates had been identified in the past [14]. A carboniferous substrate characterizes the zone, affected by faults of slate nature, with gravel colluvium and sandy clay materials on foothills as a result of previous paleo slides. Natural slopes (average angles of 12°) were also identified to be on the verge of instability [15].

The parking excavation reactivated small slides previously identified in the area [16,17] and produced new ones. The first instability signals were displayed at the end of the year, when large cracks appeared in the head of the slope and propagated afterwards along the hillside, showing cracks and longitudinal and transversal deformations, even breaking the slope toe. This type of deformation characterizes an paleo-slide failure mechanism involving an important volume of material [15,18–20]. Due to the great deformations and large unstable masses involved in the landslide, the typical solutions aimed at absolutely and instantaneously stopping the ground movement would have represented a higher cost than the total elimination of the unstable hillside. Instead, the approach was based on controlling the hillside movement evolution and implementing a series of measures to slow down the ground movements, achieving the stabilization of the slope after some years, but also ensuring no damage to any infrastructure during that time.

This paper shows and analyses the stabilization measures projected and implemented on "El Portalet" Pass. The area has been intensively investigated since 2004 by diverse authors. Classical geotechnical investigations, including boreholes and the installation of inclinometers [15], were conducted after landslide activation. Several Differential GPS campaigns as well as the use of different satellite surface monitoring techniques, including the ground-based SAR (GB-SAR) technique and the Differential Synthetic Aperture Radar Interferometry (DinSAR) technique [14,21–24], were performed from 2006 to 2010. Besides, a new geotechnical investigation was conducted to monitor the landslide evolution 10 years after the initial implementation of the stabilization measures. These data are presented for the first time in this paper. Data coming from all of those investigations are used to assess the performance and the degree of success of the stabilization measures, establishing the expected future evolution of the landslide.

2. Materials and Methods

2.1. Geographical and Geological Situation

The case study is located at "Sallent de Gallego" (Huesca, Spain). That area belongs to the high basin of the Gallego River Valley (Figure 1), in the Central Spanish Pyrenees.

The studied hillside corresponds with the southwestern slope of the spurs located between Petrusos Peak and the Old Pass of Sallent (latitude 42°48′4.83″ N; longitude 0°24′48.19″ O), with heights between 2128 m and 1848 m. The road next to which the parking lot was built, road A-136, is situated at the hillside foot, and links "El Portalet" Pass and "Sallent de Gallego" village. The Gallego River flows through lower elevations, parallel to the road. The excavation was performed at altitudes between 1735 and 1775 m. Figure 1 shows a parking excavation view, where road A-136 can also be seen.

From the geological point of view, the area under study is located in the Pyrenean Axial Zone, primarily composed of Devonian and Carboniferous materials, amongst which slates (sometimes with sandstone interlayers) are common, with some calcareous zones, all of which are affected by an intense Hercynian folding, accompanied by low-grade metamorphism. From a tectonic point of view, all of these materials are part of the Gavarnie thrust, which contains part of the southern configuration of the Pyrenean chain. Most of the hillside is covered by recent colluvial and quaternary deposits [25–27].

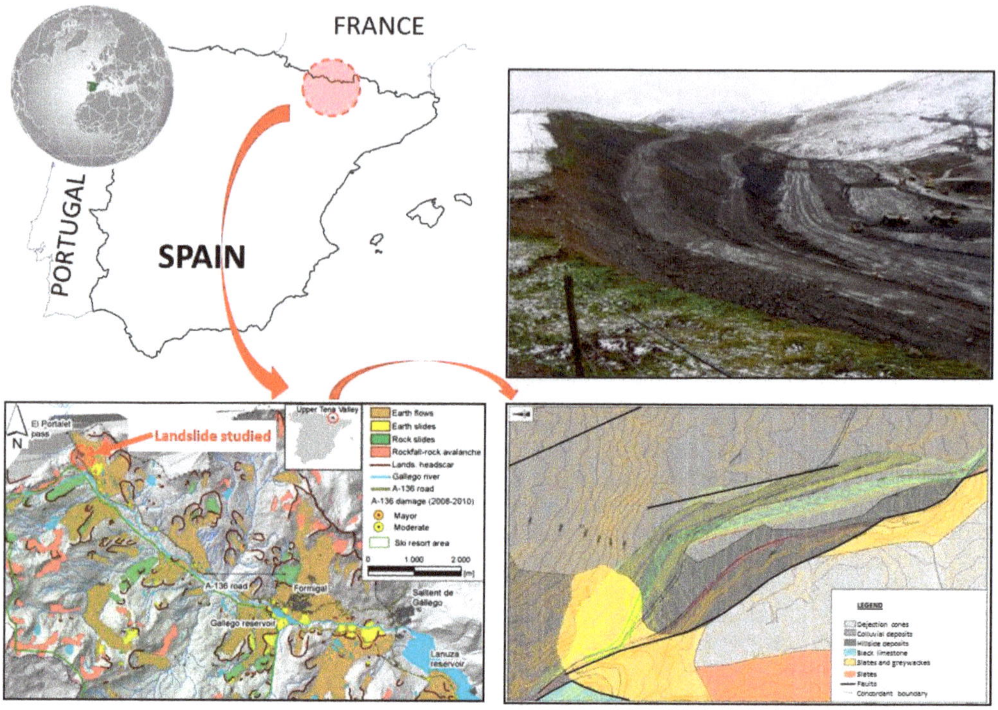

Figure 1. Geographical and geological setting of the area under study in the Central Spanish Pyrenees and view of the parking excavation works (note: bottom left image modified from [22]).

The area contains low strength rocks with intense active slope processes and with portions with low structural control. This is observed between the Inner Chains and "El Portalet" on the Gallego River basin, where slates, schists and clays have caused several slope evolution phenomena, associated with mass movements. Slumps are the most spectacular shapes, which leave a scar or tension crack on the head of the slope. Those slumps are deep and affect the substrata, also presenting deep weathering due to both their plasticity and tectonization. In some cases, Devonian slates have acted as a lubricant, their structure and faults located at the top of the slope sometimes causing an additional instability issue. The length of tension cracks can exceed 1 km, and the landslides portray the typical internal corrugation, a consequence of a mass movement.

One of these slumps was mapped by García Ruiz [16,17]. Two parallel tension cracks, which form successive steps (see Figure 1), match the southern slope of the Petrusos spurs and they also overlap with the convex surface presented by the rocky hillside. At its toe, the slumps can be found with two coalescent lobes that reach the fracture northwest of the sunken block of "El Portalet". The high density of such morphologies can be found on the Gallego head of the slope, especially between "El Portalet" and Formigal ski station in addition to the Peña Foratata southwestern slope, where superposed landslides tend to form great slumps. Most of them do not look very active, but other slumps are still moving or show a local reactivation.

This area is also located in a seismic zone, presenting a basic seismic acceleration of 0.10 g (with g being the gravitational acceleration), according to the Spanish NCSR-02 standard [28].

2.2. Previous Field Investigations

Previous geological–geotechnical investigations [15] conducted in 2005 consisted of six boreholes (location is shown in Figure 2) with depths between 24 m and 40 m. Standard Penetration Tests (SPT) were performed on each borehole when crossing soils and very weathered rocks, with a frequency of about 3 m. Disturbed and undisturbed samples were extracted from each level identified. Laboratory tests conducted on such samples included general identification tests (e.g., grain size, Atterberg limits, unit weight) and mechanical tests (uniaxial compression strength on rocks and direct CD shear tests and triaxial CU tests on soils).

Inclinometers were installed on each borehole performed. Besides, topographical landmarks were placed along the hillside to measure surface movements. Attempts to locate landmarks were made next to the existing tension cracks on the head of the slope as well as in the excavated slope. Movements between pairs of landmarks were measured every 15 days and five measurement campaigns were carried out.

In addition, an inventory of water points (e.g., springs and permanent courses) was carried out along the entire hillside, also recording the water table position in the excavated slope.

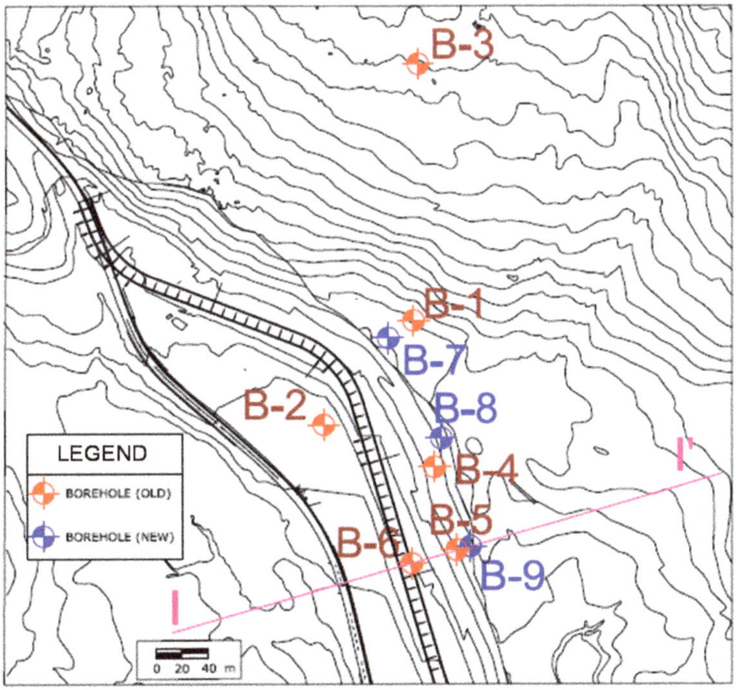

Figure 2. Location of the boreholes performed in the previous study [15] ("old" boreholes, conducted in 2005, reprinted with permission from ref. [15], Copyright Year 2005, Copyright Owner's F.J. Torrijo) and the new geotechnical investigation ("new" boreholes, conducted in 2016); profile I-I′ relates to Figure 5.

2.3. Corrective Measures

A series of corrective measures were proposed to be performed on the slope to stabilize the landslide. Those measures (Figure 3) were not aimed to stop the hillside movement immediately and absolutely but to slow it down and prevent any damage to surrounding infrastructures. Eventually, the measures projected were expected to completely stabilize the hillside and stop the movement (or reduce it to a minimal value) after some years' time. Corrective measures applied included:

- A slight modification of the slope geometry by means of the excavations of intermediate areas (benches) on the slope.
- Installation of deep drainage measures (Californian drains) as well as ditches to improve slope drainage and remove water coming from seepage and runoff; in total, 47 single drains were installed with a separation of 15 m; the horizontal slope was 6% following Serrano and Gómez [29] and Forrester [30]. Drains were executed with slotted PVC pipes with a thickness of 1.5 mm and a diameter of 75 mm.
- Implementation of slope conservation and stabilization measures such as slope profiling and re-vegetation.
- Performance of a flexible retaining wall at the slope toe; wall height was 6 m from the ground and 9.2 m from the bottom of the foundation and its width was found between 3.5 and 4.1 m; its vertical slope was 6:10.

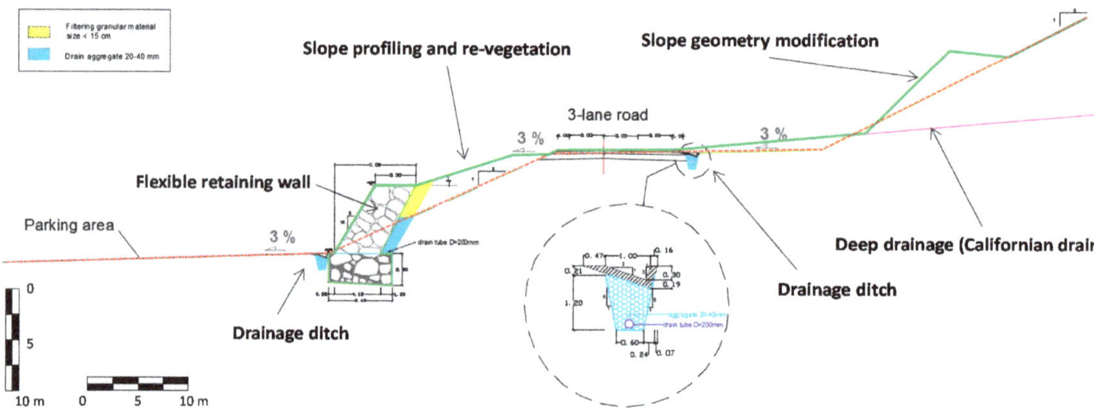

Figure 3. Stabilization measures projected; red dotted line and green line show the slope profile before and after implementing the stabilization measures (highlighted in bold), respectively.

2.4. Satellite Surveys

After conducting stabilization works on the slope, the ground-based SAR (GB-SAR) technique was used to monitor the landslide during the years 2006 and 2007 as well as to predict the slope movement evolution [14]. That technique is based on SAR interferometry, i.e., the use of consecutive pairs of SAR images for obtaining information about displacements according to the phase difference (interferogram) between two consecutive SAR images [31,32]. For monitoring the landslide movements [33], the GB-SAR sensor was installed about 600 m from the slope (target distance was between 200 and 1300 m), and worked for 47 days, with an acquisition rate of 1 image per hour, a range resolution of 1.7 m and an azimuth resolution of 0.74°.

Results of GB-SAR were validated with the results of five Differential GPS (DGPS) campaigns, where 93 points were monitored, both stable points located next to the landslide active area and points within the landslide mass. A good match between the two techniques (GB-SAR and DGPS) was found, being the maximum difference 1.5 cm.

In addition, the Differential Synthetic Aperture Radar Interferometry (DinSAR) technique was employed [21,22] to monitor different landslides of the Upper Tena Valley (located in the Central Spanish Pyrenees area), including the area under study. That technique uses microwave remote sensing and produces the measurement of surface displacement of high accuracy with a great coverage capacity [34]. The SPN approach [35], which combines both the Persistent Scatterers [35–38] and Small Baselines [39–43] methods, was used.

In total, 43 SAR images acquired by ERS-2 and ENVISAT satellites (from 2001 to 2007), 14 SAR images acquired by the TerraSAR-X satellite (2008) and 12 SAR images acquired by ALOS PALSAR (L-band) satellite (from 2006 to 2010) were processed. An auxiliary DGPS campaign was also undertaken to validate the results.

More details about this investigation can be found in [14,21–24].

2.5. New Field Investigations

In 2016, approximately 10 years after conducting the stabilization works of the slope, three new boreholes were performed (Figure 4a) at the area under study. Those boreholes attained a depth between 30 m and 45 m and were performed close to those executed in the previous field investigation (see Figure 2). On each borehole, SPT tests were performed in soils and very weathered rocks, with a frequency of about 3 m. Disturbed samples were extracted from each level found and general identification laboratory tests were carried out (grain size, Atterberg limits and unit weight).

Similar to the actions performed in the previous geological–geotechnical investigation, inclinometers were installed on the new boreholes (Figure 4b). Besides, a new inventory of water points was carried out along the entire hillside (Figure 4c).

Figure 4. New geotechnical investigation works: (**a**) Borehole performance; (**b**) Inclinometer installation; (**c**) Inventory of water points; (**d**) Example of the Quaternary material obtained in the boreholes.

2.6. Limit Equilibrium Analysis

Stability analysis of the landslide was carried out in its natural state and after implementing the corrective measures based on limit equilibrium methods. These methods consider the static mechanical laws and assume the shear strength of the soil to be totally and simultaneously developed along the sliding surface (failure surface). With such methods, the slope safety factor may be computed as the ratio between the available shear strength in the sliding surface and the needed shear strength to keep a strict equilibrium of the sliding mass.

Nowadays, the most common way of applying limit equilibrium methods is using the method of slices. This method divides the soil sliding mass into many vertical slices to solve the stability problem. The method assumes that failure of the soil is governed by the Mohr–Coulomb criterion, slices behave as rigid bodies and no stresses exist inside each slice. The equation system obtained once the equilibrium of forces is established on each slice requires assuming different simplifications and/or hypotheses to solve the system, all of which leads to having several "sub-methods". In this work, Bishop's method [44] and the Morgenstern–Price method [45] were used. The former is an approximate method that establishes the equilibrium of vertical forces and bending moments, assumes a horizontal resultant of the interslice forces and does not take into account interslice shear forces. The latter is an exact method that establishes the stability problem using the three equilibrium conditions in slices of differential thickness and assumes that the inclination of the forces between slices is proportional to a given function.

Several slope profiles were selected to compute the landslide safety factor applying the two mentioned methods. Simulations were run with and without considering the seismic acceleration (0.10 g, with g being the gravitational acceleration).

3. Results

3.1. Field Investigations and Satellite Surveys

In situ (SPT) and laboratory tests conducted on the different investigations carried out identified six geotechnical units in the area under study. Table 1 lists the average parameters of those units related to SUCS classification, Atterberg limits, unit weight, water content and mechanical parameters (cohesion and friction angle for soil materials and uniaxial compression strength for rocks).

Table 1. Geotechnical parameters of the materials found in the area under study.

Geotechnical Unit	Depth [m]	SUCS	W_P	PI	γ [kN/m^3]	W [%]	c [kPa]	φ [°]	UCS [MPa]
Colluvial deposits	2–10	GM-GC	30.5	8.4	21.0	7.0	–	29.0	–
Green sand-clay	2–6	SC	32.1	10.4	20.4	14.0	–	18.0	–
Black sand-clay	2–6	SC-GC	30.9	10.6	22.2	8.1	–	25.0	–
Fragmented calcareous rocks and slates	2–6	GC-SC	33.6	12.9	21.7	6.7	–	38.0	–
Fault breccia	–	–	–	–	23.0	6.0	20.0	25.0	–
Slate rock	–	–	–	–	27.7	–	–	–	12.5

Notation: SUCS: Soil Unified Classification System (Casagrande's classification); W_P: plastic limit; PI: plasticity index; γ: unit weight; W: water content; c: cohesion (residual); φ: friction angle (residual); UCS: uniaxial compression strength.

Rock mass of the area consisted of variably calcareous slates with calcite filled veins (Geotechnical unit "Slate rock") belonging to the "Facies Culm" from the Carboniferous period. This rock mass presented an average RQD of 60% and the RMR reached an average value of 60. According to the geological–geotechnical investigations conducted, rocky formation crops out only in the northern area, where no instabilities were detected. The rest of the slope surface is covered by quaternary deposits (Figure 4d), which include colluvial deposits, green clayey sands, black clayey sands and fragmented calcareous rocks and slates. Additionally, fault breccia materials were found in an old fault detected, though currently inactive (yet some water was seen circulating through it). Except for colluvial deposits, quaternary materials come from a roto-translational, with unidirectional movement, landslide that occurred in the area in the past. Materials appeared unstructured and remoulded due to paleo-sliding.

Data from boreholes, inclinometers and surface movements led to defining a failure surface located at depths between 5 m and 19 m (see Figure 5), with a fairly planar shape and with greater curvature in the slope toe area. The failure surface daylights on the slope and is observed on the ground surface (at both the slope head and toe) as seen in Figure 6. Surface movements in the E–SE direction, topographically recorded during the previous geotechnical investigations [15], showed small movements on the head of the slope as well as on the excavated slope. Those movements ranged between 1 cm and 3 cm, with movements in the northern area being slightly inferior to southern areas.

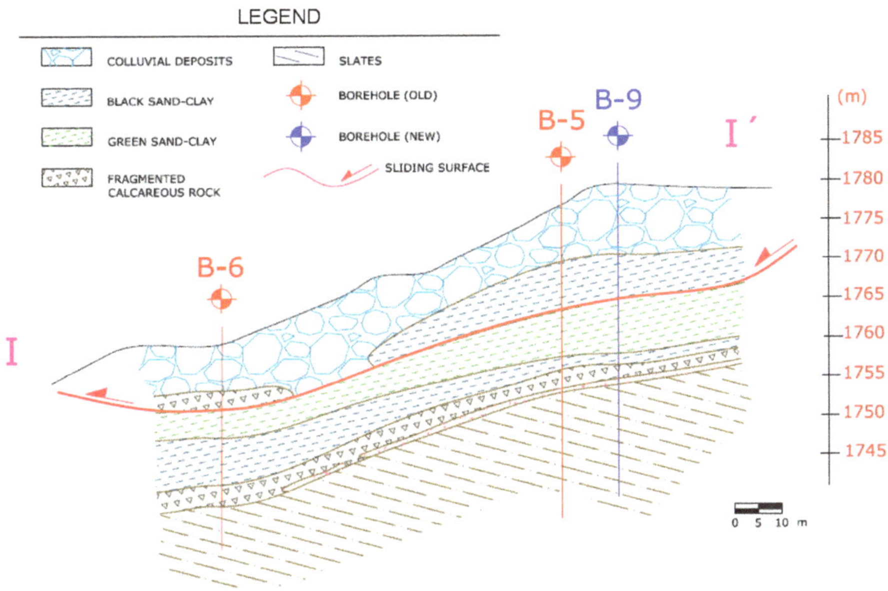

Figure 5. Geotechnical profile I-I′ (see Figure 2) of the area under study; failure surface detected is shown in red.

Figure 6. Daylight of the failure surface: (**a**) stress crack; (**b**) warped plane morphology.

Inclinometers revealed the development of the failure surface mainly through the sandy clay materials in contact with the fragmented calcareous rock. An example of inclinometer data obtained in the previous study [15] and new geotechnical investigations is shown in Figure 7 (note that those inclinometers belong to profile I-I′ shown in Figure 5; failure surface may be easily identified).

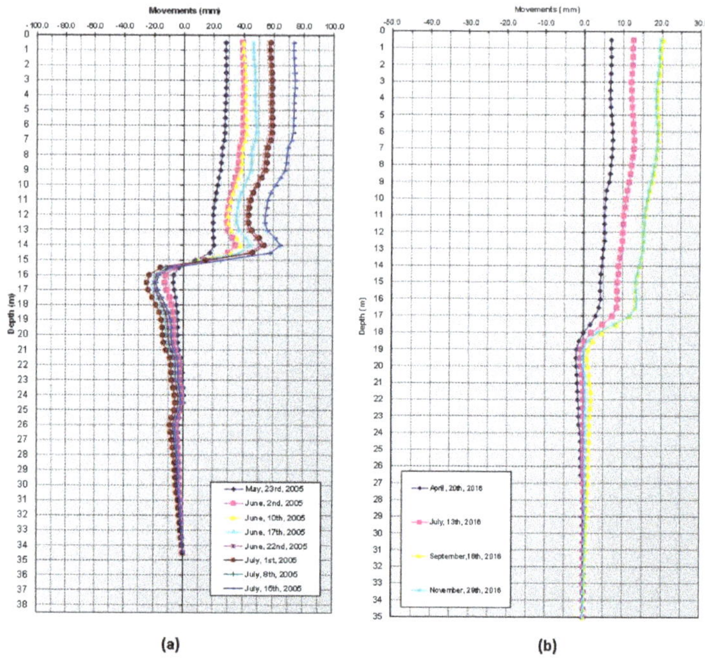

Figure 7. Inclinometers records: (**a**) Inclinometer placed in borehole B-5, reprinted with permission from ref. [15], Copyright Year 2005, Copyright Owner's F.J. Torrijo (old geotechnical investigations [15], conducted in 2005); (**b**) Inclinometer placed in borehole B-9 (new geotechnical investigations, conducted in 2016).

Topographical works developed during the previous investigation [15] detected the existence of three main landslides in the area under study (landslides, 1, 2 and 3 in Figure 8). This information was confirmed by the DinSAR data [21,22]. One of those landslides (landslide 3) was not affected by any anthropogenic activity (i.e., was not the consequence of the parking excavation) and was an extremely slow movement [46,47] with an estimated average velocity [22] of approximately 16 mm/year and affected by seasonal rainfalls. The second landslide (landslide 1) was activated by the parking excavation and corresponds to the slope movement under study. Its movement rate, estimated by the auxiliary DGPS campaign of approximately 0.1 m/year, meant it was not possible to compute its velocity with the DinSAR technique [22]. The third landslide (landslide 2) was located next to the previous one, was also activated by the parking excavation and its average velocity was estimated [22] at approximately 25 mm/year. The ALOS PALSAR differential interferometry [24] was also applied to this last landslide, obtaining similar results.

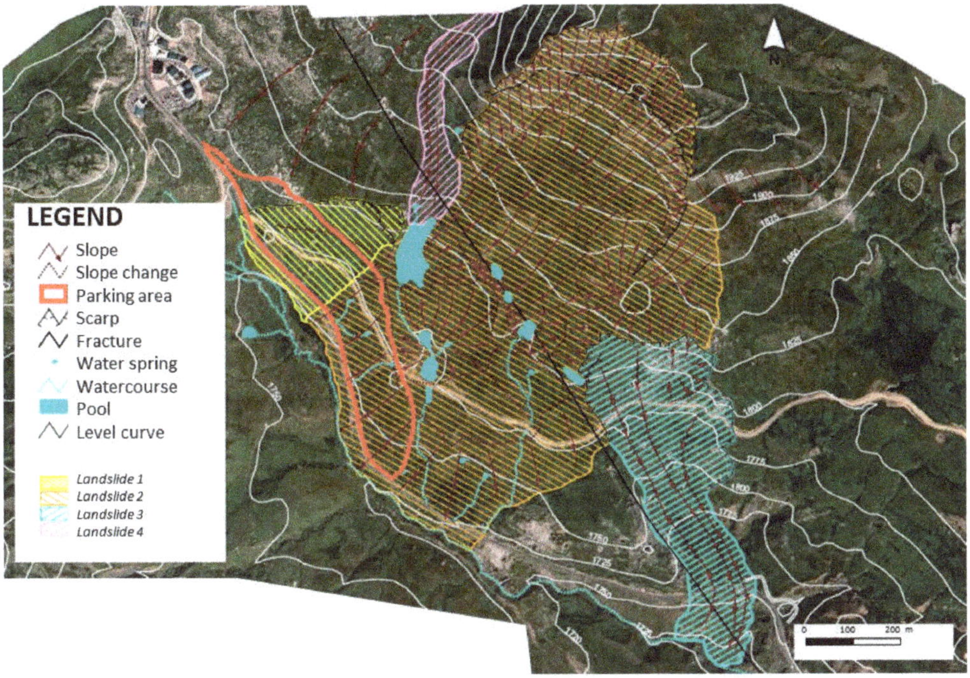

Figure 8. Main landslides detected in the area under study.

DGPS measurements [14], conducted after performing the stabilization work on the slope (i.e., landslide 2 according to Figure 8), showed that maximum displacement rates were around 1 mm/day in the summer while rates increased up to 2 mm/day in the fall. Those measurements also identified that the landslide's most active part was an area located below the main scarp. GB-SAR technique discovered that the landslide displacements were fairly linear in time, although slight differences were observed due to daily rainfall variations. Generally, an acceleration of the landslide occurs when rainfall events take place.

In fact, hydrogeological results obtained in the previous study [15] and the new geotechnical investigation indicated that surface materials were permeable and the hillside drains with abundant subterranean water flow. Continuous flow springs near the slope toe were identified, with that flow occurring through both the permeable material and the failure surface. The water table was found approximately on the failure surface, suffering oscillations with precipitations. Thus, the relationship between the landslide evolution rate and rainfall reported by the use of the GB-SAR technique [14] was confirmed due to the easy infiltration of water in the sliding mass, thanks to the high drainage capacity of the colluvial material located on the top of it.

3.2. Limit Equilibrium Analysis

Several slope profiles were analysed using limit equilibrium methods by means of the method of slices. The geotechnical parameters obtained in the field investigations and listed in Table 1 were used to conduct such analyses. The Bishop's method [44] was applied to compute the safety factor without considering the seismic action. The Morgenstern–Price method [45] was applied to compute the safety factor considering the seismic action, which was taken into account by a pseudo-static approach considering two orthogonal forces, their value based on the Spanish code [28].

Under natural conditions, once the parking area was excavated and prior to implementing any corrective measure, profile I-I′, according to Figures 2 and 5, was found to be the most unfavourable profile. Without taking into account the seismic component, the safety factor at that profile resulted in 0.86. That confirms the existence of the landslide produced by the parking excavation and the subsequent movement of the slope. It should be mentioned that this landslide was a first-time failure, not a reactivation of a previous one. When the seismic action was taken into account, a safety factor of 0.59 was obtained for the same profile (i.e., about 30% lower).

The analyses were run again introducing the projected measures and considering the final stage of the projected slope. The water table was located following the data obtained from the inventory of water points carried out in the new investigations and considering that the capacity of California drains is greater than the recharge of the aquifer (an average level was considered since the water table depends on rainfalls, see the next section). Profile I-I′ was still found to be the most unfavourable. In this case, the safety factor without taking into account the seismic component increased to 1.3; taking into account the seismic acceleration, the safety factor was 1.1 (about 15% lower). That result indicates that the slope is stable and the landslide is expected to be under control with the projected measures. Figure 9 shows the results computed by the Bishop′s method for profile I-I′ before and after implementing the stabilization measures using the software Slide v5.028 [48].

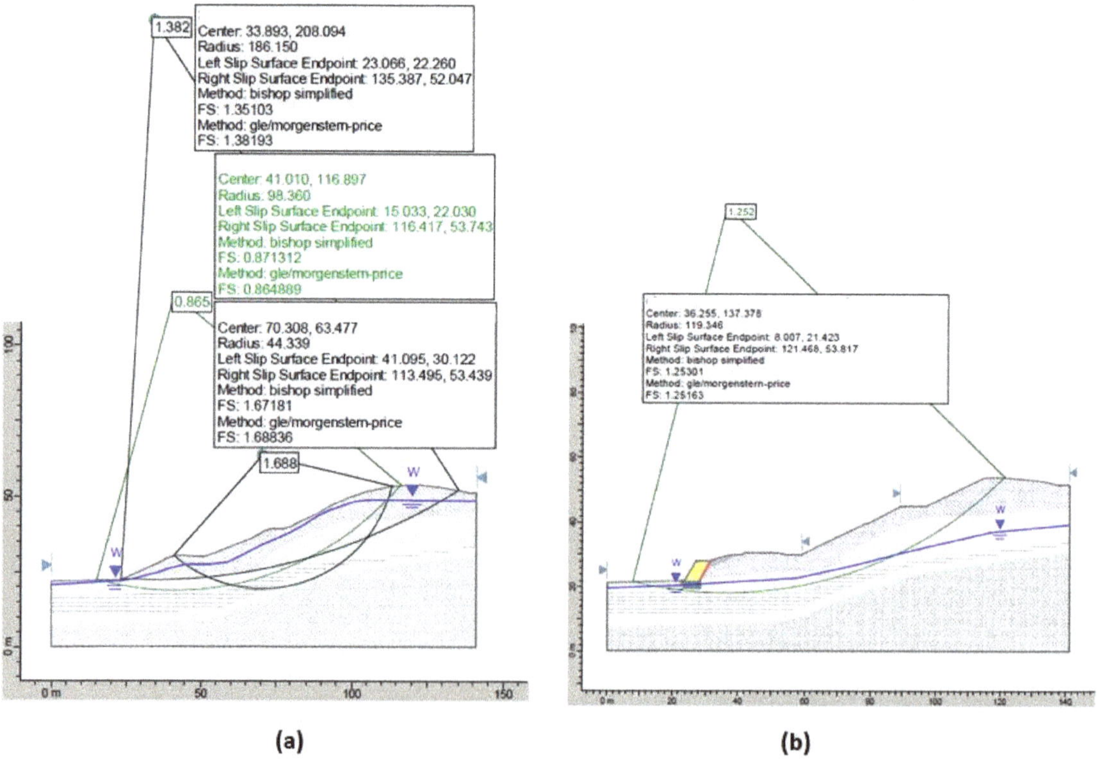

Figure 9. Safety factor analyses of profile I-I′ (see Figure 2): (**a**) slope in its natural state after the parking excavation; (**b**) slope state after implementing the stabilization measures.

It is interesting to mention that the stability analysis calculations conducted considered the whole slope and the corrective measures were designed accordingly. Considering only a part of the slope was not correct, this may have given rise to erroneous results. For instance, Figure 10 shows a stability analysis carried out by both the Bishop [44] and the Morgenstern–Price [45] methods taking into account only the area around the toe wall. The safety factor obtained was nearly 1.6 in both cases, about 23% and 45%, respectively, greater than that computed in the scenario considered in Figure 9b. This may have led to an assumption that the installation of that wall was the unique corrective measure needed, which resulted in it not being enough for controlling the hillside movement.

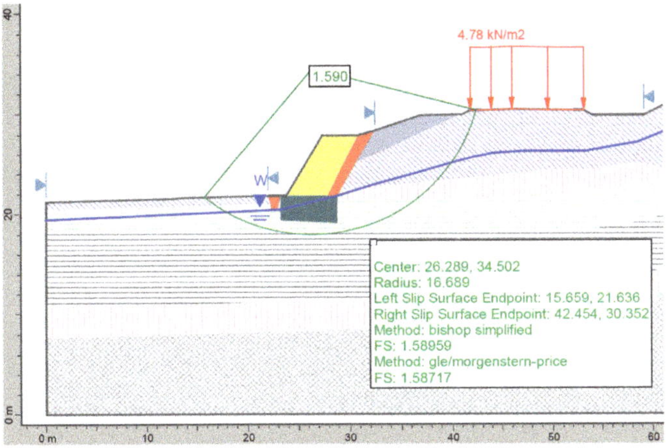

Figure 10. Stability analysis considering the projected wall alone.

3.3. Drainage Evolution

Water table lowering plays an essential role in the stability of the landslide studied. The approximate volume of the adjacent aquifer to the area under study was estimated at 0.86 Hm^3. However, based on the available data, the aquifer lateral supply was unknown, so this volume could be much higher. Although the effective porosity of the different materials traversed by the California drains was also unknown, adopting an estimated value of 5–7%, the volume of water to drain is probably between 100,000 m^3 and 200,000 m^3. Besides, the groundwater recharge of the aquifer should be considered and this does not stop, being especially significant due to the high rainfalls in the area. Recharge can be estimated at around 1900 mm per year, widely distributed throughout the year except for the summer season.

As a consequence, the water table lowering does not only depend on the drainage capacity and the efficiency of the Californian drain but also on the weather and the annual recharge (Figure 11). As a result, the drainage of the slope needs time. During that time, the slope is expected to have a safety factor lower than 1.0, which explains the movement of the slope recorded by the different investigations (in fact, the safety factor increased from the situation depicted in Figure 9a to the one given in Figure 9b). Thus, although the future solution in the long term will be stable, attaining this situation will need to wait for some time during which the slope will exhibit unstable behaviour.

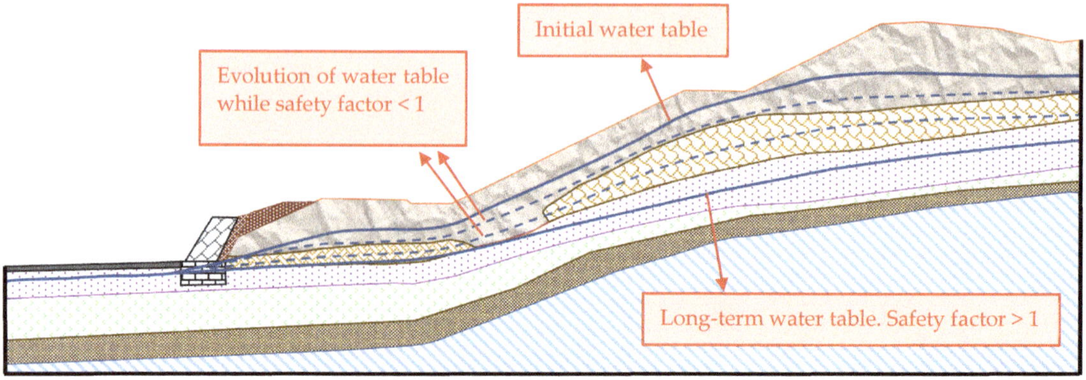

Figure 11. Expected evolution of the water table and its qualitative relationship with the safety factor.

Although the tendency of the water table is the one explained, this evolution does not take place through a uniform decay. Figure 12 shows the evolution of the lateral displacements recorded by the inclinometers located at a depth of 5 m in boreholes B5 and B9 (Figure 7). Both boreholes were located close to one another and correspond to the previous (year 2005) and new (year 2016) geotechnical investigations, respectively.

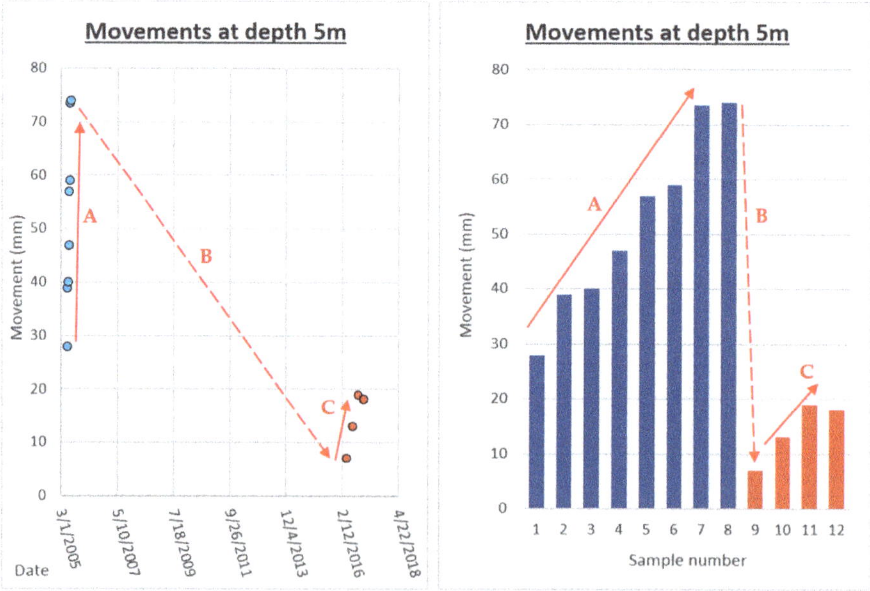

Figure 12. Evolution of the movements at a depth of 5 m. Long-term trend. The movements are represented in blue for borehole B5 (installed in year 2005) and brown for borehole B9 (installed in year 2016), Date given in Day/Month/Year. Letter A refers to the fast increment in deformations in 2005; B refers to the reduction of deformations between 2005 and 2016; C refers to the acceleration of deformations registered in 2016.

Figure 12 shows that, in 2005, the deformations were increasing and the slope was moving faster (A). Between 2005 and 2016 the deformations were greatly reduced, so the landslide was stopping (B). Finally, according to 2016 records, the slope continued to accelerate; from April to September the movement evolved from 5 to 20 mm, although from

September to November it seemed to stabilize (C). This acceleration in a given year was also observed in 2005 for the same period, April to September. That period corresponds to the spring season when precipitations normally occur in the area under study together with the snowmelt phenomenon [49]. Thus, these results confirm the influence of the water table in the landslide stability, the long-term tendency observed in the evolution of the slope movement and the long time needed to attain drainage of the ground mass (Supplementary Materials).

4. Discussion

The case study presented in this paper at "El Portalet" Pass in the Spanish Pyrenees is an example of dealing with a large landslide by slowing it and controlling its behaviour by different instrumentation and monitoring techniques. The proposed approach was fostered by the slow surface movements (about 0.3 m/year) observed during the previous geotechnical investigations and recorded in the inclinometers installed [15]. This value led to the conclusion that the hillside material had a slow type of movement [46] with both distensile and compressive areas. Therefore, permanent infrastructures (such as the parking lot) situated over the hillside were expected not to be seriously harmed by the natural movement of the slope. Nevertheless, stability of the landslide in the long term must be ensured due to both safety and economic efficiency of the parking lot and the surrounding area.

The slow movement allowed for the projection of a series of hillside stabilization measures that tended towards the reduction of the movement rate velocity, which eventually attained the total stopping of the landslide evolution. These measures included the modification of the slope geometry, the installation of drainage and the construction of a flexible retaining toe wall. Data obtained from diverse geotechnical investigations, satellite monitoring [14,15,18,19,21–24] and numerical models were used to assess the effectiveness of such measures. In addition, new geotechnical investigations, involving new boreholes and the installation of new inclinometers, were conducted, showing that the evolution of the landslide tends towards its total stabilization (no movement).

The initial design of the parking lot included a basic stability analysis of the hillside and the implementation of a rigid retaining wall to contain the ground [15]. The landslides activated once the parking area was excavated however, which showed an inefficacy of the analysis and measures. As indicated in Section 3.2, the stability analysis of the area under study should have considered the entire slope and not have been restricted to the area just adjacent to the parking excavation to avoid erroneous results and misunderstand the measures needed. A stability analysis that leads to a safety factor greater than 1.0 does not guarantee the slope to be stable, especially when involving large areas, as is the case under study. Besides, it should be noted that limit equilibrium methods do not consider deformations, so a slope may be unstable sometimes even though the safety factor is greater than 1.0, as reported in some cases found in the literature [50].

An important aspect involving a large area and also shown in this case study is the possibility that a landslide triggers other landslides in the area. The cascade effect that a potential landslide may have in other points of the slope should never be neglected. These additional landslides are difficult to predict unless a geological exploration is conducted in the area potentially affected. All in all, the stability analysis of potential landslides involving large ground masses, even though adjusted to the corresponding design requirements, should be supported on thorough geological and geotechnical surveys that enable an assessment that the analysis conducted properly fits the natural processes.

In the case under study, the influence of the water table plays a critical role in the stability of the landslide. In particular, the drainage time needed for lowering the water table was shown to be decisive in the instability of the system. Thus, establishing only a future situation and assuming this situation will be reached in a short term may give rise to errors. As shown in this work, stability takes time to be reached, an issue that should be considered in the design of slope stability measures and the subsequent monitoring process. Thus, although a large number of drains were installed, the water outlet from the aquifer did not have a high discharge ratio, so lowering the water table needs more than 10 years to be effective.

Regional climatology showed a direct influence on the water table evolution. The region under study presents high values of rain and snow precipitation as well as snowmelt events [49], all of which prevent the water table from decreasing efficiently. According to the data obtained by the inclinometers installed at the geotechnical investigations, during certain periods of the year the water table increases, as indicated by the landslide acceleration in 2016. This result is evidence that the water table does not always decrease but it may also increase in some periods throughout the year in which the water supply (due to precipitation and snowmelt events) exceeds the outlet of water through the drainage measures.

Therefore, the transition period between the construction works and the stabilization of the long-term water table may give rise to an unfavourable scenario in terms of stability. This motivates the need for monitoring of large landslides such as this case under study to establish the moment when the landslide may be considered to be under control. In this case under study, once the water table reaches the desired level considered in the stability analyses, seasonal oscillations are not expected to cause any instability in the slope, so the landslide may be considered stable and no significant movements of the ground are expected from that point.

Finally, it is interesting to mention that the cost of the stabilization measures carried out in this case study is estimated at EUR 1.5 M. This type of solution forces slope control (monitoring) and the acceptance of a certain degree of ground deformations, however, considering that the cost of the implementation of the common measures aimed to totally stop the landslide would be considerably much higher, it is clear that the engineering solution adopted was a good answer for solving the classical engineering duality cost–safety.

Supplementary Materials: A Kmz file is available at https://www.mdpi.com/article/10.3390/app11167176/s1.

Author Contributions: Conceptualization, F.J.T. and J.G.-R.; methodology, F.J.T. and J.G.-R.; project administration, G.C. and F.J.T.; software, J.G.-R.; validation, G.C. and F.J.T.; formal analysis, M.Á.E. and F.J.T.; writing—original draft preparation, J.G.-R.; writing—review and editing, G.C. and M.Á.E.; visualization, G.C. and M.Á.E.; supervision, G.C. and F.J.T. All authors have read and agreed to the published version of the manuscript.

Funding: This research did not receive any specific grant from funding agencies in the public, commercial, or not-for-profit sectors. The authors fully acknowledge the financial support provided by the Department of Geological and Geotechnical Engineering of the UPV.

Institutional Review Board Statement: Not applicable.

Informed Consent Statement: Not applicable.

Data Availability Statement: The data presented in this study are available on request from the corresponding author. The data are not publicly available due to privacy reasons.

Acknowledgments: Authors thank ICOG for providing them with the geotechnical data related to the "previous geotechnical investigation" works. Thanks are also given to IBERGEOTECNIA for its collaboration in the "new geotechnical investigation" works.

Conflicts of Interest: The authors declare no conflict of interest.

References

1. González de Vallejo, L.I.; Beltrán, F.J.; Ferrer, M. Estabilización y control de un gran deslizamiento en rocas lutíticas. In Proceedings of the III National Symposium on Slopes and Unstable Hillsides, La Coruña, Spain, 3 December 1992.
2. Jiang, Q.; Qi, Z.; Wei, W.; Zhou, C. Stability assessment of a high rock slope by strength reduction finite element method. *Bull. Eng. Geol. Environ.* **2014**, *74*, 1153–1162. [CrossRef]
3. Moreno, J.; Peña, A.; Pinto, H. Dynamic Barriers for Protections against Rocks Falls. *Rev. Construcción* **2016**, *15*, 27–37. [CrossRef]

4. Hutchinson, N.N. Assessment of the effectiveness of corrective measures in relation to geological conditions and types of slope movement. *Bull. Int. Asocc. Eng. Geol.* **1977**, *16*, 131–155. [CrossRef]
5. D'Acunto, B.; Urciuoli, G. Groundwater regime in a slope stabilized by drain trenches. *Math. Comput. Model* **2006**, *43*, 754–765. [CrossRef]
6. Cotecchia, F.; Lollino, P.; Petti, R. Efficacy of drainage trenches to stabilise deep slow landslides. *Géotech. Lett.* **2016**, *6*, 1–6. [CrossRef]
7. Conte, E.; Troncone, A. A performance-based method for the design of drainage trenches used to stabilize slopes. *Eng. Geol.* **2018**, *239*, 158–166. [CrossRef]
8. Wei, Z.L.; Wang, D.F.; Xu, H.D.; Sun, H.Y. Clarifying the effectiveness of drainage tunnels in landslide controls based on high-frequency in-site monitoring. *Bull. Eng. Geol. Environ.* **2020**, *79*, 3289–3305. [CrossRef]
9. Troncone, A.; Pugliese, L.; Lamanna, G.; Conte, E. Prediction of rainfall-induced landslide movements in the presence of stabilizing piles. *Eng. Geol.* **2021**, *288*, 106143. [CrossRef]
10. Wei, Z.L.; Lü, Q.; Sun, H.Y.; Shang, Y.Q. Estimating the rainfall threshold of a deep-seated landslide by integrating models for predicting the groundwater level and stability analysis of the slope. *Eng. Geol.* **2019**, *253*, 14–26. [CrossRef]
11. Zhi, M.M.; Shang, Y.Q.; Zhao, Y.; Lü, Q.; Sun, H.Y. Investigation and monitoring on a rainfall-induced deep-seated landslide. *Arab. J. Geosci.* **2016**, *9*, 182. [CrossRef]
12. Corominas, J.; Moya, J.; Ledesma, A.; Lloret, A.; Gili, J.A. Prediction of ground displacements and velocities from groundwater level changes at the Vallcebre landslide (eastern Pyrenees, Spain). *Landslides* **2005**, *2*, 83–96. [CrossRef]
13. Rosone, M.; Ziccarelli, M.; Ferrari, A.; Camillo, A. On the reactivation of a large landslide induced by rainfall in highly fissured clays. *Eng. Geol.* **2018**, *235*, 20–38. [CrossRef]
14. Herrera, G.; Fernández-Merodo, J.A.; Mulas, J.; Pastor, M.; Luzi, G.; Monserrat, O. A landslide forecasting model using ground based SAR data: The Portalet case study. *Eng. Geol.* **2009**, *105*, 220–230. [CrossRef]
15. Torrijo, F.J.; Sarasa, A. *Informe Sobre las Inestabilidades Detectadas en Una Ladera en las Inmediaciones del Puerto de Portalet (Sallent de Gállego, Huesca), Donde se Está Ejecutando la Construcción de un Futuro Aparcamiento*; Technical Report; Zaragoza Ilustre Colegio Oficial de Geólogos de Aragón (ICOG): Zaragoza, Spain, 2005.
16. García Ruiz, J.M. *Mapa Geomorfológico de Sallent (Huesca) a Escala 1:50,000*; Geoforma Ediciones: Logroño, Spain, 1999.
17. García Ruiz, J.M.; Chueca, J.; Julián, A. Los movimientos en masa del alto Gállego. In *Geografía Física de Aragón. Aspectos Generales y Temáticos*; Peña, J.L., Longares, L.A., Sánchez, M., Eds.; Universidad de Zaragoza e Institución Fernando el Católico: Zaragoza, Spain, 2004; pp. 141–152.
18. Torrijo, F.J.; Andrés, J.; Sarasa, A.; Bona, M.E. Estudio y propuesta de corrección de las inestabilidades detectadas en la zona de "El Portalet" (Huesca). In Proceedings of the VII National Symposium on Slopes and Unstable Hillsides, Barcelona, Spain, 27–30 October 2009.
19. Torrijo, F.J.; Rodrigo, M.; Esteban, C. Instrumentación y corrección de las inestabilidades detectadas en la zona de "El Portalet" (Huesca, España). In Proceedings of the VIII Chile Conference on Geotechnical Engineering, Santiago, Chile, 26–29 November 2014.
20. Chistaras, B.; Argyriadis, M.; Moraiti, E. Landslides in the marly slope of the Kapsali area in Kithira Island, Greece. *Bull. Eng. Geol. Environ.* **2014**, *73*, 839–844. [CrossRef]
21. Herrera, G.; Notti, D.; García-Davalillo, J.C.; Mora, O.; Cooksley, G.; Sánchez, O. Analysis with C-and X-band satellite SAR data of the Portalet landslide area. *Landslides* **2011**, *8*, 195–206. [CrossRef]
22. Herrera, G.; Gutiérrez, F.; García-Davalillo, J.C.; Guerrero, J.; Notti, D.; Galve, J.P.; Fernández-Merodo, J.A.; Cooksley, G. Multi-sensor advanced DInSAR monitoring of very slow landslides: The Tena Valley case study (Central Spanish Pyrenees). *Remote Sens. Environ.* **2013**, *128*, 31–43. [CrossRef]
23. Notti, D.; García-Davalillo, J.C.; Herrera, G.; Mora, O. Assessment of the performance of X-band satellite radar data for landslide mapping and monitoring: Upper Tena Valley case study. *Nat. Hazards Earth Syst. Sci.* **2010**, *10*, 1865–1875. [CrossRef]
24. García-Davalillo, J.C.; Herrera, G.; Notti, D.; Strozzi, T.; Álvarez-Fernández, I. DInSAR analysis of ALOS PALSAR images for the assessment of very slow landslides: The Tena Valley case study. *Landslides* **2014**, *11*, 225–246. [CrossRef]
25. Galé, C. Evolución Geoquímica, Petrográfica y de Condiciones Geodinámicas de los Magmatismos Pérmicos en los Sectores Central y Occidental del Pirineo. Ph.D. Thesis, Universidad de Zaragoza, Zaragoza, Spain, 2005.
26. Herrero, L. Caracterización de Inestabilidades de Ladera y Propuesta de Actuación en Sallent de Gállego. Master's Thesis, Universitat Politècnica de València, Valencia, Spain, 2010.
27. I.T.G.E. *Mapa Geológico de España a Escala 1:50,000*; Hoja 145; Sallent: Madrid, Spain, 1990; p. 54, 1 mapa.
28. NCSR-02. *Seismic Resistance Construction Standard*; Spanish Ministry of Public Works: Madrid, Spain, 2002.
29. Serrano, P.A.; Gómez, R. Métodos de estabilización de taludes en suelos. In *Manual de Estabilización y Revegetación de Taludes*; López Jimeno, C., Ed.; U.D. Proyectos, E.T.S.I. Minas, Universidad Politécnica de Madrid: Madrid, Spain, 1999; pp. 151–244.
30. Forrester, K. *Subsurface Drainage for Slope Stabilization*; ASCE Press: Reston, VA, USA, 2000; p. 208.
31. Bamler, R.; Hartl, P. Synthetic aperture radar interferometry. *Invere. Probl.* **1998**, *14*, R1–R154. [CrossRef]
32. Noferini, L.; Pieraccini, M.; Mecatti, D.; Macaluso, G.; Atzeni, C.; Mantovani, M.; Marcato, G.; Pasuto, A.; Silvano, S.; Tagliavini, F. Using GB-SAR technique to monitor slow moving landslide. *Eng. Geol.* **2007**, *95*, 88–98. [CrossRef]

33. Tarchi, D.; Casagli, N.; Fanti, R.; Leva, D.D.; Luzi, G.; Pasuto, A.; Pieraccini, M.; Silvano, S. Landslide monitoring by using ground-based SAR interferometry: An example of application to the Tessina landslide in Italy. *Eng. Geol.* **2003**, *68*, 15–30. [CrossRef]
34. Rosen, P.A.; Hensley, S.; Joughin, I.R.; Li, F.K.; Madsen, S.N.; Rodríguez, E. Synthetic aperture radar interferometry. *Proc. IEEE* **2000**, *88*, 333–382. [CrossRef]
35. Arnaud, A.; Adam, N.; Hanssen, R.; Inglada, J.; Duro, J.; Closa, J. ASAR ERS interferometric phase continuity. In Proceedings of the IGARSS 2003, Toulouse, France, 21–25 July 2003.
36. Ferretti, A.; Prati, C.; Rocca, F. Permanent scatterers in SAR interferometry. *IEEE Trans. Geosci. Remote* **2001**, *39*, 8–20. [CrossRef]
37. Hooper, A.; Zebker, H.; Segall, P.; Kampes, B. A new method for measuring deformation on volcanoes and other natural terrains using InSAR persistent scatterers. *Geophys. Res. Lett.* **2004**, *31*, 23. [CrossRef]
38. Werner, C.; Wegmüller, U.; Strozzi, T.; Wiesmann, A. Interferometric point target analysis for deformation mapping. In Proceedings of the IGARSS 2003, Toulouse, France, 21–25 July 2003.
39. Berardino, P.; Fornaro, G.; Lanari, R.; Sansosti, E. A new algorithm for surface deformation monitoring based on small baseline differential SAR interferograms. *IEEE Trans. Geosci. Remote* **2002**, *40*, 2375–2383. [CrossRef]
40. Lundgren, P.; Casu, F.; Manzo, M.; Pepe, A.; Berardino, P.; Sansosti, E. Gravity and magma induced spreading of Mount Etna volcano revealed by satellite radar interferometry. *Geophys. Res. Lett.* **2004**, *31*, L04602. [CrossRef]
41. Mora, O.; Mallorquí, J.J.; Broquetas, A. Linear and nonlinear terrain deformation maps from a reduced set of interferometric SAR images. *IEEE Trans. Geosci. Remote* **2003**, *41*, 2243–2253. [CrossRef]
42. Prati, C.; Ferretti, A.; Perissin, D. Recent advances on surface ground deformation measurement by means of repeated spaceborne SAR observations. *J. Geodyn.* **2010**, *49*, 161–170. [CrossRef]
43. Schmidt, D.A.; Bürgmann, R. Time-dependent land uplift and subsidence in the Santa Clara Valley, California, from a large interferometric synthetic aperture radar data set. *J. Geophys. Res. Atmos.* **2003**, *108*, 2416. [CrossRef]
44. Bishop, A.W. The use of slip circle in the stability analysis of slopes. *Geotechnique* **1955**, *5*, 7–17. [CrossRef]
45. Morgenstern, N.R.; Price, V. The analysis of the stability of general slip surfaces. *Geotechnique* **1965**, *15*, 79–93. [CrossRef]
46. Cruden, D.M.; Varnes, J. Landslide types and processes. In *Landslides Investigation and Mitigation Special Report 247*; Turner, K., Schuster, R., Eds.; Transportation Research Board: Washington, DC, USA, 1996; pp. 36–90.
47. Glade, T.; Crozier, M.J. The nature of landslide hazard impact. In *Landslide Hazard and Risk*; Glade, T., Anderson, M., Crozier, M.J., Eds.; Wiley: Chichester, UK, 2005; pp. 43–74.
48. RocScience. *Slide Manual*; RocScience, Inc.: Toronto, ON, Canada, 2005.
49. Lastrada, E.; Garzón-Roca, J.; Cobos, G.; Torrijo, F.J. A Decrease in the Regulatory Effect of Snow-Related Phenomena in Spanish Mountain Areas Due to Climate Change. *Water* **2021**, *13*, 1550. [CrossRef]
50. Garzón-Roca, J.; Torrijo, F.J.; Company, J.; Capa, V. Designing Soil-Nailed Walls Using the Amherst Wall Considering Problematic Issues during Execution and Service Life. *Int. J. Geomech. ASCE* **2019**, *19*, 05019006. [CrossRef]

applied sciences

MDPI

Article

A Novel Decomposition-Ensemble Learning Model Based on Ensemble Empirical Mode Decomposition and Recurrent Neural Network for Landslide Displacement Prediction

Xiaoxu Niu [1], Junwei Ma [1,2,*], Yankun Wang [3], Junrong Zhang [4], Hongjie Chen [5] and Huiming Tang [1,2]

1 Three Gorges Research Center for Geo-Hazards of Ministry of Education, China University of Geosciences, Wuhan 430074, China; nxx@cug.edu.cn (X.N.); tanghm@cug.edu.cn (H.T.)
2 National Observation and Research Station for Geohazards of the Three Gorges Reservoir Region, Wuhan 430074, China
3 School of Geosciences, Yangtze University, Wuhan 430100, China; ykwang@yangtzeu.edu.cn
4 Faculty of Engineering, China University of Geosciences, Wuhan 430074, China; zjr@cug.edu.cn
5 Technology R&D Center, Huaneng Lancang River Hydropower Ltd., Kunming 650214, China; chj_1018@163.com
* Correspondence: majw@cug.edu.cn

Featured Application: The proposed decomposition-ensemble learning model can be efficiently used to enhance the prediction accuracy of landslide displacement prediction and can also be extended to other difficult forecasting tasks in the geosciences with extremely complex nonlinear data characteristics.

check for updates

Citation: Niu, X.; Ma, J.; Wang, Y.; Zhang, J.; Chen, H.; Tang, H. A Novel Decomposition-Ensemble Learning Model Based on Ensemble Empirical Mode Decomposition and Recurrent Neural Network for Landslide Displacement Prediction. *Appl. Sci.* **2021**, *11*, 4684. https://doi.org/10.3390/app11104684

Academic Editors: Miguel Llorente Isidro, David Moncoulon and Ricardo Castedo

Received: 27 April 2021
Accepted: 18 May 2021
Published: 20 May 2021

Abstract: As vital comments on landslide early warning systems, accurate and reliable displacement prediction is essential and of significant importance for landslide mitigation. However, obtaining the desired prediction accuracy remains highly difficult and challenging due to the complex nonlinear characteristics of landslide monitoring data. Based on the principle of "decomposition and ensemble", a three-step decomposition-ensemble learning model integrating ensemble empirical mode decomposition (EEMD) and a recurrent neural network (RNN) was proposed for landslide displacement prediction. EEMD and kurtosis criteria were first applied for data decomposition and construction of trend and periodic components. Second, a polynomial regression model and RNN with maximal information coefficient (MIC)-based input variable selection were implemented for individual prediction of trend and periodic components independently. Finally, the predictions of trend and periodic components were aggregated into a final ensemble prediction. The experimental results from the Muyubao landslide demonstrate that the proposed EEMD-RNN decomposition-ensemble learning model is capable of increasing prediction accuracy and outperforms the traditional decomposition-ensemble learning models (including EEMD-support vector machine, and EEMD-extreme learning machine). Moreover, compared with standard RNN, the gated recurrent unit (GRU)-and long short-term memory (LSTM)-based models perform better in predicting accuracy. The EEMD-RNN decomposition-ensemble learning model is promising for landslide displacement prediction.

Keywords: landslide displacement prediction; decomposition-ensemble model; recurrent neural network (RNN); ensemble empirical mode decomposition (EEMD); maximal information coefficient (MIC)

1. Introduction

Landslides are a ubiquitous global hazard [1] posing significant threats to life and property. The statistics data show that landslide disasters affected 5 million people and caused total damage of 4.7 billion US dollars during the period from 2000 to 2020 [2].

As shown in Figure 1, China, the USA, Japan, Nepal, and India are the most landslide-prone regions [3], among which China suffers the most landslide disasters. In the past two decades, landslides have killed 3706 people and caused over 2 billion US dollars of estimated damage to China. Landslide early warning has proven to be the most effective measure for landslide mitigation [4,5], and landslide displacement prediction has been catching extensive attention from practitioners and scholars because of its significant importance in early landslide warning systems [6,7]. However, due to the inherent nonlinear characteristics of landslide monitoring data, achieving the desired prediction accuracy remains highly difficult and challenging. Therefore, it is essential to develop an effective and accurate prediction model to improve the performance of landslide displacement prediction, thus aiding landslide mitigation.

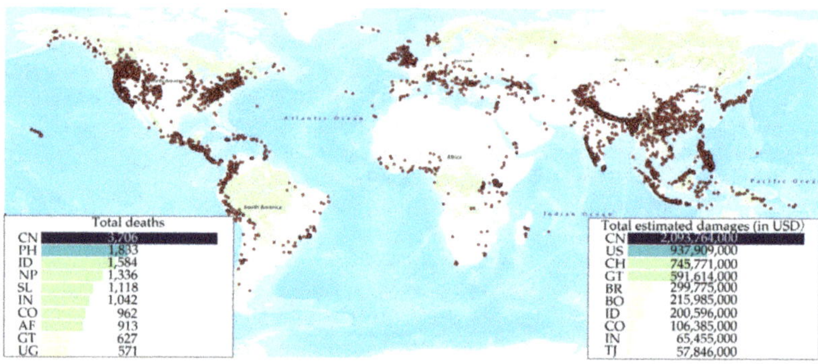

Figure 1. Spatial distribution of landslide disasters during the period from 2000 to 2020. Each dot represents a single landslide. The insets show the total deaths and total estimated damages. (Source: https://public.emdat.be/data, accessed on 2 December 2020).

A variety of landslide displacement prediction models have been proposed since the pioneering work of Saito [8]. These prediction models generally fall into two main groups: Physics-based models and data-driven models [9]. Physics-based models generally require a clear understanding of the physical processes that involve a large amount of input, sophisticated mathematical tools, and significant user expertise. Therefore, the generalization ability of physics-based models is limited [4].

Recently, data-driven models, including artificial neural networks (ANNs) [10], decision trees [4], extreme learning machines (ELMs) [11,12], support vector machines (SVMs) [13–15], quantile regression neural networks [16], random forest (RF) [17], and kernel-based ELMs and SVMs [9,18,19], have attracted attention in landslide displacement prediction. These studies have demonstrated that a data-driven model is capable of providing satisfactory predictions by recognizing movement patterns in historical monitoring data and establishing a mapping between input and output displacements without the requirement of complex physical processes. Recent applications have demonstrated the feasibility of data-driven models to capture nonlinear relationships and to model landslide dynamic processes based on historical model data; however, limitations remain.

First, in most data-driven models, the input variables that have an important influence on the accuracy of landslide displacement prediction [9] are selected based on a priori expert knowledge, trial and error, or linear cross-correlation [9,12]. Nevertheless, a priori expert knowledge of landslide systems is biased [20], or not always available, or even when available, knowledge acquisition tends to be a difficult and time-consuming process. Generally, input variable selection via trial and error is a brute-force process that is computationally expensive, especially for data-driven models with large input candidates. The most commonly used linear correlation coefficients only evaluate linear correlation and cannot reveal the nonlinear relationships that are generally involved in data-driven models. Therefore, a clear need exists for a systematic input variable selection process that does not rely on a priori expert knowledge, is computationally inexpensive, and can describe nonlinear relationships.

Second, conventional data-driven models ignore the intrinsic temporal dependency, which involves the effect of preceding actions on present actions [21,22]. Actually, measured landslide displacement data contain temporal dependencies [23,24].

The abovementioned limitations can be addressed from the following perspectives. The first is to utilize mutual information index describing nonlinear relationships by the amount of related information that is jointly owned by two or more variables [25] for input variable selection. The second solution is to recognize intrinsic temporal dependencies by deploying advanced modeling techniques. A promising solution is the recurrent neural network (RNN) [26]. The temporal dependency in monitoring data can be captured by adopting a sequential approach, thereby improving the ability to model dynamic systems. In addition, the "decomposition-ensemble" learning paradigm can also be considered a promising tool for analyzing series with complex nonlinearity characteristics and enhancing prediction accuracy [27–31]. The effectiveness of the "decomposition-ensemble" has already been confirmed in a variety of fields.

Based on the "decomposition-ensemble" principle, a novel "decomposition-ensemble" learning model integrating EEMD and RNN was proposed in this study to enhance the performance of landslide displacement prediction. The Muyubao landslide located in the Three Gorges Reservoir area was selected as a case study to verify the performance of the proposed model.

2. Study Area and Datasets

2.1. Overview of the Muyubao Landslide

The Muyubao landslide, an ancient landslide, is located in Zigui County, Hubei Province and is situated on the right bank of the Yangtze River (see Figure 2 for landslide location). The length and width of the landslide are approximately 1500 m and 1200 m, respectively. The landslide is 50 m thick on average. The landslide covers approximately 2 million m^2 in the planar area and has a volume of approximately 90 million m^3. The altitude at the toe of the landslide is 100 m, and the altitude at the crown is 520 m (see Figure 2 for the landslide geological profile). The Muyubao landslide mainly slides in a direction of 20 degrees from North. The borehole analysis reveals that the Muyubao landslide slide along a soft coal layer with an average thickness of 0.2 m. The landslide materials are distributed in two layers: The upper Quaternary deposit and the lower highly disturbed rock mass (Figure 2). The underlaid bedrock is mudstone and sandstone of the Jurassic Xiangxi Formation.

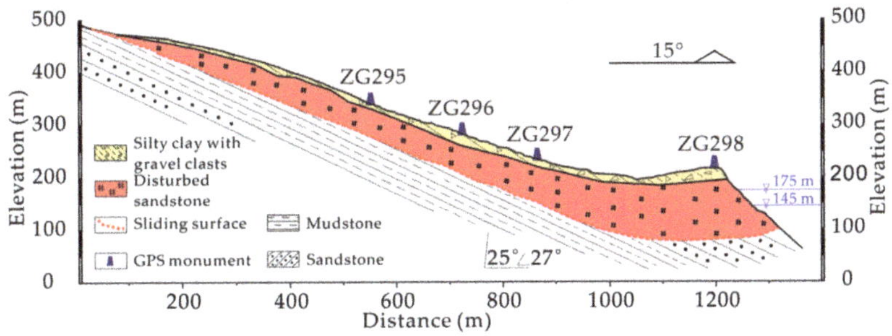

Figure 2. *Cont.*

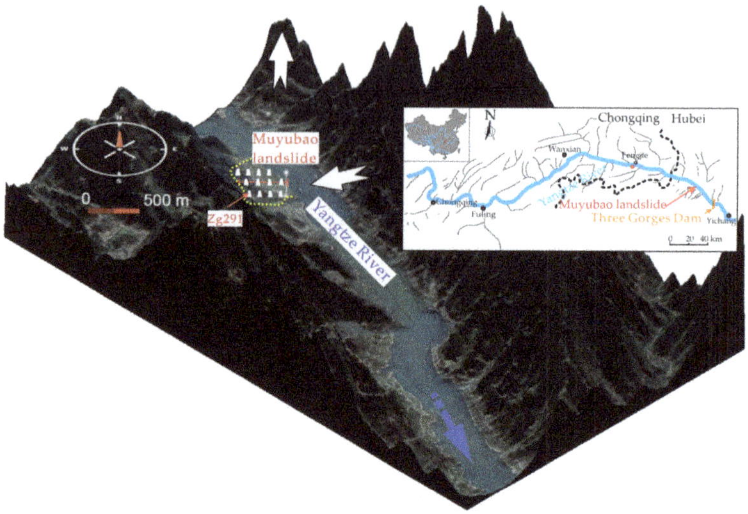

Figure 2. Location and geological profile of the Muyubao landslide, Three Gorges Reservoir area.

2.2. Data Collection

The ancient Muyubao landslide was reactivated by the impoundment of the Three Gorges Reservoir in September 2006. A landslide monitoring system consisting of twelve GPS survey monuments was installed on the landslide mass (see Figure 2 for GPS monument locations) to monitor landslide movement. Nearly 13 years of monitoring data from October 2006 to October 2018 were acquired. According to the monitoring data, the maximum landslide displacement occurred at ZG291 with a cumulative displacement of 2437.36 mm. The landslide displacement at ZG291, reservoir level in the Yangtze River, and rainfall intensity are shown in Figure 3. As shown, the Muyubao landslide exhibits step-like deformation. Sharp increments of displacement occur mainly from November to March, with the reservoir level decreasing from 175 m to 165 m.

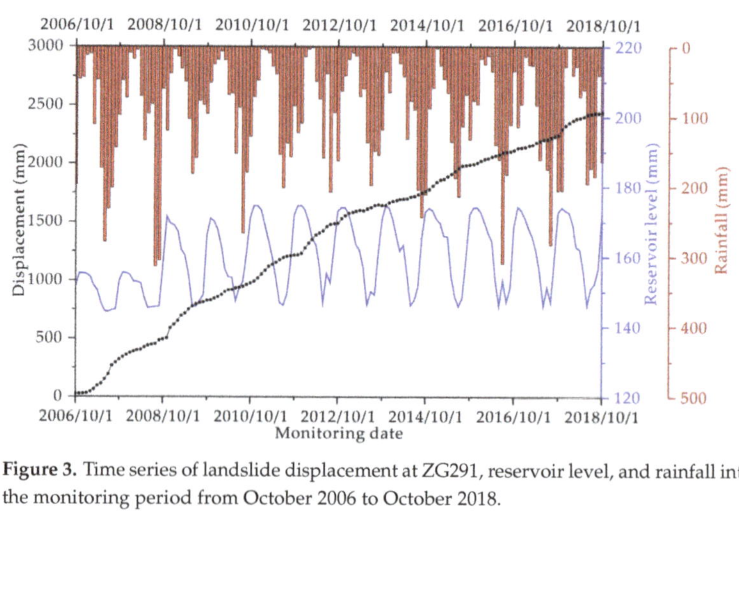

Figure 3. Time series of landslide displacement at ZG291, reservoir level, and rainfall intensity during the monitoring period from October 2006 to October 2018.

3. Methodology

3.1. Ensemble Empirical Mode Decomposition

Empirical mode decomposition (EMD) is an approach to decompose nonlinear signals into a finite number of simple components called intrinsic mode functions (IMFs). These components form a complete and nearly orthogonal basis for the original signal. The main idea of EMD is repeatedly subtracting the local mean from the original signal. EEMD was improved from EMD to overcome modal aliasing problems by adding white noise [32], and it has been widely used for the decomposition of nonlinear and nonstationary signals [33]. EEMD has the advantages of robust self-adaptability and local variation. As shown in Figure 4, the EEMD decomposition process can be briefly described as the following steps:

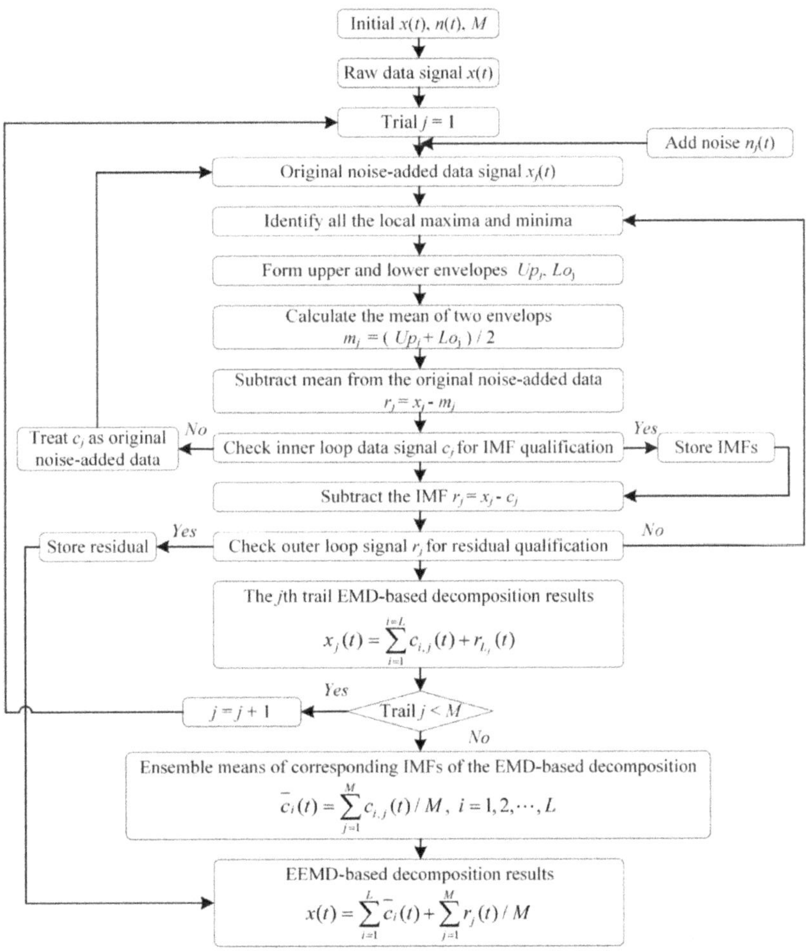

Figure 4. Schematic diagram of the EEMD decomposition process.

Add a random noise signal $n_j(t)$ to the original raw data $x(t)$ to obtain the noise-added data signal $x_j(t)$

$$x_j(t) = x(t) + n_j(t), j = 1, 2, \cdots, M \quad (1)$$

(1) Use EMD to decompose the noise data $x_j(t)$ into some IMFs:

$$x_j(t) = \sum_{i=1}^{L} c_{i,j}(t) + r_{L,j}(t)), j = 1, 2, \cdots, M \quad (2)$$

where $c_{i,j}(t)$ is the ith IMF of noise-added data $x_j(t)$ in the jth decomposition and $r_{L,j}(t)$ is the corresponding residue.

(2) Perform M trials by repeating steps (1) and (2) with diverse white noise.

(3) Calculate the mean values of the corresponding IMFs $\bar{c}_i(t)$ and residue $\bar{r}_L(t)$ as follows:

$$\bar{c}_i(t) = \sum_{j=1}^{M} c_{i,j}(t)/M \quad (3)$$

$$\bar{r}_L(t) = \sum_{j=1}^{M} r_{L,j}(t)/M \quad (4)$$

3.2. *Maximal Information Coefficient (MIC)*

Compared with the traditional statistical indexes such as the Pearson coefficient, MIC allows to detect various correlation relationships including linear, non-linear, functional, and non-functional relationships. Secondly, the MIC is designed to maintain similar results even in presence of equal levels noise of different types [34,35].

For continuous variables x and y, the MIC between x and y is described by the following formula:

$$\begin{gathered} \mathrm{MIC}(x,y) = \max\{I(x,y)/\log_2 \min\{n_x, n_y\}\} \\ \text{where} \\ I(x,y) = H(x) + H(y) - H(x,y) \\ = \sum_{i=1}^{n_x} p(x_i)\log_2 \frac{1}{p(x_i)} + \sum_{j=1}^{n_y} p(y_j)\log_2 \frac{1}{p(y_j)} - \sum_{i=1}^{n_x}\sum_{j=1}^{n_y} p(x_i y_j)\log_2 \frac{1}{p(x_i y_j)} \end{gathered} \quad (5)$$

where $P(x_i)$ presents the marginal probability of x, $P(y_j)$ presents the marginal probability of y, $P(x_i, y_j)$ presents the joint probability density function of x and y, and n_x, n_y is the number of bins of the partition of the x- and y-axis. An MIC of zero indicates that there is no dependence between the concerned variables, while MIC of one implies a stronger relationship [36]. Based on previous research, the final input variables with MICs greater than 0.1 [37,38] were selected from input candidates for model training.

3.3. *Recurrent Neural Network*

An RNN is an artificial neural network wherein adjacent hidden neurons are connected [39]. These recurrent structures of RNNs can transfer time dependence through hidden units and consider temporal correlations. There are three main types of RNNs: Standard RNN, long short-term memory (LSTM), and gated recurrent unit (GRU) (Figure 5).

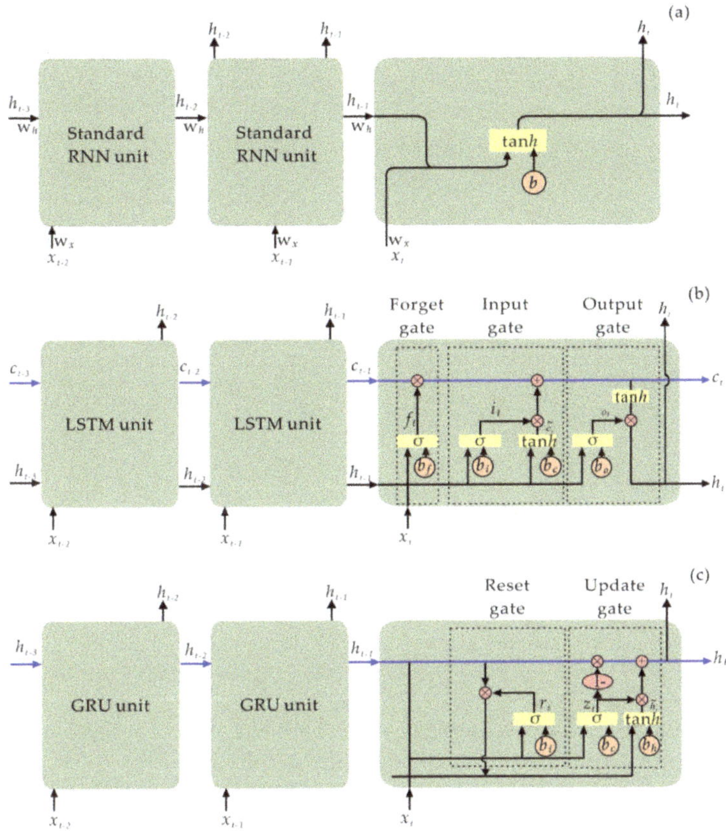

Figure 5. Basic structures of RNN units: (**a**) Standard RNN; (**b**) LSTM; (**c**) GRU.

3.3.1. Standard RNN

A standard RNN is a simple and powerful RNN. Figure 5a shows the typical structure of a standard RNN. x_t is the input vector at time step t and h_t is the hidden state of RNN cell at time step t, which is computed based on the hidden state (h_{t-1}) at the previous time step t–1 and the input vector (x_t) at the current time step t. Formally, the output of the hidden units of the standard RNN can be formulated as follows:

$$h_t = \tan\text{h}(W_x x_t + W_h h_{t-1} + b) \quad (6)$$

The final output of RNN depends on not only the input of the current time step but also the calculated of the hidden layer in the previous time step. Theoretically, RNN can take advantage of all information no matter how long the sequences are. However, according to previous studies, because of the vanishing gradient problem, standard RNNs are suitable only for short-term dependencies [39,40].

3.3.2. LSTM

LSTM was improved to overcome the gradient disappearance problem [41] in standard RNN [42]. Figure 5b shows the basic structure of LSTM. A typical LSTM cell consists of one unit state and three types of gates: Input gate (i_t), output gate (o_t), and forget gate (f_t). These three gates act as filters, serving different purposes. The input gate (i_t) determines what new information is going to be stored in the cell state (C_t). The output gate (o_t) specifies what information from the cell state (C_t) is used as output. The forget gate (f_t) determines what information will be moved away from the cell state (C_t). More formally, the outputs of the input gate (i_t), output gate (o_t), and forget gate (f_t) can be formulated as follows:

$$f_t = \sigma\left(W_{fx}x_t + W_{fh}h_{t-1} + b_f\right) \tag{7}$$

$$i_t = \sigma(W_{ix}x_t + W_{ih}h_{t-1} + b_i) \tag{8}$$

$$o_t = \sigma(W_{ox}x_t + W_{oh}h_{t-1} + b_o) \tag{9}$$

The current cell state (C_t) can be formulated as follows:

$$C_t = f_t \odot C_{t-1} + i_t \odot \widetilde{C}_t \tag{10}$$

The unit state $\widetilde{C}_t$ can be described by the following formula:

$$\widetilde{C}_t = \tan\text{h}(W_{Cx}x_t + W_{Ch}h_{t-1} + b_C) \tag{11}$$

The LSTM unit (h_t) can be formulated as follows:

$$h_t = o_t \odot \tan\text{h}(C_t) \tag{12}$$

where W_{fh}, W_{ih}, W_{oh}, and W_{Ch} are the linear correlation coefficient matrices; W_{fx}, W_{ix}, W_{ox}, and W_{Cx} are the coefficient matrices of the input variable; $\sigma(\cdot)$ denotes the sigmoid activation function; and b_f, b_i, b_o, and b_C are the bias terms of the corresponding formula.

3.3.3. GRU

GRU was developed by [43] to simplify LSTM. Figure 5c shows the basic structure of GRU. A typical GRU unit contains two types of gates: A reset gate (r_t) and an update gate (z_t). The reset gate (r_t) controls how much information from the previous state is written into the current candidate hidden layer vector $\widetilde{h}_t$. The smaller the reset gate (r_t), the less information from the previous state is written. The update gate (z_t) is used to control the degree to which the state information h_{t-1} at the previous time step $t-1$ will be brought into the current time step t. The larger the value of the update gate (z_t), the more the state information at the previous time step is brought in. The reset gate (r_t) and update gate (z_t) can be defined by the following formula:

$$r_t = \sigma(W_{rx}x_t + W_{rh}h_{t-1} + b_r) \tag{13}$$

$$z_t = \sigma(W_{zx}x_t + W_{zh}h_{t-1} + b_z) \tag{14}$$

The candidate hidden layer vector $\widetilde{h}_t$ is defined as follows:

$$\widetilde{h}_t = \tan\text{h}(W_{hx}x_t + W_{hh}(r_t \odot h_{t-1}) + b_h) \tag{15}$$

The output of the GRU unit can be formulated as follows:

$$h_t = (1 - z_t) \odot h_{t-1} + z_t \odot \widetilde{h}_t \tag{16}$$

where W_{rx}, W_{rh}, Wzx, and W_{zh} are the weight matrices; b_r and b_z are the bias terms.

3.4. *Decomposition-Ensemble Learning Model for Landslide Displacement Prediction*

Based on the principle of the "decomposition-ensemble" methodology, a three-step learning model integrating EEMD and RNN can be formulated for landslide displacement prediction. As shown in Figure 6, the proposed EEMD-RNN learning model mainly consists of the following steps: Data decomposition, individual prediction, and ensemble prediction.

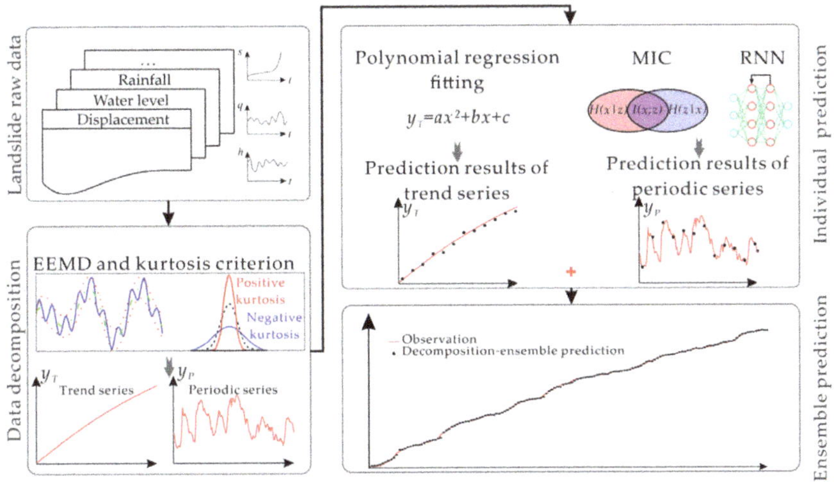

Figure 6. Overall process of the decomposition-ensemble learning model based on EEMD and RNN.

3.4.1. Data Decomposition

The data decomposition technique is useful for the accurate prediction of landslide displacement, as it can reduce the complexity and improve the interpretability of nonlinear time series. In the present study, EEMD and kurtosis criteria were applied for landslide displacement decomposition and construction of trend and periodic components for further landslide displacement prediction.

Kurtosis is a dimensionless parameter [44,45] describing the waveform peak that is formulated as follows:

$$K = \frac{1}{M}\sum_{t=1}^{M}\left[\frac{x(t)-\mu}{\sigma}\right]^4 \tag{17}$$

where M is the signal length, μ presents the average of the signal, and σ presents the standard deviation. A decomposed component with a higher kurtosis retains more deformation characteristics.

EEMD was used to decompose the landslide displacement data shown in Figure 3 into six IMFs and one residual (Figure 7). According to previous works [32], the noise added to the original signal and the maximum number of iterations were set to 0.2 and 100, respectively. The decomposed IMFs oscillate in descending order. The corresponding kurtoses for the decomposition components are listed in Table 1. The obtained kurtoses indicate that the decomposed residual term retains the overall deformation trend of the original time series with the largest kurtosis. Therefore, the residual component was treated as the main trend series for further landslide displacement prediction. The periodic series was obtained by subtracting the trend series from the original series [15]. As shown in Figure 7, the obtained trend components (y_T) and periodic components (y_P) show two characteristics: The trend components show an approximate monotonic increase in displacement with time, and the periodic components exhibit characteristics of a chaotic time series.

Table 1. The kurtoses for the decomposition components.

K_{IMF1}	K_{IMF2}	K_{IMF3}	K_{IMF4}	K_{IMF5}	K_{IMF6}	K_{Residue}
−18.318	−26.058	−46.556	−36.069	−17.362	0.137	28.819

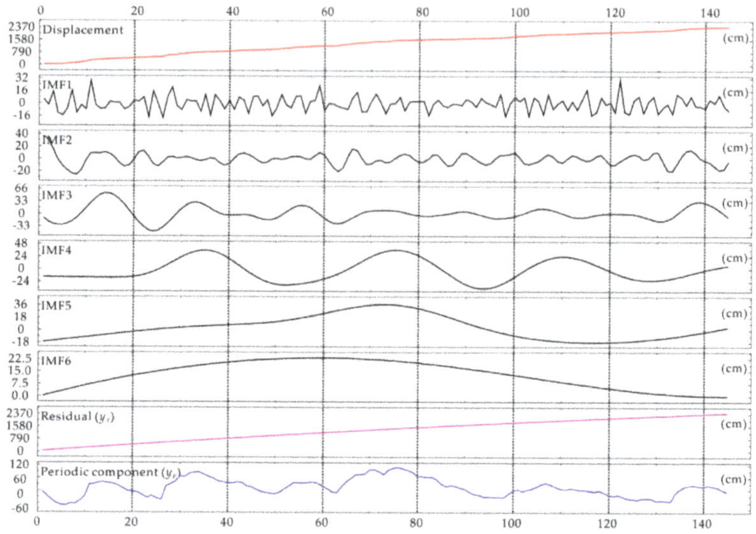

Figure 7. Decomposition results of landslide displacement using the EEMD method and time series of original landslide displacement and the trend and periodic components.

3.4.2. Individual Prediction

In the present study, 124 measurements from October 2006 to January 2017 were used as the training set for the prediction model, and 21 measurements from February 2017 to October 2018 were treated as testing data. According to previous research on landslide displacement prediction [46,47], the trend components are mainly controlled by internal geological conditions and can be perfectly predicted by polynomial regression fitting. In contrast, the periodic component is mainly controlled by external triggering factors, such as rainfall intensity and reservoir fluctuation. The major difficulty in landslide displacement prediction is accurate prediction of the periodic components. Therefore, polynomial regression fitting was treated as an individual prediction model to predict trend components. The trend component shown in Figure 7 can be fitted as follows:

$$y_T(t) = -0.0337t^2 + 21.7046t + 7.1479 \tag{18}$$

The coefficient of determination (R^2) for the trend component is 1000, which indicates a perfect model for the prediction of the trend component.

Aiming at interpreting the behaviors between input candidates and model outputs and excluding irrelevant and redundant variables to develop accurate and cost-effective prediction models [48], the RNN with MIC-based input variable selection was implemented for individual prediction of periodic components. Based on previous research related to landslide displacement prediction [49], seven commonly used variables were selected as input candidates, including three state candidates and four trigger candidates. The selected four trigger input candidates are one-month antecedent rainfall (x_1), two-month antecedent rainfall (x_2), average values of reservoir level for the current month (x_3), and reservoir fluctuation for the current month (x_4). The state candidates are displacement in the past month (x_5), displacement of landslides in the past two months (x_6), and displacement of landslides in the past three months (x_7). Pair plots and MICs between the input candidates and periodic components are shown in Figure 8. The pair plots show an approximately linear dependency between the periodic components (y_P) and state candidates (x_5, x_6, and x_7). The MICs indicate that the seven input candidates have significant dependency on the periodic components, with MIC values larger than 0.2. Therefore, seven input candidates were treated as the input for individual prediction of periodic components.

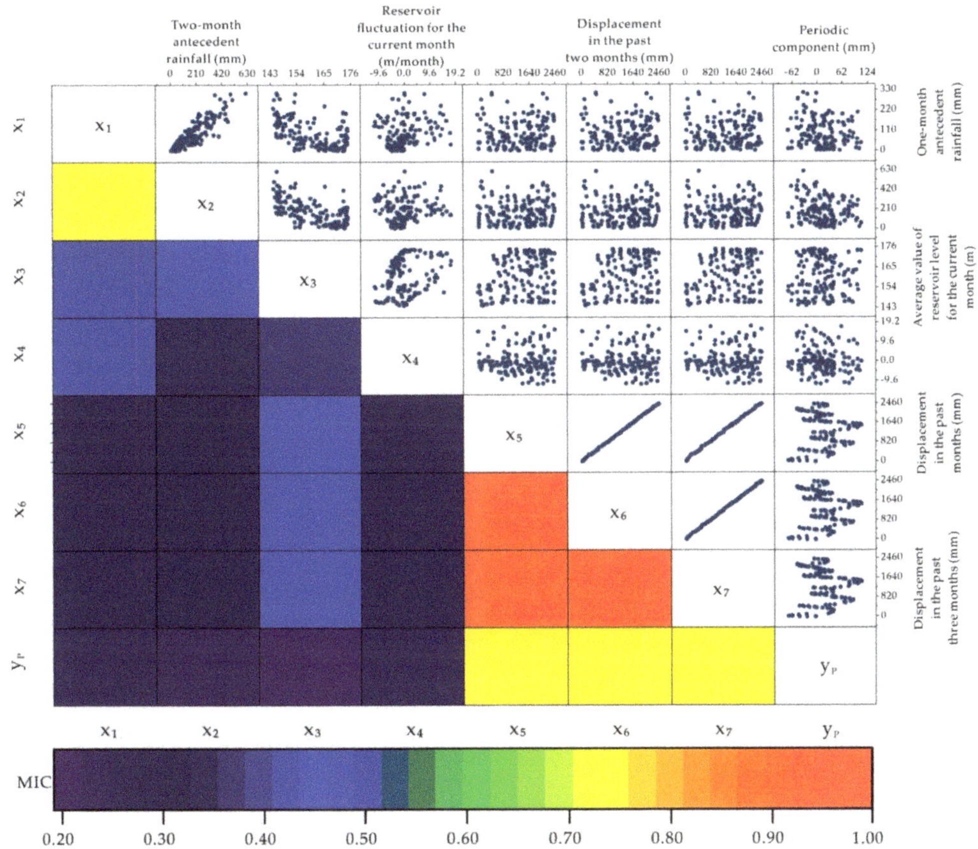

Figure 8. Pair plots and MICs between the input candidates and periodic components.

The landslide measurements were first normalized in the range of 0 to 1 by min-max feature scaling. After the outputs from the EEMD-RNN approach were renormalized, the final displacement predictions were obtained. The simple trial and error method was adopted for the parameter tuning in RNN, GRU, and LSTM networks. The results from trial-and-error analysis show that RNN, GRU, and LSTM networks with one hidden layer for landslide displacement prediction is better than using a multi-layer network. Therefore, one hidden layer with topologies of 7-50-1, 7-55-1, and 7-50-1 was set up for RNN, LSTM, and GRU in the present study. The epoch strategy referring to the process by which all data are sent into the network to complete an iterative calculation was adopted. The epoch sizes were set to 1000, 400, and 100, respectively. Moreover, learning rate scheduling was adopted for faster convergence and convergence to a better minimum [50]. The corresponding learning rate parameters for RNN, LSTM, and GRU were set to 0.6, 0.7, and 0.5, respectively. More details about the parameter settings in the comparative studies are shown in Table 2.

Table 2. Parameter settings in the comparative studies.

Model	Parameters
EMD	Maximum value of siftings as an ending standard = 10
EEMD	Maximum value of siftings as an ending standard = 10; Quantity of copies of the original signal to use as the ensemble = 200 Value of additional noise = 0.2 Maximum number of parallel threads = 1
RNN	Learning rate = 0.6 Neuronic quantity in hidden layer = 55 Maximum value of interactions = 1000
GRU	Learning rate = 0.5 Neuronic quantity in hidden layer = 50 Maximum value of interactions = 400
LSTM	Learning rate = 0.7 Neuronic quantity in hidden layer = 50 Maximum value of interactions = 1000
SVM	Penalty factor = 0.1 Kernel function parameter = 3 Tolerance of termination criterion = 0.001
ELM	Neuronic quantity in hidden layer = 20 Random seed = 1

3.4.3. Ensemble Prediction

The final ensemble predictions of landslide displacement were obtained by aggregating the predictions of trends and periodic components. A comparative analysis was conducted with the following decomposition-ensemble learning model: EEMD-based RNN, EEMD-LSTM, EEMD-GRU, EEMD-SVM, EEM-ELM, and EMD-LSTM. The parameters of the different models used in the comparative studies are listed in Table 2. The model comparative processes were performed in RStudio Version 1.2.5042 running on an Intel(R) Core (TM) i5-6300HQ CPU @ 2.3 GHz with 4 GB RAM.

3.5. Evaluation Metrics

In the present study, six evaluation metrics, namely the mean absolute error (MAE), mean square error (MSE), mean absolute percentage error (MAPE), normalized root mean square error (NRMSE), coefficient of determination (R^2), and Kling-Gupta efficiency (KGE), were applied to evaluate the model performance. These evaluation metrics are defined as follows:

$$\text{MAE} = \frac{1}{N}\left(\sum_{t=1}^{N}\left|y_{pre,t} - y_{obs,t}\right|\right) \tag{19}$$

$$\text{MSE} = \frac{1}{N}\sum_{t=1}^{N}\left(y_{pre,t} - y_{obs,t}\right)^2 \tag{20}$$

$$\text{MAPE} = \frac{1}{N}\left(\sum_{t=1}^{N}\left|\frac{y_{pre,t} - y_{obs,t}}{y_{obs,t}}\right|\right) \times 100\% \tag{21}$$

$$\text{NRMSE} = \frac{1}{\overline{y}_{obs}}\sqrt{\frac{\sum_{t=1}^{t}\left(y_{pre,t} - y_{obs,t}\right)^2}{N}} \tag{22}$$

where N is the quantity of deformation monitoring data; $y_{obs,t}$ presents the measured values of landslide displacement; $y_{pre,t}$ is predicted values of landslide displacement; $\overline{y}_{obs}$ and $\overline{y}_{pre}$ represent the mean values of observations and predictions; r is the linear relative coefficient between simulated displacement values $y_{pre,t}$ and observed displacement values $y_{obs,t}$; $\alpha = = \sigma_{y_{pre}}/\sigma_{y_{obs}}$ is a metric of the relative variability between predicted and observed displacement; and $\beta = \mu_{y_{pre}} / \mu_{y_{obs}}$ is the ratio between the average predicted displacement to the average observed displacement. The MAE is the average of the absolute errors between the predicted values and actual values, which reflects the actual predicted value error. The MSE is the expected value of the square of the difference between the predicted values and actual values, which evaluates the degree of variability in the data. The MAPE further considers the radio between error and the actual value. In general, the smaller the MAE, MSE, and MAPE values, the better the model performs. The NRMSE allows to read the errors in a more understandably way, since it is a non-dimensional parameter. The R^2 measures the linear relationship between the predicted values and actual values of a dependent variable, whereby a high value of R^2 (up to one) signposts a perfect model. The KGE values range from negative infinity to 1. It can evaluate the model performance from three perspective views: Correlation, bias, and variability [51,52]. For an ideal prediction model, the value of KGE should be as close to 1 as possible.

4. Results and Discussion

The final ensemble predictions from EEMD-RNN, EEMD-LSTM, EEMD-GRU, EEMD-SVM, EEMD-ELM, and EMD-LSTM are shown in Figure 9. The evaluation metrics, including MAE, MSE, MAPE, NRMSE, R^2, and KGE, are shown in Figure 10. As shown, satisfactory predictions were achieved, with R^2 values greater than 0.98, which demonstrate the effectiveness of the "decomposition-ensemble" learning model.

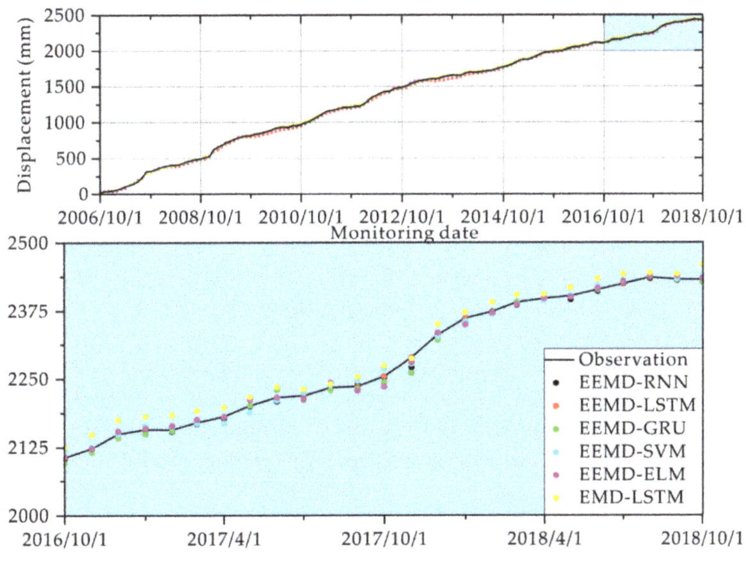

Figure 9. Time series plots of observed and predicted landslide displacement. The training set is shown with the white background, while the testing set is shown with the blue background.

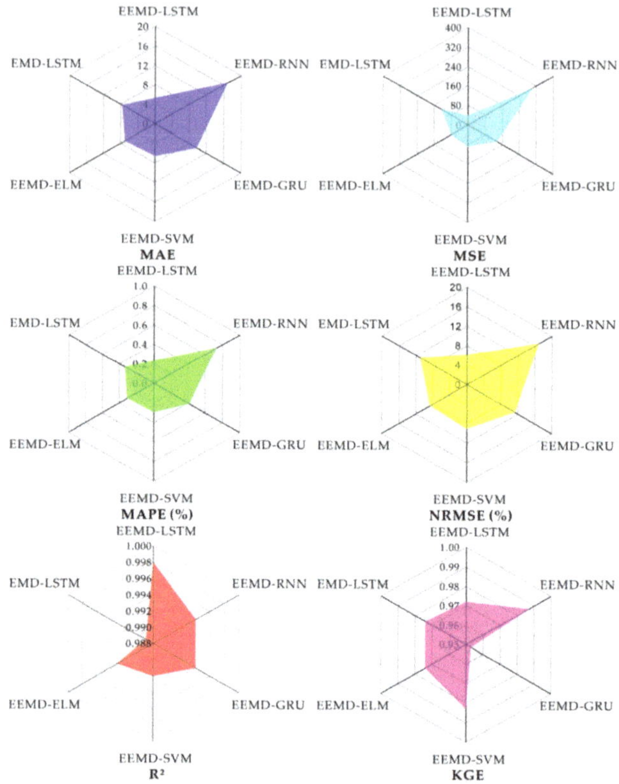

Figure 10. Comparison of model performance in terms of MAE, MSE, MAPE, NRMSE, R^2, and KGE.

4.1. Comparison of EEMD-SVM, EEMD-ELM, and EEMD-RNNs

As shown in Figure 10, in terms of correlation (R^2), there are no significant differences among the models. For standard RNN, SVM, and ELM, the values of R^2 are 0.994, 0.992, and 0.993, respectively, and the values of KGE are 0.987, 0.983, and 0.974, respectively. In terms of KGE, predictions with less bias and variability were achieved by the RNN-type network than SVM and ELM because recurrent networks provide higher nonlinearity. Moreover, landslide movements are essentially suspended during the dry season. Because static models can only learn current information and can only learn from a portion of historical data, static approaches, including ELM and SVM, provide unreasonable results. The suspended movement characteristics can be approximated well using dynamic RNN approaches through connections of adjacent hidden neurons and learning from a fully historical sequence.

4.2. Comparison of EEMD-Based Standard RNN, LSTM and GRU

As shown in Figure 10, LSTM has higher prediction accuracy than GRU, both of which are better than standard RNN. For the standard RNN, LSTM, and GRU models, the MAE values are 16.935, 5.357, and 9.8425, respectively, and the NRMSE values are 6.1, 11.3, and 17, respectively. The evaluation metrics in Figure 10 illustrate that the prediction accuracy of the LSTM and GRU models is better than that of the standard RNN model. The problem of gradient disappearance in standard RNN is the primary cause for this performance distinction. The LSTM and GRU approaches are more practicable for landslide displacement prediction because of the gated unit structures.

In this study, the GRU model consumes 47.66 s to train, while the LSTM model consumes 201.75s, an increase of nearly three times the computational cost due to the complex network structure. The comparative analysis shows that the LSTM and GRU models provide equally satisfactory performance for landslide displacement prediction, but GRU is more efficient because of its simpler network structure.

4.3. Comparison of EMD-LSTM and EEMD-LSTM

As shown in Figure 10, the R^2 and KGE of EMD-LSTM are lower than those of the EEMD-LSTM decomposition-ensemble learning model. The lower performance statistics of the EMD-LSTM decomposition-ensemble learning model are caused mainly by the mode mixing problem in EMD.

Figure 11 compares the model performance for the periodic component in terms of R^2 when varying the training data size: The model performance improves with the data capacity of the training set. The outperformance of EEMD-LSTM over EEMD-based static methods, including ELM and SVM, is not remarkable when the training dataset is smaller than 80%. This can be explained as follows: Compared to traditional models, more parameters must be tuned in the LSTM-based prediction model. Therefore, more input data are required to maintain the model performance.

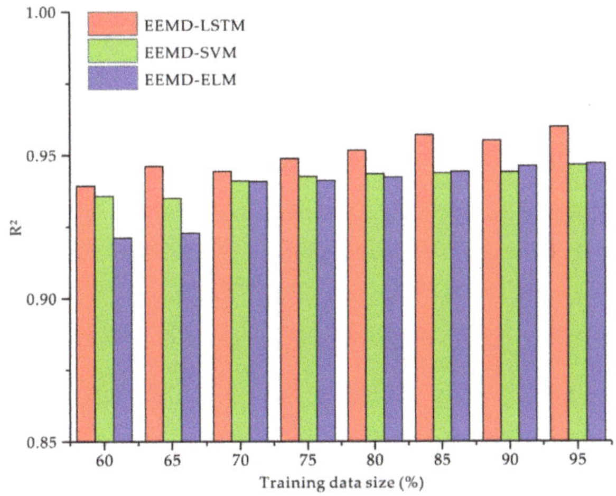

Figure 11. Comparison of model performance for periodic components in terms of R^2 when varying the training set size.

The case study from the Muyubao landslide shows that the hybrid EEMD-RNN decomposition-ensemble learning model is promising for accurate prediction of landslide displacement by combining the advantages of EEMD and RNN. The main advantages of the proposed EEMD-based RNN decomposition-ensemble learning model can be outlined as follows:

The MIC-based input variable selection is a systematic process without any a priori expert knowledge, computationally inexpensive, and capable of describing the nonlinear relationships. The performance of prediction model is able to be improved by EEMD decomposition of complicated forecasting problems into several easier ones, and the temporal dependency in complicated monitoring data is captured by adopting a RNN approach, thereby improving the ability to model dynamic systems.

Although the EEMD-RNN decomposition-ensemble learning model has potential for the accurate prediction of landslide displacement, it has inherent limitations associated with data-driven approaches, including lack of transparency and a requirement for large quantities of training data [53,54].

5. Conclusions

According to the decomposition-ensemble principle, a novel three-step decomposition-ensemble learning model integrating EEMD and RNN was proposed for landslide displacement prediction. The experimental results from the Muyubao landslide in the Three Gorges Reservoir area demonstrate that the proposed EEMD-RNN decomposition-ensemble learning model is capable of increasing prediction accuracy and outperforms traditional decomposition-ensemble learning models (including EMD-LSTM, EEMD-SVM, and EEMD-ELM) in terms of prediction accuracy. Moreover, the GRU- and LSTM-based models perform better than standard RNN by providing equally satisfactory performance in terms of predicting accuracy. Due to the simpler structure, GRU is more efficient than standard RNN and LSTM. Therefore, in practical application, EEMD-GRU learning model is more suitable for medium-term to long-term horizon displacement prediction of reservoir landslide in the Three Gorges Reservoir area. In addition to landslide displacement prediction, the proposed EEMD-RNN decomposition-ensemble learning model can also be extended to other difficult forecasting tasks in geosciences with extremely complex nonlinear data characteristics.

Author Contributions: The work was carried out in collaboration between all the authors. X.N. and J.M. guided and supervised this research; Y.W., J.Z., H.C. and H.T. performed the field investigation; X.N. wrote the original draft; and X.N. and J.M. reviewed and edited the draft. All authors have contributed to, seen, and approved the manuscript. All authors have read and agreed to the published version of the manuscript.

Funding: This research was funded by the Major Program of National Natural Science Foundation of China (Grant No. 42090055), Chongqing Geo-disaster Prevention and Control Center (Grant No. 20C0023), Science and Technology Projects of the Huaneng Lancang River Hydropower Co., Ltd. (Grant No. HNKJ18-H24), and the National Natural Science Foundation of China (Grant Nos. 41702328 and 42090055). All support is gratefully acknowledged.

Institutional Review Board Statement: Not applicable.

Informed Consent Statement: Not applicable.

Data Availability Statement: The data used in this study are available from the corresponding author upon reasonable request.

Conflicts of Interest: The authors declare no conflict of interest.

References

1. Froude, M.J.; Petley, D.N. Global fatal landslide occurrence from 2004 to 2016. *Nat. Hazards Earth Syst. Sci.* **2018**, *18*, 2161–2181. [CrossRef]
2. Cred, F. *EM-DAT. The OFDA/CRED International Disaster Database*; Université Catholique de Louvain: Brussels, Belgium, 2011; Available online: www.emdat.be (accessed on 2 December 2020).
3. Li, W.; Liu, C.; Scaioni, M.; Sun, W.; Chen, Y.; Yao, D.; Chen, S.; Hong, Y.; Zhang, K.; Cheng, G. Spatio-temporal analysis and simulation on shallow rainfall-induced landslides in China using landslide susceptibility dynamics and rainfall ID thresholds. *Sci. China Earth Sci.* **2017**, *60*, 720–732. [CrossRef]
4. Ma, J.; Tang, H.; Liu, X.; Hu, X.; Sun, M.; Song, Y. Establishment of a deformation forecasting model for a step-like landslide based on decision tree C5.0 and two-step cluster algorithms: A case study in the Three Gorges Reservoir area, China. *Landslides* **2017**, *14*, 1275–1281. [CrossRef]
5. Ma, J.; Tang, H.; Liu, X.; Wen, T.; Zhang, J.; Tan, Q.; Fan, Z. Probabilistic forecasting of landslide displacement accounting for epistemic uncertainty: A case study in the Three Gorges Reservoir area, China. *Landslides* **2018**, *15*, 1145–1153. [CrossRef]
6. Picarelli, L.; Behavior, S.; Regime, P.P.; Mechanisms, L.; Mechanics, L. *Landslides—Disaster Risk Reduction*; Springer: Cham, Switzerland, 2009.
7. Hu, X.; Wu, S.; Zhang, G.; Zheng, W.; Liu, C.; He, C.; Liu, Z.; Guo, X.; Zhang, H. Landslide displacement prediction using kinematics-based random forests method: A case study in Jinping Reservoir Area, China. *Eng. Geol.* **2021**, *283*, 105975. [CrossRef]
8. Saito, M. Forecasting the Time of Occurrence of a Slope Failure. In Proceedings of the 6th International Conference on Soil Mechanics and Foundation Engineering, Montreal, QC, Canada, 8–15 September 1965; pp. 537–541.
9. Ma, J.; Niu, X.; Tang, H.; Wang, Y.; Wen, T.; Zhang, J. Displacement prediction of a complex landslide in the Three Gorges Reservoir Area (China) using a hybrid computational intelligence approach. *Complexity* **2020**, *2020*. [CrossRef]

10. Du, J.; Yin, K.; Lacasse, S. Displacement prediction in colluvial landslides, Three Gorges Reservoir, China. *Landslides* **2013**, *10*, 203–218. [CrossRef]
11. Lian, C.; Zeng, Z.; Yao, W.; Tang, H. Extreme learning machine for the displacement prediction of landslide under rainfall and reservoir level. *Stoch. Environ. Res. Risk Assess.* **2014**, *28*, 1957–1972. [CrossRef]
12. Cao, Y.; Yin, K.; Alexander, D.E.; Zhou, C. Using an extreme learning machine to predict the displacement of step-like landslides in relation to controlling factors. *Landslides* **2016**, *13*, 725–736. [CrossRef]
13. Huang, F.; Huang, J.; Jiang, S.; Zhou, C. Landslide displacement prediction based on multivariate chaotic model and extreme learning machine. *Eng. Geol.* **2017**, *218*, 173–186. [CrossRef]
14. Miao, F.; Wu, Y.; Xie, Y.; Li, Y. Prediction of landslide displacement with step-like behavior based on multialgorithm optimization and a support vector regression model. *Landslides* **2018**, *15*, 475–488. [CrossRef]
15. Wen, T.; Tang, H.; Wang, Y.; Lin, C.; Xiong, C. Landslide displacement prediction using the GA-LSSVM model and time series analysis: A case study of Three Gorges Reservoir, China. *Nat. Hazards Earth Syst. Sci.* **2017**, *17*, 2181–2198. [CrossRef]
16. Ma, J.; Liu, X.; Niu, X.; Wang, Y.; Wen, T.; Zhang, J.; Zou, Z. Forecasting of Landslide Displacement Using a Probability-Scheme Combination Ensemble Prediction Technique. *Int. J. Environ. Res. Public Health* **2020**, *17*, 4788. [CrossRef]
17. Li, C.D.; Criss, R.E.; Fu, Z.Y.; Long, J.J.; Tan, Q.W. Evolution characteristics and displacement forecasting model of landslides with stair-step sliding surface along the Xiangxi River, three Gorges Reservoir region, China. *Eng. Geol.* **2021**, *283*, 105961. [CrossRef]
18. Han, H.; Shi, B.; Zhang, L. Prediction of landslide sharp increase displacement by SVM with considering hysteresis of groundwater change. *Eng. Geol.* **2021**, *280*, 105876. [CrossRef]
19. Zhou, C.; Yin, K.; Cao, Y.; Intrieri, E.; Ahmed, B.; Catani, F. Displacement prediction of step-like landslide by applying a novel kernel extreme learning machine method. *Landslides* **2018**, *15*, 2211–2225. [CrossRef]
20. Brandt, L.A.; Benscoter, A.M.; Harvey, R.; Speroterra, C.; Bucklin, D.; Romañach, S.S.; Watling, J.I.; Mazzotti, F.J. Comparison of climate envelope models developed using expert-selected variables versus statistical selection. *Ecol. Model.* **2017**, *345*, 10–20. [CrossRef]
21. Messinger, D.S.; Mattson, W.I.; Todd, J.T.; Gangi, D.N.; Myers, N.D.; Bahrick, L.E. Temporal dependency and the structure of early looking. *PLoS ONE* **2017**, *12*, e0169458. [CrossRef]
22. Wu, J.; Zeng, W.; Yan, F. Hierarchical Temporal Memory method for time-series-based anomaly detection. *Neurocomputing* **2018**, *273*, 535–546. [CrossRef]
23. Rahimi, M.; Shafieezadeh, A.; Wood, D.; Kubatko, E.J.; Dormady, N.C. Bayesian calibration of multi-response systems via multivariate Kriging: Methodology and geological and geotechnical case studies—ScienceDirect. *Eng. Geol.* **2019**, *260*, 105248. [CrossRef]
24. Samia, J.; Temme, A.; Bregt, A.; Wallinga, J.; Guzzetti, F.; Ardizzone, F.; Rossi, M. Do landslides follow landslides? Insights in path dependency from a multi-temporal landslide inventory. *Landslides* **2017**, *14*, 547–558. [CrossRef]
25. Liu, H.; Chen, C. Spatial air quality index prediction model based on decomposition, adaptive boosting, and three-stage feature selection: A case study in China. *J. Clean. Prod.* **2020**, *265*, 121777. [CrossRef]
26. Fan, C.; Wang, J.; Gang, W.; Li, S. Assessment of deep recurrent neural network-based strategies for short-term building energy predictions. *Appl. Energy* **2019**, *236*, 700–710. [CrossRef]
27. Zhang, X.; Wang, J. A novel decomposition-ensemble model for forecasting short-term load-time series with multiple seasonal patterns. *Appl. Soft Comput.* **2018**, *65*, 478–494. [CrossRef]
28. Yu, L.; Dai, W.; Tang, L. A novel decomposition ensemble model with extended extreme learning machine for crude oil price forecasting. *Eng. Appl. Artif. Intel.* **2016**, *47*, 110–121. [CrossRef]
29. Yu, L.; Wang, Z.; Tang, L. A decomposition–ensemble model with data-characteristic-driven reconstruction for crude oil price forecasting. *Appl. Energy* **2015**, *156*, 251–267. [CrossRef]
30. Niu, M.; Hu, Y.; Sun, S.; Liu, Y. A novel hybrid decomposition-ensemble model based on VMD and HGWO for container throughput forecasting. *Appl. Math. Model.* **2018**, *57*, 163–178. [CrossRef]
31. Zuo, G.; Luo, J.; Wang, N.; Lian, Y.; He, X. Decomposition ensemble model based on variational mode decomposition and long short-term memory for streamflow forecasting. *J. Hydrol.* **2020**, *585*, 124776. [CrossRef]
32. Wu, Z.; Huang, N.E. A study of the characteristics of white noise using the empirical mode decomposition method. *Proc. Math. Phys. Eng. Sci.* **2004**, *460*, 1597–1611. [CrossRef]
33. Li, G.; Hu, Y. An enhanced PCA-based chiller sensor fault detection method using ensemble empirical mode decomposition based denoising. *Energy Build.* **2019**, *183*, 311–324. [CrossRef]
34. Zhang, Y.; Jia, S.; Huang, H.; Qiu, J.; Zhou, C. A novel algorithm for the precise calculation of the maximal information coefficient. *Sci. Rep.* **2014**, *4*, 1–5. [CrossRef]
35. Kinney, J.B.; Atwal, G.S. Equitability, mutual information, and the maximal information coefficient. *Proc. Natl. Acad. Sci. USA* **2014**, *111*, 3354–3359. [CrossRef]
36. Reshef, D.N.; Reshef, Y.A.; Finucane, H.K.; Grossman, S.R.; McVean, G.; Turnbaugh, P.J.; Lander, E.S.; Mitzenmacher, M.; Sabeti, P.C. Detecting novel associations in large data sets. *Science* **2011**, *334*, 1518–1524. [CrossRef]
37. Lu, F.; Chen, Z.; Liu, W.; Shao, H. Modeling chlorophyll-a concentrations using an artificial neural network for precisely eco-restoring lake basin. *Ecol. Eng.* **2016**, *95*, 422–429. [CrossRef]

38. Tian, W.; Zhibo, C.; Yuqun, F.; Weinan, Q.; Zhihua, L.; Chao, Z. Sensitivity analysis of influencing factors of induced electricity based on MIC in double-circuit transmission line. *Electr. Meas. Instrum.* **2020**, 1–7. Available online: https://kns.cnki.net/kcms/detail/23.1202.TH.20201016.1612.010.html (accessed on 2 December 2020).
39. Wei, X.; Zhang, L.; Yang, H.-Q.; Zhang, L.; Yao, Y.-P. Machine learning for pore-water pressure time-series prediction: Application of recurrent neural networks. *Geosci. Front.* **2021**, *12*, 453–467. [CrossRef]
40. Pascanu, R.; Mikolov, T.; Bengio, Y.O. On the difficulty of training Recurrent Neural Networks. *JMLR.org* **2013**, *1211*, 5063.
41. Jun, M.A.; Ding, Y.; Cheng, J.; Jiang, F.; Wan, Z. A Temporal-Spatial Interpolation and Extrapolation Method Based on Geographic Long Short-Term Memory Neural Network for PM2.5. *J. Clean. Prod.* **2019**, *237*, 117729. [CrossRef]
42. Hochreiter, S.; Schmidhuber, J. Long short-term memory. *Neural Comput.* **1997**, *9*, 1735–1780. [CrossRef] [PubMed]
43. Cho, K.; Van Merriënboer, B.; Gulcehre, C.; Bahdanau, D.; Bougares, F.; Schwenk, H.; Bengio, Y. Learning phrase representations using RNN encoder-decoder for statistical machine translation. *arXiv* **2014**, arXiv:1406.1078.
44. Alam, S.S.; Bhuiyan, M.I.H. Detection of seizure and epilepsy using higher order statistics in the EMD domain. *IEEE J. Biomed. Health Inform.* **2013**, *17*, 312–318. [CrossRef]
45. Li, Y.; Xu, M.; Liang, X.; Huang, W. Application of bandwidth EMD and adaptive multiscale morphology analysis for incipient fault diagnosis of rolling bearings. *IEEE Trans. Ind. Electron.* **2017**, *64*, 6506–6517. [CrossRef]
46. Zhang, J.; Tang, H.; Tannant, D.D.; Lin, C.; Xia, D.; Liu, X.; Zhang, Y.; Ma, J. Combined forecasting model with CEEMD-LCSS reconstruction and the ABC-SVR method for landslide displacement prediction. *J. Clean. Prod.* **2021**, *293*, 126205. [CrossRef]
47. Zhang, J.; Tang, H.; Tao, W.; Ma, J.; Tan, Q.; Xia, D.; Liu, X.; Zhang, Y. A Hybrid Landslide Displacement Prediction Method Based on CEEMD and DTW-ACO-SVR—Cases Studied in the Three Gorges Reservoir Area. *Sensors* **2020**, *20*, 4287. [CrossRef] [PubMed]
48. Fang, W.; Huang, S.; Huang, Q.; Huang, G.; Meng, E.; Luan, J. Reference evapotranspiration forecasting based on local meteorological and global climate information screened by partial mutual information. *J. Hydrol.* **2018**, *561*, 764–779. [CrossRef]
49. Li, L.; Wu, Y.; Miao, F.; Xue, Y.; Huang, Y. A hybrid interval displacement forecasting model for reservoir colluvial landslides with step-like deformation characteristics considering dynamic switching of deformation states. *Stoch. Environ. Res. Risk Assess.* **2020**, 1–24. [CrossRef]
50. Senior, A.; Heigold, G.; Ranzato, M.; Yang, K. An Empirical Study of Learning Rates in Deep Neural Networks for Speech Recognition. In Proceedings of the 2013 IEEE International Conference on Acoustics, Speech and Signal Processing, Vancouver, BC, Canada, 26–31 May 2013; pp. 6724–6728. [CrossRef]
51. Chen, Y.; Yuan, H. Evaluation of nine sub-daily soil moisture model products over China using high-resolution in situ observations. *J. Hydrol.* **2020**, *588*, 125054. [CrossRef]
52. Gupta, H.V.; Kling, H.; Yilmaz, K.K.; Martinez, G.F. Decomposition of the mean squared error and NSE performance criteria: Implications for improving hydrological modelling. *J. Hydrol.* **2009**, *377*, 80–91. [CrossRef]
53. Khosravi, K.; Mao, L.; Kisi, O.; Yaseen, Z.M.; Shahid, S. Quantifying hourly suspended sediment load using data mining models: Case study of a glacierized Andean catchment in Chile. *J. Hydrol.* **2018**, *567*, 165–179. [CrossRef]
54. Xu, S.; Niu, R. Displacement prediction of Baijiabao landslide based on empirical mode decomposition and long short-term memory neural network in Three Gorges area, China. *Comput. Geosci.* **2018**, *111*, 87–96. [CrossRef]

applied sciences

MDPI

Article

Stabilization Methodology in Foundation Soils by ERT-3D Application in Estepona, South Spain

Alfonso Gutiérrez-Martín [1,*], José I. Yenes [2], Marta Fernández-Hernández [3,4] and Ricardo Castedo [4]

1 Escuela Técnica Superior de Arquitectura-Universidad de Málaga, 29071 Málaga, Spain
2 Escuela Politécnica Superior del Ejército, (ESPOL), 28002 Madrid, Spain; jyengal@et.mde.es
3 Instituto Geológico y Minero de España-IGME, 28003 Madrid, Spain; m.fernandez@igme.es or marta.fernandezh@alumnos.upm.es
4 ETSI Minas y Energía, Universidad Politécnica de Madrid, 28003 Madrid, Spain; ricardo.castedo@upm.es
* Correspondence: alfgutmar@uma.es

Featured Application: A non-invasive solution developed to preserve damaged buildings using an injection of cement grout in the soils to stabilize it.

Abstract: The paper proposes a novel methodology for the stabilization of shallow foundations, with a simplified model combined with 3D electrical resistivity tomography (ERT-3D and consolidation injections. To determine its usefulness, the method has been applied in a case located in Estepona (southern Spain). The chosen tomography model is the dipole–dipole configuration, with an optimized distance between electrodes of 0.80 m for a better visualization of the foundation subsoil; with this parameterization, a total of 72 electrodes were installed in the analyzed case. In this work, the depth of the anomaly in the building's supporting subsoil was detected ranging from 2.00 m to 3.90 m deep. The study also delineates areas of high resistivity variations (50–1000 Ω m) in the middle and eastern end of the field. These data have been validated and corroborated with a field campaign. The results of the ERT-3D monitoring are presented, once the investment data has been processed with the RES3DINV software, from the beginning to the end of the stabilization intervention. The novelty occurs with the interaction between the tomography and the foundation consolidation injections, until the final stabilization. This is a very useful methodology in case of emergency consolidation, where there is a need to minimize damage to the building. Thus, people using this combined system will be able to practically solve the initial anomalies of the subsoil that caused the damages, in a non-invasive way, considerably lowering the value of the resistivities.

Keywords: electric tomography; three-dimensional; electrodes; seat control; foundations; stabilization methodology

Citation: Gutiérrez-Martín, A.; Yenes, J.I.; Fernández-Hernández, M.; Castedo, R. Stabilization Methodology in Foundation Soils by ERT-3D Application in Estepona, South Spain. *Appl. Sci.* **2021**, *11*, 4455. https://doi.org/10.3390/app11104455

Academic Editor: Daniel Dias

Received: 13 April 2021
Accepted: 11 May 2021
Published: 13 May 2021

1. Introduction

The success of architectural structures, which are built directly on the earth's surface, depends, among other factors, on the support offered by the foundation materials bearing the structures' loads [1–6] In turn, the ability of a building's foundation to offer the necessary support for architectural structures depends on the bearing capacity of the soils underneath, and, if the upper layer has heterogeneous physical properties, it could cause spatial variability in the foundation material's strength. Spatial variability of the soil's bearing capacity puts stress on poorly supported architectural structures. Associated structural failure could occur as total, partial, or differential settlement, or even total collapse of the structure, with differential settlement in the foundations being one of the worst problems [7–10].

The global prevalence of failure and collapse, with the associated loss of life and property, has made it necessary to ensure that buildings are properly constructed [11]. It is common to find many stress-induced cracks and other defect-related issues due to the various structures and foundations of historic, public, and private buildings, and some of these problems are disasters in the making [12–14]. This necessitates proposing a non-invasive intervention methodology for the subsoil of these buildings to avoid collapse.

One of the most used non-invasive techniques for subsoil exploration is geophysics and, more specifically, electrical resistivity tomography, in both two and three dimensions [15–21], and metrological perspectives of tomography in civil engineering [22]. In recent years, this method has proven to be an efficient tool, not only to monitor wall degeneration [23], but also to detect other types of problems in building foundations, such as subsoil degradation [14,24,25]. Spatial variability in load-bearing rock/soil beneath a building puts stress on poorly supported structures, causing failure [10,26]. Therefore, a geophysical investigation offers a faster, non-invasive means of obtaining detailed, credible information about the subsurface under a building. Electric tomography can also image the ground's distribution and structural deformation, both of which offer credible information regarding the strength the rock/soil is likely to offer a building [24,27,28]. Electrical resistivity imaging has, for decades, been very effective in illuminating the subsurface, and apt at providing information about the soil's physical properties for economic, environmental, and geological engineering. As mentioned, this technique is a useful and non-invasive method to diagnose subsoil problems in shallow foundations of buildings [3,29–32], which represent 25% of the claims reported in Spain, according to recent data provided by Aseguradora Mutua de Arquitectos Superiores, a public limited company (ASEMAS S.A.). However, the three-dimensional geophysical research approach ERT 3D [33–35] is better able than the two-dimensional approach ERT 2D to characterize the subsurface and determine heterogeneity in measured rock properties along the vertical (Z) and orthogonal horizontal (X, Y) axes [36,37].Thus, the determination of the variation in the soil properties along the three orthogonal directions would allow the evaluation of the spatial variation in the strength of the foundations, imposed by the heterogeneous properties of the soil [38].

Electrical resistivity tomography is an appropriate procedure for detecting and controlling underground consolidation and stabilization, specifically the ERT-3D tomography technique [39–42]. As mentioned above, the method consists of setting out parallel lines of observation, which cover the study area, to obtain underground data; however, the key point lies in processing the data. Fortunately, important advances have been made in 3D resistive imaging and its inversion processing through applying the powerful resistance inversion software, RES3DINV [43].This is possibly the most widespread, accurate, simple, and affordable software for data inversion in electrical tomography, and, thus, it has been used in this work. Several factors have been reported to impact the variation of electrical resistivity in the subsurface. These include variation in rock type, rock fabrics, rock deformation, water saturation, different degrees of weathering, etc. [44,45]. These factors, which are known to impact spatial variation on ground electrical resistivity, are also capable of impacting variation in other physical properties of rocks [46,47].With different electrode distances, ERT-3D offers the possibility of locating these gaps/cavities and possible holes under the foundation of a building, which can cause severe deformation and settlement. Both consolidating and stabilizing soil through injections have been proven effective [48–51] and are used to mitigate or even solve differential settling. The injection material generally depends on the lithological morphology of the soil in question. Synthetic resin is frequently used as an injection fluid [52], but this paper proposes injections of cement grout as a more economical and versatile solution.

For a soil consolidation project to be successful, electrical resistivity data must be available in advance to provide initial information on the subsurface structure. In our methodology for consolidating and stabilizing buildings with shallow foundations, we will use electrical tomography as a tool that detects empty cavities, which are typical examples of anthropic (resistive) fillings with low bearing capacity in subsoils [3]. Thus, this work proposes ERT -3D to monitor the subsoil in real time, from the beginning to the final consolidation through the different injection passes into the subsoil, until its stabilization. As an example of a case of application of our combined methodology, we have chosen a historic building with serious stability problems in its foundations [5], located in Estepona, on the Spanish south coast. The subsoil was consolidated with injections of cement grout applying the electrical methodology, resulting in the preservation of the building, and avoiding its possible collapse.

2. The Model Development

The ERT method discharges an electric current into the ground and measures the potential difference at two determined points on the surface. The suitability of this method lies in the fact that irregularities in the subsoil beneath a building can be identified as contrasts or anomalies in the subsoil's electrical properties.

This method is based on Ohm's law:

$$Pa = k(\Delta V/l) \tag{1}$$

where: Pa = apparent resistivity [36]; k = geometric constant that depends only on the reciprocal positions of the current and potential electrodes; ΔV = potential difference; l = intensity of the injected current.

The apparent values of resistivity depend on the real resistivity distribution in the tested area. The true resistivity distribution can be estimated through a reverse procedure, based on minimizing an adequate function [15,39,44]. The solution to this problem is not unique. For the same set of data, a wide range of models can calculate the same apparent resistivity values. A preliminary lithological analysis is usually carried out on the subsoil's nature, to reduce the range of possible models, which can be incorporated into the reverse subroutine.

The solution method used minimizes the difference between the apparent resistivities measured and those calculated by the RES3DINV software, which uses the limited softness inversion formulation, restricting the model's change in resistivity values [15,36,43,53]. This study has used RES3DINVx64, which implements a smooth routine based on least squares and is practically the only commercial software of its type available [36,54,55]. The inversion routine used by the program is based on the smoothness-constrained least-squares method [56]. The basic smoothness-constrained least-squares method is based on the following equation.

$$(J^T J + \lambda F)\, \Delta q_k = J^T g - \lambda\, F q_{k-1} \tag{2}$$

where:

$$F = \alpha_x C_X^T C_X + \alpha_Y C_Y^T\, C_Y + \alpha_Z C_Z^T\, C_Z \tag{3}$$

J^T = Transpose of J.
J = Jacobian matrix of partial derivatives.
λ = Damping factor.
q = Disturbance vector.
k = Iteration number
g = Data mismatch vector.
αx, αy, αz = weights for roughness filters
Cx, Cy = horizontal roughness filters
Cz = vertical roughness filter

This method is advantageous because of its versatility since the damping and roughness factor filters adjust to the different types of data. This program uses the Gauss–Newton method, which recalculates the Jacobian matrix, after each interaction [57]. The interpretations of electrical tomography profiles are made using RES3DINVx64 [34,36] for resistivity and induced polarization. As mentioned above, this calculation software is based on the least-squares method with forced smoothing, modified with the Quasi-Newton optimization technique. The inversion method designs a subsoil model using rectangular prisms and determines the resistivity values for each of them, minimizing the difference between the apparent resistivity values observed [15,58]. The results of our model show an uncertainty in its acceptable magnitude. They are in a critical state with respect to their structural stability, load capacity and, under these conditions, the safety of the building's habitability. This is at an acceptable level of uncertainty associated with these increasing risks of instability [5].

2.1. Application

Some results reported in the literature and obtained in the last eleven years regard the application of the ERT-3D method to study geotechnical anomalies in the subsoils on slopes after landslides located in different geographical contexts.Such reports make us consider the ERT method as a tool and methodology very suitable to investigate these geotechnical anomalies of the subsoil during the pre-event and post-event phases of a disaster cycle [30,31,59–61] or simply a poor foundation support subsoil (anthropic fill).

In fact, during the pre-event phase, the resistivity contrasts that characterize ERT-3D allow defining the geological environment of the subsoil. They allow identifying areas of high-water content that could be responsible for reactivation events. In the post-event phase, ERT-3Denablesus to reconstruct the damaged or altered subsoil body by also providing information on the volume of the removed or altered material. This information can help to better plan future mitigation activities. Our original application not only detects these subsoil alterations in real time, but we also propose an original tool for subsequent mitigation, in case of affecting existing buildings or infrastructure in a non-invasive way and thus preventing their collapse.

One of the biggest drawbacks of 3D tomography for the investigation of shallow and disturbances in the subsurface was the fact that it did not provide continuous acquisitions over time, which made it unsuitable for the study of the dynamic nature of shallow and disturbances.

Fortunately, the development of systems for the continuous acquisition of electrical resistivity time and software for data inversion [61] are paving the way to test this method during the emergency phase, such as the application and methodology developed in the present work, where in real time we are applying the geoelectric investment data and carrying out the consolidation of the subsurface area, increasing its resistivity in the emergency case, and thus being able to recover existing constructions in the altered subsoil.

The possibility of using ERT-3D to monitor geotechnical changes and alterations in the first layers of a ground clearance-settlement area will add important information during the emergency phase. At this time, the preliminary results obtained in our methodology when applying the ERT-3D for this purpose are very encouraging. The proof of this is that we were able to recover the building that was based on the geotechnically altered substrate, where the grout cement, filler material proposed in our methodology for consolidating the subsoil, satisfactorily filled the gaps that water and air occupied in the altered substrate. Cement grout has advantages over other fillers, such as its economy, easy means of performing on site, ease of dosing, and versatility, among others.

In applying our methodology in these cases of emergencies where there is a removal of surface soil after a shallow of the subsurface that affects buildings and infrastructures [62,63], we have observed that a low resistivity zone lies in the upper area from the surface to a depth up to 4.00 m, and the zone shows a higher resistivity of greater than 400 Ω m. According to the data obtained in this work, the upper area of the soil unit consists of mostly silt soil with granule from the erosion of the bedrock. Therefore, we thought that the anomaly in the upper area is attributed to the silt soil and granule with low resistivity. This zone of low resistivity coincides with the zone of removal and presence of water, which we consolidated with our grout cement filling methodology and in real time the monitoring was carried out to verify the satisfactory level of the landfill.

The procedure followed by the authors clearly defines the altered surface material with cavities and gaps of the rocky matrix (phyllite) by lower resistivities in the bedrock (50–150 Ω m). This methodology characterizes the subsoil altered by shallow bedrock material mainly composed of (by) clay material (colluvial) with high pore-hole content and high resistivities. The modelling of the subsoil analyzed with our system was monitored with the ERT of the grout process using the 3D RESINV software [63]. Another novelty of our tool is the geological characterization of this type of shallow phenomenon, and its affectation to existing buildings on the slope that in the South of Spain, in La Cordillera Bética, occurs with recurrence [62–64]. To have a tool like this proposal to characterize and mitigate damage to existing buildings and infrastructures in a non-invasive way in this type of geological formation is an advantage.

The results after the application of the methodology in the case study showed the effectiveness of the diagnostic and intervention methodology for mitigating the serious damage suffered by the building, preventing the collapse and destruction of the building while preserving the safety of its inhabitants.

2.2. Phases: Developed Methodology

Anew methodology has been developed consisting of different phases and applied to the case study:

1. It is necessary to interpret the characteristics of the subsurface soil/rock in an electrical resistivity tomography methodology. The ERT-3D tomograms will show how subsurface materials affect the resistivity of the subsurface.
2. Using a geophysical method to boreholes will provide a detailed knowledge of the studied site, with no limits over analysis and misinterpretation.
3. This will delineate depths and thicknesses of subsurface layers and identify the stiffness and distribution of subsurface materials in the Baetic Mountain Range.
4. The electrical resistivity of metamorphic rocks in the Baetic Mountain Range (phyllite) is dependent on the degree of fracturing and the resistivity of the interstitial and pore water in the rock and its water content in such facture zones. Thus, the electrical resistivity ranges from 3000 of Ω m to a fraction of 1.00 Ω m.
5. Finally, a non-invasive system of filling of holes through the injection of cement grout in the subsoil will be applied, monitored in real time with the ERT and thus increase the resistivity of the subsoil, and with it, its consolidation.

3. Case Study

This research shows the results of applying the proposed methodology to a specific case of differential settlements in a building in anat-risk area after soil removal [5]—see Figure 1. The land movements occurred after heavy rains [64] in the 2009–2010 hydrological years, but measurements presented in this paper were taken in 2012.

The applied methodology is based on ERT-3D and consists of placing electrodes along profiles separated from each other according to the resolution, depth, and objectives to be covered. The lower the separation, the greater the resolution; the greater the separation, the greater the depth. An optimal distance of 0.80 m. has been determined in our methodology between electrodes to obtain a balance between resolution tomographic profiles and depth in accordance with the problem in question.

Prior to performing the tomographic profiles, it is recommended that preliminary geological research be carried out in the area where the buildings are located [64]. In this case, the affected buildings were located at Paraje del Arroyo, La Cala, Estepona, Malaga (Spain), which is in the south of the Baetic Mountain Range (South Iberian Peninsula). The damage occurred at coordinates 36.461094, −5.160498 (Figure 1). In the first phase, it is proposed that a granulometric analysis and a Standard Penetration Test (SPT) be executed in the geotechnical surveys. In the present case, soil consistency increased with depth (Table 1). Change in the geotechnical response of the soil occurred at an approximate depth between one and four meters, according to the SPT hits along the analyzed sample (Table 1).

Note that the numbers in bold correspond to values that are too low. The grey background switching to white marks a discontinuity and an important change of capacity and bearing resistance in the subsoil, coincident with level I of Table 2.

Figure 1. Aerial picture.Red circle: building location. April 2012.Source: geographical application Googlemaps/https://www.google.com/maps/place (accessed on 25 April 2012), coordinates 36.461094, −5.160498.

Table 1. Summary of the consistency and admissible stress of the soil according to the rotary probes. Measurements were taken at the height of the main facade of the most affected building.

Depth (m)	Hit N_{20}	Consistency	Admissible Stress (KN/m^2)
0.00–1.00	**4**	**Soft**	**30**
1.00–2.00	**3**	**Soft**	**20**
2.00–3.00	**6**	**Slightly hard**	**40**
3.00–4.00	**7**	**Slightly hard**	**50**
4.00–4.40	**10**	**Slightly hard**	**70**
4.40–6.00	19	Moderately hard	120
6.00–7.00	52	Hard	310
7.00–8.00	63	Hard	350
8.00–8.60	84	Hard	440

Table 2. Lithography from the affected area via subsoil removal. Data are from lab tests. The depth of soil affected by lack of cohesion and similarity correspond with level I.

Levels	Lithology	Depth (m)
I	Colluvial clay material	0.00–4.40
II	Modified phyllites	4.40–6.00
III	Phyllites	6.00–25.00

Based on geological and geotechnical studies, the materials extracted in the area depict the lithological levels listed in Table 2. The analyzed soils were mainly rocky and clayey, of varying thickness, which could be verified in field work and through laboratory geotechnical tests.

3.1. Damaged Building Analysis

The most common deformations in a building are related to foundation differential movements. These alterations result in structural deformations in the building and angular distortions in its foundation [3,65]. This provokes stress throughout the construction, and when this stress limit is exceeded, cracking or breakages occur.

The Spanish Building Technical Code (2006) defines the maximum angular distortion values for a building's ultimate limit of state and service, conditioning this value to the type of structure. In the service limit state, the allowable angular distortion value is L/500 for reticulated structures, and L/300 for isostatic structures and for load-bearing walls (the studied case). For L, the length in a straight line between the axes of the footings is analyzed. However, the structure's ability to assume these deformations will depend, among other factors, on the stiffness of each element comprising it, so any fixed distortion value could be conservative or give rise to non-tolerable deformations [66–69].

According to current urban regulations, the rural house examined in this study would be difficult to demolish and rebuild. Therefore, a major challenge to this research and methodology was finding a solution to recover the building without having to demolish it. The main building in this study was a rectangular two-story building, 17.15 m long and 12.05 m wide, with a surface area of approximately 206.66 m^2 per floor and a total building area of 413.32 m^2 (Figure 2).

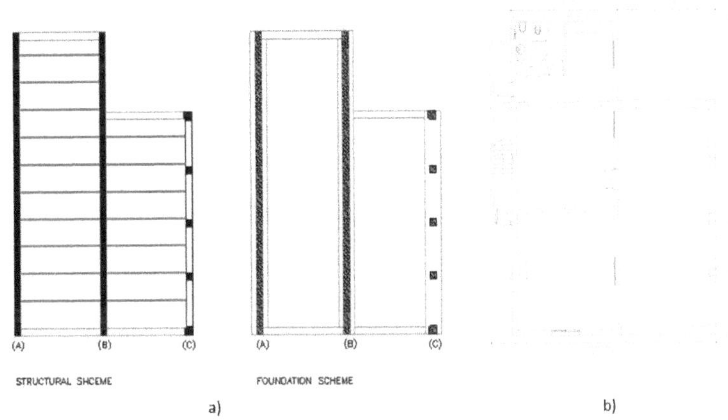

Figure 2. (**a**) Structural scheme of the damaged building, formed by three load-bearing walls, A, B, and C, and forged of semi-resistant joists, 24 cm thick. Building foundation scheme formed by three parallel, longitudinal shallow foundations (A, B, C), where the load-bearing walls rest and transmit the structure's weight. (**b**) Floor plan of the distribution of the damaged building in the slope.

This was a residential building and was used as a rural house (country holiday house); hence, guest and owner safety was paramount. The distribution comprised two building components: two floors, with an open area on the ground, and an upper, noble floor (Figure 2). Both the main building and the entrance foyer were designed in the late 1980s and completed in the early 2000s.The building's structure consisted of load-bearing walls and solid brick pilasters on the façade's main porch (vertical structure); these rested on a shallow foundation base of a reinforced concrete simple wall footing, formed by three parallel simple wall footings (A, B, C) (Figure 2). The horizontal structures of the building were24 cm thick unidirectional slabs formed of semi-resistant joists, supported in three spaces by load-bearing walls and pilasters resting on a shallow foundation composed of three longitudinal foundations in a trench.

From the previous geotechnical and structural analysis of the building illustrated in Figures 2 and 3, a problem of support in the foundation for differential seats was deduced.

Of the A, B, and C foundations, B and C had become unstable (sunk downward) with respect to A (Figure 4). This phenomenon caused the building to swing forward, which caused serious damage to its structure, as can be seen in Figure 4. The foundation of wall A also suffered a small movement in its seat (S1) (Figure 4).

Figure 3. The building's state of conservation in April 2012, after the movement of the subsoil and its shallow foundation. (C—see Figure 2 for reference) Structure of load pilasters and semi-circular arches, a structure that has suffered a decrease in its foundation with respect to the foundations of the load-bearing walls (B) and (A) on Figure 2. This has caused the building to deform and rotate forward.

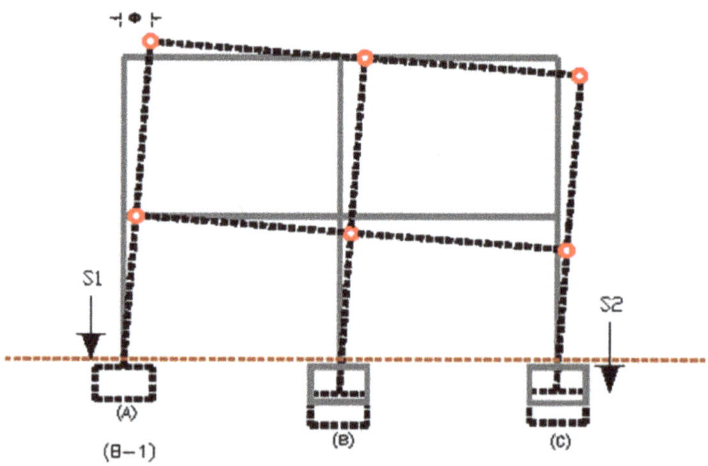

Figure 4. Cross-section scheme, presenting the likely behavior of the building's foundation and structure. There was a differential settlement of the S1 and S2 foundations due to a possible settlement-removal of the soil and the lack of bearing capacity of the resistant subsoil. Displacement occurred through the variable Φ = 95 mm, which is the depth B and C have descended, with respect to A. The seat of the foundation for wall A (S1) has lowered a depth of 25 mm, remaining within the admissible limit according to the Spanish Building Technical Code (CTE).

Due to the type of movement and structural typology, displacement occurred as shown in Figure 4. When load-bearing walls deform, they collapse, and destruction of the building becomes a major threat.

To determine the building deformation as a function of the angular distortion, it was calculated as the differential settlement, defined as the settlement difference between foundations A and C, which were the extreme foundations affected by the settlements, applying the following equation. Knowing that in this case, it was the same height that has descended B and C with respect to A:

$$\delta_{AC} = S_A - S_C \tag{4}$$

where δ_{AC} = foundation vertical differential displacement between points A and C; S_A = suffered differential settlement at point A; S_C = suffered differential settlement at point C.

The equation of angular distortion is also applied [3,70–72]:

$$\beta = \frac{\delta_{AC}}{L_{AC}} \tag{5}$$

where: β = angular distortion; L_{AC} = distance between foundation points A and C.

The Spanish Building Technical Code (2006), also known as CTE, establishes limitations on β movements, according to building structure typology. Studies regarding β began in the 1940s [73] and continue today [74–76]. The CTE limitation for the β value, for building damages, must be >L/300 for load-bearing walls. In this case, the differential settlement between foundations A and C was δ_{AC} = 95 mm and β = 0.0086, which are above the admissible limit established by the CTE, indicating imminent building collapse.

Authors also link angular distortion β values with damages suffered in enclosures (shell and internal partitions) and structural elements. The present case exceeded the limit values established by these authors, where the dimensions of cracks in enclosures must be >25 mm to be considered a serious structural risk, according to a programmed damage classification [77]. As shown in Figure 4, this building had serious damage, due to cracks in enclosures and structural elements (load-bearing walls), which were measured at greater than 50 mm. Therefore, according to this classification, the building was close to collapse. Due to the type of structure, foundation, and damages suffered, the methodology proposed in this research seems appropriate for the case. It would allow detection of zones of soil weakness where the differential settlement has been produced and consolidation of the subsoil from outside the construction, thus solving the building's issues non-invasively.

3.2. Analysis and Stabilisation Methodology

According to the data—regulations, the state of the building, the undeveloped land surrounding the building, and the necessity of keeping the building stable (i.e., avoiding imminent collapse)—a simplified model was proposed, in which ERT-3D would be used, with subsequent processing of the reversal data via the RES2DINV software [43], as seen in Figure 5. This analysis will consider the geotechnical data given in the previous sections.

The developed methodology allowed the researchers to see the reality of the soil, as well as any possible cavities in the subsoil supporting the building's foundation. These cavities were filled with grout to consolidate the soil, and then, via contrast campaigns (ERT-3D), the researchers checked to determine if the filling worked, if it moved, etc. Thus, the proposed methodology provided a complete, three-dimensional analysis of the subsoil supporting the damaged foundation, where the differential settlements occurred. Later, along with this original geophysical technique, the researchers monitored, through successive contrast campaigns, the consolidation and underpinning of the subsoil with different passes of consolidation injections. The injection material depends on the lithological morphology of the soil to be consolidated. Due to the type of subsoil in this case, as well as the material's low cost and ease of use and control, this paper proposed injections of controlled cement grout [78,79]. Synthetic resin is frequently used as an injection fluid, though it does not allow for high solicitation [52]. Three beneficial effects of injecting cement grout into the subsoil have been observed: (1) filling existing gaps, (2) soil compaction, and (3) interstitial water reduction and/or elimination.

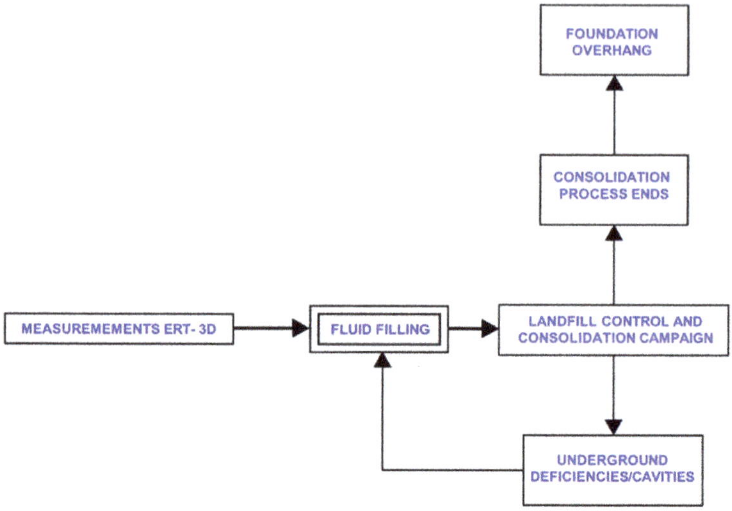

Figure 5. Block–flow diagram, showing the phases and methodology developed for intervening with the subsoil under the foundations of buildings with differential settlements.

3.3. Approach Methodology: Application

Had the proposed methodology not been applied in this case, the building could have been damaged to a critical point, making its repairs too expensive to pursue. Other, more invasive proceedings, such as underpinning through micropyles [5,80–82], were not feasible due to the state of the building and the structural typology of the load-bearing walls. However, making consolidation injections into the subsoil allowed recovery, to a certain degree, of the differential settlement suffered and rendered the building useable again.

The first necessary action involved using electrical tomography to analyze the subsoil beneath the affected foundations. In an initial campaign, this study accomplished various tests with spacings of 0.80, 1.00, and 1.50 m to identify the best resolution for checking the subsoil's state. A spacing of 0.80 m was chosen, based on another recent research [11].

The Syscal Switch 48 (V114 ++) from IRIS Instruments was used as electrical tomography equipment. It is a multi-electrode piece of equipment with an integrated computer capable of managing up to 900 electrodes, with resolution characteristics of resolution/accuracy:1 μV/0.2% [43]. The equipment power source is 250 W and 2.5 A, which generates 880 Vp-p pulses, and the manufacturer has incorporated a transmitter and receiver into the system. Among the equipment's features are a time injection adjustment, an apparent resistivity and chargeability automatic processor, a 3D real-time resistivity control, a voltage and current injection curve control, an integrated PC, and a commutation processor. The locations of the two topographies' parallel profiles (E1-E2) were shown via coordinates given by a GPS from Garmin Etrex Ventura [14], with a WG84 coordinate system. The results are listed in Table 3.

In the developed methodology, two parallel profiles (E1 and E2) were selected, covering an area of approximately 240 m^2, which included the main affected building area in accordance with Figure 6 and Table 3. The electrodes were placed in parallel.

RES3DINV—3D data inversion software for electrical imaging and induced polarization (IP) [83]—was used to automatically invert the acquired data from apparent resistivity and produce a 3D resistivity model. The two parallel profiles were executed using a dipole–dipole matrix. The dipole–dipole matrix was chosen due to its sensitivity to lateral variations of resistivity [41,84]; thus, structures such as gaps and cavities that were damaging the building's subsoil could be detected. This dipole–dipole matrix provided us with a sharper horizontal resolution of the whole, which is an important advantage in this type of building pathology. As discussed previously, an inter-electrode separation of 0.80 m was estimated as suitable for this methodology. The two profiles extended along 28 m, with 72 electrodes distributed according to the coordinates listed in Table 3 and Figure 6.

Table 3. UTM coordinates of the tomography profiles; location of the profiles according to the distribution of Figure 6.

Start UTM Coordinates of Tomography Profile E_1.		Final UTM Coordinates of Tomography Profile E_2.	
X	Y	X	Y
306,385	4,037,250	306,383	4,037,254
306,403	4,037,262	306,388	4,037,281

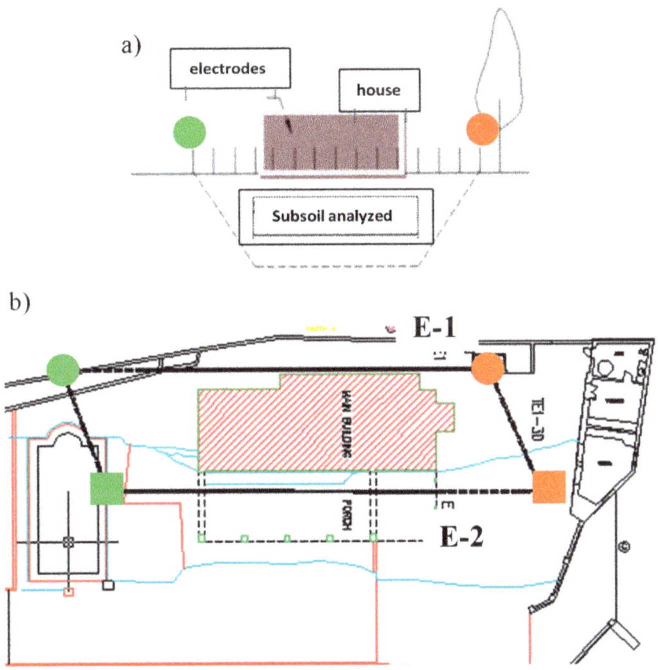

Figure 6. (**a**) Longitudinal scheme of the methodology and installation of the electrodes at 0.80 m in the building (house), as well as the subsoil analyzed with ERT -3D. The analyzed subsoil has a depth of 5.80 m, which corresponds to levels I and II of the geotechnical layers detected. (**b**) ERT -3D (TE1-3D) delimited tomography surface, covering the main building's foundation structure, within the parallel profiles proposed: E1 and E2. Circles refer to the E1 profile, while squares refer to E2; orange refers to the initial UTM coordinates, while green refers to the final ones (Table 3).

Figure 7 shows a 3D electrical resistivity model from the studied terrain and the depths reached, in this case, up to 5.80 m depth. Both profiles show resistivity values represented in a range of colors for better and easier observation of the variations in subsoil vertical and horizontal resistivities. Zones with resistivity values between 0–55 Ω m corresponded to levels of clay or colluvial material, represented in green; zones between 55–1000 Ω m, shown in orange, corresponded to fillings or colluvium removed with low compactness; and resistivity > 1000 Ω m, in red, indicated anomalies due to the presence of interstitial gaps.

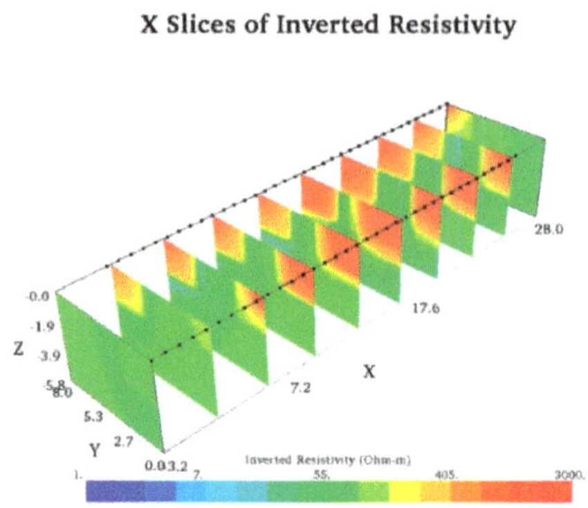

Figure 7. Advanced inversion algorithm of 3D synthetic data (RES3DINV software). Electric Ground Section ERT-3D: cross-section diagram of the ground, with the probable behavior of the soil and foundation. Cross-section scheme, with the likely behavior of the soil and foundation.

Figure 7 shows resistivity anomalies presented in slices of subsoil supporting the foundation of the damaged building. The red color shows a higher resistivity value, up to 3000 Ωm, which suggests that there are cavities causing the resistivity to rise. From the obtained resistivities, combined with the geotechnical data (SPT standards) [85,86], the physicochemical properties of geological materials can be determined.

The survey depth was established as 5.80 m, considering the previous geotechnical results of Table 1. From that depth, level III began to appear, determined by a rocky matrix of phyllite. The main anomalies were located on the main load-bearing walls, specifically A and B.

The structure of the building had been seriously damaged by the seat of its foundation and the turning of the building, specifically in the main facade between pilasters and the main bearing wall (wall B). There were also apparent interstitial anomalies in the subsoil up to 3.90 m deep in the dielectric profile E2 (wall B), and 2.00 m in the dielectric profile E1 (wall A). The anomalies and settlement in the foundation of wall B (S2) were of greater importance than the settlement at the base of wall A (S1), with a characteristic angular distortion of $\beta = 0.0086$, hence the forward inclination of the building. In summary, according to the algorithm developed in Figure 6, the first phase of the tomographical campaign was carried out, by analyzing the resistivities of the subsoil supporting the building and considering previous data, including the initial geotechnical data. Anomalies and areas with low compaction and/or gaps were identified and isolated in ERT-3D for the building's main load-bearing walls (Figure 8), so this knowledge might be acted on in the future.

Given the powerful information about the subsoil under the building's main load-bearing walls, the second phase of this methodology was proposed: consolidation injections into the subsoil for those detected anomalies (flow diagram, Figure 5). Figure 9 isolates the areas to be treated with injections via the RES3DINV investment program.

The damages observed in the analyzed building and the initial geotechnical tests confirmed the diagnosis from the ERT-3D tomography methodology. The initial ERT-3D results—that is, the application of the first phase of the flow chart (Figure 5)—showed high coincidence in the geological composition of the supporting subsoil beneath the damaged building: altered clay edges and colluvial sand from the meteorisation of the rock matrix (phyllite).

This innovative intervention technique allowed the researchers to evaluate the induced effects in the field and, in light of them, carry out possible modifications in the distribution of injection points to consolidate the subsoil. In the procedure's second phase, after the injections in the anomalies were made, the ERT-3D contrast was repeated to check the consolidation results.

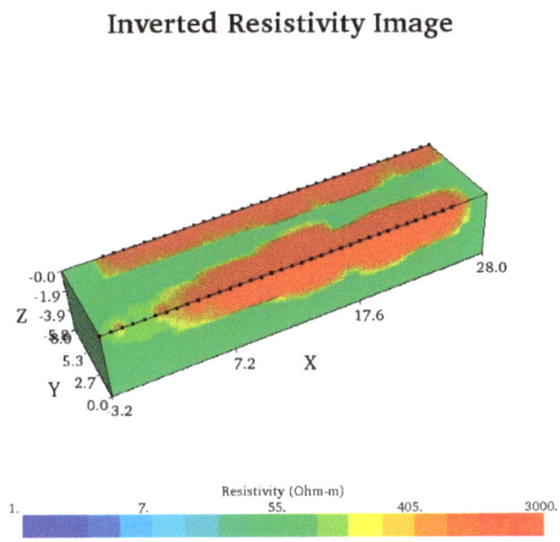

Figure 8. Volume plot rendered for the subsoil model of the building analyzed with topography included results of 3D inversion of electrical resistivity tomography over the whole survey area. On the left, view from the south, and, on the right, view from the north.

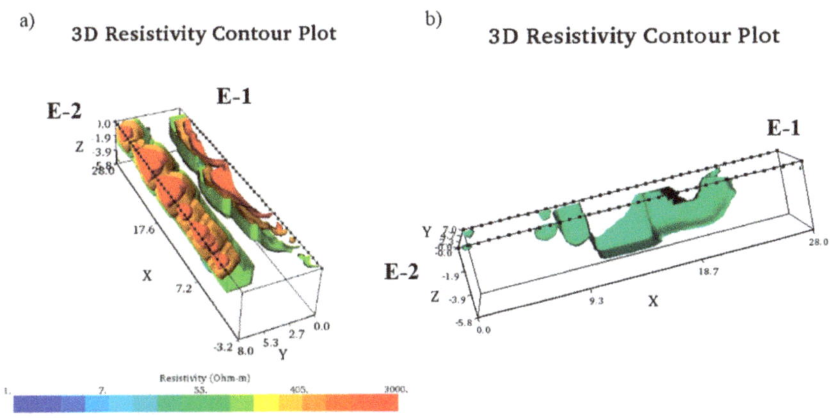

Figure 9. Example of a 3D survey using the offset dipole–dipole: (**a**) resistivity contour plot by anomalies. Scheme obtained through ERT-3D. (**b**) Resistivity contour plot by anomalies. Isolation of areas to be treated with soil consolidation injections grout using RES3DINV software.

The second phase of the proposed procedure (Figure 5) intended to obtain uniformity of chemical and physical features in the stabilized subsoil [65,70,86] until the desired goal was reached. As shown in Figure 10, the contrast tomography gave a favorable result after the cement grout injections were applied. The resistivities up to 3000 Ω m that caused distortions (cavities), affecting the stability of the building in Figure 8, were generally reduced drastically, even at values below 55 Ω m. This means the anomalies and stability issues of the building were resolved. In this way, the affected building could be recovered, and demolition avoided.

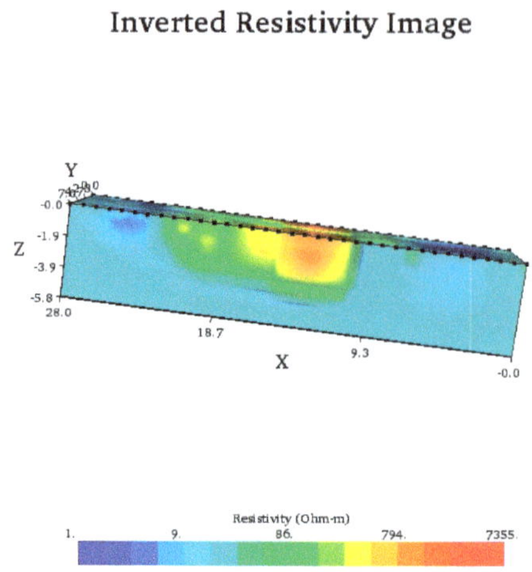

Figure 10. Results of 3D inversion of electrical resistivity tomography over the whole survey area. On the left, view from the south, and, on the right, view from the north. Contrast tomography after the consolidation injections in the subsoil.

The proposed intervention methodology is, therefore, highly effective in buildings damaged by settlements in shallow foundations, due to a deficient stress capacity of the subsoil, in this case by a removed anthropic fill. In this case study, a building on the verge of collapse, was recovered. Figure 10, showing the tomography performed after consolidation injections, indicates that up to 3000 Ω m resistivity anomaly areas have disappeared and left the building's subsoil consolidated and stabilized.

Four beneficial effects of our combined interactive methodology (tomography–injection) have been observed on subsoil where a building foundation rests:

(1) Fills existing gaps improving its bearing capacity.
(2) 100% subsoil compaction.
(3) Reduces and/or eliminates interstitial water in subsoil gaps.
(4) Recovery of severely damaged buildings with hardly any collateral effects.

4. Results and Discussion

The interpretation and adaptation of this paper's proposed methodology were satisfactory and successful. The building, on the verge of collapse, recovered its stability and was ready for use again. This procedure is promising for buildings that have reached their ultimate limit state but must be maintained either because they could not be rebuilt due to urban regulations (as in this case), or because they are listed as historic and cultural heritage buildings to be preserved at all costs.

The advantage of this proposed method is that ERT-3D allows researchers to determine if the foundation settlement and subsidence of the subsoil has been permanently eliminated after consolidation injections. Consolidating foundations through grout injections is inherently problematic because researchers do not know how much injection grout or how many injections are necessary, and they are unable to control or physically see the result, as it is underground. Therefore, an intervention methodology, such as the one presented here uniting ERT-3D monitoring with the consolidation injection process, is essential for checking the results. The advantage of this ERT-3D methodology, as well as consolidation grout injection compared to other reclining systems (e.g., micropyles), is that it is a non-invasive method for an already damaged building. This makes the system useful in extreme cases, like the one analyzed here.

In the application of our methodology, two electrical profiles were made, E1 and E2 (Figure 6), an upper part of approximately 3.90 m thick was defined for E2 (wall B) and

2.00 m for E1 (wall A) with high resistivity, ranging from 500 to up to 3000 Ω m. This section corresponded to the colluvial material (porous) that covers the metamorphic rock (phyllites). Under this lithology was the rocky matrix, formed by altered phyllites with less resistivity, of the order of 55 Ω m. The upper colluvial material, which was in contact with the underside of the building's footings, showed a significant degree of alteration, which means a high resistivity (from 1000 to up to 3000 Ω m). However, the resistivity values were quite variable within this mass of colluvial material, and this is linked to two causes: (1) lack of compaction and (2) presence of gaps and cavities in the subsoil where the foundation of the building supported, which coincided with the levels I and II altered. These data confirm the instability results of the shallow foundation of the building. The building's final state, after the application of this methodology, is shown in Figure 11.

Figure 11. Current state of the main building. Image taken after the intervention and application of the methodology. Had this methodology not been used, the only solution would have been demolition of the building.

5. Conclusions and Follow-Up

The ERT-3D technique proposed in this research proved to be versatile, fast, and cost-effective for detecting subsurface anomalies. In the present case study, the researchers were able to analyze the subsurface to a depth of 5.80 m, deep below the severely damaged building. Nevertheless, anomalies only existed at a depth of 3.90 m, below the foundation of load-bearing wall B. When the subsoil has holes, those spaces are filled with air or with water which, when evaporated, are dielectric. This means that the terrain presents a strong gradient anomaly and very high resistivity values, which causes serious damage to the foundation and structure of the building. Our methodology solves this problem, by reducing the resistivity of these initial anomalies (cavities) by means of the combined grout injection in two passes, with the control of resistivities through 3D tomography, and returning the stabilization of the foundation of the building in a non-invasive way. It has been applied to a case study that has been totally effective, where initially we had resistivities up to 3000 Ω m and it was effectively reduced with our methodology to 55 Ω m.

In the present case study, to implement the methodology, the subsoil cavities and weaknesses were first identified and located beneath the building's foundations. Secondly, a consolidation grout injection campaign was carried out, enhancing the subsoil conditions to an optimum result. The consolidation fluid injections eliminated the high value of initial subsoil resistivities (up to 3000 Ω m) and moved those resistivities to values between 55 and 1000 Ω m, filling in those cavities. Through this process, ERT-3D showed itself to be a fundamental, minimally invasive tool for research. It was useful both during the project phase and in the follow-up, underpinning work when the consolidation injections were placed under the foundations.

The limitations of our methodology increase with the depth of the altered subsoil, from 5.80 m. As has been shown in our research, it is difficult to have a reliable electrical section of resistivities to intervene later with consolidation injections. Although it would be necessary to use another geophysical methodology and another type of deep consolidation intervention using a pile or micropyle, our methodology is appropriate for emergency surface consolidations in disturbed porous subsoils. As they work safely outside the affected building, this methodology offers researchers an accurate lithological model and correctly highlights subsurface anomalies. Hence, ERT-3D allows them to determine the causes of the instability of the foundation and building and how it can be later consolidated and stabilized, in addition to monitoring the chosen solution.

Author Contributions: Conceptualization, A.G.-M. and R.C.; methodology, A.G.-M.; software, A.G.-M. and M.F.-H.; validation, A.G.-M., J.I.Y. and M.F.-H.; formal analysis, A.G.-M. and M.F.-H.; investigation, A.G.-M., J.I.Y. and R.C.; writing—original draft preparation, A.G.-M. and R.C.; writing—review and editing, R.C., A.G.-M.; visualization, R.C. and M.F.-H.; supervision, R.C. All authors have read and agreed to the published version of the manuscript.

Funding: This research received no external funding.

Data Availability Statement: Some or all data, models, or approach that support the findings of this study are available from the corresponding author upon reasonable request.

Conflicts of Interest: The authors declare no conflict of interest.

References

1. Keller, G.V.; Frischknecht, F.C. *Electrical Methods in Geophysical Prospecting*; Pergamon Press: Oxford, UK, 1966.
2. Nobahar, A.; Popescu, R. Some effects of soil heterogeneity on bearing capacity of shallow foundations. In Proceedings of the ASCE Special Conference, Houston, TX, USA, 10–13 September 2001.
3. González de Vallejo, L.; Ferrer, M.; Ortuño, L.; Oteo, C. *Ingeniería Geológica*; Prentice Hall: Madrid, Spain, 2002.
4. Ede, A.N. Building collapse in Nigeria: The trend of casualties in the last decade (2000–2010). *Int. J. Civ. Environ. Eng.* **2010**, *10*, 32–42.
5. Gutiérrez-Martín, A. El Agua de Infiltración de Lluvia, Aagente Desestabilizador de Taludes en la Provincia de Málaga. Modelos constitutivos. Ph.D. Thesis, University of Granada, Granada, Spain, 2016.
6. Olayanju, G.M.; Mogaji, A.K.; Lim, H.S.; Ojo, T.S. Foundation integrity assessment using integrated geophysical and geotechnical techniques: Case study in crystalline basement complex, southwestern Nigeria. *J. Geophys. Eng.* **2017**, *14*, 675–690. [CrossRef]
7. Fishman, Y.A. Effect of geological factors upon the pattern of failure and the stability of rock foundations of concrete dams. *Bull. Int. Assoc. Eng. Geol.* **1979**, *20*, 18–20. [CrossRef]
8. Terzaghi, K.; Peck, R.B.; Mesri, G. *Soil Mechanics in Engineering Practice*, 3rd ed.; John Wiley Sons Inc.: Hoboken, NJ, USA, 1996.
9. El-Kateb, T.; Chalaturnyk, R.; Robertson, P.K. An overview of soil heterogeneity: Quatification and implications on geotechnical field problems. *Can. Geotech. J.* **2003**, *40*, 1–15. [CrossRef]
10. Cho, S.E. Effects of spatial variability of soil properties on slope stability. *Eng. Geol.* **2007**, *92*, 97–109. [CrossRef]
11. Olarte, J.; Dashti, S.; Liel, A.B. Can ground densification improve seismic performance of the soil-foundation-structure system on liquefiable soils? *Earthq. Eng. Struct. Dyn.* **2018**, *47*, 1193–1211. [CrossRef]
12. Fagbenle, O.I.; Oluwunmi, A.O. Building failure and collapse in Nigeria: The influence of the informal sector. *J Sustain. Dev.* **2010**, *3*, 268–276.
13. Olofsson, B.; Jernberg, H.; Rosenqvist, A. Tracing leachates at waste sites using geophysical and geochemical modelling. *Environ. Earth Sci.* **2006**, *49*, 720–732. [CrossRef]
14. Osinowo, O.O.; Akanji, A.O.; Akinmosin, A. Integrated geophysical and geotechnical investigation of the failed portion of a road in Basement Complex terrain, southwestern Nigeria. *RMZ-Mater. Geoenviron.* **2011**, *58*, 143–162.
15. Loke, M.H.; Barker, R.D. Practical techniques for 3D resistivity surveys and data inversion. *Geophys. Prospect.* **1996**, *44*, 449–523. [CrossRef]
16. Ogilvy, R.; Meldrum, P.; Chambers, J.; Williams, G. The use of 3D electrical resistivity tomography to characterize waste and leachate distributions within a closed landfill, Thriplow, UK. *J. Environ. Eng. Geophys.* **2002**, *7*, 11–18. [CrossRef]
17. Aizebeokhai, A.P.; Olayinka, A.I.; Singh, V.S. Numerical evaluation of 3D geoelectrical resistivity imaging for environmental and engineering investigations using orthogonal 2D profiles. In *SEG Technical Program Expanded Abstracts 2009*; Society of Exploration Geophysicists: New York, NY, USA, 2009. [CrossRef]

18. Arango-Galván, C.; De la Torre-González, B.; Chávez-Segura, R.E.; Tejero-Andrade, A.; Cifuentes-Nava, G.; Hernández-Quintero, E. Structural pattern of subsidence in an urban area of the southeastern Mexico Basin inferred from electrical resistivity tomography. *Geofísica Int.* **2011**, *50*, 401–409. [CrossRef]
19. Cardarelli, E.; Di Filippo, G.; Tuccinardi, E. Electrical resistivity tomography to detect buried cavities in Rome: A case study. *Near Surf. Geophys.* **2006**, *4*, 387–392. [CrossRef]
20. Argote-Espino, D.; Tejero-Adrade, A.; Cifuentes-Nava, G.; Iriarte, L.; Farías, S.; Chávez, R.E.; López, F. 3D electrical propespection in the archaeological site El Pahu, Hidalgo State, Central Mexico. *J. Archaeol. Sci.* **2013**, *40*, 1213–1223. [CrossRef]
21. Chávez, R.E.; Cifuentes-Nava, G.; Tejero, A.; Hernández-Quintero, J.E.; Vargas, D. Tomografía de resistividad eléctrica 3D (ERT) especial aplicada para detectar fracturas enterradas en áreas urbanas: San Antonio Tecomal, Milpa Alta, México. *Geofísica Int.* **2014**, *53–54*, 425–434. [CrossRef]
22. Vanhove, Y.; Wang, M.; Wieczorowski, M.; Mathia, T.G. Metrological perspectives of tomography in civil engineering. In Proceedings of the 6th World Congress on Industrial Process Tomography, Beijing, China, 6–9 September 2010.
23. Schabowicz, K.; Hola, J. Nondestructive elastic-wave test of foundation slab in office building. *Mater. Trans.* **2012**, *53*, 296–302. [CrossRef]
24. Hola, J.; Schabowicz, K. State-of-the-art non-destructive methods for diagnostic testing of building structures—anticipated development trends. *Arch. Civil Mech. Eng.* **2010**, *10*, 5–18. [CrossRef]
25. Matawal, D.S. The challenges of building collapse in Nigeria. In Proceedings of the National Technical Workshop on Building Collapse in Nigeria, Akure, Nigeria, 12–14 November 2012; pp. 3–54.
26. Choi, P.; Kim, N.-H.; Lee, B.-H.; Won, M.C. Application of ultrasonic shear-wave tomography to identify horizontal crack or delamination in concrete pavement and bridge. *Constr. Build. Mater.* **2016**, *121*, 81–91. [CrossRef]
27. Palacky, G.V. Resistivity Characteristics of Geologic Targets. *Electromagn. Methods Appl. Geophys.* **1988**, *1*, 52–129. [CrossRef]
28. Putiska, M.; Nikolaj, M.; Dostal, I.; Kusnirak, D. Determination of cavities using electrical resistivity tomography. *Contrib. Geophys. Geod.* **2012**, *42*, 201–211. [CrossRef]
29. Aizebeokhai, A.P.; Olayinka, A.I.; Singh, V.S. Application of 2D and 3D geolectrical resistivity imaging for engineering site investigation in a crystalline basement terranin, southwestem Nigeria. *Environ. Earth Sci.* **2010**, *61*, 1481–1492. [CrossRef]
30. Peng, A.; Aijun, Z.; Yichuan, X.; Bo, Z.; Wenyuan, R. Experimental study on settling characteristics of thick self-weight collapsible loess in Xinjiang Ili region in China using field immersion test. *Soils Found.* **2018**, *55*, 1476–1491.
31. Shogaki, T.; Fujii, Y.; Nakagawara, Y. Deformation of the foundation and structure of Tomioka Silk Mill's East Cocoon Warehouse. *Soils Found.* **2019**, *59*, 789–800. [CrossRef]
32. Youssef, A.M.; Pradhan, B.; Sabtan, A.A.; El-Harbi, H.M. Coupling of remote sensing data aided with field investigations for geological hazards assessment in Jazan area, Kingdom of Saudi Arabia. *Environ. Earth Sci.* **2011**, *65*, 119–130. [CrossRef]
33. Loke, M.H. Electrical Imaging Surveys for Environmental and Engineering Studies, a Practical Guide to 2D and 3D Surveys. Available online: https://pages.mtu.edu/~{}ctyoung/LOKENOTE.PDF (accessed on 11 January 2021).
34. Loke, M.; Chambers, J.; Rucker, D.; Kuras, O.; Wilkinson, P. Recent developments in the direct-current geoelectrical imaging method. *J. Appl. Geophys.* **2013**, *95*, 135–156. [CrossRef]
35. Olvera-Ramírez, A.B.; Chávez, R.E.; Hernández-Quintero, J.E. Detección de zonas de fracturamiento en áreas urbanas mediante el método capacitivo de tomografía eléctrica. *Rev. Geofísica* **2014**, *64*, 129–142. [CrossRef]
36. Loke, M.H. Tutorial: 2-D and 3-D Electrical Imaging Surveys. Available online: http://personales.upv.es/jpadin/coursenotes.pdf (accessed on 12 January 2021).
37. Fernández, G.; Teixidó, T.; Peñab, J.A.; Burilloa, F.; Claros, J. Using shallow geophysical methods to characterise the monumental building at the Segeda I site (Spain). *J. Archaeol. Sci. Rep.* **2015**, *2*, 427–436. [CrossRef]
38. Badmus, B.S.; Akinyemi, O.D.; Olowofela, J.A.; Folarin, G.M. 3D electrical resistivity tomography survey for the basement of the Abeokuta terrain of Southwestern Nigeria. *J. Geol. Soc. India* **2012**, *80*, 845–854. [CrossRef]
39. Dahlin, T.; Bernstone, C. A Roll-Along Technique for 3D Resistivity Data Acquisition with Multi-Electrode Arrays. In Proceedings of the 10th EEGS Symposium on the Application of Geophysics to Engineering and Environmental Problems, Reno, NV, USA, 23–26 March 1997.
40. Day-Lewis, F.D.; White, E.A.; Johnson, C.D.; Lane, J.W.; Belaval, M. Continuous resistivity profiling to delineate submarine groundwater discharge—examples and limitations. *Lead. Edge* **2006**, *25*, 724–728. [CrossRef]
41. Clément, R.; Descloitres, M.; Günther, T.; Oxarango, L.; Morra, C.; Laurent, J.-P.; Gourc, J.-P. Improvement of electrical resistivity tomography for leachate injection monitoring. *Waste Manag.* **2010**, *30*, 452–464. [CrossRef]
42. Xiao, X.-H.; Xiao, P.-W.; Dai, F.; Li, H.-B.; Zhang, X.-B.; Zhou, J.-W. Large Deformation Characteristics and Reinforcement Measures for a Rock Pillar in the Houziyan Underground Powerhouse. *Rock Mech. Rock Eng.* **2017**, *51*, 561–578. [CrossRef]
43. Loke, M.H. RES3DMOD Ver. 2.1 M.H. Loke. Available online: www.geoelectrical.com (accessed on 24 September 2015).
44. Ward, S.H.; Fraser, D.C. Conduction of electricity in rocks. In *Mining Geophysics*; Society of Exploration Geophysicists: Houston, TX, USA, 1967; Volume 2, pp. 197–223.
45. Zonge, K.L. Electrical Parameters of Rocks as Applied to Geophysics. Ph.D. Thesis, The University of Arizona, Tucson, AZ, USA, 1972.
46. Utset, A.; Castellanos, A. Drainage effects on spatial variability of soil electrical conductivity in a vertisol. *Agric. Water Manag.* **1999**, *38*, 213–222. [CrossRef]

47. Carroll, Z.L.; Oliver, M.A. Exploring the spatial relations between soil physical properties and apparent electrical conductivity. *Geoderma* **2005**, *128*, 354–374. [CrossRef]
48. Jiménez Salas, J.A.; Justo Alpañes, J.L. *Geotecnia y Cimientos II–III*; Rueda: Madrid, Spain, 1981.
49. Ni, J.C.; Cheng, W.-C. Using fracture grouting to lift structures in clayey sand. *J. Zhejiang Univ. A* **2010**, *11*, 879–886. [CrossRef]
50. Kumari, D.; Xiang, W.-N. Review on biologically based grout material to prevent soil liquefaction for ground improvement. *Int. J. Geotech. Eng.* **2019**, *13*. [CrossRef]
51. Liu, Y.; Chang, C.-W.; Namdar, A.; She, Y.; Lin, C.-H.; Yuan, X.; Yang, Q. Stabilization of expansive soil using cementing material from rice husk ash and calcium carbide residue. *Constr. Build. Mater.* **2019**, *221*, 1–11. [CrossRef]
52. Uretek Engineering: Stabilization of Soils and Buildings by Expansive Resin Injections. Available online: http//www.uretek.es (accessed on 5 September 2019).
53. Parasnis, D.S. *Principles of Applied Geophysics*, 5th ed.; Chapman & Hall: London, UK; New York, NY, USA, 1997.
54. De Groot-Hedlin, C.; Constable, S. Occam's inversion to generate smooth, two-dimensional models form magne-totelluric data. *Geophysics* **1990**, *55*, 1613–1624. [CrossRef]
55. Sasaki, Y. Resolution of Resistivity Tomography Inferred from Numerical Simulation. *Geophys. Prospect.* **1992**, *40*, 453–463. [CrossRef]
56. Loke, M.; Acworth, I.; Dahlin, T. A comparison of smooth and blocky inversion methods in 2D electrical imaging surveys. *Explor. Geophys.* **2003**, *34*, 182–187. [CrossRef]
57. Loke, M.H.; Dhalin, T. A comparison of the Gauss–Newton and quasi-Newton methods in resistivity imaging inversion. *J. Appl. Geophys.* **2002**, *49*, 149–162. [CrossRef]
58. Chambers, J.E.; Meldrum, P.I.; Gunn, D.A.; Wilkinson, P.B.; Kuras, O.; Weller, A.L.; Ogilvy, R.D. Hydroge-ophysical Monitoring of Landslide Processes Using Automated Time-Lapse Electrical Resistivity Tomography (ALERT). In Proceedings of the Near Surface—15th European Meeting of Environmental and Engineering Geophysics, Dublin, Ireland, 7–9 September 2009.
59. Niesner, E. Subsurface resistivity changes and triggering influences detected by continuous geoelectric monitoring. *Lead. Edge* **2010**, *29*, 952–955. [CrossRef]
60. de Bari, C.; Lapenna, V.; Perrone, A.; Puglisi, C.; Sdao, F. Digital photogrammetric analysis and electrical resistivity tomography for investigating the Picerno landslide (Basilicata region, southern Italy). *Geomorphology* **2011**, *133*, 34–46. [CrossRef]
61. Loke, M.; Dahlin, T.; Rucker, D.F. Smoothness-constrained time-lapse inversion of data from 3D resistivity surveys. *Near Surf. Geophys.* **2013**, *12*, 5–24. [CrossRef]
62. Gutiérrez-Martín, A. A GIS-physically-based emergency methodology for predicting rainfall-induced shallow landslide zonation. *Geomorphology* **2020**, *359*, 107121. [CrossRef]
63. Herrada, M.A.; Gutiérrez-Martin, A.; Montanero, J.M. Modeling infiltration rates in a saturated/unsaturated soil under the free draining condition. *J. Hydrol.* **2014**, *515*, 10–15. [CrossRef]
64. Gutiérrez-Martín, A.; Herrada, M.Á.; Yenes, J.I.; Castedo, R. Development and validation of the terrain stability model for assessing landslide instability during heavy rain infiltration. *Nat. Hazards Earth Syst. Sci.* **2019**, *19*, 721–736. [CrossRef]
65. Jiménez Montoya, P.; García Meseguer, A.; Moran Cabre, F. *Hormigón Armado*; Gustavo Gili Editorial: Barcelona, Spain, 2010.
66. Burland, J.B.; Wroth, C.P. *Settlement of Buildings and Associated Damage*; Pentech Press: London, UK, 1974.
67. Burland, J.B.; Broms, B.; De Mello, V.F.B. Behaviour of foundations and structures. State of Art. Report. Session 2. In Proceedings of the Ninth Conference on Soil Mechanics and Foundation Engineering, Tokyo, Japan, 10–15 July 1977; p. 495.
68. Boscardin, M.D.; Cording, E.J. Building Response to Excavation-Induced Settlement. *J. Geotech. Eng.* **1989**, *115*, 1–21. [CrossRef]
69. Dimmock, P.S.; Mair, R.J. Effect of building stiffness on tunnelling-induced ground movement. *Tunn. Undergr. Space Technol.* **2008**, *23*, 438–450. [CrossRef]
70. Rodríguez Ortiz, J.M. *Cursoaplicado de Cimentaciones*; Servicio de Publicaciones del Colegio Oficial de Arquitectos: Madrid, Spain, 1982.
71. Negulescu, C.; Foerster, E. Parametric studies and quantitative assessment of the vulnerability of a RC frame building exposed to differential settlements. *Nat. Hazards Earth Syst. Sci.* **2010**, *10*, 1781–1792. [CrossRef]
72. Sanabria, M.P.; Guardiola-Albert, C.; Tomás, R.; Herrera, G.; Prieto, A.; Sánchez, H.; Tessitore, S. Subsidence activity maps derived from DInSAR data: Orihuela case study. *Nat. Hazards Earth Syst. Sci.* **2014**, *14*, 1341–1360. [CrossRef]
73. Meyerhof, G.G. The settlement analysis of building frames. *Struct. Eng.* **1947**, *25*, 369–409.
74. Burland, J.B. The assessment of the risk of damage to buildings due to tunnelling and excavation. In Proceedings of the Second International Conference on Soil Structure Interaction in Urban Civil Engineering, Planning and Engineering for the Cities of Tomorrow, Zurich, Switzerland, 7–8 March 2002; Volume 1.
75. Diaz, C.; Cornadó, C.; Santamaría, P.; Rosell, J.R.; Navarro, A. Actuación preventiva de diagnóstico y control de movimientos en los edificios afectados por subsidencia en el barrio de la Estación de Sallent (Barcelona). *Inf. Construcción* **2015**, *67*, 089. [CrossRef]
76. Hintze, S. Risk Analysis in Foundation Engineering with Application to Piling in Loose Friction Soils in Urban Situations. Ph.D. Thesis, Division of Soil and Rock Mechanics, Royal Institute of Technology, Stockholm, Sweden, 1994.
77. Malinin, A.; Gladkov, I.; Malinin, D. Experimental research of jet-grouting parameters in different soil conditions. In Proceedings of the Geo Shangai International Conference 2010, Shanghai, China, 3–5 June 2010. [CrossRef]
78. Argal, E.S. Modern Technologies and Problems of Ground Stabilization by Injection. *Soil Mech. Found. Eng.* **2020**, *57*, 219–226. [CrossRef]

79. Bustamante, M. *A Method for Calculating Anchors and Injected Micropiles*; Bulletin of the SEMSC: Madrid, Spain, 1986; pp. 81–82.
80. Juran, I.; Bruce, D.A.; Dimillio, A.; Benslimane, A. Micropiles: The state of practice. Part II: Design of single micropiles and groups and networks of micropiles. *Proc. Inst. Civ. Eng. Ground Improv.* **1999**, *3*, 89–110. [CrossRef]
81. Pinyol, N.M.; Alonso, E. Design of micropiles for tunnel face reinforcement: Un drainedupper bound solution. *J. Geotech. Geoenviron. Eng.* **2012**, *138*, 89–99. [CrossRef]
82. Sastry, R.G.; Chahar, S.; Singh, M. Multiple regression analysis of geoelectric imaging and geotechnical site investigation test results. *Curr. Sci.* **2018**, *114*, 1946–1952. [CrossRef]
83. Seaton, W.J.; Burbey, T.J. Aquifer characterization in the Blue Ridge physiographic province using resistivity pro-filing and borehole geophysics: Geologic analysis. *J. Environ. Eng. Geophys.* **2000**, *5*, 45–58. [CrossRef]
84. Fletcher, G.F.A. Standar penetration test: Its uses and abuses. *J. Soil Mech. Found. Div.* **1965**, *91*. [CrossRef]
85. Inazumi, S. In-Situ Ground Surveying by the NSWS Testing Machine. *Int. J. GEOMATE* **2011**, *1*, 1–9. [CrossRef]
86. Yang, S.; Ren, X.; Zhang, J. Study on embedded length of piles for slope reinforced with one row of piles. *J. Rock Mech. Geotech. Eng.* **2011**, *3*, 167–178. [CrossRef]

applied sciences

MDPI

Article

Dynamic Effect of the Earth Fissure Sites in the Yuncheng Basin, China

Ge Cao [1], Yahong Deng [1,2,*], Jiang Chang [1], You Xuan [1], Nainan He [1] and Huandong Mu [3]

1 School of Geological Engineering and Geomatics, Chang'an University, Xi'an 710054, China
2 Key Laboratory of Mine Geological Hazards Mechanism and Control, Ministry of Natural Resources, Xi'an 710054, China
3 Institute of Geotechnical Engineering, Xi'an University of Technology, Xi'an 710048, China
* Correspondence: dgdyh@chd.edu.cn

Abstract: Earth fissures are widely distributed worldwide, and the Fenwei Basin in China is one of the regions with the most significant number and scale of fissures in the world. The Yuncheng Basin is an important constituent basin of the Fenwei Basin in China, where earth fissures are densely developed and cause severe damage. In particular, the impact of earth fissures on the seismic response of the site is still unknown and is an urgent problem that needs to be solved. Based on microtremor tests, three types of typical earth fissure sites in the Yuncheng Basin were selected for field testing. Through spectrum analysis, the dynamic response characteristics of the earth fissure sites were determined. The results show that the dynamic response of the site is significantly affected by the earth fissures. The dynamic response strength of the site is the largest on both sides of the earth fissures, and it decreases and gradually stabilizes with increasing distance from the fissures. The influence range of the earth fissures on the hanging side is slightly longer than the heading side.

Keywords: earth fissure; microtremor testing; spectrum analysis; amplification effect

check for updates

Citation: Cao, G.; Deng, Y.; Chang, J.; Xuan, Y.; He, N.; Mu, H. Dynamic Effect of the Earth Fissure Sites in the Yuncheng Basin, China. *Appl. Sci.* **2023**, *13*, 9923. https://doi.org/10.3390/app13179923

Academic Editors: Miguel Llorente Isidro, Ricardo Castedo and David Moncoulon

Received: 8 August 2023
Revised: 28 August 2023
Accepted: 31 August 2023
Published: 1 September 2023

1. Introduction

Earth fissures are linear tensile fissures with a certain extension length that develop on the Earth's surface and may be accompanied by vertical fissures. They are affected by various complex factors such as internal and external forces and human activities [1]. Earth fissures have developed widely around the world and occur in the United States, Mexico, Africa, Europe, and throughout most of China [2,3]. Thus, the formation and development of earth fissures seriously affect infrastructure construction and restrict national economies [4].

The study of earth fissures began in 1929 [5,6], and humans have been studying earth fissures for a century. The first step in preventing and controlling earth fissure hazards is to identify the mechanisms of their formation and evolution. There are many factors affecting the formation of earth fissures, and the explanations of the formation mechanisms of earth fissures by domestic and foreign scholars can mainly be divided into three categories: ground subsidence induced by tectonic genesis [7,8], groundwater genesis [9–11], and structural and groundwater compound genesis [12–14]. Among them, China is one of the countries whose earth fissures have an enormous scale of development, the broadest distribution, and the most substantial phenomenon of surface damage caused by earth fissures. Since 2000, more than 5000 earth fissures have been discovered in China, mainly in the Fenwei Basin, the Hebei Plain, and the Su-Xi-Chang area (Figure 1) [15]. Compared to the other two regions, the planar spread of the earth fissures within the Fenwei Basin is the longest and most active. There have been many excellent results and consensus on the distribution pattern, activity characteristics, and genesis mechanisms of fissures in the Fenwei Basin [16–28].

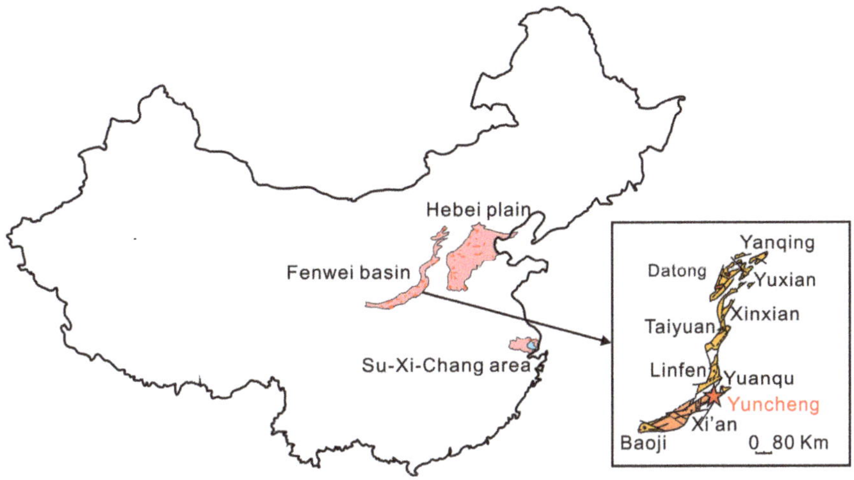

Figure 1. Geographical location of the Yuncheng Basin.

The existence and activity of earth fissures do not only cause direct damage to surrounding structures, but when an earthquake occurs, the presence of earth fissures can change and even increase the seismic response of the site, causing the structures near the earth fissures to suffer severe damage or even be destroyed. Thus far, research on the development, distribution characteristics, and formation mechanisms of earth fissures all over the world has produced results, but our understanding of the dynamic effects of earth fissure sites is insufficient. Several studies have investigated the dynamic effects of earth fissures under earthquake loads based on numerical simulations [29,30], and others have used indoor simulation experiments to explore the influence of earth fissures on the dynamic response of a site [31–35], but the existing results are few and unsystematic and are not sufficient to guide engineering practice. Therefore, in terms of the specifications for site investigation and engineering design on Xi'an ground fractures (DBJ61-6-2006), [36] only regards the seismic effects of earth fissure sites as a general site threat in accordance with the code for the seismic design of buildings (GB50011-2010) [37]. Furthermore, because there are no systematic measurement data on the seismic response of earth fissure sites, in order to study this problem it is necessary to develop new ideas and appropriate methods.

A microtremor is a kind of constant micro-movement that has no specific seismic source and can be observed at any time, and its amplitude is generally only a few microns. Microtremor studies originated in the 1950s. Kanao and Tamaka [38] quantitatively analyzed the nature of microtremor surface waves. After this, many research works were carried out on the microtremor sources, formation mechanisms, and waveforms of microtremors [39–42]. In the late 1960s, Toksoz and Lacoss used a seismic network to separate, extract, and analyze the various periodic components of microtremors [43,44] and reported on the components of the different bands of the microtremors and their possible causes. In the 1990s, Nakamura proposed a new microtremor method [45]. This horizontal-to-vertical spectral ratio (H/V) method has gradually become the focus of the analysis of site dynamics around the world. Many scholars have tried to explain the theoretical basis of this method from different angles [46,47]. In 1994, Lermo and Chavez-Garcia [48] summarized the three main types of microtremor analysis methods and concluded that the H/V method could effectively eliminate the source effect. Since then, the H/V method has been widely used in the field of engineering. Many scholars in China have also contributed to the microtremor theory and practical engineering applications [49–51]. A large number of studies have shown that microtremor movement is an efficient, economical, and convenient method for testing the dynamic characteristics of a site which contains a large amount of site soil structure information that can be used as field measurement data. Moreover, the microtremor test has initially been applied to the topic of the dynamic effect of earth fissure sites [52,53], and the study showed that the existence of earth fissures does indeed aggravate the vibration intensity of the site.

The Yuncheng Basin is an important part of the Fenwei Basin and contains the most well-developed earth fissures in China. The earth fissures in the Yuncheng Basin have caused serious disasters, and have become the most notable geological disaster in this area. Furthermore, the basin has a large population, developed agriculture, and industry, and the earth fissures not only cause direct damage to local roads and houses, but also severely restrict the construction of urban infrastructure, and thus they restrict the economic and social development of the Yuncheng Basin. The results of this study provide theoretical support for the development of seismic fortification and avoidance measures for the earth fissure sites in this area.

2. Regional Structure

From north to south, the Fenwei graben system consists of the Datong Basin, the Xinding Basin, the Taiyuan Basin, the Linfen Basin, the Yuncheng Basin, the Weihe Basin, and several other large fault basins (Figure 1). The Yuncheng Basin is located in the southwestern part of Shanxi Province and has a total area of 4885 km^2.

The Yuncheng Basin is a Cenozoic faulted basin superimposed on the multi-cyclic superimposed Ordos Basin. The sedimentary strata have been deformed by multiple tectonic movements and have experienced multiple tectonic periods, such as the Yanshanian and Himalayan movements. A representative large rift basin formed later [27]. The basic framework of the Zhongtiao mountain uplift area and the Emei platform uplift area in the basin was laid by the strong tectonic movement in the Yanshanian period, and it was mainly affected by the Himalayan movement during the Paleogene. Large-scale fault block tectonic movement occurred in this area. The tectonic setting of the basin has changed from a compressive and twisting thrust fault to a tension and twisting normal fault, and the embryonic form of the basin has been modified since then. The main structural deformations preserved today include a fault system in the Himalayan strike-slip extensional background [2].

The main forms of tectonic movement in the Yuncheng Basin are faults. The Basin was controlled by the Yanshan movement in the Mesozoic era, the Himalayan movement in the Cenozoic era, and especially by tectonic movement since the Late Cenozoic era, and therefore fault structures have been well-developed in the basin. The main active fault in the basin is the large fault on the northwest side of Zhongtiao mountain, which is also the main controlling boundary fault in the Yuncheng Basin [25]. There are eight active faults in the basin, with NE, NEE, and NNE strikes. They enclose the boundaries of the Yuncheng Basin and the secondary structural units inside the basin, and they mutually restrict and control the structural framework of the basin [27]. The Yuncheng Basin is bordered by the Emei Platform to the west, and the Zhongtiao mountains to the southeast. The NE-trending Mingtiaogang in the middle of the basin divides the Yuncheng Basin into the Sushui river plain and the Qinglong river plain. The basin is an asymmetrical sag basin that is deep in the south and shallow in the north.

The basin contains a large amount of Cenozoic strata, accounting for about 80% of the total area, and the thickness of the strata increases from the northeast to the southwest. The thickness of the Cenozoic strata in the basin is generally greater than 1000 m, and the Quaternary sediments are also more than 300 m thick. This loose, thick sedimentary layer provided the material basis for the extensive development of earth fissures.

In general, factors such as the uplift of the horst in the Yuncheng basin, the faulted basement, the over-exploitation of groundwater, and the loose, thick sedimentary layer have laid the foundation for the development of earth fissures in the basin.

3. Characteristics of the Earth Fissures in the Yuncheng Basin

The development of earth fissures is controlled by tectonic movement. The earth fissures in a basin are mainly affected by the secondary structural units. The Yuncheng Basin has been affected by the Himalayan tectonic movement since the Cenozoic era. Due to the control of the faults at the northern foot of the Zhongtiao mountain, the basement has been in a faulted extensional environment. Under the strong tectonic activity of the Zhongtiao mountain fault, the secondary structural unit in the basin was controlled, which intensified the uplift of the block and caused the formation of the Mingtiaogang earth fissures. The dip slip extension of the hanging wall normal fault in the basin caused the tension shear cracking of the surface soil layer, which laid the structural foundation for the development of earth fissures in the basin. In addition, since the 1980s, the over-exploitation of groundwater has been severe in this area, causing the groundwater level to continuously fall and the falling funnel to expand year by year, which has induced a large number of earth fissures in the Yuncheng Basin. Thus, the Yuncheng Basin has become the faulted basin with the largest number of earth fissures in Shanxi.

The earth fissures in the basin are mainly distributed on top of and on both sides of the Mingtiaogang uplift within the basin, in the Zhongtiao mountain uplift to the southeast, and in the Emei platform front edge to the north (Figure 2). The earth fissures in the Yuncheng Basin are distributed along the boundaries of the geomorphology, are concentrated along the fault zone, and are associated with land subsidence. The earth fissures predominantly have NE strikes. According to our survey, the earth fissure disasters in the basin mainly occurred before 1980 and between 1995 and 2005. A total of 119 earth fissures (belts) have developed in the study area. They mainly strike NE and are generally 100–2000 m long, with the longest reaching up to 5000 m. They are generally 0.05–0.5 m wide, and the widest can reach up to 2 m. The small earth fissures account for 45% of the earth fissures (length less than 500 m); the medium earth fissures account for 28% (length between 500 and 1000 m); and the large earth fissures account for 17% (length between 1000 and 5000 m). There is only one giant earth fissure in the basin. It is located in the salt lake district and is 10 km long and 0.3–1 m wide.

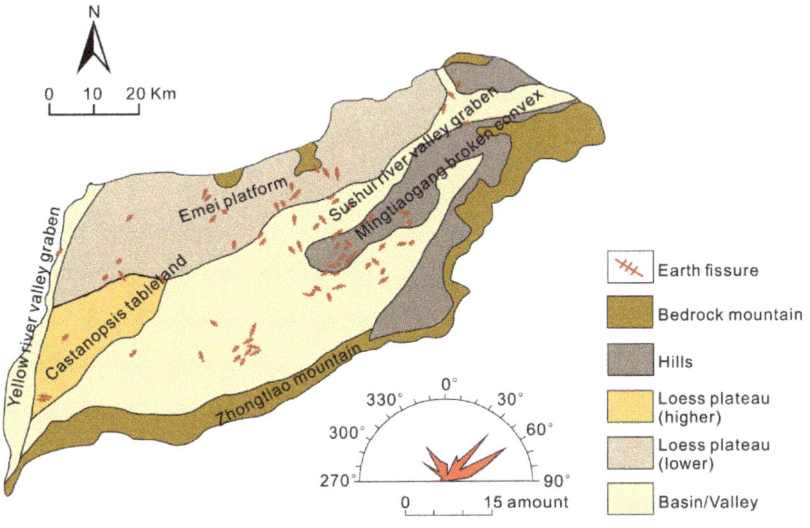

Figure 2. Structure and earth fissure distribution map of the Yuncheng Basin.

Figure 3 shows six typical earth fissures in the Yuncheng Basin and their profiles. Earth fissure F_1 (Pleistocene earth fissure on the southern margin of the Mingtiaogang uplift) is located in the salt lake district of Yuncheng. It is the only giant earth fissure in the Yuncheng Basin. The earth fissure F_1 appeared in 1975. It has a long extension, large scale, and strong continuity, and is not controlled by the pavement, buildings, or roads along the route. It is 10 km long and crosses Taocun–Banpo–Wucao–Xincao and other villages, and it is very destructive along the strike (Figure 4a,c). Earth fissure F_1 strikes NE, which is consistent with the active fault in the underlying bedrock. In a plane view, the earth fissure F_1 is longer, and its overall shape is linear. In profile, the site is dominated by loess and paleosol, with secondary fissures on either side of the main fissure symmetrically developed and approximately parallel. The vertical plane of the two sides near the surface of the main fissure is small, and the deep dislocation is large, showing the characteristic of a synsedimentary fault (Figure 3a).

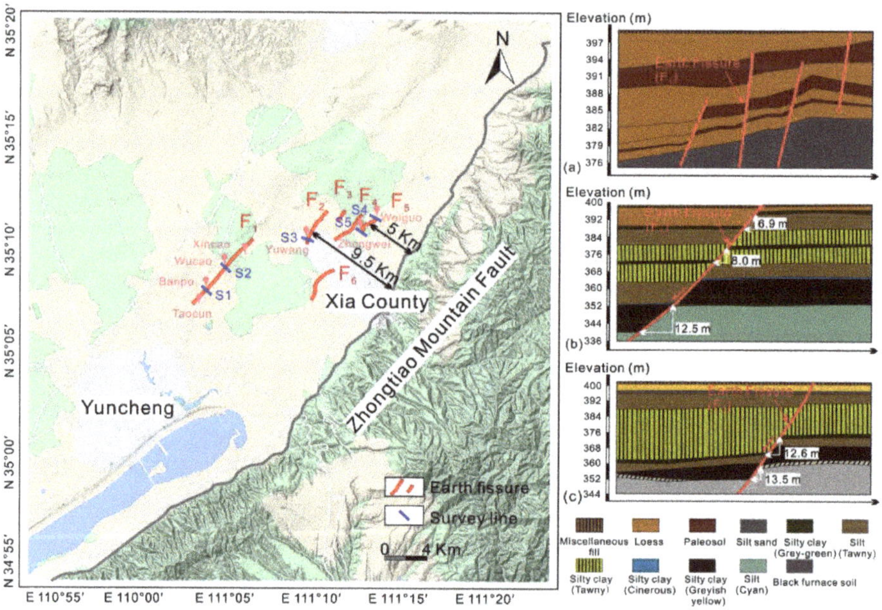

Figure 3. The locations of six typical earth fissures and survey lines in the Yuncheng Basin and their profiles. (**a**) Profile of F_1 earth fissure site, (**b**) profile of F_2 earth fissure site, and (**c**) profile of F_5 earth fissure site.

The Zhongtiao mountain piedmont fault zone is located in the southern part of Xia County. Affected by tectonic movement, the five earth fissures (F_2–F_6) developed in the Xia County area are all distributed on the northwestern side of the Zhongtiao mountain fault zone. The earth fissures are distributed in parallel rows at nearly equal intervals from southeast to northwest (Figure 3). The earth fissure F_2 is farthest from (i.e., ~9.5 km) the Zhongtiao mountains in Xia County. It has the most significant vertical surface dislocation among the five earth fissures in Xia County. The largest surface vertical dislocation occurs in the playground of Xinmiao Primary School in Yuwang, where the surface dislocation reaches 30 cm (Figure 4b,d). The earth fissure F_2 appeared in 1998 and developed rapidly from 2007 to 2008. It strikes, which is consistent with the trend of the Zhongtiao mountains, and extends for 3.9 km. Its scale is relatively large. The earth fissure F_2 passes through Yuwang and other places, and it is more destructive along the strike. As can be seen from the cross-section shown in Figure 3b, the formation is composed of interbedded silt and silty clay, and the vertical dislocation increases with increasing depth. The earth fissure F_5 is 5 km from the Zhongtiao mountain fault zone, making it the closest of the five earth fissures to the fault zone. The earth fissure F_5 starts in Yuguo and ends in Zhongwei. Its overall trend is NE55°, which is consistent with the trend of the Zhongtiao mountains. It extends for about 2.3 km, and the vertical dislocation of the ground surface is 8–20 cm. The earth fissure F_5 appeared in 2000, and it entered a period of rapid development after 2007. It is active and destructive along the strike, and it is still developing. The earth fissure is a tilt-slip tension crack with obvious horizontal extensional movement and vertical differential movement. The differential settlement on both sides of the earth fissure is obvious, and the maximum vertical dislocation is 13.5 m. As can be seen from the cross-section shown in Figure 3c, the formation is composed of silt and silty clay. The stratum is relatively weak.

Figure 4. Earth fissure hazards in the Yuncheng Basin. (**a**–**c**) earth fissure hazard, (**d**) surface dislocations caused by earth fissures.

4. Microtremor Tests and Analysis

4.1. Survey Line and Data Point Layout

Survey lines were along the previously described earth fissures (F_1, F_2, and F_5). Two survey lines (S1 and S2) were laid in Banpo and Wucao for the earth fissure F_1, only one survey line (S3) was laid in Yuwang for the earth fissure F_2, and one survey line each was laid in Yuguo and Zhongwei (lines S4 and S5, respectively) for the earth fissure F_5. As an example, the survey line layout of F_2 in Yuwang is shown schematically in Figure 5. The survey lines were perpendicular to the earth fissures. Each survey line contained 18 data acquisition points, and 9 measurement points were set on either side of the earth fissure. The survey lines were about 60 m long.

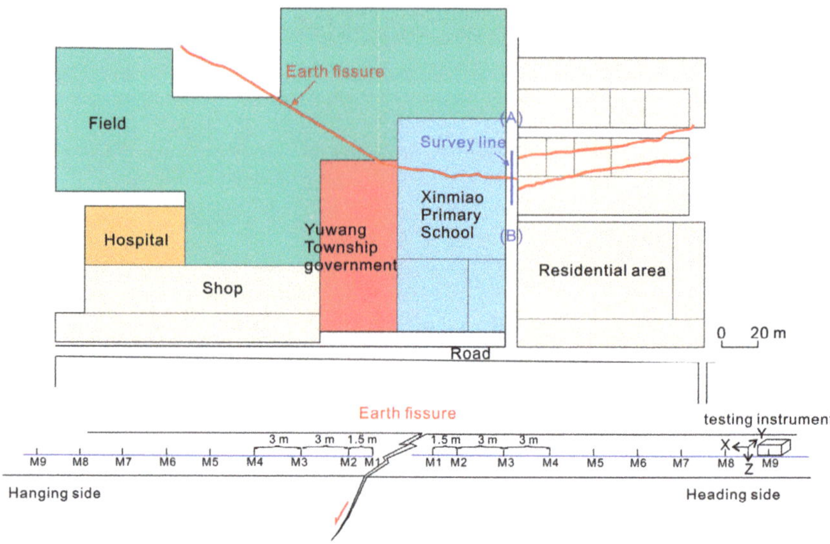

Figure 5. Schematic diagram of microtremor testing.(A represents hanging side, B represents heading side, M represents Measurement point, also in the following figures).

4.2. Equipment and Methodology

The testing instrument used for the microtremor was a high-sensitivity servo-type velocity network seismograph (CV-374AV) manufactured by Tokyo Sokushin Co., Ltd. Company (Tokyo, Janpan). The sampling frequency of the instrument was 0.1–100 Hz. It can monitor in three orthogonal directions of microtremor data at the same time. The testing instrument meets the requirements of the microtremor test.

The tests were carried out at night when it was quiet and in good weather in order to avoid obvious vibration sources. The microtremors in the X, Y, and Z directions were measured at each measurement point, and each measurement point was monitored for more than 10 min. If pedestrians or vehicles passed by during the test, these details were recorded, and the affected time period was avoided as much as possible when selecting the data.

The collected speed–time history curve from the reliable signal was intercepted and converted into the acceleration time–history curve. Intercepted data were imported into the SeismoSignal program. After the preprocessing had been completed, including filtering and baseline corrections, Fourier spectrum, response spectrum, and Arias intensity analyses were performed.

5. Microtremor Analysis of the Earth Fissure Sites

Figures 6–10 show the results of the microtremor spectrum analysis of the five survey lines. Figures 6 and 7 show the results obtained from the S1 and S2 survey lines on the F_1 earth fissure site, respectively. Figure 8 shows the result of the F_2 fissure site, and Figures 9 and 10 show the spectral results of the F_5 earth fissure site.

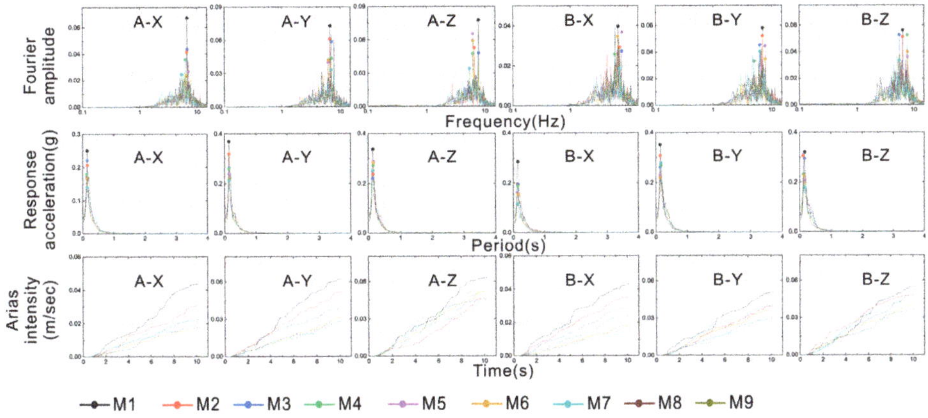

Figure 6. Fourier spectrum, response spectrum, and Arias intensity analysis results for line S1.

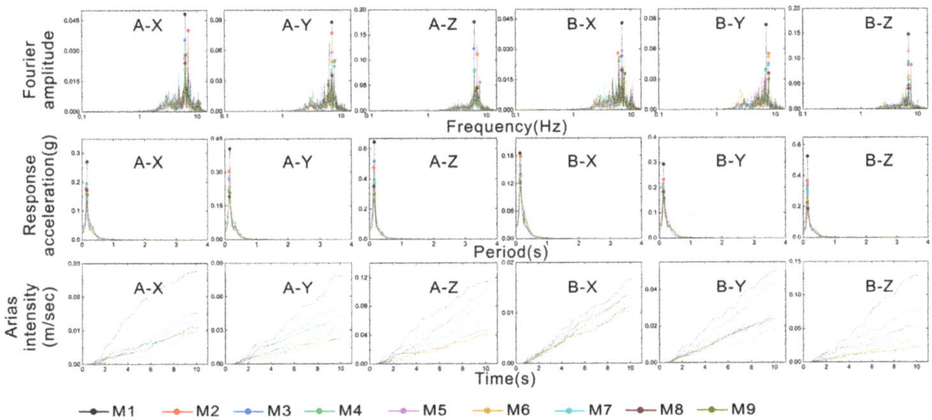

Figure 7. Fourier spectrum, response spectrum, and Arias intensity analysis results for line S2.

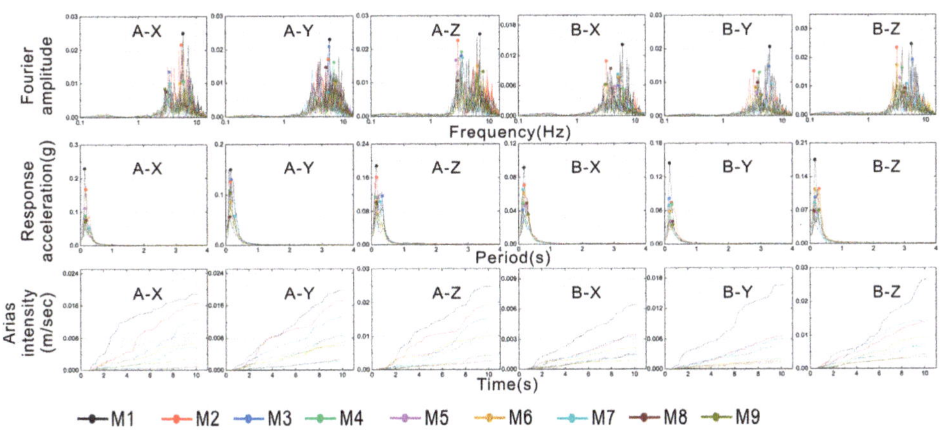

Figure 8. Fourier spectrum, response spectrum, and Arias intensity analysis results for line S3.

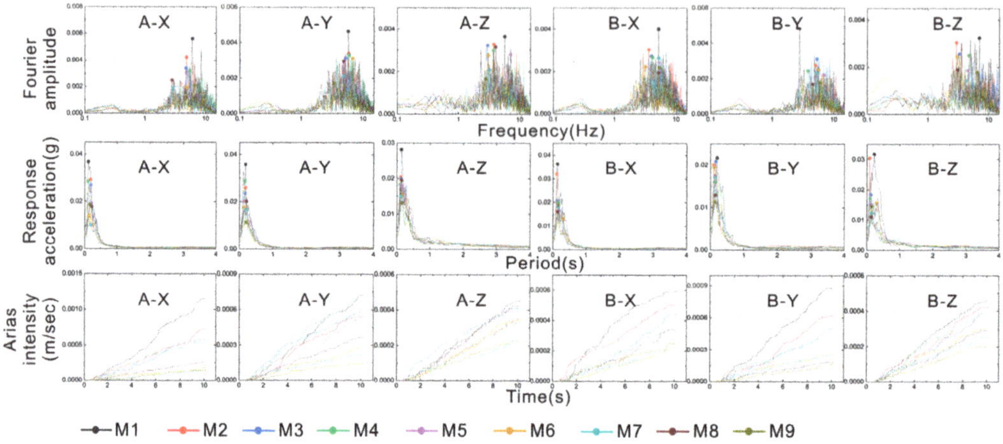

Figure 9. Fourier spectrum, response spectrum, and Arias intensity analysis results for line S4.

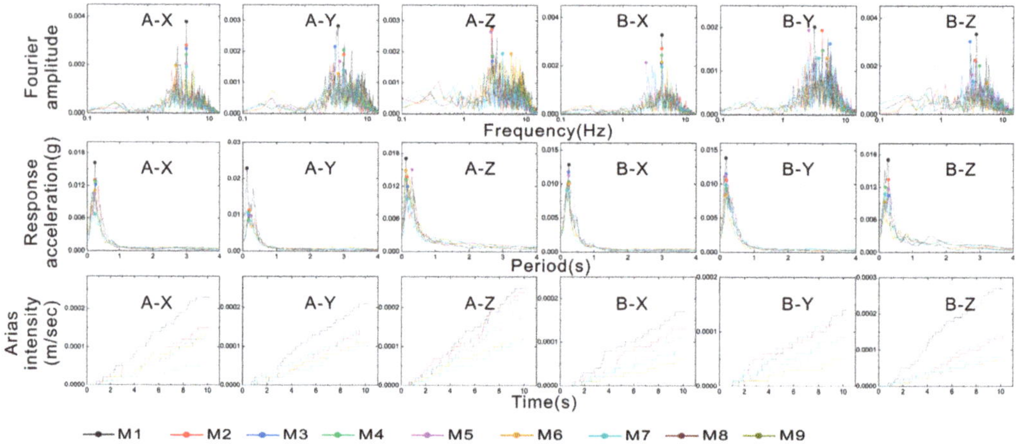

Figure 10. Fourier spectrum, response spectrum, and Arias intensity analysis results for line S5.

The Fourier spectral results show that the same earth fissure site yielded consistent Fourier spectral features, while the spectra patterns from different fissure sites were distinctive. As can be seen from Figures 6 and 7, the Fourier spectrum patterns of the F_1 earth fissure site are dominated by 'single-peak' spectra, with small spectral areas, prominent main peaks, and narrow spectral energy distribution intervals. From Figures 8–10, it can be seen the Fourier spectra of the F_2 and F_3 fissure sites are both broad and dominated by 'multi-peak' spectra, with more secondary peaks and a relatively wide range of spectral energy distribution. The Fourier spectral pattern is dependent on the fissure site, with the F_1 fissure site having relatively hard soil conditions, dominated by interbedded loess and paleosol. The F_2 and F_5 fissure sites have different site conditions, but the profiles reveal that the two sites have similar soil conditions, dominated by silty clay intercalated with silt, with a relatively weak soil layer. Therefore, the Fourier spectral patterns of the earth fissure sites F_2 and F_5 are similar, and they are significantly different from the Fourier spectral characteristics of the F_1 site.

Further analysis of the predominant frequencies revealed that the presence of earth fissures had little effect on the predominant frequencies of the site. This means that the predominant frequencies of the same survey line did not change significantly with increasing distance from the earth fissures, and the ranges of the predominant frequencies on the hanging and heading sides were basically the same. From the results of the spectral analysis, it can be concluded that the predominant frequency of the F_1 earth fissure site is 6–7 Hz, that of F_2 is 3–5 Hz, and that of F_5 is 3–5 Hz. However, because of the different soil structures of different earth fissure sites, the predominant frequencies of the different sites are different.

The response spectra of the same fissure sites were also generally consistent, with all three types of fissure sites showing a predominantly 'single-peak' response spectrum, with prominent main peaks and few or no secondary peaks. Similar to the results of the Fourier spectrum analysis, the response spectrum of the F_1 fissure site was the narrowest among all of the fissure sites, while the F_2 and F_5 fissure sites revealed a similar response spectrum with a slightly larger spectral area than that of F_1. All three fissure sites had a relatively concentrated predominant period, which was distributed in the interval of 0–0.5 s.

The Arias intensity is the curve of the energy accumulation at the measurement point over time. The difference between the energy of the different measurement points can be seen better from the final accumulated energy. Based on the Arias intensity of the three earth fissure sites, the closer the measurement point to the earth fissure, the greater the energy accumulated at the measurement point. The accumulated energies of each site reached the extreme values at measurement points A1 and B1, and as the distance increased, the energy gradually decreased and finally stabilized.

From the spectral results, it can be further found that earth fissures have no significant influence on the inherent characteristics of the site, such as the predominant frequency and the predominant period. However, when we explored the dynamic response law of the earth fissure site, the amplitude difference of each measurement point was notable. From Figures 6–10, there are significant differences in the amplitudes of the various measurement points, but it is difficult to visualize the relationship between the magnitudes of the various measurement points and the location of the earth fissure from the spectrum results. Therefore, we averaged the amplitudes of each measuring point in the X, Y, and Z directions, and obtained the relationship curve between the average amplitude of the Fourier spectrum, response spectrum, Arias intensity, and the distance from the earth fissure, as shown in Figure 11. As can be seen from the figure, although the response intensity varied from site to site, the measurement points near the earth fissures all exhibited a significant amplification effect, which gradually attenuated and stabilized with increasing distance from the earth fissure.

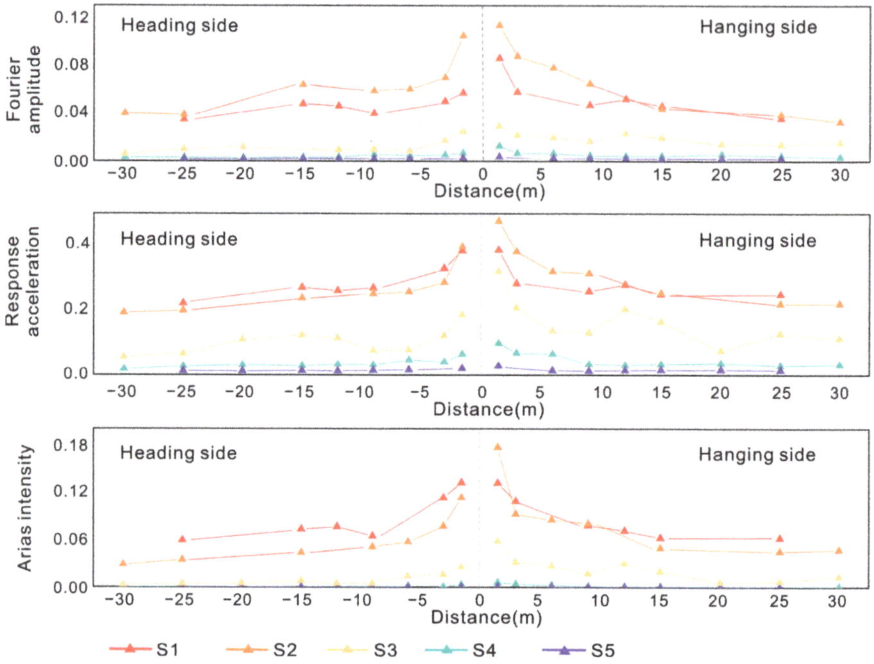

Figure 11. The amplitude attenuation curves of each site.

6. Analysis of Dynamic Effects of the Earth Fissure

6.1. Amplification Effect

In order to more intuitively reveal the degree and scope of the impact of the earth fissures on the dynamic response of the site, the concept of the amplification factor was introduced. Figure 12 shows that the amplitudes of the three measurement points on both sides of the earth fissure on each survey line were basically the same and tended to be stable. Therefore, the stable amplitude was defined as the average of the amplitudes of these three measurement points, and the amplification factor was defined as the average amplitude of each measurement point divided by the stationary amplitude. Figure 12 shows the attenuation curve and fitting curve of the amplification factor with distance. As can be seen from the figure, the earth fissure site amplification effect revealed via Fourier spectrum, response spectrum, and Arias intensity had the same attenuation mode, and the dynamic amplification response of the site was most significant when closest to the earth fissure and tended to decay as the distance from the fissure increased until reaching the area farthest away from the earth fissure, where the dynamic response of the site gradually stabilized and the amplification factor approached 1. This also indicates that the influence of the earth fissure on the dynamic response of the site had a limited range.

According to the amplification factor fitting curve (Figure 12), we can obtain the amplification factor extreme value of different analysis results. As shown in Table 1, the extreme value of the hanging and heading sides obtained by the Fourier method is 1.8–2.1. The extreme value of the amplification factor of the hanging side obtained by the response spectrum is 1.8, and that of the heading side is 1.6, while the Arisa intensity has the largest amplification factor extreme values of 3.2 for the hanging side and 2.9 for that of the heading. The amplification factors obtained by three analysis methods are different, we can choose different amplification factors according to the needs of seismic fortification. The Fourier method focuses on the inherent information of the site soil layer, which can intuitively reflect the vibration characteristics of the site. Therefore, we can select the Fourier amplification factor when considering only the dynamic characteristics of the site itself. The response spectrum can reflect the dynamic characteristics of different structural particle systems under ground motion, which is very effective for the seismic fortification of structures. The Arias intensity is a time-dependent curve, indicating the strength of the overall dynamic response of the site over a period of time. According to the amplification factor of the Arias intensity, the peak acceleration of the site under seismic action can be obtained, and the seismic fortification level of the site can be adjusted. Different methods show that the extreme value of the amplification factor in the hanging side of the same earth fissure site is slightly larger than that of the heading. Therefore, under the dynamic load, when the distance from the earth fissure is the same, the dynamic amplification response of the hanging side is stronger.

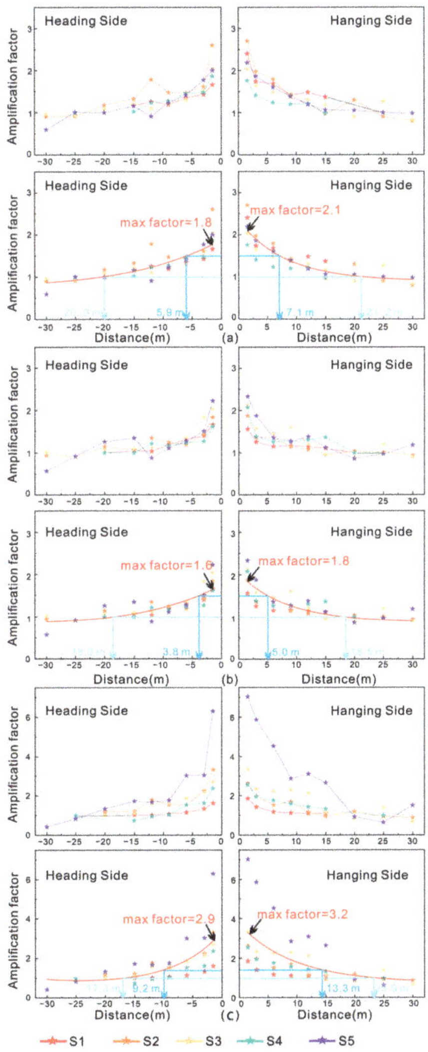

Figure 12. Amplification factor diagrams of the different analyses. (**a**) Fourier amplification factor and fitting curve, (**b**) response spectrum amplification factor and fitting curve, and (**c**) Arias intensity amplification factor and fitting curve.

Table 1. The extreme value of the amplification factor of each analysis method.

	Hanging Side	Heading Side
Fourier	2.1	1.8
Response	1.8	1.6
Arias	3.2	2.9

6.2. Range of Influence

From the amplification factor fitting curve, the range of distances corresponding to the amplification factor of 1.5 was divided into three regions for the influence of the dynamic response of earth fissure, as shown in Table 2. In future seismic fortification work, the seismic fortification levels of buildings and structures can be adjusted according to the amplification factors in different areas.

Table 2. The influence range of each analysis method.

	Hanging Side		Heading Side	
Amplification factor	1.5	1	1.5	1
Fourier	7.1 m	21.2 m	5.9 m	20.0 m
Response	5.0 m	18.5 m	3.8 m	18.0 m
Arias	13.3 m	23.9 m	9.2 m	17.3 m

When the amplification factor is greater than 1.5, it is in an area where the dynamic response of the site is amplified most intensely. All three methods have varying degrees of influence range, in which Fourier and response spectra are relatively close, the hanging side is about 5.0–7.1 m, and the heading side is 3.8–5.9 m. The area of Arias intensity is relatively large, with the hanging side at 0–13.3 m and the heading side at 0–9.2 m, and when the amplification factor is between 1.5 and 1, it is in an area where the dynamic response of the site is significantly amplified, and the buildings and structures in this area should also increase their seismic protection level accordingly. Furthermore, when the amplification factor is attenuated to 1, the site can be considered as no longer being affected by the amplification effect of the earth fissure, and therefore a zone of influence of the dynamic response can be delineated accordingly. The Fourier method yielded a hanging and heading side of approximately 20 m, while the response spectrum was about 18 m, and the Arias intensity showed an influence zone of 23.9 m for the hanging side and 17.3 m for the heading side.

As a result, the influence range of the same method was closer to both fissure sides, but the hanging side influence range was always greater than that of the heading. Therefore, from the perspective of practical engineering for seismic protection, new buildings should try to avoid the influence area. If it cannot be avoided, the seismic protection intensity of the building should be increased as much as possible according to the corresponding amplification factor. Moreover, the dynamic amplification effect of the hanging side has a wider and more extensive scope of influence than that of the heading side, in particular for the seismic fortification level of the hanging sides of the earth fissure sites.

7. Conclusions

(1) The presence of the earth fissure significantly amplifies the peaks of the direct Fourier spectra, the acceleration response spectra, and the Arias intensity within the site. In particular, the peaks of each spectrum increased dramatically in the range of 5 m immediately on either side of the fissure.

(2) The amplification effect at the fissure site follows the pattern of steep increase in the near field, slow rise in the middle field, and steady rise in the far field. Areas with amplification factors higher than 1.5 showed a steep rise in amplification, with both spectral and intensity peaks rapidly increasing to several times the original site and eventually reaching extreme values at the outcrops of the fissures. Areas with an amplification factor

between 1.5 and 1 are zones with a slow rise of the amplification effect, and the dynamic response of the site is still affected by the fissures.

(3) The amplification of the dynamic response of an earth fissure site with positive fault characteristics has a significant "hanging side effect." The amplification factor is often higher in the steep rise zone of the hanging side, and the hanging side has a further influence range.

(4) In the actual seismic fortification of the project, it is necessary to avoid the area where the amplification factor rises sharply. Alternatively, the strength of structures and foundations in the area should be increased, and additional seismic fortification measures should be taken. In areas where the amplification factor rises slowly, the seismic fortification level of the structure should also be strengthened according to the amplification factor, so that it can withstand 1.5 times the expected seismic factor.

In this study, based on the microtremor test, typical earth fissures in the Yuncheng Basin of China were taken as the research objects, and dynamic amplification patterns at earth fissure sites were systematically revealed using spectral characterization of the microtremors. We propose preliminary seismic fortification for different dynamic amplification areas. This study provides a reference for the seismic fortification of sites with similar engineering conditions and the subject of seismic amplification effect in similar sites. In addition, the numerical simulation of seismic responses of fissure sites and different structures is expected to further improve the results of the amplification effect and provide more detailed suggestions for seismic fortification under different seismic intensities.

Author Contributions: Conceptualization, Y.D.; Software, G.C.; Investigation, G.C., J.C., Y.X. and N.H.; Writing—original draft, G.C.; Writing—review & editing, G.C., Y.D. and H.M.; Supervision, Y.D.; Funding acquisition, Y.D. and H.M. All authors have read and agreed to the published version of the manuscript.

Funding: This research was funded by [National Natural Science Foundation of China] grant number [41772275], [Fundamental Research Funds for the Central Universities] grant number [300102268203], [Scientific Research Plan Projects of Shaanxi Education Department] grant number [20JK0801], [Natural Science Basic Research Program of Shaanxi Province] grant number 2022JQ-289, [Fundamental Research Funds for the Central Universities] grant number 300102262505, and [Key Research and Development projects of Shaanxi Province] grant number 2022SF-197.

Institutional Review Board Statement: Not applicable.

Informed Consent Statement: Not applicable.

Data Availability Statement: The data presented in this study are available on request from the corresponding author. The data are not publicly available due to the processed data required to reproduce these findings cannot be shared at this time as the data also forms part of an ongoing study.

Conflicts of Interest: The authors declare no conflict of interest.

References

1. Carpenter, M.C. *Earth-Fissure Movements Associated with Fluctuations in Ground-Water Levels Near the Picacho Mountains, South-Central Arizona, 1980–1984;* U.S. Geological Survey: Reston, VA, USA, 1991. [CrossRef]
2. Rojas, E.; Arzate, J.; Arroyo, M. A method to predict the group fissuring and faulting caused by regional groundwater decline. *Eng. Geol.* **2002**, *65*, 245–260. [CrossRef]
3. Laike, M.A. Environmental hazard from fissures in the main Ethiopian rift. *J. Afr. Earth Sci.* **1998**, *27*, 481–490. [CrossRef]
4. Qiao, J.W.; Peng, J.B.; Deng, Y.H.; Wen, H.; Zang, M. The study on basic characteristic of earth fissure in Linfen basin. *J. Eng. Geol.* **2015**, *23*, 856–865. [CrossRef]
5. Snider, L.C. A suggested explanation for the surface subsidence in the goose greek oil and gas field. *AAPG Bull.* **1927**, *11*, 729–745. [CrossRef]
6. Minor, H.E. Goose greek oil field. Harris county, Texas. *Adv. World Pet. Geosci.* **1925**, *9*, 286–297. [CrossRef]
7. Leonard, R.J. An earth fissure in Southern Arizona. *J. Geol.* **1929**, *37*, 765–774. [CrossRef]
8. Li, Y.S.; Li, J.Z.; Bian, J.M.; Wu, X.D. *The Ground Fissures in Xi'an*; Seismological Press: Beijing, China, 1986.
9. James, T.N.; Langer, A.M.; Kerr, P.F. Giant desiccation polygons of great basin playas. *Geol. Soc. Am. Bull.* **1968**, *79*, 69–90. [CrossRef]

10. Fletcher, J.E.; Harris, K.; Peterson, H.B.; Chandler, V.N. Piping. *Eos Trans. Am. Geophys. Union* **1954**, *35*, 258–262. [CrossRef]
11. Li, X.S. Some new view on the genetic mechanism of Xi'an ground fractures. *J. Xi'an Collage Geol.* **1994**, *16*, 75–80. [CrossRef]
12. Holzer, T.L. Faulting caused water level declines, San, Joaquin Valley, California. *Water Resour. Res.* **1980**, *16*, 1065–1070. [CrossRef]
13. Holzer, T.L. *Reconnaissance Maps of Earth Fissure and Land Subsidence, Bowie and Willecox Areas, Arizona*; Miscellaneous Field Studies Map; U.S. Geological Survey: Reston, VA, USA, 1980. [CrossRef]
14. Yang, K.Y.; Wu, C.J. *Deformation Monitoring and Research in Xi'an Ground Fissures*; Shannxi Normal University Press: Xi'an, China, 1986.
15. Wan, J.W.; Li, B.; Tan, C.X.; Zhang, P.; Feng, C. Characteristics and mechanism of earth fissures in China: A case Study of Fenhe River-Weihe river basin, Hebei plain and Suzhou-Wuxi-Changzhou plain. *Geol. Rev.* **2019**, *65*, 1383–1396. [CrossRef]
16. Deng, Y.H.; Peng, J.B.; Mu, H.D.; Li, L.; Sun, Z. Ground fissures germination mechanism of deep structure activities in Weihe basin. *J. Jilin Univ.* **2007**, *4*, 433–440. [CrossRef]
17. Deng, Y.H.; Peng, J.B.; Li, L.; Huandong, M.U. Causative relationship between basement stretching and ground fissures formation in Weihe basin. *J. Eng. Geol.* **2013**, *21*, 92–96. [CrossRef]
18. Peng, J.B.; Fan, W.; Li, X.A.; Wang, Q.L.; Feng, X.J.; Zhang, J.; Li, X.S.; Lu, Q.Z.; Hunag, Q.B.; Ma, R.Y.; et al. Some key questions in the formation of ground fissures in the Fen-Wei basin. *J. Eng. Geol.* **2007**, *4*, 433–440. [CrossRef]
19. Peng, J.B.; Li, X.A.; Fan, W.; Chen, Z.X.; Su, S.R.; Song, Y.H.; Lu, Q.Z.; Deng, Y.H.; Chen, L.W.; Sun, P. Classification and development pattern of caves in the loess plateau. *Earth Sci. Front.* **2007**, *14*, 234–244. [CrossRef]
20. Peng, J.B.; Qiao, J.W.; Leng, Y.Q.; Wang, F.; Xue, S. Distribution and mechanism of the ground fissures in Wei river basin, the origin of the silk road. *Environ. Earth Sci.* **2016**, *75*, 718. [CrossRef]
21. Wu, Q.; Jiang, Z.Q.; Li, Y.L. *Research on Ground Fissure Disaster in Shanxi Faulted Basin*; Geological Publishing House: Beijing, China, 2003.
22. Song, Y.H.; Peng, J.B.; Lu, Q.Z.; Li, Z. A type of special Shear-crush ground fissures located in Shanxi graben system. *J. Eng. Geol.* **2007**, *4*, 448–452. [CrossRef]
23. Meng, L.C.; Peng, J.B.; Lu, Q.Z.; Li, Z. Genetic and analysis on ground fissures existed in the transformer substation of Wanrong county. *Geotech. Investig. Surv.* **2010**, *38*, 1–6.
24. Xu, J.S.; Zhuang, H.D.; Tang, D.Q.; Li, Z.C. Characteristic and mechanism analysis of ground fissures in Yuncheng basin. *J. Geol. Hazards Environ. Preserv.* **2010**, *21*, 97–100. [CrossRef]
25. Chen, Y.M. Study on the characteristics of ground fissures in Shanxi fault basin. *Chin. J. Geol. Hazard Control* **2016**, *27*, 72–80+89. [CrossRef]
26. Qiao, J.W.; Peng, J.B.; Lu, Q.Z. Basic characteristic and mechanism analysis of Weiguo earth fissure in Yuncheng basin. *J. Nat. Disasters* **2017**, *26*, 110–119. [CrossRef]
27. Zhao, J.Y.; Wang, H.G.; Lu, Q.Z.; Qiao, J. Basic features of ground fissures in Yuncheng basin and emei platform. *Chin. J. Geol. Hazard Control.* **2018**, *29*, 58–67. [CrossRef]
28. He, M.; Zhai, D.L.; Qiao, J.W.; Kang, C. Basic features and mechanism of Dalyu-Baizhang ground fissure in Yuncheng basin, Shanxi province. *Chin. J. Geol. Hazard Control* **2019**, *30*, 74–83. [CrossRef]
29. Liu, C.; Peng, J.B.; Chen, L.W.; Deng, Y.; Wu, M. Finite element analysis of seismic effects on ground fissures in Xi'an. *J. Eng. Geol.* **2012**, *20*, 979–985. [CrossRef]
30. Fan, W.; Nie, Z.Q.; Chen, D.M.; Deng, L.S. Study on seismic effect of ground fissure-A case study on a ground fissure in Xianyang, Shannxi. *J. Catastrophology* **2014**, *29*, 82–88. [CrossRef]
31. Liu, N.N.; Huang, Q.B.; Men, Y.M.; Peng, J.B. Model test on dynamic response of ground with fissures under seismic loading. *Chin. J. Rock Mech. Eng.* **2014**, *33*, 1024–1031.
32. Huang, Q.B.; Gao, H.; Liu, N.N.; Ma, Y.J. Shaking table model test on seismic response of metro tunnel crossing ground fissure site. *J. Geomech.* **2018**, *24*, 785–794. [CrossRef]
33. Xiong, Z.M.; Zhang, C.; Huo, X.P.; Chen, X. Shaking table tests on acceleration response of ground fissure site. *Chin. J. Geotech. Eng.* **2018**, *40*, 520–526. [CrossRef]
34. Mu, H.D.; Deng, Y.H.; Li, R.J.; Xu, G. Dynamic response and influencing factors of ground fissure sites in Xi'an. *China Earthq. Eng. J.* **2019**, *41*, 418–427. [CrossRef]
35. Xiong, Z.M.; Zhang, C.; Chen, X. Model test on ground motion parameters of site with fissures under seismic loading. *Rock Soil Mech.* **2019**, *40*, 421–428. [CrossRef]
36. Shaanxi Provincial Construction Office. *Specification for Site Investigation and Engineering Design on Xi'an Ground Fractures (DBJ61-6-2006)*; Shanxi Provincial Construction Press: Changfeng, China, 2006.
37. Ministry of Housing and Urban Rural Development of the People's Republic of China. *Code for Seismic Design of Buildings (GB50011-2010)*; China Architecture & Building Press: Beijing, China, 2010.
38. Kanai, K.; Tamaka, T. Measurement of microtremor I. *Bull. Earthq. Res. Inst.* **1954**, 199–209.
39. Aki, K. Space and time spectra of stationary stochastic eaves, with special reference to microtremors. *Bull. Earthq. Res. Inst.* **1957**, *35*, 415–456. [CrossRef]
40. Ohta, Y.; Kagami, H.; Goto, N.; Kudo, K. Observation of 1-to 5-second microtremors and their application to earthquake engineering. Part I: Comparison with long-period accelerations at the Tokachi-Oki earthquake of 1968. *Bull. Seismol. Soc. Am.* **1978**, *68*, 767–779. [CrossRef]

41. Kagami, H.; Duke, C.M.; Liang, G.C.; Ohta, Y. Observation of 1 to 5-second microtremors and their application to earthquake engineering. *Part II. Evaluation of site effect upon seismic wave amplification due to extremely deep soil deposits. Bull. Seismol. Soc. Am.* **1982**, *72*, 987–998. [CrossRef]
42. Kagami, H.; Okada, S.; Keishi, S.; Oner, M.; Dravinski, M.; Mal, A.K. Observation of 1 to 5 second microtremors and their application to earthquake engineering. Part III. A two-dimensional study of site effects in the San Fernando valley. *Bull. Seismol. Soc. Am.* **1986**, *76*, 1801–1812. [CrossRef]
43. Lacoss, R.T.; Kelly, E.J.; Toksoz, M.N. Estimation of seismic noise structure using Arrays. *Geophysics* **1969**, *34*, 21–38. [CrossRef]
44. Toksoz, M.N.; Richard, T.L. Microseisms: Mode structures and source. *Science* **1968**, *159*, 872–873. [CrossRef] [PubMed]
45. Nakamura, Y. A method for dynamic characteristics estimation of subsurface using microtremors on the ground surface. *Q. Rep. Rtri* **1989**, *30*, 25–33.
46. Bard, P.Y. Microtremors measurements:a tool for site effect estimation? In Proceedings of the Second International Symposium on the Effects of Surface Geology on Seismic Motion, Yokohama, Japan, 1–3 December 1998; pp. 1251–1279.
47. Konno, K.; Ohmachi, T. Ground-motion characteristics estimated from spectral ratio between horizontal and vertical components of microtremor. *Bull. Seismol. Soc. Am.* **1998**, *88*, 228–241. [CrossRef]
48. Lermo, J.; Chavez-Garcia, F.J. Are microtremors useful in site response evaluation? *Bull.-Seismol. Soc. Am.* **1994**, *84*, 1350–1364. [CrossRef]
49. Guo, M.Z.; Song, Z.Q. On Nakamura method for dynamic analysis of ground characteristics with microtremor. *World Inf. Earthq. Eng.* **2000**, *2*, 88–92. [CrossRef]
50. Guo, M.Z.; Xie, L.L.; Yu, H.Y. Research pm analytic methods for microtremor mechanism. *World Earthq. Eng.* **2002**, *18*, 17–23. [CrossRef]
51. Guo, M.Z.; Ren, F.H. The microtremor methods for dynamics measurement of sites. *World Earthq. Eng.* **2005**, *21*, 139–142. [CrossRef]
52. Zhang, L.G.; Deng, Y.H.; Xue, J.; Wang, H.; Liu, Q.; Yang, Y. Preliminary research on dynamic response characteristics of microtremor at ground fissure sites-illustrated with F_6 ground fissure in Xi'an. *J. Eng. Geol.* **2017**, *25*, 784–793. [CrossRef]
53. Chang, J.; Deng, Y.H.; Xuan, Y.; Yan, Z.; Wu, W.; He, J. The dynamic response of sites with earth fissures as revealed by mcrotremor analysis—A case study in the Linfen basin, China. *Soil Dyn. Earthq. Eng.* **2020**, *132*, 106076. [CrossRef]

applied sciences

MDPI

Article

Hazard Model: Epidemic-Type Aftershock Sequence (ETAS) Model for Hungary

Péter Szabó [1,*], László Tóth [2] and Judith Cerdà-Belmonte [1]

[1] Natural Catastrophe Competence Center, Group Actuarial, UNIQA Insurance Group AG, 1029 Wien, Austria
[2] GeoRisk Earthquake Engineering, 1118 Budapest, Hungary
* Correspondence: peter.szabo@uniqa.hu

Abstract: In this article we present a space–time epidemic-type aftershock sequence (ETAS) model for the area of Hungary, motivated by the goal of its application in insurance risk models. High-quality recent instrumental data from the period 1996–2021 are used for model parameterization, including data from the recent nearby Zagreb and Petrinja event sequences. In the earthquake-triggering equations of our ETAS model, we replace the commonly used modified Omori law with the more recently proposed stretched exponential time response form, and a Gaussian space response function is applied with a variance add-on for epicenter error. After this model was tested against the observations, an appropriate overall fit for magnitudes $M \geq 3.0$ was found, which is sufficient for insurance applications, although the tests also show deviations at the $M = 2.5$ threshold. Since the data used for parameterization are dominated by Croatian earthquake sequences, we also downscale the model to regional zones via parameter adjustments. In the downscaling older historical data are incorporated for a better representation of the key events within Hungary itself. Comparison of long-term large event numbers in simulated catalogues versus historical data shows that the model fit by zone is improved by the downscaling.

Keywords: insurance hazard model; earthquake clustering; space–time ETAS model; Hungarian earthquake catalogue; 2020 Petrinja earthquake

Citation: Szabó, P.; Tóth, L.; Cerdà-Belmonte, J. Hazard Model: Epidemic-Type Aftershock Sequence (ETAS) Model for Hungary. *Appl. Sci.* **2023**, *13*, 2814. https://doi.org/10.3390/app13052814

Academic Editors: Miguel Llorente Isidro, Ricardo Castedo and David Moncoulon

Received: 19 December 2022
Revised: 16 February 2023
Accepted: 18 February 2023
Published: 22 February 2023

1. Introduction

1.1. Motivation

The origin of the study presented in this article is an ongoing effort by the UNIQA Insurance Group (www.uniqagroup.com, accessed on 5 February 2023) to build a proprietary earthquake model for Hungary. Earthquake models in insurance are used to measure the risk of a set of objects, such as buildings, civil engineering structures, and means of transport, with a focus on assessing the potential losses caused by severe earthquakes.

One key application of an earthquake model or, more generally, a natural catastrophe model in insurance is the determination of capital requirements. According to the Solvency II regulatory standards applicable in European Union member countries [1,2], the solvency capital requirement (SCR) of an insurance company shall correspond to the 200-year loss; that is, the insurer must hold their own funds to cover the risk of a loss whose probability of occurrence within a year is 0.5%. The SCR is calculated either by the Solvency II standard formula or by an internal model. For natural perils, the standard formula is a deterministic approximation of the risk based on fixed factors applied to insured amounts of the objects by geographical area. On the other hand, an internal model is based on the stochastic simulation of the peril, and its output is not just an SCR single value but a full probability distribution forecast of the loss. The use of an internal model by an insurance company for calculating the SCR is subject to case-by-case regulatory approval by the competent supervisory authorities, including a comprehensive and lengthy validation process. The development of an internal model requires a significant effort; however, the benefits include a more accurate and more detailed evaluation of the risk, and a range of model uses in risk management tasks beyond the calculation of the SCR. Another typical and important use of stochastic catastrophe models is the optimization and pricing of protection covers for the purpose of risk transfer. Reinsurance risk transfer is an instrument for risk mitigation, and it can reduce the SCR of the ceding insurer.

The general architecture of a stochastic earthquake model for insurance is the following: (1) the first fundamental module is a statistical model of earthquake occurrences, which is used to generate a synthetic event catalogue via stochastic simulation. This is followed by two equally important ones: (2) the ground motion attenuation model, translating the catalogue into event footprints, and (3) the vulnerability model, translating the event footprints into the loss of the insured objects in financial terms. All three modules require a full study of their own. In this article, only the first module is discussed, that is, a model of earthquake occurrences.

Earthquake models used in the field of insurance typically focus on the mainshocks only. This simplifies the modelling effort drastically since the mainshocks are assumed to follow a stationary process. This simplification is justified if aftershock losses are low compared to the mainshock loss. Furthermore, customer claims arising from aftershock losses are often not reported separately from the mainshock; therefore, it is a straightforward approach to model them only implicitly via conservative parameterization. Nevertheless, the recent nearby Petrinja 2020 event in Croatia and its aftershocks highlighted some properties of earthquake sequences [3], suggesting that clustering cannot necessarily be safely ignored even in a Central European model: aftershock activity after such a major event may last for months or even years, and some aftershocks (or foreshocks) are significant events on their own. Moreover, there are historical records of complex earthquake sequences (e.g., Romania 1991) lasting for several months and including multiple significant shocks [4]. Large events are still possible late in the sequence, and large aftershocks can trigger second-generation sequences, epidemic style.

1.2. Aim of the Study

The aim of this study is to build an epidemic-type aftershock sequence (ETAS) model for Hungary using recent instrumental earthquake catalogue data. ETAS methodology assumes that, on top of a constant background seismicity, every single earthquake is a potential trigger for subsequent events. Such models have been widely used to analyse earthquake clustering both for studies covering broad regions and local studies focusing on individual earthquake sequences. This study builds on existing methodologies, however, the planned implementation for insurance application is always considered in the choice of the methods.

Due to its moderate seismicity, Hungary is not an optimal target area for an ETAS model. Nonetheless, the nearby Zagreb and Petrinja event sequences in 2020–2021 produced rich data that make the parameter fitting feasible. The authors do not know about a published previous region-specific ETAS study for Hungary relying on these recent catalogue data; therefore, the results can be interesting themselves.

The data used in the study are described in Section 2. In Section 3, an overall model for Hungary is formulated and in Section 4, its results are described. In Sections 5 and 6, downscaled parameters are determined, representing local regions within the total modelled area. The motivation of parameter downscaling stems from the application of the model in insurance where local geographical outputs are needed. Section 7 includes the conclusions of the study.

1.3. Background

Since the initial formulation of the model by Ogata [5,6], ETAS methodology has been researched extensively. Works by several authors suggested different methods and algorithms for parameter estimation [6–10] and investigated the optimal forms of event-triggering functions [6,8,11,12]. The literature on ETAS also includes a discussion of important model properties. One such model property is magnitude independence, i.e., the assumption that the magnitude of an earthquake is independent of predecessor events [8,13]. The stability or criticality conditions of the event triggering, as well as the question of self-similarity, have been investigated [14,15]. Certain enhancements of the model that allow a more accurate modelling of complex processes, such as three-dimensional or finite-source ETAS models, have also been developed [9,16,17]; however, the benefits of these advanced features for modelling a moderately active region like Hungary are not immediately clear.

This study significantly relies on the work of Zhuang et al. [7,8], who developed an iterative algorithm for ETAS parameterization, based on a kernel density estimation of the background activity in combination with stochastic declustering. The authors analyse the properties of the method in regional studies covering areas of Japan and New Zealand, and they suggest a range of methods for testing the different components of the model.

ETAS parameter estimation is affected by known sensitivities and biases, which are discussed at length by Seif et al. [18], who demonstrate these effects on actual and simulated catalogues derived from Italian and South Californian data. Typical sources of parameter bias include early aftershock incompleteness in the wake of a major earthquake, the anisotropy of aftershock clusters, and the choice of the magnitude threshold. These effects also need to be considered in this Hungarian study.

2. Data

The data quality requirements of an ETAS model fitting are challenging. A consistent instrumental earthquake catalogue for Hungary and the seismograph network collecting the measurements for this catalogue have only been in place since 1996 [19], and there has been a scarcity of significant Hungarian events from 1996 until 2022. For a sufficiently rich catalogue, the 2020 Zagreband the 2020 Petrinja, Croatia events and their aftershock sequences need to be included in the study. It is relevant to note that the Petrinja mainshock also caused damage and triggered insurance claims within Hungary [20]. These two event sequences have been recorded by the Hungarian seismograph network. On the other hand, both clusters fell largely outside the core geographical window of the Hungarian earthquake catalogue; this means that the study area is extended to the periphery of the data sources where the completeness and accuracy of data is less than optimal. Furthermore, large Croatian event sequences will have a dominant effect on the model fitting.

The core geographical window of the study is the area between latitudes 45.5–49.0 N and longitudes 16.0–23.0 E. In order to include the Zagreb and Petrinja events, a margin to the west and to the south is added by defining the extended window as the area between latitudes 45.3–49.0 N and longitudes 15.7–23.0 E. The magnitude threshold of the study is set at $M_w = 2.5$.

The data available for the study include both historical and recent instrumental data belonging to several Hungarian earthquake catalogues. These catalogues have been merged and event magnitudes have been homogenized to moment magnitude M_w for this study. Hereafter, the subscript w is omitted when referring to magnitudes. The component catalogues are the following (see also Figure 1):

1. Catalogue A: Historical catalogue for the period 456 A.D. to 1995 [21], collected from several primary sources, partly instrumental and partly macroseismic. The number of events included in the extended window that reach the magnitude threshold is 2402. Due to obvious data quality limitations, these events are not used directly for ETAS parameter fitting but only as data for subsequent model downscaling.
2. Catalogue B: Instrumental catalogue covering the core window and the period 1996 to 2019, based on the recordings of the Hungarian seismograph network [19]. The number of events reaching the magnitude threshold is 654.
3. Catalogue C: Instrumental catalogue data for the period 1996 to 2010, collected and merged from international sources [22–24]. While some of the underlying sources are continuously updated, the merged dataset is only available to the end of 2010.

The number of events included in the extended window that reach the magnitude threshold is 1107, partly overlapping with catalogue B data in the core window.

4. Catalogue X: Initial event list (IEL) data covering the period 2012 to 2021 [25], based on the recordings of the Hungarian seismograph network. IEL data are an interim phase of the yearly updates of the Hungarian instrumental earthquake catalogue (catalogue B). Geographical coverage is wider than the core window by a margin of 0.2 degrees latitude and 0.3 degrees longitude in all directions. Automated preliminary epicenter determinations are post-processed for the monthly publications of the list, which may result in some coordinate shifts across the window boundary. Thus, the western and southern edge of the extended window corresponds approximately to the geographical scope of the IEL. The number of events from IEL included in the extended window that reach the magnitude threshold is 2313, fully overlapping with catalogue B data in the core window in the period 2012 to 2019.

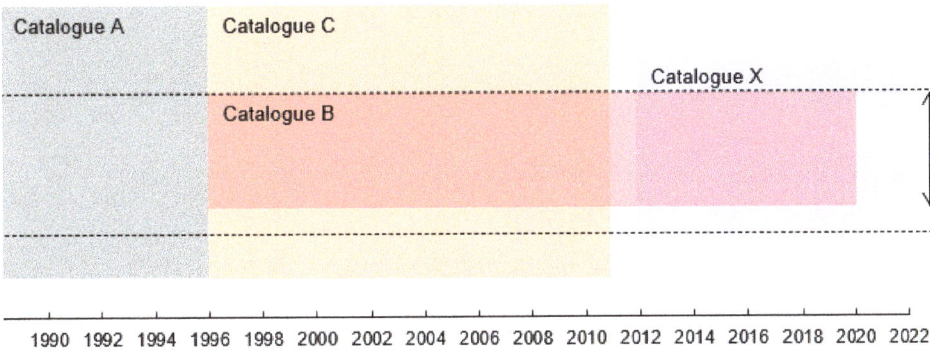

Figure 1. Schematic diagram of the coverage and overlaps of the source catalogues used in the study.

Microseismic events caused by quarry blasts and explosions had been identified and excluded from the datasets during the preparation of the catalogues prior to this study.

It is observed that the events in the Romanian area of the study show unusual magnitude–frequency patterns. This suggests an inhomogeneity of the merger of the primary sources underlying the Romanian part of catalogue C. Therefore, the affected area is excluded from the ETAS parameter fitting due to concerns of data quality. The boundary of the excluded area follows the area source zone boundaries of the SHARE project [26]. Therefore, the exclusion affects a strip within Hungary along the Romanian border.

Magnitude–frequency distributions suggest that the completeness threshold of the remaining catalogue on the core window is $M \geq 2.8$ at the beginning in 1996, which has gradually improved to $M \geq 2.5$ by 2000 and to $M \geq 2.2$ by 2013. The sensitivity contours of the Hungarian seismograph network indicate completeness for $M \geq 2.5$ on most of the modelled area in the core window with some lower-sensitivity areas on the periphery and with gradually improving coverage over time [19]. The catalogue is considered complete for $M \geq 2.5$ in the extended window only since 2018, except for a completeness gap immediately after the Petrinja 2020 mainshock: the detection of early aftershocks after a large earthquake is typically incomplete since the seismograms in this initial period are saturated with overlapping waves from multiple events [18]. Given the trade-off between completeness and the size of the catalogue, we find that the optimal choice of the threshold for this study is $M \geq 2.5$. A threshold increase would marginalize the part of the catalogue within Hungary itself and would sharply reduce the number of events for estimating the spatial distribution of the background process. For the model parameterization, completeness is assumed on the core window since 2000 and on the extended window since 2018, except for a 1-day period after the Petrinja mainshock.

After merging catalogues B, C, and X, applying the $M \geq 2.5$ threshold, removing duplications, removing events outside the completeness periods, and removing events in the excluded area, the remaining size of the target catalogue for ETAS parameter fitting is 1978. An additional 236 events reaching $M \geq 2.5$ are used as auxiliary trigger events; this includes 27 events from year 1999 and 209 events from the Petrinja completeness gap. A map of the fitting catalogue is shown in Figure 2. Another 218 events from catalogues B, C, and X outside the completeness periods are used as additional epicenter nodes for the kernel density estimation of the spatial distribution of background events.

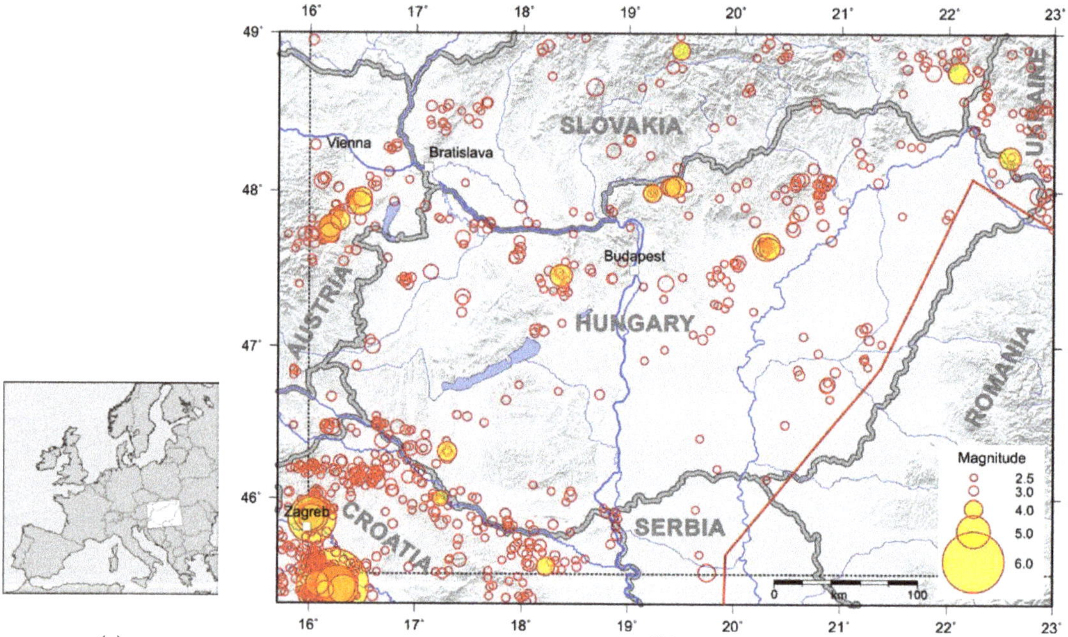

(a) (b)

Figure 2. (**a**) Location of the model area within Europe. (**b**) Plot of the earthquake catalogue for Hungary used for epidemic-type aftershock sequence (ETAS) model parameterization—target catalogue plus auxiliary events, 1999–2021. The map covers the extended window. The boundaries of the core window (dashed black line) and excluded area (red line) are marked. Events reaching $M \geq 4.0$ are highlighted.

For model calculations, geographical latitude–longitude coordinates (φ, λ) are converted to an (x, y) kilometre grid via the following transformation:

$$x = R \cdot \cos\varphi \cdot \tan(\lambda - \lambda_0), \;\; y = R \cdot (\varphi - \varphi_0), \tag{1}$$

where $R = 6\,371.01$ is the mean radius of the Earth in kilometres. This corresponds to a tangent transverse cylindrical projection whose central meridian is at λ_0 and that transforms parallels into straight lines. The point (φ_0, λ_0) is set at (47.5°, 19°), i.e., the rounded coordinates of Budapest, which are transformed into $(x, y) = (0, 0)$ in the kilometre grid.

3. Epidemic-Type Aftershock Sequence (ETAS) Modelling Methods

3.1. Model Formulation

The ETAS model describes earthquake occurrences as a non-stationary Poisson point process where every event can trigger a wave of aftershocks. In a time-only model, the rate of occurrence at time t, conditional on the process history H_t, is expressed as [5]

$$\lambda(t|H_t) = \mu + \sum_{j\,:\,t_j<t} K \cdot e^{\alpha(M_j-m_0)} \cdot g(t-t_j), \tag{2}$$

which is extended to a space–time model as [6]

$$\lambda(t,x,y|H_t) = \mu \cdot u(x,y) + \sum_{j\,:\,t_j<t} K \cdot e^{\alpha(M_j-m_0)} \cdot g(t-t_j) \cdot f(x-x_j, y-y_j|M_j), \tag{3}$$

where the first term represents background seismicity with a constant overall rate μ and area density $u(x,y)$, and the sum in the second term reflects the rate of activity triggered by events in the catalogue prior to time t. An exponential relationship is assumed between the trigger event magnitude M_j and the triggered rate of activity, where K and α are constant parameters and m_0 is the magnitude threshold. The functions g and f describe the form of the time and space response, respectively, relative to the trigger event occurrence time and epicenter. The process history $H_t = \{(t_j, x_j, y_j, M_j) : t_j < t\}$ is the catalogue of events before t, defined by their occurrence time, location, and magnitude. Depth is not included in this ETAS model as a third coordinate, and the question of the depth distribution is not discussed in this article. This is a reasonable simplification since all events in the study catalogue were shallow crustal earthquakes and all except nine of them had a focal depth of less than 20 km.

It was assumed that event magnitudes follow a truncated exponential (Gutenberg-Richter) distribution independent of the process history [8,13]. The probability density function (PDF) of magnitudes between the bounds m_0 and m_x is given by the following:

$$J(m) = \frac{\beta \cdot e^{-\beta(m-m_0)}}{1-e^{-\beta(m_x-m_0)}}, \tag{4}$$

where β is the frequency decay parameter. According to the magnitude independence assumption, an aftershock can be larger than its trigger event.

For this study, we used a stretched exponential function truncated for a minimum delay as the time response:

$$g(t) = \begin{cases} t^{q-1} \cdot \exp(-\eta t^q) & \text{if } t > \Delta t_0, \\ 0 & \text{if } t \le \Delta t_0 \end{cases} \tag{5}$$

where Δt_0 is a minimum delay parameter, and the normal distribution with an allowance for epicenter error is the space response function:

$$\begin{gathered} f(x,y|M) = \frac{1}{2\pi\sigma^2} \exp\left[-\left(x^2+y^2\right)/2\sigma^2\right], \\ \text{with } \sigma^2 = D^2 \cdot e^{\alpha(M-m_0)} + \varepsilon^2. \end{gathered} \tag{6}$$

where D is the scale parameter of the magnitude-dependent part of the variance, and ε is a parameter representing a variance add-on for epicenter error. Since both Equations (5) and (6) represent a departure from the mainstream techniques, we explain the rationale of these approaches below.

The most often used time response function is the modified Omori law [27], which is a power law function in the form of $g(t) = (t+c)^{-p}$. The stretched exponential alternative has been suggested by Mignan [12], who, besides quantitative tests showing that the stretched exponential form is better fitted to the observations from several regions than the power law, also expressed qualitative concerns about the modified Omori law. Firstly, the positive time shift constant c avoids the singularity at zero, but it is difficult to find a physical explanation for it. Secondly, a parameter value of $p \leq 1$ would mean an infinite number of triggered events unless the process is artificially truncated in time. Furthermore, due to the heavy tail of the power law function, supercritical parameter fitting outcomes, which imply an unstable process, are sometimes obtained when the modified Omori law is used in an epidemic-type model [18]. The minimum delay parameter Δt_0 in Equation (5) is motivated by experience with stochastic simulations based on the model. This truncation was introduced to avoid modelling an unrealistically high number of aftershocks immediately after the trigger event, which were never observed or identified as separate events in actual measurements since the point process model reaches its limits at very short time intervals.

Regarding the space response function, observed aftershock sequences often show anisotropic patterns. Nonetheless, isotropic functions are widely used in ETAS models as a simplification. Both the power law function, e.g., in the form $f(x, y|M) \propto \left[\left(x^2 + y^2\right)/e^{\alpha(M-m_0)} + d\right]^{-q}$ and the bivariate normal distribution have been discussed in the literature with a general preference for the power law based on goodness-of-fit diagnostics [6,8]. The Gaussian form is chosen in this study mainly for its analytical simplicity and ease of implementation in the model. Since the exponential scaling parameter α is the same in Equations (3) and (6), the space response formula assumes that the aftershock area grows in proportion to the number of events triggered, which is in line with the observed trend known as the Utsu–Seki law [6,27]. This constraint from early ETAS model variants has been challenged in the literature, raising the suggestion for a separate spatial scaling parameter γ [8,11]. However, due to the small number of major clusters available in the Hungarian catalogue, it is preferred to keep the number of fitted parameters as low as possible. Finally, for the moderately strong Hungarian events in the study period, a weak spatial scaling trend between trigger event magnitudes and aftershock radii is observed. It is assumed that the underlying effect is the epicenter error, which becomes a dominant factor at low magnitudes; hence, the motivation to include the ε parameter. We note that the formulation in Equation (6) is a simplified reflection of epicenter error when the space response function f is centred on the observed epicenters rather than the underlying ones. In a precise mathematical model, the conditional distributions of the error vectors would be neither isotropic nor independent.

When an isotropic space–time model is fitted to non-isotropic aftershock data, a typical impact is a downward bias of the α parameter; an indication of this effect is a significant gap between the α values estimated from space–time versus time-only models [18,28]. This effect is avoided by applying the constraint $\alpha = \beta$, motivated by the assumption of self-similarity: when the magnitude threshold of the model is shifted from m_0 to m_1, the K parameter is rescaled to $K_1 = K_0 \cdot e^{(\alpha-\beta)(m_1-m_0)}$ while ignoring the upper magnitude bound for simplicity; therefore, the model parameters are almost unchanged across magnitude scales if $\alpha = \beta$ except in the vicinity of the upper bound m_x. Such a model also preserves Båth's law, i.e., the observation that the average magnitude difference between a trigger event and its biggest aftershock appears to be invariant [9,14,18] (S1.2.1) [29].

The branching ratio of the model characterizing the stability of the process is defined as the average number of the first-generation descendants triggered by an event and expressed as [8,18]:

$$\varrho = \int_{m=m_0}^{m_x} \int_{t=0}^{\infty} K \cdot e^{\alpha(m-m_0)} \cdot g(t) \cdot J(m)\, dt\, dm. \tag{7}$$

The process is subcritical (stable) if $\varrho < 1$, it is critical (unstable) if $\varrho = 1$, and it is supercritical (unstable, potentially explosive) if $\varrho > 1$. In a model describing earthquake recurrence over extended periods, subcritical behaviour is assumed. Therefore, assuming a self-similar process with $\alpha = \beta$ and substituting Equations (4) and (5) in Equation (7), the branching ratio is:

$$\varrho = \frac{K}{\eta q} \cdot \frac{\beta(m_x - m_0)}{1 - e^{-\beta(m_x - m_0)}} \cdot \exp\left(-\eta \Delta t_0^q\right), \tag{8}$$

which considers the truncation of the time response function at Δt_0. There is a near-linear dependence of ϱ on the magnitude span $m_x - m_0$, where the upper bound is difficult to estimate and the lower bound is subject to expert judgement. Therefore, the branching ratio according to Equations (7) and (8) is better viewed as a model property rather than as a physical parameter of the observed earthquake catalogue.

3.2. Parameter Fitting

In this subsection the parameter fitting process for a space–time ETAS model formulated in Equations (3)–(6) is described. The study period is $[0, T]$ with $T = 8\,036$ where time is measured in days beginning at 2000-01-01 00:00 UTC. The broader study area is $A = [X_0, X_1] \times [Y_0, Y_1] = [-258,\ 313\,] \times [-245,\ 167]$, where space coordinates follow the kilometre grid defined in Equation (1). The core study area A_0 and the proper study area A_1 are narrower than A, so $A_0 \subset A_1 \subset A$ (see Figures 2 and 3). The lower and upper magnitude bounds are set at $m_0 = 2.45$ and $m_x = 6.55$, respectively; it is assumed that the magnitude values in the catalogues, rounded to multiples of 0.1, reflect the mid-points of the respective magnitude bins. In addition, the minimum aftershock delay parameter is fixed at $\Delta t_0 = 2.315{\cdot}10^{-4}$ days (equivalent to 20 s).

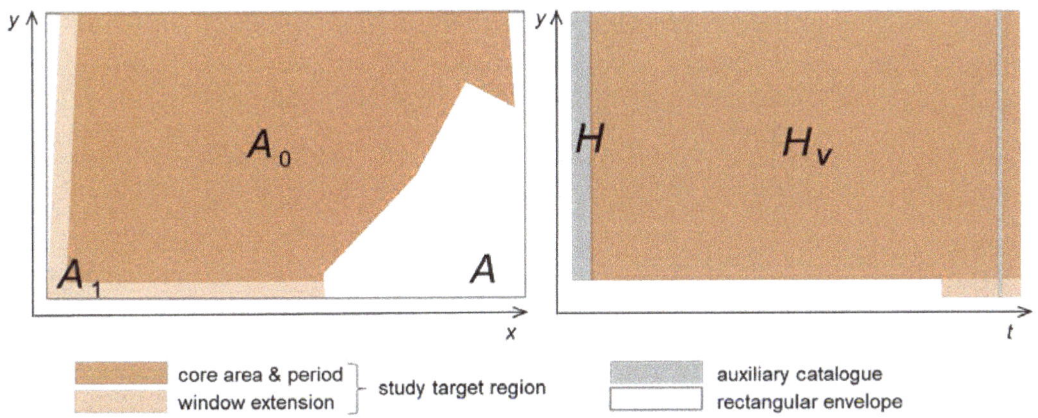

Figure 3. Schematic diagram of the space–time study target region (see also Figures 1 and 2); not to scale (the Petrinja 2020 gap is exaggerated). The auxiliary catalogue includes core window events in 1999 and the Petrinja gap events in 2020.

The maximum likelihood estimation (MLE) method seeks the parameters where the maximum of the likelihood function is reached. Expressed as a function of the parameter set $\theta = (\mu, K, \eta, q, \alpha, D, \varepsilon)$, the log-likelihood function is [6]:

$$\log L(\theta) = \sum_{i \in H_V} \log \lambda_\theta(t_i, x_i, y_i | H_{:i}) - \iiint_V \lambda_\theta(t, x, y | H_t)\, dx\, dy\, dt, \tag{9}$$

where V is the space–time target region of the study. The different event ranges considered in Equation (9) must be noted: H_V is the target catalogue, i.e., the list of all the events in V. H_V can also be a subset of the fitting catalogue H; in other words, auxiliary events from outside the target region V can be taken into account as historical triggers in the conditional occurrence rate λ_θ for Equation (9) [7,8]. Whereas H_V is assumed to be a complete catalogue, the completeness of the auxiliary catalogue $H \backslash H_V$ is not necessary. In the first term, $H_{:i} \subset H$ is a shorthand notation for the sub-catalogue of prior events that can have a triggering effect on event i. In this case, $H_{:i} = H_{t_i - \Delta t_0}$ according to the minimum delay assumption.

When implementing the maximum likelihood estimation, the complex non-rectangular form of the space–time target region, shown in Figure 3, must be considered: (1) the extended geographical window itself is not rectangular in the kilometre grid, (2) an area along the Romanian border is excluded due to concerns of data inhomogeneity, (3) the catalogue of the window extension is considered complete only from 2018, and finally, (4) a 1-day completeness gap from the target region beginning after the Petrinja 2020 mainshock is excluded. The exclusion of a time window after a large event is one of the techniques suggested in the literature to deal with the problem of aftershock incompleteness [18,28]. In fact, events falling into this time gap are used as auxiliary events, so their triggering effect is considered in the model, nonetheless.

When an estimate of the background density function $u(x,y)$ is provided, the parameters $(\mu, K, \eta, q, \alpha, D, \varepsilon)$ that maximize the log-likelihood function can be computed through non-linear optimization. In this study, the Davidon–Fletcher–Powell method [30] is implemented using custom Python code. The space integrals in the log-likelihood function in Equation (9) are approximated by integration on the rectangular broader area A; the difference is immaterial since the Gaussian kernels in the space functions quickly fall close to zero outside the study area A_1. In accordance with the self-similarity assumption, parameter α is set to be equal with the β magnitude–frequency decay parameter of the target catalogue H_V, which is estimated at $\beta = 2.602$ and is equivalent to a Gutenberg–Richter $b = 1.13$.

The function $u(x,y)$ is approximated by the kernel smoothing of a declustered epicenter catalogue, following an iterative process suggested by Zhuang et al. [7,8] after certain adaptations. An initial declustered catalogue is obtained via the window method, i.e., by removing those events from the full catalogue that fall inside a pre-defined space–time neighbourhood of a bigger event, where the window parameters suggested in [31] (pp. 173–174) are used. Starting from the initial declustering, cycles of the following steps are iterated until convergence is reached:

1. kernel smoothing for $u(x,y)$;
2. log-likelihood optimization for $(\mu, K, \eta, q, D, \varepsilon)$;
3. follow-on declustering steps based on the ETAS model.

For a better-smoothed background density estimate, during the kernel smoothing and declustering steps, the epicenter catalogue E is broader than the catalogue H used for the log-likelihood optimization. The latter is kept narrower to reduce computing time. The epicenter catalogue E includes all $M \geq 2.5$ events in the study area A_1 from catalogues B, C, and X in the years 1996–2021, giving a total number of 2 432 potential epicenter nodes before declustering. The incompleteness of E implies an inhomogeneous coverage between the core area and the extension margin, which needs to be corrected by appropriate weighting.

Declustering in an ETAS context aims to identify background events rather than mainshocks, as in the parameterization of a stationary Poisson model. Since both the background process and the trigger process contribute simultaneously to the Poisson rate at event i, declustering in an ETAS context is non-binary. A background weight $0 \leq \xi_i \leq 1$ is defined for each event in the epicenter catalogue, expressing the relative contribution of the background process to event i [7,8]:

$$\xi_i = \mu \cdot u(x_i, y_i) / \lambda(t_i, x_i, y_i | E_{:i}). \tag{10}$$

Despite its non-binary nature, the observed distribution of ξ_i is typically strongly U-shaped with modes close to 0 and 1, suggesting that background events can be well distinguished from triggered events in most cases [7,9]. As a simplification, the continuous background weights ξ_i are replaced by the binary background indicators χ_i:

$$\chi_i = \begin{cases} 0 \text{ if } \xi_i < 1/2, \\ 1 \text{ if } \xi_i \geq 1/2. \end{cases} \tag{11}$$

The conversion into binary indicators is mathematically not strictly necessary, as the equations used in the parameter fitting also work with continuous weights. The advantage of binary indicators is that they are robust, so they ensure a quick convergence of the iteration, and they also provide an unambiguous declustered catalogue of background events.

Given the background indicators χ_i, the background density function is approximated as a weighted sum of Gaussian kernels placed around the epicenter nodes [7,8]:

$$u(x,y) = \frac{1}{C}\cdot\left[\sum\nolimits_{i\in E_0}\chi_i\cdot k_d(x-x_i,y-y_i) + w\cdot\sum\nolimits_{i\in E\backslash E_0}\xi\chi_i\cdot k_d(x-x_i,y-y_i)\right], \quad (12)$$

with

$$k_d(x,y) = \frac{1}{2\pi d}\cdot\exp\left(-\frac{x^2+y^2}{2d^2}\right), \quad (13)$$

where C is a normalising constant and E_0 is the epicenter sub-catalogue corresponding to the core area A_0. The weighting w of the sum over $E\backslash E_0$ is introduced to correct the uneven coverage of the epicenter catalogue between the core area A_0 and the extension margin $A_1\backslash A_0$, and it is defined such that the corrected total weight in the two area parts is proportional to the respective background occurrence rates, estimated from the most recent declustering step of the iteration. While Zhuang et al. [7,8] suggest a variable bandwidth, a fixed bandwidth is used in this study, set by expert judgement to d = $\sqrt{50}$ = 7.071 km.

4. Modelling Results

4.1. ETAS Parameters

The convergence of the fitted parameters after each iteration of the declustering–kernel smoothing–log-likelihood optimization cycle is shown in Table 1; the preferred parameters are those from iteration 3. A plot of the modelled density of the background rate $\mu\cdot u(x,y)$ is shown in Figure 4 and the histogram of background weights ξ_i and background indicators χ_i is shown in Figure 5. The branching ratio of the model with the parameters after the last iteration is $\varrho = 0.47$. The major clusters in the study catalogue show anisotropy, which could have an impact in an alternative parameterization when the α parameter is set free. The result from space–time fitting is $\alpha = 1.79$ while a time-only fitting yields $\alpha = 2.76$—this pattern is consistent with the effects reported by Hainzl et al. [28].

Table 1. Convergence steps of the fitted ETAS parameters for the study catalogue. Parameter α is fixed through the fitting. The value μ_0 is included for information and it reflects the daily frequency of background events within the core area A_0. The preferred parameters are these from iteration 3.

Iteration	μ (1/Day)	μ_0 (1/Day)	K	η	q	α	D (km)	ε (km)	log L
1	0.0970	0.0675	0.0049	0.4178	0.2533	2.6021	0.0426	2.3464	−16,613.68
2	0.1136	0.0646	0.0049	0.4186	0.2519	2.6021	0.0418	2.3325	−16,569.50
3	0.1144	0.0644	0.0049	0.4184	0.2517	2.6021	0.0417	2.3395	−16,569.11

4.2. Parameter Ranges

The estimation of parameter errors is not straightforward, since the errors obtained from analytical solutions can be different from simulation results [32]. Here, two simple tests to derive approximate error ranges are carried out: (1) In the first test, the MLE calculation is re-run on 20 simulated catalogues, using the preferred parameters for the simulation. Each simulated catalogue includes 2214 events, i.e., the same number as the actual study catalogue H. For comparability with the actual catalogue, a fixed copy of the $M = 6.3$ Petrinja mainshock is also included at the beginning of each simulation. Table 2 shows the minimum and maximum fitted parameters from these simulations. (2) In the second test, the sensitivity of the MLE calculation to a ± 0.1 error in the magnitude of the Petrinja mainshock is reported.

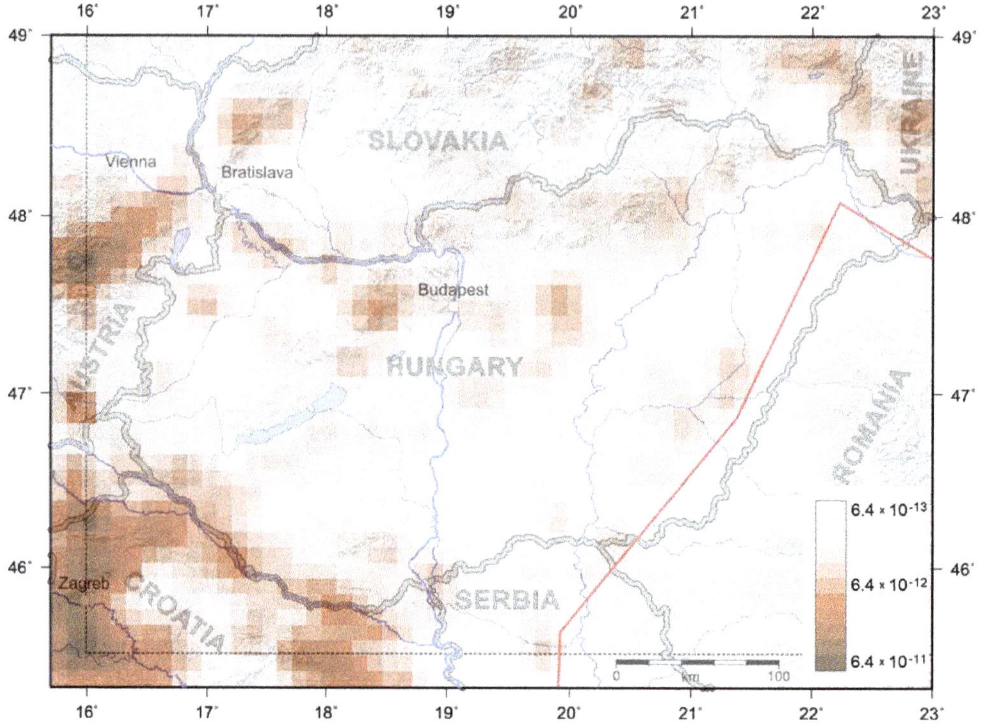

Figure 4. Modelled background density $\mu \cdot u(x, y)$ plotted at 10 km pixel size (unit is $1/(\text{day} \cdot \text{km}^2)$).

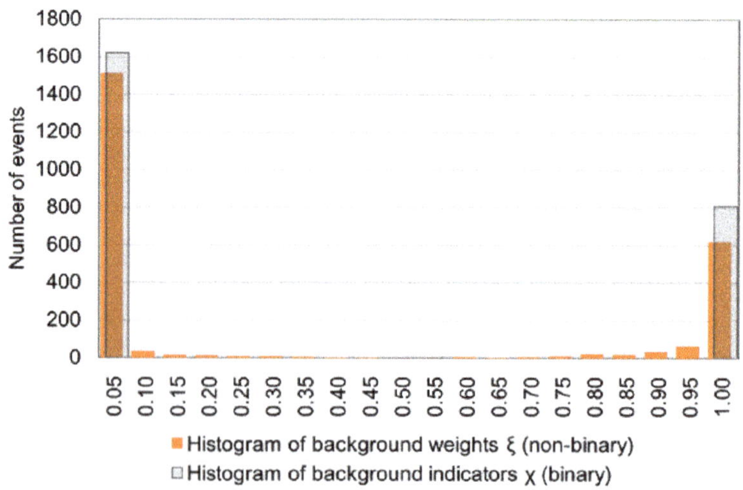

Figure 5. Histogram of background weights ξ_i over the epicenter catalogue of the study. A background weight close to 1 indicates a background event while a background weight close to zero indicates a triggered event.

Table 2. Approximate error ranges for the estimated parameters and for the branching ratio ϱ.

	μ (1/Day)	K	η	q	D (km)	ε (km)	ϱ
Parameter ranges from 20 simulations							
min	0.1031	0.0044	0.3694	0.2200	0.0389	2.2394	0.44
max	0.1284	0.0056	0.5148	0.2805	0.0455	2.4919	0.49
Parameters calculated with alternative magnitude of the Petrinja mainshock							
$M = 6.2$	0.1142	0.0052	0.4011	0.2700	0.0481	2.3035	0.49
$M = 6.4$	0.1148	0.0045	0.4401	0.2308	0.0357	2.3842	0.44

4.3. Event Accumulation Test

The cumulative number of modelled events is obtained by integrating the conditional occurrence rate function [5]:

$$\Lambda(t) = \int_0^t \iint_{A(\tau)} \lambda(\tau, x, y|H_\tau)\, dx\, dy\, d\tau, \tag{14}$$

where the notation $A(\tau)$ reflects the time-dependent area coverage of the study. Figure 6 shows the comparison of modelled and observed cumulative event numbers in the study period. The estimated number of non-observed Petrinja aftershocks is 155; this is the difference between the modelled and observed number of occurrences in the 1-day interval after the Petrinja mainshock. When the observed curve is corrected for this difference, the model captures the large Croatian earthquake sequences quite well. A weak point is the seemingly uneven observed background rate, which is not matched by the model: the underlying issue is that the southern part of the study area is unevenly covered by the patchwork of catalogues; this affects not only the southern extension margin but the southern edge of the core window too.

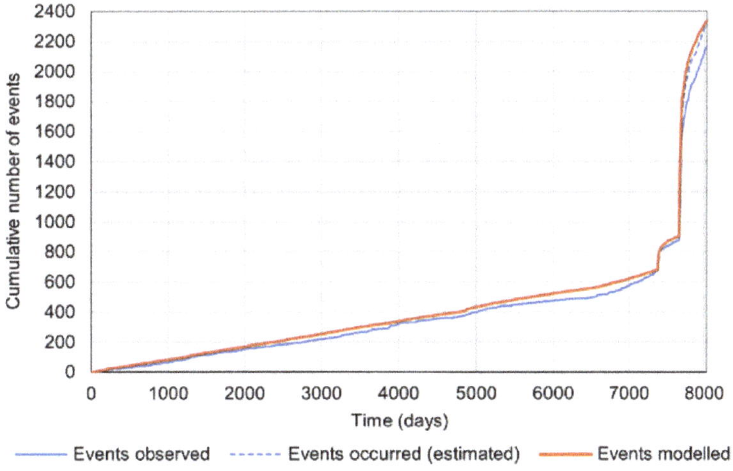

Figure 6. Modelled versus observed event accumulation in the study period 2000–2021 for $M \geq 2.5$. The two large event surges late in the period correspond to the Zagreb and Petrinja, Croatia, 2020–2021 earthquake sequences. The model estimates that 155 occurrences in the Petrinja aftershock sequence are missing from the observed catalogue. Additionally, it is worth noting the step-up of the background rate from day 6576 (1 January 2018), which is the beginning of the period when the study area covers the extended window.

4.4. Poisson Goodness-of-Fit Test

The Poisson behaviour of the background process on the core area and of the full non-declustered process on the total modelled area is tested.

In the case of the background process, the modelled occurrence rate on the core area A_0 is $\mu_0 = 0.0643$ at the magnitude threshold $M \geq 2.5$. When splitting the period into k-day intervals, the event counts by interval should follow a Poisson distribution with a mean parameter $k \cdot \mu_0$ according to the model. The outcome of chi-squared tests at a 0.95 confidence level is the following: at the magnitude threshold $M \geq 2.5$ the catalogue typically fails the test for the period 2000–2021, but it passes the test for the period 2000–2010. At the magnitude threshold $M \geq 3.0$, with the Poisson mean scaled accordingly, the catalogue passes the test for the full 2000–2021 period. This can be explained by possible data deficiencies, i.e., missing small events along the periphery after the termination of Catalogue C in 2010.

In the case of the full non-declustered process, the time transformation $t \longmapsto \Lambda(t)$, according to Equation (14), is used to convert the process into a stationary one [5]. When splitting the full period into k-unit intervals of transformed time, the event counts by interval should follow a Poisson distribution with a mean parameter k for $M \geq 2.5$. Chi-squared tests at a 0.95 confidence level fail the model at the threshold $M \geq 2.5$, but the model passes the tests at $M \geq 3.0$. This indicates that the observed process has some features at low magnitudes that this simplified model is unable to capture. Figure 7 shows some of the results of the Poisson goodness-of-fit tests.

The following three tests (i.e., distance test, time lag test, and triggering ability test) have been suggested and used with Japanese earthquake data by Zhuang et al. [8] in order to assess how well the space and time response functions of a model match the observations. The next subsections show the test results of this model with the Hungarian data.

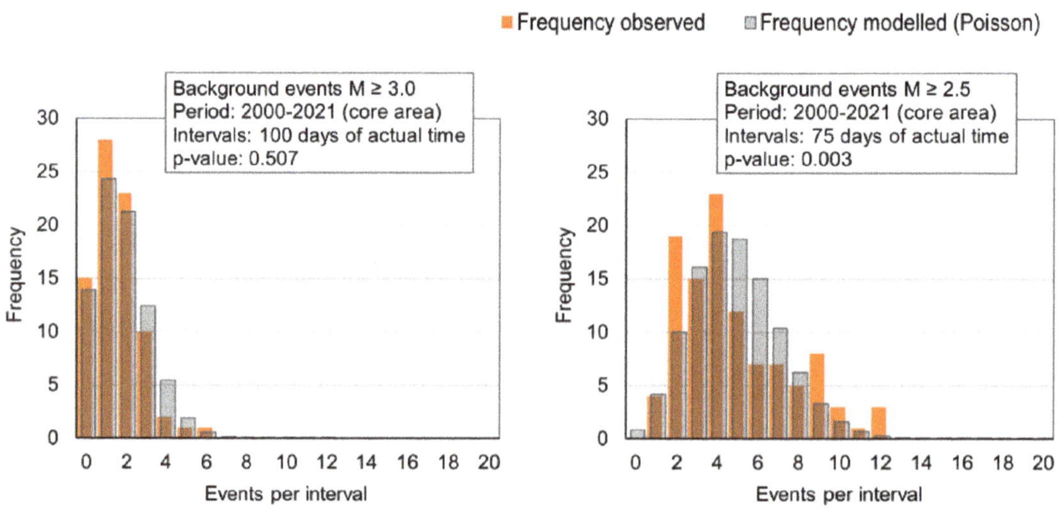

(**a**) (**b**)

Figure 7. *Cont.*

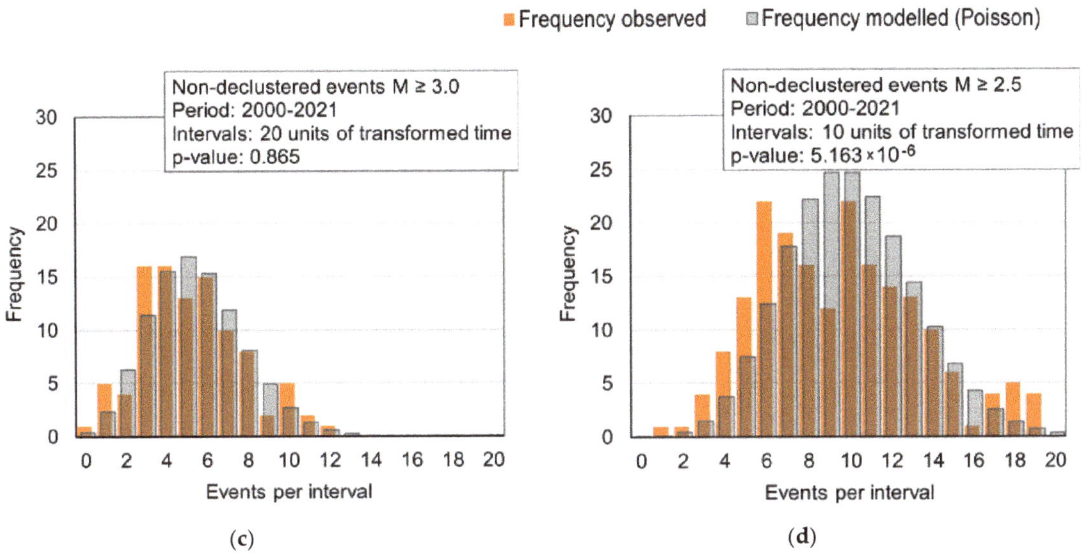

Figure 7. Examples of chi-squared Poisson goodness-of-fit tests: The distribution of event counts per interval are compared to the modelled Poisson distribution for (**a**) background events in the core area, $M \geq 3.0$; (**b**) background events in the core area, $M \geq 2.5$; (**c**) all events, $M \geq 3.0$; and (**d**) all events, $M \geq 2.5$.

4.5. Distance Test

This test compares the distribution of observed distances between trigger events and their aftershocks against the modelled space response function [8]. If an event i is triggered by an event j, the standardized distance between the two events can be defined as:

$$r_{ij} = \frac{\sqrt{(x_i - x_j)^2 + (y_i - y_j)^2}}{\sqrt{D^2 \cdot e^{\alpha(M_j - m_0)} + \varepsilon^2}}. \tag{15}$$

If the approximation of using a Gaussian space response function in the model is correct, then r_{ij} should follow a chi distribution with two degrees of freedom (or Rayleigh distribution with $\sigma = 1$). Since ancestor–descendant relationships between events cannot be identified unambiguously, the weights ζ_{ij} are introduced to express the fraction of event i, which is triggered by event j:

$$\zeta_{ij} = \frac{K \cdot e^{\alpha(M_j - m_0)} \cdot g(t_i - t_j) \cdot f(x_i - x_j, y_i - y_j | M_j)}{\lambda(t_i, x_i, y_i | H_{:i})}. \tag{16}$$

In order to build a histogram of the observed distribution of r_{ij} the sum of the ζ_{ij} values is taken in each standardized distance bin, where i runs over H_V and j runs over $H_{:i}$.

The test results are shown in Figure 8, and they show a good fit. This outcome seems to be in contrast with the results of Zhuang et al., who conclude that the Gaussian space response form fits poorly to the Japan Meteorological Agency (JMA) catalogue data, as opposed to the power law, which shows good results. However, the authors used the Rayleigh distribution with the wrong parameter for the test, that is, $\sigma = 1/\sqrt{2}$ instead of $\sigma = 1$ [8] (Equation (26), p. 7, Figure 7a, p. 10). With the correct theoretical distribution, the difference between the Gaussian model variant and the JMA data is much less than what the 2004 article suggests. In the case of this study, the additional spatial parameter ε also helps to achieve better results in this test.

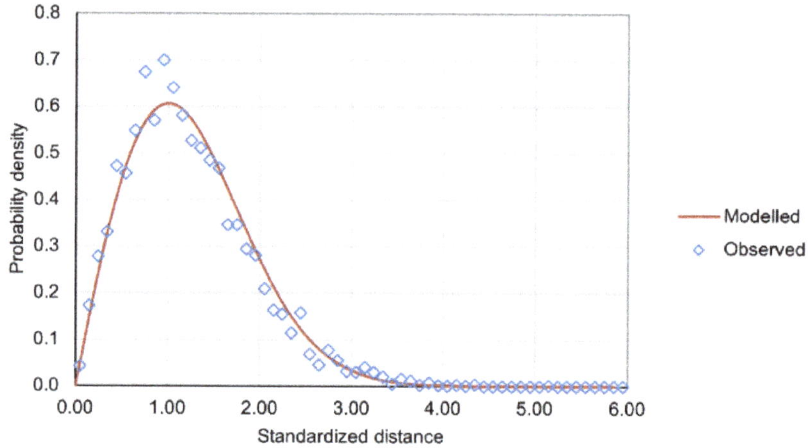

Figure 8. Distance test results: reconstruction of the distribution of the standardized triggering distance from observations versus the modelled distribution.

4.6. Time Lag Test

The purpose of this test is to analyse the fit of the modelled time response function to the observations [8]. The test method is analogous to the distance test: instead of standardized distances, the distribution of the time lag $t_{ij} = t_i - t_j$ between triggered events and their ancestors is reconstructed using the triggering weights ζ_{ij}, as defined in Equation (16). Unlike in the distance test, i is allowed to run over the full catalogue H rather than excluding the first day after the Petrinja mainshock; as a result, the incompleteness of the catalogue is also visible in the test outcome. Two views of the test output are shown in Figure 9: a plot of the probability density function (PDF) on a logarithmic scale and a complementary cumulative distribution function (CCDF) log–log plot, with the latter suggested as a more powerful diagnostic tool by Mignan [12].

The test results show the following deviations: a downward deviation of the observed PDF from the model for very short time lags, apparently due to the incompleteness of early aftershock data in the Petrinja earthquake sequence. However, the test shows this deviation only at time intervals of less than 10^{-2} days, that is, less than 15 min, whereas in the study catalogue, events with $M < 3$ are missing from the initial part of the observed Petrinja aftershock sequence for more than 10^{-1} days, that is, more than 2.4 h. The test also shows a downward deviation of the observed PDF from the model for very long time lags due to the boundary effect at the end of the study period. The latter effect also explains the downward deviation of the observed CCDF from the model.

4.7. Triggering Ability Test

This test compares the modelled and observed triggering ability of events by magnitude class [8]. The observed number of events triggered by an event j is expressed as $\sum_i \zeta_{ij}$ when using the triggering weights defined in Equation (16) while the modelled ultimate number is an exponential function of M_j, and the expected number within the study period is obtained by restricting the integration bounds of the time responses. The test results are shown in Figure 10. While the test confirms the exponential trend, it also reveals an upward deviation from the model at low magnitudes. The graph suggests an exponential trend with a lower α parameter than that of the model; the likely underlying effect is the anisotropy of aftershock clusters.

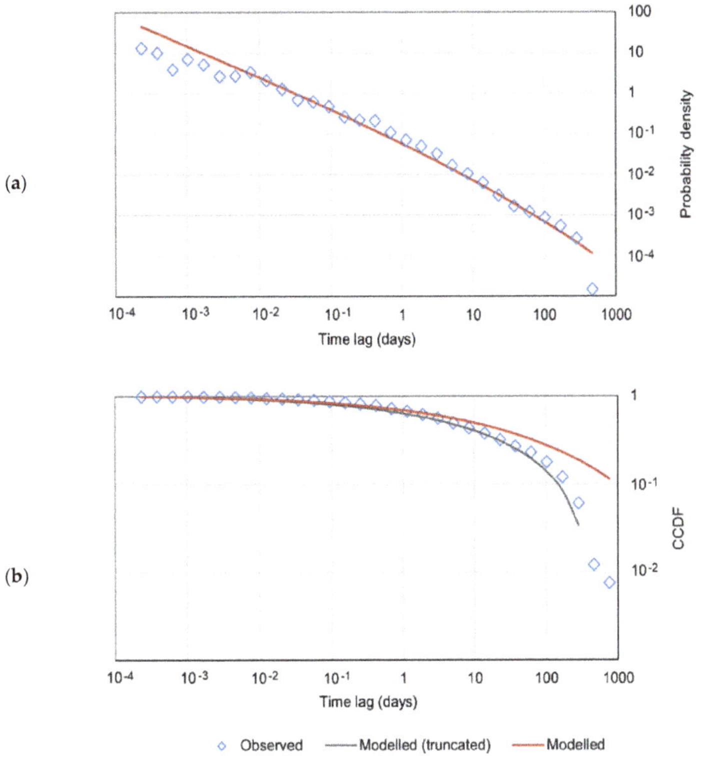

Figure 9. Time lag test results: reconstruction of the distribution of the time lag between triggered events and their ancestors from observations. (**a**) Plot of probability density. (**b**) Plot of complementary cumulative distribution function (CCDF). To check the hypothesis that the deviation of the observed curve is explained by the time boundary effect, an alternative CCDF is also shown where the time lag is truncated at 400 days.

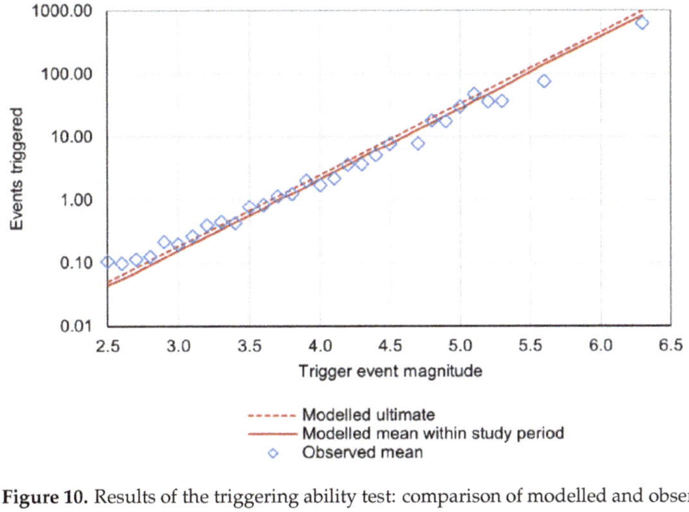

Figure 10. Results of the triggering ability test: comparison of modelled and observed triggered event numbers by trigger event magnitude.

5. Parameter Downscaling Methods

5.1. Motivation for Downscaling

By downscaling, we mean the adaptation of the model parameters to smaller zones within the study area, where a stand-alone ETAS parameter estimation is not feasible due to insufficient data. Due to the data quality requirements of an ETAS study, the parameterization presented in the previous section relies on the instrumental catalogue data from the period 1996–2021, which are dominated by the 2020–2021 Croatian earthquake sequences. On the other hand, the largest historical occurrences within Hungary itself date to earlier than the study period. From the insurance industry point of view, a Hungarian earthquake model needs a regional parameterization, covering the central area around Budapest, since this area has the densest accumulation of insured property in the country. Furthermore, for a full coverage of the country, the geographical coverage gap along the Romanian border would need to be filled.

Sections 5.2 and 5.3 describe an approach for adapting ETAS parameters to regional source zones, which incorporate historical earthquake data from a longer period. Therefore, we combined ETAS with standard techniques that have been used to parameterize stationary mainshock-only models. The zonation used in this study is based on the area sources of the SHARE project [26], with zone groupings for larger datasets. Figure 11 shows the map of the zones. The downscaling effort was restricted to the part of the zones within the core geographical window, defined between 45.5–49.0 N and 16.0–23.0 E.

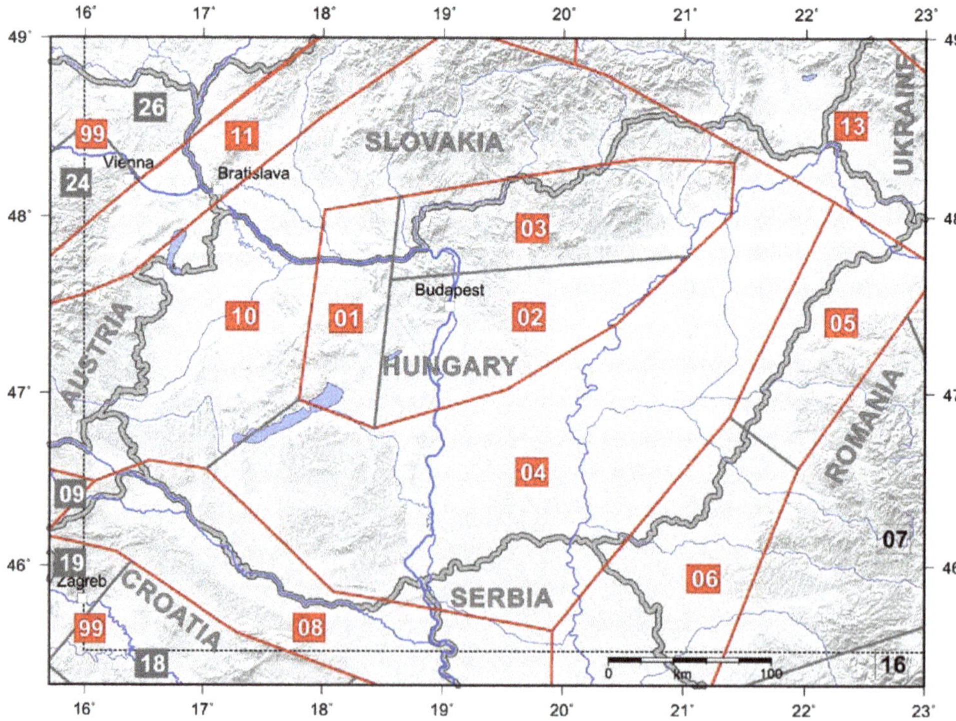

Figure 11. Source zones and zone groupings used in the model, based on the area sources of the SHARE project [26]. The core geographical window is also displayed—the downscaled model covers the part of the zones falling within the core window. Zones 01–02–03, Zones 04–10, and Zones 05–06 are grouped for the model. The non-contiguous zones with grey labels are grouped for the model as the residual Zone 99. Zones with empty labels are not modelled.

5.2. Maximum Likelihood Estimation with Variable Observation Periods

In this section, we refer to the maximum likelihood estimation (MLE) method developed by Weichert [33]. In a nutshell, the aim of the estimation is to determine recurrence parameters for mainshocks, defined as the largest earthquakes in each cluster. The accumulation of mainshocks is assumed to follow a stationary Poisson point process with a constant recurrence rate ν. It is assumed that event magnitudes are independent and follow a truncated exponential distribution with decay parameter β between the bounds m_0 and m_x, according to Equation (4). Magnitudes are rounded and grouped into the magnitude bins $[m_k - \delta, m_k + \delta)$. The length T_k of the time interval when the historical event catalogue is complete varies by the magnitude class m_k. The number of events in the respective magnitude class and completeness period is denoted by n_k. The parameters m_0, m_x, δ, and T_k are determined by prior expert judgement and are fixed in the estimation; the unknown parameters to be fitted are ν and β. Under these assumptions, the event count n_k is drawn from a Poisson distribution whose mean parameter is:

$$\lambda_k(\nu, \beta) = \nu \cdot T_k \cdot \int_{m_k-\delta}^{m_k+\delta} J(m)\, dm = \nu \cdot T_k \cdot e^{-\beta(m_k - m_0)} \cdot \frac{2\sinh(\beta\delta)}{1 - e^{-\beta(m_x - m_0)}}. \tag{17}$$

The calculus for finding the maximum of the corresponding likelihood function $L(\nu, \beta)$ leads to the equations [33]:

$$\frac{\sum_k T_k \cdot m_k \cdot \exp(-\beta m_k)}{\sum_k T_k \cdot \exp(-\beta m_k)} = \frac{\sum_k n_k \cdot m_k}{\sum_k n_k} = \overline{m}, \tag{18}$$

$$\nu = \frac{1 - e^{-\beta(m_x - m_0)}}{2\sinh(\beta\delta)} \cdot \frac{\sum_k n_k}{\sum_k T_k \cdot \exp(-\beta(m_k - m_0))}. \tag{19}$$

The recurrence parameter estimates are obtained by solving Equation (18) for β and then calculating ν from Equation (19). The strength of this MLE method is that it makes optimum use of available data by allowing magnitude-dependent observation periods. The method aims to find only temporal parameters. Its limitation is the reliance on a stationary Poisson event counting process, which normally restricts its application to declustered catalogues. Because the largest event in every cluster is selected, the estimated β is expected to be lower than the β parameter of the full population.

In our ETAS model context, the above MLE technique with variable observation periods could be straightforwardly adapted to find the (μ, β) parameters of the stationary Poisson background process while focusing on the initial event in each cluster rather than the largest one. For that, the calculation of the background weights ξ_i and indicators χ_i according to Equations (10) and (11) would need to be extended to the full historical catalogue. Such an extension is possible, with the following caveats: due to incomplete data, the process may identify some triggered earthquakes as background events. In addition, the high epicenter uncertainty of historical events, and especially the use of macroseismic epicenters, may lead to implausible results, unless Equation (10) is adjusted to take this effect into account, e.g., by the convolution of the space distributions with error functions. However, we did not elaborate in this paper on this method and its results because this study took a different approach for the downscaling of ETAS parameters, as seen in Section 5.3.

5.3. Transformed Time Estimates

To be able to perform the MLE with variable observation periods on a non-declustered catalogue, the $t \longmapsto \Lambda(t)$ transformation was used, which converts the event counts into a stationary Poisson point process with unit rate [5], where $\Lambda(t)$ is given by Equation (14).

When the event catalogue is incomplete, the part of the transformed time triggered by missing events is unknown. Given the ETAS parameters $(\mu, K, \eta, q, \alpha)$ and the magnitude–frequency decay parameter β, an estimate is provided for the transformed time interval length $\Lambda(T_2) - \Lambda(T_1)$ corresponding to the observation period $[T_1, T_2]$ where the observations are complete at magnitude threshold m_c. Here, the space–time region is defined by a geographical zone and an observation period, and, as a simplification, it is assumed that the triggering effects crossing the boundaries of this region are negligible in both directions, which is a reasonable assumption if there are no major events close to the zone and period boundaries. This simplification allows for disregarding of the space–time integration bounds of the response functions g and f and ignoring all events outside the selected region. The total transformed time is divided into the following parts: (1) Λ_0 is the transformed time coming from the background process, (2) Λ_{1+} is the transformed time triggered by the above-threshold events, and (3) Λ_{1-} is the transformed time triggered by the below-threshold events. Given that the observed below-threshold events only set a lower bound, Λ_{1-} is unknown.

$$\Lambda_0 = \mu \cdot (T_2 - T_1), \tag{20}$$

$$\Lambda_{1+} = \frac{K}{\eta q} \cdot \exp\left(-\eta \Delta t_0^q\right) \cdot \sum\nolimits_{j:t_j \in [T_1,T_2],, \ M_j \geq m_c} e^{\alpha(M_j - m_0)}, \tag{21}$$

$$\Lambda_{1-} \geq \Lambda_{1-}^{\min} = \frac{K}{\eta q} \cdot \exp\left(-\eta \Delta t_0^q\right) \cdot \sum\nolimits_{j:t_j \in [T_1,T_2],, \ M_j < m_c} e^{\alpha(M_j - m_0)}. \tag{22}$$

In Equation (22), Λ_{1-} is estimated by emulating iterations of the triggering process by tracing back every below-threshold event either to some above-threshold event or to the background:

$$\begin{aligned} \Lambda_{1-}^{\text{est}} &= (\Lambda_0 + \Lambda_{1+}) \cdot \left[p(m_c) \cdot \varrho(m_c) + p(m_c)^2 \cdot \varrho(m_c)^2 + \ldots \right] = \\ &= (\Lambda_0 + \Lambda_{1+}) \cdot \tfrac{p(m_c) \cdot \varrho(m_c)}{1 - p(m_c) \cdot \varrho(m_c)}, \end{aligned} \tag{23}$$

where $p(m_c)$ is the probability that the magnitude of an event is lower than m_c according to Equation (24), and $\varrho(m_c)$ is the average reproduction rate of events below magnitude m_c according to Equation (25). The truncation at m_c is carried out to avoid double counting with Λ_{1+}.

$$p(m_c) = \frac{1 - e^{-\beta(m_c - m_0)}}{1 - e^{-\beta(m_x - m_0)}}, \tag{24}$$

$$\varrho(m_c) = \begin{cases} \frac{K}{\eta q} \cdot \exp\left(-\eta \Delta t_0^q\right) \cdot \beta \cdot \frac{m_c - m_0}{1 - e^{-\beta(m_x - m_0)}} & \text{if } \alpha = \beta, \\ \frac{K}{\eta q} \cdot \exp\left(-\eta \Delta t_0^q\right) \cdot \frac{\beta}{\beta - \alpha} \cdot \frac{1 - e^{-(\beta - \alpha)(m_c - m_0)}}{1 - e^{-\beta(m_x - m_0)}} & \text{if } \alpha \neq \beta. \end{cases} \tag{25}$$

Considering the lower bound in (22), the total estimated transformed time is then

$$\Lambda(T_2) - \Lambda(T_1) = \Lambda_0 + \Lambda_{1+} + \max\left(\Lambda_{1-}^{\text{est}}, \ \Lambda_{1-}^{\min}\right). \tag{26}$$

The MLE framework with variable observation periods is used with the above time transformation for downscaling the ETAS parameters $(\mu, K, \eta, q, \alpha, D, \varepsilon)$ to zones. The starting values are the estimated overall parameters from Table 1. The background rate parameter μ is estimated for each zone from the 2000–2021 catalogue as $\sum \chi_i / T$, where T is the length of this period in days; because the χ_i indicators are robust, they are not recalculated with the downscaled parameters. The time response shape parameters (η, q) are left unchanged during the downscaling as most of the zone sub-catalogues have insufficient data to reparametrize them. The trigger scale parameter K is rescaled to $K \cdot r$ for each zone individually, and the α parameter is adjusted by zone using the $\alpha = \beta$ assumption. The transformed time intervals corresponding to the observation periods change with the rescaling. Starting from $r = 1$, the following iteration steps are repeated until convergence:

1. Parameter β is obtained by solving Equation (18) for the non-declustered event set in the transformed time, and α is set equal to β.

2. The scaling factor r is set such that $\nu = 1$ for the non-declustered event set in the transformed time, where ν is calculated according to Equation (19).

There are no sufficient data by zone for an independent rescaling of the space response parameters D and ε. Therefore, ε is kept unchanged while D is rescaled to $D \cdot r^{1/2}$; the latter assumption keeps the number of triggered events per unit of aftershock area approximately invariant.

6. Parameter Downscaling Results

6.1. Downscaled Parameters

The magnitude bounds are set at $m_0 = 2.45$ and $m_x = 6.55$, and the magnitude class width is $2\delta = 0.1$. For simplicity, the same upper bound is used for all zones. Considering the maximum magnitudes of the model variants as in Tóth et al. [34], $m_x = 6.55$ appears a conservative choice for the core geographical window. The largest homogenized magnitude in the historical catalogue observed within this area is $M = 6.1$ which is estimated for the Érmellék 1834 earthquake. The observation periods by magnitude class are given in Table 3.

Table 3. Observation periods by magnitude class.

Magnitude	Begin	End	T_k (Days)
$M \geq 2.5$	2000	2021	8036
$M \geq 3.0$	1995	2021	9862
$M \geq 3.5$	1970	2021	18,993
$M \geq 4.0$	1880	2021	51,865
$M \geq 4.5$	1750	2021	99,346
$M \geq 5.0$	1750	2021	99,346
$M \geq 5.5$	1700	2021	117,609
$M \geq 6.0$	1700	2021	117,609

The downscaled parameter estimates by zone are shown in Table 4. The obvious boundary effects at the Croatian edge of the core window make the transformed time estimates unreliable in Zone 99. For this zone all original parameters other than μ are kept unchanged. Because of data quality concerns, the parameters are not downscaled in the Romanian border Zone 05–06 either. As a placeholder solution for including the Romanian border area in the model, the kernel estimate of the background density $u(x, y)$ and the declustering indicators χ_i are extended to Zone 05–06 without changing the original ETAS parameters while applying a 0.5 credibility weight to all events with questionable data quality (that is, all events in zone 05–06 whose only source is Catalogue C) and restricting the estimation of μ to the years 2000–2010. Another problem area is zone 13 on the northeastern periphery, where there is a very steep frequency decay at low-end magnitudes, which is a possible indication of the presence of an inhomogeneous sub-population. Therefore, in this zone, the lowermost magnitude classes—$M < 2.7$, for the estimation of μ, and $M < 2.6$, for transformed-time MLE—are excluded from the calculation. Thereby, some low-magnitude residual activity in this zone is left unmodelled.

6.2. Sensitivity of the Downscaled Parameters

For assessing the robustness of the downscaling, the parameters are recalculated using two alternative parameterizations of the overall model from Section 4.2., i.e., with the two parameter sets calculated with alternative magnitudes of the Petrinja mainshock. The results of this calculation are shown in Table 5. Except for D, the changes in most parameters are small. One cannot conclude from these results that the errors of the downscaled parameters are small; however, the test suggests that the sensitivity of these parameters to the overall parameterization is limited and that the parameters depend more strongly on the zone catalogues.

Table 4. Preferred ETAS parameters by zone after downscaling. The event numbers used for parameter downscaling and the implied branching ratios ϱ are also shown. The downscaling approaches were not applied to the Romanian border Zone 05–06 due to data deficiencies and to the residual Zone 99 due to boundary effects. Note that the μ background rates (without Zone 05–06) do not add up to the core area total μ_0 in Table 1; the difference is due to the non-covered residual activity in Zone 13.

Zone	μ (1/Day)	K	η	q	$\alpha = \beta$	D (km)	ε (km)	N	ϱ
01–02–03	0.0101	0.0066	0.4184	0.2517	2.4406	0.0486	2.3395	191	0.60
04–10	0.0113	0.0037	0.4184	0.2517	2.8073	0.0366	2.3395	139	0.39
08	0.0174	0.0048	0.4184	0.2517	2.7404	0.0416	2.3395	233	0.49
11	0.0092	0.0054	0.4184	0.2517	2.3869	0.0438	2.3395	151	0.47
13	0.0059	0.0031	0.4184	0.2517	2.8651	0.0335	2.3395	60	0.33
05–06 *	0.0170	0.0049	0.4184	0.2517	2.6021	0.0417	2.3395	68.5 **	0.47
99 *	0.0078	0.0049	0.4184	0.2517	2.6021	0.0417	2.3395	63	0.47

* No downscaling. ** With credibility weights.

Table 5. Alternative results for the downscaled parameters, starting from two alternative parameter sets of the overall model calculated with different magnitudes of the Petrinja mainshock, $M = 6.2$ and $M = 6.4$, respectively (cf. Table 2).

Zone	Alternative	μ (1/Day)	K	η	q	$\alpha = \beta$	D (km)	ε (km)	ϱ
01–02–03	$M = 6.2$	0.0101	0.0067	0.4011	0.2700	2.4414	0.0545	2.3035	0.60
	$M = 6.4$	0.0101	0.0065	0.4401	0.2308	2.4423	0.0429	2.3842	0.60
04–10	$M = 6.2$	0.0113	0.0038	0.4011	0.2700	2.8070	0.0410	2.3035	0.39
	$M = 6.4$	0.0113	0.0037	0.4401	0.2308	2.8073	0.0323	2.3842	0.39
08	$M = 6.2$	0.0174	0.0049	0.4011	0.2700	2.7400	0.0466	2.3035	0.49
	$M = 6.4$	0.0174	0.0047	0.4401	0.2308	2.7414	0.0368	2.3842	0.49
11	$M = 6.2$	0.0092	0.0054	0.4011	0.2700	2.3870	0.0490	2.3035	0.47
	$M = 6.4$	0.0092	0.0052	0.4401	0.2308	2.3872	0.0387	2.3842	0.47
13	$M = 6.2$	0.0059	0.0032	0.4011	0.2700	2.8651	0.0376	2.3035	0.33
	$M = 6.4$	0.0059	0.0031	0.4401	0.2308	2.8680	0.0296	2.3842	0.33

6.3. Test of Long-Term Event Numbers by Zone

For this test, 500 stochastic catalogues were generated using the downscaled parameters and another 500 stochastic catalogues with the original parameters, each stochastic catalogue spanning 300 years. This provided a modelled distribution of event numbers by zone before and after downscaling, which were compared with the respective event numbers in the historical catalogue. The comparison focused on magnitudes $M \geq 4.0$, i.e., these events with a potential for insurance losses. Event numbers in the observation periods according to Table 3 were compared to these in the equivalent simulated periods by magnitude class. The test results are shown in Figure 12.

6.4. Test of Event Accumulation by Zone, 2000–2021

The comparison of modelled versus observed cumulative event numbers (cf. Figure 6) is performed in this test for each zone separately. The results are shown in Figure 13. The model shows a better fit to the observations in the central Zones (01–02–03 and 04–10) than on the periphery; this can also reflect variations of underlying data quality. The biggest deviation is observed in Zone 13 where the smallest events were excluded from the downscaling to obtain a better fit to long-term event counts in higher magnitude classes.

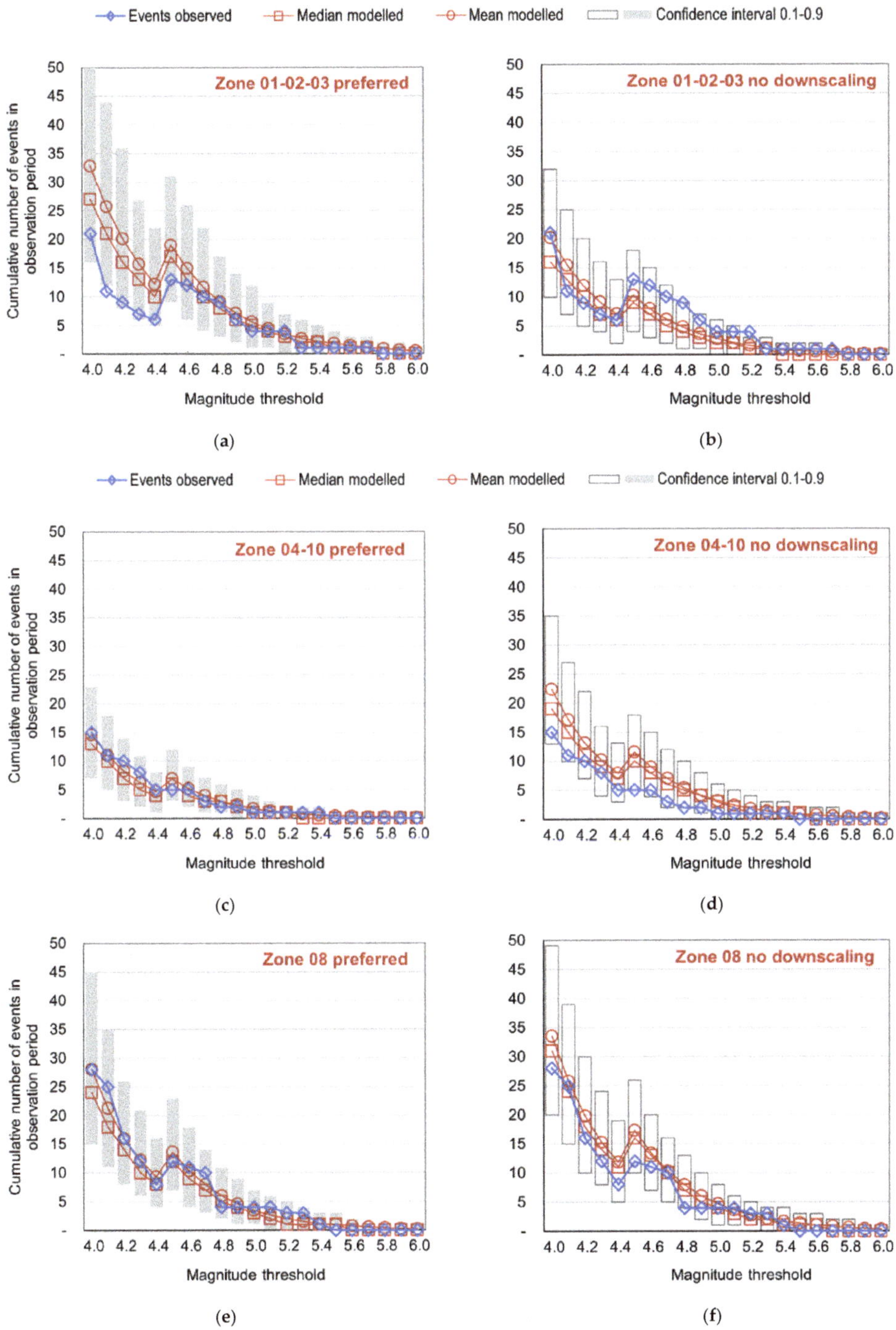

Figure 12. *Cont.*

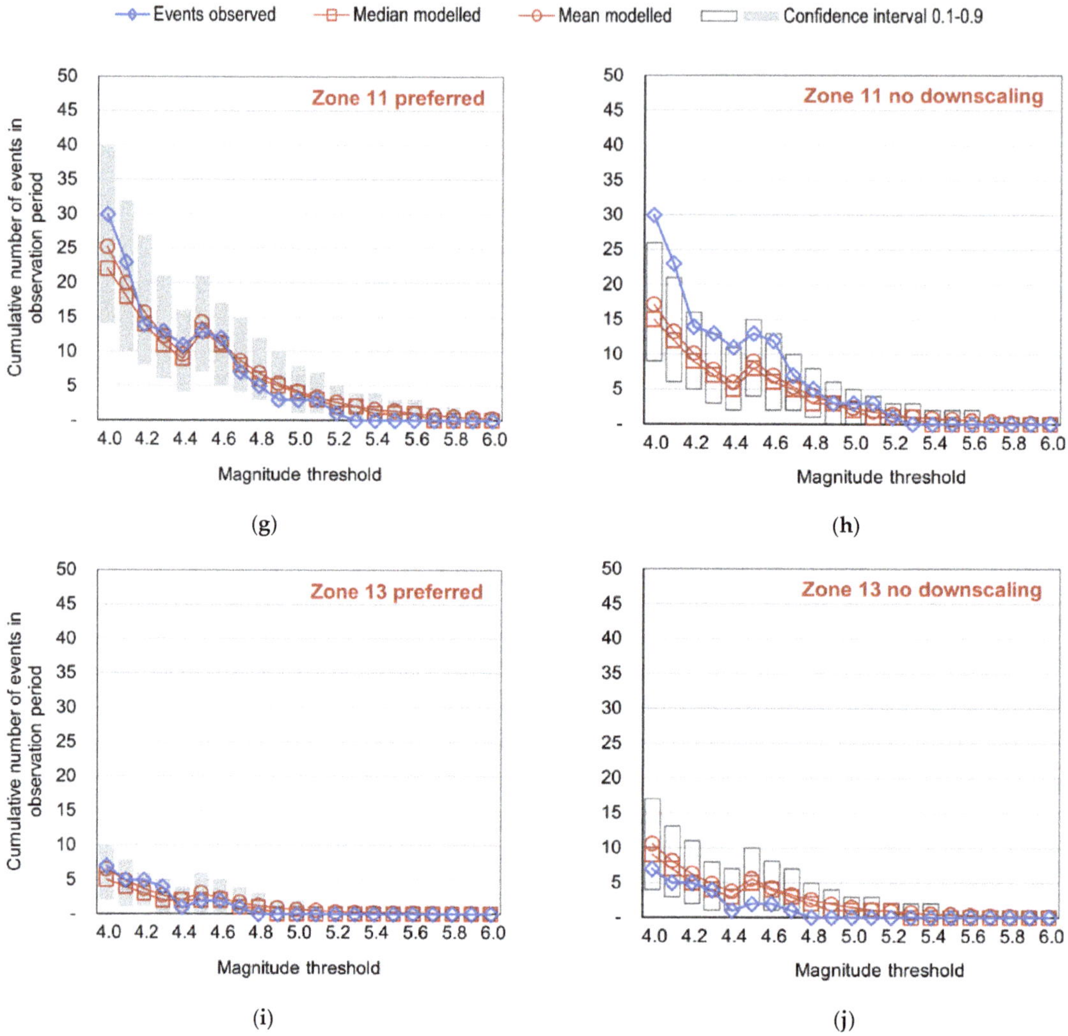

Figure 12. *Cont.*

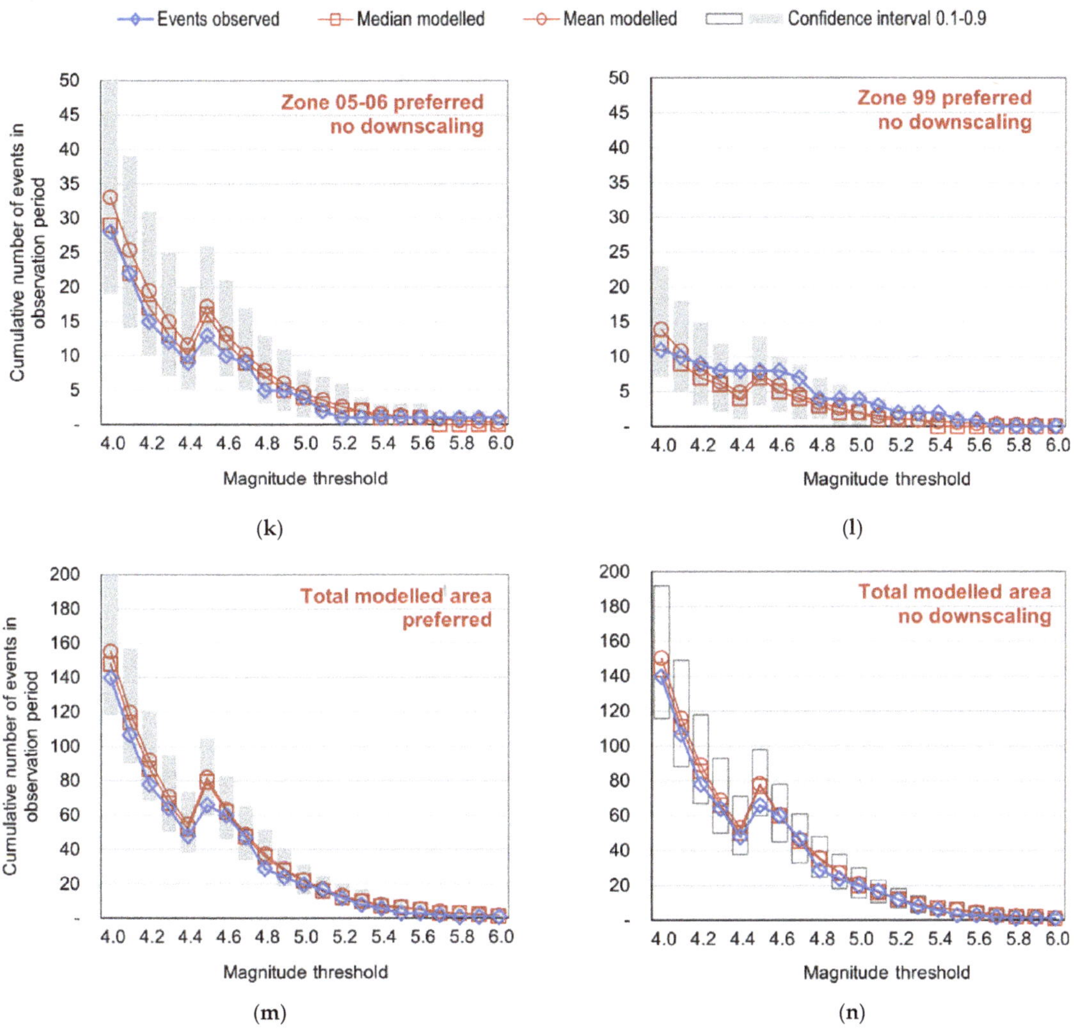

Figure 12. Comparison of long-term event counts by zone, modelled and observed. The modelled distributions were obtained by generating 500 stochastic catalogues using the downscaled parameters and another 500 stochastic catalogues with the original parameters, each one spanning 300 years. The observation periods reflecting completeness differ by magnitude class; therefore, the event numbers shown are not always decreasing with the magnitude threshold. (**a**,**c**,**e**,**g**,**i**) Zone-by-zone downscaled model results. (**b**,**d**,**f**,**h**,**j**) Results for the same zones with parameters before downscaling, for comparison. (**k**,**l**) Results for Zones 05–06 and 99 where the downscaling approach is not used. (**m**,**n**) Results for the total modelled area, with and without downscaling.

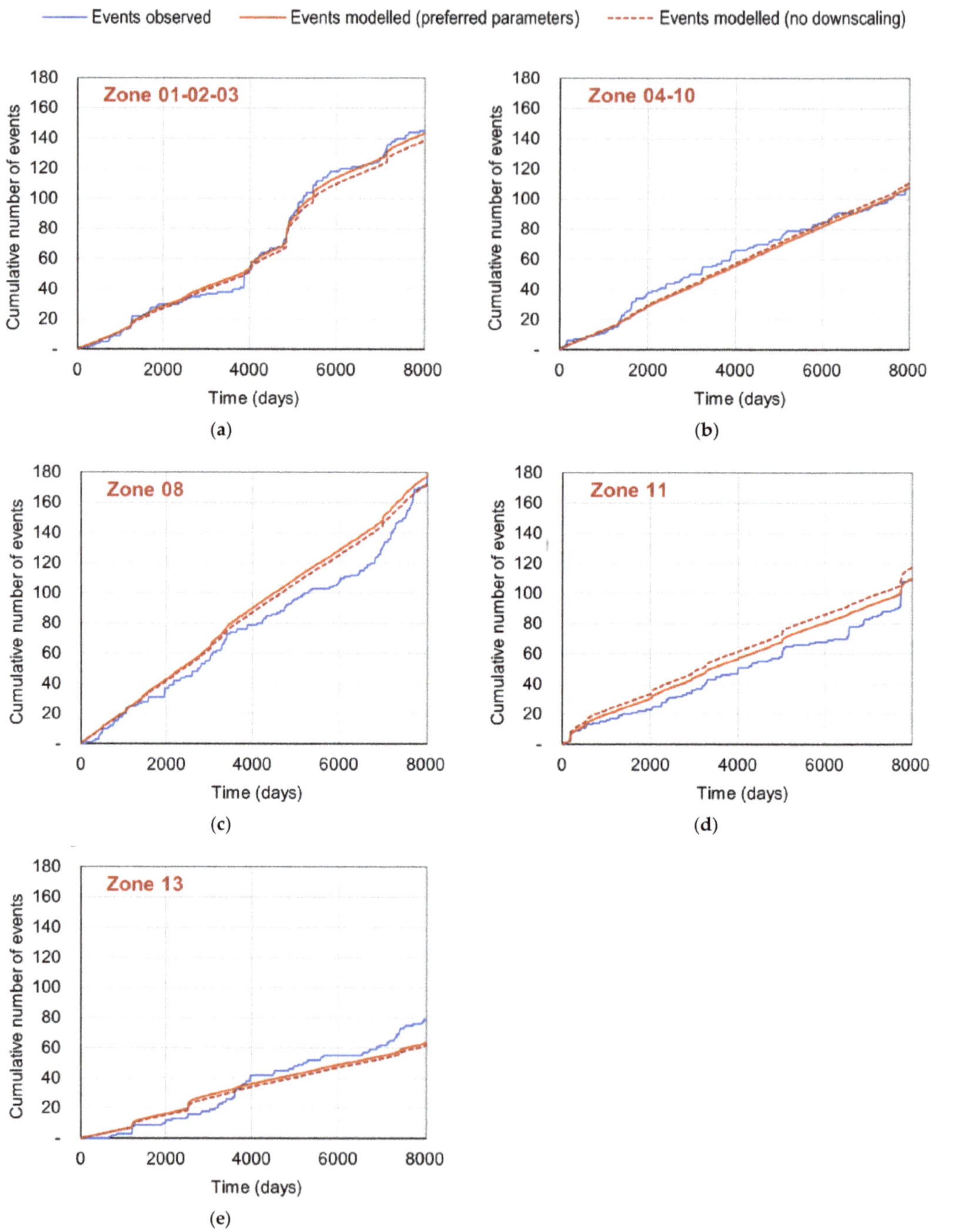

Figure 13. Modelled versus observed event accumulation by zone in the study period 2000–2021 for $M \geq 2.5$. (**a–e**) Results of the test shown for each zone. Zones 05–06 and 99 with no downscaling are omitted.

7. Discussion and Conclusions

ETAS methodology has been applied to Hungarian earthquake data of the period 1996–2021. Despite the data challenges caused by moderate seismicity, we succeeded in fitting space–time ETAS parameters to recent instrumental data of the region, which include the 2020–2021 Croatian earthquake sequences. Our conclusions from the modelling exercise are the following:

1. Poisson goodness-of-fit tests of the modelled versus the observed process show a satisfactory fit for events $M \geq 3.0$, which is appropriate for application in insurance. On the other hand, the same tests show significant deviations from the model at the magnitude threshold $M \geq 2.5$. These deviations at low magnitudes likely result from the inhomogeneities of the observed process between earthquake clusters, from data deficiencies, and possibly also from boundary effects at the edge of the modelled area or at the magnitude threshold.
2. A stretched exponential time response function is used in the model. Based on the results of the event accumulation and time lag tests, we conclude that this time response form is appropriately fitted to the observed data. Deviations from the theoretical curve in the time lag test occur at very short time periods due to early aftershock incompleteness and at long time periods due to the time bounds of the observed catalogue. The modelled branching ratio $\varrho = 0.47$ implies a stable process. A systematic comparison of the stretched exponential function against the more commonly used modified Omori law as an alternative, however, was not part of this study.
3. The model uses a Gaussian space response function with an allowance for epicenter error. The results of the standardized distance test show a good fit between the model and the observed data. The $\varepsilon = 2.34$ km value of the error parameter is consistent with the horizontal error estimates of the underlying catalogue. On the other hand, the fitted ε parameter possibly also absorbs a correction for a deviation of the data from the Gaussian space response model or from the $\gamma = \alpha$ constraint. An indication for this is that the 42-m value of the D parameter is unlikely to reflect the aftershock radius of $M = 2.5$ events by itself.
4. After the overall ETAS parameter estimation, parameter adjustments for regional zones (downscaling) were also calculated. In this step, historical catalogue data going back to 1700 were incorporated. Despite the known limitations of the historical catalogue, we find it important that, in an insurance model for a moderately active region like Hungary, the most destructive historical events are taken into account for parameterization and for model testing. Even after the inclusion of historical events, the dataset used for downscaling is still dominated by recent instrumental data which have a lower magnitude threshold of completeness. The modelled event numbers were tested against long-term historical observations in the magnitude range relevant for insurance losses. For the total modelled area, the test outcomes show a good fit both with and without downscaling, with the modelled median numbers falling slightly closer to the observations with the non-downscaled parameters. However, downscaling provides clear improvements to the results by zone. While the parameters of the overall model are subject to significant uncertainties, the tests indicate that the downscaled parameters are only moderately sensitive to the variations of the overall parameters.
5. For a Hungarian model, the central Zone 01–02–03 is especially important. In this zone, the downscaled model overestimates long-term event numbers in the 4.0–4.5 magnitude range. On the other hand, it gives better-fitting and more conservative results than the original parameterization for magnitudes $M > 4.5$, which is desirable in the case of a model for insurance applications. Zone 01–02–03 has experienced some notable event clusters in the 2000–2021 period, corresponding to the Oroszlány 2011, Tenk 2013, and Érsekvadkert-Iliny 2013–2015 events. The observed overall event accumulation in the zone is well-fitted to the downscaled model, as

shown in Figure 13a; however, this graph masks the inhomogeneities between the two 2013 event sequences, which were geographically distant but overlapped in time.

Finally, we reflect on how the results of this study contribute to the planned application in insurance. In this study, only the first module of an earthquake risk model was covered, that is, a statistical model of event occurrences. With the help of this model, synthetic earthquake catalogues can be generated, representing sequences of events that may happen in the future and possibly cause losses. Unlike in those models where only mainshocks are modelled, the synthetic catalogues of this model also include aftershock sequences. However, the subset of the mainshocks can be extracted from the catalogues if a simplified model variant is preferred. If aftershocks are modelled, then the future scenarios are not independent from the past, as large earthquakes are followed by periods of increased seismic activity. This effect can be captured by the stochastic catalogue generator of this model if recent events are added to the simulation as fixed inputs, occurring before the simulated time interval.

The synthetic catalogues mentioned above will be used as a starting point for simulating insurance losses when the full model is in place. Future work on the model will focus on developing those modules that were not discussed in this study: the ground motion attenuation and the vulnerability modules. Another possible area of future work is to extend the geographical coverage of the model to other Central European countries.

Author Contributions: Conceptualization, J.C.-B. and P.S.; Methodology, P.S.; Software, P.S.; Formal analysis, P.S.; Data curation, L.T.; Writing—original draft preparation, P.S.; Writing—review and editing, J.C.-B. and L.T.; Supervision, J.C.-B. and L.T.; Project administration, J.C.-B. All authors have read and agreed to the published version of the manuscript.

Funding: This research was fully funded by the UNIQA Insurance Group AG.

Data Availability Statement: Original datasets analysed in this study are mostly publicly available and can be found here: http://mek.niif.hu/04800/04801/ (accessed on 5 February 2023), http://www.isc.ac.uk/iscbulletin/search/ (accessed on 5 February 2023), https://earthquake.usgs.gov/earthquakes/search/ (accessed on 5 February 2023), http://www.georisk.hu/Bulletin/bulletinh.html (accessed on 5 February 2023), http://www.georisk.hu/Tajekoztato/tajekoztato.html (accessed on 5 February 2023). Other, pre-processed data presented in this study are available on request from the author: László Tóth; email: toth@georisk.hu.

Acknowledgments: The authors acknowledge the financial support of the UNIQA Insurance Group AG, Wien, Austria and especially Kurt Svoboda and Roman Schneider. The authors also thank Balázs Sághy at UNIQA Biztosító Zrt., Budapest, Hungary for all his support and encouragement given to this research, and for his help with software tools.

Conflicts of Interest: The authors declare no conflict of interest.

References

1. Directive 2009/138/EC of the European Parliament and of the Council of 25 November 2009 on the Taking-Up and Pursuit of the Business of Insurance and Reinsurance (Solvency II). Available online: https://eur-lex.europa.eu/eli/dir/2009/138/oj (accessed on 5 February 2023).
2. Commission Delegated Regulation (EU) 2015/35 of 10 October 2014 Supplementing Directive 2009/138/EC of the European Parliament and of the Council on the Taking-Up and Pursuit of the Business of Insurance and Reinsurance (Solvency II). Available online: https://eur-lex.europa.eu/eli/reg_del/2015/35/oj (accessed on 5 February 2023).
3. Petrinjski Potresi: Prosinac 2020.—Prosinac 2021.—Geofizički Odsjek. Godina Dana od Razornog Potresa Kod Petrinje. Available online: https://www.pmf.unizg.hr/geof/seizmoloska_sluzba/potresi_kod_petrinje/2020--2021 (accessed on 16 December 2022).
4. Oros, E.; Placinta, A.O.; Popa, M.; Diaconescu, M. The 1991 Seismic Crisis in the West of Romania and Its Impact on Seismic Risk and Hazard Assessment. *Environ. Eng. Manag. J.* **2019**, *19*, 609–623. [CrossRef]
5. Ogata, Y. Statistical models for earthquake occurrences and residual analysis for point processes. *J. Am. Stat. Assoc.* **1988**, *83*, 9–27. [CrossRef]
6. Ogata, Y. Space-Time Point-Process Models for Earthquake Occurrences. *Ann. Inst. Stat. Math.* **1998**, *50*, 379–402. [CrossRef]
7. Zhuang, J.; Ogata, Y.; Vere-Jones, D. Stochastic Declustering of Space-Time Earthquake Occurrences. *J. Am. Stat. Assoc.* **2002**, *97*, 369–380. [CrossRef]

8. Zhuang, J.; Ogata, Y.; Vere-Jones, D. Analyzing earthquake clustering features by using stochastic reconstruction. *J. Geophys. Res.* **2004**, *109*, B05301. [CrossRef]
9. Hardebeck, J. Appendix S–Constraining Epidemic Type Aftershock Sequence (ETAS) Parameters from the Uniform California Earthquake Rupture Forecast, Version 3 Catalog and Validating the ETAS Model for Magnitude 6.5 or Greater Earthquakes. In Uniform California Earthquake Rupture Forecast, Version 3 (UCERF3)—The Time-Independent Model, Field, E.H., Biasi, G.P., Bird, P., Dawson, T.E., Felzer, K.R., Jackson, D.D., Johnson, K.M., Jordan, T.H., Madden, C., Michael, A.J., et al., U.S. Geological Survey Open-File Report 2013–1165, 97 p., California Geological Survey Special Report 228, and Southern California Earthquake Center Publication 1792. Available online: https://pubs.usgs.gov/of/2013/1165/ (accessed on 5 February 2023).
10. Veen, A.; Schoenberg, F.P. Estimation of Space–Time Branching Process Models in Seismology Using an EM–Type Algorithm. *J. Am. Stat. Assoc.* **2008**, *103*, 614–624. [CrossRef]
11. Ogata, Y.; Zhuang, J. Space–time ETAS models and an improved extension. *Tectonophysics* **2006**, *413*, 13–23. [CrossRef]
12. Mignan, A. Modeling aftershocks as a stretched exponential relaxation. *Geophys. Res. Lett.* **2015**, *42*, 9726–9732. [CrossRef]
13. Stallone, A.; Marzocchi, W. Empirical evaluation of the magnitude-independence assumption. *Geophys. J. Int.* **2019**, *216*, 820–839. [CrossRef]
14. Turcotte, D.L.; Holliday, J.R.; Rundle, J.B. BASS, an alternative to ETAS. *Geophys. Res. Lett.* **2007**, *34*, L12303. [CrossRef]
15. Zhuang, J.; Werner, M.J.; Harte, D.S. Stability of earthquake clustering models: Criticality and branching ratios. *Phys. Rev. E* **2013**, *88*, 062109. [CrossRef]
16. Guo, Y.; Zhuang, J.; Zhou, S. A hypocentral version of the space–time ETAS model. *Geophys. J. Int.* **2015**, *203*, 366–372. [CrossRef]
17. Guo, Y.; Zhuang, J.; Zhou, S. An improved space-time ETAS model for inverting the rupture geometry from seismicity triggering. *J. Geophys. Res. Solid Earth* **2015**, *120*, 3309–3323. [CrossRef]
18. Seif, S.; Mignan, A.; Zechar, J.D.; Werner, M.J.; Wiemer, S. Estimating ETAS: The effects of truncation, missing data, and model assumptions. *J. Geophys. Res. Solid Earth* **2016**, *122*, 449–469. [CrossRef]
19. *Magyarországi Földrengések Évkönyve. Hungarian Earthquake Bulletin 1995–2019*; GeoRisk: Budapest, Hungary, 1996–2020; HU ISSN 1219-963X; Available online: http://www.georisk.hu/Bulletin/bulletinh.html (accessed on 5 February 2023).
20. MABISZ: A Decemberi Földrengések Magyar Kárbejelentéseinek a Száma Meghaladja a Horvátországit. Available online: https://mabisz.hu/a-decemberi-foldrengesek-magyar-karbejelenteseinek-a-szama-meghaladja-a-horvatorszagit/ (accessed on 16 December 2022).
21. Zsíros, T. A Kárpát-Medence Szeizmicitása és Földrengés Veszélyessége: Magyar Földrengés Katalógus (456–1995), MTA GGKI: Budapest, Hungary, 2000. Available online: https://mek.oszk.hu/04800/04801/ (accessed on 5 February 2023).
22. International Seismological Centre. On-Line Bulletin. 2022. Available online: https://doi.org/10.31905/D808B830 (accessed on 29 April 2021).
23. U.S. Geological Survey. Available online: https://earthquake.usgs.gov/earthquakes/search/ (accessed on 29 April 2021).
24. Grünthal, G.; Wahlström, R.; Stromeyer, D. The unified catalogue of earthquakes in central, northern, and northwestern Europe (CENEC)—Updated and expanded to the last millennium. *J. Seism.* **2009**, *13*, 517–541. [CrossRef]
25. GeoRisk—Földrengés Mérnöki Iroda—Earthquake Engineering. Havi Földrengés Tájékoztató (Monthly EQ list). Available online: http://www.georisk.hu/Tajekoztato/tajekoztato.html (accessed on 5 February 2023).
26. Woessner, J.; Laurentiu, D.; Giardini, D.; Crowley, H.; Cotton, F.; Grünthal, G.; Valensise, G.; Arvidsson, R.; Basili, R.; Demircioglu, M.B.; et al. The 2013 European Seismic Hazard Model: Key components and results. *Bull. Earthq. Eng.* **2015**, *13*, 3553–3596. [CrossRef]
27. Utsu, T. Aftershock and earthquake statistics (I): Some parameters which characterize an aftershock sequence and their interrelations. *J. Fac. Sci. Hokkaido Univ.* **1969**, *3*, 129–195.
28. Hainzl, S.; Christophersen, A.; Enescu, B. Impact of Earthquake Rupture Extensions on Parameter Estimations of Point-Process Models. *Bull. Seism. Soc. Am.* **2008**, *98*, 2066–2072. [CrossRef]
29. Felzer, K.R.; Becker, T.W.; Abercrombie, R.E.; Ekström, G.; Rice, J.R. Triggering of the 1999M_W7.1 Hector Mine earthquake by aftershocks of the 1992M_W7.3 Landers earthquake. *J. Geophys. Res. Atmos.* **2002**, *107*, ESE 6-1–ESE 6-13. [CrossRef]
30. Fletcher, R.; Powell, M.J.D. A Rapidly Convergent Descent Method for Minimization. *Comput. J.* **1963**, *6*, 163–168. [CrossRef]
31. Burkhard, M.; Grünthal, G. Seismic source zone characterization for the seismic hazard assessment project PEGASOS by the Expert Group 2 (EG1b). *Swiss J. Geosci.* **2009**, *102*, 149–188. [CrossRef]
32. Wang, Q.; Schoenberg, F.P.; Jackson, D.D. Standard Errors of Parameter Estimates in the ETAS Model. *Bull. Seism. Soc. Am.* **2010**, *100*, 1989–2001. [CrossRef]
33. Weichert, D.H. Estimation of the earthquake recurrence parameters for unequal observation periods for different magnitudes. *Bull. Seism. Soc. Am.* **1980**, *70*, 1337–1346. [CrossRef]
34. Tóth, L.; Győri, E.; Mónus, P.; Zsíros, T. Seismic Hazard in the Pannonian Region. In *The Adria Microplate: GPS Geodesy, Tectonics and Hazards*; Pinter, N., Ed.; Springer: Dordrecht, The Netherlands, 2006; pp. 369–384. [CrossRef]

applied sciences

MDPI

Article

The Conditional Probability of Correlating East Pacific Earthquakes with NOAA Electron Bursts

Cristiano Fidani [1,2,3]

[1] Istituto Nazionale di Geofisica e Vulcanologia, 00143 Roma, Italy; cristiano.fidani@ingv.it
[2] Central Italy Electromagnetic Network, 63847 Fermo, Italy
[3] Osservatorio Sismico "A. Bina", 63847 Perugia, Italy

Abstract: A correlation between low L-shell 30–100 keV electrons precipitating into the atmosphere and M $\geq$ 6 earthquakes in West Pacific was presented in past works where ionospheric events anticipated earthquakes by 1.5–3.5 h. This was a statistical result obtained from the Medium Energy Protons Electrons Detector on board the NOAA-15 satellite, which was analyzed for 16.5 years. The present analysis, utilizing the same database, translated into adiabatic coordinates during geomagnetic quiet periods, lead to another significant correlation regarding East Pacific strong earthquakes. This new correlation is still observed between high energy precipitating electrons detected by the NOAA-15 0° telescope and M $\geq$ 6 events of another very dangerous seismic region of the Pacific ring of fire. The particle precipitation that contributed to this correlation was characterized by electron L-shell, pitch-angle, possible disturbance altitudes, and geographical locations. This correlation occurred circa 57 h prior to the East Pacific earthquakes, according to past single cases of reports. The conditional probability corresponding to the cross-correlation peak of 0.024 per binary events reached a value of 0.011. A probability gain of 2 was calculated for earthquakes after an independent L-shell EBs detection, it is therefore applicable for future earthquake forecasting experiments. Moreover, a time-dependent probability gain approaching the correlation peak was estimated.

Keywords: electron bursts; NOAA satellite; East Pacific earthquakes; conditional probability; risk reduction

Citation: Fidani, C. The Conditional Probability of Correlating East Pacific Earthquakes with NOAA Electron Bursts. *Appl. Sci.* **2022**, *12*, 10528. https://doi.org/10.3390/app122010528

Academic Editors: Ricardo Castedo, Miguel Llorente Isidro and David Moncoulon

Received: 27 August 2022
Accepted: 15 October 2022
Published: 18 October 2022

1. Introduction

Low Earth Orbit (LEO) satellites fly at altitudes ranging between 200 and 2000 km, providing a platform of observation extending over hundreds of km, thus being able to monitor large regions struck by strong earthquakes (EQs). Likewise, given that LEO satellites are not stationary, and they circle the Earth many times every day, this platform is capable of monitoring the entire Earth's surface, assuring multiple passages on the same areas in a few days [1]. This type of monitoring uses non-seismic detectors which are mainly electromagnetic, being that the atmosphere is absent at satellite altitudes. Non-seismic phenomena observed on the ground during strong EQs are well recognized as happening before, during, and after a seismic manifestation [2]. These phenomena include fluid migration [3], the Earth's electric currents [4], atmospheric phenomena [5], and electromagnetic perturbations [6]. However, at ground level, they can be influenced by local effects, including land, atmospheric variables, and anthropogenic activities [7]. Thus, they cannot be reliably studied and related to the different phases of EQ preparation [8]. In this regard, LEO satellite observations overcome such difficulties by averaging signals on large areas and thus reducing the influence of local phenomena [9].

Electromagnetic detectors at LEO altitudes are also disturbed by atmospheric phenomena [10] and extraterrestrial perturbations mainly associated with the Sun [11]. Therefore, it has been suggested that remote sensing from the near-earth space of phenomena observed with EQs may be associated with seismic phenomena by statistical approaches [12]. An additional advantage of LEO satellites is that regarding terrestrial observations, they are able to continuously monitor multiple regions to investigate many strong seismic events over a few years for statistical analyses [13].

Statistical correlations between time series of strong earthquakes and time series of remotely collected signals have been recently obtained in several fields of electromagnetic observations [14]. A study lasting 7.5 years concerning Pc1 pulsations reported from a low-latitude station in Parkfield, California, an enhanced occurrence probability of such phenomena about 5–15 days prior to EQs, during the daytime [15]. A statistical correlation was calculated for ULF geomagnetic fluctuations, and this phenomenon anticipated moderate earthquakes by 1–5 days at Japanese ground stations [16]. Seismo-ionospheric effects on long sub-ionospheric paths have been investigated in amplitude variations of signals [17,18]. In a statistical study concerning VLF/LF wave paths, transmitter signal amplitudes revealed perturbations with a frequency excess of 3–6 days before strong seismic events occurred in 10 years [19]. Concerning the magnetometers onboard LEO satellites, shallow earthquakes with M > 5.5 have been anticipated by both magnetic field perturbation and electron density signals of $10^{-0.96+0.51 \cdot M}$ and $10^{-3.46+0.83 \cdot M}$ days, respectively, using the first 8 years of Swarm data [20]. A significant correlation between VLF wave intensity and strong EQs occurring at 0–4 h was observed in a statistical study using the micro-satellite DEMETER [21,22], where EQs were preceded by decreasing intensities. Furthermore, the largest occurrence rates of anomalies on TEC data from the global ionosphere map were correlated with strong superficial EQs, they occurred 1–5 days before the EQs [23,24]. Significant increases in electron density measurements observed on board the DEMETER satellite were correlated with moderate seismic events that occurred 10 to 6 days later [25]. A statistical analysis of NOAA POES data has evidenced Electron Bursts (EBs), which are sudden increases in electron fluxes, in the loss cone 2–3 h before the $M \geq 6$ quakes in Indonesia and the Philippines [26]. This analysis has been recently improved upon by considering the interval of 1.5–3.5 h before strong EQs in the West Pacific and resolving the ambiguity of recognizing EBs without knowing the EQ epicenters a priori [27]. The aim was to develop a methodology for EQ forecasting and verification [28].

Based on the theory of conditional probability developed in past studies [27,28] for EQ forecasting, an application for both EQ and tsunami risk reductions will be described for seismic activity in the East Pacific. Moreover, Section 2 will describe the databases used for the analysis and their relative restrictions. Section 3 will deal with the selection of EBs used to verify their correlation with East Pacific's strong EQs. This correlation is used for possible risk reduction applications discussed in Section 4, due to a longer time lag of correlated EQs. Finally, a summary of the results is reported in the Conclusions.

2. Materials and Methods

Data from the NOAA-15 polar satellite were used in this study. Particle counting rates (CRs) were produced by detectors on board the satellite that monitors protons and electrons flying in polar orbits at altitudes between 808 and 824 km at perigee and apogee, respectively, with an inclination of 98.7° and a period of 101.2′ [29]. The particle detectors, where the analysis was applied, were composed of eight solid-state telescopes called medium energy proton and electron detectors (MEPED). The latter was measuring proton and electron fluxes on 0°, 90°, and Omni-directions from 30 keV to 200 MeV [30]. MEPED data were provided every 2 s, whereas all the sets of orbital parameters were provided every 8 s; consequently, 8 s averages of the CRs were calculated, discarding unreliable negative values, and correcting for proton contamination [31]. Only the zenith telescope, 0°, was used to count electrons in this, as both the 90° and omnidirectional telescopes were investigated in the past, reporting no positive correlations [26–28]. The detected electron energies were within three ranges of 30–2700 keV, 100–2700 keV, and 300–2700 keV, so three differences in CRs were calculated to create new sets of data from 30 keV to 100 keV, 100 keV to 300 keV, and 300 keV to 2700 keV. In fact, different behaviors in particle dynamics are determined by energy, and, thus, the new sets resulted in a simpler analysis. When the mirror point of such electrons goes under 100 km, particles are sure to be stopped in the residual atmosphere. This occurs when electrons enter the loss cone, which can be determined by minimizing electron mirror point altitudes through the UNILIB libraries [32]; so, the minimum mirror altitude was also added to the data.

The International Geomagnetic Reference Field model (IGRF-13) was used to precisely determine the B-field and the L-shell at the satellite orbit. The dynamics of electrons were described using adiabatic invariants as they are more stable. Therefore, the analysis parameters were the geomagnetic field at mirror points $Bm = B/\cos^2\alpha$, where α is the pitch angle that is the difference between the electron velocity and geomagnetic field direction, and the L-shell. A 4-dimensional matrix (t; L; α; B) was filled with data relative to electron CRs where t is the time. Intervals were chosen to be of 8 s for t, 0.1 for L, 15° for α, and non-linear for B, with the shortest of these latter used where CRs were highest, to better describe the CR spatial variations and with the largest intervals used where the CRs were less frequent, so to obtain a greater number of samples [26–28]. The time period was from July 1998 to December 2014. The parameter L was restricted to a range between 1.0 and 2.2 to focus the analysis on the inner Van Allen Belts. Moreover, the α range was 180°, and the geomagnetic field range was between 16 μT and 47 μT.

The solar activity is the primary cause of precipitating electrons, it is also able to influence the inner Van Allen Belts, thus, days of moderate to high solar activity were excluded from the analysis. This was performed by neglecting data when the daily Ap index overcame a threshold. The threshold was set according to the seasons and years due to the solar cycle using the relation Ap = 11.1 + 0.8 sin [0.37(year − 1996)] + {2.1 − 0.1 sin [0.37(year − 1996)]} cos [0.0172(day − 27)]. The phase shift defined by the above relation corresponds to the minimum of the Sun's activity in 1996, whereas the 27-day modulation was dictated by the Sun's rotation. Furthermore, CRs were not included in the analysis whenever the Dst index resulted in being lower than −27 nT. This was done so as not to be influenced by any substorm activity.

Being that the CR distributions were Poissonian in every interval of (t; L; α; B), an electron CR fluctuation was considered a statistical fluctuation with a probability of less than 1% if the value of the Poisson distribution(CR fluctuation) was less than 0.01. CRs that satisfy this relation were defined as EBs. When the satellite runs a semi-orbit, all detected EBs were labeled as only one EB. Including only EBs with L-shells in a restricted interval [27], a histogram of Correlation Events (CEs) between EQs and EBs was calculated. CE was defined by the time difference between EQs and EBs, $T_{EQ} - T_{EB}$, which permitted the filling of the histogram with a bin interval of 2 h. Whenever a CE peak appeared at a certain time interval for some kind of EQs, it was tested for its significance [26] thus determining whether it was a positive correlation. The conditional probability of an EQ event following the observation of an EB event was defined for binary events by [27,28]:

$$P(EQ \mid EB) = CE/N_{EB}. \tag{1}$$

The cross-correlation Pearson coefficient for binary events called the Matthews correlation coefficient [33] can be obtained by [27,28]:

$$P(EQ \mid EB) = P(EQ) + corr(EQ,EB)\{P(EQ)[1 - P(EQ)][1 - P(EB)]/P(EB)\}^{1/2}, \tag{2}$$

and

$$corr(EQ,EB) = [CE/N_h - P(EQ)P(EB)]\{P(EQ)[1 - P(EQ)]P(EB)[1 - P(EB)]\}^{-1/2}, \tag{3}$$

where $P(EQ) = N_{EQ}/N_h$, $P(EB) = N_{EB}/N_h$, N_{EQ} was the number of considered EQs, N_{EB} was the number of considered EBs, and N_h was the number of time intervals. The corresponding ratio $P(EQ \mid EB)/P(EQ) = G$ is the probability gain of the correlation peak.

3. Results

The extraction of EBs from the NOAA-15 database of CRs starting from July 1998 to December 2014 was performed using the same procedure as in previous analyses [26–28]. Being so, the starting set of EBs to correlate to EQs remained unvaried and is reported in Figure 2 of a past work [28]. These EBs were all in the loss cone with bouncing altitudes lower than 200 km, and for 95% of cases lower than 100 km, in correspondence to the South Atlantic Anomaly (SAA). Figure 1 reports the geographical distribution of the NOAA-15 trajectories in one day, where the electron flux distribution along the trajectory is represented in colors. The black contour in the center represents the SAA. As reported in previous analysies, see, for example, Figure 1 of [26], precipitating electrons were observed far from the SAA to the west up until 170° in longitude, indicated by the cyan contour of Figure 1. The space-time distribution of the detected EBs consists of a series of consecutive events corresponding to time intervals lasting up to a few minutes. Although, one or more non-consecutive detections characterized the space-time distribution of EBs along the single semi-orbit. Moreover, the observation of one or more EBs in a semi-orbit was defined as only an EB event. Being that the EB L-shells were limited at L < 1.4 by the requirements of precipitating particles from the inner Van Allen Belts, the time length of every EBs was always found less than 12 min. For what concerns the EB time, was defined as the average time among the times of selected EBs in the semi-orbit.

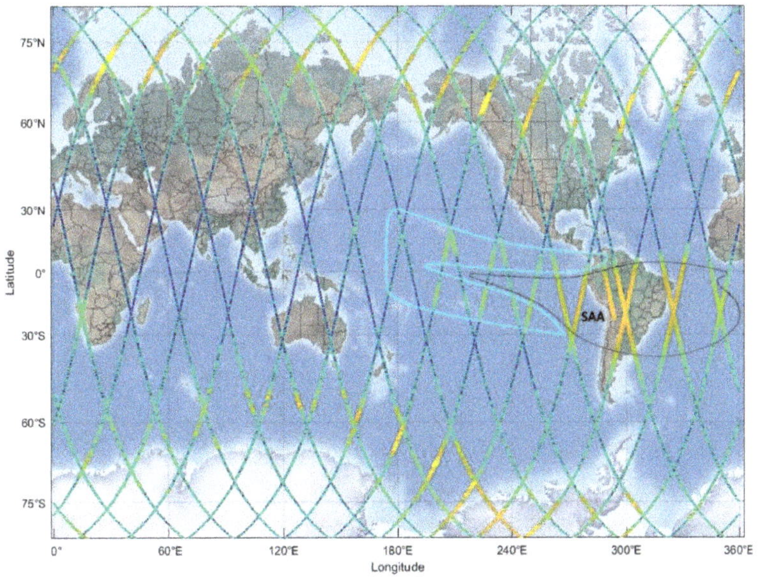

Figure 1. The set of NOAA-15 semi-orbits on 4 January 2001, with the CRs along the line orbits evidenced by colors; CR = 0 in blue, CR = 10^0 in light blue, CR = 10^1 in green, CR = 10^2 in orange, and CR = 10^3 in yellow; the SAA region is evidenced using a black contour, while the region where the satellite detect electrons in the loss cone, is evidenced by a cyan contour.

Unlike the previous study [26], which investigated strong EQs worldwide, the subsequent work [28] reported on correlations obtained with EQs restricted to the Philippines and Indonesian areas, in order to maximize the ratio between the correlation amplitude and the non-correlation amplitude. Afterward, numerous geographical areas were tested for their suitability, with the aim of maximizing this ratio, so a stronger correlation was obtained for West Pacific EQs [27]. During our study, the extension of the EQ area was pushed to the Central Pacific until it reached the East Pacific, therein another correlation peak emerged. Given this, a new analysis started for EQs in East Pacific.

A set of 6 correlation plots between 30–100 eV EBs and East Pacific M $\geq$ 6 EQs are shown in Figure 2 where the bin time interval was chosen between 2 and 6 h. A range of $\pm$3 days was considered for the time difference between EQs and EBs, where $T_{EQ} - T_{EB}$ positive highlights that EBs precede EQs, while EQs precede EBs for negative time differences. CE distributions were Poissonian as in past cases, and the average values are indicated using black dashed lines, while the red dashed lines are used for standard deviations. Significant correlation peaks appear at time differences around 57 h, which means that the EB observation anticipated the corresponding EQ. Unlike past works [26,28], where the correlation significance was represented by EQ projections at certain altitudes, here the correlation significance was calculated exclusively with respect to the EB features. In analogy with the case of West Pacific EQs [27], the L-shell interval of electrons to maximize the correlations between EBs and East Pacific EQs was obtained at $1.1 \leq L \leq 1.3$. A plot of this EB L-shell parameter compared to latitudes is shown in Figure 3. The plot reports two distinct distributions, a correlation for negative latitudes and a lack of correlation for positive ones. For what concerns electron pitch angles, they were concentrated in intervals around 67° and 117°, with $56° \leq \alpha \leq 72°$ and $108° \leq \alpha \leq 126°$. The correlations in Figure 2 were obtained for EQ depths lower than 200 km, supporting the hypothesis that only EQs close to the surface seem to be correlated to ionospheric activity.

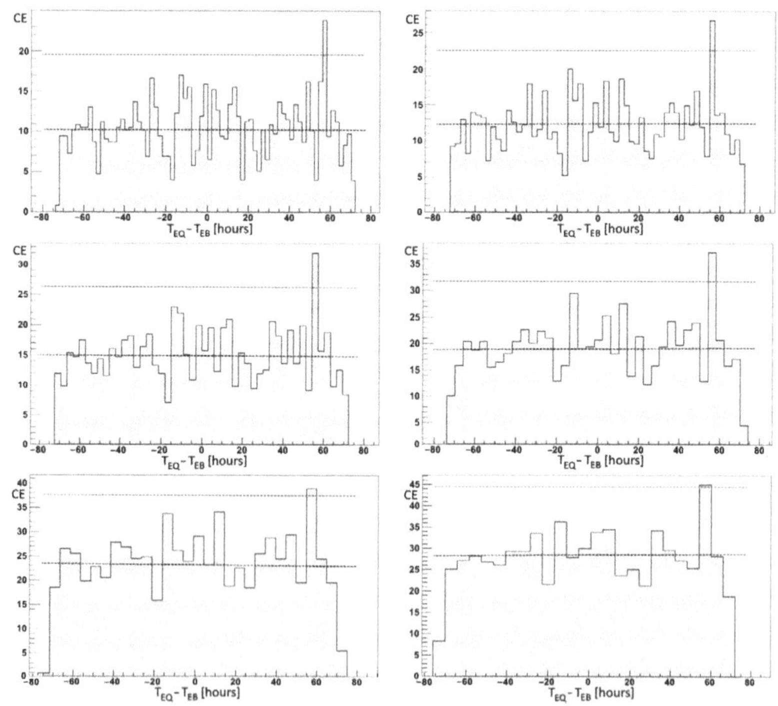

Figure 2. Correlations were obtained by filling histograms of the time difference $T_{EQ} - T_{EB}$ which is positive when EBs anticipated EQs. They are plotted for different bins of 2, 2.5, 3, 4, 5, and 6 h from top left to right; the centers of correlation peaks are 57.2, 56.05, 55.8, 56.3, 57.3, and 57.2 h, respectively; averages are evidenced by the black horizontal dotted lines, while the red ones indicated the 3 σ levels; note that border bins are not completely populated as they are partially out the range of $\pm$3 days.

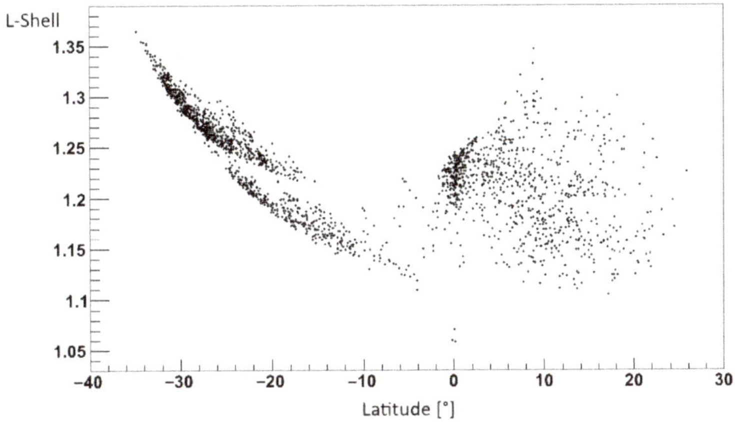

Figure 3. The L-shell parameter distribution is compared to the EB latitudes; the distribution pattern is different from those of the EB correlation with West Pacific EQs as in this case EB longitudes correspond to the SAA.

The significance of the correlation peaks was calculated for all the cases shown in Figure 2. The plot shown in Figure 4 summarizes the results regarding the number of standard deviations that define the correlation peak significance above the average values concerning the bin duration and the relative average correlated events. A statistical increase in the correlation peak significances was observed, which started to exceed 3 standard deviations for a bin correlation less than or equal to 6 h and reached a maximum of over 4 for a bin correlation of 4 h. The correlation peak significance was less than 3 for bins greater than 6 h. Concerning the number of correlated events, it decreases for short correlation bins. In Figure 4, a continuous red line indicates the number of standard deviations of 57 h correlation peaks observed with different bins, a continuous black line indicates the number of events at the peaks, whereas the dashed line indicates the averages. All the correlation peak significances overcame the threshold of 99%. As in the past publications, both correlation calculus using a randomized space and time distributions of EQs were also calculated by using the same EQ times, and the same EQ epicenters, respectively. In these randomized cases, the previously obtained correlation peaks completely disappeared.

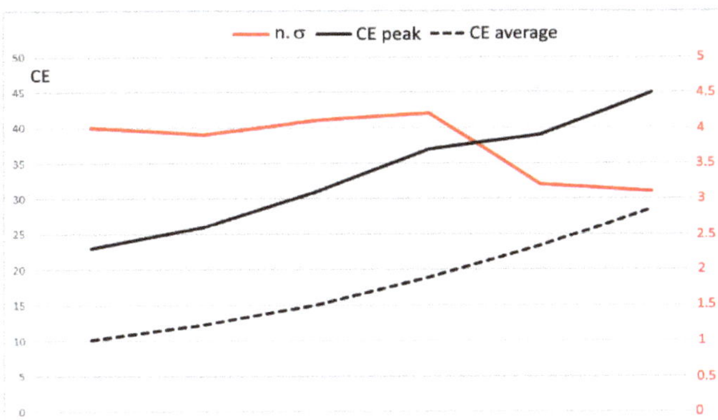

Figure 4. The correlation significance in red is compared with the correlation peaks in black and the corresponding correlation averages for the bin intervals reported in Figure 2.

The maximum number of 45 correlation events was found for the greatest time bin of 6 h, which corresponded to 45 EQs identified in the map in Figure 5. The geographical region where EQs correlated with NOAA-15 EBs is delimited by $-40°$ to $30°$ in latitude and by $245°$ to $300°$ in longitude, the yellow line in Figure 5, so included are regions with strong seismic activity, such as Mexico, Caribbean Sea, Guatemala, Honduras, Nicaragua, Costa Rica, Panama, Columbia, Ecuador, Perú, Bolivia, Chile, and a large part of South-eastern Pacific. EQ epicenters are indicated by red dots in Figure 5, about half of which are located offshore. Note that the total number of mainshocks that occurred during 16.5 years in Figure 5 yellow square was 199, they were about 1/3 of the mainshocks that stroked West Pacific in the same period. However, the 45 East Pacific mainshocks of the peak correlation are near the 44 EQs of the West Pacific correlation peak, being the East Pacific correlation bin 3 times the West Pacific correlation bin. The time distribution of the 45 considered EQs from July 1998 to December 2014 is shown in Figure 6, with their relative magnitudes. Concerning EB detection positions, the geographical region is delimited at $-35°$ and $20°$ in latitude, and at $205°$ and $295°$ in longitude, divided into two inclined belts indicated by cyan contours in Figure 5 whose inclinations are due to the asymmetry of the geomagnetic field. Finally, electron mirror points of detected EBs over the EQ epicenters are plotted in Figure 7, ranging from minimum altitudes between 100 km and 700 km.

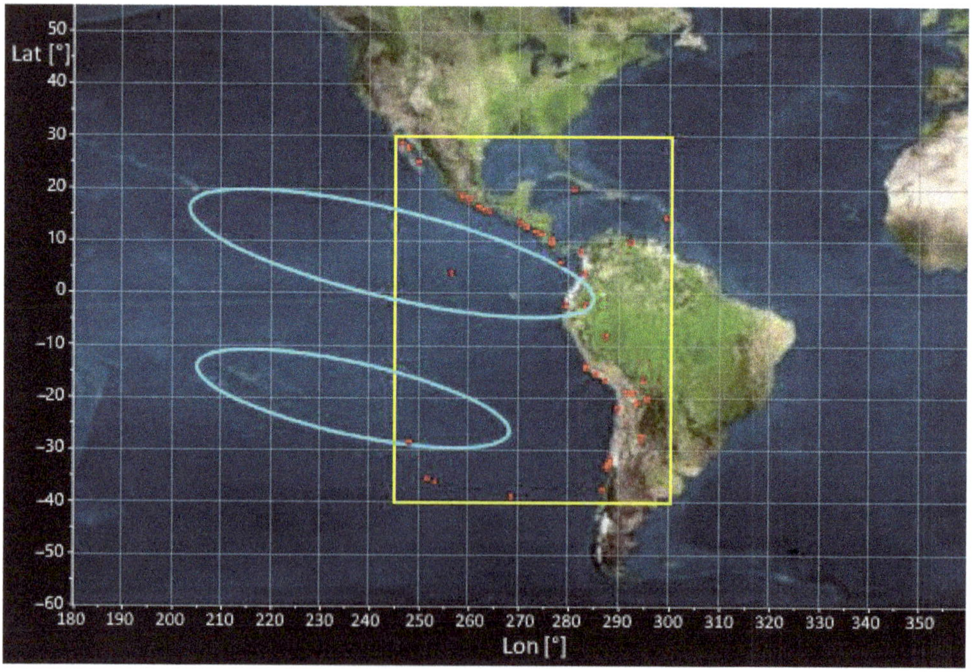

Figure 5. The very large area of EQ epicenters that contributed to the correlations in Figure 2 is delimited by yellow lines, EQ positions are reported by red dots, and cyan contours delimitate the region where the NOAA satellite detected EBs correlated with EQs.

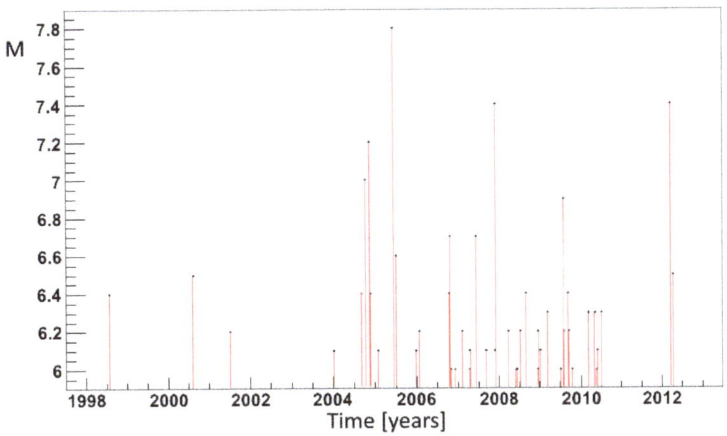

Figure 6. The EQ magnitudes reported with respect to EQ times contributed to the correlations in Figure 2.

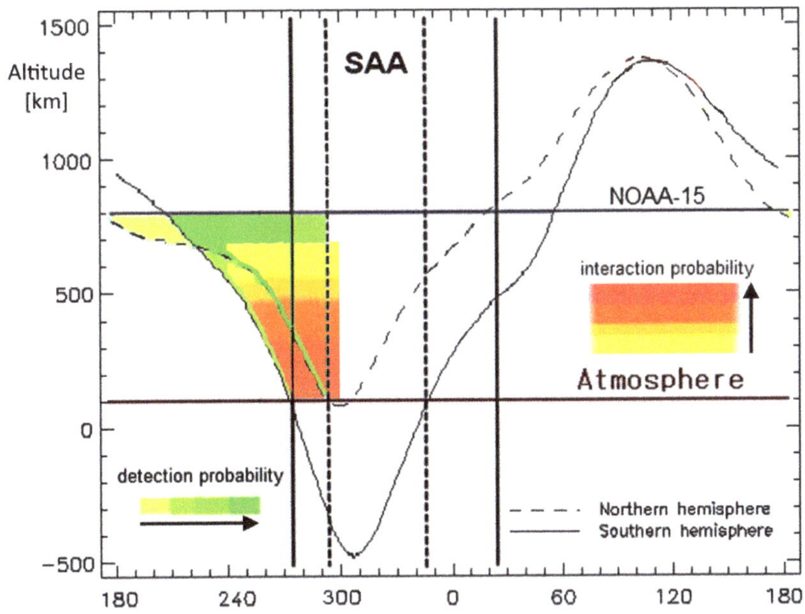

Figure 7. The electron mirror point altitudes at L = 1.2 are indicated by continuous and dashed lines, which are compared to continuous horizontal lines indicating NOAA-15 and atmosphere altitudes; green-colored areas indicate longitude regions where electrons are detectable, and the yellow to red areas indicate the probable interaction areas between EQs and EBs; arrows next to shades indicate increasing probabilities; the SAA region excluded by the analysis is delimited by vertical lines, which discriminate between the southern and the northern hemisphere using continuous and dashed lines, respectively.

4. Discussion

A comparison of the possible results obtained above for East Pacific EQs with those obtained for West Pacific EQs [27] is useful to describe the similarities and differences between them. The first striking difference was found in the correlation time difference which was a few hours for West Pacific EQs and a few days for East Pacific EQs, even if EBs anticipated EQs. The first striking similarity was found between the maximum CE number. Concerning the CE significance, calculated using the σ number, the results were slightly worse being the maximum significance of a little more than 4, observed for bin intervals of 4 h, compared to a little more than 5 for bin intervals of 2 h of the past. The CE number relative to the maximum significance was lower for the East Pacific case. The geographical positions of EBs were nearly entirely overlapping, with a slight shift to the East of around 10°, whereas the relative positions concerning the correlated EQ epicenters were completely different and non-overlapping. Altitudes of electrons calculated over the new EQ epicenters were also completely different, resulting in many hundreds of kilometers lower compared to past studies. The EQ area in the West Pacific belonged more to the Northern hemisphere than to the area of the East Pacific EQs that belonged more to the Southern hemisphere. This seems to reflect the geomagnetic field asymmetry, following the geomagnetic equator crossing the geographic equator from north to south and moving eastwards to those longitudes. The EQs are instead quite similar for the two regions, being in both cases superficial and equally distributed between land and oceanic crusts.

Concerning the adiabatic parameters of EBs, the L-shell range in this study resulted in being more extended towards lower values, and the latitude dependence was lost for positive latitudes. Moreover, the positive latitude dependence of L-shell was observed for EBs over West Pacific EQ epicenters around 125°, see Figure 2 of [27]. However, EB longitudes over East Pacific EQs are close to the SAA, where the geomagnetic field deviates from the dipolar shape. Specifically, the angle between the satellite's orbit and the geomagnetic line is small for positive latitudes and close to 90° for negative latitudes. Therefore, the satellite crosses the $1.1 \leq L \leq 1.3$ in more than 20° of positive latitude, whereas it crossed the same interval in 5° of negative latitudes. In fact, the pitch angle of positive latitude EBs is restricted to $56° \leq \alpha \leq 72°$ while the pitch angle of negative latitude EBs is restricted to $108° \leq \alpha \leq 126°$. Furthermore, two different distribution bands were observed, see Figure 3, for negative latitudes. They were separated approximately by $\alpha = 120°$, with $1.1 \leq L \leq 1.22$ in the interval $108° \leq \alpha \leq 120°$ and with $1.22 \leq L \leq 1.3$ in the interval $120° \leq \alpha \leq 126°$. The pitch angle results are divided into two intervals for both studies, coinciding exactly. Being so, the striking difference in the correlation times between the two studies seems to be more naturally associated with the major difference retrieved from the correspondent analyzed data of the two cases: which is the altitude of electrons passing above the East and West Pacific EQs. If the Lithosphere-Ionosphere interaction occurs above the future epicenters and the magnitude depends on the distance of electrons from the lithosphere; as previously proposed for magnetic pulses [6], a lower altitude of charged particles above the East Pacific means that lower magnitude magnetic pulses could be able to modify electron trajectories, even if the general behavior of magnetic pulses have not been proven decreased with the temporal remoteness of strong EQs. Case studies based upon data from the DEMETER satellite of East Pacific EQs, such as the Chile EQ M = 8.8 on 27 February 2010, and the Haiti EQ M = 7.0 on 12 January 2010, recorded similar delays with electron detections [34], also associated with VLF wave activity.

Preparedness for strong EQs and tsunamis has been developed over recent decades [35] either by triggering early warnings or by rapidly assessing the expected damages [36]. Early warnings consist of alerts seconds before the arrival of destructive seismic waves in populated regions. Such an alert may be useful in controlling the shutdown of gas pipelines and critical facilities, reducing speeds of rapid-transit vehicles, and also advising the interested populations to follow the necessary precautions. Reduced lead times limit the possible preparedness actions, thus leaving part of the population at the mercy of imminent danger. Utilizing the possible correlation results obtained above for a short-time warning in strong EQs and tsunami preparedness might contribute to reducing the impact of EQs.

A way of using such results is to define a scenario for a short-time prediction model [27]. Following the work of Console [37], the representation of the EQ prediction model would need to define the target volume where EQs occur V_T, which is a 2-d geographic space + 1-d time-space. This volume is displayed by a black contour in Figure 8, and with EQs occurrence by red stars. If an EQ occurs in the alarm volume V_A in yellow, it is a success (S), while if an EQ occurs out of the yellow volume it is a failure of prediction (F). The precursor volume V_P, which is generally different from V_T, contains the alarm events. In this case, V_P is the cyan volume of Figure 8 where EBs are detected, the time of EB observations lasting less than 12 min. When an EB is detected in V_P it is an alarm that defines a V_A. The time span of the EQ observations multiplied by the area of East Pacific where the correlations in Figure 2 are calculated is V_T. An EQ event is included in this scenario only if $M \geq 6$. Unlike West Pacific earthquakes where geographical areas of V_P and V_A were completely disjoined, areas of EQs and EBs are largely overlapping in this scenario. In this representation, V_A has the dimension T which coincides with the correlation bin dimension, it is generated by an EB detection in V_P, and covers the same area as V_T. To have the greatest significance the 4 h bin interval of 54.3–58.3 h should be chosen. The V_A dimension and shape are the same for each detected EBs in V_P; if an EQ with $M \geq 6$ occurs in V_A one success is collected. Similarly, if an EQ occurs out of the V_A, meaning that it happened before or after the 4 h between the 56.3 h following the EB, one failure is collected. Moreover, one false alarm is collected for any EB detection not followed by any EQ. Non-continuous V_T is defined as when the Sun's activity leads to discarding days or due to the NOAA-15 satellite intermittence into the detection area west of the SAA. A V_A is generated within V_T 56.3 h after an EB is detected in V_P, and the time ahead is the vertical distance in the representation of Figure 8.

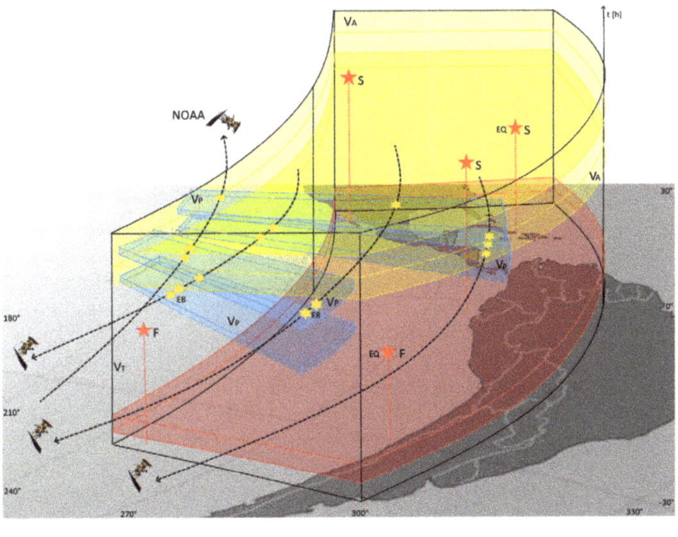

Figure 8. The scenario is the volume representation where a forecasting model can be tested, here it is delimited by the target volume V_T, the alert volume V_A, and the precursor volume V_P, the latter two being products among the geographical coordinates of EQs and EBs and the time of observations. Discontinuities in time are due to the solar activity intermittence. The alarm duration can be chosen to be 4 h in this work which corresponds to the greatest significance. V_T and V_A cover the entire East Pacific area, while V_P is restricted to two latitudinal areas being more extended on the west. In red is the possible precursor volume which hypothesizes a physical action on the ionosphere of the future epicenter lithosphere.

The V_T can be obtained by multiplying the East Pacific area by the total number of hours [27], this total number corresponds to N_h = 36,135 for 4 h intervals. The following was observed in V_T: a total number of N_{EQ} = 199 EQs with M $\geq$ 6 and depths $\leq$ 200 km, resulting in a P(EQ) = 0.0055, a total number of N_{EB} = 3371 EBs which are alarms, and a total number of N_S = 37 CEs which are successes. The success rate was calculated by N_S/N_{EB} = 0.011, which is the conditional probability (1) [27], $1 - (N_S/N_{EB}) = 0.989$ is the false alarm rate, while the alarm rate is N_S/N_{EQ} = 0.186, and the failure rate is $1 - (N_S/N_{EQ}) = 0.814$. As there were more alarms than those for the West Pacific EQs, with the success being near the same for the two cases, the number of false alarms in this study increased compared to previous research. A cross-correlation of 0.024 was calculated using the relation (3). The EQ occurrence probability of at least one target event was estimated to be P(EQ) when no EB was observed two and a half days before, while it increased to P(EQ | EB) given by (2) two and a half days after the EB observation. The probability distribution is shown in Figure 9, which indicates a probability gain near G = 2. As for the study in the West Pacific [27], days with more than one burst can be found with a frequency of about 20%, and EBs belonging to successive orbits generated a partial overlapping between two consecutive alarms. These are not considered here and will be presented in a future publication.

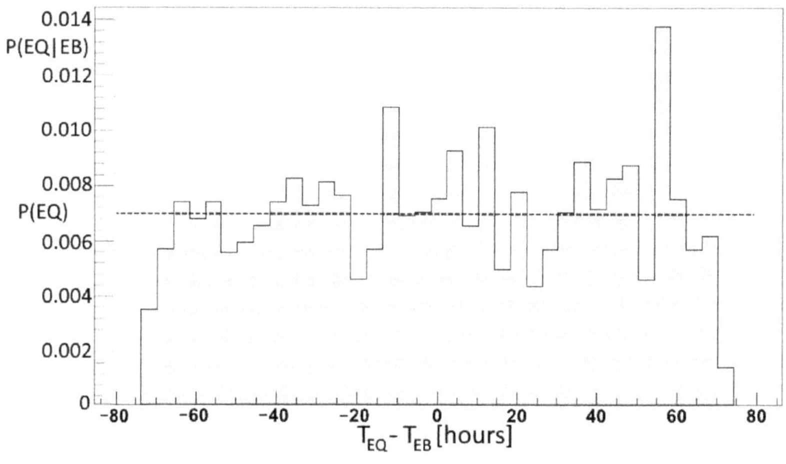

Figure 9. The conditional probability distribution P(EQ | EB) which corresponds to the correlation bin of 4 h in Figure 2, the average probability indicated by a horizontal dashed line coincides with the unconditional probability P(EQ).

An early warning system for evacuation should be based on effective EQ observations [37]. However, the geohazard risk reduction can gain valuable preparation time by adopting a probabilistic short-term warning a few hours prior, especially for tsunamis [36]. Thus, using NOAA-15 electron CR analysis is in principle achievable based on the infrastructure of antennas operating along the West Coast of the US [28], at the same longitude where EBs are detected. The system would need to be able to download NOAA data early enough so it can be analyzed in a few minutes and then compared with geomagnetic activity within a day. As EBs were detected in the same region where they anticipated West Pacific EQs [27], the probability of a strong EQ in the West Pacific cannot be neglected, which would vanish in less than 4 h. Then, if no strong EQs occur in the West Pacific, based on completely statistical evaluations of disastrous events presented in this work, an East Pacific coast forecasting can be generated using G increases in EQ-probability. Moreover, correlation bins of time lengths greater than 6 h were investigated, they concerned the cases of 7, 8, 9,10, 12, 14, and 17 h. A decrease in significance was generally observed for the corresponding correlations, where the sigma decreased up to 1.1 for the 17 h case.

Furthermore, the probability gain was calculated for each bin and reported in the plot of Figure 10 for each starting/ending time of the correlation peak. The case of 1.5 h was also added to this plot. A fit of the G distribution is reported in Figure 10, which can be interpreted as the time-dependent conditional probability of a strong EQ in the East Pacific area after an EB has been observed at time 0. Figure 10 shows that an increase in conditional probability is observed at about 40 h from the NOAA detection, around 54 h the probability magnitude overcomes the 99% significance level up to around 59 h. Finally, the probability decreases at an unconditional level around 70 h after the EB detection.

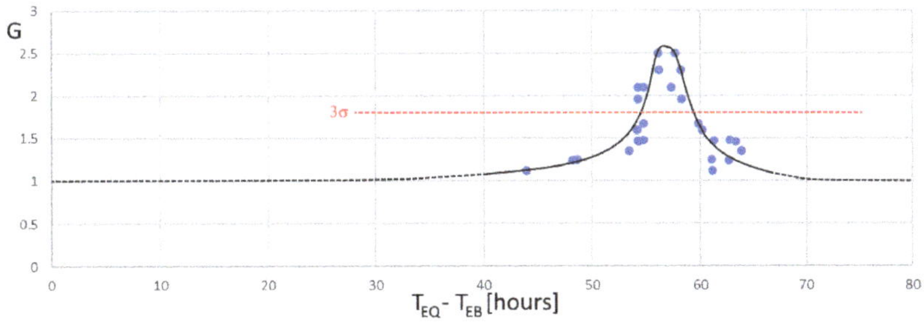

Figure 10. The probability gain approaching the correlation peak around 57 h; it was obtained by fitting the probability gain distribution from bin intervals of 1.5, 2, 2.5, 3, 4, 5, 6, 7, 8, 9, 10, 12, 14, and 17 h; the 3 sigmas limit is indicated by the red horizontal dashed.

5. Conclusions

A new statistical correlation analysis between precipitating EBs and strong EQs was carried out from the analysis of exactly 16.5 years of NOAA-15 particle data. This seemed to indicate that electrons in the loss cone were mainly observed around 57 h before main shocks with $M \geq 6$ in the East Pacific, comprising countries where seismic activity is frequently a danger. Results are in line with previous single-case observations by the DEMETER satellite. The new correlation time is significantly longer than the correlation time found for strong West Pacific EQs, with EBs occurring a few hours prior. It again supports the hypothesis that there might exist a link between ionospheric and lithospheric activities of shallow EQs whose depths are less than 200 km. As for West Pacific EQs, also this correlation happens regardless of whether the EQs occurred in the sea or on the mainland.

The parameter L-shell of the electrons was defined uniquely, as in the previous publication on West Pacific EQs [27], therein providing a precise definition of an EB to also carry out an EQ forecast in the East Pacific. The L-shell distribution in this new case was distributed differently in terms of latitude, compared to the West Pacific case, which probably reflects the asymmetry of the geomagnetic field. A probability gain G = 2 was calculated which could be used to increase the pre-alarm time of some early warnings for both strong EQs and tsunamis on the East Pacific's coasts. Despite the high number of false alarms being greater than in the West Pacific case, a time-dependent conditional probability interpretation of the process was proposed.

Funding: This research received no external funding.

Institutional Review Board Statement: Not applicable.

Informed Consent Statement: Not applicable.

Data Availability Statement: The NOAA-15 electron CRs from 1998, first accessed on 22 August 2022: http://www.ngdc.noaa.gov/stp/satellite/poes/dataaccess.html. Corrections by proton contaminations were carried out using software downloaded from the Virtual Radiation Belt Observatory, first accessed on 22 August 2022: http://virbo.org/POES#Processing. The geomagnetic field was re-evaluated together with L-shells on the NOAA-15 orbit using the International Geomagnetic Reference Field (IGRF-13) model, first accessed on 22 August 2022, and downloaded at: http://www.ngdc.noaa.gov/IAGA/vmod/igrf.html. Geomagnetic Ap indexes and Dst variations, first accessed on 22 August 2022, were downloaded at the links https://www.ngdc.noaa.gov/geomag/data.shtml and http://wdc.kugi.kyoto-u.ac.jp/dst_final/index.html, respectively. The UNILIB libraries to calculate mirror altitudes, first accessed on 22 August 2022, were downloaded at https://www.mag-unilib.eu. Finally, EQ events were first accessed on 22 August 2022, and downloaded at https://earthquake.usgs.gov/earthquakes/search/, and the database was declustered using CLUSTER 2000 software, first accessed on 22 August 2022, and downloaded at https://www.usgs.gov/media/images/cluster2000.

Acknowledgments: I would like to express my thanks to Craig J. Rodger and Janet Green from NOAA for their useful codes to subtract the proton contamination of electron channels. Additionally, I would like to express my gratitude to M. Kruglanski for the library to calculate bouncing altitudes. I would like to thank the project "Limadou Science +", and an anonymous reviewer for his improving suggestion.

Conflicts of Interest: The authors declare no conflict of interest.

References

1. Barnhart, W.D.; Hayes, G.P.; Wald, D.J. Global Earthquake Response with Imaging Geodesy: Recent Examples from the USGS NEIC. *Remote Sens.* **2019**, *11*, 1357. [CrossRef]
2. Rikitake, T. *Predictions and Precursors of Major Earthquakes*; Terra Scientific Publishing Company: Tokyo, Japan, 2003.
3. Chiodini, G.; Cardellini, C.; Di Luccio, F.; Selva, J.; Frondini, F.; Caliro, S.; Rosiello, A.; Beddini, G.; Ventura, G. Correlation between tectonic CO_2 Earth degassing and seismicity is revealed by a 10-year record in the Apennines, Italy. *Sci. Adv.* **2020**, *6*, eabc2938. [CrossRef] [PubMed]
4. Varotsos, P.A.; Sarlis, N.V.; Skordas, E.S.; Lazaridou, M.S. Electric pulses some minutes before earthquake occurrences. *Appl. Phys. Lett.* **2007**, *90*, 064104. [CrossRef]
5. Piscini, A.; De Santis, A.; Marchetti, D.; Cianchini, G. A Multi-parametric Climatological Approach to Study the 2016 Amatrice–Norcia (Central Italy) Earthquake Preparatory Phase. *Pure Appl. Geophys.* **2017**, *174*, 3673–3688. [CrossRef]
6. Fidani, C.; Orsini, M.; Iezzi, G.; Vicentini, N.; Stoppa, F. Electric and Magnetic Recordings by Chieti CIEN Station during the Intense 2016–2017 Seismic Swarms in Central Italy. *Front. Earth Sci.* **2020**, *8*, 536332. [CrossRef]
7. Parrot, M. The micro-satellite DEMETER. *J. Geodyn.* **2002**, *33*, 535–541. [CrossRef]
8. Scholz, C. *The Mechanics of Earthquakes and Faulting*, 3rd ed.; Cambridge University Press: Cambridge, UK, 2019.
9. Pulinets, S. Space technologies for short-term earthquake warning. *Adv. Space Res.* **2005**, *37*, 643–652. [CrossRef]
10. Parrot, M. Use of satellites to detect seismo-electromagnetic effects, Main phenomenological features of ionospheric precursors of strong earthquakes. *Adv. Space Res.* **1995**, *15*, 1337–1347. [CrossRef]
11. Pulinets, S.; Boyarchuk, K.A. *Ionospheric Precursors of Earthquakes*; Springer: Berlin/Heidelberg, Germany, 2004.
12. Parrot, M. Statistical analysis of the ion density measured by the satellite DEMETER in relation with the seismic activity. *Earthq. Sci.* **2011**, *24*, 513–521. [CrossRef]
13. Merayo, J.M.; Jørgensen, J.L.; Friis-Christensen, E.; Brauer, P.; Primdahl, F.; Jørgensen, P.S.; Allin, T.H.; Denver, T. The Swarm Magnetometry Package. In *Small Satellites for Earth Observation*; Sandau, R., Röser, H.P., Valenzuela, A., Eds.; Springer: Dordrecht, The Netherlands, 2008.
14. Ouzounov, D.; Pulinets, S.; Hattori, K.; Taylor, P. *Pre-Earthquake Processes: A Multidisciplinary Approach to Earthquake Prediction Studies, AGU Geophysical Monograph*; Wiley: Hoboken, NJ, USA, 2018; p. 365. [CrossRef]
15. Bortnik, J.; Cutler, J.W.; Dunson, C.; Bleier, T.E. The possible statistical relation of Pc1 pulsations to Earthquake occurrence at low latitudes. *Ann. Geophys.* **2008**, *26*, 2825–2836. [CrossRef]
16. Han, P.; Hattori, K.; Hirokawa, M.; Zhuang, J.; Chen, C.-H.; Febriani, F.; Yamaguchi, H.; Yoshino, C.; Liu, J.-Y.; Yoshida, S. Statistical analysis of ULF seismomagnetic phenomena at Kakioka, Japan, during 2001–2010. *J. Geophys. Res. Space Phys.* **2014**, *119*, 4998–5011. [CrossRef]
17. Righetti, F.; Biagi, P.F.; Maggipinto, T.; Schiavulli, L.; Ligonzo, T.; Ermini, A.; Adeline, I.; Moldovan, A.; Buyuksarac, A.; Gonzales, H.; et al. Wavelet analysis of the LF radio signals collected by the European VLF/LF network from July 2009 to April 2011. *Ann. Geophys.* **2012**, *55*, 171–180.
18. Maggipinto, T.; Biagi, P.F.; Colella, R.; Schiavulli, L.; Ligonzo, T.; Ermini, A.; Martinelli, G.; Moldovan, I.; Silva, H.; Contadakis, M.; et al. The LF radio anomaly observed before the Mw = 6.5 earthquake in Crete on October 12, 2013. *Phys. Chem. Earth Parts A/B/C* **2015**, *85–86*, 98–105. [CrossRef]
19. Hayakawa, M.; Kasahara, Y.; Nakamura, T.; Muto, F.; Horie, T.; Maekawa, S.; Hobara, Y.; Rozhnoi, A.; Solovieva, M.; Molchanov, O.A. A statistical study on the correlation between lower ionospheric perturbations as seen by subionospheric VLF/LF propagation and earthquakes. *J. Geophys. Res. Earth Surf.* **2010**, *115*, A09305. [CrossRef]

20. Marchetti, D.; De Santis, A.; Campuzano, S.A.; Zhu, K.; Soldani, M.; D'Arcangelo, S.; Orlando, M.; Wang, T.; Cianchini, G.; Di Mauro, D.; et al. Worldwide Statistical Correlation of Eight Years of Swarm Satellite Data with M5.5+ Earthquakes: New Hints about the Preseismic Phenomena from Space. *Remote Sens.* **2022**, *14*, 2649. [CrossRef]
21. Němec, F.; Santolík, O.; Parrot, M.; Berthelier, J.J. Spacecraft observations of electromagnetic perturbations connected with seismic activity. *Geophys. Res. Lett.* **2008**, *35*, L05109. [CrossRef]
22. Němec, F.; Santolik, O.; Parrot, M. Decrease of intensity of ELF/VLF waves observed in the upper ionosphere close to earthquakes: A statistical study. *J. Geophys. Res. Earth Surf.* **2009**, *114*, A4. [CrossRef]
23. Liu, J.; Chen, C.; Chen, Y.; Yang, W.; Oyama, K.; Kuo, K. A statistical study of ionospheric earthquake precursors monitored by using equatorial ionization anomaly of GPS TEC in Taiwan during 2001–2007. *J. Southeast Asian Earth Sci.* **2010**, *39*, 76–80. [CrossRef]
24. Ke, F.; Wang, Y.; Wang, X.; Qian, H.; Shi, C. Statistical analysis of seismo-ionospheric anomalies related to Ms > 5.0 earthquakes in China by GPS TEC. *J. Seism.* **2016**, *20*, 137–149. [CrossRef]
25. Yan, R.; Parrot, M.; Pinçon, J.-L. Statistical Study on Variations of the Ionospheric Ion Density Observed by DEMETER and Related to Seismic Activities. *J. Geophys. Res. Space Phys.* **2017**, *122*, 12421–12429. [CrossRef]
26. Fidani, C. Particle precipitation prior to large earthquakes of both the Sumatra and Philippine Regions: A statistical analysis. *J. Southeast Asian Earth Sci.* **2015**, *114*, 384–392. [CrossRef]
27. Fidani, C. West Pacific Earthquake Forecasting Using NOAA Electron Bursts with Independent L-Shells and Ground-Based Magnetic Correlations. *Front. Earth Sci.* **2021**, *9*, 673105. [CrossRef]
28. Fidani, C. Probability, Causality and False Alarms using Correlations between Strong Earthquakes and NOAA High Energy Electron Bursts. *Ann. Geophys.* **2020**, *62*, 543. [CrossRef]
29. Davis, G. History of the NOAA satellite program. *J. Appl. Remote Sens.* **2007**, *1*, 012504. [CrossRef]
30. Evans, D.S.; Greer, M.S. Polar Orbiting Environmental Satellite Space Environment Monitor—2: Instrument Descriptions and Archive Data Documentation. In *NOAA Technical Memorandum January*; Space Environment Center: Boulder, CO, USA, 2000; Volume 155.
31. Rodger, C.J.; Clilverd, M.A.; Green, J.C.; Lam, M.M. Use of POES SEM-2 observations to examine radiation belt dynamics and energetic electron precipitation into the atmosphere. *J. Geophys. Res. Earth Surf.* **2010**, *115*, A04202. [CrossRef]
32. Krunglanski, M. *UNILIB Reference Manual*; Belgisch Instituut Voor Ruimte—Aeronomie: Brussel, Belgium, 2002.
33. Matthews, B.W. Comparison of the predicted and observed secondary structure of T4 phage lysozyme. *Biochim. Biophys. Acta Protein Struct.* **1975**, *405*, 442–451. [CrossRef]
34. Anagnostopoulos, G.; Vassiliadis, E.; Pulinets, S. Characteristics of flux-time profiles, temporal evolution, and spatial distribution of radiation-belt electron precipitation bursts in the upper ionosphere before great and giant earthquakes. *Ann. Geophys.* **2012**, *1*, 55. [CrossRef]
35. Tucker, B.E. Reducing earthquake risk. *Science* **2013**, *341*, 1070–1072. [CrossRef]
36. Zollo, A.; Iannaccone, G.; Lancieri, M.; Cantore, L.; Convertito, V.; Emolo, A.; Festa, G.; Gallovič, F.; Vassallo, M.; Martino, C.; et al. Earthquake early warning system in southern Italy: Methodologies and performance evaluation. *Geophys. Res. Lett.* **2009**, *36*, L00B07. [CrossRef]
37. Console, R. Testing earthquake forecast hypotheses. *Tectonophysics* **2001**, *338*, 261–268. [CrossRef]

applied sciences

MDPI

Article

Stochastic Generator of Earthquakes for Mainland France

Corentin Gouache [1,*], Pierre Tinard [2] and François Bonneau [1]

1 GeoRessources, ASGA, CNRS, Université de Lorraine, 54000 Nancy, France; francois.bonneau@univ-lorraine.fr
2 Caisse Centrale de Réassurance, R&D Modélisation, 75008 Paris, France; ptinard@ccr.fr
* Correspondence: corentin.gouache@gmail.com

Abstract: Mainland France is characterized by low-to-moderate seismic activity, yet it is known that major earthquakes could strike this territory (e.g., Liguria in 1887 or Basel in 1356). Assessing this French seismic hazard is thus necessary in order to support building codes and to lead prevention actions towards the population. The Probabilistic Seismic Hazard Assessment (PSHA) is the classical approach used to estimate the seismic hazard. One way to apply PSHA is to generate synthetic earthquakes by propagating information from past seismicity and building various seismic scenarios. In this paper, we present an implementation of a stochastic generator of earthquakes and discuss its relevance to mimic the seismicity of low-to-moderate seismic areas. The proposed stochastic generator produces independent events (main shocks) and their correlated seismicity (only aftershocks). Main shocks are simulated first in time and magnitude considering all available data in the area, and then localized in space with the use of a probability map and regionalization. Aftershocks are simulated around main shocks by considering both the seismic moment ratio and distribution of the aftershock's proportion. The generator is tested with mainland France data.

Keywords: generator of earthquakes; low-to-moderate seismicity; stochastic; France

Citation: Gouache, C.; Tinard, P.; Bonneau, F. Stochastic Generator of Earthquakes for Mainland France. *Appl. Sci.* **2022**, *12*, 571. https://doi.org/10.3390/app12020571

Academic Editors: Ricardo Castedo, Miguel Llorente Isidro, David Moncoulon and Arcady Dyskin

Received: 1 October 2021
Accepted: 28 December 2021
Published: 7 January 2022

1. Introduction

Mainland France seismicity is considered as low to moderate due to its high return periods and weak maximal magnitudes. Only a few earthquakes have produced damages in this territory since the introduction of the French insurance compensation system in 1982. The six major earthquakes occurring after 1982 resulted in less than EUR 600 M of insured losses, representing approximately 1.5% of the total amount compensated with this compensation scheme, https://catastrophes-naturelles.ccr.fr/ (accessed on 20 December 2021). Nevertheless, major earthquakes could strike this territory (e.g., Liguria in 1887, Basel in 1356) and create financial losses and casualties. For example, CCR (*Caisse Centrale de Réassurance*, French insurance company) and BRGM (*Bureau de Recherches Géologiques et Minières*, French geological survey) have quantified the probable insurance losses associated with historical earthquakes if they occurred in the present day:

- Arette (1967): EUR 120–140 M [1];
- Lambesc (1909, last earthquake-related deaths): EUR 1.8–2.5 Bn [2];
- Marais vendéen (1799): EUR 1.5–2.3 Bn [3];
- Liguria (1887) localized in front of Nice [4]: EUR 9–13 Bn, https://catastrophes-naturelles.ccr.fr/documents/148935/368920/Bilan+Cat+Nat+1982-2020.pdf/22925938-97e6-4ad4-df6a-f47aed6b3135?t=1623850155584 (accessed on 20 December 2021).

It shows that estimating the seismic hazard and risk is necessary even in low-to-moderate seismic areas, since losses can be significant.

The Probabilistic Seismic Hazard Assessment (PSHA) [5] is a group of methodologies that estimates seismic hazard in a probabilistic way using seismic, tectonic and geologic data, as well as their uncertainties. It leads to a measure of probabilities of exceedance for different hazard levels (e.g., Peak Ground Acceleration) in a given site. The spatio-temporal analysis of past seismicity recorded in an area leads to distributions of the earthquake's occurrence and location that are the cornerstone of the PSHA. Traditionally, these distributions can be used in an integrative way [5] or as an input to constrain a stochastic generator of earthquakes [6]. The main advantage of the latter is to explicitly consider some synthetic seismic events that allow for the computation of the contributions of these earthquakes to the hazard evaluated at a given site. This aspect is useful in an insurance context and also makes it possible to link earthquakes to others hazards (e.g., [7] for tsunamis, [8] for liquefaction, [9] for landslides and so on). Regardless, this study only focuses on earthquakes.

The PSHA process takes into account the spatio-temporal behavior of seismicity. To carry this out, three principal methods can be involved:

- Homogeneous seismotectonic zones, hereafter called the zoning method [5]. Seismicity and tectonics are considered homogeneous within each zone. Seismic temporality is defined by computing a specific Frequency–Magnitude Distribution (FMD) in each zone. These zones are produced from various and complex data, such as seismic, geodesic and geologic ones, e.g., [10] (for France application);
- The smoothing method [11,12]. FMD is computed smoothly in space based on the past seismicity;
- The active faults method, e.g., [13] (for recent California application). Each known active fault is analyzed in order to associate it with its own FMD.

Each of these methods have drawbacks and advantages. These methods are classically used in parallel in order to take into account the advantages of each. Their application depends on the user's wishes and mainly on the studied region.

Mainland France is a low-to-moderate seismic area since it is localized on an intraplate domain, which is also called a stable region, i.e., far from tectonic plate boundaries.

The spatial segregation of the zoning method induces a drastic reduction in data. In fact, available data in each region are far smaller than those observed in the whole studied area. This finding is even more critical in low-to-moderate seismic areas, where seismic data are scarce. Using this method calls for specific care regarding data completeness and representativeness.

The smoothing method provides a good description of the seismicity but is not complete. In intraplate regions, where seismic frequencies are low, longer observation periods are needed. This is particularly true for extreme events (highest magnitudes) that may not have been seen during the period of observations. In fact, according to some authors, e.g., [14,15], stable regions present a seismicity that is more diffuse and homogeneous than past observations.

Finally, in intraplate domains, only few or even zero active faults are known. Moreover, a seismic cycle assumption could be questionable in stable regions [15]. According to these authors, intraplate seismicity is not driven by repeating tectonic loading and release energy but by transient crust deformations [15] caused by other origins than tectonics; see [16] (for a review in a French context). For this reason, the application of the active fault method is disputed. For example, for an application in the southeastern part of France [17], Martin et al. used this method as a complement to others, but only for the Provence region, where active faults are better known, e.g., [18,19].

In this paper, we propose a new stochastic generator to simulate the seismic catalogues of earthquakes that considers the specificities of the intraplate region (e.g., little data, various and complex origins of seismicity). We start by presenting the stochastic generator that simulates both main shocks (main and independent events) and aftershocks (correlated events). Then, we apply this methodology to mainland France.

2. Methodology

The proposed generator first simulates main shocks and then their associated aftershocks. A global scheme of the method is illustrated in Figure 1. In this scheme, the method is represented by the blue blocks, whereas inputs are marked as green blocks. Data used to produce these inputs are referred to in orange. Since this paper only focuses on methodology, the user is free to use their own data.

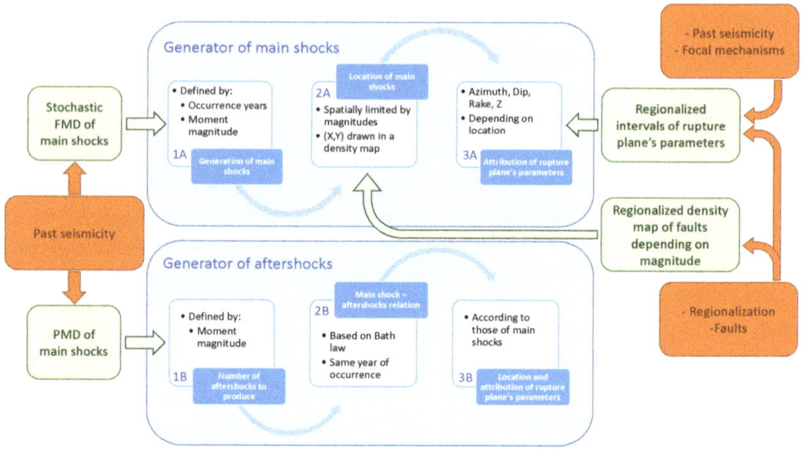

Figure 1. General workflow of earthquake generation. Blue: generator. Green: generator's inputs. Orange: data. FMD: Frequency–Magnitude Distribution. PMD: Proportion–Magnitude Distribution.

2.1. Main Shock Generation

Only few data are available in low-to-moderate seismic regions. Knowing this, instead of computing a Frequency–Magnitude Distribution (FMD) per zone or a smoothing FMD, we established the FMD of main shocks recorded on the whole territory. This has the advantage of maximizing the number of data and evaluating the probability of occurrence of an earthquake in France without any prior assumption. This follows recent studies, e.g., [14,15], that stipulate that stable region's seismicity is more diffuse and homogeneous than past observations suggest. Main shocks are thus generated through time (year) and magnitude for the whole territory. Then, the position of the main shock is moved according to its magnitude and a fault probability map.

2.1.1. Main Shock Generation in Time and Magnitude

Synthetic main shocks are first generated through time and magnitude. The time step chosen is the year, while the magnitude step is set to 0.1. We first set the number of years N_y through which the synthetic catalogue will span. Then, the generator calls for a stochastic Frequency–Magnitude Distribution (FMD) of main shocks, which corresponds to a probabilistic distribution of annual number of main shocks for each magnitude step. The annual density $\lambda_y(M)$ of main shocks is defined year after year y for each magnitude step M according to this stochastic FMD of main shocks.

Main shocks are simulated yearly following a homogeneous Poisson point process of annual density $\lambda_y(M)$:

$$n_y(M) \sim Poisson(\lambda_y(M)), \quad (1)$$

with $n_y(M)$ being the number of main shocks of magnitude $M \in [M_{min}, M_{max}]$ generated at the year $y \in [1, N_y]$.

The annual number of main shocks of magnitude equal to M at the year y ($\lambda_y(M)$ Equation (1)) is drawn in the stochastic FMD of main shocks. However, these FMDs are defined for main shocks with magnitude greater than or equal to M ($\lambda_y(\geq M)$). The annual density obtained from this kind of FMD is not $\lambda_y(M)$ but $\lambda_y(\geq M)$. Thus, $\lambda_y(M)$ is defined as:

$$\forall y \in [1, N_y], \forall M \in [M_{min}, M_{max}],$$
$$\lambda_y(M) = \begin{cases} \lambda_y(\geq M) & \text{if } M = M_{max} \\ \lambda_y(\geq M) - \lambda_y(\geq M + dM) & \text{otherwise} \end{cases} \quad (2)$$

In some cases, $\lambda_y(\geq M + dM)$ can be higher than $\lambda_y(\geq M)$, and so $\lambda_y(M)$ is set to 0.

This step is referred as 1A in Figure 1.

2.1.2. Main Shock Generation in Space

Every main shock simulated in terms of year and magnitude is then localized somewhere in the whole of mainland France.

A recent study [16] stipulates that the structural inheritance—that is, crust or mantle weakening from past tectonic—can play an important role in deformation localization in intraplate context. According to previous works and new observation, these authors estimate that 55–95% of the stable region's seismicity occurs near such weaknesses. Based on this, we suppose here that such geological objects, that give testimony to hundreds of millions years of seismicity, give more exhaustive information than past seismicity in predicting possible location of future earthquakes. In this paper, for a first approximation, we propose guiding the location of generated main shocks by using a fault probability map.

However, even in active regions around the world, fault activity could be difficult to address, e.g., [20] (for the Canterbury earthquake in New Zealand). Recent studies in China [21] or earthquakes in France [22] and Australia [23] serve as reminders for how difficult it is to define fault activity in stable regions. Since, in these regions, faults can be inactive over long periods of time [14,21], we decided to consider all faults, regardless of their presumed importance and activity.

The map of faults is converted into a density map by computing the average length of fault lines per unit area on a Cartesian grid. After that, the density values are normalized in order to obtain a sum equal to 1, which leads to a probability map. In that sense, each cell of the grid is characterized by a probability of hosting faults. Grid resolution is set by the user.

Looking at past seismicity, one would notice that earthquakes with the highest magnitudes are concentrated in the most active regions, even in intraplate domains. For example, these regions in France are the Alps, the Pyrenees and the Rhine basin. We decide to apply a magnitude threshold to drive highest magnitudes' earthquakes in areas where they have most chance to happen. To achieve this, we segregate the fault probability map by regionalization. Each region is characterized by a maximal magnitude $M_{max\,region}$. This magnitude corresponds to that of the largest earthquakes that this region can host.

Coupling regionalization with probability map leads to a set of regions defined by their own probability map and maximal magnitude $M_{max\,region}$. A main shock of magnitude M generated at year y can only be localized in eligible regions, i.e., regions where $M \leq M_{max\,region}$. Once the set of eligible regions is defined, the location of this main shock is drawn through the cumulative distribution function of the corresponding probability map (Figure 2). This procedure is a heterogeneous Poisson process whose density depends on the magnitude of the synthetic main shock to be located.

The Cartesian grid on which fault probability map is computed is used to select a cell, whose resolution depends on the user, where the main shocks' coordinates (the black point in Figure 2) are uniformly drawn. Figure 2 summarizes the localization process. Owing to this approach, spatial distribution of generated main shocks is limited by magnitude, with respect to the fault probability map that characterizes seismicity.

This step is referred as 2A in Figure 1.

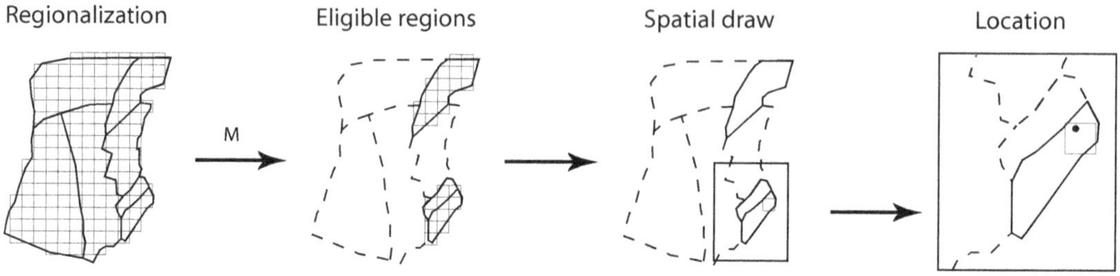

Figure 2. General workflow of main shocks localization. A probability map is defined through a Cartesian grid. Each region is characterized by their own probability map and maximal magnitude allowed. The magnitude M of the generated main shock limits the number of eligible regions. Spatial drawing is then realized only within eligible regions. Once a cell of the grid is drawn, epicentral coordinates of the main shock (black point) are uniformly drawn within this cell.

2.1.3. Rupture Plane's Parameters

Seismic hazard produced by an earthquake at a given site is computed by ground-motion prediction equations. These equations accept the definition of earthquakes as points. However, the geometry of rupture plane associated to an earthquake could play a significant role in seismic hazard computation. Knowing this, a rupture plane needs to be described for each generated earthquake.

In this study, a rupture plane is considered as a plane at a particular depth and is parametrized by a length (L) and three orientation angles: (i) azimuth (clockwise angle from north between 0° and 360°), (ii) dip (angle between 0° and 90° from the horizontal) and (iii) rake (angle between −180° and 180°) that define the components of fault movement (normal, reverse, strike-slip).

L is evaluated from moment magnitude M_w of the earthquake using the following relation, e.g., [24,25]:

$$L = 10^{(M_w - l_1)/l_2}, \tag{3}$$

with L being the rupture plane's length in km and l_1 and l_2 being two constants. For the other parameters, they are defined per region, where intervals of values are set according to data. Classically, these kinds of data are used:

- Instrumental seismic catalogue for depth;
- Focal mechanism catalogues for azimuth, dip and movement;
- Fault map for azimuth.

See Appendix A for an example of definition of these intervals for an application in mainland France.

Values for each parameter are randomly drawn in a uniform distribution when a main shock in generated in a given region.

This step is referred as 3A in Figure 1.

2.2. *Aftershock Generation*

Aftershocks could play a significant role in the seismic risk, e.g., [26], since their occurrence can destroy a building damaged by the main shock. Generating aftershocks is thus necessary in the context of seismic risk. Once again, the low amount of data makes seismic sequences difficult to observe in intraplate domains. Thus, applying the famous Epidemic Type Aftershocks-Sequence (ETAS [27]) model, which is a marked point process fond of parameters, is difficult.

Computing Frequency–Magnitude Distribution (FMD) of main shocks in order to generate main shocks through time and magnitude calls for differentiation between main shocks and aftershocks in the data analyzed. This leads to a Proportion–Magnitude Distribution (PMD) of main shocks that depicts the proportion of main shock in the studied catalogue in function of magnitude. Here, we propose generating aftershocks complementarily to main shocks by following the same PMD used to produce the main shocks. Once every main shock is generated, aftershocks are produced and related to their own main shock. This aftershock's production is carried out through three steps described below and represented in Figure 1 (1-3B).

2.2.1. Number of Aftershocks to be Produced

The first step consists of defining the number $NbAs(\geq M)$ of aftershocks with magnitude greater than or equal to M to be produced. This number is obtained for each magnitude step according to the number of main shocks $NbMs(\geq M)$ already generated and the proportion of main shocks $PropMs$ as follows:

$$NbAs(\geq M) = NbMs(\geq M) \times \left(\frac{1}{PropMs(\geq M)} - 1\right). \tag{4}$$

An example of PMD of main shocks, as well as the number of main shocks with magnitudes greater than of equal to 4 generated over 100,000 years, are shown in Figure 3. The number of aftershocks $M \geq 4$ to be produced, computed by Equation (4), is also visible in Figure 3. Thus, aftershocks are firstly associated with their magnitude.

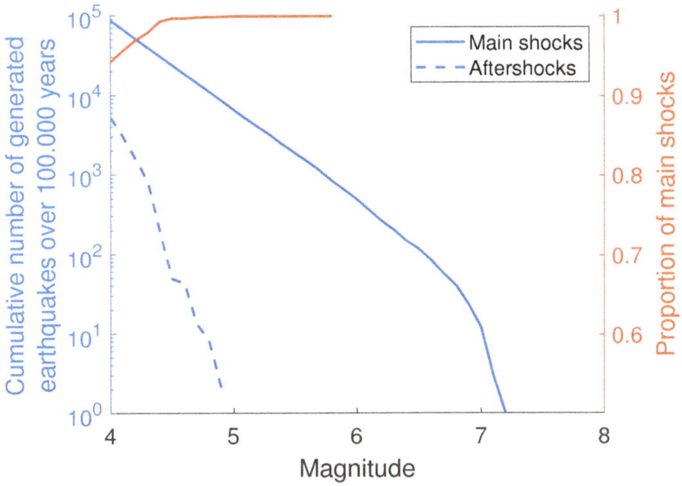

Figure 3. Example of number of aftershocks generated over 100,000 years. This number is obtained from examples of number and proportion of main shocks (Equation (4)).

2.2.2. Aftershock–Main-Shock Relation

According to Båth law [28,29] (Equation (5)), main shock is the event of the sequence with the highest magnitude. Moreover, its magnitude M_{ms} is limited from below according to the magnitudes M_{as} of its aftershocks as follows:

$$M_{ms} \geq M_{as} + \Delta M\,. \tag{5}$$

where ΔM is a constant equal to 1.2 for crustal seismicity (<50 km depth, in [28]).

In this paper, we use this convenient law to find the eligible main shocks for each aftershock. In fact, owing to the first step, the number of aftershocks to be produced is known for each magnitude step (Figure 3). Applying Equation (5) for each aftershock allows us to define the minimal magnitude $M_{ms\,min}$ of its related main shock. Its linked main shock is then randomly selected among all of the main shocks with magnitude greater than or equal to $M_{ms\,min}$. Aftershock's year of occurrence is set to the same as its main shock.

However, one needs to keep in mind that Equation (5) is an empirical model and thus not a general truth. For example, the M6.5 Amatrice (Italy) earthquake provoked an aftershock of magnitude 6.1 three months later. In order to take into account ΔM variability in the aftershock–main-shock association, the method of seismic moment ratio [30] is used in this paper. It computes this term:

$$\Delta M = -\frac{log_{10}(R)}{1.5}, \tag{6}$$

according to the R ratio equal to:

$$R = \frac{\sum M_{0\,as} - \sum M_{0\,fs}}{M_{0\,ms}} \approx 0.05, \tag{7}$$

where $\sum M_{0\,as}$ and $\sum M_{0\,fs}$ are the seismic moment, equivalent to energy, released by all of the aftershocks and foreshocks, respectively, and $M_{0\,ms}$ is the seismic moment released by their main shock. According to [30], R is equal to 0.05.

A ΔM distribution is obtained by sampling R in a Gaussian law with mean equal to 0.05 and a standard deviation of 0.0125 [30]. Such a distribution gives an average ΔM equal to 0.87 while 90% of ΔM values between 0.77 (Q5) and 1.02 (Q95). Drawing a ΔM in this distribution makes it possible to estimate $M_{ms\,min}$, the minimal magnitude of its main shock, from Equation (5).

2.2.3. Aftershock's Location and Rupture Plane's Parameters Definition

We suppose that aftershocks are localized on the same fault as the main shock. Aftershocks' epicenter (x_{as}, y_{as}) are thus placed at approximately $0.75\,L$ from main shocks' epicenter (x, y) in the direction of its azimuth (azi) to within $10°$:

$$\begin{array}{l} x_{as} \sim N(x,\, 0.75 \times L \times sin(-azi \pm 10)) \\ y_{as} \sim N(y,\, 0.75 \times L \times cos(-azi \pm 10)) \end{array}, \tag{8}$$

where L (km) is the rupture plane length associated with the main shock (Equation (3)).

Depth z_{as} (km), azimuth azi_{as} (°) and dip dip_{as} (°) of aftershock's rupture plane are sampled into normal laws. These laws have a mean that is equal to the value of the main shock (z, azi et dip) and a standard deviation depending on the parameter:

$$\begin{array}{l} z_{as} \sim N(z,\, 2.5) \geq 0 \\ azi_{as} \sim N(azi,\, 5) \geq 0 \\ dip_{as} \sim N(dip,\, 2.5) \geq 0 \end{array}. \tag{9}$$

Movement of aftershock's rupture plane (normal, reverse, strike-slip or unknown) is equal to the main shocks' rupture plane.

3. Application to Mainland France

3.1. Data Used in This Study

3.1.1. Regionalization

Even in intraplate domains, the spatial distribution of seismicity can seem heterogeneous. For example, in France, the Pyrenees and Alps are linked to a collision of European and African plates, which accentuates the tectonic features and seismicity in these areas. The SHARE project [31] has proposed segregating European territories according to tectonic features by differentiating pure Stable Continental Regions (SCR) and Oceanic Crusts (OC) from shallow active regions. This tectonic distinction is visible in Figure 4 (Figure 2 in [32]) for France. This regionalization is used in this paper in order to limit the spatial distribution of magnitudes (see Section 2.1.2 and Figure 2).

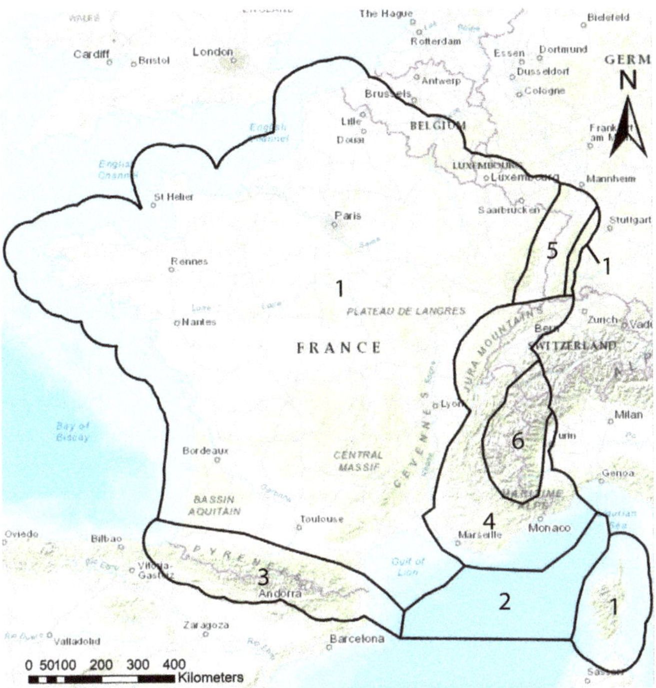

Figure 4. SHARE tectonic regionalization [32] used in this paper. Regions 1 correspond to Stable Continental Regions (SCR). Region 2 represents the Oceanic Crust (OC). Regions 3 (Pyrenees) and 4 (Alps) represent compressional active shallow regions, whereas regions 5 (Rhine Basin) and 6 (Alps) correspond to extensional ones.

3.1.2. Seismic Catalogues

Historical Catalogue

The historical French catalogue FCAT-17 [33] is composed of 4250 earthquakes felt and reported by the French population from 463 to 1964. Despite being associated with large uncertainties, such a long catalogue is useful to define the maximal magnitude of the study (M_{max}) and per region ($M_{max\,region}$). In this study, we decide to select these maximal magnitudes as the maximum recorded magnitudes in the FCAT-17 catalogue, plus its uncertainty. The strongest earthquake of this catalogue occurred in Liguria in 1887, with a moment magnitude of 6.7 ± 0.6. Thus, the maximal magnitude M_{max} is set to $6.7 + 0.6 = 7.3$. Since this earthquake is localized in region 4 (compressional Alps, Figure 4), this region is characterized by a $M_{max\,region}$ equal to 7.3. Table 1 summarizes the maximal magnitudes observed in each SHARE region.

Table 1. Maximal magnitude $M_{max\,region}$ of each region. Region numbers refer to Figure 4.

Region	1	2	3	4	5	6
$M_{max\,region}$	6.5	5.5	7	7.3	7	6.6

Consistent with the recent study [34], we consider that a seismic event with a magnitude of up to 5.5 can occur everywhere in France. As a result, the $M_{max\,region}$ is raised to 5.5 if the estimation using the catalogue leads to a smaller value, which is the case for the OC region (n°2, Table 1). Maximal magnitudes are constant and, thus, their uncertainties are not explored in this paper.

Instrumental Catalogue

This study uses both the instrumental French catalogue SIHex [35] and the RéNaSS catalogues, https://api.franceseisme.fr/fr/search (accessed on 20 December 2021). The former is composed of 37,408 earthquakes recorded between 1965 and 2009, whereas the latter regroups 25,042 seismic events indexed from 2010 to 2020. In total, the merged catalogue brings together 62,450 earthquakes that occurred during the 1965–2020 period.

Such a catalogue isn't complete since the seismometers' number and resolution have improved over time. Also, the catalogue needs to be processed in order to reveal representative and meaningful information. The classical approach to make this catalogue exhaustive is to consider cut-off magnitudes (M_c) and years (Y_c). The catalogue can be considered exhaustive on a particular territory for earthquakes with a magnitude greater than or equal to M_c only during the period Y_c-2020. For the French territory, we determine M_c and Y_c thanks to the cumulative visual method [36,37]. The couples of M_c and Y_c that we found are visible in Table 2.

Table 2. Periods of completeness in function of magnitude for the instrumental catalogue.

M_C	2–2.9	$\geq$3
Period	1970–2020	1965–2020

Figure 5 presents the spatial distribution of the 15,567 earthquakes $M_w \geq 2$ contained in the instrumental catalogue. The strongest earthquake of this catalogue occurred near Arette in 1967, with a moment magnitude of 5.2 ± 0.3.

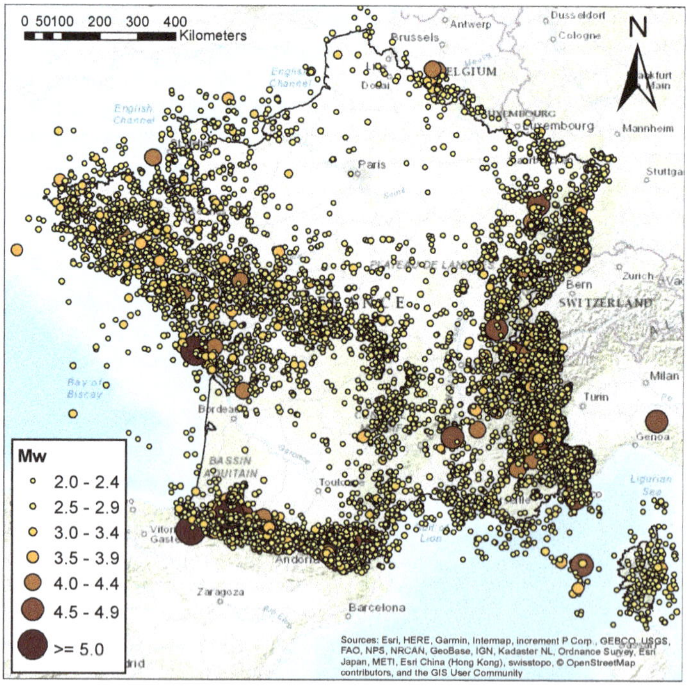

Figure 5. SIHex and ReNaSS catalogues (1965–2020) used in this study to calculate the FMD of main shocks for the whole of mainland France. Only the 15,567 earthquakes with magnitudes greater than or equal to 2 are represented.

3.1.3. Faults

For a first approximation, we propose building a spatial probability map (Figure 2) by using the CHARM database, https://infoterre.brgm.fr/page/telechargement-cartes-geologiques (accessed on 20 December 2021), produced by BRGM (Figure 6a). This database contains fault traces, i.e., intersection lines between faults and the Earth surface mapped on geological maps. The CHARM database is composed of 10,576 terrestrial fault lines covering the whole of mainland France and its vicinity (Figure 6a). As already stated in Section 2.1.2, we decide to consider all faults, regardless of their presumed importance and activity. This map of fault traces is converted into a density map by computing the average length of fault lines per unit area on a Cartesian grid of 5 km. The density map is illustrated in Figure 6b.

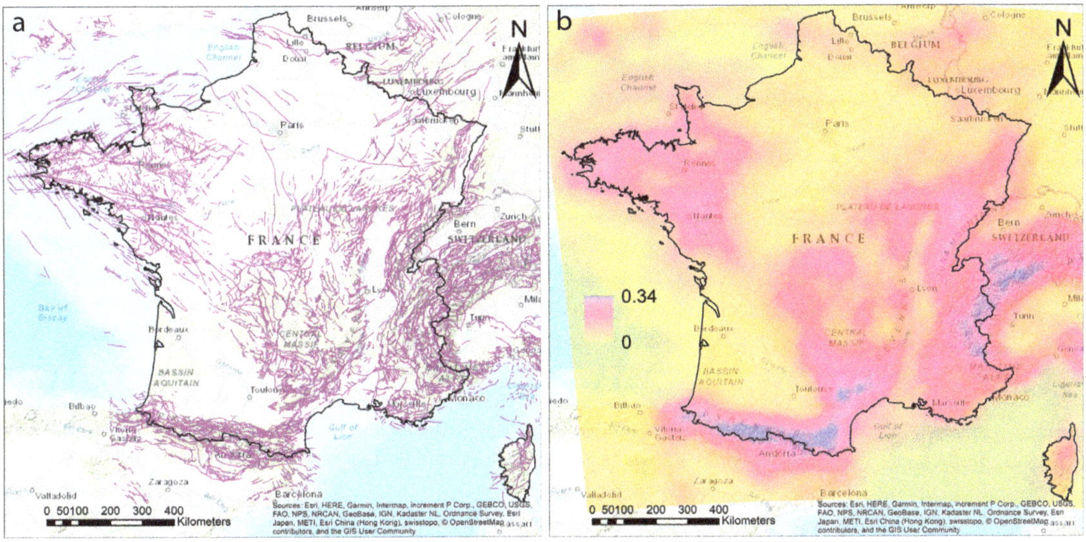

Figure 6. (**a**) Fault traces map of the French territory (CHARM database of BRGM). (**b**) Spatial density map of fault traces.

Spatial distributions of past earthquakes (Figure 5) and fault traces (Figure 6) show the same trend. However, one can remark some differences between these two maps. For example, some moderate earthquakes have been observed in the north of Bordeaux, whereas no fault trace is visible nearby. These differences are explained by:

- Unknown, e.g., [20], and/or new faults;
- The fact that the CHARM database only contains terrestrial fault traces, which means that covered, deep and/or bathymetric faults are missing;
- Very old faults that have probably been inactive for a long period of time (for example, Armorican Massif).

In order to allow for the possibility of localizing main shocks anywhere [14,15], thus avoiding specific cases, such as the one observed near Bordeaux, regions where no fault lines have been documented are associated with a minimal density value instead of 0. In this study, in the first instance, this minimal value is arbitrarily set to 1% of the maximal density value. In that sense, we consider that the least represented regions have a density that is 100 times lower than in most represented regions. Concretely, according to Figure 6b, every 5×5 km cell characterized by a density value lower than $\frac{0.34}{100} = 0.0034$ is associated with this minimal value (0.0034). The cells concerned by this density modification represent around 15% of mainland France. It mainly concerns Parisian and Aquitaine basins (SCR), as well as the Mediterranean Sea (oceanic crust). This map is then transformed to a probability map due to normalization. The final regionalized fault probability map is used as an input of the generator of earthquakes for mainland France ("*Regionalized density map of faults depending on magnitude*", Figure 1).

Faults (Figure 6a) can also be employed to define the probability density functions (pdf) of azimuth in each region. Please refer to Appendix A for an application to mainland France.

3.2. Temporal Description of the Instrumental Catalogue

3.2.1. Frequency–Magnitude Distribution

The frequency–magnitude distribution (FMD) describes the temporal occurrence of earthquakes. It consists of computing the annual number of earthquakes as a function of the magnitude. The classical way to compute FMD is to apply the GR law [38] on observed data:

$$log_{10}(N(\geq M)) = a + b \times (M - M_{min}), \tag{10}$$

where M is the earthquake's magnitude and $N(\geq M)$ is the annual number of earthquakes with a magnitude greater than or equal to the M observed in the seismic catalogue. In this study, M stands for the moment magnitude (M_w), and it is limited by a minimal magnitude M_{min} (which, here, is equal to 2).

The seismic catalogue is composed of two types of earthquakes: main shocks and their correlated seismicity (fore shocks or aftershocks). Since they are characterized by different spatial and temporal behaviors, seismologists usually analyze them separately. A declustering algorithm is classically used for that purpose, e.g., [39]. It applies spatio-temporal windows around earthquakes in a catalogue in order to find clusters. In each cluster, the earthquake with the maximal magnitude is considered as the main shock, and the others are considered as correlated events. Thus, it is possible to segregate independent events from dependent ones by conserving only main shocks and alone earthquakes (clusters with only one earthquake). Then, the population of main shocks can be studied separately through a Poisson distribution in time, space and magnitude. In this study, the G85 declustering algorithm [40] is chosen. The Proportion–Magnitude Distribution (PMD) of main shocks obtained by applying this algorithm on French instrumental seismicity is represented by blue empty dots Figure 7.

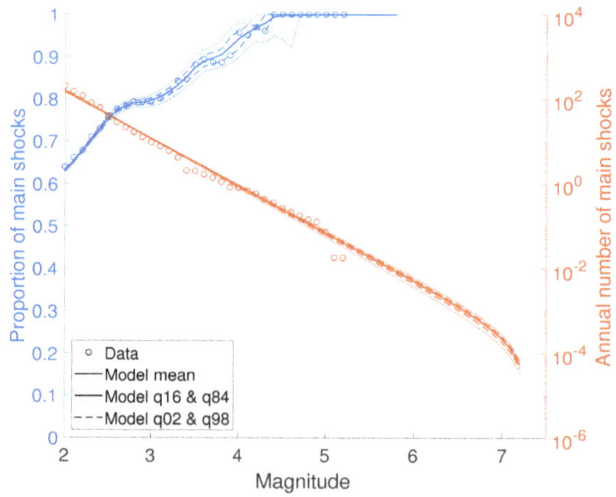

Figure 7. Proportion–Magnitude and Frequency–Magnitude Distributions (PMD and FMD) of main shocks for the whole of mainland France. The PMD is obtained by applying the G85 declustering algorithm [40] on data. A Regression of the GR law (Equation (10)) on data has been applied from $M2$ to $M5$ and has been extrapolated with the use of the truncated GR law (Equation (11)).

Orange empty dots in Figure 7 show the annual number of main shocks observed in the exhaustive and declustered catalogue as a function of their magnitude. Applying Equation (10) to these data requires a straight line distribution. Such a distribution is observed between magnitudes 2 and 5 (Figure 7). Thus, we apply Equation (10) between $M2$ and $M5$ and obtain $a = 4.41$ and $b = -1.12$.

From $M5.1$, the annual numbers of main shocks fall and do not respect the GR law (Equation (10)). This discrepancy for large magnitudes is classically attributed to a non-completeness of data for these magnitudes. Assuming that, an extrapolation of the GR law that is estimated to be between magnitudes 2 and 5 is carried out for $M \geq 5.1$.

As the total energy released by earthquakes is finite, some deviation from the GR straight line is required for largest magnitudes in order to avoid infinite distribution. A truncation is generally applied on the GR law [41]:

$$\forall M \in [M_{min} M_{max}], N(\geq M) = 10^a \times \frac{e^{-\beta(M-M_{min}} - e^{-\beta(M_{max}-M_{min})}}{1 - e^{-\beta(M_{max}-M_{min})}}, \quad (11)$$

with $\beta = b \times log(10)$. Applying Equation (11) with a and b estimated before (Equation (10)) gives the FMD of main shocks modeled from the exhaustive and declustered catalogue. This FMD is visible in Figure 7 and is characterized by a mean slope (b value) that is equal to 1.12 and an asymptotic fall of frequencies for the maximal magnitude $M_{max}7.3$.

The mean return periods (inversely equal to annual frequencies) of main shocks are summarized for six magnitude steps in Table 3. These return periods describe the average waiting time between main shocks of magnitudes greater than or equal to M. As expected, the mean return periods increase as the magnitude grows.

Table 3. Mean return periods of main shocks in function of magnitude M computed in the instrumental catalogue. These return periods are calculated from G85-GR application (Figure 7). Standard deviations are also reported. d: days; y: years.

M≥	2	3	4	5	6	7
Return periods	2.4 ± 0.1 d	31.9 ± 0.3 d	1.14 ± 0.05 y	15 ± 1 y	200 ± 24 y	4700 ± 700 y

3.2.2. Consideration of Magnitudes' Uncertainties through a Monte Carlo Scheme

Taking into account magnitudes' uncertainties in the calculation of FMD through a Monte Carlo scheme allows us to obtain a stochastic set of PMD and FMD that corresponds to the global representation of main shocks' proportion and occurrence (Figure 7). This Monte Carlo process consists of producing 1000 initial catalogues of 15,567 earthquakes with $M \geq 2$. Only the magnitude differs from one catalogue to another, since the magnitude of each earthquake is drawn in a normal law $N(\mu, \sigma)$, where μ is its magnitude and σ its magnitude's uncertainty.

In Figure 7, each magnitude step is characterized by a set of annual frequencies of main shocks. Thus, for each magnitude step, the annual frequency of main shocks can be described by a probability density function (pdf). These pdfs are visible in Figure 8 for six magnitude steps. For a given magnitude M, the annual frequency of main shocks of magnitudes greater than or equal to M is thus defined as a pdf. The pdf values are normalized by their sum. In this way, these normalized values are dimensionless and their sum converges to 1, which makes them probabilities.

These pdfs of annual frequencies as a function of magnitude are inputs of the generator of earthquakes for mainland France (*"Stochastic FMD of main shocks"*), whereas the mean PMD of main shocks constitutes another input (*"PMD of main shocks"*, Figure 1).

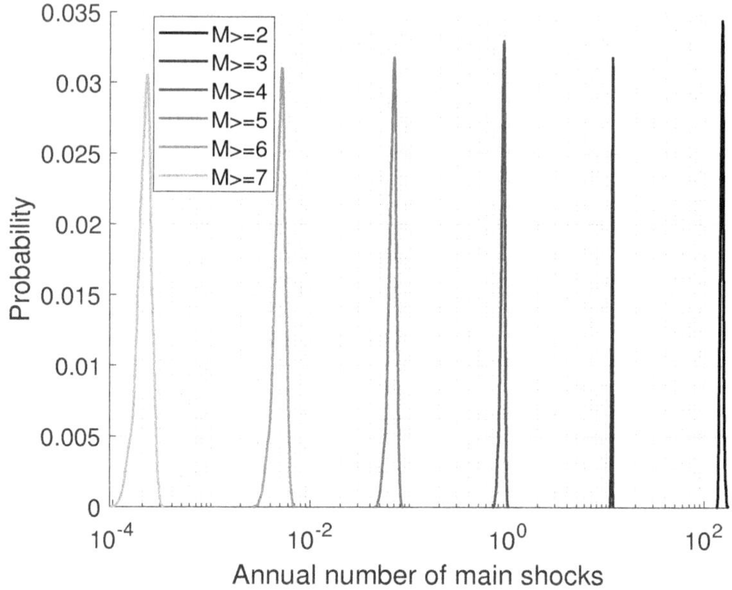

Figure 8. Probability density function of frequency of main shocks for different magnitudes according to the stochastic FMD of main shocks visible in Figure 7.

3.3. Results

This section presents 100,000 years of synthetic seismicity generated by using SHARE regionalization (Figure 4). Generated earthquakes have magnitudes greater than or equal to 4.

3.3.1. Main Shock Generation

Cumulative Seismic Moment Distribution

Cumulative seismic moments produced by both synthetic and instrumental catalogues are compared Figure 9. The seismic moment M_0 (N.m) released by an earthquake is calculated for each real or synthetic main shock directly from its moment magnitude M_w following [42]:

$$M_0 = 10^{1.5\,M_w+9.1}\,. \tag{12}$$

According to Table 2, the completeness period associated with instrumental $M \geq 4$ earthquakes is 1965–2020, which represents 56 years. Comparisons are thus made over this completeness period. Since 100,000 years have been generated, we divided the synthetic seismicity into a set of 1785 sub-catalogues of 56 years. Each of these sub-catalogues are comparable to the complete instrumental catalogue.

The final seismic moment observed in the instrumental catalogue is smaller than the final mean seismic moment generated. This slight difference is due to the fact that the return periods of $M \geq 4$ main shocks given as an input are equal to 1.14 years (Table 3) and 1.24 years, according to the initial catalogue. Thus, the initial catalogue (red line Figure 9) is composed of 45 $M \geq 4$ main shocks over 56 years, whereas the sub-catalogues of 56 years are on average composed of 49 $M \geq 4$ main shocks (black line Figure 9). Thus, the generator produces a few more $M \geq 4$ earthquakes than we observe in the initial catalogue.

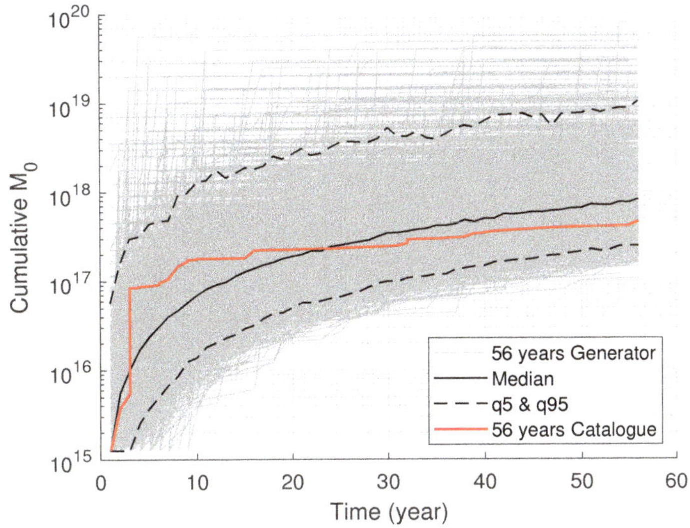

Figure 9. Cumulative seismic moment from observed and generated $M \geq 4$ seismicities over the period of completeness (Table 2). The synthetic catalogue used is 100,000 years long.

The mean distribution of generated seismic moments is, by definition, smooth, and could differ from the one observed in the instrumental catalogue, but trends are similar. Distributions visible in the synthetic sub-catalogues show variations, as well as the instrumental distribution.

Frequency-Magnitude Distribution

The Frequency–Magnitude Distribution (FMD) of generated main shocks can be observed in the 100,000 years of generated seismicity. This FMD of main shocks is compared Figure 10 with the mean one given as an input to the generator (Figure 7). The generator manages to reproduce this FMD of main shocks, especially for magnitudes smaller than 6.5. For higher magnitudes, slight variations are observed between the generator's FMD and the data's FMD. This part needs to be improved in order to avoid these slight variations.

The FMD of generated main shocks for the six SHARE regions (Figure 4) are also shown in Figure 10. Table 4 presents the b-values computed by regressing the GR law (Equation (10)) on these FMDs of main shocks, from magnitudes 4 to 5 and between *M*5 and *M*6.

Table 4. b-values computed by regressing GR law (Equation (10)) for different ranges of magnitudes on synthetic FMD of main shocks (Figure 10) obtained in each SHARE region. Region numbers refer to Figure 4.

Region	1	2	3	4	5	6	Mainland France
b(M4-5)	1.12	1.09	1.10	1.10	1.21	1.15	1.12
b(M5-6)	1.23	1.98	0.98	1.00	1.04	0.99	1.12

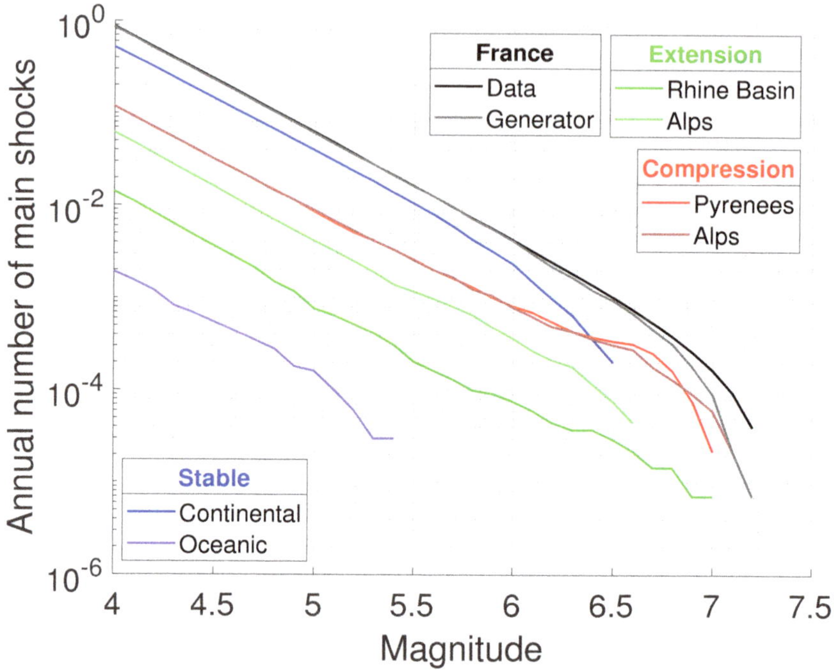

Figure 10. Mean frequency–magnitude distributions of main shocks given as input (black) and observed in a synthetic catalogue of 100,000 years for the whole of mainland France (grey) and for the SHARE regions (Figure 4).

Aside from the Rhine basin region (n°5 Table 4), every region shares practically the same slope of the FMD of generated main shocks before magnitude 5 (Figure 10): from 1.09 to 1.15. These slopes are close to the slope of the FMD of main shocks given as an input: 1.12. This is expected, since we only use this FMD as an input. Homogeneity is thus observed in the results.

Table 5 lists b-values calculated on the data in each region. Please refer to Appendix B for more details on how these values have been computed. Regardless of what is obtained from the observed (Table 5) or synthetic (b(M4-5), Table 4) seismicities, the b-values of a given region keep the same order with respect to the b-value used as the input. In fact, regions 1, 5 and 6 are characterized by both b-values being greater than or equal to the reference b-value (1.12). In parallel, regions 2, 3 and 4 are characterized by both b-values being lower than the reference b-value. Thus, variations of synthetic b-values around the initial value are consistent with b-values obtained from data.

Table 5. b-values computed by regressing GR law (Equation (10)) on FMD of main shocks observed in instrumental seismicity for each SHARE region. Region numbers refer to Figure 4. See Appendix B for more details.

Region	1	2	3	4	5	6	Mainland France
b	1.18	0.72	1.05	1.06	1.36	1.21	1.12

Moreover, for $M \geq 5.1$, slopes of FMD diverge from one region to another. Between magnitudes 5 and 6, the FMD of stable oceanic crust and the Pyrenees are, respectively, characterized by the highest and the lowest slopes: 1.98 and 0.98 (Table 4). The higher the slope, the lower the seismic activity. Thus, according to the generator, stable continental and oceanic regions are the lowest seismic areas, and the most active regions (Alps, Pyrenees and Rhine basin) are the highest seismic areas. These results are in line with our knowledge of mainland France seismicity. The use of an input as a fault probability map and only one FMD of main shocks thus leads to consistent results.

From the magnitude 5.5, the limitation of large earthquakes' spatial distribution through the regionalization seems to be efficient. In fact, when the magnitude is too high to appear in a given region, it moves to another one, which concentrates the generated main shocks in regions where high magnitudes are allowed. Thus, the limitation of large earthquakes' spatial distribution has an impact on the change in b-values, which is noted in Table 4. Moreover, a relatively pronounced asymptotic fall of frequency is noticed at the region of maximal magnitude, except for the Rhine basin region (Figure 10). This observation is also in line with our knowledge of general seismicity: without using the truncated GR law (Equation (11)), the generator seems to reproduce this asymptotic fall at the largest magnitudes. Finally, the sum of each region FMD is equal to the FMD of the whole of mainland France. This behavior is due to the use of only one FMD as an input, and brings coherence to the results.

3.3.2. Aftershock Generation

Figure 11 illustrates the number of main shocks and aftershocks per unit of magnitude generated over 100,000 years. The higher the magnitude, the higher the proportion of main shocks. This is consistent with the literature and with the Proportion–Magnitude Distribution (PMD) of main shocks given as an input of the generator (Figure 3), which is identical to the one observed in the synthetic seismicity. Thus, the generator of aftershocks is working well, since it manages to reproduce this PDM.

Figure 12 shows the number of aftershocks per main shock as a function of the magnitude of the main shock. This number is constant and equal to 1.6 on average for magnitudes of main shocks between 5.2 and 6.9.

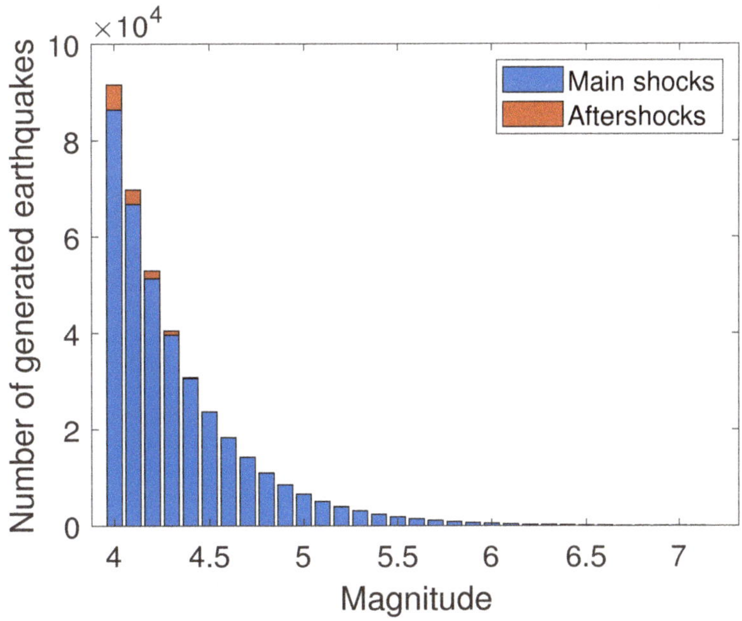

Figure 11. Number of main shocks and aftershocks generated over 100,000 years as function of magnitude.

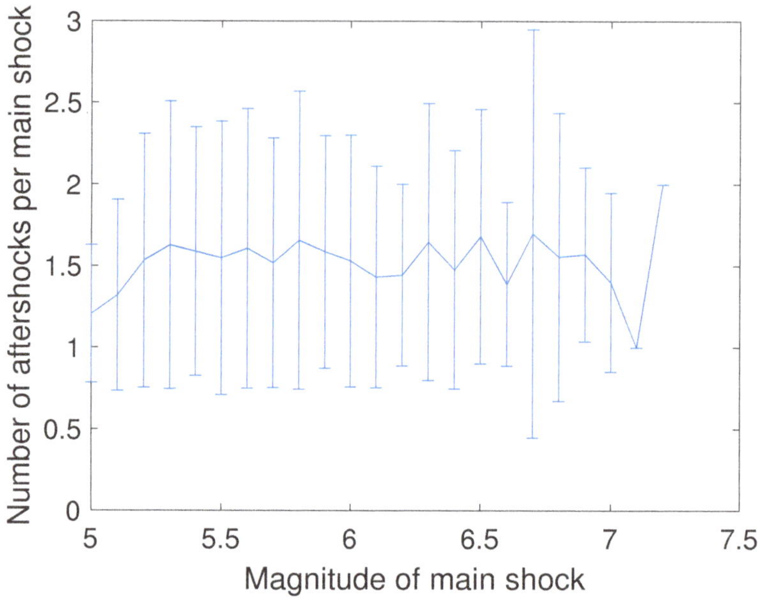

Figure 12. Number of aftershocks produced by one main shock according to magnitude of the latter. Error bars correspond to standard deviation observed in the 100,000-year synthetic catalogue.

4. Discussion

4.1. On the Use of Fault Probability Map

The use of the spatial probability of faults to guide the location of generated earthquakes is a strong choice. As already stated, it comes from both various seismic origins in stable regions and the fact that existing faults are more likely to localize further seismic events [16]. The lack of completeness and clear definition of active faults have led us to consider all faults, regardless of their alleged importance and activity. This choice is questionable but consistent with the fact that, in stable regions, faults can be inactive over long periods of time [14,21]. Results described in this paper are encouraging, since they are consistent with data given as an input and with our knowledge of French seismicity. However, the method presented in this article corresponds to a first step in a new way to generate stable seismicity, and some improvements must be achieved.

First of all, we have seen that the spatial distribution of past seismicity in mainland France is heterogeneous and that the strongest occurring earthquakes are concentrated in specific regions (Alps, Pyrenees and Rhine basin). To consider this statistical observation, we opt for a spatial limitation of magnitudes by applying regionalization. Still, we can imagine using other methodologies to generate seismicity in the most active French regions. For example, a methodology that is closer to what is conventionally carried out with FMD computed by region/zone or smoothly, e.g., [17,34,43] (for recent mainland France applications). Another possible methodology could be to compute a FMD of main shocks in the Alps, the Pyrenees and the Rhine basin and to use a spatial probability map of past seismicity. One could also use a spatial probability of faults weighted by stress rate or others tectonic and geologic information in order to address the seismic potential of faults.

Furthermore, one can imagine weighting the probability map used in this study with our knowledge of intraplate domains. For example, various origins of stable seismicity in the long term could be considered: topography potential energy, erosion and glacial isostatic adjustment since the last glaciation and so on [16]. Transient processes (e.g., fluid pore pressure increase and hydrological or sedimentary load change [15]) leading to stable seismicity could also be taken into account.

Finally, the use of a fault map calls for exhaustiveness of these faults. However, this is not the case, since the map used is only composed of terrestrial fault traces, excluding covered, deep and/or bathymetric faults. Attempting to complete this database should be one of the next steps of our work.

4.2. On the Definition of Maximal Magnitudes

As already stated, 1500 years of historic data (FCAT-17 catalogue) cannot be representative for stable seismicity. That is why we choose to use a fault map instead of an earthquake map in order to drive the synthetic earthquake's location. Thus, our definition of maximal magnitudes is paradoxical, since it is based on historical seismicity.

Analyzing analogical regions in Europe or all around the world through a Bayesian approach, e.g., [44,45], can make the definition of maximal magnitudes more robust and not constant. This approach also has the advantage of defining maximal magnitude not as a constant but across a range of values, which is interesting when investigating maximal magnitudes' uncertainties.

4.3. On the Aftershock Production

Epidemic Type Aftershocks Sequence (ETAS) models are marked point processes [27] that produce main shocks and associated aftershocks. For that purpose, they use various laws, such as the aftershocks production law ([46] from [47]):

$$K = k \times 10^{\alpha(M_{ms} - M_c)}, \tag{13}$$

where K is the number of aftershocks produced by a main shock of magnitude M_{ms}, M_c is the minimal magnitude of interest and k and α are two constants. Thus, the number of aftershocks produced by a main shock increases with M_{ms}.

In our results (Figure 12), the number of aftershocks per main shock seems to be independent to the magnitude of main shocks. This observation is mainly due to the fact that the Proportion–Magnitude Distribution (PMD) of main shocks used as an input in this study (Figure 3) is derived from the instrumental catalogue. However, as already stated, this catalogue is far from exhaustive for large magnitudes ($M > 5$). Although it is not a problem for the FMD of main shocks, since it is produced by the extrapolation of the exhaustive part (see Section 3.2), it is a problem for PMD, which is not extrapolated and is thus not exhaustive. Thus, within the whole $M \geq 4$ generated earthquakes, aftershocks represent only 6% (Figure 3), which is low, as Figure 11 illustrates.

A PMD of main shocks computed thanks to the G85 declustering algorithm [40] from the exhaustive historical catalogue (FCAT-17) should be more representative of the true distribution. According to this catalogue, aftershocks represent 15% of the whole $M \geq 4$ seismicity. Applying the PMD of main shocks estimated from the exhaustive FCAT-17 catalogue gives results that are shown in Figure 13.

These results seem more realistic, since they are based on more exhaustive data and since the number of aftershocks per main shock follows the Equation (13). Moreover, our method to generate aftershocks is non-parametric in contrast to this equation, which needs to set two parameters (k and α). This is an advantage in the context of low-to-moderate seismicity, where objective parametrization from data is limited since data are sparse.

Main shocks are generated according to the proportion of main shocks p observed in the data, whereas the number of aftershocks produced depends on the complementary of p $(1 - p)$. However, results in Figure 13 have been obtained by using instrumental data to generate main shocks and historical data to produce aftershocks. Thus, two PMDs have been used, and so the consistency between p and $1 - p$ no longer holds. A solution could be to produce both the PMD and FMD of main shocks with historical data. Nevertheless, this catalogue is characterized by large uncertainties (magnitude, time of occurrence and space) and is known to overestimate magnitudes compared to the instrumental catalogue (e.g., Figure 2c in [16]). Its use must be carried out with care.

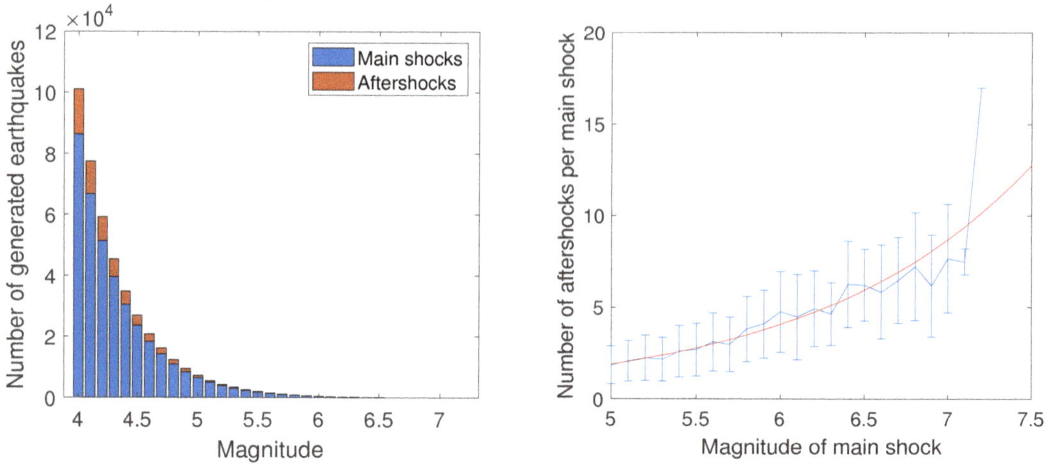

Figure 13. Number of aftershocks produced over 100,000 years by using the PMD estimated from the exhaustive historical French catalogue (FCAT-17). (**Left**): Number of main shocks and aftershocks as function of magnitude. (**Right**): Number of aftershocks produced by one main shock according to magnitude of the latter. Red line represents the regression of Equation (13) on these results: $k = 0.887$ and $\alpha = 0.331$.

5. Conclusions

In this paper, a new generator of earthquakes is proposed and applied to mainland France. Applying a stochastic generator of earthquakes to a French context is not new (see, for example, Ref. [48] for a Pyrenean application). However, the method of generating synthetic seismicity is new. Classically, two methods exist to compute a Frequency–Magnitude Distribution (FMD) of main shocks: (i) smoothly through a kernel approach [11,12] or (ii) discretely by using a zoning approach [5]. This allows us to analyze spatial and temporal behaviors of seismicity at the same time, but calls for data number reduction. In intraplate domains, i.e., far from tectonic plate boundaries, data are sparse. Thus, contrary to these classical approaches, the proposed generator uses a FMD of main shocks at a national scale in order to generate main shocks in time and magnitude in order to maximize the number of data available. These main shocks are then spatially distributed through the use of a probability map and regionalization. The former is used to guide the location of main shocks, whereas the latter allows us to limit the distribution of large earthquakes in space. Intraplate seismicity seems to be more uniformly positioned than in active regions [14,15], structural inheritance *"can play a strong role in deformation localization"* [16] and stable faults' activity is difficult to define [21–23]. For these reasons, faults, regardless of whether they are supposed as active or not, are used to produce the probability map.

Aftershocks are then produced by using the Båth law [28,29], the seismic moment ratio [30] and the Proportion–Magnitude Distribution (PMD) of main shocks. This approach defines aftershocks and main shocks differently from the famous ETAS models. Nevertheless, it remains consistent, since the numbers of main shocks and aftershocks are complementary and depend on data. Moreover, unlike ETAS models, this approach has the advantage of being non-parametric.

Temporal and energetic behaviors of generated main shocks are in line with inputs (FMD of main shocks and regionalization) and our knowledge of mainland France seismicity.

However, some improvements can be achieved, such as completing the fault map; in particular, by bathymetric faults, and better describing maximal magnitudes in each region. Moreover, we have seen that using instrumental seismicity to produce the DPM of main shocks is not representative enough. Using the FCAT-17 historical French catalogue could be a solution, but it needs to be carried out with care due to its large uncertainties and its overestimation of magnitudes. Finally, the method proposed in this paper should be applied to other stable continental regions, such as Northwestern European countries, Australia and so on in order to test its effectiveness.

Author Contributions: Conceptualization, P.T., C.G. and F.B.; methodology, C.G., P.T. and F.B.; data curation, C.G. and P.T.; writing—original draft preparation, C.G.; writing—review and editing, C.G., F.B. and P.T. All authors have read and agreed to the published version of the manuscript.

Funding: This work was performed in the frame of the RING and EXTRA& CO/ICEEL projects at Université de Lorraine. It has been funded by Caisse Centrale de Réassurance (CCR).

Institutional Review Board Statement: Not applicable.

Informed Consent Statement: Not applicable.

Data Availability Statement: Fault lines from CHARM database (BRGM).

Acknowledgments: We would like to thank the industrial and academic sponsors of the RING-GOCAD Consortium managed by ASGA for their support, especially CCR. The authors also want to thanks the anonymous reviewers for their meaningful reviews that helped considerably to improve this article.

Conflicts of Interest: The authors declare no conflict of interest.

Abbreviations

The following abbreviations are used in this manuscript:

PSHA	Probabilistic Seismic Hazard Assessment
FMD	Frequency–Magnitude Distribution
PMD	Proportion–Magnitude Distribution
SIHex	*Sismicité Instrumental de l'Hexagone* [35]
FCAT-17	French CATalogue 2017 [33]

Appendix A. Parameters of Rupture Plane

Each region is characterized by rupture plane's parameters defined by range of values. These values are defined according to seismic and tectonic data (seismicity, faults and focal mechanisms). In this study, we use the European [49] and World [50] focal mechanism catalogues. Table A1 summaries these ranges of values for depth, azimuth, dip and movement.

Table A1. Upper and lower bounds of the ranges of values defined for depth, azimuth, dip and movement used in this study. Movement: U = unknown/N = normal/S = strike-slip/R = reverse.

Region	Depth (km)		Azimuth (°)		Dip (°)		Movement
	Min	Max	Min	Max	Min	Max	
1	0	25	0	359	47	87	N & S & R
2	5	30	0	359	20	90	U
3	0	15	20	140	50	82	N
4	0	20	−10	60	45	77	S & R
5	0	20	10	70	40	82	S
6	0	15	−20	50	37	71	N

Since SHARE regions are large (Figure 4), range of values are wide: up to 25 km for depth, 120° for azimuth and 40° for dip (Table A1). All the movements (normal, strike-slip and reverse) are allowed in the stable continental region due to its wideness. Moreover, every azimuth value can be drawn in this region. Finally, since no focal mechanism is available in the oceanic crust region, all the dip (20 to 90°) and azimuth (0 to 359°) values are allowed.

Azimuth can also be described through pdf of values by deriving fault map (Figure 6a). However, these data give information only on orientation (between 0 and 180°) and not azimuth (between 0 and 360°). We thus propose to calculate pdf of orientation from fault map. Then, when an orientation $O°$ is drawn in one of these pdf, the azimuth is chosen as equal to $O°$ or $O + 180°$. Figure A1 illustrates pdf of orientation derived from fault map for each SHARE region (except the oceanic crust one).

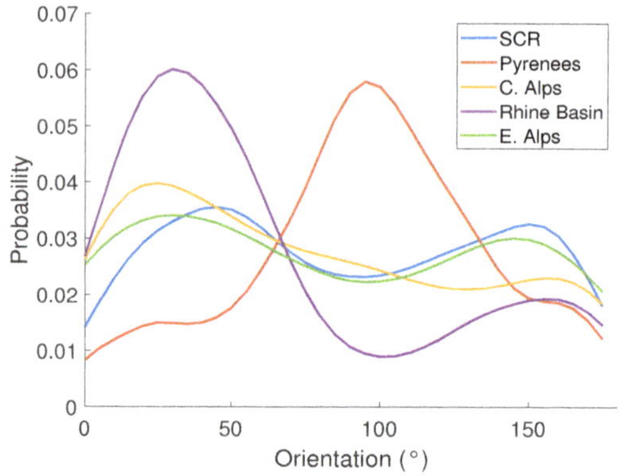

Figure A1. Pdf of orientation obtained from fault map for each SHARE region. C. and E. Alps respectively stand for Compressional and Extensional Alps (regions 4 and 6 Figure 4). No fault is localized in the oceanic crust region in the data we used thus this region isn't associated to pdf of orientation.

Appendix B. Application of GR Law in Each SHARE Regions

For some reasons explained in the text, only one Frequency-Magnitude Distribution (FMD) of main shocks, computed at national scale, is used in this paper. However, in order compare FMD obtained in each region from synthetic main shocks (Figure 10) with data, we also decide to apply GR law regression on instrumental catalogue for each region.

We first divide the initial instrumental catalogue into six sub-catalogues, one per region. Only earthquakes localized in a given region are listed in the associated sub-catalogue. Then, we explore completeness of these sub-catalogues through the cumulative visual method [36,37]. Results of this analysis are shown Table A2.

Table A2. Cut-off years obtained from instrumental catalogue in each SHARE region according to the cumulative visual method [36,37]. Region numbers refer to Figure 4.

M_C	2–2.1	2.2–2.3	2.4–2.5	2.6–2.7	2.8–2.9	3–3.1	3.2–3.3	3.4–3.5	≥3.6
Region 1	1975	1975	1970	1970	1970	1970	1970	1970	1965
Region 2	1965	1965	1965	1965	1965	1965	1965	1965	1965
Region 3	1980	1975	1975	1970	1970	1970	1970	1965	1965
Region 4	1975	1975	1975	1975	1975	1975	1975	1965	1965
Region 5	1980	1980	1980	1980	1980	1980	1980	1965	1965
Region 6	1975	1975	1975	1975	1975	1970	1965	1965	1965

Once sub-catalogues are complete, only main shocks are kept by applying G85 declustering algorithm [40]. Finally, we regress GR law on the observed frequencies of main shocks as function of magnitude. Obtained b-values for each region are summarized Table A3.

Table A3. b-values computed by regressing GR law (Equation (10)) on FMD of main shocks observed in instrumental seismicity for each SHARE region. Magnitude's ranges and number of data used for regression are also detailed. Region numbers refer to Figure 4.

Region	1	2	3	4	5	6
Magnitude's range	2–5	2–3.8	2–5.2	2–4.9	2–4.1	2–4.7
Number of data	3531	40	1962	1539	301	1614
b	1.18	0.72	1.05	1.06	1.36	1.21

One can see that the two extreme b-values, 0.72 and 1.36 (Table A3), are obtained with the lower number of data: 40 and 301 respectively. Thus, these values must be used carefully.

References

1. Rey, J.; Tinard, P. Partenariat CCR-BRGM 2014-2019. In *Travaux Menés en 2017 dans le cadre du Programme Commun sur le Risque Sismique*; Technical Report BRGM/RP-68082-FR; BRGM: Orléans, France, 2017; pp. 1–26.
2. Rey, J.; Tinard, P. Estimation financière du risque sismique à l'échelle départementale: À l'interface entre sismologie et réassurance, travaux communs CCR-BRGM (2014–2019). In Proceedings of the 9ème Colloque de l'AFPS: AFPS15, Anticiper, Limiter, Gérer les Effets des Séismes Dans Les Territoires, Marne-La-Vallée, France, 30 November–2 December 2015; pp. 1–12.
3. Rey, J.; Tinard, P. Partenariat CCR-BRGM 2014-2019. In *Travaux Menés en 2016 dans le Cadre du Programme Commun sur le Risque Sismique*; Technical Report BRGM/RP-66473-FR; BRGM: Orléans, France, 2016; pp. 1–25.
4. Mouroux, P.; Le Brun, B.; Depinois, S.; Bertrand, E.; Masure, P. *Projet Européen RISK-UE: Application à la Ville de Nice*; Technical Report BRGM/RP-53202-FR; BRGM: Orléans, France, 2004; pp. 1–137.
5. Cornell, A. Engineering Seismic Risk Analysis. *Bull. Seismol. Soc. Am.* **1968**, *58*, 1583–1606. [CrossRef]
6. Musson, R.M.W. Determination of Design Earthquakes in Seismic Hazard Analysis Through Monte Carlo Simulation. *J. Earthq. Eng.* **1999**, *3*, 463–474. [CrossRef]
7. Goda, K.; De Risi, R. Probabilistic Tsunami Loss Estimation Methodology: Stochastic Earthquake Scenario Approach. *Earthq. Spectra* **2017**, *33*, 1302–1323. [CrossRef]
8. Sitharam, T.G. The Quientessence of 25 Years of Our Contributions to Geotechnical Earthquake Engineering. *Indian Geotech. J.* **2021**, *51*, 3–49. [CrossRef]
9. Ji, J.; Wang, C.-W.; Gao, Y.; Zhang, L. Probabilistic investigation of the seismic displacement of earth slopes under stochastic ground motion: A rotational sliding block analysis. *Can. Geotech. J.* **2021**, *58*, 952–968. [CrossRef]
10. Baize, S.; Cushing, E.M.; Lemeille, F.; Jomard, H. Updated seismotectonic zoning scheme of Metropolitan France, with reference to geologic and seismotectonic data. *Bull. Société Géol. Fr.* **2013**, *184*, 225–259. [CrossRef]
11. Jacob, K.; Armbruster, J.; Barstow, N.; Horton, S. Probabilistic Ground Motion Estimates for New York: Comparison with Design Ground Motions in National and Local Codes. In Proceedings of the 5th USA National Conference on Earthquake Engineering, Chicago, IL, USA, 10–14 July 1994; pp. 119–128.
12. Woo, G. Kernel Estimation Methods for Seismic Hazard Area Source Modeling. *Bull. Seismol. Soc. Am.* **1996**, *86*, 353–362.
13. Milner, K.R.; Shaw, B.E.; Goulet, C.A.; Richards-Dinger, K.B.; Callaghan, S.; Jordan, T.H.; Dieterich, J.H.; Field, E.H. Toward Physics-Based Nonergodic PSHA: A Prototype Fully Deterministic Seismic Hazard Model for Southern California. *Bull. Seismol. Soc. Am.* **2021**, *111*, 898–915. [CrossRef]
14. Stein, S.; Liu, M.; Camelbeeck, T.; Merino, M.; Landgraf, A.; Hintersberger, E.; Kübler, S. Challenges in Assessing Seismic Hazard in Intraplate Europe. *Geol. Soc. Lond. Spec. Publ.* **2015**, *432*, 13–28. [CrossRef]
15. Calais, E.; Camelbeeck, T.; Stein, S.; Liu, M.; Craig, T.J. A New Paradigm for Large Earthquakes in Stable Continental Plate Interiors: LARGE EARTHQUAKES IN SCRs. *Geophys. Res. Lett.* **2016**, *43*, 10621–10637. [CrossRef]
16. Mazzotti, S.; Jomard, H.; Masson, F. Processes and Deformation Rates Generating Seismicity in Metropolitan France and Conterminous Western Europe. *Bull. Société Géol. Fr.* **2020**, *191*, 1–20. [CrossRef]
17. Martin, C.; Ameri, G.; Baumont, D.; Carbon, D.; Senfaute, G.; Thiry, J.-M.; Faccioli, E.; Savy, J. Probabilistic Seismic Hazard Assessment for South-Eastern France. *Bull. Earthq. Eng.* **2018**, *16*, 2477–2511. [CrossRef]
18. Guyonnet-Benaize, C. Modélisation 3D Multi-Échelle Des Structures Géologiques de La Région de La Faille de La Moyenne Durance (SE France). Ph.D. Thesis, Aix–Marseille 1, Marseille, France, 2011.
19. Guyonnet-Benaize, C.; Lamarche, J.; Hollender, F.; Viseur, S.; Münch, P.; Borgomano, J. Three-dimensional structural modeling of an active fault zone based on complex outcrop and subsurface data: The Middle Durance Fault Zone inherited from polyphase Meso-Cenozoic tectonics (southeastern France). *Tectonics* **2015**, *34*, 265–289. [CrossRef]

20. Quigley, M.; Villamor, P.; Furlong, K.; Beavan, J.; Dissen, R.; Litchfield, N.; Stahl, T.; Duffy, B.; Bilderback, E.; Noble, D.; et al. Previously Unknown Fault Shakes New Zealand's South Island. *Eos Trans. Am. Geophys. Union* **2010**, *91*, 469–470. [CrossRef]
21. Liu, M.; Stein, S.; Wang, H. 2000 Years of Migrating Earthquakes in North China: How Earthquakes in Midcontinents Differ from Those at Plate Boundaries. *Lithosphere* **2011**, *3*, 128–132. [CrossRef]
22. Ritz, J.-F.; Baize, S.; Ferry, M.; Larroque, C.; Audin, L.; Delouis, B.; Mathot, E. Surface Rupture and Shallow Fault Reactivation during the 2019 Mw 4.9 Le Teil Earthquake, France. *Commun. Earth Environ.* **2020**, *1*, 2662–4435. [CrossRef]
23. Quigley, M.; King, T.; Clark, D. The 2016 Mw 6.1 Petermann Ranges earthquake rupture, Australia: Another "one-off" stable continental region earthquake. In Proceedings of the EGU 2020, Vienna, Austria, 4–8 May 2020. [CrossRef]
24. Wells, D.L.; Coppersmith, K.J. New Empirical Relationships among Magnitude, Rupture Length, Rupture Width, Rupture Area, and Surface Displacement. *Bull. Seismol. Soc. Am.* **1994**, *84*, 974–1002.
25. Leonard, M. Self-Consistent Earthquake Fault-Scaling Relations: Update and Extension to Stable Continental Strike-Slip Faults. *Bull. Seismol. Soc. Am.* **2014**, *104*, 2953–2965. [CrossRef]
26. Iervolino, I.; Chioccarelli, E.; Giorgio, M. Aftershocks' Effect on Structural Design Actions in Italy. *Bull. Seismol. Soc. Am.* **2018**, *108*, 2209–2220. [CrossRef]
27. Ogata, Y. Statistical Models for Earthquake Occurrences and Residual Analysis for Point Processes. *J. Am. Stat. Assoc.* **1988**, *83*, 9–27. [CrossRef]
28. Richter, C.F. *Elementaty Seismology. Publisher: W.F. Freeman and Company, San Francisco, and Bailey Bros*; Swinfen Ltd.: London, UK, 1958.
29. Båth, M. Lateral Inhomogeneities of the Upper Mantle. *Tectonophysics* **1965**, *2*, 483–514. [CrossRef]
30. Zakharova, O.; Hainzl, S.; Bach, C. Seismic Moment Ratio of Aftershocks with Respect to Main Shocks: Seismic Moment Ratio of Aftershocks. *J. Geophys. Res. Solid Earth* **2013**, *118*, 5856–5864. [CrossRef]
31. Woessner, J.; Laurentiu, D.; Giardini, D.; Crowley, H.; Cotton, F.; Grünthal, G.; Valensise, G.; Arvidsson, R.; Basili, R.; Demircioglu, M.B.; et al. The SHARE Consortium. The 2013 European Seismic Hazard Model: Key components and results. *Bull. Earthq. Eng.* **2015**, *13*, 3553–3596. [CrossRef]
32. Delavaud, E.; Cotton, F.; Akkar, S.; Scherbaum, F.; Danciu, L.; Beauval, C.; Drouet, S.; Douglas, J.; Basili, R.; Sandikkaya, M.A.; et al. Toward a ground-motion logic tree for probabilistic seismic hazard assessment in Europe. *J. Seismol.* **2012**, *16*, 451–473. [CrossRef]
33. Manchuel, K.; Traversa, P.; Baumont, D.; Cara, M.; Nayman, E.; Durouchoux, C. The French Seismic CATalogue (FCAT-17). *Bull. Earthq. Eng.* **2017**, *16*, 2227–2251. [CrossRef]
34. Drouet, S.; Ameri, G.; Le Dortz, K.; Secanell, R.; Senfaute, G. A Probabilistic Seismic Hazard Map for the Metropolitan France. *Bull. Earthq. Eng.* **2020**, *18*, 1865–1898. [CrossRef]
35. Cara, M.; Cansi, Y.; Schlupp, A.; Arroucau, P.; Béthoux, N.; Beucler, E.; Bruno, S.; Calvet, M.; Chevrot, S.; Deboissy, A. SI-Hex: A New Catalogue of Instrumental Seismicity for Metropolitan France. *Bull. Société Géol. Fr.* **2015**, *186*, 3–19. [CrossRef]
36. Tinti, S.; Mulargia, F. Completeness analysis of a seismic catalog. *Ann. Geophys.* **1983**, *3*, 407–414.
37. Tinti, S.; Mulargia, F. An improved method for the analysis of the completeness of a seismic catalogue. *Lett. Nuovo C. (1971–1985)* **1985**, *42*, 21–27. [CrossRef]
38. Gutenberg, B.; Richter, C.F. Frequency of Earthquakes in California. *Bull. Seismol. Soc. Am.* **1944**, *34*, 185–188. [CrossRef]
39. Gardner, J.K.; Knopoff, L. Is Sequence of Earthquakes in Southern California, with Aftershocks Removed, Poissonian? *Bull. Seismol. Soc. Am.* **1974**, *64*, 1363–1367. [CrossRef]
40. Grünthal, G. The Up-Dated Earthquake Catalogue for the German Democratic Republic and Adjacent Areas–Statistical Data Characteristics and Conclusions for Hazard Assessment. In Proceedings of the 3rd International Symposium on the Analysis of Seismicity and Seismic Risk, Prague, Czechoslovakia, 17–22 June 1985.
41. Cornell, A.; Vanmarcke, E.H. The Major Influences on Seismic Risk. In Proceedings of the 4th World Conference on Earthquake Engineering, Santiago, Chile, 13–18 January 1969; pp. 69–83.
42. Hanks, T.C.; Kanamori, H. A Moment Magnitude Scale. *J. Geophys. Res. Solid Earth* **1979**, *84*, 2348–2350. [CrossRef]
43. Beauval, C.; Bard, P.-Y.; Danciu, L. The Influence of Source-and Ground-Motion Model Choices on Probabilistic Seismic Hazard Levels at 6 Sites in France. *Bull. Earthq. Eng.* **2020**, *18*, 4551–4580. [CrossRef]
44. Ameri, G.; Baumont, D.; Gomes, C.; Le Dortz, K.; Le, B.; Martin, C.; Secanell, R.; Vanmarcke, E.H. On the Choice of Maximum Earthquake Magnitude for Seismic Hazard Assessment in Metropolitan France–Insight from the Bayesian Approach. In Proceedings of the 9ème Colloque National de l'AFPS: AFPS15, Anticiper, Limiter, gérer les Effets des Séismes Dans les Territoires, Marne-La-Vallée, France, 30 November–2 December 2015; pp. 1–13.
45. Zentner, I. *Bayesian Estimation of the Maximum Magnitude Mmax Based on the Statistics of Extremes*; Technical Report SIGMA-2-2018-D5-004; SIGMA: St. Louis, MO, USA, 2018; pp. 1–45.
46. Helmstetter, A.; Sornette, D. Importance of Direct and Indirect Triggered Seismicity in the ETAS Model of Seismicity. *Geophys. Res. Lett.* **2003**, *30*. [CrossRef]
47. Utsu, T. Aftershocks and Earthquake Statistics(1): Some Parameters Which Characterize an Aftershock Sequence and Their Interrelations. *J. Fac. Sci. Hokkaido Univ.* **1970**, *3*, 129–195.
48. Beauval, C.; Hainzl, S.; Scherbaum, F. Probabilistic seismic hazard estimation in low-seismicity regions considering non-Poissonian seismic occurrence. *Geophys. J. Int.* **2006**, *164*, 543–550. [CrossRef]

49. Vannucci, G.; Imprescia, P.; Gasperini, P. Earthquake Mechanisms of the Mediterranean Area (EMMA) Version 3: An Improved Tool for Characterizing the Tectonic Deformation Styles in the Mediterranean. In Proceedings of the EGU 2009, Vienna, Austria, 19–24 April 2009.
50. Heidbach, O.; Barth, A.; Müller, B.; Reinecker, J.; Stephansson, O.; Tingay, M.; Zang, A. *WSM Quality Ranking Scheme, Database Description and Analysis Guidelines for Stress Indicator*; Technical Report 16-01; World Stress Map: Tokyo, Japan, 2016; pp. 1–55.

applied sciences

MDPI

Article

The Effect of the Wenchuan and Lushan Earthquakes on the Size Distribution of Earthquakes along the Longmenshan Fault

Chun Hui [1,2,3], Changxiu Cheng [1,2,3,4,*], Shi Shen [1,2,3], Peichao Gao [1,2,3], Jin Chen [1,2], Jing Yang [5] and Min Zhao [1,2,3]

1 Key Laboratory of Environmental Change and Natural Disaster, Beijing Normal University, Beijing 100875, China; 201831051038@mail.bnu.edu.cn (C.H.); shens@bnu.edu.cn (S.S.); gaopc@bnu.edu.cn (P.G.); chenjin@mail.bnu.edu.cn (J.C.); zhaom@smail.nju.edu.cn (M.Z.)

2 State Key Laboratory of Earth Surface Processes and Resource Ecology, Beijing Normal University, Beijing 100875, China

3 Center for Geodata and Analysis, Faculty of Geographical Science, Beijing Normal University, Beijing 100875, China

4 National Tibetan Plateau Data Center, Beijing 100101, China

5 College of Data Science, Taiyuan University of Technology, Taiyuan 030024, China; yangj@mail.bnu.edu.cn

* Correspondence: chengcx@bnu.edu.cn

Abstract: Changes in the stress state of faults and their surroundings is a highly plausible mechanism explaining earthquake interaction. These stress changes can impact the seismicity rate and the size distribution of earthquakes. However, the effect of large earthquakes on the earthquake size distribution along the Longmenshan fault has not been quantified. We evaluated the levels of the *b* value for the stable state before and after the large earthquakes on 12 May 2008 (Wenchuan, M_S 8.0) and 20 April 2013 (Lushan, M_S 7.0) along the Longmenshan fault. We found that after the mainshocks, the size distribution of the subsequent earthquakes shifted toward relatively larger events in the Wenchuan aftershock zone (*b* value decreased from 1.21 to 0.84), and generally remained invariable in the Lushan aftershock zone (*b* value remained at 0.76). The time required for the *b* value to return to stable states after both mainshocks was entirely consistent with the time needed by the aftershock depth images to stop visibly changing. The result of the temporal variation of *b* values shows decreasing trends for the *b* value before both large earthquakes. Our results are available for assessing the potential seismic risk of the Longmenshan fault as a reference.

Keywords: *b* value; stable state; depth; Longmenshan fault

check for updates

Citation: Hui, C.; Cheng, C.; Shen, S.; Gao, P.; Chen, J.; Yang, J.; Zhao, M. The Effect of the Wenchuan and Lushan Earthquakes on the Size Distribution of Earthquakes along the Longmenshan Fault. *Appl. Sci.* **2021**, *11*, 8534. https://doi.org/10.3390/app11188534

Academic Editors: Ricardo Castedo, Miguel Llorente Isidro and David Moncoulon

Received: 10 August 2021
Accepted: 11 September 2021
Published: 14 September 2021

1. Introduction

Following the Wenchuan M_S 8.0 earthquake on 12 May 2008, the Longmenshan fault zone was struck by the 20 April 2013 M_S 7.0 Lushan earthquake. The Longmenshan fault zone is composed of several almost parallel thrust faults, forming a boundary fault between the Sichuan Basin and Tibetan Plateau, and controls the seismicity of the Longmenshan region (Figure 1). The epicenters of the Wenchuan and Lushan earthquakes were approximately 87 km apart, and the focal mechanism of both events showed a thrust rupture [1].

According to the characteristics of the Wenchuan and Lushan earthquakes, whether the M_S 7.0 Lushan event was a strong aftershock of the M_S 8.0 Wenchuan earthquake or a new and independent event has been a topic of debate. For example, some researchers suggest that the two large earthquakes were independent events. The reasons are as follows: (1) there is no overlapping area between the Wenchuan and Lushan earthquake rupture zones [2]; (2) the Wenchuan and Lushan earthquakes were generated in different faults in the Longmenshan fault zone [3]; (3) the rupture processes of the Wenchuan and Lushan earthquakes were different, and the aftershock zones of the two events were nearly 45 km apart [4]. Alternatively, some scientists propose that the Wenchuan and Lushan earthquakes were a mainshock–aftershock sequence and note that the Lushan event struck in an area where Coulomb stress was increased due to the Wenchuan earthquake [5,6].

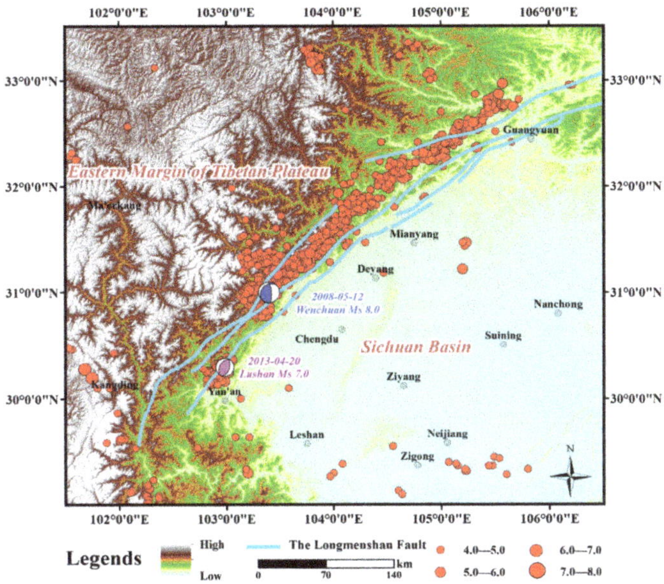

Figure 1. The topographic and tectonic map of the Longmenshan fault zone and its surrounding region. Blue beach ball represents the focal mechanism of the Wenchuan M_S 8.0 earthquake. Pink beach ball represents the focal mechanism of the Lushan M_S 7.0 earthquake. Red circles represent epicenters of earthquakes ($M_S \geq 4$) from 1 January 2000 to 1 January 2019.

The controversy over the relationship between the Wenchuan and Lushan earthquakes highlights the complexity of earthquake interaction in the Longmenshan fault zone. It is widely accepted that earthquake interactions can be understood by identifying changes in static and dynamic stress states around faults [7–9]. The most observable effect of this stress change is a significant increase in seismicity rate, which is generally considered an aftershock phenomenon [10–12]. Statistically, aftershock activity is classically described by $n(t) = K/(t + c)p$, where $n(t)$ is the number of aftershocks after time t and K, c, and p are constants that describe the aftershock productivity [13]. Ogata [14,15] described aftershock activity as a multigenerational branching process and proposed the epidemic-type aftershock sequence model, which is a stochastic point process model of self-exciting point processes.

Changes in stress can impact the seismicity rate and the frequency size distribution, which is alternatively known as the frequency-magnitude distribution (FMD) [16] or Gutenberg–Richter (G–R) law [17] and is expressed as $\log N = a - b\ M$, where N is the number of events in a given time period with magnitude greater than M, a describes the seismicity of a given seismogenic volume, and b is the slope of the FMD. Previous studies showed that b values fall within the range of 1.02 ± 0.03 on a large scale for a long time [18,19]. For regions on a smaller scale, the b values show a broad range of spatial and temporal variations. For example, the b value ranged from 0.5 to 2.5 in the Andaman–Sumatra region and California [20]. Interpretation of the variation of b values is based on several factors, including stress state [21,22], focal depth [23], faulting style [24,25], fluid pressure [26], and so on.

The earthquake size distribution generally follows a power law, with a slope of b values, which characterizes the relative occurrence of large and small events. A low b value indicates a larger proportion of large earthquakes and vice versa. Zhao et al. [27]

(2008) compared the spatial footprint of b values before and after the Wenchuan earthquake in the Longmenshan fault zone, and the results showed that the b values tend to change from lower in the southern region to higher in the northeastern region. The temporal change in b values before the Wenchuan M_S 8.0 earthquake showed a decreasing trend [28–30], and the Lushan M_S 7.0 earthquake showed similar temporal change trends in b values [31]. These studies focused on the variation trend of b values before and after the two large events along the Longmenshan fault. However, the fundamental effect of the Wenchuan and Lushan earthquakes on the size distribution of earthquakes along the Longmenshan fault has not been quantified, which limits our understanding of how the apparent stress changes in the region affect the size distribution of earthquakes.

In this study, we evaluated the spatiotemporal evolution of the b values along the Longmenshan fault in the past nearly 20 years. Moreover, we estimated the levels of the b value for the stable state before and after the Wenchuan and Lushan earthquakes and quantified the effects of the two large earthquakes on the size distribution of subsequent events at different times. In addition, the spatial evolution process of the deep seismogenic environment in the Wenchuan and Lushan aftershock zones in two and three dimensions was illustrated via spatial scanning and data fitting, which can be used to analyze the aftershock activity of the two large earthquakes.

2. Data and Postulates

The earthquake catalog we used here was documented by the regional seismic network and then verified by the China Earthquake Networks Center (CENC) along the Longmenshan fault during the period from 1 January 2000 to 1 January 2019. It is a relatively complete catalog containing the Wenchuan-Lushan earthquake sequence.

Figure 2 shows the magnitude-time distribution of earthquakes in the Wenchuan source region and Lushan source region. The locations of the earthquakes in this catalog were corrected for accuracy. Each event includes the time, location, magnitude, and depth. Homogeneity of the catalog was iterated and optimized for subsequent research and analysis.

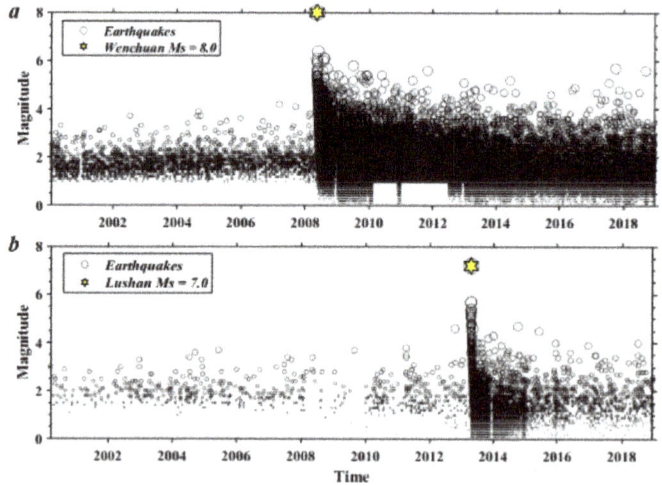

Figure 2. Magnitude versus time for events in the Longmenshan fault zone from 2000 to 2019. (**a**) Magnitude versus time for events in the M_S 8.0 Wenchuan source region; (**b**) magnitude versus time for events in the M_S 7.0 Lushan source region.

Stress change has been widely used to interpret the triggering of the mainshock–mainshock and mainshock–aftershock events [7,32–36]. However, the exact measurement of stress states is difficult; thus, a relationship between the stress state and b value has been proposed [16,24,37,38]. Schorlemmer et al. [24] demonstrated that the b value could be regarded as a stress indicator that depends inversely on differential stress. Therefore, changes in the stress state on faults lead to the variation in b values, which is followed by time-dependent recovery. In general, the larger the magnitude of the earthquake, the greater the stress changes and the larger the b value fluctuations. No event of magnitude larger than M_S 7.0 had been reported in the historical record of the Longmenshan fault zone and the catalog we used in this paper contains more than 80,000 events and only two large earthquakes greater than magnitude 6.5, that is, the Wenchuan M_S 8.0 and Lushan M_S 7.0 earthquakes. Thus, these two events are mainly responsible for the apparent changes in stress state along the Longmenshan fault zone in the past 20 years.

Therefore, there are two postulates: first, without a large earthquake perturbation, the b value will remain in a stable state with a small fluctuation range for a long time; second, after the perturbation of a large earthquake, the b value may recover to another stable state. We evaluated the levels of the b value for the stable state before and after these two large earthquakes in the study region.

3. Methods

The main methods we used include estimation methods (MaxCurvature for the estimation of the completeness magnitude; maximum likelihood estimation (MLE) for the b value estimation), test methods (Akaike information criterion (AIC) for the variation trend of b values; the nonlinearity index (NLIndex) for the linearity assessment of frequency-magnitude distribution) and kriging interpolation to describe the spatial and temporal evolution images of aftershock focal depth.

3.1. Completeness Magnitude (M_C) and b Value Estimation

The estimation of the completeness of earthquake catalogs is essential to the computation of b values, and the lowest magnitude of all earthquakes that are reliably detected in a space-time volume is defined as the completeness magnitude (M_C) [39]. The lower the M_C, the higher the detection capability. Here, we use the MaxCurvature technique, which estimates the M_C by locating the magnitude that is the highest frequency of events in the FMD. Mignan [40] showed that the MaxCurvature technique underestimates the M_C in cases involving gradually curved FMDs and postulated that this underestimation tendency arises from spatiotemporal heterogeneities within the earthquake monitoring network. Therefore, we used the corrected MaxCurvature method with a correction factor of +0.2 [41], and the uncertainties were determined by bootstrapping.

The least-squares method and maximum likelihood estimation are often used to calculate the b value, and the latter approach is considered more stable. In this work, the maximum likelihood estimation used to calculate the b value and its standard deviation [42,43]:

$$b = \frac{1}{\ln(10)(\overline{M} - M_C)} \tag{1}$$

where $\overline{M}$ is the average magnitude of earthquakes with $M \geq M_C$; M_C is the cutoff magnitude. The confidence limit of the b value is expressed as follows:

$$\sigma = \frac{b}{\sqrt{N}} \tag{2}$$

where N is the number of earthquake cases of the given sample.

3.2. Estimation of the Frequency-Magnitude Distribution (FMD) Extrapolation

The nonlinearity index (NLIndex) can be used to assess whether the extrapolation of a given high-magnitude FMD is likely an overestimate or underestimate of the probable rates for large events [44].

(1) Calculate the b value for all M_{cut} from M_{cmin} to the highest M_{cut} for which N_{min} events are still sampled.
(2) Divide the standard deviation by the largest individual b value uncertainty for each possible M_{cut}, and for each M_{cut}, this value is the NLIndex.
(3) Divide the NLIndex for each M_{cut} by the number of estimated b values to weight the result by data density. Please check Italics.
(4) Find the minimum weighted NLIndex to find the best M_C that produces the most linear FMD fit.

If NLIndex ≤ 1, the FMD is regarded as linear, and if NLIndex > 1, the FMD is not linear. The slope of M_{cut} is clearly positive or negative, respectively, indicating that the FMD overestimates or underestimates large M rates.

3.3. *Akaike Information Criterion*

To quantify the changing trend of b values, the P test was conducted for b values in two sample windows based on the Akaike information criterion (AIC) [45]. Hypothesis 1: the b values in the two sample windows are the same; Hypothesis 2: the b values in the two sample windows are different, represented as b_1 and b_2, respectively. The hypothesis of the difference of the AIC leads to the difference ΔAIC [46]:

$$\Delta AIC = -2(N_1 + N_2)\ln(N_1 + N_2) + 2N_1 \ln(N_1 + \frac{N_2 b_1}{b_2}) + 2N_2 \ln(N_2 + \frac{N_1 b_2}{b_1}) - 2 \quad (3)$$

where N_i is the number of events in the sample windows and b_i is b values in the sample windows. P_b represents the probability that the events in the two sample windows come from the same population and can be derived from the AIC as follows:

$$P_b = e^{(-\Delta AIC/2)-2} \quad (4)$$

The b value in the sample window represents a significant change when $\Delta AIC \geq 2$ ($P_b \approx 0.05$) and is highly significant when $\Delta AIC > 5$ ($P_b \approx 0.01$) [47].

3.4. *Kriging Interpolation*

Kriging interpolation is the most commonly used geostatistical approach for spatial interpolation. With this method, a semivariogram is used to express the spatial relationship of the distance between samples. This technique depends on the spatial model between samples to predict attribute values at unsampled locations [48]. As a widely used interpolation method, kriging takes into account the distance between unknown positions and the sample locations as well as the distance between sample locations, effectively reducing the interference of clustering in samples on the accuracy of the interpolated estimates [49]. We used the kriging interpolation algorithm to produce maps incorporating anisotropy and underlying trends from irregularly spaced data. The exponential semivariance model with the smallest prediction errors was chosen over the Gaussian and spherical models for the spatial interpolation of focal depth data.

4. Results and Analysis

4.1. *Completeness Magnitude (M_C) and Linearity Assessment of Frequency-Magnitude Distribution (FMD)*

As shown in Figure 3, the results of the corrected MaxCurvature method show that the M_C of the earthquake catalog used in this work is M_C = 1.5. This result is consistent with the results of previous studies on the Wenchuan earthquake zone [30,50]. Fang et al. [51] described in detail the aftershock performance and analysis of the Lushan earthquake based on the combined data from permanent and temporary seismic stations. They concluded that the minimum complete magnitude was M = 1.0. To unify the consistency of the M_C of the Wenchuan M_S 8.0 and Lushan M_S 7.0 earthquakes in the Longmenshan fault zone, we selected events with magnitudes of $M \geq M_C$ = 1.5.

We performed a linearity check on FMD, and the results are shown in Figure 4. The NLIndex (red) is shown for different cutoff magnitudes (upper inset) and the NLIndex ≤ 1 for all cut off magnitudes; thus, the linear FMD is accepted as the best M_C.

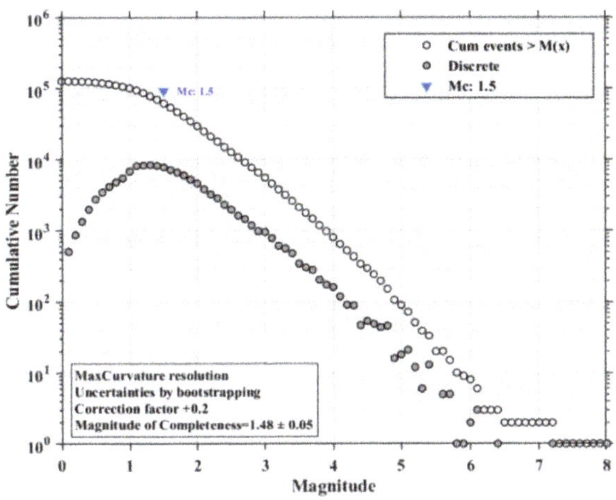

Figure 3. Frequency-magnitude distribution of the seismicity of the Wenchuan-Lushan sequence from 1 January 2000 to 1 January 2019.

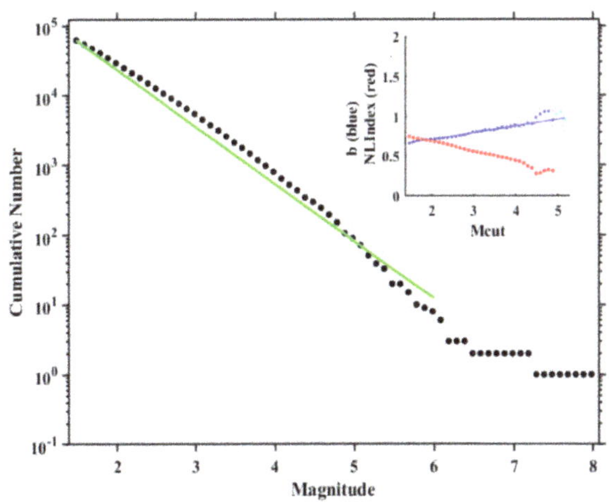

Figure 4. Frequency-magnitude distribution and NLIndex.

4.2. Time-Space Analysis of b Values

Earthquake frequency will increase immediately within a short time after a large event and may exceed the recording capacity of the seismic network. Before establishing the time-space series of b values with aftershocks, we should eliminate the events documented in the early catalog, which is somewhat heterogeneous and incomplete in small events [41]. In this work, the exclusion period depends on the magnitude of completeness over time. Therefore, we first removed the events documented in the initial catalog within two months after the Wenchuan M_S 8.0 and Lushan M_S 7.0 earthquakes, a period for which the data are highly incomplete. Then, we calculated the spatiotemporal distributions of b values before and after two large events that occurred from 2000–2019 along the Longmenshan fault by selecting events with $M \geq M_C = 1.5$ and using a time window and spatial grid to calculate the b values. In this computation, the window lengths were set to at least 500 events in the

Wenchuan aftershock zone and 200 events in the Lushan aftershock zone. Each window was moved forward by one event at a time.

Figure 5a,b display time series of the b value in source regions, and the overall change trend conforms to our postulation that b values will undergo relatively significant changes in a period of time before and after a large earthquake. Specifically, b values show a decreasing trend before the occurrence of both large earthquakes in both zones. To ensure that this trend is statistically significant, we quantitatively assessed the temporal variation in b values using the P parameter test and selected three windows before the M_S 8.0 event (W_1, W_2, and W_3) and the M_S 7.0 event (L_1, L_2, and L_3). Window selection was based on the significance of changes in b values (Figure 5a,b). The results are shown in Table 1. The b value in the sample window has significantly changed when $\Delta\text{AIC} \geq 2$ ($P_b \approx 0.05$). Table 1 shows that the b values decreased before both large earthquakes with statistically significant variations.

After the Wenchuan M_S 8.0 event, the b values in the Wenchuan aftershock zone experienced a period of dramatic fluctuation (indicated by the pink shading lasting not more than one year) before gradually stabilizing within a small fluctuation range (Figure 5a), which was similar to the range of b values in the third period (Figure 5c, b = 0.84). After the Lushan M_S 7.0 event, the b values in the Lushan aftershock zones increased rapidly and then slowly dropped to a stable state (Figure 5b), which was similar to the FMD in the first period (Figure 5d, b = 0.76). As shown in Figure 5b (red shading), the b value required less than ten months to return to a stable state.

The reference b values can be estimated for the background levels (for the period, b = 1.21 in the Wenchuan aftershock zone and b = 0.76 in the Lushan aftershock zone). When the perturbation effect of the mainshocks gradually decreases, the b values in the Lushan aftershock zone eventually return to the background level (b = 0.76), whereas those in the Wenchuan aftershock zone drop below the background level (from 1.21 to 0.84). To date, there have been no earthquakes that have significantly changed the stability of the Longmenshan fault zone since the M_S 7.0 Lushan earthquake.

The temporal distribution of earthquakes also indicates the change in stress state of faults and their surroundings. As shown in Figure 2a, before the M_S 8.0 Wenchuan earthquake, the frequency of events in the Wenchuan aftershock zone had been decreasing for a year since 2006, and only two events greater than M_S 4 occurred during the period when b values were significantly decreasing (Figure 5a). However, earthquakes that were greater than M_S 4 struck the entire Wenchuan aftershock zone after the M_S 8.0 event (Figure 1). In addition, there were no strong aftershocks above magnitude 6.5 along the faults, and only six events greater than M_S 4 occurred within two months after the mainshock. These phenomena indicate that without the continuous perturbation of strong aftershock, the b value gradually stabilized to the state mainly determined by the background earthquakes, which tended to shift to larger events following the M_S 8.0 event in the Wenchuan aftershock zone.

After the M_S 8.0 Wenchuan earthquake, the Lushan aftershock zone also experienced a "seismic quiescence" of approximately two years, and only one event greater than M_S 4 occurred before the M_S 7.0 Lushan earthquake during the period when b values were significantly decreasing (Figure 2b). Moreover, only a few events greater than M_S 4 occurred within two months after the mainshock; the subsequent events basically returned to the magnitude of the background earthquakes before the mainshock, which shows that the b values in the Lushan aftershock zone eventually returned to the background level (Figure 5b).

To analyze the spatial footprints of the changes in the b values, we divided the two study regions into $0.1° \times 0.1°$ grids, sampled the 300 events nearest to each grid node within radius of 30 km, and re-estimated the M_C in each node. For this purpose, we used a bootstrap approach to sample the events 1000 times randomly. The spatial footprints of the changes in the b values are consistent with the FMDs (Figure 5c,d). Figure 5e–g demonstrate the spatial variation in the b values throughout the Wenchuan aftershock zone. Figure 5h–j show the spatial variation in the b values in the Lushan aftershock zone.

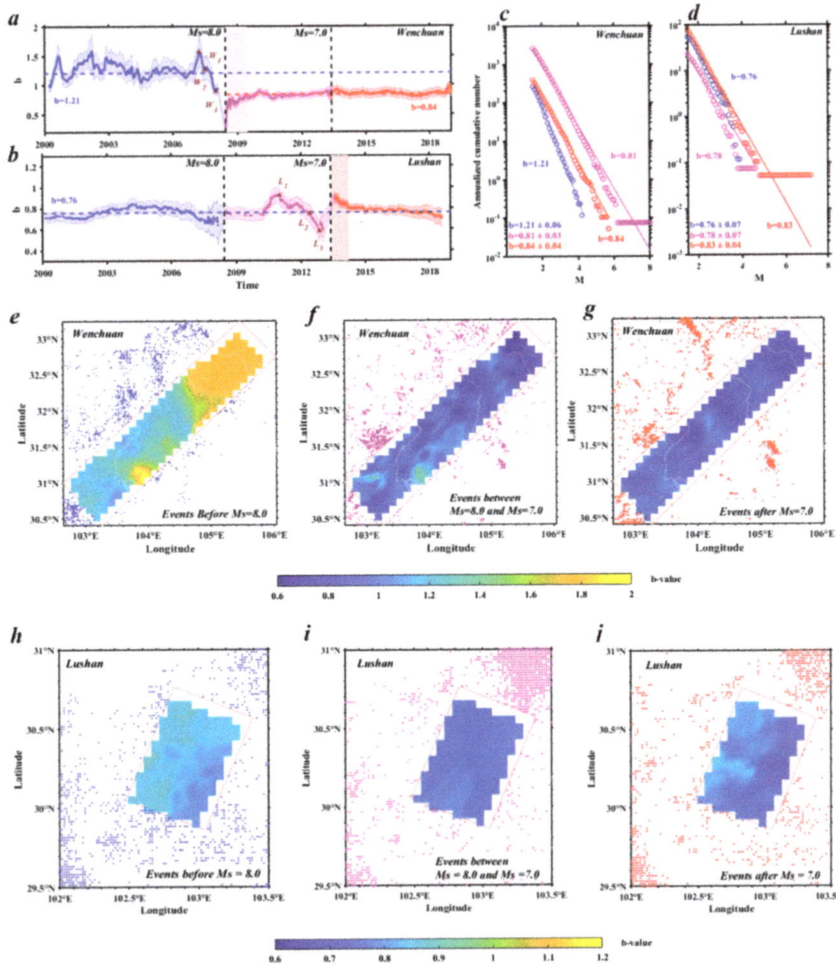

Figure 5. Temporal and spatial analysis of the b values for the Wenchuan-Lushan sequence. (**a**,**b**) Temporal variation of b values for the Wenchuan and Lushan source regions. The dashed black lines represent the times of the M_S 8.0 (Wenchuan) and M_S 7.0 (Lushan) events, and the dashed blue lines show the background b values for the Wenchuan and Lushan source regions. The dashed red line represents the b value after the time of the M_S 7.0 event (Lushan) for the Wenchuan source region. The shaded regions represent the uncertainties in the b values. (**c**,**d**) FMDs for the two aftershock zones in three different periods. (**e**–**g**) Map showing the spatial footprint of b values for the Wenchuan aftershock zone in three different periods. (**h**–**j**) Map showing the spatial footprint of b values for the Lushan aftershock zone in three different periods.

Table 1. Results of the *P* parameter test between windows.

Windows	ΔAIC	P_b
W_1 and W_2	2.6	0.03
W_2 and W_3	9.9	9.5×10^{-4}
W_1 and W_3	28.9	7.1×10^{-8}
L_1 and L_2	2.6	0.03
L_2 and L_3	4.6	0.01
L_1 and L_3	20	5.9×10^{-6}

The *b* value in the southern part of the Wenchuan source region and Lushan source region was lower than that in the northern part of the Wenchuan aftershock before the M_S 8.0 Wenchuan event (Figure 5e,h). This pattern is consistent with the characteristics of the Longmenshan fault, which is a strike-slip fault in the north and a thrust fault in the southern section. It is generally considered that the *b* value is inversely proportional to stress, and the *b* values of different types of faults are as follows: *b* (normal) > 1, *b* (strike-slip)~1, and *b* (thrust) < 1 [24,25]. The conditions changed markedly after the M_S 8.0 Wenchuan event; the *b* values decreased in the Wenchuan source region and Lushan source region (Figure 5f,i). This finding illustrates the effect of the Wenchuan earthquakes on stress change along the Longmenshan fault. Figure 5g,j illustrate the stable state of the *b* value in the Longmenshan fault zone.

4.3. Evolution of Images of Aftershock Activity Depicted by Focal Depth

Spatial scanning was performed using events with depth data in the catalog, i.e., at a step size of 0.1° for both longitude and latitude. For all earthquakes in each 0.1° × 0.1° grid point, the average depth was used as the depth value of the grid point, and then kriging interpolation was applied to all the grids. We counted at least ten events in each grid node in the Wenchuan aftershock zone as samples (as well as five events for the Lushan aftershock zone) to prevent the average depth of grid points from being affected by too few events. The contour lines of depth distribution at different periods after the mainshock were obtained and superimposed with regional faults (Figures 6 and 7).

To illustrate the evolution of the deep seismogenic environment in the Wenchuan aftershock zone. Figure 6 shows the spatial evolution of the aftershock depth at one day, one week, one month, six months, one year, and three years after the mainshock. The analysis shows that the focal depth of the Wenchuan aftershock zone spread along the direction of the fault. With Mianyang as the boundary, the aftershock zone can be divided into the southern section and the northern section. The depth distribution is generally deep in the southeast and shallow in the northwest, and the average aftershock focal depth is 10–15 km.

A comparison of the aftershock activity images depicted by the focal depth information shown in Figure 6d,e revealed that the image formed one year after the mainshock did not show a significantly different pattern in the following two years. This finding indicates that after the mainshock, the aftershock frequency tends to be stable one year later, which means that the active aftershock period of the Wenchuan M_S 8.0 was less than one year.

Figure 7 shows the spatial evolution of aftershock depths at one day, one week, five months, ten months, one year and three years after the mainshock in the Lushan aftershock zone. The analysis shows that the focal depths of the Lushan aftershock are distributed around the fault, with the fault as the boundary, with a trend of deep in the southeast and shallow in the northwest. Moreover, a comparison of the regional aftershock activity images depicted by the focal depth information in Figure 7d,e revealed that the pattern of the image formed ten months after the mainshock presented limited changes in the following one year, which indicates that the aftershock frequency tended to be stable ten months after the Lushan mainshock. Therefore, the active aftershock period of the Lushan M_S 7.0 earthquake was less than ten months.

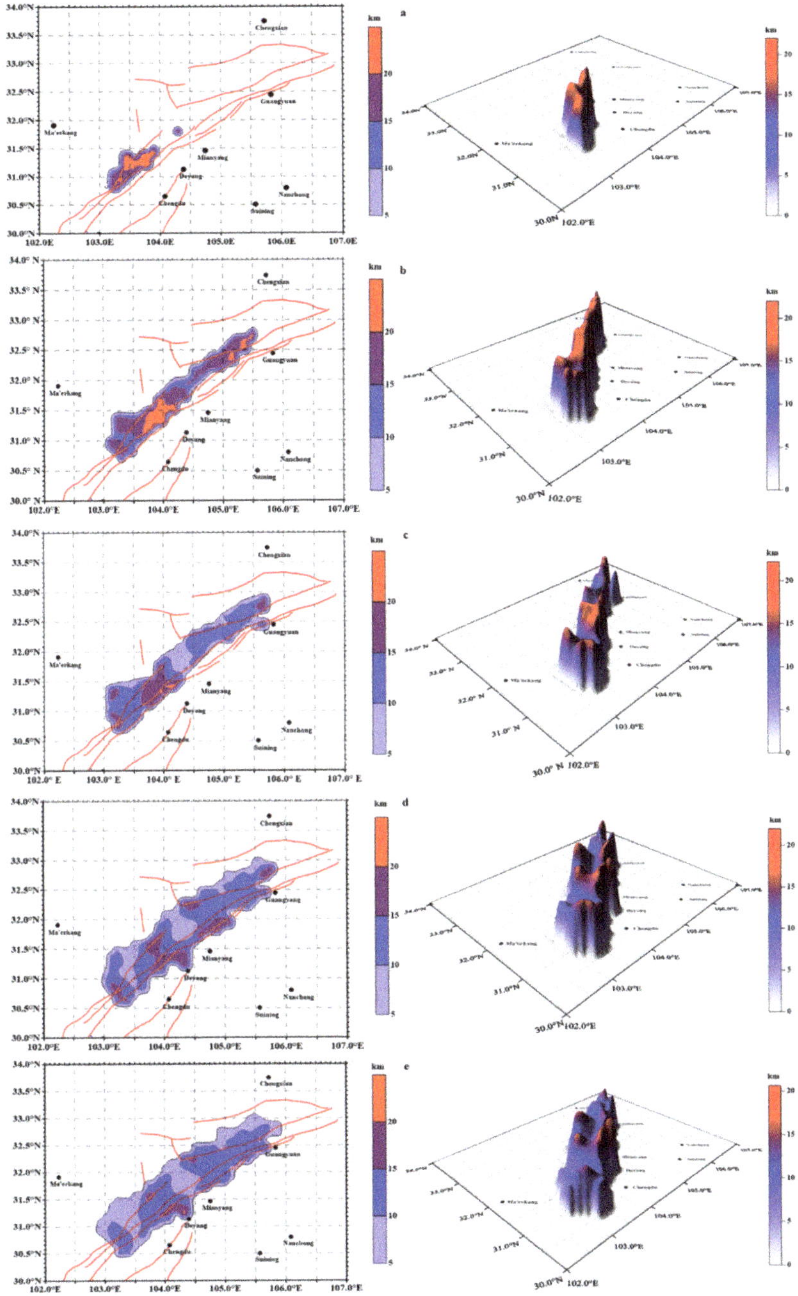

Figure 6. Spatial distribution of the focal depths of the aftershocks following the Wenchuan M_S 8.0 earthquake on the fault plane. Red lines represent the locations of faults. (**a**) One day after the mainshock; (**b**) 1 week after the mainshock; (**c**) 1 month after the mainshock; (**d**) 1 year after the mainshock; and (**e**) 3 years after the mainshock.

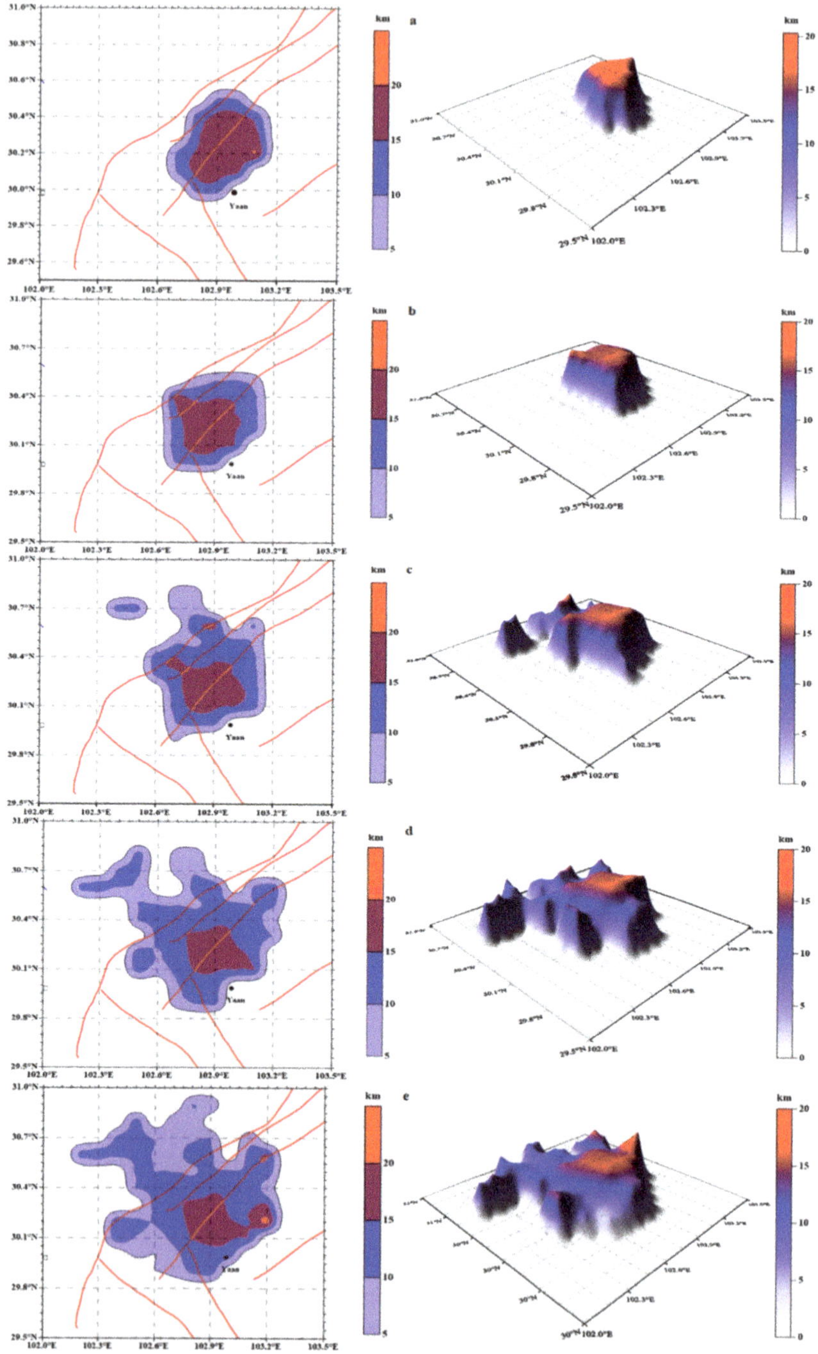

Figure 7. Spatial distribution of the focal depths of the aftershocks following the Lushan M_S 7.0 earthquake on the fault plane. Red lines represent the locations of faults. (**a**) One day after the mainshock; (**b**) 1 month after the mainshock; (**c**) 5 months after the mainshock; (**d**) 10 months after the mainshock; and (**e**) 2 years after the mainshock.

5. Discussion

Earthquake interaction can change the stress state of faults, which is reflected in both earthquake activity rate and earthquake size distribution. The relationship between stress state and b value can be used to quantify the effect of large earthquakes that can significantly change the stress states and, therefore, the earthquake size distribution. In contrast to previous studies on the Wenchuan and Lushan earthquakes, our paper focuses on a quantitative analysis and discussion of the effect of the Wenchuan M_S 8.0 and Lushan M_S 7.0 earthquakes on the size distribution of the earthquakes along the Longmenshan fault at different times.

Interpretation of the b value and its variability according to physical mechanisms has received considerable attention and discussion. In most cases, the observation of spatial and temporal b value variability can be caused by several factors: (i) Process of estimation: homogeneity of catalog and method of calculation can affect the results. All the data in this work are from the Longmenshan fault and its surroundings, and each event includes the time, location and magnitude depth. The maximum likelihood estimation and least-squares regression method [30,41–43,52–55] are used to estimate the b value and its uncertainty, but the latter is excessively affected by the largest earthquake magnitude. Marzocchi et al. [56] measured the bias on b values caused by the magnitude binning and catalog incompleteness when the b value is estimated by the maximum likelihood estimation and provided guidance to reduce the likelihood of being misled by b value variation. (ii) Stress conditions: the b value and its variation represent stress buildup and release. The differential stress is inversely dependent on the b value has been observed in laboratory experiments [57,58] as well as in the field [59,60]. The stress acting on a fault may control the variation in the b value in space and time. Parsons et al. [61] calculated the regional Coulomb stress changes on major faults surrounding the rupture resulting from the Wenchuan M_S 8.0 and showed that significant stress increased in the Lushan aftershock zone. Other studies obtained similar results [5,6,62]. The spatial variation in the b values throughout the Lushan aftershock zone decreased after the M_S 8.0 mainshock (Figure 5i), and the same effect occurred in the southern part of the Wenchuan aftershock zone after the M_S 7.0 event (Figure 5g). These findings suggest that the b value is negatively correlated with stress and indicate the effect of the earthquake interaction along the Longmenshan fault zone. (iii) Crustal tectonics: the variation in b value can be interpreted according to the tectonic characteristics, i.e., rock heterogeneity [63], focal depth [23], pore pressure [26], and fault types [24,25]. Previous studies have shown that the b value in different types of faults is b (normal) > 1, b (strike-slip)~1, and b (thrust) < 1 [26,64]. As shown in the spatial footprints in Figure 5, the b value of the southern part of the Longmenshan fault zone is lower than that in the northern part. This pattern is consistent with the tectonic characteristics of the Longmenshan fault, which is a strike-slip fault in the north and a thrust fault in the southern part [65,66].

Stress changes seem to be a key factor that affects the b value and its variation. Excepting the approach of estimation, all other factors are secondary because they are directly or indirectly affected by the stress [67]. Therefore, the observed falls in the b values shown in Figure 5a,b was interpreted as changes in the related stress conditions, which could be precursors to large earthquakes. However, these temporal variations may occur over a timescale ranging from months to years, and the timeliness and effectiveness of this variability as an indicator are difficult to guarantee. Additionally, there is usually an insufficient number of events to accurately calculate the b value before large earthquakes. Gulia and Wiemer [41] pointed out that the period following a moderate earthquake is rich in such data, with thousands of events occurring within a short period. These events may allow real-time monitoring of the evolution of b values. The authors claim that the probability of a larger earthquake following a moderate earthquake increases by several orders of magnitude if the b value remains the same or drops significantly rather than increases. However, Brodsky [68] suggested that the observed pattern revealing the changes in b values is a statistical effect rather than deterministic and that researchers need more cases to test this claim.

In general, thousands of aftershocks occur in the period following a large earthquake. Based on these abundant data, there are two typically common operational aftershock forecasting models used in aftershock hazard assessment, namely the short-term earthquake probability (STEP) model [69] and epidemic-type aftershock sequence model [14,15]. Gulia et al. [16] reported that these models forecast a high probability for a repeat of the mainshock rupture and thus substantially overestimate the aftershock hazard. This paradox can be resolved by taking into account the stress changes and their effect on the earthquake size distribution.

Our results showed that the time series of the b value in the Longmenshan fault zone after the mainshocks exhibited a period of significant fluctuation before returning to the stable state in both the Wenchuan aftershock zone (one year) and the Lushan aftershock zone (ten months). Figures 6 and 7 show that the time required for the b values to return to a stable state after both mainshocks was entirely consistent with the time required for the aftershock depth images to cease changing visibly. The spatial footprints of the changes in the b value reveal that the southern part of the Longmenshan fault zone is lower than the northern part. This finding demonstrates that the Wenchuan and Lushan events did not change the pattern of higher stress in the southern part of the Longmenshan fault zone than in the northern part. However, the most obvious change is that after the mainshock, the size distribution of the subsequent earthquakes in the Wenchuan source region shifts toward relatively larger events (lower b values).

The Longmenshan fault zone began to develop in the Late Triassic, and severe tectonic deformation occurred during the Indo-China and Himalayan movements, forming a combination of thrust and strike-slip displacement [66,70]. Previous studies did not comprehensively quantify the stable state for the Longmenshan fault zone before and after the two large events in a long time series [5,6,27,29,30,55,71]. The b value is a measurable indicator of earthquake size distribution within a specified region and period of time and is dependent on stress. With our present results, we reported the temporal and spatial variation in b values before and after two big earthquakes and fitted the source depth in time and space to quantify the stress changes and their effect on the earthquake size distribution in the Longmenshan fault zone.

6. Conclusions

Based on the tectonic characteristics and potential seismicity surrounding the aftershock zones of the Wenchuan M_S 8.0 and Lushan M_S 7.0 earthquakes, we studied the spatial and temporal variation of b values in two source regions from 2000 to 2019. In addition, the spatial evolution process of the deep seismogenic environment in the Wenchuan and Lushan aftershock zones was drawn by spatial scanning and depth data fitting.

The results depict the decreasing trends of b values before the two large earthquakes in the study region. Additionally, the b value in the Wenchuan aftershock zone took approximately one year to enter a new stable state (b values ranging from 1.21 to 0.84), while the b value in the Lushan aftershock zone took approximately ten months to return to its original stable state (b = 0.76). Moreover, the major aftershock active periods of the Wenchuan M_S 8.0 and Lushan M_S 7.0 earthquakes were less than one year and ten months, respectively, which are consistent with the time required for the b value to return to a stable state. The spatial footprints of the changes in the b values results reveal that the Wenchuan M_S 8.0 and Lushan M_S 7.0 events did not change the pattern of high b values in the north and low b values in the south along the Longmenshan fault zone.

We quantified the effect of the Wenchuan M_S 8.0 and Lushan M_S 7.0 earthquakes on the size distribution of earthquakes along the Longmenshan fault. Future studies can focus on how to quantify the effect of large earthquake size distribution across different tectonic regimes and apply the findings in potential seismic risk assessment.

Author Contributions: Conceptualization, C.H., C.C. and P.G.; Data curation, P.G. and J.Y.; Formal analysis, C.H.; Funding acquisition, S.S.; Investigation, S.S.; Methodology, S.S. and J.C.; Resources, C.C.; Software, J.C., J.Y. and M.Z.; Supervision, P.G.; Validation, P.G. and M.Z.; Visualization, S.S. and M.Z.; Writing–Original draft, C.H.; Writing–Review & editing, C.C. All authors have read and agreed to the published version of the manuscript.

Funding: This research was supported by the National Natural Science Foundation of China (Grant No. 41771537) and the Fundamental Research Funds for the Central Universities.

Data Availability Statement: Datasets for this research are available in these in-text data citation references: Shi et al. (2018), [http://doi.org/10.6038/cjg2018M0024], Li et al. (2018), [http://doi.org/10.6038/cjg2018M0129] and China Earthquake Networks Center (CENC), [http://www.cenc.ac.cn/].

Acknowledgments: The authors thank American Journal Experts (AJE) for polishing the language of this article.

Conflicts of Interest: No conflict of interest exist in the submission of this manuscript, and the manuscript is approved by all authors for publication.

References

1. Jia, K.; Zhou, S.; Zhuang, J.; Jiang, C. Possibility of the independence between the 2013 Lushan earthquake and the 2008 Wenchuan earthquake on Longmen Shan fault, Sichuan, China. *Seismol. Res. Lett.* **2014**, *85*, 60–67. [CrossRef]
2. Zhang, Y.Q.; Dong, S.W.; Hou, C.T.; Shi, J.S.; Wu, Z.H.; Li, H.L.; Sun, P.; Liu, G.; Li, J. Seismogenic structure of the April 20, 2013, Lushan Ms7 earthquake in Sichuan. *Acta Geol. Sin. Engl. Ed.* **2013**, *87*, 633–645.
3. Li, Y.Q.; Jia, D.; Wang, M.M.; Shaw, J.H.; He, J.K.; Lin, A.M.; Xiong, L.; Rao, G. Structural geometry of the source region for the 2013 Mw 6.6 Lushan earthquake: Implication for earthquake hazard assessment along the Longmen Shan. *Earth Planet. Sci. Lett.* **2014**, *390*, 275–286. [CrossRef]
4. Du, F.; Long, F.; Ruan, X.; Yi, G.X.; Wu, J. The M7.0 Lushan earthquake and the relationship with the M8.0 Wenchuan earthquake in Sichuan, China. *Chin. J. Geophys.* **2013**, *56*, 1772–1783.
5. Parsons, T.; Segou, M. Stress, distance, magnitude, and clustering influences on the success or failure of an aftershock forecast: The 2013 M 6.6 Lushan earthquake and other examples. *Seismol. Res. Lett.* **2014**, *85*, 44–51. [CrossRef]
6. Zhu, S. Is the 2013 Lushan earthquake (Mw = 6.6) a strong aftershock of the 2008 Wenchuan, China mainshock (Mw= 7.9)? *J. Geodyn.* **2016**, *99*, 16–26. [CrossRef]
7. Toda, S.; Stein, R.S.; Lin, J. Widespread seismicity excitation throughout central Japan following the 2011 M= 9.0 Tohoku earthquake and its interpretation by Coulomb stress transfer. *Geophys. Res. Lett.* **2011**, *38*. [CrossRef]
8. Wedmore, L.; Walker, J.F.; Roberts, G.P.; Sammonds, P.; McCaffrey, K.; Cowie, P. A 667 year record of coseismic and interseismic Coulomb stress changes in central Italy reveals the role of fault interaction in controlling irregular earthquake recurrence intervals. *J. Geophys. Res. Solid Earth* **2017**, *122*, 5691–5711. [CrossRef]
9. Verdecchia, A.; Pace, B.; Visini, F.; Scotti, O.; Peruzza, L.; Benedetti, L. The role of viscoelastic stress transfer in long-term earthquake cascades: Insights after the central Italy 2016–2017 seismic sequence. *Tectonics* **2018**, *37*, 3411–3428. [CrossRef]
10. Kilb, D.; Gomberg, J.; Bodin, P. Aftershock triggering by complete Coulomb stress changes. *J. Geophys. Res. Solid Earth* **2002**, *107*, ESE 2-1–ESE 2-14. [CrossRef]
11. Toda, S.; Stein, R.S.; Beroza, G.C.; Marsan, D. Aftershocks halted by static stress shadows. *Nat. Geosci.* **2012**, *5*, 410–413. [CrossRef]
12. DeVries, P.M.; Viégas, F.; Wattenberg, M.; Meade, B.J. Deep learning of aftershock patterns following large earthquakes. *Nature* **2018**, *560*, 632–634. [CrossRef] [PubMed]
13. Utsu, T.; Ogata, Y. The centenary of the Omori formula for a decay law of aftershock activity. *J. Phys. Earth* **1995**, *43*, 1–33. [CrossRef]
14. Ogata, Y. Statistical models for earthquake occurrences and residual analysis for point processes. *J. Am. Stat. Assoc.* **1988**, *83*, 9–27. [CrossRef]
15. Ogata, Y. Space-time point-process models for earthquake occurrences. *Ann. Inst. Stat. Math.* **1998**, *50*, 379–402. [CrossRef]
16. Gulia, L.; Rinaldi, A.; Tormann, T.; Vannucci, G.; Enescu, B.; Wiemer, S. The Effect of a Mainshock on the Size Distribution of the Aftershocks. *Geophys. Res. Lett.* **2018**, *45*, 13277–13287. [CrossRef]
17. Gutenberg, B.; Richter, C.F. Frequency of earthquakes in California. *Bull. Seismol. Soc. Am.* **1944**, *34*, 185–188. [CrossRef]
18. Wech, A.G.; Creager, K.C.; Houston, H.; Vidale, J.E. An earthquake-like magnitude-frequency distribution of slow slip in northern Cascadia. *Geophys. Res. Lett.* **2010**, *37*. [CrossRef]
19. El-Isa. Continuous-cyclic variations in the b-value of the earthquake frequency-magnitude distribution. *Earthquake Science* **2013**, *26*(5), 301–320. [CrossRef]
20. Nuannin, P.; Kulhanek, O.; Persson, L. Spatial and temporal *b* value anomalies preceding the devastating off coast of NW Sumatra earthquake of December 26, 2004. *Geophys. Res. Lett.* **2005**, *32*. [CrossRef]
21. Amitrano, D. Brittle-ductile transition and associated seismicity: Experimental and numerical studies and relationship with the *b* value. *J. Geophys. Res. Solid Earth* **2003**, *108*. [CrossRef]

22. Goebel, T.H.W.; Schorlemmer, D.; Becker, T.W.; Dresen, G.; Sammis, C.G. Acoustic emissions document stress changes over many seismic cycles in stick-slip experiments. *Geophys. Res. Lett.* **2013**, *40*, 2049–2054. [CrossRef]
23. Spada, M.; Tormann, T.; Wiemer, S.; Enescu, B. Generic dependence of the frequency-size distribution of earthquakes on depth and its relation to the strength profile of the crust. *Geophys. Res. Lett.* **2013**, *40*, 709–714. [CrossRef]
24. Schorlemmer, D.; Wiemer, S.; Wyss, M. Variations in earthquake-size distribution across different stress regimes. *Nature* **2005**, *437*, 539. [CrossRef]
25. Gulia, L.; Wiemer, S. The influence of tectonic regimes on the earthquake size distribution: A case study for Italy. *Geophys. Res. Lett.* **2010**, 37. [CrossRef]
26. Bachmann, C.E.; Wiemer, S.; Goertz-Allmann, B.; Woessner, J. Influence of pore-pressure on the event-size distribution of induced earthquakes. *Geophys. Res. Lett.* **2012**, *39*. [CrossRef]
27. Zhao, Y.; Wu, Z. Mapping the *b*-values along the Longmenshan fault zone before and after the 12 May 2008, Wenchuan, China, Ms 8.0 earthquake. *Nat. Hazards Earth Syst. Sci.* **2008**, *8*, 1375–1385. [CrossRef]
28. Zhang, S.; Zhou, S. Spatial and temporal variation of *b*-values in southwest China. *Pure Appl. Geophys.* **2016**, *173*, 85–96. [CrossRef]
29. Liu, Y.; Pei, S.; Pei, S. Temporal and spatial variation of *b*-value before and after Wenchuan earthquake and its tectonic implication. *Chin. J. Geophys.* **2017**, *60*, 2104–2112.
30. Shi, H.X.; Meng, L.Y.; Zhang, X.M.; Chang, Y.; Yang, Z.T.; Xie, W.Y.; Hattori, K.; Han, P. Decrease in *b* value prior to the Wenchuan earthquake (M(s) 8.0). *Chin. J. Geophys. Chin. Ed.* **2018**, *61*, 1874–1882.
31. Zhao, B.; Yang, G.; Wang, J.; Dong, F. Research on b-Values Based on Fault Buffers. *Pure Appl. Geophys.* **2020**, *177*, 71–80. [CrossRef]
32. Stein, R.S.; Barka, A.A.; Dieterich, J.H. Progressive failure on the North Anatolian fault since 1939 by earthquake stress triggering. *Geophys. J. Int.* **1997**, *128*, 594–604. [CrossRef]
33. Mallman, E.P.; Zoback, M.D. Assessing elastic Coulomb stress transfer models using seismicity rates in southern California and southwestern Japan. *J. Geophys. Res. Solid Earth* **2007**, *112*. [CrossRef]
34. Delescluse, M.; Chamot-Rooke, N.; Cattin, R.; Fleitout, L.; Trubienko, O.; Vigny, C. April 2012 intra-oceanic seismicity off Sumatra boosted by the Banda-Aceh megathrust. *Nature* **2012**, *490*, 240–244. [CrossRef]
35. Durand, V.; Bouchon, M.; Karabulut, H.; Marsan, D.; Schmittbuhl, J. Link between Coulomb stress changes and seismic activation in the eastern Marmara sea after the 1999, Izmit (Turkey), earthquake. *J. Geophys. Res. Solid Earth* **2013**, *118*, 681–688. [CrossRef]
36. Sumy, D.F.; Cochran, E.S.; Keranen, K.M.; Wei, M.; Abers, G.A. Observations of static Coulomb stress triggering of the November 2011 M5. 7 Oklahoma earthquake sequence. *J. Geophys. Res. Solid Earth* **2014**, *119*, 1904–1923. [CrossRef]
37. Zuniga, F.R.; Wyss, M. Most-and least-likely locations of large to great earthquakes along the Pacific coast of Mexico estimated from local recurrence times based on *b*-values. *Bull. Seismol. Soc. Am.* **2001**, *91*, 1717–1728. [CrossRef]
38. Nanjo, K.; Hirata, N.; Obara, K.; Kasahara, K. Decade-scale decrease in *b* value prior to the M9-class 2011 Tohoku and 2004 Sumatra quakes. *Geophys. Res. Lett.* **2012**, *39*. [CrossRef]
39. Woessner, J.; Wiemer, S. Assessing the quality of earthquake catalogues: Estimating the magnitude of completeness and its uncertainty. *Bull. Seismol. Soc. Am.* **2005**, *95*, 684–698. [CrossRef]
40. Mignan, A.; Werner, M.J.; Wiemer, S.; Chen, C.C.; Wu, Y.M. Bayesian estimation of the spatially varying completeness magnitude of earthquake catalogs. *Bull. Seismol. Soc. Am.* **2011**, *101*, 1371–1385. [CrossRef]
41. Gulia, L.; Wiemer, S. Real-time discrimination of earthquake foreshocks and aftershocks. *Nature* **2019**, *574*, 193–199. [CrossRef] [PubMed]
42. Aki, K. Maximum likelihood estimate of b in the formula log N = a-bM and its confidence limits. *Bull. Earthq. Res. Inst. Tokyo Univ.* **1965**, *43*, 237–239.
43. Utsu, T. A method for determining the value of "*b*" in a formula log n = a-bM showing the magnitude-frequency relation for earthquakes. *Geophys. Bull. Hokkaido Univ.* **1965**, *13*, 99–103.
44. Tormann, T.; Wiemer, S.; Mignan, A. Systematic survey of high-resolution *b* value imaging along Californian faults: Inference on asperities. *J. Geophys. Res. Solid Earth* **2014**, *119*, 2029–2054. [CrossRef]
45. Akaike, H. A new look at the statistical model identification. In *Selected Papers of Hirotugu Akaike*; Springer: Berlin/Heidelberg, Germany, 1974; pp. 215–222.
46. Utsu, T. *Report of the Joint Research Institute for Statistical Mathematics*; Institute for Statistical Mathematics; Tokyo, Japan, 1992.
47. Utsu, T. Representation and analysis of the earthquake size distribution: A historical review and some new approaches. *Pure Appl. Geophys.* **1999**, *155*, 509–535. [CrossRef]
48. McGrath, D.; Zhang, C.; Carton, O.T. Geostatistical analyses and hazard assessment on soil lead in Silvermines area, Ireland. *Environ. Pollut.* **2004**, *127*, 239–248. [CrossRef]
49. Ha, H.; Olson, J.R.; Bian, L.; Rogerson, P.A. Analysis of heavy metal sources in soil using kriging interpolation on principal components. *Environ. Sci. Technol.* **2014**, *48*, 4999–5007. [CrossRef]
50. Huang, Q. Seismicity changes prior to the Ms8. 0 Wenchuan earthquake in Sichuan, China. *Geophys. Res. Lett.* **2008**, *35*. [CrossRef]
51. Fang, L.; Wu, J.; Wang, W.; Du, W.; Su, J.; Wang, C.; Yang, T.; Cai, Y. Aftershock observation and analysis of the 2013 M S 7.0 Lushan earthquake. *Seismol. Res. Lett.* **2015**, *86*, 1135–1142. [CrossRef]
52. Lopez-Pineda, L. and Rebollar, C.J. Source characteristics of the Mw 6.2 loreto earthquake of 12 March 2003 that occurred in a transform fault in the middle of the Gulf of California, Mexico. *Bull. Seismol. Soc. Am.* **2005**, *95*, 419–430. [CrossRef]

53. Singh, A.P.; Mishra, O.P.; Rastogi, B.K.; Kumar, D. 3-D seismic structure of the Kachchh, Gujarat, and its implications for the earthquake hazard mitigation. *Nat. Hazards* **2011**, *57*, 83–105. [CrossRef]
54. Woessner, J.; Laurentiu, D.; Giardini, D.; Crowley, H.; Cotton, F.; Grunthal, G.; Valensise, G.; Arvidsson, R.; Basili, R.; Demircioglu, M.B.; et al. The 2013 European Seismic Hazard Model: Key components and results. *Bull. Earthq. Eng.* **2015**, *13*, 3553–3596. [CrossRef]
55. Li, L.; Adhikari, L.B.; Gang, L.; FuWang, G. Characteristics of temporal-spatial distribution of the aftershocks of the 2008 M(s) 8. 0 Wenchuan Earthquake. *Chin. J. Geophys. Chin. Ed.* **2018**, *61*, 1797–1805.
56. Marzocchi, W.; Spassiani, I.; Stallone, A.; Taroni, M. How to be fooled searching for significant variations of the *b*-value. *Geophys. J. Int.* **2020**, *220*, 1845–1856. [CrossRef]
57. Varotsos, P.; Sarlis, N.; Skordas, E.; Lazaridou, M. Seismic Electric Signals: An additional fact showing their physical interconnection with seismicity. *Tectonophysics* **2013**, *589*, 116–125. [CrossRef]
58. Sarlis, N.V.; Skordas, E.S.; Varotsos, P.A.; Nagao, T.; Kamogawa, M.; Tanaka, H.; Uyeda, S. Minimum of the order parameter fluctuations of seismicity before major earthquakes in Japan. *Proc. Natl. Acad. Sci. USA* **2013**, *110*, 13734–13738. [CrossRef]
59. Scholz, C.H. On the stress dependence of the earthquake *b* value. *Geophys. Res. Lett.* **2015**, *42*, 1399–1402. [CrossRef]
60. Rodriguez-Perez, Q.; Ramon Zuniga, F. Imaging *b*-value depth variations within the Cocos and Rivera plates at the Mexican subduction zone. *Tectonophysics* **2018**, *734*, 33–43. [CrossRef]
61. Parsons, T.; Ji, C.; Kirby, E. Stress changes from the 2008 Wenchuan earthquake and increased hazard in the Sichuan basin. *Nature* **2008**, *454*, 509–510. [CrossRef]
62. Shan, B.; Xiong, X.; Zheng, Y.; Jin, B.; Liu, C.; Xie, Z.; Hsu, H. Stress changes on major faults caused by 2013 Lushan earthquake and its relationship with 2008 Wenchuan earthquake. *Sci. China-Earth Sci.* **2013**, *56*, 1169–1176. [CrossRef]
63. Gerstenberger, M.; Wiemer, S.; Giardini, D. A systematic test of the hypothesis that the *b* value varies with depth in California. *Geophys. Res. Lett.* **2001**, *28*, 57–60. [CrossRef]
64. Beauval, C.; Scotti, O. Quantifying sensitivities of PSHA for France to earthquake catalog uncertainties, truncation of ground-motion variability, and magnitude limits. *Bull. Seismol. Soc. Am.* **2004**, *94*, 1579–1594. [CrossRef]
65. Hubbard, J.; Shaw, J.H. Uplift of the Longmen Shan and Tibetan plateau, and the 2008 Wenchuan (M = 7.9) earthquake. *Nature* **2009**, *458*, 194–197. [CrossRef] [PubMed]
66. Lei, J.; Zhao, D. Structural heterogeneity of the Longmenshan fault zone and the mechanism of the 2008 Wenchuan earthquake (Ms 8.0). *Geochem. Geophys. Geosystems* **2009**, *10*. [CrossRef]
67. El-Isa, Z.H.; Eaton, D.W. Spatiotemporal variations in the *b*-value of earthquake magnitude–frequency distributions: Classification and causes. *Tectonophysics* **2014**, *615*, 1–11. [CrossRef]
68. Brodsky, E.E. Determining whether the worst earthquake has passed. *Nature* **2019**, *574*, 185–186. [CrossRef]
69. Gerstenberger, M.C.; Wiemer, S.; Jones, L.M.; Reasenberg, P.A. Real-time forecasts of tomorrow's earthquakes in California. *Nature* **2005**, *435*, 328. [CrossRef]
70. Deng, B.; Liu, S.G.; Jansa, L.; Cao, J.X.; Cheng, Y.; Li, Z.W.; Liu, S. Sedimentary record of Late Triassic transpressional tectonics of the Longmenshan thrust belt, SW China. *J. Asian Earth Sci.* **2012**, *48*, 43–55. [CrossRef]
71. Wang, Y.; Wang, F.; Wang, M.; Shen, Z.-K.; Wan, Y. Coulomb Stress Change and Evolution Induced by the 2008 Wenchuan Earthquake and its Delayed Triggering of the 2013 M-W 6.6 Lushan Earthquake. *Seismol. Res. Lett.* **2014**, *85*, 52–59. [CrossRef]

applied sciences

MDPI

Article

Liquefaction Potential and V_{s30} Structure in the Middle-Chelif Basin, Northwestern Algeria, by Ambient Vibration Data Inversion

Abdelouahab Issaadi [1,2,*], Ahmed Saadi [1,2], Fethi Semmane [1], Abdelkrim Yelles-Chaouche [1] and Juan José Galiana-Merino [2,3]

1 Centre de Recherche en Astronomie, Astrophysique et Géophysique (CRAAG), Algiers 16340, Algeria
2 University Institute of Physics Applied to Sciences and Technologies, University of Alicante, 03080 Alicante, Spain
3 Departement of Physics, Systems Engineering and Signal Theory, University of Alicante, 03080 Alicante, Spain
* Correspondence: ia30@alu.ua.es

Abstract: The Middle-Chelif basin, in northwestern Algeria, is located in a seismically active region. In its western part lies the El-Asnam fault, a thrust fault responsible for several strong earthquakes. The most important being the El-Asnam earthquake (Ms = 7.3) of 1980. In the present study, ambient vibration data with single-station and array techniques were used to investigate the dynamic properties of the ground and to estimate the V_{s30} structure in the main cities of the basin. Soil resonance frequencies vary from 1.2 to 8.3 Hz with a maximum amplitude of 8.7 in. Collapsing behavior has also been demonstrated west of the city of El-Attaf, reflecting a strong potential for liquefaction. A V_{s30} variation map and a soil classification for each city were obtained mainly by inversion of the HVSR and Rayleigh wave dispersion curves. Finally, an empirical prediction law of V_{s30} for the Middle-Chelif basin was proposed.

Keywords: Middle-Chelif Basin; ambient vibrations; HVSR; array techniques; V_{s30}; site classification; liquefaction

check for updates

Citation: Issaadi, A.; Saadi, A.; Semmane, F.; Yelles-Chaouche, A.; Galiana-Merino, J.J. Liquefaction Potential and V_{s30} Structure in the Middle-Chelif Basin, Northwestern Algeria, by Ambient Vibration Data Inversion. *Appl. Sci.* **2022**, *12*, 8069. https://doi.org/10.3390/app12168069

Academic Editors: Miguel Llorente Isidro, Ricardo Castedo and David Moncoulon

Received: 28 June 2022
Accepted: 1 August 2022
Published: 12 August 2022

Publisher's Note: MDPI stays neutral with regard to jurisdictional claims in published maps and institutional affiliations.

1. Introduction

Northern Algeria is located in the collision zone between the African and Eurasian plates. It is characterized by moderate to high seismic activity (e.g., [1–3]), mainly concentrated in the marginal areas of the Neogene basins [4]. The Chelif Basin is located in the northwestern part of Algeria (Figure 1). It is the largest of the northern Neogene sedimentary basins and hosts an important seismic activity. The basin is mainly affected by NE-SW oriented reverse faults [4,5]. The most important is the El-Asnam fault, a reverse fault about 40 km long [6], that generated several destructive earthquakes during the last century, such as the 1934 Carnot earthquake, now El-Abadia, (M_S = 5.1, [7]); the 1954 Orléansville earthquake, now Chlef (M_S = 6.7, [7]); and the 1980 El-Asnam earthquake, now Chlef (M_S = 7.3, [6]). The latter is the largest and most destructive earthquake recorded in Algeria in the instrumental era.

The Chelif Basin is divided into three parts: the Lower-Chelif Basin, the Middle-Chelif Basin, and the Upper-Chelif Basin. The Middle-Chelif extends from Oued-Fodda in the west to Ain-Defla in the east (Figure 1). The cities of the Middle-Chelif suffered important damage during the 1980 El-Asnam earthquake. The cities of Oued-Fodda, El-Abadia, and El-Attaf were almost totally destroyed. In addition, several secondary effects of the earthquake were observed along the rupture zone [8], such as cracks, settlement, and soil liquefaction. Moreover, the coseismic uplift of the western part of the fault has obstructed the flow of the Chelif River, causing a flood which formed a natural dam, where the phenomenon of liquefaction occurred over a wide area west of El-Abadia and El-Attaf [8,9].

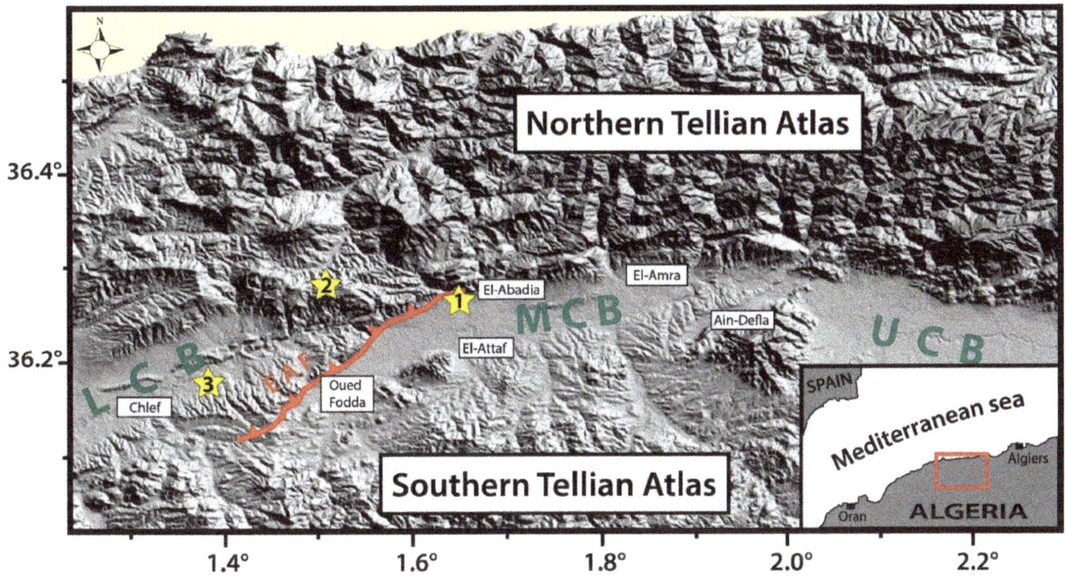

Figure 1. Location of the study area. The numbered yellow stars correspond to the epicenters of the major earthquakes in the area: (1) the 1934 Carnot (I_0 = IX) earthquake, (2) the 1954 Orléansville (*Ms* 6.7) earthquake, and (3) the 1980 El-Asnam earthquake (*Mw* 7.2). EAF = El-Asnam fault trace; UCB = Upper-Chelif Basin; MCB = Middle-Chelif Basin; LCB = Lower-Chelif Basin.

Prior to the 1980 El-Asnam earthquake, several geological and geophysical studies were conducted on the Chelif Basin (e.g., [10,11]). However, these studies were mainly concentrated in the Lower-Chelif Basin due to the presence of oil indices. Only a few studies have been carried out on the Middle-Chelif Basin [12,13]. Right after the 1980 earthquake and given the extent of its damage, the region has finally been the subject of several geological, seismological, and geophysical studies (e.g., [6,8,14,15]). The firm Woodward and Clyde Consultants [15] conducted an important seismic microzonation study in eight cities of the Lower and Middle-Chelif Basins, including El-Abadia and El-Attaf. During the investigations, several holes were drilled using Standard Penetration Tests (SPT). The study provided geotechnical, hydrogeological, landslide, and liquefaction potential maps for each of the eight cities. The Neogene formations of the Middle-Chelif were described in detail in [12], and later in [16]. Furthermore, the structural aspect of the shallow and deep lithological units was recently imaged using land gravity data [17].

In the present study, we used ambient vibrations data to characterize some geotechnical features and to estimate some dynamic properties of the soil column in the cities of El-Attaf, El-Abadia, and Ain-Defla. Ambient vibration-based techniques have been used previously in the Chelif Basin [18–21]. The aim was to determine the resonance frequencies of the ground, the shear-wave velocity structure of the sedimentary layers, and the bedrock depth, where the impedance contrasts with the sedimentary cover and may be the cause for ground shaking amplification during strong earthquakes. This work is a continuation of the ones carried out in the Middle-Chelif basin [20,21].

In the first part of this study, we used ambient vibration data recorded from single stations to estimate the ground resonance frequencies and assess the liquefaction potential in the three cities under study. In the second part, we used ambient vibration data recorded from single station and array techniques to estimate the average shear-wave velocity in the upper 30 m of the soil column (V_{s30}). Finally, a V_{s30} predictive equation for the Middle-Chelif Basin was proposed.

This study contributes to the seismic hazard assessment in northern Algeria. The results obtained in this work can be used for ground motion simulation, for the calculation of amplification factors, and for many other studies related to the reduction of seismic risk.

2. Geological Framework

The Middle-Chelif is an intra-mountainous basin structured during the Neogene [5,10,12], and located within the Tellian Atlas mountain belts (Figure 1). The depression is filled with a thick cover of Mio-Plio-Quaternary sediments. The basement is composed of hard clays and marls of Cretaceous age [12,16]. In its southern part, autochthonous formations of Jurassic to Silurian age outcrop [22], and form the Temoulga, Rouina, and Doui massifs (Figure 2). The Middle-Chelif plain is crossed from west to east by the Cheliff River, the longest in Algeria, which contributes to form the actual alluviums.

The stratigraphical column is composed of a succession of marine, continental, and lacustrine deposits. Lateral variations in facies were also reported [10,16], and further confirmed in [21], where important lateral variations in the shear-wave velocity were observed within the same formations. All these perturbations in the sedimentary cover reflect several instability periods with intense tectonic activities and different episodes of marine regression and transgression that conditioned the sedimentation process [10]. However, the geologists are not in agreement concerning the age of the first sedimentary deposits. Some authors attributed those sediments to the Lower Miocene [10,12], while others assigned them to the Middle Miocene (Serravallo-Tortonian) [16,23]. The sediments are affected by a series of normal and thrust faults, located in the margins of the Middle-Chelif Plain.

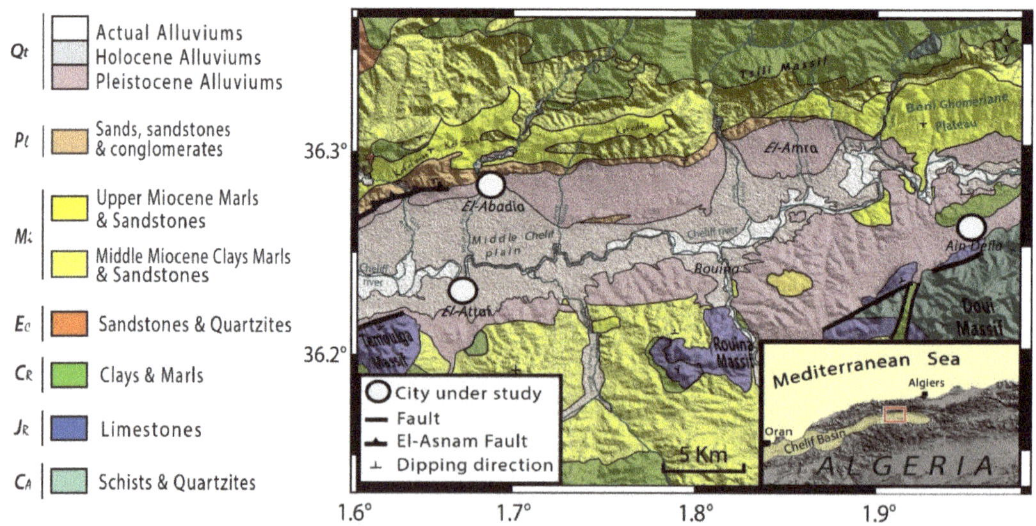

Figure 2. Geological map of the study area. Compiled and modified from [22,24].

In terms of lithology, the Miocene series occupies the major part of the sedimentary column. These formations were described and detailed in [12]. The first deposits are detritic continental series of conglomerates and marls. These sediments are overlayed by different intercalations of marls, clays, limestones, and sandstones [12,16]. A thin layer of blue marls marks the transition between the Miocene and the Pliocene sediments [10,12]. The Miocene series outcrop in succession on the hills that overlook the city of El-Abadia (Figure 2). The shear-wave velocity varies from 640 to 1450 m/s for the Miocene formations [21]. The Pliocene is divided in two stages, marine and continental, with alternations between sands, sandstones, and conglomerates [10,16]. The Quaternary deposits are continental and predominant in the Middle-Chelif Plain. They are represented by Holocene and Pleistocene alluviums.

The three cities under study are built on Quaternary alluviums of different thicknesses. The city of Ain-Defla is built at the bottom of the northern flank of the Doui massif (Figure 2), where the topmost layer of the soil is composed of old Quaternary clays and gravels (Pleistocene), reaching a maximum thickness of 60 m [21]. The old alluviums lay directly over hard Jurassic limestone and Silurian schists and quartzite [22].

The cities of El-Attaf and El-Abadia are built on stiff Quaternary soil, the topmost layer is thin (<20 m) [21], and composed of recent alluviums (Holocene), which lie over older alluviums (Pleistocene). The engineering bedrock in El-Abadia is composed of Pliocene sandstones, while in El-Attaf, it is composed of Miocene marls and sandstones [14,21]. In these cases, the engineering bedrock has been characterized as the first layer of the soil column that contains a shear-wave velocity value above 750 m/s.

The presence of important sandbanks at shallow depths (<10 m) in El-Attaf and El-Abadia may lead to liquefaction phenomena during ground shaking. The risk is weak in El-Abadia since the sandy layers are dense and the ground water level is between 15 and 30 m deep [15]. However, in El-Attaf the risk is significant since the sandy banks are loose and the ground water level is shallow (between 7 and 10 m) [15].

3. Data and Methodology

3.1. Horizontal-to-Vertical Spectral Ratio Technique (HVSR)

The HVSR (or H/V) technique [25] allows retrieval of the resonance frequency of the soil at a given site, using single station ambient vibrations measurements. The theoretical aspect of this technique consists of calculating the ratio between the amplitude spectra of the vertical and horizontal components of ambient vibrations. As a result, an HVSR curve is obtained. The HVSR frequency peak is well correlated with the soil resonance frequency.

Although this technique is very effective in estimating the soil resonance frequency [26], the scientific community is reluctant about its ability to estimate the amplification factors [27–29]. This issue is due to the contribution of different seismic waves to the wavefield. As for now, it is very difficult to quantify the ratio between body and surface waves. Bonnefoy-Claudet et al. [30] showed that in the case of high impedance contrasts between the bedrock and the sediments, the HVSR curve is mainly controlled by surface waves. The relatively low amplitude of the body waves is not sufficient to correctly estimate the amplification factor. Moreover, La Rocca et al. [31] and Benkaci et al. [32] have proven that the HVSR peak amplitude varies considerably with time. The amplitude of the frequency peak obtained from the HVSR technique in this study is interpreted as a relative indicator and not as a true amplification factor value.

Ambient vibration single-station measurements were carried out in October 2021 at 71 sites in the cities of El-Attaf (24 sites), Ain-Defla (33 sites), and El-Abadia (14 sites) (Figure 3). Some measurement points [15] were taken from a previous study [21]. Recordings were performed at night and in calm weather, as recommended by the SESAME project [33]. The acquisition time was 20 min. At some sites, the recording time was extended to 30 min due to anthropogenic noise from human activity. The equipment used for the recording was a pair of Tromino seismographs, with a sampling rate of 512 samples per second.

The HVSR technique was processed following the recommendations of the SESAME project [33]. The whole data were processed in the same way using the Geopsy software [34]. First, the signals were divided into several windows of 30 s each, tapered by a 5% cosine function. The window selection was made automatically using an anti-triggering algorithm, which allows selecting windows where the ambient noise is stationary. After that, the Fast Fourier Transform (FFT) is computed for each component (vertical and both horizontals) and the amplitude spectrum of both horizontal components is combined by an RMS (root mean square) average computation. After that, the ratio between the amplitude spectrum of both vertical and horizontal components is calculated. The HVSR curve is calculated for each window. The resulting curve is smoothed using the Konno–Ohmachi algorithm [35], with a smoothing coefficient of 40. Finally, the curves are averaged and the final HVSR curve is retrieved in the frequency range between 0.2 and 20 Hz. The resulting HVSR curve may contain one or several frequency peaks, which are directly linked to impedance contrasts at the soil column.

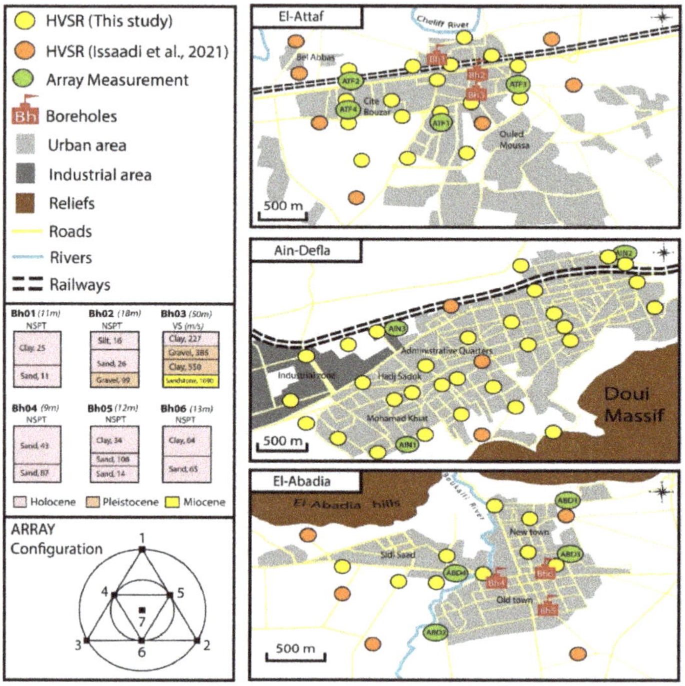

Figure 3. Distribution of the data compiled in this study. The orange circles correspond to data measured in [21]. The left panels correspond to the legend, the borehole data, and the array configuration, from top to bottom, respectively. Bh03 was taken from [14]. The rest was taken from [15].

3.2. Seismic Vulnerability Index (K_g) and Shear Strain (Υ)

The stability of structures during an earthquake depends on the behavior of the ground, which in turn depends on the dynamic properties of the soil column. Ishihara [36] established a relation between the shear strain deformation and the dynamic properties of the soil, by compiling several earthquake data, reports, and laboratory tests (Table 1). Nakamura [37] introduced a method based on the vulnerability index calculation using ambient vibration data to estimate the shear strain, for the purpose of potential earthquake damage assessment (soil liquefaction, landslides).

Table 1. Strain dependence of dynamic properties of the soil [36].

Size of Strain Υ	10^{-6}	10^{-4}	10^{-3}	10^{-2}	10^{-1}
Phenomena	Wave	Vibration	Crack	Settlement	Landslide, soil compaction, liquefaction
Dynamic properties	Elasticity		Elasto-plasticity		Collapse

The vulnerability index is calculated using the following equation [37]:

$$K_g = e \times \frac{\left(\frac{A_g^2}{F_g}\right)}{(\pi^2 \times V_b)} \tag{1}$$

where F_g is the resonance frequency of the soil, and A_g is the corresponding amplitude obtained with HVSR method. V_b is the velocity at the bedrock and e is the applied dynamic force. According to Nakamura [37], where several values of dynamic force were applied and tested with several velocity values for bedrock, the optimal results are obtained with an applied dynamic force of 60%. Thus, assuming this value and a velocity at the bedrock around 6×10^4 cm/s [37], Equation (1) can then be simplified as follows:

$$K_g = \left(\frac{A_g^2}{F_g}\right) \times 10^{-6} \tag{2}$$

The shear strain can then be calculated by multiplying the vulnerability index (K_g) with the maximum observed acceleration, or the peak ground acceleration (α_g) [37].

$$\Upsilon = K_g \times \alpha_g \tag{3}$$

where the α_g value is in cm/s^2 (Gal). The peak ground acceleration values for El-Attaf and El-Abadia cities are 410 and 520 cm/s^2, respectively, for a return period of 500 years [15]. For Ain-Defla city, a value of 250 cm/s^2 is proposed [38].

This technique is used in the present work mainly to assess the liquefaction potential in the study area. For a shear strain value $\Upsilon > 10^{-2}$, phenomena of liquefaction and landslides are likely to occur during earthquakes.

3.3. *Array Techniques*

Array-based techniques allow to obtain surface wave dispersion curves from ambient vibration records. In this study, three array techniques were used to extract the Rayleigh wave dispersion curves: the frequency-wavenumber (F-K) analysis [39–42], the spatial auto-correlation (SPAC) technique [43–45], and the extended spatial auto-correlation (ESAC) technique [44,46]. Some assumptions about the soil conditions are required before using these techniques. For example, it is assumed that the ambient vibration wavefield is essentially dominated by surface waves (especially the fundamental mode), and that the subsurface layers are homogenous and horizontally stratified, which means that in each layer the seismic waves propagate at a constant velocity [44].

The array measurement campaign was carried out in January 2021. The difficulty of finding open fields far enough from human activities inside urban areas, limited the measurement sites to 11 (four in El-Attaf and El-Abadia, and three in Ain-Defla). The measurements were made during daytime with a recording time of 40 min. The configuration of the array is triangular. Concretely, the sensor deployment consists of an equilateral triangle of 30 m on each side, with a sensor placed at each vertex, a sensor placed midway on each side, and an additional sensor placed at the center (Figure 3). This configuration provides a better coverage. According to [44], the equilateral triangle is the most efficient configuration for array techniques. The equipment used was formed by 7 SARA SS10 triaxial velocity sensors (f_0 = 1 Hz) connected to SL06 digitizers.

3.3.1. The Frequency-Wavenumber (F-K) Analysis

F-K analysis is one of the most commonly used techniques for estimating Rayleigh wave dispersion curves. It has some advantages over other techniques, such as the ability to identify the direction of ambient vibration source and the recognition of the different modes presented in the wavefield [47]. The latter is composed of a superposition of several propagated waves. The F-K technique allows estimating the velocity and the direction of approach (back-azimuth) of these waves [48].

The F-K technique is based on two fundamental assumptions: the first is that the process is stationary in time. The second is that the process is stationary in the horizontal plane and that the propagation of the wavefront is only in the vertical direction. The stationary aspect of propagated seismic waves allows the power spectral density function of the frequency wavenumber to be calculated, which contains information about the power as a function of frequency and velocity vectors of the propagated wave. There are two methods for calculating the power spectral density: the maximum likelihood method (MLM) or the high-resolution method [40,41] and the beamforming method (BFM) [42]. The BFM technique is used in this study, as it is less sensitive to errors than the MLM technique [41].

For all the array techniques, only the vertical component of the records was analyzed, as it is the one required for the estimation of the Rayleigh wave dispersion curves. The Sesarray software package [34] was used to perform the F-K analysis. First, the coordinates of each of the seven sensors were introduced in the WARANGPS software in order to calculate the array transfer function and the theoretical wavenumber limits (K_{min} and K_{max}). Then, the signals were loaded into the Geopsy software and the BFM method was applied. The signals were divided into several windows of frequency-dependent lengths (50 periods). After that, an anti-triggering algorithm was used for the window selection. The processing requires the input of the grid step and grid size parameters. The grid size corresponds to the K_{max} value, which is related to the aliasing limit. The grid step determines the maximum resolution and was chosen as $K_{min}/2$. Once all these parameters were introduced, the final processing was launched and the dispersion curve was obtained. The processing was the same for the data of the 11 arrays.

3.3.2. The SPAC and ESAC Techniques

The spatial auto-correlation and the extended spatial auto-correlation techniques are based on the assumption of a stochastic wavefield being stationary both in space and time [43].

In theory, the SPAC technique consists of calculating a single-phase velocity value at each frequency in a predefined frequency band by fitting the SPAC coefficient to a Bessel function. For a circular array, the Bessel function represents the average cross-correlation between pairs of stations as a function of their distance. Aki [43] showed that the SPAC coefficient at a given frequency has the same form as the 0th order Bessel function.

The SPAC method requires a circular array configuration with a centrally located sensor [44]. A modification of this technique has been suggested by Bettig et al. [45], which allows the SPAC technique to be applied to arrays of arbitrary configuration. The modification consists of replacing the use of fixed radius values with rings of finite thickness.

The ESAC method differs from SPAC by fixing the frequency values instead of the radius. At each frequency, the normalized transverse spectrum is fitted to the Bessel function. The inverted Bessel function that has the best fit with the normalized cross-spectra allows to obtain the phase velocity [46].

As with the F-K analysis, the SPAC technique was performed using the SESARRAY software package [34]. The first step is to define the ring parameters using the SPAC toolbox of the Geopsy software. Once the coordinates of the sensors are entered, the software will define a set of spatially distributed sensor pairs (e.g., 21 pairs for 7 sensors). Then, the sensor pairs must be included in one or more rings. For this purpose, inner and outer radii of rings that best correspond to the sensor pairs were introduced. Note that a ring can contain a minimum of two pairs. A maximum number of rings is recommended for better resolution [45,49]. Similar to the F-K analysis, the signals are divided into frequency-dependent length windows containing 50 periods. Windows are selected using the anti-triggering algorithm. Then, the analysis is launched and the spatial auto-correlation curves are obtained for each ring. The Spac2Disp software is used to display the phase velocity histograms derived from the set of the calculated spatial auto-correlation values. Then, the Rayleigh phase velocity values that best contribute to the dispersion curve are chosen within the K_{min} and K_{max} values. In this way, a final dispersion curve is obtained.

The ESAC analysis was carried out using a specific Matlab© (Natick, MA, USA) application developed by the University of Alicante [50]. For each sensor, the recorded signal is divided into non-overlapping 30 s windows. Then, the cross-spectrum is calculated and smoothed using a triangular window, in the frequency range from 0.1 to 15 Hz.

3.4. Inversion of Dispersion Curves and HVSR Curves

The inversion was carried out with the Dinver software (Sesarray package [34]) using the neighborhood algorithm [51]. In order to have a better spatial distribution for the V_{s30} values, some HVSR curves were inverted in El-Abadia (5 sites) and Ain-Defla (5 sites). Only the part around the fundamental frequency peak was considered in the inversion process [52]. For the array data inversion process, the 3 dispersion curves obtained for each array using the 3 different techniques (F-K, SPAC, and ESAC) (Figure 4) were averaged to obtain a better constrained dispersion curve with an optimized frequency range. The input parameters required for the inversion (Vp, Vs, densities, and number of layers) were taken from previous studies [14,21]. The maximum number of iterations was set to 300 iterations, and 100 models were generated at each iteration. The experimental average dispersion curve was compared to the theoretical one via a misfit value. Then, the Vs model corresponding to the minimum misfit was selected (Figure 5).

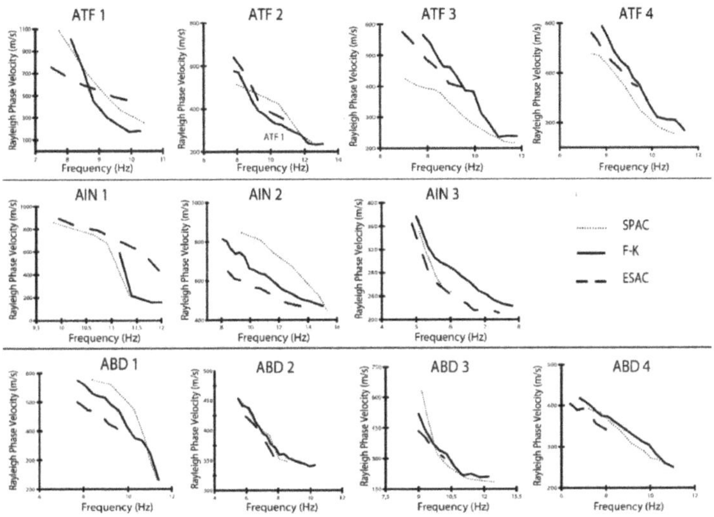

Figure 4. Rayleigh wave dispersion curves obtained from the F-K, SPAC, and ESAC methods.

3.5. Estimation of the V_{s30}

3.5.1. V_{s30} from NSPT Measurements

There are a considerable number of studies that propose equations relating shear-wave velocity to the Normalized Standard Penetration Test (NSPT). However, the equations are specific to the region under study. Sil et al. [53] compiled data from different continents and proposed empirical equations correlating NSPT values with shear-wave velocity for sands (Equation (4)), clays (Equation (5)), and for all soil types (Equation (6)):

$$V_S = 79.217 \times N^{0.3699} \tag{4}$$

$$V_S = 99.708 \times N^{0.3358} \tag{5}$$

$$V_S = 75.478 \times N^{0.3799} \tag{6}$$

where V_S is the shear-wave velocity and N-value is the number of blows in the SPT measurements. However, in the case the SPT borehole does not reach 30 m, which is the case in this study, V_{s30} can be correlated from the average velocity at depth z using the following equation from [54]:

$$Log\ V_{s30} = a + (b \times Log\ V_s z) \tag{7}$$

$V_s z$ is the velocity at depth z, and a and b are coefficients that vary with depth (see Table 2 in [54]).

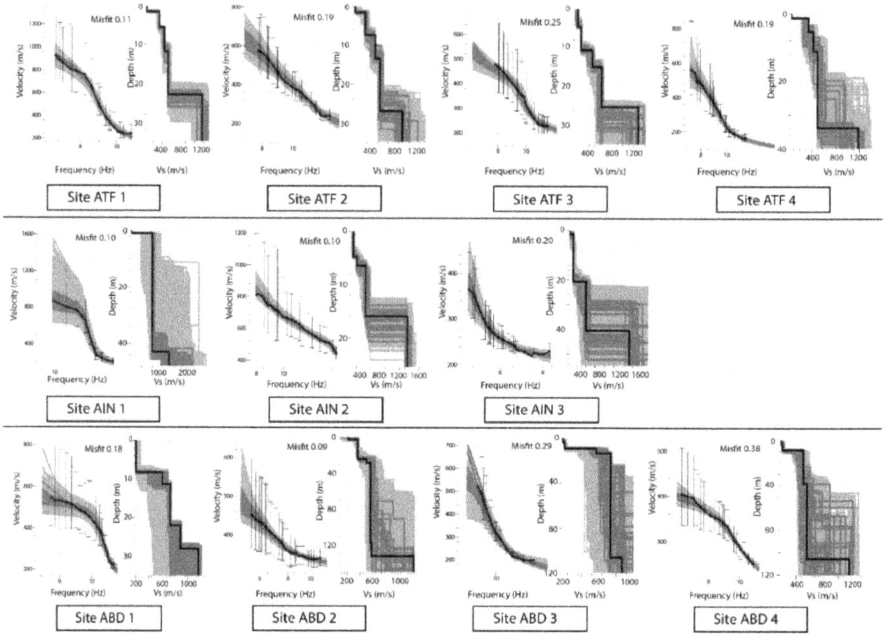

Figure 5. Results of the inversion of Rayleigh waves dispersion curves. The left panels for each site represent the dispersion curves. The right panels represent the Vs models. The black line corresponds to the best fit model, the dark grey represents models with minimum misfit + 10%. All the tested models are in light grey. The misfit is shown at the middle top of each site.

3.5.2. V_{s30} from Vs Models

The inversion of the dispersion curves allows retrieval of the Vs models. The averaged V_{s30} can be calculated from the Vs models using the following equation [55]:

$$V_{s30} = \frac{30}{\sum_{i=1}^{N} \frac{Hi}{Vi}} \quad (8)$$

H_i is the thickness and V_i is the shear-wave velocity of the layer i.

The obtained V_{s30} values were then spatially meshed using the Kriging method [56], with a linear transformation. A map of V_{s30} variation was obtained for each city. As for the classification of the sites, it was completed according to the NEHRP site classification [57] (Table 2).

Table 2. NEHRP soil classification as a function of the average shear-wave velocity to 30 m depth [57].

Title 1	Title 2	Title 3
Vs > 1500.	Hard rock	A
760 < Vs ≤ 1500	Rock	B
360 < Vs ≤ 760	Very dense soil and soft rock	C
180 < Vs ≤ 360	Stiff soil	D
Vs ≤ 180	Soft soil	E

4. Results and Discussion

4.1. Soil Resonance Frequencies and Amplitudes

The soil resonance frequencies and the corresponding amplitudes for the cities of El-Attaf, El-Abadia, and Ain-Defla, are mapped in Figure 6. In El-Attaf city, the resonance frequencies are between 1.2 and 8.3 Hz. In the major part of the city, the predominant frequencies are between 1.2 and 5 Hz. The obtained values are related to impedance contrasts in the subsoil between the Quaternary alluviums and the Miocene marls and sandstones. Near Ouled Moussa, south of the city, higher frequencies are observed (between 5 and 8.3 Hz). This increase is most likely related to the Cretaceous marls outcropping in the south [21]. Since the buildings in this area have between 1 and 5 floors, the resonance frequencies of the ground are close to the buildings' frequencies, which can be damaging for the structures during strong shaking. The predominant amplitudes of the frequency peaks vary between 2.2 and 8.7. The amplitudes are lower to the northern areas of the city. The highest amplitudes (between 5 and 8.7) are observed in "Cité Bouzar" neighborhood, in the western areas of the city.

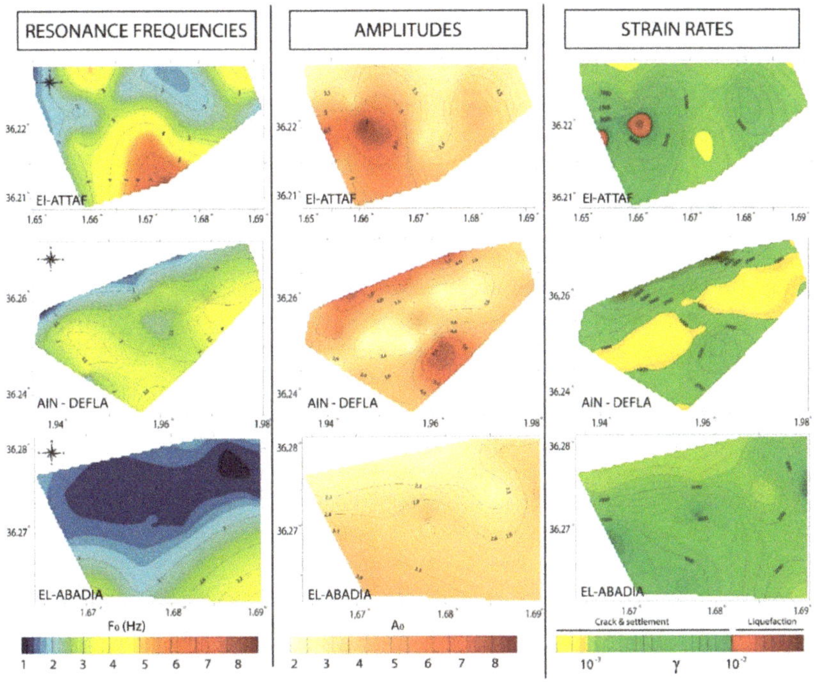

Figure 6. Results of the single−station measurements analysis.

In the El-Abadia city, the resonance frequencies vary between 1.4 and 4.2 Hz, while the corresponding amplitudes vary between 2 and 4. The obtained resonance frequencies are related to impedance contrasts between the Quaternary alluviums and the Pliocene sandstones. The amplitudes slightly increase to the southwest towards the Middle-Chelif plain, where the Quaternary stiff and soft soils are thicker. Finally, in the Ain-Defla city, the resonance frequencies vary between 1.4 and 4.5. These frequencies are related to the impedance contrasts between the Quaternary deposits and the Cretaceous-Jurassic bedrock. The increase in the resonance frequency peak from north to south is related to the presence of the Doui Massif and its hard Jurassic limestones [22] to the south of the city. The predominant amplitude of the frequency peaks varies between 2 and 7.1. The amplitudes are relatively low in the central part of the city. However, in "Mohamad Khiat" neighborhood, in the SW, the amplitudes are relatively higher (5–7.1). In the northern areas, it reaches a value of 6.8.

4.2. Shear Strain and Liquefaction Potential

The shear strain variation map for the three cities (right column in Figure 6) gives valuable information about the dynamic properties of the soils and the possible behavior during earthquakes. In El-Abadia and Ain-Defla cities, the shear strain values reflect an elasto-plastic soil behavior, where cracks and settlements may occur during strong ground shaking. However, in the central part of Ain-Defla city, the lower strain values indicate that the soil column tends towards a more elastic appearance, which is probably due to the thickening of the ancient Quaternary deposits in this zone, with the presence of very dense clays and gravels [13]. In both cities, the results show that the soils do not show any predisposition to liquefaction and landslide phenomena during earthquakes.

On the other hand, in the El-Attaf city, strain rate shows different dynamic properties of the soil, an elasto-plastic behavior in the east, and a collapse behavior in the west, more precisely in the "Cité Bouzar" neighborhood, with a soil subject to liquefaction. Piezometric measurements in "Cité Bouzar" have shown that the water table is around 7 m deep [15]. The presence of sandbanks and water table at very shallow depths increase the risk of liquefaction in the area, which was the case during the 1980 El-Asnam earthquake. Indeed, liquefaction phenomena were reported west of El-Attaf, where large sand boil formations (>6 m) were observed [9].

4.3. Dispersion Curve Inversion and Vs Models

The dispersion curves obtained using the three techniques, F-K, SPAC, and ESAC, and presented in Figure 4, are plotted within the theoretical limits of the wavenumber (Kmin, Kmax). The dispersion curves are valid between 7 and 12 Hz in El-Attaf and between 5 and 12 Hz in Ain-Defla. In the El-Abadia city, the curves are valid between 5.5 and 11 Hz. This difference is due to local site conditions. The dispersion curves are well correlated at most sites (Figure 4). At sites ATF1 and AIN1, the dispersion curves obtained with the ESAC technique tend to diverge from the other curves at high frequencies. We note that at these sites, the array was deployed on slightly sloping terrain, which might indicate that the ESAC technique could be more sensitive to slopes than the other techniques.

An average curve was calculated at each site. In this way, a better constrained dispersion curve is used for the inversion process to obtain a better consistency of the resulting Vs profiles. The results of the inversion are shown in Figure 5. In the El-Attaf city, the engineering bedrock corresponds to the Miocene marls and sandstones, with a Vs value varying between 970 and 1280 m/s. Soft and stiff Quaternary alluvium occupies the first 30 m of the soil column. In the Ain-Defla city, the bedrock Vs value varies between 1390 and 1450 m/s. The thickness of the Quaternary deposits varies between 16 and 43 m. At El-Abadia city, the bedrock Vs value is divided into two different ranges, between 870 and 900 m/s to the east (ABD1, ABD3), and between 1150 and 1200 m/s to the west (ABD2, ABD4). This difference is related to the change in bedrock composition from Late Pliocene sandstones in the east to Miocene marls and sandstones in the west. The thickness of the Quaternary layers varies between 11 and 29 m.

4.4. V_{s30} Structure and Site Classification

V_{s30} was calculated from the shear-wave velocity models and the additional SPT surveys. A map of V_{s30} variation, along with site classification, is provided for each of the three cities (Figures 7–9).

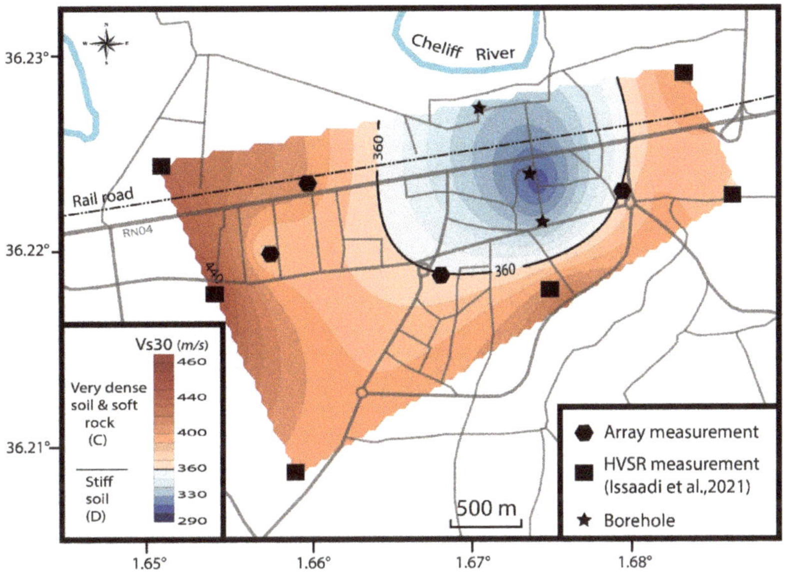

Figure 7. V_{s30} map and soil classification in the El-Attaf city [21].

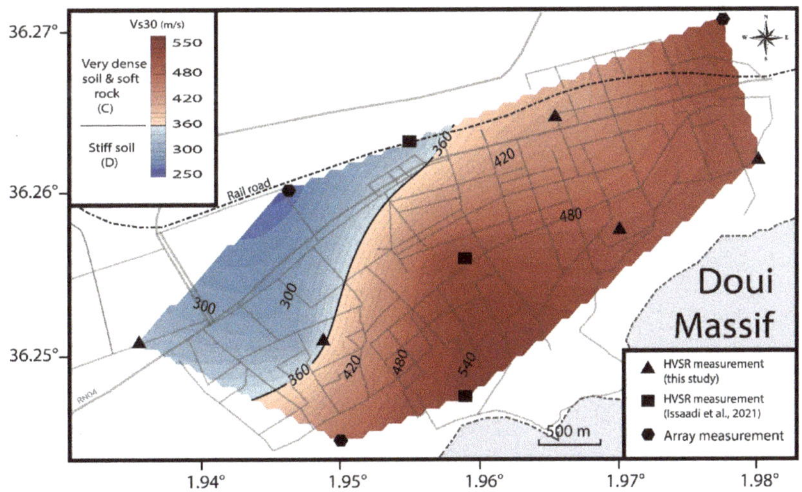

Figure 8. V_{s30} map and soil classification in the Ain-Defla city [21].

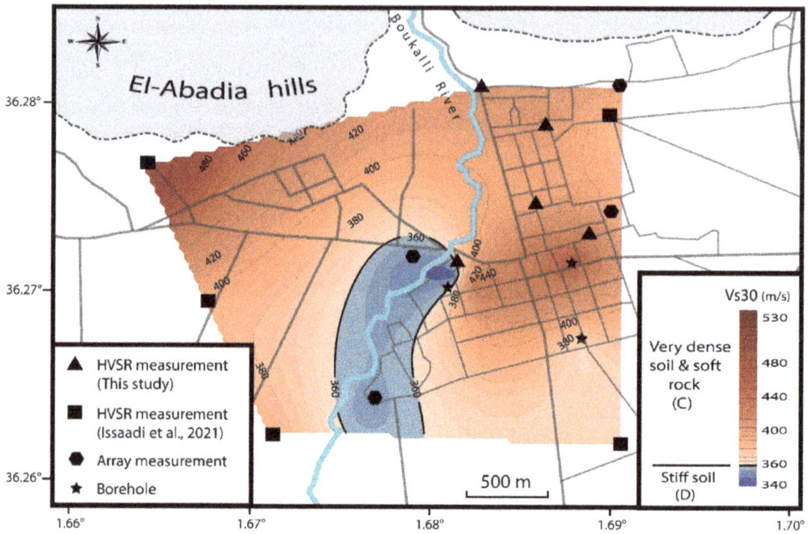

Figure 9. V_{s30} map and soil classification in the El-Abadia city [21].

In El-Attaf city, the V_{s30} values vary between 290 and 460 m/s (Figure 7). In most of the city, the soil is classified as very dense (C), except for a small area in the center where the soil is classified as stiff (D). The increase in velocity towards the west is caused by the presence of shallow Jurassic limestones, which outcrop about 1 km west of the city in the Temoulga Massif (Figure 2). In the southern part of the city, the slight increase in V_{s30} is related to the thinning of the Quaternary alluvium and the predominance of the Miocene stiff formations [21].

In the Ain-Defla city, the V_{s30} values vary between 250 and 550 m/s (Figure 8). The soil is classified as very dense and soft rock (C) in most of the city. The velocity gradually decreases towards the northwest and the soils become stiff (D). The variation in V_{s30} at Ain-Defla is mainly controlled by the ratio of Quaternary alluvium to Jurassic limestone in the first 30 m. In the south, where the city backs onto the Doui Massif, the ancient Quaternary alluvium forms a thin layer and the upper 30 m of the soil is dominated by Jurassic limestone. Moving northwest, the Quaternary deposits begin to be thicker and dominate the top 30 m of the soil column, and thus, V_{s30} values decrease and the soils are classified as stiff. In the El-Abadia city, the V_{s30} values range from 340 and 530 m/s (Figure 9). In most of the city, soils are classified as very dense and soft rock (C). The high V_{s30} values are related to the presence of Pliocene conglomerates and sandstones at shallow depths. The shear-wave velocity values decrease towards the south where the Quaternary alluvium is thicker [16]. The lowest Vs values are observed around the southern part of the Boukalli River in the city, where the upper layer is composed of present alluvium.

4.5. V_{s30} *Predictive Equation for the Middle-Chelif Basin*

The wavelength corresponding to the V_{s30} value was estimated for each average dispersion curve. The average wavelength found is $\lambda = 41$ m $\pm$ 3. After that, the V_{s30} values were correlated with V_{R41} values (Rayleigh wave velocity at $\lambda = 41$ m), and the best linear fitting was obtained (Equation (9)):

$$V_{s30} = 10171 * V_{R41} - 6719 \quad (9)$$

The regression plot, along with the residuals, is shown in Figure 10. The correlation degree is $R^2 = 0.9472$. Equation (9) was applied to dispersion curves obtained in two other cities of the Middle-Chelif Basin from a previous study [21] (Table 3). The aim was to evaluate the reliability of the V_{s30} predictive equation. The V_{s30} values were predicted within a maximum error of 5.6%, and the site classifications were correct. For the dispersion curves obtained in the present study, the V_{s30} values were predicted within a maximum error of 7.6%, and the site classifications were correct, except for ATF1 site.

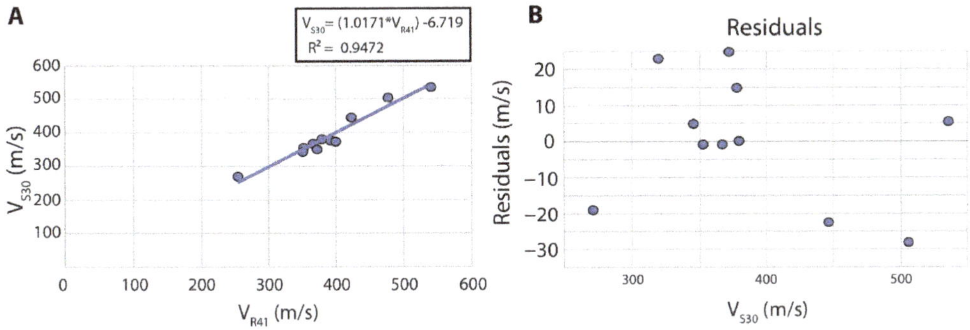

Figure 10. (**A**) Regression line (V_{S30} vs. V_{R41}) and equation given. R^2 is the correlation coefficient. (**B**) The corresponding residual values.

Table 3. Evaluation of V_{s30} predictive equation based on V_{s30} values obtained in this study, and in [21].

Site	Location.	V_{s30} (m/s)	Predicted V_{s30} (m/s)	Error (%)	Actual Site Classification	Predicted Site Classification
			This study			
ATF1	EL-Attaf	349.85	371.64	5.9	D	C
ATF2	EL-Attaf	373.97	399.10	6.3	C	C
ATF3	EL-Attaf	379.08	379.27	0.0	C	C
ATF4	EL-Attaf	367.03	366.05	0.3	C	C
AIN1	Ain-Defla	505.73	477.42	6.0	C	C
AIN2	Ain-Defla	535.85	541.50	1.0	C	C
AIN3	Ain-Defla	270.42	251.42	7.6	D	D
ABD1	El-Abadia	445.84	423.00	5.4	C	C
ABD2	El-Abadia	352.64	351.30	0.4	D	D
ABD3	El-Abadia	377.57	392.49	3.8	C	C
ABD4	El-Abadia	345.05	349.98	1.4	D	D
			Ref. [21]			
AR1	Oued-Fodda	225.87	230.85	2.2	D	D
AR2	Oued-Fodda	223.65	211.80	5.6	D	D
AR3	Oued-Fodda	449.07	457.90	1.9	C	C
AR4	Oued-Fodda	402.88	412.22	2.3	C	C
AR5	Oued-Fodda	467.75	457.90	2.2	C	C
AR7	El-Amra	454.45	469.59	3.2	C	C

In the case of the ATF1 site, the estimated and predicted V_{s30} values are close to the limit between class C and D (according to the NEHRP classification [57]).

5. Conclusions

In the present study, ambient vibration records were used to characterize the dynamic properties of the soil, and the velocity structure of its upper 30 m (V_{s30}) in the cities of El-Attaf, Ain-Defla, and El-Abadia, in the Middle-Chelif Basin. Both single-station and array-based techniques were applied. The studied cities are a good example of growing cities located in a highly seismic zone. This study improves the one carried out by the WCC (1984) by investigating the behavior of the soils and quantifying the liquefaction potential. Additionally, the V_{s30} values allowed classifying the soils of the three cities for the first time.

The HVSR technique was applied on single-station measurements to estimate the ground resonance frequencies. In the El-Attaf city, the frequencies vary between 1.2 and 8.3 Hz, and between 1.4 and 4.2 Hz in the El-Abadia city. In theAin-Defla city, the resonance frequencies vary between 1.4 and 4.5 Hz. The frequency peaks are directly related to impedance contrasts at different depths between sediments and bedrock. The corresponding amplitudes are ranging between 2 and 8.7.

The obtained resonance frequencies and the corresponding amplitudes were used to calculate the shear strain, which may give an idea about possible behavior of the soils during major earthquakes. In the El-Abadia and Ain-Defla cities, the shear strain values reflect the elasto-plastic behavior of the soil column. Cracks and settlements may occur during earthquakes, especially in El-Abadia city. In El-Attaf, the shear strain analysis also shows an elasto-plastic behavior of the soil in most of the city. Except in its western part, where a collapse behavior is observed. Consequently, the soil is subject to liquefaction.

Rayleigh wave dispersion curves were obtained from array recordings at 11 sites, using F-K, SPAC, and ESAC techniques. Shear-wave velocity models were obtained from the inversion of the mean dispersion curves. From the V_{s30} variation maps, the local soils were classified using the NEHRP chart for site classification. In the El-Attaf city, the V_{s30} values vary between 300 and 470 m/s. The soil is classified as very dense and soft rock (C) in most of the city. In the Ain-Defla city, V_{s30} values vary between 250 and 530 m/s. The soils are classified as very dense (C) in the central and eastern sides. In the west, the soils are stiff (D) due to the thickening of the Quaternary alluviums. Finally, in the El-Abadia city, the V_{s30} values vary between 340 and 530 m/s. In the major part of the city, the soil is classified as very dense and soft rock (C). In addition, a predictive equation for V_{s30} in the Middle-Chelif Basin was proposed based on the obtained dispersion curves and V_{s30} values.

The three studied cities extend to the alluvial plains of the Middle-Chelif, an area of unstable soils, which has undergone several phenomena induced by past earthquakes (e.g., soil liquefaction, landslides, cracks, settlements). A well constrained characterization of the dynamic properties of the soil, as well as the shear-wave velocity structure, allows a better understanding of the soil behavior during strong earthquakes. Therefore, this allows minimizing potential damage during potential earthquakes. The present study aims to contribute to the seismic hazard assessment in northern Algeria.

Author Contributions: A.I. acquired and processed the data and wrote the paper. A.S. acquired and prepared the data, helped with figure conceptualization. F.S. proposed the methodology, the structure, and reviewed the paper. A.Y.-C. provided the acquisition materials and supervised the paper. J.J.G.-M. provided Matlab code and reviewed the paper. All authors have read and agreed to the published version of the manuscript.

Funding: This research received no external funding.

Data Availability Statement: The data used in this study belongs to the research center in astronomy astrophysics and geophysics (CRAAG, Algeria).

Acknowledgments: The authors would like to thank the members of the CRAAG team that participated in field measurements: B. Melouk, O. Haddad, and R. Chimouni.

Conflicts of Interest: The authors declare no conflict of interest.

References

1. Ayadi, A.; Maouche, S.; Harbi, A.; Meghraoui, M.; Beldjoudi, H.; Mahsas, A.; Benouar, D.; Heddar, A.; Kherroubi, A.; Frogneux, M.; et al. Strong Algerian earthquake strikes near Capital city. *EOS* **2003**, *84*, 561–568. [CrossRef]
2. Abacha, I.; Boulahia, O.; Yelles-Chaouche, A.; Semmane, F.; Beldjoudi, H.; Bendjama, H. The 2010 Beni-Ilmane, Algeria, earthquake sequence: Statistical analysis, source parameters, and scaling relationships. *J. Seismol.* **2019**, *23*, 181–193. [CrossRef]
3. Ait Benamar, D.; Mouloud, H.; Belhai, D.; Semmane, F.; Harbi, A.; Tebbouche, M.Y.; Boukri, M.; Meziani, A.A.; Aourari, S.; Braham, M. The 17 July 2013 Hammam Melouane earthquake: Observations and analysis of geological and seismological data. *J. Iber. Geol.* **2022**, *48*, 163–180. [CrossRef]
4. Yelles-Chaouche, A.; Boudiaf, A.; Djellit, H.; Bracene, R. La tectonique active de la région nord-algérienne. *Comptes Rendus Geosci.* **2006**, *338*, 126–139. [CrossRef]

5. Meghraoui, M. Géologie des Zones Sismiques du Nord de l'Algérie: Paléosismologie, Tectonique Active et Synthèse Sismotectonique. Ph.D. Thesis, Université de Paris-Sud, Paris, France, 1988.
6. Ouyed, M.; Meghraoui, M.; Cisternas, A.; Deschamps, A.; Dorel, J.; Frechet, J.; Gaulon, R.; Hatzfeld, D.; Philip, H. Seismotectonics of the El Asnam earthquake. *Nature* **1981**, *292*, 26–31. [CrossRef]
7. Benouar, D. *Materials for the Investigation of the Seismicity of Algeria and Adjacent Regions during the Twentieth Century*; Annali Di Geofisica: Roma, Italia, 1994; pp. 459–860.
8. Philip, H.; Meghraoui, M. Structural analysis and interpretation of the surface deformations of the El Asnam earthquake of October 10, 1980. *Tectonics* **1983**, *2*, 17–49. [CrossRef]
9. Clough, W.G.; Mass, S.S.; Hamou, A.h. *Observation of Liquefaction and Landsliding as a Result of the October 10, 1980 El-Asnam, Algeria Earthquake*; Blume Earthquake Engineering Center, Departement of Civil Engineering, Stanford University: Stanford, CA, USA, 1981; p. 43.
10. Perrodon, A. Etude géologique des bassins sublittoraux de l'Algérie occidentale. *Publ. Serv. Cart. Géologique L'Algérie* **1957**, *12*, 328.
11. Repal, S.N. Le basin néogène du Chélif. In Proceedings of the XIXe Congrés de Géologie International, Alger, Algeria, 1952; p. 56. Available online: https://www.abebooks.fr/rechercher-livre/titre/le-bassin-neogene-du-chelif/auteur/repal-s-n/ (accessed on 27 June 2022).
12. Brives, A. *Les Terrains Miocenes du Bassin du Chélif et du Dahra*; Imprimerie P. Fontana & Cie: Algiers, Algeria, 1897.
13. CGG. *Geophysical Study of the Cheliff Plain*; Internal report; Compagnie Generale de Geophysique: Paris, France, 1969.
14. Talaganov, K.; Aleksovski, D.; Milutinovic, Z.; Ameur, B.; Arsovski, M.; Jancevski, J.; Andreevski, V. *Studies for Elaboration of the Code for Repair and Strengthening of Damaged Buildings in the Region of El Asnam: Engineering Geology, Geotechnical and Geophysical Characteristics of the TOWN of El Asnam and Other Sites*; Report IZIIS; Institute of Earthquake Engineering & Engineering Seismology: Skopje, North Macedonia, 1982; pp. 55–82.
15. Woodward-Clyde Consultants. *Seismic Microzonation of Ech-Chellif Region, Algeria*; Woodward-Clyde Consultants: London, UK, 1984.
16. Meghraoui, M. Etude neotectonique de la region nord-est d'El Asnam: Relation avec le seisme du 10.10. 1980. Ph.D. Thesis, Université de Paris VII, Paris, France, 1982.
17. Bendali, M.; Abtout, A.; Bouyahiaoui, B.; Boukerbout, H.; Marok, A.; Reolid, M. Interpretation of new gravity survey in the seismogenic Upper Chelif Basin (North of Algeria): Deep structure and modeling. *J. Iber. Geol.* **2022**, *48*, 205–224. [CrossRef]
18. Layadi, K.; Semmane, F.; Yelles-Chaouche, A. Site-Effects Investigation in the City of Chlef (Formerly El-Asnam), Algeria, Using Earthquake and Ambient Vibration Data. *Bull. Seismol. Soc. Am.* **2016**, *106*, 2185–2196. [CrossRef]
19. Layadi, K.; Semmane, F.; Yelles-Chaouche, A. S-wave velocity structure of Chlef City, Algeria, by inversion of Rayleigh wave ellipticity. *Near Surf. Geophys.* **2018**, *16*, 328–339. [CrossRef]
20. Issaadi, A.; Semmane, F.; Yelles-Chaouche, A.; Galiana-Merino, J.; Layadi, K. A Shear-Wave Velocity Model in the City of Oued-Fodda (Northern Algeria) from Rayleigh Wave Ellipticity Inversion. *Appl. Sci.* **2020**, *10*, 1717. [CrossRef]
21. Issaadi, A.; Semmane, F.; Yelles-Chaouche, A.; Galiana-Merino, J.J.; Mazari, A. Shallow S-wave velocity structure in the Middle-Chelif Basin, Algeria, using ambient vibration single-station and array measurements. *Appl. Sci.* **2021**, *11*, 11058. [CrossRef]
22. Kireche, O. Etude Géologique et Structurale des Massifs de la Plaine du Chéliff (Doui, Rouina, Temoulga). Ph.D. Thesis, Université d'Alger, Algiers, Algeria, 1977.
23. Thomas, G. Géodynamique d'un Bassin Intramontagneux: Le Bassin du bas Chéliff Occidental (Algérie) durant le Mio-Plio-Quaternaire. Ph.D. Thesis, Université de Pau, Pau, France, 1985.
24. Brives, A.; Jackob, M.; Ficheur, M. *Carte Géologique de Oued-Fodda 1/50.000*; Service Géologique de l'Algérie: Algiers, Algeria, 1906.
25. Nakamura, Y. A method for dynamic characteristics estimation of subsurface using microtremor on the ground surface. *Railw. Tech. Res. Inst. Q. Rep.* **1989**, *30*, 25–33.
26. Field, E.H.; Jackob, K.H. A comparison and test of various site response estimation techniques, including three that are non reference—Site dependent. *Bull. Seismol. Soc. Am.* **1995**, *85*, 1127–1143.
27. Arai, H.; Tokimasu, K. S-wave velocity profiling by inversion of microtremor H/V spectrum. *Bull. Seismol. Soc. Am.* **2004**, *94*, 53–63. [CrossRef]
28. Bonnefoy-Claudet, S.; Cornou, C.; Bard, P.Y.; Cotton, F.; Moczo, P.; Kristek, J.; Fäh, D. H/V ratio: A tool for site effects evaluation: Results from 1D noise simulations. *J. Appl. Geophys.* **2006**, *167*, 827–837. [CrossRef]
29. Rosa-Cintas, S.; Galiana-Merino, J.; Rosa-Herranz, J.; Molina, S.; Martínez-Esplá, J. Polarization analysis in the stationary wavelet packet domain: Application to HVSR method. *Soil Dyn. Earthq. Eng.* **2012**, *42*, 246–254. [CrossRef]
30. Bonnefoy-Claudet, S.; Cotton, F.; Bard, P.-Y. The nature of noise wavefield and its applications for site effects studies: A literature review. *Earth-Sci. Rev.* **2006**, *79*, 205–227. [CrossRef]
31. La Rocca, M.; Chiappetta, G.; Gervasi, A.; Festa, R.L. Non-Stability of the noise HVSR at sites near or on topographic heights. *Geophys. J. Int.* **2020**, *222*, 2162–2171. [CrossRef]
32. Benkaci, N.; El Hadi Oubaiche, J.-L.C.; Bensalem, R.; Benouar, D.; Abbes, K. Non-Stability and Non-Reproducibility of Ambient Vibration HVSR Peaks in Algiers (Algeria). *J. Earthq. Eng.* **2021**, *25*, 853–871. [CrossRef]
33. Acerra, C.; Aguacil, G.; Anastasiadis, A.; Atakan, K.; Azzara, R.; Bard, P.-Y.; Basili, R.; Bertrand, E.; Bettig, B.; Blarel, F. *Guidelines for the Implementation of the H/V Spectral Ratio Technique on Ambient Vibrations Measurements, Processing and Interpretation*; European Commission: Louxembourg, 2004.

34. Wathelet, M.; Chatelain, J.L.; Cornou, C.; Giulio, G.D.; Guillier, B.; Ohrnberger, M.; Savvaidis, A. Geopsy: A user-friendly open-source tool set for ambient vibration processing. *Seismol. Res. Lett.* **2020**, *91*, 1878–1889. [CrossRef]
35. Konno, K.; Ohmachi, T. Ground-motion characteristics estimated from spectral ratio between horizontal and vertical components of microtremor. *Bull. Seismol. Soc. Am.* **1998**, *88*, 228–241. [CrossRef]
36. Ishihara, K. *Soil Behaviour in Earthquake Geotechnics*; Oxford University Press: Oxford, UK, 1996.
37. Nakamura, Y. Seismic vulnerability indices for ground and structures using microtremor. In Proceedings of the World Congress on Railway Research in Florence, Florence, Italy, 16–19 November 1997.
38. *Regles Parasismiques Algerienne RPA99/Version 2003*; RPA99/2003. Centre National de Recherche Appliquée en Génie Parasismique: Algiers, Algeria, 2003.
39. Kelly, E.J.; Levin, M.J. *Signal Parameter Estimation for Seismometer Arrays*; Massachusetts Institute of Technologies, Lincoln Lab Lexington: Cambridge, MA, USA, 1964.
40. Capon, J.; Greenfield, R.; Kolker, R. Multidimensional maximum-likelihood processing of a large aperture seismic array. *Proc. IEEE* **1967**, *55*, 192–211. [CrossRef]
41. Capon, J. High-resolution frequency–wavenumber spectral analysis. *Proc. IEEE* **1969**, *57*, 1408–1419. [CrossRef]
42. Lacoss, R.T.; Kelly, E.J.; Toksoz, M.N. Estimation of seismic noise structure using Array. *Geophysics* **1969**, *29*, 21–38. [CrossRef]
43. Aki, K. Space and time spectra of stationary stochastic waves, with special reference to microtremors. *Bull. Earthq. Res. Inst.* **1957**, *35*, 415–456.
44. Okada, H. *The Microtremor Survey Method Geophysical Monograph Series 12*; Asten, M.W., Ed.; Society of Exploration Geophysicists: Tulsa, OK, USA, 2003.
45. Bettig, B.; Bard, P.; Scherbaum, F.; Riepl, J.; Cotton, F.; Cornou, C.; Hatzfeld, D. Analysis of dense array noise measurements using the modified spatial auto-correlation method (SPAC): Application to the Grenoble area. *Boll. Geofis. Teor. Ed Appl.* **2001**, *42*, 281–304.
46. Ling, S. An extended use of the spatial autocorrelation method for the estimation of geological structure using microtremors. In Proceedings of the 89th SEGJ (The Society Exploration Geophysicists of Japan) Conference, Nagoya, Japan, 12–14 October 1993; pp. 44–48.
47. Parolai, S.; Mucciarelli, M.; Gallipoli, M.R.; Richwalski, S.M.; Strollo, A. Comparison of empirical and numerical site responses at the Tito Test Site, Southern Italy. *Bull. Seismol. Soc. Am.* **2007**, *97*, 1413–1431. [CrossRef]
48. Rost, S. Array Seismology: Methods and Applications. *Rev. Geophys.* **2002**, *40*, 2-1–2-27. [CrossRef]
49. Köhler, A.; Ohrnberger, M.; Scherbaum, F.; Wathelet, M.; Cornou, C. Assessing the reliability of the modified three-component spatial autocorrelation technique. *Geophys. J. Int.* **2007**, *168*, 779–796. [CrossRef]
50. Galiana-Merino, J.J.; Rosa-Cintas, S.; Rosa-Herranz, J.; Garrido, J.; Peláez, J.A.; Martino, S.; Delgado, J. Array measurements adapted to the number of available sensors: Theoretical and practical approach for ESAC method. *J. Appl. Geophys.* **2016**, *128*, 68–78. [CrossRef]
51. Wathelet, M. An improved neighborhood algorithm: Parameter conditions and dynamic scaling. *Geophys. Res. Lett.* **2008**, *35*, 5. [CrossRef]
52. Fäh, D.; Kind, F.; Giardini, D. Inversion of local S-wave velocity structures from average H/V ratios, and their use for the estimation of site-effects. *J. Seismol.* **2003**, *7*, 449–467. [CrossRef]
53. Sil, A.; Haloi, J. Empirical correlations with standard penetration test (SPT)-N for estimating shear wave velocity applicable to any region. *Int. J. Geosynth. Ground Eng.* **2017**, *3*, 22. [CrossRef]
54. Boore, D.M. Estimating V_{s30}(or NEHRP site classes) from shallow velocity models (depths < 30 m). *Bull. Seismol. Soc. Am.* **2004**, *94*, 591–597.
55. Code, P. *Eurocode 8: Design of Structures for Earthquake Resistance-Part 1: General Rules, Seismic Actions and Rules for Buildings*; European Committee for Standardization: Brussels, Belgium, 2005.
56. Cressie, N. The origins of kriging. *Math. Geol.* **1990**, *22*, 239–252. [CrossRef]
57. BSS Council. *NEHRP Recommended Provisions for Seismic Regulations for New Buildings and Other Structures (FEMA 450)*; Federal Emergency Management Agency: Washington, DC, USA, 2003.

applied sciences

MDPI

Article

Shallow S-Wave Velocity Structure in the Middle-Chelif Basin, Algeria, Using Ambient Vibration Single-Station and Array Measurements

Abdelouahab Issaadi [1,2,*], Fethi Semmane [1], Abdelkrim Yelles-Chaouche [1], Juan José Galiana-Merino [2,3] and Anis Mazari [1]

1 Centre de Recherche en Astronomie Astrophysique et Géophysique (CRAAG), Algiers 16340, Algeria; fethisemmane@yahoo.fr (F.S.); kyelles@yahoo.fr (A.Y.-C.); anissmazari@gmail.com (A.M.)
2 University Institute of Physics Applied to Sciences and Technologies, University of Alicante, Crta. San Vicente del Raspeig, s/n, 03080 Alicante, Spain; jj.galiana@ua.es
3 Department of Physics, Systems Engineering, and Signal Theory, University of Alicante, Crta. San Vicente del Raspeig, s/n, 03080 Alicante, Spain
* Correspondence: issaadi.a.ouahab@gmail.com

Abstract: In order to better assess the seismic hazard in the northern region of Algeria, the shear-wave velocity structure in the Middle-Chelif Basin is estimated using ambient vibration single-station and array measurements. The Middle-Chelif Basin is located in the central part of the Chelif Basin, the largest of the Neogene sedimentary basins in northern Algeria. This basin hosts the El-Asnam fault, one of the most important active faults in the Mediterranean area. In this seismically active region, most towns and villages are built on large unconsolidated sedimentary covers. Application of the horizontal-to-vertical spectral ratio (HVSR) technique at 164 sites, and frequency–wavenumber (F–K) analysis at 7 other sites, allowed for the estimation of the ground resonance frequencies, shear-wave velocity profiles, and sedimentary cover thicknesses. The electrical resistivity tomography method was used at some sites to further constrain the thickness of the superficial sedimentary layers. The soil resonance frequencies range from 0.75 Hz to 12 Hz and the maximum frequency peak amplitude is 6.2. The structure of the estimated shear-wave velocities is presented in some places as 2D profiles to help interpret the existing faults. The ambient vibration data allowed us to estimate the maximum depth in the Middle-Chelif Basin, which is 760 m near the city of El-Abadia.

Keywords: Middle-Chelif sedimentary basin; HVSR; array measurement; frequency–wavenumber (F–K) method

check for updates

Citation: Issaadi, A.; Semmane, F.; Yelles-Chaouche, A.; Galiana-Merino, J.J.; Mazari, A. Shallow S-Wave Velocity Structure in the Middle-Chelif Basin, Algeria, Using Ambient Vibration Single-Station and Array Measurements. *Appl. Sci.* **2021**, *11*, 11058. https://doi.org/10.3390/app112211058

Academic Editors: Ricardo Castedo, Miguel Llorente Isidro and David Moncoulon

Received: 10 October 2021
Accepted: 19 November 2021
Published: 22 November 2021

1. Introduction

Northern Algeria is characterized by a series of Neogene basins (e.g., Constantine Basin, Hodna Basin, Soummam Basin, Tizi-ouzou Basin, Mitidja Basin, Medea Basin, and Chelif Basin), elongated in an E–W direction, and surrounded by the Tellian Atlas mountain belts, which act as a substratum for their sedimentary covers [1–3]. These basins host important seismic activity, mainly in their marginal zones [4]. Active tectonics in the northern part of the country and the related seismicity are due to the fact that this zone is located at the boundary between the African and Eurasian convergent plates [4,5].

In the northwestern part of Algeria, between the septentrional and meridional Tellian Atlas mountain belts, lies the Chelif Basin, a wide depression of over 450 km in length, which is the largest and the most subsident of the sublittoral basins [2,3]. The dimensions of the basin and the complexity of its hydrographic network has led several authors to divide it into three sub-basins: The Lower-, the Middle-, and the Upper-Chelif. In this study, we focus on the plain extending from Oued-Fodda to Ain-Defla (Figure 1). However, one must note that this subdivision is not unanimous within the scientific community. While some authors assign our study area to the Middle-Chelif Basin [6–8], others have assigned it to the oriental part of the Lower-Chelif Basin [2,3,9]. In this study, we consider the study area as a part of the Middle-Chelif Basin.

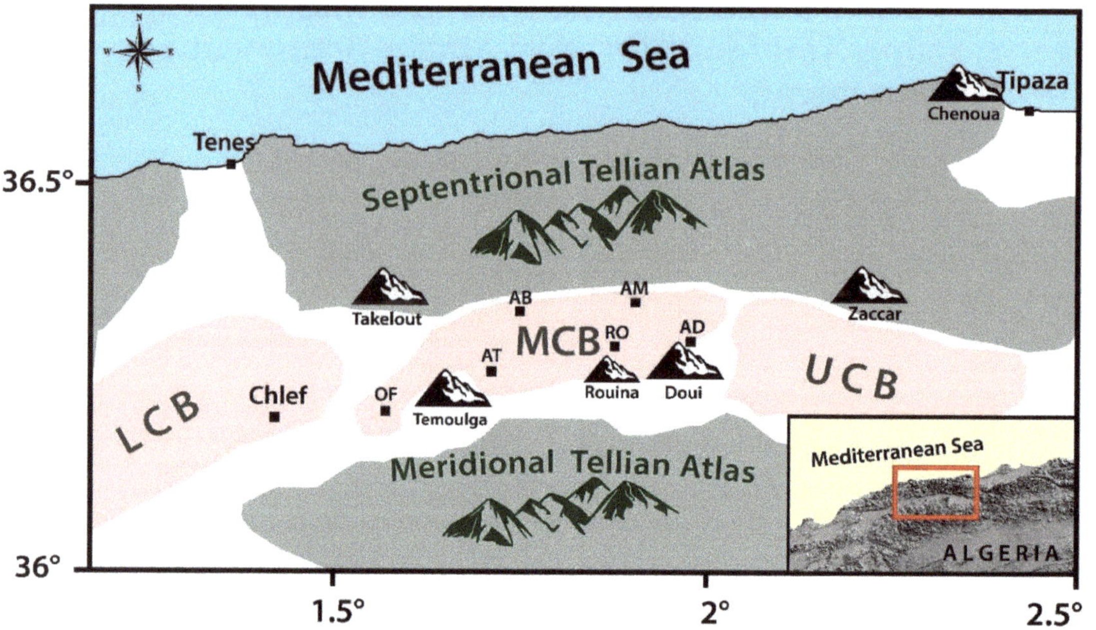

Figure 1. Situation of the Middle-Chelif Basin. LCB: Lower-Chelif Basin. MCB: Middle-Chelif Basin. UCB: Upper-Chelif Basin. OF: Oued-Fodda. AB: El-Abadia. AT: El-Attaf. AM: El-Amra. RO: Rouina. AD: Ain-Defla.

The Middle-Chelif plain is home to over 400,000 inhabitants, who are mainly concentrated on its edges, and distributed over its principal cities, which are: Ain-Defla, Rouina, El-Amra, El-Attaf, El-Abadia, and Oued-Fodda.

The region has witnessed several destructive earthquakes, such as the 1858 El-Amra earthquake (I_0 = IX, [10]); the 1934 El-Abadia earthquake (I_0 = IX, [10]); the 1954 Orléansville earthquake (Ms 6.7, [10]); and the 1980 El-Asnam earthquake (Ms 7.3, [11]).

In the last few decades, techniques based on ambient vibrations have been widely used to study sedimentary basins because of the quick implementation and the lower cost of the process, compared to boreholes or other geophysical prospecting methods (e.g., [12–16]). It is proven that soft sedimentary layers can amplify the ground shaking and prolong its duration during an earthquake, which can be harmful for buildings. Therefore, the determination of the geotechnical characteristics of the soils is primary for seismic risk and site effects assessments. One of the most important parameters for determining the geotechnical characteristics is the shear-wave velocity structure. The high-velocity contrast between soft sediments and the bedrock, along with the local geological and topographical aspects, are important factors for ground motion amplification.

The Chelif Basin has been the subject of several geological and geophysical studies. The majority of these studies have been done on the Lower-Chelif Basin, and only a few of them have included the western part of the Middle-Chelif. It was only at the end of the 1960's that the CGG (Compagnie Générale de Géophysique) carried out a geophysical prospecting campaign by electric methods in the Middle-Chelif Plain in order to study the structure of the sedimentary deposits [7].

Since the 1980 El-Asnam earthquake, and the damage it caused, the Middle-Chelif region has been the subject of a multitude of geological and geophysical studies. The first study was carried out by the Institute of Earthquake Engineering of Skopje University

(Macedonia), which resulted in the elaboration of the code for the repairing and strengthening of damaged buildings in the Chelif region. The code was based on studies of seismic hazards and the geotechnical conditions of the soil. Several seismic refraction profiles were performed in the cities of Chlef and El-Attaf [17]. The Woodward Clyde Consultants also carried out a complete geotechnical and geological study in eight cities of the Chelif region, including Oued-Fodda, El-Attaf, and El-Abadia, where several boreholes were drilled. The study outcomes provided geotechnical and hydrogeological maps for each city, along with seismic microzoning survey maps [18]. The neotectonic and paleoseismological studies carried out on the formations of the oriental Lower-Chelif Basin (e.g., [9,19]) were intended to highlight the tectonic elements and to identify geological structures and faults likely to be reactivated by generating earthquakes.

More recently, site effects investigations using earthquakes and ambient vibration data were conducted in the cities of Chlef in the Lower-Chelif Basin [20,21] and Oued-Fodda in the Middle-Chelif Basin [22]. The main objectives were to estimate the soil resonance frequencies, the shear-wave velocities in the sedimentary rocks, and the depths to seismic bedrock, where the relatively high impedance contrast with Miocene formations may amplify ground shaking during an earthquake. In the Mitidja basin, Bouchelouh [23] and Tebbouche [24] estimated and mapped the roof of the engineering bedrock using mainly ambient vibration data.

In this study, ambient vibration measurements using single-station and array techniques were used to estimate the shear-wave velocity for the sedimentary layers, and to map the bedrock structure in the Middle-Chelif Basin. In the first part of the study, we estimate the soil resonance frequencies (f_0) and the corresponding amplitudes (A_0) using the horizontal-to-vertical spectral ratio (HVSR) technique [25,26]. In the second part, we perform array measurements using the frequency–wavenumber (F–K) technique [27–30] to retrieve the surface wave dispersion curves. After that, the dispersion and HVSR curves are inverted jointly to estimate the shear-wave velocity (Vs) profiles at each site. Available borehole information and electrical resistivity profiles constrain the Vs models used for the inversion process. As a result, the bedrock model is proposed.

This study is a continuation of previous work conducted in the city of Oued-Fodda [22]. The results obtained in this work will contribute to the seismic hazard reassessment in the Chelif Basin. They can also be used for seismic hazard mitigation studies, such as strong ground motion simulation, soil liquefaction studies, and for the updating of the Algerian seismic code.

2. Geological Framework

The Neogene basins, in the western part of Algeria, stretch parallel to the Mediterranean coast and lie within the septentrional and meridional Tellian Atlas mountain belts, which belong to the southern branch of the Alpine chain [1,31]. The intramountainous Chelif Basin is a wide depression of over 450 km filled with Mio-Plio-Quaternary deposits [1–3,9,32,33]. It is formed by a succession of plains, hills, and folds, boarded by the Dahra and Boumaad Mountains in the north, and the Ouarsenis Mountains in the south. These Tellian belts were structured during the Mesozoic [34] and constitute the substratum of the Neogene deposits.

The Middle-Chelif Basin extends from the Sara-El-Maarouf anticline to the Beni-Ghomerian plateau (Figure 2). In its central part, the Quaternary plain forms a narrow strip that stretches for over 50 km. This region is affected by normal and thrust faults, including the so-called "El-Asnam Fault". The orientation of these faults is parallel to the synclinal axis of the basin. The cross-sections performed by the National Society of Petroleum [35] and the General Company of Geophysics [7] show that the synclinal axis of this basin appears to be north of the Chelif River. In the southern part of this plain, inlier terrains of Jurassic-to-Silurian age outcrops form the Doui, Rouina, and Temoulga Massifs, from east to west, respectively. These schistosity massifs are considered to be autochthonous formations [33,34]. Although the Jurassic formations are predominant in these outcrops, the bedrock is mainly Cretaceous in most of the basin area, and it is composed of Cenomanian to Senonian clays and marls [36].

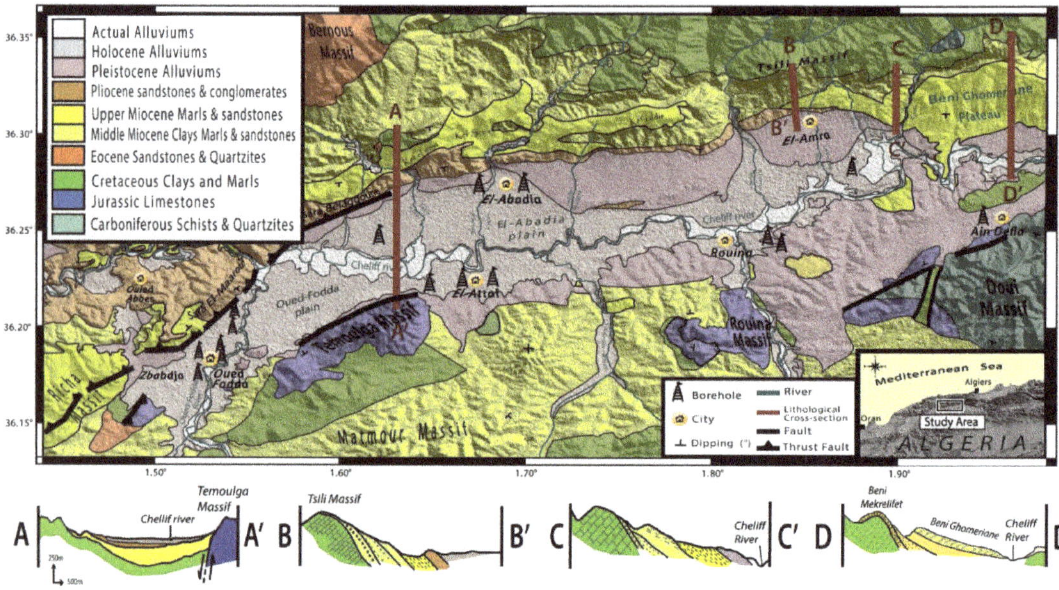

Figure 2. Geological map of the Middle-Chelif Basin. Modified and compiled from [33,36]. The lithological cross-section, AA′, is digitalized from [35]; BB′, CC′, and DD′ are from [32].

Several previous geological studies have detailed the lithological and stratigraphical aspect of the Neogene deposits in the Middle-Chelif Basin [2,9,32]. The first sedimentary deposit is a red detritic series of conglomerates, poudingues, and marls. Brives [32] and Perrodon [2] attribute these formations to the Lower Miocene, while Meghraoui [9] and Thomas [3] assign them to the Middle Miocene (Serravalo-Tortonian). These formations outcrop on the southern flank of the Tsili Massif, north of El-Amra (Figure 2). Above these formations lies an important intercalation of marls, clays, and limestones of the Lower Tortonian [9]. The boreholes and the electric surveying data used in this study [7] show that these formations can reach a considerable thickness of 400 m. This layer lies in discordance over the conglomeratic series in the northern part of the basin, and directly over the Cretaceous marls in the south. The passage between the Upper Tortonian sandstones and the Lower Tortonian marls is completed abruptly [32]. This contrast can be observed in Kef-Ensoura, north of El-Abadia, and also on the Beni-Ghomerian Plateau, where the two formations outcrop. The Messinian is represented mainly by blue marls with a maximum thickness of 50 m [9]. The marine regression in the Pliocene divided this stage into marine and continental formations [2,3,32]. The marine Pliocene is represented by sands and sandstones, while the continental is composed of red sands and conglomerates [2,9,32]. The Pliocene constitutes the principal deposits of the Sara-el-Maarouf and the Sara-Belaggoune anticlines, north of Oued-Fodda, and extends eastward by outcropping on a narrow band until the Beni-Ghomerian Plateau. However, this formation does not outcrop in the southern parts of the basin, which suggests that it disappears somewhere under the Quaternary plain. These deposits can reach a thickness of 200 m [9].

The Quaternary is predominant in the wide plain that stretches from Oued-Fodda to Ain-Defla. The Pleistocene formations surround the Holocene alluviums and form the first hills in the northern and southern margins of the plain. The Pleistocene is composed of clays, gravels, and conglomerates. The Holocene is constituted of recent alluviums. The

Chelif River crosses the Middle-Chelif Quaternary Plain from Ain-Defla to Ouled-Abbes and drags all the necessary materials that form the actual alluviums.

The geological and geotechnical data show important lateral variations in the facies and thicknesses within the sedimentary layers and the bedrock. For that, the study area was divided into three zones: The Oued-Fodda Plain, the Carnot Plain, and the Ain-Defla Region (Figure 3). The Oued-Fodda Plain lies between the Temoulga Massif and the Sara-El-Maarouf anticline in a NE–SW direction, with an average width of 3 km. The fault affecting the Temoulga Massif, east of the plain, and the El-Asnam fault in the west, suggest that this plain acts as a graben structure. The different outcrops in the area suggest a low sedimentary thickness [22]. The bedrock is mainly composed of Senonian marls.

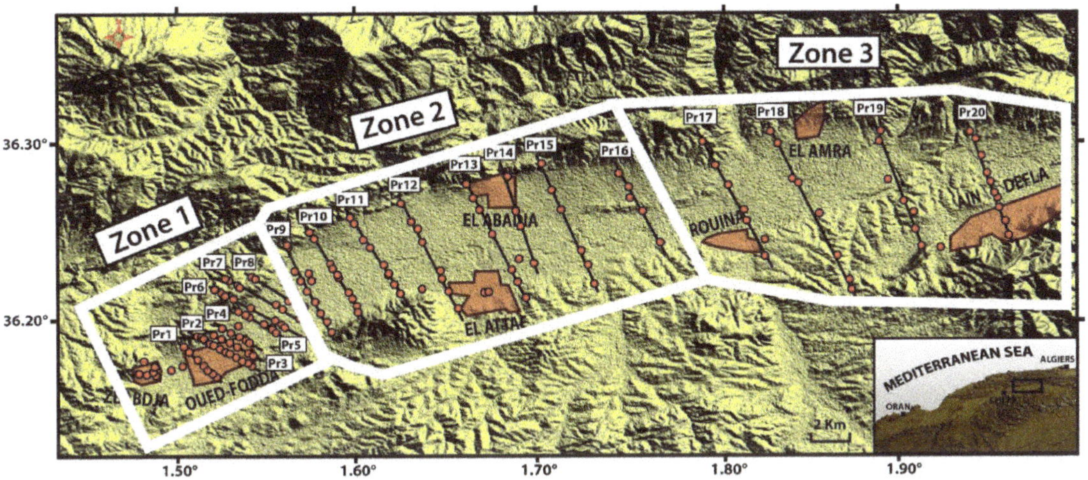

Figure 3. Zonation map of the study area. Red dots correspond to single-station measurements. Red polygons correspond to the limits of the different cities.

The Carnot Plain, also known as the El-Attaf Plain, extends from Bir-Safsaf to Rouina in an E–W direction. The main cities of the plain are El-Attaf and El-Abadia, built on the southern and northern plain margins, respectively. The average basin width of 8 km suggests that the maximum depth of the basin is likely to be observed in this region. Lateral variations in the facies are observed in the Cretaceous series, which is considered as the bedrock of the sedimentary deposits in this area. The facies change from marly to mainly composed of limestones west of El-Abadia. The geophysical data show that the sedimentary layers can reach a depth of 700–800 m from El-Abadia to Rouina [6,7,35].

3. Methodology

3.1. The Horizontal-to-Vertical Ratio Technique

The HVSR technique (also known as H/V) was initiated by Nogoshi and Igarashi [26], and developed by Nakamura [25]. It is one of the most commonly used techniques that allows for the retrieval of one of the most important parameters of site response, i.e., the resonance frequency at a given site. This technique has proven its reliability in estimating the soil resonance frequency [37], although several authors have reported that this technique is not reliable in estimating the amplification factors, especially at lower frequencies [38–41]. This inability is due to the contribution of body and surface waves, and it is the latter ones that control the HVSR curve. The low contribution of body waves is sufficient for estimating the fundamental frequency, but not enough for estimating the amplification factor [38,39,41]. Recently, LaRocca [42] has also demonstrated how the peak amplitude of the HVSR of the seismic noise may vary considerably with time.

This method consists of estimating the ratio between the Fourier amplitude spectrum of the horizontal and vertical components of the ambient noise, which provides the HVSR curve as a result. The resulting HVSR frequency peak is very well-correlated with the resonance frequency of the soil [43]. This peak corresponds to an impedance contrast in the soil column.

The low cost, the simplicity of acquisition and processing, and the reliability in estimating the soil resonance frequencies, makes this technique one of the most used for site characterizations, especially for urban areas.

3.2. Inversion of the HVSR Curves

The inversion of the HVSR curves allows for the retrieval of the shear-wave velocity values for the sedimentary layers, along with the bedrock depths of our study area.

The velocity profiles are calculated using the neighborhood algorithm [44]. As there is no unique solution for the inversion problem, several velocity models are calculated. For each model, the calculated ellipticity of the Rayleigh wave fundamental mode is compared to the HVSR curve and a misfit value is given [45]. The ellipticity curves and the corresponding velocity models with the lower misfit values are considered.

As for all the inversion processes, the nonuniqueness of the solution is a very delicate matter to deal with, and hundreds of thousands of velocity models can result from one simple HVSR curve. For that reason, some soil parameters are required to better guide and constrain the process. The main parameter is the number of the sedimentary layers and the range of thicknesses for each layer. Geological or lithological cross-sections are usually good enough for providing these two parameters. Some geotechnical parameters are also required for each layer, either as fixed values or as intervals. These parameters are the Vp, Vs, density, and Poisson's ratio. The latter is set between 0.2 and 0.5 (universal values for soils). For good and reliable results, it is necessary to have good data coverage in the study area, as well as boreholes and geophysical prospections (e.g., the seismic refraction experiment).

3.3. The Frequency–Wavenumber (F–K) Analysis

In the last decade, array-based techniques have become popular and are widely used for the analysis of ambient vibrations. The concept of configurations in arrays, with simultaneous ambient vibration recordings, has proven its reliability and efficiency through several studies and investigations [16,46,47]. One of the most applied techniques for array processing is frequency–wavenumber analysis, commonly known as F–K analysis [27–30]. This technique is performed in the frequency domain and allows for the estimation of the back azimuth and the slowness of the seismic wave sources recorded by the array [47].

The theoretical aspect of this technique is based on two main assumptions. The first one is that the wavefront propagation is on the vertical plane, and the process is stationary in the other two planes. The second assumption is that the process is stationary in time. The microtremor wavefield is a superposition of seismic waves propagated from several distant sources [28]. The F–K analysis exploits the stationary and stochastic character of these propagated waves to construct the frequency–wavenumber power spectral density function, which holds information on the power as a function of the frequency and the velocity of travelling waves. The power spectral density can be estimated using the following two following methods: The beam-forming method (BFM) [29,30], and the maximum likelihood method (MLM) or the high-resolution method (HR) [27,28,48]. In this study, we used the beam-forming method to calculate the frequency–wavenumber power spectral density and thereby retrieve the velocity and the directions of the seismic propagating waves. The HR method provides a higher resolution in the frequency–wavenumber plane than the BFM method. However, the latter is less sensitive to measurement errors since it uses less computations [27]. Rosa-Cintas [16] calculated the shear-wave velocity using both the BFM and the HR methods and the obtained dispersion curves were almost similar.

The ability of these methods to identify the directional properties of the noise wavefield is highlighted by Maresca [46].

3.4. Electrical Resistivity Surveying

The electrical methods allow for the estimation of the resistivity at the different layers that compose the subsurface structure and, therefore, for inferences of the thicknesses of these layers [49–51]. In this study, we performed two electrical prospections in order to obtain the thickness of the Quaternary alluviums. Moreover, the information available from thirteen additional electrical surveys was used.

The resistivity of a rock is the physical property that determines the ability of the rock to conduct the electric current. In a sedimentary layer, the resistivity is mainly controlled by the electrical resistivity of the fluids within this layer.

The prospecting consists of injecting an electric current into the soil column using two electrodes. After that, we measure the potential difference between two other electrodes that are planted between the injecting electrodes. This technique allows us to image the vertical succession of layers with different resistivities. One must note that the higher the resistivity is, the more solid the rock is. The depth of investigation is proportional to the interspacing between the transmitter electrodes. The constant spacing between the electrodes allows for the obtainment of a profile of the lateral and vertical variations of the resistivity.

The configuration of the measuring device is chosen according to the type of study and the expected results. There are several possible electrode configurations. The most commonly used are the Wenner, Schlumberger, and dipole–dipole configurations [49,52]. In this study, we used the dipole–dipole configuration.

4. Geotechnical Information

The set of parameters to be determined in the inversion process requires additional geotechnical information (number of layers, thicknesses, Vp, Vs, and densities). For a large area, such as the Middle-Chelif Basin, where lateral variations in the facies are quite frequent [2,9,32], a wide coverage of geotechnical data is recommended. In this study, geological and geophysical data were used and compiled for a well-guided inversion process with reliable results.

The synclinal aspect of the study area made it easier to determine the number of sedimentary layers since the oldest layers outcrop in succession on the edges of the Middle-Chelif Quaternary Plain, and dip under the alluviums [36]. The modified geological map, along with the lithological cross-sections (Figure 2), were good enough for fixing the number of layers.

A total of 16 geotechnical boreholes were also used, whose depths varied from 20 to 250 m [6,7,18]. These boreholes allowed us to fix the thicknesses of the Quaternary and Pliocene layers in some areas. In order to complete the lack of data and to obtain information on deeper layers, a total number of 13 electrical profiles were added. Nine of them were made by the General Company of Geophysics between El-Attaf and Ain-Defla [7]. The profiles are oriented in a N–S direction, with an average length of 5 km and a maximum depth of investigation of 1000 m. The remaining electrical profiles were carried out by the ANRH (Agence National de Ressource Hydraulique) in the El-Attaf region. The average length was 800 m, with an investigation depth of 150 m.

The velocity and density values for the different sedimentary rocks are provided in Talaganov [17], obtained by using the seismic refraction method in the cities of Chlef, Beni-Rached, and El-Attaf. For the seismic bedrock, a Vs value ranging between 1800 and 2600 m/s was assigned. These values were obtained by the inversion of the ambient vibration data for the cities of Chlef [20] and Oued-Fodda [22]. There are two different formations composing the bedrock, the Cretaceous flysches, clays, and marls in the Middle-Chelif Plain, and the older Jurassic limestones on its southern borders. For the limestones, a Vs value of 2800 m/s and a Vp value of 4800 m/s were assigned using local earthquake tomographic inversion results [53]. The younger marls and clays of the late Cretaceous constitute the bedrock of the sedimentary deposits in the region of Oued-Fodda since it outcrops about 2 km southwest of the city. The Vs values obtained in Issaadi [22] are, therefore, assigned to the Cretaceous marls (Upper Senonian).

A lithological and geotechnical model for the Middle-Chelif Basin was built from the compilation of the whole of the gathered data (Table 1).

Table 1. Geotechnical information used for the inversion process. Hl: Holocene. Ps: Pleistocene. As: Astian. Pl: Placenzian. Ms: Messinian. Tr: Tortonian. Sr: Seravallian. Sn: Senonian.

Per/Epoch.	Age	Rock	Thickness (m)	Vp (m/s)	Vs (m/s)	σ (KN/m^3)	References
Quaternary	Hl	Alluviums	2–20	350–600	200–350	16–18	[6,7,17,18,22]
	Ps	Clays & Conglomerates	10–100	600–1700	400–700	18–22	[6,7,17]
Pliocene	As	Sands & Conglomerates	10–80	550–2000	350–750	17–22	[7,17]
	Pl	Sands & Conglomerates	30–80	750–2400	500–900	19–23	[7,17]
Miocene	Ms	Marls	30–80	1100–2900	700–1200	20–24	[7,17]
	Tr	Limestones & Sandstones	40–150	1300–3200	900–1300	23–25	[7,17]
	Sr-Tr	Clays & Marls	100–450	1200–3400	800–1400	21–25	[7,17,35]
	Sr	Poudingues	10–60	1900–3600	1300–1500	22–25	[35]
Cretaceous	Sn	Marls, clays & Quartzites	-	2600–4600	1800–2500	24–27	[22]
Jurassique	-	Limestones	-	3000–5000	2000–2800	25–27	[53]

5. Data Acquisition and Processing

5.1. Data Acquisition

Ambient vibrations were recorded using single-station and array measurement techniques. The HVSR method was applied at 164 ambient vibration measurement points, distributed over 20 profiles (Figure 3). The measurements were made following SESAME recommendations [54]. The F–K method was applied to array measurements in seven sites. Additional measurements, using the electrical prospection technique, were carried out in the Mecheta-Nouasser and El-Amra regions. The aim was to determine the thicknesses of the Plio-Quaternary layers for the HVSR inversion. It is well-known that, in a sedimentary basin, the maximum depth to bedrock is observed beneath the youngest formations, which are generally located in the middle of the basin. Thus, ambient vibrations were recorded mainly in the Quaternary plain of the Middle-Chelif, from Zbabdja village in the west to Ain-Defla in the east (Figure 3). Moreover, in order to better image the synclinal shape of the basin, the HVSR profiles are perpendicular to its axis, varying from a WNW–ESE direction in Oued-Fodda to a NNW–SSE in the rest of the plain.

The single-station measurements were carried out in calm weather, with a recording time of 16 min away from human activities, and 26 min inside cities and villages. Ambient vibrations were recorded using Tromino seismographs, with a sampling rate of 512 samples per second. The first array measurement campaign (AR1, AR4, and AR2 in Figure 4) was carried out in February 2019, using between seven and nine Mark L22 seismographs (f_0 = 2 Hz) set in a circular configuration, with two apertures of 20 m and 50 m, respectively, at each site. The vertical components of these sensors were plugged into an Arduino-based multichannel acquisition system [55] for 30 min of simultaneous recording at each site. The second measurement campaign (AR3, AR5, AR6, and AR7) was carried out in September 2020, using nine new SS10 (f_0 = 1 Hz) triaxial velocity sensors connected to their respective

SL06 digitizers (SARA electronic instruments). The dispersion curves obtained in this way produced good resolution only for the first 50 m of the soil column. Thus, additional single-station measurement points were carried out at the center of each array, recording ambient vibrations at the same time in order to perform a joint inversion and, thus, obtain a better resolution in the shallow and deep sediments.

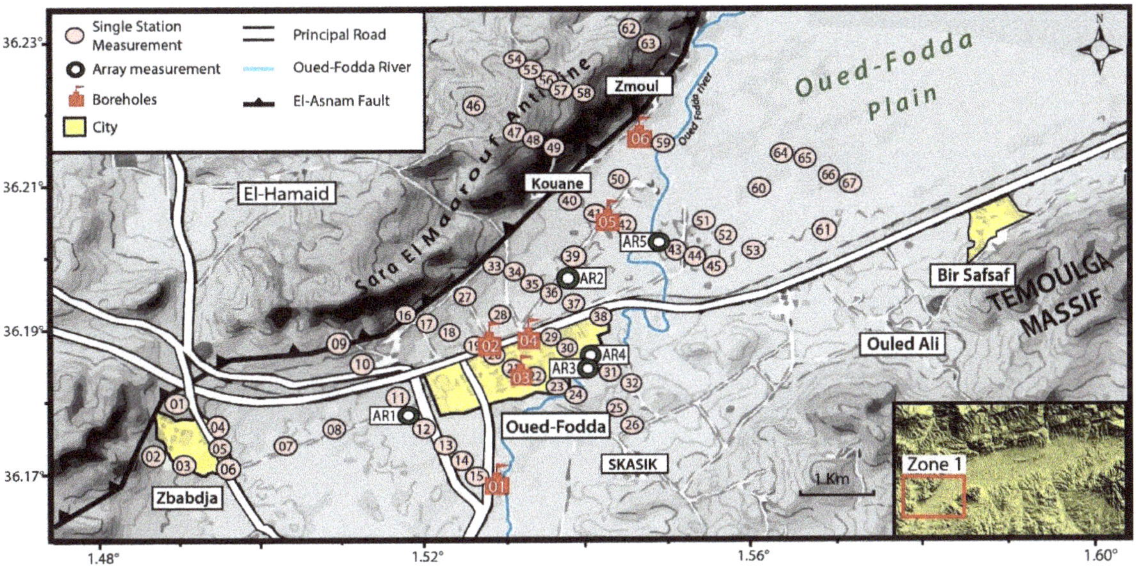

Figure 4. Compiled data in Zone 1.

For the electrical resistivity tomography survey, we used a set of 40 electrodes in dipole–dipole devices, linked to an ABEM Terrameter LS 2 resistivity meter. The interspacing between the electrodes was 10 m.

The distribution of the single-station and array measurements was made according to the available geophysical and geological data. The aim was to optimize the data coverage for the three zones.

Zone 1. The Oued-Fodda plain.

Ambient vibrations were recorded at 67 sites, including 6 in the village of Zbabdja and 9 in Oued-Fodda city (Figure 4), and 58 of these are distributed over 8 profiles oriented in a WNW–ESE direction. As an attempt to image the rupture trace of the 1980 El-Asnam earthquake in the upper sedimentary layers, the profiles, Pr 6, Pr 7, and Pr 8 (Figure 3), are crossing the surface trace around the village of Zmoul (Figure 4), with 11 measurement points on the Sara-El-Maarouf anticline, which corresponds to the overlapping block. Ambient vibrations were also recorded using the array measurement technique at 5 sites, including 2 (AR3 and AR4) within Oued-Fodda city (Figure 4).

Zone 2. The Carnot Plain.

A total set of 64 single-station measurement points were conducted in this zone, distributed over 8 profiles oriented in a NW–SE direction (Figures 3 and 5). An array measurement in a circular configuration was carried out in the city of El-Attaf, with apertures of 20 m and 40 m, along with two single-station measurement points at the same site, in order to perform a joint inversion. An electrical resistivity survey was also carried out in the Mechta-Nouasser locality (Figure 5).

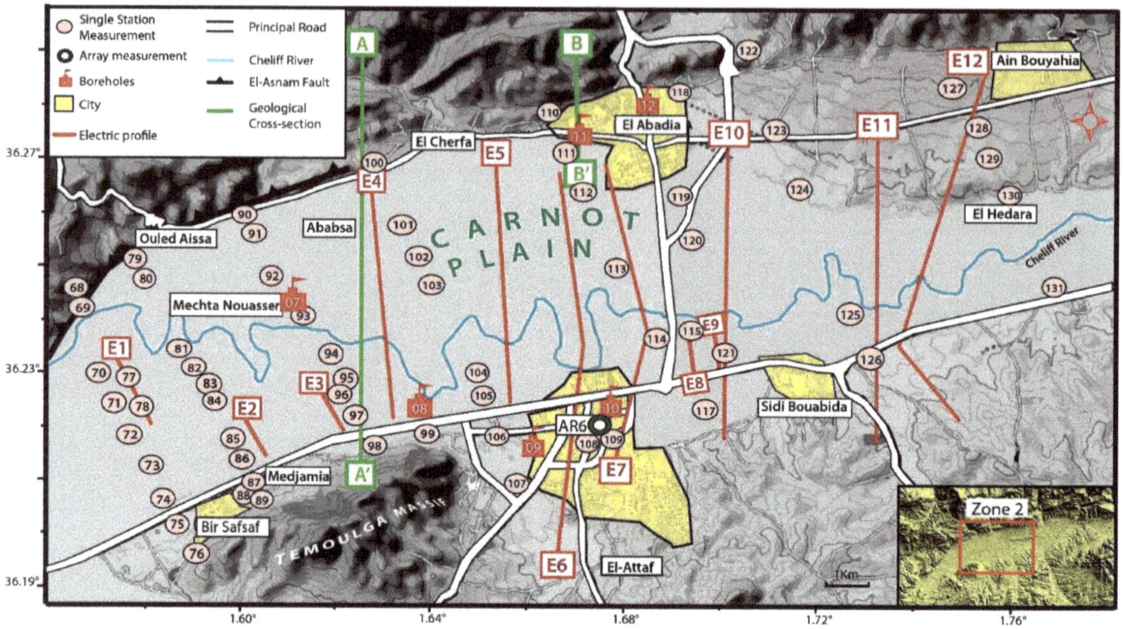

Figure 5. Compiled data in Zone 2.

Zone 3. Rouina-Ain-Defla Region.

In this zone, 33 single-station measurement points were carried out along 4 profiles, oriented NNW–SSE. Two of these profiles, Pr 17 and Pr 20 (Figures 3 and 6), are crossing the cities of Rouina and Ain-Defla.

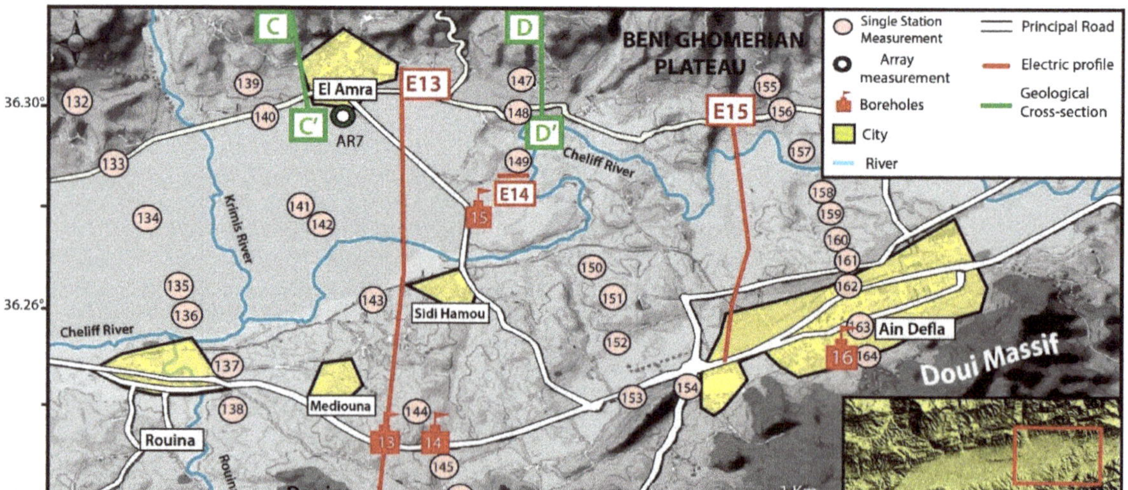

Figure 6. Compiled data in Zone 3.

In addition, an array ambient vibration measurement was carried out near the town of El-Amra, with apertures of 20 m and 50 m (AR7). Furthermore, an electrical resistivity survey (E14) was performed south of the El-Amra locality to better constrain the thickness of the shallow Quaternary layers in this area.

5.2. Data Processing

5.2.1. HVSR Technique

The HVSR technique was applied to the ambient vibration measurements using Geopsy software as a processing tool [45]. The recordings of the 164 measurement points have been processed as follows: The time series were divided into 30-s windows that were 5% cosine-tapered. The windows were selected automatically using the anti–triggering algorithm, which allows for the avoidance of transients and the selection of only windows with stationary ambient vibrations. The anti–trigger parameters were used as recommended by the SESAME project [54]. The fast Fourier transform (FFT) was computed for each window. After that, the Konno–Ohmachi algorithm was applied in order to smooth the amplitude spectra [56], with a smoothing coefficient of 40. The HVSR curve was calculated for each selected window, and an averaged HVSR curve was retrieved. Some examples of the obtained HVSR curves at each zone are shown in Figure 7.

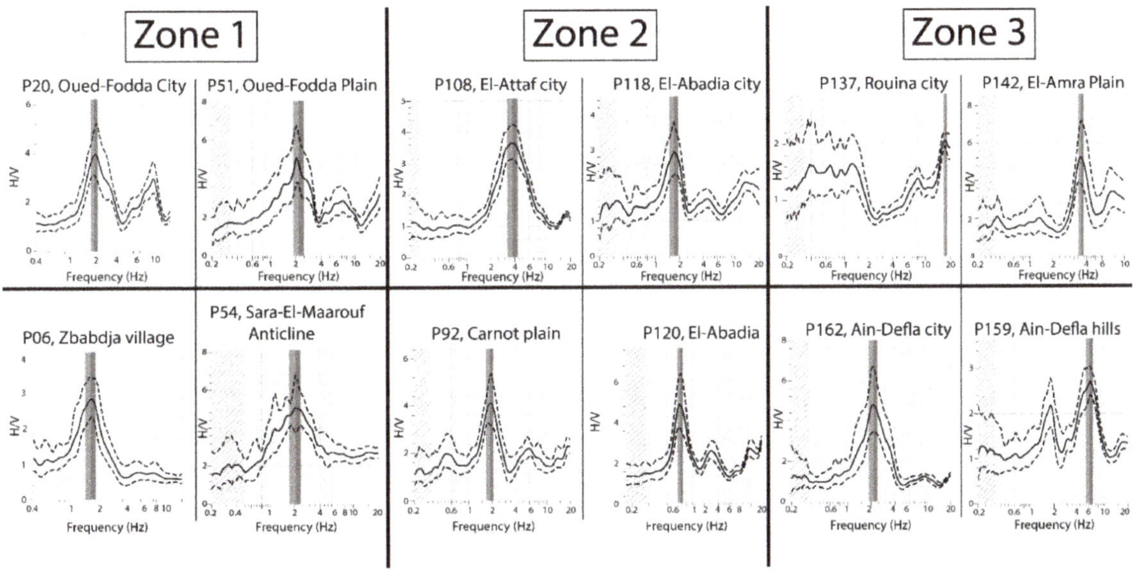

Figure 7. Some examples of the calculated HVSR curves at each zone.

5.2.2. F–K Analysis

The F–K analysis was applied to array recordings using the Sesarray package [45]. The first step consisted of calculating the array transfer function, along with the theoretical wavenumber limits (K_{min}–K_{max}) (Figure 8), from the number of receivers and the XY coordinates of each receiver of the array. The computation was carried out using WARANGPS software from the Sesarray package. After that, the beam-forming method (BFM) was applied to the vertical components using Geopsy software. The signals were divided into windows of frequency-dependent lengths, including 50 periods. For the processing, two F–K gridding parameters had to be defined: the grid step and the grid size. The grid step corresponds to the $K_{min}/2$ value, which determines the maximum resolution. The grid size corresponds to the K_{max} value, which determines the aliasing limit. As a result, a dispersion curve is obtained, i.e., the slowness as a function of the frequency.

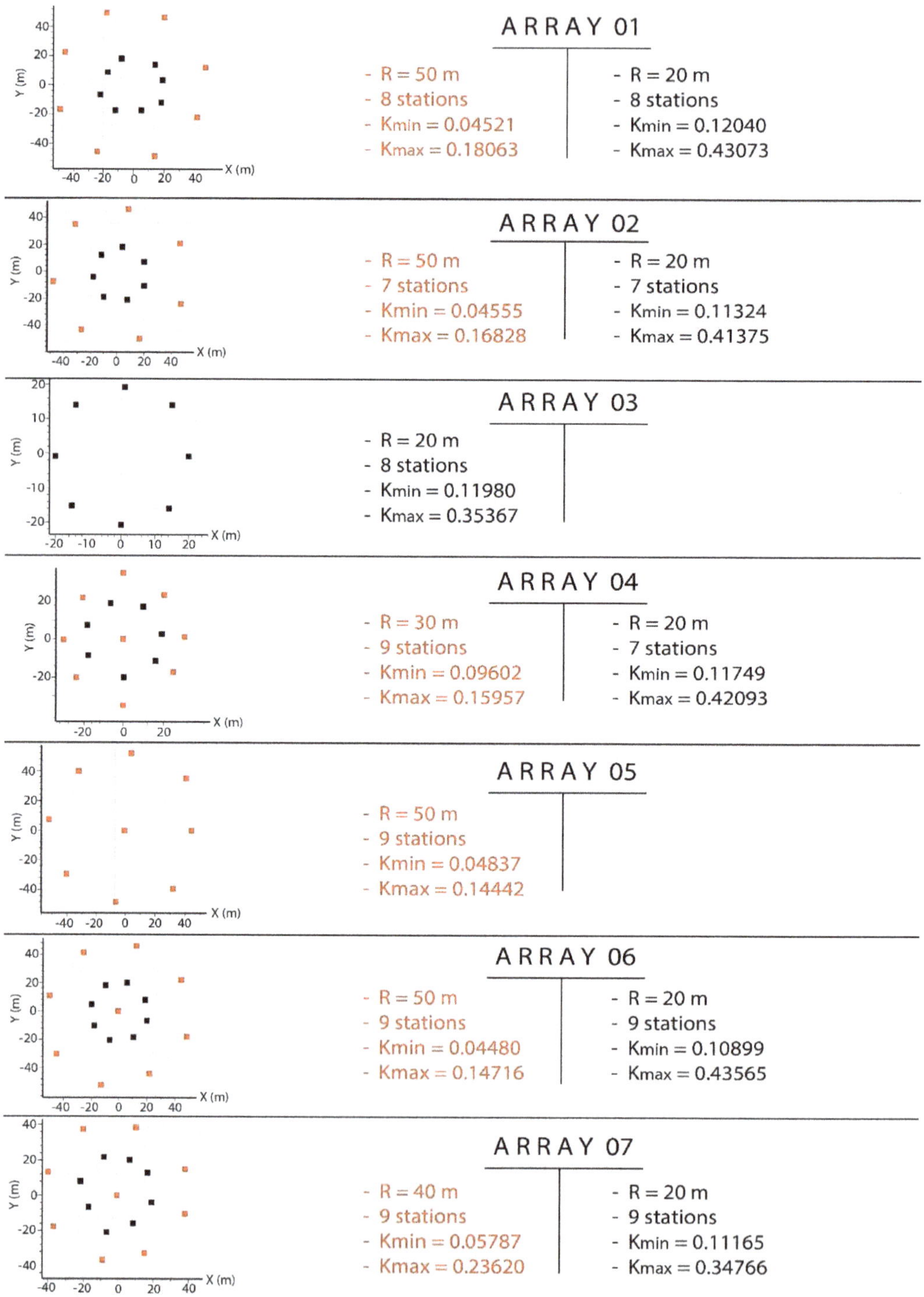

Figure 8. Theoretical wavenumbers obtained at each array recording site.

5.2.3. Inversion of HVSR and Dispersion Curves

The inversion process was carried out using the Dinver software from the Sesarray package [45]. This software uses the neighborhood algorithm to estimate the shear-wave velocity profiles from the HVSR and dispersion curves [44]. The geotechnical model built from the gathered geotechnical data (Table 1) was used as an input parameterization for the process. However, the number of layers was not constant; it varied from one site to another depending on the lithological aspect at each measurement site.

The HVSR curves were inverted entirely (0.2–20 Hz). The maximum number of iterations was fixed to 350 iterations, and 100 models were generated at each iteration. Only models with acceptable misfit values were considered. The threshold for an acceptable misfit value was 0.45 for curves with one frequency peak, and 0.6 for curves with two frequency peaks. For curves with multiple frequency peaks, misfit values under 0.8 were considered.

For the array measurements, the dispersion curves were not inverted entirely; the curves were cut and considered only within the theoretical wavenumber limits ($K_{min}/2$, K_{max}). For each dispersion curve, the corresponding HVSR curve was added in order to perform a joint inversion. Both curves were equally weighted.

6. Results and Discussions

6.1. Fundamental Frequencies and the Corresponding Amplitudes

The resonance frequencies and the corresponding amplitudes obtained from the HVSR analysis are mapped in Figure 9. The frequencies range between 0.75 and 12 Hz, while the amplitudes vary from 2 to 6.2. The obtained HVSR curves contain one frequency peak, two peaks, and multiple peaks. Curves with two frequency peaks are predominant, especially in the Quaternary plain.

In the Oued-Fodda Plain (Zone1), the predominant resonance frequencies of the soil vary between 1.5 and 3 Hz, and the corresponding amplitudes vary between 2 and 6. However, particularly in the center of Oued-Fodda city, higher frequencies, between 5 and 12 Hz, are observed. This high-frequency range is related to the outcropping bedrock in the area [22]. In the village of Zbabdja, about 3 km west of Oued-Fodda city, the frequencies are around 1.5 Hz. In terms of shape, two types of curves are observed in this zone: curves with two frequency peaks are predominant in the Oued-Fodda Plain, while curves with one peak were obtained only from recordings in the central parts of Oued-Fodda city and Zbabdja village. Second peaks at higher frequencies are related to impedance contrasts in the soil column at shallow depths [22]. These peaks are only observed at sites located on Holocene alluviums. This allowed us to deduce that this shallow impedance contrast is responsible for the second frequency peaks that are between the Holocene silt layer and the Pleistocene gravel and conglomerate layers.

In the Sara-El-Maarouf anticline, the predominant resonance frequency is around 2 Hz, while the amplitudes of the peaks range between 2.5 and 4.5. The second peaks at higher frequencies appear to be related to the impedance contrast between the Upper Pliocene sands and the Lower Pliocene sandstones.

In the Carnot Plain (Zone 2), the predominant resonance frequencies vary between 0.75 and 5 Hz. The corresponding amplitudes vary from 2.2 to 5.8. HVSR curves with one frequency peak, two peaks, and multiple peaks were observed in this area.

The variation in shape for these curves is mainly related to the variation in the lithology, the apparition of new sedimentary layers, and the disappearance of others at some places, caused by different factors that affect the sedimentation process as erosions and faults. Concerning the shapes of the peaks, they are mainly controlled by the dipping angles of the different layers [54,57].

Peaks at higher frequencies (between 8 and 12 Hz) are observed on the hills overlooking the El-Attaf city from the southwest. These peaks appear to be caused by the outcropping Cretaceous marls. In the city of El-Abadia, the resonance frequency is around 2 Hz, while in El-Attaf city, it is around 3 Hz.

Figure 9. (**A**) Fundamental frequencies obtained from the HVSR analysis. (**B**) Amplitudes of the HVSR fundamental frequency peaks. Red circles represent the cities. The blue line represents the surface trace of the El-Asnam fault. OF: Oued-Fodda. AB: El-Abadia. AT: El-Attaf. RO: Rouina. AM: El-Amra. AD: Ain-Defla.

In the third zone, the resonance frequencies range between 1.3 and 3 Hz. The corresponding amplitudes vary between 2.2 and 3.5. Two types of HVSR curves were obtained: curves with one frequency peak in the southern part, and curves with two frequency peaks in the northern part. The frequency of the second peak varies from 5 to 12 Hz and appears to be related to an impedance contrast between the Holocene and Pleistocene alluviums at shallow depths. In the city of Ain-Defla, the resonance frequency of the soil increases from 1.8 to 3 Hz, from north to south, respectively. The corresponding peak amplitudes are around 5.

6.2. F–K Analysis: Surface Wave Dispersion Curves

In Figure 10, the dispersion curves estimated from the different deployed arrays are shown. In some places, a single array was enough to obtain a reliable dispersion curve, while, in other places, two arrays of different apertures were implemented, obtaining an average dispersion curve. The frequency range is fixed by the theoretical wavenumber limits ($K_{min}/2$, K_{max}), which depend on the array configuration and aperture (Figure 8). The curves are, therefore, not very reliable outside the theoretical wavenumber limits.

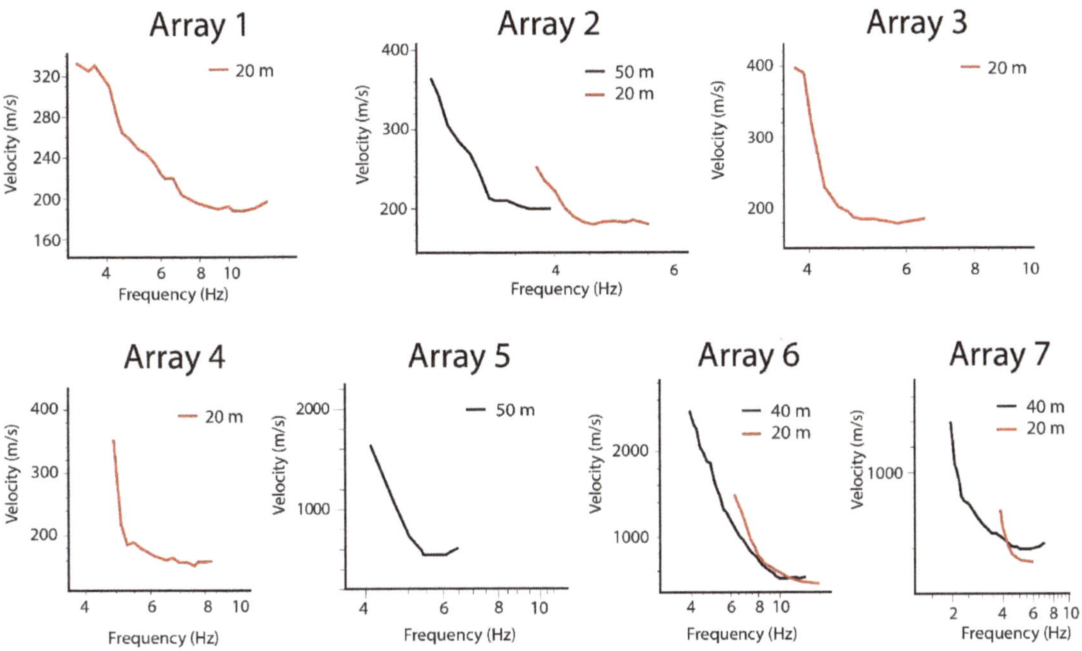

Figure 10. Surface wave dispersion curves.

In the city of Oued-Fodda and its surroundings, the analysis of the array measurements at sites AR1, AR2, AR3, AR4, and AR5 provided dispersion curves between 2.5 and 10 Hz. The dispersion curves obtained from the array recording AR6 in the city of El-Attaf are reliable in the frequency range between 5 and 10 Hz, and between 2.5 and 8 Hz in the city of El-Amra (AR7).

6.3. *Electrical Resistivity Tomography*

The electrical resistivities, obtained from measurements in Mechta-Nouasser (E1, Figure 5) and in the southeast of El-Amra (E14, Figure 6), are shown in Figure 11. In order to interpret the obtained results, we used the resistivity scale established by the CGG [7] for the Middle-Chelif Basin (left panel in Figure 11).

For the resistivity profile (E1), the depth of the investigation of 115 m allowed for the identification of the two Quaternary layers. The thicknesses of these two layers (1 and 2 in Figure 11) appeared to be about 50 m (~10 m of Holocene and ~40 m of Pleistocene alluviums). For the profile (E14), the depth of investigation was 105 m. The resistivity scale for the Middle-Chelif Basin allowed us to identify four different sedimentary layers; the two top layers belong to the Quaternary, while the underlying layers are attributed to the Miocene. The thickness of the Quaternary deposits is about 35 m (~10 m of Holocene and ~25 m of Pleistocene).

The results of the electrical resistivity measurements in Mecheta-Nouasser (E1) and El-Amra (E14) allowed us to fix the thicknesses of the Quaternary layers (Holocene and Pleistocene) for the inversion of the HVSR curves recorded at Sites 77, 78, and 149.

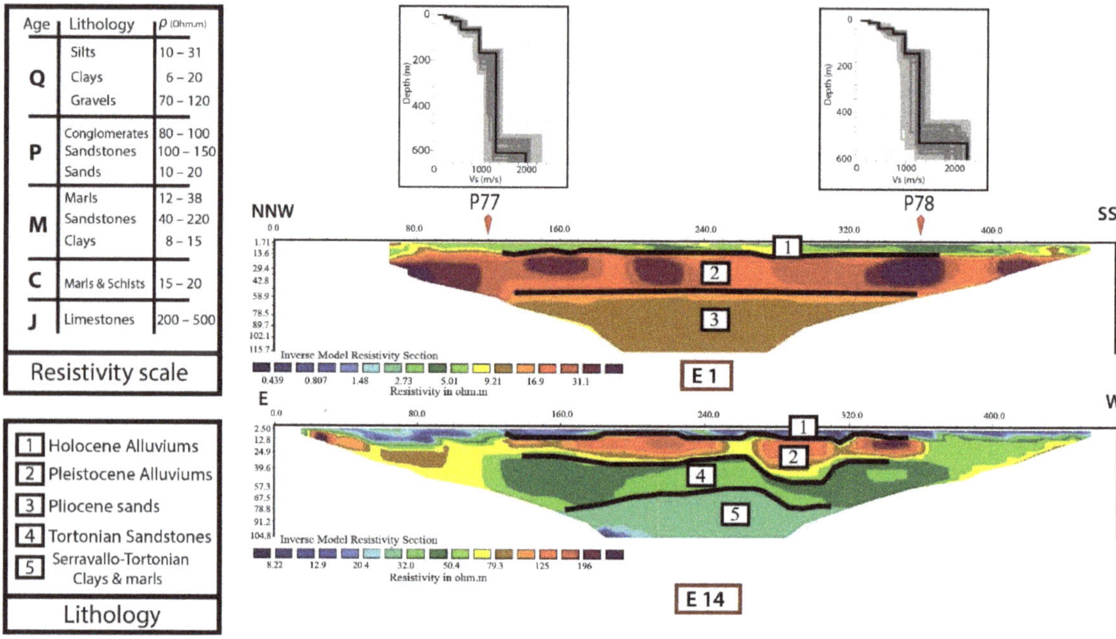

Figure 11. Resistivity profiles. The shear-wave velocity models correspond to the HVSR points, P77 and P78. The top left panel represents the resistivity scale for the Middle-Chelif Basin [7]. The bottom left panel represents the lithological units identified in the resistivity profiles.

6.4. Shear-Wave 2D Velocity Profiles

The inversion of the obtained HVSR curves allowed for the retrieval of the S-wave velocity models at each site. For sites where array measurements were also taken, the joint inversion of the obtained dispersion and HVSR curves was carried out. The number of analyzed layers, as well as the range of the respective Vs and thickness parameters, have been constrained according to previous information obtained from geological cross-sections, boreholes, and geotechnical studies (see Table 1). The aim was to obtain a result closer to the true model by minimizing the misfit value. Some examples are shown in Figure 12.

For the first zone, the eight obtained 2D velocity profiles are mapped in Figure 13. The two uppermost layers in the plain correspond to the two Quaternary stages, the Holocene and Pleistocene, mainly composed of alluviums. The obtained Vs values for the Holocene alluviums vary between 210 and 350 m/s. This range is explained by the lateral variation between clays and silts. For the Pleistocene layer, the Vs value ranges between 405 and 630 m/s. This variation is due to the alternance between grave and clay facies. The high contrast in velocity between the two Quaternary stages is responsible for the second frequency peak in the area. The thickness of the Holocene formation does not exceed 10 m, while the Pleistocene deposits reach a maximum thickness of 65 m north of Oued-Fodda city.

On the Sara-El-Maarouf anticline, in the profiles Pr6, Pr7, and Pr8 (Figure 13), the two topmost layers correspond to the Upper and Lower Pliocene. The Vs values obtained for the Upper Pliocene sands vary between 300 and 400 m/s, and between 610 and 780 m/s for the Lower Pliocene sandstones. As for the Quaternary stages down in the plain, the contrast in velocity between the two Pliocene layers is responsible for the second frequency peak in the HVSR curves obtained from recordings on the Sara-El-Maarouf anticline. These formations reach a maximum thickness of 100 m in the oriental part of the anticline. A thin

Tortonian limestone layer lies under the Pliocene layers, for which the Vs value ranges between 650 and 900 m/s. This layer dips into the northwest and is not present in the Oued-Fodda Plain. The Quaternary layers in the plain are, therefore, lying directly over a thick layer composed of Serravallo-Tortonian marls and clays. This is the thickest sedimentary layer in the Middle-Chelif Basin. It reaches a maximum thickness of 250 m in this zone. The Vs value for this layer varies between 850 and 1290 m/s. This formation is in direct contact with the Cretaceous bedrock, which is mainly composed of hard marls. The obtained shear-wave velocity of these marls varies between 1700 and 2300 m/s in this zone.

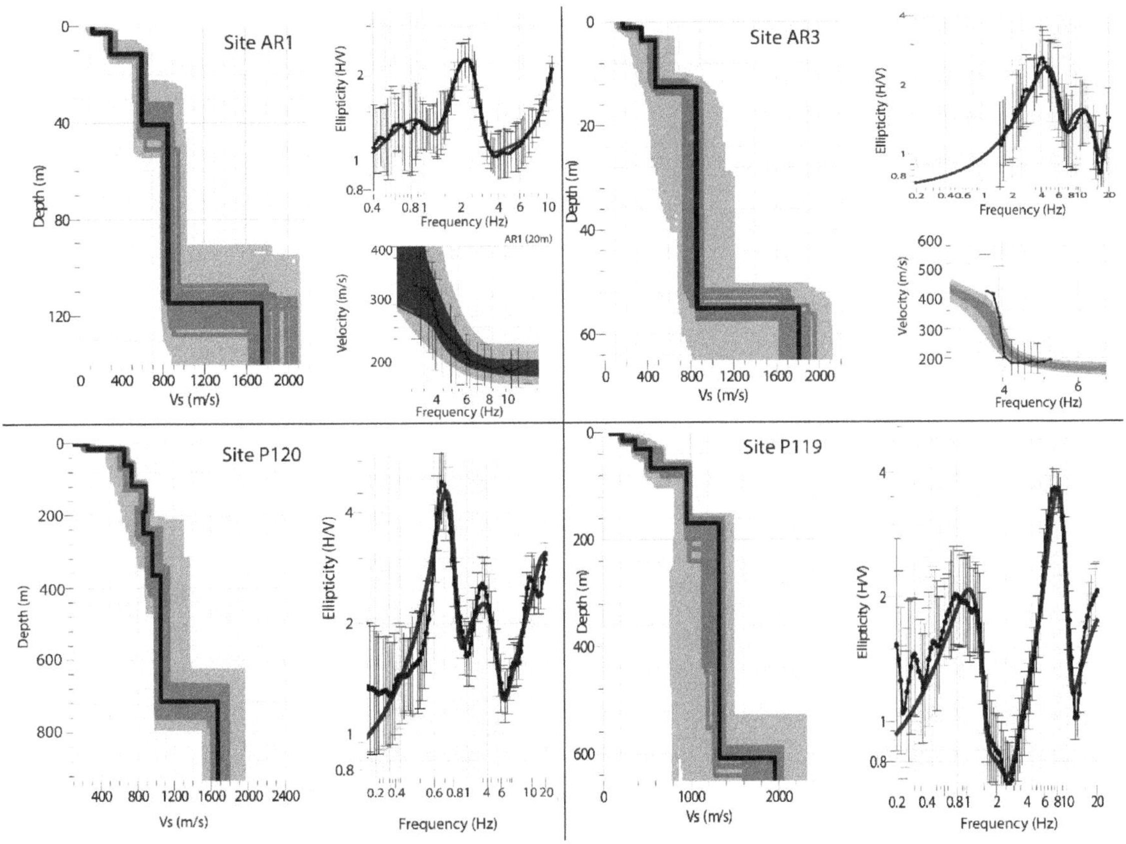

Figure 12. Examples of the inversion results. For each site, the left panels for each site represent the Vs models. The black line corresponds to the best fit model, the dark grey represents models with minimum misfit + 10%. All the tested models are in light grey. In the right panels, we show the computed fundamental mode of the Rayleigh wave ellipticity curve (dark grey curve), and the inverted part of the HVSR curve (black dotted curve). For sites AR1 and AR3, the bottom of the right panel represents the surface wave dispersion curve.

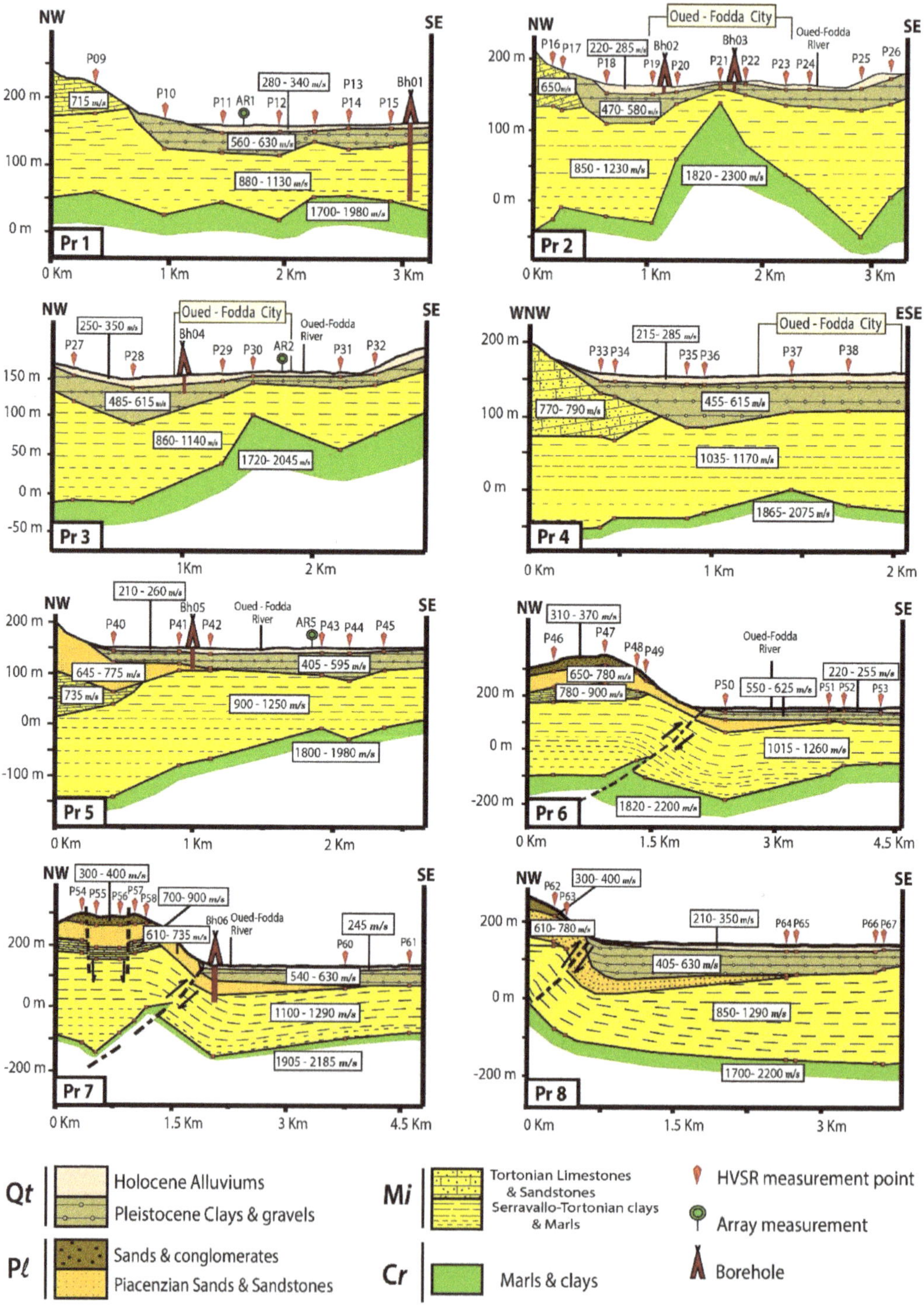

Figure 13. 2D Shear-wave velocity profiles for Zone 1. Qt: Quaternary. Pl: Pliocene. Mi: Miocene. Cr: Cretaceous.

In the profiles, Pr 6, Pr 7, and Pr 8 (Figure 13), we can see that the sediments of the plain are separated from the ones of the Sara-El-Maarouf anticline by a deep reverse fault that outcrops at the surface, the so-called "El-Asnam fault". The surface trace coordinates of the fault, along with its dipping angle calculated in Yielding [8] and Ouyed [11], were taken into consideration for the inversion process. In the profile, Pr 7, we can observe normal faults at the top of the anticline. The surface trace of these faults was observed during the HVSR measurement campaign. The important variations in the thickness of the layers between the measurement points, P54, P55, P56, and P57 (Profile Pr7 in Figure 12), allowed us to obtain an approximate idea on the fault position under the surface. The presence of normal faults at the top of the Sara-El-Maarouf anticline was highlighted before by Philip and Meghraoui [19]. This normal faulting comes as a result of the extension of the surface layers of the anticline due to the vertical slip on the El-Asnam fault [11,19].

The eight 2D velocity profiles in the Carnot Plain are mapped in Figure 14. Unlike the first zone, the synclinal shape of the basin is clearly visible in this zone. The depocenters are located north of the Chelif River, as attested by the CGG study [7]. The highest depths of the basin are observed around the city of El-Abadia, with a maximum depth of 760 m (P120 in Profile Pr14, Figure 14). Compared to the first zone, the lithostratigraphic column contains two additional layers of the Miocene age; the Messinian marls and the Serravallian poudingues. The Holocene alluviums have a Vs value between 220 and 370 m/s and reach a maximum thickness of 21 m in the middle of the plain. The Pleistocene alluviums reach a maximum thickness of 86 m, and the Vs value for this layer varies between 330 and 680 m/s. The Upper and Lower Pliocene layers that outcrop on a narrow band north of the plain, dip south under the Quaternary layers and disappear in the middle of the plain. The absence of Pliocene deposits in the south is due to marine regression during the late-Miocene early-Pliocene period, where the shorelines were located in the middle of the plain [2]. These formations are therefore thicker in the north and reach thicknesses of 145 m (P92 in Profile Pr11, Figure 14). The Vs value for the Upper Pliocene sands and conglomerates varies between 390 and 750 m/s, and between 510 and 900 m/s for the Lower Pliocene sandstones. The Pliocene deposits lie over the Messinian blue marls, the uppermost layer of the Miocene. The Miocene deposits occupy most of the sedimentary column of the Middle-Chelif Basin, reaching thicknesses of 550 m. However, there is no impedance contrast between the four Miocene layers in the area. The Vs values for the blue marls stands between 640 and 1190 m/s. For the Tortonian sandstones, they vary between 830 and 1280 m/s, and between 890 and 1380 m/s for the Serravallo-Tortonian clays and marls. For the Serravallian poudingues and conglomerates, the shear-wave velocities vary between 1190 and 1400 m/s.

For the Cretaceous bedrock, the Vs varies between 1650 and 2270 m/s. This variation is due to the lateral change of formations from Senonian clays and marls to Albian marls and limestones. In the southwestern part of the plain (Profiles Pr 9, Pr 10, and Pr 11, Figure 14), the Jurassic limestones that compose the Temoulga Massif dip vertically under the Neogene sediments, with a faulted contact between the limestones and the Cretaceous marls [35]. The calculated shear-wave velocities for the Jurassic limestones vary between 2300 and 2670 m/s.

The third zone, which covers the Rouina-Ain-Defla region, has a different local geological context. The vast Quaternary plain gives way to hills and plateaus. The four 2D velocity profiles realized for this zone are mapped in Figure 15. The lithostratigraphic column is dominated by the Tortonian sandstones and Serravallo-Tortonian clays and marls. The Vs values for sediments are in the same range as for the second zone. In the profile, Pr20 (Figure 15), we can see all the structural complexity that exists in this zone of closure.

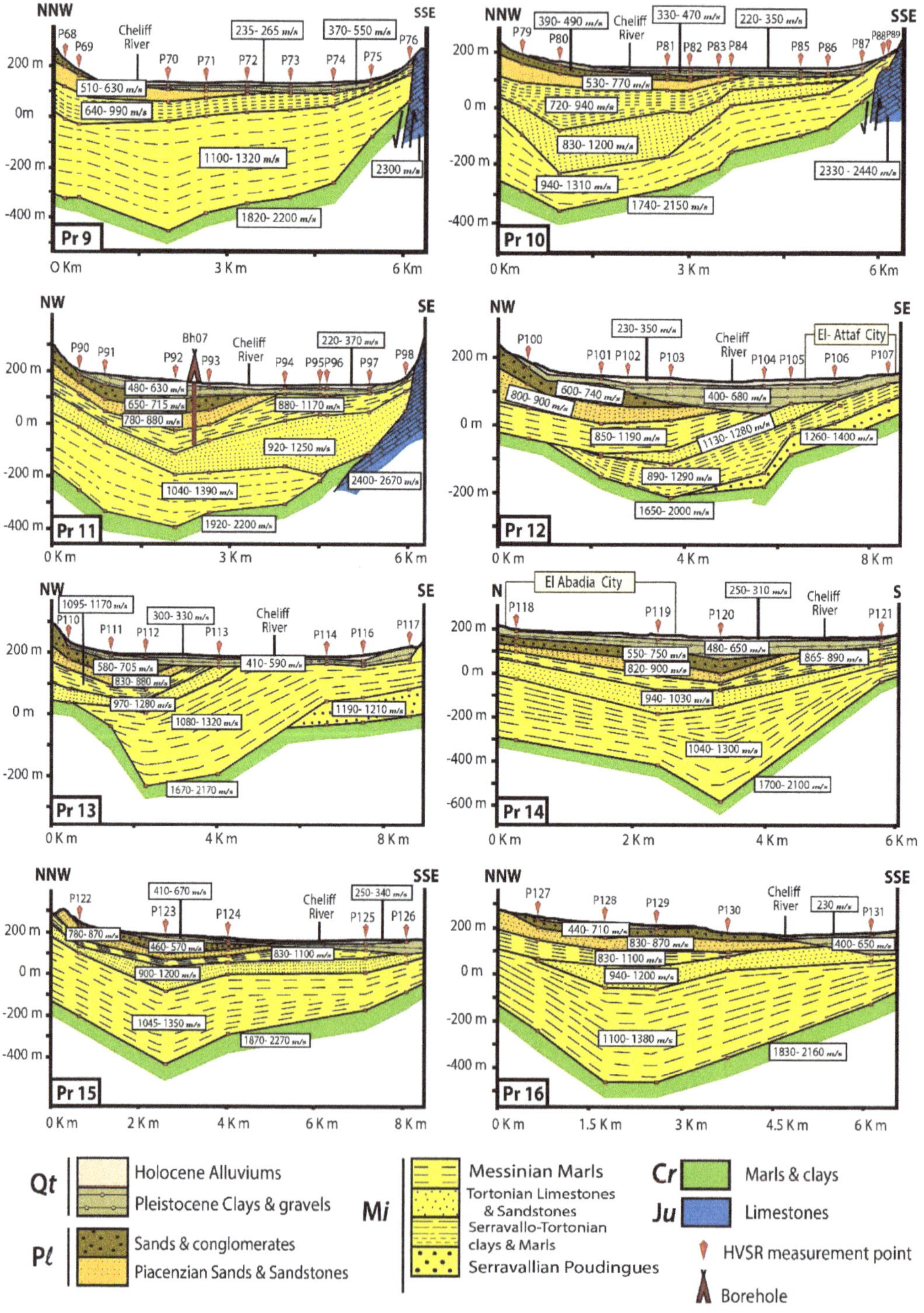

Figure 14. Shear-wave velocity profiles for Zone 2. Qt: Quaternary. Pl: Pliocene. Mi: Miocene. Cr: Cretaceous. Ju: Jurassic.

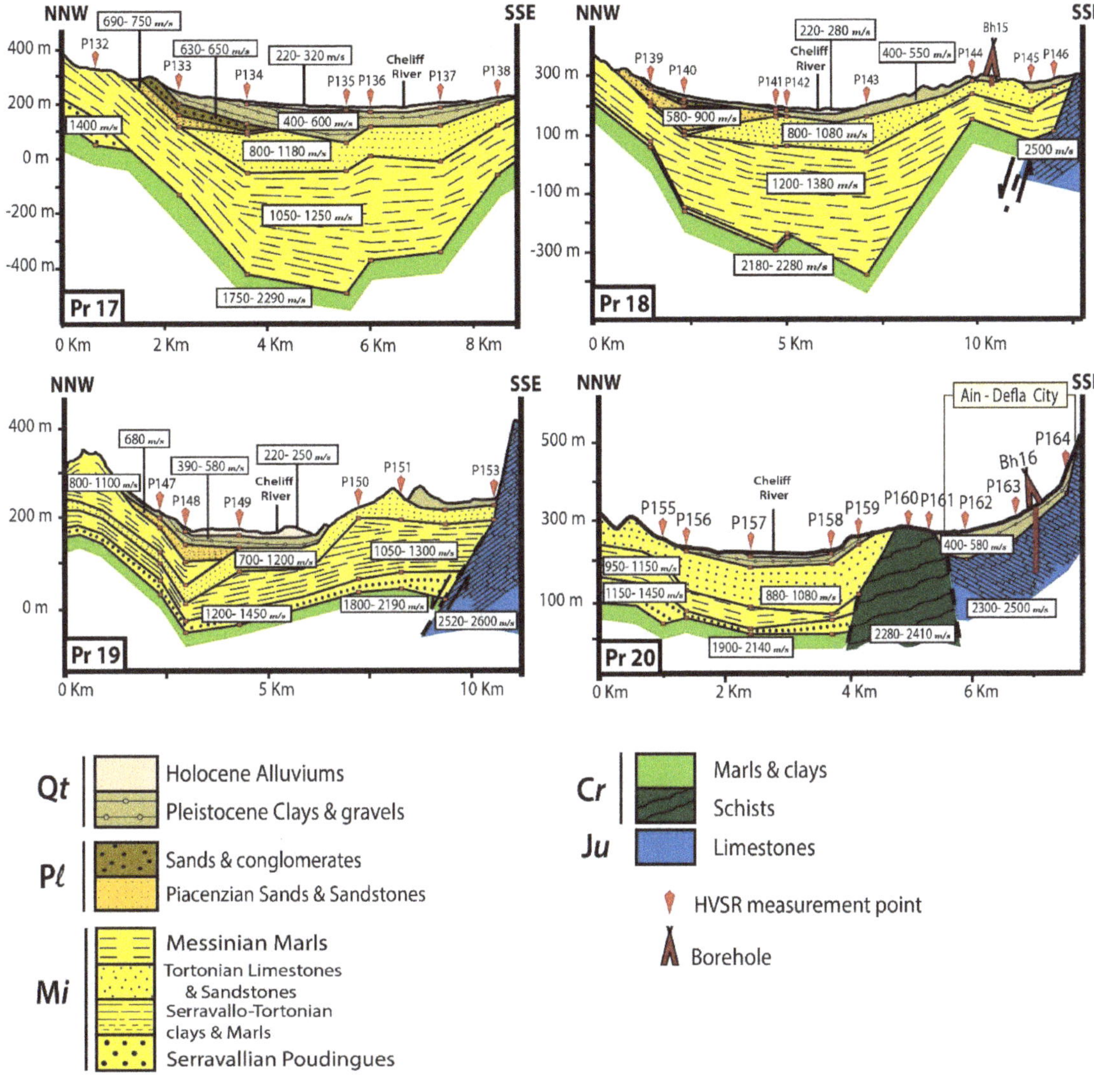

Figure 15. Shear-wave velocity profiles for Zone 3. Qt: Quaternary. Pl: Pliocene. Mi: Miocene. Cr: Cretaceous. Ju: Jurassic.

The sedimentary layers become thinner and lie over different types of bedrock. The old formations become more rugged to the south because of the presence of the epimetamorphic Doui Massif (Figure 2). North of the city of Ain-Defla (Profile Pr 20), the Albian clays and marls are separated from the Jurassic limestones by a block of Neocomian schists, which belongs to the Arib Massif that overlooks the Upper-Chelif Plain. The presence of the Neocomian schists marks the transition between the Middle and Upper Chelif Basin. The Vs values for this formation vary between 2280 and 2410 m/s.

The average Vp/Vs ratio was calculated from the 164 velocity models. The averaged ratio for the sedimentary column is 1.95. For the bedrock formations, the averaged ratio is 1.76. This value is in agreement with the 1.7 found by Bellalem [53] in a seismic tomography study.

The depths to seismic bedrock are mapped in Figure 16. We can see that the basin reaches higher depths in a narrow band stretching from the west of El-Abadia to El-Amra, and reaches a maximum depth of 760 m (P120 in Profile Pr14, Figure 14), which is very close to the 800 m measured by the CGG [7].

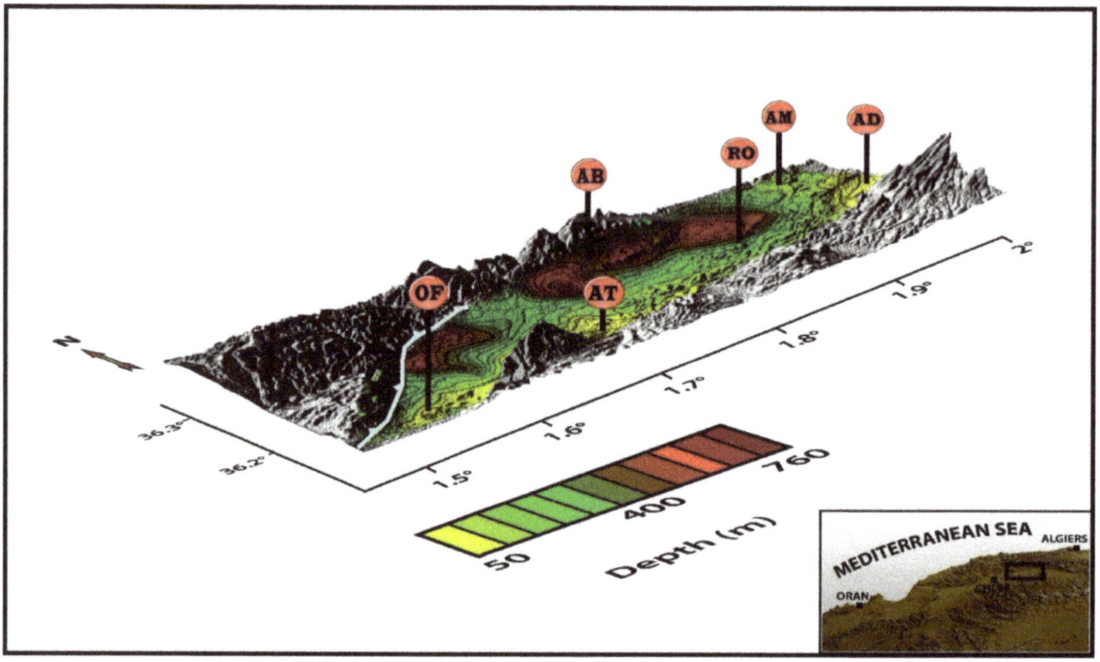

Figure 16. Bedrock depths of the Middle-Chelif Basin. Red circles represent the cities. OF: Oued-Fodda. AB: El-Abadia. AT: El-Attaf. RO: Rouina. AM: El-Amra. AD: Ain-Defla.

7. Conclusions

In the present work, ambient vibration data were used to investigate the soil characteristics and the site effects in the Middle-Chelif Basin, with both single-station and array measurements. The region is home to over 400,000 inhabitants, distributed over its six main cities: Ain-Defla, El-Amra, Rouina, El-Abadia, El-Attaf, and Oued-Fodda. Being located on a zone with moderate to high seismic activity, these cities have suffered important building damages during past earthquakes. The fact that most of them are built on soft sediment layers has played a role in amplifying the ground shaking and the lengthening of its duration.

The first tens of meters of the topsoil column play an important role in ground shaking amplification during an earthquake. In our case study, it corresponds principally to the thickness of the Quaternary layers. In the major part of the basin, these layers lie directly over the Miocene sediments, which creates an important impedance contrast. Thus, the thickness of the Quaternary layers and the corresponding Vs values had to be determined with more precision. For this purpose, seismic noise array measurements and electrical resistivity measurements were carried out. The HVSR technique was applied on ambient vibration recordings at 164 sites, and F–K analysis was applied on array recordings at 7 sites. The obtained HVSR curves allowed us to identify the soil resonance frequencies at each site. The F–K analysis allowed us to retrieve the surface wave dispersion curve at each of the 7 sites. The HVSR and dispersion curves were then jointly inverted to obtain the Vs profiles at each site.

The analysis of the HVSR curves showed the existence of different types of curves: curves with one, two, and multiple frequency peaks. The HVSR curves with two peaks are predominant in the Middle-Chelif Plain. The predominant soil resonance frequencies vary between 0.75 and 12 Hz. This wide range of frequencies is explained by the synclinal aspect of the basin, as well as the thinning of the sedimentary layers on its edges. The corresponding amplitudes vary between 2 and 6.2. However, the amplitudes obtained from the HVSR analysis may not be a reliable estimation of true amplification. The high-frequency peaks (between 5.9 and 12 Hz) appear to be related to an impedance contrast between the Holocene and Pleistocene alluviums at shallow depths.

In the Middle-Chelif Plain, the topmost layer is composed of thin Holocene alluvial deposits, for which the shear-velocity (Vs) ranges between 210 and 370 m/s. The largest thickness observed for this layer is 21 m. For the Pleistocene deposits, the Vs value varies between 330 and 680 m/s, with a maximum observed thickness of 86 m. The Pliocene deposits are present only in the northern part of the plain and on the Sara-El-Maarouf anticline in the west. The shear-wave velocity varies between 300 and 750 m/s for the Upper Pliocene sands and conglomerates, and between 510 and 900 m/s for the Lower Pliocene sandstones. The largest observed thickness for the Pliocene layers is 145 m, while the Miocene deposits occupy a major part of the sedimentary column, often composed of four distinct layers. Their thickness reaches 550 m around the city of El-Abadia. From the Messinian stiff marly formations to the Serravallian poudingues, the Vs values vary from 640 to 1450 m/s.

In the southern part of the Middle-Chelif Plain, Jurassic formations outcrop in succession on the massifs of Doui, Rouina, and Temoulga. The Jurassic limestones are dipping, generally north, and constitute the bedrock for the Neogene deposits in this part of the basin. The shear-wave velocity for these formations varies between 2300 and 2670 m/s. However, in a major part of the basin, the sedimentary deposits lie over a Cretaceous bedrock, for which the Vs values range between 1620 and 2300 m/s. The calculated Vp/Vs ratio for the sedimentary column is 1.95. For the bedrock formations, the ratio is 1.76.

The obtained velocity and resistivity profiles highlight the existence of important lateral variations in the velocity and resistivity of the whole sedimentary column. This variation is directly linked with variations in lithology (lateral change of facies), caused by different factors that have affected the sedimentation process in the Middle-Chelif Basin (marine regression and transgression episodes during the Miocene and Pliocene, erosions, faults, etc.).

The sedimentary layers show a synclinal aspect in the plain of Carnot, unlike the Oued-Fodda and Ain-Defla regions, where the underground structures are rugged, affected by different faults, folds, and outcrops. The depocenters in the Middle-Chelif Basin appear to be located in a narrow band extending from the western part of the town of El-Abadia to the northern part of the town of Rouina. The maximum observed depth to bedrock is 760 m.

The thickness of the sedimentary layers is different beneath the cities situated in the northern and southern parts of the Middle-Chelif Plain. In the city of Ain-Defla, which backs onto the northern flank of the epimetamorphic Doui Massif, the sedimentary column does not exceed 60 m. While in the cities of Oued-Fodda and El-Attaf, which were built around the western and eastern parts of the Temoulga Massif, respectively, the sedimentary column reaches a thickness of 200 m. However, the cities of Rouina, El-Abadia, and El-Amra are lying over a thick sedimentary cover that exceeds 300 m.

The cities cited in this study extend over a region where strong earthquakes with secondary effects (liquefaction, landslides, etc.) have already occurred several times before. Therefore, this study would like to be a contribution to a better assessment of the seismic hazard in the Chelif-Basin. The results obtained in this study can be used, for example, in the modeling of strong ground motions in order to update the Algerian seismic code.

Author Contributions: A.I. acquired and processed the data and wrote the paper. F.S. proposed the methodology and reviewed the paper. A.Y.-C. reviewed and supervised the paper. J.J.G.-M. helped with array processing and reviewed the paper. A.M. performed the electrical prospecting. All authors have read and agreed to the published version of the manuscript.

Funding: This study was funded by the Consellería de Participación, Transparencia, Cooperación y Calidad Democrática de la Generalitat Valenciana, and by Research Group VIGROB-116 (University of Alicante).

Institutional Review Board Statement: Not applicable.

Informed Consent Statement: Not applicable.

Data Availability Statement: The data presented in this study are available upon request from the corresponding author. The data are not publicly available because of their large size.

Acknowledgments: We would like to thank Hamai Lamine for providing the equipment for the resistivity measurements. We also thank Z.Sahari for her help with the lithological cross-sections. A special thanks to the CRAAG team that participated in field measurements: A. Belahouene, A. Sebbane, A. Saadi, B. Melouk, O. Haddad, and R. Chimouni.

Conflicts of Interest: The authors declare no conflict of interest.

References

1. Meghraoui, M. Géologie des Zones Sismiques du Nord de l'Algérie: Paléosismologie, Tectonique Active et Synthèse Sismotectonique. Ph.D. Thesis, University of Paris-Sud, Paris, France, 1988.
2. Perrodon, A. Etude géologique des bassins sublittoraux de l'Algérie occidentale. *Publ. Serv. Carte Géol. l'Algérie. NS Bull.* **1957**, *12*, 328.
3. Thomas, G. Géodynamique d'un Bassin Intramontagneux: Le Bassin du Bas Chéliff Occidental (Algérie) Durant le Mio-Plio-Quaternaire. Ph.D. Thesis, University of Pau, Pau, France, 1985.
4. Yelles-Chaouche, A.; Boudiaf, A.; Djellit, H.; Bracene, R. La tectonique active de la région nord-algérienne. *Comptes Rendus Geosci.* **2006**, *338*, 126–139. [CrossRef]
5. Serpelloni, E.; Vannucci, G.; Pondrelli, S.; Argnani, A.; Casula, G.; Anzidei, M.; Baldi, P.; Gasperini, P. Kinematics of the western Africa Eurasia plate boundary from focal mechanisms and GPS data. *Geophys. J. Int.* **2007**, *169*, 1180–1200. [CrossRef]
6. Agence Nationale des Ressources Hydroliques (ANRH). *Assessment of the Water Resources of the Middle Cheliff Alluvial Aquifer*; Agence Nationale des Ressources Hydroliques: Bir Mourad Raïs, Algeria, 2006.
7. Compagnie Generale de Geophysique (CGG). *Geophysical Study of the Cheliff Plain. Internal Report*; Compagnie Generale de Geophysique: Alger, Algeria, 1969.
8. Yielding, G.; Jackson, J.A.; King, G.C.P.; Sinvhal, H.; Vita-Finzi, C.; Wood, R.M. Relations between surface deformation, fault geometry, seismicity and rupture characteristics during the El Asnam (Algeria) earthquake of 10 October 1980. *Earth Planet Sci. Lett.* **1981**, *56*, 287–304. [CrossRef]
9. Meghraoui, M. Etude Neotectonique de la Region Nord-Est d'El ASNAM: Relation Avec le Seisme du 10.10. 1980. Ph.D. Thesis, University of Paris, Paris, France, 1982.
10. Mokrane, A.; Ait Messaoud, A.; Sebai, A.; Ayadi, A.; Bezzeghoud, M. *Les Séismes en Algérie de 1365 à 1992*; Centre de Recherche en Astronomie Astrophysique et Géophysique (CRAAG): Algiers, Algeria, 1994.
11. Ouyed, M. Le Tremblement de Terre d'El Asnam du 10 Octobre 1980: Étude des Répliques. Algérie. Ph.D. Thesis, Grenoble Alpes University, Grenoble, France, 1981.
12. Delgado, J.; Lopez-casado, C.; Estevez, A.; Giner, J.; Cuenca, A.; Molina, S. Mapping soft soils in the Segura river valley (SE Spain): A case study of microtremors as anexploration tool. *J. Appl. Geophys.* **2000**, *45*, 19–32. [CrossRef]
13. Garcia-Jerez, A.; Luzon, F.; Navarro, M.; Perez-Ruiz, J.A. Characterization of the sedimentary cover of the Zafarraya Basin, Southern Spain, by means of ambient noise. *Bull. Seismol. Soc. Am.* **2006**, *96*, 957–967. [CrossRef]
14. Luzon, F.; Al-Yuncha, Z.; Sanchez-Sesma, F. A Numerical Experiment on the Horizontal to Vertical Spectral Ratio in Flat Sedimentary Basins. *Pure Appl. Geophys.* **2001**, *158*, 2451–2461. [CrossRef]
15. Mundepi, A.; Galiana-Merino, J.; Asthana, A.L.; Rosa-Cintas, S. Soil characteristics in Doon Valley (north west Himalaya, India) by inversion of H/V spectral ratios from ambient noise measurements. *Soil Dyn. Earthq. Eng.* **2015**, *77*, 309–320. [CrossRef]
16. Rosa-Cintas, S.; Galiana-Merino, J.J.; Molina, S.; Rosa-Herranz, J.; García-Fernández, M.; Jiménez, M. Soil characterization in urban areas of the Bajo Segura Basin (Southeast Spain) using H/V, F–K and ESAC methods. *J. Appl. Geophys.* **2011**, *75*, 543–557. [CrossRef]
17. Talaganov, K.; Aleksovski, D.; Milutinovic, Z.; Ameur, B.; Arsovski, M.; Jancevski, J.; Andreevski, V. *Studies for Elaboration of the Code for Repair and Strengthening of Damaged Buildings in the Region of El Asnam: Engineering Geology, Geotechnical and Geophysical Characteristics of the Town of El Asnam and Other Sites*; IZIIS Report; Institute of Earthquake Engineering and Engineering Seismology (IZIIS): Skopje, North Macedonia, 1982; p. 82-55.

18. Woodward-Clyde-Consultants. *Seismic Microzonation of Ech-Chellif Region, Algeria*; Woodward-Clyde-Consultants: Blue Bell, PA, USA, 1984.
19. Philip, H.; Meghraoui, M. Structural analysis and interpretation of the surface deformations of the El Asnam earthquake of 10 October 1980. *Tectonics* **1983**, *2*, 17–49. [CrossRef]
20. Layadi, K.; Semmane, F.; Yelles-Chaouche, A. S-wave velocity structure of Chlef City, Algeria, by inversion of Rayleigh wave ellipticity. *Near Surf. Geophys.* **2018**, *16*, 328–339. [CrossRef]
21. Layadi, K.; Semmane, F.; Yelles-Chaouche, A. Site-Effects Investigation in the City of Chlef (Formerly El-Asnam), Algeria, Using Earthquake and Ambient Vibration Data. *Bull. Seismol. Soc. Am.* **2016**, *106*, 2185–2196. [CrossRef]
22. Issaadi, A.; Semmane, F.; Yelles-Chaouche, A.; Galiana-Merino, J.; Layadi, K. A Shear-Wave Velocity Model in the City of Oued-Fodda (Northern Algeria) from Rayleigh Wave Ellipticity Inversion. *Appl. Sci.* **2020**, *10*, 1717. [CrossRef]
23. Bouchelouh, A.; Bensalem, R.; Zaourar, N.; Machane, D.; Moulouel, H.; Oubaiche, E.H. The Miocene Roof Mapping Using Microtremor Recording and Electrical Survey Method in Blida City, Algeria. *Pure Appl. Geophys.* **2017**, *175*, 287–301. [CrossRef]
24. Tebbouche, M.; Machane, D.; Chabane, S.; Oubaiche, E.H.; Meziani, A.; Ait Benamar, D.; Moulouel, H.; Cheikh Lounis, G.; Bensalem, R.; Bendaoud, A. Imagery of the metamorphic bedrock roof of the Sahel active fault in the Sablettes (Algiers) reclaimed area by ambient vibration HVSR. *Arab. J. Geosci.* **2017**, *10*, 292. [CrossRef]
25. Nakamura, Y. A method for dynamic characteristics estimation of subsurface using microtremor on the ground surface. *Railw. Tech. Res. Inst. Q. Rep.* **1989**, *30*, 25–33.
26. Nogoshi, M.; Igarashi, T. On the amplitude characteristics of ambient noise (Part 2). *J. Seismol. Soc. Jpn.* **1971**, *24*, 26–40.
27. Capon, J. High-resolution frequency-wavenumber spectral analysis. *Proc. IEEE* **1969**, *57*, 1408–1419. [CrossRef]
28. Capon, J.; Greenfield, R.; Kolker, R. Multidimensional maximum-likelihood processing of a large aperture seismic array. *Proc. IEEE* **1967**, *55*, 192–211. [CrossRef]
29. Kelly, E.J.; Levin, M.J. *Signal Parameter Estimation for Seismometer Arrays*; Massachusetts Institute of Technologies, Lincoln Lab: Lexington, MA, USA, 1964.
30. Lacoss, R.T.; Kelly, E.J.; Toksoz, M.N. Estimation of seismic noise structure using Array. *Geophysics* **1969**, *29*, 21–38. [CrossRef]
31. Delteil, J. Tectonique de la Chaîne Alpine en Algérie D'après L'étude du Tel Oranais Oriental. Monts de la Mina, Beni Choughrane, Dahra. Ph.D. Thesis, University of Nice, Nice, France, 1974; 294p.
32. Brives, A. *Les Terrains Miocènes du Bassin du Chélif et du Dahra*; Imprimerie, P., Ed.; Fontana & Cie: Alger, Algérie, 1897.
33. Mattauer, M. *Etude Géologique de l'Ouarsenis Oriental*; Publication du Service de Cartographie d'Algérie, n°17; Service de Cartographie d'Algérie: Alger, Algeria, 1958.
34. Kireche, O. Etude Géologique et Structurale des Massifs de la Plaine du Chéliff (Doui, Rouina, Temoulga). Ph.D. Thesis, University of Algiers, Algiers, Algeria, 1977.
35. Repal, S.N. Le Basin Néogène du Chélif. In Proceedings of the XIXe Congrés de Géologie International, Service Geologique de l'Algérie, Algiers, Algeria, 8–15 September 1952; p. 56.
36. Brives, A.; Jackob, M.; Ficheur, M. *Carte Géologique De Oued-Fodda 1/50,000*; Service Géologique de l'Algérie: Algiers, Algeria, 1906.
37. Field, E.H.; Jackob, K.H. A comparison and test of various site response estimation techniques, including three that are non reference—Site dependent. *Bull. Seismol. Soc. Am.* **1995**, *85*, 1127–1143.
38. Arai, H.; Tokimasu, K. S-wave velocity profiling by inversion of microtremor H/V spectrum. *Bull. Seismol. Soc. Am.* **2004**, *94*, 53–63. [CrossRef]
39. Bonnefoy-Claudet, S.; Cornou, C.; Bard, P.Y.; Cotton, F.; Moczo, P.; Kristek, J.; Fäh, D. H/V ratio: A tool for site effects evaluation: Results from 1D noise simulations. *J. Appl. Geophys.* **2006**, *167*, 827–837. [CrossRef]
40. Lachet, C.; Bard, P.Y. Numerical and theoretical investigations on the possibilities and limitations of Nakamura's technique. *J. Phys. Earth* **1994**, *42*, 377–397. [CrossRef]
41. Rosa-Cintas, S.; Galiana-Merino, J.; Rosa-Herranz, J.; Molina, S.; Martínez-Esplá, J. Polarization analysis in the stationary wavelet packet domain: Application to HVSR method. *Soil Dyn. Earthq. Eng.* **2012**, *42*, 246–254. [CrossRef]
42. La Rocca, M.; Chiappetta, G.; Gervasi, A.; Festa, R.L. Non-Stability of the noise HVSR at sites near or on topographic heights. *Geophys. J. Int.* **2020**, *222*, 2162–2171. [CrossRef]
43. Bonnefoy-Claudet, S.; Kohler, A.; Cornou, C.; Wathelet, M.; Bard, P.Y. Effects of Love waves on microtremor H/V ratio. *Bull. Seismol. Soc. Am.* **2008**, *98*, 288–300. [CrossRef]
44. Wathelet, M. An improved neighborhood algorithm: Parameter conditions and dynamic scaling. *Geophys. Res. Lett.* **2008**, *35*, L09301. [CrossRef]
45. Wathelet, M.; Jongmans, D.; Ohrnberger, M. Surface-wave inversion using a direct search algorithm and its application to ambient vibration measurements. *Near Surf. Geophys.* **2004**, *2*, 211–221. [CrossRef]
46. Maresca, R.; Galluzzo, D.; Del-Pezzo, E. H/v spectral ratios and array techniques applied to ambient noise recorded in the Colfiorito Basin, Central Italy. *Bull. Seismol. Soc. Am.* **2006**, *96*, 490–505. [CrossRef]
47. Rost, S.; Thomas, C. Array Seismology: Methods and Applications. *Rev. Geophys.* **2002**, *40*, 1008. [CrossRef]
48. Levin, M.J. *Maximum-Likelihood Array Processing, Seismic Discrimination Semiannual Technical Summary*; Lincoln Laboratory, Massachusetts Institute of Technology: Lexington, MA, USA, 1964.
49. Allaud, L.; Martin, M. *Schlumberger, History of a Technique*; John Wiley & Sons: New York, NY, USA, 1977.

50. Loke, M.H.; Barker, R.D. Practical techniques for 3Dresistivity surveys and data inversion techniques. *Geophys. Prospect.* **1996**, *44*, 499–524. [CrossRef]
51. Schlumberger, C.; Schlumberger, M.; Doll, H.G. Electrical Coring: A Method of Determining Bottom-Hole Data by Electrical Measurements. In *Transactions of the American Institute of Mining and Metallurgical Engineers*; Paper 462; American Institute of Mining, Metallurgical, and Petroleum Engineers: New York, NY, USA, 1932; Volume 100.
52. Dahlin, T.; Loke, M.H. Resolution of 2D Wenner resistivity imaging as assessed by numerical modelling. *J. Appl. Geophys.* **1998**, *38*, 237–249. [CrossRef]
53. Bellalem, F.; Bounif, M.; Koulakov, I. P and S waves tomographic analysis of the area of El Asnam's 1980 ms 7.3 earthquake (Algeria) from its aftershock sequence. *J. Seismol.* **2015**, *19*, 253–264. [CrossRef]
54. Acerra, C.; Aguacil, G.; Anastasiadis, A.; Atakan, K.; Azzara, R.; Bard, P.-Y.; Basili, R.; Bertrand, E.; Bettig, B.; Blarel, F. *Guidelines for the Implementation of the H/V Spectral Ratio Technique on Ambient Vibrations Measurements, Processing and Interpretation*; European Commission–EVG1-CT-2000-00026 SESAME; European Commission: Brussels, Belgium, 2004.
55. Soler-Llorens, J.L.; Galiana-Merino, J.; Benabdeloued, N.; Rosa-Cintas, S.; Ortiz-Zamora, J.; Giner-Caturla, J.J. Design and implementation of an Arduino-based plug-and-play acquisition system for seismic noise measurements. *Electronics* **2019**, *8*, 1035. [CrossRef]
56. Konno, K.; Ohmachi, T. Ground-motion characteristics estimated from spectral ratio between horizontal and vertical components of microtremor. *Bull. Seismol. Soc. Am.* **1998**, *88*, 228–241. [CrossRef]
57. Woolery, E.; Street, R. Quaternary fault reactivation in the fluorspar area fault complex of western Kentucky: Evidence from shallow SH-wave reflection profiles. *Seismol. Res. Lett.* **2002**, *73*, 628–639. [CrossRef]

applied sciences

MDPI

Article

Site Response Evaluation in the Trans-Mexican Volcanic Belt Based on HVSR from Ambient Noise and Regional Seismicity

L. Francisco Pérez-Moreno [1], Quetzalcoatl Rodríguez-Pérez [2,*], F. Ramón Zúñiga [3], Jaime Horta-Rangel [1], M. de la Luz Pérez-Rea [1] and Miguel A. Pérez-Lara [1]

[1] Facultad de Ingeniería, Universidad Autónoma de Querétaro, Querétaro 76010, Mexico; lf.perezm@uaq.mx (L.F.P.-M.); horta@uaq.mx (J.H.-R.); perea@uaq.mx (M.d.l.L.P.-R.); migperez@uaq.mx (M.A.P.-L.)

[2] Consejo Nacional de Ciencia y Tecnología, Dirección Adjunta de Desarrollo Científico, Ciudad de México 03940, Mexico

[3] Centro de Geociencias, Universidad Nacional Autónoma de México, Querétaro 76230, Mexico; ramon@geociencias.unam.mx

* Correspondence: quetza@geociencias.unam.mx

Featured Application: The results obtained in this study allow an initial overview of the variation of the site response in an area for which low to moderate seismic risk is usually considered. The information presented may be used to analyze the seismic effects in the study zone associated with the geological characteristics.

Citation: Pérez-Moreno, L.F.; Rodríguez-Pérez, Q.; Zúñiga, F.R.; Horta-Rangel, J.; Pérez-Rea, M.d.l.L.; Pérez-Lara, M.A. Site Response Evaluation in the Trans-Mexican Volcanic Belt Based on HVSR from Ambient Noise and Regional Seismicity. *Appl. Sci.* **2021**, *11*, 6126. https://doi.org/10.3390/app11136126

Academic Editors: Miguel Llorente Isidro, David Moncoulon, Ricardo Castedo and Filippos Vallianatos

Received: 28 April 2021
Accepted: 28 June 2021
Published: 30 June 2021

Abstract: The Trans-Mexican Volcanic Belt (TMVB), located in central Mexico, is an area for which low to moderate seismic risk is considered. This is based on the limited instrumental data available, even though large historical earthquakes have damaged some urban centers in the past. However, site effects is an aspect that must be considered in estimating risk, because there are some instances of important amplifications that have been documented with serious effects. In this work, ambient noise and earthquake records from 90 seismic permanent and temporary stations are used to analyze site response in the TMVB. The results obtained show a heterogeneous range in the value of the fundamental frequency. When possible, a comparison was made of the results obtained from ambient noise and earthquake records. In almost all these comparisons, no significant differences were observed in terms of the fundamental frequency. However, there were some stations with a flat average HVSR ambient noise curve that contradicted earthquake data results, which showed peaks at some frequencies. Our results are a first step towards categorizing the different site responses in the TMVB but in order to provide finer details, it is necessary to improve the actual monitoring conditions.

Keywords: site effects; spectral ratio; trans-mexican volcanic belt

1. Introduction

Within the context of seismic engineering, the study of site response involves changes in variables related to the seismic intensity, in terms of amplitude, duration, and frequency content. These changes depend on the geological features at the measurement site, and usually, lead to larger amplitudes on soil sites than on hard rock.

Throughout history it is possible to find documented cases in which site effects have played a decisive role in observed damage after significant earthquakes, such as in San Francisco (1906), Mexico City (1985 and 2017), and Kobe (1995). Thus, in the seismic design of buildings and civil infrastructure it is important to estimate the maximum expected intensities at the site.

The evaluation of changes in amplitude of ground motion of a particular site is commonly made with respect to a reference position. The most popular and reliable way to do this is through the Standard Spectral Ratio (SSR) technique [1]. This method is based on the ratio of the Fourier amplitude spectra observed during an earthquake at soil $A_i(f)$ to reference site $A_j(f)$. The Fourier spectra of ground motion $A(f)$ can be expressed as the multiplication of source $S(f)$, path $P(f)$, and site $H(f)$ terms in the frequency domain:

$$A(f) = S(f) \times P(f) \times H(f) \tag{1}$$

where f is the frequency.

In the SSR technique, the reference site is required to have the most homogeneous geological conditions as it is assumed to be free of anomalous amplifications. The two sites should be close to each other in comparison with the distance to the source. Considering that the analysis is performed for stations which register the same earthquake:

$$SSR(f) = \frac{|A_i(f)|}{|A_j(f)|} = \frac{|H_i(f)|}{|H_j(f)|}. \tag{2}$$

However, it is not always possible to find records of the same event with a good signal-to-noise (S/R) ratio in two stations that meet these conditions. Thus, the Horizontal-to-Vertical Spectral Ratio (HVSR) technique [2] is used as an alternative to studying the site response. In this method, the ratio between the mean of the horizontal components $A_{Hi}(f)$ and the vertical component $A_{Vi}(f)$, recorded in a single station, is calculated:

$$HVSR(f) = \frac{|A_{Hi}(f)|}{|A_{Vi}(f)|}. \tag{3}$$

This technique can be used to estimate the fundamental resonant frequency (f_0) of a site, but is unreliable for determining its transfer function [3].

In the literature, several studies can be found in which the site response is analyzed in areas around the world, based on these empirical techniques and numerical models [4–8]. The study of this phenomenon remains current and it is an important issue to be considered in the estimation of seismic risk and damage prevention, even in areas of low seismicity [9].

The Trans-Mexican Volcanic Belt (TMVB), located in central Mexico (Figure 1), is an example of those areas where the site response might represent a potential risk despite the low frequency of earthquakes. It is a volcanic arc that covers Cretaceous and Cenozoic magmatic provinces [10]. Seismicity in this region is related to extensional faults in the crust and normal faults in the subducted Cocos plate. This zone is located between 18°30′ and 21°30′, and extends from the coast of the Pacific Ocean to the Gulf of Mexico. It has an approximate length of 1000 km and a variable width of 90–230 km.

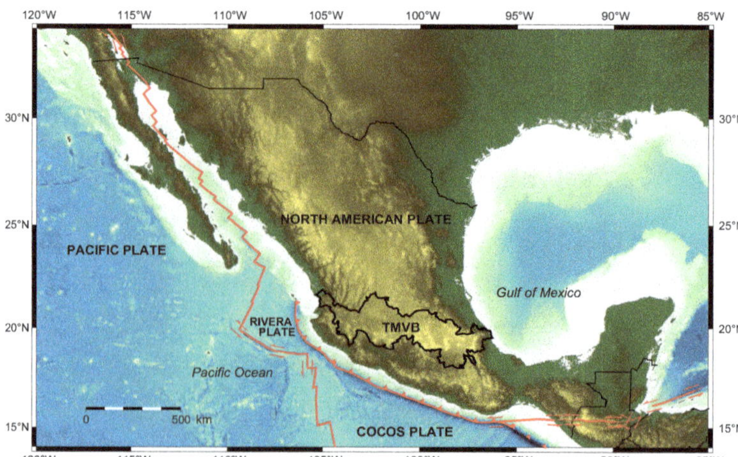

Figure 1. Location of the Trans-Mexican Volcanic Belt and interaction of tectonic plates in Mexico.

This area presents low seismicity, compared to the Pacific Ocean coast, where the Cocos plate subducts under the North American plate. Nevertheless, there are important recent and historical occurrences of destructive events [11,12]. Examples of these is the earthquake that occurred in Acambay on 12 November 1912 (Mw 6.9, H = 33 km) [13], and more recently in Puebla on 19 September 2017 (Mw 7.1, H = 51.2 km) [14]. This aspect is particularly important, in view that the TMVB comprises some of the largest cities in Mexico with significant industrial development and population growth in the last years.

It may be difficult to find a zone in the TMVB with negligible site effects, due to the variability in geological properties [10]. Several articles have been published on this topic related to Mexico, mainly focused on the behavior of site effects in Mexico City.

Ordaz et al. [15] found that stations located in the hill zone in the Valley of Mexico present amplifications that are 10 times higher than those predicted from ground motion attenuation equations. In a later study, García et al. [16] compared amplification responses at some of the stations analyzed by [15]. The records analyzed corresponded to inslab earthquakes with an epicenter outside the TMVB, and they observed similar behavior despite the source type and location.

Singh et al. [17] briefly discuss HVSR curves for 2 stations crossing the TMVB, obtained from records of 9 shallow coastal earthquakes on the Pacific Ocean coast. Likewise, Lozano et al. [18] studied the influence of source characteristics on the site effects in the Valley of Mexico from 36 interplate and inslab Mexican earthquakes and 12 teleseismic events originated in South America. Their results agree with those obtained by [16] and concluded that observed site effects were independent of the characteristics and location of the source.

Clemente-Chávez et al. [19] presented the first study of site effects in the TMVB outside the Valley of Mexico, based on local shallow seismicity. They analyzed HVSR curves obtained from 22 earthquakes recorded by 25 seismological stations located in the study zone. The events studied have depths $H < 10$ km, and magnitudes between 3.6 and 4.3. Average values for the fundamental frequency f_0 and amplification factors obtained from the HVSR curves were reported. Their results were compared with previous studies and they attributed the differences found to the location of the source. However, due to the limited amount of data employed, some of their conclusions were supported only with one or two records. Furthermore, it should be considered that HVSR curves do not unequivocally represent the actual site amplification.

Considering the low density of permanent seismological stations in the TMVB [20], the number of records that can be analyzed with the SSR technique is very limited. So, the use of ambient noise records represents a feasible option to include additional data for the analysis of site response in terms of f_0. In this paper, we analyze HVSR curves obtained from ambient noise and regional instrumental records related to 121 crustal and inslab events with an epicenter inside the TMVB. Data considered were recorded by 90 seismological stations belonging to seven seismic networks.

The results of our study may be useful in disaster prevention and in estimating the behavior of future buildings in the study area. The information presented may also be used to complement studies examining the damage observed to existing infrastructure after the occurrence of a seismic event [21,22]. However, it is important to consider that a more accurate and reliable analysis can be carried out throughout microzonation studies, so our results only give a general overview.

2. Data and Methods

To perform this study, we compiled a catalog of earthquakes with epicenter inside the area of the TMVB, based on the seismicity catalogs of the Servicio Sismológico Nacional [23], the U.S. Geological Survey [24], and the International Seismological Centre (ISC) [25]. Avoiding duplications and magnitudes reported as "non-calculable", 2113 earthquakes with magnitude 1.5–7.8 and depth 1–191 km, were selected.

Based on this information, instrumental records were searched in seven permanent and temporary seismological networks in Mexico (Table 1), with broadband stations located inside the TMVB. It should be considered that over time there have been significant changes in instrumental density in this zone [20]. Additionally, the stations that belong to permanent networks are widely dispersed, and the availability of continuous data is limited to the last 30 years.

Table 1. Seismological networks that were analyzed in this study. The continuous data availability is variable depending on each station.

Network	Period
SSN	2002–present
CGEO	2003–present
GEOSCOPE	1990–present
MASE	2004–2007
CODEX	2006–2008
MARS	2006–2007
GECO	2013–2018

Data were gathered from stations with velocity sensors belonging to Servicio Sismológico Nacional (SSN) [23], Centro de Geociencias de la Universidad Nacional Autónoma de México (CGEO), GEOSCOPE [26], and the temporary networks: The MesoAmerican Subduction Experiment (MASE) [27,28], Colima Deep Seismic Experiment (CODEX) [29], Mapping the Rivera Subduction Zone (MARS) [30], and Geometry of the Cocos Plate (GECO) [31].

The stations belonging to SSN and CGEO are equipped with STS-2 broadband seismometers with sampling rates of 80 and 100 Hz. Only the ACIG station has a Trillium 120 seismometer with a sampling rate of 20 Hz. In the case of the GEOSCOPE network, the UNM station has a velocity sensor STS-1 20 samples/s. During their period of operation, the GECO network stations were equipped with Reftek 151B-60 and Guralp 40T sensors. The temporary networks MASE and CODEX had CMG 3T, and 40T seismometers, respectively, with a sampling rate of 100 Hz. The MARS network was equipped with CMG 3T sensors of 40 sps.

Only records that met the following criteria were considered for the analyses: (1) Station and epicenter located inside the TMVB; (2) signal-to-noise ratio (S/R) > 2.0; and (3) clear arrival of P and S waves, based on visual inspection.

Considering these restrictions, a total of 352 records related to 24 inslab and 97 crustal earthquakes were chosen (Figure 2). Such events have magnitudes between 2.0 and 7.1, a hypocentral depth of $1 \leq H \leq 105$ km, and occurred between 1993 and 2019 in the TMVB. The main characteristics of the selected earthquakes are listed in Table 2.

In Figure 2 it can be observed that in the central zone there is the lowest number of records. As mentioned by Zúñiga et al. [20], the other regions present higher seismic activity due to active fault systems, the subduction zone on the Pacific Ocean coast, and the occurrence of inslab earthquakes in the central-eastern sector.

The number of records selected for each station was variable, and in most of them, the data with the characteristics sought was considerably limited. An HVSR analysis was performed at all stations with at least 10 earthquake records. This threshold was considered to include as many stations as possible, trying not to compromise the reliability of the results. It is important to note that despite the large number of seismological stations belonging to MASE, MARS, and CODEX networks, the suitable records were limited due to the characteristics of the earthquakes during their short period of operation.

The processing of records was performed by means of the package Geopsy [32]. For each analyzed record, the mean and the trend were removed. The corresponding Fourier Acceleration Spectra (FAS) were smoothed using the Konno–Omachi function [33], considering a b-value of 20. Window lengths that included 95% of the total energy were taken, beginning from the S-wave arrival.

Table 2. A database of crustal and inslab earthquakes with records that met the considered criteria. The value of magnitude is the one reported by SSN.

Date dd/mm/yyyy	Origin Time hh:mm:ss	Magnitude	Depth (km)	Latitude (°)	Longitude (°)
08/03/1993	14:37:41	3.3	11	19.25	−98.93
02/11/1994	15:31:21	3.7	7	19.41	−98.90
27/03/1996	08:05:15	3.4	8	19.77	−99.01
17/04/1996	11:19:10	3.5	10	19.21	−98.94
06/09/1996	06:55:00	4.0	5	19.21	−98.36
03/07/1997	19:39:32	2.8	7	19.25	−99.44
25/12/1997	01:29:10	4.1	11	18.97	−98.59
15/03/1998	02:01:49	4.2	5	19.50	−100.26
26/10/1998	02:35:10	3.0	6	19.06	−99.17
26/10/1998	07:50:06	3.3	4	19.06	−99.20
31/12/1998	06:44:05	3.8	2	18.98	−98.65
04/01/1999	23:21:59	3.5	3	20.11	−98.98
19/05/1999	20:23:06	3.1	5	19.19	−98.96
22/05/1999	06:57:26	3.1	4	19.18	−98.97
07/06/1999	00:24:14	3.3	5	19.17	−98.96
15/06/1999	20:42:04	7.0	63	18.13	−97.54
20/06/1999	10:25:21	4.0	3	19.31	−98.48
07/12/1999	13:53:49	3.4	3	19.22	−98.92
04/03/2000	20:55:23	4.1	4	18.84	−98.57
12/03/2000	03:44:48	4.1	5	20.10	−99.29
06/07/2000	08:36:24	3.3	8	19.18	−98.93
06/07/2000	13:15:34	3.4	5	19.19	−98.96
21/07/2000	06:13:39	6.0	48	18.09	−98.97
14/10/2000	02:44:48	3.1	12	19.36	−99.19
15/10/2000	07:50:43	2.7	5	19.39	−99.11
01/03/2001	16:26:42	3.5	46	18.98	−99.82
13/06/2001	03:57:43	3.2	12	19.27	−99.42
14/09/2001	17:13:15	2.9	14	19.31	−99.30
15/11/2001	22:18:22	3.5	4	19.56	−99.16
09/05/2002	14:25:43	3.7	19	19.49	−99.01
04/02/2003	10:59:03	4.1	2	18.92	−98.51
15/12/2003	10:39:38	4.0	4	20.35	−99.07
13/09/2004	20:58:34	3.1	11	19.42	−99.17
16/04/2005	22:55:25	3.8	16	19.43	−99.00
12/05/2005	08:06:25	3.4	37	19.20	−98.97
07/08/2005	03:25:09	4.0	7	19.74	−98.61
16/10/2005	14:12:36	3.5	14	19.30	−99.20
23/11/2005	23:11:26	3.6	20	19.35	−98.94
24/01/2006	12:59:38	3.7	5	20.34	−99.22
28/02/2006	23:58:49	3.4	2	19.35	−99.02
12/03/2006	01:41:32	3.6	2	19.17	−98.95
12/03/2006	01:47:21	3.4	5	19.18	−98.96
05/05/2006	15:24:06	3.5	1	19.17	−98.95
27/06/2006	15:40:10	4.0	16	19.26	−102.32
14/07/2006	05:15:28	3.7	72	20.19	−98.87
18/12/2006	03:38:31	3.9	44	20.92	−104.47
25/01/2007	14:38:22	3.7	30	19.18	−99.16
30/01/2007	19:00:10	3.7	9	20.35	−103.87
14/05/2007	08:23:27	4.4	12	21.36	−104.75
30/05/2007	20:42:28	4.0	3	19.21	−99.45
22/06/2007	13:36:52	4.2	25	19.16	−96.96
20/10/2007	19:25:54	3.7	75	19.47	−102.20
20/10/2007	02:37:27	3.9	20	19.95	−101.97
21/10/2007	11:29:53	3.7	93	19.53	−102.17
23/03/2009	02:53:36	3.9	46	18.92	−97.01
22/05/2009	19:24:18	5.7	62	18.11	−98.46
29/11/2009	06:34:12	4.0	5	19.35	−103.76
17/04/2010	07:03:29	4.0	2	20.45	−99.04

Table 2. *Cont.*

Date dd/mm/yyyy	Origin Time hh:mm:ss	Magnitude	Depth (km)	Latitude (°)	Longitude (°)
18/04/2010	18:26:25	4.0	79	19.27	−97.46
18/05/2010	05:29:09	4.0	5	20.35	−98.92
18/05/2010	05:32:37	3.4	10	20.37	−98.96
03/10/2010	23:07:11	4.0	6	19.48	−103.52
19/12/2010	02:46:45	4.1	16	19.03	−97.21
28/09/2011	10:51:18	4.1	23	19.76	−96.64
18/05/2012	03:07:58	4.4	1.6	20.26	−103.44
05/10/2012	00:14:57	4.2	16.1	19.45	−102.26
08/02/2013	01:24:58	3.8	6.5	20.12	−100.48
21/06/2013	09:41:33	4.0	2	18.52	−98.74
18/09/2013	00:26:15	3.3	5	19.62	−96.79
14/12/2013	22:33:40	4.1	104.6	19.45	−103.71
24/12/2013	15:27:07	3.6	3	19.52	−103.80
16/01/2014	07:23:57	3.4	9	19.27	−97.26
09/05/2014	07:50:49	3.7	3	18.98	−97.27
19/07/2014	11:57:23	3.5	8	19.17	−98.97
04/08/2014	21:10:03	4.0	59.2	18.18	−97.83
08/11/2014	02:46:50	3.6	5	19.51	−96.78
01/12/2014	08:50:06	3.4	2	19.35	−99.22
06/05/2015	04:06:46	3.7	10	19.16	−97.17
09/08/2015	15:17:17	3.7	5	18.96	−98.21
16/08/2015	14:29:17	3.4	20	19.65	−97.55
25/10/2015	02:52:43	3.7	77.3	19.26	−97.30
07/11/2015	02:29:50	4.1	5	19.73	−96.70
26/12/2015	14:08:36	4.0	5	19.55	−102.11
16/01/2016	06:28:46	3.1	42.6	19.65	−96.80
27/01/2016	21:22:44	4.1	69.5	18.28	−97.33
28/01/2016	11:41:49	3.4	44	19.45	−96.72
08/02/2016	21:16:07	4.6	6.9	19.66	−97.35
13/03/2016	19:13:50	3.7	55.3	19.31	−97.03
01/04/2016	22:34:33	3.7	58.6	18.27	−97.78
01/04/2016	05:06:51	3.1	5	19.63	−96.64
01/04/2016	05:39:39	3.4	4.1	19.68	−96.73
11/05/2016	22:35:20	4.8	8	20.81	−103.52
15/05/2016	01:43:48	3.9	5	20.82	−103.47
11/06/2016	19:06:14	3.9	12.5	18.85	−97.19
16/06/2016	00:56:29	3.3	10	19.95	−96.77
24/06/2016	00:22:54	3.4	20	19.23	−97.40
27/06/2016	19:12:25	3.4	10	19.95	−96.76
12/07/2016	17:33:29	3.5	5	20.77	−103.45
19/07/2016	04:34:43	3.9	56.7	18.29	−97.38
28/07/2016	18:21:06	3.9	5	20.73	−103.45
08/08/2016	15:57:08	3.8	20	19.44	−96.72
29/08/2016	16:24:17	3.8	3	18.95	−98.58
26/09/2016	03:44:44	3.6	10.1	19.51	−96.45
15/10/2016	01:51:48	4.1	51.3	18.87	−96.99
26/02/2017	02:11:29	4.0	23.7	19.88	−96.85
01/06/2017	13:36:59	3.8	5	19.24	−97.04
09/07/2017	09:03:33	3.5	30.7	19.69	−96.81
04/08/2017	01:07:36	3.5	10	19.50	−96.43
10/09/2017	02:54:12	2.7	9.4	19.31	−99.18
19/09/2017	18:14:40	7.1	51.2	18.33	−98.68
29/09/2017	17:02:23	3.8	50.7	18.35	−98.66
01/11/2017	20:48:04	4.3	52.8	18.25	−98.60
06/11/2017	09:38:09	3.1	3	19.19	−98.44
09/11/2017	10:12:07	2.0	8.5	19.38	−99.19
18/11/2017	07:21:54	3.6	32.4	18.64	−98.48
24/12/2017	12:25:12	3.4	3.1	18.82	−98.60
08/06/2018	18:09:34	3.9	11.5	19.95	−96.75
30/07/2018	22:15:54	3.9	24.6	19.27	−97.34
30/03/2019	22:09:22	3.9	5	19.44	−96.71
14/07/2019	19:37:06	3.8	37	19.21	−97.23
23/07/2019	19:03:20	4.0	32	19.27	−97.43

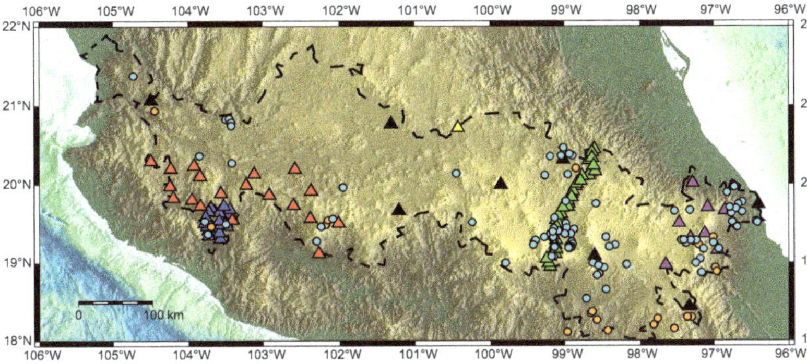

Figure 2. Location of epicenters and seismological stations analyzed in this study. Cyan circles: Crustal earthquakes; orange circles: Inslab earthquakes. Red triangles: MARS temporary network stations; blue triangles: CODEX temporary network stations; green triangles: MASE temporary network stations; purple triangles: GECO temporary network stations; black triangles: SSN permanent network stations; yellow triangle: CGEO permanent station; and white triangle: GEOSCOPE permanent station. Dashed line: TMVB.

Due to the lack of earthquake records, an HVSR analysis was performed in all the stations using ambient noise data. Although the energy of ambient noise does not compare with that of an earthquake, this second analysis allowed us to increase the number of stations analyzed and to compare the results obtained with both procedures. For each station, a database of 30 records of one-hour duration was collected on random dates within its period of operation. In all ambient noise records, the mean and trend were removed and the same smoothing process was applied as in the earthquake records. For the HVSR analyses, the records were divided into 60 windows of 1 min and the root mean square of the horizontal components was calculated. SESAME criteria [34] were considered for the analyses and classification of the curves obtained. Finally, the results were correlated with the geological characteristics of the study area.

3. Results

Due to the large extension of the study area, the stations analyzed were grouped into four zones. This division was based on the distribution of stations and on geologic information given by [10]. Most of the average curves shown correspond to the HVSR analyses based on ambient noise data, between 0.1 and 25 Hz. Some of the considered stations have a Nyquist frequency of 10 and 20 Hz, so not all curves are displayed in the same frequency range. When possible, a comparison is made with results from earthquake records. Regarding geology, information from the Geologic Map of North America [35] was used.

3.1. Western TMVB

In the westernmost part of the TMVB, there are three main fault systems: Tepic-Zacoalco, Colima, and Chapala [10]. Regarding significant earthquakes, there is a historical account of the occurrence of a damaging event during the 16th century near the town of Ameca, in the state of Jalisco. For this event, a magnitude Mw 7.2 $\pm$ 0.3 was estimated from a rupture vs. magnitude scaling relation [36,37]. Another earthquake occurred on 2 October 1847, with a mean magnitude of 5.7 $\pm$ 0.4 which has been associated to the faults of the Chapala Graben. It affected several towns and produced important damages to buildings. An intensity of IX has been estimated for this event [38].

Later, on 11 February 1875 an earthquake near the city of Guadalajara caused great destruction in nearby areas including dozens of deaths [39].

Figure 3 shows the seismological stations analyzed in the Western sector of the TMVB. As can be observed, there is only one station belonging to a permanent seismological network. The other temporary stations are mostly concentrated in the south of this sector.

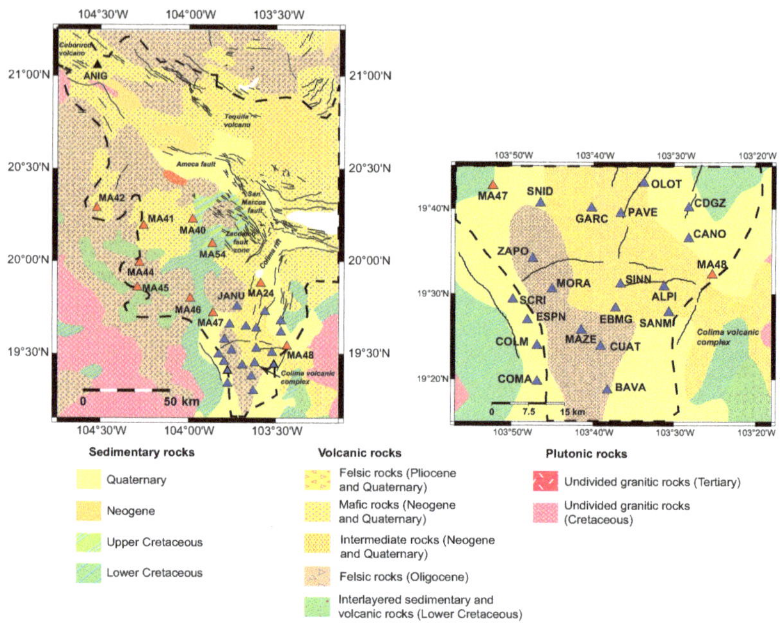

Figure 3. Geological units in the western sector of the TMVB [35]. Red triangles: MARS temporary network stations; blue triangles: CODEX temporary network stations; and black triangle: SSN permanent network station. Continuous line: Main fault systems in the area; dashed line: TMVB.

In this sector, there are a variety of geological formations ranging from the Lower Cretaceous to the Quaternary [35]. Most of the stations analyzed are located in the vicinity of the Colima volcanic complex. Figure 4 shows some of the average HVSR curves obtained in this zone, in which only the ANIG station could be analyzed with earthquake records.

In some curves there can be observed flat responses or complex shapes with low amplitudes at different frequencies: MA47, MORA, MAZE, CUAT, ZAPO, MA24, OLOT, SNID, SINN, EMBG (Figure 4), SCRI, and COMA (Figure 5). Station MA47 is located on sedimentary rocks from the Lower Cretaceous. Its HVSR curve has a flat shape and no significant impedance contrast is identified. In contrast, stations MORA, MAZE, CUAT, and ZAPO were deployed on Oligocene felsic rocks and near their location there are volcanic and sedimentary rocks from the Neogene and Quaternary. The site response could be attributed to lateral heterogeneity and the presence of sedimentary rocks from the Cretaceous. Stations MA24, OLOT, SINN, and EBMG were located on mafic and intermediate rocks from the Neogene and Quaternary. MA24 and OLOT were relatively close to each other, but a resonant frequency can not be identified in either of them. As for SINN and EBMG, the corresponding curves have a flat shape, indicating hard rock sites. Stations SCRI and COMA (Figure 5) were deployed on Quaternary sedimentary rocks, and in both of them a predominantly flat shape is also observed.

There is another group of stations in which the HVSR curves present multiple peaks, with amplitudes similar to what could be identified as the clearest peak. Station MA54 were located on sedimentary rocks from the Upper Cretaceous. The peak around 6 Hz could be attributed to a thin layer of sediments and the amplitude between 0.6 and 2 Hz to the contact with formations from the Lower Cretaceous. Meanwhile, MA46 were deployed on Oligocene felsic rocks. The observed peaks above 2 Hz could indicate multiple layers of sediments of varying thickness above the bedrock. Station MA42 was on mafic and intermediate rocks from the Neogene and Quaternary. The shape of the HVSR curve corresponding to this station could be attributed to variations in the topography or a deep interface with Oligocene formations. SNID (Figure 4) and MA48 (Figure 5) were deployed on Quaternary sedimentary rocks. The amplitudes in SNID could be associated to lateral heterogeneities by means of its location near Neogene formations. Likewise, some of the multiple peaks in MA48 could be spurious or related to multiple layers of sediments at variable depth.

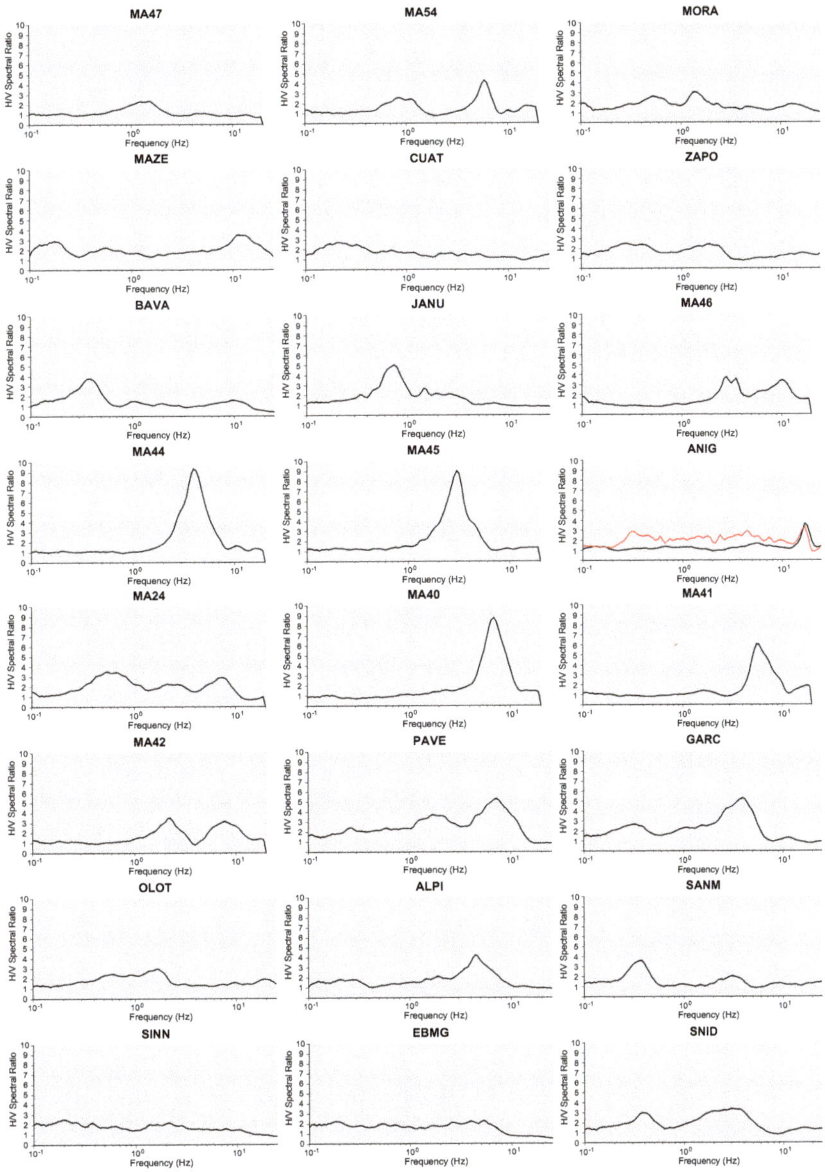

Figure 4. Average HVSR curves for stations located in the western sector of the TMVB. Blackline: HVSR results using ambient noise records; redline: HVSR results from earthquake records.

In the remaining stations, clear peaks of different amplitude can be observed, revealing different impedance contrasts between sediment layers and the bedrock: BAVA, JANU, MA44, MA45, ANIG, MA40, MA41, PAVE, GARC, ALPI, SANM (Figure 4), CDGZ, COLM, CANO, and ESPN (Figure 5). This is also indicative of possible amplifications in ground motion during the occurrence of seismic events. In the case of stations BAVA and JANU, located on Oligocene felsic rocks, the observed peaks at low frequencies could be associated to thick sedimentary deposits. High clear peaks at 3 and 4 Hz are observed at stations MA44 and MA45. These may be produced by strong impedance contrasts with interlayered sedimentary and volcanic rocks from the Lower Cretaceous.

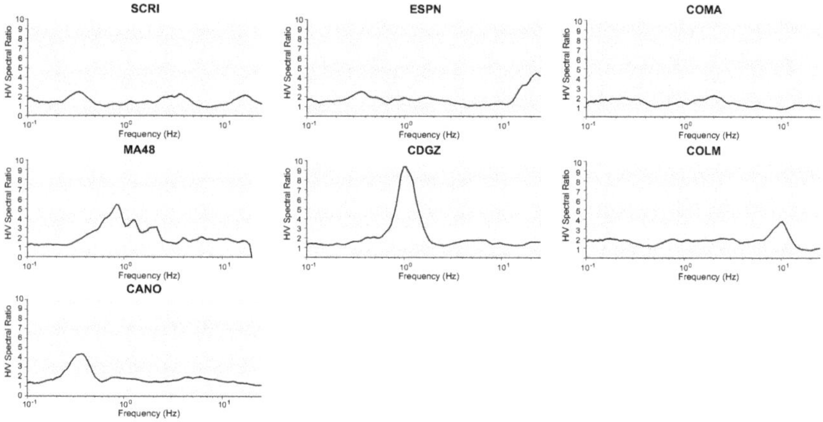

Figure 5. Average HVSR curves for stations located in the western sector of the TMVB.

The stations located on mafic and intermediate rocks from the Neogene and Quaternary have a heterogeneous response (ANIG, MA40, MA41, PAVE, GARC, ALPI, and SANM) (Figure 4). In general, a clear peak is observed above 5 Hz, which could be attributed to impedance contrast with thin layers of sediment. It should be noted that the average HVSR curve obtained with earthquake data at ANIG station does not vary significantly in its shape to that obtained from ambient noise. There is a peak above 10 Hz that could have been produced by the presence of thin volcanic layers.

Regarding the average HVSR curves of stations on Quaternary sedimentary rocks (ESPN, CDGZ, COLM, and CANO) there are varied shapes (Figure 5). In ESPN, there is a peak above 10 Hz which may require the analysis of a wider frequency bandwidth. The clear peak in CANO could be associated to contact with Neogene and Quaternary rocks. As for CDGZ, there is a high peak near 1 Hz, attributable to impedance contrast with layers of the Lower Cretaceous or thick layers of sediments. Likewise, in COLM there is a small peak around 10 Hz, probably due to the presence of shallow unconsolidated sediments.

3.2. Central TMVB

The Central sector is the one with the lowest instrumental density. There is a group of nine temporary stations concentrated on the southwest of the sector, and only two permanent stations that are considerably separated in the east (Figure 6). There were few earthquake records that met the established restrictions. For this reason, the results presented here only correspond to the analysis of ambient noise data.

Most of the stations in this sector were deployed on Neogene and Quaternary mafic rocks. In MA15, MA18, MA27, MA29, and MOIG, there are HVSR curves with flat shapes and no significant impedance contrast is observed, while stations MA20 and MA28 do not have clear peaks (Figure 7).

Station MA21 has a clear peak at 1 Hz and additional unclear peaks above this frequency. In this case, the shape of the HVSR curve could be related to the calderas that predominate in the area or lateral variations with sedimentary rocks from the Lower Cretaceous. Furthermore, the curve corresponding to the IGIG station shows peaks at frequencies beyond 10 Hz. This permanent station is located on Felsic rocks from Neogene, and its response could be related to thin layers of sediments.

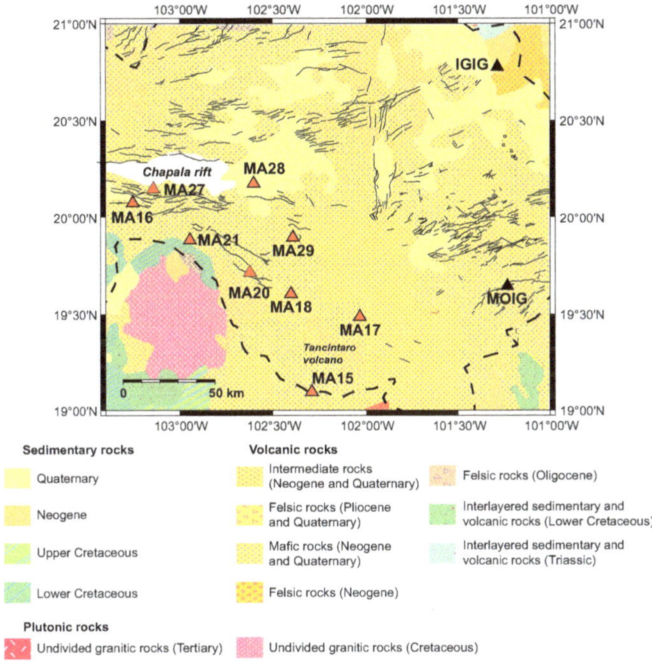

Figure 6. Geological units in the central sector of the TMVB [35]. Red triangles: MARS temporary network stations; black triangles: SSN permanent network stations. Continuous line: Main fault systems in the area; dashed line: TMVB.

In stations MA16 and MA17, there are clear peaks at 1.5 and 5 Hz, respectively. This could be attributed to impedance contrasts with layers of sediments of different thickness.

3.3. Central-Eastern TMVB

In the central-eastern zone of the TMVB is the Acambay fault system, covering some faults considered as active [40]. In the same area, there are the Taxco-San Miguel de Allende and Chapala-Tula systems, with evidence of activity in recent years. In this sector, a large number of relevant earthquakes can be mentioned over time, highlighting the following.

In the 18th century, there are reports of crustal earthquakes related to the Venta de Bravo fault. The first occurred in November 1734, followed by more than 30 strong and small events between November 1734 and March 1735 [39]. The largest crustal earthquake in the TMVB during the 20th century took place on 12 November 1912, in Acambay [13]. It had a magnitude of Mw 6.9 with a maximum intensity of IX in the epicentral area. Then, a series of 90 earthquakes were registered between February and June 1979 in a region comprising Maravatio and Mexico state. The mainshock occurred on 22 February 1979 with a magnitude Mw 5.3 and maximum intensity of VIII [40].

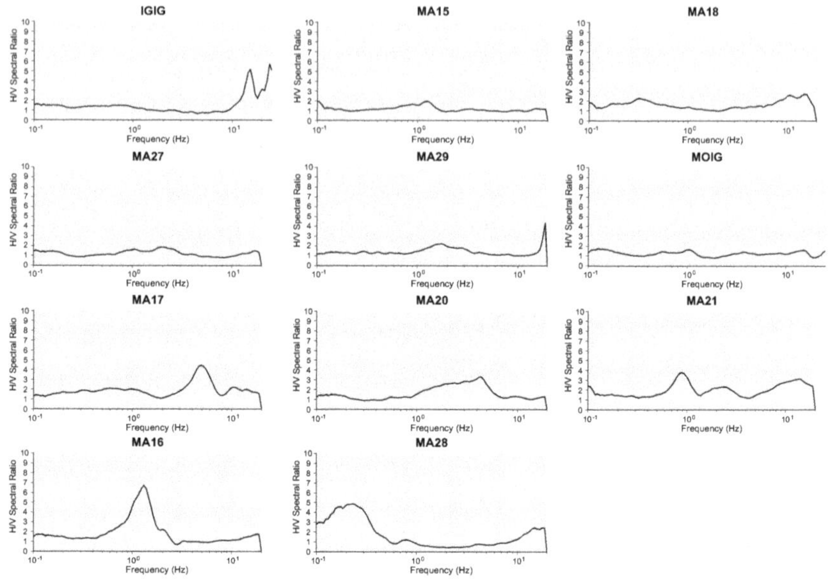

Figure 7. Average HVSR curves for stations located in the central sector of the TMVB.

Figure 8 shows the seismological stations analyzed in this sector. In this case, it was possible to analyze nine stations with earthquake data and to compare the results with the use of ambient noise. Particularly noteworthy are the stations belonging to the MASE temporary network, deployed between 2004 and 2007. Despite being numerous, due to the characteristics of the events that occurred during this period, only a few stations had earthquake records with the characteristics sought.

In this sector, there are some stations located on Neogene and Quaternary geological units (Figure 8). In JRQG, ECID, SNLU, COAC, PTRP, ESTA, and MIXC there are mostly flat HVSR curves, and no clear peaks are identified. In DHIG, CUIG, UNM, SABI, and PSIQ, the shape of the average curves obtained from ambient noise and earthquake records do not vary significantly (Figure 9). Small peaks can be observed at different frequencies that show no significant impedance contrasts or possible amplifications. As a result of this, the locations of these stations and those mentioned above may be considered as hard rock sites. In the HVSR curve corresponding to station TEPE (Figure 10), it is not possible to point out some value as the resonance frequency. This shape could be produced by lateral heterogeneity due to the vicinity of some Quaternary mafic rocks.

In some stations located on Quaternary volcanic rocks, there are almost HVSR curves of a flat shape (ACIG, PACH, MIMO, SAPE, SAPA, and TOSU). On the other hand, stations CUCE and CUNO have HVSR curves in which it is not possible to clearly identify any resonant frequency and could be considered as flat (Figure 10).

In station SALU, deployed on Quaternary sedimentary rocks, there are multiple peaks below 1 Hz and between 3 and 10 Hz (Figure 9). This could be due to the influence of topography or underlying geological units. The curve corresponding to station CHIC, located on Quaternary volcanic rocks, has multiple peaks at high frequencies, but only the one at 20 Hz can be considered as clear, indicating shallow layers of sediments. Additionally, in stations VEGU, ARBO, and VLAD, there are not very well-defined peaks between 1 and 10 Hz (Figure 10). These could be attributed to small impedance contrasts at different depths or lateral heterogeneity due to the presence of Neogene sedimentary rocks.

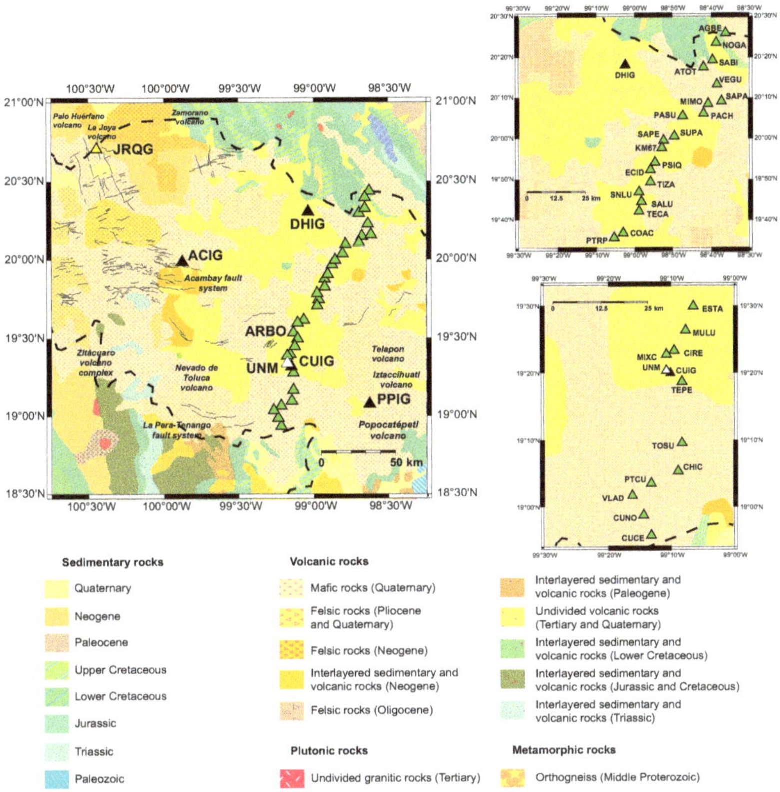

Figure 8. Geological units in the central-eastern sector of the TMVB [35]. Green triangles: MASE temporary network stations; black triangles: SSN permanent network stations; yellow triangle: CGEO permanent station; and white triangle: GEOSCOPE permanent station. Continuous line: Main fault systems in the area; dashed line: TMVB.

There are some stations in which clear peaks of varied amplitude can be identified. In the stations TECA, TIZA, and PASU, located on Quaternary sedimentary rocks, there are small peaks of a wide shape but low amplitude (Figure 9). This could indicate the presence of deposits of different stiffness and width, which by consolidation time have no significant contrasts with the underlying bedrock. The MULU and CIRE stations are special cases since the average HVSR curves present very high peaks at frequencies below 1 Hz. These responses can be explained by the fact that these temporary stations were located in the lakebed zone of Mexico City. In this area, there are reports of devastating effects during the earthquakes of 1985 and 2017.

At some of the remaining stations deployed on Quaternary volcanic rocks, it was possible to perform the analyses with ambient noise and earthquake data (NOGA, AGBE, KM67, and PPIG). The shape of the average HVSR curves obtained did not differ significantly and the amplitude only increased when earthquake records were used (Figure 10). A particularly special case is that of station KM67, where the HVSR curve obtained using ambient noise data is flat. However, when doing analysis with earthquake data, a clear peak is observed, revealing a significant impedance contrast. At this site, the energy associated with the seismic movement may have produced reverberations in underlying Cretaceous layers. In the case of the PPIG station, located at the Popocatépetl volcano, there is a peak above 10 Hz using ambient noise and earthquake data. Likewise, a peak below 1 Hz can be observed in the analysis with earthquake records. In ATOT, SUPA, and

PTCU there are clear peaks between 1 and 3 Hz that may be due to impedance contrasts with Cretaceous layers.

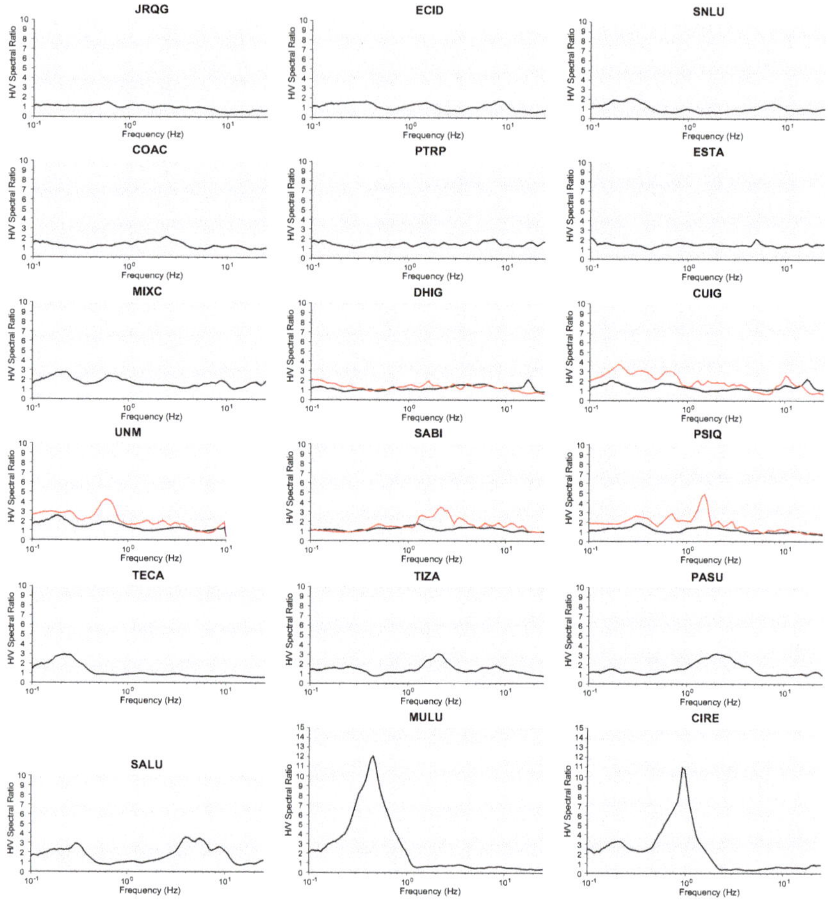

Figure 9. Average HVSR curves for some stations located in the central-eastern sector of the TMVB. Blackline: HVSR results using ambient noise records; redline: HVSR results using earthquake records.

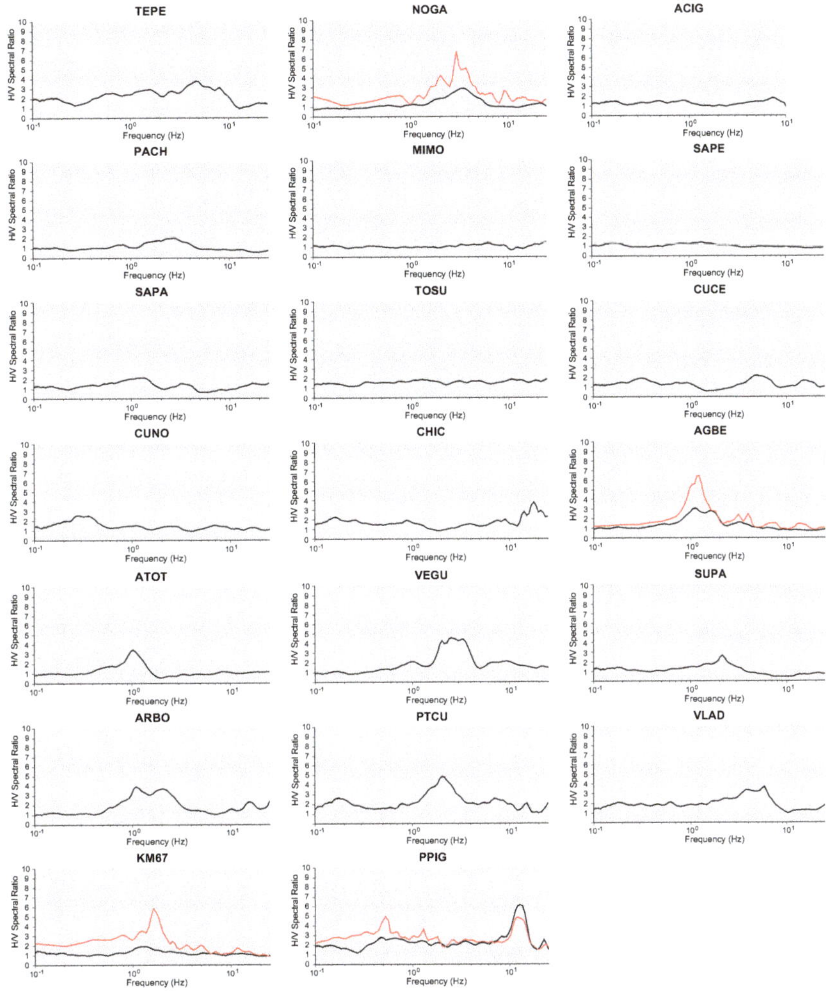

Figure 10. Average HVSR curves for some stations located in the central-eastern sector of the TMVB. Blackline: HVSR results using ambient noise records; redline: HVSR results using earthquake records.

3.4. Eastern TMVB

Regarding relevant historic earthquakes in this sector, it can be mentioned the one occurred in 1546 that caused important damage in the city of Jalapa and nearby towns. There is no specific date reported for this event nor wider descriptions in the historical records, but it is known that the first Catholic church built in America was destroyed [41]. Later, on 4 January 1920 an earthquake with magnitude Mw 6.4 occurred in the city of Jalapa. This event caused structural damage in nearby towns and it seconds the 1985 Michoacán earthquake in terms of fatalities.

Figure 11 shows the seismological stations analyzed in this sector. Most of them belong to the temporary network GECO and only two to the permanent network of the SSN.

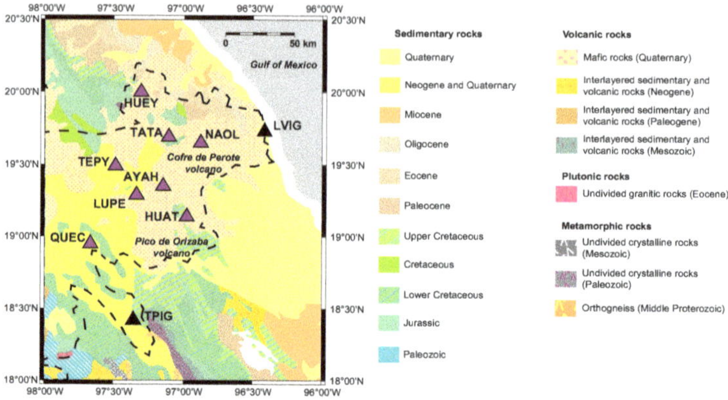

Figure 11. Geological units in the eastern sector of the TMVB [35]. Purple triangles: GECO temporary network stations; black triangles: SSN permanent network stations. Dashed line: TMVB.

In this sector, it was possible to analyze almost half of the stations with earthquake data (Figure 12). In the case of the LVIG station, no significant differences are found between the curves obtained using ambient noise and the ones obtained through earthquake data. As well as in this station, in AYAH and TPIG, no contrast of impedances is observed, and the shape of the curves is almost flat, indicating hard rock sites.

Moreover, TEPY was located on Quaternary mafic rocks, while QUEC and LUPE on Neogene and Quaternary sedimentary rocks. In these stations, there are multiple peaks. For station TEPY it is not possible to identify a value for the resonant frequency. As for QUEC and LUPE, there is a peak near 10 Hz, which may indicate the presence of thin deposits of sediments. Likewise, the peaks at lower frequencies could be due to the influence of sedimentary rocks from the Neogene and Upper Cretaceous.

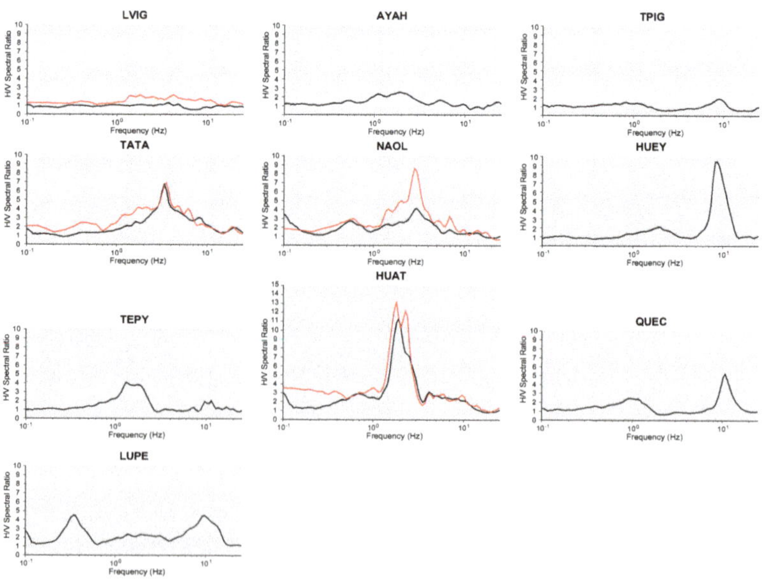

Figure 12. Average HVSR curves for stations located in the Eastern sector of the TMVB. Blackline: HVSR results using ambient noise records; redline: HVSR results using earthquake records.

Stations TATA, NAOL, HUEY, and HUAT are on Quaternary mafic rocks and have varied fundamental frequencies. The observed clear peaks could be attributed to impedance contrasts related to layers of sediments with varied thickness. In TATA and HUAT, there are visible peaks between 1 and 10 Hz. The shape of these curves does not vary significantly when using earthquake and ambient noise data. In the case of the NAOL station, the amplitude of the peak is increased but its location in the considered frequency range does not change.

4. Conclusions

From the results obtained in this study, it is observed that the site response in terms of the fundamental frequency f_0 had a wide variety throughout the TMVB and there was no visible correlation between the shape of the HVSR curves and extent of any particular geological unit. In almost 46% of the stations analyzed, the response was mostly flat or it was not possible to identify clear peaks. In most of these cases, no significant impedance contrasts were observed, so they could be considered as hard rock sites. There was another small group of stations distributed in the four sectors, in which multiple peaks were observed in the HVSR curves. Some of these peaks could be identified as spurious or attributable to the influence of underlying layers of varying stiffness and thickness.

Additionally, in approximately 36% of the analyzed sites there were clear peaks of varied amplitude in the HVSR curves. In these cases, it was possible to identify the value of the resonance frequency, which was mostly between 1 and 10 Hz. Similarly, in some sites there were peaks in frequencies lower than 1 Hz, which should not be underestimated considering the modern trend of tall buildings in the populated areas of the TMVB.

To date, most of the seismicity studies in Mexico have focused on earthquakes originated on the Pacific Ocean coast. Except for Mexico City, due to its geotechnical properties, the amplitude of these events usually reaches the TMVB considerably attenuated. For this reason, this study focused on analyzing what happens in this zone with regional earthquakes. In this regard, it is important to note that there were sites with high peaks in the four sectors analyzed, not only in the central-eastern sector where Mexico City is located.

In some stations, it was possible to make a comparison of the results obtained from ambient noise and earthquake records. In most of them, there were no significant differences in the shape of the average HVSR curves. In the analyses carried out with seismic records, no evidence of non-linear behavior was identified in the site response with respect to the ambient noise results. This could be attributed to the fact that most of the events considered are of low magnitude. In general, there are few major earthquakes with both epicenter and instrumental records inside the TMVB. However, there were some sites with a flat HVSR ambient noise curve in which new peaks appeared when analyzing earthquake data. Considering that the energy associated with ambient noise is not comparable to that associated with a seismic event, it is possible that several sites considered as hard rock may have significant amplifications during the occurrence of an earthquake.

Based on these observations, we consider it is necessary to increase the density of the permanent seismic instrumentation in the TMVB. This will allow to have greater coverage, better quality records, and the possibility of carrying out more analyses using earthquake data, including the SSR technique. A greater permanent seismic instrumentation would allow obtaining isofrequency maps, which could be used in soil-structure interaction analyses and as a basis for estimating seismic hazards and risk assessment in the TMVB. This would also allow to better delimit the risk of populated zones based on local seismicity, considering the significant growth that the cities in the area have had in recent years.

Author Contributions: Conceptualization, L.F.P.-M., Q.R.-P. and F.R.Z.; methodology, L.F.P.-M.; formal analysis, L.F.P.-M.; investigation, L.F.P.-M. and M.A.P.-L.; writing—original draft preparation, L.F.P.-M.; writing—review and editing, Q.R.-P., F.R.Z., J.H.-R. and M.d.l.L.P.-R.; supervision, J.H.-R., M.d.l.L.P.-R. and M.A.P.-L. All authors have read and agreed to the published version of the manuscript.

Funding: This research received no external funding. The APC was funded by L.F.P.-M.

Institutional Review Board Statement: Not applicable.

Informed Consent Statement: Not applicable.

Acknowledgments: Special thanks to the Mexican National Council for Science and Technology (CONACYT) for the support provided to L. Francisco Pérez-Moreno during the realization of his Ph.D. studies and to Quetzalcoatl Rodríguez-Pérez in the Cátedras program (project 1126). SSN data was obtained by the Servicio Sismológico Nacional (México). Station maintenance, data acquisition, and distribution is thanks to its personnel. The authors thank Antonio de J. Mendoza Carvajal and Jorge A. Real Pérez for the maintenance and operation of the Geometry of the Cocos Plate (GECO) array. This network was financed by PAPIIT (projects IN110913, IN105816, and IN106119) and CONACYT (project 177676). We are also particularly grateful to Xyoli Pérez-Campos for providing access to the network data. We appreciate the contributions of the four anonymous reviewers of the initial version of this paper. Their valuable comments allowed us to improve this study.

Conflicts of Interest: The authors declare no conflict of interest.

References

1. Borcherdt, R.D. Effects of local geology on ground motion near San Francisco bay. *Bull. Seismol. Soc. Am.* **1970**, *60*, 29–61.
2. Nakamura, Y. A method for dynamic characteristics estimation of subsurface using microtremor on the ground surface. *Railw. Tech. Res. Inst. Q. Rep.* **1989**, *30*, 25–33.
3. Perron, V.; Gélis, C.; Froment, B.; Hollender, F.; Bard, P.Y.; Cultrera, G.; Cushing, E.M. Can broad-band earthquake site responses be predicted by the ambient noise spectral ratio? Insight from observations at two sedimentary basins. *Geophys. J. Int.* **2018**, *215*, 1442–1454. [CrossRef]
4. Gosar, A. Study on the applicability of the microtremor HVSR method to support seismic microzonation in the town of Idrija (W Slovenia). *Nat. Hazards Earth Syst. Sci.* **2017**, *17*, 925–937. [CrossRef]
5. Madiai, C.; Facciorusso, J.; Gargini, E. Numerical Modeling of Seismic Site Effects in a Shallow Alluvial Basin of the Northern Apennines (Italy). *Bull. Seismol. Soc. Am.* **2017**, *107*, 2094–2105. [CrossRef]
6. Stanko, D.; Gülerce, Z.; Markušić, S.; Šalić, R. Evaluation of the site amplification factors estimated by equivalent linear site response analysis using time series and random vibration theory based approaches. *Soil Dyn. Earthq. Eng.* **2019**, *117*, 16–29. [CrossRef]
7. Sun, Q.; Guo, X.; Dias, D. Evaluation of the seismic site response in randomized velocity profiles using a statistical model with Monte Carlo simulations. *Comput. Geotech.* **2020**, *120*, 103442. [CrossRef]
8. Falcone, G.; Romagnoli, G.; Naso, G.; Mori, F.; Peronace, E.; Moscatelli, M. Effect of bedrock stiffness and thickness on numerical simulation of seismic site response. Italian case studies. *Soil Dyn. Earthq. Eng.* **2020**, *139*, 106361. [CrossRef]
9. Singh, S.K.; Quintanar-Robles, L.; Arroyo, D.; Cruz-Atienza, V.C.; Espíndola, V.H.; Bello-Segura, D.I.; Ordaz, M. Lessons from a Small Local Earthquake (Mw 3.2) That Produced the Highest Acceleration Ever Recorded in Mexico City. *Seismol. Res. Lett.* **2020**, *91*, 3391–3406. [CrossRef]
10. Ferrari, L.; Orozco-Esquivel, T.; Vlad, C.; Marina, M. The dynamic history of the Trans-Mexican Volcanic Belt and the Mexico subduction zone. *Tectonophysics* **2012**, *522–523*, 122–149. [CrossRef]
11. Suárez, G.; Caballero-Jiménez, G.V.; Novelo-Casanova, D.A. Active Crustal Deformation in the Trans-Mexican Volcanic Belt as Evidenced by Historical Earthquakes During the Last 450 Years. *Tectonics* **2019**, *38*, 3544–3562. [CrossRef]
12. Suárez, G.; Ruiz-Barón, D.; Chico-Hernández, C.; Zúñiga, F.R. Catalog of Preinstrumental Earthquakes in Central Mexico: Epicentral and Magnitude Estimations Based on Macroseismic Data. *Bull. Seismol. Soc. Am.* **2020**, *110*, 3021–3036. [CrossRef]
13. Suter, M.; Carrillo-Martínez, M.; Quintero, O. Macroseismic Study of Shallow Earthquakes in the Central and Eastern Parts of the Trans-Mexican Volcanic Belt, Mexico. *Bull. Seismol. Soc. Am.* **1996**, *86*, 1952–1963.
14. Sahakian, V.J.; Melgar, D.; Quintanar, L.; Ramírez-Guzmán, L.; Pérez-Campos, X.; Baltay, A. Ground motions from the 7 and 19 September 2017 Tehuantepec and Puebla-Morelos, Mexico, earthquakes. *Bull. Seismol. Soc. Am.* **2018**, *108*, 3300–3312. [CrossRef]
15. Ordaz, M.; Singh, S.K. Source spectra and spectral attenuation of seismic waves from Mexican earthquakes, and evidence of amplification in the hill zone of Mexico City. *Bull. Seismol. Soc. Am.* **1992**, *82*, 24–43.
16. García, D.; Singh, S.K.; Herráiz, M.; Pacheco, J.F.; Ordaz, M. Inslab earthquakes of Central Mexico: Q, source spectra, and stress drop. *Bull. Seismol. Soc. Am.* **2004**, *94*, 789–802. [CrossRef]
17. Singh, S.K.; Iglesias-Mendoza, A.; García, D.; Pacheco, J.F.; Ordaz, M. Q of Lg waves in the central Mexican Volcanic Belt. *Bull. Seismol. Soc. Am.* **2007**, *97*, 1259–1266. [CrossRef]
18. Lozano, L.; Herráiz, M.; Singh, S.K. Site effect study in central Mexico using H/V and SSR techniques: Independence of seismic site effects on source characteristics. *Soil Dyn. Earthq. Eng.* **2009**, *29*, 504–516. [CrossRef]
19. Clemente-Chávez, A.; Zúñiga, F.R.; Lermo, J.; Figueroa-Soto, A.; Valdés-González, C.; Montiel, M.; Chávez, O.; Arroyo, M. On the behavior of site effects in central Mexico (the Mexican volcanic belt-MVB), based on records of shallow earthquakes that occurred in the zone between 1998 and 2011. *Nat. Hazards Earth Syst. Sci.* **2014**, *14*, 1391–1406. [CrossRef]

20. Zúñiga, F.R.; Lacan, P.; Rodríguez-Pérez, Q.; Márquez-Ramírez, V.H. Temporal and spatial evolution of instrumented seismicity in the Trans-Mexican Volcanic Belt. *J. S. Am. Earth Sci.* **2020**, *98*, 102390. [CrossRef]
21. Ruggieri, S.; Tosto, C.; Rosati, G.; Uva, G.; Ferro, G.A. Seismic Vulnerability Analysis of Masonry Churches in Piemonte after 2003 Valle Scrivia Earthquake: Post-event Screening and Situation 17 Years Later. *Int. J. Archit. Herit.* **2020**, 1–29. [CrossRef]
22. Godínez-Domínguez, E.A.; Tena-Colunga, A.; Pérez-Rocha, L.E.; Archundia-Aranda, H.I.; Gómez-Bernal, A.; Ruiz-Torres, R.P.; Escamilla-Cruz, J.L. The September 7, 2017 Tehuantepec, Mexico, earthquake: Damage assessment in masonry structures for housing. *Int. J. Disaster Risk Reduct.* **2021**, *56*, 1–29. [CrossRef]
23. Catálogo de Sismos. Available online: http://www2.ssn.unam.mx:8080/catalogo/ (accessed on 16 October 2020).
24. Search Earthquake Catalog. Available online: https://earthquake.usgs.gov/earthquakes/search/ (accessed on 16 October 2020).
25. ISC Bulletin: Event Catalogue Search. Available online: http://www.isc.ac.uk/iscbulletin/search/catalogue/ (accessed on 16 October 2020).
26. GEOSCOPE. *French Global Network of Broad Band Seismic Stations*; Institut de Physique du Globe de Paris (IPGP) and Ecole Et Observatoire Des Sciences De La Terre De Strasbourg (EOST). [CrossRef]
27. Meso America Subduction Experiment. Caltech. Dataset. [CrossRef]
28. Pérez-Campos, X. MASE: Undergraduate Research and Outreach as Part of a Large Project. *Seismol. Res. Lett.* **2008**, *92*, 232–236. [CrossRef]
29. West, M. The Colima Deep Seismic Experiment: Imaging the Magmatic Root of Colima Volcano. *Int. Fed. Digit. Seismogr. Netw.* **2006**.
30. Yang, T.; Grand, S.P.; Wilson, D.; Guzman-Speziale, M.; Gomez-Gonzalez, J.M.; Dominguez-Reyes, T. Seismic structure beneath the Rivera subduction zone from finite-frequency seismic tomography. *J. Geophys. Res. Solid Earth* **2009**, *114*, B01302. [CrossRef]
31. Rodríguez-Domínguez, M.; Pérez-Campos, X.; Montealegre-Cázares, C.; Clayton, R.W.; Cabral-Cano, E. Crustal structure variations in south-central Mexico from receiver functions. *Geophys. J. Int.* **2019**, *219*, 2174–2186. [CrossRef]
32. Wathelet, M.; Chatelain, J.L.; Cornou, C.; Giulio, G.D.; Guillier, B.; Ohrnberger, M.; Savvaidis, A. Geopsy: A user-friendly open-source tool set for ambient vibration processing. *Seismol. Res. Lett.* **2020**, *91*, 1878–1889. [CrossRef]
33. Konno, K.; Ohmachi, T. Ground-motion characteristics estimated from spectral ratio between horizontal and vertical components of microtremor. *Bull. Seismol. Soc. Am.* **1998**, *88*, 228–241.
34. SESAME Team. *Guidelines for the Implementation of the H/V Spectral Ratio Technique on Ambient Vibrations: Measurements, Processing and Interpretation*; SESAME European Research Project. 2004. Available online: http://sesame.geopsy.org/Papers/HV_User_Guidelines.pdf (accessed on 15 August 2020).
35. Reed, J.C.; Wheeler, J.O.; Tucholke, B.E. *Decade of North American Geology Geologic Map of North America—Perspectives and Explanation*; The Geological Society of America: Boulder, CO, USA, 2015.
36. Suter, M. The A.D. 1567 Mw 7.2 Ameca, Jalisco, Earthquake (Western Trans-Mexican Volcanic Belt): Surface Rupture Parameters, Seismogeological Effects, and Macroseismic Intensities from Historical Source. *Bull. Seismol. Soc. Am.* **2015**, *105*, 646–656. [CrossRef]
37. Suter, M. The 1563 MI 8 Puerto de la Navidad Subduction-Zone and 1567 Mw 7.2 Ameca Crustal Earthquakes (Western Mexico): New Insights from Sixteenth-Century Sources. *Seismol. Res. Lett.* **2019**, *90*, 366–375. [CrossRef]
38. Suter, M. The 2 October 1847 MI 5.7 Chapala Graben Triggered Earthquake (Trans-Mexican Volcanic Belt, West-Central Mexico): Macroseismic Observations and Hazard Implications. *Seismol. Res. Lett.* **2017**, *89*, 35–46. [CrossRef]
39. Lacan, P.; Aguirre-Díaz, G.; Audin, L.; Perea, H.; Baize, S.; Zúñiga, F.R.; Ortuño, M. Sedimentary evidence of historical and prehistorical earthquakes along the Venta de Bravo Fault System, Acambay Graben (Central Mexico). *Sediment. Geol.* **2018**, *365*, 62–77. [CrossRef]
40. Rodríguez-Pérez, Q.; Zúñiga, F.R. Seismicity characterization of the Maravatío-Acambay and Actopan regions, central Mexico. *J. S. Am. Earth Sci.* **2017**, *76*, 264–275. [CrossRef]
41. Suárez, G.; Novelo-Casanova, D.A. A Pioneering Aftershock Study of the Destructive 4 January 1920 Jalapa, Mexico, Earthquake. *Seismol. Res. Lett.* **2018**, *89*, 1894–1899. [CrossRef]

applied sciences

MDPI

Article

Influence of Rainfall Intensity and Slope on the Slope Erosion of Longling Completely Weathered Granite

Haojun Tian [1,2] and Zhigang Kong [1,2,*]

1 Faculty of Land Resources Engineering, Kunming University of Science and Technology, Kunming 650093, China
2 MNR Key Laboratory of Geohazard Forecast and Geoecological Restoration in Plateau Mountainous Area, Kunming 650093, China
* Correspondence: zhigangkong@kust.edu.cn

Abstract: Serious slope erosion occurs in the distribution areas of fully weathered granites, and rainfall intensity and slope gradient are important factors affecting slope erosion. In this study, we investigate the erosion characteristics of Longling completely weather granites with a focus on the effects of rainfall intensity and slope gradient. Based on an indoor 60-min simulated rainfall test, we selected four slope gradients (10°, 20°, 30°, and 40°) and three rainfall intensities (50, 80, and 110 mm/h) for evaluation. A total of 12 groups of tests were conducted to analyze the erosion and surface hydrodynamic characteristics of the completely weathered granite slope. The results indicate a significant positive correlation between rainfall intensity and slope gradient, and the correlation between rainfall intensity and flow velocity became stronger as the slope gradient increased. The peak sediment yield rate represents the moment at which the change in slope shape is maximized. After the peak appears, the slope will no longer undergo great deformation, and the sediment yield rate will decrease and then become stable. Finally, rainfall intensity and slope gradient, which are the two key factors that determine slope flow velocity, are described using a binary function. The findings provide a reference for the study of slope erosion in completely weathered granites.

Keywords: completely weathered granite; runoff; sediment production; slope erosion; flow velocity hydrodynamic characteristics

check for updates

Citation: Tian, H.; Kong, Z. Influence of Rainfall Intensity and Slope on the Slope Erosion of Longling Completely Weathered Granite. *Appl. Sci.* **2023**, *13*, 5295. https://doi.org/10.3390/app13095295

Academic Editors: Miguel Llorente Isidro, Ricardo Castedo and David Moncoulon

Received: 2 March 2023
Revised: 16 April 2023
Accepted: 20 April 2023
Published: 23 April 2023

1. Introduction

Slope erosion is the result of an interactive process between slope water flow and soil, leading to erosion, transportation, and removal of soil from the slope surface under the influence of water. This phenomenon results in soil fertility decline, soil erosion, and slope damage. Numerous studies have demonstrated that the hydraulic characteristics of slopes are the primary factors that influence slope erosion [1] with rainfall intensity and slope gradient being the most crucial [2,3]. Changes in these factors can significantly alter slope runoff affecting sediment transport capacity and erosion rate [4,5]. Additionally, the effect of slope runoff on erosion varies depending on soil type. Therefore, understanding the hydraulic characteristics of slope erosion is crucial for preventing and managing slope erosion.

To better understand the hydraulic characteristics of slope runoff, scholars have utilized indoor rainfall simulation tests to investigate the mechanisms [6] of slope erosion by varying rainfall intensity and slope gradient [7,8]. Extensive research has indicated a positive correlation between slope runoff velocity and both rainfall intensity and slope gradient [9,10]. As rainfall intensity increases, the water content in slope runoff also increases, while an increase in slope gradient accelerates the runoff's speed under gravity, leading to heightened impact and erosion on the soil surface. Dong et al. [11] studied the soil accumulation on highway slopes and found that sediment yield increases with increasing rainfall intensity. Niu et al. [12] conducted indoor experiments on different slope ratios and initial seepage fields under rainfall infiltration to investigate the failure mechanism and pattern of fly ash dam slopes under rainfall influence. Yan et al. [13] conducted simulated rainfall experiments under various typical underlying surface conditions to study the runoff characteristics and mechanisms under different rainfall intensities and durations. In addition, researchers have also studied thin-layer water flow on slopes [14–16]. Yang et al. [17] conducted artificial rainfall simulation experiments under varying slope gradients and rainfall intensities to determine the velocity law of thin-layer water flow. Various hydraulic parameters, including Reynolds number, Froude number, Manning coefficient, and hydraulic power, have been applied to slope erosion research [18,19]. Yuan et al. [20] studied the soil erosion and hydraulic characteristics of the loess slope in Beijing and found that mean flow velocity is the most closely related hydraulic parameter to sediment concentration in runoff.

Slope erosion, a vital aspect of soil erosion, has garnered significant attention in China. Nevertheless, the patterns of erosion differ considerably among various soil types. Presently, research on slope erosion of diverse soil types focuses mainly on the Northwest loess region [21,22] and southern red soil region [23,24]. In the loess region, the loose soil texture and high porosity render it vulnerable to wind and water erosion [25,26], leading to severe slope erosion. Zhao et al. [27] calculated the slope, aspect, and channel network of loess slopes in different evolution stages and analyzed the relationship between variation characteristics and erosion and evolution processes of the slopes. In the southern red soil region, the red soil in the Jiangnan hilly area and Nanling Mountains are the primary research objects [28,29]. Red soil is characterized by high fertility, loose structure, and sensitivity to erosion [30,31], making it susceptible to rainfall erosion. Feng et al. [32] studied the impacts of rainfall intensity, slope gradient, and surface cover on the erosion process of granite red soil slopes and concluded that the impact of slope gradient on sediment yield increases with increasing rainfall intensity. However, in granite regions, erosion characteristics and mechanisms vary due to differences in granite weathering degree, resulting in limited applicability of slope management measures [33–35].

The northeast of Longling County in Yunnan Province is a distribution area of completely weathered granite and also an operating area of oil and gas pipelines. The oil and gas pipelines pass through the distribution area of completely weathered granite. To lay the pipelines, grooves need to be dug and then the completely weathered granite excavated from the grooves is backfilled. However, the backfill soil containing completely weathered granite has high sand content, low clay content, loose structure, poor physical and mechanical parameters, and low fertility. As a result, the working slope surface is exposed, leading to poor corrosion resistance. Under heavy rainfall conditions, the working slope is prone to erosion by slope runoff, resulting in slope collapse, surface water and soil loss, and other disasters. These problems result in pipeline exposure and suspension, which substantially impact the normal functioning of pipelines. However, traditional management methods have often proved insufficient, emphasizing the urgency of responding to pipeline safety challenges and ensuring the continuous operation of pipelines. It is crucial to conduct focused research on the completely weathered granite area, which involves determining the erosion characteristics and patterns of completely weathered granite slopes at a certain slope gradient and rainfall intensity. In addition, it is necessary to determine the spatiotemporal variation properties of flow velocity and hydraulic laws under different circumstances. This will ultimately provide a theoretical basis for soil erosion management of completely weathered granite slopes.

2. Materials and Methods

2.1. Experimental Design

According to the analysis of rainfall data in the Longling area, three rainfall intensity levels were selected for the experiments: 50, 80, and 110 mm/h (50, 80, and 110 mm/h are the average value of the maximum single-point rainstorm in the past 10 years, the calculated single-point rainstorm value in 10 years and the single-point rainstorm value in 100 years. As well, the rainfall device can be set at a level of 20–240 mm/h). Under each of these three rainfall intensity levels, four slope gradients (10°, 20°, 30°, and 40°) were experimentally evaluated resulting in a total of 12 test configurations. During each experiment, the erosion process of the simulated slope surface was monitored in real-time using a digital pan-tilt camera until the slope was damaged. The soil moisture contents and bulk densities were measured by drying methods before filling. The soil dry densities were controlled between 1.3 and 1.5 g/cm^3, and the soil moisture contents were approximately 10% (Near natural state values). The test soil was naturally dried and passed through a 10 mm sieve to remove stones and weeds. The filling thickness was 50 cm, paved, and compacted to reduce the influence of the boundary effect. After the filling was complete, the soil surface was scraped with a wooden board to create a certain surface roughness. The soil bulk density of the slope was measured using the ring knife method to verify that it reached the experimental design level. Each experiment lasted 1 h.

2.2. Test Materials and Devices

The soil used in this experiment was trench backfill soil taken from the completely weathered granite distribution area of the Longling section of the China-Myanmar oil and gas pipeline in Longling County, Baoshan City, Yunnan Province. The sampling time was July 2020.

For simulated rainfall experiments, a down-spraying ZYJY-DZ02 system was used. The rainfall source was 5.0 m, and the rainfall source was pure, non-polluted water. The adjustable-slope steel groove used in the simulated rainfall experiments was 3 m in length, 0.5 m in width, and 0.5 m in height. A V-shaped collecting port was placed at the tail of the soil bin to collect runoff and sediment samples generated by the simulated rainfall. The rainfall intensity could be controlled within the range of 20–270 mm/h using a switch in the control room, and the rainfall uniformity remained above 85%. The rainfall amount and duration could be controlled using a combination of the pressure pump and nozzle. Thus, the experimental rainfall system met the requirements for our experiments. The test device is shown in Figure 1.

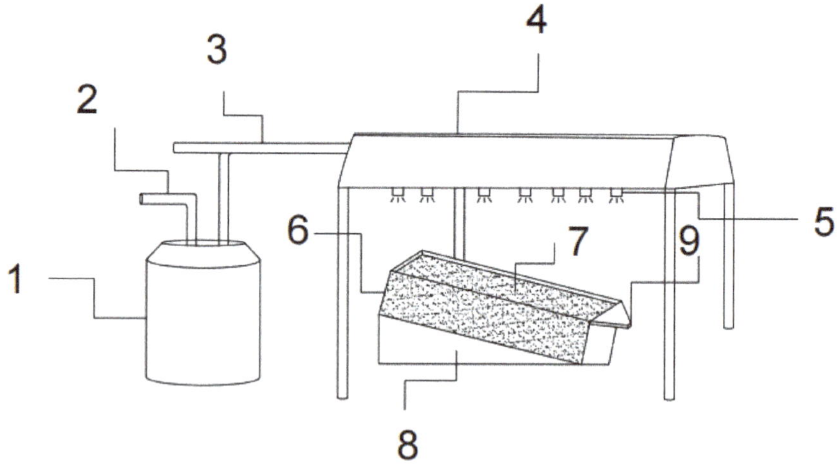

Figure 1. Schematic diagram of the system for indoor simulated rainfall experiments. 1—Water storage tank; 2—Water supply pipe; 3—Pressure stabilizing water pump, flow meter, and water valve; 4—Rainfall simulator; 5—Rainfall nozzle; 6—Test bed; 7—Completely weathered granite for testing; 8—Slope adjustment device; 9—Triangular weir.

2.3. Test Procedure

The experiments were conducted in the Soil and Water Conservation Laboratory of Kunming University of Science and Technology from August to October 2020. After the rain begins, the initial runoff time of each test was recorded, and the erosion process was observed. Each rainfall event lasted 60 min, The flow rate was measured every 3 min and a 1 L container was used to collect runoff sediment samples every 2 min (for a total of 30 samples in each test). After the slope flow became stable, a tracer method based on red ink was used to measure the slope runoff velocity. The entire test process was recorded using a high-speed camera, and the flow velocity was calculated from the recorded images. Measurements of flow velocity should consider different slope sections. The 3 m soil trough was measured from the top of the slope to the foot of the slope at 0.5–1.5 m and 1.5–2.5 m from the top of the slope. After the rainfall test, the volume of the sample in the container was measured. After the sample was allowed to stand for 24 h to settle, the supernatant was poured off, and all the sediment in the container was transferred to a disposable paper cup for drying to determine the dry weight of the sediment.

2.4. Test Data Analysis

The hydrodynamic parameters of the slope runoff were determined as follows.

(1) The runoff yield rate N (mL/s) was calculated as

$$N = \frac{q}{t},$$

where q is the runoff (mL) in a single simulated rainfall event, and t is the single sampling time (s).

(2) The sediment yield rate M (g/s) was calculated as

$$M = \frac{M'}{t},$$

where M' is the sediment yield in a single sampling time (g), and t is the single sampling time (min).

(3) The Reynolds number Re was calculated as

$$Re = \frac{vR}{\mu},$$

where v is the flow rate (m/s), R is the hydraulic radius (m), and μ is the viscosity coefficient of water flow (Based on the field test, the water temperature is 16–18 °C, μ = 1.061–1.115).

(4) The Froude number Fr was calculated as

$$Fr = \frac{v}{(gh)^{0.5}},$$

where g is the acceleration due to gravity (m/s^2) (Based on the test site, g = 9.79), and h is the water depth (m).

(5) The resistance coefficient f was calculated as

$$f = \frac{8ghJ}{v^2},$$

where g is the acceleration due to gravity (m/s^2), h is the water depth (m), and J is the hydraulic break (J = $sin\beta$, where β is the slope gradient).

(6) The Manning coefficient n was calculated as

$$n = \frac{R^{2/3}S^{1/2}}{V},$$

where R is the hydraulic radius (m), S is the slope gradient (radians), and V is the average velocity (m/s).

3. Results and Analysis

This section may be divided into subheadings. It should provide a concise and precise description of the experimental results, their interpretation, as well as the experimental conclusions that can be drawn.

3.1. Process of Runoff Formation

As shown in Figure 2, a higher intensity of precipitation results in a more pronounced surge in the runoff rate. Taking a slope gradient of 10° as an instance, the peak runoff rates achieve at precipitation intensities of 50, 80, and 110 mm/h are 16.3, 31.2, and 38.5 mL/s, respectively, with corresponding average runoff rates of 13.6, 27.7, and 32.7 mL/s. Notably, the augmentation of both peak and average runoff rates is substantial when the precipitation intensity increases from 50 to 80 mm/h. However, this enhancement is not linear, and the peak runoff rate experiences only a 23% surge, while the average runoff rate increases by less than 20% as the precipitation intensity elevates from 80 to 110 mm/h.

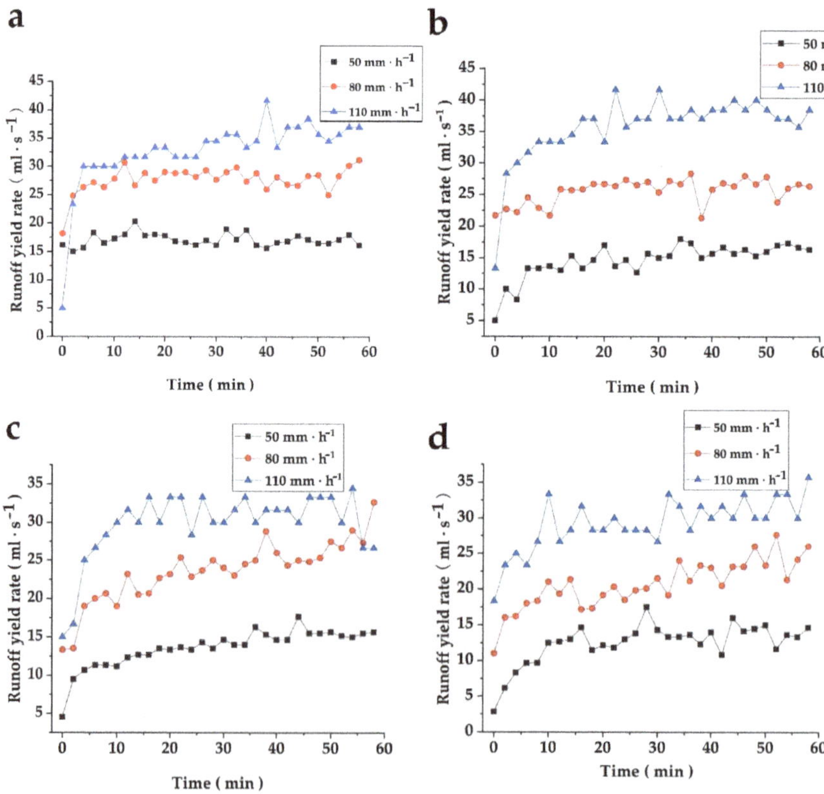

Figure 2. Variation in runoff rate with time under different rainfall intensities for the same slope gradients. (**a**) Slope is 10°. (**b**) Slope is 20°. (**c**) Slope is 30°. (**d**) Slope is 40°.

It is also worth mentioning that, for a fixed precipitation intensity, the impact of slope gradient on the runoff rate is relatively minor. Specifically, when the precipitation intensity is 110 mm/h, the average runoff rate undergoes a modest 19% escalation as the slope gradient rises from 20° to 30°. Subsequently, as the slope gradient increases from 30° to 40°, the minimum increase in average runoff rate (8%) is observed. Consequently, it can be inferred that the rise in runoff rate with increasing slope gradient diminishes substantially once the slope gradient exceeds 30°.

For a given slope gradient, the runoff rates associated with distinct rainfall intensities exhibit a descending order of 110 mm/h > 80 mm/h > 50 mm/h. Specifically, the runoff rate initially rises as rainfall intensity grows, followed by a stabilization of the runoff rate. This phenomenon is attributed to the low soil water content during the initial rainfall stage, leading to the transformation of rainfall into seepage. As the soil water content gradually increases, the slope water content approaches saturation, and the rainfall is converted into slope runoff.

Compared to the slope gradient, rainfall intensity has a more significant effect on the runoff rate. Specifically, the escalation of rainfall intensity leads to an augmented amount of precipitation on the slope, leading to a corresponding increase in the runoff volume. In contrast, a rise in the slope gradient alters the slope's shape, resulting in greater downward acceleration and flow velocity of the water traveling down the slope.

3.2. Sediment Production Process

As shown in Figure 3, an evident rise in the sediment yield rate occurs with an increase in the rainfall intensity for the same slope gradient. For instance, at a slope gradient of 40°, the peak sediment yield rates attained at rainfall intensities of 50, 80, and 110 mm/h are 0.81, 1.47, and 3.07 g/s, respectively, with corresponding average sediment yield rates of 0.24, 0.81, and 2.25 g/s, respectively. When the rainfall intensity elevates from 50 to 80 mm/h, the peak sediment yield rate experiences an 81% surge, and the average sediment yield rate increases by 235%. Similarly, when the rainfall intensity grows from 80 to 110 mm/h, the peak sediment yield rate undergoes a 107% rise, and the average sediment yield rate increases by 178%.

Under uniform rainfall intensity, a minor elevation in the slope gradient leads to a slight rise in the sediment yield rate. The most significant increment in peak sediment yield rate with increasing slope is detected under the rainfall intensity of 110 mm/h, with a substantial surge observed as the slope gradient escalates from 30° to 40°. Conversely, the most trivial rise in peak sediment yield rate occurs at the rainfall intensity of 80 mm/h, with only a modest increase detected as the slope gradient increases from 10° to 20°. The maximal increment in the average sediment yield rate transpires when the rainfall intensity is 110 mm/h, with a notable rise observed as the slope gradient increases. When the slope gradient increases from 30° to 40°, the average sediment yield rate undergoes a substantial escalation of 269%. In contrast, the most minor augmentation in average sediment yield rate is recorded when the rainfall intensity is 80 mm/h; with only a modest increment of 8% detected as the slope gradient elevates from 10° to 20°. For insignificant slope gradients and rainfall intensities, the effect of slope gradient on sediment yield rate is not apparent. However, as the slope gradient and rainfall intensity increase, the sediment yield rate also rises, with the slightest growth rate observed under the slope gradient of 20° and the rainfall intensity of 80 mm/h. Based on these observations, it can be speculated that a critical point exists at or near these conditions.

Elevating either the rainfall intensity or slope gradient induces an augmentation in sediment yield to a certain extent. Raising the slope gradient diminishes the stability of the slope. Meanwhile, alterations in rainfall intensity modify the intensity and erosive force of slope flow, with augmented runoff leading to heightened sediment yield. The simultaneous effect of increased slope and rainfall intensity accelerates the slope erosion process, giving rise to rills and depressions on the slope surface and collapses in areas with the most severe erosion. Augmentations in both rainfall intensity and slope gradient also abbreviate the time taken to reach the peak sediment yield.

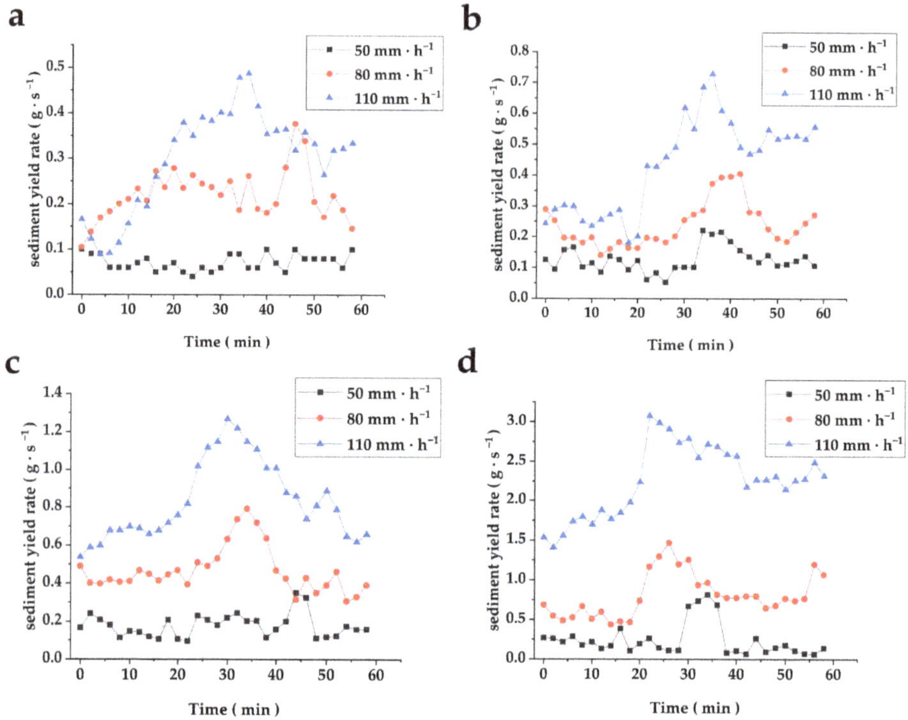

Figure 3. Changes in sediment yield rate with time under different rainfall intensities for the same slope. (**a**) Slope is 10°. (**b**) Slope is 20°. (**c**) Slope is 30°. (**d**) Slope is 40°.

Under the same slope gradient, the sediment yield rate initially rises with the intensifying rainfall intensity, subsequently declines, and ultimately stabilizes. This occurs because as the slope surface experiences erosion, forming rills, the quantity of sediment transported escalates. When the slope morphology is disrupted, a sudden collapse triggers a massive volume of sediment to be carried away by the slope flow, reaching the peak sediment yield rate. Following the initial significant damage, the slope morphology rapidly achieves stability; during this phase, the slope flow is inadequate to inflict secondary damage on the slope, causing the sediment yield rate to diminish and stabilize.

3.3. *Spatial and Temporal Differences in Flow Velocity*

As shown in Figure 4, Under the examined slope gradients, the escalation in rainfall intensity leads to substantial growth in flow rate. Using the 40° slope gradient as an example, with rainfall intensities of 50, 80, and 110 mm/h, the mean flow velocities are 0.15, 0.18, and 0.26 m/s, while the peak flow velocities are 0.17, 0.23, and 0.32 m/s, respectively. As the rainfall intensity rises from 50 to 80 mm/h, the average flow velocity experiences a 23% enhancement, and the peak velocity shows a 35% expansion. When the rainfall intensity climbs from 80 to 110 mm/h, the average flow velocity exhibits a 35% increment, and the peak velocity demonstrates a 40% amplification. Hence, the upsurge in rainfall intensity has a notable impact on the flow velocity.

The maximum flow velocity reaches 0.32 m/s, while the minimum flow velocity is a mere 0.09 m/s. The timing of the peak flow velocity does not coincide with the timing of runoff and sediment production. This is due to the formation of rills on the slope during erosion by overland flow, which alters the slope's shape, rendering it uneven and causing the overland flow to deviate from uniform laminar flow. The impact of rills on overland flow is multifaceted. Runoff within rills can continuously affect their inner walls, making them smoother, thereby diminishing resistance and augmenting flow velocity. Simultaneously, rills may also evolve along the cross-section, raising the height difference between the rill and the original slope, consequently generating a height drop in the slope flow and diminishing the flow velocity.

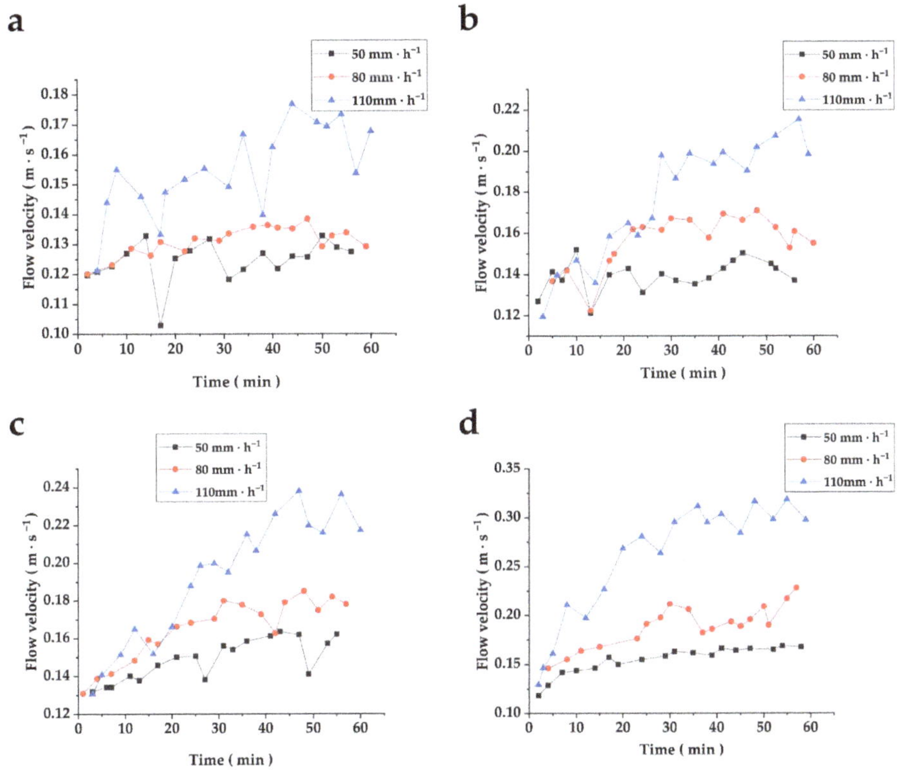

Figure 4. Comparison of slope flow velocity under different slope gradients. (**a**) Slope is 10°. (**b**) Slope is 20°. (**c**) Slope is 30°. (**d**) Slope is 40°.

During simulated rainfall, the flow velocity on the slope exhibits fluctuations with a general upward tendency. The flow velocity demonstrates a significant positive correlation with both slope gradient and rainfall intensity. As the flow velocity escalates, the slope flow intensifies the erosion of the slope, and the continuous erosion of the slope surface by runoff leads to the emergence of rills on the slope surface. Rills predominantly appear in the middle section of the slope, while their presence in the upper and lower sections is scarce. This can be ascribed to the flow velocity and slope strength. The flow velocity in the upper part of the slope is comparatively low, and the volume of runoff is relatively small; as a result, the erosive force is insufficient to erode the slope. The flow velocity in the middle section is relatively high, and the volume of runoff is larger. In this scenario, the erosive force on the slope is relatively significant, culminating in the formation of rills of varying sizes in the middle section of the slope under the action of erosion from runoff. Although the flow velocity in the lower slope section is relatively substantial, the slope strength exceeds that of other sections. Consequently, the volume of runoff in the lower section of the slope is larger than in the upper section but smaller than in the middle section.

3.4. Hydraulic Characteristics of Slope Runoff

As shown in Table 1, *Re* elevates with the intensification of slope gradient and rainfall intensity. The value of *Re* yields a result lower than 500; therefore, despite the evident rise in *Re*, the flow of the slope runoff remains laminar. The mean value of Fr exceeds 1 signifying rapid overland flow that escalates with the augmentation of slope gradient and rainfall intensity. The resistance coefficient f diminishes with the amplification of slope gradient and rainfall intensity, which also elucidates why enhancing the slope gradient or rainfall intensity leads to greater slope runoff velocity. The Manning coefficient n reflects the roughness of the slope surface. The greater the slope gradient, the greater the rainfall intensity, and the erosion of slope runoff, the easier it is to make the slope surface appear depressed and promote the formation of rills. The more complex the slope morphology, the greater the n.

Table 1. Slope runoff hydraulic parameters under different slope gradients and rainfall intensities.

Rainfall Intensity ($mm{\cdot}h^{-1}$)	Gradient (°)	Average Discharge Velocity ($m{\cdot}s^{-1}$)	Reynolds Number (*Re*)	Froude Number (*Fr*)	Manning Coefficient (*f*)	Resistance Coefficient (*n*)
50	10	0.124	43.421	1.143	0.017	0.427
	20	0.140	76.094	1.453	0.025	0.397
	30	0.149	108.447	1.773	0.025	0.341
	40	0.155	207.893	2.036	0.030	0.292
80	10	0.131	46.240	1.445	0.016	0.402
	20	0.156	93.630	1.756	0.021	0.387
	30	0.165	125.573	1.990	0.025	0.312
	40	0.189	249.569	2.236	0.028	0.265
110	10	0.155	50.124	1.729	0.014	0.383
	20	0.177	138.843	2.018	0.017	0.356
	30	0.192	205.076	2.482	0.017	0.271
	40	0.226	302.162	2.980	0.021	0.192

3.5. Slope Flow Velocity, Rainfall Intensity, and Slope Gradient Fitting Equation

An intensification of precipitation strength results in an expansion of surface water flow volume, subsequently accelerating the flow velocity of slope runoff. An increase in slope magnitude signifies a greater inclination of the terrain, which, under the influence of gravitational forces, amplifies the energy of water traversing the slope and further accelerates the flow velocity of slope runoff. The combined action and mutual constraints of these factors ultimately impact the velocity of water flow along the slope. To delve deeper into the impact of rainfall intensity and slope gradient on slope flow velocity, the rainfall intensity and slope gradient tangent values were denoted as X and Y, respectively. A binary fitting analysis of the rainfall intensity and slope gradient tangent value was subsequently conducted. The fitted surface is illustrated in Figure 5, and the fitting equation is as follows:

$$v = 0.057\ln X + 0.033\ln Y - 0.053 \quad (R^2 = 0.908)$$

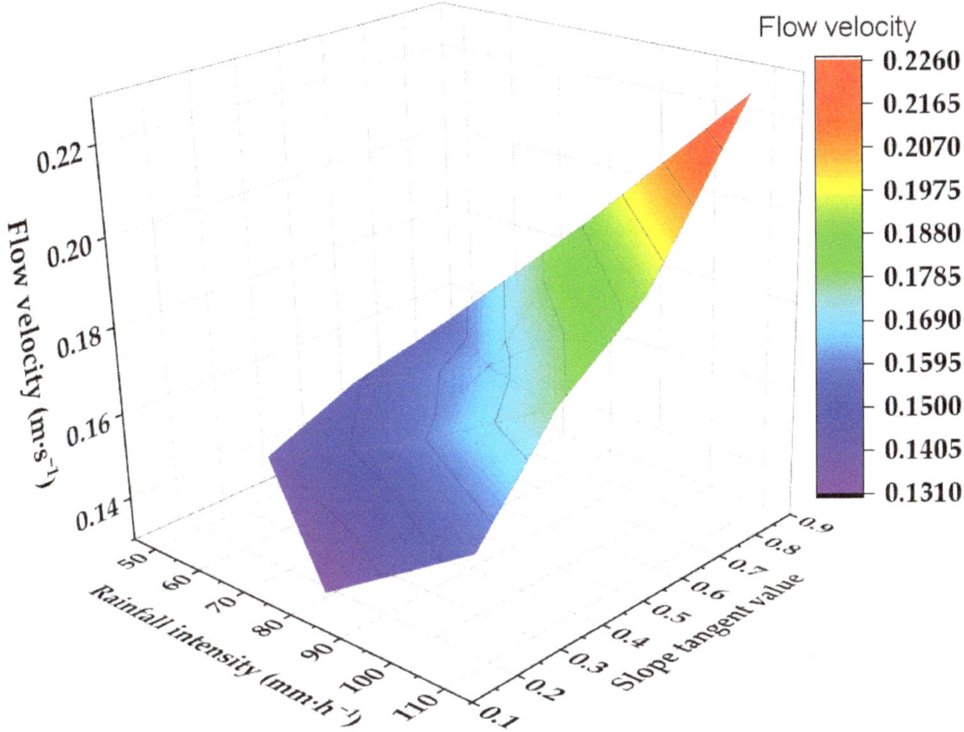

Figure 5. Relationships among slope flow velocity, rainfall intensity, and slope gradient tangent value.

4. Conclusions

(1) There is a significant positive correlation between rainfall intensity and slope gradient, and as the slope gradient increases, the correlation between rainfall intensity and flow velocity becomes stronger.

(2) The appearance of the peak sediment yield rate represents the maximum temporal extent of changing slope morphology. After the peak value appears, the secondary deformation of the slope will not occur in a short time, and the sediment yield rate decreases and then becomes stable.

(3) As two important factors of slope velocity, rainfall intensity, and slope gradient can be described and predicted using binary functions.

Author Contributions: Conceptualization, H.T.; methodology, H.T.; software, H.T.; validation, Z.K.; formal analysis, H.T.; investigation, Z.K.; resources, Z.K.; data curation, H.T.; writing—original draft preparation, H.T.; writing—review and editing, Z.K.; visualization, Z.K.; supervision, Z.K.; project administration, Z.K.; funding acquisition, Z.K. All authors have read and agreed to the published version of the manuscript.

Funding: This research was funded by Key Research and Development Plan of Yunnan Province: The Technology of the Comprehensive Risk Assessment of the Earthquake Catastrophe and the Disaster Chains in Yunnan and Its Application, (Grant No. 202203AC100003); the scientific and technological development project of Southwest Pipeline Co., Ltd. (Chengdu, China), National Pipe Network Group Research on Hydraulic Protection and Soil and Water Conservation of Oil and Gas Pipelines through Fully Weathered Granite Area (Grant No. 2018016); and science and technology development project of China Hydropower Foundation Co., Ltd. (Tianjin, China). Evaluation of rapid excavation of slope cut-off wall in complex geological background area and treatment technology of mud and water inrush in tunnel engineering (Grant No. 2022530103001936).

Institutional Review Board Statement: Not applicable.

Informed Consent Statement: Not applicable.

Data Availability Statement: The data presented in this study are available on request from the corresponding author.

Conflicts of Interest: The authors declare no conflict of interest.

References

1. Zhang, L.; Li, Z.; Wang, S. lmpact of runoff regimes on sediment yield and sediment flow behavior at slope scale. *Trans. Chin. Soc. Agric. Eng.* **2015**, *31*, 124–131.
2. Wu, X.; Yang, Y.; Ye, Z. Stability of rock-soil slopes under rainfall infiltration and its influencing factors. *J. Water Resour. Water Eng.* **2022**, *33*, 189–199.
3. Ji, Z.; Zhang, L. Spatial-Temporal Variation of Distribution of Rainfall Erosivity in the Poyang Lake Basin. *Yellow River* **2019**, *41*, 81–84.
4. Zhang, Z.; Hong, B.; Huang, Y. Study on the relationship between runoff characteristics and soil erosion of laterite slope under rainfall. *Mt. Res.* **2017**, *35*, 535–542.
5. Luo, R.; Zhang, G.; Cao, Y. Progress in the research of hydrodynamic characteristics of sediment-laden overland flow. *Prog. Geogr.* **2009**, *28*, 567–574.
6. Ke, Q.; Zhang, K. A review on the erosion-related researches by simulated rainfall experiments in China. *Sci. Soil Water Conserv.* **2018**, *16*, 134–143.
7. Zhang, R.; Zhang, L.; FU, X. Research on Relationships between Sediment Yield and Hydraulics Parameters on Slope. *J. Soil Water Conserv.* **2017**, *31*, 81–86.
8. Zheng, J.; She, D.; Xu, C. Evolution process of erosion ril on saline soil slope in coastal reclamation with different rainfall intensities and slope gradients. *J. Hohai Univ. (Nat. Sci.)* **2015**, *43*, 313–318.
9. Li, J.; Wang, Y.; Zhang, C. Characteristic of Flow Velocity and Sediment Yielding on Engineering Accumulations Slope with Three Soil Textures. *J. Yangtze River Sci. Res. Inst.* **2019**, *36*, 28–35.
10. El-kateb, H.; Zhang, H.; Zhang, P.; Mosandl, R. Soil erosion and surface runoff on different vegetation covers and slope gradients: A field experiment in Southern Shaanxi Province, China. *Catena* **2013**, *105*, 1–10. [CrossRef]
11. Dong, J.; Zhang, K.; Guo, Z. Runoff and soil erosion from highway construction spoil deposits: A rainfall simulation study. *Transp. Res. Part D Transp. Environ.* **2012**, *17*, 8–14. [CrossRef]
12. Niu, H.-K.; Li, Q.; Zhang, L.-T.; Li, X.; Wang, J.-T. Experimental Study on Failure Mechanism and Mode of Fly-Ash Dam Slope Triggered by Rainfall Infiltration. *Appl. Sci.* **2022**, *12*, 9404. [CrossRef]
13. Yan, C.; Hu, H.; Xu, X. Changing pattern of runoff coefficients in urban underlving surfaces under simulated rainfall conditions. *Sci. Soil Water Conserv.* **2022**, *20*, 24–30.
14. Zhang, K.; Wang, G.; Sun, X. Experiment on hydrodynamic characteristics of thin layer water on slope. *Trans. Chin. Soc. Agric. Eng.* **2014**, *30*, 182–189.
15. Li, G.; Abrahams, A.D.; Atkinson, J.F. Correction factors in theDetermination of mean velocity of overland flow. *Earth Surf. Process. Landf.* **1996**, *21*, 509–515. [CrossRef]
16. Zhuang, X. Spatial distribution of sheet flow velocity along slope under simulated rainfall conditions. *Geoderma* **2018**, *321*, 1–7. [CrossRef]
17. Yang, D.; Gao, P.; Liu, X. Study on the Velocity of Shallow Water Flow on Slope. *J. Soil Water Conserv.* **2019**, *33*, 72–80.
18. Hao, H.; Du, Y.; Cao, D. Experimental study of flow hydraulic characteristics and soilerosion on red soil hilly slope. *Sci. Soil Water Conserv.* **2018**, *16*, 1–8.
19. PAN, L.; Li, R.; LI, Q. Effects of Straw Mulching on Runoff and Sediment Characteristics ofSloping Farmland in the Karst Area of Western Guizhou. *J. Soil Water Conserv.* **2021**, *35*, 9–16.
20. Yuan, H.; Xin, Z.; Jiang, Q. Slope Erosion and Its Hydrodynamic Characteristic of CinnamonSoil Under Continuous Rainfall. *J. Soil Water Conserv.* **2020**, *34*, 14–20, 30.
21. He, J.; Li, X.; Cai, Q. Experimental study on soil erosion characteristics and spatial variation of runoff velocity on typical loess slope. *Geogr. Res.* **2022**, *41*, 1327–1337.
22. Cao, L.; Zhang, K.; Dai, H. Modeling Interrill Erosion on Unpaved Roads in the Loess Plateau of China. *Land Degrad. Dev.* **2015**, *26*, 825–832. [CrossRef]
23. Tian, P.; Qiu, H.; Feng, Y. Effects of Rainfall ntensity and Slope Gradient on Runoff and SedimentProduction and Erosion Dynamic Process on Red Soil Slope. *Res. Soil Water Conserv.* **2020**, *27*, 1–8.
24. Liang, Z.; Zhuo, M.; Guo, T. Hydrodynamic characteristics of slope surface flow under different rainfall intensities and slopes. *Ecol. Environ. Sci.* **2015**, *24*, 638–642.
25. Ji, L.; Wang, W.; Kang, H. Differences in hydraulic erosion processes of the earth and eath-rock Lou soil engineering accumulation in the Loess Region. *Chin. J. Appl. Ecol.* **2020**, *31*, 1587–1598.

26. Cai, L.; Tang, G.; Xiong, L. An Analysis on Fractal Characeeristics of Typical Landform Patterns in Northern Shaanxi Loess Plateau Based on DEM. *Bull. Soil Water Conserv.* **2014**, *34*, 141–144, 329.
27. Zhao, W.; Zhang, H.; Jiang, H. Variation characteristics of loess erosion hillside under indoor artificial rainfall conditions. *Arid Land Geogr.* **2019**, *42*, 867–875.
28. Zhang, X.; Li, X.; Lin, J. Characteristics of soil aggregates and organic carbon in eroded gully in red soil region of Southern China. *Trans. Chin. Soc. Agric. Eng.* **2020**, *36*, 115–123.
29. Zhou, M.; Xiao, H.; Nie, X. Analysis and Prospect of Soil Organic Carbon Research Process in Recent 30 Years at Home and Abroad. *Res. Soil Water Conserv.* **2020**, *27*, 391–400.
30. Jiang, F.; Ye, Y.; Guo, X. Spatial Distribution of Soil Available Phosphorus and lts Driving Factors Under Different Erosion Dearees in South China. *Soils* **2018**, *50*, 1013–1021.
31. Shi, Z.; Yang, J.; Li, Z. Soil Conservation in the Hilly Red Soil Region of Southern China. *J. Soil Water Conserv.* **2018**, *32*, 6–9.
32. Feng, X.; Zha, X.; Huang, S. Processes and characteristics of erosion on hillslopes ofgranite red soil under simulated rainfall. *Sci. Soil Water Conserv.* **2014**, *12*, 19–23.
33. Pan, S.; Chen, S.; Yang, H. Specimen size effect on strength and stress-strain properties of strongly weathered granite in coastal zone. *Geotech. Investig. Surv.* **2021**, *49*, 6–10.
34. Zhao, N.; Dong, S.; Chen, X. Experimental lnvestigation on the Dynamic Mechanical Properties of Weakly Weathered Granite. *J. China Three Gorges Univ. (Nat. Sci.)* **2022**, *44*, 62–70.
35. Deng, L.; Zhang, L.; Fan, X. Simulation Study on Runoff Processes and Erosion Rate on the Weathered Granite Slope in Southern China. *J. Soil Water Conserv.* **2018**, *34*, 143–150.

applied sciences

MDPI

Article

Time-Dependence of the Mechanical Behavior of Loess after Dry-Wet Cycles

Kai Liu [1], Tianfeng Gu [1,*], Xingang Wang [1,2,*] and Jiading Wang [1]

1 State Key Laboratory of Continental Dynamics, Department of Geology, Northwest University, Xi'an 710069, China; liukaii@stumail.nwu.edu.cn (K.L.); wangjddz@gmail.com (J.W.)
2 State Key Laboratory of Geohazard Prevention and Geoenvironment Protection, Chengdu University of Technology, Chengdu 610059, China
* Correspondence: gutf@nwu.edu.cn (T.G.); xgwang@nwu.edu.cn (X.W.)

Abstract: The structure, time-dependent mechanical deformation, and strength characteristics of loess, which is loose and porous with well-developed vertical joints, are greatly affected by the dry-wet cycles, which are attributed to periodic artificial irrigation, rainfall, and water evaporation. To better understand the creep characteristics of loess under the effect of dry-wet cycles, Q_2 loess samples obtained from the South Jingyang County, China, were subjected to different dry-wet cycles (0, 5, 10, 15, 20) and sheared in triaxial creep tests. The experimental results revealed that: firstly, the maximum value of the deviatoric stress corresponding to creep failure gradually decreases with an increase in the dry-wet cycles. Secondly, the long-term strength of the loess after dry-wet cycles were obtained through the Isochronous Curve Method. It is found that the long-term strength and the number of dry-wet cycles showed an exponential decreasing relationship. In addition, the creep damage mechanism of loess due to dry-wet cycles is proposed. This study may provide the basis for understanding the mechanical behavior of the loess under the effect of dry-wet cycles, as well as guidelines for the prevention and prediction of loess landslide stability.

Keywords: loess; dry-wet cycles; triaxial creep test; creep behavior; strain-time; long-term strength

check for updates

Citation: Liu, K.; Gu, T.; Wang, X.; Wang, J. Time-Dependence of the Mechanical Behavior of Loess after Dry-Wet Cycles. *Appl. Sci.* **2022**, *12*, 1212. https://doi.org/10.3390/app12031212

Academic Editors: Ricardo Castedo, Miguel Llorente Isidro and David Moncoulon

Received: 15 December 2021
Accepted: 23 January 2022
Published: 24 January 2022

1. Introduction

Loess is a weakly cemented loose sediment with a special material composition, which is widely distributed in the Loess Plateau of China (Figure 1) [1,2]. Creep phenomenon is common in loess-covered areas, and the creep characteristics of loess pose an important impact on the landslide formation and the construction of engineering structures [3–5]. The creep properties of loess vary due to the change of the environment. For example, unsaturated loess soils are usually subjected to wetting because of the natural rainfall or artificial irrigation, and loess soils would then be subjected to a drying process, i.e., water evaporation, as rainfall or irrigation stops. Such a process could be called a dry-wet cycle [6,7] and leads to the following consequences: the loss of soluble salts in the loess [8–10], the deterioration of the mechanical properties of the loess [11,12], and the change of shear strength, deformation, and permeability of the loess [13–15]. Therefore, the creep deformation of the loess was aggravated, which will contribute to the formation of disasters such as landslides.

The influence of dry-wet cycles on the time-dependent behavior of rocks has been considered in some research [16–19], while in the last few years increasing attention has been paid to experimental research into the behavior of loess soils. It is acknowledged that the reduction in the shear strength of the loess due to the dry-wet cycles is obvious [3,20,21]. Malusis et al. [22] reported that the hydraulic conductivity of the soil gradually increases as the number of dry-wet cycles increases. Until now, extensive attempts have been made on the physical and mechanical properties of loess after dry-wet cycles by using the uniaxial tensile shear test [23], direct shear test [24,25], the uniaxial compressive test [26], conventional triaxial test [7,27], the soil-water characteristic curve (SWCC) of loess [28], etc. For example, Yuan et al. [23] carried out an experimental study on the tensile strength of the intact loess samples under dry-wet cycles and concluded that the repeated dry-wet cycle destroys the original structure of the intact loess and causes its tensile strength to disappear. Furthermore, Mu et al. [24] investigated the shear characteristics of loess samples under dry-wet cycles through direct shear tests and pointed out that the shear strength of loess gradually decreased as the number of dry-wet cycles increase. By conducting experiments, Li et al. [26] reported that the unconfined compressive strength, elastic modulus, and cohesive force of the loess specimen decreased with an increase in the number of dry-wet cycles, while the vertical compressive strain and failure deformation increased. Hu et al. [27] conducted triaxial tests considering three influence factors such as dry density, dry-wet cycle amplitude, and the lower bound water content of dry-wet cycles and established a compacted loess deterioration model (CLDM). Wang et al. [29] explored the effect of the dry-wet cycle on the collapsible deformation characteristics of compacted loess, and the results revealed that the compressive strain at all levels of deviatoric stress of specimens with different initial density gradually increases with the increase of the number of dry-wet cycles. The higher the initial compaction degree, the more significant effect of the dry-wet cycle on the mechanical behavior of compacted loess.

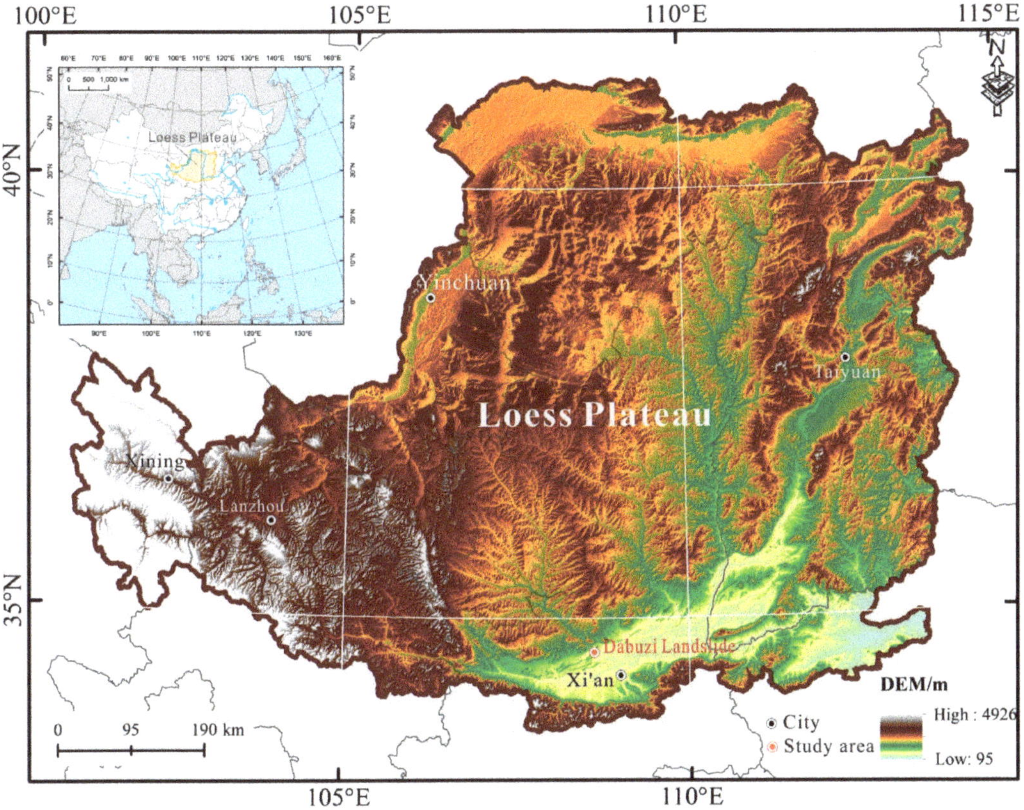

Figure 1. The study area.

The studies above have been focused on the variation in conventional mechanical properties such as shear strength, uniaxial tensile strength, uniaxial compressive strength, and triaxial shear strength of the loess after the dry-wet cycles. However, very little attention has been paid to the damaging effect of the dry-wet cycles on the creep properties of loess. Additionally, a few experimental results have shown that the creep strains of loess soils developed after dry-wet cycles are not negligible [30] and can reach a critical value, eventually triggering loess landslides. In this backdrop, typical loess samples obtained from South Jingyang, China, were used to conduct triaxial creep tests under different numbers of dry-wet cycles to gain full insight into the time-dependent characteristics of loess soils and the deterioration of the long-term strength caused by dry-wet cycles. Furthermore, the damage mechanism of the creep characteristics of loess due to the dry-wet cycles is explored. This study would be beneficial for understanding the time-dependence of mechanical behaviors of loess soils as well as the long-term stability of loess slopes.

2. Triaxial Creep Tests

2.1. Sampling Site and Dry-Wet Cycle Process

The loess samples used in test were collected from the back scalp of the Dabuzi landslide in Jingyang County, Shaanxi Province (Figure 1). The sampling site was about 20 m from the edge of the platform, and the collected samples were identified as Lishi loess (Q2) according to previous studies [31]. The loess in this area have suffered from different degrees of dry-wet cycles as a result of the long-term repeat rise and fall of the groundwater level due to the farmland irrigation (Figure 2a) [32]. Therefore, the mechanical properties of loess deteriorates, resulting in the occurrence of the Dabuzi landslide (Figure 2b). All loess samples were firstly sealed and kept in iron buckets. Then, samples were brought back to the laboratory and cut immediately to a cylindrical specimen with a diameter of 61.8 mm and height of 125 mm. The procedures based on the ASTM standards were conducted to determine the index properties, namely Atterberg limits [33], specific gravity and density [34], and moisture content [35]. The basic physical index properties of loess are listed in Table 1.

Figure 2. The sampling site. (**a**) Irrigation canal, (**b**) landslide boundary.

Table 1. The physical index properties of loess.

Moisture Content (%)	Dry Density (g/cm^3)	Density (g/cm^3)	Specific Gravity	Void Ratio
16	1.503	1.858	2.711	0.768

A two-step process for preparation of the sample was used with each dry-wet cycle. The first step is sample saturation, which is briefly described as follows. The specimens were saturated by the vacuum saturation method. Filter papers and porous stones were firstly placed on the top and bottom of the soil sample. Then, the soil sample was fixed in a saturator chamber, as shown in Figure 3a, and then the saturator chamber with the soil sample was tightened with screw nuts and placed in a vacuum container (Figure 3b). After that, the soil specimen was submerged in de-aired water inside a vacuum container for at least 24 h to achieve a degree of saturation greater than 97% [36,37]. The drying phase was followed by wetting. After saturation, the soil sample was taken out by loosening the screw nuts and removing the filter papers and porous stones. Then, the soil sample sealed by the saturator chamber was placed in a drying oven at about 105 °C for 48 h to fully remove its water content [38,39]. Next, the soil sample was taken out and cooled down. This process, consisting of saturation and drying phases, is called one dry-wet cycle in this study. After the last dry-wet cycle, the corresponding water required to obtain the target water content was dripped on the top surface of the soil sample, and then the soil sample was wrapped with a plastic wrap and placed in a humidifier for 48 h to obtain a homogeneous sample in room temperature [40,41]. In this study, samples with different dry-wet cycles (0, 5, 10, 15, and 20) were prepared and the sample after 10 dry-wet cycles was shown in Figure 3c.

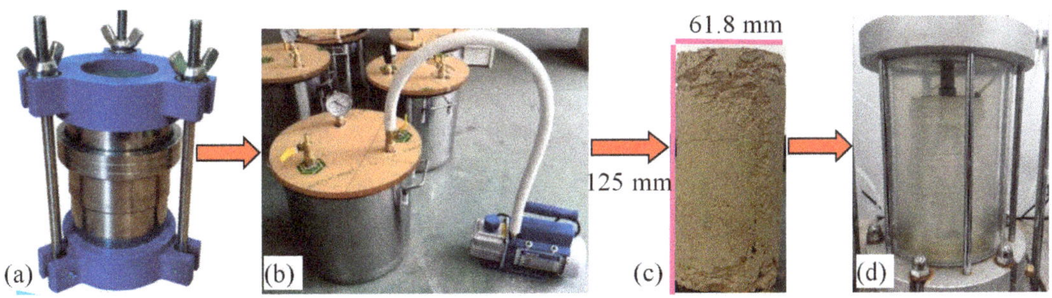

Figure 3. Specimen with the dry-wet cycles. (**a**) Frame saturator chamber, (**b**) vacuum container, (**c**) specimen after dry-wet cycles, (**d**) triaxial chamber.

2.2. Testing Apparatus

The FSR-20 triaxial creep apparatus for the testing cylinder of soil samples with a height of 125 mm and diameter of 61.8 mm in the State Key Laboratory of Continental Dynamics in China (see Figures 3c and 4) was utilized in this study. As shown in Figure 4, the apparatus consists of several parts, such as an axial loading system, pore water pressure controlling system, confining pressure controlling system, matric suction controlling system, data measurement, and a collect system. By using a pneumatic servo control system, the apparatus is capable of providing not only constant shear stress but also constant air pressure for a long time with high accuracy [3,42]. Creep tests can be carried out for confining pressure up to 1 MPa, pore water pressure up to 500 kPa, axial load up to 2 MPa, and pore gas pressure up to 500 kPa. The tests may be run either undrained or consolidated undrained with or without pore water pressure measurement, or drained condition. In this study, drained creep tests were conducted on loess samples.

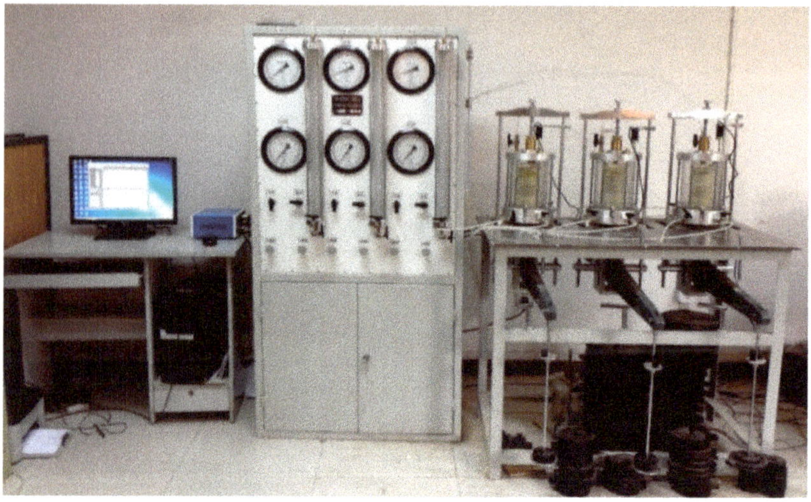

Figure 4. FSR-20 triaxial creep apparatus.

2.3. Testing Scheme

According to the basic physical index properties listed in Table 1, samples with a moisture content of 16% were tested by using multi-level loading method [4,43]. Considering the density (ρ = 1858 kg/m^3) and the thickness (20 m) of the overlying loess, the confining pressure of 200 kPa was chosen for the triaxial creep test. The multi-level loading method used in this study is described as follows: the first level of deviatoric stress was applied to the specimen until the stability was attained when the deformation of the specimen was less than 0.01 mm in 24 h. To find an optimum time during which the deformation of the specimen was less than 0.01 mm, a series of preliminary tests were conducted, where 24 h, 36 h, and 48 h were used for specimens with the same testing conditions. The creep shear test results show that for the cases of 24 h, 36 h, and 48 h, the strength and strain-time curves of loess have no obvious difference, indicating that 24 h is enough for obtaining the stability state of the loess. Therefore, 24 h was selected here from the point of view of time saving. Then, the second level of deviatoric stress was applied to the specimen until the stability state was achieved. The specimen was subjected to shearing until creep failure occurred at some level of deviatoric stress. In this study, the failure occurred in the specimen when the strain of specimen reached 20% following the previous study concluded by Xie et al. [44]. The samples after different dry-wet cycles were subjected to consolidation and shearing in the conventional triaxial compression test to calculate the instantaneous failure strength. Then, the loading scheme of the triaxial creep test (Table 2) was determined according to the instantaneous failure strength.

Table 2. Loading scheme of the triaxial creep test.

The Number of Dry-Wet Cycles	Deviatoric Stress/kPa
0	25, 50, 62.5, 75, 87.5, 100, 150, 250, 350
5	50, 100, 150, 200, 250, 300
10	50, 125, 200, 250, 300
15	50, 100, 175, 250
20	50, 100, 175, 250

3. Test Results

Based on the experimental data collected, the typical creep strain curves for loess with a different number of dry-wet cycles is plotted in Figures 5–9. After that, according to Boltzmann's linear superposition principle [45,46], the creep curves corresponding to each level of deviatoric stress was obtained by using "the coordinate translation method" [47,48]. Figures 5–9 show the whole process curve of creep for loess and creep curves of loess with a different number of dry-wet cycles. In Figures 5–9, q represents the deviatoric stress, ε is strain, w is the moisture content, and t is time.

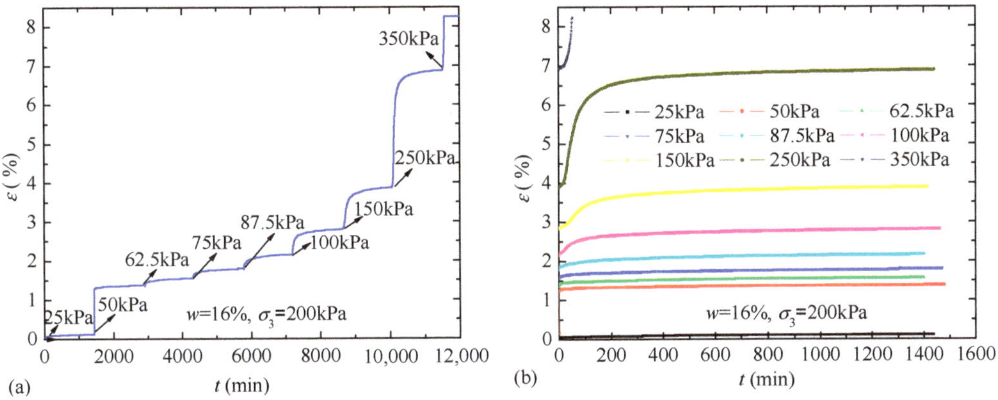

Figure 5. (**a**) The whole process curve of creep for loess without dry-wet cycles, (**b**) creep curves of the loess without dry-wet cycles.

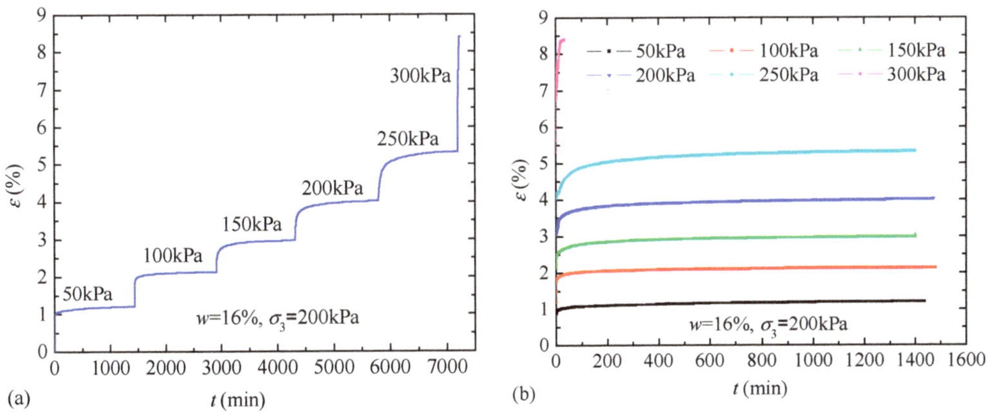

Figure 6. (**a**) The whole process curve of creep for loess with 5 dry-wet cycles, (**b**) creep curves of loess with 5 dry-wet cycles.

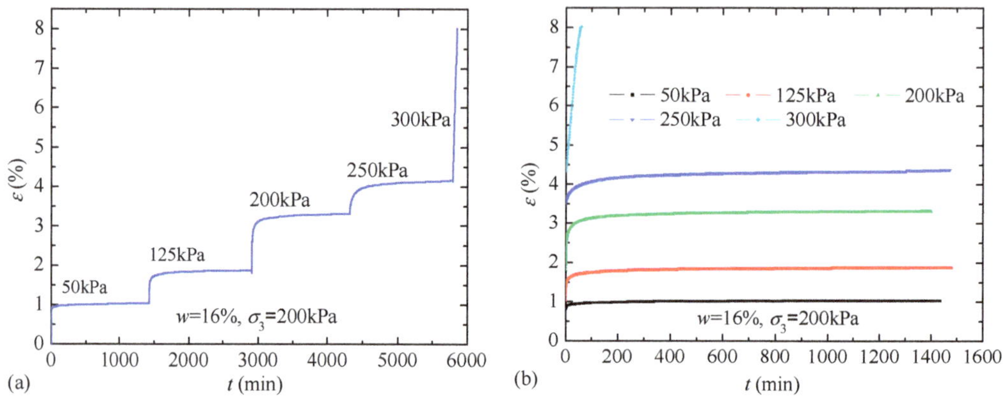

Figure 7. (**a**) The whole process curve of creep for loess with 10 dry-wet cycles, (**b**) creep curves of loess with 10 dry-wet cycles.

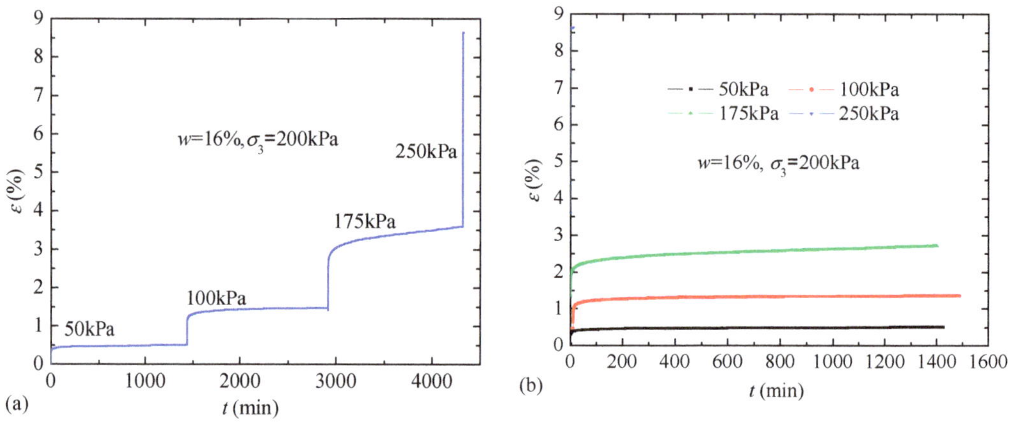

Figure 8. (**a**) The whole process curve of creep for loess with 15 dry-wet cycles, (**b**) creep curves of loess with 15 dry-wet cycles.

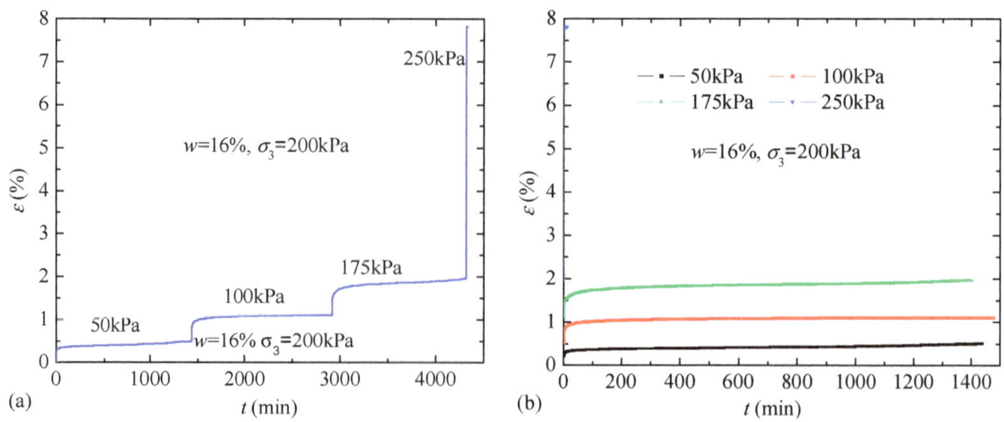

Figure 9. (**a**) The whole process curve of creep for loess with 20 dry-wet cycles, (**b**) creep curves of loess with 20 dry-wet cycles.

It can be observed from Figures 5–9 that the characteristics of the whole process curve of creep for loess and creep curves are as follows: (i) The specimen with different dry-wet cycles underwent three typical creep stages, namely the decelerating creep stage, the steady state stage, and accelerate creep stage. With the same number of dry-wet cycles, the creep curve tends to be flat with an increase of time when a small load is applied, that is, the steady state stage. However, with a higher loading, the creep curve of the stable phase increases linearly with time. (ii) The slope of the curve corresponding to the steady state stage increases with an increase of loading. However, sudden failures could be observed within the loess specimen once the load exceeds a certain value. For example, when the deviator stress of the sample with 20 dry-wet cycles increased to 250 kPa, the deformation dramatically increases. (iii) With the increase of the number of dry-wet cycles, the maximum value of the deviatoric stress corresponding to the creep failure gradually decreases, indicating that the creep strength of the sample is deteriorated by the dry-wet cycles.

4. Discussion

4.1. The Long-Term Strength of Loess Samples with Different Dry-Wet Cycles

The long-term strength of rock and soil is an important parameter for assessing the stability of landslides [49,50]. At present, the main methods to obtain long-term strength through creep test curves include the isochronous curve method [3,51], which uses the creep test curve and the Boltzmann superposition principle [52] and states that the stress response of a system to a time dependent shearing deformation may be written as the sum of responses to a sequence of step strain perturbations in the past [53], to obtain the stress-strain isochronous curves corresponding to the different deviatoric stress at the same time [54]. Figure 10a–e show the isochronous curves corresponding to the creep curves of the samples after five kinds of dry-wet cycles (n = 0, 5, 10, 15, 20). Based on the isochronous curves, the long-term strength (q_L) was obtained.

The experimental data listed in Table 3 was plotted in Figure 11. One can observe that there is a clear correlation between the long-term strength and the number of dry-wet cycles. The overall relationship between the long-term strength of loess and the number of dry-wet cycles can be described by the following equation:

$$q_L = 123.50e^{(-n/9.98)} + 127.33,\ n \leq 20, \tag{1}$$

where q_L is the long-term strength measured in kPa and n is the number of the dry-wet cycles.

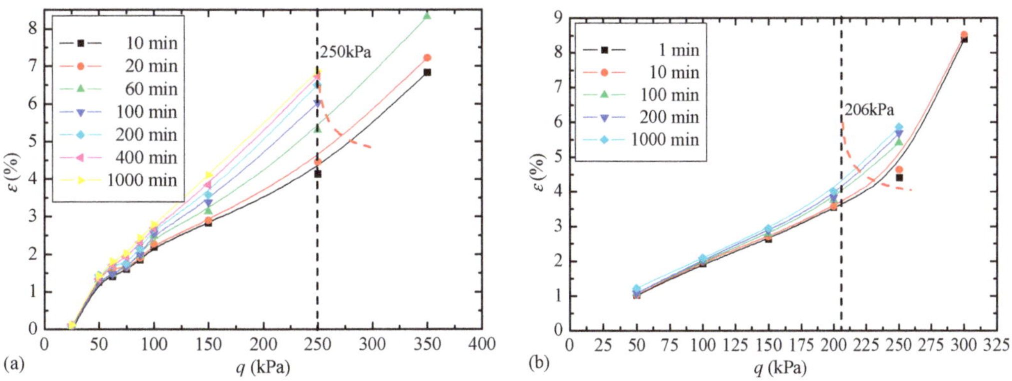

Figure 10. *Cont.*

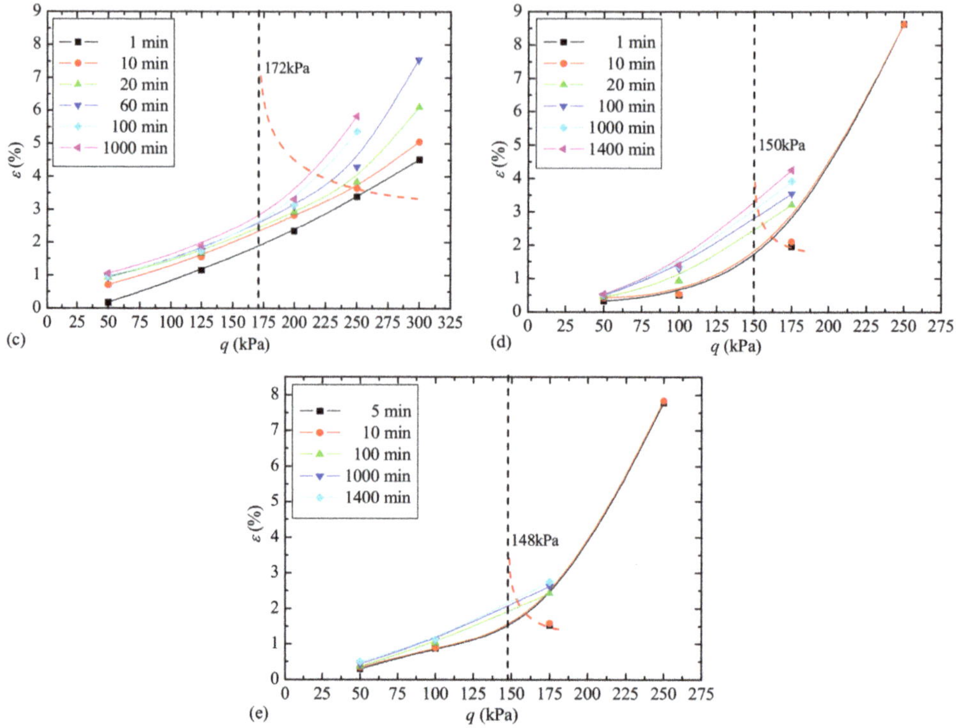

Figure 10. Creep isochronous curve after different dry-wet cycles. (**a**) $n = 0$, (**b**) $n = 5$, (**c**) $n = 10$, (**d**) $n = 15$, (**e**) $n = 20$.

Table 3. Long-term strength after different dry-wet cycles (qL).

n	0	5	10	15	20
q_L	250	206	172	150	148

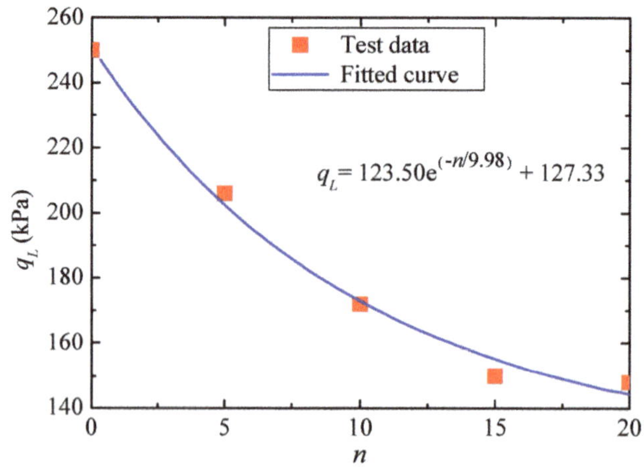

Figure 11. The fitting curve of the long-term strength after different dry-wet cycles.

It is clear that the long-term strength of loess decreases when increasing the dry-wet cycles. However, the reducing magnitude of long-term strength gradually decreases with the increase of the number of dry-wet cycles, until it reaches a constant value. That is to say, the long-term strength of the loess and the number of dry-wet cycles ($n \leq 20$) shows an exponential decreasing relationship (see Equation (1)).

4.2. Creep Damage Mechanism of Loess Samples Due to the Dry-Wet Cycles

The creep damage mechanism of the loess samples due to the dry-wet cycles was explained from the perspective of macro pictures of the loess samples after creep failure and the change of long-term strength with dry-wet cycles. Figure 12 shows loess samples after experiencing creep failure. One can observe that the obvious failure surface did not appear in the specimen, whereas the bending and bulging occurred on the side of the specimen. Furthermore, as the number of dry-wet cycles increases, the phenomenon of bending and bulging of the specimen becomes more noticeable. Such a phenomenon can be explained as follows: under the effect of a dry-wet cycle, the microstructure of loess is destroyed due to the repeated migration and loss of soluble salts between soil particles in the loess sample [8,10]. Moreover, the dry-wet cycles also promote the development and expansion of joint fractures in the loess sample [7], resulting in the worsening of the creep mechanical properties (see Table 3 and Figure 11). As the number of dry-wet cycles further increased, the dissolved salts in the loess sample gradually dissolved in water, which results in the coarse particles being further dispersed and disintegrated and the total pore volume of the sample being increased [25]. Therefore, the creep curves of the loess specimen show a strain-softening pattern, and the bending or lateral swelling occurred in the specimen (see Figure 12, n = 5 and n = 10). When the number of dry-wet cycles reaches a certain value, structural stability (micro- and macro-structure stability) within the loess soils was achieved. Consequently, the long-term strength of loess samples tends to be stable (see Figure 10, n = 15 and n = 20), causing the creep mechanical properties to be consistent with each other.

Figure 12. Specimen after creep failure.

4.3. Limitations of the Experimental Test in This Study

Concerning the influence of dry-wet cycles, the experimental results herein can be used for the determination of numerical simulation parameters and for better evaluating the long-term stability of loess slopes. Therefore, the study would be beneficial to understanding the time-dependence of the mechanical behaviors of loess soils as well as the long-term stability of loess slopes, which provide the basis for understanding the mechanical behavior of the loess under the effect of dry-wet cycles, as well as guidelines for the prevention and prediction of loess landslide stability. However, there are some potential limitations in this study that could be addressed in future research. First, the loess samples were sheared under a saturated condition with a maximum value of 20 dry-wet cycles, whereas the triaxial creep tests were not conducted on unsaturated loess samples with a wide range of dry-wet cycles. Therefore, the conclusions and theoretical formulas obtained in this study have certain restrictions. Secondly, the water used in this study is distilled water, which is unrealistic due to the fact that the water in the filed contains many mixtures due to solute transport and other chemical effects [10,55–57]. Finally, the influence of dry-wet cycle on loess in the field is very complicated due to the influence of rainfall, irrigation water, evaporation, and other factors [8,9,11,12], which could not be simply simulated by dry-wet cycles. On the other side, it is difficult to estimate the actual number of dry-wet cycles that the loess experienced in the field. Thus, there are some difficulties in applying the experimental results in this study to engineering practice. All limitations above would be considered in future research.

5. Conclusions

To assist in the understanding of the time-dependent behaviors and the long-term stability of loess slopes after dry-wet cycles, the mechanical properties of intact loess samples were investigated through triaxial creep tests. The following conclusions can be drawn according to the obtained experimental results:

(i) With the same number of dry-wet cycles, the strain-time curve of the loess samples shows a similar trend, where the strain eventually reaches a certain value with an increase of time when a small load is applied, whereas the creep curve of the stable phase increases linearly with time when the loess specimen is subjected to a higher loading. As the number of dry-wet cycles increases, the maximum value of the deviatoric stress corresponding to the creep failure gradually decreases, indicating that the deterioration of triaxial compressive strength is attributed to the dry-wet cycles.

(ii) The long-term strength of loess samples with different dry-wet cycles was obtained by using the Isochronous Curve Method, and the long-term strength of loess samples and the number of dry-wet cycles showed an exponential decreasing relationship.

(iii) The creep damage mechanism of loess samples due to dry-wet cycles can be explained as follows: the dry-wet cycle causes repeated migration and loss of soluble salts between soil particles in the loess sample, which results in damage of the sample's microstructure. Consequently, the development and expansion of joint fractures within the loess samples were also aggravated. The combined effects of microstructural damage and expansion of joint fractures causes a greater deterioration in the creep properties of loess soils.

Author Contributions: Conceptualization, X.W. and T.G.; methodology, K.L.; software, X.W. and K.L.; writing—original draft preparation, K.L.; writing—review and editing, X.W. and T.G.; visualization, X.W.; supervision, J.W.; funding acquisition, X.W. All authors have read and agreed to the published version of the manuscript.

Funding: This research was supported by the National Natural Science Foundation of China (No. 41902268) and the Opening Fund of State Key Laboratory of Geohazard Prevention and Geoenvironment Protection (Chengdu University of Technology) (No. SKLGP2020K016).

Institutional Review Board Statement: Not applicable.

Informed Consent Statement: Not applicable.

Data Availability Statement: Not applicable.

Acknowledgments: We thanks the editors, and the two anonymous reviewers for their constructive comments, which has helped us improve the quality of the manuscript.

Conflicts of Interest: The authors declare no conflict of interest.

References

1. Derbyshire, E.; Dijkstra, T.; Smalley, I.; Li, Y. Failure mechanisms in loess and the effects of moisture content changes on remoulded strength. *Quat. Int.* **1994**, *24*, 5–15. [CrossRef]
2. Lian, B.; Peng, J.; Zhan, H.; Huang, Q.; Wang, X.; Sheng, H. Formation mechanism analysis of irrigation-induced retrogressive loess Landslides. *Catena* **2020**, *195*, 104441. [CrossRef]
3. Wang, X.; Wang, J.; Zhan, H.; Li, P.; Qiu, H.; Hu, S. Moisture content effect on the creep behavior of loess for the catastrophic Baqiao landslide. *Catena* **2020**, *187*, 104371. [CrossRef]
4. Lian, B.; Wang, X.; Peng, J.; Huang, Q. Shear rate effect on the residual strength characteristics of saturated loess in naturally drained ring shear tests. *Nat. Hazards Earth Syst. Sci.* **2020**, *20*, 2843–2856. [CrossRef]
5. Yates, K.; Fenton, C.H.; Bell, D.H. A review of the geotechnical characteristics of loess and loess-derived soils from Canterbury, South Island, New Zealand. *Eng. Geol.* **2018**, *236*, 11–21. [CrossRef]
6. Pan, P.; Shang, Y.-Q.; Lü, Q.; Yu, Y. Periodic recurrence and scale-expansion mechanism of loess landslides caused by groundwater seepage and erosion. *Bull. Int. Assoc. Eng. Geol.* **2017**, *78*, 1143–1155. [CrossRef]
7. Xu, J.; Li, Y.; Wang, S.; Wang, Q.; Ding, J. Shear strength and mesoscopic character of undisturbed loess with sodium sulfate after dry-wet cycling. *Bull. Eng. Geol. Environ.* **2020**, *79*, 1523–1541. [CrossRef]
8. Fan, X.; Xu, Q.; Scaringi, G.; Li, S.; Peng, D. A chemo-mechanical insight into the failure mechanism of frequently occurred landslides in the Loess Plateau, Gansu Province, China. *Eng. Geol.* **2017**, *228*, 337–345. [CrossRef]
9. Zhang, F.; Wang, G. Effect of irrigation-induced densification on the post-failure behavior of loess flowslides occurring on the Hei-fangtai area, Gansu, China. *Eng. Geol.* **2017**, *236*, 111–118. [CrossRef]
10. Zhang, F.; Wang, G.; Kamai, T.; Chen, W.; Zhang, D.; Yang, J. Undrained shear behavior of loess saturated with different concen-trations of sodium chloride solution. *Eng. Geol.* **2013**, *155*, 69–79. [CrossRef]
11. Wen, B.-P.; Yan, Y.-J. Influence of structure on shear characteristics of the unsaturated loess in Lanzhou, China. *Eng. Geol.* **2014**, *168*, 46–58. [CrossRef]
12. Xu, L.; Coop, M.R. Influence of structure on the behavior of a saturated clayey loess. *Can. Geotech. J.* **2016**, *53*, 1026–1037. [CrossRef]
13. Billard, A.; Muxart, T.; Derbyshire, E.; Wang, J.T.; Dijkstra, T. Landsliding and land use in the loess of Gansu Province, China. *Supplementband* **1993**, *87*, 117–131.
14. Carey, J.M.; McSaveney, M.J.; Petley, D.N. Dynamic liquefaction of shear zones in intact loess during simulated earthquake loading. *Landslides* **2016**, *14*, 789–804. [CrossRef]
15. Sorbino, G.; Nicotera, M.V. Unsaturated soil mechanics in rainfall-induced flow landslides. *Eng. Geol.* **2012**, *165*, 105–132. [CrossRef]
16. Liu, X.; Jin, M.; Li, D.; Zhang, L. Strength deterioration of a Shaly sandstone under dry–wet cycles: A case study from the Three Gorges Reservoir in China. *Bull. Eng. Geol. Environ.* **2017**, *77*, 1607–1621. [CrossRef]
17. Sawatsubashi, M.; Kiyota, T.; Katagiri, T. Effect of initial water content and shear stress on immersion-induced creep deformation and strength characteristics of gravelly mudstone. *Soils Found.* **2021**, *61*, 1223–1234. [CrossRef]
18. Wang, X.-G.; Lian, B.-Q.; Wang, J.-D.; Feng, W.-K.; Gu, T.-F. Creep damage properties of sandstone under dry-wet cycles. *J. Mt. Sci.* **2020**, *17*, 3112–3122. [CrossRef]
19. Zhang, F.; Jiang, A.; Yang, X. Shear creep experiments and modeling of granite under dry-wet cycling. *Bull. Eng. Geol. Environ.* **2021**, *80*, 5897–5908. [CrossRef]
20. Khan, M.A.; Hossain, M.S.; Samir, S.; Aramoon, A. Impact of Wet-Dry Cycles on the Shear Strength of High Plastic Clay Based on Direct Shear Testing. *Geotech. Front.* **2017**, *2017*, 615–622. [CrossRef]
21. Tang, C.; Shi, B.; Liu, C.; Zhao, L.; Wang, B. Influencing factors of geometrical structure of surface shrinkage cracks in clayey soils. *Eng. Geol.* **2008**, *101*, 204–217. [CrossRef]
22. Malusis, M.A.; Yeom, S.; Evans, J.C. Hydraulic conductivity of model soil–bentonite backfills subjected to wet–dry cycling. *Can. Geotech. J.* **2011**, *48*, 1198–1211. [CrossRef]
23. Yuan, Z.; Tang, C.; Yang, P.; Shu, A.; Yang, L. Experimental studies of compressive strength of undisturbed loess in drying-wetting cycle. *J. Eng. Geol.* **2018**, *26*, 155–161. (In Chinese)
24. Mu, H.; Deng, Y.; Rongjian, L.I. Experimental study on strength characteristics of loess at ground fissures in xi'an under action of dry and wet cycle. *J. Eng. Geol.* **2018**, *26*, 1132–1137. (In Chinese)
25. Yan, C.; Zhang, Z.; Jing, Y. Characteristics of strength and pore distribution of lime-flyash loess under freeze-thaw cycles and dry-wet cycles. *Arab. J. Geosci.* **2017**, *10*, 544. [CrossRef]
26. Li, G.; Wang, F.; Ma, W.; Fortier, R.; Mu, Y.; Mao, Y.; Hou, X. Variations in strength and deformation of compacted loess exposed to wetting-drying and freeze-thaw cycles. *Cold Reg. Sci. Technol.* **2018**, *151*, 159–167. [CrossRef]
27. Hu, C.; Yuan, Y.; Wang, X.; Mei, Y.; Liu, Z. Experimental study on strength deterioration model of compacted loess under wet-ting-drying cycles. *Chin. J. Rock Mech. Eng.* **2018**, *37*, 2804–2818. (In Chinese)
28. Zhao, T.; Wang, J. Soil-water characteristic curve for unsaturated loess soil considering density and wetting-drying cycle effects. *J. Cent. South Univ. Sci. Technol.* **2012**, *43*, 2445–2453.
29. Wang, F.; Li, G.; Mu, Y.; Zhang, P.; Fan, S. Experimental study of deformation characteristics of compacted loess subjected to dry-ing-wetting cycle. *Rock Soil Mech.* **2016**, *37*, 2306–2312.

30. Xu, J.; Li, Y.; Ren, C.; Lan, W. Damage of saline intact loess after dry-wet and its interpretation based on SEM and NMR. *Soils Found.* **2020**, *60*, 911–928. [CrossRef]
31. Wei, T.; Fan, W.; Yuan, W.; Wei, Y.-N.; Yu, B. Three-dimensional pore network characterization of loess and paleosol stratigraphy from South Jingyang Plateau, China. *Environ. Earth Sci.* **2019**, *78*, 333. [CrossRef]
32. Duan, Z.; Cheng, W.-C.; Peng, J.-B.; Wang, Q.-Y.; Chen, W. Investigation into the triggering mechanism of loess landslides in the south Jingyang platform, Shaanxi province. *Bull. Eng. Geol. Environ.* **2019**, *78*, 4919–4930. [CrossRef]
33. *D4318-10;* Standard Test Methods for Liquid Limit, Plastic Limit, and Plasticity Index of Soils. ASTM International: West Conshohocken, PA, USA, 2010.
34. *D 854;* Standard Test Methods for Specific Gravity of Soil Solids by Water Pycnometer. ASTM International: West Conshohocken, PA, USA, 2014.
35. *D 2216;* Standard Test Methods for Laboratory Determination of Water (Moisture) Content of Soil and Rock by Mass. ASTM International: West Conshohocken, PA, USA, 2010.
36. *D 6836;* Standard Test Methods for Determination of the Soil Water Characteristic Curve for Desorption Using a Hanging Column, Pressure Extractor, Chilled Mirror Hygrometer, and/or Centrifuge. ASTM International: West Conshohocken, PA, USA, 2002.
37. Sadeghi, H.; Hossen, S.B.; Chiu, A.C.; Cheng, Q.; Ng, C. Water retention curves of intact and re-compacted loess at different net stresses. *Jpn. Geotech. Soc. Spéc. Publ.* **2016**, *2*, 221–225. [CrossRef]
38. Guo, S.; Wu, J.; Dang, T.; Liu, W.; Li, Y.; Wei, W.; Syers, J.K. Impacts of fertilizer practices on environmental risk of nitrate in semiarid farmlands in the Loess Plateau of China. *Plant Soil* **2010**, *330*, 1–13. [CrossRef]
39. Xu, P.; Zhang, Q.; Qian, H.; Yang, F.; Zheng, L. Investigating the mechanism of pH effect on saturated permeability of remolded loess. *Eng. Geol.* **2021**, *284*, 105978. [CrossRef]
40. Hu, W.; Cheng, W.-C.; Wen, S.; Rahman, M. Effects of chemical contamination on microscale structural characteristics of intact loess and resultant macroscale mechanical properties. *Catena* **2021**, *203*, 105361. [CrossRef]
41. Nan, J.; Peng, J.; Zhu, F.; Ma, P.; Liu, R.; Leng, Y.; Meng, Z. Shear behavior and microstructural variation in loess from the Yan'an area, China. *Eng. Geol.* **2021**, *280*, 105964. [CrossRef]
42. Chen, X.; DU, S.; Zhang, D.; Li, B.; Ji, T. Triaxial creep properties of fiber reinforced expansive soil. *J. Eng. Geol.* **2017**, *25*, 80–87.
43. Yang, S.-Q.; Hu, B. Creep and Long-Term Permeability of a Red Sandstone Subjected to Cyclic Loading After Thermal Treatments. *Rock Mech. Rock Eng.* **2018**, *51*, 2981–3004. [CrossRef]
44. Xie, X.; Qi, S.; Zhao, F.; Wang, D. Creep behavior and the microstructural evolution of loess-like soil from Xi'an area, China. *Eng. Geol.* **2018**, *236*, 43–59. [CrossRef]
45. Shukla, A.; Joshi, Y.M. Boltzmann superposition principle for a time-dependent soft material: Assessment under creep flow field. *Rheol. Acta* **2017**, *56*, 927–940. [CrossRef]
46. Wang, J.; Wang, X.; Zhan, H.; Qiu, H.; Hu, S. A new superlinear viscoplastic shear model for accelerated rheological deformation. *Comput. Geotech.* **2019**, *114*, 103132. [CrossRef]
47. Betten, J. *Creep Mechanics;* Springer Science & Business Media: Berlin/Heidelberg, Germany, 2008.
48. Vinogradov, V.; Milton, G. The total creep of viscoelastic composites under hydrostatic or antiplane loading. *J. Mech. Phys. Solids* **2005**, *53*, 1248–1279. [CrossRef]
49. Al-Zubaydi, A.H.; Al-Atalla, M.A.; Al-Kiki, I.M. Long term strength and durability of clayey soil stabilized with lime. *Eng. Technol. J.* **2011**, *29*, 725–735.
50. Fatahi, B.; Le, T.M.; Le, M.Q.; Khabbaz, H. Soil creep effects on ground lateral deformation and pore water pressure under em-bankments. *Geomech. Geoengin.* **2013**, *8*, 107–124. [CrossRef]
51. Liu, L.; Xu, W. Experimental Researches on Long-Term Strength of Granite Gneiss. *Adv. Mater. Sci. Eng.* **2015**, *2015*, 187616. [CrossRef]
52. Kowalewski, Z.L. Assessment of the Multiaxial Creep Data Based on the Isochronous Creep Surface Concept. In *IUTAM Symposium on Creep in Structures;* Springer: Dordrecht, The Netherlands, 2001; pp. 401–410. [CrossRef]
53. Metri, V.; Briels, W. Brownian dynamics investigation of the Boltzmann superposition principle for orthogonal superposition rheology. *J. Chem. Phys.* **2019**, *150*, 014903. [CrossRef]
54. Cong, L.; Hu, X. Triaxial rheological property of sandstone under low confining pressure. *Eng. Geol.* **2017**, *231*, 45–55. [CrossRef]
55. Jury, W.A. Simulation of solute transport using a transfer function model. *Water Resour. Res.* **1982**, *18*, 363–368. [CrossRef]
56. Kong, R.; Gao, L.; Zhao, W.; Zhao, B. Multi-scale analysis of geotechnical and physicochemical changes in loess caused by nano-SiO2 pile migration. *Front. Earth Sci.* **2021**, *9*, 820. [CrossRef]
57. Yao, R.-J.; Yang, J.-S.; Zhang, T.-J.; Hong, L.-Z.; Wang, M.-W.; Yu, S.-P.; Wang, X.-P. Studies on soil water and salt balances and scenarios simulation using SaltMod in a coastal reclaimed farming area of eastern China. *Agric. Water Manag.* **2014**, *131*, 115–123. [CrossRef]

applied sciences

MDPI

Article

Debris Flow Characteristics in Flume Experiments Considering Berm Installation

Hyungjoon Chang, Kukhyun Ryou * and Hojin Lee

School of Civil Engineering, Chungbuk National University, Cheongju 28644, Korea; param79@cbnu.ac.kr (H.C.); hojinlee@cbnu.ac.kr (H.L.)
* Correspondence: rgh0126@naver.com; Tel.: +82-10-4240-9941

Abstract: This study was conducted to identify the characteristics and mobility of debris flows and analyze the performance of a berm as a debris flow mitigation measure. The debris flow velocity, flow depth, Froude number, flow resistance coefficients, and mobility ratio were accordingly determined using the results of flume tests. To analyze the influence of the berm, the results for a straight channel test without a berm were compared with those for a single-berm channel test. The debris flow velocity was observed to increase with increasing channel slope and decreasing volumetric concentration of sediment, whereas the mobility ratio was observed to increase with increasing channel slope and volumetric concentration of sediment. In addition, it was confirmed that the installation of a berm significantly decreased the debris flow velocity and mobility ratio. This indicates that a berm is an effective method for reducing damage to areas downstream of a debris flow by decreasing its potential mobility. By identifying the effects of berms on debris flow characteristics according to the channel slope and volumetric concentration of sediment, this study supports the development of berms to serve as debris flow damage mitigation measures.

Keywords: debris flow; berm; mitigation measure; volumetric concentration

check for updates

Citation: Chang, H.; Ryou, K.; Lee, H. Debris Flow Characteristics in Flume Experiments Considering Berm Installation. *Appl. Sci.* **2021**, *11*, 2336. https://doi.org/10.3390/app11052336

Academic Editor: Ricardo Castedo

Received: 2 February 2021
Accepted: 2 March 2021
Published: 6 March 2021

1. Introduction

The frequency of torrential rainfall has increased worldwide due to climate change caused by global warming, in turn increasing the occurrence of forest soil sediment disasters, such as landslides and debris flows [1]. Indeed, in South Korea, the occurrence of forest soil sediment disasters has increased due to the increasing frequency of torrential rainfall. The average annual area affected by the occurrence of forest soil sediment disasters in Korea was 290 ha for the period from 1980 to 1999, but significantly increased to 469 ha for the period from 2000 to 2019. In particular, a large-scale debris flow occurred in July 2011 at Mt. Woomyeon, located in an urban area, causing serious human casualties and property damage. This incident increased public interest in debris flow safety and expanded debris flow policy-making from mountainous areas to include urban areas as well [2]. It is known that debris flows can be initiated by various factors such as rainfall, snowmelt, typhoons, volcanoes, and earthquakes [3], but rainfall has been identified as the main cause of the debris flows that have occurred in Korea [4,5].

Debris flow, a type of mass movement, is a mixture composed of water and sediments of various particle sizes from clay to boulders. It is a dynamic phenomenon that moves downhill under the influence of gravity and may cause human casualties and property damage along the way [3,6–9]. Field monitoring of debris flows is difficult because they occur irregularly and exist for only a short duration of time [7,10,11]. Furthermore, it is difficult to accurately predict and respond to debris flows because they increase in scale by absorbing the sediment and water along their movement paths through strong erosive force [7,11–13]. Notably, the debris flow is distinguished from other mass movements because it is a sediment–water mixture, and thus can transfer momentum under the influence of grain friction, grain collision, and viscous fluid dynamics [14]. In addition, momentum transfer caused by collisions between grains in a debris flow leads to excess pore water pressure, which reduces the shear resistance of the debris flow and results in very high mobility [14]. Thus, a debris flow can produce a peak discharge dozens of times higher than that of a flood in the same watershed [15,16], and its flow characteristics may vary depending on the relative contents of water and sediment as well as the size and type of particles [3]. Therefore, it is necessary to identify the flow and deposition characteristics of debris flows and establish mitigation measures to predict and reduce the damage they cause.

Debris flow mitigation measures can be mainly divided into structural and non-structural measures. Structural mitigation measures directly reduce debris flow damage using structures installed in the path of a debris flow; erosion control dams, flexible debris flow barriers, and berms can all be considered appropriate structures for this purpose. Non-structural mitigation measures indirectly limit debris flow damage using methods such as debris flow forecasting and warning systems or land use regulations. Figure 1 shows examples of real world slopes with multiple berms. This study evaluated the use of a berm, a structural debris flow mitigation measure consisting of a small step installed on a slope that effectively disperses rainwater, reduces the momentum of debris flows, and reduces the riverbed erosion. Berms can be easily constructed at a low cost compared to other debris flow mitigation measures. Therefore, several studies applying berms as a debris flow mitigation measure have been recently conducted. VanDine [17] introduced a method of applying berms to lateral, deflection, and terminal walls serving as open debris flow control structures, and suggested major considerations for structural design. Prochaska et al. [18] pointed out that existing analysis methods for debris basins and deflection berms were not sufficient for the prediction of debris flow volume or the subsequent design of the berm geometry, impact load, and outlet, and accordingly suggested solutions and guidelines. Kim and Lee [19] performed numerical simulations by applying the finite difference method based on the mass conservation and momentum conservation equations to evaluate the behavior and mechanism of debris flow in response to the installation of berms on a slope. Sharma et al. [20] calculated the axial force, horizontal displacement, and safety factor using the finite element method to identify the stability of a slope according to the soil cohesion, the angle of internal friction, and the width and height of the berm installed upon it.

(a) (b)

Figure 1. Examples of using berms in real world slopes: (**a**) Slopes with berms in coal mine areas (Dr Ajay Kumar Singh/Shutterstock.com); (**b**) slopes with berms in Peru (lialina/Shutterstock.com).

To effectively design debris flow mitigation measures, it is necessary to first calculate the main debris flow parameters, such as the potential debris volume, impact force, flow velocity, peak discharge, mobility ratio, runout distance, and total travel distance [16,21]. In particular, the flow velocity must be considered in the design of any debris flow mitigation measure because it directly influences the impact force and deposition characteristics, and is affected by the channel topography [3,15,22,23]. In the countries where studies on debris flow characteristics have been actively conducted, the debris flow process and impact force can be estimated relatively accurately [10,21,24]. Domestic studies on debris flows in South

Korea, however, have been limited to trend analyses for simple experimental conditions. Numerical analysis, field observations, and flume experiments have all been previously used for research into the behavior and mechanism of debris flows. Numerical analyses provide relatively high accuracy, but it is difficult to obtain the necessary parameters through field observations, and as such, the observational data required to verify the results of such analyses are insufficient [2]. Thus, while field observations and numerical analyses are suitable for developing and testing methods to predict the behavior of debris flows, flume experiments are more suitable for conducting research on the behavior of debris flows under controlled conditions to develop related prediction equations [8]. Indeed, flume experiments are the only available method for identifying the flow and deposition characteristics of debris flows through repeatable experiments with reproducible results [25]. For these reasons, flume experiments have been mainly used in the past to identify the behaviors of debris flows.

In this study, flume experiments were therefore performed to analyze debris flow characteristics and mobility ratios according to channel slope, volumetric concentration of sediment, and presence of a berm. The flow velocity, flow depth, Froude number, and mobility ratio of the debris flows were then calculated based on the experimental observations. In addition, flow resistance coefficients were calculated by substituting the observed flow velocity and flow depth into the debris flow velocity estimation equations.

2. Methods for Assessing Debris Flow Characteristics

2.1. Flow Velocity

The debris flow velocity must be considered in any debris flow risk assessment and subsequent design of mitigation measures, because it significantly affects the impact force. It can be directly measured through field monitoring or flume experiments, or calculated using flow velocity estimation equations. The basic formula suggested in previous studies for estimating the flow velocity of a debris flow is [16,26–28]:

$$v = Ch^a\alpha^b \tag{1}$$

where v is the flow velocity, C is the flow resistance coefficient, h is the flow depth, α is the channel slope, and a and b are exponential factors defined according to the flow characteristics.

Table 1 defines Equations (2) through (6), which have been proposed by Hungr et al. [15], Rickenmann [16,29], Takahashi [22], Koch [26], and Lo [27]. These equations have been proposed to calculate the flow velocity for different flow types through field observations, flume experiments, and numerical analyses based on Equation (1). In Table 1, ρ is the mixed density of debris flow (kg·m^{-3}), g is the gravitational acceleration (9.81 m·s^{-2}), k is the cross-sectional shape factor (3 for wide rectangular channels, 5 for trapezoidal channels, and 8 for semicircular channels), μ is the apparent dynamic viscosity of the debris flow (Pa·s), ξ is the lumped coefficient depending on the volumetric concentration of sediment (m$^{-1/2}$·s^{-1}), n is the Manning coefficient (m$^{-1/3}$·s), C_1 is the Chezy coefficient (m$^{1/2}$·s^{-1}), and C_2 is the empirical coefficient proposed by Koch [26] (m$^{0.78}$·s^{-1}). In this study, the flow resistance coefficients μ, ξ, n, C_1, and C_2 were calculated by substituting the experimentally observed debris flow velocity and flow depth into Equations (2) through (6).

The various coefficients in Equations (2) through (6) are empirical constants known to provide results that are reasonably consistent with field observations [27]. Hungr et al. [15] suggested that Equation (2) is suitable to estimate the debris flow velocities because a laminar flow regime is formed near the peak of debris flow surge. The variable k in Equation (2) has a different value depending on the cross-sectional shape of the channel [16]. Rickenmann [16], and Eu et al. [2], used a value of 3 as k, which is valid for a wide rectangular channel. This is because the debris flow depth is generally smaller than the width. Equation (3) is based on Bagnold's theory [30] of dilatant grain shearing in the inertial regime. Takahashi [22] estimated the debris flow velocity based on Equation (3) and suggested that it is suitable for stony debris flows in Japan. Equation (4) is known as Manning's formula, and Mizuyama [31] proposed using it for estimating the debris flow velocity. Equation (5) is known as Chezy's formula, and was initially introduced to estimate the velocity of snow avalanches; Rickenmann [29] used it to estimate the debris flow velocity. Equation (6) was proposed by Koch [26] through numerical analyses. Koch [26] confirmed that the Newtonian turbulent, Voellmy, and empirical models, which have smaller exponents (a and b in Equation (1)), are closer to observed values in the field than other models.

Table 1. Equations used to estimate the mean velocity of debris flow according to flow type.

Flow Type	Equation	Reference
Newtonian laminar flow	$v = \frac{\rho g h^2 \alpha}{k\mu}$ (2)	[15]
Dilatant grain shearing flow	$v = \frac{2}{3}\xi h^{3/2}\alpha^{1/2}$ (3)	[15,22]
Newtonian turbulent flow	$v = \frac{1}{n}h^{2/3}\alpha^{1/2}$ (4)	[15,27]
Voellmy flow	$v = C_1 h^{1/2}\alpha^{1/2}$ (5)	[16,26,29]
Empirical equation	$v = C_2 h^{0.22}\alpha^{0.33}$ (6)	[26]

2.2. Froude Number

Debris flows are dominated by the influence of gravity. Therefore, similarity must be applied using the Froude number, which represents the ratio of inertial force to gravity. The Froude number is expressed as

$$F_r = \frac{v}{\sqrt{gh}} \tag{7}$$

where F_r is the Froude number.

2.3. Mobility Ratio

The debris flow mobility ratio is a dimensionless number, generally known as the Heim coefficient, obtained by dividing the total drop between the initial occurrence and final deposition points of the debris flow by the total travel distance, which is defined as the horizontal distance between the two points. This concept was introduced by Heim [32] to analyze rock avalanches, and its applicability was expanded when Iverson [14] applied it to the analysis of debris flows. The debris flow mobility ratio can be used to predict the runout distance and flow velocity [33], and many studies have used this ratio because it accurately reflects the risk and potential mobility of debris flows [8,11,14,33,34].

3. Flume Experiments

3.1. Experimental Setup

Figure 2 shows the experimental setups and main parameters of the flume experiments performed in this study. Here, H is the total drop between the initial occurrence and final deposition points of the debris flow, L is the total travel distance. The setup for the flume experiments was composed of a sample box, a channel, and a deposition plane. The sample box was 0.2 m long, 0.15 m wide, and 0.3 m high, and installed at the top of the channel. The sediment–water mixture was supplied to the channel by manually lifting the gate of the sample box. The channel was made of steel in consideration of the strong erosion that accompanies debris flow, and had a length of 1.3–1.9 m, a width of 0.15 m, and a height of 0.3 m (Table 2). The channel was fabricated in separate upper and lower slope segments so that a 0.6-m long berm could be installed between them. At the outlet of the channel, a 1.5-m long, 1.0-m wide deposition plane was installed to analyze the debris flow deposition characteristics. The deposition plane was composed of 0.05-m grids in the longitudinal and lateral directions.

Table 2. Dimensions of the channels for each test type.

Test Type	Length (m)	Width (m)	Height (m)
Straight channel test	1.3	0.15	0.3
Single-berm channel test	1.9	0.15	0.3

To observe the flow and deposition characteristics of the debris flow, cameras capable of capturing video at 60 fps were installed at the front and on a side of the flume. The front camera was used to measure the mean velocity of the head of the debris flow from the top of the channel to the outlet, and the side camera was used to measure the maximum depth of the debris flow at a point 0.1 m upstream of the channel outlet. The debris flow mobility ratio was calculated upon the completion of sediment–water mixture deposition by observing the total drop H and total travel distance L, as indicated in Figure 2.

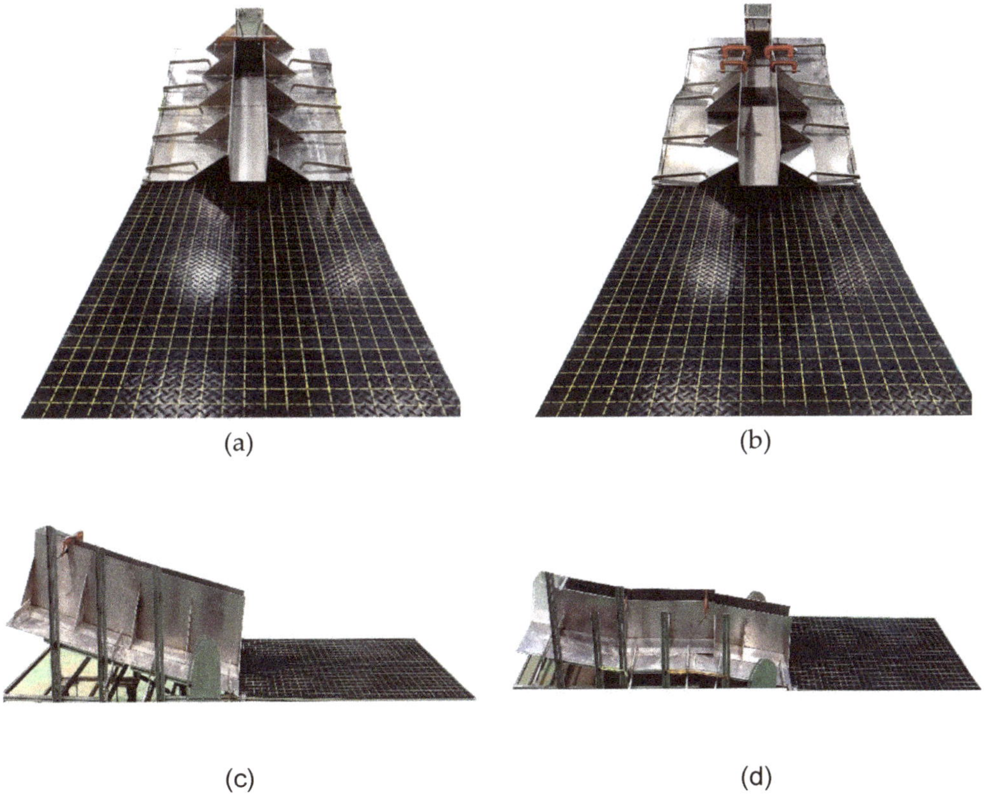

Figure 2. *Cont.*

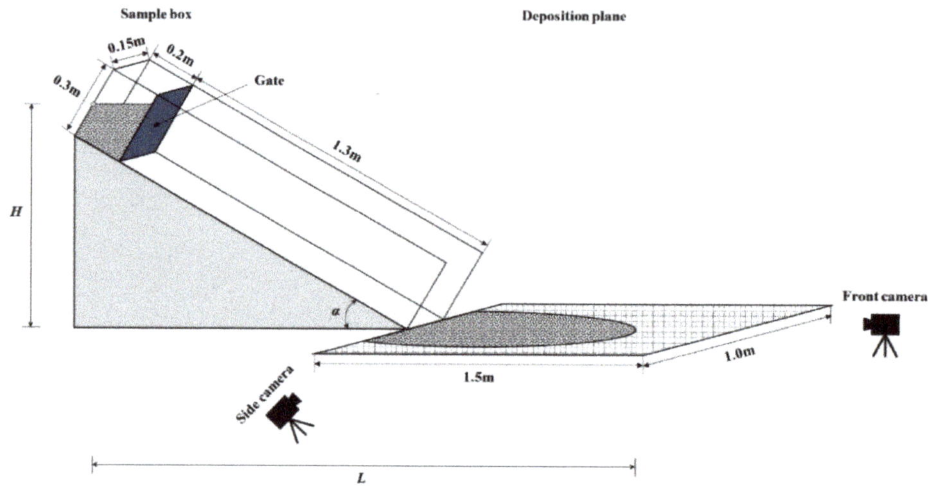

(e)

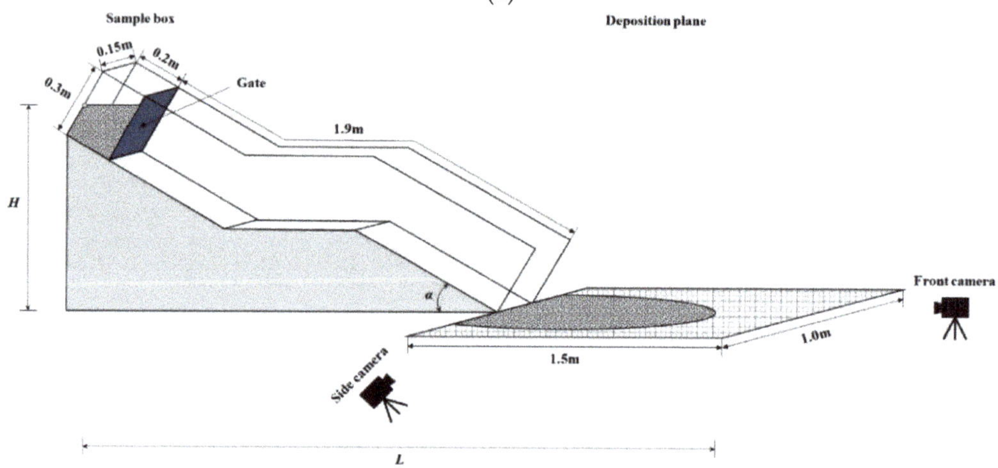

(f)

Figure 2. Experimental setups for flume experiments: (**a**) Front image of the straight channel test; (**b**) front image of the single-berm channel test; (**c**) side image of the straight channel test; (**d**) side image of the single-berm channel test; (**e**) schematic diagram of the straight channel test; and (**f**) schematic diagram of the single-berm channel test.

3.2. Experimental Conditions

Debris flows are a sediment–water mixture, thus its viscosity is difficult to measure using only a general soil test [1]. Takahashi [9] indirectly estimated the viscosity of debris flows using the volumetric concentration of sediment, and confirmed that the viscosity of debris flows increases as the volumetric concentration of sediment increases. As the viscosity of debris flows increases, the shear resistance and bed friction of the debris flows increase. This leads to a decrease in the momentum of the debris flows, which changes the debris flow characteristics. Therefore, it is necessary to compose various experimental conditions for the viscosity of debris flows. The volumetric concentration of sediment can be obtained by dividing the volume of sediment by the volume of the sediment–water mixture as follows:

$$C_V = \frac{V_S}{(V_S + V_W)} = \frac{V_S}{V_{total}} \quad (8)$$

where C_V is the volumetric concentration of sediment, V_S is the volume of sediment, V_W is the volume of water, and V_{total} is the total volume of the sediment–water mixture.

In this study, the viscosity of the debris flow was also considered using C_V. Table 3 shows the experimental cases evaluated in this study considering the experiment type (straight channel test or single-berm channel test), α (10–25° in 5° increments), and C_V (0.40–0.60 in increments of 0.05). Consequently, a total of 40 experimental conditions were evaluated in this study, each of which was repeated five times.

Table 3. Experimental conditions used in this study.

	Straight Channel Test				Single-Berm Channel Test			
Channel Slope α (°)	**10**	**15**	**20**	**25**	**10**	**15**	**20**	**25**
Volumetric concentration C_V	0.40	0.40	0.40	0.40	0.40	0.40	0.40	0.40
	0.45	0.45	0.45	0.45	0.45	0.45	0.45	0.45
	0.50	0.50	0.50	0.50	0.50	0.50	0.50	0.50
	0.55	0.55	0.55	0.55	0.55	0.55	0.55	0.55
	0.60	0.60	0.60	0.60	0.60	0.60	0.60	0.60

3.3. Sample Properties

In this study, a sediment–water mixture was used to reproduce the conditions of a debris flow. Table 4 shows the particle composition of the sediment as determined through a sieve analysis, with reference to previous studies [35,36], and Figure 3 shows the particle-size distribution curve of the sediment. The weight ratios of each particle size range were 25% for 4.75–9.50 mm, 25% for 2.00–4.75 mm, and 50% for 2.00 mm or less. In this study, C_V was set to be 0.40–0.60, based on previous studies with similar experimental setup scales [1,8,11,12]. Table 5 shows the weight of sediment and water required for each experiment when C_V was adjusted within the 0.40–0.60 range, where W is the weight of the sediment-water mixture. The same volume of sediment–water mixture (4500 cm^3) was used in the straight channel test and single-berm channel test. As C_V increased from 0.40 to 0.60, the density of the sediment–water mixture linearly increased from 1578 to 1867 $kg{\cdot}m^{-3}$ (Table 5).

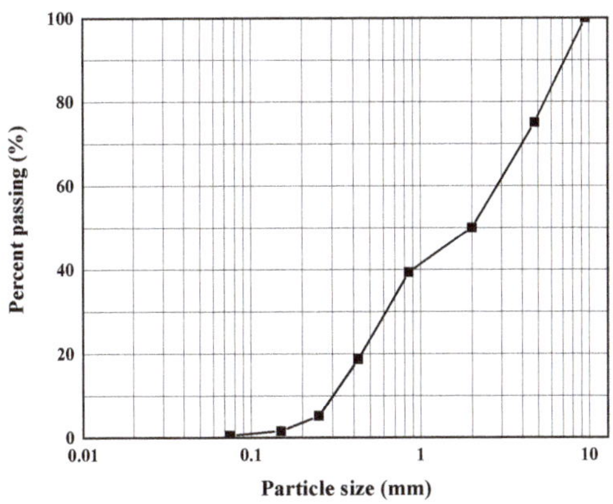

Figure 3. Grain-size distribution curve of the mixed sediment used in the debris flow.

Table 4. Weight ratios of the sediment–water mixture by particle size.

Particle Size (mm)	Weight Ratio (%)
4.750–9.500	25.00
2.000–4.750	25.00
0.850–2.000	10.67
0.425–0.850	20.63
0.250–0.425	13.57
0.150–0.250	3.63
0.075–0.150	1.00
<0.075	0.50

Table 5. Weight of debris flow samples according to volumetric concentration of sediment C_V.

C_V	<2 mm (kg_f)	2–4.75 mm (kg_f)	4.75–9.5 mm (kg_f)	Water (kg_f)	W (kg_f)	ρ ($kg{\cdot}m^{-3}$)
0.40	2.200	1.100	1.100	2.700	7.100	1578
0.45	2.475	1.238	1.238	2.475	7.425	1650
0.50	2.750	1.375	1.375	2.250	7.750	1722
0.55	3.025	1.513	1.513	2.025	8.075	1794
0.60	3.300	1.650	1.650	1.800	8.400	1867

3.4. Experimental Method

Figures 4 and 5 shows examples and a flow chart of flume experiments performed in this study, respectively. Experimental videos were separately presented in Videos S1 and S2. Flume experiments were first planned by determining the setup and conditions for each experiment, as defined in Table 3. Then, the effects of α, C_V, and berm installation were examined for each of the 40 defined test conditions in the following sequence. Sediment and water were added to the sample box to meet the required weight and mixed well. The gate was then rapidly removed from the sample box to supply the sediment–water mixture. Finally, the flow velocity, flow depth, and mobility ratio of the resulting debris flow were observed as described in Section 3.1, and the flow resistance coefficient and Froude number were calculated accordingly.

(a) (b)

Figure 4. Examples of flume experiments conducted in this study: (**a**) Straight channel test (α was 25° and C_V was 0.50); and (**b**) single-berm channel test (α was 25° and C_V was 0.50).

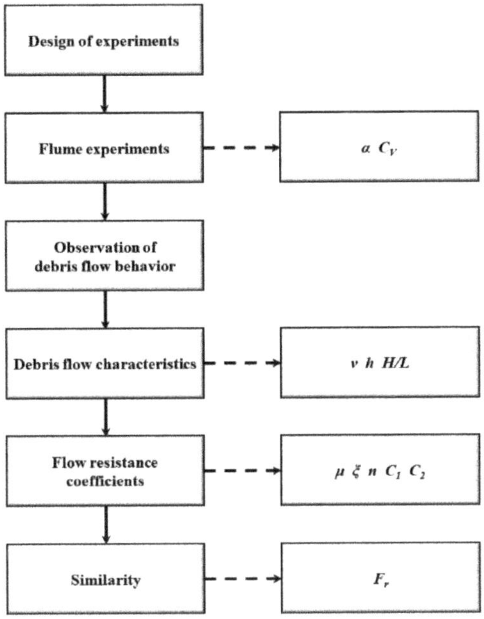

Figure 5. Flow chart of flume experiments.

4. Results and Analysis

Table 6 shows the debris flow velocity, flow depth, Froude number, total travel distance, and mobility ratio observed through the flume experiments, where N is the experiment number. In the case of the straight channel test, the development of debris flow was insufficient to observe the flow depth when α was 10° and C_V was 0.60; however, debris flow was sufficiently developed when α was 15° or greater. In the single-berm channel test, debris flow stopped in the channel when C_V was 0.60, regardless of α, so these values are not reported in the analysis below. In addition, when α was 20° or less and C_V was between 0.50 and 0.55, the development of debris flow was insufficient to observe the flow depth in some experiments.

Table 6. Observed data from flume experiments.

Test Type	N	α (°)	C_V	v (m·s^{-1})	h (m)	F_r	H (m)	L (m)	H/L
Straight channel	1	10	0.40	1.826	0.0198	4.14	0.391	2.459	0.159
	2	10	0.45	1.697	0.0160	4.28	0.391	2.340	0.167
	3	10	0.50	1.559	0.0108	4.79	0.391	2.194	0.178
	4	10	0.55	1.324	0.0058	5.55	0.391	2.105	0.186
	5	10	0.60	1.169	-	-	0.391	1.780	0.220
	6	15	0.40	1.929	0.0184	4.54	0.507	2.630	0.193
	7	15	0.45	1.811	0.0150	4.72	0.507	2.541	0.200
	8	15	0.50	1.697	0.0104	5.31	0.507	2.339	0.217
	9	15	0.55	1.621	0.0050	7.32	0.507	2.098	0.242
	10	15	0.60	1.231	0.0038	6.38	0.507	1.763	0.288
	11	20	0.40	2.265	0.0168	5.58	0.620	2.661	0.233
	12	20	0.45	2.083	0.0142	5.58	0.620	2.560	0.242

Table 6. *Cont.*

Test Type	N	α (°)	C_V	v (m·s^{-1})	h (m)	F_r	H (m)	L (m)	H/L
	13	20	0.50	1.912	0.0102	6.04	0.620	2.417	0.256
	14	20	0.55	1.806	0.0048	8.32	0.620	2.176	0.285
	15	20	0.60	1.629	0.0034	8.92	0.620	1.765	0.351
	16	25	0.40	2.453	0.0154	6.31	0.728	2.739	0.266
	17	25	0.45	2.347	0.0104	7.35	0.728	2.613	0.278
	18	25	0.50	2.181	0.0094	7.18	0.728	2.429	0.300
	19	25	0.55	2.070	0.0045	9.85	0.728	2.187	0.333
	20	25	0.60	1.912	0.0032	10.79	0.728	1.817	0.400
Single-berm channel	21	10	0.40	1.594	0.0125	4.55	0.391	2.907	0.134
	22	10	0.45	1.383	0.0079	4.97	0.391	2.587	0.151
	23	10	0.50	1.152	-	-	0.391	2.349	0.166
	24	10	0.55	0.870	-	-	0.391	2.255	0.173
	25	10	0.60	-	-	-	-	-	-
	26	15	0.40	1.746	0.0106	5.41	0.507	2.969	0.171
	27	15	0.45	1.527	0.0078	5.52	0.507	2.791	0.182
	28	15	0.50	1.435	0.0030	8.36	0.507	2.602	0.195
	29	15	0.55	1.153	-	-	0.507	2.252	0.225
	30	15	0.60	-	-	-	-	-	-
	31	20	0.40	2.026	0.0104	6.34	0.620	3.089	0.201
	32	20	0.45	1.855	0.0074	6.88	0.620	2.873	0.216
	33	20	0.50	1.678	0.0054	7.29	0.620	2.617	0.237
	34	20	0.55	1.319	-	-	0.620	2.280	0.272
	35	20	0.60	-	-	-	-	-	-
	36	25	0.40	2.375	0.0178	5.68	0.728	3.159	0.230
	37	25	0.45	2.184	0.0100	6.97	0.728	2.938	0.248
	38	25	0.50	1.967	0.0080	7.02	0.728	2.647	0.275
	39	25	0.55	1.699	0.0030	9.90	0.728	2.380	0.306
	40	25	0.60	-	-	-	-	-	-

4.1. Debris Flow Characteristics

4.1.1. Flow Velocity

Figure 6 shows the flow velocities observed through the flume experiments and their changes due to the installation of the berm. Figure 6a,b show the debris flow velocity according to α and C_V, respectively. Figure 6c,d show the percent change in debris flow velocity according to α and C_V, respectively, after the berm was installed. The experimental results indicated that the debris flow velocity increased as α increased or C_V decreased (Figure 6a,b). In the case of the straight channel test, the incremental increase in flow velocity with increasing α was similar for C_V values less than 0.55, but increased noticeably with α at a higher C_V. The incremental decrease in flow velocity with increasing C_V was similar, regardless of α. In the case of the single-berm channel test, the incremental increase in flow velocity with increasing α was found to be similar, regardless of C_V. The incremental decrease in flow velocity with increasing C_V was similar when α was 15° or greater, but increased noticeably with C_V when α was less than 15°. In addition, it was confirmed that the installation of the berm on the slope reduced the debris flow velocity by 3.2–34.3% (Figure 6c,d). The average decreases in debris flow velocity were found to be 22.9%, 17.4%, 15.2%, and 9.5% for a C_V of 0.40–0.55 and α of 10°, 15°, 20°, and 25°, respectively, and 9.0%, 13.0%, 15.9%, and 27.0% for an α of 10°–25° and C_V of 0.40, 0.45, 0.50, and 0.55, respectively. This confirmed that the installation of the berm was more effective in reducing the debris flow velocity at smaller α values and larger C_V values.

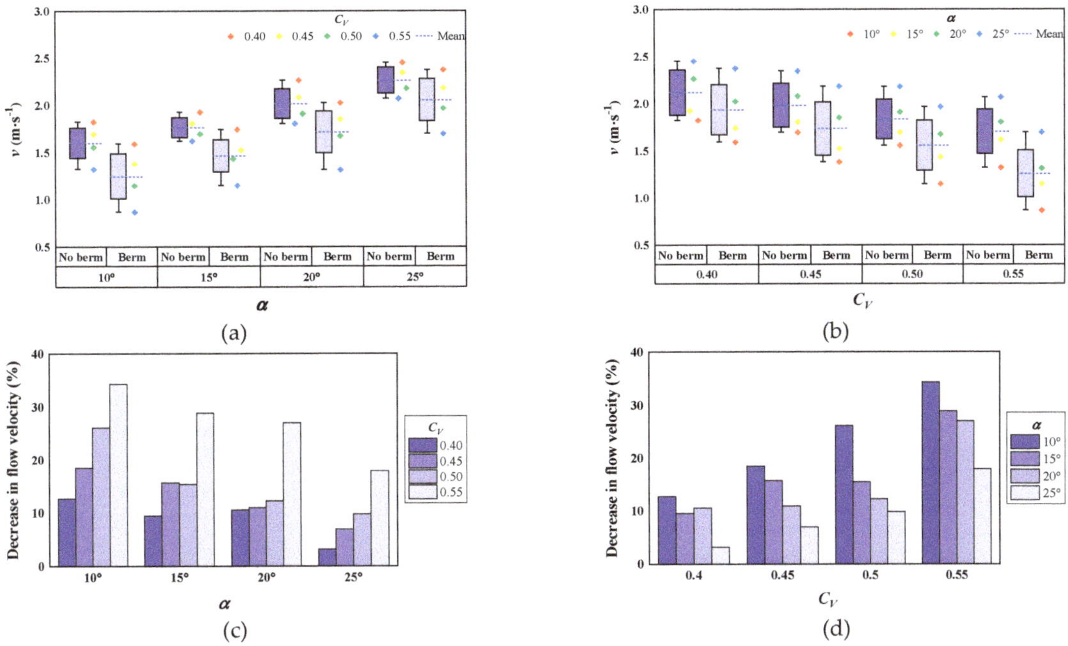

Figure 6. (**a**) Flow velocity according to α (at C_V = 0.40–0.55); (**b**) flow velocity according to C_V (at α = 10°–25°); (**c**) percent decrease in flow velocity due to berm according to α (at C_V = 0.40–0.55); (**d**) percent decrease in flow velocity due to berm according to C_V (at α = 10°–25°).

4.1.2. Flow Depth

Figure 7 shows the flow depths observed through the flume experiments and their changes due to the installation of the berm. Figure 7a,b show the debris flow depth according to α and C_V, respectively. Figure 7c,d show the percent change in debris flow depth according to α and C_V, respectively, after the berm was installed. In the case of the straight channel test, the debris flow depth decreased as α increased or C_V increased (Figure 7a,b). The decrease in flow depth with increasing α slowed when C_V was 0.50 or greater, and the decrease in flow depth with increasing C_V slowed when α was 15° or greater. In the case of the single-berm channel test, the debris flow depth decreased as C_V increased (Figure 7b). The decrease in flow depth with increasing α changed when C_V was 0.50, and the flow depth suddenly increased when α was 25° and C_V was 0.45 or less (Figure 7a). The installation of the berm was found to decrease the debris flow depth by an average of 3.9–71.2%, even though the flow depth increased by 15.6% when α was 25° and C_V was 0.40 (Figure 7c,d). The average decreases in debris flow depth were 43.7%, 53.8%, 44.4%, and 1.1% for a C_V of 0.40–0.50 and α values of 10°, 15°, 20°, and 25°, respectively, and 25.4%, 37.6%, and 44.4% for an α of 10°–25° and C_V values of 0.40, 0.45, and 0.50, respectively. This confirmed that the installation of the berm was more effective in reducing the debris flow depth at larger C_V values.

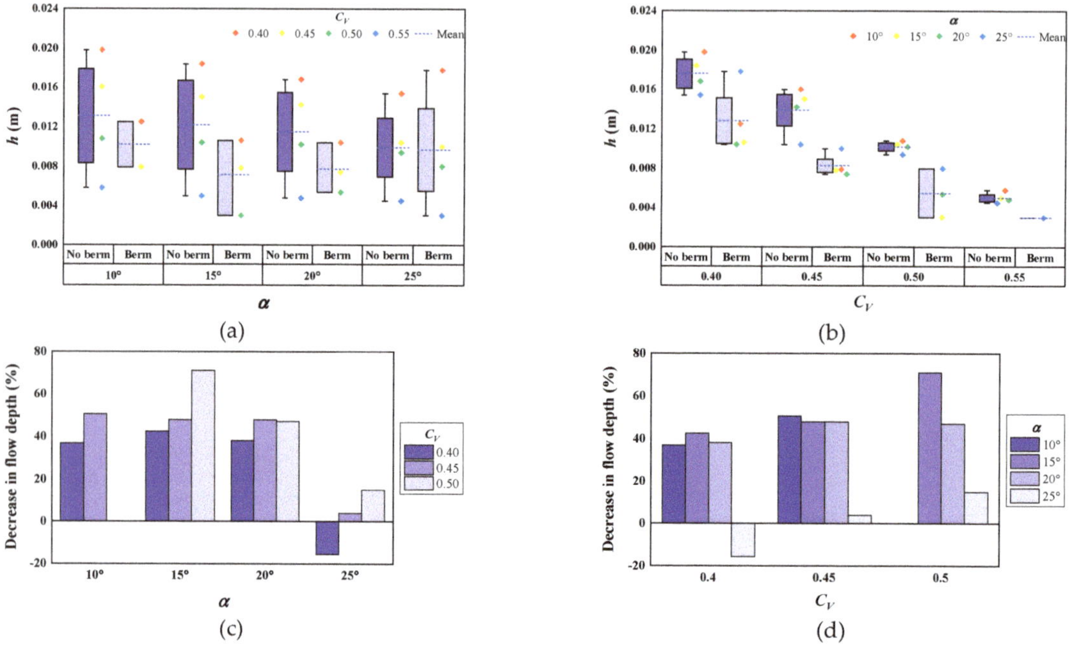

Figure 7. (**a**) Flow depth according to α (at C_V = 0.40–0.55); (**b**) flow depth according to C_V (at α = 10°–25°); (**c**) percent decrease in flow depth due to berm according to α (at C_V = 0.40–0.50); (**d**) percent decrease in flow depth due to berm according to C_V (at α = 10°–25°).

4.1.3. Froude Number

Figure 8 shows the Froude numbers calculated through the flume experiment observations and the change in Froude number due to the installation of the berm. Figure 8a,b show the Froude number according to α and C_V, respectively. Figure 8c,d show the change in the Froude number according to α and C_V, respectively, when the berm was installed. In the case of the straight channel test, the Froude number increased as α increased (Figure 8a). With the berm installed, the Froude number decreased by 2.1–9.7% when α was 25° and C_V was 0.50 or less (Figure 8c,d); however, when α was less than 25°, the Froude number increased by 9.7–58.0% due to the installation of the berm (Figure 8c,d). The average incremental increases in the Froude number were found to be 12.8%, 30.8%, and 19.5% for a C_V of 0.40–0.50 and α values of 10°, 15°, and 20°, respectively, and 8.4%, 12.3%, and 25.6% for an α of 10°–25° and C_V values of 0.40, 0.45, and 0.50, respectively. In other words, the average increase in the Froude number due to the installation of the berm increased as C_V increased.

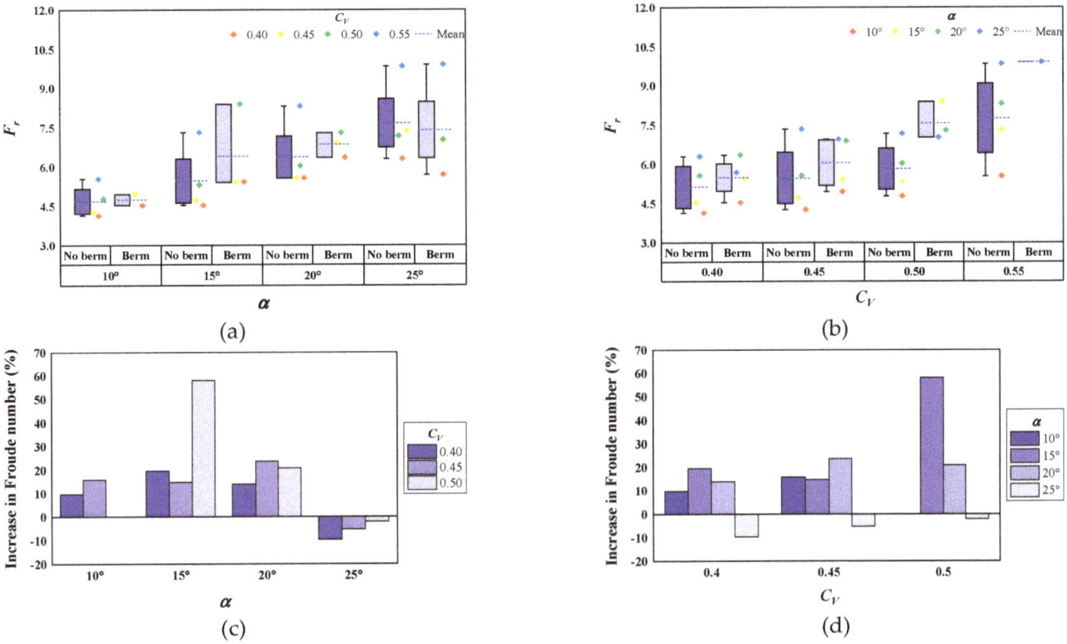

Figure 8. (**a**) Froude number according to α (at C_V = 0.40–0.55); (**b**) Froude number according to C_V (at α = 10°–25°); (**c**) percent increase in Froude number due to berm according to α (at C_V = 0.40–0.50); (**d**) percent increase in Froude number due to berm according to C_V (at α = 10°–25°).

4.1.4. Flow Resistance Coefficients

Table 7 shows the ranges, mean values, and standard deviations of the flow resistance coefficients μ, ξ, n, C_1, and C_2 calculated using Equations (2)–(6) based on the experimental results in this study. Figure 9 shows the experimentally obtained flow resistance coefficient values according to α and C_V. The experimental results showed that α had little influence on any flow resistance coefficients (Figure 9a,c,e,g,i), but C_V clearly affected all flow resistance coefficients except C_2 (Figure 9b,d,f,h,j): As C_V increased, the values of μ and n decreased, whereas the values of ξ and C_1 increased. When the berm was installed, the values of ξ decreased by 4.3% when α was 25° and C_V was 0.40. Except for this one observation, each flow resistance coefficient exhibited a consistent change pattern due to the installation of the berm. The installation of the berm decreased μ by 8.6–93.8%, increased ξ by up to 563%, decreased n by 7.0–57.2%, increased C_1 by 10.6–91.4%, and increased C_2 by 2.9–26.4%.

Table 7. Flow resistance coefficients calculated in this study.

Flow Resistance Coefficient	Range	Mean	Standard Deviation
μ (Pa·s)	0.006–0.243	0.089	0.075
ξ ($m^{-1/2}·s^{-1}$)	2240–30,764	9152	7514
n ($m^{-1/3}·s$)	0.006–0.019	0.012	0.003
C_1 ($m^{1/2}·s^{-1}$)	27.47–61.53	38.87	8.132
C_2 ($m^{0.78}·s^{-1}$)	6.48–9.05	8.00	0.639

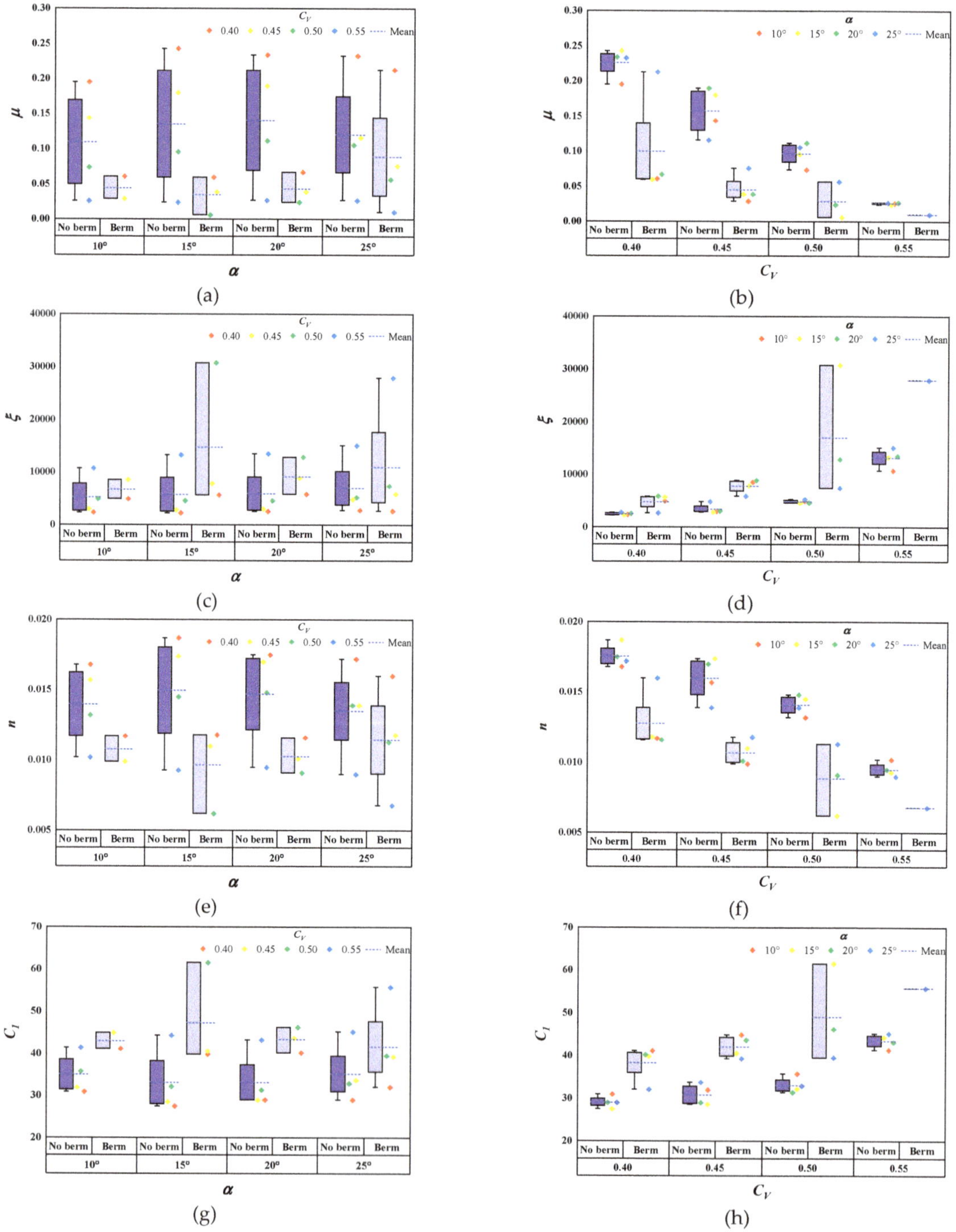

Figure 9. *Cont.*

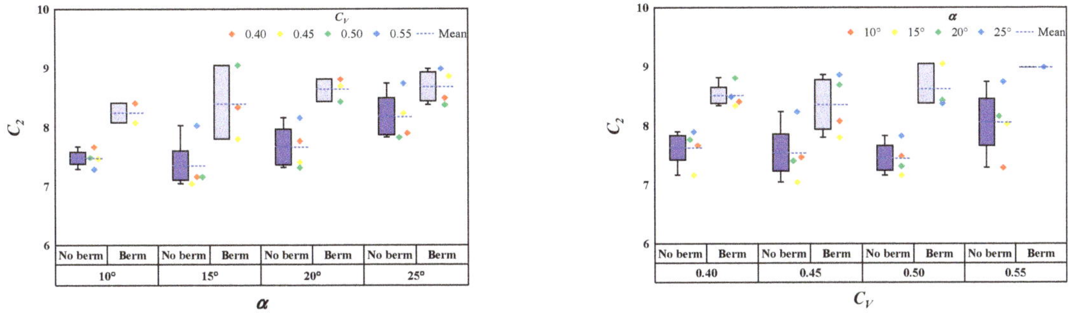

(i) (j)

Figure 9. (**a**) μ according to α (at C_V = 0.40–0.55); (**b**) μ according to C_V (at α = 10°–25°); (**c**) ξ according to α (at C_V = 0.40–0.55); (**d**) ξ according to C_V (at α = 10°–25°); (**e**) n according to α (at C_V = 0.40–0.55); (**f**) n according to C_V (at α = 10°–25°); (**g**) C_1 according to α (at C_V = 0.40–0.55); (**h**) C_1 according to C_V (at α = 10°–25°); (**i**) C_2 according to α (at C_V = 0.40–0.55); (**j**) C_2 according to C_V (at α = 10°–25°).

4.1.5. Mobility Ratio

Figure 10 shows the debris flow mobility ratio observed through the flume experiments and the percent decrease in mobility ratio due to the installation of the berm. Figure 10a,b show the debris flow mobility ratio according to α and C_V, respectively. Figure 10c,d show the percent decrease in the debris flow mobility ratio according to α and C_V, respectively, when the berm was installed. The experimental results confirmed that the debris flow mobility ratio increased as α or C_V increased (Figure 10a,b). In the case of the straight channel test, the incremental increases in the mobility ratio due to the increase in α were similar when C_V was less than 0.55, but larger when it was 0.55 or greater. The incremental increases in mobility ratio due to the increase in C_V were similar when α was 15° or greater, but considerably smaller when it was less than 15°. In the single-berm channel test, the incremental increases in the mobility ratio due to the increase in α were similar when C_V was less than 0.55, but larger when it was 0.55 or higher. The incremental increases in mobility ratio due to the increase in C_V were similar when α was 20° or greater, but relatively smaller when it was less than 20°. In addition, the debris flow mobility ratio was observed to decrease by 4.6–15.7% due to the installation of the berm on the slope (Figure 10c,d). The average reductions in the debris flow mobility ratio were found to be 9.8%, 9.4%, 9.1%, and 10.2% for a C_V of 0.40–0.55 and α values of 10°, 15°, 20°, and 25°, respectively, and 13.6%, 10.0%, 8.2%, and 6.7% for an α of 10°–25° and C_V values of 0.40, 0.45, 0.50, and 0.55, respectively. In other words, the installation of the berm was more effective in reducing the debris flow mobility ratio at lower C_V values.

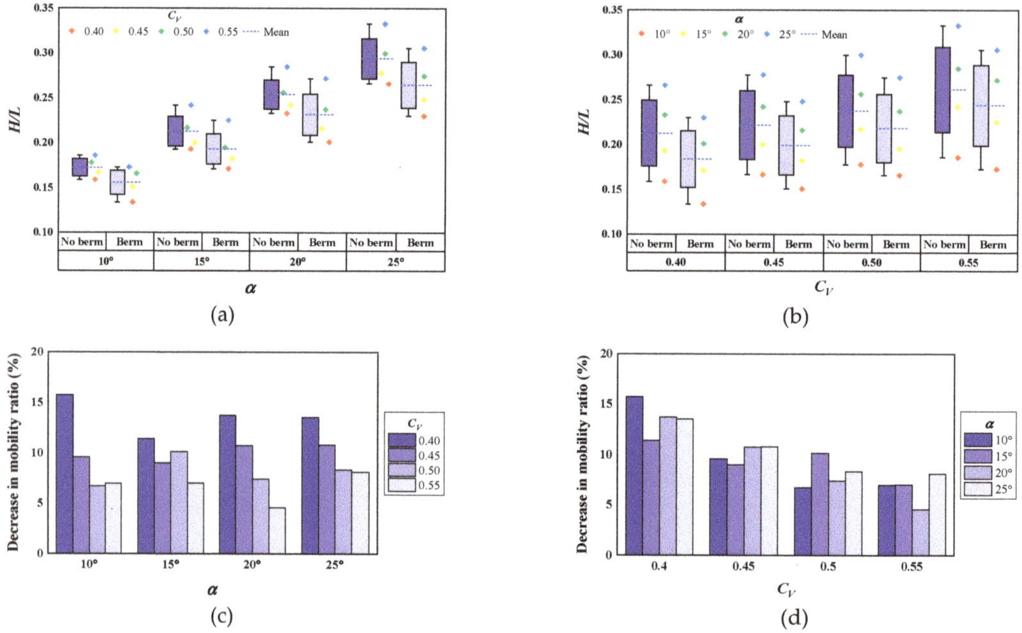

Figure 10. (**a**) Mobility ratio according to α (at C_V = 0.40–0.55); (**b**) mobility ratio according to C_V (at α = 10°–25°); (**c**) percent decrease in mobility ratio due to berm according to α (at C_V = 0.40–0.55); (**d**) percent decrease in mobility ratio due to berm according to C_V (at α = 10°–25°).

5. Discussion

Debris flows are affected by the characteristics of the sediment–water mixture (magnitude, C_V, and grain-size distribution) and channel shape (α, width, and curvature) [3,6,9]. In this study, various α and C_V values were therefore evaluated in flume tests along with changes in channel shape due to the installation of a berm in the middle of the channel. The experimental results showed that the development of a flow was influenced to occur at α of 15° (Table 6), which has been mentioned as the slope that causes debris flow in previous studies [3,9]. Takahashi [9] mentioned that the debris flow pattern suddenly changes when its C_V is less than 0.55 because of active particle separation, and that a debris flow cannot reach the outlet of its channel when the C_V exceeds 0.58. In this study, the development of the debris flow was also observed to differ around a C_V of 0.55; at a C_V of 0.60 in the single-berm channel test, the debris flow stopped in the channel before reaching the outlet (Table 6). In the straight channel test, however, it appears that α had a larger influence on the flow. Thus, debris flow occurred when α was 15° or higher, even when C_V exceeded 0.58.

As α increased, the debris flow velocity and mobility ratio were both observed to increase (Figures 6a and 10a). The changes in the flow velocity and mobility ratio differed around an α of 15°. The flow depth consistently decreased as α increased when no berm was installed, but it suddenly increased at an α of 25° when the berm was installed (Figure 7a). This appears to be because the debris flow moved along the steeply sloped channel with considerable momentum, and then the flow suddenly changed under the influence of the channel cross section geometry where the berm was installed. The installation of the berm caused the Froude number to increase when α was less than 25°, but it decreased when it was 25° (Figure 8c). In other words, it was confirmed that the 15° slope known to cause debris flow indeed affects the debris flow velocity and mobility ratio, and that an α of 25° affects the flow depth and Froude number in single-berm channel.

As C_V increased, the debris flow velocity and flow depth decreased (Figures 6b and 7b), but the mobility ratio increased (Figure 10b). A C_V of 0.50 marked important the changes in the development of debris flow depth and Froude number, and a C_V of 0.55 marked important changes in the development of the debris flow velocity and mobility ratio. At a C_V of 0.60, the debris flow did not reach the outlet at any α when the berm was installed, but it reached the outlet if α was 15° or higher when no berm was installed. Furthermore, it was confirmed that the C_V dominated the flow resistance coefficients μ, ξ, n, and C_1, while C_2 was not affected by the C_V, likely because it was defined using a numerical analysis, unlike the other flow resistance coefficients, which were defined using the experimental results. Similarly, Rickenmann [16] mentioned that C_2 is appropriate for the numerical analysis of unsteady debris flows.

The Froude numbers obtained in this study ranged from 4.14 to 10.79. This range is similar to those reported in studies conducted using an experimental setup with a channel length of 2 m or less [2,35]. The Froude numbers of actual debris flows have been determined to range from 0.36 to 7.56 in previous studies [3,15,23,37–39], and the Froude numbers of debris flows produced using flume experiments have been reported to range from 0.6 to 12.44 [2,12,35,40–43]. Since debris flow is affected by various conditions, the Froude number varies depending on the characteristics of the target debris flow. In general, the Froude numbers of actual debris flows are higher than those obtained in flume experiments. In the case of an actual debris flow, it appears that the Froude number decreases because the flow depth increases with the large amount of sediment absorbed into the flow by the riverbed erosion during the movement process [7,13].

In this study, the use of a berm was considered as a debris flow mitigation measure. The installation of the berm in the channel was observed to reduce the debris flow velocity, depth, and mobility ratio by up to 34.3%, 71.2%, and 15.7%, respectively (Figure 6c,d; Figure 7c,d; and Figure 10c,d, respectively). This indicates that the berm effectively decreased the potential kinetic energy and mobility of the debris flow moving downstream under the influence of gravity. However, it is important to note that various experimental berm conditions were not considered in this study due to laboratory test limitations. The effects of berms on debris flow characteristics can be more effectively identified if additional experimental berm conditions such as the length, location, and back slope are considered.

6. Conclusions

In South Korea, where mountainous areas account for more than 63% of the land, an increasing amount of debris flow has occurred each year due to rapidly increasing torrential rainfall. However, related studies on debris flow and the preparation of mitigation measures are currently insufficient. Therefore, this study was conducted using a straight channel test (without a berm) and a single-berm channel test to determine the effects of channel slope, volumetric concentration of sediment, and berm installation on the resulting flow velocity, flow depth, Froude number, flow resistance coefficients, and mobility ratio.

The experimental results showed that the debris flow velocity and mobility ratio increased, but the debris flow depth decreased as the channel slope increased. In addition, as the volumetric concentration of sediment increased, the debris flow velocity and depth both decreased, whereas the mobility ratio increased. When the berm was installed on the channel slope, the debris flow velocity, depth, and mobility ratio all significantly decreased, indicating that the installation of a berm on a slope can effectively decrease the spread of debris flow in downstream areas. In this study, the Froude number exhibited a range similar to those determined in previous studies at similar experimental scales.

The results of this study provide a useful understanding of the effects of channel slope and volumetric concentration of sediment on debris flow characteristics. They also provide details describing the effects of berm installation, which are required to design adequate debris flow damage reduction measures. In future studies, the down-channel depositions will be further analyzed to derive the correlation between the flow characteristics and deposition characteristics.

Author Contributions: Conceptualization and methodology, H.L.; formal analysis, H.C.; investigation, K.R.; writing-original draft preparation, K.R. and H.C. All authors have read and agreed to the published version of the manuscript.

Funding: This research was supported by Basic Science Research Program through the National Research Foundation of Korea (NRF) funded by the Ministry of Education (NRF-2017R1D1A3B03035477 & NRF-2019R1A6A3A01096145).

Institutional Review Board Statement: Not applicable.

Informed Consent Statement: Not applicable.

Data Availability Statement: The data presented in this study are not available without the author's agreement. To use the data, a request should be made to the corresponding author.

Conflicts of Interest: The authors declare no conflict of interest. The funders had no role in the design of the study; in the collection, analyses, or interpretation of data; in the writing of the manuscript, or in the decision to publish the results.

References

1. Eu, S.; Im, S.; Kim, D.; Chun, K.W. Flow and deposition characteristics of sediment mixture in debris flow flume experiments. *For. Sci. Technol.* **2017**, *13*, 61–65. [CrossRef]
2. Eu, S.; Im, S.; Kim, D. Development of debris flow impact force models based on flume experiments for design criteria of soil erosion control dam. *Adv. Civ. Eng.* **2019**, *2019*, 1–8. [CrossRef]
3. Costa, J.E. Physical geomorphology of debris flows. In *Developments and Applications of Geomorphology*; Costa, J.E., Fleisher, P.J., Eds.; Springer: Berlin/Heidelberg, Germany, 1984; pp. 268–317.
4. Kang, H.; Kim, Y. The physical vulnerability of different types of building structure to debris flow events. *Nat. Hazards* **2016**, *80*, 1475–1493. [CrossRef]
5. Eu, S.; Im, S. Examining the impact force of debris flow in a check dam from small-flume experiments. In Proceedings of the 7th International Conference on Debris-Flow Hazards Mitigation, Golden, CO, USA, 10–13 June 2019.
6. Coussot, P.; Meunier, M. Recognition, classification and mechanical description of debris flows. *Earth Sci. Rev.* **1996**, *40*, 209–227. [CrossRef]
7. Archetti, R.; Lamberti, A. Assessment of risk due to debris flow events. *Nat. Hazards Rev.* **2003**, *4*, 115–125. [CrossRef]
8. D'Agostino, V.; Cesca, M.; Marchi, L. Field and laboratory investigations of runout distances of debris flows in the Dolomites (Eastern Italian Alps). *Geomorphology* **2010**, *115*, 294–304. [CrossRef]
9. Takahashi, T.; Das, D.K. *Debris Flow: Mechanics, Prediction and Countermeasures*, 2nd ed.; CRC Press: Leiden, The Netherlands, 2014; p. 572.
10. Hübl, J.; Suda, J.; Proske, D.; Kaitna, R.; Scheidl, C. Debris flow impact estimation. In Proceedings of the International Symposium on Water Management and Hydraulic Engineering, Ohrid, Macedonia, 1–5 September 2009.
11. de Haas, T.; Braat, L.; Leuven, J.R.F.W.; Lokhorst, I.R.; Kleinhans, M.G. Effects of debris flow composition on runout, depositional mechanisms, and deposit morphology in laboratory experiments. *J. Geophys. Res. Earth Surf.* **2015**, *120*, 1949–1972. [CrossRef]
12. Cui, P.; Zeng, C.; Lei, Y. Experimental analysis on the impact force of viscous debris flow. *Earth Surf. Process. Landf.* **2015**, *40*, 1644–1655. [CrossRef]
13. Frank, F.; McArdell, B.W.; Huggel, C.; Vieli, A. The importance of entrainment and bulking on debris flow runout modeling: Examples from the Swiss Alps. *Nat. Hazards Earth Syst. Sci.* **2015**, *15*, 2569–2583. [CrossRef]
14. Iverson, R.M. The physics of debris flows. *Rev. Geophys.* **1997**, *35*, 245–296. [CrossRef]
15. Hungr, O.; Morgan, G.C.; Kellerhals, R. Quantitative analysis of debris torrent hazards for design of remedial measures. *Can. Geotech. J.* **1984**, *21*, 663–677. [CrossRef]
16. Rickenmann, D. Empirical relationships for debris flows. *Nat. Hazards* **1999**, *19*, 47–77. [CrossRef]
17. VanDine, D.F. *Debris Flow Control Structures for Forest Engineering*; BC Ministry of Forests: Victoria, BC, Canada, 1996; p. 68.
18. Prochaska, A.B.; Santi, P.M.; Higgins, J.D. Debris basin and deflection berm design for fire-related debris-flow mitigation. *Environ. Eng. Geosci.* **2008**, *14*, 297–313. [CrossRef]
19. Kim, S.; Lee, H. A study of the debris flow activity on the one-stepped channel slope. *Soil Water Res.* **2015**, *10*, 32–39. [CrossRef]
20. Sharma, A.; Raju, P.T.; Sreedhar, V.; Mahiyar, H. Slope stability analysis of steep reinforced soil slopes using finite element method. In *Geotechnical Applications*; Anirudhan, I.V., Maji, V.B., Eds.; Springer: Singapore, 2019; pp. 163–171.
21. Proske, D.; Suda, J.; Hübl, J. Debris flow impact estimation for breakers. *Georisk Assess. Manag. Risk Eng. Syst. Geohazards* **2011**, *5*, 143–155. [CrossRef]
22. Takahashi, T. Debris flow. *Annu. Rev. Fluid Mech.* **1981**, *13*, 57–77. [CrossRef]
23. Hu, K.; Tian, M.; Li, Y. Influence of flow width on mean velocity of debris flows in wide open channel. *J. Hydraul. Eng.* **2013**, *139*, 65–69. [CrossRef]

24. Scheidl, C.; Chiari, M.; Kaitna, R.; Müllegger, M.; Krawtschuk, A.; Zimmermann, T.; Proske, D. Analysing debris-flow impact models, based on a small scale modelling approach. *Surv. Geophys.* **2013**, *34*, 121–140. [CrossRef]
25. Iverson, R.M. Scaling and design of landslide and debris-flow experiments. *Geomorphology* **2015**, *244*, 9–20. [CrossRef]
26. Koch, T. Testing various constitutive equations for debris flow modelling. In *Hydrology, Water Resources and Ecology in Headwaters*; Kovar, K., Tappeiner, U., Peters, N.E., Craig, R.G., Eds.; IAHS Press: Wallingford, UK, 1998; pp. 249–257.
27. Lo, D.O.K. *Review of Natural Terrain Landslide Debris-Resisting Barrier Design: Geo Report No. 104*; Geotechnical Engineering Office, Civil Engineering Department, The Government of Hong Kong Special Administrative Region: Kowloon, Hong Kong, China, 2000; p. 92.
28. Prochaska, A.B.; Santi, P.M.; Higgins, J.D.; Cannon, S.H. A study of methods to estimate debris flow velocity. *Landslides* **2008**, *5*, 431–444. [CrossRef]
29. Rickenmann, D. Debris flows 1987 in Switzerland: Modelling and fluvial sediment transport. In *Hydrology in Mountainous Regions II*; Sinniger, R.O., Monbaron, M., Eds.; IAHS: Lausanne, Switzerland, 1990; pp. 371–378.
30. Bagnold, R.A. *Experiments on a Gravity-Free Dispersion of Large Solid Spheres in a Newtonian Fluid under Shear*; Proceedings of the Royal Society: London, UK, 1954.
31. Mizuyama, T.; Ishikawa, Y. *Technical Standard for the Measures against Debris Flow (Draft)*; Sabo (Erosion Control) Division, Sabo Department, Public Works Research Institute, Ministry of Construction: Tokyo, Japan, 1988; p. 48.
32. Heim, A. Der Bergsturz von Elm. *Z. Der Dtsch. Geol. Ges.* **1882**, *34*, 74–115.
33. Toyos, G.; Dorta, D.O.; Oppenheimer, C.; Pareschi, M.T.; Sulpizio, R.; Zanchetta, G. GIS-assisted modelling for debris flow hazard assessment based on the events of May 1998 in the area of Sarno, Southern Italy: Part I. Maximum run-out. *Earth Surf. Process. Landf.* **2007**, *32*, 1491–1502. [CrossRef]
34. Bathurst, J.C.; Burton, A.; Ward, T.J. Debris flow run-out and landslide sediment delivery model tests. *J. Hydraul. Eng.* **1997**, *123*, 410–419. [CrossRef]
35. Fairfield, G. Assessing the dynamic influences of slope angle and sediment composition on debris flow behaviour: An experimental approach. Master's Thesis, Durham University, Durham, UK, 2011.
36. Im, S.; Eu, S.; Kim, D. Understanding debris flow characteristics using flume experiments. In *Advancing Culture of Living with Landslides*; Mikos, M., Tiwari, B., Yin, Y., Sassa, K., Eds.; Springer: Cham, Switzerland, 2017; pp. 357–361.
37. Jordan, R.P. Debris flows in the southern Coast Mountains, British Columbia: Dynamic behaviour and physical properties. Ph.D. Thesis, The University of British Columbia, Vancouver, BC, Canada, 1995.
38. Hürlimann, M.; Rickenmann, D.; Graf, C. Field and monitoring data of debris-flow events in the Swiss Alps. *Can. Geotech. J.* **2003**, *40*, 161–175. [CrossRef]
39. Li, Y.; Liu, J.; Su, F.; Xie, J.; Wang, B. Relationship between grain composition and debris flow characteristics: A case study of the Jiangjia Gully in China. *Landslides* **2015**, *12*, 19–28. [CrossRef]
40. Bugnion, L.; McArdell, B.W.; Bartelt, P.; Wendeler, C. Measurements of hillslope debris flow impact pressure on obstacles. *Landslides* **2012**, *9*, 179–187. [CrossRef]
41. Scheidl, C.; McArdell, B.W.; Rickenmann, D. Debris-flow velocities and superelevation in a curved laboratory channel. *Can. Geotech. J.* **2015**, *52*, 305–317. [CrossRef]
42. Wendeler, C.; Volkwein, A. Laboratory tests for the optimization of mesh size for flexible debris-flow barriers. *Nat. Hazards Earth Syst. Sci.* **2015**, *15*, 2597–2604. [CrossRef]
43. Jiang, R.; Fei., W.; Zhou, H.; Huo, M.; Zhou, J.; Wang, J.; Wu, J. Experimental and numerical study on the load and deformation mechanism of a flexible net barrier under debris flow impact. *Bull. Eng. Geol. Environ.* **2020**, *79*, 2213–2233. [CrossRef]

applied sciences

MDPI

Article

Modeling the Impact of Extreme Droughts on Agriculture under Current and Future Climate Conditions Using a Spatialized Climatic Index

Dorothée Kapsambelis [1,2,*], David Moncoulon [1], Martine Veysseire [3], Jean-Michel Soubeyroux [4] and Jean Cordier [2]

1 Caisse Centrale de Réassurance (CCR) Department R&D Modeling Cat & Agriculture, 157 Boulevard Haussmann, 75008 Paris, France; dmoncoulon@ccr.fr
2 Agrocampus Ouest, Economie et Gestion, SMART-LERECO, 65 rue de Saint-Brieuc, 35000 Rennes, France; jean.cordier@agrocampus-ouest.fr
3 Meteo-France, Direction des Services Méteorologiques, 42 Avenue Gaspard-Coriolis, CEDEX 1, 31057 Toulouse, France; martine.veysseire@meteo.fr
4 Meteo-France, Direction de la Climatologie et des Services Climatiques, 42 Avenue Gaspard-Coriolis, CEDEX 1, 31057 Toulouse, France; jean-michel.soubeyroux@meteo.fr
* Correspondence: dkapsambelis@ccr.fr; Tel.: +33-1-44-35-32-72

Abstract: Extreme droughts have a strong impact on agricultural production. In France, the 2003 drought generated records of yield losses at a national scale for grassland (more than 30%) and for cereals (more than 10% for soft winter wheat and winter barley). These extreme events raise the question of farm resilience in the future. Studying them makes it possible to adapt risk management policy to climate change. Therefore, the objective of this paper was to analyze the frequency and the intensity of extreme drought in 2050 and their impact on crop yield losses (grassland and cereals) in France. We used the DOWKI (Drought and Overwhelmed Water Key Indicator) meteorological index based on a cumulative water anomaly, which can explain droughts and their consequences on agricultural yield losses at a departmental scale. Then, using the ARPEGE-Climat Model developed by Meteo-France, DOWKI was projected in 2050 and grassland, soft winter wheat, and winter barley yield losses were simulated. The results compare the frequency and intensity of extreme droughts between the climate in 2000 and 2050. Our results show that the frequency of extreme droughts (at least as intense as in 2003) doubled in 2050. In addition, the yield losses due to 10-year droughts increased by 35% for grassland and by more than 70% for cereals.

Keywords: extreme droughts; climate change; modeling; crop yield losses; crop insurance

Citation: Kapsambelis, D.; Moncoulon, D.; Veysseire, M.; Soubeyroux, J.-M.; Cordier, J. Modeling the Impact of Extreme Droughts on Agriculture under Current and Future Climate Conditions Using a Spatialized Climatic Index. *Appl. Sci.* **2022**, *12*, 2481. https://doi.org/10.3390/app12052481

Academic Editor: Joao Carlos Andrade dos Santos

Received: 16 December 2021
Accepted: 23 February 2022
Published: 27 February 2022

Publisher's Note: MDPI stays neutral with regard to jurisdictional claims in published maps and institutional affiliations.

1. Introduction

1.1. Consequences of Climate Change on Agriculture

Agriculture in France represents an important economic activity (leading producer in the European Union). In 2014, of the EUR 373 billion of gross agricultural products (GAP) produced in the European Union, France produced EUR 67 billion, representing 18% of the GAP [1,2]. France is the main producer of wheat and cattle in the European Union, and these two activities cover a large part of its utilized agricultural land [3]. In 2003, a severe drought caused a massive decrease in agricultural production and income (30% of the production was lost [4]), despite the rise in prices that some crops experienced [5]. Grassland yields were also greatly reduced and public support (via the Calamity Fund System) was necessary to allow farmers to get through the year, especially in the milk production community. These elements indicate that despite technological progress, crop production remains highly dependent on water resources and climatic conditions. In this context, increasing our scientific knowledge of the intensity and frequency of these extreme droughts is necessary to evaluate their impact on agricultural production.

There is a global consensus in the scientific community that the climate will be very different in the middle of the 21st century [6–9]. The main cause of climate change is the increase in atmospheric concentrations of several greenhouses gases as a result of human activity [10,11]. Several models are used in the community to study different scenarios of climate change and its consequences on agriculture. The following studies highlight several important points:

- Annual temperatures are expected to increase [12,13]. As a consequence, the crop-growing period will be shorter and the grain protein concentration will decrease [14,15]. In addition, a high frequency of severe drought events during summer periods is to be expected [7,11,16]. These events may cause crop yield reduction or stagnation (depending on areas and species) [17,18]. Some studies show that in the Mediterranean region, yields could greatly decrease compared to historical yield trends [2,19].
- A decrease in annual precipitation, especially in the Mediterranean region, such as the South of France, will contribute to an increase in winter droughts [11].
- Many studies highlight that cereal crops (in particular wheat and corn, which are the most studied crops) will experience a decrease in yield due to an increased number of days above 30 °C [12,13,18,20–23].
- Other studies underline that high temperatures have direct and indirect negative impacts on dairy production [24,25] by affecting animal health and grassland yields, in particular.

Finally, several studies demonstrate that heat waves like the 2003 one, which particularly affected crop yields, will be more frequent in the future [2,6,26–28].

Extreme drought events are difficult to study because there are by definition rare events that occur very infrequently, so an archive of historical data may contain just a few extreme events [29]. The IPCC defines the concept of "extreme" as "the occurrence of a value of a weather or climate variable above (or below) a threshold value near the upper (or lower) ends ('tails') of the range of observed values of the variable" [8]. Thus, the notion of an extreme drought event is dependent on the value of the climatic index chosen to characterize the climate. Therefore, the link between the climate index and the impact on agriculture (yield losses) has to be explicit. However, annual yield losses are due to a set of phenomena (diseases, climatic events, changes in cropping practices) and it is not easy to assess the weight of one phenomenon independently. The best-known drought indicators are:

1. Standardized Precipitation Index (SPI) developed by [30] and based on precipitation data to determine the exact period and duration of meteorological droughts. The values of the index are fitted to a log-logistic distribution for standardization.
2. Standardized Precipitation Evaporation Index (SPEI) developed by [31] and based on the difference between precipitation and reference evapotranspiration (ET0).

These indicators characterize meteorological and hydrological droughts, which are not necessarily similar to agricultural droughts. Although many indices have been developed to analyze the evolution of droughts, the direct relation between these indices and crop yield has not been frequently investigated. Some studies have been conducted in China, in the United States, and in Europe comparing the SPEI, PDSI (Palmer–Drought Severity Index), and SPI indices for the detection of agricultural droughts [32,33]. The best correlations are obtained with SPEI. In addition, in Canada, a study was carried out to analyze the correlation between grassland losses linked to droughts and certain agro-climatic indices like PDSI and SPI. The results indicate that the coefficients of determination remained very low with all indices [34]. Other drought indices have been developed for specific territory, such as ARID to study the link between water stress and plant growth in the United States [35]. Thus, some studies show that an index developed with the parameter of precipitation alone, like SPI, is not sufficient to explain the variability in crop production due to drought, particularly for extreme events like the one in 2003 [36,37], because this drought was characterized by an increase in evapotranspiration rates [38]. At the French country scale, one indicator used to analyze the effect of the climate change on agriculture is the Standardized Soil Water Index (SSWI) [39]. This indicator represents the useful water reserve of the soil, or water availability for plants. Water deficit and temperature are parameters commonly used to study the climate effect on agricultural crops [40,41].

1.2. Objectives of This Study

In the paper, we propose an evaluation of the frequency and intensity of extreme droughts in current climate and future climate conditions (year 2050). Our methodology is based on a simple drought index [42] correlated to crop yield losses that can be projected into the future using a global climate model—for instance, the ARPEGE-Climat Model from Meteo-France.

Based on this method, the objectives are to:

- Analyze the climate and its evolution between the years 2000 and 2050;
- Detect extreme events and analyze their frequency and intensity;
- Simulate their consequences on crop yield losses.

The model is applied to three crop categories: grasslands, soft winter wheat, and winter barley.

This study aims to provide insight on the following issues:

1. Do the historic records contain the most extreme events and, if not, how can we characterize them?
2. What is the probability of occurrence in the current climate conditions?
3. Will the intensity and/or frequency of these events evolve in the future?

2. Materials and Methods

2.1. Modeling Extreme Droughts and Their Consequences on Yield Losses

2.1.1. DOWKI Computation on the SAFRAN Reanalysis

We used the DOWKI, which characterizes extreme events of drought and excess water. DOWKI is (1) simple to compute, (2) purely meteorological, and (3) independent of crop categories. It can be compared to yield losses for several types of crops and on large areas. DOWKI is a cumulative efficient rain anomaly, computed on a 10-day time step, between the current year value and the historical average. It is computed for the growing period of a given crop, and starts at 0 on 1 January. It is expressed in mm. Its equation for drought event characterization is as follows:

$$ERNC_{i,n} = \left[(P_{i-1,n} - PET_{i-1,n}) - \left(\overline{P_{i-1,P} - PET_{i-1,P}}\right)\right] + \left[(P_{i,n} - PET_{i,n}) - \left(\overline{P_{i,P} - PET_{i,P}}\right)\right]$$

$$DOWKI_{drought\ c,\ n} = \min ERNC_{i0 \to ij,c,n}$$

where $ERNC_{i,n}$ is the cumulative rain anomaly computed in decade i for year n. P is the precipitation and PET is the potential evapotranspiration. $\left(\overline{P_{i-1,P} - PET_{i-1,P}}\right)$ represents the average of the difference between P and PET computed for all i-1 10-day periods in period P (historical period). DOWKI is an annual value and is computed by taking the minimum of the values of $ERNC$ for any 10-day period between i_0 (first 10-day period) and i_j (final 10-day period).

In [42], DOWKI was computed for representative meteorological stations at the departmental scale to match with available yield loss data, and we showed model uncertainties as we simulated all the crops losses. One notable limitation is the climate measure at a single point over the department. On the one hand, this measure is not necessarily representative of the climate over the whole territory of the department. On the other hand, the crop parcels were not necessarily located at the climate measuring point (meteorological station point). In this second case precisely, this would mean that we measured the hazard but not the agricultural risk. In this paper, we computed DOWKI on the SAFRAN-Grid—8 km × 8 km over the French metropolitan area for two reasons:

1. To reduce the uncertainties of the model;

2. To match with the output scale of the ARPEGE-Climat model using a quantile–quantile downscaling method on the SAFRAN daily reanalysis data (1981–2010 for rainfall and 1989–2018 for potential evapotranspiration).

After computing DOWKI on the SAFRAN grid, we crossed the SAFRAN reanalysis grid with the Graphic Plot Register (GPR) shown in Figure 1. To compute an index value by department, we calculated the DOWKI average values in each cell of the department where the crop was present. This methodology allowed us to measure climate risk specifically on crop production since we integrated the hazard parameter (DOWKI value) and crop vulnerability (the DOWKI computation corresponds to on the crop vulnerability period and the agricultural parcel location).

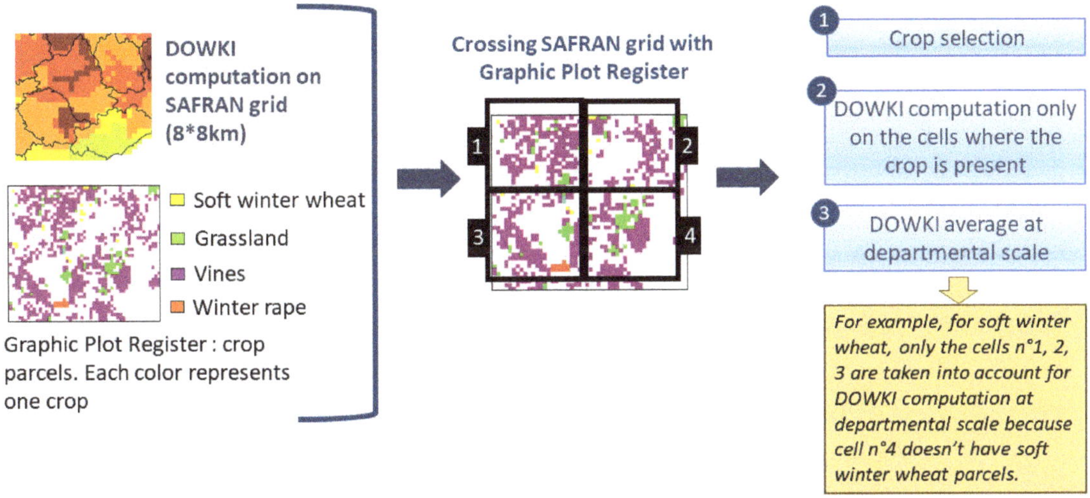

Figure 1. Crossing the SAFRAN-grid with Graphic Plot Register for DOWKI computation at the departmental scale.

2.1.2. Computing Yield Losses

We used the AGRESTE database (https://agreste.agriculture.gouv.fr, accessed date: 5 March 2019), which refers to yield by crop and department in the historical period (1989–2018 for soft winter wheat and winter barley and 2000–2018 for grassland) with one value by year and by crop produced and declared on a given surface. Yield losses for the n-th year were computed by comparing the annual yield with a yield reference defined by the Olympic average over 5 years. This methodology is used in agricultural public policies like crop insurance [43]. The crop yield loss computation using the Olympic average is presented here:

$$Yield\ loss_{c,n} = \frac{Yield_{c,n} - (\sum_{n-5}^{n-1} yield_c - Max(\sum_{n-5}^{n-1} yield_c) - Min(\sum_{n-5}^{n-1} yield_c))}{(\sum_{n-5}^{n-1} yield_c - Max(\sum_{n-5}^{n-1} yield_c) - Min(\sum_{n-5}^{n-1} yield_c))}$$

In which c is the culture and n is the year.

Figure 2 represents yield losses for soft winter wheat, winter barley, and grassland on the French farm scale for the historical period 2000–2018. Over this period, yields were affected by several events:

- The most significant soft winter wheat and winter barley yield losses were registered for the 2016 excess water event (27% and 17%, respectively) and the 2003 drought (14% and 16%, respectively).

- The most important grassland yield losses were registered for the 2003 drought (32%) and the 2011 drought (21%).

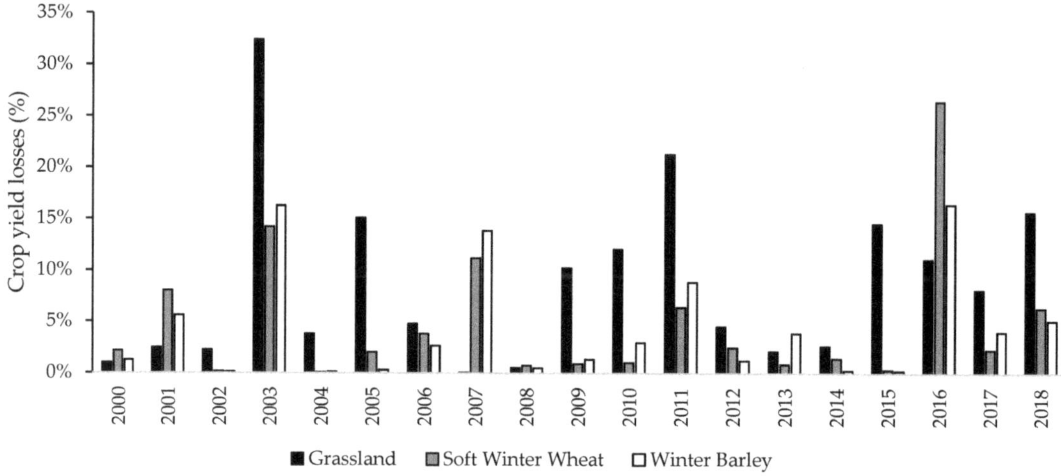

Figure 2. Crop yield losses (%) for grassland, soft winter wheat, and winter barley at the national scale computed over the historical period 2000–2018 with the AGRESTE database.

Cereals seem to be sensitive to two natural extreme hazards—excess water and droughts—and grassland specifically to droughts. Over the historical period, there were two extreme drought events: in 2003 and in 2011. These two years saw largescale severe droughts, the worst droughts in 30 years. On average, these two events caused 25% crop losses for grassland and 10% crop losses for soft winter wheat. In order to characterize the extreme events in the future, we used these two extreme droughts as a reference.

2.1.3. Analyzing the Link between DOWKI and Yield Losses

DOWKI values and yield losses were computed for each department over the historical period 2000–2018. This calibration matrix of 1800 values was used to study the statistical relationships between the index value and the yield losses.

The index values were classified using 50 mm steps. For each class we calculated the number of yield loss values exceeding 0% and the average yield loss value.

The parameters of the model were:

- The period over which the annual value of the index was computed. This period corresponded to the vulnerability period of the crop and was different for each crop;
- The extreme event threshold at the departmental scale;
- The minimum cultivation area to be taken into account to rule out small areas in which yields are very volatile.

These parameters were optimized using an experimental design, which consisted of computing a high number of calibration processes with different values for each parameter. The size of the experimental design was (p^n), with p being the number of parameters (here, p = 4) and n being the number of values for each parameters (here, n = 10). In our case study, the number of calibration processes was p^n = 10,000. The experimental design was evaluated by analyzing the following parameters:

- Average error at the national scale over the entire historical period;
- Average error at the departmental scale over the entire historical period.

This experimental design allowed us to select the best parameters by minimizing both errors. The best parameters are presented in Table 1. A specific experimental design with the same 4 parameters was run for each crop studied. Thus, a selection of the best parameters for grassland, soft winter wheat, and winter barley was made. The following table gives these parameter values.

Table 1. Best parameters used for the calibration of the model by crop.

Parameters	Annual Index Period Computation	Extreme Event Threshold	Minimum Cultivation Area
Grassland	3rd 10 days in April–2nd 10 days in September	−200 mm	2×10^4
Winter barley	3rd 10 days in April–1st 10 days in August	−200 mm	2×10^4
Soft winter wheat North of France	3rd 10 days in April–1st 10 days in August	−200 mm	9×10^4
Soft winter wheat South of France	3rd 10 days in April–1st 10 days in August	−200 mm	1.1×10^4

For soft winter wheat, two climatic regions were defined—the North and South of France—to improve the calibration results.

2.2. *Modelling Climate Scenarios with ARPEGE-Climat*

2.2.1. General Methodology

Unlike most climatic projections, here the ARPEGE-Climat was not used to simulate a continuous period between 2000 and 2050 but to simulate a 400-year-long time series with year 2000 climate forcing and with year 2050 climate forcing under an RCP 8.5 scenario. The objective was to collect a large panel of possible meteorological situations, but not ones that necessarily occurred, for these two target years. These 400 years had to be interpreted as possible realizations of a given targeted year. With 400 possible realizations for the year 2000 and the year 2050, we had at our disposal a large series of data. Then, it was possible to analyze extreme events and to estimate probabilities of occurrence.

Meteorological data such as precipitations and potential evapotranspiration were the outputs of this model for the climate in 2000 and the climate in 2050. The results were analyzed on an 8 km × 8 km grid for the whole French territory.

2.2.2. Targeting the Year 2050

The year 2050 was chosen for this study as the target year for our climatic projections. This mid-term target year, 30 years in the future, will allow us to analyze the consequences of climate change on crop production and support public policy decisions. Under the financial context, insurers are able to make projections of their market in 2050. The target year 2100—widely used by climatologists to study the impact of climate change—is too far in the future to make serious hypotheses on the evolution of agriculture, landscapes, economy, and risk management policies.

2.2.3. Choice of RCP 8.5

The Representative Concentration Pathway 8.5 scenario (RCP 8.5) is characterized by increasing greenhouse gas concentration levels (>1370 eq-CO_2 in 2100). This scenario is the most extreme and corresponds to a radiative forcing of +5 W/m^2 in 2050 (only +4 W/m^2 for RCP 4.5) [7]. The RCP 8.5 scenario represents a "pessimistic" or "conservative" vision of what the climate could be like in 2050. In this scenario, the energy demand is high, with the highest greenhouse gas emissions, corresponding to a high population and modest technological improvements. In France, RCP 8.5 corresponded to a temperature increase of 2.2 °C in 2050 compared to the 1976–2005 period, and a temperature increase of 1.7 °C for RCP 4.5 in 2050 [44,45]. According to the IPCC, RCP 8.5 corresponds today to historical paths since 1992.

2.2.4. ARPEGE-Climat Model Description and Parameterization

The numerical model ARPEGE is a global and spectral general circulation model developed for an "operational numerical weather forecast" by Meteo-France in collaboration with the ECMWF (European Centre for Medium-Range Weather Forecasts). ARPEGE-Climat became the atmospheric part of the CNRM earth-modelling system, which couples different components of the climate system (atmosphere, ocean, land surface, sea ice). The ARPEGE grid can be tilted and stretched by changing the position of the pole and by increasing the horizontal resolution over an area of interest. This zoom ability allows regional climate to be studied with ARPEGE-Climat.

In our case, ARPEGE-Climat had the pole in Germany (9.97° E, 50.00° N). The spatial resolution over Europe was about 20 km. The time step of the model was 600 s (10 min).

The exchanges between atmosphere and soil were taken into account by the specific SVAT (Soil Vegetation Atmosphere Transfer) module SURFEX (V7) implemented in ARPEGE-Climat.

The climate forcing allowed the climate to be kept stationary using fixed parameters:

- Fixed greenhouse gases concentrations accorded with the choice of Representative Concentration Pathway (RCP) for the fixed year (here, 2000 or 2050);
- Stationary sea-surface temperature series (adapted to each RCP with a quantile mapping method);
- Fixed stratospheric ozone concentrations;
- Fixed aerosol concentrations.

2.2.5. Model Outputs

The archive held model outputs over Europe and North Africa at stretched and tilted grid points in ARPEGE at an hourly step time for 36 near-surface parameters and at a 3-hour time step for 5 altitude parameters at 9 different levels. Then, data were generally interpolated on user-specific grids.

2.2.6. Downscaling and Post-Processing

We needed precipitation and potential evapotranspiration for metropolitan France. Precipitation could be directly extracted and interpolated on the 8 km × 8 km SAFRAN grid. Potential evapotranspiration was computed at a daily step time according to the Penman–Monteith formula, with 2 m temperature, sea level pressure, 2 m specific humidity, 10 m wind speed, surface downwards global short-wave radiation, and surface long-wave radiation. These parameters were retrieved and interpolated on the 8 km × 8 km SAFRAN grid.

The imperfections of the models induced biases in the outputs and downscaling, and interpolation is not a perfect method. Therefore, we removed the biases with 30 years of the climatic reference database SAFRAN (SIM2 reanalysis).

The precipitation and the parameters used to compute the potential evapotranspiration were generally corrected with the quantile mapping method. A specific method was developed at Meteo-France for global radiation. Last, potential evapotranspiration was corrected with the quantile-mapping method.

2.3. Uncertainty Analysis

The simulation results using a model chain like ours carried important uncertainties that needed to be evaluated and taken into account in the confidence interval of the results. Different uncertainties were contained in our model chain: climatic model uncertainties, index uncertainties, and damage model uncertainties.

As seen in its definition above, the DOWKI index computation is deterministic with no addition of uncertainties between the input data (P and PET) and index value. The hazard uncertainties were thus contained in the values of P and PET provided by the ARPEGE-Climat values and the downscaling process. To evaluate these uncertainties contained in the input data, we relied on two hypotheses:

- The simulation of 400 repeats of the same target year (years 2000 and 2050) will include a large part of the climatic uncertainty;
- The comparison between the target year 2000 produced by ARPEGE-Climat and the historical reanalysis of SAFRAN in the 2000–2018 period will complete this analysis.

The most important uncertainty lay in the crop yield loss simulations using DOWKI values and the damage model. As seen during the calibration process, false positives and false negatives induced model errors. We decided to take this uncertainty into account in the confidence interval by simulating each climate year in the ARPEGE-Climat model 100 times: For each of the 100 repeats of the same year, a yield loss value was randomly chosen within the index class at the department scale. This method allowed the confidence interval (for example, quantiles 10 and 90) to be estimated for each year and department.

3. Results

3.1. Historical Reanalysis

The relationship between the DOWKI values and yield losses for grassland is illustrated in Figure 3. The damage model is the statistical relation between climatic index and yield losses at the department scale. It is a combination of two predictive models:

- Prediction of the occurrence of a claim, i.e., yield loss exceeding 5% at the department scale;
- Prediction of the yield loss value at the department scale.

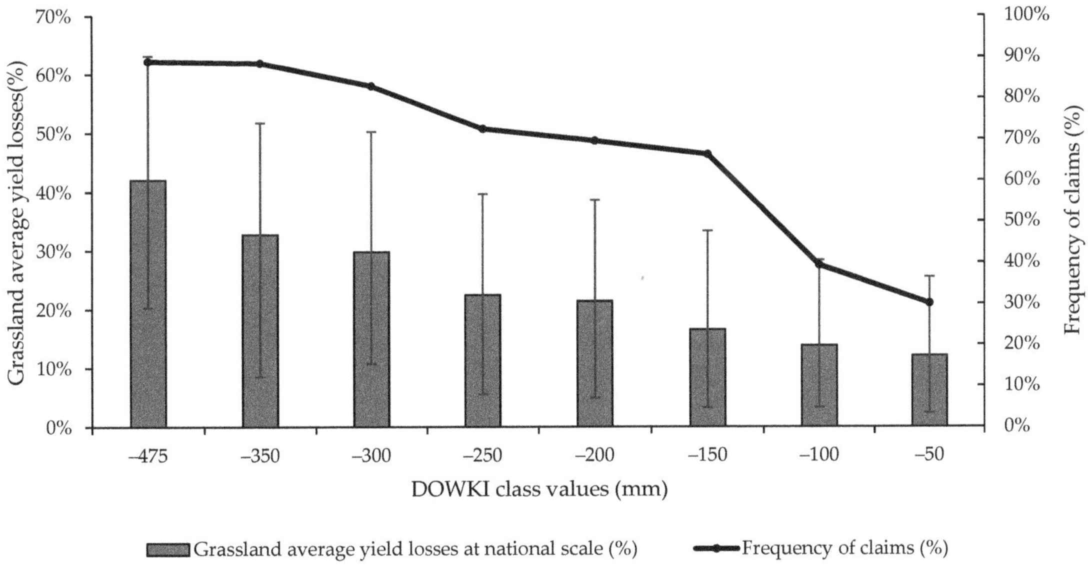

Figure 3. Damage function for grassland yield loss simulations: frequency of claims, percentiles 10 and 90, and average of yield losses according to the DOWKI values.

When a DOWKI value exceeded −475 mm, the departmental yield loss was equal to 42% with a probability of occurrence close to 90%. For DOWKI values close to zero, the probability of claims was significantly lower (~30%), as was the yield loss value (10%).

The calibration generated false positives and false negatives not explained by our index. False positives are departments where the index indicated, for a given year, an intense drought but without consistent yield loss. A false negative, on the contrary, is a case where high yield loss could not be explained by the index value. Several hypotheses were formulated to explain these errors (Table 2).

Table 2. False positives and false negatives.

False Positives	False Negatives
Adaptation of agricultural practices: modifying the sowing period or harvest period, choice of varieties	Development and propagation of disease
Protection measures (irrigation)	Combination of several climatic events, including droughts

To validate our damage model, back testing was performed by comparing, at the national scale, the observed yield losses and the simulated yield losses (Figure 4).

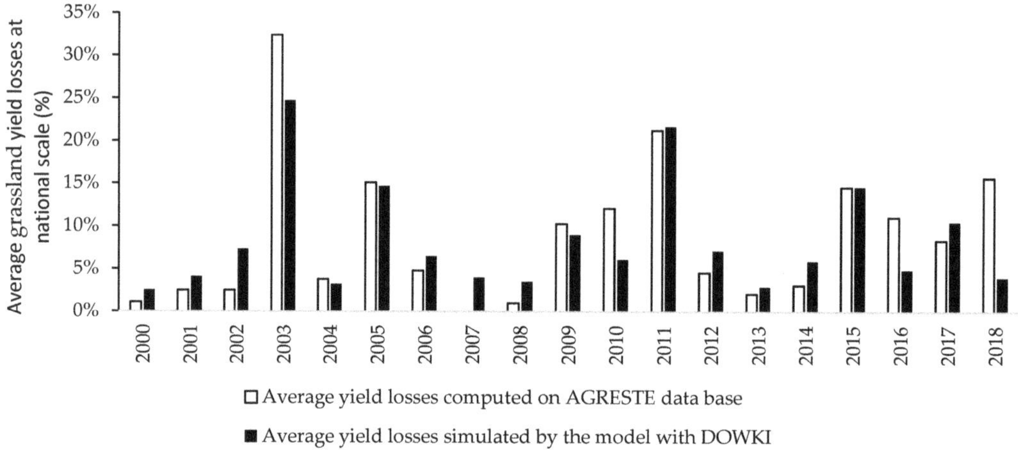

Figure 4. Average grassland yield losses at the national scale (%) computed in the AGRESTE database and simulated by the model with DOWKI values.

The back-testing relative error at the national scale was 5.5% for grasslands (14.6% for soft winter wheat and 20.4% for winter barley). The 2011 and 2003 intense droughts were explained by the model with an underestimation of 24% (2003) and overestimation of 2% (2011), but the highest simulated yield losses remained, as expected. The lowest yield losses at the national scale (years 2000, 2001, 2002, and 2008) were overestimated by the model, but the simulated yield losses were still the lowest in the distribution. The two droughts in 2003 and 2011 were characterized by a lack of precipitation and an augmentation of evapotranspiration rates. In addition, for the 2003 drought, record extreme temperatures were experienced during the summer. The main difference between these two droughts is that they did not begin at the same period of the year. The 2011 drought was a spring drought and the extreme values of DOWKI were computed in June. For the 2003 drought, extreme values of DOWKI were computed in August. The DOWKI values were more extreme for grassland than for cereals because the drought lasted all of August and the vulnerability period of cereals is shorter.

The most difficult issue with these model results is the case of the drought in 2018: High yield losses due to an extreme drought that occurred in the northeast region of France were not detected by our model. This was due to multiannual drought cycles. The DOWKI index value was initialized at 0 on 1 January of each simulated year, whereas in 2017, the soils were abnormally dry in December.

As shown in Figure 4, the back testing of the model showed its capacity to simulate extreme drought events and predict the national yield loss.

3.2. Agro-Climatic Model Results in 2000 and 2050

3.2.1. Comparison of DOWKI Distributions between 2000 and 2050

Using ARPEGE-Climat, two event sets of 400 years (2000 climate and 2050 climate) were computed. The first issue was to determine whether these distributions were significantly different. The numerous repeats in each target year allowed us to use a statistical test to answer this first question.

We compared the distributions of the annual average national scale DOWKI values with a Wilcoxon–Mann–Whitney non-parametric test commonly used to compare medians of two samples that do not follow a Gaussian distribution. The test rejected the null hypothesis that the two distributions were samples from continuous distributions with equal medians. The p-value was equal to 0.027.

3.2.2. National Analysis

In this first approach, we analyzed the frequency of extreme droughts with an intensity equal or superior to 2003 and 2011 at the national scale. In the current climate distribution, 29 drought events were identified. A quick estimation of the return period of these extreme droughts in current climate was 13 years. This first result is consistent with the 30 years of available historical data (1989–2018), with two extreme droughts (2003 and 2011), giving an empirical return period of 15 years. In the 2050 scenario, 57 extreme drought events were identified with a return period of seven years.

On average, these droughts affected 81.7% of utilized agricultural land (UAL) in 2000 and 86.1% in 2050. All these events were systemic, with a minimum of 61.8% (2000 climate) and 52.7% (2050 climate) of UAL affected by drought.

The annual DOWKI value at the national scale decreased by 40% (DOWKI was equal to −78 mm for the climate in 2000 and −110 mm for the climate in 2050) when comparing the 2000 and 2050 distributions.

At the national scale, the effect of climate change on the frequency of extreme droughts, considering 2003 as the reference, will increase significantly (+100%) between 2020 and 2050, according to the ARPEGE-Climat simulations. These events will remain systemic in the climate in 2050, with at least 50% of the UAL affected by drought.

Beyond the average, Figure 5 illustrates the yield losses (%) at the national scale for soft winter wheat (a), winter barley (b), and grassland (c) with respect to their return period in the current climate and in 2050.

A lot of information can be extracted from Figure 5. When integrating the model uncertainties (percentile 10–90), the empirical cumulative distribution functions (ECDF curves) between the 2000 climate and the 2050 climate did not overlap over 10 years, showing a significant increase in yield losses between 2000 and 2050 for all return periods.

First, the annual average loss will increase in 2050 by:

- 40% for grasslands;
- 47% for winter barley;
- 45% for soft winter wheat.

The yield losses due to 10-year droughts will increase by:

- 35% for grassland;
- 75% for soft winter wheat;
- 79% for winter barley (Table 3).

The results show a more important yield loss increase for cereals than grassland (Table 3) due to 10-year droughts. The whole of France would be impacted by a significant increase in risk in 2050, but the evolution would be even more significant in the northern half of France, where straw cereals are cultivated. Indeed, we analyzed the DOWKI values of 10-year droughts between the 2000 climate and the 2050 climate: A critical increase in the water balance anomaly (30–50%) was registered in the North of France, particularly where cereals are cultivated [46]. In the South of France, for 10-year droughts an increase in the water balance anomaly of 10–30% was recorded [46].

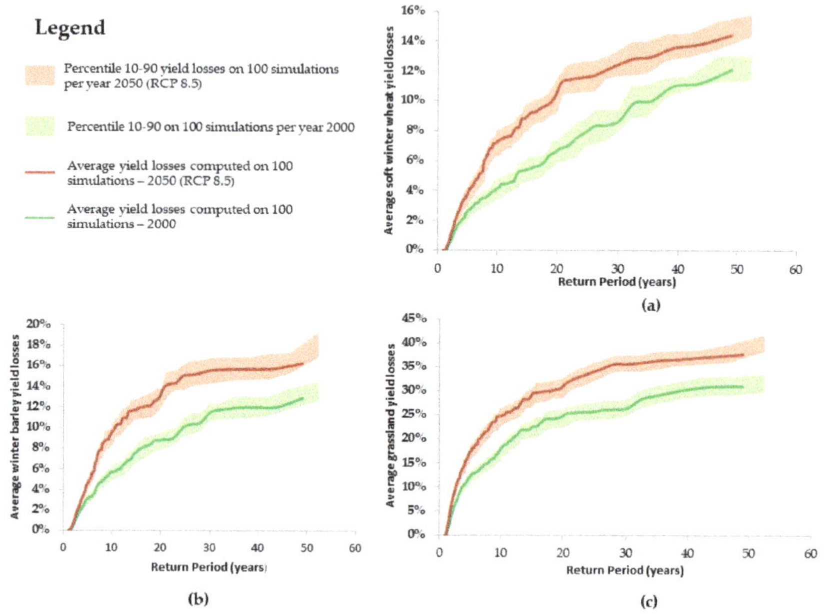

Figure 5. Average yield losses and percentile 10–90 on 100 simulations for the year 2000 and the year 2050 (RCP 8.5) for (**a**) soft winter wheat, (**b**) winter barley, and (**c**) grassland.

Table 3. Average yield losses at the national scale for soft winter wheat, winter barley, and grassland for 10-year droughts for the climate in 2000 and the climate in 2050.

10-Year Droughts	Average Yield Losses at the National Scale—Climate in 2000	Average Yield Losses at the National Scale—Climate in 2050
Soft winter wheat	4.2%	7.4%
Winter barley	5.3%	9.5%
Grassland	18.5%	25.0%

The return period of the highest losses (50-year return period at current climate) will become every 19.6 years for grasslands and winter barley and every 26.1 years for soft winter wheat.

In terms of output and income losses, these droughts will affect the agricultural economy with a loss of:

- 4.5 million metric tons and a deficit of EUR 745 million (with an average price of EUR 168/T) for soft winter wheat;
- 1 million metric tons and a deficit of EUR 163 million (with an average price of EUR 164/T) for winter barley;
- 8.8 million metric tons of dry matter and an indirect deficit of EUR 1.3 billion (with an average price estimated at EUR 150/T of dry matter).

3.2.3. Regional Analysis

Were the evolutions highlighted at the national scale consistent with a geographical study at the departmental scale?

We analyzed the intensity and frequency of extreme droughts at the local level using DOWKI values computed at an 8 km × 8 km scale and crop yield losses simulated at the departmental level.

The results presented in Figure 6a illustrate that the increase in the water deficit will be more significant on average in the South of France (southwest and Mediterranean region). Overall, we observed a worsening water deficit of 30% to 50% throughout France and above 50% in the south.

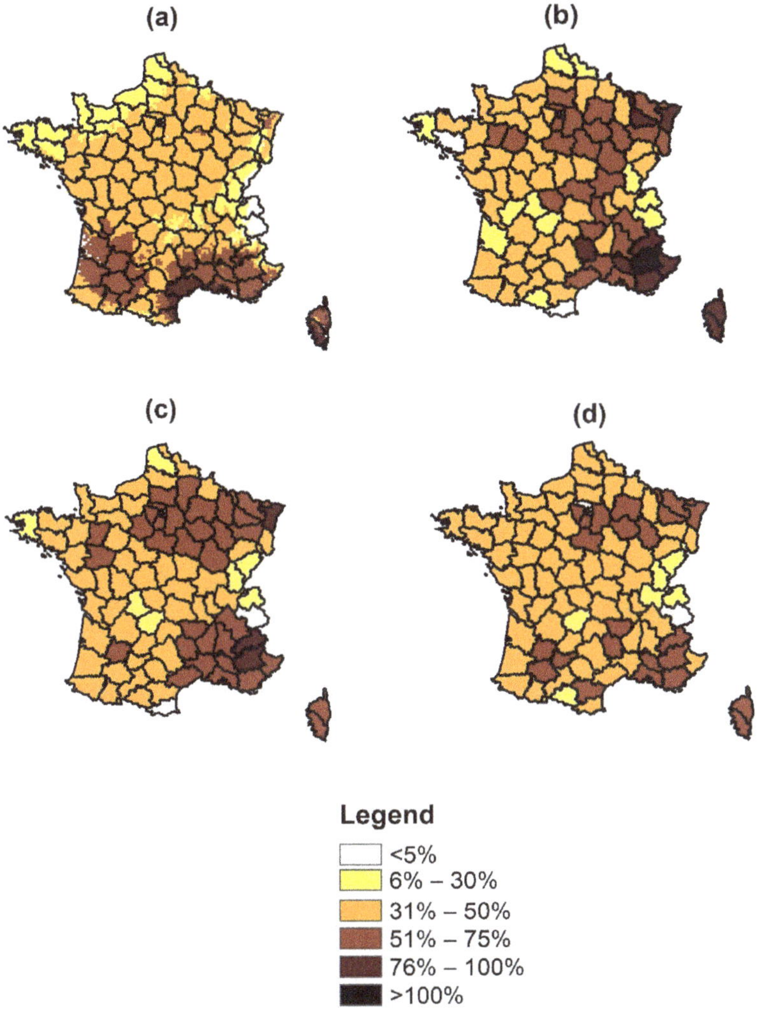

Figure 6. Annual average evolution between the climate in 2000 and in 2050 of (**a**) DOWKI values, (**b**) yield losses of soft winter wheat, (**c**) yield losses of winter barley, and (**d**) grassland.

Brittany, Normandy, and the coastal northern regions showed the lowest evolution of drought index (<30%). In these areas, yield losses for grassland will increase by 30 to 50% on average.

The translation of the DOWKI values in terms of yield losses showed the following results: the northeastern and southeastern parts of France will incur high yield loss increases for straw cereals. Depending on the department, Figure 6b,c shows that the yield losses will increase by 30% to 100% in 2050 for straw cereals. For grasslands, the whole of France will be affected by a significant increase in yield losses of between 30% and 75%.

4. Discussion

4.1. Comparison of the Results with Others Studies

This study shows that significant droughts from the recent past generated high yield losses at the national scale with a systemic impact on the French territory. More extreme events were computed under the current climate in terms of hazard and yield losses. Their probability of occurrence was estimated by our model to be 13 years. Our results show that the frequency of these extreme events will increase in the future to a return period of seven years.

These results are consistent with the ClimSec project [39]; with the study of IPCC [7,44], which focuses on extreme events; and with other European studies using EUROCORDEX models [47–50]. Climatic projections indicate that droughts will have a severity never before registered in terms of spatial extension and intensity. Other studies point out that the frequency of extreme drought events will strongly increase in the future, leading to a crop yield decrease, including grasslands under the RCP 8.5 scenario in all French territory [26]. In addition, studies focusing on specific countries and analyzing the evolution of droughts using climatic indices show an increase in severe drought in Greece [51], and a decrease in wheat yield due to drought severity [52] and an increase in drought frequency and severity in Spain [53], as well as in others areas like in China [54,55] and the United States [56].

Many studies show that the Mediterranean region appears to be very exposed to droughts in the future [57,58]). Indeed, the different models used at the regional scale (RCM models) to measure the impact of climate change on drought events agree that the droughts will be more intense in southern Europe, especially in the Mediterranean region [7,57,59]. Extreme heat wave studies show that the Mediterranean region will therefore probably record a cumulative water deficit anomaly, but this is, however, more widely throughout France, where the evolution between the climate in 2000 and in 2050 will be the most marked [60–63].

These extreme events are the most worrying for the sustainability of agricultural production systems because they generate very significant losses at the country level, affecting food security. For example, the extreme drought in 2011 was responsible for losses of more than USD 1 billion for animal production in United States [64]. In the European Union, losses due to the 2003 drought are estimated at EUR 13 billion, including EUR 4 billion for France [65]. Nowadays it is well documented that in many rural areas, small farms do not have the financial capacity to cope with systemic climate shocks [66]. In the future, climate change will increase extreme drought frequency [8,67], which raises the question of the resilience of farm income. The improvement of risk knowledge supports the assessment of the risk management systems currently in place and their sustainability in the context of climate change.

4.2. Limits of This Study

The first limit to this work is the use of a single climate model. It was important for the authors to question the reliability of this climate model. The specificity of our approach was to simulate 400 years of steady-state climate under the conditions of the years 2000 and 2050. Was the variability of other CORDEX-Drias models contained in these 2 × 400 years event sets? CORDEX-Drias simulations between 1985 and 2005 (current climate) and 2040–2060 (climate in 2050) for six different models were compared with ARPEGE-Climat. It appears that the current climate, future climate, and evolution ratio of the six models at the French scale were included in ARPEGE-Climat 400-year outputs, as shown in Figures A1 and A2 in Appendix A. After this validation was complete, it was obvious that obtaining extreme event values was tougher when mixing a short-scale event set from six different models than with the use of ARPEGE-Climat model. This study highlights the relevance of using large-scale event sets to represent the variability of climate, especially for extreme values.

Another limitation is the computation of crop yield losses using the Olympic average. This method allowed us to integrate a certain variability of yields over time. However, the crop yield loss computed was annual and was a sum of different factors, and this explains, in part, the errors in the model. Moreover, crop yields are not stable over time and many authors have shown that the cereal yield in France increased until 1996 and then stagnated or decreased [68,69]. However, the results contrast depending on geographical locations. Many studies have been done to eliminate bias introduced by non-climatic factors in the computation of yield losses [70,71], and it would be interesting to apply this kind of methodology. However, other factors may arise the same year, such as several climatic events. This was the case in 2003. A significant frost occurred in the central region of France, which contains 20% of the cultivated area for soft winter wheat [72]. The effects of frost accumulated with those of drought, which partly explains the significant crop losses and our difficulty in simulating it using a drought index.

Finally, this study was conducted with other elements being equal by definition. It did not take into account agricultural adaptation to climate change. The cultivated area for each crop modeled was the same in 2050. Our methodology was to project yield losses based on the relation between index values and historical yield losses. Therefore, cultivation of resistant varieties to extreme drought were not included in these results.

5. Conclusions

This paper analyzed the intensity and frequency of extreme agricultural droughts in 2050. For this purpose, the analysis focused on three crops: soft winter wheat, winter barley, and grassland. A new meteorological index was developed, which represents a cumulative water anomaly and is correlated to the yield losses. The model created simulated the crop yield losses at the departmental scale from the index values. Then, the index was projected in 2050 using the ARPEGE-Climat model from Meteo-France. The results compared the intensity and frequency of extreme droughts between the climate in 2000 and the climate in 2050 and show that the yield losses due to 10-year drought increased by 35% for grassland and by more than 70% for cereals.

Within the frameworks of both (1) the new CAP program (2023–2027) and (2) the French risk management scheme reform, these numbers are useful to alert and inform political stakeholders to the consequences of climate change, at the national and regional scale, on grassland and cereals. Our results show that to calibrate a risk management scheme and to be able to estimate the national farm exposure in the mid-term, the evolution of climatic extremes has to be taken into account. Insurers, reinsurers, public funds, and farmers (individually and globally) are exposed at different levels to the increase in climatic events in the next 30 years.

Insurance and reinsurance solvability is linked to the capacity to face extreme losses and, by definition, to the capacity to model the frequency/intensity curve. Nevertheless, as shown in this paper, this frequency/intensity curve cannot be considered stationary over the next 30 years. Under this condition, pricing treaties to allow loss balance in the mid-term have to integrate a mix between current and future losses.

The next step will be to integrate risk management scenarios in our model and to estimate the losses for the different stakeholders. A public–private partnership is a promising route to face systemic extreme events when insurance mutualization is to be reconsidered. Today, in France, the crop insurance diffusion rate is 30% for cereals and less than 2% for grasslands [73]. In this respect, after the occurrence of an extreme event at the national scale, the State must intervene to support farmers' resiliency. A significant increase in the diffusion rates is one way to achieve sustainable agriculture in the context of increasing risks.

Agriculture has always been able to adapt to the changing climate. However, considering pessimistic scenarios like RCP 8.5 and the fast increase in extreme droughts, risk management policies must support national agricultural production during the adaptation period.

Author Contributions: Validation, J.C. and J.-M.S.; writing—original draft, D.K., D.M. and M.V. All authors have read and agreed to the published version of the manuscript.

Funding: This research received no external funding.

Institutional Review Board Statement: Not applicable.

Informed Consent Statement: Not applicable.

Data Availability Statement: Data used for yield are available in https://agreste.agriculture.gouv.fr (accessed on 20 January 2022).

Acknowledgments: The authors want to thank Elizabeth Harader-Coustau for helping with the revision of the manuscript and Lea Boittin for the English corrections. We also want to thank our anonymous reviewers for their valued advice on the structure of the paper and the addition of complementary studies.

Conflicts of Interest: The authors declare no conflict of interest.

Appendix A

Using a single climatic model can generate bias in the results. This section presents the multi-model study. To analyze it with an objective approach, the data from five climatic models were downloaded (IPSL-CM5A, CNRM-CERFACS-ALADIN, NCC, MPI, MOHC-HadGEM2) using:

- The years 1985–2005 for the current climate;
- The years 2040–2060 for the future climate according to RCP 8.5.

The parameters we analyzed were the annual average DOWKI values for the French territory and the evolution of the annual average between 2000 and 2050 for each model.

To compare the DRIAS models with ARPEGE on the same basis, we randomly chose a set of 20 years in the current climate and 20 years in the future climate 100 times in the ARPEGE event set.

The distribution of the 100 values for ARPEGE for the annual average values and the evolution of the annual average values were compared. As shown in Figures A1 and A2 below, we can see that the 400 years of ARPEGE simulations contained the annual average values of the five models and their evolution.

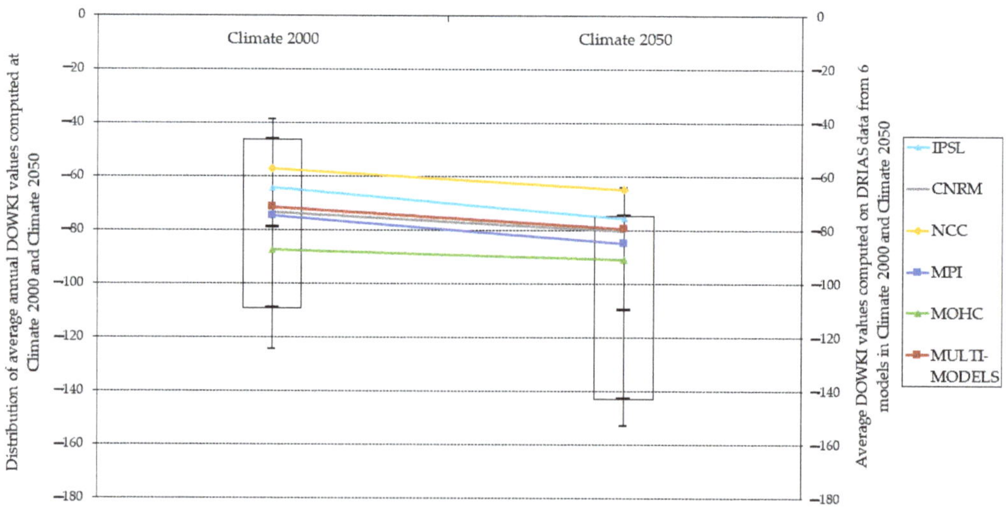

Figure A1. Distribution of the 100 average annual DOWKI values from ARPEGE in black. The limits of the box plot represent percentiles 10–90 and the error bars represent percentiles 5–95. The average is also represented by a bar in the boxplot. The average annual DOWKI values computed in 5 models are represented in color. The two distributions are computed for the climate in 2000 and the climate in 2050.

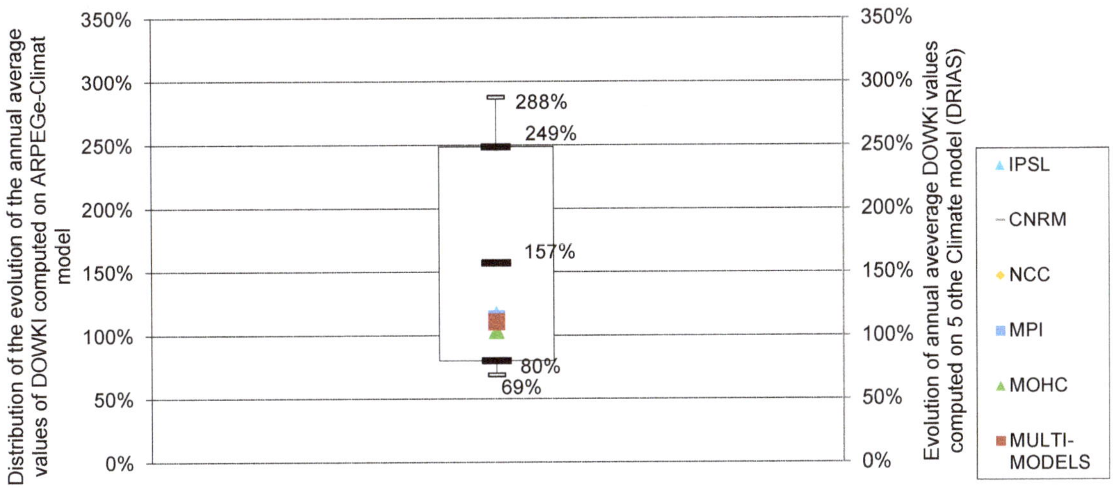

Figure A2. Distribution of the evolution between the climate in 2000 and the climate in 2050 of the annual average values of DOWKI from ARPEGE-Climat in black. The limits of the boxplot represent percentiles 10–90 and the error bars represent percentiles 5–95. The average of the 100 values is also represented by a bar in the boxplot. The evolution of the average annual DOWKI values between the climate in 2000 and the climate in 2050 computed in 5 other climate models are represented in color.

We can thus consider that ARPEGE, with a long-range simulation of 400 years, takes into account more uncertainties than the five DRIAS models.

References

1. Ministère de l'Agriculture et de l'Alimentation. *AGRESTE-Chiffres et Données. Statistique Agricole Annuelle 2018–2019*; Ministère de l'Agriculture et de l'Alimentation: Paris, France, 2020; p. 64.
2. Bindi, M.; Olesen, J. The Responses of Agriculture in Europe to Climate Change. *Reg. Environ. Chang.* **2010**, *11*, 151–158. [CrossRef]
3. Ministère de l'Agriculture et de l'Alimentation. *Mémento de La Statistique Agricole 2020 | Agreste, La Statistique Agricole*; Ministère de l'Agriculture et de l'Alimentation: Paris, France, 2020.
4. COPA COGECA. *Assessment of the Impact of the Heat Wave and Drought of the Summer 2003 on Agriculture and Forestry*; Committee of Agricultural Organisations in the European Union and General Committee for Agricultural cooperation in the European Union: Brussels, Belgium, 2003; pp. 1–15.
5. Amigues, J.P.; Debaeke, P.; Itier, B.; Lemaire, G.; Seguin, B.; Tardieu, F.; Thomas, A. *Sécheresse et Agriculture. Réduire La Vulnérabilité de L'agriculture à Un Risque Accru Du Manque d'eau*; INRA: Paris, France, 2006; p. 72.
6. Ciais, P.; Reichstein, M.; Viovy, N.; Granier, A.; Ogée, J.; Allard, V.; Aubinet, M.; Buchmann, N.; Bernhofer, C.; Carrara, A. Europe-Wide Reduction in Primary Productivity Caused by the Heat and Drought in 2003. *Nature* **2005**, *437*, 529–533. [CrossRef] [PubMed]
7. IPCC. *Climate Change 2014: Impacts, Adaptation, and Vulnerability. Part A: Global and Sectoral Aspects*; Cambridge University Press Cambridge: Cambridge, UK, 2014; p. 1132.
8. IPCC. *Managing the Risks of Extreme Events and Disasters to Advance Climate Change Adaptation*; Cambridge University Press: Cambridge, UK, 2012; p. 582.
9. IMFREX. *Impact Des Changements Anthropiques Sur La Fréquence Des Phénomènes Extrêmes de Vent de Température et de Précipitations*, Rapport final. 2002; 49.
10. IPCC. *Climate Change 2007: Impacts, Adaptation and Vulnerability. Contribution of Working Group II to the Fourth Assessment Report of the Intergovernmental Panel on Climate Change*; Cambridge University Press: Cambridge, UK; New York, NY, USA, 2007; p. 976.
11. Vert, J.; Schaller, N.; Villien, C.; Portet, F.; Mahé, T.; Sergent, A.-S.; Alexandre, S.; Nathalie, B.; Cloppet, E.; Dhôte, J.-F.; et al. *Agriculture, Forêt et Climat: Vers des stratégies d'adaptation Centre d'étude et de prospectives*; Ministère de l'Agriculture et de l'Alimentation: Paris, France, 2013.
12. Asseng, S.; Foster, I.; Turner, N.C. The Impact of Temperature Variability on Wheat Yields. *Glob. Chang. Biol.* **2011**, *17*, 997–1012. [CrossRef]
13. Alexandrov, V.; Eitzinger, J.; Cajic, V.; Oberforster, M. Potential Impact of Climate Change on Selected Agricultural Crops in North-Eastern Austria. *Glob. Chang. Biol.* **2002**, *8*, 372–389. [CrossRef]

14. Asseng, S.; Milroy, S.P. Simulation of Environmental and Genetic Effects on Grain Protein Concentration in Wheat. *Eur. J. Agron.* **2006**, *25*, 119–128. [CrossRef]
15. Spiertz, J.H.J. The Influence of Temperature and Light Intensity on Grain Growth in Relation to the Carbohydrate and Nitrogen Economy of the Wheat Plant. *Neth. J. Agric. Sci.* **1977**, *25*, 182–197. [CrossRef]
16. Vicente-Serrano, S.M.; Lopez-Moreno, J.-I.; Beguería, S.; Lorenzo-Lacruz, J.; Sanchez-Lorenzo, A.; García-Ruiz, J.M.; Azorin-Molina, C.; Morán-Tejeda, E.; Revuelto, J.; Trigo, R. Evidence of Increasing Drought Severity Caused by Temperature Rise in Southern Europe. *Environ. Res. Lett.* **2014**, *9*, 044001. [CrossRef]
17. Brisson, N.; Levrault, F. *Livre Vert du Projet CLIMATOR: 2007–2010: Changement Climatique, Agriculture et Forêt en France: Simulations D'impacts sur les Principales Espèces*; ADEME: Angers, France, 2012; ISBN 978-2-35838-278-6.
18. Porter, J.R.; Gawith, M. Temperatures and the Growth and Development of Wheat: A Review. *Eur. J. Agron.* **1999**, *10*, 23–36. [CrossRef]
19. Ciscar, J.-C.; Iglesias, A.; Feyen, L.; Szabo, L.; Van Regemorter, D.; Amelung, B.; Nicholls, R.; Watkiss, P.; Christensen, O.B.; Dankers, R.; et al. Physical and Economic Consequences of Climate Change in Europe. *Proc. Natl. Acad. Sci. USA* **2011**, *108*, 2678–2683. [CrossRef] [PubMed]
20. Schlenker, W.; Roberts, M.J. Nonlinear Temperature Effects Indicate Severe Damages to U.S. Crop Yields under Climate Change. *Proc. Natl. Acad. Sci. USA* **2009**, *106*, 15594–15598. [CrossRef] [PubMed]
21. Semenov, M. Simulation of Extreme Weather Events by a Stochastic Weather Generator. *Clim. Res.* **2008**, *35*, 203–212. [CrossRef]
22. Teixeira, E.I.; Fischer, G.; van Velthuizen, H.; Walter, C.; Ewert, F. Global Hot-Spots of Heat Stress on Agricultural Crops Due to Climate Change. *Agric. For. Meteorol.* **2013**, *170*, 206–215. [CrossRef]
23. Challinor, A.J.; Watson, J.; Lobell, D.B.; Howden, S.M.; Smith, D.R.; Chhetri, N. A Meta-Analysis of Crop Yield under Climate Change and Adaptation. *Nat. Clim. Chang.* **2014**, *4*, 287–291. [CrossRef]
24. André, G.; Engel, B.; Berentsen, P.B.M.; Vellinga, T.V.; Oude Lansink, A.G.J.M. Quantifying the Effect of Heat Stress on Daily Milk Yield and Monitoring Dynamic Changes Using an Adaptive Dynamic Model. *J. Dairy Sci.* **2011**, *94*, 4502–4513. [CrossRef] [PubMed]
25. Dikmen, S.; Hansen, P.J. Is the Temperature-Humidity Index the Best Indicator of Heat Stress in Lactating Dairy Cows in a Subtropical Environment? *J. Dairy Sci.* **2009**, *92*, 109–116. [CrossRef]
26. Jacobs, C.; Berglund, M.; Kurnik, B.; Dworak, T.; Marras, S.; Mereu, V.; Michetti, M. *Climate Change Adaptation in the Agriculture Sector in Europe*; European Environment Agency (EEA): Copenhagen, Denmark, 2019.
27. Fink, A.H.; Brücher, T.; Krüger, A.; Leckebusch, G.C.; Pinto, J.G.; Ulbrich, U. The 2003 European Summer Heatwaves and Drought -Synoptic Diagnosis and Impacts: European Heatwave—Impacts. *Weather* **2004**, *59*, 209–216. [CrossRef]
28. Van der Velde, M.; Tubiello, F.N.; Vrieling, A.; Bouraoui, F. Impacts of Extreme Weather on Wheat and Maize in France: Evaluating Regional Crop Simulations against Observed Data. *Clim. Change* **2012**, *113*, 751–765. [CrossRef]
29. Tank, A.M.G.K.; Können, G.P. Trends in Indices of Daily Temperature and Precipitation Extremes in Europe, 1946–1999. *J. Clim.* **2003**, *16*, 3665–3680. [CrossRef]
30. McKee, T.B.; Doesken, N.J.; Kleist, J. The Relationship of Drought Frequency and Duration to Time Scales. In Proceedings of the Eighth Conference on Applied Climatology, Anaheim, California, 17–22 January 1993; American Meteorological Society: Boston, MA, USA, 1993; pp. 179–184.
31. Vicente-Serrano, S.; Beguería, S.; Lopez-Moreno, J. A Multiscalar Drought Index Sensitive to Global Warming: The Standardized Precipitation Evapotranspiration Index. *J. Clim.* **2010**, *23*, 1696–1718. [CrossRef]
32. LIU, X.; Zhu, X.; Pan, Y.; Li, S.; Liu, Y.; Ma, Y. Agricultural Drought Monitoring: Progress, Challenges, and Prospects. *J. Geogr. Sci.* **2016**, *26*, 750–767. [CrossRef]
33. Tian, L.; Yuan, S.; Quiring, S.M. Evaluation of Six Indices for Monitoring Agricultural Drought in the South-Central United States. *Agric. For. Meteorol.* **2018**, *249*, 107–119. [CrossRef]
34. Quiring, S.M.; Papakryiakou, T.N. An Evaluation of Agricultural Drought Indices for the Canadian Prairies. *Agric. For. Meteorol.* **2003**, *118*, 49–62. [CrossRef]
35. Woli, P.; Jones, J.W.; Ingram, K.T.; Fraisse, C.W. Agricultural Reference Index for Drought (ARID). *Agron. J.* **2012**, *104*, 287–300. [CrossRef]
36. Beguería, S.; Vicente-Serrano, S.M.; Reig, F.; Latorre, B. Standardized Precipitation Evapotranspiration Index (SPEI) Revisited: Parameter Fitting, Evapotranspiration Models, Tools, Datasets and Drought Monitoring. *Int. J. Climatol.* **2014**, *34*, 3001–3023. [CrossRef]
37. Potop, V.; Možný, M.; Soukup, J. Drought Evolution at Various Time Scales in the Lowland Regions and Their Impact on Vegetable Crops in the Czech Republic. *Agric. For. Meteorol.* **2012**, *156*, 121–133. [CrossRef]
38. Rebetez, M.; Mayer, H.; Dupont, O.; Schindler, D.; Gartner, K.; Kropp, J.P.; Menzel, A. Heat and Drought 2003 in Europe: A Climate Synthetis. *Ann. For. Sci.* **2006**, *63*, 569–577. [CrossRef]
39. Vidal, J.-P.; Martin, E.; Franchistéguy, L.; Habets, F.; Soubeyroux, J.-M.; Blanchard, M.; Baillon, M. Multilevel and Multiscale Drought Reanalysis over France with the Safran-Isba-Modcou Hydrometeorological Suite. *Hydrol. Earth Syst. Sci. Discuss.* **2010**, *14*, 459–478. [CrossRef]
40. Gate, P. Le Blé Face Au Changement Climatique. *Perspect. Agric.* **2007**, *336*, 20–56.

41. Ben-Ari, T.; Adrian, J.; Klein, T.; Calanca, P.; Van der Velde, M.; Makowski, D. Identifying Indicators of Extreme Wheat and Maize Yield Losses. *Agric. For. Meteorol.* **2016**, *220*, 130–140. [CrossRef]
42. Kapsambelis, D.; Moncoulon, D.; Cordier, J. An Innovative Damage Model for Crop Insurance, Combining Two Hazards into a Single Climatic Index. *Climate* **2019**, *7*, 125. [CrossRef]
43. Ministère de l'Agriculture, de l'Agroalimentaire et de la Foret, Ministère de l'Economie et des Finances. *Cahier Des Charges Applicable Aux Entreprises D'assurance Pour La Prise En Charge Partielle de Primes et Cotisations D'assurance Récolte 2016. Pris En Application Des Articles 1, 2 et 10 Du Décret Fixant Pour Les Années 2016–2020 Les Modalités D'application de L'article L. 361-4 Du Code Rural et de La Pêche Maritime En Vue de Favoriser Le Développement de L'assurance Contre Certains Risques Agricoles*; Ministère de l'Agriculture et de l'Alimentation: Paris, France, 2016.
44. IPCC. *Climate Change 2021: The Physical Science Basis. Contribution of Working Group I to the Sixth Assessment Report of the Intergovernmental Panel on Climate Change*; Cambridge University Press: Cambridge, UK, 2021.
45. IPCC. *Regional Fact Sheet—Europe*; Cambridge University Press: Cambridge, UK; New York, NY, USA, 2021.
46. Kapsambelis, D.; Moncoulon, D.; Cordier, J. Modélisation de l'impact des sécheresses et excès d'eau sur les productions végétales à horizon 2050. In *Rapport Scientifique CCR 2021*; CCR: Paris, France, 2021; pp. 62–66.
47. Forzieri, G.; Feyen, L.; Russo, S.; Vousdoukas, M.; Alfieri, L.; Outten, S.; Migliavacca, M.; Bianchi, A.; Rojas, R.; Cid, A. Multi-Hazard Assessment in Europe under Climate Change. *Clim. Chang.* **2016**, *137*, 105–119. [CrossRef]
48. Jacob, D.; Petersen, J.; Eggert, B.; Alias, A.; Christensen, O.B.; Bouwer, L.M.; Braun, A.; Colette, A.; Déqué, M.; Georgievski, G.; et al. EURO-CORDEX: New High-Resolution Climate Change Projections for European Impact Research. *Reg. Env. Chang.* **2014**, *14*, 563–578. [CrossRef]
49. Russo, S.; Dosio, A.; Graversen, R.G.; Sillmann, J.; Carrao, H.; Dunbar, M.B.; Singleton, A.; Montagna, P.; Barbola, P.; Vogt, J.V. Magnitude of Extreme Heat Waves in Present Climate and Their Projection in a Warming World. *J. Geophys. Res. Atmos.* **2014**, *119*, 12500–12512. [CrossRef]
50. Spinoni, J.; Vogt, J.V.; Naumann, G.; Barbosa, P.; Dosio, A. Will Drought Events Become More Frequent and Severe in Europe? *Int. J. Climatol.* **2018**, *38*, 1718–1736. [CrossRef]
51. Loukas, A.; Vasiliades, L.; Tzabiras, J. Climate Change Effects on Drought Severity. *Adv. Geosci.* **2008**, *17*, 23–29. [CrossRef]
52. Mavromatis, T. Drought Index Evaluation for Assessing Future Wheat Production in Greece. *Int. J. Climatol.* **2007**, *27*, 911–924. [CrossRef]
53. Gaitán, E.; Monjo, R.; Pórtoles, J.; Pino-Otín, M.R. Impact of Climate Change on Drought in Aragon (NE Spain). *Sci. Total Environ.* **2020**, *740*, 140094. [CrossRef] [PubMed]
54. Zhang, Q.; Zhang, J. Drought Hazard Assessment in Typical Corn Cultivated Areas of China at Present and Potential Climate Change. *Nat. Hazards J. Int. Soc. Prev. Mitig. Nat. Hazards* **2016**, *81*, 1323–1331. [CrossRef]
55. Sun, C.X.; Huang, G.; Fan, Y.; Zhou, X.; Lu, C.; Wang, X.Q. Drought Occurring with Hot Extremes: Changes under Future Climate Change on Loess Plateau, China. *Earth's Future* **2019**, *7*, 587–604. [CrossRef]
56. Strzepek, K.; Yohe, G.; Neumann, J.; Boehlert, B. Characterizing Changes in Drought Risk for the United States from Climate Change. *Environ. Res. Lett.* **2010**, *5*, 044012. [CrossRef]
57. Beniston, M.; Stephenson, D.B.; Christensen, O.B.; Ferro, C.A.; Frei, C.; Goyette, S.; Halsnaes, K.; Holt, T.; Jylhä, K.; Koffi, B. Future Extreme Events in European Climate: An Exploration of Regional Climate Model Projections. *Clim. Chang.* **2007**, *81*, 71–95. [CrossRef]
58. Blenkinsop, S.; Fowler, H.J. Changes in European Drought Characteristics Projected by the PRUDENCE Regional Climate Models. *Int. J. Climatol.* **2007**, *27*, 1595–1610. [CrossRef]
59. Vidal, J.P.; Martin, E.; Kitova, N.; Najac, J.; Soubeyroux, J.-M. Evolution of Spatio-Temporal Drought Characteristics: Validation, Projections and Effect of Adaptation Scenarios. *Hydrol. Earth Syst. Sci.* **2012**, *16*, 2935–2955. [CrossRef]
60. Deque, M. Frequency of Precipitation and Temperature Extremes over France in an Anthropogenic Scenario: Model Results and Statistical Correction According to Observed Values. *Glob. Planet. Chang.* **2007**, *57*, 16–26. [CrossRef]
61. Planton, S.; Déqué, M.; Chauvin, F.; Terray, L. Expected Impacts of Climate Change on Extreme Climate Events. *Geoscience* **2008**, *340*, 564–574. [CrossRef]
62. Soubeyroux, J.M.; Vidal, J.-P.; Najac, J.; Kitova, N.; Blanchard, M.; Dandin, P.; Martin, E.; Pagé, C.; Habets, F. *Projet ClimSec. Impact du Changement Climatique en FRANCE sur la Sécheresse et l'eau du Sol. Rapport Final*; irstea: Gennes, France, 2011; p. 73.
63. Soubeyroux, J.-M.; Kitova, N.; Blanchard, M.; Vidal, J.P.; Martin, E.; Dandin, P. Caractérisation Des Sècheresses Des Sols En France et Changement Climatique: Résultats et Applications Du Projet ClimSec. *La Météorologie* **2012**, *78*, 21–30. [CrossRef]
64. Hatfield, J.; Takle, G.; Grotjahn, R.; Holden, P.; Izaurralde, R.C.; Mader, T.; Marshall, E.; Liverman, D.; Melillo, J.M.; Richmond, T.C.; et al. *Chapter 6: Agriculture. Climate Change Impacts in the United States: The Third National Climate Assessment*; Melillo, J.M., Richmond, Terese (T.C.), Yohe, G.W., Eds.; U.S. Global Change Research Program: Sydney, NSW, Australia, 2014. [CrossRef]
65. Létard, V.; Flandre, H.; Lepeltier, S. *La France et les Français Face à la Canicule: Les Leçons D'une Crise*; Rapport d'information au Sénat: Paris, France, 2004.
66. Dercon, S. Growth and Shocks: Evidence from Rural Ethiopia. *J. Dev. Econ.* **2004**, *74*, 309–329. [CrossRef]
67. Christidis, N.; Stott, P.A.; Brown, S.; Hegerl, G.C.; Caesar, J. Detection of Changes in Temperature Extremes during the Second Half of the 20th Century. *Geophys. Res. Lett.* **2005**, 32. [CrossRef]

68. Brisson, N.; Gate, P.; Gouache, D.; Charmet, G.; Oury, F.-X.; Huard, F. Why Are Wheat Yields Stagnating in Europe? A Comprehensive Data Analysis for France. *Filed Crops Res.* **2010**, *119*, 201–212.
69. Peltonen-Sainio, P.; Jauhiainen, L.; Laurila, I.P. Cereal Yield Trends in Northern European Conditions: Changes in Yield Potential and Its Realisation. *Field Crops Res.* **2009**, *110*, 85–90. [CrossRef]
70. Harsányi, E.; Bashir, B.; Alsilibe, F.; Alsafadi, K.; Alsalman, A.; Széles, A.; Bácskai, I.; Juhász, C.; Ratonyi, T.; Mohammed, S. Impact of Agricultural Drought on Sunflower Production across Hungary. *Atmosphere* **2021**, *12*, 1339. [CrossRef]
71. Mokhtar, A.; He, H.; Alsafadi, K.; Mohammed, S.; Ayantobo, O.; Elbeltagi, A.; Abdelwahab, O.; Hongfei, Z.; Quan, Y.; Abdo, H.; et al. Assessment of the Effects of Spatiotemporal Characteristics of Drought on Crop Yields in Southwest China. *Int. J. Climatol.* **2021**, 1–20. [CrossRef]
72. Ministère de l'Agriculture et de l'Alimentation. Conjoncture Agreste 2003. Ministère de l'Agriculture et de l'Alimentation: Paris, France, 2003; p. 3.
73. CCR. *Note de Synthèse Sur Le Suivi de l'aide à l'assurance Multirisques Climatique Pour l'exercice 2019 Par CCR Pour Le Ministère de l'Agriculture et de l'Alimentation*; CCR: Paris, France, 2003; p. 43.

applied sciences

MDPI

Article

Modelling Fire Risk Exposure for France Using Machine Learning

Baptiste Gualdi [1], Emma Binet-Stéphan [1], André Bahabi [1], Roxane Marchal [2,*] and David Moncoulon [2]

1 EURIA EURo Institut d'Actuariat Brest, 29200 Brest, France; baptiste0310015361u.gualdi@etudiant.univ-brest.fr (B.G.); emma.binetste@gmail.com (E.B.-S.); andre.bah@outlook.com (A.B.)

2 Caisse Centrale de Réassurance, Department R&D Cat and Agriculture Modelling, 75008 Paris, France; dmoncoulon@ccr.fr

* Correspondence: rmarchal@ccr.fr; Tel.: +33-1-44-35-31-00

Abstract: Wildfires generating damage to assets are extremely rare in France. The peril is not covered by the French natural catastrophes insurance scheme (law of 13 July 1982). In the context of the changing climate, Caisse Centrale de Réassurance—the French state-owned reinsurance company involved in the Nat Cat insurance scheme—decided to develop its knowledge on the national exposure of France to wildfire risks. Current and future forest fires events have to be anticipated in case one of the events threatens buildings. The present work introduces the development of a catastrophe loss risk model (Cat model) for forest fires for the French metropolitan area. Cat models are the tools used by the (re)insurance sector to assess their portfolios' exposure to natural disasters. The open-source national Promethée database focusing on the South of France for the period 1973–2019 was used as training data for the development of the hazard unit using machine learning-based methods. As a result, we observed an extension of the exposure to wildfire in northern areas, namely Landes, Pays-de-la-Loire, and Bretagne, under the RCP 4.5 scenario. The work highlighted the need to understand the multi-peril exposure of the French country and the related economic damage. This is the first study of this kind performed by a reinsurance company in collaboration with a scholarly institute, in this case EURIA Brest.

Keywords: forest fires; Cat model; climate change; disaster risk; machine learning; R programming language

Citation: Gualdi, B.; Binet-Stéphan, E.; Bahabi, A.; Marchal, R.; Moncoulon, D. Modelling Fire Risk Exposure for France Using Machine Learning. *Appl. Sci.* **2022**, *12*, 1635. https://doi.org/10.3390/app12031635

Academic Editor: Jason K. Levy

Received: 7 January 2022
Accepted: 1 February 2022
Published: 4 February 2022

1. Introduction

In the world, large forest fire events are generating significant damage to natural ecosystems, human lives, and critical infrastructures [1]. In the last few years, large events occurred especially in the United States and in Australia [2]. In 2017 and 2018, in California, wildfire events were estimated, respectively, to have caused $12 bn in damages for the Tubbs Fire and the CampFire. It has been estimated that wildfire caused $150 bn damage globally, with $27.7 bn for direct losses to buildings and houses, or 20% of the total [3–5]. Between 2011 and 2020, the average annual loss for the USA was $4.7 bn for forest fires [3]. More recently, we had in mind last year's Black Summer in Australia, with the sad images of koalas and kangaroos burnt by the flames; in addition to the 10,000 people displaced, 25 people died, 5.5 million ha were destroyed, and 2448 homes were destroyed [6]. Those wildfires generated colossal economic losses. Periods of long and intense droughts elevated fire risk prediction, which is especially the case in Canada and the Western USA [7]. Nowadays, in early July 2021, the world watches, helplessly, the heat wave hitting Lytton (Canada), which recorded temperatures of 49.6 °C, with flames destroying the city [8], as well as the large events in Greece and Turkey due to the greatest heat wave in thirty four years [9].

Modelling wildfire is a complex task, as several parameters have to be defined (fire propagation, fuel, wind speed, terrain type, smoke, prevention actions and building susceptibility). This leads to the development of detailed models to assess the fire propagation, ignition and dynamic, as exemplified by the well-known models of the literature (FlamMap, FSPro, FARSITe, FIRETACTIC, FEPS, HYPSLIT, PHOENIX, and Minimum Travel Time) [1,4,10–14]. The insurance sector also developed models at the asset level, modelling the roofs, walls, and windows which were the most susceptible to burning with destruction functions [3,11,15–17]. In comparison to the US or Australia, southern Europe records less burnt surface, such as the 7.4 million ha burnt between 2000 and 2018 [18]. For France, from 1982 to 2017, 12 events were recorded for a total of 350,000 ha burnt in the Mediterranean area (EM-DAT data 2021, https://public.emdat.be/, (accessed on 2 August 2021)). Then, comparing these elements to the Promethée database (Promethée data, https://www.promethee.com/, (accessed on 12 October 2020)), France records an accumulation of small events (approximately 2000 events per year) with small-to-medium surfaces, with strong spatial and temporal fluctuations (approximately 7.3 ha burnt per year). At the end of August 2021, the South of France recorded a large wildfire event of 8100 ha in the natural park Plaine des Maures. The event was extreme in terms of its propagation speed of up to 8 km/h, destroying a dozen houses in Val de Gilly Grimaud. It is the worst wildfire event occurring in France since 2–3 September 2003, when flames destroyed—in the same location—20,000 ha of forest, with three lives lost. It was demonstrated that the 1994 fire protection was successful [19]. On the contrary, for other countries, as explained by [20], there is a lack of international coordinated safety procedure for wildfires.

Nevertheless, considering climate change, it is important to anticipate the future exposure. Indeed, France is the fourth European country in terms of forest cover, with 17 million ha of forest, which means an increase of exposure in the next few years. Climate change affects the frequency of wildfires due to anomalous maximum temperatures, lower humidity, higher maximum wind speed, and fewer rainy days [5,21–23]. A study projected the fire danger due to climate change in Southern France [24]. In 2019, a lot of kermes oak trees died due to the heatwave, as the lethal temperature was reached with temperatures of more than 60 °C measured (French Ministry of Ecology, 2021, https://www.ecologie.gouv.fr/prevention-des-feux-foret, (accessed on 7 September 2021)). There is a clear need to develop a model to identify the exposed areas, and to protect them from significant losses (to the ecosystem and to the economy). Then, taking into account those elements, CCR experts in natural disaster modelling covered by the Nat Cat scheme raised interest in forest fires. In collaboration with EURIA Brest, we developed a Cat model from scratch within seven months, from data collection and hazard modelling through machine learning to exposure and damage estimates. A Cat model is the tool of the (re)insurance sector to estimate the consequences of natural disasters on their portfolios. It is composed of three submodules: hazard, vulnerability and damage units (Figure 1). We aimed to test the ability of machine learning-based methods to model the fire hazard, namely burnt surface and occurrence [25]. The hazards themselves, such as fire smoke and earth imagery, were not the target of this study. The outputs of the hazard and vulnerability units are combined into the damage unit in order to provide estimates of the amount of loss due to the natural events. A special interest was taken in the wildland–urban interface in order to consider the increasing number of houses in the French littoral at-risk areas [26]. Due to data availability, the RCP 4.5 IPCC scenario was used [27]. The model was developed using the R programming language. We combined meteorological data from spatially synchronized Safran daily weather data, building location BD TOPO IGN®, and insured values at the department scale.

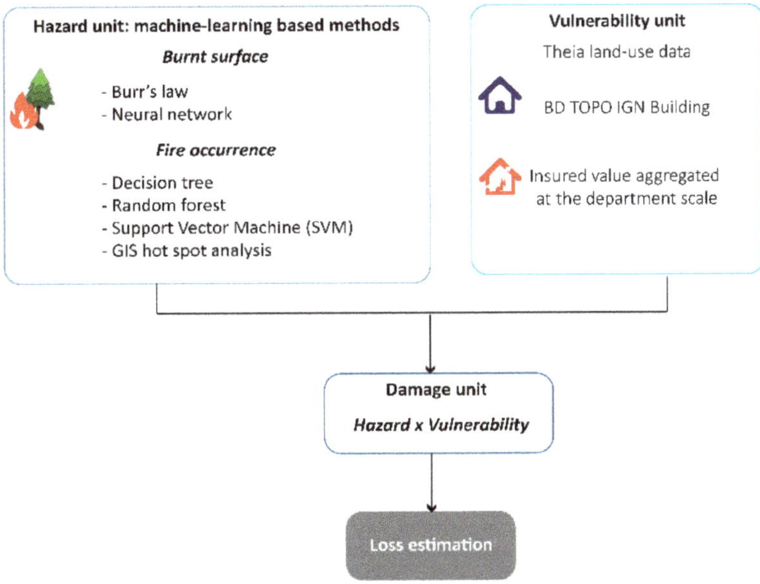

Figure 1. Cat model structure used by the (re)insurance sector to estimate the amount of loss due to natural disasters, in this case wildfire events. For the study, machine learning-based methods were integrated into the model, namely into the hazard unit. The latest was based on the use of machine-learning methods to define the number of fire events and the burnt surface. Once the best model for each of the variables was defined, the outputs were combined with the damage unit. The vulnerability unit gathers all of the information about the insured portfolios with building locations and insured values. Then, the damage unit (risk assessment), allows the loss estimation for a wildfire event.

Cat modelling allows a probabilistic assessment of wildfire risk, examining key locations in order to determine the potential property losses; the model calculates risk by looking at a range of factors, in this case simplistic factors, in order to define the first exposure of France. The aim is to estimate whether the probability of fire occurrence and burnt surface per DFCI (DFCI, https://www.geonov.fr/smartdata/carroyage/, (accessed on 12 October 2020)) mesh according to the Safran data meteorological conditions of the mesh. This study was designed to demonstrate learning for residential exposure.

The paper is structured as follows. Section 2 relates the data collection process and the implemented machine learning methods for Cat model development. Section 3 presents the results of the Cat models from current to future exposure with potential losses. The paper ends with the discussions and conclusion.

2. Materials and Methods

2.1. Fire and Meteorological Data as the Input Data for a Machine Learning-Based Hazard Unit

The historical patterns of wildfires in Southern France were based on the Promethée dataset from 1973 to 2019. The database contains, for each DFCI mesh, several metrics (Table 1).

The DFCI geographical mesh system is used in France by actors for fire prevention from the 100-km to the 2-km resolution (Figure 2).

Table 1. Metrics of the Promethée database used for the study.

Metrics	Details
Date	Date of occurrence of the wildfire
Number	Id of the fire
Type of fire	Unfilled variable
Department	Localization of the fire
INSEE ID	French ID for community
Community	Name of the community
DFCI mesh	Id of the DFCI mesh
Alert	The data and hour of the first fire alert
Origin of the alert	Policemen, population, aerial etc.
Burnt surface	In m^2
Max_burnt_surf	Maximal burnt surface for each DFCI mesh

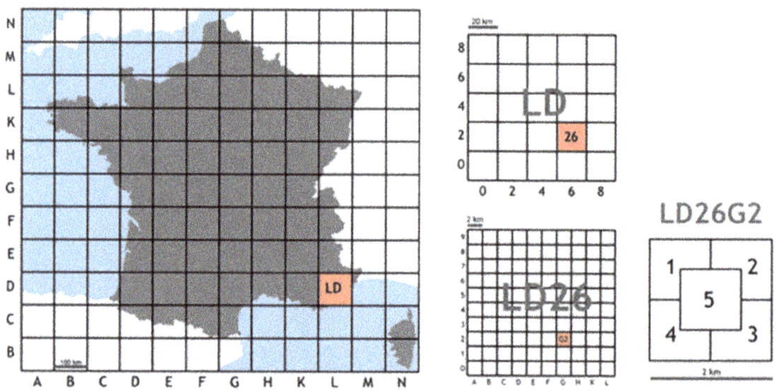

Figure 2. DFCI mesh, from the national to the local scale (https://www.data.gouv.fr/fr/datasets/carroyage-dfci-2-km/, (accessed on 12 October 2020)). The mesh has a value of 100 km with a letter code, exemplified with "LD". Then, for the 20 km resolution, two figures are added—e.g., "LD26"—and for the 2 km a letter and a figure are added—e.g., "LD26G2"—providing a unique code for each mesh.

The Fire Weather Index (FWI) system was developed by the Canadian Forest Fire Danger Rating System (CFFDRS) in the seventies [28]. This indicator is used worldwide as a trustable indicator for the study of climate change effects in fire exposure [29]. The indicator is available on the EFFIS Copernicus website throughout Europe. The FWI is calculated daily by Météo-France for France via Arpège-Climat 4.6 over the period of reference: 1959–2007. The FWI data was downloaded from Drias's Météo-France platform and for the period 1973–2007. The FWI is only available on a seasonal average from March to November. The higher the FWI is, the greater the probability of wildfire is. The winter season is not studied in this article.

The Safran data provide information about the temperature, humidity, wind and precipitation. The data were downloaded from the Drias's Météo-France platform. The resolution is 8 km $\times$ 8 km; for different RCP scenarios, a total of 8602 points cover France. The evolution of the critical meteorological parameters was calculated for 1973–2005, and for the horizon 2050 RCP 4.5 seasonality of the parameters and a 20 $\times$ 20 km^2 analysis. The study focuses on a seasonal timescale in order to highlight the variation of the meteorological metrics. Climsec Météo-France data are available as a seasonal average for the entire year. We assume that the fire event within a DFCI mesh is uniquely determined by the mesh's meteorological conditions.

EURO-CORDEX (Coordinated Downscaling Experiment) data are available daily by Safran point. The data were reanalyzed in order to obtain them for the same seasonality as the FWI and Climsec data (Table 2).

Table 2. EURO-CORDEX and Climsec data.

Metrics	Details
TASMIN	Daily minimal temperature at 2 m (altitude)
TASMAX	Daily maximal temperature at 2 m
TAS	Daily averaged temperature at 2 m
PR	Daily precipitations (mm)
SFCWIND	Wind speed at 10 m (altitude) (m/s)
SPI	Meteorological drought for 3 months
SSWI	Soil wetness index for 3 months

The meteorological Safran data were overlaid with the DFCI mesh. The resolution is 20 km × 20 km. The meteorological database is thus based on 1467 meshes (Appendix A).

2.2. *Machine Learning-Based Methods for the Development of the Hazard Unit*

Artificial intelligence and machine learning methods have been used in wildfire science since the 1990s. Within the sets of available tools suggested in Jain et al. [30], we focused only on the following: (i) decision trees, (ii) support vector machines, and (iii) artificial neural networks.

We used machine learning-based methods for the development of the hazard unit from the historical data collected in the Prométhée database. We focused only on the definition of the burnt surface and the occurrence of fire events. The meteorological data were integrated as indicators in order to assess their consequences on the area covered by the fire, and on the occurrence.

2.2.1. Burnt Surface

The first approach was to predict the burnt surface in each DFCI mesh, and to validate it for the real historical data. The training data were the total burnt surfaces and the maximal burnt surface. In order to solve the issue of extreme events and fires with low intensity, we focused only on the fires of 1 ha and 100 ha. The first tested method was the adjustment according to a law. Indeed, this makes it possible to study the data as a drawing of a random variable X, the law of which was known but the parameters of which were not. In order to do this, we must choose a known law of X that seems to be close to the distribution of our data. Then, by a method of optimization of the parameters of this law, we find the parameters that maximize the likelihood between the data and the density of this law. We can predict future data by randomly drawing this law as many times as necessary in a statistical approach. Linear regression establishes a linear relation between an explained variable and one or more explicative variables. The model was defined as follows, with Y being the explained variable, Xp being the p explicative variable, and the error and β being the parameters of the model.

We focused on the least-squares method, which minimizes the square deviation and the estimated regression. The the R packages used were fitdistr, MASS and fitdistrplus. After different tests, the most appropriate law was Burr's law.

The second process was to test the neural network. A neural network is composed of neurons distributed in several layers: input neurons, neurons in different hidden layers, and output neurons. The input data and our hidden layers will modulate these data by different weights and biases which provide output neurons and a value [31–33]. Then, the square error of the prediction was calculated by comparing the differences between the data and the predicted value. The neural network was then modified in order to minimize this error. Thus, by repeating this operation the neural network obtains accurate predictions while avoiding overlearning. The package used was nnet. The network length was 10, with 40 iterations.

2.2.2. Fire Occurrence

The objective was to predict the probability of a DFCI mesh being affected by a fire event, and to validate it with the real historical data. The training data were the number of fires recoded as a Boolean variable, in order to predict the occurrence of at least one fire event for the summer period. Three machine learning-based methods were tested, and are detailed below. In addition to the machine learning models developed with R, we tested the ArcGis® GIS-based machine learning hot-spot analysis.

A decision tree is a decision support tool that takes the form of a tree. At each node, a decision is taken according to a parameter, and we descend in the tree to a new node until we arrive at a leaf [34]. In a decision tree, in the case of a classification tree, the leaves contain qualitative variables (labels); in the case of a regression tree, the leaves contain quantitative variables. The package used was rpart. The maximal length of the trees was 3, with at least 50 individuals per terminal node. Between 1973 and 2005, during the summer season, 64% of the meshes recorded at least one fire; thus, the learning database will gather 50% of the database.

In a Random Forest, the main advantage of decision trees is their readability and speed of execution [35]. The package used was randomForest; 500 trees were chosen, with a height of five, and with at least 10 individuals per terminal node.

A Support Vector Machine (SVM) is a supervised learning technique. If the points (target values) are linearly separable in the space of explanatory variables, the SVM will search for the hyperplane boundary (the decision boundary) [36]. However, the points may be not separated by a hyperplane, and it is then possible to reconsider the problem in a higher dimensional space [37]. In order to deform the original space, we apply a kernel function; in this new space, it is then likely that there is a linear separation. The package used was e1017, and the duration of the calculation was a few minutes.

The space–time path is useful to visualize and understand the relationship between time and geography data. The geographical data are represented along the x and y axis, and the cube's height represents the time on the z axis [38]. The Promethée dataset fits perfectly with the ArcGis® geovisualization tool, as there is information on the location and on the time series.

In order to perform the analysis, the Space Time Cube tool (ArcGIS Pro 2.8 online support: https://pro.arcgis.com/fr/pro-app/latest/tool-reference/space-time-pattern-mining/learnmorecreatecube.htm, (accessed on 10 March 2021)) was used based on Promethée data, in order to define the hot spots whilst considering their evolution over the timeline. The point data per year and community were integrated within the tool, and were aggregated considering space and time. The fire events were recapitulated within a hexagonal grid at yearly time steps, and then the spatial model provided the evolution in the time under a NetCDF format. We did not define the interval distance between the points, because they are the centroid of the community. Then, the Emerging Hot Spot Analysis tool was used to read the NetCDF file. It analyzes the area in which the fire events are statistically emergent or reduced in the area. The temporal interval is defined as a year.

2.3. *Vulnerability and Damage Modelling*

For the development of the vulnerability unit, information about the portfolio exposure is required. First, the land-use type is needed; we used the Theia data at a 100-m resolution. Theia defined 23 land-use types; the data was re-categorized into 4 categories (Table 3). We assume the land use to be constant for the horizon 2050, as well as the number of buildings. The land-use data were overlaid with the DFCI meshes.

Table 3. New classification of the THEIA land-use data.

New Classification	THEIA Land Use Types
Urban	Dense buildings; diffuse buildings; industrial and commercial areas; roads
Agriculture	Colza; cereals; protein crops; soybeans; sunflowers; maize; rice; tubers/roots; grasslands; orchards; vineyards;
Forest	Wilderness deciduous forest; coniferous forest; grasslands; woody moors;
Other surfaces	Mineral; beaches; dunes; glaciers/snow; water;

Secondly, building location data were obtained from the vector building dataset BD TOPO IGN®. In this study, we only focused on the individual residential building, for which we have more detailed insurance data. For the damage model, we assume that when a fire crosses three departments with different building densities and insured values, that the three are proportionally damaged (Figure 3).

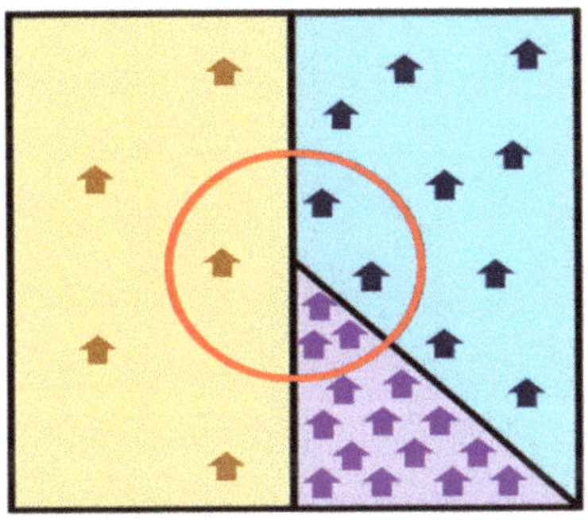

Figure 3. When a fire (in red) touches, proportionally, the department within the DFCI mesh, the density of the houses is applied, and we obtain the number of houses burnt and the damage costs.

In terms of insured damages, wildfires may have important consequences for the buildings. They can totally damage the infrastructure. Nevertheless, wildfire damage is not covered within the Nat Cat scheme, and events destroying residential assets are extremely rare in France. In French newspapers, there is information about the costs for firemen, but not for the insured losses. Thus, the wildfire-related insured damages are not available. In order to bridge the gap, we used confidential CCR portfolio data aggregated at the department scale. The leaflet R package was used to map the data. We consider, in the model, that if the fire touches a house it is totally destroyed within the damage function. As there are no data on fire-related claims, we considered the insured values as the claims. This is contrary to Australia or the USA, where buildings are destroyed and the destruction functions are then calibrated [11]. The price of a m^2 of building per department was downloaded (https://www.meilleursagents.com/prix-immobilier/ (accessed on 1 December 2021)). The fictive model applies the maximal historical burnt surface of each DFCI mesh since 1973 to the existing urban areas. Then, we count the number of damaged assets, i_{house}, to which we apply the costs of a square meter of house per department, *P€M2*, and the insured values of the house and furniture at the department scale, *P€fur*.

$$Fori_{house} = \sum P€M2 + P€fur$$

Third, the Wildland–Urban Interface (WUI), developed by the USA, describes areas where wildfires and urban areas interact, generating a potential loss of properties and life [10,39]. The WUI types intermix, and interface areas were applied to the entire French scale at the DFCI mesh of 20 × 20 km.

Concerning the damage model, we assume the application of the maximal surface of burnt areas of each DFCI mesh on the urban surface of the same mesh; then, by using the departmental insured values, we are able to calculate the potential damages.

3. Results

3.1. Statistical Analysis

The statistical analysis of the data highlights the variations of the total burnt areas and fire occurrence per season per year in the Promethée area. The seasonal variability is very important; it is correlated to the fact that if a lot of areas are burnt during the year n − 1, the probability of fire is decreased for the year 0 or year +1 (Figure 4).

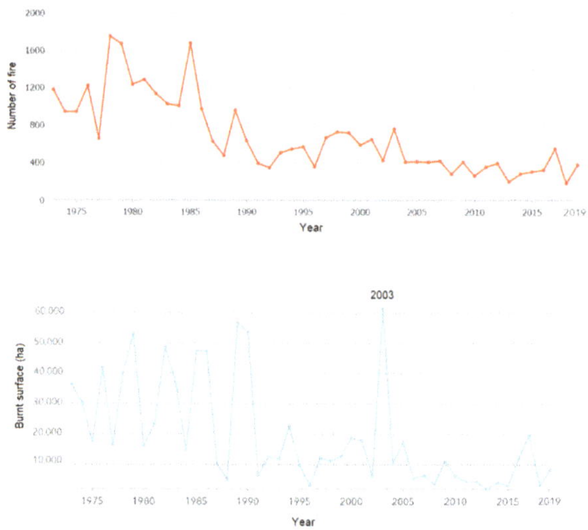

Figure 4. The occurrence and surface area of fire events tend to decrease over time, especially since the 1990s, with the reinforcement of the preventive measures. In terms of burnt surface, the year 2003 is highlighted, and is well known regarding the high intensity of the heat wave.

A large majority of the DFCI mesh has no or only two fire departures. On the contrary, some of the mesh has more than 100 fire departures over the entire studied period and over the years (Figure 5).

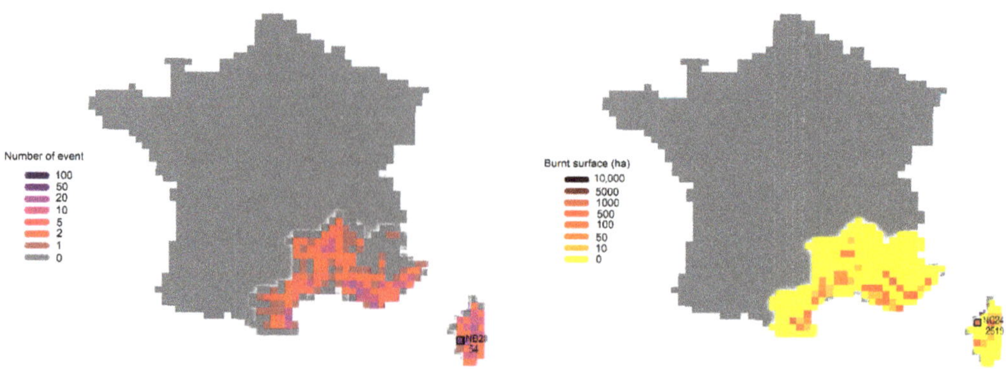

Figure 5. The two maps exemplify the diversity of the fire occurrence (**left**) and burnt surface (**right**) according to each DFCI mesh for the summer period in 2005.

For a daily analysis, the link between a high FWI and fire occurrence is important; nevertheless, when considering the average value of the FWI over three months, the link is not insured. The variability of the data poses an issue of extreme values, and adds complexity for the implementation of machine learning-based methods (see Section 3.2). In order to cope with the issue, the correlation matrix allows us to better understand the relationships and interdependence between the metrics. The number of fires is positively correlated with the FWI (0.31), the number of days without precipitation (0.21) and the mean temperature (0.29). The number of fires is, on the contrary, negatively correlated with the precipitation; in particular, the negative values are between −0.09 and −0.17. This matrix also highlights that the SPI and SSWI indexes are not correlated with other variables, and especially with the targeted variables NBFEUX and SURFTOT (values equal to 0). On the contrary, the variables representing the same data have an important coefficient of relation; for example, the variable for temperature TASQ50 and TASQ90 with a correlation of 0.96 (Figure 6).

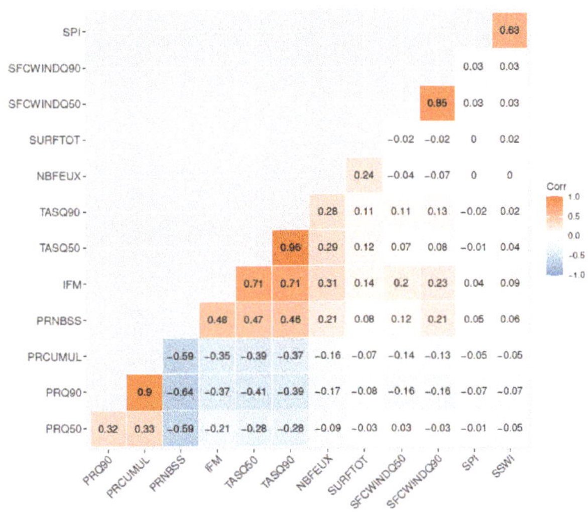

Figure 6. The correlation matrix highlights the coefficient of correlation for the different possible pairs of variables. There is a positive correlation between the FWI and the Q50/Q90 of the daily temperature (0.71). We observed a correlation between the number of days without precipitation and the number of fires (0.21).

3.2. Hazard Unit

The results below show the comparison of the estimates from the machine learning-based methods between the real and modelled scenarios in order to predict the burnt surface and the probability of fire. We also integrated the projections at horizon 2050 under the RCP 4.5 IPCC Scenario.

3.2.1. Burnt Surface

For this purpose, we started with the statistical approach based on Burr's law. We obtained a similar distribution for the majority of the low-intensity fires, and for some of the extreme events. The quality of the simulation was determined using a quantile–quantile diagram. Figure 5 shows a soft Burr's law overestimating the area of the burnt surface according to the seasons. The average error was 0.7. The densities are coherent with a very high probability of low fire intensity and a very low probability of extreme fire. The higher the fire, is the lower the probability is (Figure 7). It is a soft Burr's law. We assumed an interesting simulation, and we simulated with this law several thousand numbers in order to calculate the average burnt surface.

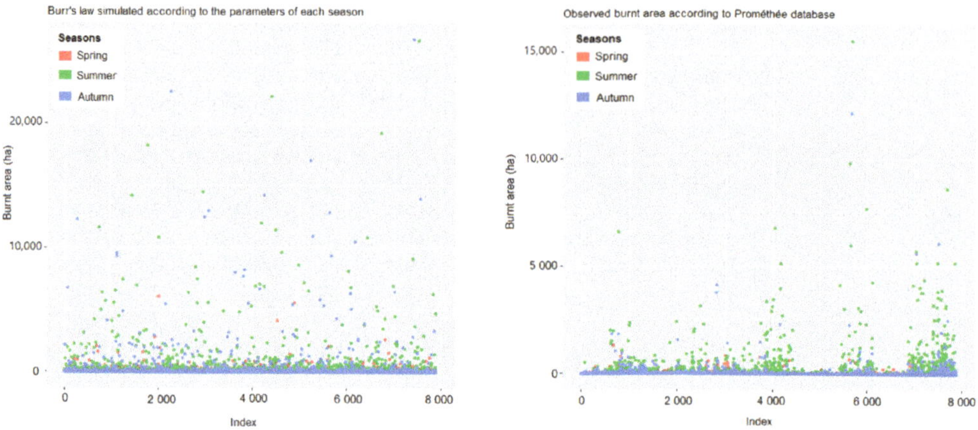

(a) (b)

Figure 7. Burnt surface simulated with Burr's law (**a**), and compared to the observed data from the Prométhée database (**b**).

In a second analysis, the neural network was tested for the estimation of the burnt surface. Figure 8 highlights the average absolute difference between our predicted values and the real ones. The average error on the learning base is in blue, and the error on the test base is given in red. The error on the test base is 15.6 ha. The use of the neural network highlights the need to focus on the prediction of the occurrence or absence of fire events.

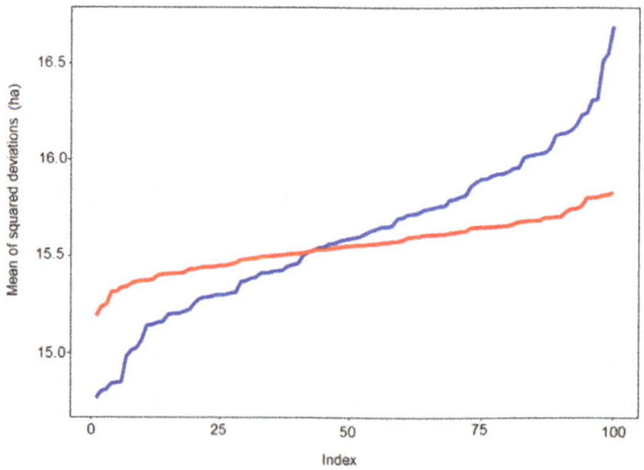

Figure 8. Comparison of the squared deviation for the neutral network based on the Prométhée database from 1973 to 2005 concerning the burnt surface.

Regarding the training data available and the machine learning outputs, we assume that Burr's Law provides better results that the neural network concerning the estimation of the burnt surface.

3.2.2. Fire Occurrence

In order to model the fire occurrence, we tested three machine learning-based methods and one GIS-based approach.

First, we tried the decision tree. At the least, the learning base will contain 50% of the lines. We still obtained a high number of false negatives and badly classified results compared to the other machine learning methods (Figure 9). We could use a deeper tree, but the result wouldn't change much, and the tree would have too many leaves to be interpretable. The main variables are practically always the same, no matter which sample we take. The confusion matrix associated to the tree based on the learning base reveals 22.03% badly classified results; 40.01% false positives and 12.38% false negatives. The results are similar for the validation base, with 24.50% being badly classified, 43.56% false positives and 13.48% false negatives. Based on this method, the false negatives are too important and the results are insufficient.

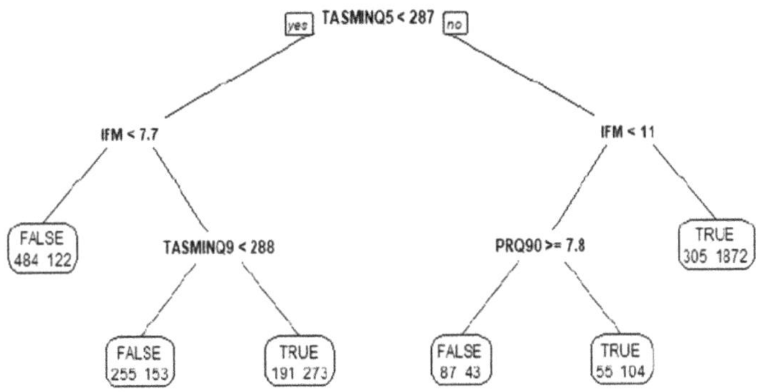

Figure 9. Extract of the decision tree developed for the study.

Next, we tested the Support Vector Machine (SVM). The objective was to split linearly the true and false results within the space of meteorological indexes (of dimension 14). The function svm of the package e1071 allows the realization of SVM by choosing among four nodal functions: linear, polynomial, radial or sigmoid. After different tests, the radial nodal function provides better results. Figure 10 provides the error rate on the learning base, and tests for a variation of the γ parameter. Among the three presented methods and related results, the three are similar in terms of performance. The more interesting rates to consider in the context of fire occurrence are the false negative and badly classified results, at around 14% (i.e., the number of fires that the model has not predicted). We assume that the models have good performance (Figure 10).

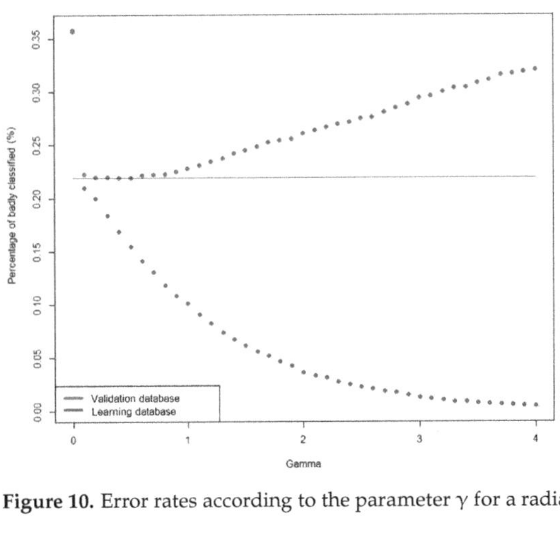

Figure 10. Error rates according to the parameter γ for a radial nodal.

Then, in the continuity of the tests, we tested the random forest method. The implementation of the random forest of the learning base does not highlight false negatives and positives from phenomena of overlearning. Nevertheless, the random forest provides similar results from a dozen trees; the rate of bad classification was around 25% for a random forest predicting the occurrence of fires larger than 10 ha, the FWI, and the Q50 of the daily minimal temperature.

We also tried to estimate the prediction of a burnt area of more than 10 ha; the best model was a random forest (Figure 11). The results are coherent and applicable to the historical datasets and for the 2050 future climate.

(a) (b)

Figure 11. The figure compares (**a**) the modelled prediction of fires of 10 ha for the summer of 2005 using a random forest, and observations from the same year (**b**).

We observe that the results of the projection are different according to the method of projection, as the models are calibrated on the Promethée area, and are projected to the entire country (Figure 12). Each model has at least 20% badly classified results and 40% false positives.

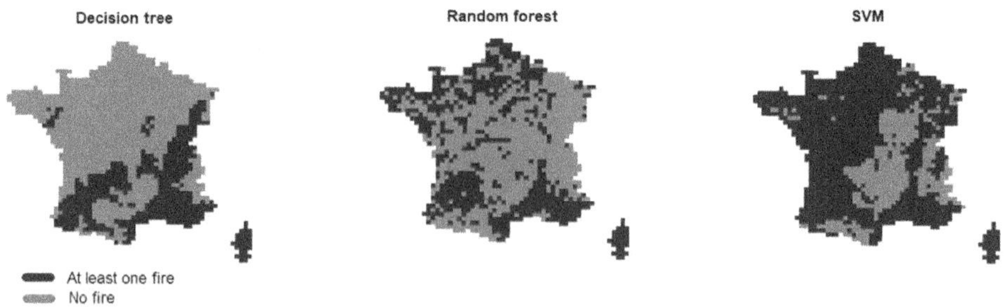

(a) (b) (c)

Figure 12. Comparison of the results from the three machine learning tools rgarding the probability of at least one fire within the 2 km DFCI mesh during summer 2050 RCP 4.5. (**a**) The decision tree projection is the more moderated model; we observed the exposure of Landes, Rhône valley and Vosges. (**b**) The random forest's projection highlights the Mediterranean and Corse areas' exposure, with an increasing exposure of the Rhône valley, Landes, Bretagne, Nord-Pas-de-Calais and Pays-de-Loire; this model seems to be the more coherent one. (**c**) SVM's projection is the more pessimistic model, generating fire events in the large majority of the territory; the results are more related to the RCP 8.5 pessimistic scenario.

The results are highly coherent for the Mediterranean area. We observe some similarities, especially for increasing exposure of Bretagne, Pays de la Loire, Centre Val de Loire, and the Atlantic coast (Landes forested areas), and an increasing exposure of the mediterranean area (Occitanie and Provence Alpes Côte d'Azur) [40].

Finally, the statistically significant hotspots and cold spots are represented on the map. The red areas indicate that, throughout the time, there is an aggregation of a high number of forest fires. The blue areas highlight a smaller number of fire events. Each hexagon is classified according to the timescale. The geographical analysis underlines the exposure of the areas nearby Béziers and Perpignan, which are areas of oscillating hot spots. The rest of the Mediterranean area is exposed in the same manner without a strong evolution in time (oscillating cold spots). For North Corse, at least 90% of the temporal intervals were statistically significant hotspots (Figure 13).

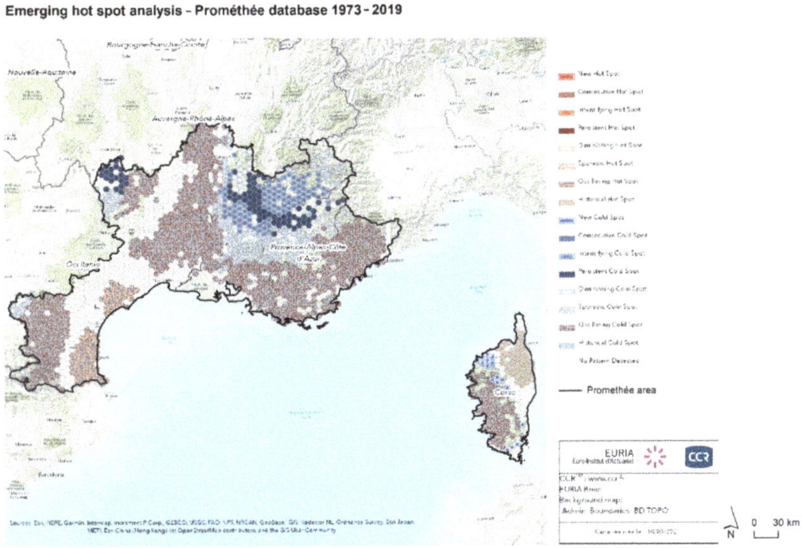

Figure 13. Emerging hot spot analysis for the Promethée dataset 1973–2019.

3.3. *Damage Model*

The R code provides, for each year, a table with the potential costs of a fire for the DFCI. Within this model, we assume that the burnt surface is entirely inhabited. Under the hypothesis of the RCP 4.5 at horizon 2050 using a random forest model predicting the occurrence or not of fire greater than 10 ha, we estimate that the damage will be, in 2050, around 35 M€ on average for residential insured areas only. The evolution of the insured values is not considered, nor are the land-use changes. The spatial repartition of the future areas exposed to wildfire events further north in France highlights the increase of the economic exposure (Figure 10). This evolution can be compared to the results of Moncoulon et al. [26] on geotechnical drought and shrinking swelling clay (SSC), in that they reveal an increasing exposure of the southern communities, as well as those is the Atlantic area.

4. Discussion

Despite the many limitations associated with simulation modelling and machine learning-based methods in these experiments discussed above, the outputs from this work provide useful information on the exposure of France to wildfires. This work introduced the foundation of the Cat model for the assessment of forest fire exposure and projection to horizon 2050. We acknowledge the limitations in the use of the model for the prediction of building exposure considering the spatial resolution chosen. First and foremost, our predictions were only applied to the RCP 4.5 scenarios. This study focused on wildfire disasters from an asset context, and we recognize that wildfires also threaten human lives and ecosystems, as well as cascading impacts on floods, landslides and potable water after severe fire events. Furthermore, the model could consider not only the physical damage but also the business interruption or critical infrastructure issues (highways, secondary roads, etc.). Furthermore, we did not consider the currently implemented fire prevention programs. We considered the exposure to be constant in the 2050 horizon. As the issue of population growth in littoral areas will be higher in the future, it will increase the exposure to fire events and greatly influence the WUI [26,41,42]. Regarding the damage model's development, wildfire-related claims data could support the determination of the damage functions; nonetheless, based on our knowledge, that kind of data is not available in France. Downscaling the model could also make the results and damage estimates more precise.

Although climate change is considered in the optimistic scenario RCP 4.5, the results of the future exposure are significant to start to raise and develop a risk culture in the future exposed areas (Bretagne, Alsace, etc.), and for the maintenance of the currently well-structured prevention processes in Southern France. There have been numerous studies that focused on the asset level and considered the fire conditions, landscape and properties' structures. Here, in order to obtain a global vision of the exposure of the entire area of France, we used simplistic models. Obtaining new data is challenging, as it requires waiting for future wildfires and potentially generating large losses, and ensuring that data of sufficient quality are collected. The relatively small number of fire events in terms of number or burned acres, and the very low number of burnt assets in France means that the applicability of the model developed overestimates the number, and the surface is not easily calibrated on the training data. Nevertheless, it offers a new visibility of France's exposure to that kind of natural disaster for the next few years. It provides elements for discussion on the issue of the underwriting of fire risk within the Nat Cat scheme.

5. Conclusions

This model synthesizes information for the French insurance sector, and contributes to understanding and reducing wildfire losses. CCR and Euria Brest developed a first-of-its-kind France Cat model projecting changes to wildfire potential under the RCP4.5 scenario at a granularity of about 20×20 km. Finally, the best model for burnt-surface prediction is Burr's law, and the random forest for the fire occurrence.

In the future, this model could be combined with GIS analysis (within distance to vegetation, slopes, and fuel type) and with satellite imagery analysis in order to make the exposure analysis more precise.

New machine learning and remote sensing data could be used to develop specific damage curves for household buildings for the vulnerability models, such as those created for hurricane damage. This is the first time, as far as we know, that a reinsurance company developed, with an institute, a prototype model that links machine learning and insurance data, and applied these models to the estimation of the expected financial loss from wildfires. Likewise, the model could be applied to other countries, as well as the pessimistic scenario RCP 8.5. The different improvements will open the door to explore a wide range of exposure management in order to reduce the climate change impact, and to support the community for preventive measures. Strong decisions have to be taken in order to avoid making 2021 the last coolest year of the rest of our lives.

This paper offers new visibility for the improvement of the preparedness in future potentially affected areas. We hope that this work will support future potentially exposed areas to integrate the analysis within their disaster risk prevention and resilience plans. Robust cat models can improve the accuracy of the predictions of the locations of the greatest risks to assets, and could provide an indication of the implementation of preventive measures. The proposed methodology could serve as a reference for wildfire risk assessment, and can be replicated elsewhere.

Author Contributions: Conceptualization, D.M. and R.M.; methodology, B.G., E.B.-S. and A.B.; software, B.G., E.B.-S. and A.B.; validation, D.M. and R.M.; formal analysis, B.G., E.B.-S. and A.B.; investigation, B.G., E.B.-S. and A.B.; resources, B.G., E.B.-S. and A.B.; data curation, B.G., E.B.-S. and A.B.; writing—original draft preparation, B.G., E.B.-S., A.B. and R.M.; writing—review and editing, R.M.; visualization, B.G., E.B.-S. and A.B.; supervision, D.M. All authors have read and agreed to the published version of the manuscript.

Funding: This research received no external funding.

Institutional Review Board Statement: Not applicable.

Informed Consent Statement: Not applicable.

Conflicts of Interest: The authors declare no conflict of interest.

Appendix A

Table A1. Developed metrics from the overall database to be integrated within the machine learning tools.

Metrics	Details
NBFEUX	Number of fire departures
NBFEUX0_5HA	Number of fire >0.5 ha
NBFEUX1HA	Number of fire >1 ha
NBFEUX10HA	Number of fire >10 ha
NBFEUX100HA	Number of fire >100 ha
SURFQ50	Quantile 50 of the fire surface (ha)
SURFQ90	Quantile 90 of the fire surface (ha)
SURFMAX	Maximal surface of the fire (ha)
SURFTOT	Total burnt area (ha)
INBFEUX	Binear code: fire/no fire; whatever the burnt surface
INBFEUX0_5HA	Binear code: fire/no fire; burnt surface >0.5 ha
INBFEUX1HA	Binear code: fire/no fire; burnt surface >1 ha
INBFEUX10HA	Binear code: fire/no fire; burnt surface >10 ha
INBFEUX100HA	Binear code: fire/no fire; burnt surface >100 ha
TASMINQ50	Quantile 50 over 3 months of the minimal temperature
TASMINQ90	Quantile 90 over 3 months of the minimal temperature
TASMAXQ50	Quantile 50 over 3 months of the maximal temperature
TASMAXQ90	Quantile 90 over 3 months of the maximal temperature
TASQ50	Quantile 50 over 3 months of the mean temperature
TASQ90	Quantile 90 over 3 months of the mean temperature
PRQ50	Quantile 50 over 3 months of the daily precipitation
PRQ90	Quantile 90 over 3 months of the daily precipitation
PRCUMUL	Total precipitation over 3 months
PRNBSS	Number of days without precipitation
SFCWINQ50	Quantile 50 over 3 months of the wind speed
SFCWINQ90	Quantile 90 over 3 months of the wind speed
SPI	Meteorological drought for 3 months
SSWI	Soil wetness index for 3 months
FWI	Seasonal average from March to November

References

1. Sfetsos, A.; Giroud, F.; Clemencau, A.; Varela, V.; Freissinet, C.; Lecroart, J.; Vlachogiannis, D.; Politi, N.; Karozis, S.; Gkotsis, I.; et al. Assessing the Effects of Forest Fires on Interconnected Critical Infrastructures under Climate Change. Evidence from South France. *Infrastructures* **2021**, *6*, 16. [CrossRef]
2. Li, S.; Banerjee, T. Spatial and temporal pattern of wildfires in California from 2000 to 2019. *Sci. Rep.* **2021**, *11*, 1–17. [CrossRef]
3. Risk Management Solutions (RMS) Fundamentals of Wildfire Modeling. Virtual Online Meeting "Exceedance 2021", May 2021. Available online: https://www.rms.com/events/exceedance-2021 (accessed on 1 December 2021).

4. Ager, A.A.; Day, M.A.; Alcasena, F.J.; Evers, C.R.; Short, K.C.; Grenfell, I. Predicting Paradise: Modeling future wildfire disasters in the western US. *Sci. Total Environ.* **2021**, *784*, 147057. [CrossRef]
5. Wang, D.; Guan, D.; Zhu, S.; Mac Kinnon, M.; Geng, G.; Zhang, Q.; Zheng, H.; Lei, T.; Shao, S.; Gong, P.; et al. Economic footprint of California wildfires in 2018. *Nat. Sustain.* **2020**, *4*, 252–260. [CrossRef]
6. Whittaker, J.; Haynes, K.; Wilkinson, C.; Tofa, M.; Samson, S.; Dilworth, T.; Collins, J.; Tait, L. *Lessons from Black Summer: How People Experienced the 2019/20 NSW Fire Season; Hazard Note*; Bushfire and Natural Hazards CRC: Merlboune, VIC, Australia, 2021; p. 4.
7. Phillips, C. How Wildfires Affect Climate Change—And Vice Versa. 2021. Available online: https://theconversation.com/how-wildfires-affect-climate-change-and-vice-versa-158688 (accessed on 1 December 2021).
8. Canicule historique au Canada le village aux 49,6 °C détruit par le feu. Le Monde.fr 2021. Available online: https://www.lemonde.fr/climat/article/2021/07/02/canicule-historique-au-canada-le-village-aux-49-6-c-detruit-par-un-incendie_6086626_1652612.html#:~{}:text=Canada-,Canicule%20historique%20au%20Canada%20%3A%20le%20village%20aux%2049%2C6%20%C2%B0,ce%20record%20mardi%2029%20juin.&text=A%20260%20kilom%C3%A8tres%20de%20Vancouver,du%20nord%2Douest%20du%20Canada. (accessed on 1 December 2021).
9. En Turquie et en Grèce, une canicule et des feux d'une ampleur historique. *Le Monde.fr* **2021**.
10. Mahmoud, H.; Chulahwat, A. A Probabilistic Cellular Automata Framework for Assessing the Impact of WUI Fires on Communities. *Procedia Eng.* **2017**, *198*, 1111–1122. [CrossRef]
11. Duff, T.J.; Penman, T.D. Determining the likelihood of asset destruction during wildfires: Modelling house destruction with fire simulator outputs and local-scale landscape properties. *Saf. Sci.* **2021**, *139*, 105196. [CrossRef]
12. Salis, M.; Arca, B.; Del Giudice, L.; Palaiologou, P.; Alcasena-Urdiroz, F.; Ager, A.; Fiori, M.; Pellizzaro, G.; Scarpa, C.; Schirru, M.; et al. Application of simulation modeling for wildfire exposure and transmission assessment in Sardinia, Italy. *Int. J. Disaster Risk Reduct.* **2021**, *58*, 102189. [CrossRef]
13. Jarry, M.; Campet, P. *Modélisation d'un Feu de Forêt*; Master pro de didactique de mathématiques Mémoire de modélisation; Université Paris 7: Paris, France, 2010; p. 28.
14. De Gennaro, M. Modélisation de La Propagation Des Grands Incendies de Forêts et Élaboration d'un Outil Opérationnel d'aide à La Lutte Tactique. Ph.D. Thesis, Aix-Marseille University, Marseille, France, 2017.
15. Badal-Valero, E.; Coll-Serrano, V.; Segura-Gisbert, J. Fire Risk Sub-Module Assessment under Solvency II. Calculating the Highest Risk Exposure. *Mathematics* **2021**, *9*, 1279. [CrossRef]
16. *IFRC Shelter Safety Handbook Some Important Information on How to Build Safer*; IFRC: Geneva, Switzerland, 2011; p. 62.
17. *The Zurich Flood Resilience Alliance PERC A Burning Issue: Insights for Resilience from Three Wildfire Events*; Institute for Social and Environmental Transition–International: Boulder, CO, USA, 2020; p. 28.
18. San-Miguel-Ayanz, J.; Durrant, T.; Boca, R.; Libertà, G.; Branco, A.; de Rigo, D.; Ferrari, D.; Maianti, P.; Vivancos, T.A.; Oom, D.; et al. *Forest Fires in Europe, Middle East and North Africa 2017 2018*; European Commission: Brussels, Belgium, 2019; ISBN 978-92-76-11234-1.
19. Evin, G.; Curt, T.; Eckert, N. Has fire policy decreased the return period of the largest wildfire events in France? A Bayesian assessment based on extreme value theory. *Nat. Hazards Earth Syst. Sci.* **2018**, *18*, 2641–2651. [CrossRef]
20. Vuorio, A.; Kovanen, P.T.; Budowle, B.; Sajantila, A.; Palo, J.U.; Stoop, J. Wildfire-Related Catastrophes: The Need for a Modern International Safety Investigation Procedure. *Front. Clim.* **2021**, *3*. [CrossRef]
21. Jolly, W.M.; Cochrane, M.A.; Freeborn, P.H.; Holden, Z.A.; Brown, T.J.; Williamson, G.J.; Bowman, D.M.J.S. Climate-induced variations in global wildfire danger from 1979 to 2013. *Nat. Commun.* **2015**, *6*, 7537. [CrossRef]
22. Norman, B.; Newman, P.; Steffen, W. Apocalypse now: Australian bushfires and the future of urban settlements. *npj Urban Sustain.* **2021**, *1*, 1–9. [CrossRef]
23. Xu, R.; Yu, P.; Abramson, M.J.; Johnston, F.H.; Samet, J.M.; Bell, M.L.; Haines, A.; Ebi, K.L.; Li, S.; Guo, Y. Wildfires, Global Climate Change, and Human Health. *N. Engl. J. Med.* **2020**, *383*, 2173–2181. [CrossRef]
24. Varela, V.; Vlachogiannis, D.; Sfetsos, A.; Karozis, S.; Politi, N.; Giroud, F. Projection of Forest Fire Danger due to Climate Change in the French Mediterranean Region. *Sustainability* **2019**, *11*, 4284. [CrossRef]
25. Murphy, K.P. *Machine Learning: A Probabilistic Perspective*; MIT Press: Cambridge, MA, USA, 2012; ISBN 978-0-262-30432-0.
26. Moncoulon, D.; Desarthe, J.; Naulin, J.-P.; Onfroy, T.; Tinard, P.; Wang, Z.-X.; Hajji, C.; Veysseire, M.; Dequé, M.; Régimbeau, F. *Conséquences du changement climatique sur le coût des catastrophes naturelles en France à l'horizon 2050*; Caisse Centrale de Réassurance & Météo France: Paris, France, 2018; p. 31.
27. Core Writing Team; Pachauri, R.K.; Meyer, L.A. (Eds.) *IPCC Climate Change 2014: Impacts, Adaptation, and Vulnerability. Part A: Global and Sectoral Aspects. Contribution of Working Group II to the Fifth Assessment Report of the Intergovernmental Panel on Climate Change*; Cambridge University Press: Cambridge, UK; New York, NY, USA, 2014.
28. Van Wagner, C.E. Development and Structure of the Canadian Forest Fire Weather Index System. *Can. For. Serv.* **1987**, *48*. Available online: http://citeseerx.ist.psu.edu/viewdoc/summary?doi=10.1.1.460.3231 (accessed on 1 December 2021)
29. Bergeron, Y.; Flannigan, M.D. Predicting the effects of climate change on fire frequency in the southeastern Canadian boreal forest. *Water, Air, Soil Pollut.* **1995**, *82*, 437–444. [CrossRef]
30. Jain, P.; Coogan, S.C.P.; Subramanian, S.G.; Crowley, M.; Taylor, S.W.; Flannigan, M.D. A review of machine learning applications in wildfire science and management. *Environ. Rev.* **2020**, *28*, 478–505. [CrossRef]

31. Touzet, C. LES RESEAUX DE NEURONES ARTIFICIELS, INTRODUCTION AU CONNEXIONNISME; Collection de l'EERIE; Marseille, France, EC2, 1992. Available online: https://hal-amu.archives-ouvertes.fr/hal-01338010/document (accessed on 1 December 2021).
32. Zhang, W.; Li, H.; Li, Y.; Liu, H.; Chen, Y.; Ding, X. Application of deep learning algorithms in geotechnical engineering: A short critical review. *Artif. Intell. Rev.* **2021**, *54*, 5633–5673. [CrossRef]
33. Zhang, W.; Phoon, K.-K. Editorial for Advances and applications of deep learning and soft computing in geotechnical underground engineering. *J. Rock Mech. Geotech. Eng.* **2022**. [CrossRef]
34. Breiman, L.; Friedman, J.H.; Olshen, R.A.; Stone, C.J. *Classification and Regression Trees*; Routledge: Abingdon, UK, 2017; ISBN 9781315139470.
35. Breiman, L. Statistical Modeling: The Two Cultures (with comments and a rejoinder by the author). *Stat. Sci.* **2001**, *16*, 199–231. [CrossRef]
36. St, J.P.O.E.D.; Dumais, S.; Osuna, E.; Platt, J.; Schölkopf, B. Support vector machines. *IEEE Intell. Syst. Appl.* **1998**, *13*, 18–28. [CrossRef]
37. Wang, J.; Chen, Q.; Chen, Y. RBF Kernel Based Support Vector Machine with Universal Approximation and Its Application. In *Advances in Neural Networks—ISNN 2004*; Yin, F.L., Wang, J., Guo, C., Eds.; Lecture Notes in Computer Science; Springer: Berlin/Heidelberg, Germany, 2004; Volume 3173.
38. Kraak, M. The Space-Time Cube Revisited from a Geovisualization Perspective. In Proceedings of the 21st International Cartographic Conference (ICC), Durban, South Africa, 10–16 August 2003.
39. Westhaver, A.; Taylor, S.W. *Developing a Method for Conducting Wildland/Urban Interface Fire Case Study Research*; Institute for Catastrophic Loss Reduction: Toronto, ON, Canada, 2020; p. 86.
40. Chatry, C.; Le Quentrec, M.; Laurens, D.; Le Gallou, J.-Y.; Lafitte, J.-J.; Creuchet, B. *Changement Climatique et Extension Des Zones Sensibles Aux Feux de Forêts*; Ministère de l'Alimentation, de l'Agriculture et de la Pêche, Ministère de l'Intérieur, de l'Outre-mer et des collectivités territoriales; Conseil Général de l'Environnement et du Développement Durable: Paris, France, 2010; p. 190.
41. Wu, C.; Venevsky, S.; Sitch, S.; Mercado, L.M.; Huntingford, C.; Staver, A.C. Historical and future global burned area with changing climate and human demography. *One Earth* **2021**, *4*, 517–530. [CrossRef]
42. Wu, Z.; Wei, R.; Chu, Z.; Liu, Q. Real-time rock mass condition prediction with TBM tunneling big data using a novel rock–machine mutual feedback perception method. *J. Rock Mech. Geotech. Eng.* **2021**, *13*, 1311–1325. [CrossRef]

applied sciences

MDPI

Article

The Manning's Roughness Coefficient Calibration Method to Improve Flood Hazard Analysis in the Absence of River Bathymetric Data: Application to the Urban Historical Zamora City Centre in Spain

Julio Garrote [1,*], Miguel González-Jiménez [2], Carolina Guardiola-Albert [2] and Andrés Díez-Herrero [2]

[1] Department of Geodynamics, Stratigraphy, and Paleontology, Faculty of Geology, Complutense University of Madrid, 28040 Madrid, Spain

[2] Geological Hazards Division, Geological Survey of Spain (IGME, CSIC), 28003 Madrid, Spain; miguel.gonzalez@igme.es (M.G.-J.); c.guardiola@igme.es (C.G.-A.); andres.diez@igme.es (A.D.-H.)

* Correspondence: juliog@ucm.es; Tel.: +34-91-394-4850

Featured Application: The methodology proposed in this manuscript makes it possible to improve the estimation of flood zones and their flow depth values in situations where there are no available bathymetric data of the channel (or they are scarce and do not allow for its shape reconstruction). It could improve flood risk assessment too.

Citation: Garrote, J.; González-Jiménez, M.; Guardiola-Albert, C.; Díez-Herrero, A. The Manning's Roughness Coefficient Calibration Method to Improve Flood Hazard Analysis in the Absence of River Bathymetric Data: Application to the Urban Historical Zamora City Centre in Spain. *Appl. Sci.* **2021**, *11*, 9267. https://doi.org/10.3390/app11199267

Academic Editors: Ricardo Castedo, Miguel Llorente Isidro and David Moncoulon

Received: 2 August 2021
Accepted: 2 October 2021
Published: 6 October 2021

Publisher's Note: MDPI stays neutral with regard to jurisdictional claims in published maps and institutional affiliations.

Abstract: The accurate estimation of flood risk depends on, among other factors, a correct delineation of the floodable area and its associated hydrodynamic parameters. This characterization becomes fundamental in the flood hazard analyses that are carried out in urban areas. To achieve this objective, it is necessary to have a correct characterization of the topography, both inside the riverbed (bathymetry) and outside it. Outside the riverbed, the LiDAR data led to an important improvement, but not so inside the riverbed. To overcome these deficiencies, different models with simplified bathymetry or modified inflow hydrographs were used. Here, we present a model that is based upon the calibration of the Manning's n value inside the riverbed. The use of abnormally low Manning's n values made it possible to reproduce both the extent of the flooded area and the flow depth value within it (outside the riverbed) in an acceptable manner. The reduction in the average error in the flow depth value from 50–75 cm (models without bathymetry and "natural" Manning's n values) to only about 10 cm (models without bathymetry and "calibrated" Manning's n values), was propagated towards a reduction in the estimation of direct flood damage, which fell from 25–30% to about 5%.

Keywords: flood risk; cultural heritage sites; bathymetric data; manning's roughness coefficient; hydrodynamic modelling

1. Introduction

Floods are probably the most frequently recurring natural phenomena affecting society (humans and goods) in terms of space and time, regardless of their geographical location or socioeconomic development, as shown by the data collected by the International Disasters Database for the period 1900–2018 (CRED, 2020). This is the main reason why flood risk management has become an essential tool from both a social and economic perspective, with the objective of reducing losses associated with both factors. Furthermore, urban historical centers are characterized by the presence of multiple types of cultural heritage sites, which give it a priceless value upon the consideration of the impossibility of the restoration or recovery of the possible damage caused by natural hazards, in this specific case, by river flooding. This abovementioned duality, but also mainly the irreversibility, of flood damage on cultural heritage (which cannot be reproduced once it has been destroyed) makes the prediction and assessment of the flood risks that may affect these sites a critical task for their preservation. Given this situation, the first natural disaster management strategies that included cultural heritage among its objectives began to be developed in the 1990s. Among these initiatives, one could highlight the "Carta del Rischio" [1], which has been developed by the Italian Central Institute for Restoration since 1992, or the "Noah's Ark" project of the European Union [2], launched in 2002. At specific sites, only a few works provide applications to case studies at different scales (from regional to local scales) [3–7].

After assessing direct tangible damage (i.e., the direct damage resulting from the physical contact of floodwater with property and its contents), it was found that the economic flood losses have been increasing throughout the past half-century. In the last decade (2008–2018), the economic losses associated with floods exceeded 35 billion USD [8] and, within this period, the flood losses exceeded 19 billion USD in 2012 alone [9–11].

The design, sizing, implementation and effectiveness of flood damage mitigation measures require rigorous risk analyses [12]. Within the aspects that a flood risk analysis includes, the greatest technical efforts are usually concentrated on flood hazard assessments. Flood vulnerability analysis is the other determining factor regarding the correct estimation of risk, mainly from the use of magnitude–damage models (e.g., [13–19]). Since the most common methods of flood hazard analysis are based on hydrological and hydraulic models [20], the greatest uncertainties and sources of error may come from the input data to these models. In the case of hydraulic or hydrodynamic models, the geometry of the channel (topography and bathymetry) [21–24], the boundary conditions (roughness, flow regime, etc. [25]) and the Manning's roughness coefficient (e.g., [26–28]) are determinants of the goodness of the model and results. In short, the effectiveness of risk mitigation may depend on the correct estimation of parameters such as the roughness of the terrain surface and the detailed bathymetric characterization of the riverbed.

In addition to the key points mentioned above, the urban character of cities' historical centers must be taken into account. This latter characteristic determines the preferred flow direction of floods inside the urban areas, turning the streets into improvised river channels. To achieve all these objectives, it is essential to have topographic data that are capable of reproducing the geometry and variability of the terrain [22,24,29,30], with the information coming from LiDAR (light detection and ranging) sensors being the most widely used today to derive DEMs (digital elevation models). However, most LiDAR data are not capable of reproducing the river channel morphology due to its inability to penetrate water bodies. Airborne LiDAR sensors (one of the possible sources of LiDAR data) are usually not capable of penetrating water bodies (turbid, turbulent or deep streams and rivers), but they constitute the most common LiDAR data due to their cost-effectiveness for wide areas relative to other LiDAR sources. In fact, as is discussed by Kinzel et al. [31], the available bathymetric LiDAR techniques are usually not designed for shallow waters and are not optimized for providing the spatial resolution necessary for mapping small-to-medium-sized rivers.

However, an accurate representation of river bathymetry (bed topography) plays a critical role in multiple hydrologic and hydraulic applications, including but not limited to flood modelling [23]. To solve these limitations, the acquisition of topographic data through ground surveys and subsequent combination with subaerial LiDAR data may be an efficient solution [32–34]. However, this approach has only been used for short river reaches. When the river reach length increases, so do the logistical and cost considerations. Therefore, under the scenario of long river reach study areas, it is common to use alternative methods or models for estimating bathymetry for use in hydraulic or hydrologic analyses. These alternative models assumed a general simplification of river bathymetric geometry, both from the use of simple geometric shapes (triangular, trapezoidal or parabola for a river cross-section) or from other geomorphological and hydraulic methods. The former approach is more frequent [23,35–38]. From a hydraulic perspective, the horizontally divided channel method (HDCM) was previously used [39,40] to solve the absence of bathymetric data. The HDCM separately considered the flow above and below the bank top and is a better option [39] for dividing the flow than the vertically divided channel method (VDCM) when the floodplain roughness is significantly greater than the channel roughness. When the flow is horizontally divided, the lower part of the flow fits with the bankfull flow (taken from the dominant discharge concept), which has a return period of 1-2 years according to the field observations of Wolman and Miller [41]. However, this bankfull flow return period is dependent on local meteorological characteristics, thus it ranges from about 2 years in the north of Spain to 5 or more years in the southern and southeastern parts of the country. Once the bankfull flow is defined, it is detracted from the inflow hydrograph. In the same way, Chone et al. [40] used LiDAR topography to subtract the flow rate at the LiDAR date from the inflow hydrograph as a solution for the absence of river bathymetric data.

Whatever the proposed methodology, model calibration (when possible) is one of the main tasks that will ensure the quality of the results obtained. To carry out this calibration process, the availability of field data related to the event to be reproduced is essential. When this information is available, model calibration by varying the value of the Manning coefficient is one of the most frequent approaches, which is described by Ardıçlıoğlu and Kuriqui [42]. However, valid information is not always available for the calibration of hydraulic models. This situation is more frequent when we do not try to model a specific event, but rather intend to model a design flow event that has a low frequency of occurrence (high return period). In these cases, and in many others, the lack of available information to calibrate the hydraulic models is compensated by the consideration of a benchmark model against which the results obtained in the rest of the models will be compared [43]. In general, this control model is defined by the availability of more or better data for it.

Several studies have already shown that incorporating bathymetry provides more accurate hydraulic simulations and flooding area estimations, which improve flood hazard analysis at the same time as flood risk assessments. Therefore, the goal of this study was not to reinforce these past findings but to try to open up another approach to improve flood hazard assessment in floodplains, that is, improving flood hazard mapping and main flow variables (flow depth) outside the main river channel (that is, into the flood plains) through the use of calibrated but not "natural" values of Manning's roughness coefficient. In other words, we looked for a Manning value or range of values that compensated for the effect that the absence of bathymetric data would have on the hydraulic modelling results. This new approach was supported by the assumption that most flood-exposed elements (people and goods) are located out of the river channel and not inside it; therefore, the flow parameters (depth and velocity) in the river channel are not really key points for flood hazard assessment in most cases. Under these assumptions, the use of two LiDAR DEMs (with and without bathymetric data) for the Douro River reach in Zamora city (Spain) allowed us to calibrate a Manning's roughness coefficient for the main river channel (which is consistent with the value obtained previously from three cross-sections of the river) so that the flooded area and the flow depth in the floodplain converged with the results obtained in the control (or benchmark) model. The hydrodynamic model using the calibrated value show very similar flooded areas, as well as flow parameters (both for the 500-year and 100-year return period peak flows); thus, it can be used for flood hazard and risk assessment. Furthermore, the results of this new approach were compared with results from previous methodological approaches, such as the modified inflow hydrographs. Moreover, the approach proposed here can be replicated for any kind of river and it is cost-saving, as it does not require large ground surveys.

2. Study Area and Data Description

This study focused on comparing the flood area extension over different hydrodynamic models, where Manning's roughness coefficient was changed in a calibration process. The study area comprises the Douro River reach in Zamora city (Spain), where most of the historical urban center is located on the right side of Douro River, but the urban center located on the left side is developed over lower-lying and flood-prone areas. The Douro River reach in Zamora (Figure 1A) is 7 km long with a mean width of 225 m, and is characterized by a relatively deep, sand-clayey channel with some fully vegetated sandy bars and with several meanders and a flat floodplain.

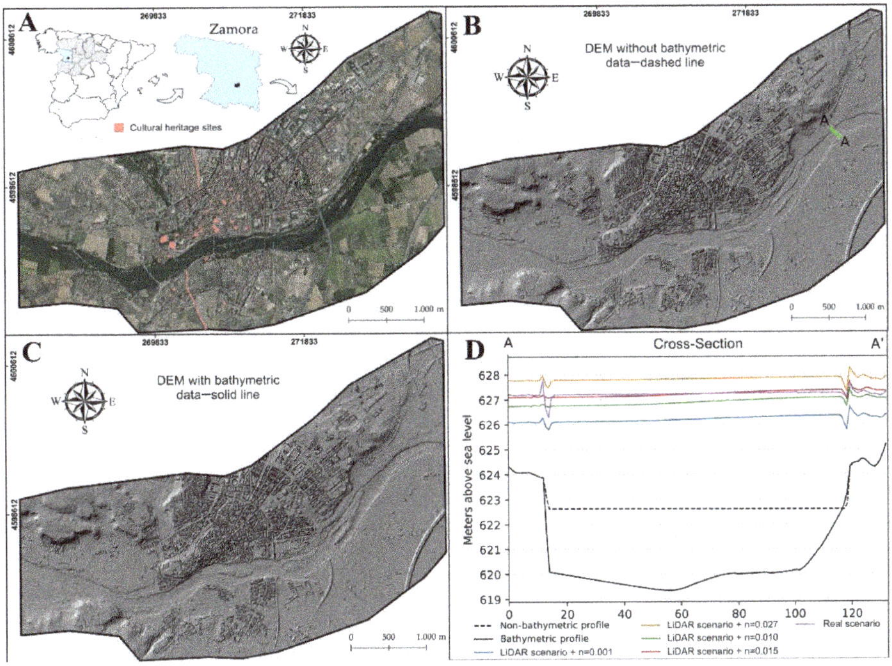

Figure 1. The study area and cultural heritage sites location map (**A**). The digital elevation model (DEM) from LiDAR data (**B**) not showing bathymetry, which was included from echo sounders with a D-GPS (**C**). Example of a cross-section flow depth calibration (**D**).

Topographic LiDAR data [44] are publicly available in the form of point cloud files, which are filtered and classified following the American Society for Photogrammetry and Remote Sensing (ASPRS) standard classification. The vertical accuracy of the LiDAR data used in this study is reported to be lower than 20 cm [44]. The terrain and building types of the LiDAR classification were used to derive the different DEMs. Due to the absence of LiDAR data in the main river channel (since water has low reflectance), it was modelled using a second-order polynomial interpolation of multiple LiDAR cross-sections. Each of these sections presented a height equal to the minimum LiDAR point value and had a length that was restricted to channel boundaries. Although a spline and a third-order polynomial interpolation were tried, the second-order one was chosen because it provided a smooth and downstream-constant slope surface. Bathymetry data were obtained in the form of cross-section measurements that were obtained from boat-mounted echo sounders with D-GPS (differential global positioning system) data acquisition. Both sets of topographic data were combined into a 1 m spatial resolution to create a final DEM (Figure 1C), which was considered to be the "real scenario" (or benchmark scenario). On the other hand, a LiDAR point cloud was also interpolated in the absence of bathymetric data into a 1 m spatial resolution DEM (Figure 1B), which was considered to be the "LiDAR scenario".

The peak flow values that were used for the hydrodynamic models were obtained from a streamflow gauge at the upstream location of the Douro River reach. A generalized extreme values (GEV) distribution was fitted to the annual maximum flow time series, and the 500-year return period peak flow value (2274 $m^3 s^{-1}$) was selected for our analysis upon the assumption that this value can be considered as a low-frequency or "extraordinary" flood event. To test the efficiency and validity of the proposal, a second peak flow value was selected: in this case, the peak flow value associated with a return period of 100 years (1872 $m^3 s^{-1}$). The reason for not selecting a lower peak flow value, therefore one that was more contrasted with the 500-year return period, was the need for the hydraulic model to include the overflow of the main channel in the flood plain. This need was based on the premise that the usefulness of our proposal is only fulfilled in the floodplain, and therefore, in both cases (different peak flows), the floodplain should be affected.

Finally, PNOA ortho-photographs from 2017 (0.25 m of spatial resolution) were used in combination with a field survey to define the value of Manning's roughness coefficient (Manning's n); this map was created for the whole study area at a scale of 1:4.000. It contains ten different roughness units, each of which was assigned a unique Manning's n according to previously proposed tables [45,46]. For the "real scenario", a Manning's roughness coefficient value of 0.027 was assigned for the Douro River's main channel. All other Manning's roughness coefficient values related to each terrain surface unit were calculated and were kept constant throughout the calibration process.

3. Methodology

Figure 2 shows the full methodological approach that was used to define the best calibrated Manning's n value in the river channel, which allowed us to map the flooded area and hydraulic conditions in the absence of bathymetric data.

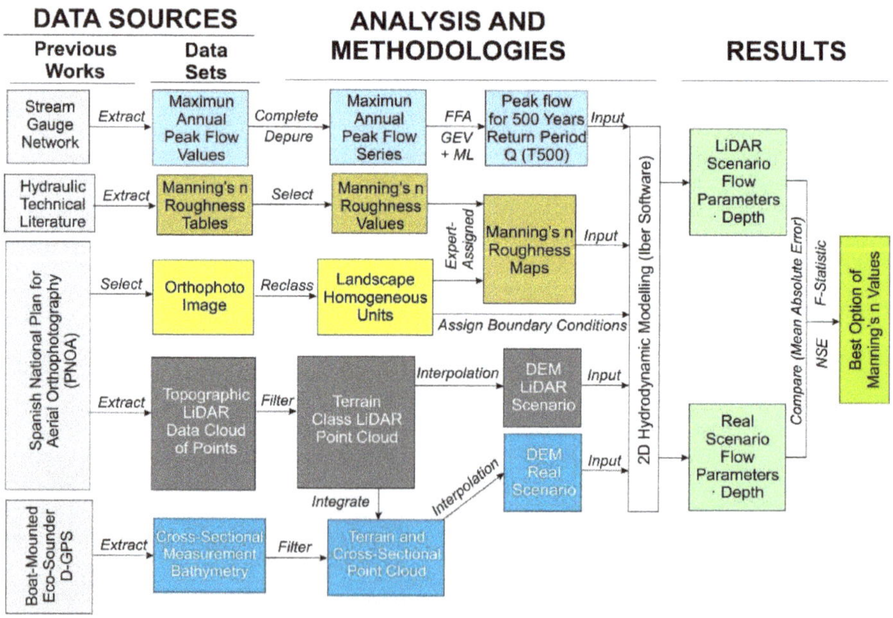

Figure 2. Flow chart of the full methodological approach that was used in the Manning's n value calibration process, from data sources until the calibrated vs. control scenarios comparison.

The entire methodology and analysis that was carried out to obtain the calibrated Manning's n value that best reproduced (in the absence of bathymetric data) the hydraulic conditions and the flooded area in the floodplain associated with a flood event relative to the benchmark model (availability of bathymetric data) was based on a simple premise: the flow through the channel can be expressed in a simplified form from the following equation:

$$Q = A * V, \quad (1)$$

where Q is the total flow ($m^3 s^{-1}$), A is the cross-sectional area (m^2) and V is the mean flow velocity ($m s^{-1}$).

Starting from this equation, if the absence of bathymetric data meant a decrease in the cross-sectional area of the channel, an increase in the flow velocity of the water in the channel could compensate for the reduction in area and thus allow the flow rate that was circulating through the channel to remain constant. This approach is shown graphically in Figure 3.

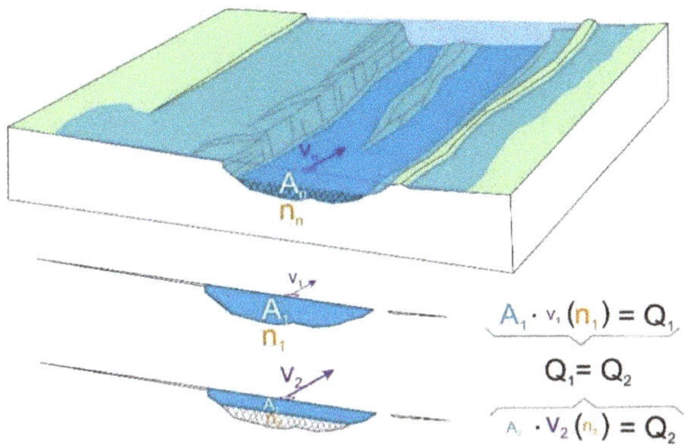

Figure 3. Basic conceptual diagram for the calibration of the parameter n, assuming that the flow rate (Q) is a function of the cross-sectional area of the channel (A_1 and A_2) and the mean flow velocity (V_1 and V_2, which, in turn, is dependent on the Manning's *n* values (n_1 and n_2)). The size of the symbols identifying the variables is directly related to the magnitude of the variables.

3.1. Comparison of Bathymetric Representation

Comparisons of MDE differences and flow depth differences as input and output parameters of the hydrodynamic modelling were carried out. The first of the comparison processes allowed us to quantify the spatially distributed cross-sectional area reduction due to the non-availability of LiDAR data on submerged areas of the main channel. The second one allowed us to spatially assess the distributed differences in flow depth value, as well as to select the optimal model ("LiDAR scenario" plus "best Manning's *n* value") to simulate the flood hazard levels relative to the results of the original, or control, model.

The errors in the bathymetric representation for the "LiDAR scenario" and "real scenario" were compared both using an arithmetic subtraction between the two models (Figure 4A) to obtain a difference value for each pixel of the model and by calculating the mean absolute error (MAE), as shown in Equation (2):

$$MAE = \frac{\sum_{i=1}^{n} |H_L - H_r|}{n_c} \quad (2)$$

where H_L is the elevation of the ith cell for the "LiDAR scenario" DEM, H_r is the elevation of the ith cell for the "real scenario" DEM and n_c is the number of cells used for the analysis.

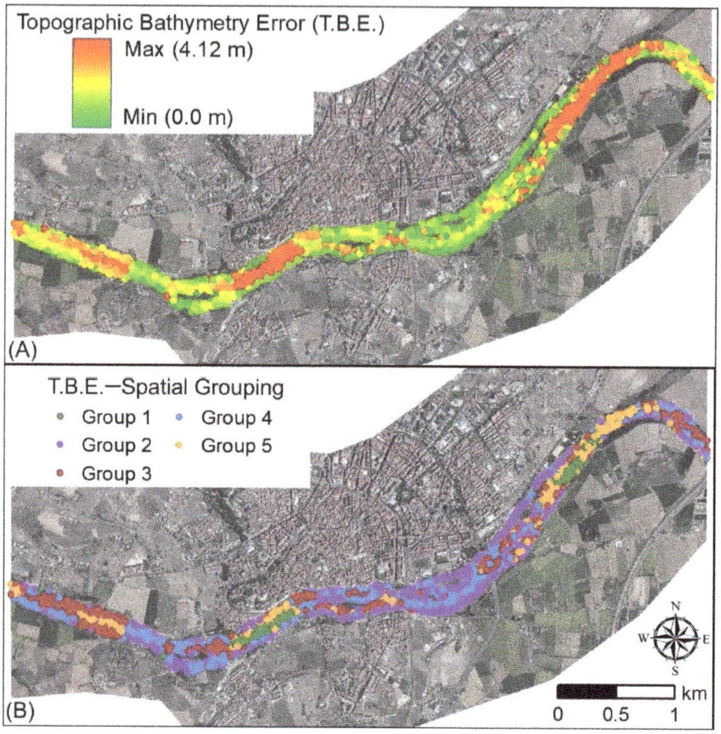

Figure 4. Topographic differences between the "real scenario" and the "LiDAR scenario" (**A**), where green colors show lower differences and red colors show higher topographic differences. A grouping analysis (**B**) of the errors without spatial constraints showed a non-uniform pattern.

3.2. *Hydraulic Modelling and Calibration Process*

The hydrodynamic modelling for all configurations was performed using 2D Iber software [47], which is a two-dimensional mathematical unsteady flow model that is used to simulate the free surface flow in rivers and estuaries and employs a high-resolution finite volume as a numerical method to solve the depth-averaged 2D shallow water equations, also known as Saint Venant equations. Iber software requires a geometric description of the channel in the form of a 3D mesh for performing hydraulic computations. Only the hydraulic modelling of peak flows was carried out, as these peak flows were related to maximum flooded area extension.

The calibration process was carried out in a similar way to a sensitivity analysis of the hydrodynamic model to changes in the Manning's n parameter. The extreme values of the Manning's n for the Duero riverbed were, on the one hand, a minimum value of 0.001, and on the other hand, a maximum value of 0.027 (similar to the value used in the "real scenario"). No higher values of the Manning's n were considered on the basis that, with the same value of surface roughness parameter and a decrease in the cross-sectional area of the channel (due to the absence of bathymetric data in the "LiDAR scenario"), both the flooded area and the flow depth values should be clearly higher than the one obtained in the "real scenario", which here was considered the benchmark model. To complement and complete the analysis, two other scenarios were considered for comparison with the control model. On the one hand, a model with a variable and spatially distributed Manning's n value was considered (which was obtained from the previous models using the Manning's n value that showed the best fit for each of the pixels within the river channel, as is shown in Figure 5). Second, the HDCM model, as applied by Chone et al. [40], was considered by subtracting the flow value at the date of acquisition of the LiDAR topography from the peak flow value of the 500- and 100-year return period.

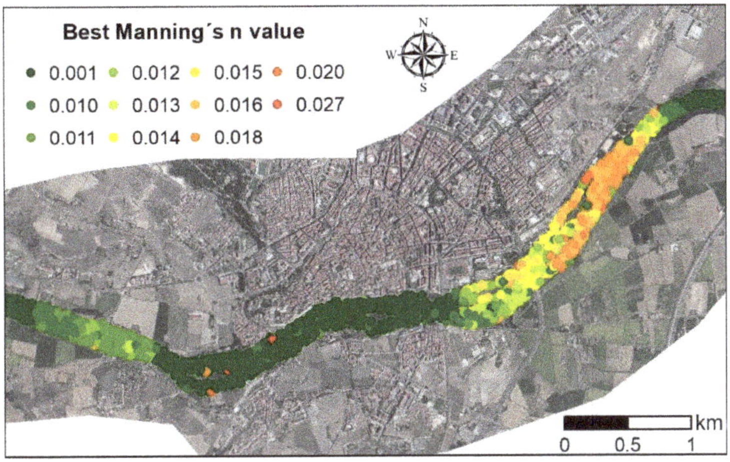

Figure 5. Map showing the spatial distribution of the Manning's n value best fit option along the Douro River channel in Zamora city, Spain. The Manning's n value best fit was found from the relationship between the channel bathymetry differences ("real scenario"–"LiDAR scenario") and the flow depth differences.

As the present approach is only useful for floodplain areas and not for river channels, the data collected for the calibration process were located in the Douro river floodplain. The results of different Manning's n values inside the river channel were also compared to the benchmark model to achieve a spatially distributed Manning model where Manning's n value inside the river channel varied for each model cell. Manning's n value is dependent on which model (with a constant value of Manning within the channel) performed best in the comparison against the control (or benchmark) model.

3.3. Comparison of Hydrodynamic Model Outputs

To evaluate the suitability of the different hydrodynamic models that were generated in the calibration process, the flow depth parameter was selected since it was the most commonly used parameter for flood risk analysis through magnitude–damage models as a flood vulnerability approach. The hydraulic outputs (flow depth) generated by the Iber software were exported into geo-referenced ASCII raster files (*.asc files) for each of the "LiDAR scenario + Manning's n value" models, and further outputs were compared to the benchmark model generated using the "real scenario" topography in the GIS software ESRI® ArcGIS 10.6.1. Spatially distributed flow depth differences between calibrated models and the control model, as well as random sample creation and geo-statistical analysis input parameter calculations, were also generated inside the ESRI® ArcGIS environment.

The analysis of different results started with descriptive statistics of the flow depth value differences between the "LiDAR scenario" models (with different Manning's n values) and the "real scenario" in the set of random samples (nearly 7000 random points for the 500-year return period peak flow and around 3600 random points for the 100-year return period). The mean, median, mode, standard deviation, variance and Nash–Sutcliffe efficiency index were calculated and compared for analysis purposes. Box plots were used to explore the spatial distribution behavior of different levels of depth errors. First, error location points were classified depending on the distance to the river bank with classes of 100 m width. Then, a boxplot was graphed for each spatial class. Distance bands to the river were computed using the Near ArcMap tool [48]. The Matplotlib library [49] from Phyton software was used to perform a descriptive and graphical statistical analysis.

The Nash–Sutcliffe efficiency (NSE) index was used as given in Equation (3):

$$\mathrm{NSE} = 1 - \frac{\sum_{i=1}^{n}(\mathrm{D_m} - \mathrm{D_o})^2}{\sum_{i=1}^{n}(\mathrm{D_o} - \overline{\mathrm{D}}_\mathrm{o})^2} \quad (3)$$

where $\mathrm{D_m}$ is the flow depth value of the modelled "LiDAR scenario", $\mathrm{D_o}$ is the flow depth value in the "real scenario" and $\overline{\mathrm{D}}_\mathrm{o}$ is the mean flow depth value for the "real scenario".

Furthermore, the flood inundation extent that resulted from each calibrated Manning's n model was also compared with the "real scenario" model using the F-statistic, as given in Equation (4):

$$F\ statistic = \frac{A_{Lr}}{A_L + A_r - A_{Lr}} \times 100 \quad (4)$$

where A_L is the observed inundation area (inundation area of the reference model in this case, "real scenario"), A_r is the modelled inundation area ("LiDAR scenario") and A_{Lr} is the area that is common to both the "real scenario" and the "LiDAR scenario" inundation maps.

The F-statistic index, which was previously used in [23,29,50–52], allowed us to compare floods throughout the inundation extent resulting from each hydrodynamic model, where the hydrodynamic models differed was in the bathymetry data (as previously used) or in the Manning's *n* value at the main river reach (as was used here). A value of 100 meant a perfect match between the observed and predicted areas of inundation, and a lower F indicated a discrepancy between the two.

4. Results and Discussion

4.1. "Real Scenario" vs. "LiDAR Scenario" Bathymetric Differences

The use of the MAE index on a random sample of 3000 points (located in the river channel, independent of the samples obtained in the floodplain) made it possible to estimate the error in the bathymetry data between the "real scenario" and the "LiDAR scenario". The MAE index adopted a value of 1.57 m. This was not surprising if we consider the type of riverbed that the Douro River had in the study area under analysis. However, when transferring the MAE index value to the average width of the river's cross-section, we obtained a reduction in the cross-sectional area of 353.25 m^2. This reduction in the cross-section could cause a reduction in the flow capacity of the channel of 353.25–529.87 $m^3\ s^{-1}$ at times when the flow was medium–low and its average speed could be considered to be between 1–1.5 $m\ s^{-1}$. At times of flooding, when the average speed of the water could rise to values of 3 $m\ s^{-1}$ or more, the reduction in the flow capacity in the channel could rise to values of 1059 $m^3\ s^{-1}$ or even more.

The absolute error in bathymetric data between the "real scenario" and the "LiDAR scenario" ranged from 0–4.115 m at the Douro River reach, and it showed a clear trend from the river banks to its centerline (Figure 4A). At the same time, a more irregular pattern was observed in the downstream direction, where the application of a grouping analysis over the spatial distribution of errors in bathymetry did not show clear trends when no spatial constraints were used (compare with spatial constraints, such as K-nearest neighbors or Delaunay triangulation techniques, which show artificial grouping due to the requirements of spatial constraints between points in the same group). When no spatial constraints were used, the grouping of the random sample points could be related to the "riffle and pools" longitudinal profile shape (Figure 4B) of natural rivers.

All previous data pointed out the importance of bathymetric data for the correct hydrodynamic modelling of the flow event flooding area, as well as the uncertainty of flood parameter results when using LiDAR data without a bathymetric data improvement. In this sense, Cook and Merwade [29] previously pointed out this problem when they showed a reduction of between 5 and 20% in flooded areas when bathymetric data were combined with LiDAR data, with this reduction range being based on the channel type and hydrodynamic model (1D or 2D). In the same way, Saleh et al. [53] highlight that, in the context of determining water levels at a specific location, the difference in bathymetry could be very important. Furthermore, they also noted how this difference may be more evident when we use DEMs or remote sensing techniques to identify river geometry at the regional scale since it is difficult to obtain an accurate bathymetric river representation. Other assessments [23] pointed towards the same conclusion, but there is no quantification of the dependence between bathymetric errors and flooded area differences.

4.2. Manning's n Value Calibration

The Manning's n value calibration process was carried out first for the 500-year return period peak flow, and it was done in two steps. First, focusing on a cost-saving approach that would allow its application in places with very different socio-economic development, the calibration was carried out based on the results obtained in a series of cross-sections of the Douro River. At the location of each cross-section, a hydrodynamic model was generated and a best fit Manning's n value model was defined (Figure 1D). A non-homogeneous Manning's n value for each cross-section was found, which pointed towards the assumed hydrodynamic simplification when we use a homogeneous surface roughness coefficient for the whole river reach. This point is discussed later.

From the different best fit Manning's n values of each cross-section, a mean value of $n = 0.015$ was obtained. However, given the non-homogeneous results obtained in the previous step, a second calibration phase was carried out.

In this second phase, the hydrodynamic model was extended throughout the study area. The latter aspect favored the achievement of multiple objectives, such as the rapid comparison of results in the location of the river cross-sections that were used in the first calibration phase. However, it also allowed for statistical and geostatistical analyses of the results to be carried out to achieve a better understanding of them in such a way that one could finally obtain a selection, based on scientific criteria, of the hydrodynamic model based on the "LiDAR scenario" topography and the best value (or range of values) of Manning's n coefficient.

As a complement to the above, the results obtained from two other models were analyzed. On the one hand, use was made of a model that presented a variable spatial distribution in the value of the roughness coefficient in the channel (obtained from the roughness values that offered a better adjustment relative to the control model, i.e., the "real scenario", for each point in the channel). On the other hand, a model with a roughness coefficient value of 0.027 was considered (similar to the control model), albeit with a reduced peak flow value (from 2274 to 2114 $m^3 s^{-1}$) depending on the measured flow rate at the date of acquisition of the LiDAR data.

The calibration, optimization and validation of the parameters in the models require having observational elements of calibration, such as the existence of a gauging station with its available stage–discharge curve in the case of the hydraulic models [54,55]; the existence of stage water plaques of historical floods with a known peak flow value; or other types of high-water marks, such as paleo-hydrological evidence (slackwater deposits or dendro-geomorphological floods [56]). In addition, it is not sufficient to have a single element for calibration because different combinations of parameters can give convergent results at one point; rather, multiple points are needed to adjust them, and these are not always available [28]. In fact, as pointed out by Hawker et al. [43], the availability of observational elements of calibration is not possible in many cases.

However, the readjustment of compensation between parameters, which is the strategy followed in this work, made it possible to compensate for the deficiencies in bathymetry by readjusting the roughness in a relatively simple way that was applicable to any section or situation, whether or not there were calibration elements.

In short, the methodological approach adopted, although it can be complemented with other approaches, is an innovative strategy that can substantially improve hydrodynamic models, hazard analyses, risk assessments and, finally, the effectiveness of flood risk mitigation measures. Evidently, this was all done in places where there was no bathymetric information and where such information could not be obtained, and only for the floodplain, not for the river channel.

4.3. Flow Depth Models Analysis and Optimal Model Selection

The hydrodynamic output parameter of flow depth, which was derived from each Manning's n value calibration model that was constructed upon the topography of the "LiDAR scenario" model, was compared to the control model, i.e., the so-called "real scenario". First, the global results were analyzed both as flow depth differences and as flood area extensions by using descriptive statistics parameters, such as mean, median, variance or standard deviation, and with the use of the Nash–Sutcliffe efficiency index and the F-statistic index. All these statistical indexes are shown in Table 1.

Table 1. Statistical data about Manning's n value calibration process, considering both hydrodynamic outputs, namely, flow depth and flooding area.

	Mean	Median	Mode	Standard Deviation	Variance	F-Statistic	NSE Index
LiDAR Scenario + n = 0.001	0.231	0.417	0.100	−0.110	0.174	81.76	0.9976
LiDAR Scenario + n = 0.010	0.024	0.326	0.026	0.050	0.106	91.94	**0.9998**
LiDAR Scenario + n = 0.011	**0.000**	0.311	0.010	0.050	0.096	92.30	0.9976
LiDAR Scenario + n = 0.012	−0.029	0.298	**−0.007**	0.050	0.089	92.68	0.9860
LiDAR Scenario + n = 0.013	−0.061	0.287	−0.027	0.050	0.082	93.19	0.9944
LiDAR Scenario + n = 0.014	−0.094	0.280	−0.045	0.050	0.079	93.46	0.9727
LiDAR Scenario + n = 0.015	−0.130	−0.073	0.030	0.269	0.073	**93.92**	0.9994
LiDAR Scenario + n = 0.016	−0.169	−0.106	0.040	0.263	0.069	93.34	0.9985
LiDAR Scenario + n = 0.018	−0.251	0.257	−0.191	−0.182	0.066	91.11	0.9953
LiDAR Scenario + n = 0.020	−0.332	0.254	−0.314	−0.312	0.065	89.10	0.9920
LiDAR Scenario + n = 0.027	−0.641	0.260	−0.660	−0.750	0.068	84.19	0.9236
LiDAR Scenario + ndistrib	0.010	**0.005**	0.050	0.316	0.100	93.05	0.9981
real Scenario + Q reduced + n = 0.027	−0.393	−0.407	−0.416	0.258	0.067	87.65	0.9894

From the analysis of the simulation points as a function of the flow depth values, it was found that the differences between the "real scenario" and the "LiDAR scenario (Manning's n value = 0.011)" had the lowest mean deviation, with a perfect average fit (0.000); the lowest median deviation results were found for the "LiDAR scenario (spatially distributed Manning's n value)", which slightly underestimated the flow depth (0.005); the lowest mode value was related to the scenario with Manning's n value equal to 0.012; the variance of the deviation had a random behavior.

The Nash–Sutcliffe efficiency index was calculated as one minus the ratio of the error variance of the modelled hydrodynamic output divided by the variance of the observed hydrodynamic output. In our assessment, the output was the flow depth parameter.

The NSE index showed very similar values for almost all of the hydrodynamic models. The best value was associated with the hydrodynamic model with a Manning's n value of 0.010, but the differences relative to other models (like models with n = 0.015, 0.016 or a spatially distributed n value) were not significant. All models showed high values of the NSE; therefore, the error variance of the modelled hydrodynamic outputs was much lower than the actual variance of the "real scenario" hydrodynamic model flow depth parameter. From the results of the NSE index, it was difficult to make the selection of the best Manning's calibrated hydrodynamic model.

Just as the results of the flow depth parameter given by the NSE index were not fully significant to select the best calibrated hydrodynamic model option, the results from the

F-statistic index (Table 1) produced the same result. The F-statistic index showed many close values for the hydrodynamic models, with Manning's n value ranging from 0.010 to 0.018 (always above a value of 90), with the best value of the F-statistic index linked to the Manning's n value of 0.015. The F-statistic improvement from Manning's n value of 0.010 to 0.015 was only about 1.5%.

As a conclusion, we could say that the analysis of the flow depth results from all calibrated models not giving us a strong criterion for the better option selection means that all models using Manning's n values ranging from 0.010 to 0.016 may be suitable to reproduce the conditions shown by the control model, i.e., the "real scenario" hydrodynamic model. The results shown by these models were similar in quality to those obtained by the use of the variable and spatially distributed Manning n value model, and they were better than the result shown by the HDCM approach. Based on the results obtained up to this point, a second phase of flow depth values analysis was necessary and a geo-statistical approach was carried out, where the distance to the main river channel was taken into account for the flow depth analysis.

When the results were analyzed as a function of the distance to the riverbanks (Figure 6), less deviation was also observed for the LiDAR scenario models with Manning's n values ranging from 0.011 to 0.015 (the model with an n value of 0.016 began to show a slight deviation towards flow depth overestimation). From the best-calibrated models, those linked to a Manning's n value equal to 0.013 or 0.014 probably showed the lowest dispersion against the zero error line.

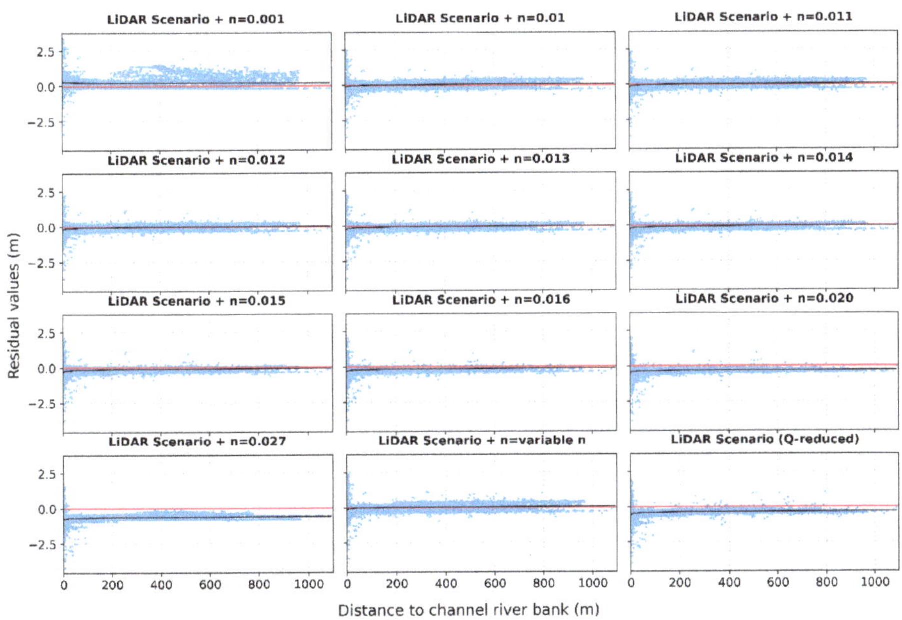

Figure 6. Scatter plot graphical representation of the relationship between the variable flow depth and distance to the riverbank for different Manning's n value calibrated models. The red line shows an error equal to zero and the black line shows a logarithmic fitting to the point values.

If the analysis of flow depth values was carried out in intervals of distance from the riverbanks (Figure 7), the scenarios that best adjusted the intervals of maximum–minimum deviation were also those ranging from Manning's n value equal to 0.010 to 0.014. The "LiDAR scenario (Manning's n value = 0.010)" showed the best results for shorter distances from the riverbank, with a very good fit for distances from 0 to 150 m. The "LiDAR scenario

(Manning's n value = 0.012)" gave the best results for distances from 150–300 m off the riverbank. Moreover, the "LiDAR scenario (Manning's n value = 0.013)" showed a good fit for all distances up to 300 m. However, for distances over 300 m, the "LiDAR scenario (Manning's n value = 0.014)" probably showed the best fit from the range of different Manning's n values calibrated. For this model, most of the error in the flow depth value lay within an interval of ± 10–12 cm, although this error increased to 15–20 cm when we got close to the riverbank. All these models substantially improved the results obtained in the model in which the natural value of Manning's parameter n was maintained.

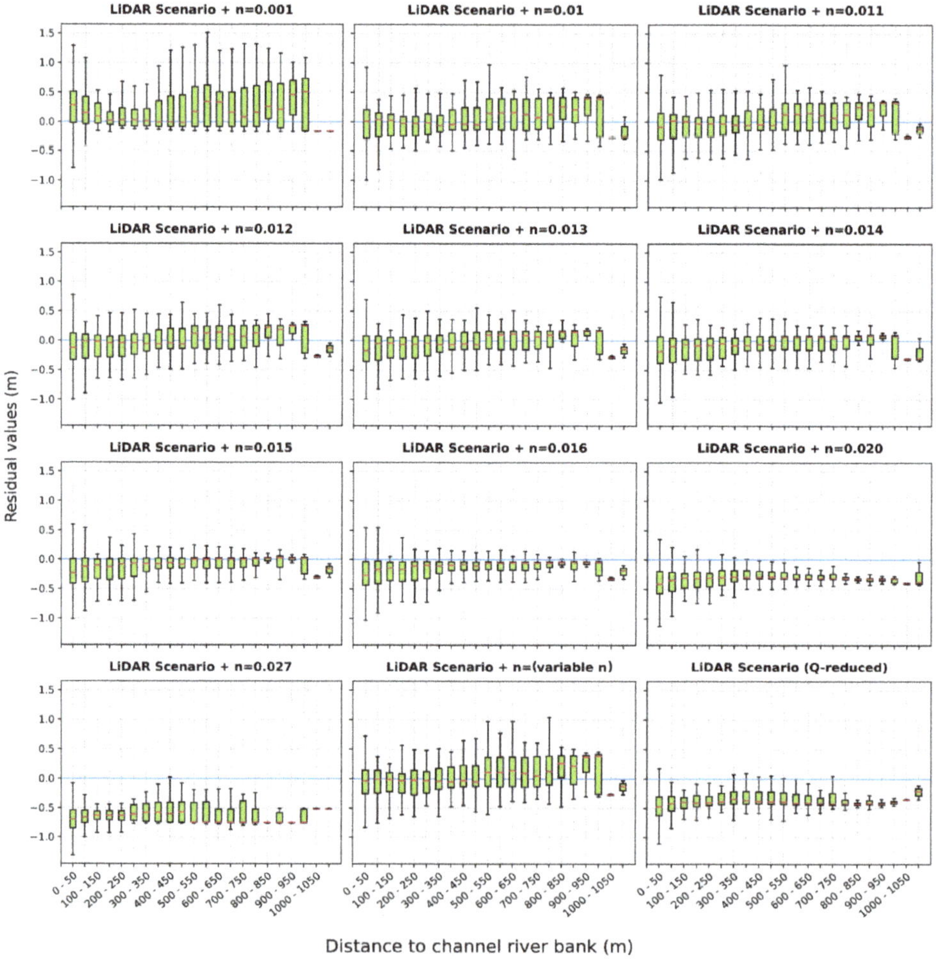

Figure 7. Box plot graphics showing the relationship between the flow depth errors and the distance to the riverbank (within 50 m intervals). Positive errors were related to the underestimation of the flow depth value by the calibrated model, and vice versa.

When we analyzed the results that were associated with the spatially distributed Manning's parameter n model and those associated with the HDCM model, we observed that in both cases, the results were better than those associated with the model that preserved the natural value of Manning's n parameter. However, in neither case did the results approach the best of the models with a constant and calibrated Manning's n value. Thus, the HDCM model was one of the worst performers in this study. The bad performance of the HDCM model may have been due to the large difference between flow rates on the date of the LiDAR data and the 500-year return period peak flow, as well as the likely significant differences in flow velocity in each case, where the higher flow velocities would require less of a channel cross-sectional area (Figure 3). On the other hand, the model with a spatially distributed Manning's n value provided a very good fit with the control model ("real scenario") of up to about 500 m distance from the channel; however, at further distances, it underestimated the flow depth more than the models with a constant Manning's n parameter and values between 0.013 and 0.015. Therefore, if the risk is to be assessed at a short distance because this is where the exposed and vulnerable elements are located (farms, transport infrastructure, etc.), the scenario "LiDAR scenario (Manning's n value = 0.011)" or the spatial distributed Manning's n value model are of interest, while if risk analysis is to be carried out for elements distant from the riverbed (homes and towns far from the river but within a flood zone), the scenario "LiDAR scenario (Manning's n value = 0.012 to 0.015)" can be used. This gives rise to an interesting discussion on the need to use different roughness indices depending on the flow rate and its return period, as some authors have already pointed out (but in the opposite direction to these results [55]). This variation in the parameters and indices to be used in hydrological and hydraulic models depending on the magnitude of the event has already been described extensively in the scientific-technical literature for other parameters, such as initial abstractions (curve number) as a function of precipitation intensity.

The coefficient of water bottom friction was investigated extensively and is known to depend on the particle sizes of materials on the river bed. There have been many studies on friction parameter estimation, especially on a relationship between estimated Manning's coefficients and river bed conditions. These range from the classical tables and lists [57,58], to present-day estimations using fractals and connectivity [59,60] from remote sensing information [61], as well as including visual guides [45] and technical determination procedures [62,63]; all of these methods can be grouped in two kinds of approaches: (i) grain size–roughness relationships for different river bottom patches or polygons and (ii) micro-topographical analyses of bathymetrical data.

The first group is used in technical reports and studies of large river reaches for hydrodynamic modelling and civil engineering; the second group is usual in scientific detailed studies of small river channels for geomorphological and ecological analysis. Both approaches are necessary and complementary because they depend on the scale and objectives of roughness estimation. However, it is very important to quantify predictive uncertainty in the hydrodynamic modelling of shallow water flow in response to uncertainty in friction parameterization [64].

However, in any case, a fully spatially distributed Manning's coefficient based upon the physical characteristics of terrain (mainly the riverbed characteristics) was not achieved due to the complex distribution and the high degree of spatial variability in the physical characteristics of the terrain (grain-size and micro-topography distributions) and vegetation. This objective is already feasible for floodplains [61] by using UAVs, but not for the submerged areas of river channels.

Based on the whole scope of the results of our analysis, the use of a spatially distributed Manning's n value became necessary. Manning's n value is usually calibrated within a river reach using a uniform value for all reaches, although the use of non-uniform values was pointed out by previous studies [65] that used a different Manning's n value for each river reach at the Lower Tapi River (India) to get the best-fit calibrated HEC-RAS model. In this sense, Attari and Hosseini [66] showed a methodological framework for the automatic river segmentation into different river reaches that were fitted with a non-uniform Manning's n value. Both approaches were utilized prior to the use of a non-uniform value for the roughness coefficient along a sequential river reach segmentation, but they did not use a real spatially distributed Manning's n value. Although a more complete study of spatially distributed values of Manning's n parameter would be necessary and convenient, the approximation to its use that was carried out in the present study did not show significantly better results than those of the other models considered.

In our methodological approach, we used the 500-year return period peak flow to develop the methodological framework, while the 100-year return period peak flow was used as the test model. The statistical results of the test model (Table S1 in Supplementary Materials) showed a slight difference from the 500-year return period peak flow model. A similar dependence on the statistic used was observed relative to the hydraulic model, which offered better results compared to the control model. The geostatistical analysis of results for the test model, considering the distance from the riverbank, showed very similar trends (Figures S1 and S2) to those related to the 500-year return period. Therefore, the scatter plot of Figure S1 shows that the best fit with the control (or benchmark) model was linked to Manning's n value in the range of 0.014–0.016. The results of the box plot (Figure S2), which were the same as for 500-year return period models, showed differences in the best fit that was linked to the distance to the riverbank. As discussed above, the best fit for shorter distances was obtained with a lower Manning's n value (about 0.011), while for distances equal to or greater than 500 m, the model that offered the best results was possibly the one with a Manning's n value of 0.016. As for the 500-year return period, the test model linked to the HDCM approach showed an overestimation of the flow depth values for all distances to the riverbank. For all hydraulic models related to the 100-year return period, an increase in uncertainty could be observed at further distances from the riverbank, which was due to the smaller flooding area at the floodplain (and the consequent lower number of sample points at further distances) than the one related to the 500-year return period models.

In general, we observed that the best-fitting Manning's n value increased slightly from the methodological developed model (500-year return period peak flow) to the methodological test model (100-year return period peak flow). However, taking this into account, the optimal range of Manning's n value from 0.014 to 0.016 could be defined. Within this range of Manning's n value, the flow depth errors in the river floodplain were drastically reduced from the model without bathymetry data and a "normal" (0.027 in the study site) Manning's n value.

The results from the two peak flows considered in the present assessment point towards the validation of our methodological approach, and the usefulness of using a calibrated and reduced values of Manning's n coefficient where the topography of the riverbed is not available and its acquisition lies outside the economic budget of flood risk managers.

4.4. Local Results at Cultural Heritage Sites in Zamora (Spain)

Beyond the results of the overall study area (with or without riverbank distance dependence), some control points (Figure 8) that are linked to different housing types in Zamora city were used for the result quality analysis of the Manning's-n-value-calibrated models. Four control points were selected, representing different types of buildings in the vicinity of the city of Zamora. Checkpoints 1 and 3 represented buildings in an urban environment with a high building density, checkpoint 2 corresponded to a cultural heritage site (chapel) and, finally, checkpoint 4 corresponded to a house that was isolated among agricultural fields.

In all cases, the absence of bathymetric data in the peak flow hydraulic modelling implied a flow depth overestimation ranging from 50 to 75 cm, where no process of artificial modification of the Manning's n value (calibrated model with $n = 0.027$) was carried out. These flow depth errors could be drastically reduced through the Manning's n value calibration process, as shown in Figure 8. However, as was also observed previously, the results obtained at the four control points did not show complete homogeneity. Again, the range of values between 0.010 and 0.014 seemed to show the best results overall. However, depending on the spatial location of the control point, it was observed that values lower than 0.010 could also give optimal results, but values higher than 0.014 did not. The selection of only four locations may not be representative for the whole study area, and the previously exposed results (where the best fit was obtained for the Manning's n value ranging between 0.014 to 0.016) had greater confidence, but they were illustrative of the overestimation of flow depth in the absence of bathymetric data without some kind of calibration process.

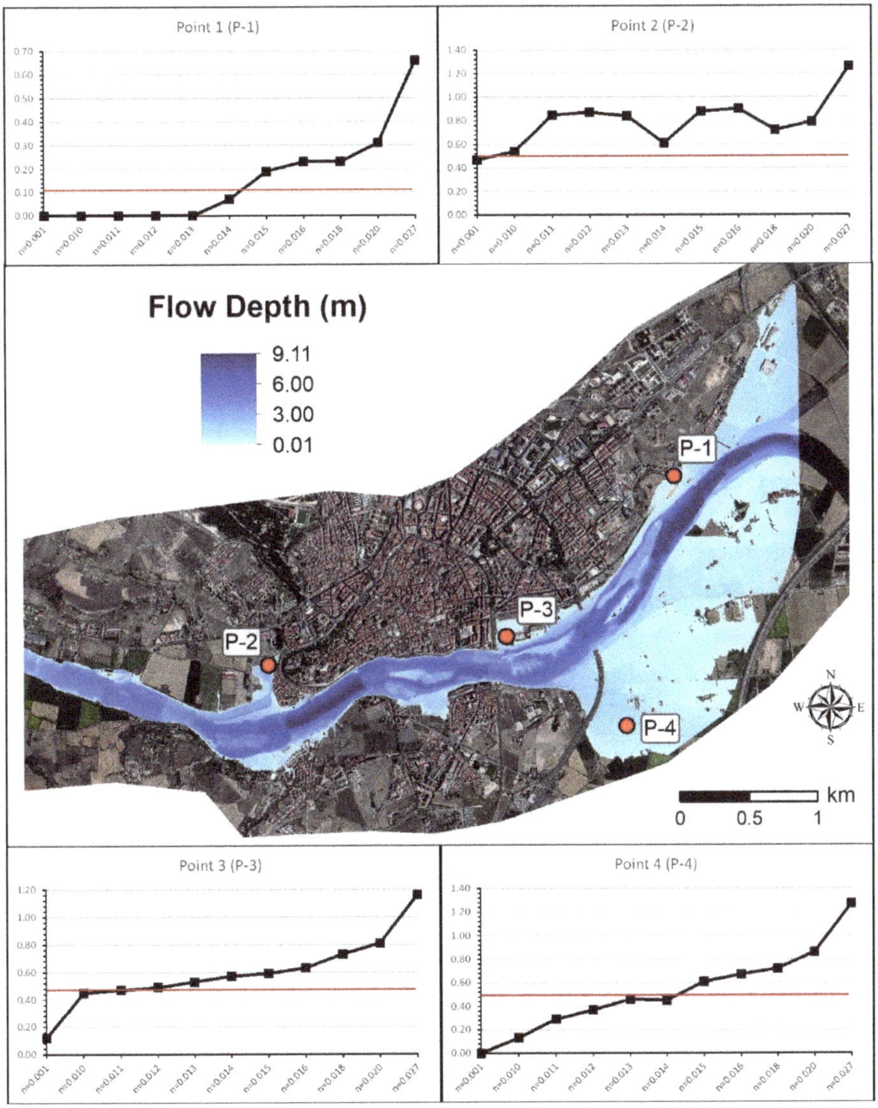

Figure 8. Local Manning's n value calibration results at control points in Zamora city. The upper and lower graphs show flow depth values for each calibrated model, where the red line shows the flow depth value for the control model, i.e., the "real scenario".

By transferring these differences in flow depth to the analysis of direct economic damage caused by floods, they can lead to significant differences in damage estimates. Thus, based on a widely used magnitude–damage model [15], we found that a difference in the flow depth of 50–75 cm could lead to a variation in the direct damage estimates of 25–30%. From the same model, we could estimate that, if we associated an average error of 10 cm with the calibrated model with $n = 0.014$, the error in the damage estimate would be around 6%. These percentages of economic direct damage can vary depending on the non-linearity of the distribution of flood damage associated with the flow depth value, as well as the damage model used. Wagenaar et al. [67] pointed out that the resulting uncertainties in estimated damage (due to different models) are in the order of magnitude of a factor of 2 to 5.

Furthermore, from the results obtained, it can be seen that the use of a reduced or lower (relative to that which is naturally associated with the characteristics of the riverbed) Manning's n value between 0.011 and 0.016 could lead to an error in the estimation of the flow depth that was no more than 25 cm, which, transferred to the estimation of direct damage, meant an approximate damage value error of 12%. Therefore, the use of an artificially lower Manning's n value could reduce the error in estimating flood damage by half. Therefore, this can be an interesting starting point for the improvement of flood damage estimates in areas without bathymetric data availability (even taking into account the fact that obtaining bathymetric data will always be the best option to achieve the best results). Furthermore, it could serve to carry out hazard analyses and, therefore, more personalized risk analyses depending on the elements at risk to be analyzed and their distances from the riverbanks.

However, from a practical point of view, this would introduce complexity into the systematic production of hazard and risk mapping, such as those of FEMA [68] in the USA, SNCZI [69] in Spain, or the Flood Factor [70], also in the USA. At the same time, it would give dynamism and ease of updating to large-scale local studies, which are optimal for urban areas or vulnerable infrastructures (large dams, nuclear power stations, industrial complexes, etc.).

5. Conclusions

The present manuscript shows a new approach for improving flood hazard maps where bathymetric data are not available (or they are scarce, such as a few cross-sections for the whole river reach). The proposed solution to this unavailability of bathymetric data is valid for river floodplains but not inside the river main channel. Unlike the approaches based on the generation of simplified bathymetric shapes, or the hydraulic corrections (HDCM model), the proposal of this manuscript was based on the calibration of the Manning's n value (roughness surface index). This calibration point toward the use of abnormally low values for a natural channel sought to compensate for the reduction in the channel cross-section area using an increase of the flow velocity in the channel itself. The main conclusions that can be derived from this study are as follows:

- The results obtained show a clear improvement when compared to the direct use of LiDAR topographic information combined with a Manning's n value according to the characteristics of the river channel studied. The results from the methodological developed model (500-year return period peak flow) and the methodological testing model (100-year return period peak flow) converged toward a range of Manning's n values from 0.014 to 0.016, which is far from the 0.027 that is selected based on the Douro River reach characteristics. A slight uncertainty in the best range of the Manning's n value was seen depending on the magnitude of the peak flow rate used. The magnitude of the peak flow rate and the optimal Manning's value show an inverse relationship.
- For the case of the Douro River in Zamora, the results of the hydrodynamic modelling under the "LiDAR scenario" and the "natural" Manning's n value conditions ($n = 0.027$) caused average errors of 50–75 cm in the flow depth estimation. By calibrating the Manning's n value, these average errors could be reduced to a value close to 10 cm.
- By transferring the errors in flow depth to the estimation of direct damage due to floods (based on the widely used USACE damage magnitude model), we achieved a

reduction in the error in the percentage of damage from values of 25–30% to errors close to 5%.

- The results of the calibration of the Manning's n value for the 500-year return period peak flow showed that the best fit varied according to the distance from the riverbank such that we could select different values of n (within a range between 0.011 and 0.014) depending on whether the area of greatest interest (in the hazard assessment) was close to the channel (lower value of n) or far from it (higher value of n). In any case, this implied the use of values of approximately half those used in real conditions. In the case of the 100-year return period peak flow, the best option of Manning's value for areas far from the riverbank went up to 0.016.
- A uniform Manning's n value (for the river channel) was used both for the control "real scenario" and for each Manning's n value calibrated model. However, a spatially distributed Manning's n value was used too, and the results, although good, did not significantly improve on those obtained in the constant value models. In fact, it was observed that this model offered better results for distances up to 550 m from the riverbed, but at greater distances, the results worsen.
- A hydraulic approach, namely, the HDCM, was also used, but the results were far from satisfactory. In fact, the results associated with this model were among the worst of those obtained in the present study. This situation calls into question the usefulness of this approach as a solution to the absence of bathymetric data in cases where the flow rates (on the date of acquisition of the topographic data, and those associated with the study) are very different.
- The present work is the first contribution to a methodological framework that should be improved by applying it in other areas where the river characteristics (river slope, channel typology, sinuosity, percentage of reduction of the channel cross-section, etc.) are different from those shown in the present work. In this way, the objective of having a range of Manning's n values depending on the specific characteristics of the study area could be achieved. This approach could represent an interesting scientific-technical innovation in the analysis of flood risks. Furthermore, in the present state, the use of "not natural lower Manning's n value" was shown to be an optimal option for the improvement of flood damage estimates in urban areas where there is no availability of bathymetric data.

Supplementary Materials: The following are available online at https://www.mdpi.com/article/10.3390/app11199267/s1, Figure S1: Scatter plot of the 100-year return period residual values for calibrated models, Figure S2: Box plot of the 100-year return period residual values for calibrated models, Table S1: Statistical results for the 100-year return period hydraulic models.

Author Contributions: All authors have made a significant contribution to the final version of the paper. Conceptualization, J.G.; Hydrodynamic modelling, J.G. and M.G.-J.; Statistical analysis, J.G., M.G.-J. and C.G.-A.; Writing—original draft, J.G. and A.D.-H.; Writing—review and editing, all authors. All authors have read and agreed to the published version of the manuscript.

Funding: This research was funded by the project DRAINAGE, CGL2017–83546–C3–R (MINE-ICO/AEI/FEDER, UE); it is specifically part of the assessments carried out under the subproject DRAINAGE–3–R (CGL2017–83546–C3–3–R). The participation of M.G.-J. is supported by the National System of Youth Guarantee (activity with reference PEJ2018-002477) co-financed under the Youth Employment Operational Program, with financial resources from the Youth Employment Initiative (YEI) and the European Social Fund (ESF).

Data Availability Statement: The data presented in this study are available on request from the corresponding author. The data are not publicly available due to their file size, which is greater than 4 Gb.

Acknowledgments: The authors would like to thank to Gerardo Benito (MNCN, CSIC) and the Duero River Water Authority (CHD) who provided us with the bathymetric data from Zamora city and its surroundings.

Conflicts of Interest: The authors declare no conflict of interest.

Abbreviations

DEM	Digital elevation model
FEMA	Federal Emergency Management Agency
HDCM	Horizontally divided channel method
LiDAR	Light detection and ranging
MAE	Mean absolute error
NSE	Nash–Sutcliffe efficiency index
PNOA	Plan Nacional Orto-fotografía Aérea (Aerial Ortho-photography National Plan)
SNCZI	Sistema Nacional de Cartografía de Zonas Inundables (Flood Prone Areas Mapping National Plan)
VDCM	Vertically divided channel method

References

1. Istituto Centrale per il Restauro, Carta del Rischio, versione 2.1.1, Direzione Generale Sicurezza del Patrimonio Culturale (MIBACT). Available online: http://www.cartadelrischio.beniculturali.it/webgis/ (accessed on 30 October 2020).
2. Consiglio Nazionale Delle Ricerche, NOAH's Ark project, CORDIS EU research results. Available online: https://cordis.europa.eu/project/id/501837 (accessed on 30 October 2020).
3. Lanza, S.G. Flood hazard threat on cultural heritage in the town of Genoa (Italy). *J. Cult. Herit.* **2003**, *4*, 159–167. [CrossRef]
4. Mazzanti, M. Valuing cultural heritage in a multi-attribute framework microeconomic perspectives and policy implications. *J. Socio-Econ.* **2003**, *32*, 549–569. [CrossRef]
5. Ortiz, R.; Ortiz, P.; Martín, J.M.; Vázquez, M.A. A new approach to the assessment of flooding and dampness hazards in cultural heritage, applied to the historic centre of Seville (Spain). *Sci. Total Environ.* **2016**, *551–552*, 546–555. [CrossRef]
6. Arrighi, C.; Brugioni, M.; Castelli, F.; Franceschini, S.; Mazzanti, B. Flood risk assessment in art cities: The exemplary case of Florence (Italy). *J. Flood Risk Manag.* **2018**, *11*, S616–S631. [CrossRef]
7. Figueiredo, R.; Romão, X.; Paupério, E. Flood risk assessment of cultural heritage at large spatial scales: Framework and application to mainland Portugal. *J. Cult. Herit.* **2020**, *43*, 163–174. [CrossRef]
8. CRED. The International Disaster Database [online], Centre for Research on the Epidemiology of Disasters. Available online: http://emdat.be/emdat_db/ (accessed on 10 April 2021).
9. Munich RE. NatCatSERVICE Database. Available online: https://natcatservice.munichre.com/ (accessed on 17 July 2020).
10. Ward, P.J.; Jongman, B.; Sperna Weiland, F.C.; Bouwman, A.; Van Beek, R.; Bierkens, M.; Ligtvoet, W.; Winsemius, H.C. Assessing flood risk at the global scale: Model setup, results, and sensitivity. *Environ. Res. Lett.* **2013**, *8*, 044019. [CrossRef]
11. Visser, H.; Bouwman, A.; Petersen, A.; Ligtvoet, W. *A Statistical Study of Weather–Related Disasters: Past, Present and Future*; PBL Netherlands Environmental Assessment Agency: The Hague, The Netherlands, 2012.
12. Merz, B.; Kreibich, H.; Schwarze, R.; Thieken, A. Assessment of economic flood damage. *Nat Hazards Earth Syst. Sci.* **2010**, *10*, 1697–1724. [CrossRef]
13. USACE. *Economic Guidance Memorandum (EGM) 04-01, Generic Depth-Damage Relationships for Residential Structures with Basements*; USACE: Washington, DC, USA, 2003.
14. Garrote, J.; Alvarenga, F.M.; Díez-Herrero, A. Quantification of flash flood economic risk using ultra-detailed stage–damage functions and 2-D hydraulic models. *J. Hydrol.* **2016**, *541*, 611–625. [CrossRef]
15. Huizinga, J.; De Moel, H.; Szewczyk, W. *Global Flood Depth–Damage Functions: Methodology and the Database with Guidelines*; Joint Research Centre: Luxembourg, 2017.
16. Deniz, D.; Arneson, E.E.; Liel, A.B.; Dashti, S.; Javernick-Will, A.N. Flood loss models for residential buildings, based on the 2013 Colorado floods. *Nat. Hazards* **2017**, *85*, 977–1003. [CrossRef]
17. Zabret, K.; Hozjan, U.; Kryžanowsky, A.; Brilly, M.; Vidmar, A. Development of model for the estimation of direct flood damage including the movable property. *J. Flood Risk Manag.* **2018**, *11*, S527–S540. [CrossRef]
18. Schoppa, L.; Sieg, T.; Vogel, K.; Zöller, G.; Kreibich, H. Probabilistic flood loss models for companies. *Water Resour. Res.* **2020**, *56*, e2020WR027649. [CrossRef]
19. Paprotny, D.; Kreibich, H.; Morales-Nápoles, O.; Castellarin, A.; Carisi, F.; Schröter, K. Exposure and vulnerability estimation for modelling flood losses to commercial assets in Europe. *Sci. Total Environ.* **2020**, *737*, 140011. [CrossRef]
20. Díez-Herrero, A.; Lain-Huerta, L.; Llorente-Isidro, M. *A Handbook on Flood Hazard Mapping Methodologies*; Publications of the Geological Survey of Spain (IGME): Madrid, Spain, 2009.
21. Casas, A.; Benito, G.; Thorndycraft, V.R.; Rico, M. The topographic data source of digital terrain models as a key element in the accuracy of hydraulic flood modelling. *Earth Surf. Proc. Land.* **2006**, *31*, 444–456. [CrossRef]
22. Bermúdez, M.; Zischg, A.P. Sensitivity of flood loss estimates to building representation and flow depth attribution methods in micro–scale flood modelling. *Nat. Hazards* **2018**, *92*, 1633. [CrossRef]
23. Dey, S.; Saksena, S.; Merwade, V. Assessing the effect of different bathymetric models on hydraulic simulation of rivers in data sparse regions. *J. Hydrol.* **2019**, *575*, 838–851. [CrossRef]

24. Arrighi, C.; Campo, L. Effects of digital terrain model uncertainties on high resolution urban flood damage assessment. *J. Flood Risk Manag.* **2019**, *12*, e12530. [CrossRef]
25. Vieux, B.E. Hydraulic Roughness. In *Distributed Hydrologic Modeling Using GIS*; Vieux, B.E., Ed.; Springer: Heidelberg, Germany, 2016; pp. 101–114. [CrossRef]
26. Boyer, M. Estimating the Manning coefficient from an average bed roughness in open channels. *Eos. Trans. Am. Geophys. Union* **1954**, *35*, 957–961. [CrossRef]
27. Kidson, R.; Richards, K.; Carling, P. Hydraulic model calibration for extreme floods in bedrock-confined channels: Case study from northern Thailand. *Hydrol. Process.* **2006**, *20*, 329–344. [CrossRef]
28. Ballesteros, J.; Bodoque, J.; Díez-Herrero, A.; Sanchez-Silva, M.; Stoffel, M. Calibration of floodplain roughness and estimation of flood discharge based on treering evidence and hydraulic modelling. *J. Hydrol.* **2011**, *403*, 103–115. [CrossRef]
29. Cook, A.; Merwade, V. Effect of topographic data, geometric configuration and modeling approach on flood inundation mapping. *J. Hydrol.* **2009**, *377*, 131–142. [CrossRef]
30. Boettle, M.; Kropp, J.P.; Reiber, L.; Roithmeier, O.; Rybski, D.; Walther, C. About the influence of elevation model quality and small-scale damage functions on flood damage estimation. *Nat. Hazards Earth Syst. Sci.* **2011**, *11*, 3327–3334. [CrossRef]
31. Kinzel, P.J.; Legleiter, C.J.; Nelson, J.M. Mapping River Bathymetry with a Small Footprint Green LiDAR: Applications and Challenges. *J. Am. Water Resour. Ass. (JAWRA)* **2012**, *49*, 183–204. [CrossRef]
32. Allouis, T.; Bailly, J.S.; Pastol, Y.; LeRoux, C. Comparison of LiDAR Waveform Processing Methods for Very Shallow Water Bathymetry Using Raman, Near-Infrared and Green Signals. *Earth Surf. Proc. Land.* **2010**, *35*, 640–650. [CrossRef]
33. Bangen, S.G.; Wheaton, J.M.; Bouwes, N.; Bouwes, B.; Jordan, C. A methodological intercomparison of topographic survey techniques for characterizing wadeable streams and rivers. *Geomorphology* **2014**, *206*, 343–361. [CrossRef]
34. Hostache, R.; Matgen, P.; Giustarini, L.; Teferle, F.N.; Tailliez, C.; Iffly, J.; Corato, G. A drifting GPS buoy for retrieving effective riverbed bathymetry. *J. Hydrol.* **2015**, *520*, 397–406. [CrossRef]
35. Fonstad, M.A.; Marcus, W.A. Remote sensing of stream depths with hydraulically assisted bathymetry (HAB) models. *Geomorphology* **2005**, *72*, 320–339. [CrossRef]
36. Trigg, M.A.; Wilson, M.D.; Bates, P.D.; Horritt, M.S.; Alsdorf, D.E.; Forsberg, B.R.; Vega, M.C. Amazon flood wave hydraulics. *J. Hydrol.* **2009**, *374*, 92–105. [CrossRef]
37. Gichamo, T.Z.; Popescu, I.; Jonoski, A.; Solomatine, D. River cross-section extraction from the ASTER global DEM for flood modeling. *Environ. Model. Softw.* **2012**, *31*, 37–46. [CrossRef]
38. Grimaldi, S.; Li, Y.; Walker, J.P.; Pauwels, V.R.N. Effective representation of river geometry in hydraulic flood forecast models. *Water Resour. Res.* **2018**, *54*, 1031–1057. [CrossRef]
39. Bradbrook, K.F.; Lane, S.N.; Waller, S.G.; Bates, P.D. Two dimensional diffusion wave modelling of flood inundation using a simplified channel representation. *Int. J. River Basin Manag.* **2004**, *2*, 211–223. [CrossRef]
40. Choné, G.; Biron, P.M.; Buffin-Bélanger, T. Flood hazard mapping techniques with LiDAR in the absence of river bathymetry data. In *E3S Web of Conferences, Proceedings of the River Flow 2018—Ninth International Conference on Fluvial Hydraulics, Lyon, France, 5–8 September 2018*; EDP Science: Les Ulis, France, 2018; Abstract 06005.
41. Wolman, M.G.; Miller, J.P. Magnitude and Frequency of Forces in Geomorphic Processes. *J. Geol.* **1960**, *68*, 54–74. [CrossRef]
42. Ardıçlıoğlu, M.; Kuriqi, A. Calibration of channel roughness in intermittent rivers using HEC-RAS model: Case of Sarimsakli creek, Turkey. *SN Appl. Sci.* **2019**, *1*, 1080. [CrossRef]
43. Hawker, L.; Bates, P.; Neal, J.; Rougier, J. Perspectives on Digital Elevation Model (DEM) Simulation for Flood Modeling in the Absence of a High-Accuracy Open Access Global DEM. *Front. Earth Sci.* **2018**, *6*, 233. [CrossRef]
44. Instituto Geográfico Nacional, LiDAR PNOA. 2014. Available online: https://pnoa.ign.es/el-proyecto-pnoa-lidar (accessed on 30 October 2020).
45. Arcement, G.J.; Schneider, V.R. *Guide for Selecting Manning's Roughness Coefficients for Natural Channels and Flood Plains*; USGS: Denver, CO, USA, 1989.
46. Brunner, G.W. *HEC-RAS, River Analysis System Hydraulic Reference Manual, Version 4.1*; U.S. Army Corps of Engineers: Davis, CA, USA, 2010.
47. Bladé, E.; Cea, L.; Corestein, G.; Escolano, E.; Puertas, J.; Vázquez-Cendón, E.; Dolz, J.; Coll, A. Iber: Herramienta de simulación numérica del flujo en ríos. *Rev. Intern. Met. Num. Calc. Dis. Ingen.* **2014**, *30*, 1–10. [CrossRef]
48. ESRI. ArcMap. 2011. Available online: https://desktop.arcgis.com/es/arcmap/ (accessed on 30 October 2020).
49. Hunter, J.D. Matplotlib: A 2D Graphics Environment. *Comput. Sci. Eng.* **2007**, *9*, 90–95. [CrossRef]
50. Bates, P.D.; De Roo, A.P.J. A simple raster-based model for flood inundation simulation. *J. Hydrol.* **2000**, *236*, 54–77. [CrossRef]
51. Horritt, M.S.; Bates, P.D. Evaluation of 1D and 2D numerical models for predicting river flood inundation. *J. Hydrol.* **2002**, *268*, 87–99. [CrossRef]
52. Tayefi, V.; Lane, S.N.; Hardy, R.J.; Yu, D. A comparison of one- and two dimensional approaches to modelling flood inundation over complex upland floodplains. *Hydrol. Process.* **2007**, *21*, 3190–3202. [CrossRef]
53. Saleh, F.; Ducharne, A.; Flipo, N.; Oudin, L.; Ledoux, E. Impact of river bed morphology on discharge and water levels simulated by a 1D Saint—Venant hydraulic model at regional scale. *J. Hydrol.* **2012**, *476*, 1–9. [CrossRef]
54. Casas, A.; Lane, S.; Hardy, R.J.; Benito, G.; Whiting, P.J. Reconstruction of subgrid-scale topographic variability and its effect upon the spatial structure of three-dimensional river flow. *Water Resour. Res.* **2010**, *46*, W03519. [CrossRef]

55. Casas, A.; Stuart, L.; Dapeng, Y.; Benito, G. A method for parameterising roughness and topographic sub-grid scale effects in hydraulic modelling from LiDAR data. *Hydrol. Earth Syst. Sci.* **2010**, *14*, 1567–1579. [CrossRef]
56. Benito, G.; Díez-Herrero, A. Palaeoflood Hydrology: Reconstructing Rare Events and Extreme Flood Discharges. In *Hydro-Meteorological Hazards, Risks, and Disasters*; Paron, P., Di Baldassarre, G., Eds.; Elsevier: Amsterdam, The Netherlands, 2015; pp. 65–104.
57. Cowan, W.L. Estimating hydraulic roughness coefficients. *Agr. Eng.* **1956**, *37*, 473–475.
58. Chow, V.T. *Open-Channel Hydraulics*; McGraw-Hill Book Co.: New York, NY, USA, 1959; 680p.
59. Ercan, A.; Kavvas, M.L. Scaling and self-similarity in two-dimensional hydrodynamics. *Chaos* **2015**, *25*, 075404. [CrossRef] [PubMed]
60. Larsen, L.G.; Ma, J.; Kaplan, D. How Important Is Connectivity for Surface Water Fluxes? A Generalized Expression for Flow Through Heterogeneous Landscapes. *Geophys. Res. Lett.* **2017**, *44*, 10349–10358. [CrossRef]
61. Prior, E.M.; Aquilina, C.A.; Czuba, J.A.; Pingel, T.J.; Hession, W.C. Estimating Floodplain Vegetative Roughness Using Drone-Based Laser Scanning and Structure from Motion Photogrammetry. *Remote Sens.* **2021**, *13*, 2616. [CrossRef]
62. Limerinos, J.T. *Determination of the Manning Coefficient from Measured Bed Roughness in Natural Channels*; United States Geological Survey: Washington, DC, USA, 1970; 53p.
63. Simons, D.B.; Richardson, E.V. *Resistance to Flow in Alluvial Channels*; United States Geological Survey: Washington, DC, USA, 1966; 70p.
64. Horritt, M.S. A linearized approach to flow resistance uncertainty in a 2-D finite volume model of flood flow. *J. Hydrol.* **2006**, *316*, 13–27. [CrossRef]
65. Timbadiya, P.V.; Patel, P.L.; Porey, P.D. Calibration of HEC-RAS model on prediction of flood for lower Tapi River, India. *J. Water Resour. Prot.* **2011**, *3*, 805–811. [CrossRef]
66. Attari, M.; Hosseini, S.M. A simple innovative method for calibration of Manning's roughness coefficient in rivers using a similarity concept. *J. Hydrol.* **2019**, *575*, 810–823. [CrossRef]
67. Wagenaar, D.J.; de Bruijn, K.M.; Bouwer, L.M.; de Moel, H. Uncertainty in flood damage estimates and its potential effect on investment decisions. *Nat. Hazards Earth Syst. Sci.* **2016**, *16*, 1–14. [CrossRef]
68. FEMA. Federal Emergency Management Agency, Flood Map, Flood Insurance Rate Map (FIRM). 2009. Available online: https://floodpartners.com/fema-flood-map/ (accessed on 30 October 2020).
69. Dirección General del Agua, Sistema Nacional de Cartografía de Zonas Inundables (SNCZI). 2013. Available online: https://sig.mapama.gob.es/snczi/index.html?herramienta=DPHZI (accessed on 25 October 2020).
70. First Street Foundation, Flood Factor. 2019. Available online: https://floodfactor.com/ (accessed on 27 October 2020).

applied sciences

MDPI

Article

Detecting Areas Vulnerable to Flooding Using Hydrological-Topographic Factors and Logistic Regression

Jae-Yeong Lee * and Ji-Sung Kim

Korea Institute of Civil Engineering and Building Technology, 283 Goyangdae-ro, Ilsanseo-gu, Goyang 10223, Korea; jisungk@kict.re.kr
* Correspondence: jaeyeonglee@kict.re.kr

Abstract: As a result of rapid urbanization and population movement, flooding in urban areas has become one of the most common types of natural disaster, causing huge losses of both life and property. To mitigate and prevent the damage caused by the recent increase in floods, a number of measures are required, such as installing flood prevention facilities, or specially managing areas vulnerable to flooding. In this study, we presented a technique for determining areas susceptible to flooding using hydrological-topographic characteristics for the purpose of managing flood vulnerable areas. To begin, we collected digital topographic maps and stormwater drainage system data regarding the study area. Using the collected data, surface, locational, and resistant factors were analyzed. In addition, the maximum 1-h rainfall data were collected as an inducing factor and assigned to all grids through spatial interpolation. Next, a logistic regression analysis was performed by inputting hydrological-topographic factors and historical inundation trace maps for each grid as independent and dependent variables, respectively, through which a model for calculating the flood vulnerability of the study area was established. The performance of the model was evaluated by analyzing the receiver operating characteristics (ROC) curve of flood vulnerability and inundation trace maps, and it was found to be improved when the rainfall that changes according to flood events was also considered. The method presented in this study can be used not only to reasonably and efficiently select target sites for flood prevention facilities, but also to pre-detect areas vulnerable to flooding by using real-time rainfall forecasting.

Keywords: flood vulnerability; spatial analysis; logistic regression; ROC analysis; flood detection

Citation: Lee, J.-Y.; Kim, J.-S. Detecting Areas Vulnerable to Flooding Using Hydrological-Topographic Factors and Logistic Regression. *Appl. Sci.* **2021**, *11*, 5652. https://doi.org/10.3390/app11125652

Academic Editor: Ricardo Castedo

Received: 13 May 2021
Accepted: 10 June 2021
Published: 18 June 2021

1. Introduction

Floods can have several causes, and result mainly from hydro-meteorological causes such as typhoons and localized torrential downpours. Recently, changes in atmospheric flow caused by global warming and climate change have brought about major meteorological problems. In particular, in Northeast Asian regions such as Korea, China, and Japan, atmospheric flow stagnated due to the abnormal high temperatures in the polar regions in the summer of 2020. This led to the longest rainy season ever, causing huge losses.

Other causes of flooding include a decrease in the rainwater storage effect of forests due to reckless development, and an increase in impervious areas due to urbanization. In Seoul, Korea, as the Gangnam region began to be developed in earnest after the 1970s, the low-lying areas were newly developed for residential purposes and lost their rainwater storage function [1]. In such densely populated urban areas, the occurrence of flooding will increase further because low-lying areas will be developed to resolve the inadequacy of the supply of housing compared to demand.

To reduce the loss caused by frequent floods in recent years, central and local governments have established measures to prevent such flood damage. However, budget limitations mean that not all areas vulnerable to flooding can be refurbished with flood prevention facilities. For this reason, it is important to prioritize relatively more vulnerable areas, and in some cases, information on vulnerable areas should be provided to residents.

Providing such information not only improves the ability of residents to cope with floods through education and training, but also has the effect of restraining the development of relevant areas.

The advantage in using physically based models is their high capability for prognosis and forecasting, while their disadvantage is the high input data demand [2]. For this reason, techniques for identifying flood vulnerable areas using topographic factors have been suggested in various ways by previous studies. The determination of flood vulnerable areas is one of the representative non-structural measures in flood defense, and should be performed reasonably through hydrological and topographic analysis of rainfall-runoff. As such, techniques for determining flood vulnerable areas have been studied by researchers in a number of fields including hydrology, topography, and soil science. Dehortin et al. [3] laid the foundation for calibrating or evaluating surface runoff susceptibility mappings through on-site monitoring that measures surface runoff. Lagadec et al. [4] presented the indicateur du ruissellement intense pluvial (IRIP) technique that maps the characteristics of surfaces that are susceptible to generation, transferal, and accumulation of surface rainfall-runoff. Lee et al. [5] compared the detection rates of flood vulnerability based on topographic factors using general data such as advanced spaceborne thermal emission and reflection radiometer (ASTER) and shuttle radar topography mission (STRM). Lee and Kim [6] analyzed the correlation between topographic factors considering rainfall-runoff characteristics, as well as actual inundation trace data.

Flood vulnerability has been estimated using the physical characteristics of the surfaces on which rainfall-runoff are likely to accumulate, such as lowlands and gentle slopes; more recently, studies have been performed that attempt to use machine learning to calculate flood vulnerability. Logistic regression, a field of machine learning, can suggest vulnerability in the study area in a probabilistic manner through binary classification of past data (yes or no) by connecting topographic factors and natural disasters such as floods and landslides [7–9]. In addition, studies on estimating flood vulnerability using other machine learning techniques are also being conducted by many researchers. Among those, studies using random forests [10,11] and principal component analysis (PCA) [12,13] have been actively conducted. In addition to studies that applied a single technique, studies which compare or connect several techniques have also been conducted. Pradhan and Lee [14] compared and proposed methods of detecting landslide-prone areas with logistic regression and artificial neural network (ANN). Lee et al. [15] compared flood vulnerability estimated using random forests and boosted trees with topographic factors as input data. Li et al. [16] used logistic regression, Naive Bayes, AdaBoost, and random forests to estimate flood vulnerability around the world, and compared the detection capabilities for each model. To reduce the dimensions of various topographic factors, studies on applying logistic regression after PCA [17–19] have also been conducted.

KICT [20] stated that it was necessary to establish special measures for areas prone to flooding and strengthen flood forecast warning systems, in order to respond to floods. Shin and Park [21] mentioned that the floods that occurred in Seoul in 2010 and 2011 had a high spatial correlation, and that they occurred repeatedly in the same area. In particular, it was analyzed that one-third of the areas which flooded in 2011 were areas that had previously suffered from floods [21]. On this basis, this study confirmed that flood vulnerable areas should be determined through an analysis of the topographical causes of areas where floods frequently occur in Seoul, the study area, and intensively managed.

A variety of approaches have been conducted to identify flood vulnerable areas, and the most representative of them is the method using the numerical models [22–24]. This method is to designate an expected flooding area by calculating the hydraulic-hydrological characteristics of rainfall-runoff for a hypothetical scenario precipitation with a numerical model. Although numerical models showed great capabilities for predicting a diverse range of flooding scenarios, they often require various types of hydro-geomorphological monitoring datasets, requiring intensive computation, which prohibits short-term prediction [25]. Previous studies have suggested data-based techniques for determining flood vulnerable areas using hydrological-topographic factors due to the efficiency of data collection and analysis. However, these methods only calculate the flood vulnerability at the planning level, and do not detect floods for various actual events. To supplement this, in this study, a logistic regression model estimating flood vulnerability that changes according to rainfall was developed and the detection performance was evaluated with a new event.

In spatial data-based flood vulnerability analysis, it is important to select input data that can affect floods and collect data. The input data were selected by referring to the topographical factors mainly used in the previous studies [7–19] introduced above (slope, elevation, topographic wetness index, curvature, stream power index, distance from river, in order of most use). Meanwhile, in Korea, hydrological-topographic data can be easily obtained through the websites [26–29] of government agencies. These data can be regarded as reliable data because they are produced with strict quality control.

The purpose of this study is to develop a technique for determining flood vulnerable areas in order to reduce the damage caused by flooding. As shown in Figure 1, this technique can calculate flood vulnerability by estimating logistic regression coefficients taking into account the hydrological-topographic factors in the study area. This methodology can map flood vulnerable areas suitable for each flood event by changing the values according to the rainfall situation. With this, if real-time rainfall forecasting is used, flooding can be predicted.

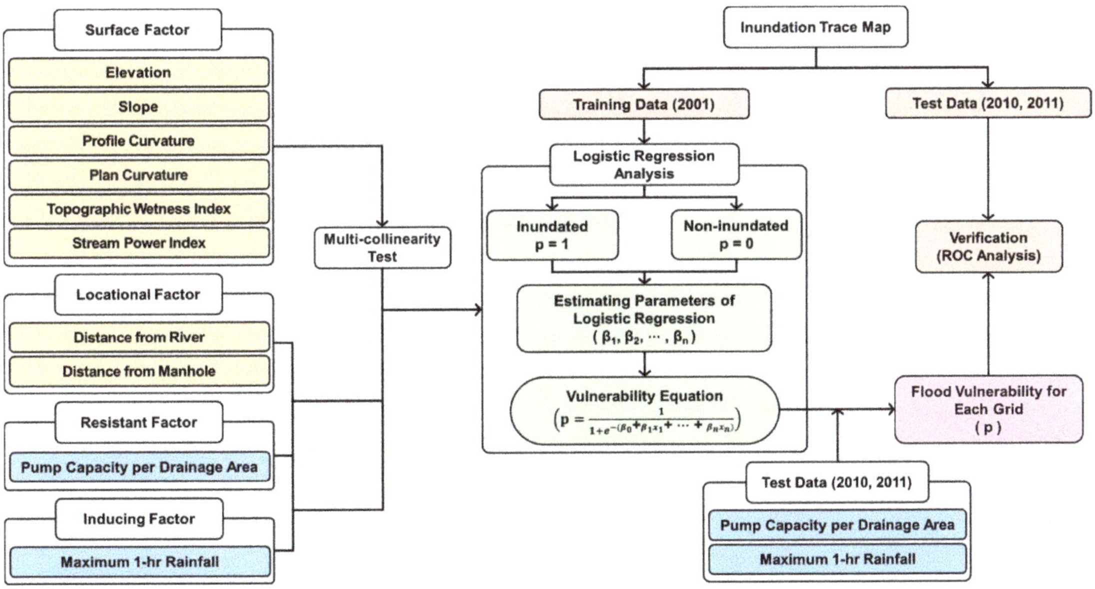

Figure 1. Flow chart of this study.

2. Study Area and Materials

2.1. Seoul Metropolitan City

Seoul metropolitan city (SMC), the capital city of South Korea, has seen continued population growth with the progress of industrialization and urbanization since the 1960s. As a result, this city is not only a densely populated region with more than 10 million people, which is 20% of the total population of the country, on an area of 605 km^2, but also shows a concentration of capital in highly dense office regions such as Gwanghwamun and Gangnam. In this environment, severe flooding occurred in 2010 and 2011, causing great damage to life and property in Seoul. The flood that occurred on 21 September 2010 flooded 17,905 households and injured one person. The flood of 27 July 2011 inundated

14,809 households, causing 19 deaths and 41 injuries [21]. With flood damage occurring every year since then, the city of Seoul has been striving to prevent it by increasing the design frequency of drainage pipes and installing pump stations.

In this study, inundation trace maps generated in 2001 [26] were used to develop a logistic regression model to calculate flood vulnerability for each grid. The inundation trace maps for 2010 and 2011 were used to evaluate the performance of the developed regression model. Figure 2 shows the extent of the study area and the traces of flooding in 2001.

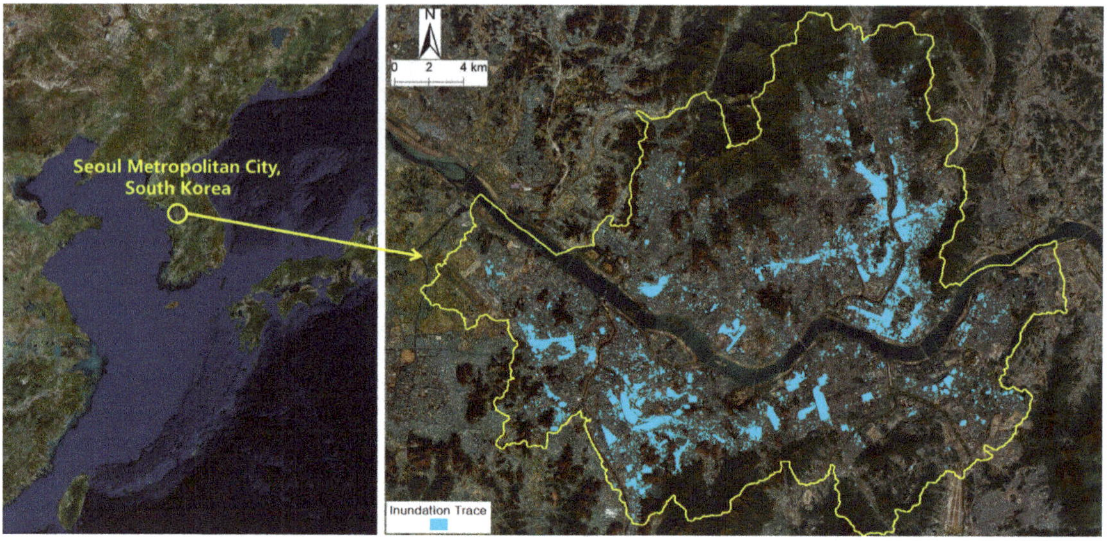

Figure 2. Inundation traces that occurred in 2001 in Seoul, the study area.

2.2. Hydrological-Topographic Factors

Hydrological-topographic factors were classified into three topographic factors (surface, locational, and resistant) and one hydrological factor (inducing factor). Elevation, slope, profile curvature, plan curvature, topographic wetness index (TWI), and stream power index (SPI) were considered for surface factors, which are the characteristics of runoff moving on the surface by gravity. For locational factors, distance from river and manhole were considered to indicate the range affected by catchment runoff due to natural factors (river) and artificial factors (manhole). As a resistant factor, pump capacity per drainage area was analyzed to consider the effect of drainage pumps installed to protect against urban flooding. The maximum 1-h rainfall was used as an inducing factor, which is an external factor that can directly affect the occurrence of floods.

2.2.1. Surface Factors

The characteristics of surfaces that are vulnerable to flooding are typically lowlands, gentle slopes, and concave terrains, and can be estimated through spatial analysis. In this study, a digital topographic map drawn to a scale of 1:5000 (2018) provided by the NGII [27] was used to calculate the topographic factors of the study area. The digital topographic map was converted into a 30 × 30 m digital elevation model (DEM) through spatial analysis because the contour lines and elevation points were composed in a vector form. Raster calculations were performed with this DEM (elevation) to calculate five surface factors including slope, profile curvature, plan curvature, topographic wetness index, and stream power index (Figure 3).

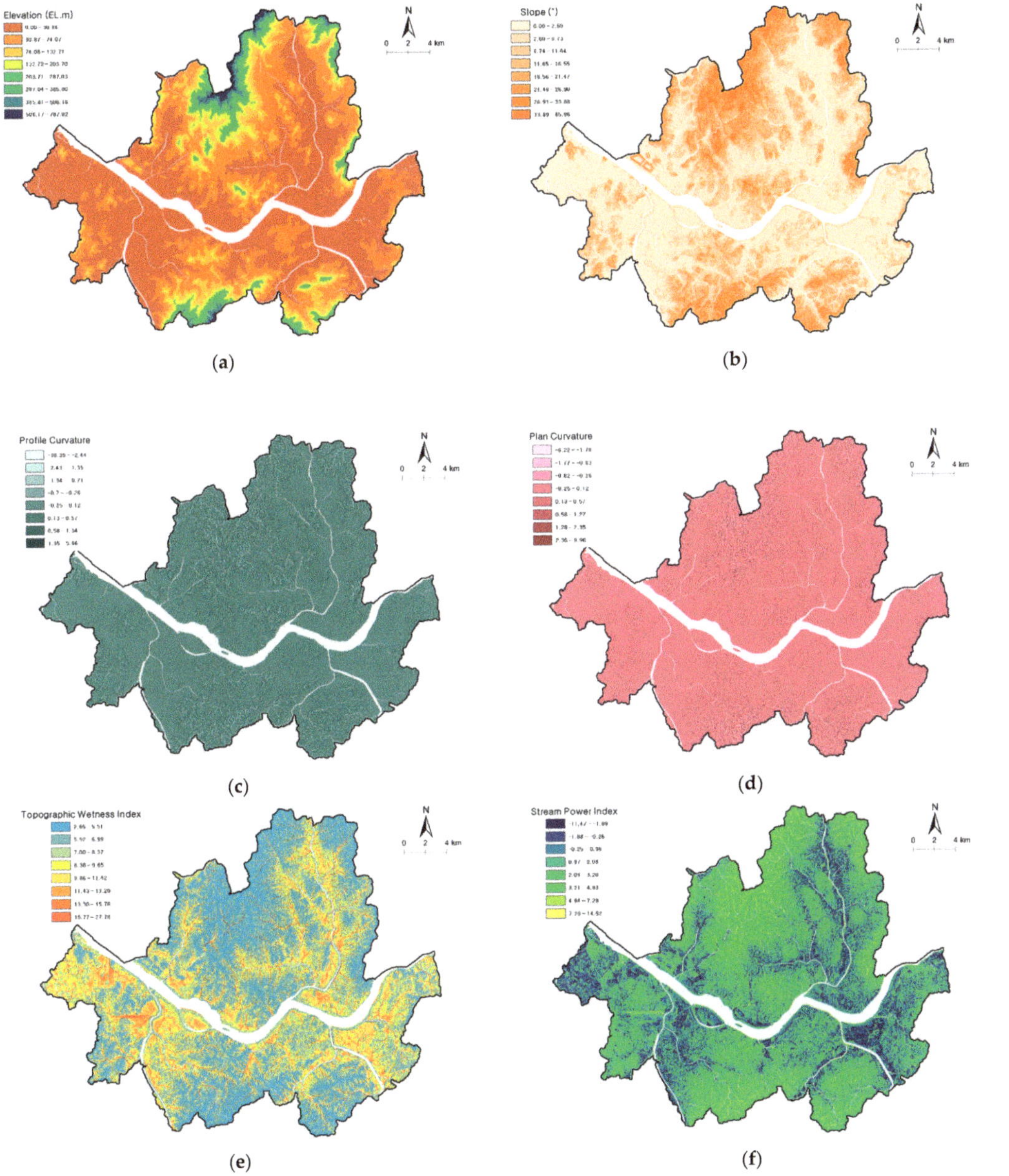

Figure 3. Topographic factors: (**a**) Elevation; (**b**) Slope; (**c**) Profile curvature; (**d**) Plan curvature; (**e**) Topographic wetness index (TWI); (**f**) Stream power index (SPI).

Elevation is the most representative factor explaining the characteristics of a surface that is prone to flooding; more lowlands means the area is more vulnerable to flooding. Since the flow velocity is slow in areas with gentle slopes, the runoff from rainfall accumulates and causes a flood. Curvature can be calculated as the second derivative of the surface, and can be classified into profile curvature and plan curvature, respectively, depending on whether it is calculated in a direction parallel to or perpendicular to the slope. Profile curvature is the curvature in the downward direction of the slope, and flooding is likely to occur in a concave terrain (positive). Plan curvature is the curvature in the horizontal direction of the slope, and runoff is likely to accumulate in a valley (negative). The topographic wetness index (TWI) was derived from the study of Beven and Kirkby [30] and can be calculated as shown in Equation (1). The TWI means that the gentler the slope (θ) of the target grid and the larger the basin area (a) of the upstream region, the higher the potential wetness index of the region. The stream power index (SPI), which was proposed by Moore et al. [31], represents the degree of sediment movement and erosion from surface runoff, and is calculated as shown in Equation (2).

$$TWI = \ln(a/\tan\theta) \tag{1}$$

$$SPI = \ln(a \times \tan\theta) \tag{2}$$

2.2.2. Locational Factors

Runoff from rainfall that reaches the ground flows from high to low along the slope by gravity. In natural basins, rainfall runoff gathers to form a river, while in urban areas such runoff is concentrated to manholes through a drainage pipe network. Therefore, areas near rivers or manholes are likely to be vulnerable to flooding when localized torrential downpours exceeding the capacity occur. To calculate the distance from the river and manhole, the location of the river and manholes was calculated for each grid using a digital topographic map (Figure 4a,b).

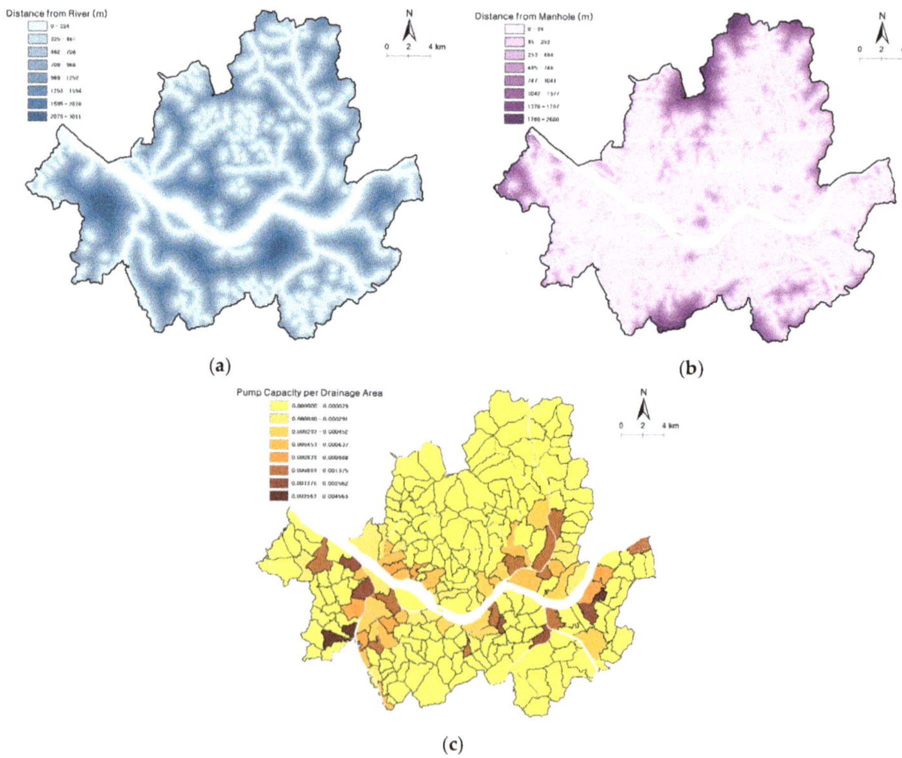

Figure 4. Locational and resistant factors: (**a**) Distance from river; (**b**) Distance from manhole; (**c**) Pump capacity per drainage area.

2.2.3. Resistant Factor

In urban areas, drainage pumping stations, which are representative facilities to reduce flood damage in lowlands during localized torrential rain, are installed [32]. In this study, statistical data from the Ministry of Environment (ME) [33] were collected to investigate the location and specifications of drainage pumping stations in the study area. On the other hand, since the specific time of the establishment of drainage pumping stations and that of the increase in the capacity could be not confirmed, the year-end statistical data of a year before the flood event, which was applied to the development and verification of this model, were used. It was found that 91 pumping stations were operated in Seoul in 2000, and the total pumping capacity was 118,196 m^3/min. In addition, there were 239 drainage sections in Seoul, and each drainage pumping station was designed to fit the area of the drainage section where the facility was located. Therefore, to reflect this, pumping capacity (C, m^3/min) was divided by the area (A, m^2) of the drainage section to calculate pumping capacity per drainage area (P), as shown in Equation (3) (Figure 4c).

$$P = C/A \tag{3}$$

2.2.4. Inducing Factor

Recently, the frequency of localized torrential rains has been increasing due to climate change [34]. In Seoul, which was affected by this, the number of occurrences of more than 30 mm/h of rainfall increased by 2.3 times throughout the year compared to before 1990, and that of more than 50 mm/h of rainfall increased by 5.3 times [35]. In addition, Son et al. [35] analyzed that rainfall of 75.0 mm and 15.5 mm/h was observed at the Seodaemun (412) and Dobong (406) observatories in Seoul at 14:00 on 21 September 2010, respectively, showing an approximately 5-fold difference between the two observatories. As such, in terms of the temporal distribution of rainfall, the occurrence frequency of concentrated torrential rains (30 mm/h or more) increases, and the spatial distribution also shows a large deviation due to localized heavy rains. Therefore, it was confirmed that topographic and hydrological factors should be connected when estimating flood vulnerability in this study.

Inundation damage in Seoul resulted mainly from inland floods, which occurred in urban lowlands or were caused by rainfall that could overwhelm the drainage infrastructure, rather than fluvial floods [21]. In Korea, when designing drainage pipes to protect against flood, the rainfall duration and the return period generally considered are 1 h and 10–30 years, respectively [36]. Therefore, in this study, maximum 1-h rainfall was used as an inducing factor that causes urban flooding.

Korea Meteorological Administration (KMA) [29] provides various types of observation data, such as automated synoptic observing system (ASOS) and automated weather system (AWS), as shown in Table 1. ASOS is installed in the location of the former KMA to perform observation tasks such as observing weather phenomena and data sharing through international cooperation, and AWS is installed in places where observation by a human operator is difficult, such as on mountains or islands, to monitor localized severe weather phenomena in real time [37]. There were a total of 32 rainfall observatories located in and near Seoul, as shown in Table 1, but to secure the reliability of the data required to develop a regression model, it was necessary to select data in consideration of missing observations, and the opening/closure of such observatories. The data from the Gangseo (404) and Gwangjin (413) observatories were excluded because missing data were found at the time of the occurrence of maximum 1-h rainfall observed at a nearby observatory in 2001. Those from Bukaksan (422), Guro (423), Gangbuk (424), and Namhyeon (425) observatories were excluded because they opened after 2001. The selected data were interpolated using the inverse distance weighting (IDW) method to assign the rainfall at the point where the observatories were located to all the relevant grids (Figure 5).

On the other hand, inundation trace maps provided information on flooded areas, but did not provide information on the date and time of flooding. If information on the time of flooding is not available, such data cannot be linked to rainfall data. Therefore, in this study, the maximum 1-h rainfall occurring in July 2001 was used as an independent variable for logistic regression. After that, to evaluate the performance of the regression equation, flood vulnerability was estimated using the maximum 1-h rainfall in September 2010 and July 2011, and compared with the inundation trace maps.

Table 1. Observatories in Seoul and maximum 1-h. rainfall.

Stn. No.	Stn. Name	Type	Start Date of Obs.	Max. 1-h. Rainfall (mm)		
				July 2001	September 2010	July 2011
108	Seoul	ASOS	1 October 1907	90.0	71.0	60.5
116	Gwanak 1	Radar	10 November 1993	96.5	32.0	90.0
400	Gangnam	AWS	4 December 1994	100.0	71.5	71.0
401	Seocho		4 December 1994	94.5	78.5	85.5
402	Gangdong		4 December 1994	82.5	79.0	56.5
403	Songpa		5 December 1994	86.0	65.0	91.5
404	Gangseo		5 December 1994	Missed	100.5	64.5
405	Yangcheon		5 December 1994	78.0	71.5	61.0
406	Dobong		6 December 1994	83.5	31.0	67.5
407	Nowon		6 December 1994	86.0	35.5	61.0
408	Dongdaemun		7 December 1994	93.5	65.5	69.0
409	Jungnang		7 December 1994	95.0	73.0	64.5
410	KMA		15 December 1999	111.0	81.5	55.0
411	Mapo		8 December 1994	100.5	85.0	63.5
412	Seodaemun		8 December 1994	77.0	81.5	52.5
413	Gwangjin		8 December 1994	Missed	74.0	56.0
414	Seongbuk		9 December 1994	82.5	53.0	59.0
415	Yongsan		9 December 1994	100.5	68.0	49.0
416	Eunpyeong		19 December 1994	76.0	54.0	68.5
417	Geumcheon		3 January 1998	108.0	98.0	62.0
418	Hangang		10 December 1994	96.0	69.5	48.0
419	Junggu		10 December 1994	96.5	73.5	47.0
420	Bukhansan		20 January 1994	83.5	6.5	Closed
421	Seongdong		22 August 2000	86.5	75.0	52.5
422	Bukaksan		3 November 2010	Not opened	Not opened	61.5
423	Guro		28 December 2001	Not opened	78.0	46.0
424	Gangbuk		28 December 2001	Not opened	33.5	67.0
425	Namhyeon		3 November 2010	Not opened	Not opened	112.5
509	Gwanak 2		14 September 1992	97.5	95.5	110.0
510	Yeongdeungpo		27 June 1989	94.0	72.0	65.0
540	Goyang		15 November 1992	54.0	39.0	101.5
541	Namyangju		21 November 1992	72.5	26.5	64.0

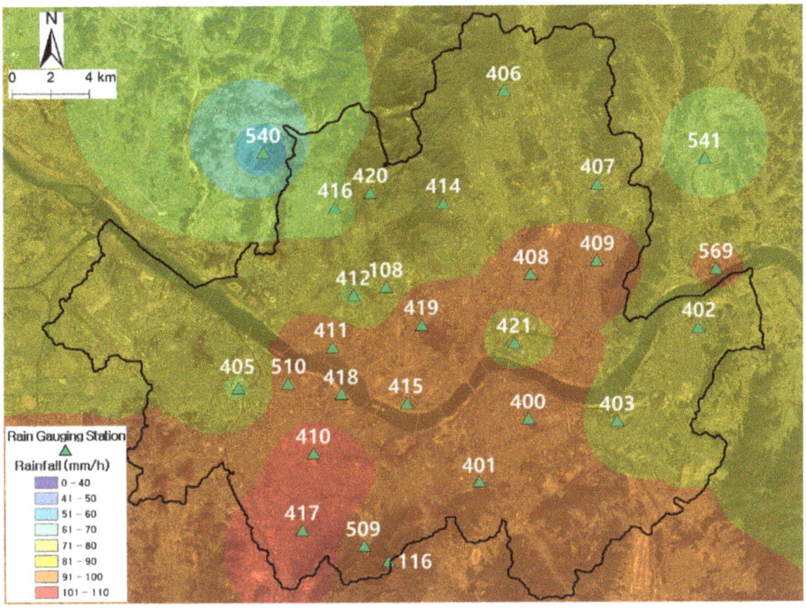

Figure 5. Rainfall interpolation result (2001).

3. Methodology

3.1. Multi-Collinearity Test

Multi-collinearity problems can cause when there is a correlation between two or more variables in a regression model. This problem can cause the calculations to be false, and the logistic parameters are incorrect and or inexact [38]. As the surface factors used in this study, five independent variables (slope, profile curvature, plan curvature, TWI, and SPI) calculated from elevation were used. Applying variables derived from one raster data to a regression model may cause multi-collinearity problems [17]. Therefore, the determination of multi-collinearity is an important step in detecting flood vulnerability using a logistic regression model. The variance inflation factor (*VIF*), one of the indicators used to determine multi-collinearity, can be calculated using the coefficient of determination (R^2) as in Equation (4).

$$VIF = \frac{1}{1 - R^2} \tag{4}$$

Lin [39] stated that variables can be judged to have multi-collinearity when VIF is greater than 10. Table 1 shows that there is no multi-collinearity problem as the VIF values for the six independent variables of the surface factors ranged from 1.099 to 2.679 (Table 2). Therefore, it was confirmed that six surface factors can be used as independent variables in logistic regression analysis to calculate flood vulnerability.

Table 2. Evaluation of vulnerability according to area under the curve (AUC).

Elevation	Slope	Profile Curvature	Plan Curvature	TWI	SPI
1.552	2.679	1.099	1.114	1.477	1.259

3.2. Logistic Regression

Logistic regression is a probability model proposed by Cox [40], which is used for classification and prediction by expressing the relationship between dependent variables and independent variables as a regression equation. It was mainly proposed to classify events in which the dependent variable follows a binomial distribution, such as the relationship between test scores and whether they pass the exam, or patient health status and whether they have a disease.

Odds ratio (*OR*) was introduced to utilize logistic regression for binary classification. OR represents the ratio of the probability, p, that an event will occur, and the probability, $1 - p$, that it will not occur, and it is calculated as follows.

$$OR = \frac{p}{1-p} \tag{5}$$

In addition, the problem of binary classification is that a linear regression analysis cannot be performed, because the dependent variable is represented as "0" or "1", and thus the range is different from the independent variable having a continuous distribution. Accordingly, the dependent variable is adjusted to $(-\infty, \infty)$ in the range of [0, 1] through the logit transformation that applies the logarithm to OR. This can be expressed using the following equation.

$$\text{Logit}(OR) = \log\left(\frac{p}{1-p}\right) = Y = \beta_0 + \beta_1 x_1 + \cdots + \beta_n x_n \tag{6}$$

In this study, for the calculation of the regression coefficient (β_n) of Equation (6), the occurrence of flooding events (Y) for all grids in the study area and 10 hydrological-topographic factors ($x_1 \sim x_{10}$) were used. In addition, the maximum likelihood estimator is used to determine regression coefficients including the constant term [41].

Next, a logistic function is used to calculate the flooding probability for target grids using the calculated regression coefficients. The logistic function can be calculated as follows by using the inverse function relation in Equation (6).

$$e^{\beta_0 + \beta_1 x_1 + \cdots + \beta_n x_n} = \frac{p}{1-p} \tag{7}$$

$$(1-p)e^{\beta_0 + \beta_1 x_1 + \cdots + \beta_n x_n} = p \tag{8}$$

$$e^{\beta_0 + \beta_1 x_1 + \cdots + \beta_n x_n} = p\left(1 + e^{\beta_0 + \beta_1 x_1 + \cdots + \beta_n x_n}\right) \tag{9}$$

$$p = \frac{1}{1 + e^{-(\beta_0 + \beta_1 x_1 + \cdots + \beta_n x_n)}} \tag{10}$$

The probability of flooding p can be obtained by inputting the hydrological-topographic factor for a target grid to Equation (10). This flooding probability (p) corresponds to flood vulnerability in this study. The flood vulnerability estimated through the logistic regression has the range [0, 1], and the closer to 1, the higher the probability of flood occurrence.

3.3. 2 × 2 Confusion Matrix and ROC Analysis

In this study, a receiver operating characteristics (ROC) analysis was conducted using a 2 × 2 confusion matrix to check the extent to which the areas with high flood vulnerability calculated using the logistic regression model were consistent with the inundation trace maps. The 2 × 2 confusion matrix and ROC analysis have been mainly used in the medical field, including the performance evaluation of reagents that discriminate negative from positive patients in the diagnostic test of COVID-19, which has been spreading around the world in recent years. This technique has recently been extended and applied to the fields of machine learning and object recognition to evaluate the classification accuracy of artificial intelligence [42,43]. ROC analysis allows us to determine whether a test method is useful by showing a curve for the degree to which an event is detected for each test method [44,45]. To draw this curve, four components of a 2 × 2 confusion matrix should be used.

As shown in Table 3, the 2 × 2 confusion matrix can be composed of 4 different combinations depending on whether the flood vulnerable area and inundation traces on the map coincide. If the flood vulnerable area and inundation traces coincide, it can be expressed as true positives (*TP*) or true negatives (*TN*); otherwise, it is expressed as false positives (*FP*) or false negatives (*FN*). For the plot of the ROC curve, the x-axis is calculated as 1-specificity, showing specificity which is the ratio of accurately predicted areas (TN) among the areas where no actual flooding occurred (FP + TN). The y-axis of the graph shows sensitivity, which is the ratio of the areas selected as flood vulnerable areas (TP) among the flooded areas (TP + FN). When expressed as an equation, specificity and sensitivity can be expressed as Equations (11) and (12), respectively, and the range of values is [0, 1] [45].

$$Specificity = \frac{TN}{FP + TN} \tag{11}$$

$$Sensitivity = \frac{TP}{TP + FN} \tag{12}$$

Table 3. Overview of 2 × 2 confusion matrix.

		Inundation Trace	
		Inundated Area	**Non-Inundated Area**
Flood vulnerability	**Vulnerable area**	True positives (TP)	False positives (FP)
	Non-vulnerable area	False negatives (FN)	True negatives (TN)

In ROC analysis, the performance of a test method can be evaluated by calculating the area under the curve (AUC). It can be evaluated that the closer the AUC is to 1, the better the detection performance is, while the closer the AUC is to 0, the worse the detection performance is. According to Ying et al. [46] and Simundic [47], the AUC can be evaluated as shown in Table 4. In addition, if the ROC curve is distributed below the diagonal with a slope of 1 and the AUC is calculated to be 0.5 or less, it means that the test method is not useful.

Table 4. Evaluation of vulnerability according to area under the curve (AUC).

AUC	Evaluation
0.9–1.0	Excellent
0.8–0.9	Very good
0.7–0.8	Good
0.6–0.7	Sufficient
0.5–0.6	Bad
<0.5	Test not useful

4. Results and Discussion

4.1. Logistic Regression Modeling

For the analysis, the city of Seoul was divided using a grid into 648,174 30 m × 30 m squares excluding rivers, and 47,065 of these were found to have inundation traces. Through this process, the grid where flooding had occurred and those where it had not were classified as 1 and 0, respectively, and these values were entered into Y of Equation (5). In addition, the hydrological-topographic factors for each grid were required in the logistic regression equation to estimate the flood vulnerability. This study intends to provide information on the changes in vulnerability according to rainfall, rather than calculating unchanged flood vulnerability for each grid by considering only the topographic factors. Therefore, two logistic regression models were developed and their performance was compared: an equation that used only topographic factors (T) as independent variables, and one that also included hydrological factors (TR). As a result, the logistic regression coefficients and constant terms of the two equations were determined, as shown in Equations (14) and (15), respectively.

$$z = \beta_0 + \beta_1 x_1 + \cdots + \beta_n x_n \tag{13}$$

$$\begin{aligned} z_T = {} & -4.394 - 1.391 \times Elevation - 0.120 \times Slope \\ & +0.049 \times Profile\ Curvature + 0.070 \times Plan\ Curvature \\ & +0.335 \times TWI - 0.147 \times SPI + 0.240 \times Distance\ from\ River \\ & -5.746 \times Distance\ from\ Manhole \\ & -0.093 \times Pump\ Capacity\ per\ Area \end{aligned} \tag{14}$$

$$\begin{aligned} z_{TR} = {} & -4.486 - 1.323 \times Elevation - 0.206 \times Slope \\ & +0.074 \times Profile\ Curvature + 0.101 \times Plan\ Curvature \\ & +0.374 \times TWI - 0.163 \times SPI + 0.253 \times Distance\ from\ River \\ & -5.610 \times Distance\ from\ Manhole \\ & -0.081 \times Pump\ Capacity\ per\ Area + 0.503 \times Rainfall \end{aligned} \tag{15}$$

With the data for 2001, the flood vulnerability was calculated using the hydrological-topographic factors and the determined regression coefficients, for all grid in the study area (Figure 6). In the figure, a darker color indicates that the area is more vulnerable, while a lighter color indicates that the area is less vulnerable. The flood vulnerability was represented by classifying the probability in the range [0, 1] into five classes using the natural breaks method. The idea of the natural breaks method is to minimize variance among objects within the chosen subsets, and maximize variance between the subsets [48]. The five classes included very high (1.00–0.50), high (0.50–0.34), medium (0.34–0.22), low (0.22–0.13), and very low (0.13–0.02). In addition, as areas with a probability of less than 2% were not evaluated to be vulnerable, a vulnerability level was not assigned to these.

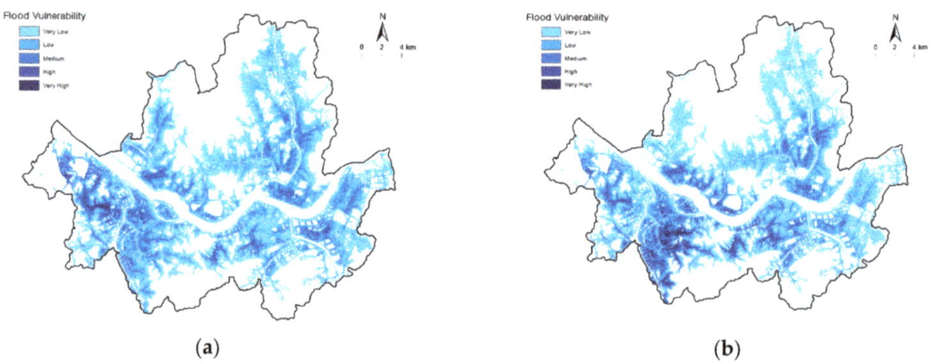

Figure 6. Results of logistic regression (2001): (**a**) Topographic data; (**b**) Topographic and hydrological data.

Flood vulnerability, which was calculated with two logistic regression equations, was divided into vulnerability considering only topographic factors (Figure 6a) and one that also considered maximum 1-h rainfall, a hydrological factor (Figure 6b). From the difference between Figure 6a,b, it can be seen that the vulnerability varies by region according to the spatial distribution of the maximum 1-h rainfall. When considering the hydrological factor, the area with very high-intensity rainfall of 100–110 mm/h (the area marked in red in the southwest in Figure 5) was more susceptible to flooding than when only topographic factors were considered. On the other hand, the area with rainfall of 60–70 mm/h (the area represented in green in the northwest in Figure 5) was found to be less vulnerable.

An ROC analysis was conducted to quantitatively confirm whether the flood vulnerable areas determined by the technique proposed in this study and those where floods occurred in the past coincided. To plot the ROC curve with 10 points, the flood vulnerability of target areas was divided into 10 equal parts using quartiles. If many floods occurred in areas with high vulnerability in the ROC curve (lower side of the x-axis), the sensitivity would increase, and in particular would increase sharply at the beginning of the curve. Consequentially, as the AUC increases, it can be evaluated that the technique of this study detects flooded areas well. The ROC curve is shown in Figure 7.

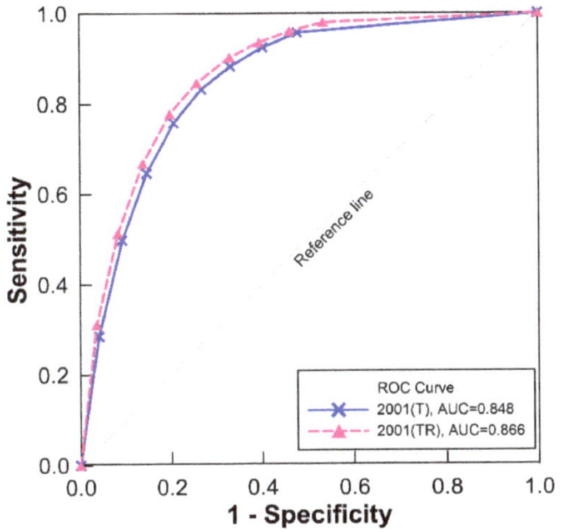

Figure 7. ROC curves of flood vulnerability and inundation traces.

Through ROC analysis, it was found that the AUC of flood vulnerability considering only topographic factors and that including rainfall were 0.848 and 0.866, respectively, and both were evaluated as "very good" as shown in Table 4. Further, the precision was calculated to confirm the rate at which flood occurrence was detected for the flood vulnerability, which was classified into five classes. This can be obtained by using the number of samples classified as positive in a 2 × 2 confusion matrix as shown in Equation (16), and the range is [0, 1] (perfect value is 1) [49,50]. The precision for each class was calculated as shown in Figure 8 and, in both cases, it was found that floods were detected at a rate of more than 50% in the very high class, and more than 40% in the high class.

$$Precision = \frac{TP}{TP + FP} \tag{16}$$

Based on the results of this analysis, it is considered that the logistic regression model detects flood occurrence well in the study area. Although the inputting of the hydrological factor did not make a distinct difference, it can be assumed that the vulnerability will change and the detection rate will improve if new rainfall data are input, even without topographic changes. To confirm this, the performance of the logistic regression equation was evaluated using the maximum 1-h rainfall and inundation trace maps in 2010 and 2011.

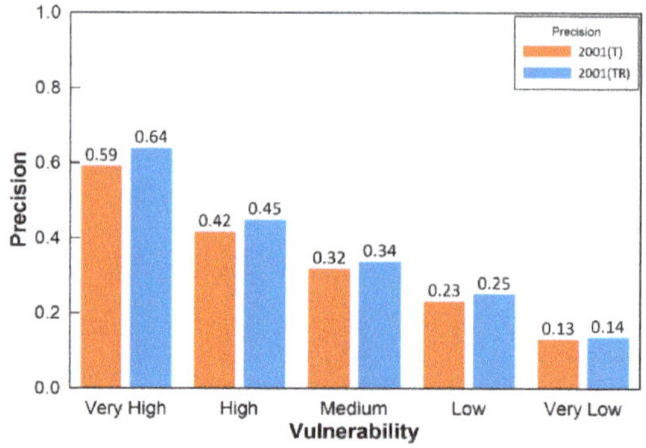

Figure 8. Proportion of flood occurrence in vulnerable areas.

4.2. Mapping Vulnerable Areas in Other Flood Events

To evaluate the performance of the logistic regression model developed in this study, inundation trace maps and the maximum 1-h rainfall data were collected for floods that occurred in September 2010 and July 2011. Of the rainfall observatories, the Bukaksan (422) and Namhyeon (425) observatories were excluded because they had not yet opened in 2010, and the Bukhansan (420) observatories, which were closed in 2011, were also excluded. The rainfall of the observatories was interpolated using the IDW method as shown in Figure 9. In 2010, very high-intensity rainfall of around 100 mm/h occurred in the west of the study area, while relatively low-intensity rainfall of around 60 mm/h was recorded in the north. Overall, in 2011, it rained less than in 2010, except in the south with 100 mm/h of rainfall.

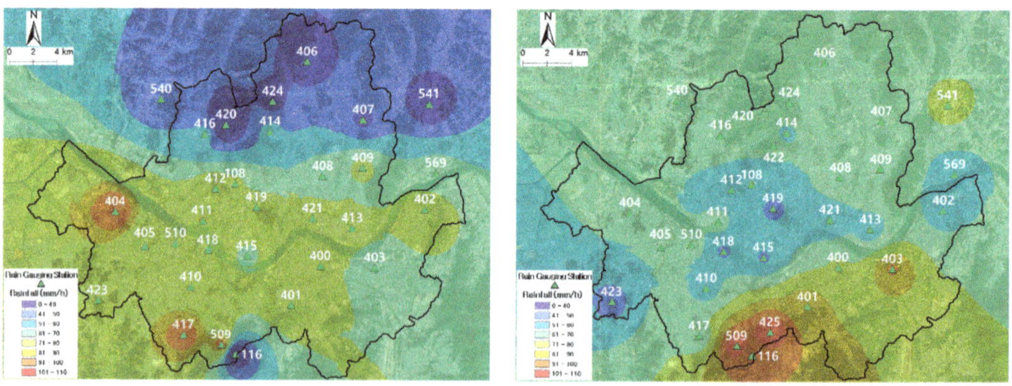

(a) (b)

Figure 9. Rainfall interpolation results by year: (**a**) September 2010; (**b**) July 2011.

In addition, the year-end statistical data for 2009 and 2010 [51,52] were used to consider drainage pumping stations that were newly built or increased in capacity after flooding in 2001. In 2009 and 2010, the number of pumping stations was increased from 91 in 2000 to 111, and the capacity also was increased to 155,313 m^3/min (2009) and 161,279 m^3/min (2010). The changed pump capacity per drainage area is shown in Figure 10.

In the logistic regression model developed above with data for 2001, the same values as those in 2001 were used for surface and locational factors, and values for 2010 and 2011 were entered for resistant and inducing factors, respectively. The flood vulnerability that was recalculated by inputting pumping capacity and rainfall to this model is shown in Figure 11. Interestingly, as low rainfall was input compared to that used in the development of the model, the vulnerability in 2010 and 2011 was decreased significantly compared to in 2001. In addition, in 2010, the high-intensity rainfall in the west increased the vulnerability, and the low-intensity rainfall in the north decreased the vulnerability. In 2011, most areas were calculated to have low vulnerability, except for the increase in vulnerability in some areas due to high-intensity rainfall in the south.

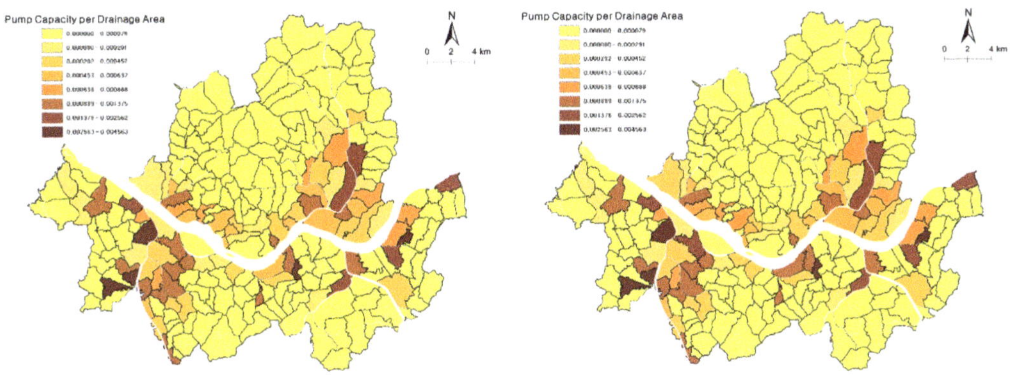

(a) (b)

Figure 10. Density of drainage pump capacity by year: (**a**) September 2010; (**b**) July 2011.

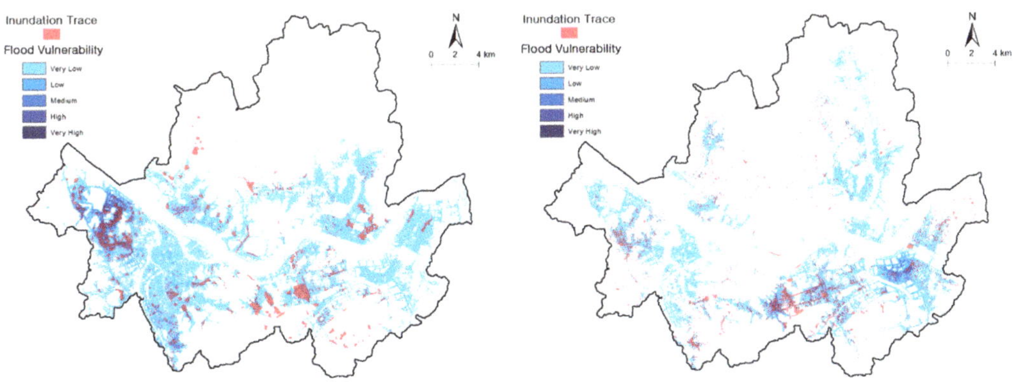

(a) (b)

Figure 11. Results of the selection for flood vulnerable areas: (**a**) September 2010; (**b**) July 2011.

An ROC analysis was conducted to quantitatively analyze the extent to which calculated flood vulnerability in 2010 and 2011 actually detects floods. The ROC curves for 2010 and 2011 are shown in Figure 12. In both cases, it was found that the measure of flood vulnerability (AUC = 0.861, 0.815) that considered the hydrological factors together detected flood occurrence better than that (AUC = 0.841, 0.766), which considered only the topographic factors. The detection rate was calculated as shown in Figure 13. In 2010, among vulnerable areas considering rainfall, flooding occurred at a rate of 66% (T, 57%) in the very high class; 54% (T, 31%) in the high; 47% (T, 18%) in the medium; 33% (T, 12%) in the low; and 11% (T, 7%) in the very low. In 2011, floods occurred at a rate of 36% (T, 17%) in the very high class; 41% (T, 12%) in the high; 31% (T, 9%) in the medium; 28% (T, 6%) in the low; and 9% (T, 4%) in very low. Through ROC analysis and precision, it was found that the model for calculating flood vulnerability that only considers topographic factors has a disadvantage of overestimating vulnerable areas, but that the detection rate could be improved by up to over four times (in the low class in 2011) when the rainfall was also considered.

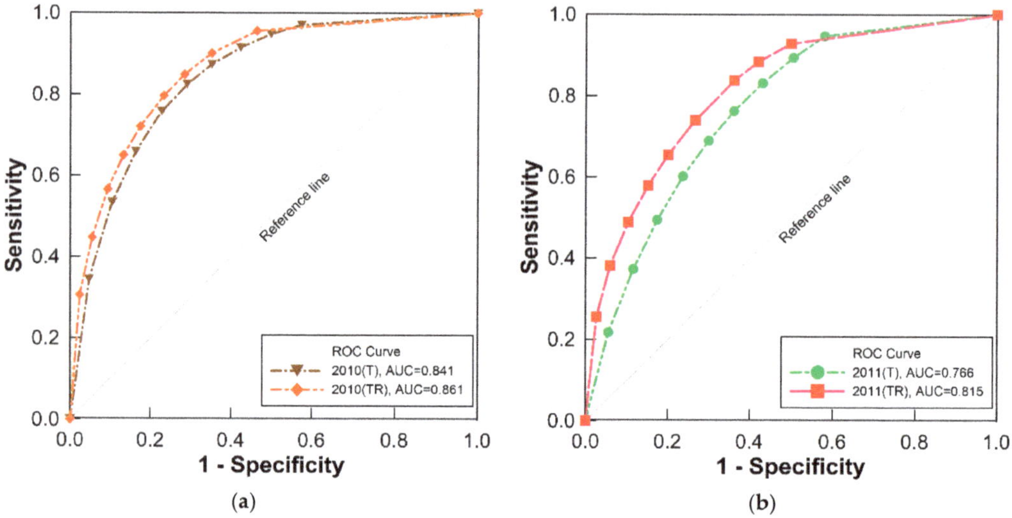

Figure 12. ROC curves of flood vulnerability and inundation traces for performance evaluation: (**a**) September 2010; (**b**) July 2011.

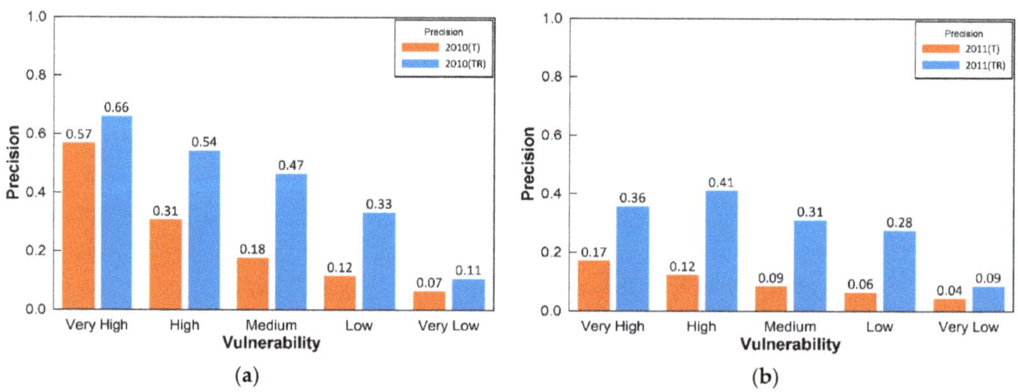

Figure 13. Proportion of flood occurrence in vulnerable areas for performance evaluation: (**a**) September 2010; (**b**) July 2011.

4.3. Discussion

This study proposed a technique for calculating the flood vulnerability that changes according to the rainfall situation using hydrological-topographic factors. Lee et al. [5] suggested that studies using globally available data, such as SRTM and ASTER, are needed so that they can be used even in areas where data are insufficient in flood vulnerability analysis. In addition, they said that it is necessary to develop a technique that can evaluate flood vulnerability in a simple but scientific way that can be applied to areas where there are no data on hydrological observations or poor quality. Against this background, in this study, topographic data that can be used anywhere was used as an independent variable of the logistic regression model, and data on soil or land use, which may not be available depending on the region, were not added. However, reviewing previous studies, 89% of the floods that occurred in Seoul in 2011 occurred in areas with an impermeability rate of 70% or higher [21]. Further, it was analyzed that 52.1% of the study area consisted of roads, residential, and commercial areas, and 89.4% of floods occurred in these areas. It would be good if soil impermeability or land use is added as an independent variable through future research.

In this study, flood vulnerability was calculated using hydrological-topographic factors and compared with historical inundation trace data. As a result, there were some cases where flooding occurred in areas with a calculated vulnerability lower than 0.5, and other cases where flooding did not occur even in areas above 0.5. However, it remains uncertain whether areas with relatively low vulnerability are safer. It is true that areas with high vulnerability require intensive management due to their high probability of flooding, but even areas with low vulnerability should be managed with constant attention to reduce flood damage. Since floods occur for very complicated causes, it may be difficult to detect them using only the factors used in this study. Kim et al. [53] proposed an optimal input data selection method by combining total rainfall, rainfall of various durations, kurtosis, and skewness to predict urban flooding using a deep neural network. If the characteristics of rainfall, such as various durations, kurtosis, and skewness, are considered as an inducing factor, the detection accuracy for flood vulnerable areas can be improved.

The flood vulnerability, calculated using hydrological-topographic factors, did not take into account the density and importance of the population and capital in the study area. If a flood occurs in a densely populated area, it is difficult for many people to evacuate all at once, so even if the area has a low vulnerability, it is necessary to pay close attention to the area. Similarly, in areas where major social overhead capital (SOC) facilities are located, such as power plants or water supply/wastewater treatment facilities, great damage can be caused to surrounding areas when floods occur. Rehman et al. [54] reviewed scholarly articles related to flood vulnerability from 1990 to 2018, noting that flood vulnerability is being analyzed in social, environmental, and economic contexts, and presented a list of indicators that can be used for future research. In this regard, it is necessary to distinguish the vulnerability of regions with high socio-economic vulnerability from the criteria for calculating flood vulnerability ratings for other regions.

Another point to be improved in the methodology of this study is that the vulnerability of the entire study area was calculated using one logistic regression equation. Accordingly, there is a limitation in that areas with the same hydrological-topographic factors are likely to be determined as vulnerable areas even though flood damage did not occur there. These areas may not actually be flooded due to flood protection facilities such as a retarding basin, or drainage pipe networks that have already been expanded to handle the greater amount of rainfall through local government management. Based on this, a good direction for a future study would be to develop a logistic regression equation for each drainage section rather than only one equation for the entire study area. It is considered that the disaster prevention performance of the drainage section can be reflected indirectly through such equations without using a physical model.

5. Conclusions

In this study, we proposed a technique to detect flood vulnerable areas by simultaneously considering topographic and hydrological factors to reduce damage caused by flooding. To estimate the vulnerability to flooding of the study area, a logistic regression model was developed using historical inundation trace data, and hydrological-topographic factors based on the grid system. The conclusions obtained through this study are as follows.

(1) A logistic regression model was established by dividing into a model that only considered topographic factors (T) and one that included hydrological factors (TR), and the results were compared. When comparing the two models, it was found that the estimated result was different due to the influence of rainfall. In addition,

according to the results of ROC analysis and precision calculation, it was found that the method of estimating the flood vulnerability that included the hydrological factor was relatively better for detecting the flood occurrence pattern.

(2) Flood events in 2010 and 2011 were applied to evaluate the performance of the developed logistic regression model. Through ROC analysis with inundation trace maps, it was found that the AUC improved from 0.841 (T) to 0.861 (TR) in 2010, and from 0.766 (T) to 0.815 (TR) in 2011, indicating that the model including hydrological factors was better for detecting flooding patterns. In addition, according to the actual flood occurrence rates calculated in the vulnerable areas determined in consideration of the rainfall, the detection rate was significantly improved compared to the approach that considered only topographical factors, which overestimated vulnerable areas.

(3) There were some cases in which flooding occurred in areas with low vulnerability, while in other cases, high vulnerability areas saw no floods occur. The cause of a flood is very complicated because the factors influence the events in a complex manner through interaction. Therefore, constant attention and management is required even for areas with low vulnerability. In particular, areas with many residents or important SOC facilities should be managed separately.

(4) The technique for determining flood vulnerable areas proposed in this study enables an analysis to be performed quickly and conveniently because the topographic factors are fixed in the logistic regression equation and only new rainfall is inputted to it. If real-time rainfall forecasting is available, it will allow us to quickly and easily predict areas that are likely to be flooded. In addition, through further research, if a regression model for each drainage section is developed by subdividing the characteristics of flood damage factors, it is expected that the detection rates will improve and more reliable flood information will be provided.

Author Contributions: Conceptualization, J.-Y.L. and J.-S.K.; methodology, J.-Y.L. and J.-S.K.; software and validation, J.-Y.L.; writing—original draft preparation, J.-Y.L.; writing—review and editing, J.-S.K. All authors have read and agreed to the published version of the manuscript.

Funding: This study was supported by a Grant (127568) from the Water Management Research Program funded by Ministry of Environment of Korean government.

Institutional Review Board Statement: Not applicable.

Informed Consent Statement: Not applicable.

Acknowledgments: This study was supported by a Grant (127568) from the Water Management Research Program funded by Ministry of Environment of Korean government.

Conflicts of Interest: The authors declare no conflict of interest.

References

1. Koh, T.G.; Lee, W.Y. A Study on Urban Planning Technique for Flood Reduction in the Lowlands. *Seoul Stud.* **2012**, *13*, 287–300.
2. Balica, S.F.; Popescu, I.; Beevers, L.; Wright, N.G. Parametric and Physically Based Modelling Techniques for Flood Risk and Vulnerability Assessment: A Comparison. Environ. *Modeling Softw.* **2013**, *41*, 84–92. [CrossRef]
3. Dehotin, J.; Breil, P.; Braud, I.; Lavenne, A.D.; Lagouy, M.; Sarrazin, B. Detecting Surface Runoff Location in a Small Catchment using Distributed and Simple Observation Method. *J. Hydrol.* **2015**, *525*, 113–129. [CrossRef]
4. Lagadec, L.R.; Patrice, P.; Braud, I.; Charzelle, B.; Moulin, L.; Dehotin, J.; Breil, P. Description and Evaluation of a Surface Runoff Susceptibility Mapping Method. *J. Hydrol.* **2016**, *541*, 405–509. [CrossRef]
5. Lee, K.S.; Lee, D.E.; Jung, S.H.; Lee, G.H. Analysis of Large-scale Flood Inundation Area using Optimal Topographic Factors. *J. Korean Water Resour. Assoc.* **2018**, *51*, 481–490.
6. Lee, J.Y.; Kim, J.S. ROC Analysis of Topographic Factors in Flood Vulnerable Area Considering Surface Runoff Characteristics. *Eco. Resilient Infra.* **2020**, *7*, 327–335.
7. Tehrany, M.S.; Lee, M.J.; Pradhan, B.; Jebur, M.N.; Lee, S.R. Flood Susceptibility Mapping using Integrated Bivariate and Multivariate Statistical Models. *Environ. Earth Sci.* **2014**, *72*, 4001–4015. [CrossRef]
8. Lee, S.M.; Lee, M.J. Susceptibility Mapping of Umyeonsan using Logistic Regression Model and Post-validation through Field Investigation. *Korean J. Remote Sens.* **2017**, *33*, 1047–1060.

9. Jung, M.K.; Kim, J.G.; Uranchimeg, S.; Kwon, H.H. The Probabilistic Estimation of Inundation Region using a Multiple Logistic Regression. *J. Korea Water Resour. Assoc.* **2020**, *53*, 121–129.
10. Wang, Z.; Lai, C.; Chen, X.; Yang, B.; Zhao, S.; Bai, X. Flood Hazard Risk Assessment Model Based on Random Forest. *J. Hydrol.* **2015**, *527*, 1130–1141. [CrossRef]
11. Rizeei, H.M.; Pradhan, B.; Saharkhiz, M.A. An Integrated Fluvial and Flash Pluvial Model using 2D High-resolution Sub-grid and Particle Swarm Optimization-based Random Forest Approaches in GIS. *Complex Int. Syst.* **2019**, *5*, 283–302. [CrossRef]
12. Kalantari, Z.; Nickman, A.; Lyon, S.W.; Olofsson, B.; Folkeson, L. A Method for Mapping Flood Hazard along Roads. *J. Envron. Manag.* **2014**, *133*, 69–77. [CrossRef] [PubMed]
13. Khatun, S.; Pal, S. Categorization of Morphometric Surface through Morphometric Diversity Analysis in Kushkarani River Basin in Eastern India. *Asian J. Phys. Chem. Sci.* **2017**, *2*, 1–19. [CrossRef]
14. Pradhan, B.; Lee, S.R. Delineation of Landslide Hazard Areas on Penang Island, Malaysia, by using Frequency Ratio, Logistic Regression, and Artificial Neural Network Models. Environ. *Earth Sci.* **2010**, *60*, 1037–1054. [CrossRef]
15. Lee, S.M.; Kim, J.C.; Jung, H.S.; Lee, M.J.; Lee, S.R. Spatial Prediction of Flood Susceptibility using Random-forest and Boosted-tree Models in Seoul Metropolitan City, Korea. *Geom. Nat. Hazard Risk* **2017**, *8*, 1185–1203. [CrossRef]
16. Li, X.; Yan, D.; Wang, K.; Weng, B.; Qin, T.; Liu, S. Flood Risk Assessment of Global Watersheds Based on Multiple Machine Learning Models. *Water* **2019**, *11*, 1654. [CrossRef]
17. Nandi, A.; Mandal, A.; Wilson, M.; Smith, D. Flood Hazard Mapping in Jamaica using Principal Component Analysis and Logistic Regression, Environ. *Earth Sci.* **2016**, *75*, 465. [CrossRef]
18. Melly, B.L.; Schael, D.M.; Nick, R.M.; Gama, P.T. Mapping Ephemeral Wetlands: Manual Digitisation and Logistic Regression Modelling in Nelson Mandela Bay Municipality, South Africa. *Wetl. Ecol. Manag.* **2017**, *25*, 313–330. [CrossRef]
19. Sun, X.; Chen, J.; Bao, Y.; Han, X.; Zhan, J.; Peng, W. Landslide Susceptibility Mapping using Logistic Regression Analysis along the Jinsha River and Its Tributaries Close to Derong and Deqin County, Southwestern China. *Int. J. Geo-Inf.* **2018**, *7*, 438. [CrossRef]
20. Korea Institute of Construction Technology (KICT). *Investigation Report on Heavy Rainfall in the Metropolitan Region in July 2011*; KICT: Goyang, Korea, 2011.
21. Shin, S.Y.; Park, C.Y. Analyzing Relationships between Land Use Characteristics and Flood Damage Area: The Case of Seoul. *Korea Spat. Plan. Rev.* **2014**, *81*, 3–20.
22. Zhao, G.; Xu, Z.; Pang, B.; Tu, T.; Xu, L.; Du, L. An Enhanced Inundation Method for Urban Flood Hazard Mapping at the Large Catchment Scale. *J. Hydrol.* **2019**, *571*, 873–882. [CrossRef]
23. Kim, B.H.; Sanders, B.F.; Schubert, J.E.; Famiglietti, J.S. Mesh Type Tradeoffs in 2D Hydrodynamic Modeling of Flood with a Godunov-based Flood Solver. *Adv. Water Resour.* **2014**, *68*, 42–61. [CrossRef]
24. Liang, Q.; Du, G.; Hall, J.W.; Borthwick, A.G.L. Flood Inundation Modeling with an Adaptive Quadtree Grid Shallow Water Equation Solver. *J. Hydraul. Eng.* **2008**, *134*, 1603–1610. [CrossRef]
25. Mosavi, A.; Ozturk, P.; Chau, K. Flood Prediction using Machine Learning Models: Literature Review. *Water* **2018**, *10*, 1536. [CrossRef]
26. Seoul Open Data Plaza. Available online: https://data.seoul.go.kr (accessed on 14 October 2020).
27. National Geographic Information Institute (NGII). Available online: http://map.ngii.go.kr (accessed on 2 January 2019).
28. National Spatial Data Infrastructure Portal. Available online: http://data.nsdi.go.kr (accessed on 14 October 2020).
29. Weather Data Open Portal. Available online: https://data.kma.go.kr (accessed on 1 February 2021).
30. Beven, K.J.; Kirkby, M.J. A Physically Based Variable Contributing Area Model of Basin Hydrology. *Hydrol. Sci. J.* **1979**, *24*, 43–69. [CrossRef]
31. Moore, I.D.; Grason, R.B.; Ladson, A.R. Digital Terrain Modelling: A Review of Hydrological, Geomorphological, and Biological Applications. *Hydrol. Proc.* **1991**, *5*, 3–30. [CrossRef]
32. Song, Y.H.; Park, M.J.; Lee, J.H. Analysis of Urban Inundation Reduction Effect by Early Operation of Drainage Pumping Station. *J. Korean Soc. Hazard Mitig.* **2014**, *14*, 267–276. [CrossRef]
33. Ministry of Environment (ME). *Statistics of Sewerage 2000*; ME: Seoul, Korea, 2000.
34. Jeong, I.J.; Kim, S.K.; Choi, J.H.; Jung, S.G. A Study on Element Derived for Disaster Recovery Support Priorities. *J. Korean Soc. Hazard Mitig.* **2016**, *16*, 67–71. [CrossRef]
35. Son, A.L.; Han, K.Y.; Bae, S.H. Temporal and Spatial Characteristics Analysis of Rainfall in Seoul. *J. Korean Soc. Hazard Mitig.* **2013**, *13*, 83–95. [CrossRef]
36. Ministry of Environment (ME). *A Study on the Adequacy of Probability Years when Calculating Design Rainfall for Public Sewer Facilities*; ME: Sejong, Korea, 2018.
37. Hyun, J.H.; Park, H.S.; Chung, G.H. Effects of the Difference between ASOS and AWS Data on Runoff Characteristics. *J. Korean Soc. Hazard Mitig.* **2019**, *19*, 443–449. [CrossRef]
38. Al-Juaidi, A.E.M.; Nassar, A.M.; Al-Juaidi, O.E.M. Evaluation of Flood Susceptibility Mapping using Logistic Regression and GIS Conditioning Factors. *Arab. J. Geosci.* **2018**, *11*, 765. [CrossRef]
39. Lin, F.J. Solving Multicollinearity in the Process of Fitting Regression Model using the Nested Estimate Procedure. *Qual. Quant.* **2008**, *42*, 417–426. [CrossRef]
40. Cox, D.R. The Regression Analysis of Binary Sequences. *J. R. Stat. Soc.* **1958**, *20*, 215–242. [CrossRef]

41. Lopez, A.V.; Morales, C. Flash Flood Forecasting in Sao Paulo using a Binary Logistic Regression Model. *Atm.* **2020**, *11*, 473.
42. Bae, H.B.; Kwon, O.S. Untact Face Recognition System Based on Super-resolution in Low-resolution Images. *J. Korea Multimed. Soc.* **2020**, *23*, 412–420.
43. Lee, S.M.; Park, K.D.; Kim, I.K. Comparison of Machine Learning Algorithms for Chl-a Prediction in the Middle of Nakdong River (Focusing on Water Quality and Quantity Factors). *J. Korean Soc. Water Wastewater* **2020**, *34*, 277–288. [CrossRef]
44. DeLong, E.R.; DeLong, D.M.; Clarke-Pearson, D.L. Comparing the Areas under Two or More Correlated Receiver Operating Characteristic. *Biometrics* **1988**, *44*, 837–845. [CrossRef] [PubMed]
45. Metz, C.E. Basic Principles of ROC Analysis. *Semin. Nucl. Med.* **1978**, *8*, 283–298. [CrossRef]
46. Ying, G.S.; Maguire, M.; Quinn, G.; Kulp, M.T.; Cyert, L. Vision in Preschoolers Study Group. ROC Analysis of the Accuracy of Noncycloplegic Retinoscopy, Retinomax Autorefractor, and SureSight Vision Screener for Preschool Vision Screening. *Invest. Ophthal. Vis. Sci.* **2011**, *52*, 9658–9664. [CrossRef]
47. Simundic, A.M. Diagnostic Accuracy-part1, Basic Concepts: Sensitivity and Specificity, ROC Analysis, STARD Statement. *Point Care J. Near-Patient Testing Tech.* **2012**, *11*, 6–8. [CrossRef]
48. Jenks, G. The Data Model Concept in Statistical Mapping. *Int. Yearb. Cartogr.* **1967**, *7*, 186–190.
49. Razali, N.; Ismail, S.; Mustapha, A. Machine Learning Approach for Flood Risks Prediction. *IAES Int. J. Art. Int.* **2020**, *9*, 73–80. [CrossRef]
50. Chapi, K.; Singh, V.P.; Shirzadi, A.; Shahabi, H.; Bui, D.T.; Pham, B.T.; Khosravi, K. A Novel Hybrid Artificial Intelligence Approach for Flood Susceptibility Assessment. *Environ. Modeling Softw.* **2017**, *95*, 229–245. [CrossRef]
51. Ministry of Environment (ME). *Statistics of Sewerage 2009*; ME: Seoul, Korea, 2009.
52. Ministry of Environment (ME). *Statistics of Sewerage 2010*; ME: Seoul, Korea, 2010.
53. Kim, H.I.; Lee, J.Y.; Han, K.Y.; Cho, J.W. Applying Observed Rainfall and Deep Neural Network for Urban Flood Analysis. *J. Korean Soc. Hazard Mitig.* **2020**, *20*, 339–350. [CrossRef]
54. Rehnam, S.; Sahana, M.; Hong, H.; Sajjad, H.; Ahmed, B.B. A Systematic Review on Approaches and Methods used for Flood Vulnerability Assessment: Framework for Future Research. *Nat. Hazard* **2019**, *97*, 975–998.

applied sciences

MDPI

Article

Marine Gas Hydrate Geohazard Assessment on the European Continental Margins. The Impact of Critical Knowledge Gaps

Ricardo León *, Miguel Llorente and Carmen Julia Giménez-Moreno

Geological Survey of Spain (IGME), Department of Research and Prospective Geoscience, Rios Rosas 23, 28003 Madrid, Spain; m.llorente@igme.es (M.L.); j.gimenez@igme.es (C.J.G.-M.)
* Correspondence: r.leon@igme.es

Featured Application: Results can be used by policy makers and companies for planning seafloor activities. The susceptibility assessment and the analysis of its reliability can be applied in other regional works of geohazard assessment.

Abstract: This paper presents a geohazard assessment along the European continental margins and adjacent areas. This assessment is understood in the framework of the seafloor's susceptibility to (i.e., likelihood of) being affected by the presence of hydrate deposits and the subsequent hazardous dissociation processes (liquefaction, explosion, collapse, crater-like depressions or submarine landslides). Geological and geophysical evidence and indicators of marine gas hydrates in the theoretical gas hydrate stability zone (GHSZ) were taken into account as the main factors controlling the susceptibility calculation. Svalbald, the Barents Sea, the mid-Norwegian margin-northwest British Islands, the Gulf of Cádiz, the eastern Mediterranean and the Black Sea have the highest susceptibility. Seafloor areas outside the theoretical GHSZ were excluded from this geohazard assessment. The uncertainty analysis of the susceptibility inference shows extensive seafloor areas with no data and a very low density of data that are defined as critical knowledge gaps.

Keywords: gas hydrates; European margins; geohazard assessment

Citation: León, R.; Llorente, M.; Giménez-Moreno, C.J. Marine Gas Hydrate Geohazard Assessment on the European Continental Margins. The Impact of Critical Knowledge Gaps. *Appl. Sci.* **2021**, *11*, 2865. https://doi.org/10.3390/app11062865

Academic Editor: Paraskevi Nomikou

Received: 2 March 2021
Accepted: 19 March 2021
Published: 23 March 2021

1. Introduction

Marine gas hydrates are crystalline solids forming ice-like marine deposits. They are composed of water molecules surrounding light hydrocarbon gases, such as methane (the most common), ethane and propane, in cage-like lattices [1]. They are common in shallow marine sediments (<1000 m bsf) below 350 mwd under high pressure, relatively low temperature and high hydrocarbon gas saturation in pore water conditions [2–4]. Bacterial methanogenesis, thermogenesis [5] and serpentinised oceanic crust [6,7] are the main source of CH_4 in continental margin sediment.

Marine gas hydrate is considered an important geohazard feature [8]. Three main environmental parameters control the nucleation and dissociation of marine methane hydrates: seafloor temperature, geothermal gradient and pressure [4]. Depressurization due to drops in sea level and warming of bottom water is the natural main scenario where hydrate dissociation can take place, driving large-scale natural gas release with potentially profound impacts, generating landslides, pockmarks, collapses, seafloor explosions and gas release [3,9]. These processes have also been hypothesized in the geological record [10]. However, under stable pressure/temperature conditions inside the gas hydrate stability zone (GHSZ), hydrocarbon seepage (such as pockmarks and gas flares) is likely to occur along fluid migration pathways of deep hydrocarbon reservoirs [11]. Nevertheless, these pressure-temperature conditions of shallow sediments may be modified by human activity on deep-water infrastructure such as wellheads, pipelines, production facilities, seabed anchors, cable touchdown areas on the seabed and catenaries in the water column [12].

Geological event inventories are a useful tool for regional risk analysis [13]. Global gas hydrate inventories have been used to make predictions about global hydrate volumes [14,15] and related risks [16], and to make impact projections based on future warm scenarios [17]. In the near future, marine gas hydrates will become a severe geohazard because of the unfavourable consequences of global warming on the marine gas hydrate stability field [11,17]. They will thus trigger seafloor instabilities in gas hydrate areas that are currently stable [17], discharging marine methane from shallow near-shore environments (0–50 m) to the atmosphere [18], lowering pH and causing geochemical changes in the water column due to aerobic oxidation [19]. However, moderate methane submarine emissions are absorbed by fragile chemosynthetic ecosystems that prosper in the vicinity of venting gas seeps [20,21]

Evidence of marine methane hydrates has been reported in eight main regions along the European continental margins, and in adjacent areas such as offshore Greenland and Svalbard, the Norwegian margin, offshore the northern British Islands, the southern Iberian and northwest African margins (the Gulf of Cádiz and Alborán Sea), and the Black, Marmara and eastern Mediterranean seas [22]. However, hydrate-related data (geological, geophysical and oceanographic) are not homogeneously covered in the whole extent of the European continental margins. This issue is especially important for obtaining hydrate-related predictions (e.g., creating a predictive—and quick—static and continuous model for the hydrate stability field along the whole of the European continental margins).

This paper presents, for the first time on the whole of the European margins and adjacent areas, a geohazard assessment (susceptibility analysis) of the presence of marine gas hydrates. It also assesses the main knowledge gaps of hydrate-related information with a pan-European scope, and analyses their impact on the uncertainty of susceptibility inference. Susceptibility is understood as the likelihood of the seafloor to be affected by the presence of hydrate deposits.

2. Geological Setting

The study area offers a wide view of the European margins from Macaronesia (SW corner: 24°15′ N; 36°10′ W) to the Black Sea and the Barents Sea (NE corner: 60°40′ E, 90° N). This hydrate framework covers three great domains on the European continental margins: the Arctic, the northeast Atlantic Sea and the south European Alpine Belt (Figure 1).

Hydrate systems in the Arctic province are located on the west Greenland, Svalbard and western Barents Sea margins. The east Greenland and west Svalbard margins were created during the Cretaceous to Paleogene continental rifting [23]. East Greenland is bordered by a wide continental shelf and deep basins. Gas seepage, bottom-simulating reflector (BSR) levels and pore water anomalies related to offshore Mesozoic sedimentary basins have been reported, associated with thermogenic gas migration along fractures [24]. The west Svalbard margin is composed of glacigenic debris flows, turbidites, hemipelagic sediments and contourites [25], where gas flares, seepages and BSR levels take place [26]. Hydrates are mainly represented in mud volcanoes in south Svalbard (e.g., Håkon–Mosby mud volcano—HM in Figure 1; [27]). Hydrocarbon gases are mainly thermogenic and locally biogenic [28], but abiogenic [29] contributions have been reported.

The Barents Sea and northern Norwegian margins are composed of a complex structure of sedimentary basins (e.g., Møre and Vøring; MB and VB in Figure 1, respectively) and structural highs resulting from the Cenozoic rifting [30]. The sedimentary basins are filled mainly with fine-grained hemipelagic sediments (Miocene–Pliocene) and glacigenic debris flows and contourites (Plio-Pleistocene) [31]. The Storegga Slide (SS in Figure 1) affected a huge sediment volume (mainly of the Møre Basin) as a response to climatic variability, 8200 ya [32]. In the Barents Sea, hydrate indicators are mainly BSR levels, gas chimneys and seepage pipes associated with vertical fluid flow systems and shallow gas.

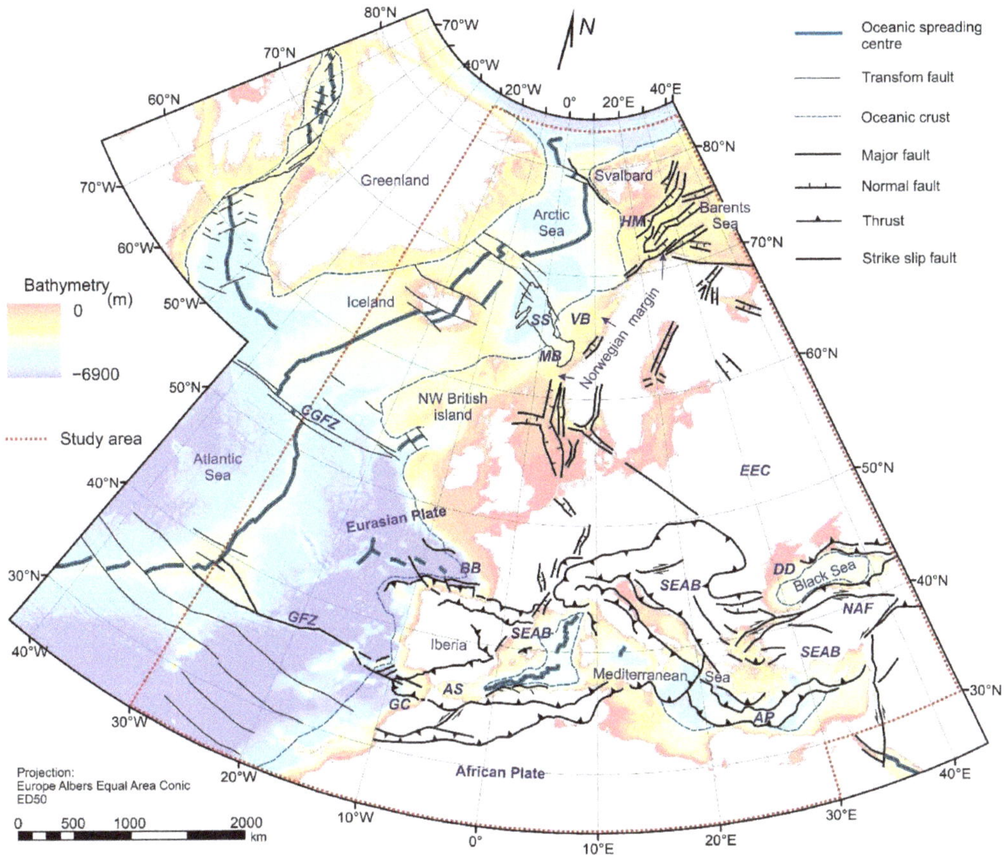

Figure 1. Main tectonic structures and geological domains on the European continental margins and adjacent areas. Location of the study area. HM, Håkon–Mosby mud volcano; VB, Vøring Basin; MB, Møre Basin; SS, Storegga Slide; CGFZ, Charlie–Gibbs fracture zone; EEC, East European craton; Biscay Bay; GFZ, Gloria Fracture Zone; GC, Gulf of Cádiz; AS, Alborán Sea; SAEB, south European Alpine Belt, NAF, North Anatolian Fault; DD, Danube delta fan. (Taken from [23,33,34]).

However, no hydrates have been obtained. The nature of gases is mostly thermogenic, migrating through faults and fractures [35]. On the mid-Norwegian margin, hydrate samples were recovered during the TTR-16 cruise [36], and the indicators are BSR levels, bright spots, gas chimneys and pockmarks [37], all of them circumscribed to the header of the Storegga Slide. Here, gases have a microbial origin with thermogenic components [38].

The northwest British Islands margins (Figure 1) are composed of basins with thick Cenozoic series [39]. Several seepages and gas chimneys are present on the upper slope, controlled by unconformities and fractures and sourced from the Upper Carboniferous and Middle and Upper Jurassic successions [40]. No hydrates have been recovered.

The hydrate province of the south European Alpine Belt (SEAB in Figure 1) is located on the southern Iberian and northwest African margins, in the eastern Mediterranean and in the Black Sea. These areas are located in the context of the Alpine orogeny, owing to the convergence between the African and Eurasian plates [41]. On the southern Iberian and northwest African margins and in the Gulf of Cádiz and Alborán Sea (GC and AS in Figure 1), hydrate samples have been recovered in mud volcanoes [42,43], associated with other hydrocarbon fluid flow structures such as pockmarks and hydrocarbon-derived authigenic carbonate (HDAC) [44]. This fluid flow is controlled by fractures linked to a deep-rooted mud diapirism in the allochthonous unit of the Gulf of Cádiz (AUGC in Figure 1; [45]). Hydrocarbon gases have three origins: thermogenic, mixed thermogenic/biogenic on the subsurface [46] and abiogenic [7].

In the eastern Mediterranean Sea, the main seabed fluid flow areas are the accretionary complex and the Nile delta (AP and ND in Figure 1, respectively), where multiple mud volcanoes, pockmark fields and broad degassing areas with chemosynthetic fauna and authigenic carbonates are present [47]. Hydrates have only been observed in mud volcanoes along the accretionary complex [48]. The potential sources for hydrocarbon are related to late Messinian and Miocene to recent sapropels [49]. In mud volcanoes, gas has a thermogenic signature, while in pockmarks, the signature is predominantly microbial methane [50]. In the Marmara Sea (MS in Figure 1), a pull-apart basin onshore of the North Anatolian Fault (NAF in Figure 1) [51], hydrates of thermogenic origin related to seismic indicators (e.g., bright spots and transparent and chaotic zones) [52], gas flares, mud volcanoes and pockmarks [53] have been acquired.

The Black Sea (2212 mwd) is an extensional, mostly anoxic back-arc basin that contains the largest hydrogen sulphide and methane reservoirs in the world [22]. It is composed of two sub-basins, the eastern and western ones [54]. The evidence and indicators of marine gas hydrates are BSR levels, seismic blanking, bright reflections, pockmarks and mud volcanoes. The northwestern part is dominated by organic-rich delta fan complexes (the Danube [DD in Figure 1] and Dniepr rivers), where gas flares (microbial origin) and BSR levels have been observed linked to their canyon and levee systems [55]. Gas hydrates have been recovered in mud volcanoes related to gas chimneys, active faulting and diapirism in deep areas of the western basin, as well as in shallow (upper-middle slope) areas in the southern and eastern parts [55,56]. In these areas, gas shows a mixed thermogenic and microbial composition in the subsurface.

3. Data Source and Methods

This study used the hydrate-related GIS database of the GARAH project 731166, GeoERA-GE-1, H2020 Environment (https://geoera.eu/projects/garah4/; accessed on 20 December 2020). This GIS database (GARAH*ydrates*; [57]) is INSPIRE-compliant and stores hydrate-related geological, geophysical and oceanographic information (Figure 2). It is the result of a data collection from two main groups of data: (i) data of a pan-European scope from free public databases and project results, such as EMODnet, PERGAMON and MIGRATE; and (ii) data of a regional scope from scientific organizations. The source of the data, owner and person/institution of contact is stored in each database record.

The majority of the geological and geophysical evidence and indicators of marine gas hydrates of GARAH*ydrates* came from the results of Work Package 1 of the MIGRATE COST action—ES1405 (https://www.migrate-cost.eu/; data given by MIGRATE COST Action to GARAH project on 31 January 2019) led by the University of Southampton and the National Oceanographic Centre, in which 21 organizations from 15 countries were involved. The MIGRATE database contains 1892 records (vector and raster) and stores information regarding direct and indirect evidence of gas hydrates. The data on direct evidence of gas hydrates are from samples described in publications. The data on indirect evidence include seismic indicators such as BSR levels and areas, gas chimneys, high reflectivity areas and velocity anomalies. Other gas hydrate information includes seabed features (gas seepages areas), heat flow data, sediment thickness models, pore water anomalies, theoretical models of the base of the GHSZ, and relief and bathymetry models.

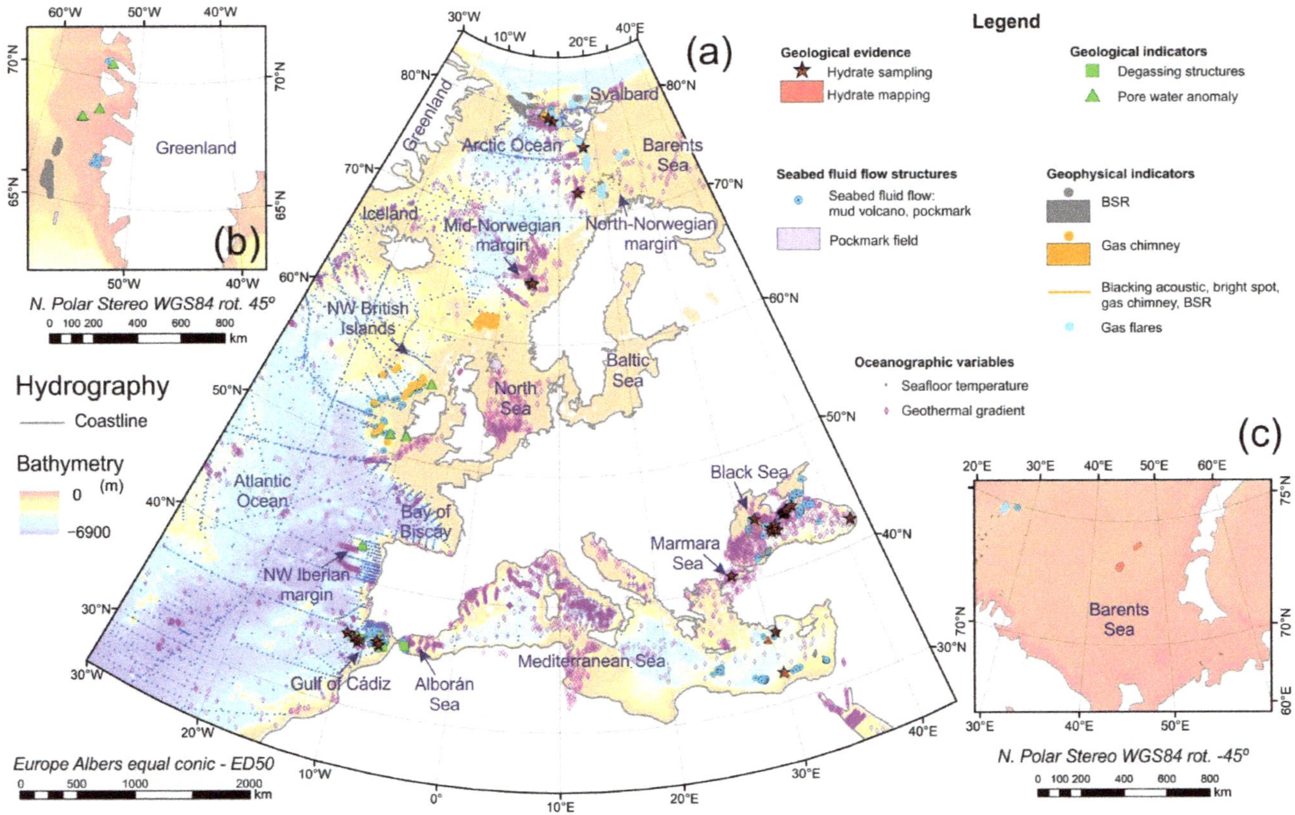

Figure 2. Geological, geophysical and oceanographic data sets used in the present paper. Marine gas hydrate evidence and indicators, and oceanographic variables stored in the GARAH*ydrates* data base. (**a**) Study area. (**b**) West Greenland. (**c**) Barents Sea.

EMODnet Geology marine minerals (https://www.emodnet-geology.eu/map-viewer/?p=marine_minerals; accessed on 1 January 2019) supplied 28 records in polygon shapefile format regarding marine gas hydrate evidence in the Gulf of Cádiz, the Barents Sea and the Black Sea. The PERGAMON database was developed by the geological surveys of Spain (IGME) and Ireland (GSI) in 2011 and 2012 in the framework of the PERGAMON COST action—ES0902 (https://www.cost.eu/actions/ES0902/; data given by PERGAMON Cost Action to GARAH project on 1 January 2019). This database supplied seafloor temperature data and theoretical models of the thickness of the GHSZ in the Arctic Sea.

The geothermal gradient data were obtained from the global heat flow database of the International Heat Flow Commission (website: http://engineering.und.edu/geology-and-geological-engineering/globe-heat-flow-database/index.cfm; accessed on 31 March 2020). The data were downloaded with the ODV application (https://odv.awi.de/; accessed on 1 February 2019). Seafloor temperature is a composite dataset developed by the Geological Survey of Spain (IGME) using CTD data downloaded from the World Ocean Database (https://www.ncei.noaa.gov/products/world-ocean-database; accessed on 1 February 2019) and the British Oceanographic Data Centre (https://www.bodc.ac.uk/; data given by British Geological Survey to GARAH project on 30 November 2018). Finally, bathymetry was obtained from three sources: the EMODnet Bathymetry portal (https://www.emodnet-bathymetry.eu/; accessed on 31 March 2020), IBCAO (https://www.gebco.net/data_and_products/gridded_bathymetry_data/arctic_ocean/; accessed on 1 February 2019) and GEBCO (https://www.gebco.net/; accessed on 31 March 2020).

Several records were added or updated using regional data from scientific organizations: (i) British Geological Survey (BGS) technical reports for geophysical indicators in the north of the British Islands [58–60]; (ii) the BGS 250k map series/MCA Civil Hydrography Prog data for pockmark mapping; (iii) marine gas hydrate evidence in the Black Sea from SRDE-Geoinform of Ukraine; and (iv) regional models of the base of the GHSZ for CH_4 and/or CO_2 in the Biscay Bay from the Bureau de Recherches Géologiques et Minières (BRGM).

The thickness of the GHSZ was taken from Núñez-Varela [61]. The hydrate stability field has been calculated with the CSMHYD program [62] for a standard composition of biogenic gas [63] and a salinity assumption of 36 psi for all the study area.

4. Results

4.1. Hydrate-Related Information Stored in GARAHydrates

The hydrate-related information is structured in four levels inside GARA*Hydrates*: (i) geological and geochemical evidence and indicators, (ii) geophysical indicators, (iii) seabed fluid flow structures, and (iv) oceanographic variables (Figure 2). Four types of items describe the information: location items, property metadata, geo-descriptors and references/comments (Table 1). Location items describe the geographical location (coordinates, geological setting, etc.). Property-reference metadata store the owner of the data and contact information. Geo-descriptors describe the geological, geochemical and geophysical characteristics of the evidence or indicator. Finally, references/comments store bibliographic references and other comments of interest of each item of evidence or indicator.

The level of information "geological and geochemical evidence and indicators" stores evidence (e.g., crystals of gas hydrates) and indicators (e.g., degassing structures and pore water anomalies) of gas hydrates acquired by direct sampling. The level "geophysical indicators" stores seismic or electric features of gas hydrate presence in the sediment column, such as high resistivity, BSR levels, bright spots, acoustic blanking facies and gas chimneys. The level "seabed fluid flow structures" stores structures related to fluid migration in areas where evidence or indicators of marine gas hydrates have been observed. Finally, the level "oceanographic variables" stores information about seafloor temperature, geothermal gradient and bathymetry.

Table 1. Description of the attributes (items) of the GARAH*ydrates* GIS database. NN, not null; LV, list of values.

Set of Items	Field Name	Description
Principal key	ID_IndiNa	Unique identification code for each record
Location items	Lat_DD	Latitude in decimal degrees (WGS84)
	Long_DD	Longitude in decimal degrees (WGS84)
	WaterDepth	Seafloor depth (metres water depth)
	GeoSettin	Geographical/geological Setting—list of values
	LocalSite	Local site where the evidence is located
Property metadata	Data_Sourc	Institution/company if owner of data. Project, database or publication where data were collected
	Cruise	Oceanographic cruise where data were recovered or observed
	CName	Contact name
	Email	Contact Email
Geo-descriptors	E_I	E = direct evidence; I = indirect indicator—constraints: NN
	FF_Type	Type of evidence or indicator—constraints: NN; LV: FF_Type
	Descripti	Description of the evidence—free text
	Sedi_Type	Sediment type—LV
	D_Indi_mtp	Depth of the top of the evidence below seabed in metres
	D_Indi_mbt	Depth of the bottom of the evidence below seabed in metres
	Size	Size (volume, km^2, tons, etc.)
References and comments	DOI	DOI of main data publication
	Reference	References to data. Author, Year and Title. Link to PDF in data repository
	Comments	Comments—free text

More than 136,000 records of hydrate samples, seismic indicators and seabed fluid flow structures have been stored (Table 2). In west Greenland, no hydrates were recovered; there are only six indirect items of evidence or indicators, such as pore water anomalies of chloride. However, BSR levels (~9400 km^2) and gas flares and seepages (~3500 km^2) have been mapped. West Greenland–Svalbard–Barents Sea is, together with the Black Sea, one of the most extensive hydrate regions in the study area, with more than 2100 km^2 and 26,300 km^2 of mapped hydrates and BSR levels (58), respectively, as well as numerous gas flares and seepages. Although the mid-Norwegian margin is a very localized hydrate region, it is the only site where hydrates have been recovered on the northwest Atlantic European margins. Here, gas hydrates are linked to the old slumped slope (Storegga), BSR levels and seepage area. On the southern Iberian and northwest African margins and in the Mediterranean Sea, hydrates have only been recovered on mud volcanoes (in the Gulf of Cádiz and eastern Mediterranean Sea). A high number of hydrate samples have been recovered in these regions as a result of many oceanographic cruises by scientific groups. The Black Sea is the hydrate region that has most gas hydrate surfaces mapped (more than 3650 km^2), with more than 80 mud volcanoes on the seafloor, 91% of them inside the GHSZ.

Finally, three information layers are stored in the oceanographic variable group: seafloor temperature (5896 records), geothermal gradient (4332 records) and bathymetry (composite raster dataset with a cell size ~100×100 m).

Table 2. Geological and geophysical evidence and indicators of marine gas hydrates stored in GARAH*ydrates*.

Geological Settings	Geological Evidence				Geophysical Indicators									
	Hydrates		Degassing Structures	Pore Water Anomalies	BSRs		High Res.	Gas Chimneys		Acoustic Blanking	Gas Flares	Seabed Structures		
	Samples	Km^2/Levels			Loc. Sites	Km^2/Levels		Loc. Sites	Km^2/Levels			Pocks-Seeps Loc. Sites	Pocks-Seeps km^2/Levels	Mud Volcanoes
West Greenland				6		9410/2					5		3565/1	2 mud diapir
East Greenland—Svalbald—Barents Sea	4	2163/4 *E.* Barents S.			14	26,316/58		16	3110/95		65	1007		1
Mid-Norwegian margin	6					4278/9						76		
northeast Atlantic				8	139			24		139		36	58,273/42	
Southern Iberian and northwest African margins	22 Gulf of Cádiz		10		1			7		4		44	233/1 Alborán	63
Eastern Mediterranean Sea	2					2360/4						24		59
Marmara Sea	1											512		2
Black Sea	23	3655/7			15	15,058/14	4					31		102

4.2. Susceptibility Assessment

The presence of gas hydrates in marine sediments is a geohazard that has not yet been evaluated in the whole of the European continental margins. This work uses the database of marine gas hydrate evidence and indicators developed in the GARAH project to make a pan-European assessment of hydrate presence on its continental margins. This assessment was carried out in two steps: analysis and weighting of factors and susceptibility calculation.

4.2.1. Analysis and Weighting of Factors

Several factors were taken into account in this assessment: marine gas hydrate evidence, seismic indicators, seabed fluid flow structures and thickness of the GHSZ.

Evidence of marine gas hydrates is the ground truth of where hydrate exists in the seafloor and/or sub-seafloor. This layer establishes seafloor areas with a moderate–high likelihood of occurrence of dissociation processes in the seafloor or sub-seafloor. The magnitude of the processes will depend on the quantity of hydrate in the sedimentary column and its type within the sediment (massive, in layers, disseminated, etc.).

Seismic indicators show seafloor areas where hydrates could exist. Marine gas hydrates have not been recovered, but there is a moderate to high likelihood of them occurring. In areas where marine gas hydrates have been recovered by direct sampling (geological evidence), seismic indicators allow hydrate occurrence to be inferred.

Seafloor geological structures related to hydrocarbon fluid migration, such as pockmarks, gas flares, mud volcanoes and HDAC are directly linked to deep hydrocarbon reservoirs. The occurrence of these structures reveals a free fluid leakage from the sedimentary structure to the water column. These processes can occur inside the GHSZ, in some cases due to preferential fluid migration from deep reservoirs, and in other cases, as a result of hydrate dissociation processes. This layer of information will thus establish a wide spectrum of the susceptibility of presence of marine gas hydrates and their hazardousness. It will range from low or void in areas outside the GHSZ to moderate to high in areas inside the GHSZ.

The thickness of the GHSZ establishes the theoretical seafloor area where the occurrence of hydrates is physically possible under optimal gas saturation and salinity conditions. Seafloor areas inside the GHSZ will be considered potential areas to be affected by dissociation processes. In addition, the intersection between the base of the GHSZ and the seafloor will be considered a potential strip of the high likelihood of fluid leakage and dissociation processes. The three oceanographic variables taken into account in the thickness calculation were seafloor temperature, geothermal gradient and bathymetry.

Each geological and geophysical item of evidence and indicator was weighted according to the confidence/certainty of finding hydrates at the site. The maximum weight (or confidence) was given to recovered samples of gas hydrates or evidence of hydrate dissociation, such as degassing or liquidation structures in gravity cores. Seismic indicators of the presence of gas hydrates or hydrocarbon seabed fluid flow such as BSRs, acoustic blanking, amplitude anomalies and the presence of geological structures of seabed fluid flow in the vicinity of the GHSZ were weighted with a lower value, between 0.8 and 0.9, based on expert criteria (Table 3).

Regarding the theoretical GHSZ, the seafloor was weighted in three categories. Seafloor areas outside the theoretical GHSZ were excluded as not likely to be affected by hydrate dissociation processes. On the other hand, any location inside the GHSZ was selected as theoretically likely to suffer dissociation processes. A strip at the up-dip limit of the GHSZ (50 m in thickness) was a critical area for these dissociation processes (Figure 3).

Table 3. Weights given to each hydrate-related item of evidence or indicator for the development of the density map of evidence/indicators.

Evidence and Indicator	Description	Weight
Gas hydrate samples	Crystals or aggregates of gas hydrates observed in gravity cores.	1
Degassing structures	Bubbles and/or vacuoles (porosity) in sediment liquefactions observed in gravity core samples.	1
Pore water anomalies	Chemical and isotopic pore water anomalies that are caused by hydrate dissociation (e.g., downward chlorinity decrease combined with δ18O increase).	0.9
BSRs	Only bottom-simulating reflectors generated by the impedance contrast between the gas-hydrated sediment above and a free gas layer below. Opal BSRs are excluded.	0.9
High resistivity	Anomalous high electrical resistivity in logs due to the presence of massive hydrates.	0.9
Velocity anomalies	Anomalous seismic propagation velocity in the sediment due to the presence of both gas hydrates and free gas.	0.8
Acoustic blanking facies	Zones devoid of reflections in seismic profiles because of the presence of free gas in the sediment.	0.8
Dim spots	Local low-amplitude seismic attribute anomalies that may indicate the presence of hydrocarbons.	0.8
Bright spots	Seismic amplitude or high-amplitude anomalies that may indicate the presence of hydrocarbons.	0.8
Gas chimneys	Areas of poor data quality or push-downs caused by subsurface leakage of gas from a poorly sealed hydrocarbon accumulation.	0.8
Seabed features	Geomorphological features related to seabed fluid flow.	0.8
Gas seepage: pockmarks, mud volcanoes, gryphons	Steady or episodic, slow or rapid, visible or invisible flow of gaseous hydrocarbons from subsurface sources to the Earth's surface. Pockmarks are craters in the seabed caused by fluids erupting and streaming through the sediments. Mud volcanoes are positive cone-shaped reliefs created by the extrusion of mud, water and gases, mainly hydrocarbon fluids.	0.8
Mud diapirs	Positive, cone-shaped reliefs created by intrusion inside the sediment column of mud, water and gases, mainly hydrocarbon fluids.	0.8
HDAC	Hydrocarbon-derived authigenic carbonate formed as a consequence of the anaerobic oxidation of methane by consortia of microbes.	0.8
Gas flares	Acoustic artefacts in the water column caused by gas bubbles.	0.8

4.2.2. Susceptibility Calculation

The proposed methodology analyses the geological hazard by means of the susceptibility assessment. The term "susceptibility" is employed here to define the likelihood of occurrence of hydrates in the sediment column, and subsequently the likelihood of them being affected by dissociation processes resulting from natural or human-induced activities (liquefaction, explosions, collapse, crater-like depressions or submarine landslides). Susceptibility assessment is applied as the first step in a pan-European risk assessment, owing in particular to (i) the regional scope of the assessment (the European continental margins and adjacent areas) and (ii) the current state of European gas hydrate-related information characterized by intensively studied areas with a high density of high-quality data and wide areas of critical knowledge gaps with no data.

The baseline scenario (the initial hypothesis) is that gas hydrate occurrence is only possible in seafloor areas where pressure (bathymetry) and seafloor temperature conditions are inside the theoretical GHSZ. In this zone, the occurrence of gas hydrates is directly related to the presence of evidence (direct samples of hydrates) or indicators of it (e.g., pore water and velocity anomalies, BSRs and gas chimneys), as well as the occurrence of hydrocarbon fluid flow structures. Finally, the likelihood of the seafloor being affected by gas hydrate dissociation processes will be great at the base of the GHSZ and in the vicinity of gas hydrate evidence and indicators.

In order to prove this initial hypothesis, a susceptibility assessment was carried out through map algebra in a GIS environment from a density map of evidence and indicators and the pan-European map of the GHSZ on the seafloor.

The first step for the development of the density map was to create a lattice of evidence and indicators at a resolution of the work scale (5 × 5 km), which will be the final resolution of the susceptibility assessment map. In this lattice, each geographical feature of evidence and indicators was weighted according to Table 3.

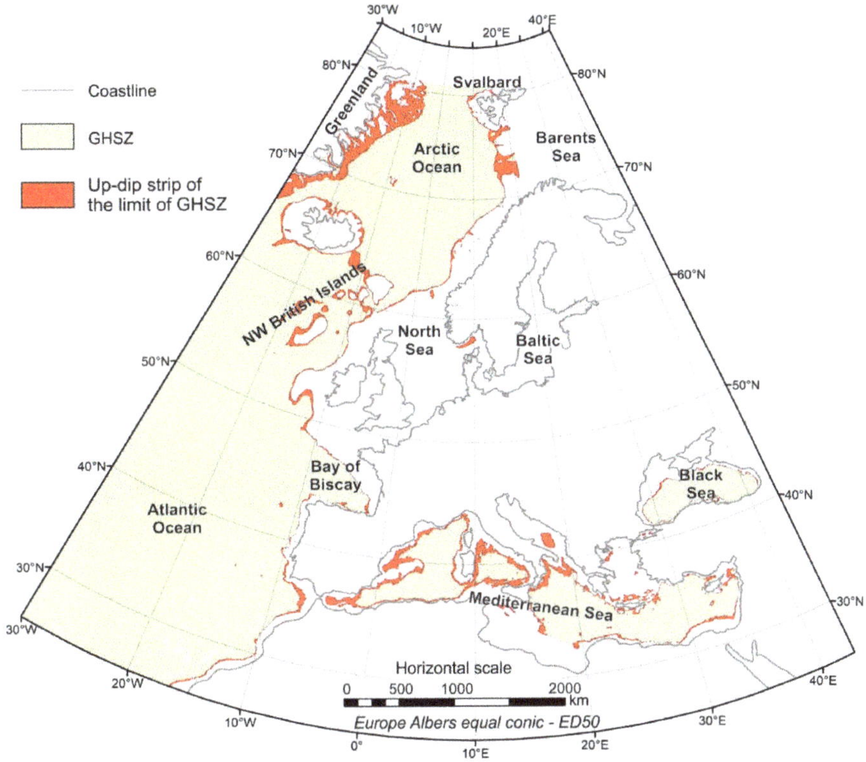

Figure 3. Theoretical up-dip limit of the GHSZ in the study area (modified from Núñez-Varela, 2020). In the volcanic area of Macaronesia (the Canary and Madeira islands), the up-dip limit has been eliminated because of the absence of hydrocarbon reservoirs on the flanks of the islands.

A hypothesis was then established that considered the database of ground evidence for sites that have been sampled, but occurrences of gas hydrates might not be restricted to these point locations. If a given pixel were located between a ground evidence and an indicator, the likeliness of that pixel containing gas hydrates would be greater than that of a pixel located far from either. Given the discrete nature of the features described within the database and the relative concept of this likeliness, a regionalization technique was applied following a smoothed saturated algorithm of kernel density. Here, the weighting of the features represents an abstract concept of the confidence of having gas hydrates. This technique consists of a kernel density estimation, which fits a smoothly tapered surface to each point or polyline. The search radius (default option) was calculated on the basis of the spatial configuration and the number of input points. This approach corrects for spatial outliers (input points that are very far from the rest), so they will not make the search radius unreasonably large. These values of the weighted density map were then normalized from zero to one (Figure 4) to remove the per area output of the model, providing a relative likeliness of between 0 and 1, where 0 means that there is very little (but not no) confidence of finding significant amounts of gas hydrates at that location and 1 means that there is certainty of finding gas hydrates at that location.

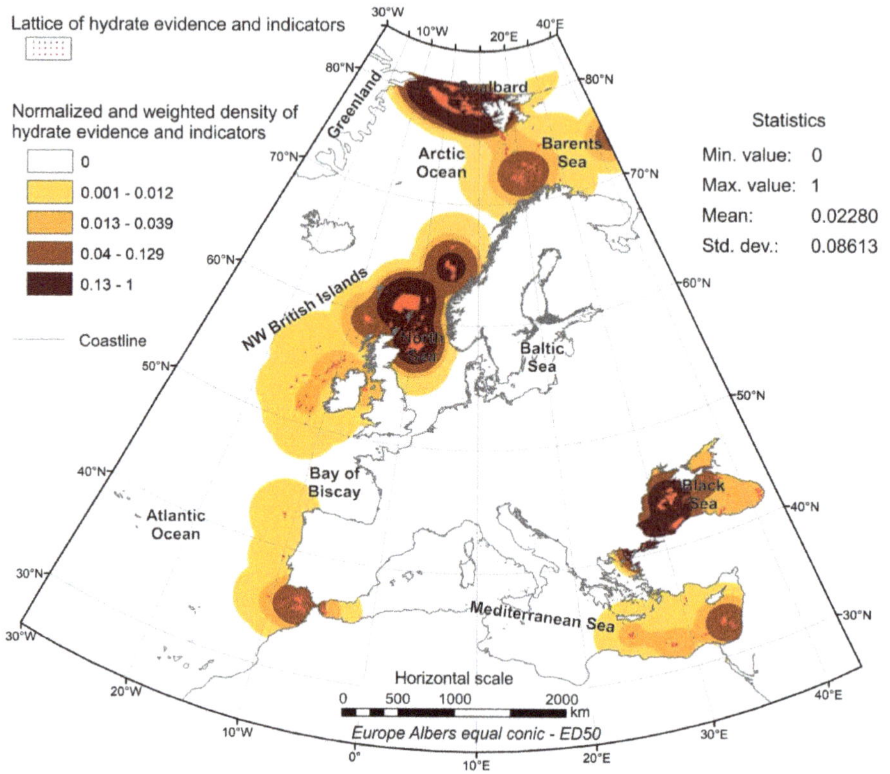

Figure 4. Normalized (zero to one) and weighted density map of hydrate evidence and indicators. Lattice of hydrate evidence and indicators overlapped, red dots. Density map developed with the "kernel density" algorithm of ArcGIS®. Parameters: population field, weight (taken from Table 3); cell size, 5000; method, geodesic.

Regarding the weighted map of the theoretical GHSZ (Figure 3), the up-dip limit of the GHSZ in the vicinity of the low-latitude volcanic islands in the Atlantic Ocean (i.e., the Azores, Madeira and the Canary Islands) was not taken into account because of the absence of hydrocarbon reservoirs at these geological sites. Finally, this map was weighted in relation to the mean value of the normalized density map of evidence and indicators (mean = 0.00228; Figure 4). Thus, according to its likelihood of being affected by dissociation processes, the GHSZ on the seafloor was weighted using expert criteria. The strip of up-dip of the GHSZ was weighted with 0.00228 and the rest of the GHSZ with 0.00114 (half the likelihood). Seafloor areas outside the GHSZ were given a value of zero.

The susceptibility assessment was performed by map algebra, taking into account the control maps of density of hydrate evidence and indicators and the weighted map of the GHSZ on the seafloor. Specifically, the final map (Figure 5) was conceived as a segmentation in three levels by quantiles resulting from the addition of the above control maps:

$$Sc = \delta_{ei} + GHSZ_w \quad (1)$$

where Sc is the susceptibility map; δ_{ei} is the normalized weighted density map of hydrate evidence and indicators; and $GHSZ_w$ is the weighted map of the GHSZ on the seafloor. The final Sc value was masked with the positive values of the GHSZ map. Seafloor areas outside the GHSZ have a susceptibility value of zero.

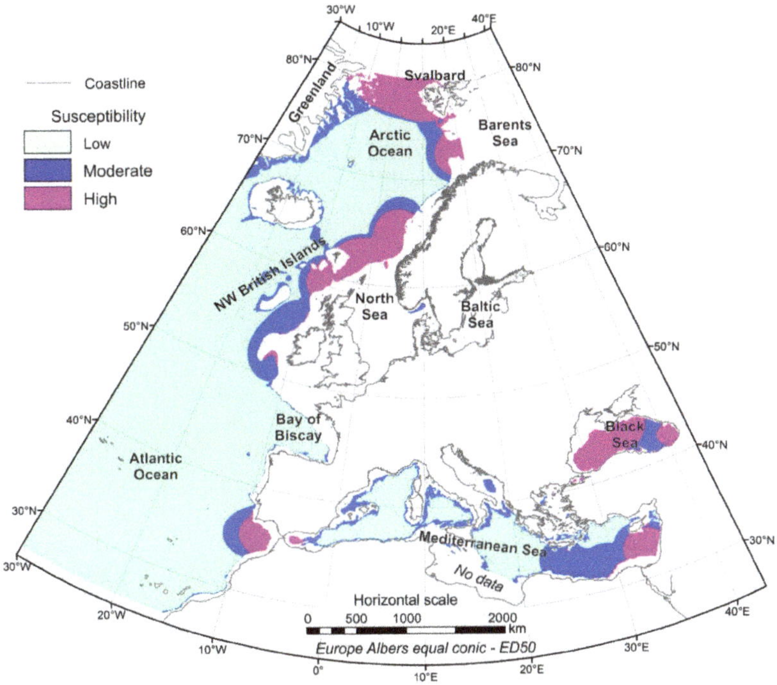

Figure 5. Susceptibility assessment of the seafloor to the presence of hydrates on the European continental margins and adjacent areas.

Susceptibility values were segmented into three levels by quantiles: low from 0.009 to 0.0129, middle from 0.0129 to 0.0325 and high from 0.0325 to 1.009. The susceptibility assessment shows seven areas with high values: Svalbard, the northern Norwegian margin—Barents Sea, the continental slope of the mid-Norwegian margin and the North Sea, the Gulf of Cádiz, the eastern Mediterranean and the Black Sea. Moderate values are located on the west Greenland continental shelf, near the northwest British Islands and on the continental slope of the western and northern Mediterranean Sea.

5. Discussion

This section analyses the pan-European database of susceptibility of presence of marine gas hydrates from a geohazard point of view, considering the impact of spatial data distribution on the uncertainty value and on the identification of critical knowledge gaps.

5.1. Hydrate-Related Knowledge Gaps

Nucleation and dissociation of marine methane hydrates are directly controlled by the presence of hydrocarbon gas solubility in the sediment pore water and three environmental parameters: seafloor temperature, geothermal gradient and pressure (water depth) [3]. However, free public information about these key parameters shows a non-homogeneous continuity along the European continental margins. This issue is especially critical for understanding the behaviour of the GHSZ or making predictions or calculations on it. In particular, it is essential for assessments related to geohazards and risks, assessments of the abundance of sediment-hosted gas hydrates, and assessments of the role of CO_2-rich hydrates in the geological storage of CO_2. The issue is also of broad interest to the scientific community: petroleum geologists, biologists and ecologist working on vulnerable ecosystems, researchers on natural hazards and tsunamis, civil engineers and policy makers.

We therefore assessed the critical knowledge gaps in the geological and geophysical evidence and indicators and the oceanographic variables taken into account in the calculation of the GHSZ thickness. This assessment was carried out along density maps of seafloor temperature, geothermal gradient and hydrate evidence and indicators. Owing to the regional (pan-European) scope, an area unit was established at a sedimentary basin scale of 100,000 km^2 (a searching radius of ca. 178.4 km) and a surface resolution of 5 km × 5 km.

Evidence of marine methane hydrates has been reported in eight main regions along the European continental margins (Figure 2): offshore Greenland and Svalbard, the Norwegian margin, offshore the northern British Islands, the southern Iberian and northwest African margins (the Gulf of Cádiz and Alborán Sea), and the Black, Marmara and eastern Mediterranean seas. These areas show a high density of high-quality data resulting from several scientific oceanographic cruises and an intensive scientific survey. Outside the limits of these areas, the lack of evidence and indicators is obvious. In our opinion, the reliability of this lack of evidence is controversial (Figure 6). Although the majority of European continental margins have been prospected for the oil industry, some deep ocean basin areas have not. Therefore, some areas with a lack of evidence (possibly located in deep ocean basins) could be treated as information gaps resulting from a lack of prospection or scientific fluid flow research.

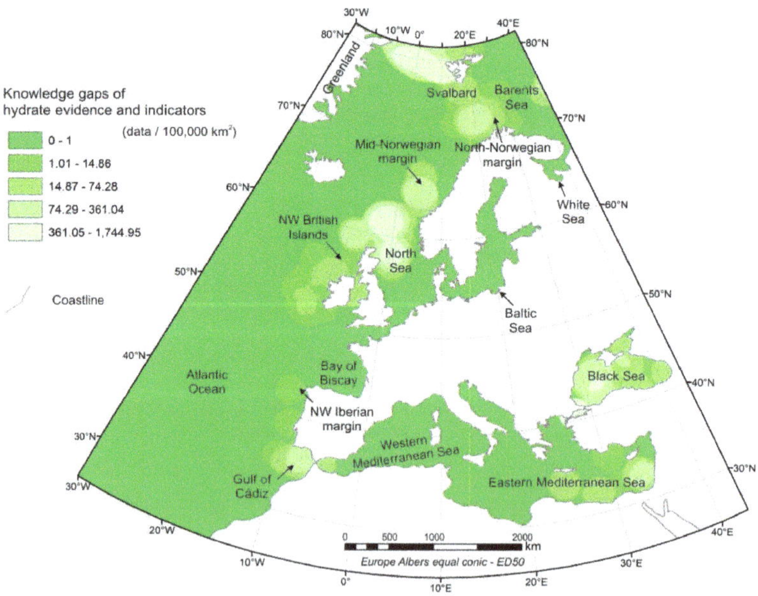

Figure 6. Knowledge gap assessment of hydrate evidence and indicators. Density map developed with the "point density" algorithm of ArcGIS®. Pixel value, number of items of evidence and indicators of hydrates per 100,000 km^2. Parameters: population field, none; cell size, 5000; radius, 178,415 metres; areal units, square kilometres; method, geodesic.

Seafloor temperature and marine geothermal data have a heterogeneous distribution. Marine geothermal data appear to be concentrated with high density in some of the above-mentioned eight main regions with hydrate evidence surveyed by scientific cruises (Figure 7a). On the other hand, seafloor temperature data, the most sensitive variable in the theoretical calculation of the base of the GHSZ [63], are especially concentrated in the Black Sea and on the eastern Atlantic continental shelf (Figure 7b). For the two above datasets, areas with less than 1 record per 100,000 km^2 were selected as knowledge gaps (KG in Figure 7). These knowledge gaps are especially critical (i) in areas where direct hydrate samples have been recovered, (ii) in the vicinity of the up-dip limit of the GHSZ, and (iii) in areas where seabed fluid flow structures have been detected. The critical knowledge gaps for geothermal gradient data are east of Greenland, Svalbard–Barents Sea, the White Sea, northwest of the British Islands and the southeastern Mediterranean Sea; and for seafloor temperature they are east Greenland, the western Barents and White seas, the northern Black Sea and the southeastern Mediterranean Sea (north of Libya). In addition, because of the critical need to understand geothermal gradient in areas that have a relatively high spatial variance, high-resolution coverage is critical, in particular in order to assess the potential for uncertainty predictions for similar areas with no data.

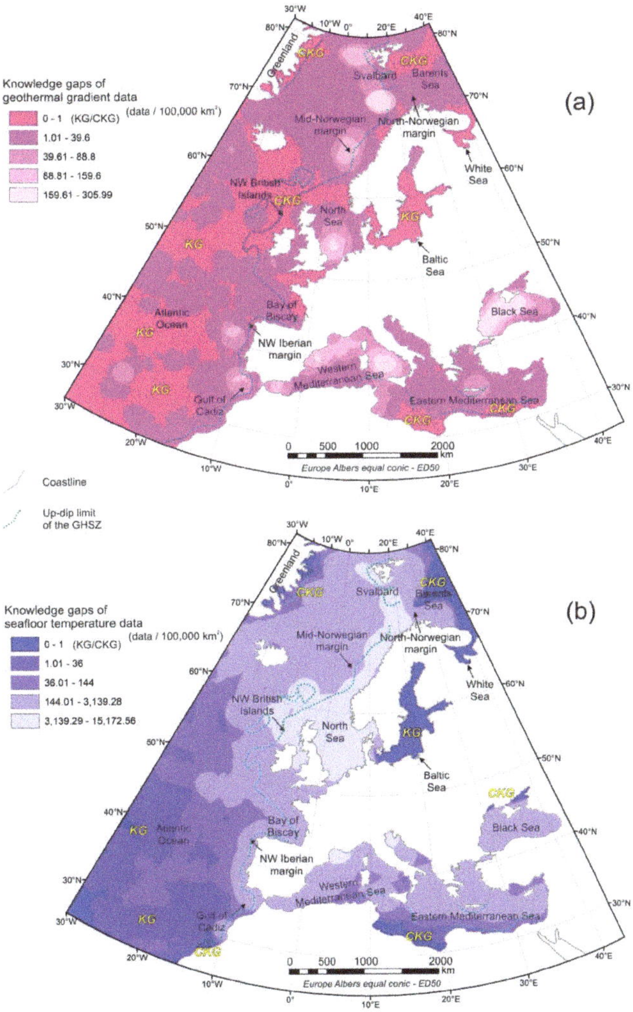

Figure 7. (**a**) Knowledge gap analysis of geothermal gradient data. KG, knowledge gap of geothermal gradient data; CKG, critical knowledge gap of geothermal gradient data. (**b**) Knowledge gap analysis of seafloor temperature data. KG, knowledge gap of seafloor temperature data; CKG, critical knowledge gap of seafloor temperature data. Density maps developed with the "point density" algorithm of ArcGIS®. Pixel value, number of data per 100,000 km^2. Parameters: population field, none; cell size, 5000; radius, 178,415 metres; areal units, square kilometres; method, geodesic.

In general, the public bathymetry data collected (EMODnet Bathymetry and IBCAO) have a quite acceptable quality and have been very useful for the objectives of this hydrate-related pan-European study. The original grid has a cell size of 100 × 100 m and the inference was calculated with a cell size of 5 × 5 km.

The EMODnet Bathymetry dataset has a quality value stored for each pixel/bathymetry datum calculated from derived bathymetric parameters: minimum water depth in metres to the lowest astronomical tide (LAT), average water depth in metres to the LAT, maximum water depth in metres to the LAT, standard deviation of water depth in metres, number of values used for interpolation over the grid cell, the interpolation flag (identification of extrapolated cells), average water depth smoothed by means of a spline function in metres to the LAT, an indicator of the offsets between the average and smoothed water depth as a % of the water depth, and a reference to the prevailing source of data with metadata. Unfortunately, this information (per pixel) is not available on the web portal. Consequently, quantifiable targets for calculating knowledge gaps are not available. However, through a visual analysis, areas with poor accuracy or lack of data were selected, especially on the North African Mediterranean margins (e.g., north of Libya). These areas were classified as bathymetry knowledge gaps (KG in Figure 8b,c). However, owing to the resolution of the inference (5 × 5 km), no critical gaps were determined for the bathymetric data.

5.2. Reliability of the Susceptibility Assessment

In areas where hydrate evidence and indicators have been reported, the assumption of a salinity of 36 psi for the theoretical GHSZ calculated by Núñez-Varela [61] along the study area involves an error in the thickness calculation of ±2–10 m in the Arctic region (33–35 psi), ±5 m in the Mediterranean Sea (38 psi) and ±30 m in the Black Sea (17 psi) [64]. Nevertheless, considering the regional scale of the susceptibility assessment (cell size of 5 × 5 km), these errors are acceptable in the thickness calculation because they lay inside the vertical precision of the assessment. The error made by the difference in salinity in the theoretical GHSZ is lower than the vertical precision of this calculation due to the variation of the bathymetry each 5 km along the continental slope. In particular, ±15–120 m in the Arctic region, and ±100–200 m in both the Mediterranean and Black Sea. Regarding the Black Sea, although errors coming from salinity are still acceptable, the susceptibility resulting should be taken with caution, as values could be higher.

In order to assess the reliability of the susceptibility inference, a qualitative value of uncertainty (very high, high, middle, low and very low) was established as a function of the data density taken into account in the susceptibility calculation (Figure 9). The reliability (u) is thus equal to the sum of the density maps of geothermal gradient (ρ_{gr}), seafloor temperature (ρ_{st}) and hydrate evidence and indicators (ρ_{hy}):

$$u = \rho_{gr} + \rho_{st} + \rho_{hy} \tag{2}$$

Five levels of reliability were established. The reliability is considered "very low" with values from 0 to 36 data per 100,000 km^2, approximately less than ca. 1 datum per 50 km in mean; and "low", "middle", "high" and "very high" from 36 to 144, from 144 to 648, from 648 to 3149, and from 3149 to 15,218 data per 100,000 km^2, respectively. These levels were defined by the geometrical segmentation of u-value, except "very low" and "low", which were defined by expert criteria. Very low reliability areas were catalogued as global knowledge gaps (KG in Figure 9) that are critical (CKG) in the vicinity of the up-dip limit of the GHSZ and hydrocarbon seabed fluid flow structures.

Areas located in the proximity of the continental shelf and intensively surveyed by oceanographic cruises show the most reliable results in the susceptibility assessment. By contrast, areas distant from the coastline (e.g., the mid-Atlantic Ocean) and areas that are inaccessible because of the presence of icebergs (e.g., east Greenland) or political issues (e.g., north of Libya) have very high uncertainty.

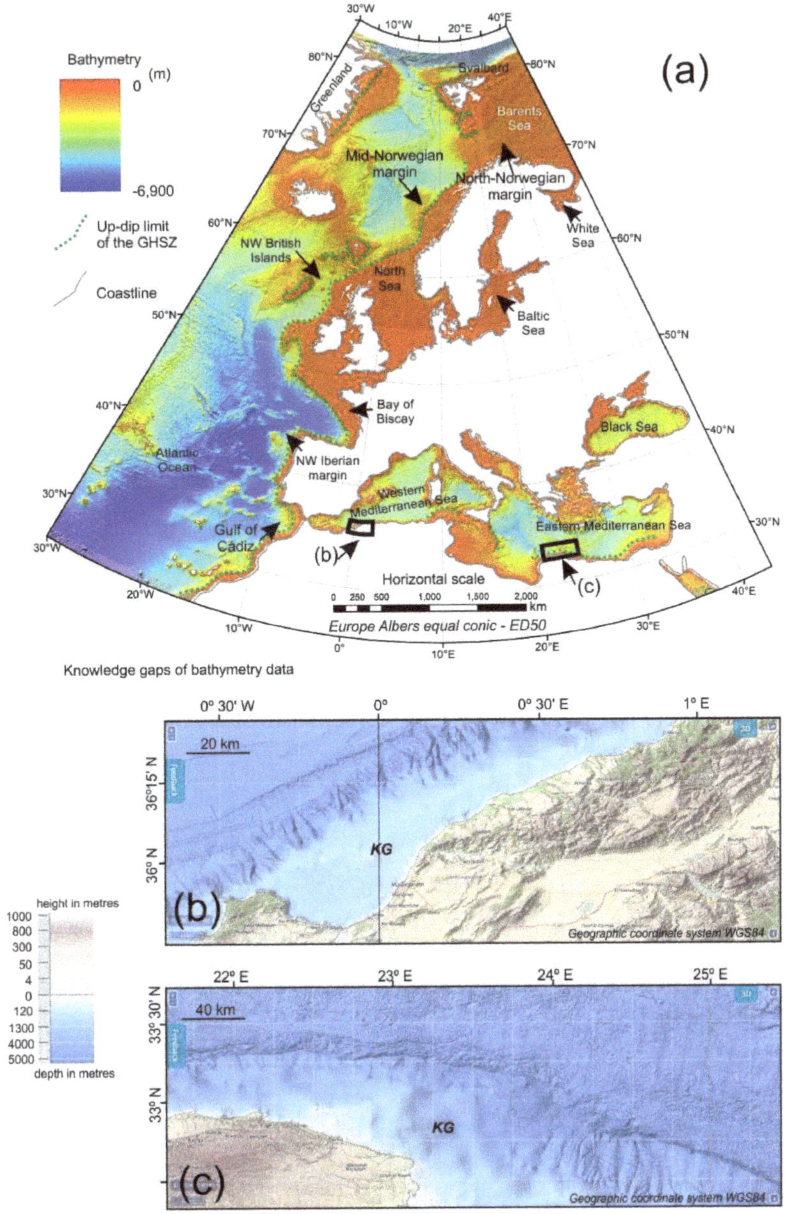

Figure 8. Knowledge gaps of bathymetry data. (**a**) EMODnet Bathymetry mosaic in the study area (cell size ca. 200 × 200 m). (**b**) Detail of knowledge gap on the Algerian margin. (**c**) Detail of knowledge gap on the Libyan margin. Details downloaded from the EMODnet Bathymetry web portal (https://portal.emodnet-bathymetry.eu/; accessed on 1 January 2021).

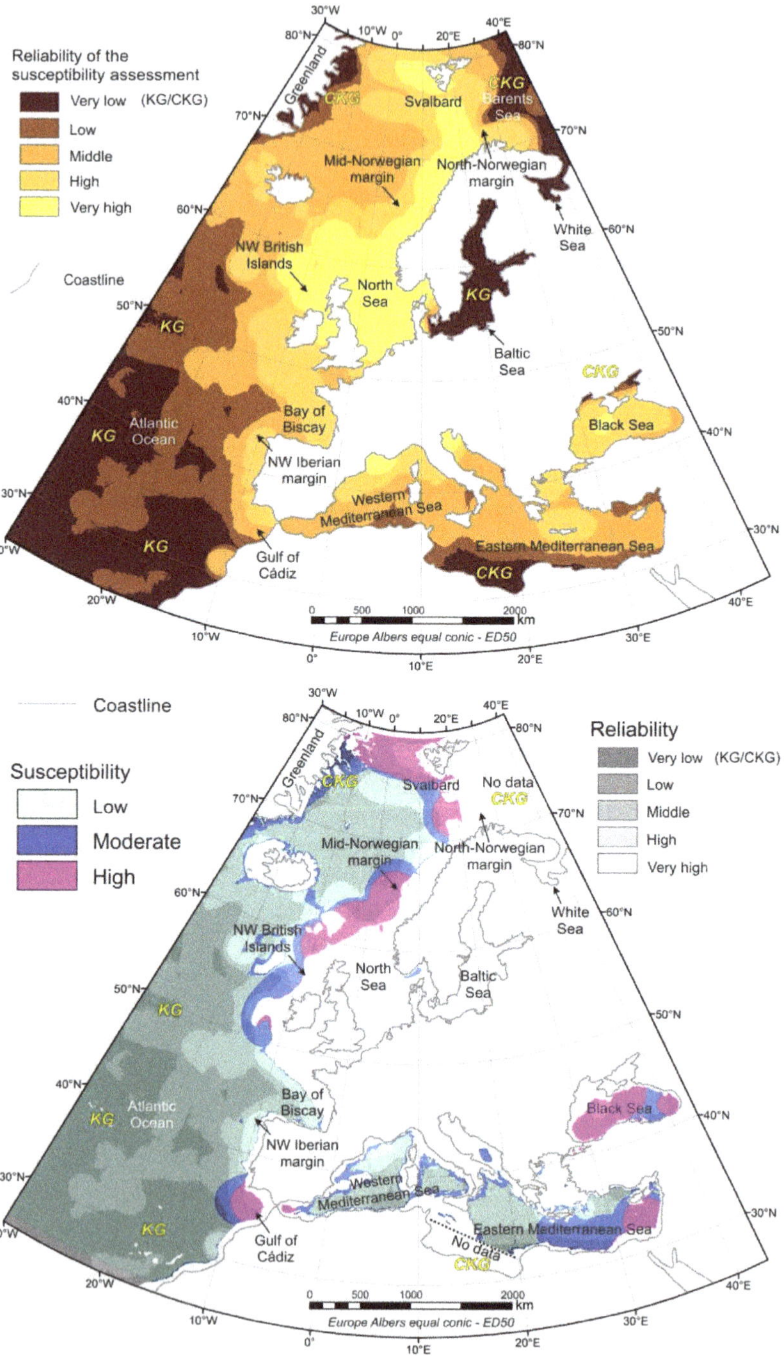

Figure 9. Reliability of the susceptibility assessment of hydrate presence on the European continental margins. (**a**) Reliability prediction in the inference of the susceptibility of the seafloor to the presence of hydrates. KG, global knowledge gap; CKG, global critical knowledge gap. (**b**) Reliability and susceptibility assessments overposted.

5.3. Spatial Significance of the Susceptibility Assessment and the Impact of Knowledge gaps

Owing to the methodology applied, the hydrate evidence knowledge gaps are directly related to low values of susceptibility. These knowledge gaps may have two possible causes: (i) the catalogue is incomplete, these areas have been poorly surveyed, no records have been recovered, but hydrates may exist and subsequently a high susceptibility may be potentially latent; and (ii) there are no data because there is no evidence of hydrates. In order to solve this uncertainty, two concepts are added in the susceptibility assessment: the up-dip limit of the GHSZ and the presence of hydrocarbon seabed fluid flow structures. Particularly, examples of this situation are the east Greenland shelf, the Irish margin, the western Iberian margin and the western Mediterranean Sea, where no hydrates have been recovered but hydrocarbon seabed fluid flow structures and seismic indicators (e.g., on the Irish margin) have been observed.

High susceptibility values are located in areas with a high density of evidence and indicators. The majority of gas hydrate evidence stored in the database was recovered in focused seabed fluid flow structures such as mud volcanoes or pockmarks. This is especially significant on the southern European margins in the Gulf of Cádiz and the eastern Mediterranean and Black seas. In these cases, gas hydrates are circumscribed to the feeder systems of the hydrocarbon fluid migration structures, which, subject to certain exceptions, do not exceed 0.1 to 1 km and 4 km in diameter for pockmarks and mud volcanoes, respectively. In these areas, there is therefore no continuous spatial variation in the presence of hydrates. Gas hydrates appear with a located distribution (nugget effect?) and focused inside the hydrocarbon fluid flow structures where fluid migration is mainly controlled by faults [45,65]. However, the presence of hydrocarbon fluid flow structures shows a continuous spatial variation in fluid leakage areas. In these areas, the density map shows areas where hydrate-bearing fluid flow structures are more probable and, subsequently, the likelihood of the seafloor suffering gas hydrate dissociation processes as a result of natural or human activities could also be high. Finally, although the susceptibility could be high in mud volcano fields, for instance, the real risk or magnitude of dissociation processes will be low because of the typology or internal structure of hydrates inside the sediment. In mud volcanoes, hydrates constitute small (millimetres or centimetres) crystals or aggregates, and their real volume is low.

Moderate susceptibility values seem to be controlled by the GHSZ and in particular by the optimal theoretical environmental conditions for hydrate presence on the continental shelves of the Arctic region and Mediterranean Sea. In our opinion, the presence of moderate values on the eastern continental shelf of Greenland and their absence on the western Norwegian shelf is directly related to the presence of cooler bottom water masses on the eastern continental shelf of Greenland and the subsequent influence on the theoretical GHSZ. Although no hydrates have been recovered in the Mediterranean Sea, owing to the particular seafloor temperature/pressure conditions (bathymetry) on the continental slope, this area has a slightly elevated likelihood of occurrence of hydrate dissociation processes in the hypothetical presence of hydrocarbon gases in the sediment column.

The future global warming scenario projected by the scientific community [66–68] increases the susceptibility assessed by the direct effect on the gas hydrate stability of the ocean temperature increase [69] and the isostatic rebound in polar and sub-polar areas [9]. This direct effect (seafloor temperature increase and effective seafloor pressure decrease) will have a high impact on the eastern Greenland shelf, the northwest Norwegian margin, Svalbard and the Barents Sea, and subsequently the susceptibility in these areas will increase greatly. Furthermore, future changes in the thermohaline circulation [67,68] could have dramatic effects at high latitudes on the seafloor temperature and subsequently on hydrate stability.

6. Conclusions

Geological and geophysical evidence and indicators of the presence of marine gas hydrates and the oceanographic variables controlling the GHSZ show a heterogeneous distribution and knowledge gaps (areas with <1 records per 100 km^2) along the European continental margins. Some of these knowledge gaps have been classified as critical: (i) for geothermal gradient, data on the east of Greenland, Svalbard, the northern Norwegian margin, the southern Barents Sea and the White Sea, the north of the British Islands, the Gulf of Cádiz, the Bay of Biscay, the north-western Iberian margin and the southwestern Mediterranean Sea; and (ii) for seafloor temperature, data on east of Greenland, the western Barents Sea and White Sea, the northern Black Sea and the south-eastern Mediterranean Sea.

The susceptibility assessment of occurrence of hydrate dissociation processes on the seafloor shows high values in Svalbard, the northern Norwegian margin—Barents Sea, the continental slope of the mid-Norwegian margin and the North Sea, the Gulf of Cádiz and the eastern Mediterranean and the Black Sea. Moderate values are observed on the continental shelf of western Greenland, the northwest of the British Islands and the continental slope of the western and northern Mediterranean Sea.

Author Contributions: Conceptualization and writing—original draft preparation, R.L. and C.J.G.-M.; methodology, validation, formal analysis and writing—review and editing, R.L., M.L. and C.J.G.-M. All authors have read and agreed to the published version of the manuscript.

Funding: This research was funded by the European Union's Horizon 2020 research and innovation programme under grant agreement No 731166, GARAH project (GeoERA- GeoE.171.002 GE-1), EMODnet Bathymetry—High Resolution Seabed Mapping (EASME/EMFF/2018/007).

Data Availability Statement: Data used in this paper are available in a public and permanent repository (https://data.mendeley.com/datasets/vbt6hspgpn/draft?preview=1, (accessed on 20 March 2020)) with doi:10.17632/vbt6hspgpn.1.

Acknowledgments: We thank C. Guardiola, A. Lounds, the WP3 team of the GARAH project (P. Mata, C. Rochelle, A. Burnol, T. Nielsen, J. Hopper, I. Reguera, M. Stewart, S. Cervel and U. Larsen), and the MIGRATE COST Action coordination and WP1 teams (K. Wallmann, S. Bünz, T. Minshull, H. Marín-Moreno, J. Schicks., J. Bialas, G. Cifci, M. Giustiniani, J. Hopper, V. Magalhaes, Y. Makovsky, M.D. Max, T. Nielsen, S. Okay, I. Ostrovsky, N. O'Neill, A. Plaza-Faverola, C. Rochelle, S. Roy, K. Schwalenberg, K. Senger, S. Vadakkepuliyambatta, A. Vasilev).

Conflicts of Interest: The authors declare no conflict of interest. The funders had no role in the design of the study; in the collection, analysis or interpretation of data; in the writing of the manuscript; or in the decision to publish the results.

References

1. Paull, C.K.; Dallimore, S.R.; Caress, D.W.; Gwiazda, R.; Melling, H.; Riedel, M.; Jin, Y.K.; Hong, J.K.; Kim, Y.-G.; Graves, D.; et al. Active mud volcanoes on the continental slope of the Canadian Beaufort Sea. *Geochem. Geophys. Geosystems* **2015**, *16*, 3160–3181. [CrossRef]
2. Boswell, R.; Yamamoto, K.; Lee, S.-R.; Collett, T.; Kumar, P.; Dallimore, S. Chapter 8—Methane Hydrates. In *Future Energy*, 2nd ed.; Letcher, T.M., Ed.; Elsevier: Boston, MA, USA, 2014; pp. 159–178; ISBN 978-0-08-099424-6.
3. Sloan, E.D. Fundamental principles and applications of natural gas hydrates. *Nature* **2003**, *426*, 353–359. [CrossRef]
4. You, K.; Flemings, P.B.; Malinverno, A.; Collett, T.S.; Darnell, K. Mechanisms of Methane Hydrate Formation in Geological Systems. *Rev. Geophys.* **2019**, *57*, 1146–1196. [CrossRef]
5. Dickens, G.R. The potential volume of oceanic methane hydrates with variable external conditions. *Org. Geochem.* **2001**, *32*, 1179–1193. [CrossRef]
6. Fryer, P.; Wheat, C.G.; Mottl, M.J. Mariana blueschist mud volcanism: Implications for conditions within the subduction zone. *Geology* **1999**, *27*, 103–106. [CrossRef]
7. Hensen, C.; Scholz, F.; Nuzzo, M.; Valadares, V.; Gràcia, E.; Terrinha, P.; Liebetrau, V.; Kaul, N.; Silva, S.; Martínez-Loriente, S.; et al. Strike-slip faults mediate the rise of crustal-derived fluids and mud volcanism in the deep sea. *Geology* **2015**, *43*, 339–342. [CrossRef]
8. Ruppel, C.D.; Kessler, J.D. The interaction of climate change and methane hydrates. *Rev. Geophys.* **2017**, *55*, 126–168. [CrossRef]
9. Wallmann, K.; Riedel, M.; Hong, W.L.; Patton, H.; Hubbard, A.; Pape, T.; Hsu, C.W.; Schmidt, C.; Johnson, J.E.; Torres, M.E.; et al. Gas hydrate dissociation off Svalbard induced by isostatic rebound rather than global warming. *Nat. Commun.* **2018**, *9*, 83. [CrossRef]
10. Argentino, C.; Conti, S.; Fioroni, C.; Fontana, D. Evidences for Paleo-Gas Hydrate Occurrence: What We Can Infer for the Miocene of the Northern Apennines (Italy). *Geosciences* **2019**, *9*, 134. [CrossRef]

11. Serov, P.; Vadakkepuliyambatta, S.; Mienert, J.; Patton, H.; Portnov, A.; Silyakova, A.; Panieri, G.; Carroll, M.L.; Carroll, J.; Andreassen, K.; et al. Postglacial response of Arctic Ocean gas hydrates to climatic amelioration. *Proc. Natl. Acad. Sci. USA* **2017**, *114*, 6215–6220. [CrossRef] [PubMed]
12. McConnell, D.R.; Zhang, Z.; Boswell, R. Review of progress in evaluating gas hydrate drilling hazards. *Mar. Pet. Geol.* **2012**, *34*, 209–223. [CrossRef]
13. Hervás, J.; Bobrowsky, P. Mapping: Inventories, Susceptibility, Hazard and Risk. In *Landslides—Disaster Risk Reduction*; Sassa, K., Canuti, P., Eds.; Springer: Berlin/Heidelberg, Germany, 2009; pp. 321–349; ISBN 978-3-540-69970-5.
14. Kvenvolden, K.A.; Lorenson, T.D. The Global Occurrence of Natural Gas Hydrate. In *Natural Gas Hydrates: Occurrence, Distribution, and Detection*; American Geophysical Union (AGU): Washington, DC, USA, 2001; pp. 3–18; ISBN 978-1-118-66841-2.
15. Wallmann, K.; Pinero, E.; Burwicz, E.; Haeckel, M.; Hensen, C.; Dale, A.; Ruepke, L. The Global Inventory of Methane Hydrate in Marine Sediments: A Theoretical Approach. *Energies* **2012**, *5*, 2449–2498. [CrossRef]
16. Liu, F.; Che, Y.; Xu, S.; Chen, Q.; Tan, L. Managing Information of Gas Hydrate Reservoirs of the South China Sea in an Integrated GIS Database. In *Proceedings of the GeoShanghai 2018 International Conference: Geoenvironment and Geohazard*; Farid, A., Chen, H., Eds.; Springer: Singapore, 2018; pp. 37–44.
17. Maslin, M.; Owen, M.; Betts, R.; Day, S.; Dunkley Jones, T.; Ridgwell, A. Gas hydrates: Past and future geohazard? *Philos. Trans. R. Soc. Math. Phys. Eng. Sci.* **2010**, *368*, 2369–2393. [CrossRef] [PubMed]
18. Weber, T.; Wiseman, N.A.; Kock, A. Global ocean methane emissions dominated by shallow coastal waters. *Nat. Commun.* **2019**, *10*, 4584. [CrossRef] [PubMed]
19. Reeburgh, W.S. Oceanic Methane Biogeochemistry. *Chem. Rev.* **2007**, *107*, 486–513. [CrossRef] [PubMed]
20. Bernardino, A.F.; Levin, L.A.; Thurber, A.R.; Smith, C.R. Comparative Composition, Diversity and Trophic Ecology of Sediment Macrofauna at Vents, Seeps and Organic Falls. *PLoS ONE* **2012**, *7*, e33515. [CrossRef]
21. Ristova, P.P.; Wenzhöfer, F.; Ramette, A.; Felden, J.; Boetius, A. Spatial scales of bacterial community diversity at cold seeps (Eastern Mediterranean Sea). *ISME J.* **2015**, *9*, 1306–1318. [CrossRef]
22. Minshull, T.A.; Marín-Moreno, H.; Betlem, P.; Bialas, J.; Bünz, S.; Burwicz, E.; Cameselle, A.L.; Cifci, G.; Giustiniani, M.; Hillman, J.I.T.; et al. Hydrate occurrence in Europe: A review of available evidence. *Mar. Pet. Geol.* **2020**, *111*, 735–764. [CrossRef]
23. Oakey, G.N.; Chalmers, J.A. A new model for the Paleogene motion of Greenland relative to North America: Plate reconstructions of the Davis Strait and Nares Strait regions between Canada and Greenland. *J. Geophys. Res. Solid Earth* **2012**, *117*. [CrossRef]
24. Nielsen, T.; Laier, T.; Kuijpers, A.; Rasmussen, T.L.; Mikkelsen, N.E.; Nørgård-Pedersen, N. Fluid flow and methane occurrences in the Disko Bugt area offshore West Greenland: Indications for gas hydrates? *Geo-Mar. Lett.* **2014**, *34*, 511–523. [CrossRef]
25. Mattingsdal, R.; Knies, J.; Andreassen, K.; Fabian, K.; Husum, K.; Grøsfjeld, K.; De Schepper, S. A new 6 Myr stratigraphic framework for the Atlantic–Arctic Gateway. *Quat. Sci. Rev.* **2014**, *92*, 170–178. [CrossRef]
26. Dumke, I.; Burwicz, E.B.; Berndt, C.; Klaeschen, D.; Feseker, T.; Geissler, W.H.; Sarkar, S. Gas hydrate distribution and hydrocarbon maturation north of the Knipovich Ridge, western Svalbard margin. *J. Geophys. Res. Solid Earth* **2016**, *121*, 1405–1424. [CrossRef]
27. Vogt, P.R.; Cherkashev, G.A.; Ginsburg, G.D.; Ivanov, M.K.; Crane, K.; Lein, A.; Sundvor, E.; Pimenov, N.; Egorov, A.V. *EOS, Transaction, American Geophysican Union-Wiley Online Library*; Wiley: Hoboken, NJ, USA, 1997; pp. 556–557.
28. Plaza-Faverola, A.; Vadakkepuliyambatta, S.; Hong, W.-L.; Mienert, J.; Bünz, S.; Chand, S.; Greinert, J. Bottom-simulating reflector dynamics at Arctic thermogenic gas provinces: An example from Vestnesa Ridge, offshore west Svalbard. *J. Geophys. Res. Solid Earth* **2017**, *122*, 4089–4105. [CrossRef]
29. Johnson, J.E.; Mienert, J.; Plaza-Faverola, A.; Vadakkepuliyambatta, S.; Knies, J.; Bünz, S.; Andreassen, K.; Ferré, B. Abiotic methane from ultraslow-spreading ridges can charge Arctic gas hydrates. *Geology* **2015**, *43*, 371–374. [CrossRef]
30. Doré, A.G. Barents Sea Geology, Petroleum Resources and Commercial Potential. *Arctic* **1995**, *48*, 207–221. [CrossRef]
31. Rise, L.; Ottesen, D.; Berg, K.; Lundin, E. Large-scale development of the mid-Norwegian margin during the last 3 million years. In *Ormen Lange—An Integrated Study for Safe Field Development in the Storegga Submarine Area*; Solheim, A., Bryn, P., Berg, K., Sejrup, H.P., Mienert, J., Eds.; Elsevier: Oxford, UK, 2005; pp. 33–44; ISBN 978-0-08-044694-3.
32. Bryn, P.; Berg, K.; Forsberg, C.F.; Solheim, A.; Kvalstad, T.J. Explaining the Storegga Slide. *Mar. Pet. Geol.* **2005**, *22*, 11–19. [CrossRef]
33. Asch, K. The 1:5 Million International Map of Europe and Adjacent Areas—IGME 5000 2005, Bundesanstalt für Geowissenschaften und Rohstoffe (BGR). Available online: https://download.bgr.de/bgr/Geologie/IGME5000/pdf/igme5000.zip (accessed on 20 March 2020).
34. Woudloper. *Tectonic Map of Southern Europe, North Africa and the Middle East, Showing Tectonic Structures of the Western Alpine Mountain Belt*. 2009. Available online: https://commons.wikimedia.org/wiki/File:Tectonic_map_Mediterranean_EN.svg (accessed on 20 March 2020).
35. Tasianas, A.; Bünz, S.; Bellwald, B.; Hammer, Ø.; Planke, S.; Lebedeva-Ivanova, N.; Krassakis, P. High-resolution 3D seismic study of pockmarks and shallow fluid flow systems at the Snøhvit hydrocarbon field in the SW Barents Sea. *Mar. Geol.* **2018**, *403*, 247–261. [CrossRef]
36. Akhmetzhanov, A.M.; Kenyon, N.H.; Ivanov, M.K.; Westbrook, G.; Mazzini, A. *Deep-Water Depositional Systems and Cold Seeps of the Western Mediterranean, Gulf of Cadiz and Norwegian Continental Margins*; Preliminary results of investigations during the TTR-16 cruise of RV Professor Logachev, May–July 2006; United Nations Educational, Scientific and Cultural Organisation: Paris, France, 2008.

37. Bünz, S.; Mienert, J.; Berndt, C. Geological controls on the Storegga gas-hydrate system of the mid-Norwegian continental margin. *Earth Planet. Sci. Lett.* **2003**, *209*, 291–307. [CrossRef]
38. Vaular, E.N.; Barth, T.; Haflidason, H. The geochemical characteristics of the hydrate-bound gases from the Nyegga pockmark field, Norwegian Sea. *Org. Geochem.* **2010**, *41*, 437–444. [CrossRef]
39. Shannon, P.M.; Moore, J.G.; Jacob, A.W.B.; Makris, J. Cretaceous and Tertiary basin development west of Ireland. *Geol. Soc. Lond. Pet. Geol. Conf. Ser.* **1993**, *4*, 1057–1066. [CrossRef]
40. Rensbergen, P.V.; Rabaute, A.; Colpaert, A.; Ghislain, T.S.; Mathijs, M.; Bruggeman, A. Fluid migration and fluid seepage in the Connemara Field, Porcupine Basin interpreted from industrial 3D seismic and well data combined with high-resolution site survey data. *Int. J. Earth Sci.* **2007**, *96*, 185–197. [CrossRef]
41. Moores, E.M.; Fairbridge, R.W. (Eds.) *Encyclopedia of European and Asian Regional Geology*; Encyclopedia of Earth Sciences Series; Springer: Dordrecht, The Netherlands, 1997; ISBN 978-0-412-74040-4.
42. Kenyon, N.H.; Ivanov, M.K.; Akhmetzhanov, A.M.; Akhmanov, G.G. *Interdisciplinary Approaches to Geoscience on the North East Atlantic Margin and Mid-Atlantic Ridge: Preliminary Results of Investigations During the TTR-10 Cruise of RV Professor Logachev, July–September 2000*; IOC Technical Series; Unesco, United Nations Educational: Paris, France, 2001.
43. Kenyon, N.H.; Ivanov, M.K.; Akhmetzhanov, A.M.; Akhmanov, G.G. *Multidisciplinary Study of Geological Processes on the North East Atlantic and Western Mediterranean Margins: Preliminary Results of Geological and Geophysical Investigations During the TTR-9 Cruise of R/V Professor Logachev June–July, 1999*; IOC Technical Series; Unesco, United Nations Educational: Paris, France, 2000.
44. León, R.; Somoza, L.; Medialdea, T.; Maestro, A.; Díaz-del-Río, V.; Fernández-Puga, M. del C. Classification of sea-floor features associated with methane seeps along the Gulf of Cádiz continental margin. *Deep Sea Res. Part II Top. Stud. Oceanogr.* **2006**, *53*, 1464–1481. [CrossRef]
45. Medialdea, T.; Somoza, L.; Pinheiro, L.M.; Fernández-Puga, M.C.; Vázquez, J.T.; León, R.; Ivanov, M.K.; Magalhaes, V.; Díaz-del-Río, V.; Vegas, R. Tectonics and mud volcano development in the Gulf of Cádiz. *Mar. Geol.* **2009**, *261*, 48–63. [CrossRef]
46. Stadnitskaia, A.; Ivanov, M.K.; Blinova, V.; Kreulen, R.; van Weering, T.C.E. Molecular and carbon isotopic variability of hydrocarbon gases from mud volcanoes in the Gulf of Cadiz, NE Atlantic. *Mar. Pet. Geol.* **2006**, *23*, 281–296. [CrossRef]
47. Mascle, J.; Mary, F.; Praeg, D.; Brosolo, L.; Camera, L.; Ceramicola, S.; Dupré, S. Distribution and geological control of mud volcanoes and other fluid/free gas seepage features in the Mediterranean Sea and nearby Gulf of Cadiz. *Geo-Mar. Lett.* **2014**, *34*, 89–110. [CrossRef]
48. De Lange, G.J.; Brumsack, H.-J. The occurrence of gas hydrates in Eastern Mediterranean mud dome structures as indicated by pore-water composition. *Geol. Soc. Lond. Spec. Publ.* **1998**, *137*, 167–175. [CrossRef]
49. Merey, Ş.; Longinos, S.N. Does the Mediterranean Sea have potential for producing gas hydrates? *J. Nat. Gas Sci. Eng.* **2018**, *55*, 113–134. [CrossRef]
50. Römer, M.; Sahling, H.; Pape, T.; dos Santos Ferreira, C.; Wenzhöfer, F.; Boetius, A.; Bohrmann, G. Methane fluxes and carbonate deposits at a cold seep area of the Central Nile Deep Sea Fan, Eastern Mediterranean Sea. *Mar. Geol.* **2014**, *347*, 27–42. [CrossRef]
51. Armijo, R.; Meyer, B.; Hubert, A.; Barka, A. Westward propagation of the North Anatolian fault into the northern Aegean: Timing and kinematics. *Geology* **1999**, *27*, 267–270. [CrossRef]
52. Bourry, C.; Chazallon, B.; Charlou, J.L.; Pierre Donval, J.; Ruffine, L.; Henry, P.; Geli, L.; Çagatay, M.N.; İnan, S.; Moreau, M. Free gas and gas hydrates from the Sea of Marmara, Turkey: Chemical and structural characterization. *Chem. Geol.* **2009**, *264*, 197–206. [CrossRef]
53. Sarıtaş, H.; Çifçi, G.; Géli, L.; Thomas, Y.; Marsset, B.; Henry, P.; Grall, C.; Rochat, A. Gas occurrence and shallow conduit systems in the Western Sea of Marmara: A review and new acoustic evidence. *Geo-Mar. Lett.* **2018**, *38*, 385–402. [CrossRef]
54. Nikishin, A.M.; Okay, A.I.; Tüysüz, O.; Demirer, A.; Amelin, N.; Petrov, E. The Black Sea basins structure and history: New model based on new deep penetration regional seismic data. Part 1: Basins structure and fill. *Mar. Pet. Geol.* **2015**, *59*, 638–655. [CrossRef]
55. Egorov, V.N.; Artemov, Y.G.; Gulin, S.B.; Gennadiy, P.G. Methane seeps in the Black Sea: Discovery, quantification and environmental assessment. *J. Black SeaMediterranean Environ.* **2011**, *17*, 171–185.
56. Shnyukov, E.; Yanko-Hombach, V. *Mud Volcanoes of the Black Sea Region and Their Environmental Significance.*; Springer Nature: Cham, Switzerland, 2020; ISBN 978-3-030-40315-7.
57. Leon, R.; Rochelle, C.; Burnol, A.; Gimenez-Moreno, C.J.; Nielsen, T.; Hopper, J.; Reguera, I.; Mata, P.; Stewart, M.; Cervel, S. A pan-European GIS focused on gas hydrates: A research base-line in geohazards and geological storage of CO_2. In Proceedings of the Copernicus Meetings, EGU General Assembly 2020, Online, 4–8 May 2020. EGU2020-4861. [CrossRef]
58. Holmes, R. *Estimated Base of Methane Hydrate and Observations from Analogue Air-Gun Profiles West of Shetland*; British Geological Survey Technical Report WB/97/5C; British Geological Survey: Edinburgh, UK, 1997.
59. Holmes, R.; Finlayson, K.; Griffiths, M.A.; Andresen, P.C. *Regional Shallow Gas Study West of Shetland*; Report 95276/BGS Technical Report WB/96/45C; Hydrosearch Associates Ltd.; British Geological Survey: Edinburgh, UK, 1996.
60. Long, D. *Hydrates and Potential Existence in the Faeroe-Shetland Channel*; BGS Technical Report WB/96/34C; British Geological Survey: Edinburgh, UK, 1996.
61. Núñez-Varela, E. Cálculo del campo teórico de la zona de estabilidad de los hidratos de gas natural biogénico en los márgenes continentales europeos. *Open_igme* **2020**, *VII*, 85.
62. Sloan, E.D. *Clathrate Hydrates of Natural Gases*, 2nd ed.; Dekker: New York, NY, USA, 1998.

63. León, R.; Somoza, L.; Giménez-Moreno, C.J.; Dabrio, C.J.; Ercilla, G.; Praeg, D.; Díaz-del-Río, V.; Gómez-Delgado, M. A predictive numerical model for potential mapping of the gas hydrate stability zone in the Gulf of Cadiz. *Mar. Pet. Geol.* **2009**, *26*, 1564–1579. [CrossRef]
64. Sloan, E.D.; Koh, C.A. *Clathrate Hydrates of Natural Gases*; CRC Press, Taylor & Francis Group: Boca Raton, FL, USA; London, UK; New York, NY, USA, 2007; ISBN 978-1-4200-0849-4.
65. Argentino, C.; Conti, S.; Crutchley, G.J.; Fioroni, C.; Fontana, D.; Johnson, J.E. Methane-derived authigenic carbonates on accretionary ridges: Miocene case studies in the northern Apennines (Italy) compared with modern submarine counterparts. *Mar. Pet. Geol.* **2019**, *102*, 860–872. [CrossRef]
66. Gillett, N.P. Weighting climate model projections using observational constraints. *Philos. Trans. R. Soc. Math. Phys. Eng. Sci.* **2015**, *373*, 20140425. [CrossRef] [PubMed]
67. Schmittner, A.; Latif, M.; Schneider, B. Model projections of the North Atlantic thermohaline circulation for the 21st century assessed by observations. *Geophys. Res. Lett.* **2005**, *32*. [CrossRef]
68. Vellinga, M.; Wood, R.A. Global Climatic Impacts of a Collapse of the Atlantic Thermohaline Circulation. *Clim. Change* **2002**, *54*, 251–267. [CrossRef]
69. Ketzer, M.; Praeg, D.; Rodrigues, L.F.; Augustin, A.; Pivel, M.A.G.; Rahmati-Abkenar, M.; Miller, D.J.; Viana, A.R.; Cupertino, J.A. Gas hydrate dissociation linked to contemporary ocean warming in the southern hemisphere. *Nat. Commun.* **2020**, *11*, 3788. [CrossRef]

applied sciences

MDPI

Review

Recent Advancement in Assessment and Control of Structures under Multi-Hazard

Matin Jami [1], Rajesh Rupakhety [1,*], Said Elias [2], Bjarni Bessason [3] and Jonas Th. Snæbjörnsson [4]

1 Earthquake Engineering Research Centre, Faculty of Civil and Environmental Engineering, School of Engineering and Natural Science, University of Iceland, Austurvegur 2a, 800 Selfoss, Iceland; amj38@hi.is
2 Department of Construction Management and Engineering (CME), Faculty of Engineering Technology (ET), University of Twente (UTWENTE), 7522 NB Enschede, The Netherlands; elias.rahimi@utwente.nl
3 Faculty of Civil and Environmental Engineering, School of Engineering and Natural Science, University of Iceland, 102 Reykjavík, Iceland; bb@hi.is
4 Department of Engineering, School of Technology, Reykjavik University, 101 Reykjavík, Iceland; jonasthor@ru.is
* Correspondence: rajesh@hi.is; Tel.: +354-5254129

Abstract: This review presents an up-to-date account of research in multi-hazard assessment and vibration control of engineering structures. A general discussion of the importance of multi-hazard consideration in structural engineering, as well as recent advances in this area, is presented as a background. In terms of performance assessment and vibration control, various hazards are considered with an emphasis on seismic and wind loads. Although multi-hazard problems in civil engineering structures are generally discussed to some extent, the emphasis is placed on buildings, bridges, and wind turbine towers. The scientific literature in this area is vast with rapidly growing innovations. The literature is, therefore, classified by the structure type, and then, subsequently, by the hazard. Main contributions and conclusions from the reported studies are presented in summarized tables intended to provide readers with a quick reference and convenient navigation to related publications for further research. Finally, a summary of the literature review is provided with some insights on knowledge gaps and research needs.

Keywords: multi-hazard; earthquake; wind; flood; hazards; hurricane; mitigation; resilience; risk assessment; bridge; building; wind turbine; control system

check for updates

Citation: Jami, M.; Rupakhety, R.; Elias, S.; Bessason, B.; Snæbjörnsson, J.T. Recent Advancement in Assessment and Control of Structures under Multi-Hazard. *Appl. Sci.* **2022**, *12*, 5118. https://doi.org/10.3390/app12105118

Academic Editors: Miguel Llorente Isidro, Ricardo Castedo and David Moncoulon

Received: 21 March 2022
Accepted: 13 May 2022
Published: 19 May 2022

1. Introduction

Natural hazards, such as earthquakes and wind forces, pose a challenge for human safety and comfort. Forces generated by these natural processes can damage, or even collapse, vulnerable civil engineering structures. The risk to lives and properties posed by natural hazards increases with urbanization, where large cities and metropolitan areas get more and more densely populated. Increasing urbanization and shortage of land results in the need to build taller and more complex structures which can be more vulnerable to lateral forces created by wind and earthquakes. Effects of natural hazards on civil engineering structures is, therefore, an important field of research.

Between 1998 and 2017, natural disasters affected 4.4 billion people worldwide, caused 1.3 million casualties [1], and resulted in economic loss of 2900 billion USD. During this 20-year period, floods, storms, and earthquakes were the most frequent hazards, accounting for 43.4, 28.2, and 7.8% of all natural disasters, respectively. Although floods were the most frequent hazard during this time, earthquakes and storms have been the deadliest and the costliest, respectively. Floods and earthquakes, combined, killed nearly one million people and resulted in an economic loss of almost 2000 billion USD during this 20-year period. The frequencies, casualties, and economic losses, caused by different types of natural hazards between 1998 and 2017, are shown in Figure 1. The numbers in Figure 1, which are based on CRED report [1], clearly show that earthquakes and storms are the most damaging natural hazards. It is interesting to note that earthquakes have killed more people than all other hazards combined.

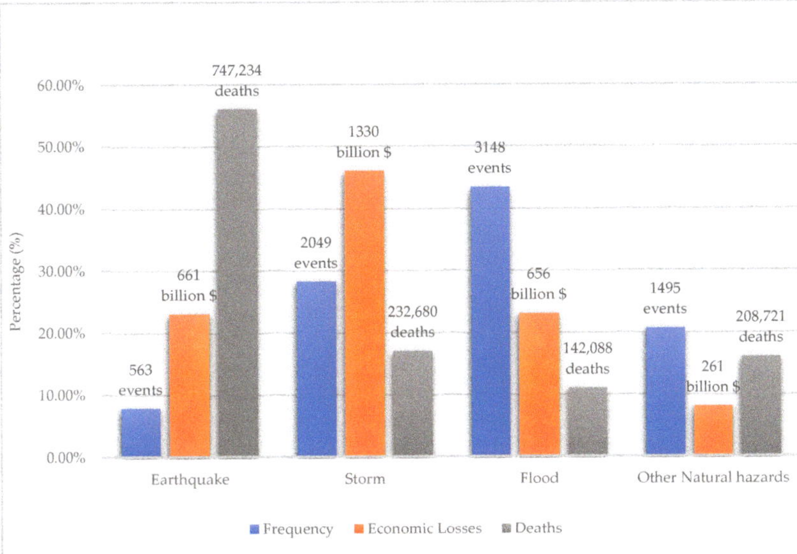

Figure 1. Frequencies of different natural hazards and their effects during 1998–2017 (based on CRED report, [1]).

Different natural processes affect structures and people in different ways. While the simultaneous occurrence of two different types of damaging hazards, such as strong wind and earthquake, is rare, some natural processes can induce secondary hazards. For example, fire and landslides are known to occur after strong earthquakes (see, for example, Ravankah et al. [2]). Moreover, a structure may be exposed to different types of natural hazards, albeit not simultaneously, during its lifetime. Therefore, it needs to be resistant to forces and damage mechanisms imposed by more than one natural process. Structures optimally designed for actions from one type of natural hazard may not necessarily be well equipped to deal with actions from all types of hazards. This leads to the need for hazard mapping, considering different types of natural processes and their interdependencies.

Consideration of multiple hazards in urban development is gaining popularity in the research community. For example, Bathrellos et al. [3] studied probabilities of incidence of floods, landslides, and earthquakes, in a specific area in Northeastern Greece, to map multiple hazards and identify areas suitable for urban development. Hicks et al. [4] explore disaster risk reduction from a multi-hazard perspective. Regional multi-hazard mapping for urban development is gaining popularity in research (e.g., [5]). Vulnerability and design of structures against multiple hazards is also gaining popularity in research. As an example, Aly [6], as well as Aly and Abburu [7], discuss some fundamental differences between wind and earthquake-resistant designs of high-rise buildings. A review of studies on the vulnerability of buildings subjected to wind and earthquake forces is presented by Indirli et al. [8]. A framework for life-cycle loss estimation in tall buildings subjected to wind and seismic forces is presented by Venanzi et al. [9]. Civil engineering infrastructure, such as dams, bridges, roads etc., are lifelines of modern society. Although multi-hazard risk assessment of infrastructure is challenging [10,11], it is an important tool to improve their safety and operability following natural disasters, which is instrumental for social resilience. Various factors affecting costs and performance of infrastructure in a multi-hazard environment is discussed in Ettouney and Alampalli [12]. Performance and fragilities of special structures, such as dams and floodwalls exposed to multiple hazards, are studied in Ardebili and Saouma [13] and Bodda [14].

Natural events, such as wind and earthquakes, impose dynamic forces on buildings and other civil engineering structures. Damage caused by such forces depends on the dynamic properties of the structure, as well as the characteristics of the wind forces and ground motion. In most cases, structural damage is a result of excessive vibration. Vibration control, which refers to reducing oscillations of structures exposed to dynamic forces, can, therefore, be used as a protective measure. Vibration control makes use of active, passive, or hybrid secondary devices that are installed on the structure and designed/tuned/actuated for optimal reduction in structural responses such as displacement, acceleration, etc. Base-isolation, for example, has been a popular and effective protection against earthquakes (see, for example, [15–19]). Tuned mass dampers (TMD) and other supplemental damping devices of different designs and configurations have also been known to effectively reduce wind and earthquake-induced vibrations of different types of structures (see, for example, [20–22]). Vibration control systems can provide an alternative protection for existing structures where retrofitting or strengthening is considered too costly or not feasible, due to factors such as aesthetics, cultural aspects, etc. Control devices that are effective against the forces generated by one type of natural hazard might not be effective against other hazards. For example, base isolation systems, which are effective for seismic protection of structures, might cause an adverse response during strong wind [23]. Due to the uncertainties in the amplitude and frequency content of dynamic forces, induced by wind and earthquakes, and their relationship with the properties of the affected structures, consideration of a multi-hazard scenario is especially important when designing vibration control systems.

This work is an attempt to bring together and synthesize valuable information and conclusions presented in a vast body of research literature on multi-hazard effects and control of civil engineering structures. The work is based on a review and synthesis of published literature. Relevant studies were searched through scholarly databases such as Web of Science, Google Scholar, and Scopus. The keywords used for searching were "multi-hazard", "vibration-control", "seismic control", "tuned mass dampers", "seismic fragility", and "life-cycle assessment". The search results were then narrowed first by scanning the titles of articles to include only those that indicated relevance in the multi-hazard problem, addressing one or more of the criteria: (a) hazard mapping/quantification, (b) performance assessment, (c) design and/or optimization, (c) fragility assessment, (d) life-cycle and/or cost-benefit analysis, and (e) vibration control. This resulted in more than 400 articles. The Abstract and Conclusion sections of these articles were then studied to further filter out studies that did not address the multi-hazard problem. This resulted in 220 articles. The references listed in these articles were then checked to search for more relevant articles. Special attention was given to state-of-the art review studies. References listed in studies were checked in detail to search for additional relevant articles. In total, 263 articles were studied and are referenced to in this work. Among these, there are 210 journal articles, 17 books/reports, 11 theses, 14 book chapters, and 11 conference papers. These include 18 state-of-the art reviews.

Initial thematic development of the work was, first and foremost, based on the keywords listed in these articles. The keywords in these articles were extracted, and their frequencies were counted. In total, 699 unique keywords were found. Multi-word keywords were then replaced by a single word (called, here, a reduced keyword) that is representative of the scope of the work. For example, "vibration control" was reduced to "control", "risk assessment" was reduced to "risk", and so on. In some cases, such as "wind turbines", both words were retained. This resulted in 117 keywords. Similar keywords were then grouped together to identify themes/scopes. For example, "seismic", "ground motion", and "earthquakes" were placed under the theme of "Seismic". The number of occurrences of these themes were then counted, and the themes were ranked. Frequency distribution of the most frequent themes is presented in Figure 2. Some reduced keywords with a low frequency of occurrence are therefore not considered useful in creating an overall theme of the subject being studied and are not shown in the figure. Control is the most frequent theme, and multi-hazard is the third most frequent theme. Seismic and wind loads are the most frequently considered hazards. In terms of structures, bridges, buildings, and wind turbines are frequent themes, while only a few (less than 10) occurrences of other infrastructure and lifelines were encountered. This thematic distribution of the studied articles was used to prepare the main structure of this paper, which is schematically presented in Figure 3.

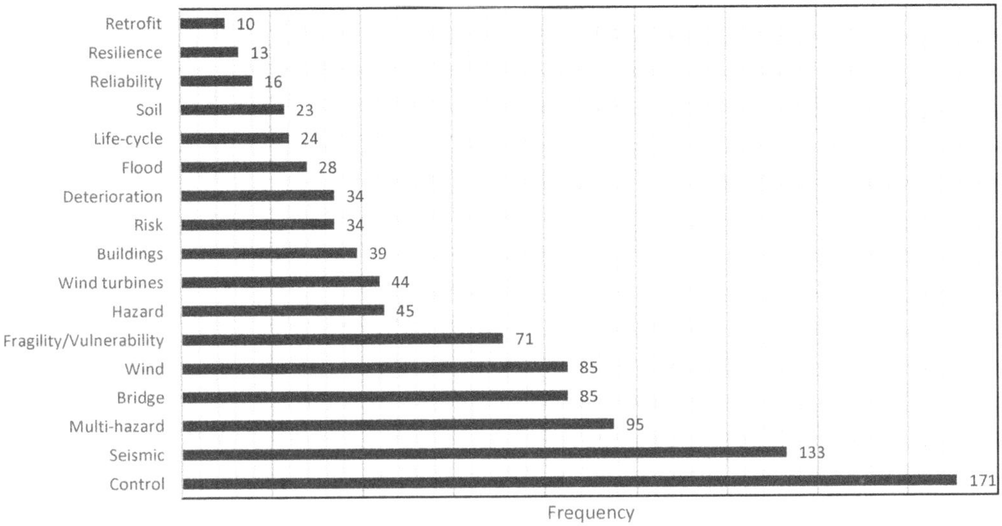

Figure 2. Frequency distribution of the most relevant themes extracted from reduced keywords.

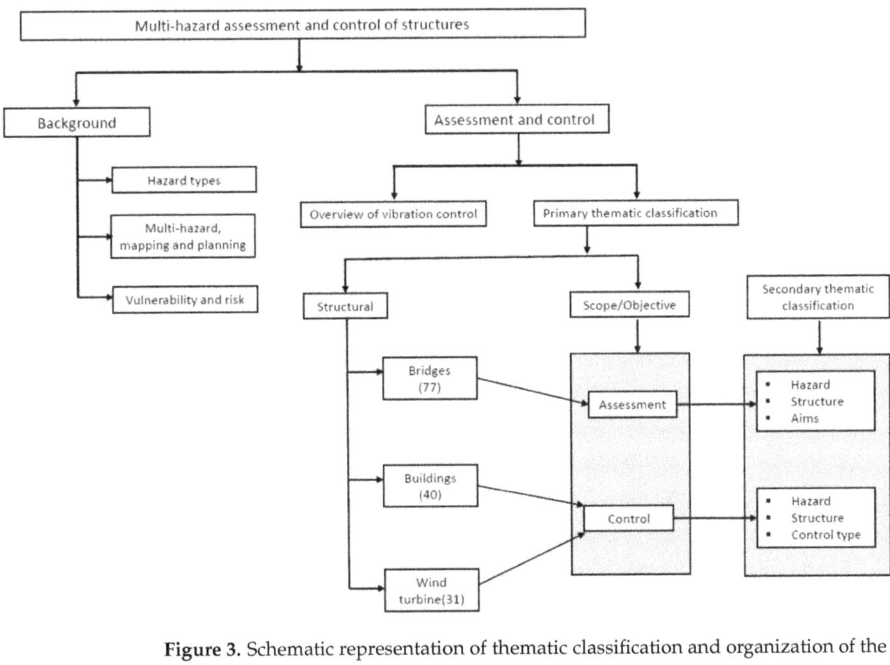

Figure 3. Schematic representation of thematic classification and organization of the paper.

Background information about different types of hazards, multi-hazard scenarios, and associated vulnerability and risk is provided in Section 2. This section is not a state-of-the-art review of these topics but rather background information for the rest of the paper (see Figure 3). The main part of the paper, which is the state-of-the-art review part of the paper, is briefly termed as assessment and control (see Figure 3). The review is based on the themes encountered in the studied literature. Topics such as fragility/vulnerability assessment, life-cycle assessment, multi-hazard assessment, and reliability assessment are covered under the assessment theme, while the control theme mainly deals with vibration suppression. As vibration control is the most dominant theme of the studied papers (see Figure 2), a brief literature review of different control systems is provided in Section 3. The review is primarily classified by two themes: namely, structure and scope of work. Bridges, buildings, and wind turbines are covered in Sections 4–6, respectively. Each of these sections is sub-divided into assessment and control sub-sections. The literature on bridges is dominated by the assessment theme, which is classified into secondary themes such as hazard type, type of bridge, and the main aims of the study. The literature on buildings and wind turbines contains several studies of multi-hazard vibration control. For each of these structures, the studies reviewed here are sub-classified into secondary themes of hazard, type of building/wind turbine, and the type of control device.

2. Risk: Hazard, Exposure, and Vulnerability

Risk related to disasters (disaster risks) can include loss of lives, disrupted economy, damages to the environment, etc. Risk is linked to the combination of hazard, physical exposure, and vulnerability of the infrastructure. The roles of each of these factors are briefly reviewed in the following sections.

2.1. Hazard

The definition of "hazard" in a broader sense is "any external or internal process or event that might degrade the performance of the system on hand" [12]. The United Nations General Assembly [24] defines hazard as "a process, phenomenon or human activity that may cause loss of life, injury or other health impacts, property damage, social and economic disruption or environmental degradation".

Natural events, such as storms, earthquakes, or floods, are well-known hazards with widespread potential to turn into disasters. While these events are mostly sudden and occur in a relatively short time window, slower processes, such as fatigue, corrosion, ageing, etc., can also impact structural performance over their life span.

Among the three elements that constitute disaster risk, hazard is the one that is mostly beyond human control. Nevertheless, a proper understanding of the occurrence frequency, spatio-temporal distribution, and intensity of the hazard is important for disaster risk reduction. Recent advances in sensing technology: data collection, processing, storage, sharing capabilities; and modelling/computational tools have improved our understanding how different hazards affect civil engineering structures. Hazards can be of different types. For example, they can be natural events, such as earthquakes, or man-made ones, such as explosions.

Different classifications of hazard have been proposed for multi-hazard studies. For example, Ettouney and Alampalli [12] discuss the classification of hazards based on Temporal, Frequency, and Newtonian characteristics. Temporal characterization distinguishes between simultaneous occurrence, segregation in time, and cascading effects. Frequency characterization distinguishes continuous processes, such as corrosion, from intermittent processes such as earthquakes. Intermittent processes can be further classified as frequent, intermediate, or rare. Newtonian characterization is another useful approach for hazard classification that is generally used in design codes. In design codes, hazards are generally quantified in terms of loads, such as wind load, earthquake load, etc. The impact of these loads can be quantified by different metrics, such as stress, deformation, etc., and are evaluated based on Newtonian mechanics. Such hazards have been termed as Newtonian [12]. Other processes, such as corrosion, wear and tear, fatigue, etc., are termed as non-Newtonian [12]. Natural hazards can also be classified based on their origin and the geo-atmospheric processes associated with them, such as

1. Biophysical (wildfire).
2. Atmospheric (wind or thunderstorms, lightning, hail, snow, and climate change).
3. Hydrological (drought and flood).
4. Shallow Earth Processes (erosion, subsidence and uplift, and mass movement).
5. Geophysical (volcanic eruption, tsunami, landslide, earthquake, and snow avalanche).

The idea that a structure needs to resist different types of hazards during its service life is well-established in civil engineering. For example, design codes and standards have provisions for different types of actions such as dead load, live load, wind load, seismic load, etc. Simultaneous occurrence of multiple actions is addressed in design codes through load combinations. Such recognition of multiple actions and load combinations does not encompass the real extent of multi-hazard effects and interactions. Multi-hazard generally refers to the concept where two or more hazards interact through structural performance. A multi-hazard interaction, for example, can impact risk due to a hazard when a decision regarding structural exposure and vulnerability against frequency, location, and amplitude of another hazard is made. For example, a change in design wind load and/or structural capacity can impact structural vulnerability to earthquake forces. Multi-hazard interactions may result in common or conflicting design solutions. For example, provision of structural ductility is beneficial for both blast and seismic loads.

Padgett and Kameshwar [25] present a comprehensive classification of multi-hazard combinations for bridges. Although their classification was intended for bridges, it can be generalized for most civil engineering structures, as is presented in Figure 4. Classification of hazard, according to Figure 4, helps to understand potential interactions between different hazards through their effects on structures.

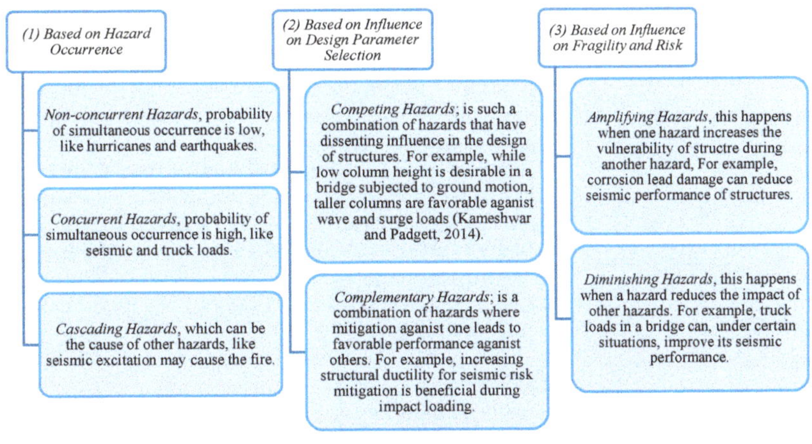

Figure 4. Classification of multi-hazard combinations [26] (modified from Padgett and Kameshwar [25]).

Multi-hazard consideration is important for structural safety and reliability. Duthinh and Simiu [27] present an interesting point regarding the traditional practice of treating different hazards independently and designing structural components based on the more demanding hazards. Taking an example of wind and earthquakes, they show that the ASCE Standard 7 provisions are not risk-consistent in the sense that, in regions affected by both strong wind and earthquakes, risks of exceedance of limit states can be up to twice as high as those in regions where only one of these hazards dominates. Kappes et al. [28] discuss the challenges of analyzing multi-hazard risk and existing frameworks to address those challenges. Zaghi et al. [29] presents the limitations of modern design codes in adequately addressing multi-hazard risk and emphasizing the need for common nomenclature for multi-hazard design. They also mention several problems and challenges in the multi-hazard design of structures. Different aspects of multi-hazard approaches, to mitigate risk to civil engineering infrastructure, are further discussed in Gardoni and Lafave [30]. The implications of considering potential multi-hazard effects in the life cycle cost analysis of an infrastructure is addressed by Jalayer et al. [31] and Fereshtehnejad and Shafieezadeh [32].

2.2. *Exposure, Multi-Hazard Mapping and Planning*

In the context of disaster risk, the UNDRR (United Nations Office for Disaster Risk Reduction) defines exposure as "the situation of people, infrastructure, housing, production capacities and other tangible human assets located in hazard-prone areas" [33]. Exposure is a necessary factor for disaster risk. Exposure is one of the risk determinants that can be controlled, to some extent, by proper planning. Such decisions are, however, not feasible in cases of existing risk: for example, large cities already built-in hazardous space. Reducing exposure to multi-hazards is more challenging than if only a single hazard is considered. Urban planning, land-use policy-making, environmental protection decisions, etc., need to rely on, and benefit from, multi-hazard considerations.

Local and regional scale mapping of different types of hazards is essential in multi-hazard considerations when analyzing exposure. Although significant advancements have been made in mapping individual hazards, mapping multi-hazard is challenging due to the differences in their physical phenomena, measures of frequency/amplitude, impact on structures, etc.

Barua et al. [34] present a multi-hazard map for different districts of Bangladesh based on local historical disaster database and comparison of scenario hazard scales with those in other countries. Their study includes earthquakes, tornadoes, floods, and cyclones, which are combined through a weighing scheme. Pourghasemi et al. [35] present multi-hazard mapping of Fars Province in southern Iran. They consider floods, fires, and landslides. They test two different machine learning algorithms in predicting distribution of these hazards based on historical data, and make use of different conditioning factors such as aspect, elevation, drainage, annual mean rainfall, etc. They highlight the importance of multi-hazard mapping in land-use planning, sustainable development, and watershed management in the study region. A similar study for the western region of Iran is presented in Pourghasemi et al. [36].

For multi-hazard mapping of relatively small areas, the Analytical Hierarchy Process (AHP) has been proposed as a suitable method (see, for example, [3]). It is a class of Multi-Criteria Decision Analysis (MCA) and relies on the connection between influencing factors and hazards rather than statistics from historical databases. In this sense, the method is subjective in assigning intensities and weights of different hazards. Some examples of AHP application in multi-hazard mapping can be found in Bathrellos et al. [3], Karaman [37], and Khatakho et al. [5].

2.3. *Vulnerability and Risk*

Vulnerability lies within the characteristics or properties of the elements (structures) at risk, making them susceptible to impacts of hazards. The UNGA [24] defines vulnerability as "the conditions determined by physical, social, economic and environmental factors or processes which increase the susceptibility of an individual, a community, assets or systems to the impacts of hazards". The concept of vulnerability is used in a broad sense and with different meanings in different fields. Vulnerability is the risk determinant that is the most feasible one to manage/control/reduce through human action or interference. Vulnerability reduction is, therefore, one of the most effective forms of risk reduction. However, the quantification of vulnerability of civil engineering structures even to individual hazards, such as earthquakes, is a challenging task with many uncertainties (see, for example, [38–40]). In a multi-hazard scenario, the overall vulnerability can be different from the vulnerability to a single hazard, which makes the definition of vulnerability especially challenging. Its complexity is due to the variations in structural material types, geometries, environments, exposure to hazards, usage, age, maintenance, and many other factors. Quantification of structural vulnerability to different types of hazards is a popular and growing research field. Structural vulnerability is mostly expressed in terms of fragility or vulnerability curves, which quantify, in a probabilistic sense, the chance of exceeding undesired states of damage conditioned to a given intensity of hazard. On a larger geographical scale, vulnerability classification of structures relies on general information about the structures, their usage, and exposure to hazards. Such classifications are commonly used for buildings. This method of vulnerability assessment was used by Nassirpour et al. [41] to rank school infrastructure in the Philippines, considering flood, wind, and earthquake hazards. A vulnerability assessment methodology for building, subjected to both single and multi-hazards, was presented by Schwarz et al. [42]. By following the principles of the European Macroseismic Scale 1998 (EMS-98, [43]), they developed vulnerability tables for different hazards (wind, flood, and earthquake). A framework to create multi-dimensional vulnerability models from vulnerability tables was also presented and applied in a few test cities in Germany. Gautam and Dong [44] present multi-hazard damage to structures in central Nepal caused by the 2015 Gorkha Earthquake and the 2017 Chhatiune Khola flash flood. A conceptual model for multi-hazard assessment of the vulnerability of historic buildings is presented by Ayala et al. [45] (2006) with an example application considering English parish churches. A comprehensive review of single and multi-hazard vulnerability and risk in historic urban areas is presented by Julia and Ferreira [46]. They also present interesting examples of the use of multi-hazard risk analysis in historic urban areas.

An interesting methodology for the risk evaluation of offshore structures subjected to ocean waves, wind, and ground motion is proposed by Bhartia and Vanmarcke [47]. They consider failure probabilities under short-term loads, as well as overall risks, due to loads of different intensities. Their results show that limit states (of failure), structural characteristics, as well as features of different types of loads interact in a complex way, controlling the relative importance of different hazards. Aggravating effects in a multi-hazard scenario are clearly demonstrated in their case-study example of ambient (ever-present wind over the sea) and seismic loads. An outline for identification of different hazards and subsequent risk assessment has been introduced by the United States Federal Emergency Management Agency, 1997, [48]. Ciurean et al. [49] present a comprehensive report of recent developments in multi-hazard processes and risks related to research, policy, and industry.

3. Vibration Control of Structures

Most of the structural damage caused by natural forces can be attributed to excessive vibrations. For static loads, vulnerability reduction can be achieved at the design stage by increasing stiffness and/or strength of structural elements. For existing structures, retrofitting strategies also aim to improve strength and/or stiffness of the structural elements. Similar strategies can also be used for dynamic forces such as wind and earthquakes, but newer and potentially more cost-effective solutions emerging in the scientific research are percolating to practical applications. These new solutions are not necessarily about increasing structural stiffness and/or strength. They fundamentally rely on changing the dynamic properties of the structures to make them less vulnerable to natural forces. This can, contrary to retrofitting in the traditional sense, even make the overall structure more flexible. A notable example is the well-established base-isolation technology for reducing earthquake-induced vibrations of buildings and bridges. Vibration reduction, also known as vibration control, makes use of different types of devices installed on the structure to reduce vibrations caused by different types of forces. While base-isolation, supplemental damping, and bracing systems to increase lateral stiffness and ductility have been researched and used for a long time, newer control strategies that rely mainly on dynamic devices installed on the primary structure are emerging. Depending on their mode of operation and need for external energy and/or internal feedback mechanism, vibration control devices can be broadly classified as passive, active, semi-active, or hybrid systems. The following list gives a few examples of these different types of control systems

1. Passive: energy dissipation, base isolation, tuned mass dampers (TMD), and tuned liquid dampers.
2. Active: adaptive control, active bracing, and active mass damping.
3. Semi-active: semi-active energy dissipation, semi-active isolation, and semi-active mass damping.
4. Hybrid: hybrid bracing and hybrid mass damping.

Detailed definitions and the basic theory behind these different classes of structural control devices can be found in the seminal work of Housner et al. [50]. Most of these control concepts have been extensively investigated, and many of these systems are already installed in different types of structures.

Passive control devices are the ones most frequently used, as they don't require external energy supply. Tuned mass dampers (TMDs), fluid viscous dampers (FVD), tuned liquid dampers (TLD), and seismic base isolation (BI) are the most popular passive control systems.

An early state-of-the art review of seismically base isolated buildings is presented by Kelly [51], Buckle and Mayes [52], and Jangid and Datta [17]. They discuss different types of base isolation systems and summarize findings of the contemporary literature about their performance in seismic response control in addition to presenting a parametric study of crucial design parameters for optimal reduction in seismic performance. Patil and Reddy [19] present a state-of-the-art review of base isolation systems in seismic response mitigation. They focus on design code provisions for isolated structures and discuss effects of soft-soil and near-fault ground motions. Kunde and Jangid [18] present a state-of-the art review of seismically isolated bridges and identify some knowledge gaps in the contemporary literature. Soong and Spencer [21] discuss different types of supplemental energy dissipation systems, including passive and active dampers for structural control. They provide an informative timeline of the development of these control technologies and describe the state-of-the-art review in the context of seismic-resistant design and the retrofitting of structures. A review on the behavior of structures with passive control systems exposed to seismic loads is presented by Buckle [52]. This study discusses the advantages of passive systems in a seismic design and provides several examples of their successful applications. It also highlights limitations of passive control, considering uncertainties in seismic forces and limit states induced by unexpectedly demanding events, and points towards the need for better practical guidelines in their design and implementation.

A comprehensive review on the response control of structures by TMDs is reported by Elias and Matsagar [22]. They review different configurations of dampers, involving one or more tuned masses, installed at different locations of the structure. They report that the findings in the literature support effectiveness of TMDs in reducing wind and earthquake-induced vibrations of certain types of structures. They also identify potential obstacles, such as robustness and reliability, across different levels of loading, especially those that exceed the yield limit, causing inelastic deformations in the structure.

A state-of-the-art review of different types of structural control systems was presented by Saeed et al. [53]. Their review includes different control technologies that can be classified as active, semi-active, passive, or hybrid. They conclude that control systems have a huge potential and importance in modern structures.

Symans and Constantinou [54] present a detailed review of semi-active control systems for the seismic protection of structures and conclude that different solutions, such as stiffness control devices, electrorheological dampers, friction control dampers, fluid viscous dampers, etc., have the potential of practical feasibility in full-scale structures. Spencer and Nagarajaiah [55] also report on the state of the art of semi-active technologies for structural vibration control. They report that smart damping devices, such as Magnetorheological (MR) dampers, appear to combine desirable features of both passive and active control solutions, and they offer a viable control solution against wind and earthquake forces.

The literature on control of structures against wind or seismic forces is vast. As explained above, the state-of-the-art and recent findings, in the structural control of different kinds, have been presented in many works. As an example, one of the first comprehensive state-of-the-art reviews of structural control systems was published more than two decades ago by Housner et al. [50]. Although performance of different control schemes in a single hazard scenario, such as wind or an earthquake, is well-known and summarized in many works, structural control in multi-hazard scenario is an emerging field of research. While some interesting research has been published in this field, there is a lack of an overview of the state-of-the-art, ongoing progress, and future directions. Most of the literature in this regard is on seismic and/or wind-induced response reduction in bridges, buildings, and wind turbines. These topics are dealt with separately in the following sections.

4. Multi-Hazard Assessment and Control of Bridges

Bridges are lifelines of modern society. They are vulnerable to different hazards, as evidenced by several failures in the past. On 19 August 2016, a suspension railway bridge in Tolten-Chile collapsed due to train-induced vibrations [56]. On 29 August, during hurricane Katrina, the Twin Spans Bridge connecting New Orleans to Slidell, Louisiana, United States, suffered extensive damage [57]. On 21 July 2003, Kinzua Bridge in Pennsylvania, United States, was hit by a tornado with 100 mph (45 m/s) winds and collapsed [58]. On 14 January 2003, Sgt. Aubrey Cosens VC Memorial Bridge in Ontario, Canada collapsed [59] due to fatigue-induced failure of the steel hanger rods supporting the deck. On 17 January 1995, a bridge on Hanshin Expressway in Kobe, Japan collapsed during the Kobe Earthquake [60]. During the Loma Prieta earthquake in 1989, two famous bridges (Cypress Street Viaduct and San Francisco—Oakland Bay Bridge) in California, USA were heavily damaged [61,62]. The collapse of these two bridges killed forty-one persons.

Safety and reliability of bridges are controlled by a diverse set of factors related to the structural form, function, maintenance, and the hazards they are exposed to. Multi-hazard consideration is, therefore, emerging as an important topic in bridge design and safety assessment. Some of the recent advancements in this field are discussed in the following.

4.1. Multi-Hazard Assessment of Bridges

To understand the consequences of multi-hazard effects on the safety/reliability of bridges, a wide range of experimental and analytical studies have been conducted and reported in the literature. One of the most studied scenarios is the interaction of earthquakes with other actions: for example, traffic-load. The interaction between these hazards, when they occur concurrently, can be amplifying or diminishing (see, for example, [63,64]). Cascading effects might also be observed when bridges, partly damaged by earthquakes, are exposed to traffic (see, for example, [65,66]). Ground shaking and liquefaction induced by earthquakes can have complex interactions in bridge response, both amplifying and diminishing (see, for example, [67,68]). Another multi-hazard scenario for bridges is the simultaneous occurrence of high waves and hurricane surge (see, for example, [69–73]). Another scenario is foundation scour due to floods, which may increase the seismic vulnerability of bridges [74–81].

Aging and corrosion of bridge elements causes structural deterioration that can amplify the effect of other hazards, such as earthquakes or wind forces. The effect of deterioration caused by seismic and traffic loads on a reinforced concrete bridge is addressed by Deco and Frangopol [82], Choe et al. [83], Kumar et al. [84], Choe et al. [85], Gardoni and Rosowsky [86], Choine et al. [87], Rokneddin et al. [88], and Biondini et al. [89]. Long-span bridges are especially sensitive to wind forces, but they can also be affected by seismic excitation. A framework for the assessment of vulnerability of long-span bridges subjected to multi-hazards (seismic and wind excitation) is presented by Martin et al. [90]. A summary of recent advances in wind effects on long-span bridges, with a multi-hazard perspective, is presented in Chen et al. [91]. Studies on the multi-hazard effects and performance of bridges is summarized in Table 1.

Table 1. A summary of published works on bridges subjected to multi-hazard.

Type of Hazards	Reference	Type of Bridge Structure	Aims	Main Contribution/Conclusion
		Multi-Hazard in General		
	Ettouney et al. [10]	Different type of bridges	Theoretical formulation	A general theory and application to structural analysis, design, life cycle costing, risk assessment, and health monitoring
		Earthquake and Wind		
	Martina et al. [90]	suspension bridges	Fragility analysis	Fragility surfaces for bridges exposed to multiple extreme events
		Earthquake and Corrosion		
	Choe et al. [85],	RC bridges	Fragility assessment	Fragility increment functions of corroding bridge columns and their application in life cycle cost and risk analysis
	Choe et al. [83]	RC bridges	Fragility assessment	Probabilistic models of seismic demand and fragility of corroding bridges, and their application in reliability analysis
	Kumar et al. [84]	RC bridges	Life-cycle cost assessment	Probabilistic model of life cycle cost considering cumulative damage, case studies highlighting dominance of cumulative seismic damage in reducing reliability
	Gardoni and Rosowsky [86]	RC bridges	Fragility assessment	Fragility increment functions and an example application
	Ghosh and Padgett [92]	Multi-span continuous highway bridges,	Fragility assessment	Time-dependent seismic fragility curves considering ageing and deterioration
	Simon et al. [93]	RC bridges	Fragility assessment	Losses in strength and stiffness due to corrosion have marginal effects on seismic fragilities of the case-study bridge
	Alipour, et al. [94]	Highway RC bridges	Fragility assessment	Time-dependent fragility models, and life cycle cost assessment
	Ghosh and Padgett [95]	Highway RC bridges	Loss assessment	Probabilistic framework for loss assessment using component-level cost estimates, case studies highlighting the impacts of deterioration
	Sung and Su [96]	RC bridge columns	Capacity, fragility, and loss estimation	Capacity models of deteriorated columns and resulting time-dependent fragility curves
	Zhong et al. [97]	RC bridges	Fragility assessment	Component and system level fragility curves, and a case study application
	Ou et al. [98]	RC bridges	Long-term performance assessment	Case studies of several bridges highlighting the need to increase design PGA to ensure adequate performance through the design life

Table 1. *Cont.*

Type of Hazards	Reference	Type of Bridge Structure	Aims	Main Contribution/Conclusion
	Rokneddin et al. [88]	RC bridges	Reliability assessment	Time-dependent fragility models of selected bridge classes, and an algorithm for reliability assessment, case studies highlight the need for accounting network-level importance in retrofit programs
	Akiyama and Frangopol [99]	RC bridges	Life cycle reliability assessment	Modelling spatial variability of rebar corrosion using X-ray, and a computational framework to incorporate corrosion hazard in seismic reliability
	Biondini et al. [89]	RC bridges	Life cycle performance assessment	Probabilistic modelling of capacities of corroding critical sections, case study highlighting undesirable effect of corrosion in seismic performance
	Ni et al. [100]	RC bridges	Modelling impact of corrosion on seismic performance	A new constitutive model for corroded reinforcing steel, and fragility curves at various time intervals
	Shekhar et al. [101]	Highway bridges	Study of realistic corrosion models and their impacts on life-cycle cost.	Framework for seismic life cycle costs from generic corrosion measures, case study demonstrating relevance of pitting versus uniform corrosion model
	Ghosh and Sood [102]	Highway bridges	Assessment of time-dependent capacities and more realistic degradation models in seismic fragility	A methodology for seismic fragility assessment incorporating pitting corrosion models and time-dependent capacity models, predictive equations for seismic reliability assessment over the service life
	Thanapol et al. [103]	RC bridges	Incorporation of spatial distribution of corrosion in seismic reliability assessment	Models to estimate steel weight loss at critical sections using spatially variable corrosion images using X-ray technology, followed by an illustrative case study
	Rao et al. [104]	RC bridges	Fragility assessment	Framework for seismic vulnerability assessment of deteriorating RC columns
	Alipour and Shafei [105]	Highway bridges	Network resilience assessment	Computational framework for risk assessment which emphasis on effect of ageing in seismic resilience
	Ghosh et al. [106], Rokneddin et al. [107],	Highway bridge network	Network reliability analysis	Methodology to estimate bridge fragilities using deterioration parameters from instrumented bridges in the network, and an example application
Earthquake and Floods				
	Dong and Frangopol [78]	Highway bridges	Life-cycle performance assessment	A framework for time-variant loss and resilience assessment, and effect of climate change

Table 1. *Cont.*

Type of Hazards	Reference	Type of Bridge Structure	Aims	Main Contribution/Conclusion
		Earthquake, Floods, and Ground-Failures		
	Gehl and D'Ayala [108]	RC bridges	Fragility assessment	A Bayesian Networks based fragility surfaces
		Earthquake and Scour		
	Wang et al. [109]	RC bridge	Estimation of load factors	Risk-consistent load factors based on case studies
	Wang et al. [77]	Bridge with pile foundation	Performance assessment	Experimental evidence for effects of scour depth on failure mechanisms of piers and piles
	Guo et al. [110]	RC bridges	Study of the effect of time-dependent scour hazard on seismic vulnerability	Fragility surfaces and time-dependent loss estimates with two case study bridges
	Alipour et al. [111]	RC bridges	Investigation of scour-load modification factors in seismic assessment	Reliability-based load and resistance factors
	Yilmaz [112]	Highway bridges	Risk and reliability assessment	Framework for risk assessment and uncertainty analysis
	Chandrasekaran and Banerjee [113]	RC Bridge	Optimal retrofitting	An approach for retrofit optimization, case study results showing that column jacketing is effective
	Guo and Chen [114]	RC bridges	Lifecycle assessment	A framework for lifecycle assessment, case study highlighting the need for a time-sensitive assessment
	Han et al. [74]	High-rise pile cap foundation	Performance assessment	Seismic capacity of foundation is significantly affected by scour depth
		Earthquake, Scour, and Corrosion		
	Dong et al. [115]	RC bridges	Sustainability assessment	A framework for time-varying sustainability, and an illustrative application
	Asadi et al. [116]	RC bridge	Performance assessment	A framework for performance assessment and life cycle cost analysis
		Earthquake, Scour, Corrosion, and Traffic-Load		
	Deco and Frangopol [117]	Highway bridges	Risk assessment	Framework for time-varying total risk and effect of structural redundancy
		Earthquake, Scour, Corrosion, Wind, Traffic-Load and Liquefaction		
	Banerjee et al. [118]	Highway bridges	Review	Summary of state-of-the-art in different aspects of resilience: loss assessment, recovery actions, and maintenance
		Earthquakes, Corrosion, Surge, and Wave		
	Kameshwar, and Padgett [119]	RC Coastal bridges	Lifecycle risk assessment and design	Object oriented consequence-based framework for lifecycle risk assessment considering structural deterioration

Table 1. *Cont.*

Type of Hazards	Reference	Type of Bridge Structure	Aims	Main Contribution/Conclusion
Earthquakes, Wind, Tsunami, Flood, Surge, and Wave				
	Gidaris et al. [120]	Highway bridges	Review	Summary of the state-of-the art in fragility and restoration models
Earthquakes, Blast and Fire				
	Echevarria, et al. [121]	Concrete-filled fiber reinforced polymer tube (CFFT) bridge columns	Experimental investigation of lightly reinforced CFFT columns	Experimentally validated design equations and a formulation for displacement-based seismic design including fire protection provisions
Earthquake and traffic loads				
	Sun et al. [122]	Bridges in Southeast Coastal areas of China	Study combinations of seismic and truck loads	Probabilistic methodology to combine earthquake and truck load
	Ghosh et al. [64]	Highway bridges	Studying effect of traffic loads on seismic reliability	Framework for joint fragility assessment, fragility surfaces, and case study application
Earthquake, Traffics-Load, and Deterioration				
	Deco and Frangopol [82]	Bridges in general	Life-cycle risk assessment	Probabilistic framework for life cycle risk assessment of spatially distributed group of bridges
Earthquakes and High water				
	Nikellis et al. [123]	Generic	Risk assessment	An analysis of risk metrics, stakeholder perceptions, and impact on retrofit strategies
Traffic Load and Wind				
	Chen et al. [91]	Long span bridges	Review	Summary of recent advances.
Wave and Storm Surge				
	Ataei, and Padgett [70]	Costal RC bridges	Capacity assessment	Probabilistic approach to global limit state capacities
	Ataei et al. [71]	Costal RC bridges	Fragility assessment	Development of surrogate models and uncertainty analysis

4.2. Multi-Hazard Vibration Control of Bridges

Although the literature on multi-hazard vulnerability and the risk assessment of bridges is vast, retrofitting bridges for multi-hazard protection is an emerging research topic that is gaining interest. Chandrasekaran and Banerjee [113] consider three different retrofit strategies to enhance bridge performance under the multi-hazard. Wang et al. [76] note that increasing foundation stiffness can be more beneficial than increasing foundation depth in reducing seismic vulnerability of bridges subjected to scour. Sung and Su [96] use time-dependent fragility curves to estimate the total direct costs of neutralized RC bridges as a function of ground motion intensity and service time and propose it as a tool to time retrofit campaigns. Benefits of the base isolation system, as a control/retrofit solution for increasing the reliability of steel bridges subjected to ground shaking and liquefaction hazards, is demonstrated in Wang et al. [124].

To the best of our knowledge, multi-hazard considerations in vibration control of bridges has not been reported in the literature yet.

5. Multi-Hazard Assessment and Vibration Control of Buildings

This section provides a review of studies related to building response to multi-hazard, with emphasis on wind and seismic forces. Building response to seismic forces is controlled by various factors, such as amplitude, duration, and frequency content of ground shaking. It also depends on the characteristics of the building itself and the underlying soil properties. Larger earthquakes produce ground motion with more energy at lower frequencies than smaller earthquakes. Large earthquakes are hazardous to all buildings, particularly to those that have natural frequencies close to the dominant frequency of ground shaking. Such a phenomenon has been observed in ground shaking and building response during past earthquakes (see, for example, [125,126]).

Wind loads contain energy at lower frequencies than seismic ground motions. Low to mid-rise buildings with relatively low vibration frequencies are therefore more susceptible to dynamic vibrations caused by seismic forces than those due to wind. Wind forces on such structures could, nevertheless, have undesired effects on components such as roofs, windows, chimneys, etc. Damage to light and improperly anchored roofs in low-rise buildings during strong wind is, therefore, of concern. In super tall buildings, wind generally induces stronger displacement response than earthquakes. Seismic loads, however, might excite higher vibration modes of such structures, resulting in high floor accelerations. This implies low inter-story drift and, therefore, lower risk of structural damage, but high floor acceleration can be critical for non-structural components [6,9,127,128]. From a structural point of view, wind loads are, therefore, critical for flexible structures, while seismic loads are more demanding on stiff structures. Occupant comfort and safety is another consideration when it comes to the response of buildings to wind and earthquake loads. Large floor accelerations can cause discomfort to occupants and may pose a safety threat due to moving objects. Floor accelerations in tall buildings are typically higher during moderate to strong ground shaking than during strong wind. Strong seismic loading is, however, typically less frequent than strong wind. From a serviceability point of view, wind action is, therefore, more critical for occupant comfort. Multi-hazard effects in buildings also need to be looked at from a life-cycle cost perspective and accumulation of damage due to multiple events: for example, wind response of a structure partially damaged by an earthquake or vice versa. Damage accumulation and fatigue due to repeated loading from frequent actions, such as moderate to strong wind, is also an important consideration.

5.1. Multi-Hazard Assessment of Buildings

Huang [129] provides a comprehensive account of the dynamic responses of high-rise buildings under multiple hazards. It presents performance assessment methods and case study investigations using high-rise buildings in Hong Kong. Various factors, such as seismic source-to-site distance, recurrence periods, ground shaking amplitude, building height, damping ratios, properties of wind forces, etc., were considered in the analysis. The results show that seismic loads result in a higher floor acceleration response and lateral forces but weaker torsional forces and a lower displacement response compared to wind forces. The height of the buildings was also found to be an important parameter, with wind response being more sensitive to variation in height than seismic response. The results also showed that wind response is more strongly influenced by the level of damping of the building than seismic response.

Chen [127], and Rasigha and Neeladharan [128] report differences in the seismic and wind responses of mid-rise to high-rise buildings. Aly [6], as well as Aly and Abburu [7] present the responses of tall buildings subjected to wind and seismic forces. In these assessments, two tall buildings (76-story and 54-story) were considered for finite element analysis. They found that ground motions excite higher vibration modes in buildings, resulting in lower inter-story drift than wind forces, but higher floor accelerations last for a shorter time. Wind actions are, therefore, critical from an occupant comfort and serviceability consideration. Tall structures designed for strong wind may possess an adequate capacity against moderate ground shaking, but they might suffer non-structural losses due to high floor accelerations. A framework for life-cycle loss estimation, of non-structural damage in tall buildings under wind and seismic loads, is presented by Venanzi et al. [9]. Their framework assumes that damaged structures are restored to their original condition after each hazardous event. Hazardous events are not simultaneous, and small maintenance costs are neglected. Their results show that for drift-dependent damages, wind forces are costlier than seismic forces. Seismic forces are costlier, in terms of non-structural damage, due to high floor accelerations. These observations are consistent with results reported in other studies, [6,7]. Antoun [130] studied the performance of a 74-story building located in Miami to evaluate the expected losses associated with a multi-hazard (wind and earthquake forces). Performance-based approaches were used for earthquake, wind, and hurricane forces. Monetary losses corresponding to structural and non-structural damage, as well as occupant discomfort, was estimated. They report that losses due to façade damage are dominant for high probabilities of exceedance, whereas structural damage becomes dominant at lower probabilities of exceedance.

Zhang et al. [131] proposed a framework for the damage risk assessment of high-rise buildings exposed to wind and seismic forces acting separately and concurrently. They used recorded earthquake and wind data, over a period of about 47 years, to estimate hazard curves for wind and seismic forces as well as copula-based bi-hazard surfaces. They then performed multi-hazard fragility assessment and estimated damage probabilities for separate and concurrent hazard models. Their results show that damage probability due to bi-hazards dominates the total damage probability in most damage states. They highlight the need for multi-hazard considerations in the design and evaluation of tall structures subjected to wind and seismic forces. Damage risk assessment and cost-benefit analysis of mitigation strategies, in residential buildings subjected to hurricane and seismic forces, are discussed in Li [132], giving a comprehensive overview of factors that are important in risk assessment, as well as their roles and impacts in hazard mitigation. The risk-cost-benefit framework, based on life-cycle and scenario-case analyses presented by Li [132], incorporate probabilistic modelling of hazards, structural fragility, and expected costs during different service intervals.

Multi-hazard consideration in performance-based engineering and performance-based design criteria, addressing wind and seismic forces, has been researched extensively in the literature. Chiu and Chock [133] present one of the first applications of the performance-based engineering approach in a multi-hazard scenario.

A probabilistic framework, for the multi-hazard risk assessment of reinforced concrete buildings subjected to seismic and blast loads, is discussed in Asprone et al. [134]. Annual risk of structural collapse, considering seismic action and progressive collapse due to blast forces, is formulated in this study. They conclude that the Monte Carlo (MC) simulation is suitable for calculating probability of progressive collapse, as well as for identifying critical blast scenarios.

Multi-hazard performance of different structural elements, such as columns, frames, plates, walls, etc., have been reported by many researchers. Resistance capacity of precast segmental columns, subjected to impact and cyclic loading, is investigated experimentally by Zhang et al. [135]. They found that, compared to monolithic columns, segmental columns (precast segments joined together, often with pre-stressed tendons) possess better ductility and sustain lower residual drift under cyclic loading. Under impact loading, segmental columns were found to have better self-centering capacity. They showed that shear resistance of such columns can be significantly improved by introducing concrete shear keys, but it comes at some cost related to stress-concentration and potential damage to concrete segments.

Rachel [136] presents a methodology for the resilience assessment of buildings subjected to seismic, wind, fire, and various post-earthquake scenarios. The results of this study showed that post-earthquake fire resilience in moment frame buildings is independent of seismic damage if frame connections are intact. The results also showed that multi-hazard resilience of moment resisting frame buildings can be improved by strengthening and/or fire-proofing gravity columns. Shin [137] presents multi-hazard performance evaluation matrices for retrofitted non-ductile reinforced concrete buildings subjected to seismic and blast loads.

Unnikrishnan and Barbato [138] investigated multi-hazard interaction on the performance of low-rise wood-frame buildings. Chulahawat and Mahmoud [139] present an algorithm to optimize building systems, with suspended floor slabs subjected to wind and seismic hazards, and observe that tall buildings with such systems are effectively optimized for both wind and seismic forces without a significant trade-off on performance to individual hazards.

5.2. Multi-Hazard Vibration Control of Buildings

Vibration control of buildings subjected to wind or seismic forces has been extensively researched. Vibration control of buildings in multi-hazard scenarios is, on the other hand, not as extensively studied. Some important studies in this area are summarized in Table 2. Performance assessment of control devices, their optimization, and life-cycle cost analysis are the main issues that have been addressed in these studies. Wind and earthquake forces are the most considered hazard in these studies. Most of these studies present traditional control systems such as passive TMDs, passive energy dissipation devices, viscous fluid dampers, multiple tuned passive TMDs, etc. Some recent advances in this area include inerter-based TMDs (Djerouni et al., [140]; Djerouni et al. [141]; Djerouni et al. [142]; Marian and Giaralis [143]), glass curtain wall TMDs (Bedon and Amadio [144]), and sliding floor isolators (Chulahwat and Mahmoud [139]; Mahmoud and Chulahwat [145]).

Table 2. A summary of published works on vibration control of the building subjected to multi-hazard.

Reference	Hazards	Structure	Control System	Main Contribution/Conclusion
Cao et al. [146]	Wind, Blast and Earthquake.	5- and 39-story benchmark buildings	Semi-active friction damper	A new controller called input space dependent controller (ISDC) which is more effective than sliding mode controller.
Mahmoud and Chulahwat [145] and Chulahqat and Mahmoud [139]	Wind and Earthquake	7- and 10-story steel frame buildings.	Sliding floor isolation.	A modified covariance matrix adaptation evolution strategy (CMA-ES) algorithm
Dogruel and Dargush [147]	Wind and Earthquake	16-story steel frame building.	Passive Energy Dissipation (PED).	Methodology for optimal life-cycle cost estimation, and optimal design of retrofitting
Shalom et al. [148]	Wind and Earthquake	76-storey building.	Multiple Tuned Mass Dampers (MTMDs)	Life-cycle cost-based optimization framework
Roy and Matsagar [149,150]	Wind and Earthquake	9-, 20-, and 25-story steel frame buildings.	PED	Optimal retrofits for earthquakes result in undesirable effects on wind response, and vice versa; damper performance is sensitive to site-specific hazard
Chapain and Aly [151]	Wind and Earthquake	76-story building	Viscous Fluid Dampers (VFDs)	VFDs are effective in multi-hazard control
Elias and Matsagar [152]	Wind and Earthquake	76-story and 20-story building	TMD	Optimally placed and tuned TMDs are effective in multi-hazard control
Elias et al. [153]	Wind and Earthquake	76-story building	MTMDs	MTMDs with equal stiffness are better than those with equal masses.
Bedon and Amadio [144]	Earthquake and blast	4-storey steel frame building	Glass curtain walls as passive TMD	Glass curtain walls can be utilized as distributed TMD for vibration mitigation
Gong [154]	Wind and Earthquake	5-,9-, and 20-story buildings	Variable Friction Cladding Connection (VFCC)	Experimental and analytical evidence demonstrating effectiveness of VFCC.

6. Multi-Hazard Assessment and Control of Wind Turbines

The tall and slender geometry of wind turbine towers and the large top mass of the turbine and the rotors make wind turbines sensitive to both wind and seismic excitation.

Wind and seismic loading have been the two most common environmental actions considered for research on the performance assessment of wind turbine towers. For offshore turbines, wave loading is also an important factor.

6.1. Multi-Hazard Assessment of Wind Turbines

Maryam [155], as well as Maryam and Gardoni [156] highlight the importance of multi-hazard consideration in site-selection and design of wind turbines. They present a multi-hazard probabilistic framework to evaluate the structural reliability of offshore wind turbines. Considering wind and seismic action, their results show that annual probabilities of failure are higher when seismic action is considered. Comparing two identical wind turbines, one in the Gulf of Mexico and the other off the coast of California, they conclude that, although the latter location is more favorable in terms of power production, annual probabilities of failure are higher due to higher seismicity. Avossa et al. [157] present a Monte Carlo simulation-based framework for the estimation of multi-hazard fragility curves of wind turbine structures. They provide an example application of the framework, to derive failure probabilities of a prototype wind turbine structure, conditioned on wind velocity and peak ground acceleration for different operational states of the turbine. Their results show that aerodynamic damping plays an important role in the seismic fragility. Fragility in an operational state, for seismic action in the fore-aft direction, increases with wind speed up to the rated wind speed, after which it starts to decrease. When the rotor is operating at the rated condition or is parked, the probability of failure is larger than 50% and the peak ground acceleration exceeds about 70% of the acceleration of gravity. Campo and Estrada [158] present similar conclusions regarding the importance of aerodynamic damping, stating that, while wind action is more damaging at the operational state, seismic action can be more threatening when the rotor is parked. Katsanos et al. [159] report, for a 5 MW offshore wind turbine, that seismic action contributes more than wind and wave action to structural demands such as base moment and tower-top displacement. They also report on the fragility of sensitive equipment located in the nacelle, which are found to be prone to severe damage at moderate ground shaking intensity. Zuo et al. [160] investigated the fragility of a prototype 5 MW offshore wind turbine structure subjected to aerodynamic forces and wave loading. They considered different operational states of the rotor and derived fragility curves for both the supporting tower and the rotor blades. Their results show that, when the wind speed is between the cut-in and cut-out range, exceedance probability of the blade failure is much higher than that of the tower failure. They also highlight the impact of aerodynamic damping in reducing wind-induced vibrations of the tower. Zuo et al. [161] studied the effect of soil structure interaction (SSI) on the 5 MW offshore prototype model. Their results show that the fore-aft displacement demand on the tower is significantly affected by SSI. Asareh et al. [162] investigated the fragility of a 5 MW wind turbine prototype subjected to wind and seismic action. Their results show that failure due to exceedance of tower-tip displacement and rotation is more likely than yielding or buckling of the tower.

6.2. Multi-Hazard Vibration Control of Wind Turbines

Vibration control of wind turbine structures, subjected to the combined actions of wind, waves, and earthquake ground motions, is extensively reported in the literature. These studies are mostly aimed at the optimization and performance assessment of control systems. A summary of relevant studies on the vibration control of wind turbine structures subjected to multi-hazard is presented in Table 3.

Table 3. A summary of published works on vibration control of wind turbine structures subjected to multi-hazard.

Reference	Hazard	Structure	Control System	Main Contribution/Conclusion
Xie et al. [163]	Wind and Waves	Offshore, 5 MW, barge-type floating	TMD on the platform	Modelling of drivetrain dynamics, optimization of TMD parameters, simulation results confirm the importance of drivetrain dynamics and that TMDs are effective in vibration suppression
He et al. [164]	Wind and Waves	Offshore, 5 MW, barge-type floating	TMD in the nacelle	TMDs are effective in reducing standard deviation of tower-top displacements, with upto 50% reduction for TMD mass ratio of 2%
Stewart and Lackner [165]	Wind and Waves	Offshore, 5 MW, monopile	TMD in the nacelle	Misalignment of wind and wave load significantly increases base moment in the side-side direction, which can be reduced by 40% with a TMD
Zhao et al. [166]	Wind, wave and seismic	Offshore, scaled model, monopile	TMD in the nacelle	Shake table tests for modal identification and estimation of aerodynamic damping, results show that TMDs are effective in reducing seismic response, effectiveness increases with rotation speed of blades
Sun and Jahangiri [167]	Misaligned wind-wave and seismic	Offshore, 5 MW, monopile	Pendulum tuned TMD(PTMD) in the nacelle	PTMDs are slightly more effective than two linear TMDs with equivalent mass and their stroke is smaller
Sun and Jahangiri [168]	Misaligned wind-wave	Offshore, 5 MW, monopile	PTMD in the nacelle	Increase in fatigue life due to PTMD is 50% higher than that due to dual linear TMDs.
Sun et al. [169]	Misaligned wind-wave and Seismic load	Offshore, 5 MW, monopile	PTMD in the nacelle	PTMD is more robust than dual linear TMDs, and with a mass ratio of 2% reduction in short-term fatigue damage is reduced by up to 90%.
Hu et al. [170]	Wind and waves	Offshore, 5 MW, barge-type floating	Tune Mass Damper Inerter (TMDI) in the nacelle	TMDI are more effective than TMD but there is a trade-off between fore-aft load control and device stroke; performance superior to TMD can be achieved for comparable device stroke
Zuo et al. [171]	Wind and Waves	Offshore, 5 MW, monopile	MTMD	MTMDs are efficient and robust in reducing the out-of-plane vibration of blades and the tower in parked and operational conditions.
Zuo et al. [172]	Wind, Waves and Earthquake	Offshore, 5 MW, monopile	MTMD	Multi-mode control using MTMDs are more efficient than STMDs in multi-hazard scenarios
Hussan et al. [173]	Wind, Waves and Earthquake	Offshore 5 MW with standard jacket foundation	MTMD	SSI plays important role in MTMD performance, often over-estimates it
Altay et al. [174]	Wind and Earthquake	Onshore 5 MW	TMD and Tuned Liquid Column Damper TLCD	Resonant tower vibrations at lower wind speeds are effectively reduced by TMDs and TLCDs, transient tower vibrations at higher wind speeds are less effectively reduced, only nominal control of seismic-induced vibrations
Colwell and Basu [175]	Wind and Waves	Offshore, monopile	TLCD	TLCDs are effective in reducing peak response and increasing fatigue life
Dezvareh et al. [176]	Wind and Waves	Offshore, 5 MW, jacket type	Tuned Liquid Column Gas Dampers (TLCGD)	Effective in reducing nacelle displacement and acceleration protecting the tower structure and acceleration-sensitive nacelle devices

Table 3. *Cont.*

Reference	Hazard	Structure	Control System	Main Contribution/Conclusion
Bargi et al. [177]	Wind, Waves and Earthquake	Offshore, 5 MW, jacket tupe	Tuned Liquid Column Gas Dampers (TLCGD)	Nacelle acceleration is better controlled under wind-wave excitation while nacelle displacement is better controlled under seismic load, heavier devices are more efficient but less robust against detuning
Sun [178]	Wind, Waves and Earthquake	Offshore, 5 MW, monopile	Semi-active TMD (S-TMD)	Semi-active TMDs are more efficient than passive TMDs
Hemmati and Oterkus [179]	Wind, Waves and Earthquake	Offshore, 5 MW, monopile	S-TMD	S-TMD provide better control than passive TMDs with as much as 4 times lower mass
Rezaee and Aly [180]	Wind, Wave, Earthquake,	Land based and offshore, 5 MW	MR damper-used as S-TMD and with outer bracing	The dampers are efficient in reducing strong vibrations and its duration during seismic loading. While displacement control starts early during ground shaking, acceleration control lags behind by a few seconds
Xie and Aly [181]	Wind and Earthquake	Various	TMD, TLD, VD, A-TMD, S-TMD, and TLCD,	A state-of-the-art review for evaluating the performance of the various types of control systems
Rezaee and Aly [182]	Wind and Waves	On-land 5 MW	TMD, TLCD, VD, and tuned sloshing damper (TSD)	Comparative study of different dampers show that VDs are the most robust, and that TSDs are effective at a wider range of frequencies
Zhao et al. [183]	Wind and Earthquake	On-land, 1.5 MW	Scissor-Jack Braced Viscous Damper (VD-SJB)	VD-SJB is effective and practical in reducing vibrations, seismic-vibration reduction in fore-aft direction is lower in operating condition than in parked condition
Zuo et al. [184]	Wind, Waves and Earthquake	Various	Various	State-of-the-art review
Rahman et al. [185]	Wind, Waves and Earthquake	Various	Various	Literature review

7. Concluding Remarks

This work is an attempt to summarize a vast body of research literature on multi-hazard effects on structures and their vibration mitigation measures. Aspects such as performance assessment, fragility modelling, life-cycle cost assessment, and vibration control in a multi-hazard scenario are covered. The main emphasis is on wind and seismic actions on major infrastructure, such as bridges, buildings, and wind turbine towers. Understanding of multi-hazard scenarios in a probabilistic sense and mapping them out for engineering design is an evolving field. At a local scale, multi-hazard mapping using the Analytical Hierarchy Process (AHP) is gaining popularity. Multi-hazard mapping at regional scales remains a challenging task, demanding more research on unifying frameworks that standardize and unify existing probabilistic hazard assessment methods used for different natural actions. Some recent advances in multi-hazard vulnerability of buildings include multi-dimensional vulnerability modelling. Fragility modelling in a multi-hazard scenario is still a growing field of knowledge, with many unresolved questions. Some examples of such unresolved issues relate to: (i) definition of intensity measures of hazards that might interact with each other, resulting in overall effects that are of different nature than those due to individual hazards; (ii) definition of joint probabilities of exceedances of intensities of different types of hazards; (iii) lack of empirical data on actual damage recorded in multi-hazard scenario, etc.

Multi-hazard assessment of bridges is a widely studied topic. Most studies in this area focus on seismic loads and corrosion. Other effects, such as wind loads, scour, traffic loads, etc., in conjunction with seismic loads, have also been reported. Published literature on bridges subjected to seismic loads and corrosion focus on the fragility assessment. Multi-hazard effects in such assessments are generally modelled through fragility increment functions, damage accumulation, and time-dependent fragility curves. Probabilistic load and resistance factors, for different hazards affecting bridges, is another widely reported research theme. In most cases, such fragility models are intended for a life-cycle cost and risk analysis. While multi-hazard fragility of individual bridges is widely reported, there are only a few deals with bridge networks. More research is needed in capacity modelling, risk metrics, stakeholder perceptions, and load combinations.

The literature on multi-hazard effects on buildings is dominated by wind and earthquake loads. Performance-based engineering frameworks, progressive damage and collapse modelling, resilience assessment, multi-hazard performance evaluation metrices, etc., are some of the recent advances in damage risk and life cycle cost analysis of buildings. A variety of control systems such as passive TMDs, passive energy dissipation devices, viscous fluid dampers, multiple tuned passive TMDs, have been investigated in control of buildings subjected to wind and seismic forces. Some recent advances in this area seem promising and practically appealing: for example, sliding floor isolators, glass curtain wall TMDs, and variable friction cladding connections [VFCC]. While the literature on the vibration control of buildings subjected to wind or seismic action is vast, relatively few studies have addressed their simultaneous occurrence. More research is needed on probabilistic treatment of multi-hazard load cases, robustness of control devices against uncertainties in structural properties, as well as loading the feasibility of control from a life-cycle perspective.

Seismic and wind forces are the two most considered environmental actions in the performance assessment and vibration control of wind turbine structures. Wave action, and effects of wave/wind misalignment in offshore wind turbines is also widely researched. Multi-hazard probabilistic framework for reliability assessment of offshore wind turbines is relatively well-established. Monte Carlo simulation-based frameworks for multi-hazard fragility assessment are recently emerging. For land-based wind turbines, several studies have highlighted the role of aerodynamic damping in response to combined action of wind and earthquakes. Most of the published work on vibration control of wind turbines focuses on structural fragility of the supporting tower. Offshore wind turbines subjected to wind and waves is the most investigated scenario. The most reported control device is passive TMD placed on the nacelle, although use of TMD on the platform of a barge-type floating turbine has also been reported to be effective. Most of the studies conclude that control devices are effective in reducing multi-hazard fragility. Recent advances in multi-hazard control of wind turbines include interesting innovations such as braced viscous dampers, and semi-active control systems. Effect of ground motion variability on control performance is an area that needs to be studied better. Control performance against impulsive loads caused by, for example, near-fault ground motions (see, for example, Rupakhety et al. [186]; Elias et al. [187]; Sigurðsson et al. [188], Jami et al. [189]) also need to be investigated better. In addition, fragility of rotor blades and effects of drivetrain dynamics need more attention.

In most of structural vibration control studies reported in the literature, the structure is assumed to remain elastic, which may not be realistic in extreme loading conditions. Inelastic deformations of the structure can result in de-tuning of the control device resulting in lower performance. Control optimization and performance assessment of yielding structures subjected to multi-hazard scenarios, as well as damage accumulation due to multiple hazards occurring over the useful life of a structure, need to be investigated and better understood to facilitate practical applications of control systems in actual engineering projects.

Author Contributions: M.J.: Conceptualization, research, literature study, writing, and review; R.R.: methodology, writing, and review; S.E.: review; B.B.: writing, and review; J.T.S.: writing and review. All authors have read and agreed to the published version of the manuscript.

Funding: Matin Jami is supported by a doctoral grant from the University of Iceland. RR acknowledges support from the University of Iceland Research Fund.

Institutional Review Board Statement: Not applicable.

Informed Consent Statement: Not applicable.

Data Availability Statement: Not applicable.

Conflicts of Interest: The authors declare no conflict of interest.

References

1. Wallemarq, P.; Below, R.; McClean, D. *UNISDR and CRED Report: Economic Losses, Poverty & Disasters (1998–2017)*; United Nations Office for Disaster Risk Reduction: Geneva, Switzerland, 2018; Available online: https://www.cred.be/sites/default/files/CRED_Economic_Losses_10oct.pdf (accessed on 12 May 2022).
2. Ravankhah, M.; Schmidt, M.; Will, T. Multi-hazard disaster risk identification for World Cultural Heritage sites in seismic zones. *J. Cult. Herit. Manag. Sustain. Dev.* **2017**, *7*, 272–289. [CrossRef]
3. Bathrellos, G.D.; Skilodimou, H.D.; Chousianitis, K.; Youssef, A.M.; Pradhan, B. Suitability estimation for urban development using multi-hazard assessment map. *Sci. Total Environ.* **2017**, *575*, 119–134. [CrossRef] [PubMed]
4. Hicks, A.; Barclay, J.; Chilvers, J.; Armijos, M.T.; Oven, K.; Simmons, P.; Haklay, M. Global mapping of citizen science projects for disaster risk reduction. Citizen Science: Reducing Risk and Building Resilience to Natural Hazards. *Front. Earth Sci.* **2020**, *7*, 226. [CrossRef]
5. Khatakho, R.; Gautam, D.; Aryal, K.R.; Pandey, V.P.; Rupakhety, R.; Lamichhane, S.; Liu, Y.C.; Abdouli, K.; Talchabhadel, R.; Adhikari, R. Multi-hazard risk assessment of Kathmandu Valley, Nepal. *Sustainability* **2021**, *13*, 5369. [CrossRef]
6. Aly, A.M. Design of buildings for wind and earthquake. In Proceedings of the World Congress on Advances in Civil, Environmental, and Materials Research (ACEM'14), Busan, Korea, 24–28 August 2014.
7. Aly, A.M.; Abburu, S. On the design of high-rise buildings for multihazard: Fundamental differences between wind and earthquake demand. *Shock. Vib.* **2015**, *2015*, 148681. [CrossRef]
8. Indirli, M.; Valpreda, E.; Panza, G.; Romanelli, F.; Lanzoni, L.; Teston, S.; Rossi, G. Natural multi-hazard and building vulnerability assessment in the historical centers: The examples of San Giuliano di Puglia (Italy) and Valparaiso (Chile). In Proceedings of the 7th European Commission Conference "SAUVEUR", Heritage, Prague, 31 May 2006.
9. Venanzi, I.; Lavan, O.; Ierimonti, L.; Fabrizi, S. Multi-hazard loss analysis of tall buildings under wind and seismic loads. *Struct. Infrastruct. Eng.* **2018**, *14*, 1295–1311. [CrossRef]
10. Ettouney, M.M.; Alampalliz, S.; Agrawal, A.K. Theory of multihazards for bridge structures. *Bridge Struct.* **2005**, *1*, 281–291. [CrossRef]
11. García, H.J. J. Multi-Hazard Risk Assessment: An Interdependency Approach. Ph.D. Thesis, The University of British Columbia, Vancouver, BC, Canada, 2010.
12. Ettouney, M.M.; Alampalli, S. *Multihazard Considerations in Civil Infrastructure*; Taylor & Francis Group: Abingdon, UK, 2017; IRBS: 9781482208320.
13. Ardebili, M.A.H.; Saouma, V.E. Single and multi-hazard capacity functions for concrete dams. *Soil Dyn. Earthq. Eng.* **2017**, *101*, 234–249. [CrossRef]
14. Bodda, S.S. Multi-Hazard Risk Assessment of a Flood Defense Structure. Master's Theses, North Carolina State University, Raleigh, NC, USA, 2018.
15. Bessason, B.; Haflidason, E. Recorded and Numerical Strong Motion Response of a Base-Isolated Bridge. *Earthq. Spectra* **2004**, *20*, 309–332. [CrossRef]
16. Jónsson, M.H.; Bessason, B.; Haflidason, E. Earthquake response of a base-isolated bridge subjected to strong near-fault ground motion. *Soil Dyn. Earthq. Eng.* **2010**, *30*, 447–455. [CrossRef]
17. Jangid, R.S.; Datta, T.K. ; Datta, T.K. Seismic behavior of base isolated buildings: A state-of-the art-review. *Proc. ICE Struct. Build. Inst. Civ. Eng.* **1995**, *110*, 186–203. [CrossRef]
18. Kunde, M.C.; Jangid, R.S. Seismic behavior of isolated bridges: A state-of-the-art review. *Electron. J. Struct. Eng.* **2003**, *3*, 140–170.
19. Patil, S.J.; Reddy, G.R. State of art review—Base isolation systems for structures. *Int. J. Emerg. Technol. Adv. Eng.* **2012**, *2*, 438–453.
20. Buckle, I.G. Passive control of structures for seismic loads. In Proceedings of the 12th world conference on earthquake engineering, Auckland, New Zealand, 30 January–4 February 2000; pp. 2825–2838.
21. Soong, T.T.; Spencer, B.F. Supplemental energy dissipation: State-of-the-art and state-of-the-practice. *Eng. Struct.* **2002**, *24*, 243–259. [CrossRef]
22. Elias, S.; Matsagar, V. Research developments in vibration control of structures using passive tuned mass dampers. *Annu. Rev. Control* **2017**, *44*, 129–156. [CrossRef]
23. Henderson, P.; Novak, M. Response of base-isolated buildings to wind loading. *Earthq. Eng. Struct. Dyn.* **1989**, *18*, 1201–1217. [CrossRef]
24. Available online: https://reliefweb.int/report/world/hazard-definition-and-classification-review (accessed on 12 May 2022).
25. Padgett, J.E.; Kameshwar, S. Supporting Life Cycle Management of Bridges through Multi-Hazard Reliability and Risk Assessment. In *Multi-Hazard Approaches to Civil Infrastructure Engineering*; Springer: Berlin/Heidelberg, Germany, 2016; pp. 41–58.

26. Kameshwar, S.; Padgett, J.E. Multi-hazard risk assessment of highway bridges subjected to earthquake and hurricane hazards. *Eng. Struct.* **2014**, *78*, 154–166. [CrossRef]
27. Duthinh, D.; Simiu, E. Safety of structures in strong winds and earthquakes: Multihazard considerations. *J. Struct. Eng.* **2010**, *136*, 330–333. [CrossRef]
28. Kappes, M.S.; Keiler, M.; von Elverfeldt, K.; Glade, T. Challenges of analyzing multi-hazard risk: A review. *Nat. Hazards* **2012**, *64*, 1925–1958. [CrossRef]
29. Zaghi, A.E.; Padgett, J.E.; Bruneau, M.; Barbato, M.; Li, Y.; Mitrani-Reiser, J.; McBride, A. Establishing common nomenclature, characterizing the problem, and identifying future opportunities in multihazard design. *J. Struct. Eng.* **2016**, *142*, H2516001. [CrossRef]
30. Gardoni, P.; LaFave, J.M. Multi-Hazard Approaches to Civil Infrastructure Engineering: Mitigating Risks and Promoting Resilence. In *Multi-Hazard Approaches to Civil Infrastructure Engineering*; Springer: Berlin/Heidelberg, Germany, 2016; pp. 3–12.
31. Jalayer, F.; Asprone, D.; Prota, A.; Manfredi, G. Multi-hazard upgrade decision making for critical infrastructure based on life-cycle cost criteria. *Earthq. Eng. Struct. Dyn.* **2011**, *40*, 1163–1179. [CrossRef]
32. Fereshtehnejad, E.; Shafieezadeh, A. A multi-type multi-occurrence hazard lifecycle cost analysis framework for infrastructure management decision making. *Eng. Struct.* **2018**, *167*, 504–517. [CrossRef]
33. Available online: https://www.un-spider.org/risks-and-disasters/disaster-risk-management#:~{}:text=Exposure%20is%20defined%20as%20%E2%80%9Cthe,of%20assets%20in%20an%20area (accessed on 12 May 2022).
34. Barua, U.; Akhter, M.S.; Ansary, M.A. District-wise multi-hazard zoning of Bangladesh. *Nat. Hazards* **2016**, *82*, 1895–1918. [CrossRef]
35. Pourghasemi, H.R.; Gayen, A.; Edalat, M.; Zarafshar, M.; Tiefenbacher, J.P. Is multi-hazard mapping effective in assessing natural hazards and integrated watershed management? *Geosci. Front.* **2020**, *11*, 1203–1217. [CrossRef]
36. Pourghasemi, H.R.; Gayen, A.; Panahi, M.; Rezaie, F.; Blaschke, T. Multi-hazard probability assessment and mapping in Iran. *Sci. Total Environ.* **2019**, *692*, 556–571. [CrossRef]
37. Karaman, H. Integrated Multi-Hazard Map Creation by Using AHP and GIS. In *Recent Advances on Environmental and Life Science*; Geomatics Engineering Department, Istanbul Technical University: Istanbul, Turkey, 2015.
38. Bessason, B.; Rupakhety, R.; Bjarnason, J.Ö. Comparison and modelling of building losses in South Iceland caused by different size earthquakes. *J. Build. Eng.* **2022**, *46*, 103806. [CrossRef]
39. Thorvaldsdottir, S.; Bessason, B.; Rupakhety, R. Non-structural earthquake risk management for residential buildings. *Ann. Geophys.* **2021**, *64*, SE323. [CrossRef]
40. Ferreira, M.; Meroni, F.; Azzaro, R.; Musacchio, G.; Rupakhety, R.; Bessason, B.; Thorvaldsdottir, S.; Lopes, M.; Oliveira, C.S.; Solarino, S. What scientific information on non-structural elements seismic risk people need to know? Part 1: Compiling an inventory on damage to non-structural elements. *Ann. Geophys.* 2021. [CrossRef]
41. Nassirpour, A.; Galasso, C.; D'ayala, D. Multi-hazard physical vulnerability prioritization of school infrastructure in the Philippines. In Proceedings of the 11th US National Conference on Earthquake Engineering (11NCEE), Los Angeles, CA, USA, 25–29 June 2018; pp. 25–29.
42. Schwarz, J.; Maiwald, H.; Beinersdorf, S.; Kaufmann, C. Evaluation of the vulnerability of existing building stocks under single and Multi-Hazard impact. In Proceedings of the 16th European Conference on Earthquake Engineering (ECEE), Thessaloniki, Greece, 18–21 June 2018.
43. Grünthal, G.; Musson, R.; Schwarz, J.; Stucchi, M. *European Macroseismic Scale 1998*; Cahiers du Centre Européen de Geodynamique et de Seismologie: Luxembourg, 1998; Volume 15.
44. Gautam, D.; Dong, Y. Multi-hazard vulnerability of structures and lifelines due to the 2015 Gorkha earthquake and 2017 central Nepal flash flood. *J. Build. Eng.* **2018**, *17*, 196–201. [CrossRef]
45. D'Ayala, D.; Copping, A.; Wang, H. A conceptual model for multi-hazard assessment of the vulnerability of historic buildings. In Proceedings of the Fifth International Conference, Structural Analysis of Historical Constructions: Possibilities of Numerical and Experimental Techniques, New Delhi, India, 9–11 November 2006; pp. 121–140.
46. Julià, P.B.; Ferreira, T.M. From single-to multi-hazard vulnerability and risk in Historic Urban Areas: A literature review. *Nat. Hazards* **2021**, *108*, 93–128. [CrossRef]
47. Bhartia, B.K.; Vanmarcke, E.H. *Multi-Hazard Risk Analysis: Case of a Simple Offshore Structure*; National Center for Earthquake Engineering Research: Taipei, Taiwan, 1988; p. 80.
48. United States Federal Emergency Management Agency. *Multi-Hazard Identification and Risk Assessment: The Cornerstone of the National Mitigation Strategy*; Federal Emergency Management Agency: Washington, DC, USA, 1997.
49. Ciurean, R.; Gill, J.; Reeves, H.J.; O'Grady, S.; Aldridge, T. Review of Multi-Hazard Research and Risk Assessments. 2018. Available online: https://nora.nerc.ac.uk/id/eprint/524399/ (accessed on 12 May 2022).
50. Housner, G.W.; Bergman, L.A.; Caughey, T.K.; Chassiakos, A.G.; Claus, R.O.; Masri, S.F.; Skelton, R.E.; Soong, T.T.; Spencer, B.F.; Yao, J.T. Structural control: Past, present and future. *J. Eng. Mech. Am. Soc. Civ. Eng. ASCE* **1997**, *123*, 897–971. [CrossRef]
51. Kelly, J.M. Aseismic base isolation: Review and bibliography. *Soil Dyn. Earthq. Eng.* **1986**, *5*, 202–216. [CrossRef]
52. Buckle, I.G.; Mayes, R.L. Seismic isolation: History, application, and performance—A world view. *Earthq. Spectra* **1990**, *6*, 161–201. [CrossRef]
53. Saeed, T.E.; Nikolakopoulos, G.; Jonasson, J.-E.; Hedlund, H. A state-of-the-art review of structural control systems. *J. Vib. Control* **2013**, *21*, 919–937. [CrossRef]

54. Symans, M.D.; Constantinou, M.C. Semi-active control systems for seismic protection of structures: A state-of-the-art review. *Eng. Struct.* **1999**, *21*, 469–487. [CrossRef]
55. Spencer, B.F.; Nagarajaiah, S. State of the art of structural control. *J. Struct. Eng. Am. Soc. Civ. Eng. ASCE* **2003**, *129*, 845–856. [CrossRef]
56. List of Bridge Failures. Wikipedia. Available online: https://en.wikipedia.org/wiki/List_of_bridge_failures (accessed on 12 May 2022).
57. Okeil, A.M.; Cai, C.S. Survey of short-and medium-span bridge damage induced by Hurricane Katrina. *J. Bridge Eng.* **2008**, *13*, 377–387. [CrossRef]
58. Leech, T. The collapse of the Kinzua Viaduct: A combination of design oversight and material fatigue left a century-old railroad bridge vulnerable to an F-1 tornado. *Am. Sci.* **2005**, *93*, 348–354. [CrossRef]
59. Biezma, M.V.; Schanack, F. Collapse of steel bridges. *J. Perform. Constr. Facil.* **2007**, *21*, 398–405. [CrossRef]
60. Mylonakis, G.; Syngros, C.; Gazetas, G.; Tazoh, T. The role of soil in the collapse of 18 piers of Hanshin Expressway in the Kobe earthquake. *Earthq. Eng. Struct. Dyn.* **2006**, *35*, 547–575. [CrossRef]
61. Hanks, T.C.; Brady, A.G. The Loma Prieta earthquake, ground motion, and damage in Oakland, Treasure Island, and San Francisco. *Bull. Seismol. Soc. Am.* **1991**, *81*, 2019–2047. [CrossRef]
62. Nims, D.K. Collapse of the Cypress Street Viaduct as a result of the Loma Prieta Earthquake. No. UCB/EERC-89/16. 1991. Available online: https://trid.trb.org/view/337611 (accessed on 12 May 2022).
63. Kartal, H.; Soyluk, K. Dynamic Behaviour of a Cable-Stayed Bridge under Earthquake and Traffic Loads. In Proceedings of the 10th International Congress on Advances in Civil Engineering, Middle East Technical University, Ankara, Turkey, 17–19 October 2012.
64. Ghosh, J.; Caprani, C.C.; Padgett, J.E. Influence of traffic loading on the seismic reliability assessment of highway bridge structures. *J. Bridge Eng.* **2014**, *19*, 04013009. [CrossRef]
65. Franchin, P.; Pinto, P.E. Allowing traffic over mainshockdamaged bridges. *J. Earthq. Eng.* **2009**, *13*, 585–599. [CrossRef]
66. Terzić, V.; Stojadinović, B. Traffic Load Capacity of a Bridge Damaged in an Earthquake. 2009. Available online: https://www.caee.ca/10CCEEpdf/2010EQConf-001221.pdf (accessed on 12 May 2022).
67. Wang, Z.; Due~nas-Osorio, L.; Padgett, J.E. Seismic response of a bridge–soil–foundation system under the combined effect of vertical and horizontal ground motions. *Earthq. Eng. Struct. Dyn.* **2013**, *42*, 545–564. [CrossRef]
68. Moshirabadi, S.; Soltani, M.; Maekawa, K. Seismic interaction of underground RC ducts and neighboring bridge piers in liquefiable soil foundation. *Acta Geotech.* **2015**, *10*, 761–780. [CrossRef]
69. Ataei, N.; Stearns, M.; Padgett, J.E. Response sensitivity for probabilistic damage assessment of coastal bridges under surge and wave loading. Transportation Research Record. *J. Transp. Res. Board* **2010**, *2202*, 93–101. [CrossRef]
70. Ataei, N.; Padgett, J.E. Limit state capacities for global performance assessment of bridges exposed to hurricane surge and wave. *Struct. Saf.* **2013**, *41*, 73–81. [CrossRef]
71. Ataei, N.; Padgett, J.E. Fragility surrogate models for coastal bridges in hurricane prone zones. *Eng. Struct.* **2015**, *103*, 203–213. [CrossRef]
72. Ataei, N.; Padgett, J.E. Influential fluid–structure interaction modelling parameters on the response of bridges vulnerable to coastal storms. *Struct. Infrastruct. Eng.* **2015**, *11*, 321–333. [CrossRef]
73. Anarde, K.A.; Kameshwar, S.; Irza, J.N.; Nittrouer, J.A.; LorenzoTrueba, J.; Padgett, J.E.; Sebastian, A.; Bedient, P.B. Impacts of hurricane storm surge on infrastructure vulnerability for an evolving coastal landscape. *Nat. Hazards Rev.* **2018**, *19*, 04017020. [CrossRef]
74. Han, Z.; Ye, A.; Fan, L. Effects of riverbed scour on seismic performance of high-rise pile cap foundation. *Earthq. Eng. Eng. Vib.* **2010**, *9*, 533–543. [CrossRef]
75. Banerjee, S.; Ganesh Prasad, G. Seismic risk assessment of reinforced concrete bridges in flood-prone regions. *Struct. Infrastruct. Eng.* **2013**, *9*, 952–968. [CrossRef]
76. Wang, Z.; Due~nas-Osorio, L.; Padgett, J.E. Influence of scour effects on the seismic response of reinforced concrete bridges. *Eng. Struct.* **2014**, *76*, 202–214. [CrossRef]
77. Wang, S.C.; Liu, K.Y.; Chen, C.H.; Chang, K.C. Experimental investigation on seismic behavior of scoured bridge pier with pile foundation. *Earthq. Eng. Struct. Dyn.* **2015**, *44*, 849–864. [CrossRef]
78. Dong, Y.; Frangopol, D.M. Probabilistic time-dependent multihazard life-cycle assessment and resilience of bridges considering climate change. *J. Perform. Constr. Facil.* **2016**, *30*, 04016034. [CrossRef]
79. Yilmaz, T.; Banerjee, S. Impact spectrum of flood hazard on seismic vulnerability of bridges. *Struct. Eng. Mech.* **2018**, *66*, 515–529.
80. Yilmaz, T.; Banerjee, S.; Johnson, P.A. Performance of two real-life California bridges under regional natural hazards. *J. Bridge Eng.* **2016**, *21*, 04015063. [CrossRef]
81. Yilmaz, T.; Banerjee, S.; Johnson, P.A. Uncertainty in risk of highway bridges assessed for integrated seismic and flood hazards. *Struct. Infrastruct. Eng.* **2018**, *14*, 1182–1196. [CrossRef]
82. Deco, A.; Frangopol, D.M. Life-cycle risk assessment of spatially distributed aging bridges under seismic and traffic hazards. *Earthq. Spectra* **2013**, *29*, 127–153. [CrossRef]
83. Choe, D.E.; Gardoni, P.; Rosowsky, D.; Haukaas, T. Seismic fragility estimates for reinforced concrete bridges subject to corrosion. *Struct. Saf.* **2009**, *31*, 275–283. [CrossRef]
84. Kumar, R.; Gardoni, P.; Sanchez-Silva, M. Effect of cumulative seismic damage and corrosion on the life-cycle cost of reinforced concrete bridges. *Earthq. Eng. Struct. Dyn.* **2009**, *38*, 887–905. [CrossRef]

85. Choe, D.E.; Gardoni, P.; Rosowsky, D. Fragility increment functions for deteriorating reinforced concrete bridge columns. *J. Eng. Mech.* **2010**, *136*, 969–978. [CrossRef]
86. Gardoni, P.; Rosowsky, D. Seismic fragility increment functions for deteriorating reinforced concrete bridges. *Struct. Infrastruct. Eng.* **2011**, *7*, 869–879. [CrossRef]
87. Choine, M.N.; O'Connor, A.; Padgett, J.E. A seismic reliability assessment of reinforced concrete integral bridges subject to corrosion. *Key Eng. Mater.* **2013**, *569*, 366–373. [CrossRef]
88. Rokneddin, K.; Ghosh, J.; Due~nas-Osorio, L.; Padgett, J.E. Bridge retrofit prioritization for ageing transportation networks subject to seismic hazards. *Struct. Infrastruct. Eng.* **2013**, *9*, 1050–1066. [CrossRef]
89. Biondini, F.; Camnasio, E.; Palermo, A. Lifetime seismic performance of concrete bridges exposed to corrosion. *Struct. Infrastruct. Eng.* **2014**, *10*, 880–900. [CrossRef]
90. Martin, J.; Alipour, A.; Sarkar, P. Fragility surfaces for multi-hazard analysis of suspension bridges under earthquakes and microbursts. *Eng. Struct.* **2019**, *197*, 109169. [CrossRef]
91. Chen, S.; Zhou, Y.; Wu, J.; Chen, F.; Hou, G. Research of long-span bridge and traffic system subjected to winds: A system and multi-hazard perspective. *Int. J. Transp. Sci. Technol.* **2017**, *6*, 184–195. [CrossRef]
92. Ghosh, J.; Padgett, J.E. Aging considerations in the development of time-dependent seismic fragility curves. *J. Struct. Eng.* **2010**, *136*, 1497–1511. [CrossRef]
93. Simon, J.; Bracci, J.M.; Gardoni, P. Seismic response and fragility of deteriorated reinforced concrete bridges. *J. Struct. Eng.* **2010**, *136*, 1273–1281. [CrossRef]
94. Alipour, A.; Shafei, B.; Shinozuka, M. Performance evaluation of deteriorating highway bridges located in high seismic areas. *J. Bridge Eng.* **2011**, *16*, 597–611. [CrossRef]
95. Ghosh, J.; Padgett, J.E. Probabilistic seismic loss assessment of aging bridges using a component-level cost estimation approach. *Earthq. Eng. Struct. Dyn.* **2011**, *40*, 1743–1761. [CrossRef]
96. Sung, Y.C.; Su, C.K. Time-dependent seismic fragility curves on optimal retrofitting of neutralised reinforced concrete bridges. *Struct. Infrastruct. Eng.* **2011**, *7*, 797–805. [CrossRef]
97. Zhong, J.; Gardoni, P.; Rosowsky, D. Seismic fragility estimates for corroding reinforced concrete bridges. *Struct. Infrastruct. Eng.* **2012**, *8*, 55–69. [CrossRef]
98. Ou, Y.C.; Fan, H.D.; Nguyen, N.D. Long-term seismic performance of reinforced concrete bridges under steel reinforcement corrosion due to chloride attack. *Earthq. Eng. Struct. Dyn.* **2013**, *42*, 2113–2127. [CrossRef]
99. Akiyama, M.; Frangopol, D.M. Long-term seismic performance of RC structures in an aggressive environment: Emphasis on bridge piers. *Struct. Infrastruct. Eng.* **2014**, *10*, 865–879. [CrossRef]
100. Ni Choine, M.; Kashani, M.M.; Lowes, L.N.; O'Connor, A.; Crewe, A.J.; Alexander, N.A.; Padgett, J.E. Nonlinear dynamic analysis and seismic fragility assessment of a corrosion damaged integral bridge. *Int. J. Struct. Integr.* **2016**, *7*, 227–239. [CrossRef]
101. Shekhar, S.; Ghosh, J.; Padgett, J.E. Seismic life-cycle cost analysis of ageing highway bridges under chloride exposure conditions: Modelling and recommendations. *Struct. Infrastruct. Eng.* **2018**, *14*, 941–966. [CrossRef]
102. Ghosh, J.; Sood, P. Consideration of time-evolving capacity distributions and improved degradation models for seismic fragility assessment of aging highway bridges. *Reliab. Eng. Syst. Saf.* **2016**, *154*, 197–218. [CrossRef]
103. Thanapol, Y.; Akiyama, M.; Frangopol, D.M. Updating the seismic reliability of existing RC structures in a marine environment by incorporating the spatial steel corrosion distribution: Application to bridge piers. *J. Bridge Eng.* **2016**, *21*, 04016031. [CrossRef]
104. Rao, A.S.; Lepech, M.D.; Kiremidjian, A. Development of time-dependent fragility functions for deteriorating reinforced concrete bridge piers. *Struct. Infrastruct. Eng.* **2017**, *13*, 67–83. [CrossRef]
105. Alipour, A.; Shafei, B. Seismic resilience of transportation networks with deteriorating components. *J. Struct. Eng.* **2016**, *142*, C4015015. [CrossRef]
106. Ghosh, J.; Rokneddin, K.; Padgett, J.E.; Dueñas-Osorio, L. Seismic reliability assessment of aging highway bridge networks with field instrumentation data and correlated failures, I: Methodology. *Earthq. Spectra* **2014**, *30*, 795–817. [CrossRef]
107. Rokneddin, K.; Ghosh, J.; Due~nas-Osorio, L.; Padgett, J.E. Seismic reliability assessment of aging highway bridge networks with field instrumentation data and correlated failures, II: Application. *Earthq. Spectra* **2014**, *30*, 819–843. [CrossRef]
108. Gehl, P.; D'Ayala, D. Development of Bayesian Networks for the multi-hazard fragility assessment of bridge systems. *Struct. Saf.* **2016**, *60*, 37–46. [CrossRef]
109. Wang, Z.; Padgett, J.E.; Due~nas-Osorio, L. Risk-consistent calibration of load factors for the design of reinforced concrete bridges under the combined effects of earthquake and scour hazards. *Eng. Struct.* **2014**, *79*, 86–95. [CrossRef]
110. Guo, X.; Wu, Y.; Guo, Y. Time-dependent seismic fragility analysis of bridge systems under scour hazard and earthquake loads. *Eng. Struct.* **2016**, *121*, 52–60. [CrossRef]
111. Alipour, A.; Shafei, B.; Shinozuka, M. Reliability-based calibration of load and resistance factors for design of RC bridges under multiple extreme events: Scour and earthquake. *J. Bridge Eng.* **2013**, *18*, 362–371. [CrossRef]
112. Yilmaz, T. Risk assessment of highway bridges under multi-hazard effect of flood-induced scour and earthquake. Ph.D. Dissertation, The Pennsylvania State University, State College, PA, USA, 2015.
113. Chandrasekaran, S.; Banerjee, S. Retrofit optimization for resilience enhancement of bridges under multihazard scenario. *J. Struct. Eng.* **2016**, *142*, C4015012. [CrossRef]

114. Guo, X.; Chen, Z. Lifecycle multihazard framework for assessing flood scour and earthquake effects on bridge failure. *ASCE-ASME J. Risk Uncertain. Eng. Syst. Part A Civ. Eng.* **2016**, *2*, C4015004. [CrossRef]
115. Dong, Y.; Frangopol, D.M.; Saydam, D. Time-variant sustainability assessment of seismically vulnerable bridges subjected to multi-hazard. *Earthq. Eng. Struct. Dyn.* **2013**, *42*, 1451–1467. [CrossRef]
116. Asadi, P.; Hamed Ranjkesh, S.; Zeinal Hamedani, A. Life-Cycle Cost Analysis (LCCA) of RC Bridges Subjected to Multi-hazard. *Sci. Iran.* **2020**, *28*, 629–644.
117. Deco, A.; Frangopol, D.M. Risk assessment of highway bridges under multi-hazard. *J. Risk Res.* **2011**, *14*, 1057–1089. [CrossRef]
118. Banerjee, S.; Vishwanath, B.S.; Devendiran, D.K. Multihazard resilience of highway bridges and bridge networks: A review. *Struct. Infrastruct. Eng.* **2019**, *15*, 1694–1714. [CrossRef]
119. Kameshwar, S.; Padgett, J.E. Towards risk based multi-hazard resistant design of bridges. *ASCE Struct. Congr.* **2014**, *2014*, 2318–2329.
120. Gidaris, I.; Padgett, J.E.; Barbosa, A.R.; Chen, S.; Cox, D.; Webb, B.; Cerato, A. Multi-hazard fragility and restoration models of highway bridges for regional risk and resilience assessment in the United States: State-of-the-art review. *J. Struct. Eng.* **2017**, *143*, 04016188. [CrossRef]
121. Echevarria, A.; Zaghi, A.E.; Christenson, R.; Accorsi, M. CFFT bridge columns for multihazard resilience. *J. Struct. Eng.* **2016**, *142*, C4015002. [CrossRef]
122. Sun, D.; Chen, B.; Sun, B. Study based on bridge health monitoring system on multihazard load combinations of earthquake and truck loads for bridge design in the southeast coastal areas of China. *Shock Vib.* **2015**, *2015*, 829380. [CrossRef]
123. Nikellis, A.; Sett, K. Multihazard Risk Assessment and Cost–Benefit Analysis of a Bridge–Roadway–Levee System. *J. Struct. Eng.* **2020**, *146*, 04020050. [CrossRef]
124. Wang, Z.; Due~nas-Osorio, L.; Padgett, J.E. Influence of soilstructure interaction and liquefaction on the isolation efficiency of a typical multispan continuous steel girder bridge. *J. Bridge Eng.* **2014**, *19*, A4014001. [CrossRef]
125. Takewaki, I.; Murakami, S.; Fujita, K.; Yoshitomi, S.; Tsuji, M. The 2011 off the Pacific coast of Tohoku earthquake and response of high-rise buildings under long-period ground motions. *Soil Dyn. Earthq. Eng.* **2011**, *31*, 1511–1528. [CrossRef]
126. Takewaki, I.; Fujita, K.; Yoshitomi, S. Uncertainties in long-period ground motion and its impact on building structural design: Case study of the 2011 Tohoku (Japan) earthquake. *Eng. Struct.* **2013**, *49*, 119–134. [CrossRef]
127. Chen, E. Multi-Hazard Design of Mid-to High-Rise Structures. Master's Theses, University of Illinois Urbana-Champaign, Champaign, IL, USA, 2012.
128. Rasigha, J.; Neeladharan, C. Design of Structures to Resist Multi-hazard. *Int. J. Innov. Res. Sci. Eng. Technol.* **2016**, *5*, 1–6.
129. Huang, M. *High-Rise Buildings under Multi-Hazard Environment*; Springer: Berlin/Heidelberg, Germany, 2017.
130. Antoun, J.E.K. Multi-Hazard Performance Assessment of High-Rise Buildings. Master's Theses, Master of Science in Civil Engineering, The Department of Civil & Environmental Engineering, Louisiana State University, Baton Rouge, LA, USA, 2017.
131. Zheng, X.W.; Li, H.N.; Yang, Y.B.; Li, G.; Huo, L.S.; Liu, Y. Damage risk assessment of a high-rise building against multihazard of earthquake and strong wind with recorded data. *Eng. Struct.* **2019**, *200*, 109697. [CrossRef]
132. Li, Y.; Asce, M. Assessment of damage risks to residential buildings and cost–benefit of mitigation strategies considering hurricane and earthquake hazards. *J. Perform. Constr. Facil.* **2012**, *26*, 7–16. [CrossRef]
133. Chiu, G.L.F.; Chock, G.Y.K. Multihazard performance based objective design for managing natural hazards damage. In Proceedings of the IFAC, Control in Natural Disasters, Tokyo, Japan, 21–22 September 1998.
134. Asprone, D.; Jalayer, F.; Prota, A.; Manfredi, G. Proposal of a probabilistic model for multi-hazard risk assessment of structures in seismic zones subjected to blast for the limit state of collapse. *Struct. Saf.* **2010**, *32*, 25–34. [CrossRef]
135. Zhang, X.; Hao, H.; Li, C. Multi-hazard resistance capacity of precast segmental columns under impact and cyclic loading. *Int. J. Prot. Struct.* **2018**, *9*, 24–43. [CrossRef]
136. Chicchi, R.A. Multi-Hazard Resilience of Steel MFR Buildings. Ph.D. Dissertation, Purdue University, West Lafayette, IN, USA, 2017.
137. Shin, J. Multi-Hazard Performance Criteria for Non-Ductile Reinforced Concrete Frame Buildings Retrofitted with an FRP Column Jacketing System. Ph.D. Dissertation, Georgia Institute of Technology, Atlanta, GA, USA, 2017.
138. Unnikrishnan, V.U.; Barbato, M. Multihazard interaction effects on the performance of low-rise wood-frame housing in hurricane-prone regions. *J. Struct. Eng.* **2017**, *143*, 04017076. [CrossRef]
139. Chulahwat, A.; Mahmoud, H. A combinatorial optimization approach for multi-hazard design of building systems with suspended floor slabs under wind and seismic hazards. *Eng. Struct.* **2017**, *137*, 268–284. [CrossRef]
140. Djerouni, S.; Elias, S.; Abdeddaim, M.; Rupakhety, R. Optimal design and performance assessment of multiple tuned mass damper inerters to mitigate seismic pounding of adjacent buildings. *J. Build. Eng.* **2022**, *48*, 103994. [CrossRef]
141. Djerouni, S.; Abdeddaim, M.; Elias, S.; Rupakhety, R. Optimum double mass tuned damper inerter for control of structure subjected to ground motions. *J. Build. Eng.* **2021**, *44*, 103259. [CrossRef]
142. Djerouni, S.; Ounis, A.; Elias, S.; Abdeddaim, M.; Rupakhety, R. Optimization and performance assessment of tuned mass damper inerter systems for control of buildings subjected to pulse-like ground motions. *Structures* **2022**, *38*, 139–156. [CrossRef]
143. Marian, L.; Giaralis, A. Optimal design of a novel tuned mass-damper–inerter (TMDI) passive vibration control configuration for stochastically support-excited structural systems. *Probabilistic Eng. Mech.* **2014**, *38*, 156–164. [CrossRef]
144. Bedon, C.; Amadio, C. Numerical assessment of vibration control systems for multi-hazard design and mitigation of glass curtain walls. *J. Build. Eng.* **2018**, *15*, 1–13. [CrossRef]

145. Mahmoud, H.; Chulahwat, A. Multi-Hazard Multi-Objective Optimization of Building Systems with Isolated Floors under Seismic and Wind Demands. In *Multi-Hazard Approaches to Civil Infrastructure Engineering*; Springer: Berlin/Heidelberg, Germany, 2016; pp. 141–164.
146. Cao, L.; Laflamme, S.; Hong, J.; Dodson, J. Input space dependent controller for civil structures exposed to multi-hazard excitations. *Eng. Struct.* **2018**, *166*, 286–301. [CrossRef]
147. Dogruel, S.; Dargush, G.F. A framework for multi-hazard design and retrofit of passively damped structures. In Proceedings of the Architectural Engineering Conference (AEI), Denver, CO, USA, 24–27 September 2008; pp. 1–12.
148. Kleingesinds, S.; Lavan, O.; Venanzi, I. Life-cycle cost-based optimization of MTMDs for tall buildings under multi-hazard. *Struct. Infrastruct. Eng.* **2020**, *17*, 921–940. [CrossRef]
149. Roy, T.; Matsagar, V. Effectiveness of passive response control devices in buildings under earthquake and wind during design life. *Struct. Infrastruct. Eng.* **2019**, *15*, 252–268. [CrossRef]
150. Roy, T.; Matsagar, V. Probabilistic assessment of steel buildings installed with passive control devices under multi-hazard scenario of earthquake and wind. *Struct. Saf.* **2020**, *85*, 101955. [CrossRef]
151. Chapain, S.; Aly, A.M. Vibration attenuation in high-rise buildings to achieve system-level performance under multi-hazard. *Eng. Struct.* **2019**, *197*, 109352. [CrossRef]
152. Elias, S.; Matsagar, V. Optimum Tuned Mass Damper for Wind and Earthquake Response Control of High-Rise Building. In *Advances in Structural Engineering*; Springer: New Delhi, India, 2015; pp. 1475–1487.
153. Elias, S.; Rupakhety, R.; Olafsson, S. Analysis of a Benchmark Building Installed with Tuned Mass Dampers under Wind and Earthquake Loads. *Shock Vib.* **2019**, *2019*, 7091819. [CrossRef]
154. Gong, Y. Variable Friction Cladding Connection for Multi-hazard Mitigation. Ph.D. Dissertation, Iowa State University, Ames, IA, USA, 2019.
155. Maryam, M.R. Multi-Hazard Reliability Assessment of Offshore Wind Turbines. Ph.D. Thesis, Texas A&M University, College Station, TX, USA, 2012.
156. Mardfekri, M.; Gardoni, P. Multi-hazard reliability assessment of offshore wind turbines. *Wind Energy* **2015**, *18*, 1433–1450. [CrossRef]
157. Avossa, A.M.; Demartino, C.; Contestabile, P.; Ricciardelli, F.; Vicinanza, D. Some results on the vulnerability assessment of HAWTs subjected to wind and seismic actions. *Sustainability* **2017**, *9*, 1525. [CrossRef]
158. del Campo, J.O.M.; Estrada, A.P. Multi-hazard fragility analysis for a wind turbine support structure: An application to the Southwest of Mexico. *Eng. Struct.* **2012**, *209*, 109929. [CrossRef]
159. Katsanos, E.I.; Sanz, A.A.; Georgakis, C.T.; Thöns, S. Multi-hazard response analysis of a 5MW offshore wind turbine. *Procedia Eng.* **2017**, *199*, 3206–3211. [CrossRef]
160. Zuo, H.; Bi, K.; Hao, H.; Xin, Y.; Li, J.; Li, C. Fragility analyses of offshore wind turbines subjected to aerodynamic and sea wave loadings. *Renew. Energy* **2020**, *160*, 1269–1282. [CrossRef]
161. Zuo, H.; Bi, K.; Hao, H. Dynamic analyses of operating offshore wind turbines including soil-structure interaction. *Eng. Struct.* **2018**, *157*, 42–62. [CrossRef]
162. Asareh, M.A.; Schonberg, W.; Volz, J. Fragility analysis of a 5-MW NREL wind turbine considering aero-elastic and seismic interaction using finite element method. *Finite Elem. Anal. Des.* **2016**, *120*, 57–67. [CrossRef]
163. Xie, S.; Jin, X.; He, J. Structural vibration control for the offshore floating wind turbine including drivetrain dynamics analysis. *J. Renew. Sustain. Energy* **2019**, *11*, 023304. [CrossRef]
164. He, E.M.; Hu, Y.Q.; Zhang, Y. Optimization design of tuned mass damper for vibration suppression of a barge-type offshore floating wind turbine. *Proc. Inst. Mech. Eng. Part M J. Eng. Marit. Environ.* **2017**, *231*, 302–315. [CrossRef]
165. Stewart, G.M.; Lackner, M.A. The impact of passive tuned mass dampers and wind-wave misalignment on offshore wind turbine loads. *Eng. Struct.* **2014**, *73*, 54–61. [CrossRef]
166. Zhao, B.; Gao, H.; Wang, Z.; Lu, Z. Shaking table test on vibration control effects of a monopile offshore wind turbine with a tuned mass damper. *Wind Energy* **2018**, *21*, 1309–1328. [CrossRef]
167. Sun, C.; Jahangiri, V. Bi-directional vibration control of offshore wind turbines using a 3D pendulum tuned mass damper. *Mech. Syst. Signal Process.* **2018**, *105*, 338–360. [CrossRef]
168. Sun, C.; Jahangiri, V. Fatigue damage mitigation of offshore wind turbines under real wind and wave conditions. *Eng. Struct.* **2019**, *178*, 472–483. [CrossRef]
169. Sun, C.; Jahangiri, V.; Sun, H. Performance of a 3D pendulum tuned mass damper in offshore wind turbines under multiple hazards and system variations. *Smart Struct. Syst.* **2019**, *24*, 53–65.
170. Hu, Y.; Wang, J.; Chen, M.Z.; Li, Z.; Sun, Y. Load mitigation for a barge-type floating offshore wind turbine via inerter-based passive structural control. *Eng. Struct.* **2018**, *177*, 198–209. [CrossRef]
171. Zuo, H.; Bi, K.; Hao, H. Mitigation of tower and out-of-plane blade vibrations of offshore monopile wind turbines by using multiple tuned mass dampers. *Struct. Infrastruct. Eng.* **2019**, *15*, 269–284. [CrossRef]
172. Zuo, H.; Bi, K.; Hao, H. Using multiple tuned mass dampers to control offshore wind turbine vibrations under multiple hazards. *Eng. Struct.* **2017**, *141*, 303–315. [CrossRef]
173. Hussan, M.; Sharmin, F.; Kim, D. Multiple tuned mass damper based vibration mitigation of offshore wind turbine considering soil–structure interaction. *China Ocean Eng.* **2017**, *31*, 476–486. [CrossRef]

174. Altay, O.; Taddei, F.; Butenweg, C.; Klinkel, S. Vibration Mitigation of Wind Turbine Towers with Tuned Mass Dampers. In *Wind Turbine Control and Monitoring*; Springer: Berlin/Heidelberg, Germany, 2014; pp. 337–373.
175. Colwell, S.; Basu, B. Tuned liquid column dampers in offshore wind turbines for structural control. *Eng. Struct.* **2009**, *31*, 358–368. [CrossRef]
176. Dezvareh, R.; Bargi, K.; Mousavi, S.A. Control of wind/wave-induced vibrations of jacket-type offshore wind turbines through tuned liquid column gas dampers. *Struct. Infrastruct. Eng.* **2016**, *12*, 312–326. [CrossRef]
177. Bargi, K.; Dezvareh, R.; Mousavi, S.A. Contribution of tuned liquid column gas dampers to the performance of offshore wind turbines under wind, wave, and seismic excitations. *Earthq. Eng. Eng. Vib.* **2016**, *15*, 551–561. [CrossRef]
178. Sun, C. Semi-active control of monopile offshore wind turbines under multi-hazards. *Mech. Syst. Signal Process.* **2018**, *99*, 285–305. [CrossRef]
179. Hemmati, A.; Oterkus, E. Semi-active structural control of offshore wind turbines considering damage development. *J. Mar. Sci. Eng.* **2018**, *6*, 102. [CrossRef]
180. Rezaee, M.; Aly, A.M. Vibration control in wind turbines to achieve desired system-level performance under single and multiple hazard loadings. *Struct. Control Health Monit.* **2018**, *25*, e2261. [CrossRef]
181. Xie, F.; Aly, A.M. Structural control and vibration issues in wind turbines: A review. *Eng. Struct.* **2020**, *210*, 110087. [CrossRef]
182. Rezaee, M.; Aly, A.M. Vibration control in wind turbines for performance enhancement: A comparative study. *Wind. Struct.* **2016**, *22*, 107–131. [CrossRef]
183. Zhao, Z.; Dai, K.; Lalonde, E.R.; Meng, J.; Li, B.; Ding, Z.; Bitsuamlak, G. Studies on application of scissor-jack braced viscous damper system in wind turbines under seismic and wind loads. *Eng. Struct.* **2019**, *196*, 109294. [CrossRef]
184. Zuo, H.; Bi, K.; Hao, H. A state-of-the-art review on the vibration mitigation of wind turbines. *Renew. Sustain. Energy Rev.* **2020**, *121*, 109710. [CrossRef]
185. Rahman, M.; Ong, Z.C.; Chong, W.T.; Julai, S.; Khoo, S.Y. Performance enhancement of wind turbine systems with vibration control: A review. *Renew. Sustain. Energy Rev.* **2015**, *51*, 43–54. [CrossRef]
186. Rupakhety, R.; Sigurdsson, S.U.; Papageorgiou, A.S.; Sigbjörnsson, R. Quantification of ground-motion parameters and response spectra in the near-fault region. *Bull. Earthq. Eng.* **2011**, *9*, 893–930. [CrossRef]
187. Elias, S.; Rupakhety, R.; Ólafsson, S. Tuned mass dampers for response reduction of a reinforced concrete chimney under near-fault pulse-like ground motions. *Front. Built Environ.* **2020**, *6*, 92. [CrossRef]
188. Sigurðsson, G.Ö.; Rupakhety, R.; Rahimi, S.E.; Olafsson, S. Effect of pulse-like near-fault ground motions on utility-scale land-based wind turbines. *Bull. Earthq. Eng.* **2020**, *18*, 953–968. [CrossRef]
189. Jami, M.; Elias, S.; Rupakhety, R.; De Domenico, D.; Falsone, G.; Ricciardi, G. Are Bridges Safe Under Near-Fault Pulse-Type Ground Motions Considering the Vertical Component? In *International Conference of the European Association on Quality Control of Bridges and Structures*; Springer: Cham, Switzerland, 2021; pp. 1207–1215.

applied sciences

MDPI

Technical Note

Debris Flow Classification and Risk Assessment Based on Combination Weighting Method and Cluster Analysis: A Case Study of Debris Flow Clusters in Longmenshan Town, Pengzhou, China

Yuanzheng Li [1,2], Junhui Shen [1,2,*], Meng Huang [1,2] and Zhanghai Peng [1,2]

1 State Key Laboratory of Geohazard Prevention and Geoenvironment Protection, Chengdu University of Technology, Chengdu 610059, China; lyzgeo@stu.cdut.edu.cn (Y.L.); huangmeng@stu.cdut.edu.cn (M.H.); pzh@stu.cdut.edu.cn (Z.P.)

2 College of Environment and Civil Engineering, Chengdu University of Technology, Chengdu 610059, China

* Correspondence: shenjunhui@cdut.edu.cn

Abstract: Debris flows can damage infrastructure and threaten human life and property safety, especially in tourist attractions. Therefore, it is crucial to classify and evaluate the risk of debris flows. This article takes 14 debris flows in Longmenshan Town, Pengzhou, Sichuan, China, as the research object. Based on on-site geological surveys, combined with drone images and multiple remote sensing images, the essential characteristics of each debris flow are comprehensively determined. A total of nine factors are used as the primary indicators affecting the risk of debris flow: drainage density, roundness, the average gradient of the main channel, maximum elevation difference, bending coefficient of the main channel, the loose-material supply length ratio, vegetation area ratio, population density, and loose-material volume of unit area. The subjective weights of each indicator are obtained using the Analytic Hierarchy Process, while the objective weights are obtained using the CRITIC method. Based on this, the distance function is introduced to couple the subjective and objective weights, determine each indicator's combined weights, and obtain the integrated evaluation score values of different debris flow hazards. Considering the integrated evaluation score of debris flow, cluster analysis was used to classify 14 debris flows and cluster effectiveness indicators were introduced to determine the effectiveness of debris flow classification. A quantitative standard for the risk of debris flow within the study area was proposed, and finally, a risk assessment of debris flow in the study area was made. Comparing the results of this paper with the gray correlation method, the coupled synergistic method, and the geological field survey results, proves that the proposed method is feasible and provides a reasonable scientific basis for the study of the hazard assessment of regional debris flow clusters and other related issues within the scope of the Jianjiang River basin and other areas.

Keywords: combination weighting method; cluster analysis; optimization; classification of debris flows; risk assessment; Longmenshan Town

Citation: Li, Y.; Shen, J.; Huang, M.; Peng, Z. Debris Flow Classification and Risk Assessment Based on Combination Weighting Method and Cluster Analysis: A Case Study of Debris Flow Clusters in Longmenshan Town, Pengzhou, China. *Appl. Sci.* **2023**, *13*, 7551. https://doi.org/10.3390/app13137551

Academic Editors: Ricardo Castedo, Miguel Llorente Isidro and David Moncoulon

Received: 22 May 2023
Revised: 19 June 2023
Accepted: 24 June 2023
Published: 26 June 2023

1. Introduction

Debris flow is a common geological hazard widely distributed in mountainous areas. It is a debris flow composed of water, rock, soil, and steam, and its formation process is highly complex. Its eruption is sudden and short-lived [1–3]. Due to the characteristics of high density, strong fluidity, and fast flow velocity, debris flows are highly destructive. In recent years, frequent debris flows have significantly harmed human life, property, the economy, and the environment, especially in mountainous areas' seismic and geological active zones [4–6]. The CMLR (Chinese Ministry of Land and Resources) reports that thousands of disasters occur yearly in China, and mountainous hazards threaten 74 million people. Specifically, during the decade between 2001 and 2010, mountain hazards caused 9933 deaths and missing persons, excluding approximately 25,000 deaths caused by landslides, collapses, and mudslides during the Wenchuan earthquake [7]. Especially for the debris flow group located in the active seismic zone and scenic area, because of the dense population and numerous road networks, if the debris flow disaster occurs, it will cause incalculable losses, so it is essential to classify debris flow in particular areas and assess the risk of debris flow in special areas [8,9].

In recent years, research on the risk assessment of debris flows has been divided into numerical simulation [10], empirical analysis [11], and artificial intelligence [12]. Among them, numerical simulation can not only simulate the movement process of debris flows but also calculate flow velocity and movement distance. However, the parameters required for numerical simulation calculation are generally difficult to obtain. The simulation process is also relatively complex, making the obtained results not necessarily consistent with the actual situation. Moreover, simulating and calculating each debris flow is impractical for regional debris flow clusters [13,14]. The empirical analysis mainly depends on the geological engineering scientists' on-site survey results. Human subjectivity plays a leading role. Different researchers may have different evaluation criteria for the same debris flow risk, leading to the need for empirical analysis methods to combine other methods to determine the debris flow risk comprehensively [15].

With the development of artificial intelligence, extension theory [16], artificial neural networks [17], Bayesian theory [18], genetic algorithms [19], evidence weight method [20], grey correlation method, and other methods have been widely used in debris flow risk assessment [21]. When most algorithms are applied to debris flow risk assessment, complex risk assessment indicators need to be selected, indicator weights need to be calculated, and the final evaluation model needs to be determined. However, things could still be improved in the current method of determining the weight of evaluation indicators. Researchers often pay more attention to the influence of objective indicators or subjective factors, and it is necessary to consider the collective impact of the two comprehensively. Some evaluation models require manual scoring and determination of grading boundaries, leading to certain deficiencies in the evaluation results of debris flow risk. Therefore, understanding how to combine the influence of subjective and objective weights and establish a scientific evaluation model based on this has practical significance for the risk assessment of debris flows.

Through an on-site geological survey, combined with UAV images and remote sensing images, 14 debris flows were found in Longmenshan Town, Pengzhou City, Sichuan Province, China, distributed on both sides of Baishui River and Jianjiang River, with 5 on both sides of Jianjiang River, and 9 on both sides of Baishui River. Considering the rapid and sudden occurrence of debris flows and many surrounding villages and tourists, it is essential to classify and evaluate the risk of debris flows in this area.

2. Study Area

Pengzhou City is located in the northwest of Sichuan Basin and the northwest edge of Chengdu Plain, 36 km away from the urban area of Chengdu, spanning 103°40′~104°10′ east longitude and 30°54′~31°26′ north latitude. The city covers an area of 1421 km^2. Longmenshan Town is located in the north of Pengzhou, upstream of Jianjiang River, 55 km away from the Pengzhou urban area, connecting Shifang City in the east, Cifeng Town in the south, Dujiangyan Irrigation Project in the west, and Wenchuan County in the north. The research area is located in the core area of the Longmenshan Fault, with the Yingxiu Beichuan Fault passing through the right bank of the Baishui River and the Guanxian Anxian Fault passing through the downstream Xiaoyudong Town, which belongs to a controlled structure. The base tectonic layer in the study area is the Huangshuihe Group stratum, and the main rock types of the group is the "Pengguan Complex", which has experienced many strong orogenies (Himalayan movement, Indosinian movement, Chengjiang movement). Finally, neotectonics led to the formation of typical mountain canyon geomorphic features. Longmenshan Town has a humid subtropical climate, with the highest temperature of 24.8 °C in summer and the coldest temperature of 5.2 °C in January in winter. The average annual rainfall for many years is 932.5 mm. There is a significant amount of rainfall in summer, primarily rainstorms. Due to the humid subtropical climate, the plants in the study area are mainly subtropical alpine forest vegetation. Before the 2008 earthquake, the vegetation coverage rate exceeded about 60%. The "Longmen Mountain National Scenic Area" is located here and is a famous tourist resort. After the Wenchuan earthquake, geological disasters frequently occurred in the area, with different numbers and scales of debris flow occurring in 2008, 2009, 2010, 2012, and 2022, causing significant property losses and casualties. The distribution of multiple remote sensing images and debris flows in the study area is shown in the following figure (Figures 1–3).

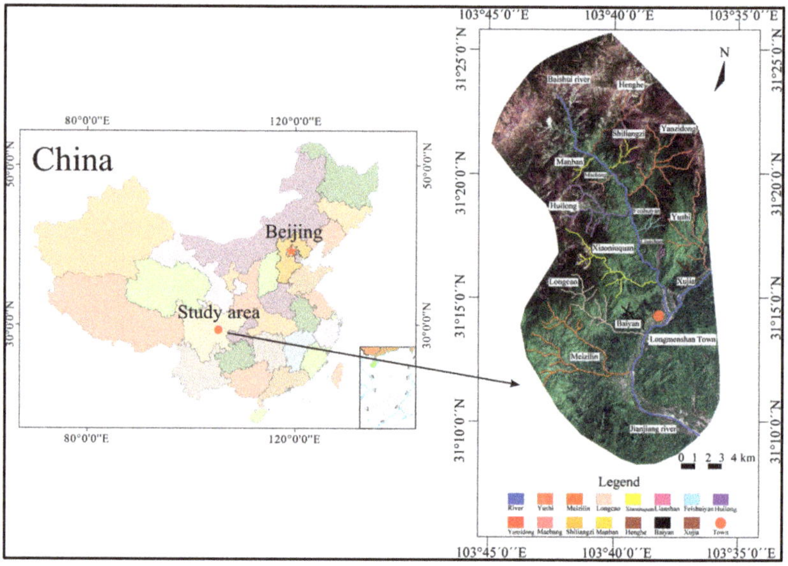

Figure 1. Remote sensing image of GF-6 and distribution of 14 debris flow gullies in the research area.

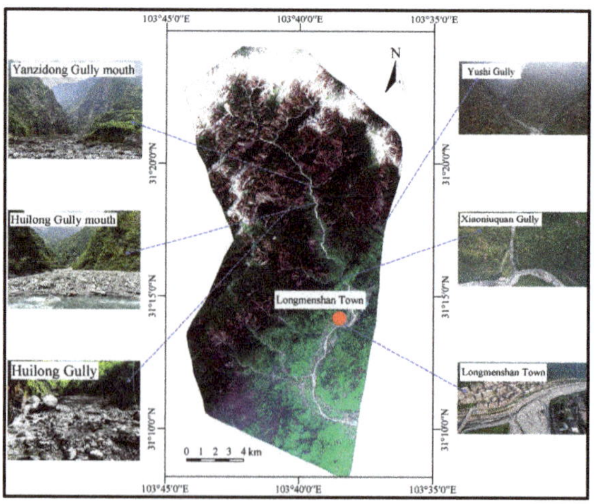

Figure 2. Remote sensing images of the research area (Pleiades).

Figure 3. Remote sensing images of the research area (GF-2).

3. Materials and Methods

The calculation of the weight of debris flow indicators mainly adopts a single subjective and objective weighting method. The subjective weighting method obtains weights based on individuals' subjective experiences, such as AHP. This has unique advantages in determining the weights of various indicators at different levels in an extensive system and can fully utilize the expert experience in the corresponding field. The objective weighting rule entirely relies on the laws of the data, such as the CRITIC method, which can reflect the relative importance of various factors. Combining the AHP and CRITIC methods can reflect researchers' intuitive understanding of debris flow in the geological field survey stage and consider the regularity of objective data, making the weight obtained more scientific.

The classification of debris flow is a straightforward guide to prevent the occurrence of debris flow disasters, and the classification method is well established [22]. However, because there are many classification methods, the same debris flow has different classification results under different criteria. The affinity propagation cluster analysis method, which applies to analyzing various geostatistical data, is applied to classify debris flows in this paper [23,24]. The final result of debris flow hazard evaluation is obtained based on the calculation results of debris flow index weights combined with the classification of debris flow hazard. The risk assessment process is shown in Figure 4:

3.1. Indicator Selection

The selection of risk assessment indicators for debris flow mainly considers the primary conditions for forming and developing debris flow disasters. From the quantitative evaluation requirements perspective, specific indicators need to reflect the debris flow risk. Geological conditions, material conditions, and trigger conditions play a crucial role in the distribution and activity of debris flows. When selecting debris flow indicators, it is necessary to consider the scientific, representative, comprehensive, and regional differences between the indicators. Table 1shows the indicators selected by researchers worldwide in recent years in the study of debris flow risk assessment. Table 1 shows specific differences in the selection of evaluation indicators in different regions. Considering that rainfall within the study area is the same, it is difficult to accurately obtain the debris flow frequency and soil particle size. At the same time, the lithology and fault length of the strata can be reflected to a certain extent by the amount of material sources. The study area is located in a scenic area with a relatively dense population. Therefore, based on on-site geological surveys, combined with the analysis results of drone images and multiple remote sensing images, there are nine specific factors that are important and closely related to the occurrence of debris flow in the research area and can be used as essential indicators for debris flow risk assessment. These are drainage density, roundness, average gradient of main channel, maximum elevation difference, Bengding coefficient of main channel, loose-material supply length ratio, vegetation area ratio, population density, and loose-material volume of unit area. This article utilizes multiple remote sensing images, digital elevation models (DEM), drone stereo aerial photography, and field investigations (Figures 1–3 and 5) (Table 2) to ultimately obtain the size of various risk assessment indicators for debris flow in the study area through calculations and depictions.

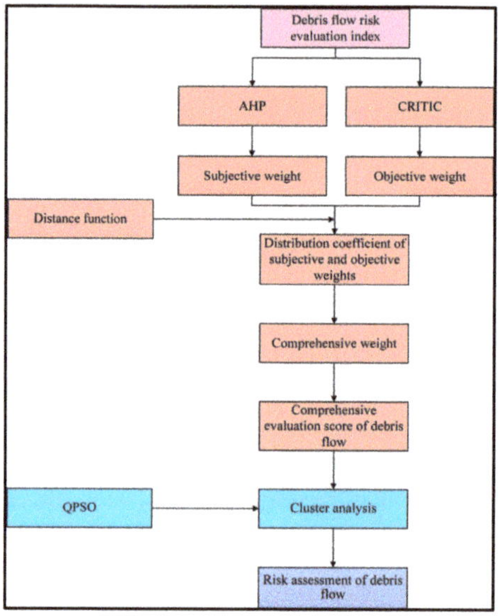

Figure 4. Debris Flow risk assessment process.

Table 1. Factors frequently used in risk assessment of debris flow.

Factors	[25]	[26]	[27]	[28]	[29]	[30]	[31]	[32]	[33]	Time
Rainfall intensity		√	√					√		3
Daily rainfall			√						√	2
Cumulative rainfall		√	√					√		3
Main channel length	√	√	√				√	√	√	6
Gully slope angle	√	√	√			√	√	√		6
Drainage density	√	√	√				√	√	√	6
Soil particle size		√	√					√		3
Basin area	√	√	√	√	√	√	√	√	√	9
Average gradient of main channel	√			√	√	√			√	5
Slope direction					√					1
Vegetation coverage	√			√		√	√			4
Loose material volume	√						√			2
Population density	√							√		2
Maximum elevation difference	√				√	√	√	√		5
Bengding coefficient of main channel	√				√			√		3
Fault length						√				1
Frequency						√	√		√	3

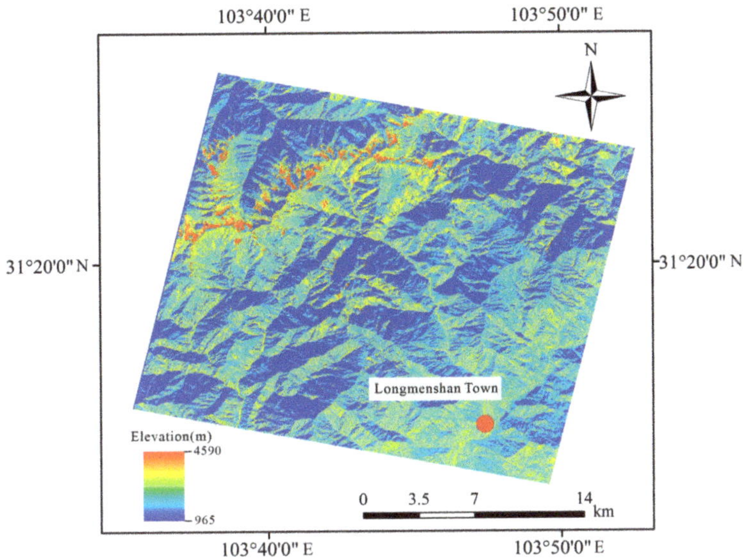

Figure 5. Research area digital elevation model (DEM).

Table 2. Data Description.

Data Type	Date	Resolution	Source
Remote sensing image (Figure 1)	2020.9	2 m	GF-6
Remote sensing image (Figure 2)	2020.8	0.5 m	Pleiades
Remote sensing image (Figure 3)	2020.11	0.8 m	GF-2
DEM (Figure 5)	2020.11	0.8 m	GF-2

Drainage density (F1) (km/km^2): The ratio of the total length of gullies developed within the debris flow basin area to the basin area, comprehensively reflecting the engineering geological conditions within the watershed. This value is calculated using ArcGIS geometry from remote sensing images.

Roundness (F2) (km/km^2): This refers to the ratio of the length of the main gully of a debris flow to the basin area. In general, at different stages of debris flow development, the flat form of valleys varies, and the degree of danger also varies. This value is calculated using ArcGIS geometry from remote sensing images.

Average gradient of main channel (F3) (°): This is the ratio of the maximum elevation difference of the main channel to its linear length. The larger the value, the better the hydrodynamic condition is. The value is obtained through DEM.

Maximum elevation difference (F4) (m): The difference between the highest and lowest elevations in the basin provides kinetic conditions for the occurrence of debris flow disasters. The value is obtained through DEM.

Bengding coefficient of main channel (F5): This refers to the ratio of the main channel length to its linear length, which reflects the degree of channel blockage. The size of the bending coefficient is positively correlated with the blockage coefficient and is related to the flow rate and scale of the debris flow. This value is calculated using ArcGIS geometry from remote sensing images.

The loose-material supply length ratio (F6) (%): This refers to the ratio of loose-material length along a channel to total channel length, which reflects the successive supplied sediments. This value is obtained through on-site geological surveys and remote sensing images.

Vegetation area ratio (F7) (%): Low vegetation coverage can cause severe soil erosion in the basin. The value is obtained through drone aerial photography and remote sensing images, and the vegetation coverage is estimated based on the depth of the color. The lighter the color, the lower the vegetation coverage, and this is corrected through drone aerial photography.

Population density (F8) (number of people per km^2): With the development of the economy and technology, human activities have become one of the essential factors affecting mudslides, and population density can reflect the intensity of human activities. This value is estimated based on the number of buildings using remote sensing images and confirmed through on-site investigations.

Loose-material volume of unit area (F9) ($\times 10^4$ m^3/km^2): The ratio of the source quantity of a single debris flow to the basin area. The source of materials in the ditch is one of the basic factors that cause debris flow disasters, and the size of the unit area loose material volume is directly proportional to the risk of debris flow. The value is obtained by combining a Laser rangefinder with a remote-sensing image, and the thickness is obtained by combining field estimation and drilling data.

Table 3 shows the evaluation index values of debris flow risk in the study area.

Table 3. Evaluation index values of debris flow risk in the research area.

Number	Debris Flow	F1	F2	F3	F4	F5	F6	F7	F8	F9
1	Xiaoniuquan	1.11	0.66	16.20	2350	1.1711	39.46	70	5	119.5
2	Lianshan	2.15	1.36	33.56	1430	1.1250	11.06	60	5	25.50
3	Feishuiyan	2.20	1.52	33.66	1420	1.1620	12.32	60	5	21.33
4	Huilong	1.25	0.49	20.09	2130	1.1535	38.72	70	5	122.70
5	Yanzidong	1.19	0.58	14.28	2245	1.1816	40.90	70	5	113.20
6	Shiliangzi	1.47	0.78	30.26	1500	1.2731	21.66	55	1	35.16
7	Machang	1.92	1.33	35.58	1430	1.1947	19.71	55	1	38.99
8	Manban	1.73	1.40	29.02	1400	1.0588	22.72	55	1	36.47
9	Henghe	0.80	0.50	18.53	2137	1.1615	39.13	50	1	135.30
10	Yushi	0.94	0.45	15.57	2500	1.2020	44.05	80	150	259.34
11	Longcao	1.80	0.75	16.62	2300	1.2160	42.45	80	150	277.80
12	Meizilin	0.80	0.33	15.59	1700	1.2429	43.22	80	150	252.57
13	Xujia	1.70	1.22	18.48	1130	1.1236	14.66	60	50	35.16
14	Baiyan	2.53	2.19	33.98	1320	1.1841	13.45	60	30	24.86

3.2. *Combination Weighting Method*

3.2.1. CRITIC Method

The CRITIC method is an objective weighting method that reflects the discreteness and factor conflict between samples through standard deviation and correlation coefficient. The size of the standard deviation is directly proportional to the degree of discreteness and factor weight, and the size of the correlation coefficient is also directly proportional to the conflict between factors. The larger the correlation coefficient, the smaller the weight [34]. The CRITIC method takes into account sample information and factor correlation. Also, it utilizes the coefficient of variation to make the dispersion reflected by standard deviation more realistic, with significant advantages [35–37]. The specific calculation steps are as follows:

(1) Assuming m samples containing n indicators, construct the original data matrix using the indicators:

$$X = \begin{bmatrix} \alpha_{11} & \alpha_{12} & \cdots & \alpha_{1n} \\ \alpha_{21} & \alpha_{22} & \cdots & \alpha_{2n} \\ \cdots & \cdots & \cdots & \cdots \\ \alpha_{m1} & \alpha_{m2} & \cdots & \alpha_{mn} \end{bmatrix} \quad (1)$$

(2) Normalization of indicators:

$$q_{ij} = \frac{\alpha_{ij} - \min_j(\alpha_{ij})}{\max_j(\alpha_{ij}) - \min_j(\alpha_{ij})} \quad (2)$$

(3) Calculate coefficient of variation:

$$\overline{\alpha_j} = \frac{\sum_{i=1}^{m} \alpha_{ij}}{m} \quad (3)$$

$$S_j = \sqrt{\frac{1}{m}\sum_{i=1}^{m} (\alpha_{ij} - \overline{\alpha_j})^2} \quad (4)$$

$$\mu_j = \frac{S_j}{\overline{\alpha_j}} \quad (5)$$

In the formula, $\overline{\alpha_j}$ is the average value of each indicator; S_j is the standard deviation; μ_j is the coefficient of variation.

(4) Calculate the correlation coefficient matrix:

$$\kappa_{ij} = \frac{cov(y_k, y_l)}{(s_k s_j)}, k = 1,2,\ldots,n; l = 1,2,\ldots,n \quad (6)$$

In the formula, κ_{ij} represents the correlation coefficient between indicators, and $cov(y_k, y_l)$ represents the covariance between indicators.

(5) Calculation of indicator information quantity:

$$\omega_j = \mu_j \sum_{i=1}^{n} (1 - \kappa_{ij}), j = 1,2,\ldots,n \quad (7)$$

The weights of each indicator are:

$$y_j = \frac{\omega_j}{\sum_{i=1}^{n} \omega_j} j = 1,2,\ldots,n \quad (8)$$

3.2.2. Analytic Hierarchy Process (AHP)

The Analytic Hierarchy Process (AHP) was proposed by renowned mathematician Saaty and is a simple and feasible decision-making method with significant subjectivity [38]. The main advantage lies in the ability to determine the weights of various indicators at different levels in the system, which has unique advantages and is simple and convenient to calculate; therefore, it is widely used [39,40]. The specific calculation steps are as follows:

(1) Establish a tiered hierarchical structure model. The hierarchical structure is generally divided into three layers: target layer, criterion layer, and scheme layer.

(2) Establish the judgement matrix. For different factors at the same level, establish a judgment matrix by comparing their impact on the target factors. The formula for constructing the judgment matrix is as follows:

$$A = (a_{ij})_{n \times n}, a_{ij} > 0, \quad a_{ij} = \frac{1}{a_{ij}}, (i, j = 1,2,\ldots n) \quad (9)$$

Among them, a_{ij} is the ratio of the influence degree of elements B_i and B_j, usually represented by a scoring method of one to nine, as shown in the Table 4 below.

Table 4. Definition of comparative importance.

1	Two decision factors (e.g., indicators) are equally important
3	Two decision factors (e.g., indicators) are equally important
5	Two decision factors (e.g., indicators) are equally important
7	One decision factor is very strongly more important
9	One decision factor is extremely more important
2, 4, 6, 8	Intermediate values
Reciprocals	If ij is the judgement value when i is compared to j, then $U_{ji} = 1/U_{ji}$ is the judgement value when j is compared to i

(3) Calculate consistency indicator CI. By calculating the eigenvalues and eigenvectors of the judgment matrix, it can be represented as:

$$CI = \frac{\lambda_{\max} - n}{n - 1} \tag{10}$$

Among λ_{max} is the eigenvalue of matrix A.
(4) Calculate consistency ratio CR

$$CR = \frac{CI}{RI} \tag{11}$$

Among them, RI is the consistency indicator of the judgment matrix (Table 5). When $CR < 0.1$, it is judged that the matrix meets the consistency requirement. Otherwise, it is considered that the matrix does not meet the consistency requirement, and further adjustments are needed until the consistency check is met.

Table 5. The random average consistency index.

n	1	2	3	4	5	6	7	8	9	10	11	12
RI	0	0	0.52	0.89	1.12	1.26	1.36	1.41	1.46	1.49	1.52	1.54

3.2.3. Combination Weighting Rule

The Analytic Hierarchy Process determines the judgment matrix mainly based on the subjective experience of experts. In contrast, the evaluation process of the CRITIC method relies entirely on the own laws of objective data. In order to reflect both the researchers' intuitive understanding of debris flow in the field geological investigation stage and to take into account the regularity of objective data, the degree of difference between the weights obtained by the two methods is consistent with the degree of difference between their corresponding distribution coefficients, this paper introduces a distance function, which couples the weights obtained by the two methods together to determine the index weights comprehensively.

Suppose the weight vector obtained by the Analytic Hierarchy Process method is $\omega_i{}^c$, the weight vector obtained by the CRITIC method is $\omega_i{}^y$, and the distance function between them is denoted as $d(\omega_i{}^c, \omega_i{}^y)$ [33]:

$$d(\omega_i{}^c, \omega_i{}^y) = \left[\frac{1}{2}\sum_{i=1}^{n}(\omega_i{}^c - \omega_i{}^y)^2\right]^{\frac{1}{2}} \tag{12}$$

Assuming that the combined weights are ${\omega_i}^z$ and the linear weighting method is used to obtain ${\omega_i}^z$, and assuming that the distribution coefficients of the two weights are a and b, respectively, then ${\omega_i}^z$ can be expressed as:

$$
{\omega_i}^z = a{\omega_i}^c + b{\omega_i}^y \tag{13}
$$

To ensure a consistent degree of variation in the magnitude of the weights and the distribution coefficients, make the distance function and the distribution coefficients the same:

$$
d({\omega_i}^c, {\omega_i}^y)^2 = (a-b)^2 \tag{14}
$$

The two weight assignment coefficients also have to satisfy Equation (15). Combining Equations (14) and (15) yields the assignment coefficients a and b. Substituting a and b into Equation (13) yields ${\omega_i}^z$.

$$
a + b = 1 \tag{15}
$$

3.3. Cluster Analysis

As an essential method for studying classification problems, cluster analysis groups similar things together as much as possible and separates things that are more different (Figure 6). The fundamental laws within things can be more clearly recognized by cluster analysis, which plays an essential role in several scientific fields [41–46]. There are certain drawbacks to the traditional clustering method: (1) artificially determining the number of groups and (2) artificially selecting the clustering center. However, the result of such division is mainly affected by human factors, which makes the final division unreliable [47–51].

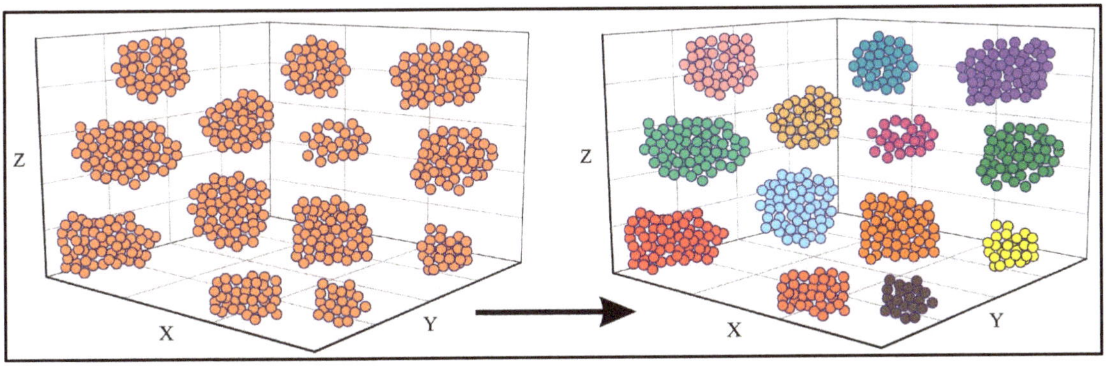

Figure 6. Cluster Analysis.

The affinity propagation clustering algorithm overcomes the shortcomings of traditional clustering analysis. Its principle is to achieve efficient and accurate data clustering by iteratively transmitting attraction and attribution information between data points in a given data set. It is a clustering algorithm based on the "information transfer" between data points. It takes the similarity between a pair of data points as input and exchanges accurate and valuable information between data points until an optimal set of class representative points and clustering are gradually formed. The main advantages include: (1) the number of clusters and the cluster center can be obtained by calculation, which does not need to be specified manually; (2) each data point can be used as a potential cluster center; (3) the clustering results are unique; (4) the starting condition of the algorithm is the input correlation matrix, and (5) there is no requirement for the symmetry of the matrix. The specific algorithm flow is as follows [52,53].

(1) Calculate data point correlation matrix.

(2) Determining the size of p (Preference) and the number of iterations.
(3) Calculate the responsibility information and the availability information between monitoring points.
(4) Update the responsibility information and availability information.
(5) Calculate the cluster center.
(6) The maximum number of cycles is reached, and the final result is obtained.

In the study of debris flow classification, traditional clustering analysis methods require the manual determination of classification numbers, which is subjective. However, using the affinity propagation clustering algorithm to classify debris flows can be completed without specifying the number of classifications and clustering centers in advance. The calculation results are unique and reasonable. However, an essential parameter in the affinity propagation algorithm is the reference p-value, which refers to the reliability of using data points as clustering centers. The size of the reference p-value directly affects the clustering results, and its size is directly proportional to the number of clusters. Improper selection of p-values can lead to poor clustering results [54]. In general, when there is no prior knowledge, the p-value is set as the median of the similarity matrix and remains unchanged during the clustering process. However, selecting the median of the similarity matrix may not necessarily result in the optimal result [55]. Considering that the basis of algorithm startup is a correlation matrix, and different p-values input in the calculation process will also lead to different calculation results, therefore, determine the correlation between debris flows before calculation, and use quantum particle swarm optimization (QPSO) [56] to optimize the p-value, find the p-value under the optimization condition of the objective function, and obtain the classification result of debris flows under the optimization p-value condition.

3.3.1. Correlation Calculation

They assumed that for any two debris flows, i and j, each has z evaluation indicators. The k_{th} evaluation indicator of the two debris flows can be expressed as i_k and j_k. The evaluation indicators of debris flow i and j can be combined into a set of data pairs (i_k, j_k) ($1 \leq k \leq z$). For any two data pairs (i_k, j_k) and (i_l, j_l) in a set, when $i_k > i_l$ and $j_k > j_l$, or $i_k < i_l$ and $j_k < j_l$, the data pair is said to be consistent; when $i_k > i_l$ and $j_k < j_l$, or $i_k < i_l$ and $j_k > j_l$, the data pair is said to be inconsistent; when $i_k = i_l$ and $j_k = j_l$, this data pair is neither consistent nor inconsistent. If correlation analysis is conducted on the evaluation indicators between two debris flows, the correlation between any two evaluation indicators can be expressed as [57]:

$$\tau_{ij} = \frac{2C}{\frac{1}{2}z(z-1)} - 1 = \frac{4C}{z(z-1)} - 1 \tag{16}$$

Among them, C is the number of identical order pairs. The value range of τ_{ij} is $[-1, +1]$, and when $\tau_{ij} = 1$, it indicates that the two debris flows have the exact level correlation; when $\tau_{ij} = -1$, it indicates that two debris flows have opposite level correlations; when $\tau_{ij} = 0$, it indicates that the two debris flows are independent of each other. For the risk assessment indicators of debris flow, due to the different dimensions of each indicator, correlation calculation can not only eliminate the impact of different dimensions of each evaluation indicator but also serve as a prerequisite for establishing a correlation matrix [58].

3.3.2. Classification and Risk Assessment of Debris Flow

Assuming there are n debris flows in total, for a particular debris flow i, it is necessary to calculate the correlation between debris flow i and $n-1$ debris flow other than debris flow i and establish a debris flow correlation matrix S_{ij}:

$$S_{ij} = \begin{bmatrix} \tau_{11} & \tau_{12} & \cdots & \tau_{1n} \\ \tau_{21} & \tau_{22} & \cdots & \tau_{2n} \\ \cdots & \cdots & \cdots & \cdots \\ \tau_{n1} & \tau_{n2} & \cdots & \tau_{nn} \end{bmatrix} \tag{17}$$

Based on the correlation matrix of debris flow, affinity propagation clustering analysis is conducted because selecting different p-values will result in different classification results. However, too many or too few classification numbers do not match debris flow's actual risk classification results. Therefore, the optimal classification results need to be determined through the clustering effectiveness function. Among many clustering indicators, the Silhouette indicator is widely used, which can not only reflect the intra-class tightness and inter-class separateness of clustering results but also assess the optimal number of clusters and evaluate the quality of clustering, so the Silhouette indicator is chosen to judge the optimal debris flow classification.

Suppose there is a data set with n data points, which is divided into K clusters C_i ($i = 1, 2, \ldots K$). $a(t)$ denotes the average dissimilarity of data point t in cluster C_j to all other data points within C_j, and $d(t, C_i)$ is the average dissimilarity of data point t in C_j to all data points in another class C_i, then $b(t) = \min\{d(t, C_i\}$, where $i = 1, 2, \ldots K, i \neq j$. Therefore, the Silhouette index of a data point is [59–61]:

$$S(t) = \frac{b(t) - a(t)}{\max\{a(t), b(t)\}} \tag{18}$$

The average $S(t)$ value $S_{avg}(C_i)$ of all data points in cluster C_i can be obtained from $S(t)$, which reflects the compactness and separation of cluster C_i. The average $S(t)$ value S_{avg} of all data points in a dataset can reflect the quality of clustering results. The larger S_{avg}, the better the clustering quality, and the optimal number of clusters must correspond to the maximum S_{avg} value. The formula is as follows:

$$S_{avg} = \frac{\sum_{t=1}^{n} S(t)}{n} \tag{19}$$

According to the above description, when the quantum particle swarm optimization (QPSO) algorithm is used to optimize the p-value, assuming that the number of p-values to be optimized is N, the p-values to be optimized are $P_1, P_2, P_3, \ldots P_N$ and the total number of variables to be solved are N, which can be transformed into an N-dimensional optimization problem [62,63]. The optimization calculation process is as follows:

Step 1: Calculate the correlation matrix S_{ij} of debris flow.

Step 2: Initialization. The qubit phase plays the role of the random initial population, which is in the range of $[0, 2\pi]$ calculated by random number function. Then, combined with the upper and lower limits of p-values variables, by solving the solution space transformation formula, the probability amplitude is converted to the variable space.

Step 3: The correlation matrix S_{ij} and p-value variables were used for affinity propagation clustering calculation, and the Silhouette index value was obtained from the output classification results. The average Silhouette index value is taken as the fitness value of the QPSO algorithm. Considering the comprehensive evaluation of classification results, the average Silhouette index value (S_{avg}) is selected as the objective function.

Step 4: Nonlinear adjustment of inertia weight.

Step 5: Update particle state (update qubit depression angle and qubit probability amplitude)

Step 6: Adaptive adjustment of mutation operator and mutation processing. The quantum nongate is used to mutate particles.

Step 7: The correlation matrix S_{ij} and p-value variables were used for affinity propagation clustering calculation, and the average value of the Silhouette index was calculated. If the average value of Silhouette index meets the stopping condition, or if the number of iterations has reached the maximum, the clustering results under the optimized p-value will be output; otherwise, return to Step 3 to continue the cyclic calculation.

Figure 7 shows the flowchart of the debris flow classification algorithm:

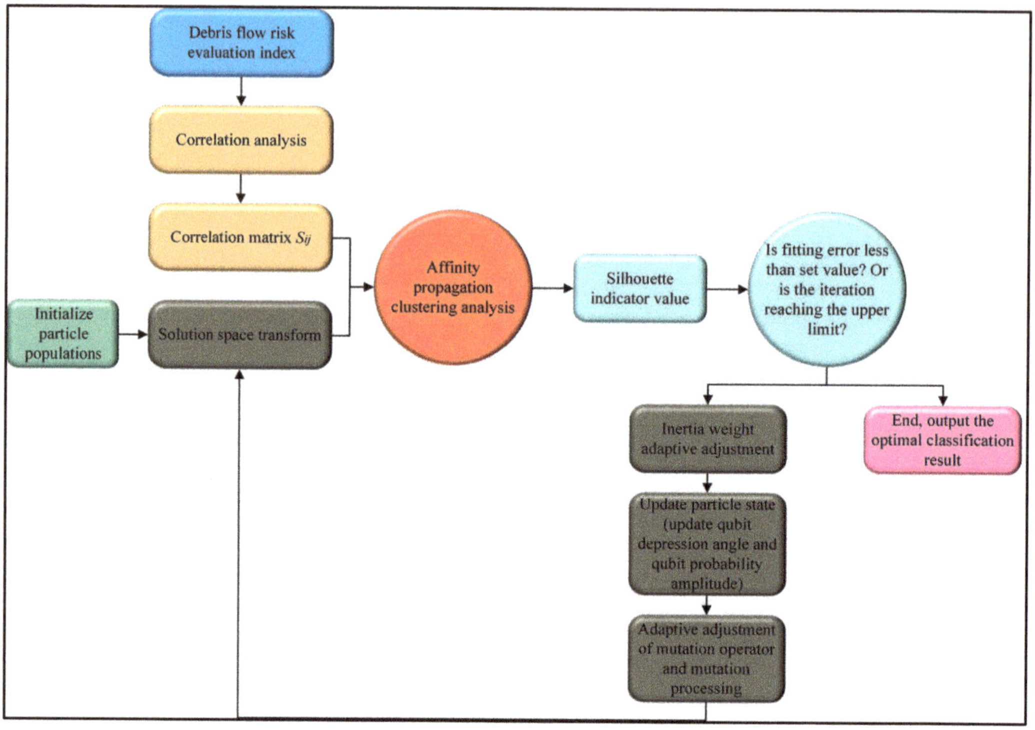

Figure 7. Flow chart of debris flow classification algorithm.

Based on the clustering analysis method proposed above, debris flows can be divided into different types. However, the clustering analysis results only classify debris flows into different categories, and another judgment is needed regarding which risk level corresponds to different categories of debris flows.

This article calculates the synthetic evaluation score (D_i) for each debris flow risk based on the combined weight values obtained by the combination weighting method (Equation (20)) [33]. The larger D_i, the greater the probability of the debris flow occurring and the more dangerous it is. Select indicators and synthetic evaluation scores to calculate the correlation between debris flows. Based on correlation calculation (τ_{ij}), establish a correlation matrix (S_{ij}) and perform cluster analysis. Based on the clustering results and synthetic evaluation scores, classify the risk of debris flow.

$$D_i = \sum_{j=1}^{n} \omega_z(j) q_{ij} \tag{20}$$

4. Classification and Risk Assessment Results of Debris Flow in the Study Area

4.1. Weight Calculation

4.1.1. Results of the AHP

Based on the selection of evaluation indicators for debris flow, a hierarchical structure model for evaluating the risk of debris flow groups in Longmenshan Town is constructed (Figure 8).

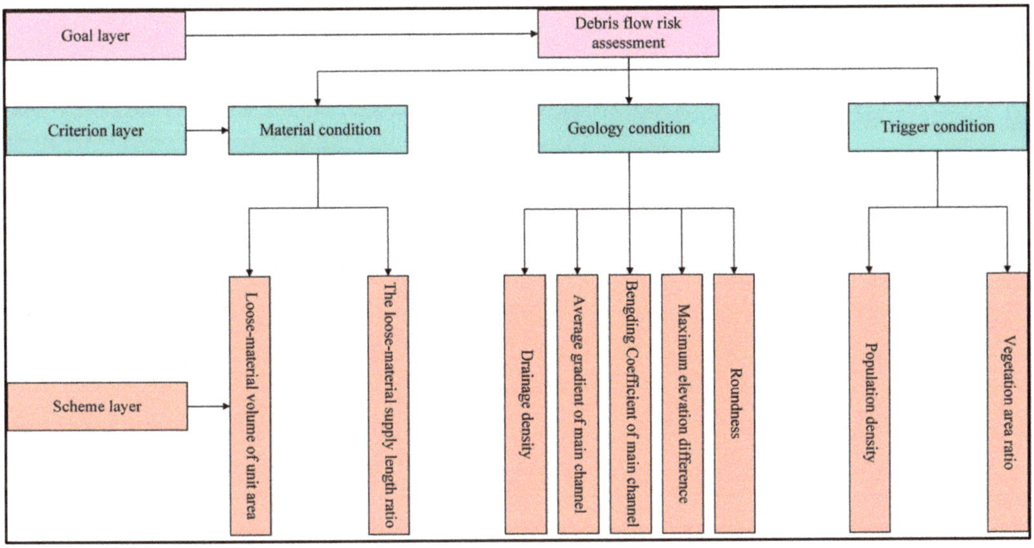

Figure 8. Hierarchical structure for debris flow risk assessment.

According to the hierarchical structure model of debris flow evaluation indicators, each evaluation indicator is graded using the 1–9 scale method, a judgment matrix is constructed (Tables 6–9), and consistency testing is conducted. Finally, the weights of each evaluation indicator are obtained (Table 10). During the evaluation process, a total of 15 experts were selected for scoring, all of whom were from the Sichuan Province Sudden Major Geological Disaster Expert Database.

Table 6. Criterion layer judgment matrix for goal layer.

Criterion Level	Material Condition	Geology Condition	Trigger Condition	*CI*	*RI*	*CR*
Material condition	1	1/3	2	0.0268	0.52	0.0516
Geology condition	3	1	3			
Trigger condition	1/2	1/3	1			

Table 7. Criterion layer judgment matrix for geology condition.

Geology Condition	F1	F3	F5	F4	F2	*CI*	*RI*	*CR*
F1	1	1/2	3	1/3	2	0.0709	1.12	0.0633
F3	2	1	3	1/4	3			
F5	1/3	1/3	1	1/4	2			
F4	3	4	4	1	4			
F2	1/2	1/3	1/2	1/4	1			

Table 8. Criterion layer judgment matrix for material condition.

Material Condition	F9	F6	CI	*RI*	*CR*
F9	1	3	0	0	0
F6	1/3	1			

Table 9. Criterion layer judgment matrix for trigger condition.

Trigger Condition	F8	F7	CI	RI	CR
F8	1	2	0	0	0
F7	1/2	1			

Table 10. The weighted values of the factors obtained by AHP.

Evaluation Index	F1	F2	F3	F4	F5	F6	F7	F8	F9
Weight	0.1	0.04	0.13	0.28	0.05	0.06	0.05	0.10	0.19

4.1.2. Results of CRITIC Method

Using the data in Table 4, establish the original evaluation index matrix using the CRITIC method, normalize the indicators, calculate the information content, and finally use Equation (8) to obtain the objective weight value (Table 11).

Table 11. CRITIC method evaluation index weight.

Evaluation Index	F1	F2	F3	F4	F5	F6	F7	F8	F9
Amount of information	1.07	0.83	1.09	1.04	1.32	0.80	1.05	1.67	0.80
Weight	0.11	0.09	0.11	0.11	0.14	0.08	0.11	0.17	0.08

4.1.3. Results of the Combination Weighting Method

Based on the weight results obtained by the Analytic Hierarchy Process and CRITIC method, the combined weights of each evaluation index are calculated using Equations (12)–(15) (Table 12). According to Equation (12), this is calculated as:

$$d({\omega_i}^c, {\omega_i}^y) = \left[\frac{1}{2}\sum_{i=1}^{n} ({\omega_i}^c - {\omega_i}^y)^2\right]^{\frac{1}{2}} = 0.1746 \quad (21)$$

Table 12. Combination weighting method for evaluating indicator weight results.

Evaluation Index	F1	F2	F3	F4	F5	F6	F7	F8	F9
AHP	0.10	0.04	0.13	0.28	0.05	0.06	0.05	0.10	0.19
CRITIC	0.11	0.09	0.11	0.11	0.14	0.08	0.11	0.17	0.08
Combination weighting method	0.10	0.06	0.12	0.21	0.09	0.07	0.07	0.13	0.14

By combining Equations (13) and (14), it can be concluded that:

$$\begin{aligned} a - b &= 0.1746 \\ a + b &= 1 \end{aligned} \quad (22)$$

Solved:

$$\begin{aligned} a &= 0.5873 \\ b &= 0.4127 \end{aligned} \quad (23)$$

Therefore, the combination weight value obtained by the combination weighting method can be expressed as:

$${\omega_i}^z = 0.5873{\omega_i}^c + 0.4127{\omega_i}^y \quad (24)$$

4.2. Classification Results of Debris Flow

Considering the synthetic weight values obtained by the combination weighting method, the synthetic evaluation score (D_i) for each debris flow risk degree is calculated (Table 13).

Table 13. Synthetic evaluation score (D_i) of debris flow in the research area.

Number	Debris Flow	D_i	Number	Debris Flow	D_i
1	Xiaoniuquan	0.5359	8	Manban	0.2113
2	Lianshan	0.2660	9	Henghe	0.3963
3	Feishuiyan	0.2692	10	Yushi	0.7948
4	Huilong	0.4727	11	Longcao	0.8301
5	Yanzidong	0.5210	12	Meizilin	0.6718
6	Shiliangzi	0.2716	13	Xujia	0.2856
7	Machang	0.2397	14	Baiyan	0.3064

This study selected 9 evaluation index values (F1~F9) from 14 debris flows in the research area as well as the synthetic evaluation score of each debris flow (D_i). Using these 10 indicators, Equation (16) was used to calculate the correlation between the 14 debris flows τ_{ij}. Based on the correlation calculation, the correlation matrix of debris flow in the study area was established according to Equation (17) (Figure 9).

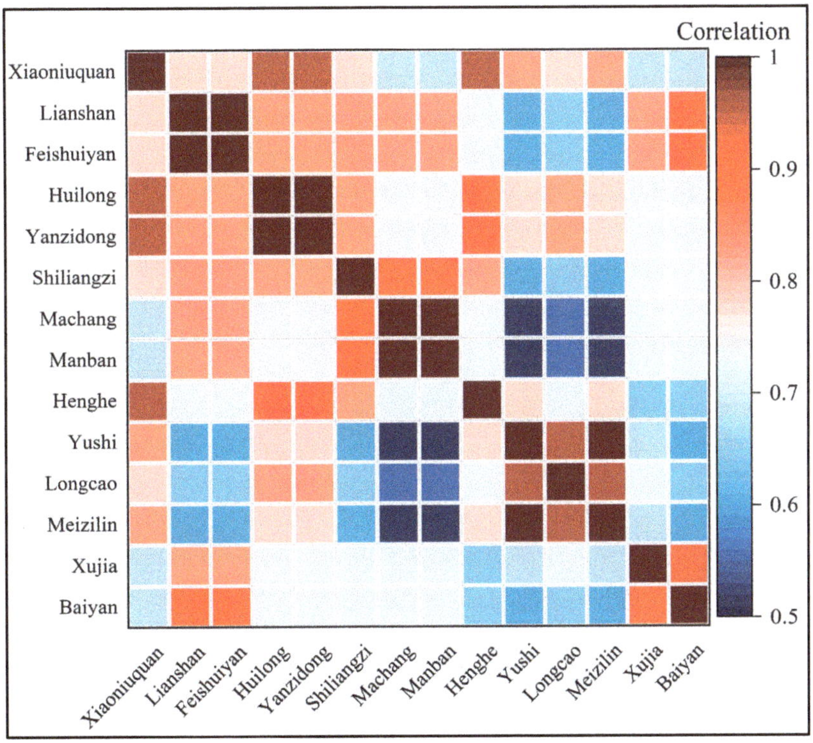

Figure 9. Calculation results of debris flow correlation in the study area.

According to the previous description of S_{avg}, the larger the calculated S_{avg} value, the more scientific and reasonable the classification results are. The correlation matrix shown in Figure 9 is used as the basis for starting the algorithm, and the QPSO-optimized affinity propagation clustering is used to compute the matrix. The calculation results showed that, when the debris flow is divided into four types, the S_{avg} index value is the highest, reaching 0.66. Therefore, the classification result is the optimal number of classifications (Table 14).

Table 14. The optimal classification results of debris flow in the research area.

Classification	Number of Categories	Debris Flow
I	3	Longcao, Meizilin, Yushi
II	4	Xiaoniuquan, Yanzidong, Huilong, Henghe
III	3	Shiliangzi, Baiyan, Xujia
IV	4	Feishuiyan, Lianshan, Machang, Manban

4.3. Risk Assessment Based on Classification Results

Based on the above clustering analysis results, debris flows are divided into four categories. However, the clustering analysis results only classify debris flows into different categories and which risk level corresponds to each of the four categories of debris flows. The clustering results are not provided, and another judgment is needed. Table 15 shows that, among Class I debris flows, Longcao Gully has the highest synthetic evaluation score of 0.8301, while Meizilin Gully has the lowest synthetic evaluation score of 0.6718. Among Class II debris flows, the synthetic evaluation score of Xiaoniuquan Gully debris flow is the highest at 0.5359, while the synthetic evaluation score of Henghe Gully is the lowest at 0.3963. Among Class III debris flows, Baiyan Gully has the highest synthetic evaluation score of 0.3064, while Shiliangzi Gully has the lowest synthetic evaluation score of 0.2716. Among Class IV debris flows, Feishui Rock Gully has the highest synthetic evaluation score of 0.2692, while Manban Gully has the lowest synthetic evaluation score of 0.2113. Therefore, based on the calculation results, the Class I debris flow in the study area is classified as extremely dangerous ($0.6718 \le D_i \le 0.8301$), the Class II debris flow is classified as highly dangerous ($0.3963 \le D_i \le 0.5359$), the Class III debris flow is classified as moderately dangerous ($0.2716 \le D_i \le 0.3064$), and the Class IV debris flow is classified as low-risk ($0.2113 \le D_i \le 0.2692$).

Table 15. Debris flow risk assessment in study area.

Classification	Number of Categories	Debris Flow	$max\ (D_i)$	$min\ (D_i)$	Debris Flow Risk Degree
I	3	Longcao, Meizilin, Yushi	0.8301	0.6718	extreme risk
II	4	Xiaoniuquan, Yanzidong, Huilong, Henghe	0.5359	0.3963	high risk
III	3	Shiliangzi, Baiyan, Xujia	0.2692	0.2113	moderate risk
IV	4	Feishuiyan, Lianshan, Machang, Manban	0.3064	0.2716	low risk

This article compared and analyzed the debris flow risk assessment results with those obtained by the grey correlation and collaborative coupling methods (Table 16). Table 16 shows that the results obtained in this article are consistent with those obtained by the coupling synergy method, generally one level higher than the risk obtained by the grey correlation method. The results obtained by the grey correlation method are mainly medium to low, with a small amount being hazardous, and there is no extreme risk. Compared to the results obtained in this article, they tend to be conservative. The results of the grey correlation method indicate that the possibility of debris flow outbreaks in the study area is minimal. However, the results obtained by the grey correlation method do not match the actual situation of multiple debris flows that have already erupted in the research area. For example, in July 2009, a debris flow broke out in Yushi Gully, depositing some houses (Figures 10 and 11). In 2012, a significant debris flow disaster occurred on August 18th in Longmenshan Town [64]. In August 2022, a debris flow disaster broke out in Longcao Gully, and on-site investigations also showed that there were multiple stages of debris flow accumulation in most debris flow channels (Figure 12). The evaluation criteria of the grey correlation method are significantly too low, so the evaluation results obtained in this article can better reflect the actual situation of debris flow outbreaks in the study area.

Table 16. Comparative analysis of risk assessment results using different methods.

Number	Debris Flow	Results of This Article	Results of Grey Correlation Method	Results of Synergistic Coupling Method
1	Xiaoniuquan	High risk degree	Moderate risk degree	High risk degree
2	Lianshan	Low risk degree	Low risk degree	Low risk degree
3	Feishuiyan	Low risk degree	Low risk degree	Low risk degree
4	Huilong	High risk degree	Moderate risk degree	High risk degree
5	Yanzidong	High risk degree	Moderate risk degree	High risk degree
6	Shiliangzi	Moderate risk degree	Low risk degree	Moderate risk degree
7	Machang	Low risk degree	Low risk degree	Low risk degree
8	Manban	Low risk degree	Low risk degree	Low risk degree
9	Henghe	High risk degree	Moderate risk degree	High risk degree
10	Yushi	Extreme risk degree	High risk degree	Extreme risk degree
11	Longcao	Extreme risk degree	High risk degree	Extreme risk degree
12	Meizilin	Extremely risk degree	High risk degree	Extreme risk degree
13	Xujia	Moderate risk degree	Low risk degree	Moderate risk degree
14	Baiyan	Moderate risk degree	Moderate risk degree	Extreme risk degree

Figure 10. Debris flow broke out in Yushi Gully (2009).

Figure 11. A Debris flow broke out in Yushi Gully, destroying houses (2009).

Figure 12. Multi-phase Debris flow buildup in Huilong Gully.

5. Discussion

The accuracy of classification and risk assessment of debris flows is crucial for preventing and controlling debris flows. According to different classification standards, the same debris flow can belong to different categories simultaneously. The traditional classification standards for debris flows have a certain lag in preventing and controlling current debris flows [25]. For regional debris flows, different regions of debris flows have different impact factors, which need to be comprehensively selected based on the actual situation of different research regions [65]. At the same time, there are also areas for improvement in calculating the weight of the selected influencing factors. For current debris flow risk assessment models, most require manual determination of risk classification standards, which results in the inability of debris flow risk assessment results to escape the influence of human subjectivity [33]. This article proposes associating the combination weighting method with affinity propagation clustering analysis to obtain combination weights scientifically and, based on this, use clustering analysis to obtain accurate classification standards.

In this article, the reasonable and correct selection of debris flow evaluation indicators and the calculation of indicator weights are prerequisites for using cluster analysis methods for debris flow classification and risk assessment [66]. This article uses various methods, such as on-site geological surveys, multiple remote sensing images, and drone images, to select nine influencing factors based on the essential characteristics of the debris flow clusters in Longmenshan Town. These influencing factors reflect the various geological, material, and trigger conditions of the debris flows within the study area and calculate the weight of the selected factors. The CRITIC method is an objective weight calculation method, but objective methods cannot reasonably exclude singular data in data processing, which may result in incorrect results. The AHP can fully utilize the experience of experts in the corresponding field to calculate weights, which is a subjective method. However, different experts judge different factors, which can cause incorrect results. Therefore, using the combination weighting method to combine the advantages of the two methods can obtain more scientific indicator weights, which is superior to applying a single method.

Compared with traditional clustering analysis, affinity propagation clustering analysis does not require manually specifying the number of clusters and cluster centers, and the clustering results are unique, with obvious advantages. When using affinity propagation clustering analysis to classify and evaluate the risk of debris flows, optimizing the affinity propagation clustering algorithm, further optimizing the algorithm performance, scientifically obtaining the classification results of debris flows, quantifying the classification standards, and correctly evaluating the risk of debris flows. However, this classification and evaluation method has limitations: (1) it does not apply to the risk assessment of individual debris flows; (2) when the number of debris flows is small, this method cannot be applied; and (3) a certain number of evaluation indicators need to be selected, and this method cannot be applied when the number of evaluation indicators is too small.

As the results of this article are based on the primary data of 14 debris flows in Longmenshan Town, Pengzhou City, China, with significant regional significance, considering the different development characteristics of debris flows in different regions, when applying this method to debris flows in other regions, the differences in regional conditions should be considered.

6. Conclusions

(1) Based on on-site geological surveys, drone images, and multiple remote sensing images, 9 debris flow risk assessment indicators were selected from 14 debris flows in Longmenshan Town, Pengzhou, China. Each indicator's subjective and objective weights were calculated using hierarchical analysis and CRITIC methods, and the two weights were coupled to obtain the synthetic weights of the evaluation indicators. Based on this, the synthetic evaluation score D_i was calculated for each debris flow so that the obtained synthetic evaluation score could scientifically reflect the risk level of each debris flow gully.

(2) This study conducted cluster analysis of 14 debris flows, established the classification model, and classified the debris flows in the study area into four categories. By combining the classification results with synthetic evaluation scores, it was ultimately determined that, among the 14 debris flows in the study area, 3 were extremely dangerous ($0.6718 \leq D_i \leq 0.8301$), 4 were highly dangerous ($0.3963 \leq D_i \leq 0.5359$), 3 were moderately dangerous ($0.2716 \leq D_i \leq 0.3064$), and 4 were low-risk ($0.2113 \leq D_i \leq 0.2692$).

Author Contributions: Y.L.: Investigation, Methodology, Data curation, Visualization, Writing—original draft. J.S.: Funding acquisition, Data curation. M.H.: Investigation. Z.P.: Investigation. All authors have read and agreed to the published version of the manuscript.

Funding: This research was funded by The National Natural Science Foundation of China (No. 41572308).

Institutional Review Board Statement: Not applicable.

Informed Consent Statement: Not applicable.

Data Availability Statement: The data presented in this study are available on request from the corresponding authors.

Conflicts of Interest: The authors declare no conflict of interest.

References

1. Chen, N.S.; Yue, Z.Q.; Cui, P.; Li, Z.L. A rational method for estimating maximum discharge of a landslide-induced debris flow: A case study from Southwestern China. *Geomorphology* **2007**, *84*, 44–58. [CrossRef]
2. Jomelli, V.; Pavlova, I.; Eckert, N.; Grancher, D.; Brunstein, D. A new hierarchical Bayesian approach to analyse environmental and climatic influences on debris flow occurrence. *Geomorphology* **2015**, *250*, 407–421. [CrossRef]
3. Cardona, F.G.; Giraldo, E.A.; Arango, M.I.; Mergili, M. Regional and detailed multi-hazard assessment of debris-flow processes in the Colombian Andes. In Proceedings of the EGU General Assembly 2021, Online, 19–30 April 2021. [CrossRef]
4. Cabral, V.; Reis, F.; Veloso, V.; Ogura, A.; Zarfl, C. A multi-step hazard assessment for debris-flow prone areas influenced by hydroclimatic events. *Eng. Geol.* **2023**, *313*, 106961. [CrossRef]
5. Joshi, M.; Rajappan, S.; Rajan, P.P.; Mathai, J.; Sankar, G.; Nandakumar, V.; Kumar, V.A. Weathering controlled landslide in Deccan traps: Insight from Mahabaleshwar, Maharashtra. *J. Geol. Soc. India* **2018**, *92*, 555–561. [CrossRef]
6. Joshi, M. Co-seismic landslides in the Sikkim Himalaya during the 2011 Sikkim Earthquake: Lesson learned from the past and inference for the future. *Geol. J.* **2022**, *57*, 5039–5060. [CrossRef]
7. Cui, P. Progress and prospects in research on mountain hazards in China. *Prog. Geogr.* **2014**, *33*, 145–152. [CrossRef]
8. Zou, Q.; Cui, P.; He, J.; Lei, Y.; Li, S. Regional risk assessment of debris flows in China-An HRU-based approach. *Geomorphology* **2019**, *340*, 84–102. [CrossRef]
9. Zhou, W.; Tang, C.; Van Asch, T.W.J.; Chang, M. A rapid method to identify the potential of debris flow development induced by rainfall in the catchments of the Wenchuan earthquake area. *Landslides* **2016**, *13*, 1243–1259. [CrossRef]

10. Wu, Y.H.; Liu, K.F.; Chen, Y.C. Comparison between FLO-2D and Debris-2D on the application of assessment of granular debris flow hazards with case study. *J. Mt. Sci.* **2013**, *10*, 293–304. [CrossRef]
11. Nie, Y.P.; Li, X.Z.; Zhou, W.; Xu, R.C. Dynamic hazard assessment of group-occurring debris flows based on a coupled model. *Nat. Hazards* **2021**, *106*, 2635–2661. [CrossRef]
12. Zhao, Y.; Meng, X.M.; Qi, T.J.; Qing, F.; Xiong, M.Q.; Li, Y.J.; Guo, P.; Chen, G. AI-based identification of low-frequency debris flow catchments in the Bailong River basin, China. *Geomorphology* **2020**, *359*, 107125. [CrossRef]
13. Beguería, S.; Van Asch, T.W.; Malet, J.P.; Gröndahl, S. A GIS-based numerical model for simulating the kinematics of mud and debris fows over complex terrain. *Nat. Hazard Earth Syst. Sci.* **2009**, *9*, 1897–1909. [CrossRef]
14. Ouyang, C.J.; He, S.M.; Xu, Q.; Luo, Y.; Zhang, W.C. A MacCormack-TVD fnite diference method to simulate the mass fow in mountainous terrain with variable computational domain. *Comput. Geosci.* **2013**, *52*, 1–10. [CrossRef]
15. Shi, M.Y.; Chen, J.P.; Song, Y.; Zhang, W.; Song, S.Y.; Zhang, X.D. Assessing debris flow susceptibility in Heshigten Banner, Inner Mongolia, China, using principal component analysis and an improved fuzzy C-means algorithm. *Bull. Eng. Geol. Environ.* **2016**, *75*, 909–922. [CrossRef]
16. Niu, C.C.; Wang, Q.; Chen, J.P.; Wang, K.; Zhang, W.; Zhou, F.J. Debris-flow hazard assessment based on stepwise discriminant analysis and extension theory. *Q. J. Eng. Geol. Hydrogeol.* **2014**, *47*, 211–222. [CrossRef]
17. Lee, D.H.; Cheon, E.; Lim, H.H.; Choi, S.K.; Kim, Y.T.; Lee, S.R. An artificial neural network model to predict debris-flow volumes caused by extreme rainfall in the central region of South Korea. *Eng. Geol.* **2021**, *281*, 105979. [CrossRef]
18. Liang, W.J.; Zhuang, D.F.; Jiang, D.; Pan, J.J.; Ren, H.Y. Assessment of debris flow hazards using a Bayesian Network. *Geomorphology* **2012**, *171*, 94–100. [CrossRef]
19. Iovine, G.; D'Ambrosio, D.; Di Gregorio, S. Applying genetic algorithms for calibrating a hexagonal cellular automata model for the simulation of debris flows characterised by strong inertial effects. *Geomorphology* **2005**, *66*, 287–303. [CrossRef]
20. Chen, X.; Chen, H.; You, Y.; Chen, X.; Liu, J. Weights-of-evidence method based on GIS for assessing susceptibility to debris flows in Kangding County, Sichuan Province, China. *Environ. Earth Sci.* **2016**, *75*, 1–16. [CrossRef]
21. Wei, Z.; Shang, Y.; Zhao, Y.; Pan, P.; Jiang, Y. Rainfall threshold for initiation of channelized debris flows in a small catchment based on in-site measurement. *Eng. Geol.* **2017**, *217*, 23–34. [CrossRef]
22. Gu, X.B.; Shao, J.L.; Wu, S.T.; Wu, Q.H.; Bai, H. The risk assessment of debris flow hazards in zhouqu based on the projection pursuit classification model. *Geotech. Geol. Eng.* **2022**, *40*, 1267–1279. [CrossRef]
23. Frey, B.J.; Dueck, D. Clustering by passing messages between data points. *Science* **2007**, *315*, 972–976. [CrossRef] [PubMed]
24. Pokharel, B.; Althuwaynee, O.F.; Aydda, A.; Kim, S.W.; Lim, S.; Park, H.J. Spatial clustering and modelling for landslide susceptibility mapping in the north of the Kathmandu Valley, Nepal. *Landslides* **2021**, *18*, 1403–1419. [CrossRef]
25. Liang, Z.; Wang, C.M.; Han, S.L.; Ullan Jan Khan, K.; Liu, Y. Classification and susceptibility assessment of debris flow based on a semi-quantitative method combination of the fuzzy C-means algorithm, factor analysis and efficacy coefficient. *Nat. Hazards Earth Syst. Sci.* **2020**, *20*, 1287–1304. [CrossRef]
26. Chang, T.C.; Chao, R.J. Application of back-propagation networks in debris flow prediction. *Eng. Geol.* **2006**, *85*, 270–280. [CrossRef]
27. Chang, T.C. Risk degree of debris flow applying neural networks. *Nat. Hazards* **2007**, *42*, 209–224. [CrossRef]
28. Lu, G.Y.; Chiu, L.S.; Wong, D.W. Vulnerability assessment of rainfall-induced debris flows in Taiwan. *Nat. Hazards* **2007**, *43*, 223–244. [CrossRef]
29. Tunusluoglu, M.C.; Gokceoglu, C.; Nefeslioglu, H.A.; Sonmez, H. Extraction of potential debris source areas by logistic regression technique: A case study from Barla, Besparmak and Kapi mountains (NW Taurids, Turkey). *Environ. Geol.* **2008**, *54*, 9–22. [CrossRef]
30. Lin, P.S.; Lin, J.Y.; Hung, J.C.; Yang, M.D. Assessing debris-flow hazard in a watershed in Taiwan. *Eng. Geol.* **2002**, *66*, 295–313. [CrossRef]
31. Meng, F.Q.; Li, G.J.; Li, M.; Ma, J.Q.; Wang, Q. Application of stepwise discriminant analysis to screening evaluation factors of debris flow. *Rock Soil Mech.* **2010**, *31*, 2925–2929. [CrossRef]
32. Chang, T.C.; Chien, Y.H. The application of genetic algorithm in debris flows prediction. *Environ. Geol.* **2007**, *53*, 339–347. [CrossRef]
33. Zhang, C.; Wang, Q.; Chen, J.P.; Gu, F.G.; Zhang, W. Evaluation of debris flow risk in Jinsha River based on combined weight process. *Rock Soil Mech.* **2011**, *32*, 831–836. [CrossRef]
34. Diakoulaki, D.; Mavrotas, G.; Papayannakis, L. Determining objective weights in multiple criteria problems: The critic method. *Comput. Oper. Res.* **1995**, *22*, 763–770. [CrossRef]
35. Nakamura, Y.; Mori, T.; Sato, M.A.; Ishii, S. Reinforcement learning for a biped robot based on a CPG-actor-critic method. *Neural Netw.* **2007**, *20*, 723–735. [CrossRef]
36. Pan, B.; Liu, S.; Xie, Z.; Shao, Y.; Li, X.; Ge, R. Evaluating operational features of three unconventional intersections under heavy traffic based on CRITIC method. *Sustainability* **2021**, *13*, 4098. [CrossRef]
37. Wang, Y.; Jiang, Y.Y. Air quality evaluation based on improved CRITIC weighting method and fuzzy optimization method. *Stat. Decis.* **2017**, *17*, 83–87. [CrossRef]
38. Saaty, T.L. A scaling method for priorities in hierarchical structures. *J. Math. Psychol.* **1977**, *15*, 234–281. [CrossRef]

39. de FSM Russo, R.; Camanho, R. Criteria in AHP: A systematic review of literature. *Procedia Comput. Sci.* **2015**, *55*, 1123–1132. [CrossRef]
40. Saaty, T.L. Decision making—The analytic hierarchy and network processes (AHP/ANP). *J. Syst. Sci. Syst. Eng.* **2004**, *13*, 1–35. [CrossRef]
41. Frades, I.; Matthiesen, R. Overview on techniques in cluster analysis. In *Bioinformatics Methods in Clinical Research*; Humana Press: Totowa, NJ, USA, 2010; pp. 81–107. [CrossRef]
42. Blashfield, R.K.; Aldenderfer, M.S. The literature on cluster analysis. *Multivar. Behav. Res.* **1978**, *13*, 271–295. [CrossRef]
43. Monnet, J.; Broucke, M. The use of a cluster analysis in a Ménard pressuremeter survey. *Proc. Inst. Civ. Eng.-Geotech. Eng.* **2012**, *165*, 367–377. [CrossRef]
44. Jimenez, R. Fuzzy spectral clustering for identification of rock discontinuity sets. *Rock Mech. Rock Eng.* **2008**, *41*, 929. [CrossRef]
45. Saeidi, O.; Torabi, S.R.; Ataei, M. Prediction of the rock mass diggability index by using fuzzy clustering-based, ANN and multiple regression methods. *Rock Mech. Rock Eng.* **2014**, *47*, 717–732. [CrossRef]
46. Xu, L.M.; Chen, J.P.; Wang, Q.; Zhou, F.J. Fuzzy C-means cluster analysis based on mutative scale chaos optimization algorithm for the grouping of discontinuity sets. *Rock Mech. Rock Eng.* **2013**, *46*, 189–198. [CrossRef]
47. Hammah, R.E.; Curran, J.H. On distance measures for the fuzzy K-means algorithm for joint data. *Rock Mech. Rock Eng.* **1999**, *32*, 1–27. [CrossRef]
48. Tokhmechi, B.; Memarian, H.; Moshiri, B.; Rasouli, V.; Noubari, H.A. Investigating the validity of conventional joint set clustering methods. *Eng. Geol.* **2011**, *118*, 75–81. [CrossRef]
49. Hammah, R.E.; Curran, J.H. Validity measures for the fuzzy cluster analysis of orientation. *Trans. Pattern Anal. Mach. Intell.* **2000**, *22*, 1467–1472. [CrossRef]
50. Hartigan, J.; Wong, M. Algorithm AS 136: A K-means clustering algorithm. *Appl. Stat.* **1979**, *28*, 100–108. [CrossRef]
51. Steinley, D. Means clustering: A half-century synthesis. *Br. J. Math. Stat. Psychol.* **2006**, *59*, 1–34. [CrossRef]
52. Wang, J.; Gao, Y.; Wang, K.; Sangaiah, A.K.; Lim, S.J. An affinity propagation-based self-adaptive clustering method for wireless sensor networks. *Sensors* **2019**, *19*, 2579. [CrossRef]
53. Bodenhofer, U.; Kothmeier, A.; Hochreiter, S. APCluster: An R package for affinity propagation clustering. *Bioinformatics* **2011**, *27*, 2463–2464. [CrossRef] [PubMed]
54. Shang, F.; Jiao, L.C.; Shi, J.; Wang, F.; Gong, M. Fast affinity propagation clustering: A multilevel approach. *Pattern Recognit.* **2012**, *45*, 474–486. [CrossRef]
55. Li, P.; Ji, H.F.; Wang, B.L.; Huang, Z.Y.; Li, H.Q. Adjustable preference affinity propagation clustering. *Pattern Recognit. Lett.* **2017**, *85*, 72–78. [CrossRef]
56. Omkar, S.N.; Khandelwal, R.; Ananth, T.V.S.; Naik, G.; Gopalakrishnan, S. Narayana Quantum behaved Particle Swarm Optimization (QPSO) for multi-objective design optimization of composite structures. *Expert Syst. Appl.* **2009**, *36*, 11312–11322. [CrossRef]
57. Kendall, M.G. A new measure of rank correlation. *Biometrika* **1938**, *30*, 81–93. [CrossRef]
58. Puth, M.T.; Neuhäuser, M.; Ruxton, G.D. Effective use of Spearman's and Kendall's correlation coefficients for association between two measured traits. *Anim. Behav.* **2015**, *102*, 77–84. [CrossRef]
59. Struyf, A.; Hubert, M.; Rousseeuw, P.J. Integrating robust clustering techniques in S-PLUS. *Comput. Stat. Data Anal.* **1997**, *26*, 17–37. [CrossRef]
60. Mamat, A.R.; Mohamed, F.S.; Mohamed, M.A.; Rawi, N.M.; Awang, M.I. Silhouette index for determining optimal k-means clustering on images in different color models. *Int. J. Eng. Technol.* **2018**, *7*, 105–109. [CrossRef]
61. Li, X.; Wang, Z.; Peng, K.; Liu, Z.X. Ant colony ATTA clustering algorithm of rock mass structural plane in groups. *J. Cent. South Univ.* **2014**, *21*, 709–714. [CrossRef]
62. Sun, J.; Wu, X.; Palade, V.; Fang, W.; Lai, C.H.; Xu, W. Convergence analysis and improvements of quantum-behaved particle swarm optimization. *Inf. Sci.* **2012**, *193*, 81–103. [CrossRef]
63. Islam, A.R.M.T.; Saha, A.; Ghose, B.; Pal, S.C.; Chowdhuri, I.; Mallick, J. Landslide susceptibility modeling in a complex mountainous region of Sikkim Himalaya using new hybrid data mining approach. *Geocarto Int.* **2021**, *37*, 9021–9046. [CrossRef]
64. Huang, X.; Tang, C. Formation and activation of catastrophic debris flows in Baishui River basin, Sichuan Province, China. *Landslides* **2014**, *11*, 955–967. [CrossRef]
65. Chang, M.; Tang, C.; Van Asch, T.W.; Cai, F. Hazard assessment of debris flows in the Wenchuan earthquake-stricken area, South West China. *Landslides* **2017**, *14*, 1783–1792. [CrossRef]
66. Zhang, W.; Chen, J.P.; Wang, Q.; An, Y.K.; Qian, X.; Xiang, L.J.; He, L.X. Susceptibility analysis of large-scale debris flows based on combination weighting and extension methods. *Nat. Hazards* **2013**, *66*, 1073–1100. [CrossRef]

MDPI
St. Alban-Anlage 66
4052 Basel
Switzerland
www.mdpi.com

Applied Sciences Editorial Office
E-mail: applsci@mdpi.com
www.mdpi.com/journal/applsci

www.ingramcontent.com/pod-product-compliance
Lightning Source LLC
Chambersburg PA
CBHW061452220926
51850CB00004B/71
9783036590394